BIG IDEAS MATH®

Course 1

A Common Core Curriculum

CALIFORNIA TEACHING EDITION

Ron Larson
Laurie Boswell

BIG IDEAS LEARNING®
Erie, Pennsylvania
BigIdeasLearning.com

BIG IDEAS LEARNING®

Big Ideas Learning, LLC
1762 Norcross Road
Erie, PA 16510-3838
USA

For product information and customer support, contact Big Ideas Learning at **1-877-552-7766** or visit us at ***BigIdeasLearning.com***.

About the Cover
The cover images on the *Big Ideas Math* series illustrate the advancements in aviation from the hot-air balloon to spacecraft. This progression symbolizes the launch of a student's successful journey in mathematics. The sunrise in the background is representative of the dawn of the Common Core era in math education, while the cradle signifies the balanced instruction that is a pillar of the *Big Ideas Math* series.

Printed in the U.S.A.

ISBN 13: 978-1-60840-676-0
ISBN 10: 1-60840-676-8

4 5 6 7 8 9 10 WEB 17 16 15 14

AUTHORS

Ron Larson is a professor of mathematics at Penn State Erie, The Behrend College, where he has taught since receiving his Ph.D. in mathematics from the University of Colorado. Dr. Larson is well known as the lead author of a comprehensive program for mathematics that spans middle school, high school, and college courses. His high school and Advanced Placement books are published by Houghton Mifflin Harcourt. Ron's numerous professional activities keep him in constant touch with the needs of students, teachers, and supervisors. Ron and Laurie Boswell began writing together in 1992. Since that time, they have authored over two dozen textbooks. In their collaboration, Ron is primarily responsible for the pupil edition and Laurie is primarily responsible for the teaching edition of the text.

Laurie Boswell is the Head of School and a mathematics teacher at the Riverside School in Lyndonville, Vermont. Dr. Boswell received her Ed.D. from the University of Vermont in 2010. She is a recipient of the Presidential Award for Excellence in Mathematics Teaching. Laurie has taught math to students at all levels, elementary through college. In addition, Laurie was a Tandy Technology Scholar, and served on the NCTM Board of Directors from 2002 to 2005. She currently serves on the board of NCSM, and is a popular national speaker. Along with Ron, Laurie has co-authored numerous math programs.

ABOUT THE BOOK

The revised *Big Ideas Math* series uses the same research-based strategy of a balanced approach to instruction that made the first *Big Ideas Math* series so successful. This approach opens doors to abstract thought, reasoning, and inquiry as students persevere to answer the Essential Questions that introduce each section. The foundation of the program is the Common Core Standards for Mathematical Content and Standards for Mathematical Practice. Students are subtly introduced to "Habits of Mind" that help them internalize concepts for a greater depth of understanding. These habits serve students well not only in mathematics, but across all curricula throughout their academic careers.

The *Big Ideas Math* series exposes students to highly motivating and relevant problems. Woven throughout the series are the depth and rigor students need to prepare for career-readiness and other college-level courses. In addition, the *Big Ideas Math* series prepares students to meet the challenge of the new Common Core testing.

We consider the *Big Ideas Math* series to be the crowning jewel of 30 years of achievement in writing educational materials.

Ron Larson   Laurie Boswell

TEACHER REVIEWERS

- Lisa Amspacher
  Milton Hershey School
  Hershey, PA
- Mary Ballerina
  Orange County Public Schools
  Orlando, FL
- Lisa Bubello
  School District of Palm Beach County
  Lake Worth, FL
- Sam Coffman
  North East School District
  North East, PA
- Kristen Karbon
  Troy School District
  Rochester Hills, MI
- Laurie Mallis
  Westglades Middle School
  Coral Springs, FL
- Dave Morris
  Union City Area School District
  Union City, PA
- Bonnie Pendergast
  Tolleson Union High School District
  Tolleson, AZ
- Valerie Sullivan
  Lamoille South Supervisory Union
  Morrisville, VT
- Becky Walker
  Appleton Area School District
  Appleton, WI
- Zena Wiltshire
  Dade County Public Schools
  Miami, FL

STUDENT REVIEWERS

- Mike Carter
- Matthew Cauley
- Amelia Davis
- Wisdom Dowds
- John Flatley
- Nick Ganger
- Hannah Iadeluca
- Paige Lavine
- Emma Louie
- David Nichols
- Mikala Parnell
- Jordan Pashupathi
- Stephen Piglowski
- Robby Quinn
- Michael Rawlings
- Garrett Sample
- Andrew Samuels
- Addie Sedelmyer
- Tyler Steffy
- Erin Taylor
- Reid Wilson

CONSULTANTS

- Patsy Davis
  Educational Consultant
  Knoxville, Tennessee

- Bob Fulenwider
  Mathematics Consultant
  Bakersfield, California

- Linda Hall
  Mathematics Assessment Consultant
  Norman, Oklahoma

- Ryan Keating
  Special Education Advisor
  Gilbert, Arizona

- Michael McDowell
  Project-Based Instruction Specialist
  Fairfax, California

- Sean McKeighan
  Interdisciplinary Advisor
  Norman, Oklahoma

- Bonnie Spence
  Differentiated Instruction Consultant
  Missoula, Montana

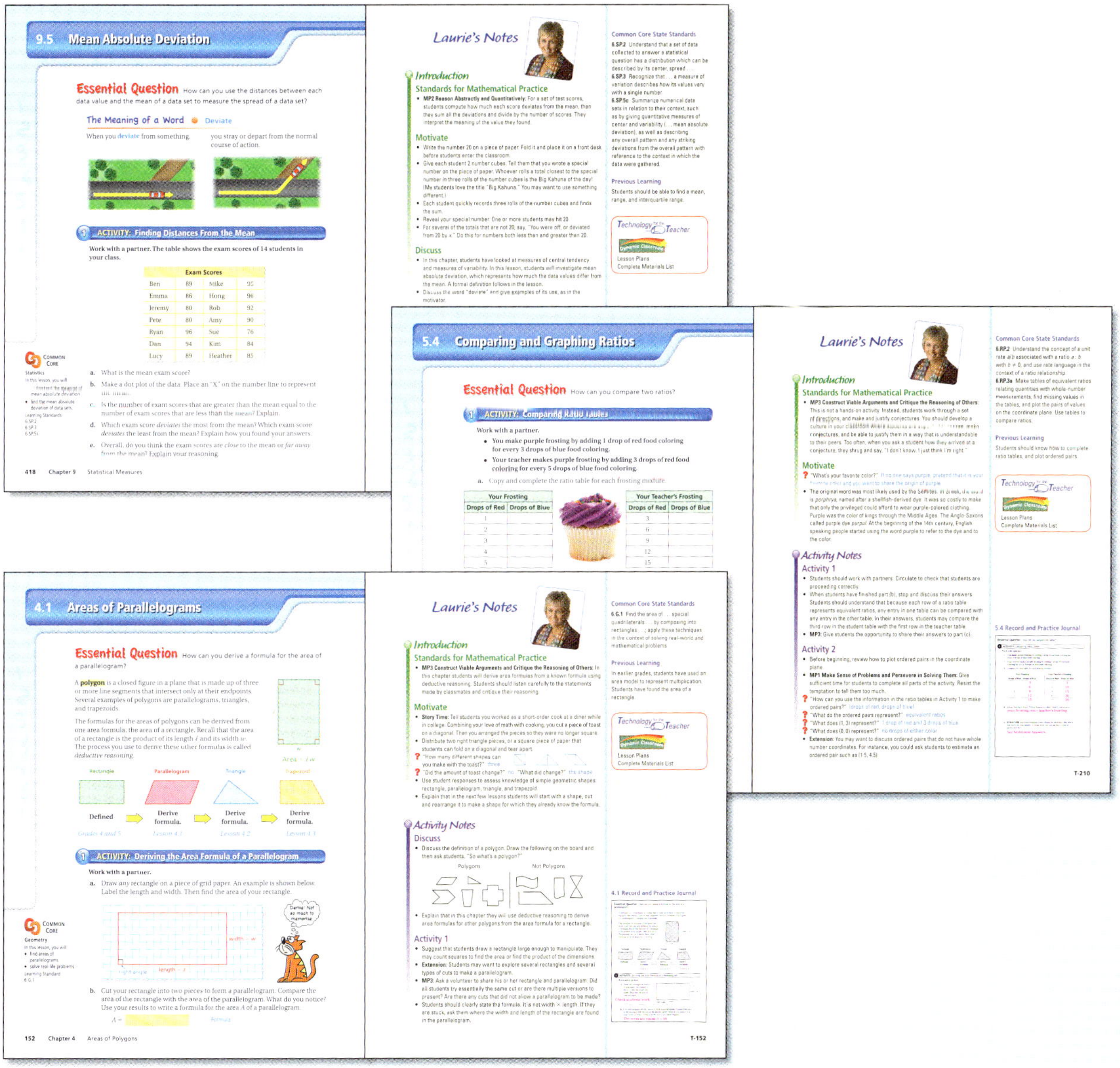

1

Numerical Expressions and Factors

“I like talking about math, and working with a partner allows me to do that.”

Fractions and Decimals

3

“I like that the Essential Question helps me begin thinking about the lesson.”

Algebraic Expressions and Properties

4 Areas of Polygons

"I really enjoy the projects at the end of the book because they help connect the math to other subjects, like science or art."

5

“I like Newton and Descartes! The cartoons are funny and I like that they model the math that we are learning.”

Ratios and Rates

Integers and the Coordinate Plane

“I really like the Big Ideas Math website! The online resources are a huge help when I get stuck or need extra help.”

7

" I like the real-life application exercises because they show me how I can use the math in my own life. "

Equations and Inequalities

8 Surface Area and Volume

“I like playing the games in the Game Closet! They are a fun way to practice concepts we are learning in class.”

9

"With the BigIdeasMath.com website I don't have to worry if I forget my book or my workbook at school."

Statistical Measures

10

Data Displays

“I like that I can print the Glossary Flashcards to review before a test.”

Appendix A: My Big Ideas Projects

PROGRAM OVERVIEW

Print

Also available online and in digital format

Pupil Edition

Also available in eReader format

Teaching Edition

Record and Practice Journal: English and Spanish

- Fair Game Review
- Activity Recording Journal
- Extra Practice Worksheets
- Activity Manipulatives
- Glossary

Assessment Book

- Pre-Course Test
- Quizzes
- Chapter Tests
- Standards Assessment
- Alternative Assessment
- End-of-Course Tests

Resources by Chapter

- Family and Community Involvement: English and Spanish
- Start Thinking! and Warm Up
- Extra Practice
- Enrichment and Extension
- Puzzle Time
- Technology Connection
- Projects with Rubrics

INTRODUCING...
My Dear Aunt Sally
A Common Core app for web, phone, tablet, and mobile devices
mydearauntsally.com

Technology

Student Resources at *BigIdeasMath.com*

Dynamic Student Edition

- Textbook (English and Spanish Audio)
- Record and Practice Journal
- Interactive Manipulatives
- Lesson Tutorials
- Vocabulary (English and Spanish Audio)
- Skills Review Handbook
- Basic Skills Handbook
- Game Closet

Teacher Resources at *BigIdeasMath.com*

Teach Your Lesson

- Dynamic Classroom
  - Whiteboard Classroom Presentations
  - Interactive Manipulatives
  - Support for Mathematical Practices
  - Answer Presentation Tool
- Multi-Language Glossary
- Teaching Edition
- Vocabulary Flash Cards
- Worked-Out Solutions

Plan Your Lesson

- Editable Resources
  - Lesson Plans
  - Assessment Book
  - Resources by Chapter
- Math Tool Paper
- Pacing Guides
- Project Rubrics

Response to Intervention

- Differentiating the Lesson
- Game Closet
- Lesson Tutorials
- Skills Review Handbook
- Basic Skills Handbook

Additional Support for Common Core State Standards

- Common Core State Standards
- Performance Tasks by Standard

DVDs

- Dynamic Assessment Resources
  - ExamView® Assessment Suite
  - Online Testing
  - Self-Grading Homework, Quizzes, and Tests
  - Report Generating
- Dynamic Teaching Resources
- Dynamic Student Edition

eInstruction EXAMVIEW® ASSESSMENT SUITE

SCOPE AND SEQUENCE

Regular Pathway

Grade 6

Ratios and Proportional Relationships	– Understand Ratio Concepts; Use Ratio Reasoning
The Number System	– Perform Fraction and Decimal Operations; Understand Rational Numbers
Expressions and Equations	– Write, Interpret, and Use Expressions, Equations, and Inequalities
Geometry	– Solve Problems Involving Area, Surface Area, and Volume
Statistics and Probability	– Summarize and Describe Distributions; Understand Variability

Grade 7

Ratios and Proportional Relationships	– Analyze Proportional Relationships
The Number System	– Perform Rational Number Operations
Expressions and Equations	– Generate Equivalent Expressions; Solve Problems Using Linear Equations and Inequalities
Geometry	– Understand Geometric Relationships; Solve Problems Involving Angles, Surface Area, and Volume
Statistics and Probability	– Analyze and Compare Populations; Find Probabilities of Events

Grade 8

The Number System	– Approximate Real Numbers; Perform Real Number Operations
Expressions and Equations	– Use Radicals and Integer Exponents; Connect Proportional Relationships and Lines; Solve Systems of Linear Equations
Functions	– Define, Evaluate, and Compare Functions; Model Relationships
Geometry	– Understand Congruence and Similarity; Apply the Pythagorean Theorem; Apply Volume Formulas
Statistics and Probability	– Analyze Bivariate Data

REACH ALGEBRA IN 8TH GRADE

Advanced Pathway

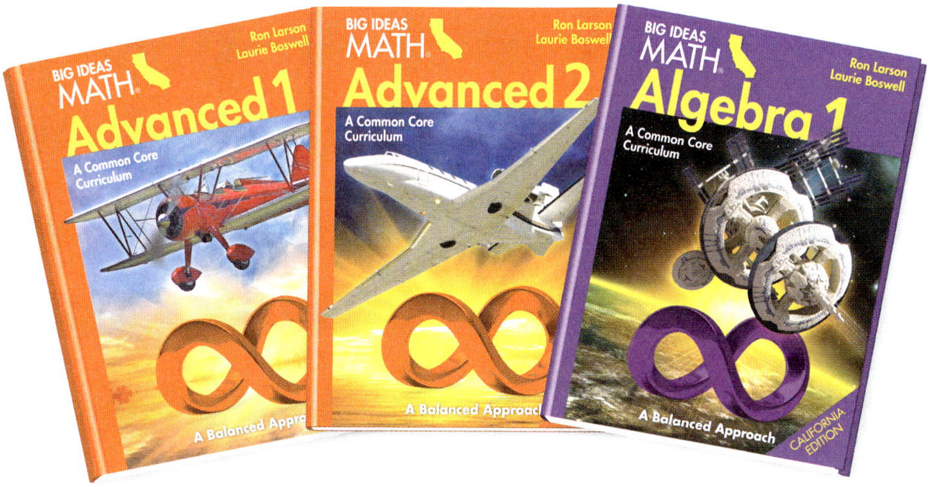

Start Acceleration in 6th Grade

Compacted Pathway

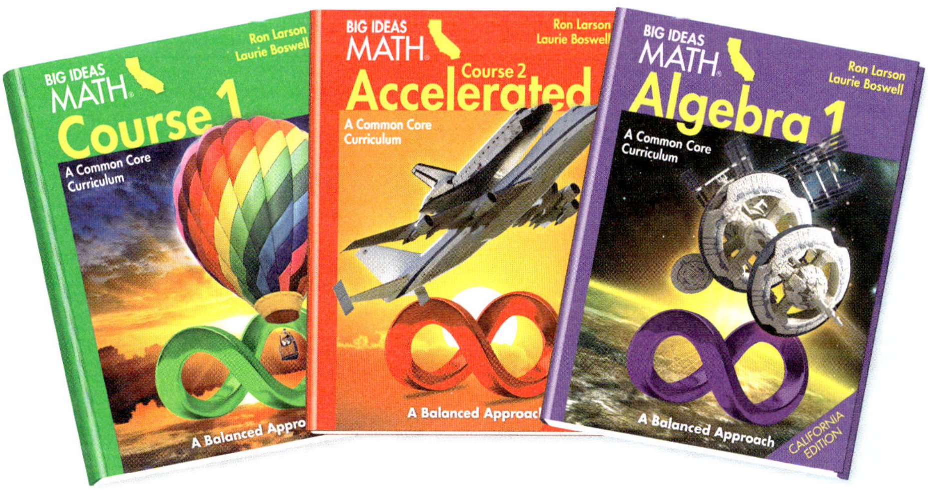

Start Acceleration in 7th Grade

COMMON CORE STATE STANDARDS TO BOOK CORRELATION

After a standard is introduced, it is revisited many times in subsequent activities, lessons, and exercises.

Domain: Ratios and Proportional Relationships

Understand ratio concepts and use ratio reasoning to solve problems.

6.RP.1 Understand the concept of a ratio and use ratio language to describe a ratio relationship between two quantities.

- **Section 5.1** *(pp. 190–195)* Ratios
- **Section 5.2** *(pp. 196–203)* Ratio Tables
- **Section 5.3** *(pp. 204–209)* Rates
- **Section 5.4** *(pp. 210–215)* Comparing and Graphing Ratios

6.RP.2 Understand the concept of a unit rate a/b associated with a ratio $a:b$ with $b \neq 0$, and use rate language in the context of a ratio relationship.

- **Section 5.3** *(pp. 204–209)* Rates
- **Section 5.4** *(pp. 210–215)* Comparing and Graphing Ratios

6.RP.3 Use ratio and rate reasoning to solve real-world and mathematical problems.

a. Make tables of equivalent ratios relating quantities with whole-number measurements, find missing values in the tables, and plot the pairs of values on the coordinate plane. Use tables to compare ratios.

- **Section 5.2** *(pp. 196–203)* Ratio Tables
- **Section 5.3** *(pp. 204–209)* Rates
- **Section 5.4** *(pp. 210–215)* Comparing and Graphing Ratios
- **Section 7.4** *(pp. 314–321)* Writing Equations in Two Variables

b. Solve unit rate problems including those involving unit pricing and constant speed.

- **Section 5.3** *(pp. 204–209)* Rates
- **Section 5.4** *(pp. 210–215)* Comparing and Graphing Ratios

c. Find a percent of a quantity as a rate per 100; solve problems involving finding the whole, given a part and the percent.

- **Section 5.5** *(pp. 218–223)* Percents
- **Section 5.6** *(pp. 224–231)* Solving Percent Problems

d. Use ratio reasoning to convert measurement units; manipulate and transform units appropriately when multiplying or dividing quantities.

- **Section 5.7** *(pp. 232–237)* Converting Measures

Domain: The Number System

Apply and extend previous understandings of multiplication and division to divide fractions by fractions.

6.NS.1 Interpret and compute quotients of fractions, and solve word problems involving division of fractions by fractions.

- **Section 2.1** *(pp. 54–61)* Multiplying Fractions
- **Section 2.2** *(pp. 62–69)* Dividing Fractions
- **Section 2.3** *(pp. 70–75)* Dividing Mixed Numbers

Compute fluently with multi-digit numbers and find common factors and multiples.

6.NS.2 Fluently divide multi-digit numbers using the standard algorithm.

- **Section 1.1** *(pp. 2–9)* Whole Number Operations

6.NS.3 Fluently add, subtract, multiply, and divide multi-digit decimals using the standard algorithm for each operation.

- **Section 2.4** *(pp. 78–83)* Adding and Subtracting Decimals
- **Section 2.5** *(pp. 84–91)* Multiplying Decimals
- **Section 2.6** *(pp. 92–99)* Dividing Decimals

6.NS.4 Find the greatest common factor of two whole numbers less than or equal to 100 and the least common multiple of two whole numbers less than or equal to 12. Use the distributive property to express a sum of two whole numbers 1–100 with a common factor as a multiple of a sum of two whole numbers with no common factor.

- **Section 1.4** *(pp. 24–29)* Prime Factorization
- **Section 1.5** *(pp. 30–35)* Greatest Common Factor
- **Section 1.6** *(pp. 36–41)* Least Common Multiple
- **Extension 1.6** *(pp. 42–43)* Adding and Subtraction Fractions
- **Section 3.4** *(pp. 132–139)* The Distributive Property
- **Extension 3.4** *(pp. 140–141)* Factoring Expressions

Apply and extend previous understandings of numbers to the system of rational numbers.

6.NS.5 Understand that positive and negative numbers are used together to describe quantities having opposite directions or values; use positive and negative numbers to represent quantities in real-world contexts, explaining the meaning of 0 in each situation.

- **Section 6.1** *(pp. 248–253)* Integers
- **Section 6.3** *(pp. 260–265)* Fractions and Decimals on the Number Line

6.NS.6 Understand a rational number as a point on the number line. Extend number line diagrams and coordinate axes familiar from previous grades to represent points on the line and in the plane with negative number coordinates.

a. Recognize opposite signs of numbers as indicating locations on opposite sides of 0 on the number line; recognize that the opposite of the opposite of a number is the number itself, and that 0 is its own opposite.

- **Section 6.1** *(pp. 248–253)* Integers
- **Section 6.3** *(pp. 260–265)* Fractions and Decimals on the Number Line

b. Understand signs of numbers in ordered pairs as indicating locations in quadrants of the coordinate plane; recognize that when two ordered pairs differ only by signs, the locations of the points are related by reflections across one or both axes.

- **Section 6.5** *(pp. 274–281)* The Coordinate Plane
- **Extension 6.5** *(pp. 282–283)* Reflecting Points in the Coordinate Plane

c. Find and position integers and other rational numbers on a horizontal or vertical number line diagram; find and position pairs of integers and other rational numbers on a coordinate plane.

- **Section 6.1** *(pp. 248–253)* Integers
- **Section 6.2** *(pp. 254–259)* Comparing and Ordering Integers
- **Section 6.3** *(pp. 260–265)* Fractions and Decimals on the Number Line
- **Section 6.4** *(pp. 268–273)* Absolute Value
- **Section 6.5** *(pp. 274–281)* The Coordinate Plane
- **Extension 6.5** *(pp. 282–283)* Reflecting Points in the Coordinate Plane

6.NS.7 Understand ordering and absolute value of rational numbers.

a. Interpret statements of inequality as statements about the relative position of two numbers on a number line diagram.

- **Section 6.2** *(pp. 254–259)* Comparing and Ordering Integers
- **Section 6.3** *(pp. 260–265)* Fractions and Decimals on the Number Line
- **Section 6.4** *(pp. 268–273)* Absolute Value

b. Write, interpret, and explain statements of order for rational numbers in real-world contexts.

- **Section 6.2** *(pp. 254–259)* Comparing and Ordering Integers
- **Section 6.3** *(pp. 260–265)* Fractions and Decimals on the Number Line
- **Section 6.4** *(pp. 268–273)* Absolute Value

c. Understand the absolute value of a rational number as its distance from 0 on the number line; interpret absolute value as magnitude for a positive or negative quantity in a real-world situation.

- **Section 6.4** *(pp. 268–273)* Absolute Value

d. Distinguish comparisons of absolute value from statements about order.

- **Section 6.4** *(pp. 268–273)* Absolute Value

6.NS.8 Solve real-world and mathematical problems by graphing points in all four quadrants of the coordinate plane. Include use of coordinates and absolute value to find distance between points with the same first coordinate or the same second coordinate.

- **Section 6.5** *(pp. 274–281)* The Coordinate Plane

Domain: Expressions and Equations

Apply and extend previous understandings of arithmetic to algebraic expressions.

6.EE.1 Write and evaluate numerical expressions involving whole-number exponents.

- **Section 1.2** *(pp. 10–15)* Powers and Exponents
- **Section 1.3** *(pp. 16–21)* Order of Operations

6.EE.2 Write, read, and evaluate expressions in which letters stand for numbers.

a. Write expressions that record operations with numbers and with letters standing for numbers.

- **Section 3.2** *(pp. 118–123)* Writing Expressions

b. Identify parts of an expression using mathematical terms (sum, term, product, factor, quotient, coefficient); view one or more parts of an expression as a single entity.

- **Section 1.5** *(pp. 30–35)* Greatest Common Factor
- **Section 3.4** *(pp. 132–139)* The Distributive Property
- **Extension 3.4** *(pp. 140–141)* Factoring Expressions

c. Evaluate expressions at specific values of their variables. Include expressions that arise from formulas used in real-world problems. Perform arithmetic operations, including those involving whole-number exponents, in the conventional order when there are no parentheses to specify a particular order (Order of Operations).

- **Section 3.1** *(pp. 110–117)* Algebraic Expressions

6.EE.3 Apply the properties of operations to general equivalent expressions.

- **Section 3.3** *(pp. 126–131)* Properties of Addition and Multiplication
- **Section 3.4** *(pp. 132–139)* The Distributive Property
- **Extension 3.4** *(pp. 140–141)* Factoring Expressions

6.EE.4 Identify when two expressions are equivalent.

- **Section 3.3** *(pp. 126–131)* Properties of Addition and Multiplication
- **Section 3.4** *(pp. 132–139)* The Distributive Property
- **Extension 3.4** *(pp. 140–141)* Factoring Expressions

Reason about and solve one-variable equations and inequalities.

6.EE.5 Understand solving an equation or inequality as a process of answering a question: which values from a specified set, if any, make the equation or inequality true? Use substitution to determine whether a given number in a specified set makes an equation or inequality true.

- **Section 7.2** *(pp. 300–307)* Solving Equations Using Addition or Subtraction
- **Section 7.3** *(pp. 308–313)* Solving Equations Using Multiplication or Division
- **Section 7.5** *(pp. 324–331)* Writing and Graphing Inequalities
- **Section 7.6** *(pp. 332–337)* Solving Inequalities Using Addition or Subtraction
- **Section 7.7** *(pp. 338–343)* Solving Inequalities Using Multiplication or Division

6.EE.6 Use variables to represent numbers and write expressions when solving a real-world or mathematical problem; understand that a variable can represent an unknown number, or, depending on the purpose at hand, any number in a specified set.

- **Section 3.2** *(pp. 118–123)* Writing Expressions
- **Section 3.3** *(pp. 126–131)* Properties of Addition and Multiplication
- **Section 3.4** *(pp. 132–141)* The Distributive Property
- **Section 7.1** *(pp. 294–299)* Writing Equations in One Variable

6.EE.7 Solve real-world and mathematical problems by writing and solving equations of the form $x + p = q$ and $px = q$ for cases in which p, q, and x are all nonnegative rational numbers.

- **Section 7.1** *(pp. 294–299)* Writing Equations in One Variable
- **Section 7.2** *(pp. 300–307)* Solving Equations Using Addition or Subtraction
- **Section 7.3** *(pp. 308–313)* Solving Equations Using Multiplication or Division

6.EE.8 Write an inequality of the form $x > c$ or $x < c$ to represent a constraint or condition in a real-world or mathematical problem. Recognize that inequalities of the form $x > c$ or $x < c$ have infinitely many solutions; represent solutions of such inequalities on number line diagrams.

- **Section 7.5** *(pp. 324–331)* Writing and Graphing Inequalities
- **Section 7.6** *(pp. 332–337)* Solving Inequalities Using Addition or Subtraction
- **Section 7.7** *(pp. 338–343)* Solving Inequalities Using Multiplication or Division

Represent and analyze quantitative relationships between dependent and independent variables.

6.EE.9 Use variables to represent two quantities in a real-world problem that change in relationship to one another; write an equation to express one quantity thought of as the dependent variable, in terms of the other quantity, thought of as the independent variable. Analyze the relationship between the dependent and independent variables using graphs and tables, and relate these to the equation.

- **Section 7.4** *(pp. 314–321)* Writing Equations in Two Variables

Domain: Geometry

Solve real-world and mathematical problems involving area, surface area, and volume.

6.G.1 Find the area of right triangles, other triangles, special quadrilaterals, and polygons by composing into rectangles or decomposing into triangles and other shapes; apply these techniques in the context of solving real-world and mathematical problems.

- **Section 4.1** *(pp. 152–157)* Areas of Parallelograms
- **Section 4.2** *(pp. 158–163)* Areas of Triangles
- **Section 4.3** *(pp. 166–171)* Areas of Trapezoids
- **Extension 4.3** *(pp. 172–173)* Areas of Composite Figures

6.G.2 Find the volume of a right rectangular prism with fractional edge lengths by packing it with unit cubes of the appropriate unit fraction edge lengths, and show that the volume is the same as would be found by multiplying the edge length of the prism. Apply the formulas $V = \ell wh$ and $V = bh$ to find volumes of right rectangular prisms with fractional edge lengths in the context of solving real-world and mathematical problems.

- **Section 8.4** *(pp. 374–379)* Volumes of Rectangular Prisms

6.G.3 Draw polygons in the coordinate plane given coordinates for the vertices; use coordinates to find the length of a side joining points with the same first coordinate or the same second coordinate. Apply these techniques in the context of solving real-world and mathematical problems.

- **Section 4.4** *(pp. 174–179)* Polygons in the Coordinate Plane

6.G.4 Represent three-dimensional figures using nets made up of rectangles and triangles, and use the nets to find the surface area of these figures. Apply these techniques in the context of solving real-world and mathematical problems.

- **Section 8.1** *(pp. 354–359)* Three-Dimensional Figures
- **Section 8.2** *(pp. 360–365)* Surface Areas of Prisms
- **Section 8.3** *(pp. 368–373)* Surface Areas of Pyramids

Domain: Statistics and Probability

Develop understanding of statistical variability.

6.SP.1 Recognize a statistical question as one that anticipates variability in the data related to the question and accounts for it in the answers.

- **Section 9.1** *(pp. 390–395)* Introduction to Statistics

6.SP.2 Understand that a set of data collected to answer a statistical question has a distribution which can be described by its center, spread, and overall shape.

- **Section 9.1** *(pp. 390–395)* Introduction to Statistics
- **Section 9.2** *(pp. 396–401)* Mean
- **Section 9.3** *(pp. 402–409)* Measures of Center
- **Section 9.4** *(pp. 412–417)* Measures of Variation
- **Section 9.5** *(pp. 418–423)* Mean Absolute Deviation
- **Section 10.2** *(pp. 440–447)* Histograms
- **Section 10.3** *(pp. 450–455)* Shapes of Distributions
- **Section 10.4** *(pp. 458–465)* Box-and-Whisker Plots

6.SP.3 Recognize that a measure of center for a numerical data set summarizes all of its values with a single number, while a measurc of variation describes how its values vary with a single number.

- **Section 9.2** *(pp. 396–401)* Mean
- **Section 9.3** *(pp. 402–409)* Measures of Center
- **Section 9.4** *(pp. 412–417)* Measures of Variation
- **Section 9.5** *(pp. 418–423)* Mean Absolute Deviation

Summarize and describe distributions.

6.SP.4 Display numerical data in plots on a number line, including dot plots, histograms, and box plots.

- **Section 9.1** *(pp. 390–395)* Introduction to Statistics
- **Section 10.1** *(pp. 434–439)* Stem-and-Leaf Plots
- **Section 10.2** *(pp. 440–447)* Histograms
- **Section 10.3** *(pp. 450–455)* Shapes of Distributions
- **Section 10.4** *(pp. 458–465)* Box-and-Whisker Plots

6.SP.5 Summarize numerical data sets in relation to their context, such as by:

a. Reporting the number of observations.

- **Section 9.1** *(pp. 390–395)* Introduction to Statistics
- **Section 9.2** *(pp. 396–401)* Mean
- **Section 9.5** *(pp. 418–423)* Mean Absolute Deviation

b. Describing the nature of the attribute under investigation, including how it was measured and its units of measurement.

- **Section 9.1** *(pp. 390–395)* Introduction to Statistics

c. Giving quantitative measures of center (median and/or mean) and variability (interquartile range and/or mean absolute deviation), as well as describing any overall pattern and any striking deviations from the overall pattern with reference to the context in which the data were gathered.

- **Section 9.2** *(pp. 396–401)* Mean
- **Section 9.3** *(pp. 402–409)* Measures of Center
- **Section 9.4** *(pp. 412–417)* Measures of Variation
- **Section 9.5** *(pp. 418–423)* Mean Absolute Deviation
- **Section 10.4** *(pp. 458–465)* Box-and-Whisker Plots

d. Relating the choice of measures of center and variability to the shape of the data distribution and the context in which the data were gathered.

- **Extension 10.3** *(pp. 456–457)* Choosing Appropriate Measures

BOOK TO COMMON CORE STATE STANDARDS CORRELATION

Chapter 1

Numerical Expressions and Factors

The Number System

6.NS.2
6.NS.4

Expressions and Equations

6.EE.1
6.EE.2b

Chapter 2

Fractions and Decimals

The Number System

6.NS.1
6.NS.3

Chapter 3

Algebraic Expressions and Properties

The Number System

6.NS.4

Expressions and Equations

6.EE.2a–c
6.EE.3
6.EE.4
6.EE.6

Chapter 4

Areas of Polygons

Geometry

6.G.1
6.G.3

Chapter 5

Ratios and Rates

Ratios and Proportional Relationships

6.RP.1
6.RP.2
6.RP.3a–d

Chapter 6

Integers and the Coordinate Plane

The Number System

6.NS.5
6.NS.6a–c
6.NS.7a–d
6.NS.8

Chapter 7

Equations and Inequalities

Ratios and Proportional Relationships

6.RP.3a

Expressions and Equations

6.EE.5
6.EE.6
6.EE.7
6.EE.8
6.EE.9

Chapter 8

Surface Area and Volume

Geometry

6.G.2
6.G.4

Chapter 9

Statistical Measures

Statistics and Probability

6.SP.1
6.SP.2
6.SP.3
6.SP.4
6.SP.5a–c

Chapter 10

Data Displays

Statistics and Probability

6.SP.2
6.SP.4
6.SP.5c–d

EMBEDDED MATHEMATICAL PRACTICES THROUGH A BALANCED APPROACH TO INSTRUCTION

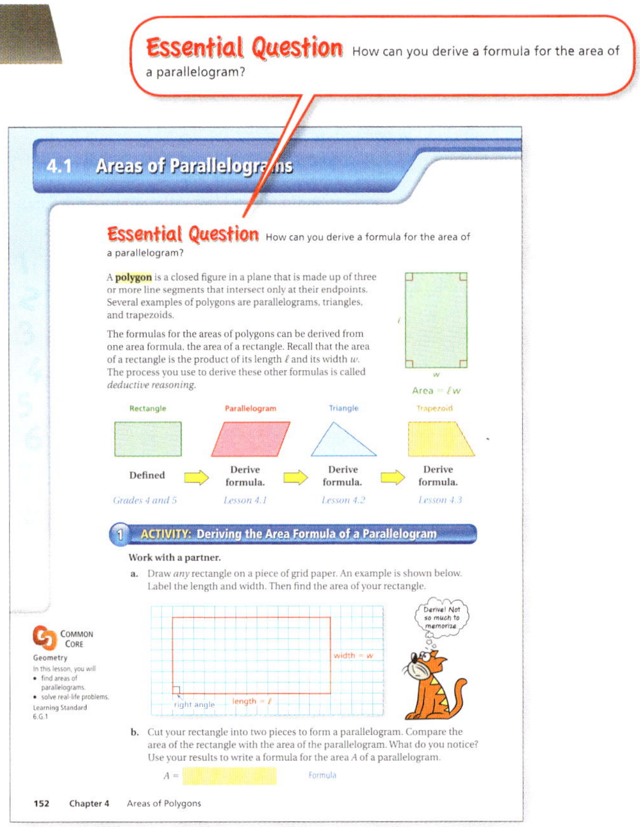
Essential Question How can you derive a formula for the area of a parallelogram?

4.1 Areas of Parallelograms

Essential Question How can you derive a formula for the area of a parallelogram?

A **polygon** is a closed figure in a plane that is made up of three or more line segments that intersect only at their endpoints. Several examples of polygons are parallelograms, triangles, and trapezoids.

The formulas for the areas of polygons can be derived from one area formula, the area of a rectangle. Recall that the area of a rectangle is the product of its length ℓ and its width w. The process you use to derive these other formulas is called *deductive reasoning*.

Area = ℓw

Rectangle	Parallelogram	Triangle	Trapezoid
Defined	Derive formula.	Derive formula.	Derive formula.
Grades 4 and 5	*Lesson 4.1*	*Lesson 4.2*	*Lesson 4.3*

1 ACTIVITY: Deriving the Area Formula of a Parallelogram

Work with a partner.

a. Draw *any* rectangle on a piece of grid paper. An example is shown below. Label the length and width. Then find the area of your rectangle.

width = w; right angle; length = ℓ

Derive! Not so much to memorize

b. Cut your rectangle into two pieces to form a parallelogram. Compare the area of the rectangle with the area of the parallelogram. What do you notice? Use your results to write a formula for the area A of a parallelogram.

$A =$ ____ Formula

COMMON CORE
Geometry
In this lesson, you will
- find areas of parallelograms.
- solve real-life problems.

Learning Standard 6.G.1

152 Chapter 4 Areas of Polygons

The Common Core State Standards require students to do more than memorize how to solve problems. The standards define skills and knowledge that young people need to succeed academically in credit-bearing, college entry courses and in workforce training programs. Mastering the skills reflected in the Common Core State Standards prepares students for college and career.

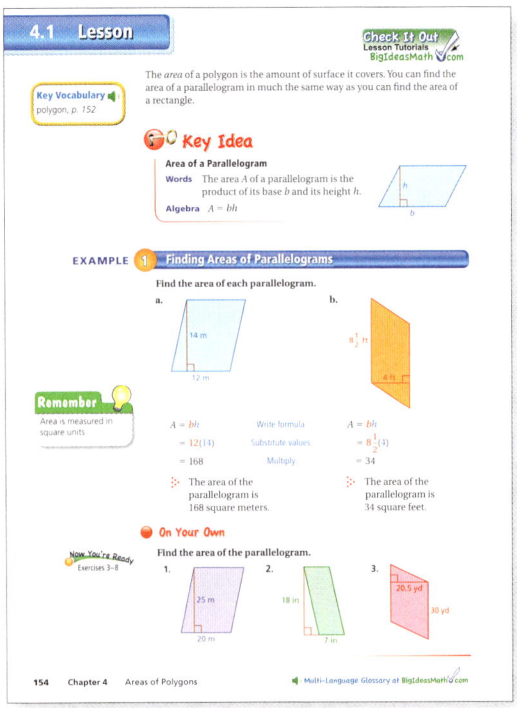
4.1 Lesson

Check It Out Lesson Tutorials BigIdeasMath.com

Key Vocabulary
polygon, *p. 152*

The *area* of a polygon is the amount of surface it covers. You can find the area of a parallelogram in much the same way as you can find the area of a rectangle.

Key Idea

Area of a Parallelogram

Words The area A of a parallelogram is the product of its base b and its height h.

Algebra $A = bh$

EXAMPLE 1 Finding Areas of Parallelograms

Find the area of each parallelogram.

a. 14 m, 12 m

$A = bh$ Write formula.
$= 12(14)$ Substitute values.
$= 168$ Multiply.

The area of the parallelogram is 168 square meters.

b. $8\frac{1}{2}$ ft, 4 ft

$A = bh$
$= 8\frac{1}{2}(4)$
$= 34$

The area of the parallelogram is 34 square feet.

Remember
Area is measured in square units.

On Your Own

Now You're Ready Exercises 3–8

Find the area of the parallelogram.

1. 25 m, 20 m
2. 18 in, 7 in
3. 20.5 yd, 30 yd

154 Chapter 4 Areas of Polygons

Multi-Language Glossary at BigIdeasMath.com

Research shows that students benefit from a program that includes equal exposure to discovery and direct instruction. By beginning each lesson with an inquiry-based activity, *Big Ideas Math* allows students to explore, question, explain, and persevere as they seek to answer Essential Questions that encourage abstract thought.

These student-directed activities are followed by direct instruction lessons, allowing for procedural fluency, modeling, and the opportunity to use clear, precise mathematical language.

DAILY SUPPORT FOR TEACHERS IN MANAGING THE COMMON CORE STATE STANDARDS

Standards for Mathematical Practice

- **MP3 Construct Viable Arguments and Critique the Reasoning of Others:** In this chapter students will derive area formulas from a known formula using deductive reasoning. Students should listen carefully to the statements made by classmates and critique their reasoning.

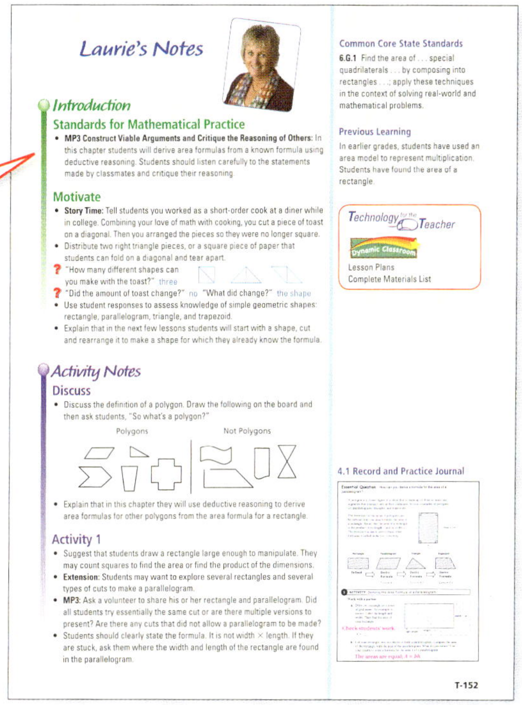

Laurie's Notes

Introduction

Standards for Mathematical Practice

- MP3 Construct Viable Arguments and Critique the Reasoning of Others: In this chapter students will derive area formulas from a known formula using deductive reasoning. Students should listen carefully to the statements made by classmates and critique their reasoning.

Motivate

- Story Time: Tell students you worked as a short-order cook at a diner while in college. Combining your love of math with cooking, you cut a piece of toast on a diagonal. Then you arranged the pieces so they were no longer square.
- Distribute two right triangle pieces, or a square piece of paper that students can fold on a diagonal and tear apart.
- "How many different shapes can you make with the toast?" three
- "Did the amount of toast change?" no "What did change?" the shape
- Use student responses to assess knowledge of simple geometric shapes: rectangle, parallelogram, triangle, and trapezoid.
- Explain that in the next few lessons students will start with a shape, cut and rearrange it to make a shape for which they already know the formula.

Activity Notes

Discuss

- Discuss the definition of a polygon. Draw the following on the board and then ask students, "So what's a polygon?"

Polygons Not Polygons

- Explain that in this chapter they will use deductive reasoning to derive area formulas for other polygons from the area formula for a rectangle.

Activity 1

- Suggest that students draw a rectangle large enough to manipulate. They may count squares to find the area or find the product of the dimensions.
- Extension: Students may want to explore several rectangles and several types of cuts to make a parallelogram.
- MP3: Ask a volunteer to share his or her rectangle and parallelogram. Did all students try essentially the same cut or are there multiple versions to present? Are there any cuts that did not allow a parallelogram to be made?
- Students should clearly state the formula. It is not width × length. If they are stuck, ask them where the width and length of the rectangle are found in the parallelogram.

Common Core State Standards

6.G.1 Find the area of . . . special quadrilaterals . . . by composing into rectangles . . . ; apply these techniques in the context of solving real-world and mathematical problems.

Previous Learning

In earlier grades, students have used an area model to represent multiplication. Students have found the area of a rectangle.

Technology for the Teacher

Dynamic Classroom

Lesson Plans

Complete Materials List

4.1 Record and Practice Journal

T-152

The *Big Ideas Math* Teaching Edition is unique in its organization. It provides teachers with complete support as they teach the program. Using side-by-side pages, teachers have access to the full student page as they teach. Throughout the book Laurie Boswell shares insights into the Common Core State Standards and the Standards for Mathematical Practice.

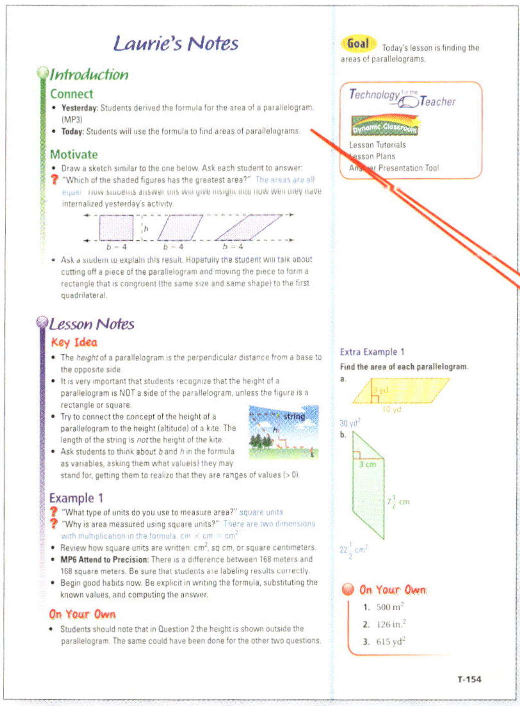

Laurie's Notes

Introduction

Connect

- Yesterday: Students derived the formula for the area of a parallelogram. (MP3)
- Today: Students will use the formula to find areas of parallelograms.

Motivate

- Draw a sketch similar to the one below. Ask each student to answer:
- "Which of the shaded figures has the greatest area?" The areas are all equal. How students answer this will give insight into how well they have internalized yesterday's activity.
- Ask a student to explain this result. Hopefully the student will talk about cutting off a piece of the parallelogram and moving the piece to form a rectangle that is congruent (the same size and same shape) to the first quadrilateral.

Lesson Notes

Key Idea

- The *height* of a parallelogram is the perpendicular distance from a base to the opposite side.
- It is very important that students recognize that the height of a parallelogram is NOT a side of the parallelogram, unless the figure is a rectangle or square.
- Try to connect the concept of the height of a parallelogram to the height (altitude) of a kite. The length of the string is *not* the height of the kite.
- Ask students to think about *b* and *h* in the formula as variables, asking them what value(s) they may stand for, getting them to realize that they are ranges of values (> 0).

Example 1

- "What type of units do you use to measure area?" square units
- "Why is area measured using square units?" There are two dimensions with multiplication in the formula. cm × cm = cm²
- Review how square units are written: cm², sq cm, or square centimeters.
- MP6 Attend to Precision: There is a difference between 168 meters and 168 square meters. Be sure that students are labeling results correctly.
- Begin good habits now. Be explicit in writing the formula, substituting the known values, and computing the answer.

On Your Own

- Students should note that in Question 2 the height is shown outside the parallelogram. The same could have been done for the other two questions.

Goal Today's lesson is finding the areas of parallelograms.

Technology for the Teacher

Dynamic Classroom

Lesson Tutorials

Lesson Plans

Answer Presentation Tool

Extra Example 1

Find the area of each parallelogram.

a. 30 yd²

b. $22\frac{1}{2}$ cm²

On Your Own

1. 500 m²
2. 126 in.²
3. 615 yd²

T-154

Each student page is accompanied by a support page. Laurie includes connections to previous learning, support for the Mathematical Practices, and closure opportunities.

Connect

- **Yesterday:** Students derived the formula for the area of a parallelogram. (MP3)
- **Today:** Students will use the formula to find areas of parallelograms.

The Teaching Edition also provides Differentiated Instruction, Response to Intervention, and English Language Learner support.

FOCUS, COHERENCE, AND RIGOR

Focus

The *Big Ideas Math* program is based on the Common Core State Standards and Standards for Mathematical Practice. There are significantly fewer standards than in the past, allowing for extended instructional time on each. The program's balanced approach, blending discovery with direct instruction, helps students move from concrete to representative understanding and then to abstract representation and the ability to apply the concepts in real life.

Coherence

Learning mathematics is cumulative. In the *Big Ideas Math* program, each grade level builds on the concepts students learned in earlier grades, without repetition. This learning creates a fabric of knowledge that is woven tighter and tighter throughout students' schooling.

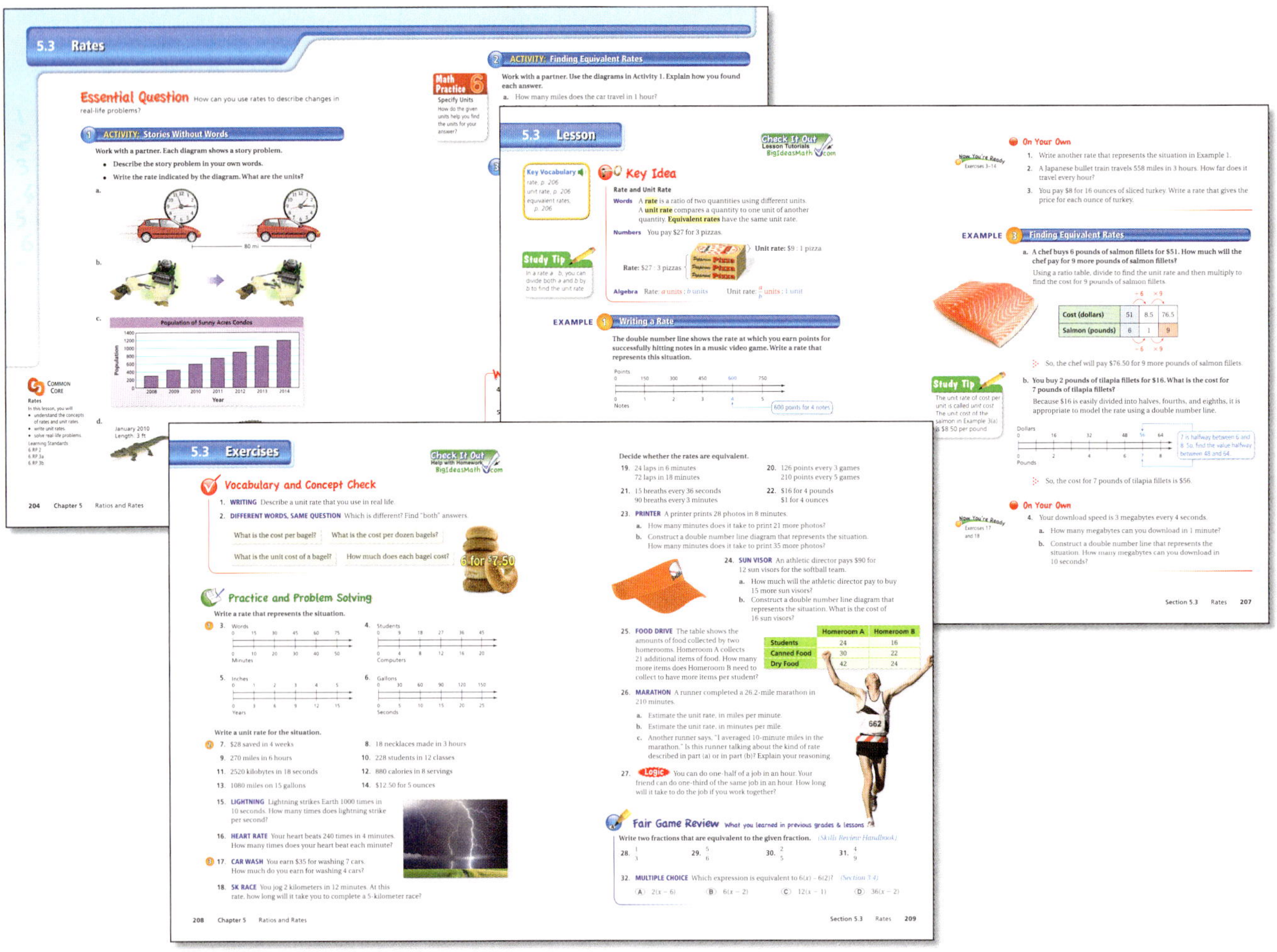

Rigor

The *Big Ideas Math* program meets the three aspects of rigor defined by the Criteria for Evaluating Mathematics Instructional Material: 1. Conceptual understanding; 2. Procedural skill and fluency; and 3. Applications through a balanced approach of instruction.

Conceptual Understanding

The program begins with an Essential Question to set the expectations for conceptual understanding.

> **Essential Question** How can you use rates to describe changes in real-life problems?

Procedural Fluency

In *Big Ideas Math* students have the opportunity to attain procedural fluency through clear stepped-out examples and rich problem sets.

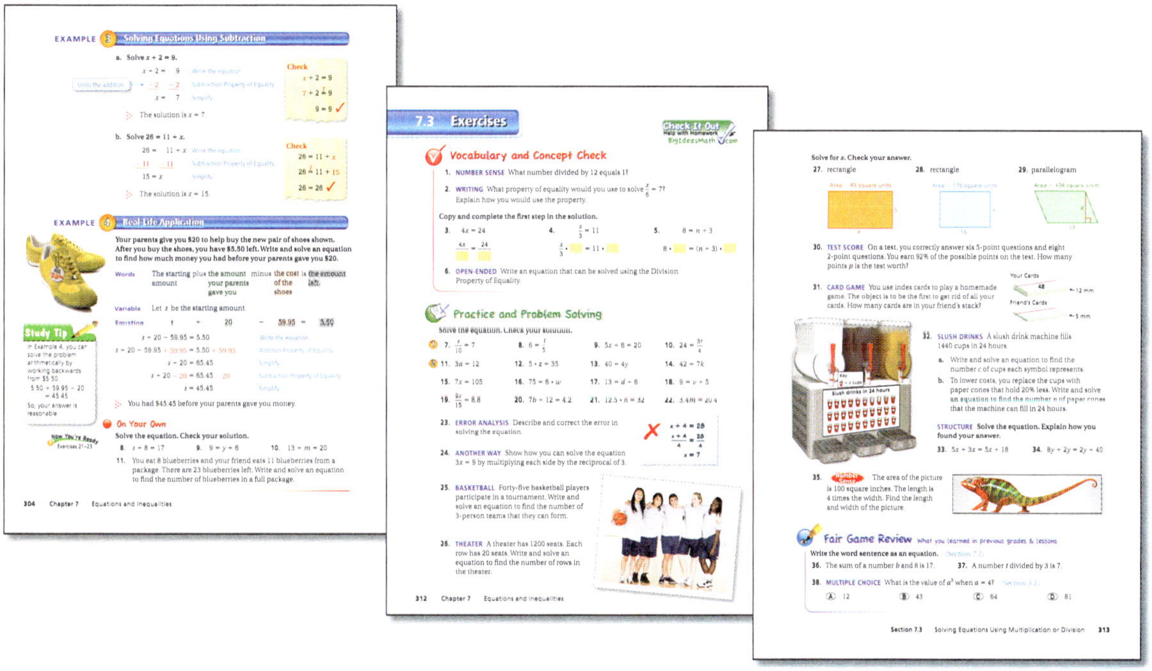

Applications

The Mathematical Practices are embedded throughout the *Big Ideas Math* program. Robust real-life applications in the lessons and exercise sets provide students the opportunity to use the practices and to develop an appreciation for mathematics as it applies to their lives.

Big Ideas Math offers a true balanced approach through student-directed activities to develop conceptual understanding and direct instruction lessons that offer processes and procedures for fluency and real-life applications throughout.

CUSTOMIZED INSTRUCTION

Print Option

The print Teaching Edition provides teachers with help through Laurie's Notes and other features that help manage the classroom. The Chapter Resource Book, Skills Review Handbook, and Assessment Book complete a teaching array that makes it easy for the teacher to differentiate, assess, and teach.

Digital Option

Teachers can use 21st century technology tools found throughout the program to provide exciting ways to stimulate learning. These tools provide innovative electronic activities, timely feedback, and measures for accurate assessment.

Blended Option

Teachers will find that using the blended option provides them a multitude of ways to teach, differentiate, and assess. Teachers and students can customize their teaching and learning by blending the power of creative technology tools with the accessibility of print resources. Rich content and the combination of creative print and online resources allows for an engaging and challenging approach to teaching the Common Core State Standards.

The Dynamic Classroom

Regardless of the option you choose for your students, *The Dynamic Classroom* will be one of your most valued tools in the *Big Ideas Math* program. This powerful tool can be used with interactive whiteboards and includes the following:

- Chapter Openers
- Start Thinking!
- Warm Ups
- Record and Practice Journal pages
- Virtual Manipulatives
- *On Your Own* Exercises
- Extra Examples
- Mini Assessments
- Closure Activities
- Graphic Organizers
- Mathematical Practices
- Answer Presentation Tool

PERSONALIZED LEARNING

The *Big Ideas Math* program offers teachers and students many ways to personalize and enrich the learning experience of all levels of learners.

Lesson Tutorials Online

Two- to three-minute lesson tutorials provide colorful visuals and audio support for every example in the textbook. The Lesson Tutorials are valuable for students who miss a class, need a second explanation, or just need some help with a homework assignment. Parents can also use the tutorials to stay connected or to provide additional help at home.

The Dynamic Student Edition

This unique tool provides students with 21st century learning tools making it easier to directly interact with the underlying mathematics. From the dynamic student textbook to engaging tutorials, students use electronic manipulatives, flashcards, and games to enhance their learning and understanding of math.

Differentiated Instruction

Through print and digital resources, the *Big Ideas Math* program completely supports the 3-Tier Response to Intervention model. Using research-based instructional strategies, teachers can reach, challenge and motivate each student with germane, high quality instruction targeted to individual needs.

CONTINUOUS PREPARATION FOR

In the Textbook

Activities

2 ACTIVITY: Multiplying Decimals Using an Area Model

Work with a partner. Use an area model to find the product. Explain your reasoning.

a. $0.8 \cdot 0.5$

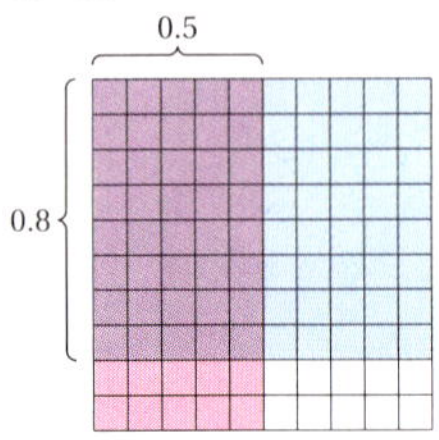

The new Common Core assessments will require a higher level of thinking. In the Activities, students are asked to explain their reasoning.

Exercises

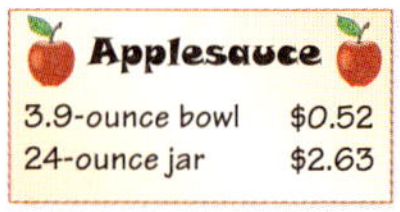

59. **PRECISION** A store sells applesauce in two sizes.
    a. How many *bowls* of applesauce fit in a *jar*? Round your answer to the nearest hundredth.
    b. Explain two ways to find the better buy.
    c. What is the better buy?

The new Common Core assessments will ask for multiple representations. The *Big Ideas Math* program provides students opportunities to use multiple approaches to solve problems.

Quizzes and Tests

20. **PHOTOGRAPHY** A motocross rider is in the air for 2.5 seconds. Your camera can take a picture every 0.125 second. Your friend's camera can take a picture every 0.15 second.
    a. How many times faster is your camera than your friend's camera?
    b. How many more pictures can you take while the rider is in the air?

The assessments in the textbook include problems that extend concepts.

Standards Assessment

10. Which number is equivalent to $5.139 - 2.64$? (6.NS.3)
    F. 2.499 H. 3.519
    G. 2.599 I. 3.599

11. Which expression is equivalent to $\frac{4}{9} \div \frac{5}{7}$? (6.NS.1)
    A. $\frac{20}{63}$ C. $\frac{45}{28}$
    B. $\frac{28}{45}$ D. $\frac{63}{20}$

12. Which of the following expressions is equivalent to a perfect square? (6.EE.1)
    F. $3 + 2^2 \times 7$ H. $(80 \div 4) \div 4$
    G. $34 + 18 \div 3^2$ I. $3^2 + 6 \times 5 \div 3$

13. You are filling baskets using 18 green eggs, 36 red eggs, and 54 blue eggs. What is the greatest number of baskets that you can fill so that the baskets are identical and there are no eggs left over? (6.NS.4)
    A. 3 C. 9
    B. 6 D. 18

14. A walkway was built using identical concrete blocks. (6.NS.1)

Think Solve Explain

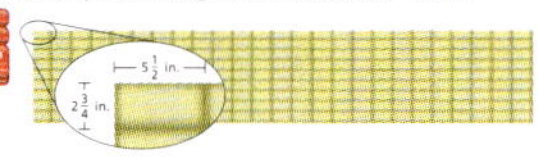

Part A How much longer, in inches, is the length of the walkway than the width of the walkway? Show your work and explain your reasoning.

Part B How many times longer is the length of the walkway than the width of the walkway? Show your work and explain your reasoning.

Answers

10. F
11. B
12. G
13. D
14. *Part A* $112\frac{3}{4}$ inches
    Part B $5\frac{5}{9}$ times as long

Item Analysis (continued)

10. F. Correct answer
    G. The student does not subtract correctly in the tenths place.
    H. The student subtracts the 3 from the 4 in the hundredths place, instead of subtracting 4 from 13, and does not borrow from the ones place.
    I. The student does not subtract correctly in the tenths place and ones place.
11. A. The student incorrectly calculates the expression by multiplying the numerators and multiplying the denominators.
    B. Correct answer
    C. The student incorrectly calculates the expression by inverting the dividend instead of the divisor.
    D. The student incorrectly calculates the expression by inverting both the divisor and the dividend.
12. F. The student does not use the order of operations correctly.
    G. Correct answer
    H. The student does not use the order of operations correctly.
    I. The student does not use the order of operations correctly.
13. A. The student does not find the greatest common factor.
    B. The student finds the product of the two common prime factors of the three numbers.
    C. The student does not find the greatest common factor.
    D. Correct answer
14. **4 points** The student demonstrates a thorough understanding of multiplying and dividing mixed numbers and whole numbers. The student calculates the length and width of the walkway correctly and gets answers of $112\frac{3}{4}$ inches for Part A and $5\frac{5}{9}$ times as long for Part B. The student shows accurate, complete work for both parts and provides clear and complete explanations.

    3 points The student demonstrates an understanding of multiplying and dividing mixed numbers and whole numbers, but the student's work and explanations demonstrate an essential but less than thorough understanding.

    2 points The student demonstrates a partial understanding of multiplying and dividing mixed numbers and whole numbers. The student's work and explanations demonstrate a lack of essential understanding.

    1 point The student demonstrates very limited understanding of multiplying and dividing mixed numbers and whole numbers. The student's response is incomplete and exhibits many flaws.

    0 points The student provided no response, a completely incorrect or incomprehensible response, or a response that demonstrates insufficient understanding of multiplying and dividing mixed numbers and whole numbers.

The Standards Assessments include questions in multiple formats to help prepare students for the Common Core assessments. A detailed Item Analysis for each question in the Pupil Edition is available in the Teaching Edition.

COMMON CORE ASSESSMENT

In the Technology

Online Test Practice

2. You are going to make a cement patio with a garden in the space shown at the left. The dimensions of the garden are shown at the right.

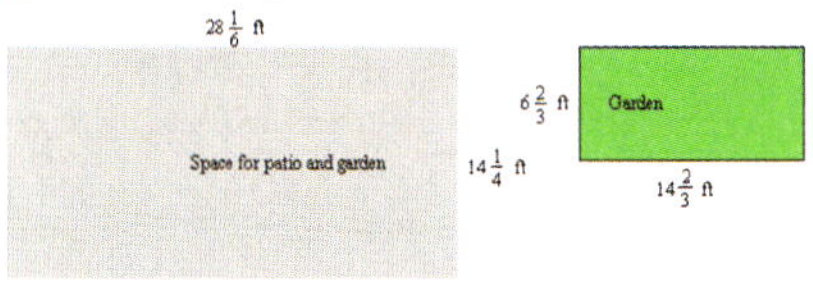

a. Draw a diagram that shows your layout of the garden and patio.
b. Describe two methods for finding the area of the cement patio.
c. Find the area of the cement patio. Which method did you use, and why?

Students will find self-grading quizzes and tests on the student side of the website.

Performance Tasks

CCSS 6.NS.1 Amusement Park

The table shows the portions of time you spent on several activities during a day at an amusement park.

	Portion of your time
Going on rides	$\frac{1}{6}$
Waiting in line for rides	$\frac{3}{10}$
Playing games	$\frac{1}{12}$
Swimming in wave pool	$\frac{3}{10}$

1. How many times more is the portion of time spent going on rides than the portion of time spent playing games? Justify your answer.

2. How many times more is the portion of time spent waiting in line for rides than the portion of time going on rides? Justify your answer.

3. The portion of time spent swimming in the wave pool is two times more than the portion of time spent having lunch. What portion of your time at the amusement park was spent having lunch?

4. If "having lunch" is the only other activity not listed in the table, what other method could you have used to find the answer in Exercise 3?

Performance Tasks for every Common Core Standard are available at *BigIdeasMath.com.*

Resources

Chapter 2 Alternative Assessment

1. A student learning to divide mixed numbers completed the following homework problem. Describe how you would help this student learn from the mistake made.

$$\times \quad 6\frac{3}{8} \div 2\frac{1}{2} = 3\frac{3}{4}$$

2. Jared is planning a road trip. The cost of the regular grade of gasoline is \$3.79 per gallon.

a. Find the cost of filling the 23-gallon tank of his car.

b. Jared drives 421.6 miles on one tank of gasoline. Calculate his fuel efficiency in miles per gallon. Round to the nearest tenth of a mile per gallon.

c. At the next gas station, Jared considers using a higher grade of gasoline with a cost of \$3.99 per gallon. How much more would he pay to fill his tank using this gasoline instead of the regular grade?

d. Toward the end of his trip, Jared realizes he only has \$35.76 left in his wallet. How many more gallons of the regular grade of gasoline can he purchase? Round to the nearest tenth of a gallon.

e. Jared is 160 miles from home. Use the car's fuel efficiency calculated in part (b) to determine if Jared will reach home before running out of gasoline.

f. His entire road trip was 1500 miles. Assuming his car had the same fuel efficiency and he always bought the regular grade of gasoline, calculate the cost of fuel for his trip.

Other resources include the Assessment Book, Resources by Chapter, and Additional Support at *BigIdeasMath.com*.

ExamView® Test Generator

1. **Extended Response** Your class has a plant that needs light. To reach the sunlight from the window, you stack books under the plant. The table shows the types of books in the classroom.

Book	Thickness
Earth science	$\frac{7}{8}$ in.
Biology	$\frac{5}{4}$ in.
Physical science	$\frac{3}{2}$ in.

a. You use 2 physical science books and 2 Earth science books. How tall is the stack?
b. You increase the stack to 3 physical science books, 4 Earth science books, and 3 biology books. How tall is this stack?
c. There are only two more of each book available. Can you make a stack that is at least 32 inches tall? If not, what is the least number of Earth science books you need borrow from another classroom? Explain how you found your answer.

ANS:
a. $4\frac{3}{4}$ in.
b. $11\frac{3}{4}$ in.
c. No; You have 6 Earth science books, 5 biology books, and 5 physical science books with a total height of 19 inches. You need at least 32 − 19 or 13 more inches. Using *Guess, Check, and Revise* you need to borrow 15 Earth science books.

PTS: 1 DIF: Level 3 NAT: 6.NS.1
KEY: multiply | fractions | whole numbers | application MSC: Analysis

The *Big Ideas Math* assessment package includes enhanced ExamView® testing.

PACING GUIDE FOR COURSE 1

Chapters 1–10 **154 Days**

Scavenger Hunt (1 Day)

Chapter 1 (17 Days)

Chapter Opener	1 Day
Activity 1.1	1 Day
Lesson 1.1	1 Day
Activity 1.2	1 Day
Lesson 1.2	1 Day
Activity 1.3	1 Day
Lesson 1.3	1 Day
Study Help/Quiz	1 Day
Activity 1.4	1 Day
Lesson 1.4	1 Day
Activity 1.5	1 Day
Lesson 1.5	1 Day
Activity 1.6	1 Day
Lesson 1.6	1 Day
Extension 1.6	1 Day
Chapter Review/Chapter Tests	2 Days

Chapter 2 (20 Days)

Chapter Opener	1 Day
Activity 2.1	1 Day
Lesson 2.1	2 Days
Activity 2.2	1 Day
Lesson 2.2	2 Days
Activity 2.3	1 Day
Lesson 2.3	1 Day
Study Help/Quiz	1 Day
Activity 2.4	1 Day
Lesson 2.4	1 Day
Activity 2.5	1 Day
Lesson 2.5	2 Days
Activity 2.6	1 Day
Lesson 2.6	2 Days
Chapter Review/Chapter Tests	2 Days

Chapter 3 (13 Days)

Chapter Opener	1 Day
Activity 3.1	1 Day
Lesson 3.1	1 Day
Activity 3.2	1 Day
Lesson 3.2	1 Day
Study Help/Quiz	1 Day
Activity 3.3	1 Day
Lesson 3.3	1 Day
Activity 3.4	1 Day
Lesson 3.4	1 Day
Extension 3.4	1 Day
Chapter Review/Chapter Tests	2 Days

Chapter 4 (13 Days)

Chapter Opener	1 Day
Activity 4.1	1 Day
Lesson 4.1	1 Day
Activity 4.2	1 Day
Lesson 4.2	1 Day
Study Help/Quiz	1 Day
Activity 4.3	1 Day
Lesson 4.3	1 Day
Extension 4.3	1 Day
Activity 4.4	1 Day
Lesson 4.4	1 Day
Chapter Review/Chapter Tests	2 Days

Chapter 5 (18 Days)

Chapter Opener	1 Day
Activity 5.1	1 Day
Lesson 5.1	1 Day
Activity 5.2	1 Day
Lesson 5.2	1 Day
Activity 5.3	1 Day
Lesson 5.3	1 Day
Activity 5.4	1 Day
Lesson 5.4	1 Day
Study Help/Quiz	1 Day

See *Big Ideas Math Advanced 1 Teaching Edition* for the Pacing Guide for Advanced 1.

Activity 5.5	1 Day
Lesson 5.5	1 Day
Activity 5.6	1 Day
Lesson 5.6	1 Day
Activity 5.7	1 Day
Lesson 5.7	1 Day
Chapter Review/Chapter Tests	2 Days

Chapter 6 (15 Days)

Chapter Opener	1 Day
Activity 6.1	1 Day
Lesson 6.1	1 Day
Activity 6.2	1 Day
Lesson 6.2	1 Day
Activity 6.3	1 Day
Lesson 6.3	1 Day
Study Help/Quiz	1 Day
Activity 6.4	1 Day
Lesson 6.4	1 Day
Activity 6.5	1 Day
Lesson 6.5	1 Day
Extension 6.5	1 Day
Chapter Review/Chapter Tests	2 Days

Chapter 7 (18 Days)

Chapter Opener	1 Day
Activity 7.1	1 Day
Lesson 7.1	1 Day
Activity 7.2	1 Day
Lesson 7.2	1 Day
Activity 7.3	1 Day
Lesson 7.3	1 Day
Activity 7.4	1 Day
Lesson 7.4	1 Day
Study Help/Quiz	1 Day
Activity 7.5	1 Day
Lesson 7.5	1 Day
Activity 7.6	1 Day
Lesson 7.6	1 Day
Activity 7.7	1 Day
Lesson 7.7	1 Day
Chapter Review/Chapter Tests	2 Days

Chapter 8 (12 Days)

Chapter Opener	1 Day
Activity 8.1	1 Day
Lesson 8.1	1 Day
Activity 8.2	1 Day
Lesson 8.2	1 Day
Study Help/Quiz	1 Day
Activity 8.3	1 Day
Lesson 8.3	1 Day
Activity 8.4	1 Day
Lesson 8.4	1 Day
Chapter Review/Chapter Tests	2 Days

Chapter 9 (14 Days)

Chapter Opener	1 Day
Activity 9.1	1 Day
Lesson 9.1	1 Day
Activity 9.2	1 Day
Lesson 9.2	1 Day
Activity 9.3	1 Day
Lesson 9.3	1 Day
Study Help/Quiz	1 Day
Activity 9.4	1 Day
Lesson 9.4	1 Day
Activity 9.5	1 Day
Lesson 9.5	1 Day
Chapter Review/Chapter Tests	2 Days

Chapter 10 (13 Days)

Chapter Opener	1 Day
Activity 10.1	1 Day
Lesson 10.1	1 Day
Activity 10.2	1 Day
Lesson 10.2	1 Day
Study Help/Quiz	1 Day
Activity 10.3	1 Day
Lesson 10.3	1 Day
Extension 10.3	1 Day
Activity 10.4	1 Day
Lesson 10.4	1 Day
Chapter Review/Chapter Tests	2 Days

Common Core State Standards for Mathematical Practice

Make sense of problems and persevere in solving them.

- Multiple representations are presented to help students move from concrete to representative and into abstract thinking
- *Essential Questions* help students focus and analyze
- *In Your Own Words* provide opportunities for students to look for meaning and entry points to a problem

Reason abstractly and quantitatively.

- Visual problem solving models help students create a coherent representation of the problem
- Opportunities for students to decontextualize and contextualize problems are presented in every lesson

Construct viable arguments and critique the reasoning of others.

- *Error Analysis*; *Different Words, Same Question*; and *Which One Doesn't Belong* features provide students the opportunity to construct arguments and critique the reasoning of others
- *Inductive Reasoning* activities help students make conjectures and build a logical progression of statements to explore their conjecture

Model with mathematics.

- Real-life situations are translated into diagrams, tables, equations, and graphs to help students analyze relations and to draw conclusions
- Real-life problems are provided to help students learn to apply the mathematics that they are learning to everyday life

Use appropriate tools strategically.

- *Graphic Organizers* support the thought process of what, when, and how to solve problems
- A variety of tool papers, such as graph paper, number lines, and manipulatives, are available as students consider how to approach a problem
- Opportunities to use the web, graphing calculators, and spreadsheets support student learning

Attend to precision.

- *On Your Own* questions encourage students to formulate consistent and appropriate reasoning
- Cooperative learning opportunities support precise communication

Look for and make use of structure.

- *Inductive Reasoning* activities provide students the opportunity to see patterns and structure in mathematics
- Real-world problems help students use the structure of mathematics to break down and solve more difficult problems

Look for and express regularity in repeated reasoning.

- Opportunities are provided to help students make generalizations
- Students are continually encouraged to check for reasonableness in their solutions

Go to *BigIdeasMath.com* for more information on the Common Core State Standards for Mathematical Practice.

Common Core State Standards for Mathematical Content for Grade 6

Chapter Coverage for Standards

1 2 3 4 **5** 6 7 8 9 10

Domain Ratios and Proportional Relationships

- Understand ratio concepts and use ratio reasoning to solve problems.

1 **2** 3 4 5 **6** 7 8 9 10

Domain The Number System

- Apply and extend previous understandings of multiplication and division to divide fractions by fractions.
- Compute fluently with multi-digit numbers and find common factors and multiples.
- Apply and extend previous understandings of numbers to the system of rational numbers.

1 2 **3** 4 5 6 **7** 8 9 10

Domain Expressions and Equations

- Apply and extend previous understandings of arithmetic to algebraic expressions.
- Reason about and solve one-variable equations and inequalities.
- Represent and analyze quantitative relationships between dependent and independent variables.

1 2 3 **4** 5 6 7 **8** 9 10

Domain Geometry

- Solve real-world and mathematical problems involving area, surface area, and volume.

1 2 3 4 5 6 7 8 **9** **10**

Domain Statistics and Probability

- Develop understanding of statistical variability.
- Summarize and describe distributions.

Go to *BigIdeasMath.com* for more information on the Common Core State Standards for Mathematical Content.

How to Use Your Math Book

- Read the **Essential Question** in the activity.

  Discussion the **Math Practice** question with your partner.

  Work with a partner to decide **What Is Your Answer?**

  Now you are ready to do the **Practice** problems.

- Find the **Key Vocabulary** words, **highlighted in yellow**.

  Read their definitions. Study the concepts in each **Key Idea**.

  If you forget a definition, you can look it up online in the

  Multi-Language Glossary at BigIdeasMath.com.

- After you study each **EXAMPLE**, do the exercises in the **On Your Own**.

  Now You're Ready to do the exercises that correspond to the example.

  As you study, look for a **Study Tip** or a **Common Error**.

- The exercises are divided into 3 parts.

  Vocabulary and Concept Check

  Practice and Problem Solving

  Fair Game Review

  If an exercise has a 1 next to it, look back at Example 1 for help with that exercise.

  More help is available at **Check It Out Lesson Tutorials BigIdeasMath.com**.

- To help study for your test, use the following.

  Quiz **Study Help**

  Chapter Review **Chapter Test**

SCAVENGER HUNT

Use this *Scavenger Hunt* to find where things are in **Chapter 1**.

Newton

Find an **Essential Question**.

Find a **Math Practice**.

Find a **Practice**.

Find the definition of a **numerical expression**.

Use the index in the back of the book.

Find the **Key Idea** Order of Operations.

Find an **EXAMPLE** of greatest common factor.

Find an **On Your Own** about $6 + 2^4 - 1$.

Find **Now You're Ready** Exercises 12–20.

Find a **Remember** about prime and composite numbers.

Find a **Vocabulary** question about expressions.

Find a **Problem Solving** question about a **FIELD TRIP**.

Find the answer in the back of the book.

Which examples **?** help with Exercise 6 on page 20?

Find 2 **Quiz**zes.

How many questions are on the **Chapter Test**?

Descartes

1 Numerical Expressions and Factors

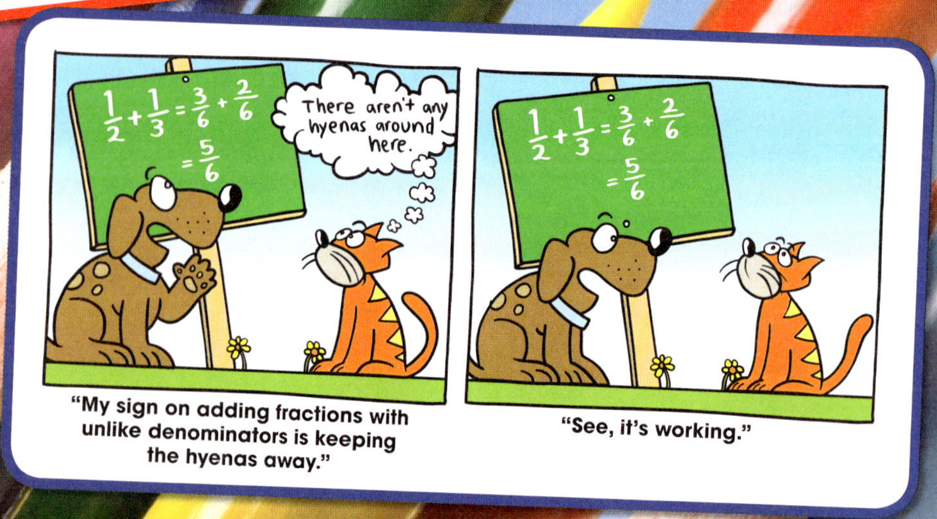

Common Core Progression

4th Grade
• Fluently add and subtract. • Find factors and multiples from 1–100.
5th Grade
• Fluently multiply. • Multiply and divide by powers of 10. • Evaluate expressions with whole number exponents with powers of 10. • Use parenthesis, brackets, or braces in numerical expressions.
6th Grade
• Fluently divide. • Write and evaluate with whole number exponents. • Find the prime factorization of a number. • Find the GCF of two whole numbers. • Find the LCM of two whole numbers.

Chapter Summary

Section	Common Core State Standard	
1.1	Learning	6.NS.2 ★
1.2	Preparing for	6.EE.1
1.3	Learning	6.EE.1 ★
1.4	Preparing for	6.NS.4
1.5	Learning	6.NS.4, 6.EE.2b
1.6	Learning	6.NS.4
★ Teaching is complete. Standard can be assessed.		

Pacing Guide for Chapter 1

Chapter Opener	1 Day
Section 1 Activity Lesson	 1 Day 1 Day
Section 2 Activity Lesson	 1 Day 1 Day
Section 3 Activity Lesson	 1 Day 1 Day
Study Help / Quiz	1 Day
Section 4 Activity Lesson	 1 Day 1 Day
Section 5 Activity Lesson	 1 Day 1 Day
Section 6 Activity Lesson Extension	 1 Day 1 Day 1 Day
Chapter Review/ Chapter Tests	2 Days
Total Chapter 1	17 Days
Year-to-Date	18 Days

Technology for the Teacher

BigIdeasMath.com
Chapter at a Glance
Complete Materials List
Parent Letters: English and Spanish

Common Core State Standards

4.OA.4 ... Determine whether a given whole number in the range 1–100 is prime or composite.

4.NF.3c Add and subtract mixed numbers with like denominators.

Additional Topics for Review

- Dividing Whole Numbers
- Order of Operations (no exponents)
- Multiples of Whole Numbers

Try It Yourself

1. prime **2.** composite

3. prime **4.** prime

5. composite **6.** composite

7. prime **8.** composite

9. composite

10. $\frac{62}{9}$ or $6\frac{8}{9}$ **11.** $\frac{106}{11}$ or $9\frac{7}{11}$

12. $\frac{33}{4}$ or $8\frac{1}{4}$ **13.** $\frac{58}{13}$ or $4\frac{6}{13}$

14. $\frac{7}{2}$ or $3\frac{1}{2}$ **15.** $\frac{4}{3}$ or $1\frac{1}{3}$

Record and Practice Journal Fair Game Review

1. composite **2.** prime

3. prime **4.** composite

5. prime **6.** composite

7. composite **8.** prime

9. prime **10.** composite

11. yes **12.** no

13. $2\frac{4}{5}$ **14.** $5\frac{5}{7}$

15. $10\frac{7}{9}$ **16.** $8\frac{10}{11}$

17. $2\frac{1}{2}$ **18.** $1\frac{1}{2}$

19. $\frac{3}{5}$ **20.** $3\frac{1}{2}$

21. $4\frac{1}{2}$ c

Math Background Notes

Vocabulary Review

- Prime Number
- Composite Number
- Mixed Number
- Improper Fraction

Identifying Prime and Composite Numbers

- Students should be familiar with factor pairs, prime numbers, and composite numbers.
- Review the definition of a prime number. A *prime number* is a whole number greater than 1 whose only factors are 1 and itself.
- You may want to review factoring with students. To find the factors of a number, try to divide the number by prime numbers that are less than the given number.
- **Common error:** Students may remember the divisibility rules for 2, 3, and 5 but forget to check for divisibility for larger prime numbers. Remind students to keep checking all of the prime numbers less than the given number before deciding whether the number is prime or composite.

Adding and Subtracting Mixed Numbers with Like Denominators

- Remind students that adding and subtracting fractions always requires a common denominator.
- Students should be familiar with mixed numbers and improper fractions, although these vocabulary terms might be new to them.
- Remind students that mixed numbers and improper fractions represent the same number in two different ways. Both represent parts greater than or equal to one whole.
- When two fractions already share a common denominator, simply add or subtract their numerators and keep the common denominator.
- Students may want to add or subtract the whole numbers first, then the fractional parts. Although this method may have worked for students before, be sure to model the correct steps to ensure that students have a good foundation for adding and subtracting mixed numbers with unlike denominators.

Reteaching and Enrichment Strategies

If students need help...	If students got it...
Record and Practice Journal • Fair Game Review Skills Review Handbook Lesson Tutorials	Game Closet at *BigIdeasMath.com* Start the next section

What You Learned Before

"Because 6 is composite, you can have 2 piles of 3 blocks, 3 piles of 2 blocks, or..."

Identifying Prime and Composite Numbers (4.OA.4)

Example 1 **Determine whether 26 is prime or composite.**
Because the factors of 26 are 1, 2, 13, and 26, it is composite.

Example 2 **Determine whether 37 is prime or composite.**
Because the only factors of 37 are 1 and 37, it is prime.

Try It Yourself

Determine whether the number is prime or composite.

1. 5 **2.** 14 **3.** 17

4. 23 **5.** 28 **6.** 33

7. 43 **8.** 57 **9.** 64

Adding and Subtracting Mixed Numbers with Like Denominators (4.NF.3c)

Example 3 **Find $2\frac{3}{5} + 4\frac{1}{5}$.**

$$2\frac{3}{5} + 4\frac{1}{5} = \frac{2 \cdot 5 + 3}{5} + \frac{4 \cdot 5 + 1}{5}$$ Rewrite the mixed numbers as improper fractions.

$$= \frac{13}{5} + \frac{21}{5}$$ Simplify.

$$= \frac{13 + 21}{5}$$ Add the numerators.

$$= \frac{34}{5}, \text{ or } 6\frac{4}{5}$$ Simplify.

Try It Yourself

Add or subtract.

10. $4\frac{1}{9} + 2\frac{7}{9}$ **11.** $6\frac{1}{11} + 3\frac{6}{11}$ **12.** $3\frac{7}{8} + 4\frac{3}{8}$

13. $5\frac{8}{13} - 1\frac{2}{13}$ **14.** $7\frac{1}{4} - 3\frac{3}{4}$ **15.** $4\frac{1}{6} - 2\frac{5}{6}$

1.1 Whole Number Operations

Essential Question How do you know which operation to choose when solving a real-life problem?

1 ACTIVITY: Choosing an Operation

Work with a partner. The double bar graph shows the history of a citywide cleanup day.

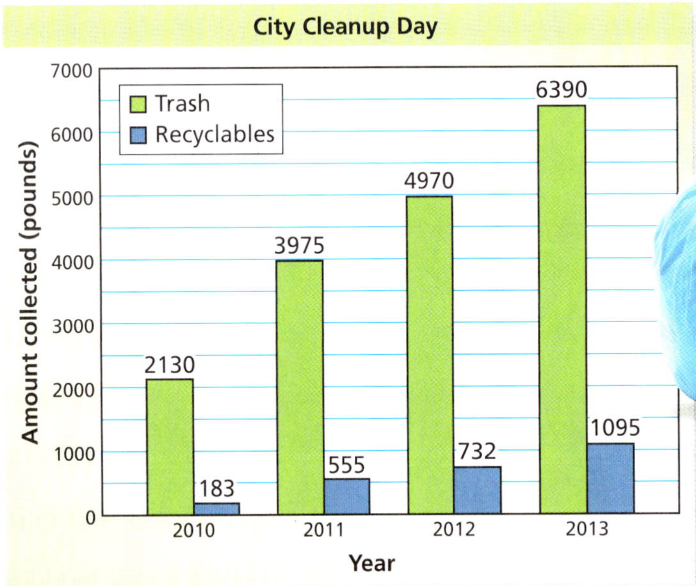

- **Copy each question below.**
- **Underline a key word or phrase that helps you know which operation to use to answer the question. State the operation. Why do you think the key word or phrase indicates the operation you chose?**
- **Write an expression you can use to answer the question.**
- **Find the value of your expression.**

a. What is the total amount of trash collected from 2010 to 2013?

b. How many more pounds of recyclables were collected in 2013 than in 2010?

c. How many times more recyclables were collected in 2012 than in 2010?

d. The amount of trash collected in 2014 is estimated to be twice the amount collected in 2011. What is that amount?

COMMON CORE

Whole Numbers

In this lesson, you will

- determine which operation to perform.
- divide multi-digit numbers.

Learning Standard
6.NS.2

Laurie's Notes

Introduction

Standards for Mathematical Practice

- **MP1a Make Sense of Problems** and **MP3a Construct Viable Arguments:** From the first day, you want to establish a norm in your classroom that each student will discuss mathematical problems with a partner and learn to form arguments based upon stated assumptions, definitions, and previously established results. Students need time to think, discuss, and evaluate their reasoning. Assure students that this will be the case in math this year.

Motivate

- Share some recycling facts with students in the form of, did you know. . .
  - An average American uses 465 trees worth of paper during his or her lifetime.
  - Each person creates about 4.7 pounds of waste every single day.
  - Paper plus cardboard combined make up 73% of the materials in landfills.
  - Over 25 billion styrofoam cups are thrown away in the United States each year.
  - Approximately 350,000 aluminum cans are made in a minute.
- If there is a recycling program at your school, take time to discuss the importance of the program.

Activity Notes

Activity 1

- As a warm-up, you may want to have students briefly describe the amount of debris collected in the city for the last four years to make sure students understand how to read the double bar graph.
- In this activity, students need to read carefully, looking for words that suggest an operation to be performed.
- **Common Error:** The word "times" does not immediately imply that multiplication is to be done. In part (c), "times" is used, and it is a division problem. You could point out that in part (d), the problem could be rephrased as "two times the amount of trash..." to help students see the difference.
- As you circulate, do an informal assessment of whole number operations. Are there students unsure about the process?
- Ask volunteers to share their work at the board.
- **Extension:** Discuss recycling efforts in your town.

Common Core State Standards

6.NS.2 Fluently divide multi-digit numbers using the standard algorithm.

Previous Learning

Students need to be familiar with basic computation facts involving whole numbers.

Technology for the Teacher

Dynamic Classroom

Lesson Plans
Complete Materials List

1.1 Record and Practice Journal

Essential Question How do you know which operation to choose when solving a real-life problem?

1 ACTIVITY: Choosing an Operation

Work with a partner. The double bar graph shows the history of a citywide cleanup day.

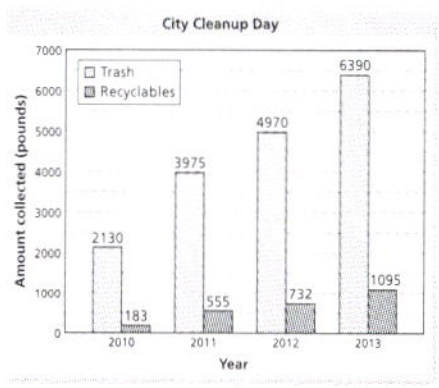

- **Underline a key word or phrase that helps you know which operation to use to answer each question below. State the operation. Why do you think the key word or phrase indicates the operation you chose?**
- **Write an expression you can use to answer the question.**
- **Find the value of your expression.**

a. What is the total amount of trash collected from 2010 to 2013?
addition; 2130 + 3975 + 4970 + 6390; 17,465 pounds

b. How many more pounds of recyclables were collected in 2013 than in 2010?
subtraction; 1095 − 183; 912 pounds

Differentiated Instruction

Kinesthetic

Provide counters to students to use with the magic square in Question 5. Students determine the number of counters needed by modeling the numbers in the completed diagonal. Next, they find a row or column with two given numbers and model those numbers with the counters. The remaining counters represent the missing number in that row or column. Students continue the process until all the numbers are found.

1.1 Record and Practice Journal

c. How many times more recyclables were collected in 2012 than in 2010?

division; 732 ÷ 183; 4 times more

d. The amount of trash collected in 2014 is estimated to be twice the amount collected in 2011. What is that amount?

multiplication; 3975 × 2; 7950 pounds

2 ACTIVITY: Checking Answers

Work with a partner.

a. Explain how you can use estimation to check the reasonableness of the value of your expression in Activity 1(a).

Sample answer: **Round each number in the sum to the nearest thousand, then add.**

b. Explain how you can use addition to check the value of your expression in Activity 1(b).

Sample answer: **Find the sum of the difference, 912, and the number being subtracted, 183. The sum is equal to the greater number in the expression.**

c. Explain how you can use estimation to check the reasonableness of the value of your expression in Activity 1(c).

Sample answer: **Round each number in the quotient 732 ÷ 183 using compatible numbers.**

d. Use mental math to check the value of your expression in Activity 1(d). Describe your strategy.

Sample answer: **Multiply the thousands, then hundreds, then tens, and then ones. Then add the results.**

3 ACTIVITY: Using Estimation

Work with a partner. Use the map. Explain how you found each answer.

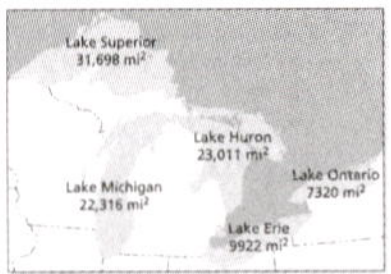

a. Which two lakes have a combined area of about 33,000 square miles?

Lake Huron and Lake Erie

b. Which lake covers an area about three times greater than the area of Lake Erie?

Lake Superior

c. Which lake covers an area that is about 16,000 square miles greater than the area of Lake Ontario?

Lake Huron

d. Estimate the total area covered by the Great Lakes.

Sample answer: **about 94,000 mi^2**

What Is Your Answer?

4. **IN YOUR OWN WORDS** How do you know which operation to choose when solving a real-life problem?

Sample answer: **Look for key words or phrases in the problem that indicate which operation to use to solve.**

5. In a *magic square*, the sum of the numbers in each row, column, and diagonal is the same and each number from 1 to 9 is used only once. Complete the magic square. Explain how you found the missing numbers.

4	9	2
3	5	7
8	1	6

Laurie's Notes

Activity 2

- **MP8 Look for and Express Regularity in Repeated Reasoning:** As students work through problems, they should always be asking themselves if the results they are getting seem reasonable.
- In this activity, students look back at Activity 1. They use estimation and mental math to check the reasonableness of their answers. Throughout the year, encourage your students to check their answers in all of their work.
- Ask volunteers to share their answers.

Activity 3

? "Can anyone name the Great Lakes?" The acronym *HOMES* can help students remember. Huron, Ontario, Michigan, Erie, Superior

? "Have any of you visited one or more of the Great Lakes?

? "Do you know which is the smallest? Ontario the largest?" Superior

- In this activity, students will use the information provided on the map to answer questions. Remind students they should begin by using estimation.
- When students share their answers, listen for their explanations of how they found each answer. For instance, in part (a), if the combined area is 33,000 square miles, Lake Superior, with an area of almost 32,000 square miles, is not going to be one of the lakes. Students might then explain that they compared the total areas of Erie and Michigan with the total areas of Erie and Huron.

What Is Your Answer?

- **Big Idea:** In Question 4, listen for the big idea, namely that key words or phrases, along with the context, help in deciding which operation is needed to answer the question.
- In Question 5, students should be able to describe their processes of finding the missing numbers, not just what the missing numbers are.

Closure

- **Exit Ticket:**
  - Forests are being cut at a rate of 100 acres per minute. How many acres per hour is this? 6000 acres per hour
  - Recycling 15 trees worth of paper reduces air pollutants by about 177,000 tons. This is about how many tons per tree? about 11,800 tons per tree

2 ACTIVITY: Checking Answers

Math Practice 6

Communicate Precisely

What key words should you use so that your partner understands your explanation?

Work with a partner.

a. Explain how you can use estimation to check the reasonableness of the value of your expression in Activity 1(a).

b. Explain how you can use addition to check the value of your expression in Activity 1(b).

c. Explain how you can use estimation to check the reasonableness of the value of your expression in Activity 1(c).

d. Use mental math to check the value of your expression in Activity 1(d). Describe your strategy.

3 ACTIVITY: Using Estimation

Work with a partner. Use the map. Explain how you found each answer.

a. Which two lakes have a combined area of about 33,000 square miles?

b. Which lake covers an area about three times greater than the area of Lake Erie?

c. Which lake covers an area that is about 16,000 square miles greater than the area of Lake Ontario?

d. Estimate the total area covered by the Great Lakes.

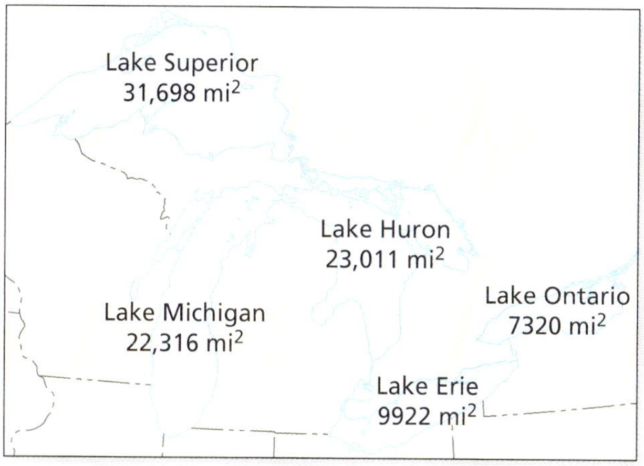

What Is Your Answer?

4. **IN YOUR OWN WORDS** How do you know which operation to choose when solving a real-life problem?

5. In a *magic square*, the sum of the numbers in each row, column, and diagonal is the same and each number from 1 to 9 is used only once. Complete the magic square. Explain how you found the missing numbers.

?	9	2
?	5	?
8	?	?

Practice Use what you learned about choosing operations to complete Exercises 8–11 on page 7.

1.1 Lesson

Check It Out
Lesson Tutorials
BigIdeasMath.com

Recall the four basic operations: addition, subtraction, multiplication, and division.

Operation	Words	Algebra
Addition	the *sum* of	$a + b$
Subtraction	the *difference* of	$a - b$
Multiplication	the *product* of	$a \times b$ $a \cdot b$
Division	the *quotient* of	$a \div b$ $\frac{a}{b}$ $b\overline{)a}$

EXAMPLE 1 Adding and Subtracting Whole Numbers

The bar graph shows the attendance at a three-day art festival.

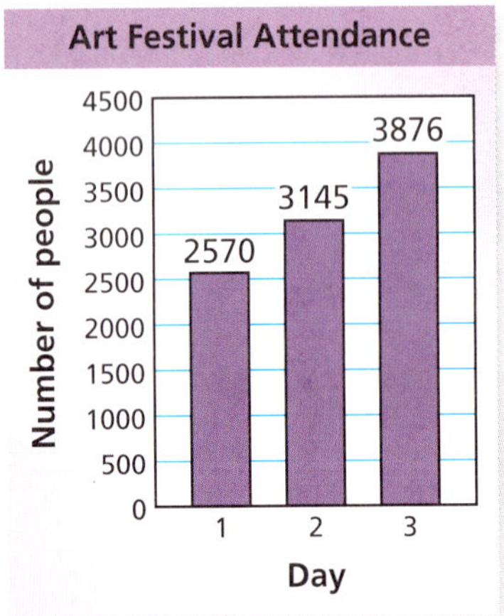

a. What is the total attendance for the art festival?

You want to find the total attendance for the three days. In this case, the phrase *total attendance* indicates you need to find the sum of the daily attendances.

Line up the numbers by their place values, then add.

$$\begin{array}{r} \scriptstyle 111 \\ 2570 \\ 3145 \\ +\ 3876 \\ \hline 9591 \end{array}$$

∴ The total attendance is 9591 people.

b. What is the increase in attendance from Day 1 to Day 2?

You want to find how many more people attended on Day 2 than on Day 1. In this case, the phrase *how many more* indicates you need to find the difference of the attendances on Day 2 and Day 1.

Line up the numbers by their place values, then subtract.

$$\begin{array}{r} \scriptstyle 10 \\ \scriptstyle 2\,\not{0}\,14 \\ \not{3}\not{1}\not{4}5 \\ -\ 2570 \\ \hline 575 \end{array}$$

∴ The increase in attendance from Day 1 to Day 2 is 575 people.

EXAMPLE 2 Multiplying Whole Numbers

A school lunch contains 12 chicken nuggets. Ninety-five students buy the lunch. What is the total number of chicken nuggets served?

You want to find the total number of chicken nuggets in 95 groups of 12 chicken nuggets. The phrase *95 groups of 12* indicates you need to find the product of 95 and 12.

$$\begin{array}{r} 12 \\ \times\ 95 \\ \hline 60 \\ 108 \\ \hline 1140 \end{array}$$

- 60: Multiply 12 by the ones digit, 5.
- 108: Multiply 12 by the tens digit, 9.
- 1140: Add.

∴ There were 1140 chicken nuggets served.

Study Tip

In Example 2, you can use estimation to check the reasonableness of your answer.

$12 \times 95 \approx 12 \times 100$
$= 1200$

Because $1200 \approx 1140$, the answer is reasonable.

Laurie's Notes

Introduction

Connect

- **Yesterday:** Students worked with partners to solve problems involving operations with whole numbers. (MP1a, MP3a, MP8)
- **Today:** Students will perform computations with whole numbers and review how to check solutions.

Motivate

? "Who likes roller coasters?"
- If possible, search the Internet for a video of a roller coaster ride (Skyrush at Hershey Park, PA) and show it to your students.
- Explain that today they will be doing a problem about an amusement park ride. There is a lot of math used at an amusement park.

Lesson Notes

Discuss

- Review the key words and phrases from yesterday that students suggested to help them recognize the operations needed to answer the questions.
- Discuss the different representations of operations using variables, as shown in the table.

? "It is less common to use the × to represent multiplication when using variables. Can you guess why?" The × symbol may be confused with the variable *x*.

Example 1

- Yesterday students read a double bar graph. Today a single bar graph provides information. Reviewing how to read a bar graph helps all students.

? "How can we find the total attendance for three days? Explain." Add the daily attendances together. The phrase "total attendance for three days" means you need to add to find the total.

? "About how much did the attendance increase from Day 1 to Day 2? Listen for about 500 to 600 people.

- When part (b) is done, ask students how to check their answers. They should be familiar with adding 575 to 2570 upward to get 3145.

? **Extension:** "What was the trend in attendance at the art show?" *Sample answer:* The increase in the number of people increased each day.

Example 2

? "What is a reasonable estimate for this problem? Explain." 1200; 12 × 95 is going to be a little less than 12 × 100 = 1200.

? "Does it matter which number is written on top?" no

- Talk through the multiplication problem. If time permits, repeat the problem by putting the 95 on top and 12 below.

? "Is our answer reasonable? Explain." Yes; 1140 is close to the estimate of 1200.

Goal Today's lesson is performing operations with whole numbers.

Technology for the Teacher

Dynamic Classroom

Lesson Tutorials
Lesson Plans
Answer Presentation Tool

Extra Example 1

Use the graph in Example 1.

a. What is the total attendance for Day 2 and Day 3 of the art festival? 7021 people

b. What is the increase in attendance from Day 2 to Day 3? 731 people

Extra Example 2

Construction paper packs from Paper Company A have 85 sheets of paper per pack. Twenty-three teachers each decide to buy a pack. What is the total number of sheets purchased? 1955 sheets of paper

On Your Own

1. 2427 2. 113

3. 4956

Extra Example 3

You make 18 equal payments for a video game system with games. You pay a total of $468. How much is each payment? $26

English Language Learners

Vocabulary

Encourage English language learners to keep a vocabulary notebook. They should include key vocabulary words as well as any other words or phrases with which they are not familiar. Accompany each word or phrase with its definition or a description.

Laurie's Notes

On Your Own

- **Neighbor Check:** Have students work independently and then have their neighbors check their work. Have students discuss any discrepancies.
- Explain to students that working independently and then checking with a neighbor will occur frequently this year.

? "Did any of you use mental math to do Question 2? Explain." Listen for an approach of subtracting 800 and then adding 1.
$912 - 799 = 912 - 800 + 1 = 112 + 1 = 113$

Example 3

- The last two examples review whole number division, an operation that some students may still need work on.
- **Teaching Tip:** When you write the problem $24\overline{)840}$ on the board, use one hand to cover the 40 in the dividend and ask, "Can 24 be divided into 8?" no Now move your hand to reveal 84 in the dividend and ask, "Can 24 be divided into 84?" yes This technique helps students to focus on the process.
- Ask students about each step in the division process. Note the downward arrow, which my students find helpful in working through the steps.

? "How do you check an answer to a division problem?" Multiply the answer (quotient) by the divisor and you should get the dividend.

On Your Own

Now You're Ready
Exercises 12–20

Find the value of the expression. Use estimation to check your answer.

1. $1745 + 682$ **2.** $912 - 799$ **3.** 42×118

EXAMPLE 3 Dividing Whole Numbers: No Remainder

You make 24 equal payments for a go-kart. You pay a total of $840. How much is each payment?

You want to find the number of groups of 24 in $840. The phrase *groups of 24 in $840* indicates you need to find the quotient of 840 and 24.

Use long division to find the quotient.
Decide where to write the first digit of the quotient.

$$\begin{array}{r} ? \\ 24\overline{)840} \end{array}$$ Do not use the hundreds place because 24 is greater than 8.

$$\begin{array}{r} ? \\ 24\overline{)840} \end{array}$$ Use the tens place because 24 is less than 84.

So, divide the tens and write the first digit of the quotient in the tens place.

$$\begin{array}{r} 3 \\ 24\overline{)840} \\ -72 \\ \hline 12 \end{array}$$

Divide 84 by 24: There are three groups of 24 in 84.
Multiply 3 and 24.
Subtract 72 from 84.

Next, bring down the 0 and divide the ones.

$$\begin{array}{r} 35 \\ 24\overline{)840} \\ -72\downarrow \\ \hline 120 \\ -120 \\ \hline 0 \end{array}$$

Divide 120 by 24: There are five groups of 24 in 120.
Multiply 5 and 24.
Subtract 120 from 120.

Remember

$\dfrac{\text{dividend}}{\text{divisor}} = \text{quotient}$

So, quotient × divisor = dividend.

Check Find the product of the quotient and the divisor.

$$\begin{array}{rl} 35 & \text{quotient} \\ \times\ 24 & \text{divisor} \\ \hline 140 & \\ 70 & \\ \hline 840 & \text{dividend} ✓ \end{array}$$

The quotient of 840 and 24 is 35.

So, each payment is $35.

On Your Own

Now You're Ready
Exercises 21–23

Find the value of the expression. Use estimation to check your answer.

4. $234 \div 9$ 5. $\dfrac{986}{58}$ 6. $840 \div 105$

7. Find the quotient of 9920 and 320.

When you use long division to divide whole numbers and you obtain a remainder, you can write the quotient as a mixed number using the rule

$$\text{dividend} \div \text{divisor} = \text{quotient} + \frac{\text{remainder}}{\text{divisor}}.$$

EXAMPLE 4 Real-Life Application

A 301-foot-high swing at an amusement park can take 64 people on each ride. A total of 8983 people ride the swing today. All the rides are full except for the last ride. How many rides are given? How many people are on the last ride?

To find the number of rides given, you need to find the number of groups of 64 people in 8983 people. The phrase *groups of 64 people in 8983 people* indicates you need to find the quotient of 8983 and 64.

Divide the place-value positions from left to right.

$$\begin{array}{r l}
140\ \text{R}23 & \\
64\overline{)8983} & \text{There is one group of 64 in 89.} \\
-\ 64 & \\
258 & \text{There are four groups of 64 in 258.} \\
-\ 256 & \\
23 & \text{There are no groups of 64 in 23.} \\
-\ 0 & \\
23 & \text{The remainder is 23.}
\end{array}$$

Do not stop here. You must write a 0 in the ones place of the quotient.

The quotient is $140\frac{23}{64}$. This indicates 140 groups of 64, with 23 remaining.

So, 141 rides are given, with 23 people on the last ride.

On Your Own

Now You're Ready
Exercises 24–26

Find the value of the expression. Use estimation to check your answer.

8. $\dfrac{6096}{30}$ 9. $45{,}691 \div 28$ 10. $3215 \div 430$

11. **WHAT IF?** In Example 4, 9038 people ride the swing. What is the least number of rides possible?

Laurie's Notes

On Your Own

- **Neighbor Check:** Have students work independently and then have their neighbors check their work. Have students discuss any discrepancies.
- Note that Question 5 is represented as a fraction. Students should be comfortable with different representations of division.

Example 4

- The last example involves whole number division with a remainder.
- Review how to write a quotient with a remainder as a mixed number. Use a simple problem such as $14 \div 4 = 3 + \frac{2}{4} = 3\frac{1}{2}$.
- Ask a volunteer to read the problem.
- Probe student understanding by asking how they could solve the problem if there were a few times when the line was empty, and rides occurred that were partially full. Can they find the total number of rides? Students would need to realize that long division will not help answer the question.

? "How can we find the number of rides given?" Divide total number of riders by 64 because there are 64 riders on each ride.

- **MP6 Attend to Precision:** Be sure to discuss the 0 needed in the ones place of the quotient. If students have made an estimate to begin with, they will recognize that 140 makes more sense than 14.

? "23 is less than a whole group of 64. What does the remainder mean and how can we write the remainder?" The remainder of 23 is the number of people on the last ride. Write 23 out of 64, which is $\frac{23}{64}$.

On Your Own

- Notice that Question 8 has a 0 in the quotient. Be sure to have students work this problem.

Closure

- The theoretical ride capacity for Skyrush (at Hershey Park, PA) is 1350 riders per hour. There are 2 trains with 32 riders each. How many rides is this per hour? about 21 rides per train per hour

On Your Own

4. 26
5. 17
6. 8
7. 31

Extra Example 4

A record-breaking rollercoaster at an amusement park can take 28 people on each ride. A total of 24,539 people ride the rollercoaster today. All of the rides are full except the last ride. How many rides are given? How many people are on the last ride? 877 rides; 11 people

On Your Own

8. $203\frac{1}{5}$
9. $1631\frac{23}{28}$
10. $7\frac{41}{86}$
11. 142 rides

Vocabulary and Concept Check

1. addition
2. multiplication
3. division
4. subtraction
5. addition
6. subtraction
7. a. dividend
   b. quotient
   c. divisor

Practice and Problem Solving

8. $2118 + 3391 + 4785 + 6354$; 16,648 people
9. $4785 - 3391$; 1394 more people
10. $6354 \div 2118$; 3 times more people
11. 4785×2; 9570 people
12. 3091
13. 7081
14. 5847
15. 2462
16. 4436
17. 433
18. 3108
19. 6944
20. 98,884
21. 31
22. 7
23. 60
24. $105\frac{4}{61}$
25. $47\frac{110}{173}$
26. $209\frac{13}{32}$

Assignment Guide and Homework Check

Level	Day 1 Activity Assignment	Day 2 Lesson Assignment	Homework Check
Basic	8–11, 47–51	1–7, 13–23 odd, 27, 38	13, 15, 19, 21, 38
Average	8–11, 47–51	1–7, 13, 15, 19, 21, 28–38 even	13, 15, 19, 21, 38
Advanced	8–11, 47–51	1–7, 28–46 even	30, 36, 38, 42, 44

For Your Information

- **Exercise 42** Remind students that there are 128 fluid ounces in a gallon.

Common Errors

- **Exercises 12–17** Students may set up the problem incorrectly using the vertical method of adding or subtracting. Remind them to line up the numbers by place value, then add or subtract.
- **Exercises 18–20** Students may incorrectly line up the multiplication of the tens digit. Remind them that the right-most digit should line up in the tens-digit place.
- **Exercises 23, 24, and 26** Students may forget to use 0 as a place holder in the quotient. Remind them to estimate the answers to check their results.

1.1 Record and Practice Journal

Find the value of the expression. Use estimation to check your answer.

1. $5947 + 2001$ **7948**
2. $2587 + 1654$ **4241**
3. $5684 + 3118$ **8802**
4. $1596 - 302$ **1294**
5. $9564 - 7581$ **1983**
6. $7094 - 989$ **6105**
7. $851 \div 37$ **23**
8. $\frac{612}{68}$ **9**
9. $8970 \div 345$ **26**
10. $\frac{5424}{52}$ **104 R16 or $104\frac{4}{13}$**
11. $8549 \div 198$ **43 R35 or $43\frac{35}{198}$**
12. $74{,}386 \div 874$ **85 R96 or $85\frac{48}{437}$**
13. Your family is traveling 345 miles to an amusement park. You have already traveled 131 miles. How many more miles must you travel to the amusement park?
    214 miles

1.1 Exercises

Check It Out
Help with Homework
BigIdeasMath.com

Vocabulary and Concept Check

VOCABULARY **Determine which operation the word or phrase represents.**

1. sum
2. times
3. the quotient of
4. decreased by
5. total of
6. minus
7. **VOCABULARY** Use the division problem shown to tell whether the number is the divisor, dividend, or quotient.

$$\begin{array}{r} 26 \\ 34\overline{)884} \end{array}$$

   a. 884 **b.** 26 **c.** 34

Practice and Problem Solving

The bar graph shows the attendance at a food festival. Write an expression you can use to answer the question. Then find the value of your expression.

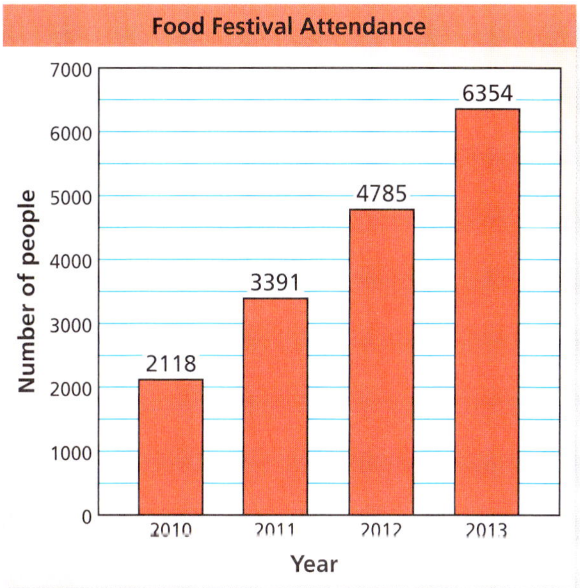

8. What is the total attendance at the food festival from 2010 to 2013?
9. How many more people attended the food festival in 2012 than in 2011?
10. How many times more people attended the food festival in 2013 than in 2010?
11. The festival projects that the total attendance for 2014 will be twice the attendance in 2012. What is the projected attendance for 2014?

Find the value of the expression. Use estimation to check your answer.

(1) **12.** $2219 + 872$

13. $\begin{array}{r} 5351 \\ +\ 1730 \\ \hline \end{array}$

14. $3968 + 1879$

15. $7694 - 5232$

16. $9165 - 4729$

17. $\begin{array}{r} 2416 \\ -\ 1983 \\ \hline \end{array}$

(2) **18.** $\begin{array}{r} 84 \\ \times\ 37 \\ \hline \end{array}$

19. 124×56

20. 419×236

(3) **21.** $837 \div 27$

22. $\dfrac{588}{84}$

23. $7440 \div 124$

(4) **24.** $6409 \div 61$

25. $8241 \div 173$

26. $\dfrac{33{,}505}{160}$

ERROR ANALYSIS **Describe and correct the error in finding the value of the expression.**

27.

$$\begin{array}{r} 39 \\ \times\ 17 \\ \hline 273 \\ 39 \\ \hline 312 \end{array}$$

28.

$$\begin{array}{r} 19 \\ 12\overline{)1308} \\ -12 \\ \hline 108 \\ -108 \\ \hline 0 \end{array}$$

Determine the operation you would use to solve the problem. Do not answer the question.

29. Gymnastic lessons cost $30 per week. How much will 18 weeks of gymnastic lessons cost?

30. The scores on your first two tests were 82 and 93. By how many points did your score improve?

31. You are setting up tables for a banquet for 150 guests. Each table seats 12 people. What is the minimum number of tables you will need?

32. A store has 15 boxes of peaches. Each box contains 45 peaches. How many peaches does the store have?

33. Two shirts cost $18 and $25. What is the total cost of the shirts?

34. A gardener works for 14 hours during a week and charges $168. How much does the gardener charge for each hour?

Find the perimeter and area of the rectangle.

35.

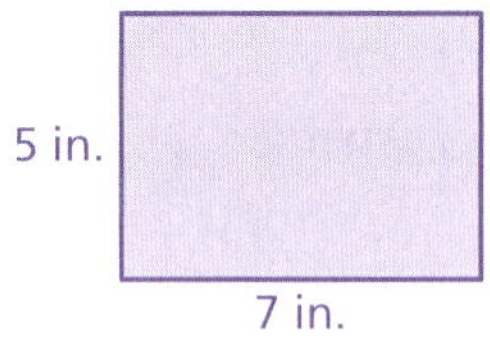

36.

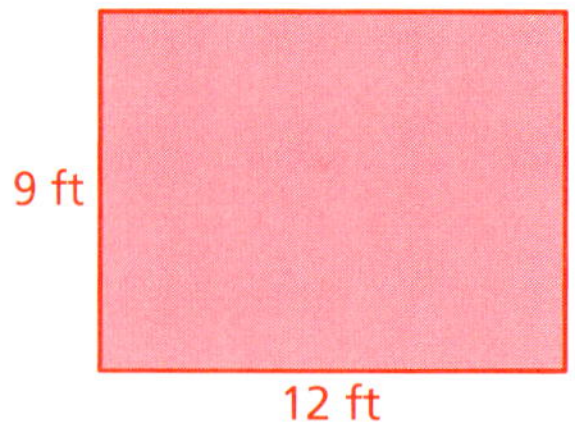

37.

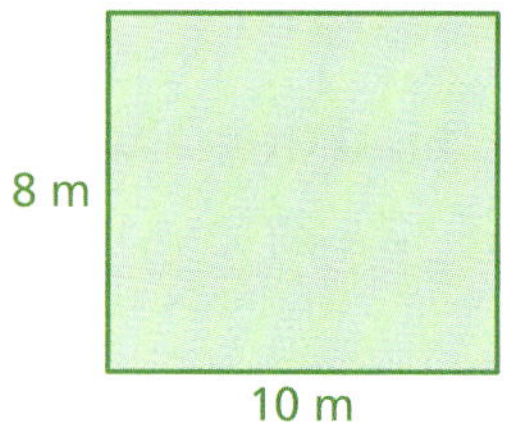

38. **BOX OFFICE** The number of tickets sold for the opening weekend of a movie is 879,575. The movie was shown in 755 theaters across the nation. What was the average number of tickets sold at each theater?

39. **LOGIC** You find that the product of 93 and 6 is 558. How can you use addition to check your answer? How can you use division to check your answer?

40. **NUMBER SENSE** Without calculating, decide which is greater: $3999 \div 129$ or $3834 \div 142$. Explain.

Common Errors

- **Exercises 29–34** Students may struggle with identifying the operation to use. Have them review their notes from the activity and lesson on which phrases indicate which operations.
- **Exercises 35–36** Students may not give a complete answer. Remind them the units of measure should be included with the numerical answer.

Practice and Problem Solving

27. The partial product 39 should be moved to the left so that the 3 is under the 2 and the 9 is under the 7. The answer should be 663.

28. Two digits were brought down instead of one after subtracting 12 from 13. The answer should be 109.

29. multiplication

30. subtraction

31. division

32. multiplication

33. addition

34. division

35. 24 in.; 35 in.^2

36. 42 ft; 108 ft^2

37. 36 m; 80 m^2

38. 1165 tickets

39. You can use addition to check your answer by adding 93 to itself 6 times. You can use division to check your answer by dividing 558 by 93 or by 6.

40. 3999 ÷ 129; *Sample answer:* The first quotient 3999 ÷ 129 has a greater dividend and a lesser divisor than the second quotient 3834 ÷ 142. So, the first quotient will be greater than the second quotient.

English Language Learners

Vocabulary

Have English language learners dedicate four pages of their notebooks to the four basic math operations. On each page, students list words or phrases that indicate the operation. For instance:

Addition	
the sum of	total cost
total attendance	combined

As students work through the problems in the book, they should add any new phrases to the pages of their notebooks.

Practice and Problem Solving

41. no; If the remainder is greater than the divisor, then the quotient should be increased until the remainder is less than the divisor or equal to zero.

42. 64 c

43. 46 tokens

44. See *Taking Math Deeper.*

45. **a.** \$424

b. $\frac{3}{4}$ qt, or $\frac{3}{16}$ gal

46. $36{,}000 \div 900 = 40$

Fair Game Review

47–50.

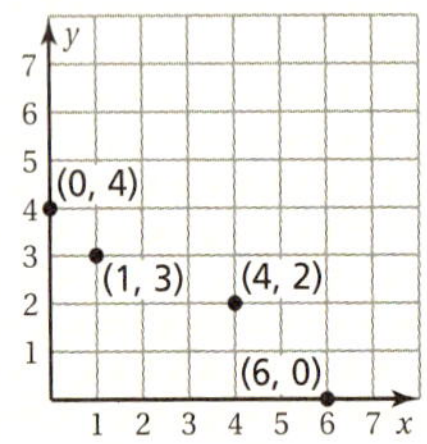

51. A

Mini-Assessment

Find the value of the expression. Check your answer using estimation.

1. $1347 + 914$ 2261

2. $2538 - 1979$ 559

3. 17×223 3791

4. $374 \div 17$ 22

5. There are 240 students and 16 rooms. How many students will be in each class if there is an equal number of students in each class? 15 students

Taking Math Deeper

Exercise 44

This problem involves several calculations. One way to keep track of the calculations is to draw diagrams.

1 Begin by finding the number of books you can display in one bookcase.

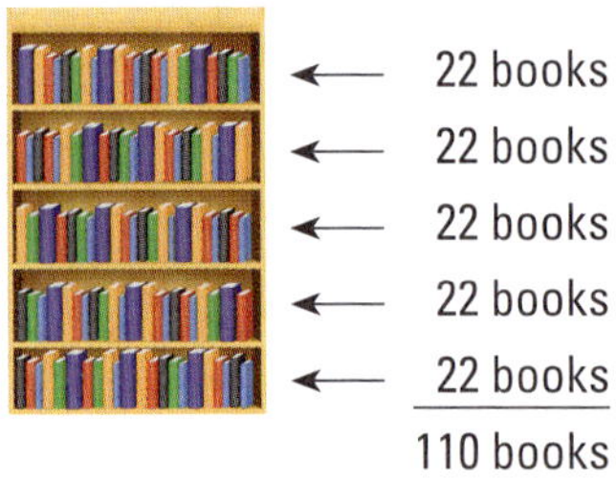

One bookcase can display up to 110 books.

2 Next, find the number of bookcases needed to display 943 books.

a. Eight bookcases can display up to 880 books and 9 bookcases can display up to 990 books. So, you need to borrow 9 bookcases.

3 The first 8 bookcases are full and the 9th bookcase will display $943 - 880 = 63$ books. Use a diagram to show how you fill the 9th bookcase, starting at the top.

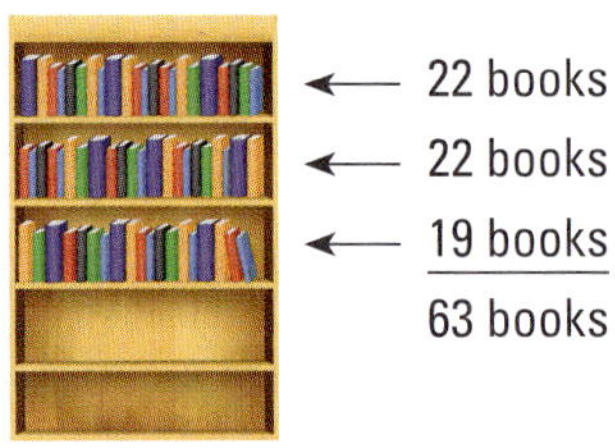

b. There are 19 books on the third shelf.

Reteaching and Enrichment Strategies

If students need help. . .	If students got it. . .
Resources by Chapter • Practice A and Practice B • Puzzle Time Record and Practice Journal Practice Differentiating the Lesson Lesson Tutorials Skills Review Handbook	Resources by Chapter • Enrichment and Extension • Technology Connection Start the next section

41. **REASONING** In a division problem, can the remainder be greater than the divisor? Explain.

42. **WATER COOLER** You change the water jug on the water cooler. How many cups can be completely filled before you need to change the water jug again?

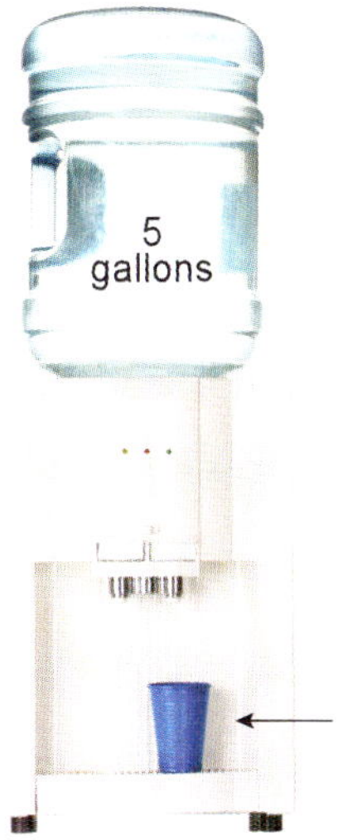

43. **ARCADE** You have $9, one of your friends has $10, and two of your other friends each have $13. You combine your money to buy arcade tokens. You use a coupon to buy 8 tokens for $1. The cost of the remaining tokens is four for $1. You and your friends share the tokens evenly. How many tokens does each person get?

44. **BOOK SALE** You borrow bookcases like the one shown to display 943 books at a book sale. You plan to put 22 books on each shelf. No books will be on top of the bookcases.

   a. How many bookcases must you borrow to display all the books?

   b. You fill the shelves of each bookcase in order, starting with the top shelf. How many books are on the third shelf of the last bookcase?

45. **MODELING** The siding of a house is 2250 square feet. The siding needs two coats of paint. The table shows information about the paint.

Can Size	Cost	Coverage
1 quart	$18	80 square feet
1 gallon	$29	320 square feet

   a. What is the minimum cost of the paint needed to complete the job?

   b. How much paint is left over?

46. Critical Thinking Use the digits 3, 4, 6, and 9 to complete the division problem. Use each digit once.

$$\square\square{,}000 \div \square 00 = \square 0$$

Fair Game Review What you learned in previous grades & lessons

Plot the ordered pair in a coordinate plane. *(Skills Review Handbook)*

47. (1, 3) 48. (0, 4) 49. (6, 0) 50. (4, 2)

51. **MULTIPLE CHOICE** Which of the following numbers is *not* prime? *(Skills Review Handbook)*

Ⓐ 1 Ⓑ 2 Ⓒ 3 Ⓓ 5

1.2 Powers and Exponents

Essential Question How can you use repeated factors in real-life situations?

As I was going to St. Ives
I met a man with seven wives
Each wife had seven sacks
Each sack had seven cats
Each cat had seven kits
Kits, cats, sacks, wives
How many were going to St. Ives? Nursery Rhyme, 1730

1 ACTIVITY: Analyzing a Math Poem

Work with a partner. Here is a "St. Ives" poem written by two students. Answer the question in the poem.

As I was walking into town
I met a ringmaster with five clowns
Each clown had five magicians
Each magician had five bunnies
Each bunny had five fleas
Fleas, bunnies, magicians, clowns
How many were going into town?

Common Core
Numerical Expressions
In this lesson, you will
- write expressions as powers.
- find values of powers.

Preparing for Standard 6.EE.1

Number of clowns: $5 =$ ______

Number of magicians: $5 \times 5 =$ ______

Number of bunnies: $5 \times 5 \times 5 =$ ______

Number of fleas: $5 \times 5 \times 5 \times 5 =$ ______

So, the number of fleas, bunnies, magicians, and clowns is ______. Explain how you found your answer.

Laurie's Notes

Introduction

Standards for Mathematical Practice

- **MP5 Use Appropriate Tools Strategically:** When students are asked to evaluate a power, they should recognize which tool (mental math, paper and pencil, or a calculator) is appropriate.

Motivate

- ? "Have you heard the St. Ives nursery rhyme?" St. Ives is an old village on the southwest tip of Cornwall, England. Another old village, Jamestown, VA, was settled in 1607. "By how many years did the settling of Jamestown pre-date the nursery rhyme?" 123 years
- Tell students that today's activity connects repeated patterns in a nursery rhyme to a repeated mathematical operation.

Demonstrate

- Ask volunteers to act out the rhyme. You can substitute "person" for "man" and "friends" for "wives" if necessary. Each wife (friend) should hold 7 pieces of paper. This helps to build a visual model of $1 \times 7 \times 7$, or 49.
- ? "If 7 triangles are drawn on each piece of paper, how many triangles are there?" $1 \times 7 \times 7 \times 7 = 343$
- Discuss the idea that this pattern could continue. For instance, there could be 7 dots in every triangle and you want to know how many dots there are. The activity today looks at a way to record repeated multiplication.

Activity Notes

Activity 1

- The poem in this activity is similar in style to the St. Ives poem. The careful reader might suggest that only one person for sure is going to town and the rest (clowns, magicians, bunnies, and fleas) were along the road and not headed to town. Moving or stationary, ask the students to determine how many there were.
- Ask a volunteer to share how they found the answer to the question of how many clowns, magicians, bunnies, and fleas there were.
- ? "What would a visual model for this problem look like?" Answers vary, but listen for a tree diagram model or perhaps 5 circles all containing 5 squares with 5 triangles within each square, and so on.
- ? "How is this student poem similar to the St. Ives poem?" *Sample answer:* You can be sure that one is going to town.
- ? "Does anyone recall what the word *factor* means? Can you give an example?" Students may say factors are numbers multiplied together. In the problem $3 \times 4 = 12$, 3 and 4 are factors, and 12 is the product.
- ? "Could a problem have more than two factors? Explain." yes; $3 \times 4 \times 5 = 60$; There are 3 factors.
- ? "How could this student poem be expanded to keep the pattern going?" Answers vary; *Sample answer:* Each flea could have bitten 5 people, or each magician could have had 5 assistants.

Common Core State Standards

6.EE.1 Write and evaluate numerical expressions involving whole-number exponents.

Previous Learning

Students need to be familiar with basic computation facts as well as the meanings of *factor* and *product*.

Technology for the Teacher

Dynamic Classroom

Lesson Plans
Complete Materials List

1.2 Record and Practice Journal

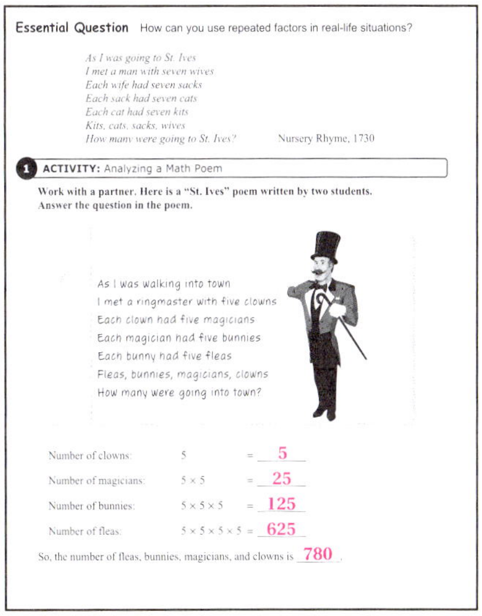

Essential Question How can you use repeated factors in real-life situations?

As I was going to St. Ives
I met a man with seven wives
Each wife had seven sacks
Each sack had seven cats
Each cat had seven kits
Kits, cats, sacks, wives
How many were going to St. Ives? Nursery Rhyme, 1730

1 ACTIVITY: Analyzing a Math Poem

Work with a partner. Here is a "St. Ives" poem written by two students. Answer the question in the poem.

As I was walking into town
I met a ringmaster with five clowns
Each clown had five magicians
Each magician had five bunnies
Each bunny had five fleas
Fleas, bunnies, magicians, clowns
How many were going into town?

Number of clowns: $5 = 5$

Number of magicians: $5 \times 5 = 25$

Number of bunnies: $5 \times 5 \times 5 = 125$

Number of fleas: $5 \times 5 \times 5 \times 5 = 625$

So, the number of fleas, bunnies, magicians, and clowns is 780.

Differentiated Instruction

Visual

Write the first three rows of power pyramids for 2, 4, 5, and 10 on the board.

2	**4**
2×2	4×4
$2 \times 2 \times 2$	$4 \times 4 \times 4$
$2 \times 2 \times 2 \times 2$	$4 \times 4 \times 4 \times 4$
5	**10**
5×5	10×10
$5 \times 5 \times 5$	$10 \times 10 \times 10$
$5 \times 5 \times 5 \times 5$	$10 \times 10 \times 10 \times 10$

Have students write each product as a power and then find the value of each power. Ask, "What is the product of 4 factors of 2? What is the product of 2 factors of 4? What is the product of 3 factors of 5? What is the product of 4 factors of 10?" 16; 16; 125; 10,000

1.2 Record and Practice Journal

2 ACTIVITY: Writing Repeated Factors

Work with a partner. Complete the table.

Repeated Factors	Using an Exponent	Value
a. 4×4	4^2	16
b. 6×6	6^2	36
c. $10 \times 10 \times 10$	10^3	1000
d. $100 \times 100 \times 100$	100^3	1,000,000
e. $3 \times 3 \times 3 \times 3$	3^4	81
f. $4 \times 4 \times 4 \times 4 \times 4$	4^5	1024
g. $2 \times 2 \times 2 \times 2 \times 2 \times 2$	2^6	64

h. In your own words, describe what the two numbers in the expression 3^5 mean.

3 is the base number, or the number being multiplied. 5 is the exponent and determines how many times the base number is used as a factor.

3 ACTIVITY: Writing and Analyzing a Math Poem

Work with a partner.

a. Write your own "St. Ives" poem.

Check students' work.

b. Draw pictures for your poem.

Check students' work.

c. Answer the question in your poem.

Check students' work.

d. Show how you can use exponents to write your answer.

Check students' work.

What Is Your Answer?

4. **IN YOUR OWN WORDS** How can you use repeated factors in real-life situations? Give an example.

Sample answer: **Real-life situations use repeated factors when something is multiplied by the same amount each time.**

5. **STRUCTURE** Use exponents to complete the table. Describe the pattern.

10	100	1000	10,000	100,000	1,000,000
10^1	10^2	10^3	10^4	10^5	10^6

Add a zero, increase the exponent by one.

Laurie's Notes

Activity 2

- Explain that in the notation 4^2, the 2 is called an *exponent*. Some students may be familiar with the notation and vocabulary. It will be made formal in the lesson. Discuss the sample with the class before students work with their partners.
- Observe strategies students use for finding the values in the last column. Mental math strategies are important. For instance, in finding the product $3 \times 3 \times 3 \times 3$, many students will find 3×3, then 9×3, and then use paper and pencil to find 27×3. Show students that they can also do the following:

$$3 \times 3 \times 3 \times 3$$
$$9 \times 9$$
$$81$$

- **MP7 Look for and Make Use of Structure:** Understanding how to represent repeated factors using an exponent requires students to recognize the pattern or structure of the expression. Taking the next step of evaluating the expression also involves understanding the structure and the underlying algebraic properties that allow 3^4 to be evaluated as described above.
- Ask students to describe their strategies for performing the computations.
- In part (h), students may not have precise language to describe what 3^5 means, which is perfectly acceptable at this stage.
- **Common Error:** In describing what 3^5 means, you are not multiplying 3 five times. There are actually only four multiplications performed. The number 3 is written five times, meaning there are five *factors* of 3.

Activity 3

- Give students sufficient time to write and illustrate their poems.
- Have students share their poems aloud. If possible, use a document camera to share their illustrations.

What Is Your Answer?

- In Question 5, point out that for a number with an exponent of 1, the value is the number. There is only 1 factor, which is the number 10.
- In Question 5, students should recognize that the exponent and the number of zeros in the answer are the same. These numbers represent the base 10 place values.

Closure

- You have a dozen boxes of a dozen donuts each, and each donut has a dozen chocolate chips. Use an exponent to write an expression for the total number of chocolate chips. 12^3

2 ACTIVITY: Writing Repeated Factors

Math Practice 8

Repeat Calculations

What patterns do you notice with each problem? How does this help you write exponents?

Work with a partner. Copy and complete the table.

Repeated Factors	Using an Exponent	Value
a. 4×4		
b. 6×6		
c. $10 \times 10 \times 10$		
d. $100 \times 100 \times 100$		
e. $3 \times 3 \times 3 \times 3$		
f. $4 \times 4 \times 4 \times 4 \times 4$		
g. $2 \times 2 \times 2 \times 2 \times 2 \times 2$		

h. In your own words, describe what the two numbers in the expression 3^5 mean.

3 ACTIVITY: Writing and Analyzing a Math Poem

Work with a partner.

a. Write your own "St. Ives" poem.

b. Draw pictures for your poem.

c. Answer the question in your poem.

d. Show how you can use exponents to write your answer.

What Is Your Answer?

4. IN YOUR OWN WORDS How can you use repeated factors in real-life situations? Give an example.

5. STRUCTURE Use exponents to complete the table. Describe the pattern.

10	100	1000	10,000	100,000	1,000,000
10^1	10^2				

Practice Use what you learned about exponents to complete Exercises 4–6 on page 14.

1.2 Lesson

Check It Out
Lesson Tutorials
BigIdeasMath.com

Key Vocabulary
power, *p. 12*
base, *p. 12*
exponent, *p. 12*
perfect square, *p. 13*

A **power** is a product of repeated factors. The **base** of a power is the repeated factor. The **exponent** of a power indicates the number of times the base is used as a factor.

base → 3; exponent → 4

$3^4 = 3 \cdot 3 \cdot 3 \cdot 3$

power — 3 is used as a factor 4 times.

Power	Words
3^2	Three *squared*, or three to the second
3^3	Three *cubed*, or three to the third
3^4	Three to the fourth

EXAMPLE 1 Writing Expressions as Powers

Math Practice 5
Choose Tools
Why are calculators more efficient when finding the values of expressions involving exponents?

Write each product as a power.

a. $4 \cdot 4 \cdot 4 \cdot 4 \cdot 4$

Because 4 is used as a factor 5 times, its exponent is 5.

So, $4 \cdot 4 \cdot 4 \cdot 4 \cdot 4 = 4^5$.

b. $12 \times 12 \times 12$

Because 12 is used as a factor 3 times, its exponent is 3.

So, $12 \times 12 \times 12 = 12^3$.

On Your Own

Now You're Ready
Exercises 4–12

Write the product as a power.

1. $6 \cdot 6 \cdot 6 \cdot 6 \cdot 6 \cdot 6$

2. $15 \times 15 \times 15 \times 15$

EXAMPLE 2 Finding Values of Powers

Find the value of each power.

a. 7^2

$7^2 = 7 \cdot 7$ — Write as repeated multiplication.

$= 49$ — Simplify.

b. 5^3

$5^3 = 5 \cdot 5 \cdot 5$

$= 125$

Multi-Language Glossary at BigIdeasMath.com

Laurie's Notes

Introduction

Connect

- **Yesterday:** Students explored how to write a product of repeated factors using an exponent. (MP5, MP7)
- **Today:** Students will use formal language to describe a power and look at the specific case of perfect squares.

Motivate

- Hold up a chess or checkers board.
- ? "How many small squares are on each edge?" 8 "How many of each color are on each edge?" 4 "How many small squares are there in all?" 64 "How many are there of each color?" 32 "If you placed two paper clips on each small square, how many paper clips would you need?" 128
- Ask students to think about all of the answers: 4, 8, 32, 64, and 128. Do they have any observations? If necessary, ask them to think about repeatedly multiplying 2 by itself.
- Introduce the vocabulary words *power*, *base*, and *exponent*.
- Explain that the more common way to refer to a power such as 3^2 is "three squared," although "three to the second" is also correct.
- **Connection:** Discuss with students the common attribute shared by the two-dimensional *square* and the three dimensional *cube*. All sides/edges are the same length. In the power 5^3, all of the factors are the same, namely a factor of 5.

Words of Wisdom

- A common error many students make is to multiply the exponent by the base. For example, $3^2 = 9$, not 6. The simpler the problem, the more often they seem to make this error.

Lesson Notes

Example 1

- Note that two different representations are used for multiplication (• and ×).
- Discuss with students the need to be careful. The dot may be mistaken as a decimal point (2.2 versus 2 • 2) and the × may be mistaken as a variable.
- ? "Are there other representations of multiplication you are familiar with?" Students may mention the use of parentheses. $3(4) = 12$

On Your Own

- **Neighbor Check:** Have students work independently and then have their neighbors check their work. Have students discuss any discrepancies.

Example 2

- Students should be able to evaluate some powers using mental math.
- In order to practice vocabulary, ask a volunteer to read the problem and the answer. You should hear, "7 squared is 49" and "5 cubed is 125."

Goal Today's lesson is writing and finding values of **powers.**

Technology for the Teacher

Dynamic Classroom

Lesson Tutorials
Lesson Plans
Answer Presentation Tool

Extra Example 1

Write each product as a power.

a. 6 • 6 • 6 • 6 6^4

b. 14 × 14 × 14 × 14 × 14 14^5

On Your Own

1. 6^6 **2.** 15^4

Extra Example 2

Find the value of each power.

a. 8^3 512

b. 5^4 625

Extra Example 3

Determine whether each number is a perfect square.

a. 50 not a perfect square

b. 9 perfect square

On Your Own

3. 216 **4.** 81

5. 81 **6.** 324

7. perfect square

8. not a perfect square

9. not a perfect square

10. perfect square

Extra Example 4

A baseball diamond is a square with a side length of 90 feet. What is the area of a baseball diamond? 8100 ft^2

On Your Own

11. 576 $in.^2$, 4 ft^2

English Language Learners

Vocabulary

English language learners may stumble over the ordinal number of the exponent. Have students write this table in their notebooks for reference.

Number	Ordinal
1	first
2	second
3	third
4	fourth
5	fifth
6	sixth
7	seventh
8	eighth
9	ninth
10	tenth

Laurie's Notes

Example 3

- As an introduction to the definition of a perfect square, write the following sequence on the board: 1, 4, 9, 16, 25, 36, . . .

? "What are the next three numbers in the sequence?" 49, 64, 81 "What is the pattern?" Square the whole numbers in order. Students will often answer that you add 3, add 5, add 7, and so on. This is also a correct pattern, so you may need to probe further to get students to recognize that the numbers in the sequence are squares of whole numbers.

- **MP5 Use Appropriate Tools Strategically:** Define *perfect square*. Give students a pile of square tiles and ask them to make a larger square. How many tiles were used? The number will always be a perfect square. The square tiles are a tool to help students visualize square numbers.

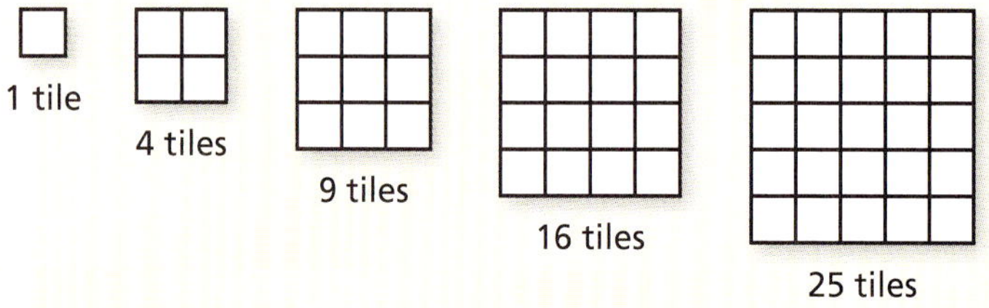

- If you give students 12 tiles and ask them to use all of the tiles to make a square, they will not be able to do so because 12 is not a perfect square.
- Work through each problem.

On Your Own

- Check to see if students used a calculator, paper and pencil, or mental math to answer Questions 3–6. Some students may find 18^2 (= 324) by finding the value of $(20 \times 18) - (2 \times 18)$.

Example 4

- Explain that a verbal model is an equation. It states in words the formula or process that will be used to solve a problem.

? "How do you find the area of a square?" square the side length "Could you find the area of the chess or checkers board? Explain." yes; If you know, or can measure, a side length, square it to find the area.

On Your Own

- If you have another square object in your classroom, have students measure it and find its area.

Closure

? "What powers of 2 are perfect squares?" those with an even exponent: $2^2 = 4$, $2^4 = 16$, $2^6 = 64$, etc.

The square of a whole number is a **perfect square**.

EXAMPLE 3 Identifying Perfect Squares

Determine whether each number is a perfect square.

a. 64

Because $8^2 = 64$, 64 is a perfect square.

b. 20

No whole number squared equals 20. So, 20 is not a perfect square.

On Your Own

Now You're Ready
Exercises 14–21 and 25–32

Find the value of the power.

3. 6^3 **4.** 9^2 **5.** 3^4 **6.** 18^2

Determine whether the number is a perfect square.

7. 25 **8.** 2 **9.** 99 **10.** 100

Remember

The *area* of a figure is the amount of surface it covers. Area is measured in square units.

The area of a square is equal to its side length squared.

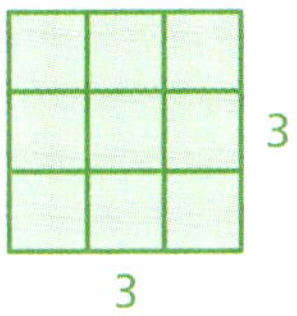

$\text{Area} = 3^2 = 9$ square units

EXAMPLE 4 Real-Life Application

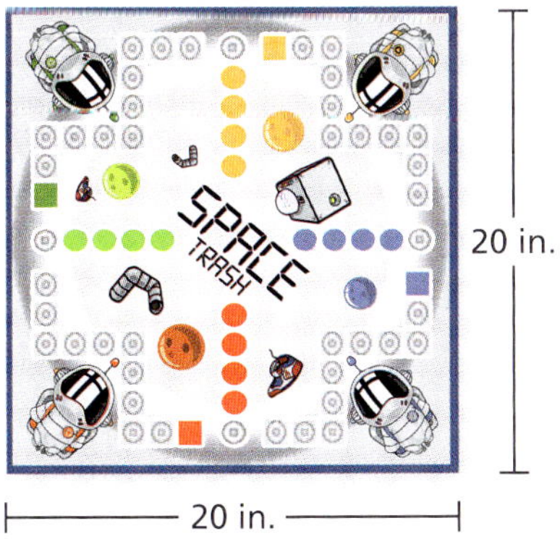

A game board is a square with a side length of 20 inches. What is the area of the game board?

Use a verbal model to solve the problem.

$$\text{area of game board} = (\text{side length})^2$$
$$= 20^2 \quad \text{Substitute 20 for side length.}$$
$$= 400 \quad \text{Multiply.}$$

The area of the game board is 400 square inches.

On Your Own

11. What is the area of the square traffic sign in square inches? in square feet?

1.2 Exercises

Check It Out
Help with Homework
BigIdeasMath.com

Vocabulary and Concept Check

1. **VOCABULARY** How are exponents and powers different?

2. **VOCABULARY** Is 10 a perfect square? Is 100 a perfect square? Explain.

3. **WHICH ONE DOESN'T BELONG?** Which one does *not* belong with the other three? Explain your reasoning.

$2^4 = 2 \times 2 \times 2 \times 2$	$3 + 3 + 3 + 3 = 3(4)$	$3^2 = 3 \times 3$	$5 \cdot 5 \cdot 5 = 5^3$

Practice and Problem Solving

Write the product as a power.

1 **4.** 9×9 **5.** 13×13 **6.** $15 \times 15 \times 15$

7. $2 \cdot 2 \cdot 2 \cdot 2 \cdot 2$ **8.** $14 \times 14 \times 14$ **9.** $8 \cdot 8 \cdot 8 \cdot 8$

10. $11 \times 11 \times 11 \times 11 \times 11$ **11.** $7 \cdot 7 \cdot 7 \cdot 7 \cdot 7 \cdot 7$ **12.** $16 \cdot 16 \cdot 16 \cdot 16$

13. ERROR ANALYSIS Describe and correct the error in writing the product as a power.

✗ $4 \cdot 4 \cdot 4 = 3^4$

Find the value of the power.

2 **14.** 5^2 **15.** 4^3 **16.** 2^5 **17.** 14^2

Use a calculator to find the value of the power.

18. 7^6 **19.** 4^8 **20.** 12^4 **21.** 17^5

22. ERROR ANALYSIS Describe and correct the error in finding the value of the power.

✗ $8^3 = 8 \cdot 3 = 24$

23. POPULATION The population of Virginia is about 8×10^6. About how many people live in Virginia?

24. FIGURINES The smallest figurine in a gift shop is 2 inches tall. The height of each figurine is twice the height of the previous figurine. Write a power to represent the height of the tallest figurine. Then find the height.

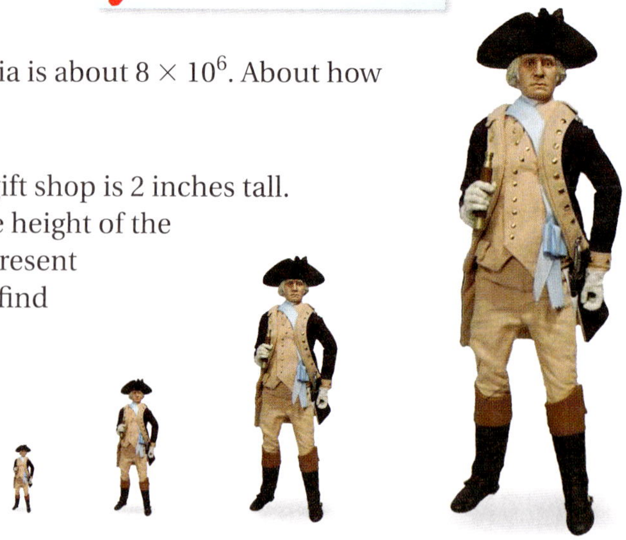

Assignment Guide and Homework Check

Level	Day 1 Activity Assignment	Day 2 Lesson Assignment	Homework Check
Basic	4–6, 40–44	1–3, 7, 13–21 odd, 22, 23, 27, 31	7, 15, 21, 22, 31
Average	4–6, 40–44	1–3, 7–13 odd, 14–24 even, 25–39 odd	7, 16, 22, 31, 37
Advanced	4–6, 40–44	1–3, 8–12 even, 13, 14–24 even, 25–33 odd, 34–39	10, 16, 22, 31, 37

Common Errors

- **Exercises 4–12** Students may miscount the number of factors. Remind them to be careful when counting the number of factors, and that the number of factors is the exponent.
- **Exercises 14–21** Students may make the same mistake that is illustrated in Exercise 22, that is, they may write the exponent as a factor. Again, remind them that the exponent is the number of times the base is used as a factor. You may want to demonstrate this point with a couple of quick examples. For instance, using Exercise 14, point out that $5^2 = 5 \times 5 = 25$ but $5 \times 2 = 10$.
- **Exercise 23** Students may write 8×10^6 as $(8 \times 10)^6$ and come up with a population of 262,144,000,000. Remind students to check the reasonableness of their answers.

1.2 Record and Practice Journal

Write the product as a power.

1. $5 \times 5 \times 5$ — 5^3
2. 13×13 — 13^2
3. $8 \bullet 8 \bullet 8 \bullet 8 \bullet 8 \bullet 8$ — 8^6
4. $12 \bullet 12 \bullet 12 \bullet 12 \bullet 12$ — 12^5
5. $10 \bullet 10 \bullet 10 \bullet 10$ — 10^4
6. $17 \times 17 \times 17$ — 17^3

Find the value of the power.

7. 4^4 — 256
8. 9^3 — 729
9. 24^2 — 576

Determine whether the number is a perfect square.

10. 47 — no
11. 16 — yes
12. 121 — yes

13. You complete 3 centimeters of a necklace in an hour. Each hour after the first, you triple the length of the necklace. Write an expression using exponents for the length of the necklace after 3 hours. Then find the length.

3^3; 27 cm

Vocabulary and Concept Check

1. An exponent indicates the number of times the base is used as a factor. A power is the entire expression (base and exponent). A power is a product of repeated factors.
2. no; yes; 10 is not the square of a whole number, so it is not a perfect square. $100 = 10^2$, so it is a perfect square.
3. $3 + 3 + 3 + 3 = 3(4)$ does not belong because it shows a product as a sum of repeated addends, whereas the other three show powers as products of repeated factors.

Practice and Problem Solving

4. 9^2
5. 13^2
6. 15^3
7. 2^5
8. 14^3
9. 8^4
10. 11^5
11. 7^6
12. 16^4
13. The base is written as the exponent and the exponent is written as the base. $4 \bullet 4 \bullet 4 = 4^3$
14. 25
15. 64
16. 32
17. 196
18. 117,649
19. 65,536
20. 20,736
21. 1,419,857
22. The exponent is written as a factor, but it should have been used to indicate the number of times the base is used as a factor. $8^3 = 8 \bullet 8 \bullet 8 = 512$
23. 8,000,000 people
24. 2^4; 16 in.

Practice and Problem Solving

25. not a perfect square

26. perfect square

27. perfect square

28. not a perfect square

29. perfect square

30. not a perfect square

31. not a perfect square

32. perfect square

33. 40,000 cm^2

34. See *Taking Math Deeper.*

35. 8 squares

36. **a.** 9 by 9 tile arrangement, 10 by 10 tile arrangement, or 11 by 11 tile arrangement

b. in the 9 by 9 arrangement: 44 tiles; in the 10 by 10 arrangement: 25 tiles; in the 11 by 11 arrangement: 4 tiles

37. See Additional Answers.

38. 7 and 8

39. 13 blocks; add $7^2 - 6^2$ blocks; 19 blocks; add $10^2 - 9^2$ blocks; 39 blocks; add $20^2 - 19^2$ blocks

Fair Game Review

40. 84 **41.** 165

42. 8 **43.** 7

44. C

Mini-Assessment

Write the product as a power. Then, find the value of the power.

1. $3 \times 3 \times 3 \times 3 \times 3$ 3^5; 243
2. 2×2 2^2; 4
3. $5 \cdot 5 \cdot 5 \cdot 5$ 5^4; 625
4. $16 \cdot 16$ 16^2; 256
5. The number of students in a school is about 11^3. About how many students are in the school? 1331 students

Taking Math Deeper

Exercise 34

This open-ended question provides an opportunity for answers at different levels.

1 Whole-Number Bases (Use *Guess, Check, and Revise.*)

Exponent 2: Students can discover that $11^2 = 121$.
Exponent 3: Students can discover that $5^3 = 125$.
Exponents 4, 5, and 6: Students can discover that none of the powers fall between 120 and 130.
Exponent 7: Students can discover that $2^7 = 128$.

2 Decimal Bases

Although the lesson is restricted to whole-number bases, it is still reasonable that some students will wonder about the following.
"If $11^2 = 121$, isn't it reasonable that 11.1^2 is just slightly more than 121?"
This type of investigative thinking can introduce the concept of square roots, which students will study in Grade 8.

3 Extension

After students have found the three answers with whole-number bases, ask them whether there are more than three. They can justify their answer using a table.

Exponent \ Base	2	3	4	5	6	7	8	9	10	11	12
2	4	9	16	25	36	49	64	81	100	121	144
3	8	27	64	125	216						
4	16	81	256								
5	32	243									
6	64	729									
7	128										
8	256										

Reteaching and Enrichment Strategies

If students need help. . .	If students got it. . .
Resources by Chapter • Practice A and Practice B • Puzzle Time Record and Practice Journal Practice Differentiating the Lesson Lesson Tutorials Skills Review Handbook	Resources by Chapter • Enrichment and Extension • Technology Connection Start the next section

Determine whether the number is a perfect square.

3 **25.** 8 **26.** 4 **27.** 81 **28.** 44

29. 49 **30.** 125 **31.** 150 **32.** 144

33. **PAINTING** A square painting measures 2 meters on each side. What is the area of the painting in square centimeters?

34. **NUMBER SENSE** Write three powers that have values greater than 120 and less than 130.

35. **CHECKERS** A checkers board has 64 squares. How many squares are in each row?

36. **PATIO** A landscaper has 125 tiles to build a square patio. The patio must have an area of at least 80 square feet.

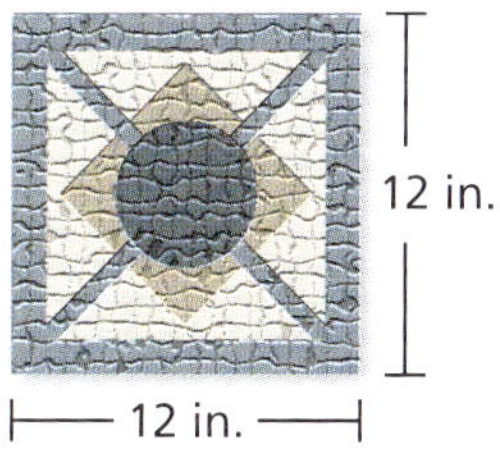

a. What are the possible arrangements for the patio?

b. How many tiles are not used in each arrangement?

37. **PATTERNS** Copy and complete the table. Describe what happens to the value of the power as the exponent decreases. Use this pattern to find the value of 4^0.

Power	4^6	4^5	4^4	4^3	4^2	4^1
Value	4096	1024				

38. **REASONING** Consider the equation $56 = \square^2$. The missing number is between what two whole numbers?

39. Repeated Reasoning How many blocks do you need to add to Square 6 to get Square 7? to Square 9 to get Square 10? to Square 19 to get Square 20? Explain.

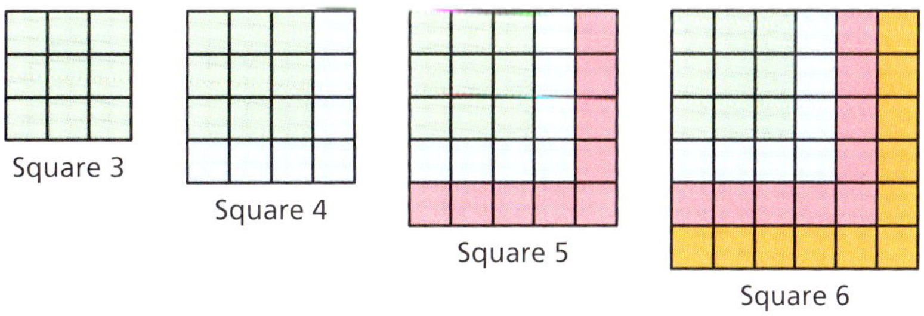

Fair Game Review What you learned in previous grades & lessons

Find the value of the expression. *(Skills Review Handbook)*

40. 6×14 **41.** 11×15 **42.** $56 \div 7$ **43.** $112 \div 16$

44. **MULTIPLE CHOICE** You buy a box of sugar-free gum that has 12 packs. Each pack has 5 pieces. Which expression represents the total number of pieces of gum? *(Skills Review Handbook)*

(A) $12 + 5$ (B) $12 - 5$ (C) 12×5 (D) $12 \div 5$

1.3 Order of Operations

Essential Question What is the effect of inserting parentheses into a numerical expression?

1 ACTIVITY: Comparing Different Orders

Work with a partner. Find the value of the expression by using different orders of operations. Are your answers the same? (Circle *yes* or *no*.)

a. Add, then multiply. $3 + 4 \times 2 =$ ____ Multiply, then add. $3 + 4 \times 2 =$ ____ Same? Yes No

b. Add, then subtract. $5 + 3 - 1 =$ ____ Subtract, then add. $5 + 3 - 1 =$ ____ Same? Yes No

c. Divide, then multiply. $12 \div 3 \cdot 2 =$ ____ Multiply, then divide. $12 \div 3 \cdot 2 =$ ____ Same? Yes No

d. Divide, then add. $16 \div 4 + 4 =$ ____ Add, then divide. $16 \div 4 + 4 =$ ____ Same? Yes No

e. Multiply, then subtract. $8 \times 4 - 2 =$ ____ Subtract, then multiply. $8 \times 4 - 2 =$ ____ Same? Yes No

f. Multiply, then divide. $8 \cdot 4 \div 2 =$ ____ Divide, then multiply. $8 \cdot 4 \div 2 =$ ____ Same? Yes No

g. Subtract, then add. $13 - 4 + 6 =$ ____ Add, then subtract. $13 - 4 + 6 =$ ____ Same? Yes No

h. Multiply, then add. $1 \times 2 + 3 =$ ____ Add, then multiply. $1 \times 2 + 3 =$ ____ Same? Yes No

COMMON CORE

Numerical Expressions

In this lesson, you will

- evaluate numerical expressions with whole-number exponents.

Learning Standard 6.EE.1

Laurie's Notes

Introduction

Standards for Mathematical Practice

- **MP6 Attend to Precision:** When working with even simple expressions such as $8 - 2 + 4$, the order of operations must be used. Students may incorrectly reason that the order in which operations are performed does not matter when it is a *simple* problem.

Motivate

- **Model:** Place two calculators, one scientific and one non-scientific, under a document camera or give the calculators to a student. Press the same sequence of keys, $20 - 8 \times 2$ and then enter (or $=$), on each calculator.

❓ "What is the answer on each calculator?" The scientific calculator should display 4 and the non-scientific calculator should display 24.

❓ "Is it okay to have different answers to the same question?" No. The correct answer is 4. To obtain the correct answer on the non-scientific calculator, perform the multiplication separately and then subtract the result from 20.

Discuss

- The non-scientific calculator performs the operations in order from left to right. The scientific calculator uses the agreed upon *order of operations* and performs the multiplication before the subtraction.

Activity Notes

Activity 1

- Introduce the activity and have students work with partners to complete it.

❓ When students have finished, ask, "Did the order in which you performed the operations matter?" Order mattered in all but parts (b), (f), and (h).

- Ask students whether they were able to predict when the order was going to matter and when it was not. They may not be able to predict at this point.
- If time permits, have students write expressions involving two different operations where the order matters and where the order does not matter. Students will recognize that it is more challenging to write expressions where the order does not matter.

Common Core State Standards

6.EE.1 Write and evaluate numerical expressions involving whole-number exponents.

Previous Learning

In Grade 5, students evaluated numerical expressions using the order of operations (no exponents).

Technology for the Teacher

Dynamic Classroom

Lesson Plans
Complete Materials List

1.3 Record and Practice Journal

Essential Question What is the effect of inserting parentheses into a numerical expression?

1 ACTIVITY: Comparing Different Orders

Work with a partner. Find the value of the expression by using different orders of operations. Are your answers the same? (Circle *yes* or *no*.)

a.	Add, then multiply. $3 + 4 \times 2 = $ **14**	Multiply, then add. $3 + 4 \times 2 = $ **11**	Same? Yes (No)
b.	Add, then subtract. $5 + 3 - 1 = $ **7**	Subtract, then add. $5 + 3 - 1 = $ **7**	Same? (Yes) No
c.	Divide, then multiply. $12 \div 3 \bullet 2 = $ **8**	Multiply, then divide. $12 \div 3 \bullet 2 = $ **2**	Same? Yes (No)
d.	Divide, then add. $16 \div 4 + 4 = $ **8**	Add, then divide. $16 \div 4 + 4 = $ **2**	Same? Yes (No)
e.	Multiply, then subtract. $8 \times 4 - 2 = $ **30**	Subtract, then multiply. $8 \times 4 - 2 = $ **16**	Same? Yes (No)
f.	Multiply, then divide. $8 \bullet 4 \div 2 = $ **16**	Divide, then multiply. $8 \bullet 4 \div 2 = $ **16**	Same? (Yes) No

Differentiated Instruction

Kinesthetic

Students should understand the order of operations before parentheses (symbols of grouping) are introduced. Write

$5 + 10 \div 5 \times 3 - 1$

on the board and have students evaluate the expression.

$5 + 10 \div 5 \times 3 - 1 = 10$

Then ask a student to add parentheses so that addition is performed first and have students evaluate the new expression.

$(5 + 10) \div 5 \times 3 - 1 = 8$

Next ask a student to add parentheses so that addition and multiplication are performed first and have students evaluate the new expression.

$(5 + 10) \div (5 \times 3) - 1 = 0$

1.3 Record and Practice Journal

g. Subtract, then add. $13 - 4 + 6 =$ **15** Add, then subtract. $13 - 4 + 6 =$ **3** Same? Yes (No)

h. Multiply, then add. $1 \times 2 + 3 =$ **5** Add, then multiply. $1 \times 2 + 3 =$ **5** Same? (Yes) No

2 ACTIVITY: Using Parentheses

Work with a partner. Use all the symbols and numbers to write an expression that has the given value.

	Symbols and Numbers	*Value*	*Expression*
a.	( ), +, ÷, 3, 4, 5	3	$(4 + 5) \div 3$
b.	( ), −, ×, 2, 5, 8	11	$(8 \times 2) - 5$
c.	( ), ×, ÷, 4, 4, 16	16	$16 \times (4 \div 4)$ or $(16 \times 4) \div 4$
d.	( ), −, ÷, 3, 8, 11	1	$(11 - 3) \div 8$ or $(11 - 8) \div 3$
e.	( ), +, ×, 2, 5, 10	70	$(2 + 5) \times 10$

3 ACTIVITY: Reviewing Fractions and Decimals

Work with a partner. Evaluate the expression.

a. $\frac{3}{4} - \left(\frac{1}{4} + \frac{1}{2}\right)$ **0**

b. $\left(\frac{5}{6} - \frac{1}{6}\right) - \frac{1}{12}$ **$\frac{7}{12}$**

c. $7.4 - (3.5 - 3.1)$ **7**

d. $10.4 - (8.6 + 0.9)$ **0.9**

e. $(\$7.23 + \$2.32) - \$5.40$ **$4.15**

f. $\$124.60 - (\$72.41 + \$5.67)$ **$46.52**

What Is Your Answer?

4. In an expression with two or more operations, why is it necessary to agree on an order of operations? Give examples to support your explanation.
It is necessary so that everyone will get the same answer. *Sample answer:* Without an order of operations, $7 + 4 \times 5$ could be 55 or 27.

5. **IN YOUR OWN WORDS** What is the effect of inserting parentheses into a numerical expression?
changes the order of operations

Laurie's Notes

Activity 2

- Explain that when parentheses are used in an expression, the operation(s) within the parentheses are performed first.
- **MP2 Reason Abstractly and Quantitatively:** As students work these problems, they need to reason quantitatively about the ending value and consider what numbers and operations they have to work with. For example, the first problem involves the operations of addition and division. Students should reason that the sum of two of the numbers must be divisible by the third number. Trial and error is not the first strategy—reasoning is.
- Explain the logic puzzle activity. The numbers may be used in any order.
- Have students share their results when all have finished.

Activity 3

- This activity gives students an opportunity to review prior work with fractions and decimals.
- Circulate as students are working these problems. Listen for student understanding and recall of finding sums and differences of fractions and decimals. Make note of which students may need a review of these skills.
- Have students share their results when all have finished.

What Is Your Answer?

- Students should be comfortable with the idea that when evaluating an expression with two or more operations, performing operations in a particular order is necessary.

Closure

- **Writing Prompt:** To evaluate $4 - 3 \times (2 + 1)$, . . . Perform the addition in parentheses, multiply, then subtract.

2 ACTIVITY: Using Parentheses

Work with a partner. Use all the symbols and numbers to write an expression that has the given value.

	Symbols and Numbers	*Value*	*Expression*
a.	( ), +, ÷, 3, 4, 5	3	
b.	( ), −, ×, 2, 5, 8	11	
c.	( ), ×, ÷, 4, 4, 16	16	
d.	( ), −, ÷, 3, 8, 11	1	
e.	( ), +, ×, 2, 5, 10	70	

3 ACTIVITY: Reviewing Fractions and Decimals

Math Practice 2

Use Operations

How do you know which operation to perform first?

Work with a partner. Evaluate the expression.

a. $\frac{3}{4} - \left(\frac{1}{4} + \frac{1}{2}\right) =$

b. $\left(\frac{5}{6} - \frac{1}{6}\right) - \frac{1}{12} =$

c. $7.4 - (3.5 - 3.1) =$

d. $10.4 - (8.6 + 0.9) =$

e. $(\$7.23 + \$2.32) - \$5.40 =$

f. $\$124.60 - (\$72.41 + \$5.67) =$

What Is Your Answer?

4. In an expression with two or more operations, why is it necessary to agree on an order of operations? Give examples to support your explanation.

5. **IN YOUR OWN WORDS** What is the effect of inserting parentheses into a numerical expression?

Practice Use what you learned about the order of operations to complete Exercises 3–5 on page 20.

1.3 Lesson

Check It Out
Lesson Tutorials
BigIdeasMath.com

Key Vocabulary
numerical expression, *p. 18*
evaluate, *p. 18*
order of operations, *p. 18*

A **numerical expression** is an expression that contains only numbers and operations. To **evaluate**, or find the value of, a numerical expression, use a set of rules called the **order of operations**.

Key Idea

Order of Operations

1. Perform operations in **P**arentheses.
2. Evaluate numbers with **E**xponents.
3. **M**ultiply or **D**ivide from left to right.
4. **A**dd or **S**ubtract from left to right.

EXAMPLE 1 Using Order of Operations

a. Evaluate $12 - 2 \times 4$.

$12 - 2 \times 4 = 12 - 8$ — Multiply 2 and 4.

$= 4$ — Subtract 8 from 12.

b. Evaluate $7 + 60 \div (3 \times 5)$.

$7 + 60 \div (3 \times 5) = 7 + 60 \div 15$ — Perform operation in parentheses.

$= 7 + 4$ — Divide 60 by 15.

$= 11$ — Add 7 and 4.

EXAMPLE 2 Using Order of Operations with Exponents

Evaluate $30 \div (7 + 2^3) \times 6$.

Evaluate the power in parentheses first.

$30 \div (7 + 2^3) \times 6 = 30 \div (7 + 8) \times 6$ — Evaluate 2^3.

$= 30 \div 15 \times 6$ — Perform operation in parentheses.

$= 2 \times 6$ — Divide 30 by 15.

$= 12$ — Multiply 2 and 6.

Study Tip

Remember to multiply and divide from left to right. In Example 2, you should divide before multiplying because the division symbol comes first when reading from left to right.

On Your Own

Now You're Ready
Exercises 6–14

Evaluate the expression.

1. $7 \cdot 5 + 3$
2. $(28 - 20) \div 4$
3. $6 \times 15 - 10 \div 2$
4. $6 + 2^4 - 1$
5. $4 \cdot 3^2 + 18 - 9$
6. $16 + (5^2 - 7) \div 3$

Laurie's Notes

Introduction

Connect

- **Yesterday:** Students explored performing the operations in a numerical expression using different orders. (MP2, MP6)
- **Today:** Students will use the order of operations to evaluate a numerical expression.

Motivate

- Write "double" on a sheet of paper and "add 10" on another sheet of paper.
- Have each student choose his or her favorite number and record it. Hold up the two sheets of paper, one at a time, and tell students to perform the operation on the number. For instance, if a student chooses 4:

  double, then add 10 ⟶ 18

  add 10, then double ⟶ 28
- Share with students that for expressions such as $2 \times 4 + 10$ *or* $10 + 4 \times 2$, the multiplication is performed first!

Words of Wisdom

- A common error is that students may forget the "left to right" portion of the rule. For instance, $24 - 10 + 6 = 20$ (not 8).
- You may introduce the common acronym PEMDAS as a memory tool for the order of operations. Remind them again, however, that the left-to-right rule is an important part of the order of operations.

Lesson Notes

Key Idea

- **FYI:** Students should recall the order of operations from Grade 5. It is being extended now to include the use of exponents in numerical expressions.

Example 1

- ? "How many operations are in the expression in part (a)?" 2 "What operation should be performed first?" multiplication
- ? "How many operations are in the expression in part (b)?" 3 "What operation should be performed first?" multiplication "Why?" Perform operations within parentheses first.

Example 2

- ? "How many operations are in this expression?" 4
- Ask a volunteer to explain the order in which the expression should be evaluated.
- **Common Error:** Students will say that 2^3 is 6 because they multiplied the base and the exponent.
- Refer to the Study Tip to reinforce the left-to-right rule.

On Your Own

- Each student should work independently before checking his or her work with a neighbor. Encourage students to show their work instead of trying to evaluate the expressions in their heads.

Goal Today's lesson is using the **order of operations** to **evaluate numerical expressions.**

Technology for the Teacher

Dynamic Classroom

Lesson Tutorials
Lesson Plans
Answer Presentation Tool

English Language Learners

Pair Activity

Encourage English language learners to verbalize the process of evaluating using the order of operations. Give each student an expression to evaluate. Include expressions containing exponents and parentheses. After evaluating the expression, the student should explain his or her solution to a partner.

Extra Example 1

a. Evaluate $16 + 5 \times 2$. 26

b. Evaluate $47 - 5 \times (32 \div 4)$. 7

Extra Example 2

Evaluate $15 \times (12 - 3^2) \div 9$. 5

On Your Own

1. 38	**2.** 2
3. 85	**4.** 21
5. 45	**6.** 22

Extra Example 3

a. Evaluate $10 - 2(1 + 3)$. 2

b. Evaluate $7 + 5(8 - 6) \times 2^3$. 87

Extra Example 4

A group of people visit a museum.

Age	Number of People	Admission Price per Person
65 and older	1	$8
13–64	2	$12
12 and under	4	$4

What is the total admission price? $48

On Your Own

7. 4 **8.** 0

9. 8

10. Cost of 10 spheres + Cost of 6 paint bottles + Cost of 9 rods

$10 \cdot 2 + 6 \cdot 3 + 9 \cdot 1$
$= 20 + 18 + 9 = 47$

Your total cost increases by $3 to $47. You would need one more sphere which costs $2 and one more rod which costs $1.

Laurie's Notes

Example 3

- Discuss different ways multiplication can be represented. Share with students that a new way to represent multiplication is to use parentheses. All of the following represent 3 times 4:

  3×4 $3 \cdot 4$ $3(4)$ $(3)(4)$

- Explain that $3(2 + 7)$ is the same as $3 \times (2 + 7)$. A number written next to the parentheses implies multiplication.
- Work through each part of the example. Before you begin each part, ask a volunteer to identify the operations in the expression.

Example 4

- Ask a volunteer to read the problem.
- Do not skip the step of writing the verbal model because it explains in words how to solve the problem.

? "Would the answer be different if the operations were performed from left to right?" yes

- **FYI:** In this contextual problem, students understand almost intuitively that the 3 products must be found before adding. Strip away the context and write the expression $9 \cdot 2 + 6 \cdot 3 + 8 \cdot 1$ on the board and students are apt to perform the operations left to right without regard to the order of operations.
- **MP6 Attend to Precision:** Attending to precision, with or without a context involved, is what we want to develop in all students.

On Your Own

- Have students identify aloud to a partner the different operations they see in each exercise.
- As they write the solution, have students say the steps aloud.
- Ask volunteers to show their solutions on the board.

Closure

Exit Ticket: Evaluate each expression.

a. $18 + 4 \times 10$ 58 **b.** $12 \div 6 \times 2$ 4 **c.** $5^2 - 20 + 3(24 - 18)$ 23

The symbols $\times$ and $\cdot$ are used to indicate multiplication. You can also use parentheses to indicate multiplication. For example, $3(2 + 7)$ is the same as $3 \times (2 + 7)$.

EXAMPLE 3 Using Order of Operations

a. Evaluate $9 + 7(5 - 2)$.

$9 + 7(5 - 2) = 9 + 7(3)$	Perform operation in parentheses.
$= 9 + 21$	Multiply 7 and 3.
$= 30$	Add 9 and 21.

b. Evaluate $15 - 4(6 + 1) \div 2^2$.

$15 - 4(6 + 1) \div 2^2 = 15 - 4(7) \div 2^2$	Perform operation in parentheses.
$= 15 - 4(7) \div 4$	Evaluate 2^2.
$= 15 - 28 \div 4$	Multiply 4 and 7.
$= 15 - 7$	Divide 28 by 4.
$= 8$	Subtract 7 from 15.

EXAMPLE 4 Real-Life Application

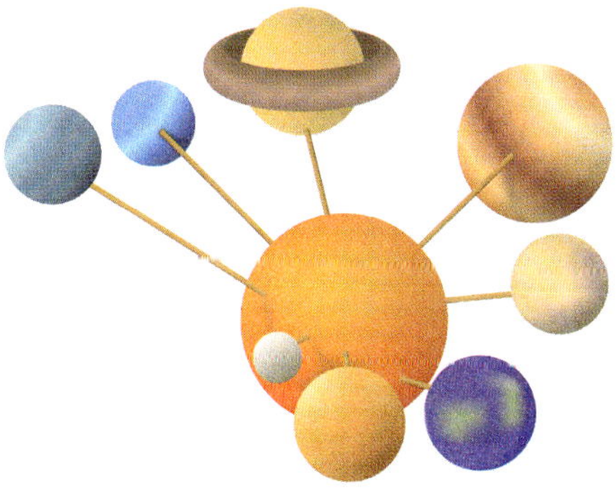

You buy foam spheres, paint bottles, and wooden rods to construct a model of our solar system. What is your total cost?

Item	Quantity	Cost per Item
Spheres	9	\$2
Paint	6	\$3
Rods	8	\$1

Use a verbal model to solve the problem.

cost of 9 spheres + cost of 6 paint bottles + cost of 8 rods

$9 \cdot 2 \quad + \quad 6 \cdot 3 \quad + \quad 8 \cdot 1$

$9 \cdot 2 + 6 \cdot 3 + 8 \cdot 1 = 18 + 18 + 8$	Multiply.
$= 44$	Add.

Your total cost is \$44.

On Your Own

Now You're Ready
Exercises 18–23

Evaluate the expression.

7. $50 + 6(12 \div 4) - 8^2$ **8.** $5^2 - 5(10 - 5)$ **9.** $\dfrac{8(3 + 4)}{7}$

10. WHAT IF? In Example 4, you add the dwarf planet Pluto to your model. Use a verbal model to find your total cost assuming you do not need more paint. Explain.

1.3 Exercises

Check It Out
Help with Homework
BigIdeasMath.com

Vocabulary and Concept Check

1. **WRITING** Why does $12 - 8 \div 2 = 8$, but $(12 - 8) \div 2 = 2$?
2. **REASONING** Describe the steps in evaluating the expression $8 \div (6 - 4) + 3^2$.

Practice and Problem Solving

Find the value of the expression.

3. $(4 \times 15) - 3$
4. $10 - (7 + 1)$
5. $18 \div (6 + 3)$

Evaluate the expression.

1 2
6. $5 + 18 \div 6$
7. $(11 - 3) \div 2 + 1$
8. $45 \div 9 \times 12$
9. $6^2 - 3 \cdot 4$
10. $42 \div (15 - 2^3)$
11. $4^2 \cdot 2 + 8 \cdot 7$
12. $3^2 + 12 \div (6 - 3) \times 8$
13. $(10 + 4) \div (26 - 19)$
14. $(5^2 - 4) \cdot 2 - 18$

ERROR ANALYSIS **Describe and correct the error in evaluating the expression.**

15. ✗ $9 + 2 \times 3 = 11 \times 3$
$= 33$

16. ✗ $19 - 6 + 12 = 19 - 18$
$= 1$

17. **POETRY** You need to read 20 poems in 5 days for an English project. Each poem is 2 pages long. Evaluate the expression $20 \times 2 \div 5$ to find how many pages you need to read each day.

Evaluate the expression.

3
18. $9^2 - 8(6 + 2)$
19. $(3 - 1)^3 + 7(6) - 5^2$
20. $8\left(1\frac{1}{6} + \frac{5}{6}\right) \div 4$
21. $7^2 - 2\left(\frac{11}{8} - \frac{3}{8}\right)$
22. $8(7.3 + 3.7) - 14 \div 2$
23. $2^4(5.2 - 3.2) \div 4$

24. **MONEY** You have four \$10 bills and eighteen \$5 bills in your piggy bank. How much money do you have?

25. **THEATER** Before a show, there are 8 people in a theater. Five groups of 4 people enter, and then three groups of 2 people leave. Evaluate the expression $8 + 5(4) - 3(2)$ to find how many people are in the theater.

4(\$10) + 18(\$5)

Assignment Guide and Homework Check

Level	Day 1 Activity Assignment	Day 2 Lesson Assignment	Homework Check
Basic	3–5, 34–38	1, 2, 7–23 odd, 24–28 even	9, 15, 21, 24, 26
Average	3–5, 34–38	1, 2, 7–15 odd, 19–25 odd, 26–32 even	9, 15, 21, 26, 32
Advanced	3–5, 34–38	1, 2, 6–14 even, 15–25 odd, 26, 28–33	12, 15, 19, 28, 32

Common Errors

- **Exercises 8 and 12** Students may want to evaluate multiplication before division in an expression (or a part of an expression) that contains division before multiplication from left to right. Remind students that multiplication and division are performed from left to right.
- **Exercises 18–23** Students may forget that parentheses can be used to indicate multiplication. Remind them of the various ways to indicate multiplication.
- **Exercises 26–28** Students may be unsure of how to evaluate these expressions. Remind them that the fraction bar means division. If necessary, have students rewrite the expressions using division symbols in place of the fraction bars. Be sure they insert additional parentheses if necessary.
- **Exercise 32** Students may not see the implication that the two groups are working *simultaneously* and they may end up with a time of 88 minutes instead of 44 minutes. Point out that the two groups are working at the same time, so together they are able to clean up 400 yards in 5 minutes.

1.3 Record and Practice Journal

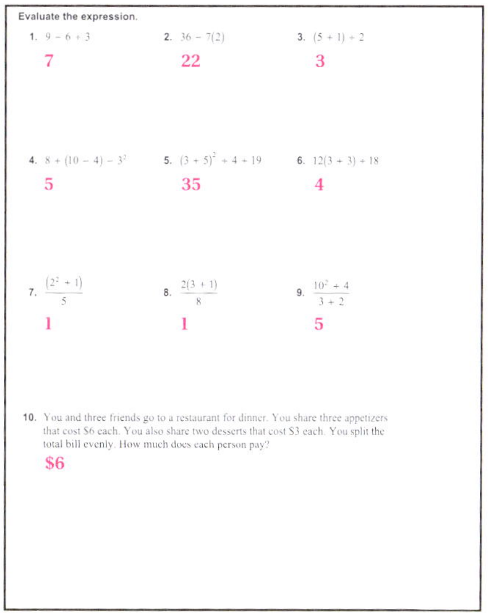
Evaluate the expression.

1. $9 - 6 \div 3$ **7**
2. $36 - 7(2)$ **22**
3. $(5 + 1) \div 2$ **3**
4. $8 + (10 - 4) - 3^2$ **5**
5. $(3 + 5)^2 \div 4 + 19$ **35**
6. $12(3 + 3) \div 18$ **4**
7. $\frac{(2^2 + 1)}{5}$ **1**
8. $\frac{2(3 + 1)}{8}$ **1**
9. $\frac{10^2 \div 4}{3 + 2}$ **5**
10. You and three friends go to a restaurant for dinner. You share three appetizers that cost \$6 each. You also share two desserts that cost \$3 each. You split the total bill evenly. How much does each person pay? **\$6**

Vocabulary and Concept Check

1. Using the order of operations for $12 - 8 \div 2$, you divide 8 by 2 and then subtract the result from 12. Using the order of operations for $(12 - 8) \div 2$, you subtract 8 from 12 and then divide by 2.
2. As illustrated below, perform the operation in parentheses, evaluate 3^2, divide 8 by 2, add 4 and 9.

$$8 \div (6 - 4) + 3^2$$
$$= 8 \div 2 + 3^2$$
$$= 8 \div 2 + 9$$
$$= 4 + 9$$
$$= 13$$

Practice and Problem Solving

3. 57 **4.** 2

5. 2 **6.** 8

7. 5 **8.** 60

9. 24 **10.** 6

11. 88 **12.** 41

13. 2 **14.** 24

15. Addition was performed before multiplication.
$9 + 2 \times 3 = 9 + 6 = 15$

16. Addition was performed before subtraction. These operations should be performed in order from left to right.

$$19 - 6 + 12 = 13 + 12$$
$$= 25$$

17. 8 pages **18.** 17

19. 25 **20.** 4

21. 47 **22.** 81

23. 8 **24.** \$130

25. 22 people

Practice and Problem Solving

26. 12
27. 1
28. 3
29. \$34
30. See *Taking Math Deeper.*
31. \$23; Add the prices of the items you buy. Then subtract the amount of the gift card from the total.
32. 44 min; Two miles is equivalent to 3520 yards. Each group can clean $200 \div 5 = 40$ yards each minute, so together the two groups can clean 80 yards each minute. So, it takes $3520 \div 80 = 44$ minutes to clean 2 miles.
33. **a.** $27 \div 3 + 5 \times 2 = 19$
    b. *Sample answer:* $9^2 + 11 - 8 \times 4 \div 1 = 60$
    c. $5 \times 6 - 15 + 9 = 24$
    d. $14 \times 2 \div 7 - 3 + 9 = 10$

Fair Game Review

34. 5.7
35. 6.1
36. 9.6
37. 0.9
38. D

Mini-Assessment

Evaluate the expression.

1. $4 + 12 \div 3$ 8
2. $20 - 4 \cdot 2^2$ 4
3. $(6^2 - 3) \times (2 + 8)$ 330
4. $4^3 \div 2 - (7 - 5)^2$ 28
5. You have four \$1 bills, three \$5 bills and two \$10 bills in your wallet. How much money do you have in your wallet? \$39

Taking Math Deeper

Exercise 30

As it is, this problem is straightforward. It has several simple solutions. For instance, $1 \times 100 + 1 - 1 \div 1 = 100$. You can make the problem more challenging by restricting the numbers that can be used. Here are some examples.

1 Use only fours.

Restrict the problem so that students can only use the digit 4. With this restriction, can you write an expression that is equal to 100 and still use all four operations (and no parentheses)?

$$\frac{444}{4} - \frac{44}{4} + 4 \times 4 - 4 \times 4 = 111 - 11 + 16 - 16 = 100$$

2 Use only threes.

Restrict the problem so that students can only use the digit 3. Notice that the pattern shown in part (1) can be used for *any* digit.

$$\frac{333}{3} - \frac{33}{3} + 3 \times 3 - 3 \times 3 = 111 - 11 + 9 - 9 = 100$$

3 Here is another way to make the problem more challenging.

(1) Use only 3's.
(2) Use each operation at most once.

$$33 \times 3 + 3 \div 3 = 99 + 1 = 100$$

Reteaching and Enrichment Strategies

If students need help. . .	If students got it. . .
Resources by Chapter • Practice A and Practice B • Puzzle Time Record and Practice Journal Practice Differentiating the Lesson Lesson Tutorials Skills Review Handbook	Resources by Chapter • Enrichment and Extension • Technology Connection Start the next section

Evaluate the expression.

26. $\dfrac{6(3+5)}{4}$ **27.** $\dfrac{12^2 - 4(6) + 1}{11^2}$ **28.** $\dfrac{26 \div 2 + 5}{3^2 - 3}$

29. FIELD TRIP Eighty students are going on a field trip to a history museum. The total cost includes

- 2 bus rentals and
- $10 per student for lunch.

What is the total cost per student?

30. OPEN-ENDED Use all four operations without parentheses to write an expression that has a value of 100.

31. SHOPPING You buy 6 notebooks, 10 folders, 1 pack of pencils, and 1 lunch box for school. After using a $10 gift card, how much do you owe? Explain how you solved the problem.

32. LITTER CLEANUP Two groups collect litter along the side of a road. It takes each group 5 minutes to clean up a 200-yard section. How long does it take to clean up 2 *miles*? Explain how you solved the problem.

33. Number Sense Copy each statement. Insert +, −, ×, or ÷ symbols to make each statement true.

a. $27 \square 3 \square 5 \square 2 = 19$ **b.** $9^2 \square 11 \square 8 \square 4 \square 1 = 60$

c. $5 \square 6 \square 15 \square 9 = 24$ **d.** $14 \square 2 \square 7 \square 3 \square 9 = 10$

Fair Game Review What you learned in previous grades & lessons

Add or subtract. *(Skills Review Handbook)*

34. 5.2 + 0.5 **35.** 8 − 1.9 **36.** 12.6 − 3 **37.** 0.7 + 0.2

38. MULTIPLE CHOICE You are making two recipes. One recipe calls for $2\frac{1}{3}$ cups of flour. The other recipe calls for $1\frac{1}{4}$ cups of flour. How much flour do you need to make both recipes? *(Skills Review Handbook)*

Ⓐ $1\frac{1}{12}$ cups Ⓑ $3\frac{1}{12}$ cups Ⓒ $3\frac{2}{7}$ cups Ⓓ $3\frac{7}{12}$ cups

1 Study Help

Check It Out
Graphic Organizer
BigIdeasMath.com

You can use an **information frame** to help you organize and remember concepts. Here is an example of an information frame for powers.

Words:
A *power* is a product of repeated factors. The *base* of a power is the common factor. The *exponent* of a power indicates the number of times the base is used as a factor.

Numbers:
$5^2 = 5 \cdot 5$

Powers

Numbers:
$4^3 = 4 \cdot 4 \cdot 4$

Example:
Find the value of the power.
$2^4 = 2 \cdot 2 \cdot 2 \cdot 2 = 16$

On Your Own

Make information frames to help you study these topics.

1. adding whole numbers
2. subtracting whole numbers
3. multiplying whole numbers
4. dividing whole numbers
5. order of operations

After you complete this chapter, make information frames for the following topics.

6. prime factorization
7. greatest common factor (GCF)
8. least common multiple (LCM)
9. least common denominator (LCD)

"Dear Mom, I am sending you an information frame card for Mother's Day!"

Sample Answers

1.

Real-Life Application Example:

On a video game, Jacob got 1685 points and earned two bonuses worth 193 and 270 points. What is his total score?

Answer: 2148 points

Key Words:

the sum of,
the total of

Adding whole numbers

Numbers:

```
  12
 1685
  193
+ 270
 2148
```

Estimate and Check:

$1685 + 193 + 270 \approx 1700 + 200 + 300$
$= 2200$

Because $2148 \approx 2200$, the answer is reasonable.

2.

Real-Life Application Example:

On the way to school, the temperature was 58°F. On the way home, it was 72°F. How much did the temperature increase?

Answer: 14°F

Key Words:

the difference of,
how many more,
how many less,
the change in

Subtracting whole numbers

Numbers:

```
 6 12
  72
- 58
  14
```

Estimate and Check:

$72 - 58 \approx 70 - 60$
$= 10$

Because $14 \approx 10$, the answer is reasonable.

3.

Real-Life Application Example:

How many cartons of milk should the cafeteria manager at Riverdale Middle School order so that all of the 136 students can get one carton per day in February (28 days)?

Answer: 3808 cartons

Key Words:

the product of,
total of equal-size groups

Multiplying whole numbers

Numbers:

```
   1
   24
  136
 x 28
 1088
+272
 3808
```

Estimate and Check:

$136 \times 28 \approx 140 \times 30$
$= 4200$

Because $3808 \approx 4200$, the answer is reasonable.

4–5. Available at *BigIdeasMath.com*.

List of Organizers

Available at *BigIdeasMath.com*

Comparison Chart
Concept Circle
Definition (Idea) and Example Chart
Example and Non-Example Chart
Formula Triangle
Four Square
Information Frame
Information Wheel
Notetaking Organizer
Process Diagram
Summary Triangle
Word Magnet
Y Chart

About this Organizer

An **Information Frame** can be used to help students organize and remember concepts. Students write the topic in the middle rectangle. Then students write related concepts in the spaces around the rectangle. Related concepts can include *Words, Numbers, Algebra, Example, Definition, Non-Example, Visual, Procedure, Details,* and *Vocabulary.* Students can place their information frames on note cards to use as a quick study reference.

Technology for the Teacher

Editable Graphic Organizer

Answers

1. 8161
2. 2703
3. 13,524
4. 9
5. 27
6. 121
7. perfect square
8. not a perfect square
9. 9
10. 3
11. 6
12. 4
13. 16 seats
14. 225 in.^{2}
15. 3 mi
16. $38

Technology for the Teacher

Online Assessment
Assessment Book
ExamView® Assessment Suite

Alternative Quiz Ideas

100% Quiz	Math Log
Error Notebook	Notebook Quiz
Group Quiz	**Partner Quiz**
Homework Quiz	Pass the Paper

Partner Quiz

- Partner quizzes are to be completed by students working in pairs. Student pairs can be selected by the teacher, by students, through a random process, or any way that works for your class.
- Students are permitted to use their notebooks and other appropriate materials.
- Each pair submits a draft of the quiz for teacher feedback. Then they revise their work, and turn it in for a grade.
- When the pair is finished they can submit one paper, or each can submit their own.
- Teachers can give feedback in a variety of ways. It is important that the teacher does not reteach or provide the solution. The teacher can tell students which questions they have answered correctly, if they are on the right track, or if they need to rethink a problem.

Reteaching and Enrichment Strategies

If students need help. . .	If students got it. . .
Resources by Chapter • Practice A and Practice B • Puzzle Time Lesson Tutorials *BigIdeasMath.com*	Resources by Chapter • Enrichment and Extension • Technology Connection Game Closet at *BigIdeasMath.com* Start the next section

1.1–1.3 Quiz

Check It Out
Progress Check
BigIdeasMath.com

Find the value of the expression. Use estimation to check your answer. *(Section 1.1)*

1. $4265 + 3896$	**2.** $5327 - 2624$
3. 276×49	**4.** $648 \div 72$

Find the value of the power. *(Section 1.2)*

5. 3^3	**6.** 11^2

Determine whether the number is a perfect square. *(Section 1.2)*

7. 36	**8.** 15

Evaluate the expression. *(Section 1.3)*

9. $6 + 21 \div 7$	**10.** $\dfrac{4(12 - 3)}{12}$
11. $16 \div 2^3 + 6 - 2$	**12.** $2 \times 14 \div (3^2 - 2)$

13. AUDITORIUM An auditorium has a total of 592 seats. There are 37 rows of seats, and each row has the same number of seats. How many seats are there in a single row? *(Section 1.1)*

14. SOFTBALL The bases on a softball field are square. What is the area of each base? *(Section 1.2)*

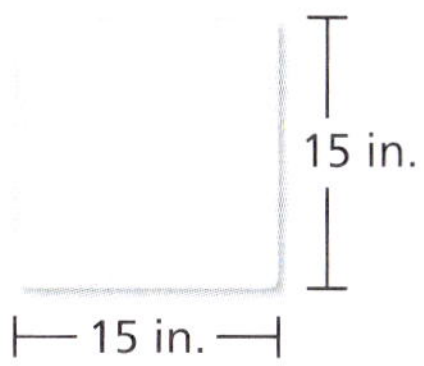

15. DUATHLON In an 18-mile duathlon, you run, then bike 12 miles, and then run again. The two runs are the same distance. Find the distance of each run. *(Section 1.3)*

16. AMUSEMENT PARK Tickets for an amusement park cost \$10 for adults and \$6 for children. Find the total cost for 2 adults and 3 children. *(Section 1.3)*

1.4 Prime Factorization

Essential Question Without dividing, how can you tell when a number is divisible by another number?

1 ACTIVITY: Finding Divisibility Rules for 2, 3, 5, and 10

Work with a partner. Copy the set of numbers (1–50) as shown.

1	2	3	4	5	6	7	8	9	10
11	12	13	14	15	16	17	18	19	20
21	22	23	24	25	26	27	28	29	30
31	32	33	34	35	36	37	38	39	40
41	42	43	44	45	46	47	48	49	50

a. Highlight all the numbers that are divisible by 2.

b. Put a box around the numbers that are divisible by 3.

c. Underline the numbers that are divisible by 5.

d. Circle the numbers that are divisible by 10.

e. **STRUCTURE** In parts (a)–(d), what patterns do you notice? Write four rules to determine when a number is divisible by 2, 3, 5, and 10.

Common Core

Common Factors and Multiples

In this lesson, you will
- use divisibility rules to find prime factorizations of numbers.

Preparing for Standard 6.NS.4

2 ACTIVITY: Finding Divisibility Rules for 6 and 9

Work with a partner.

a. List ten numbers that are divisible by 6. Write a rule to determine when a number is divisible by 6. Use a calculator to check your rule with large numbers.

b. List ten numbers that are divisible by 9. Write a rule to determine when a number is divisible by 9. Use a calculator to check your rule with large numbers.

Laurie's Notes

Introduction

Standards for Mathematical Practice

- **MP4 Model with Mathematics** and **MP2 Reason Abstractly and Quantitatively:** Writing a factor pair is the first step in finding the prime factorization of a number. The possible factor pairs are related to the dimensions of the possible rectangles with a fixed area.

Motivate

- ? Write the number 2520 on the board. "What do you think is special about 2520?" Students may state that it is even, or it ends in 0.
- ? "Because 2520 ends in 0, what number divides into it evenly?" 10
- ? "Because 2520 is an even number, what number divides into it evenly?" 2
- ? "What other numbers do you think divide into 2520 evenly?"
- Students may guess correctly that 2520 is the least number divisible by all of the numbers 1–10.

Discuss

- Review definition of *divisible.* A number is divisible by another number if the second number is a factor of the first number.

Activity Notes

Activity 1

- Tell students that they will be looking for patterns today. If they do not have a highlighter, suggest that they lightly shade over the number.
- When students have finished, ask volunteers to share their observations.
- Students will likely note that all of the numbers divisible by 2 appear in columns, as do the numbers divisible by 5 and by 10. This is a function of how the table was set up. This would not be true if the width of the first row was 9. In this case, the numbers divisible by 3 would be in column.
- You want students to focus on the numbers and not the position of the numbers in the table.
- The divisibility rule for 3 will likely be the most challenging. Students will see that the boxed numbers appear on a diagonal. Prompt with leading questions as needed.
- Check to see that all four rules are stated correctly.

Activity 2

- Help students to think about rules for 6 and 9. You might ask, "If a number is divisible by 10, what other numbers is it also divisible by? Explain."
- **MP3a Construct Viable Arguments:** Listen for students to give viable arguments of why a number divisible by 10 is also divisible by 2 and 5. This will help them think about the divisibility rule for 6.
- Check to see that the rules are stated correctly.

Common Core State Standards

6.NS.4 Find the greatest common factor of two whole numbers less than or equal to 100 and the least common multiple of two whole numbers less than or equal to 12. Use the distributive property to express a sum of two whole numbers 1–100 with a common factor as a multiple of a sum of two whole numbers with no common factor.

Previous Learning

Students need to be familiar with dividing whole numbers and with the meanings of *prime* and *composite numbers.*

Technology for the Teacher

Dynamic Classroom

Lesson Plans
Complete Materials List

1.4 Record and Practice Journal

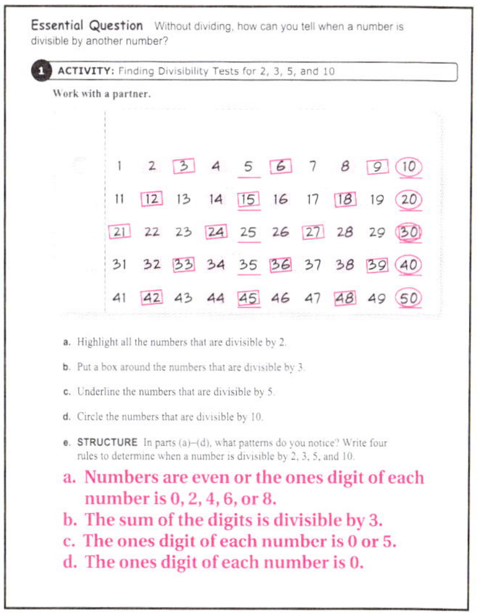

Essential Question Without dividing, how can you tell when a number is divisible by another number?

1 ACTIVITY: Finding Divisibility Tests for 2, 3, 5, and 10

Work with a partner.

1	2	3	4	5	6	7	8	9	10
11	12	13	14	15	16	17	18	19	20
21	22	23	24	25	26	27	28	29	30
31	32	33	34	35	36	37	38	39	40
41	42	43	44	45	46	47	48	49	50

a. Highlight all the numbers that are divisible by 2.

b. Put a box around the numbers that are divisible by 3.

c. Underline the numbers that are divisible by 5.

d. Circle the numbers that are divisible by 10.

e. **STRUCTURE** In parts (a)–(d), what patterns do you notice? Write four rules to determine when a number is divisible by 2, 3, 5, and 10.

a. Numbers are even or the ones digit of each number is 0, 2, 4, 6, or 8.
b. The sum of the digits is divisible by 3.
c. The ones digit of each number is 0 or 5.
d. The ones digit of each number is 0.

English Language Learners

Graphic Organizer

Have students organize the divisibility rules in a table. In the first column, list the divisor. In the second column, list the rule. In the third column, list examples of numbers that are divisible by the divisor. Being organized saves time, allowing for more time to work on language skills.

1.4 Record and Practice Journal

2 ACTIVITY: Finding Divisibility Rules for 6 and 9

Work with a partner.

a. List ten numbers that are divisible by 6. Write a rule to determine when a number is divisible by 6. Use a calculator to check your rule with large numbers.

Sample answer: **6, 12, 18, 24, 30, 36, 42, 48, 54, 60; The numbers are even and divisible by 3.**

b. List ten numbers that are divisible by 9. Write a rule to determine when a number is divisible by 9. Use a calculator to check your rule with large numbers.

Sample answer: **9, 18, 27, 36, 45, 54, 63, 72, 81, 90; The sum of the digits is divisible by 9.**

3 ACTIVITY: Rewriting a Number Using 2s, 3s, and 5s

Work with three other students. Use the following rules and only the prime factors 2, 3, and 5 to write each number on the next page as a product.

- Your group should have four sets of cards: a set with all 2s, a set with all 3s, a set with all 5s, and a set of blank cards. Each person gets one set of cards.*
- Begin by choosing two cards to represent the given number as a product of two factors. The person with the blank cards writes any factors that are not 2, 3, or 5.
- Use the cards again to represent any number written on a blank card as a product of two factors. Continue until you have represented each handwritten card as a product of two prime factors.
- You may use only one blank card for each step.

*Cut-outs are available in the back of the Record and Practice Journal.

a. 108

$2 \cdot 2 \cdot 3 \cdot 3 \cdot 3$

b. 80

$2 \cdot 2 \cdot 2 \cdot 2 \cdot 5$

c. 162

$2 \cdot 3 \cdot 3 \cdot 3 \cdot 3$

d. 300

$2 \cdot 2 \cdot 3 \cdot 5 \cdot 5$

e. Compare your results with those of other groups. Are your steps the same for each number? Is your final answer the same for each number?

The steps could differ among groups, but final results should be the same.

What Is Your Answer?

4. **IN YOUR OWN WORDS** Without dividing, how can you tell when a number is divisible by another number? Give examples to support your explanation.

Use divisibility rules.

5. Explain how you can use your divisibility rules from Activities 1 and 2 to help with Activity 3.

They can help you determine whether 2, 3, or 5 are factors of each number in Activity 3.

Laurie's Notes

Activity 3

- Give time for students to form groups of 4.
- Hand out the cards and ask a volunteer to read the directions. Be clear that the cards students have are all 2s, all 3s, all 5s, or all blank cards. No student should have a mix of numbers or blank cards. In the example shown, note that the number 5 is not used. Students should not expect to use all of their cards.
- The last step is important, writing the number as the product of all the factors they found.
- If time is a constraint, you might consider doing this activity at the board by taping numbers to the board.
- Discuss the results of the activity. The results should be the same for all groups, but the steps they took in getting there likely will vary. For instance, in part (d) the first step might be 2×150, 3×100, or 5×60.
- You could suggest additional numbers for students to factor if time allows.

What Is Your Answer?

- Listen for an answer involving the divisibility rules.
- **Extension:** Students may ask about divisibility rules for other numbers such as 4 and 8. A number is divisible by 4 if the last two digits are divisible by 4. A number is divisible by 8 if the last three digits are divisible by 8.

Closure

- Write a 3-digit number that is divisible by 3 and 10. Explain why you are correct. *Sample answer:* 510; The sum of the digits is divisible by 3, and the number ends in 0.
- Write a 3-digit number that is divisible by 2 and 3 but not 5. Explain why you are correct. *Sample answer:* 768; The number is even, the sum of the digits is divisible by 3, and the number does not end in 5 or 0.

3 ACTIVITY: Rewriting a Number Using 2s, 3s, and 5s

Work with three other students. Use the following rules and only the prime factors 2, 3, and 5 to write each number below as a product.

- Your group should have four sets of cards: a set with all 2s, a set with all 3s, a set with all 5s, and a set of blank cards. Each person gets one set of cards.
- Begin by choosing two cards to represent the given number as a product of two factors. The person with the blank cards writes any factors that are not 2, 3, or 5.
- Use the cards again to represent any number written on a blank card as a product of two factors. Continue until you have represented each handwritten card as a product of two prime factors.
- You may use only one blank card for each step.

a. **Sample:** 108

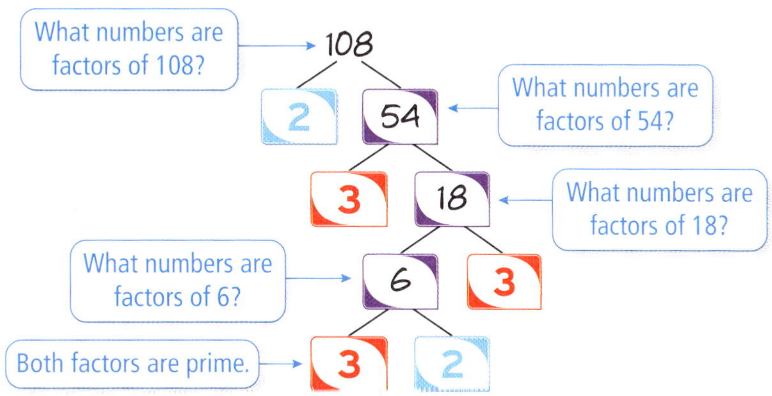

∴ $108 = 2 \cdot 3 \cdot 3 \cdot 3 \cdot 2$

b. 80 **c.** 162 **d.** 300

e. Compare your results with those of other groups. Are your steps the same for each number? Is your final answer the same for each number?

Math Practice 4

Interpret Results

How do you know your answer makes sense?

What Is Your Answer?

4. **IN YOUR OWN WORDS** Without dividing, how can you tell when a number is divisible by another number? Give examples to support your explanation.

5. Explain how you can use your divisibility rules from Activities 1 and 2 to help with Activity 3.

Practice Use what you learned about divisibility rules to complete Exercises 4–7 on page 28.

Check It Out
Lesson Tutorials
BigIdeasMath.com

Because 2 is factor of 10 and $2 \cdot 5 = 10$, 5 is also a factor of 10. The pair 2, 5 is called a **factor pair** of 10.

Key Vocabulary
factor pair, *p. 26*
prime factorization, *p. 26*
factor tree, *p. 26*

EXAMPLE 1 Finding Factor Pairs

The brass section of a marching band has 30 members. The band director arranges the brass section in rows. Each row has the same number of members. How many possible arrangements are there?

Use the factor pairs of 30 to find the number of arrangements.

$30 = 1 \cdot 30$	There could be 1 row of 30 or 30 rows of 1.
$30 = 2 \cdot 15$	There could be 2 rows of 15 or 15 rows of 2.
$30 = 3 \cdot 10$	There could be 3 rows of 10 or 10 rows of 3.
$30 = 5 \cdot 6$	There could be 5 rows of 6 or 6 rows of 5.
$30 = 6 \cdot 5$	The factors 5 and 6 are already listed.

∴ There are 8 possible arrangements: 1 row of 30, 30 rows of 1, 2 rows of 15, 15 rows of 2, 3 rows of 10, 10 rows of 3, 5 rows of 6, or 6 rows of 5.

Study Tip
When making an organized list of factor pairs, stop finding pairs when the factors begin to repeat.

On Your Own

Now You're Ready
Exercises 8–15

List the factor pairs of the number.

1. 18 **2.** 24 **3.** 51

4. WHAT IF? The woodwinds section of the marching band has 38 members. Which has more possible arrangements, the brass section or the woodwinds section? Explain.

Key Idea

Prime Factorization

The **prime factorization** of a composite number is the number written as a product of its prime factors.

You can use factor pairs and a **factor tree** to help find the prime factorization of a number. The factor tree is complete when only prime factors appear in the product. A factor tree for 60 is shown.

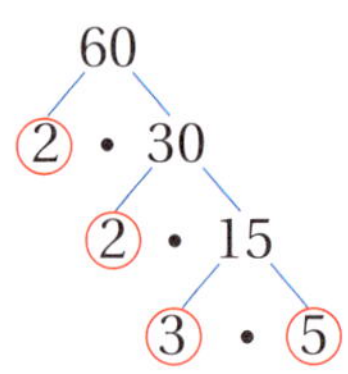

$60 = 2 \cdot 2 \cdot 3 \cdot 5$, or $2^2 \cdot 3 \cdot 5$

Remember
A *prime number* is a whole number greater than 1 with exactly two factors, 1 and itself. A *composite number* is a whole number greater than 1 with factors other than 1 and itself.

Multi-Language Glossary at BigIdeasMath.com

Laurie's Notes

Introduction

Connect

- **Yesterday:** Students wrote divisibility rules for 2, 3, 5, 6, 9, and 10. (MP2, MP3a, MP4)
- **Today:** Students will use the divisibility rules to help them write the prime factorization of numbers.

Motivate

? "Do you know someone who has ordered an item over the Internet?" Most students will say yes.

- In order to make the transactions safe, credit card numbers are encrypted.
- To encrypt a number, a company makes available (in a public key) a large number (400 digits) that is the product of two very large prime numbers.
- The sender uses that number to encrypt the message to the company.
- Only the company knows the factors needed to decrypt the message.
- While it is easy to multiply two very large prime numbers, it is nearly impossible to find the two original primes in a timely manner. There is no fast algorithm for prime factorization.
- Today students will learn how to find the prime factorization of a number.

Lesson Notes

Example 1

- Define factor pairs and give an example.
- Work through the example as shown, pointing out that when the factors begin to repeat, you are done, as explained in the Study Tip.
- **Visual model:** You could use 30 square tiles and ask how to form a rectangle using all of the tiles. The dimensions of the rectangles are the factor pairs found in this example.
- In the context of the problem, 5 • 6 is a factor pair and it can be interpreted two ways: 5 rows of 6 or 6 rows of 5. It is still one factor pair.

On Your Own

- **Neighbor Check:** Have students work independently, and then have their neighbors check their work. Have students discuss any discrepancies.
- **MP2 Reason Abstractly and Quantitatively:** Note even though 38 > 30, 38 has fewer factor pairs than 30. Students should reason that the size of the number is not what determines the number of factor pairs.

Key Idea

? "What is a composite number?" a number greater than one that has more than two factors

- Write the definition of prime factorization.
- Work through the example shown, using the vocabulary: factors, prime, and composite.

Goal Today's lesson is writing the **prime factorization** of a number.

Technology for the Teacher

Dynamic Classroom

Lesson Tutorials
Lesson Plans
Answer Presentation Tool

Extra Example 1

There are 40 members in the book club at school. Each member sits at a desk. Each row needs to have the same number of desks. How many possible arrangements are there? 8

On Your Own

1. 1, 18; 2, 9; 3, 6
2. 1, 24; 2, 12; 3, 8; 4, 6
3. 1, 51; 3, 17
4. brass section; The number 38 has only two factor pairs: 1, 38 and 2, 19. So, there are only 4 possible arrangements with the woodwinds section: 1 row of 38, 38 rows of 1, 2 rows of 19, or 19 rows of 2.

Extra Example 2

Write the prime factorization of 45.
$3^2 \cdot 5$

Extra Example 3

What is the greatest perfect square that is a factor of 675? 225

On Your Own

5. $2^2 \cdot 5$
6. $2^3 \cdot 11$
7. $2 \cdot 3^2 \cdot 5$
8. $2 \cdot 3 \cdot 7 \cdot 11$
9. 36; $396 = 2 \cdot 2 \cdot 3 \cdot 3 \cdot 11$ and $2 \cdot 3 \cdot 2 \cdot 3 = 6 \cdot 6 = 36$

Differentiated Instruction

Kinesthetic

Have students work in groups to find the prime factorization of 1575 (Example 3) using a factor tree. Give each group a different pair of factors, such as 3×525, 5×315, 7×225, 9×175, and so on. Have students compare their results. They should see that the prime factorization is the same, even when starting with different factor pairs.

Laurie's Notes

Example 2

- Ask students to name a factor pair for the number 48. This should give at least two different ways in which the problem can be started.
- Work through two different versions of a prime factor tree for 48. As the Study Tip states, it is important for students to understand that the steps may be different, but the final factorization will be the same.
- ? "Why is $2 \cdot 2 \cdot 3 \cdot 2 \cdot 2$ the same as $3 \cdot 2 \cdot 2 \cdot 2 \cdot 2$?" Commutative Property of Multiplication
- ? "Why isn't 1 listed in the prime factorization of a number?" 1 is neither prime nor composite.

Example 3

- Students will need to find the prime factorization of 1575 in order to answer the question.

On Your Own

- **Common Error:** In Question 5, students may begin to write a factor pair of 10 and 10. It is easy for them to forget that factor pairs must multiply to give the answer, not add.
- In Question 8, students should be encouraged to use their divisibility rules to find the largest factor they can, using mental math, versus starting with $2 \cdot 231$.

Closure

- **Exit Ticket:** The class ended and a student didn't finish finding the prime factorization. The student's first step was 8×24.
  a) What was the original number? 192
  b) Finish finding the prime factorization. $2 \times 2 \times 2 \times 2 \times 2 \times 2 \times 3$

EXAMPLE 2 Writing a Prime Factorization

Write the prime factorization of 48.

Choose any factor pair of 48 to begin the factor tree.

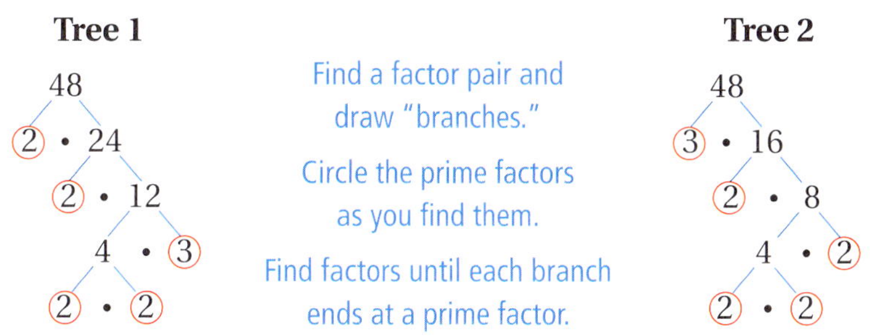

$48 = 2 \cdot 2 \cdot 3 \cdot 2 \cdot 2$ $\qquad$ $48 = 3 \cdot 2 \cdot 2 \cdot 2 \cdot 2$

The prime factorization of 48 is $2 \cdot 2 \cdot 2 \cdot 2 \cdot 3$, or $2^4 \cdot 3$.

Study Tip

Notice that beginning with different factor pairs results in the same prime factorization. Every composite number has only one prime factorization.

EXAMPLE 3 Using a Prime Factorization

What is the greatest perfect square that is a factor of 1575?

Because 1575 has many factors, it is not efficient to list all of its factors and check for perfect squares. Use the prime factorization of 1575 to find any perfect squares that are factors.

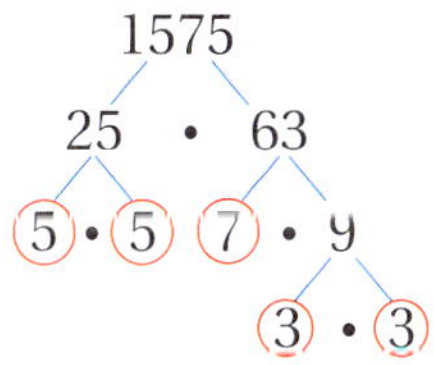

$1575 = 3 \cdot 3 \cdot 5 \cdot 5 \cdot 7$

The prime factorization shows that 1575 has three factors other than 1 that are perfect squares.

$3 \cdot 3 = 9$ $\qquad$ $5 \cdot 5 = 25$ $\qquad$ $(3 \cdot 5) \cdot (3 \cdot 5) = 15 \cdot 15 = 225$

So, the greatest perfect square that is a factor of 1575 is 225.

On Your Own

Now You're Ready
Exercises 16–23 and 29–32

Write the prime factorization of the number.

5. 20 $\qquad$ **6.** 88 $\qquad$ **7.** 90 $\qquad$ **8.** 462

9. What is the greatest perfect square that is a factor of 396? Explain.

1.4 Exercises

Check It Out
Help with Homework
BigIdeasMath.com

Vocabulary and Concept Check

1. **VOCABULARY** What is the prime factorization of a number?
2. **VOCABULARY** How can you use a factor tree to help you write the prime factorization of a number?
3. **WHICH ONE DOESN'T BELONG?** Which factor pair does not belong with the other three? Explain your reasoning.

   2, 28   4, 14   6, 9   7, 8

Practice and Problem Solving

Use divisibility rules to determine whether the number is divisible by 2, 3, 5, 6, 9, and 10. Use a calculator to check your answer.

4. 1044   5. 1485   6. 1620   7. 1709

List the factor pairs of the number.

1   8. 15   9. 22   10. 34   11. 39

12. 45   13. 54   14. 59   15. 61

Write the prime factorization of the number.

2   16. 16   17. 25   18. 30   19. 26

20. 84   21. 54   22. 65   23. 77

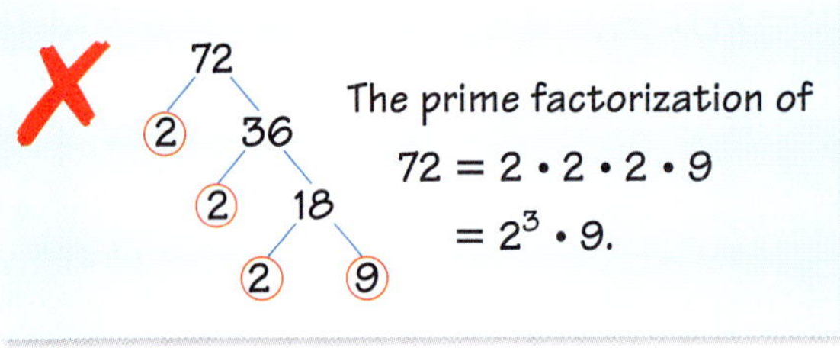

24. **ERROR ANALYSIS** Describe and correct the error in writing the prime factorization.

25. **FACTOR RAINBOW** You can use a factor rainbow to check whether a list of factors is correct. To create a factor rainbow, list the factors of a number in order from least to greatest. Then draw arches that link the factor pairs. For perfect squares, there is no connecting arch in the middle. So, just circle the middle number. A factor rainbow for 12 is shown. Create factor rainbows for 6, 24, 36, and 48.

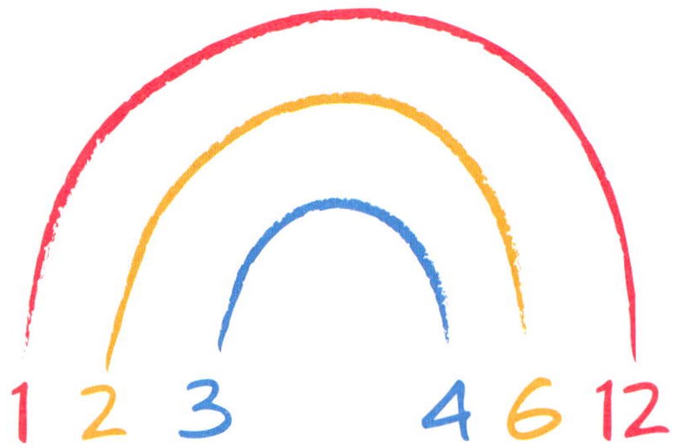

Assignment Guide and Homework Check

Level	Day 1 Activity Assignment	Day 2 Lesson Assignment	Homework Check
Basic	4–7, 40–44	1–3, 9–21 odd, 24, 25	9, 13, 17, 19, 21
Average	4–7, 40–44	1–3, 11–23 odd, 24, 26, 27, 29, 33, 35	11, 19, 21, 27, 35
Advanced	4–7, 40–44	1–3, 14–32 even, 33–36, 39	14, 22, 26, 35, 36

Common Errors

- **Exercises 16–23** Students may think that smaller numbers have fewer factors. Remind students to keep finding factors until all factors are prime numbers.
- **Exercises 26–28** Students may incorrectly find the value of a power. Remind them that the exponent is the number of times the base is a factor. That is, $3^2 = 3 \cdot 3$, not $3 \cdot 2$.

1.4 Record and Practice Journal

List the factor pairs of the number.

1. 6 — $1 \cdot 6, 2 \cdot 3$
2. 7 — $1 \cdot 7$
3. 10 — $1 \cdot 10, 2 \cdot 5$
4. 16 — $1 \cdot 16, 2 \cdot 8, 4 \cdot 4$
5. 35 — $1 \cdot 35, 5 \cdot 7$
6. 55 — $1 \cdot 55, 5 \cdot 11$

Write the prime factorization of the number.

7. 9 — 3^2
8. 24 — $2^3 \cdot 3$
9. 40 — $2^3 \cdot 5$
10. 44 — $2^2 \cdot 11$
11. 50 — $2 \cdot 5^2$
12. 65 — $5 \cdot 13$
13. A fitness instructor arranges 30 people into rows. Each row has the same number of people.
    a. Can the instructor arrange the people into rows of 6? yes
    b. Can the instructor arrange the people into rows of 9? no

Vocabulary and Concept Check

1. The prime factorization of a composite number is the number written as a product of its prime factors.
2. First, find a factor pair and draw "branches." Next, circle the prime factors as you find them. Then, find factors until each branch ends at a prime factor.
3. 6, 9 does not belong because it is a factor pair of 54 and the others are factor pairs of 56.

Practice and Problem Solving

4. 2, 3, 6, 9
5. 3, 5, 9
6. 2, 3, 5, 6, 9, 10
7. None, 1709 is a prime number.
8. 1, 15; 3, 5
9. 1, 22; 2, 11
10. 1, 34; 2, 17
11. 1, 39; 3, 13
12. 1, 45; 3, 15; 5, 9
13. 1, 54; 2, 27; 3, 18; 6, 9
14. 1, 59
15. 1, 61
16. $2 \cdot 2 \cdot 2 \cdot 2$ or 2^4
17. $5 \cdot 5$ or 5^2
18. $2 \cdot 3 \cdot 5$
19. $2 \cdot 13$
20. $2 \cdot 2 \cdot 3 \cdot 7$ or $2^2 \cdot 3 \cdot 7$
21. $2 \cdot 3 \cdot 3 \cdot 3$ or $2 \cdot 3^3$
22. $5 \cdot 13$
23. $7 \cdot 11$
24. 9 is not prime, it is equal to $3 \cdot 3$.
    $72 = 2 \cdot 2 \cdot 2 \cdot 3 \cdot 3 = 2^3 \cdot 3^2$
25. See Additional Answers.

Practice and Problem Solving

26. 180 **27.** 1575

28. 12,584 **29.** 4

30. 25 **31.** 36

32. 1

33. yes; 2 is a prime number because it only has 1 and itself as factors. The rest of the even whole numbers have 2 as a factor.

34. composite; The total number of players on the baseball team is equal to the number in each group times the number of groups, so it must be composite.

35. See *Taking Math Deeper*.

36. 6

37. cupcake table; Because 60 has more factors than 75, there are more rectangular arrangements.

38. 26 yd

39. See Additional Answers.

Fair Game Review

40. 145 **41.** 357

42. 2395 **43.** 1248

44. B

Mini-Assessment

1. List the factor pairs of 36.
   1, 36; 2, 18; 3, 12; 4, 9; 6, 6
2. Write the prime factorization of 42.
   $2 \cdot 3 \cdot 7$
3. Write the prime factorization of 56.
   $2^3 \cdot 7$
4. What is the greatest perfect square that is a factor of 192? 64

Taking Math Deeper

Exercise 35

One way to solve this problem is to try separating manipulatives into equal groups, and see if any are left over.

1. Find 36 objects to represent each of the 36 students. You can use pennies.

2. Each group should have *at least* 4 students but *no more than* 8 students. So, see if you can separate the 36 pennies into equal groups of 4, 5, 6, 7, and 8 with none left over.

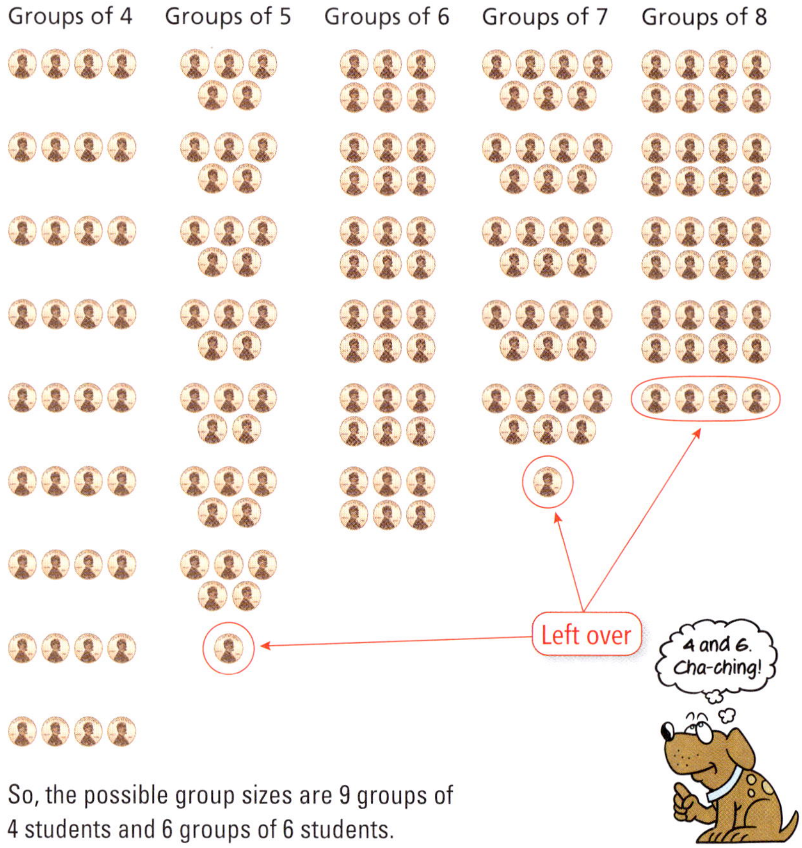

3. So, the possible group sizes are 9 groups of 4 students and 6 groups of 6 students.

Project

Research the meaning of the word *simulation*. Explain how the solution above is a simulation. Explain why simulations are useful.

Reteaching and Enrichment Strategies

If students need help. . .	If students got it. . .
Resources by Chapter • Practice A and Practice B • Puzzle Time Record and Practice Journal Practice Differentiating the Lesson Lesson Tutorials Skills Review Handbook	Resources by Chapter • Enrichment and Extension • Technology Connection Start the next section

Find the number represented by the prime factorization.

26. $2^2 \cdot 3^2 \cdot 5$ **27.** $3^2 \cdot 5^2 \cdot 7$ **28.** $2^3 \cdot 11^2 \cdot 13$

Find the greatest perfect square that is a factor of the number.

3 **29.** 244 **30.** 650 **31.** 756 **32.** 1290

33. CRITICAL THINKING Is 2 the only even prime number? Explain.

34. BASEBALL The coach of a baseball team separates the players into groups for drills. Each group has the same number of players. Is the total number of players on the baseball team *prime* or *composite*? Explain.

35. SCAVENGER HUNT A teacher divides 36 students into equal groups for a scavenger hunt. Each group should have at least 4 students but no more than 8 students. What are the possible group sizes?

36. PERFECT NUMBERS A *perfect number* is a number that equals the sum of its factors, not including itself. For example, the factors of 28 are 1, 2, 4, 7, 14, and 28. Because $1 + 2 + 4 + 7 + 14 = 28$, 28 is a perfect number. What are the perfect numbers between 1 and 28?

37. BAKE SALE One table at a bake sale has 75 oatmeal cookies. Another table has 60 lemon cupcakes. Which table allows for more rectangular arrangements when all the cookies and cupcakes are displayed? Explain.

38. MODELING The stage manager of a school play creates a rectangular acting area of 42 square yards. String lights will outline the acting area. To the nearest whole number, how many yards of string lights does the manager need to enclose this area?

Rectangular Prism

Volume = 40 cubic inches

39. **Volume** The volume of a rectangular prism can be found using the formula *volume* = *length* × *width* × *height*. Using only whole number dimensions, how many different prisms are possible? Explain.

Fair Game Review What you learned in previous grades & lessons

Find the difference. *(Skills Review Handbook)*

40. 192 − 47 **41.** 451 − 94 **42.** 3210 − 815 **43.** 4752 − 3504

44. MULTIPLE CHOICE You buy 168 pears. There are 28 pears in each bag. How many bags of pears do you buy? *(Skills Review Handbook)*

Ⓐ 5 Ⓑ 6 Ⓒ 7 Ⓓ 28

1.5 Greatest Common Factor

Essential Question How can you find the greatest common factor of two numbers?

A **Venn diagram** uses circles to describe relationships between two or more sets. The Venn diagram shows the names of students enrolled in two activities. Students enrolled in both activities are represented by the overlap of the two circles.

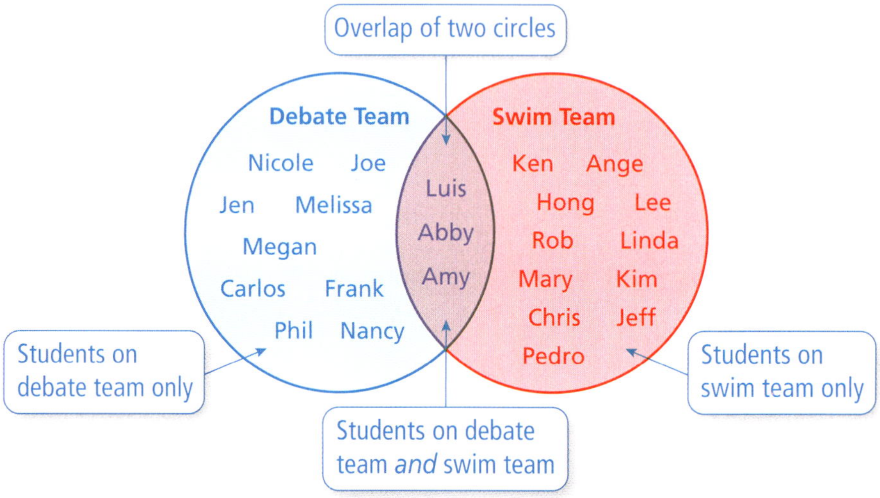

1 ACTIVITY: Identifying Common Factors

Work with a partner. Copy and complete the Venn diagram. Identify the *common factors* of the two numbers.

a. 36 and 48

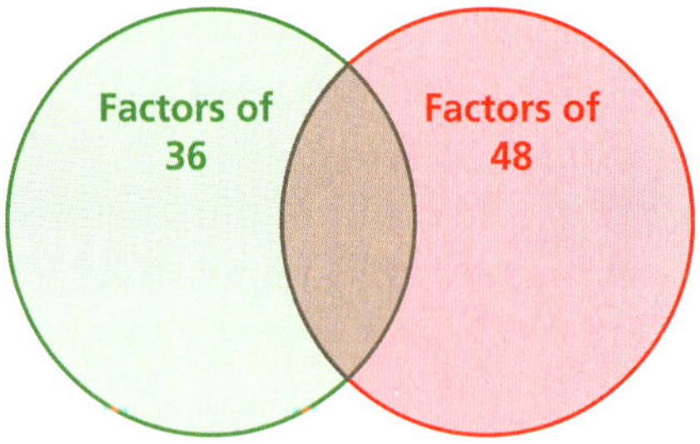

b. 16 and 56

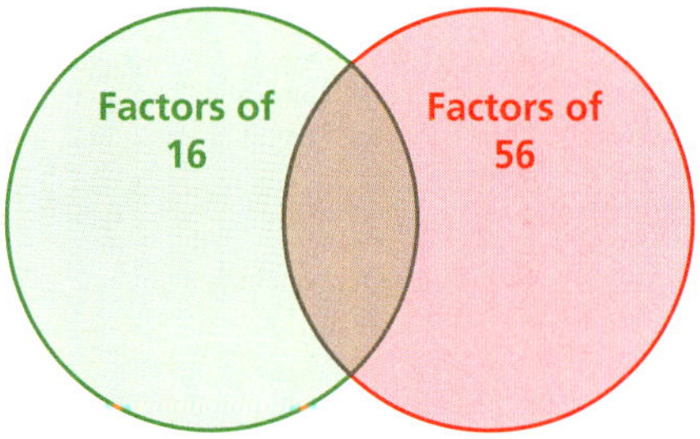

c. 30 and 75

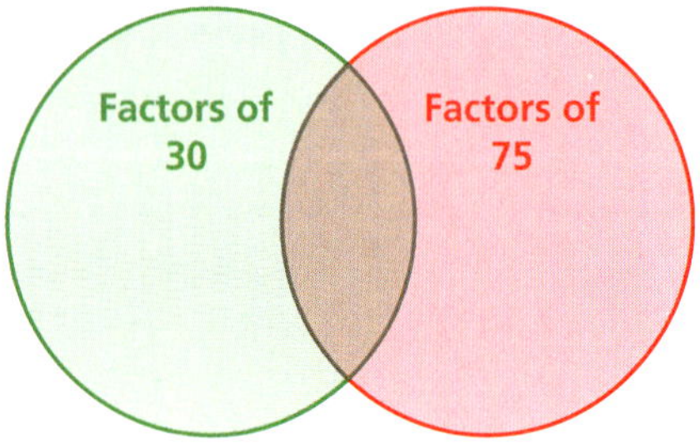

d. 54 and 90

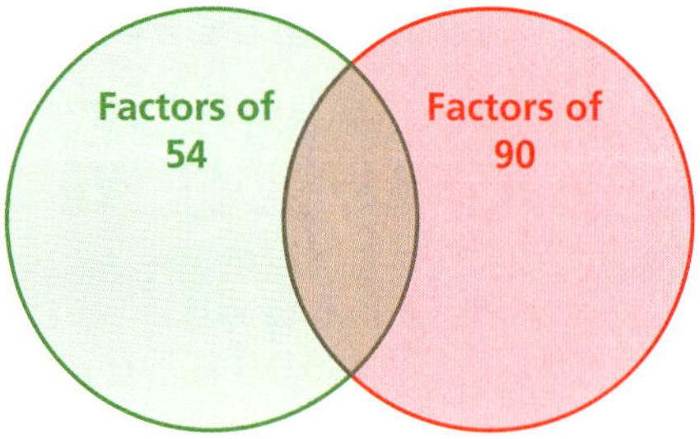

e. Look at the Venn diagrams in parts (a)–(d). Explain how to identify the *greatest common factor* of each pair of numbers. Then circle it in each diagram.

COMMON CORE

Common Factors

In this lesson, you will
- use diagrams to identify common factors.
- find greatest common factors.

Learning Standards
6.NS.4
6.EE.2b

Laurie's Notes

Introduction

Standards for Mathematical Practice

- **MP5 Use Appropriate Tools Strategically:** Oftentimes there are different strategies or methods for solving a problem. A mathematically proficient student is able to assess which method will be more efficient and why. The listing method is more efficient for the problem GCF(16, 20) while the prime factorization method is more efficient for the problem GCF(120, 32, 48).

Motivate

- Before class, make 10 index cards where 5 of them have a red square drawn, 3 have a green circle, and 2 have a red circle. Distribute the cards to 10 volunteers.
- Place two loops of yarn on the floor in the front of the room. Begin each task with students outside of the loops.
- Task 1: Ask the students holding a square to stand in one loop and students with a circle to stand in the other loop.
- Task 2: Ask the students holding a red shape to stand in one loop and students with a green shape to stand in the other loop.
- Task 3: Ask the students holding a square to stand in one loop and students with a red shape to stand in the other loop.
- For the third task, students should see that the loops must intersect. Ask where each set of students is standing, including the green circles.

Discuss

- Discuss the definition of Venn diagrams and relate them to the Motivate activity just completed. Venn diagrams are named for the English mathematician, John Venn (1834–1923), who first used the circles to show relationships between different sets of elements.

Activity Notes

Activity 1

? "What are the factors of 24?" 1, 2, 3, 4, 6, 8, 12, 24

- Introduce the activity. Have students work in pairs while you circulate to see that they are recording the factors correctly.
- Remind students of the divisibility rules that will help them determine the factors of the given numbers.

? "Were there any problems that did not have any factors in common? Explain." No, they all have at least the number 1 in common.

- **FYI:** Two numbers that have only 1 as a common factor are called *relatively prime*.
- **Connection:** Writing the factors in a Venn diagram is similar to the list method of finding the greatest common factor. The Venn diagram method allows students to see that there can be several common factors, but only one greatest common factor.

Common Core State Standards

6.NS.4 Find the greatest common factor of two whole numbers less than or equal to 100 . . ..

6.EE.2b Identify parts of an expression using mathematical terms (sum, term, product, factor, quotient, coefficient); . . .

Extending the Standards

The GCF standard specifies numbers from 1–100, so finding common factors could be handled solely by listing the factors. If you wish, the examples and exercises in this text can be worked this way. However, we chose to show the prime factorization method also.

Previous Learning

Students need to be familiar with finding factors of a number and finding the prime factorization of a number.

Technology for the Teacher

Dynamic Classroom

Lesson Plans
Complete Materials List

1.5 Record and Practice Journal

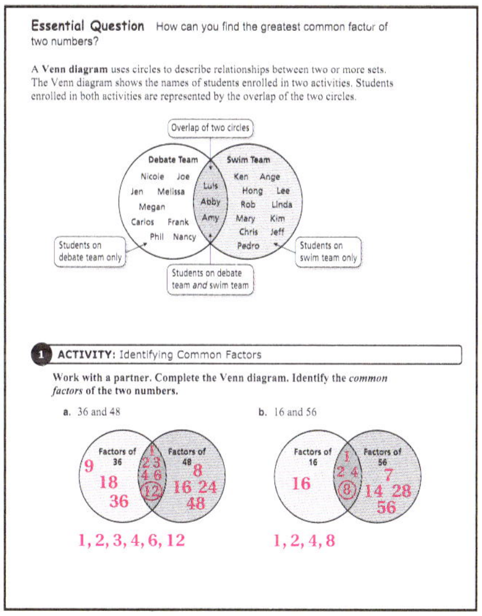

Essential Question How can you find the greatest common factor of two numbers?

A **Venn diagram** uses circles to describe relationships between two or more sets. The Venn diagram shows the names of students enrolled in two activities. Students enrolled in both activities are represented by the overlap of the two circles.

1 ACTIVITY: Identifying Common Factors

Work with a partner. Complete the Venn diagram. Identify the *common factors* of the two numbers.

a. 36 and 48

1, 2, 3, 4, 6, 12

b. 16 and 56

1, 2, 4, 8

Differentiated Instruction

Visual

Show students the following method for finding the GCF of 18, 24, and 42.

1. List the factors of each number in ascending order from left to right.
2. Look at the least number in the set, in this case 18.
3. Decide whether the greatest factor, 18, is the GCF. If not, then cross it out.
4. Look at the factor to the left, 9. Decide whether this factor is the GCF. If not, then cross it out.
5. Continue working right to left, until you find the GCF, which in this case is 6.

Factors of 18: 1, 2, 3, (6), ~~9~~, ~~18~~

Factors of 24: 1, 2, 3, 4, (6), 8, 12, 24

Factors of 42: 1, 2, 3, (6), 7, 14, 21, 28

1.5 Record and Practice Journal

c. 30 and 75

d. 54 and 90

1, 3, 5, 15

1, 2, 3, 6, 9, 18

e. Look at the Venn diagrams in parts (a)–(d). Explain how to identify the *greatest common factor* of each pair of numbers. Then circle it in each diagram.

It is the greatest of the common factors.

2 ACTIVITY: Interpreting a Venn Diagram of Prime Factors

Work with a partner. The Venn diagram represents the prime factorization of two numbers. Identify the two numbers. Explain your reasoning.

a.

b.

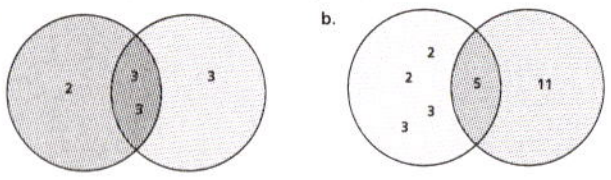

18 and 27

55 and 180

3 ACTIVITY: Identifying Common Prime Factors

Work with a partner.

a. Write the prime factorizations of 36 and 48. Use the results to complete the Venn diagram.

$36 = 2^2 \cdot 3^2$

$48 = 2^4 \cdot 3$

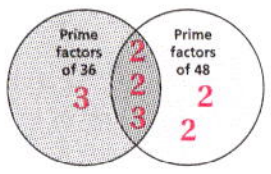

b. Repeat part (a) for the remaining number pairs in Activity 1.

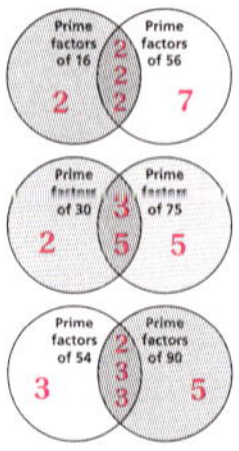

$16 = 2^4$

$56 = 2^3 \cdot 7$

$30 = 2 \cdot 3 \cdot 5$

$75 = 3 \cdot 5^2$

$54 = 2 \cdot 3^3$

$90 = 2 \cdot 3^2 \cdot 5$

c. **STRUCTURE** Compare the numbers in the overlap of the Venn diagrams to your results in Activity 1. What conjecture can you make about the relationship between these numbers and your results in Activity 1?

The product of the numbers in the overlap is equal to the greatest common factor of the numbers.

What Is Your Answer?

4. **IN YOUR OWN WORDS** How can you find the greatest common factor of two numbers? Give examples to support your explanation.

See Additional Answers.

5. Can you think of another way to find the greatest common factor of two or more numbers? Explain. *Sample answer:* Make a list of the factors of each number. Identify the common factors and the greatest common factor.

Laurie's Notes

Activity 2

- **MP1a Make Sense of Problems** and **MP5:** In this activity, students must make sense of the structure of the Venn diagram and how it is used as a visual tool to represent the factors of two numbers. Each number in the intersection is a factor of both original numbers.
- Ask volunteers to share their reasoning about each problem.
- **Extension:** Draw the Venn diagrams shown and ask students to explain how the two problems are different from those in the activity.

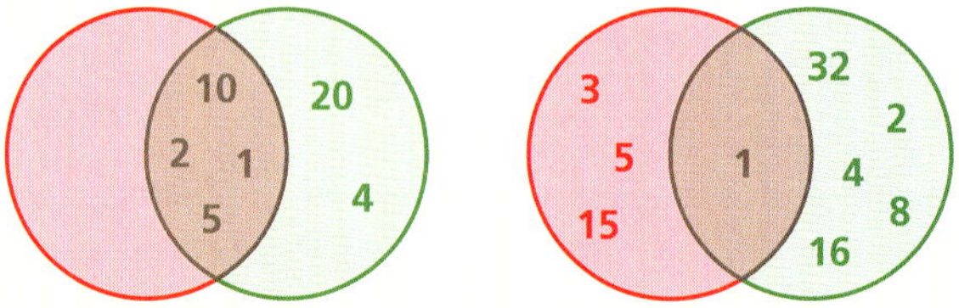

Activity 3

- Discuss the directions with students. Be sure they understand that they are using the prime factorization of each number and *not* the factors of the number. Review the definition of prime factor as needed.
- If students are having difficulty, model the first example as a whole class and then have partners work the remaining three problems.
- **MP7 Look for and Make Use of Structure:** Listen to the students' explanations to part (c). The greatest common factor can be found by determining which prime factors the numbers have in common.

What Is Your Answer?

- For Question 4, you are listening for the product of the prime factors the numbers have in common. Students should also describe the method used in Activity 1.

Closure

- **Exit Ticket:** What is the greatest common factor of 16 and 48? 16
  Name two numbers whose greatest common factor is 8. *Sample answer:* 32 and 56

2 ACTIVITY: Interpreting a Venn Diagram of Prime Factors

Work with a partner. The Venn diagram represents the prime factorization of two numbers. Identify the two numbers. Explain your reasoning.

a.

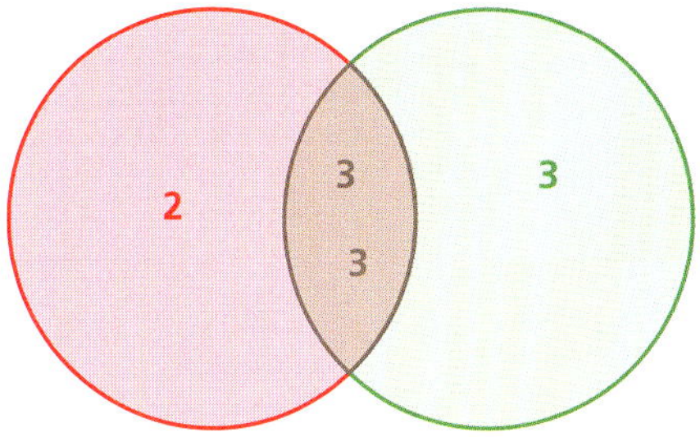

b.

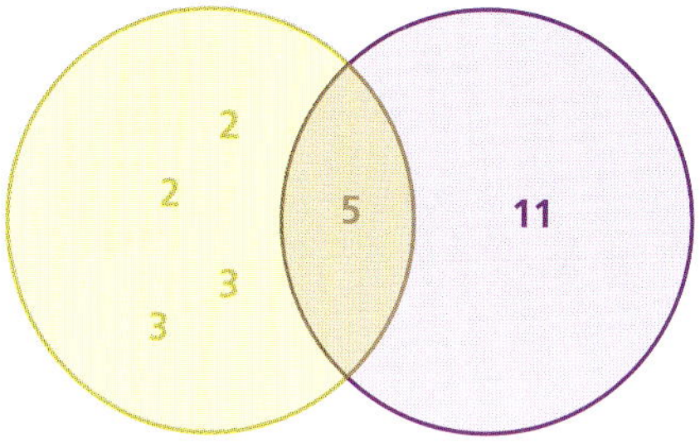

3 ACTIVITY: Identifying Common Prime Factors

Math Practice 1

Interpret a Solution

What does the diagram of the resulting prime factorization mean?

Work with a partner.

a. Write the prime factorizations of 36 and 48. Use the results to complete the Venn diagram.

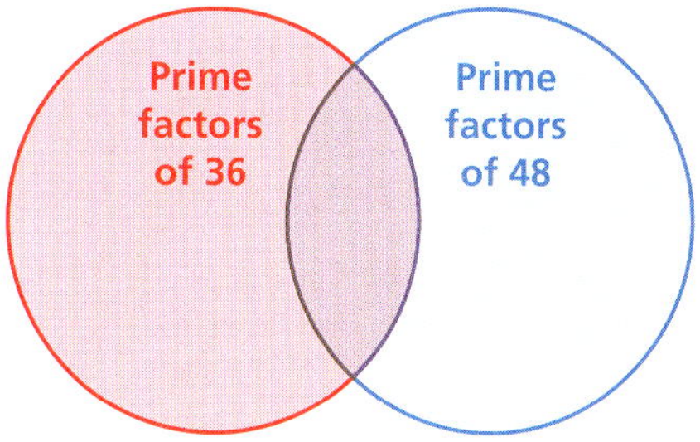

b. Repeat part (a) for the remaining number pairs in Activity 1.

c. **STRUCTURE** Compare the numbers in the overlap of the Venn diagrams to your results in Activity 1. What conjecture can you make about the relationship between these numbers and your results in Activity 1?

What Is Your Answer?

4. **IN YOUR OWN WORDS** How can you find the greatest common factor of two numbers? Give examples to support your explanation.

5. Can you think of another way to find the greatest common factor of two numbers? Explain.

Practice Use what you learned about greatest common factors to complete Exercises 4–6 on page 34.

1.5 Lesson

Check It Out
Lesson Tutorials
BigIdeasMath.com

Factors that are shared by two or more numbers are called **common factors**. The greatest of the common factors is called the **greatest common factor** (GCF). One way to find the GCF of two or more numbers is by listing factors.

> **Key Vocabulary**
> Venn diagram, *p. 30*
> common factors, *p. 32*
> greatest common factor, *p. 32*

EXAMPLE 1 Finding the GCF Using Lists of Factors

Find the GCF of 24 and 40.

List the factors of each number.

Factors of 24: (1), (2), 3, (4), 6, (8), 12, 24

Factors of 40: (1), (2), (4), 5, (8), 10, 20, 40

Circle the common factors.

The common factors of 24 and 40 are 1, 2, 4, and 8. The greatest of these common factors is 8.

So, the GCF of 24 and 40 is 8.

Another way to find the GCF of two or more numbers is by using prime factors. The GCF is the product of the common prime factors of the numbers.

EXAMPLE 2 Finding the GCF Using Prime Factorizations

Find the GCF of 12 and 56.

Make a factor tree for each number.

> **Study Tip**
> Examples 1 and 2 show two different methods for finding the GCF. After solving with one method, you can use the other method to check your answer.

Write the prime factorization of each number.

$12 = (2) \cdot (2) \cdot 3$

$56 = (2) \cdot (2) \cdot 2 \cdot 7$

Circle the common prime factors.

$2 \cdot 2 = 4$ Find the product of the common prime factors.

So, the GCF of 12 and 56 is 4.

On Your Own

Now You're Ready
Exercises 7–18

Find the GCF of the numbers using lists of factors.

1. 8, 36 **2.** 18, 72 **3.** 14, 28, 49

Find the GCF of the numbers using prime factorizations.

4. 20, 45 **5.** 32, 90 **6.** 45, 75, 120

Multi-Language Glossary at BigIdeasMath.com

Laurie's Notes

Introduction

Connect

- **Yesterday:** Students used Venn diagrams to find common factors of two numbers. (MP1a, MP5, MP7)
- **Today:** Students will learn two methods for finding the greatest common factor of two numbers.

Motivate

? Write LOL on the board and ask, "What does this mean?" Most will say "Laugh Out Loud" or "Laughing Out Loud." There are other interpretations.

- Write a list of common acronyms on the board. Ask students to identify as many as they can. They could work in pairs.

OJ	Orange juice	BLT	Bacon, lettuce, and tomato
YTD	Year to date	PIN	Personal identification number
TBA	To be announced	EST	Eastern Standard Time

? "What does the acronym GCF mean?" greatest common factor

Lesson Notes

Example 1

- The first method of finding the GCF is similar to the first activity yesterday.

? "What are the factors of 24?" 1, 2, 3, 4, 6, 8, 12, 24

? "What are the factors of 40?" 1, 2, 4, 5, 8, 10, 20, 40

? "What factors appear in both lists?" 1, 2, 4, 8

- **FYI:** Students should not just say, "The greatest common factor is 8." The complete answer is, "The greatest common factor *of 12 and 40* is 8."

Example 2

- The second method is similar to the third activity yesterday.

? "What prime factors do 12 and 56 have in common?" two factors of 2

- The greatest common factor of two (or more) numbers is the product of the prime factors that they have in common. The greatest common factor of 12 and 56 is 2 • 2 or 4.
- The solution can be checked by using the first method of listing factors.

? "Does it matter which method you use and why?" no; Each method will give you the correct answer.

? "How do you decide which method to use?" If the numbers are relatively small, use the listing method. If the numbers are relatively large, use the prime factorization method.

On Your Own

- Students should be encouraged to think and reason first. In Question 2, 72 is divisible by 18, so the GCF is 18.
- Ask volunteers to share their work at the board or document camera to check solutions.

Goal Today's lesson is finding the **greatest common factor (GCF)** of two or more numbers.

Technology for the Teacher

Dynamic Classroom

Lesson Tutorials
Lesson Plans
Answer Presentation Tool

Extra Example 1

Find the GCF of 36 and 54 using lists of factors. 18

Extra Example 2

Find the GCF of 40 and 48 using prime factorization. 8

On Your Own

1. 4	**2.** 18
3. 7	**4.** 5
5. 2	**6.** 15

Laurie's Notes

Extra Example 3

Which pair of numbers, 24 and 36, or 36 and 54, has a GCF of 12? 24 and 36

Example 3

- Read the problem.
- **MP2 Reason Abstractly and Quantitatively:** Asking students questions where they need to explain constraints or parameters helps them to focus on and develop their reasoning skills.
- ? "Why can't a greater number be a factor of a lesser number?" The greater number can't divide into the lesser number evenly.
- **Big Idea:** The GCF of two numbers will always be less than or equal to the lesser of the two original numbers.
- Work through the remainder of the problem as shown.

Extra Example 4

You are arranging your collection of DVDs into stacks. You have 16 drama, 32 action, and 40 comedy DVDs. You want all of the stacks to be the same height. What is the greatest number of DVDs you can have in a stack to make them the same height, without any left over? 8

Example 4

- Ask a volunteer to read the problem. Discuss the list of gifts. Discuss what it means to have identical groups of the gifts in each piñata.
- ? "How could you distribute the 18 bottles of nail polish?" Put 1 in 18 different piñatas, put 2 in nine different piñatas, put 3 in six different piñatas, and so on.
- Repeat the question for the earrings and hair bows.
- **MP3a Construct Viable Arguments:** It is important to ask students to justify their answers and communicate their reasoning to others.
- ? "Which method for finding the GCF should be used and why?" Prime factorization, because there are 3 numbers and finding the prime factorization of each is fairly quick.
- **Extension:** Ask what will be in each of the six piñatas.

On Your Own

7. *Sample answer:* 10, 20
8. no; The prime factorization of 30 is $2 \cdot 3 \cdot 5$, so the GCF is still $2 \cdot 3 = 6$

On Your Own

- **Reasoning:** In Question 7, students are given the GCF and they need to find the two original numbers. Write the various solutions on the board and look for similarities in the answers.

Closure

- **Writing prompt:** To find the GCF of 16 and 28, I would... *Sample answer:* List the factors of each number. Then identify the greatest common factor.

  Factors of 16: (1), (2), (4), 8, 16

  Factors of 28: (1), (2), (4), 7, 14, 28

  The greatest common factor of 16 and 28 is 4.

English Language Learners

Abbreviations

The abbreviation GCF is pronounced *gee see ef.* Other languages may pronounce the letters differently causing students to become confused. When speaking or writing, use both forms, ". . . GCF, or greatest common factor…," to reinforce the pronunciation and meaning. This is also true with the abbreviations LCM and GCD.

EXAMPLE 3 Finding Two Numbers with a Given GCF

Which pair of numbers has a GCF of 15?

Ⓐ 10, 15   Ⓑ 30, 60   Ⓒ 21, 45   Ⓓ 45, 75

The number 15 cannot be a factor of the lesser number 10. So, you can eliminate Statement A.

The number 15 cannot be a factor of a number that does not have a 0 or 5 in the ones place. So, you can eliminate Statement C.

List the factors for Statements B and D. Then identify the GCF for each.

Choice B: Factors of 30: 1, 2, 3, 5, 6, 10, 15, 30

Factors of 60: 1, 2, 3, 4, 5, 6, 10, 12, 15, 20, 30, 60

The GCF of 30 and 60 is 30.

Choice D: Factors of 45: 1, 3, 5, 9, 15, 45

Factors of 75: 1, 3, 5, 15, 25, 75

The GCF of 45 and 75 is 15.

The correct answer is Ⓓ.

EXAMPLE 4 Real-Life Application

You are filling piñatas for your sister's birthday party. The list shows the gifts you are putting into the piñatas. You want identical groups of gifts in each piñata with no gifts left over. What is the greatest number of piñatas you can make?

The GCF of the numbers of gifts represents the greatest number of identical groups of gifts you can make with no gifts left over. So, to find the number of piñatas, find the GCF.

$$18 = 2 \cdot 3 \cdot 3$$
$$24 = 2 \cdot 3 \cdot 2 \cdot 2$$
$$42 = 2 \cdot 3 \cdot 7$$

$2 \cdot 3 = 6$ Find the product of the common prime factors.

The GCF of 18, 24, and 42 is 6.

So, you can make at most 6 piñatas.

On Your Own

Now You're Ready
Exercises 23–25

7. Write a pair of numbers whose greatest common factor is 10.
8. **WHAT IF?** In Example 4, you add 6 more pairs of earrings. Does this change your answer? Explain your reasoning.

1.5 Exercises

Check It Out
Help with Homework
BigIdeasMath.com

Vocabulary and Concept Check

1. **VOCABULARY** What is the greatest common factor (GCF) of two numbers?
2. **WRITING** Describe how to find the GCF of two numbers by using prime factorization.
3. **DIFFERENT WORDS, SAME QUESTION** Which is different? Find "both" answers.

What is the greatest common factor of 24 and 32?

What is the greatest common divisor of 24 and 32?

What is the greatest prime factor of 24 and 32?

What is the product of the common prime factors of 24 and 32?

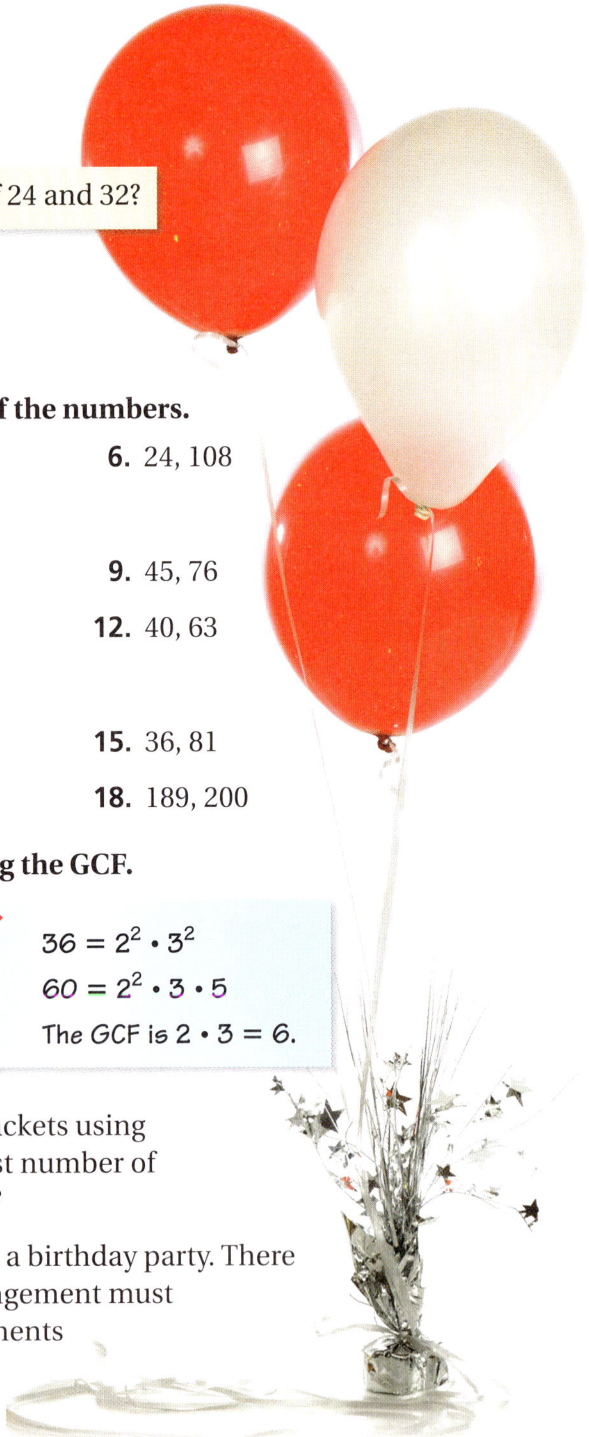

Practice and Problem Solving

Use a Venn diagram to find the greatest common factor of the numbers.

4. 12, 30
5. 32, 54
6. 24, 108

Find the GCF of the numbers using lists of factors.

1 7. 6, 15
8. 14, 84
9. 45, 76
10. 39, 65
11. 51, 85
12. 40, 63

Find the GCF of the numbers using prime factorizations.

2 13. 45, 60
14. 27, 63
15. 36, 81
16. 72, 84
17. 61, 73
18. 189, 200

ERROR ANALYSIS **Describe and correct the error in finding the GCF.**

19. ✗

$42 = 2 \cdot 3 \cdot 7$

$154 = 2 \cdot 7 \cdot 11$

The GCF is 7.

20. ✗

$36 = 2^2 \cdot 3^2$

$60 = 2^2 \cdot 3 \cdot 5$

The GCF is $2 \cdot 3 = 6$.

21. **CLASSROOM** A teacher is making identical activity packets using 92 crayons and 23 sheets of paper. What is the greatest number of packets the teacher can make with no items left over?

22. **BALLOONS** You are making balloon arrangements for a birthday party. There are 16 white balloons and 24 red balloons. Each arrangement must be identical. What is the greatest number of arrangements you can make using every balloon?

Assignment Guide and Homework Check

Level	Day 1 Activity Assignment	Day 2 Lesson Assignment	Homework Check
Basic	4–6, 35–39	1–3, 7–21 odd, 22, 23	7, 11, 15, 21, 23
Average	4–6, 35–39	1–3, 10–15, 20, 22, 23–33 odd	10, 15, 22, 23, 31
Advanced	4–6, 35–39	1–3, 10–34 even	10, 16, 22, 24, 30

For Your Information

- **Exercise 3** In the *Different Words, Same Question* exercise, three of the four choices pose the same question using different words. The remaining choice poses a different question. So there are two answers.

Common Errors

- **Exercise 3** Students may get confused with the use of the words "divisor" and "factor." Explain that the greatest common divisor (GCD) is another name for GCF. It is the greatest common factor, that can be divided evenly into the given numbers.
- **Exercises 4–6** Some students may struggle using a Venn diagram. Make sure they understand that the intersection of any (or all) circles can be used to find the GCF of the numbers represented by the circles.
- **Exercises 23–25** Some students may struggle if they use a Venn diagram for three sets. Tell them that the same rules apply for sets of three as with a set of two. Make sure they understand that the intersection of any (or all) circles can be used to find the GCF of the numbers represented by the circles.

1.5 Record and Practice Journal

Find the GCF of the numbers using lists of factors.

1. 9, 15 — 3
2. 11, 19 — 1
3. 8, 28 — 4
4. 60, 70 — 10
5. 40, 56 — 8
6. 35, 72 — 1

Find the GCF of the numbers using prime factorizations.

7. 4, 10 — 2
8. 5, 11 — 1
9. 6, 8 — 2
10. 14, 42 — 14
11. 45, 63 — 9
12. 60, 90 — 30
13. You are making identical gift bags using 24 candles and 36 bottles of lotion. What is the greatest number of gift bags you can make with no items left over?
    12 gift bags

Vocabulary and Concept Check

1. The GCF is the greatest factor that is shared by the two numbers.
2. First, find the prime factorization of both numbers. Next, identify common prime factors. Then, find the product of the common prime factors.
3. What is the greatest prime factor of 24 and 32?; 2; 8

Practice and Problem Solving

4. 6
5. 2
6. 12
7. 3
8. 14
9. 1
10. 13
11. 17
12. 1
13. 15
14. 9
15. 9
16. 12
17. 1
18. 1
19. 7 is the greatest common *prime* factor. The GCF is $2 \cdot 7 = 14$.
20. Not all of the common prime factors are included. The GCF is $2^2 \cdot 3 = 12$.
21. 23 packets
22. 8 arrangements

Practice and Problem Solving

23. 7

24. 6

25. 14

26. *Sample answer:* 16, 32, and 48; Multiply 16 by 1, 2, and 3.

27. *Sample answer:* Prime factorization because it is tedious to find all the factors of large numbers.

28. sometimes

29. always

30. never

31. 12; 6 red, 5 pink, and 4 yellow

32. See Additional Answers.

33. **a.** Because 73 is a prime number and the GCF of the three numbers is 1.

b. 18; The GCF of 54 and 36 is 18. 18 divides evenly into 72 leaving one banana left over.

34. See *Taking Math Deeper*.

Fair Game Review

35. Commutative Property of Addition

36. Associative Property of Addition

37. Commutative Property of Multiplication

38. Associative Property of Multiplication

39. B

Mini-Assessment

Find the GCF of the numbers.

1. 8, 20 4

2. 35, 56 7

3. 18, 45 9

4. 12, 54, 84 6

Taking Math Deeper

Exercise 34

This problem is conceptually challenging. *Stop and think* about what information each sentence reveals.

1 *One-foot-by-one-foot tiles cover the floor of each room.*

- A whole number of one-foot-by-one-foot tiles covers each floor. There are no partial tiles.
- The number of tiles along the length and the number of tiles along the width are factors of the number of tiles in the room.

It is also helpful to connect the tiles to a room's length, width, and area.

- The numbers of tiles along the length and width of each room represent the length (in feet) and width (in feet) of each room.
- The number of tiles in the room represents the area (in square feet).

2 *Describe how the greatest possible length of the adjoining wall is related to the total number of tiles in each room.*

- Each room has the same number of tiles along the adjoining wall. Therefore, the numbers of tiles in the rooms must have a common factor.

So, the **greatest** possible length of the adjoining wall is the **greatest** common factor of the numbers of tiles in the rooms.

3 One possible diagram is shown using two reasonable room sizes of 80 square feet and 104 square feet. The GCF of 80 and 104 is 8, meaning that the greatest possible length of the adjoining wall is 8 feet, which is also reasonable.

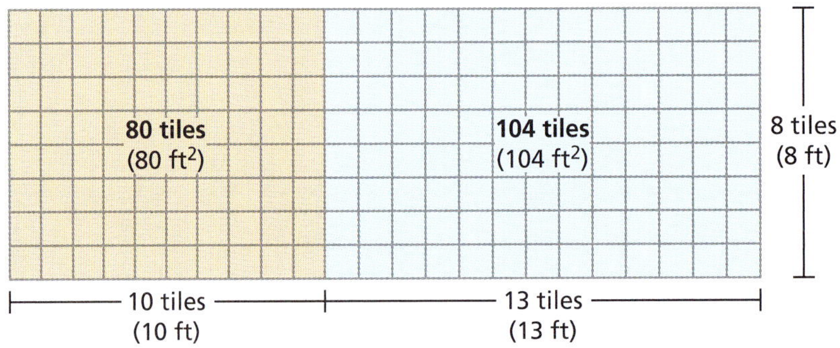

Reteaching and Enrichment Strategies

If students need help. . .	If students got it. . .
Resources by Chapter • Practice A and Practice B • Puzzle Time Record and Practice Journal Practice Differentiating the Lesson Lesson Tutorials Skills Review Handbook	Resources by Chapter • Enrichment and Extension • Technology Connection Start the next section

Find the GCF of the numbers.

4 **23.** 35, 56, 63 **24.** 30, 60, 78 **25.** 42, 70, 84

26. **OPEN-ENDED** Write a set of three numbers that have a GCF of 16. What procedure did you use to find your answer?

27. **REASONING** You need to find the GCF of 256 and 400. Would you rather list their factors or use their prime factorizations? Explain.

CRITICAL THINKING **Tell whether the statement is *always*, *sometimes*, or *never* true.**

28. The GCF of two even numbers is 2.

29. The GCF of two prime numbers is 1.

30. When one number is a multiple of another, the GCF of the numbers is the greater of the numbers.

31. **BOUQUETS** A florist is making identical bouquets using 72 red roses, 60 pink roses, and 48 yellow roses. What is the greatest number of bouquets that the florist can make if no roses are left over? How many of each color are in each bouquet?

32. **VENN DIAGRAM** Consider the numbers 252, 270, and 300.

a. Create a Venn diagram using the prime factors of the numbers.

b. Use the Venn diagram to find the GCF of 252, 270, and 300.

c. What is the GCF of 252 and 270? 252 and 300? Explain how you found your answer.

33. **FRUIT BASKETS** You are making fruit baskets using 54 apples, 36 oranges, and 73 bananas.

a. Explain why you cannot make identical fruit baskets without leftover fruit.

b. What is the greatest number of identical fruit baskets you can make with the least amount of fruit left over? Explain how you found your answer.

34. Problem Solving Two rectangular, adjacent rooms share a wall. One-foot-by-one-foot tiles cover the floor of each room. Describe how the greatest possible length of the adjoining wall is related to the total number of tiles in each room. Draw a diagram that represents one possibility.

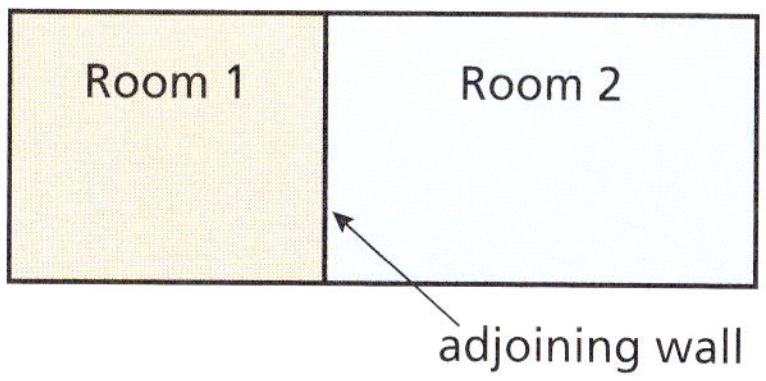

Fair Game Review What you learned in previous grades & lessons

Tell which property is being illustrated. *(Skills Review Handbook)*

35. $13 + (29 + 7) = 13 + (7 + 29)$ **36.** $13 + (7 + 29) = (13 + 7) + 29$

37. $(6 \times 37) \times 5 = (37 \times 6) \times 5$ **38.** $(37 \times 6) \times 5 = 37 \times (6 \times 5)$

39. **MULTIPLE CHOICE** In what order should you perform the operations in the expression $4 \times 3 - 12 \div 2 + 5$? *(Section 1.3)*

Ⓐ $\times, -, \div, +$ Ⓑ $\times, \div, -, +$ Ⓒ $\times, \div, +, -$ Ⓓ $\times, +, -, \div$

1.6 Least Common Multiple

Essential Question How can you find the least common multiple of two numbers?

1 ACTIVITY: Identifying Common Multiples

Work with a partner. Using the first several multiples of each number, copy and complete the Venn diagram. Identify any *common multiples* of the two numbers.

a. 8 and 12

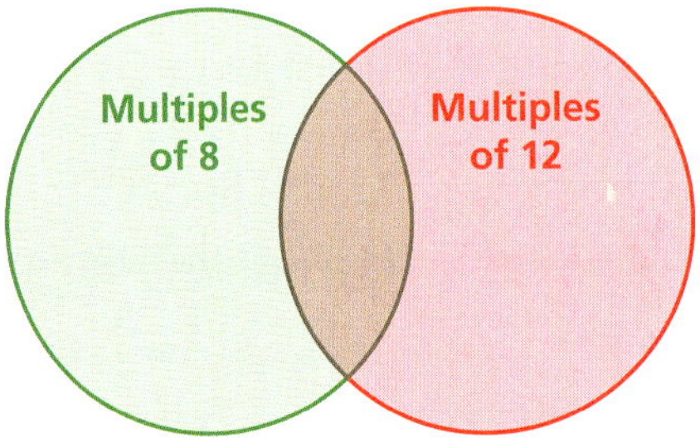

b. 4 and 14

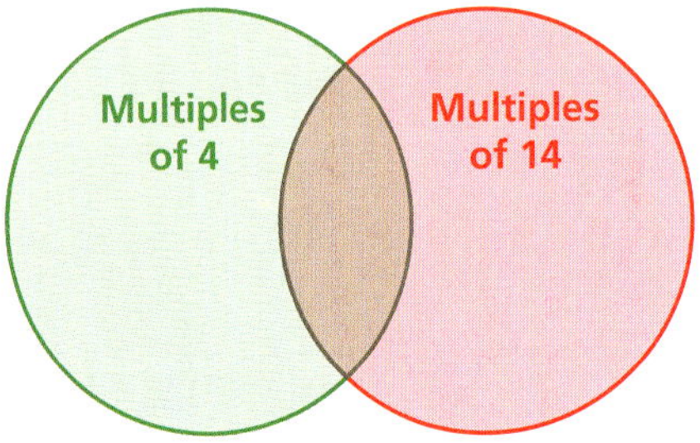

c. 10 and 15

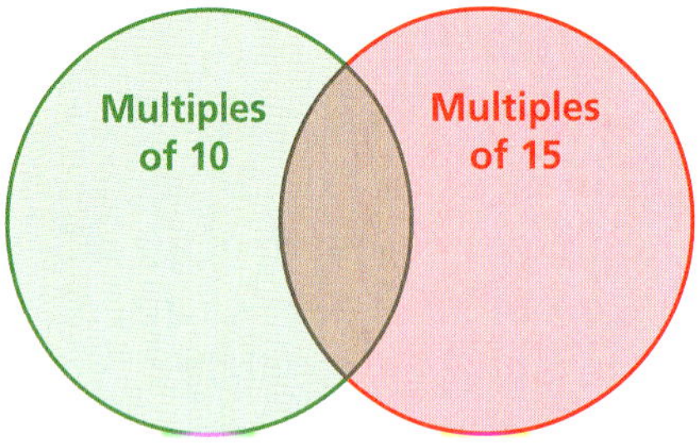

d. 20 and 35

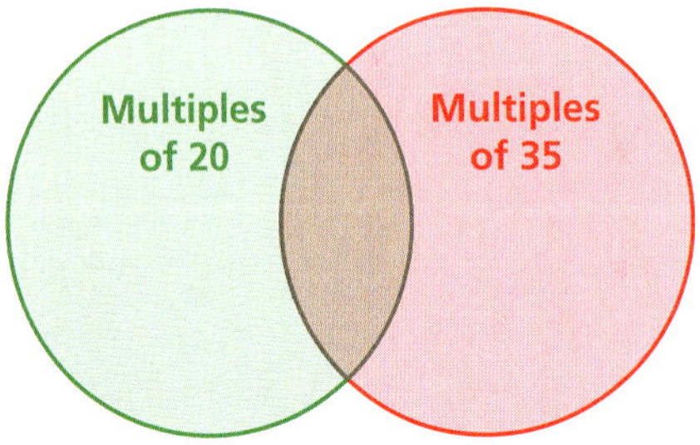

e. Look at the Venn diagrams in parts (a)–(d). Explain how to identify the *least common multiple* of each pair of numbers. Then circle it in each diagram.

COMMON CORE

Common Multiples

In this lesson, you will
- use diagrams to identify common multiples.
- find least common multiples.

Learning Standard
6.NS.4

Laurie's Notes

Introduction

Standards for Mathematical Practice

- **MP5 Use Appropriate Tools Strategically:** This section is similar to the last in that students are presented two methods for finding the least common multiple of two numbers. A mathematically proficient student is able to assess which method will be more efficient and why. The listing method is more efficient for the problem LCM(8, 12), while the prime factorization method is more efficient for the problem LCM(24, 32).

For Your Information

- A key usage of multiples in CCSS is recognizing that if a number x is a multiple of two numbers a and b, then x is a multiple of $a \cdot b$. Students will see this in Chapter 3.

Motivate

- Pose this puzzle to students: There are forty dogs. Fourteen are Golden Labs and 29 are wearing collars. If five dogs are Golden Labs and are wearing collars, how many dogs are not Golden Labs and not wearing collars? 2; Golden Labs without collars: 9, With collars and not Golden Labs: 24
- If students are stuck, ask them what tool they have used to represent the relationship between different sets of elements.
- To get students started, you could present the following Venn diagram.

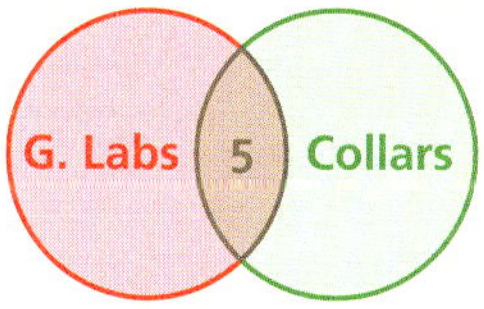

- This problem will help students remember that there is an exterior of the Venn diagram.

Activity Notes

Activity 1

? "What are the first five multiples of 4?" 4, 8, 12, 16, 20

- Introduce the activity. Students may ask how many multiples they need to list. There is no set answer, however they should list at least the first 5 or 6 multiples.
- Have each student work with his or her partner while you circulate around the room to see that they are recording the multiples correctly.
- Accuracy is important. If students list a multiple incorrectly, all subsequent multiples are likely going to be incorrect.

? "Were there any problems that had more than one multiple in common? Explain." Yes Students may not have extended their lists enough to see other multiples.

Common Core State Standards

6.NS.4 Find . . . the least common multiple of two whole numbers less than or equal to 12. . . .

Extending the Standards

The LCM standard specifies numbers from 1–12, so finding common multiples could be handled solely by listing the multiples. If you wish, the examples and exercises in this text can be worked this way. However, we chose to show the prime factorization method also.

Previous Learning

Students need to be familiar with finding multiples of a number and finding the prime factorization of a number.

Technology for the Teacher

Dynamic Classroom

Lesson Plans
Complete Materials List

1.6 Record and Practice Journal

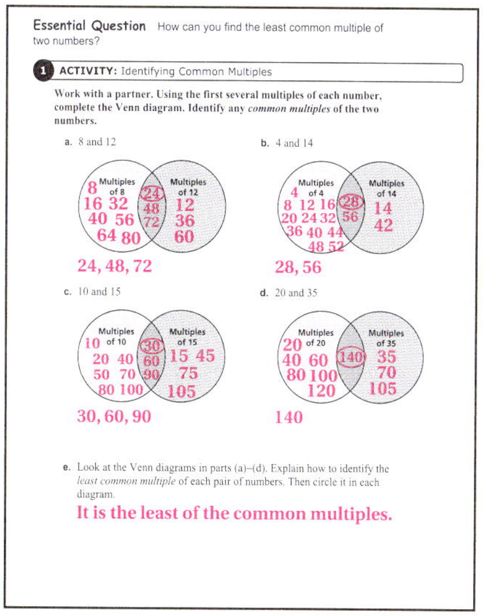

Essential Question How can you find the least common multiple of two numbers?

1 ACTIVITY: Identifying Common Multiples

Work with a partner. Using the first several multiples of each number, complete the Venn diagram. Identify any *common multiples* of the two numbers.

a. 8 and 12

Multiples of 8: 8 16 32 40 56 64 80 | 24 48 72 | Multiples of 12: 12 36 60

24, 48, 72

b. 4 and 14

Multiples of 4: 4 8 12 16 20 24 32 36 40 44 48 52 | 28 56 | Multiples of 14: 14 42

28, 56

c. 10 and 15

Multiples of 10: 10 20 40 50 70 80 100 | 30 60 90 | Multiples of 15: 15 45 75 105

30, 60, 90

d. 20 and 35

Multiples of 20: 20 40 60 80 100 120 | 140 | Multiples of 35: 35 70 105

140

e. Look at the Venn diagrams in parts (a)–(d). Explain how to identify the *least common multiple* of each pair of numbers. Then circle it in each diagram.

It is the least of the common multiples.

English Language Learners

Vocabulary

In the lesson, be sure that students can distinguish between *common multiple* and *least common multiple.* In the extension, students need to distinguish between *denominator, common denominator,* and *least common denominator.*

1.6 Record and Practice Journal

2 ACTIVITY: Interpreting a Venn Diagram of Prime Factors

Work with a partner

a. Write the prime factorizations of 8 and 12. Use the results to complete the Venn diagram.

$8 = 2^3, 12 = 2^2 \cdot 3$

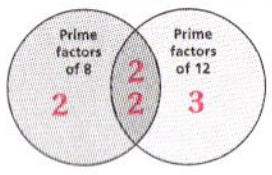

b. Repeat part (a) for the remaining number pairs in Activity 1.

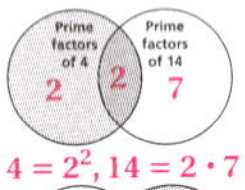

$4 = 2^2, 14 = 2 \cdot 7$

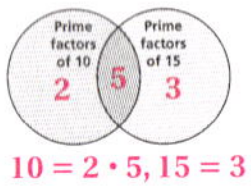

$10 = 2 \cdot 5, 15 = 3 \cdot 5$

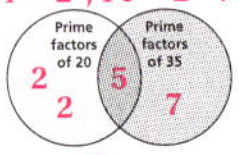

$20 = 2^2 \cdot 5, 35 = 5 \cdot 7$

c. **STRUCTURE** Compare the numbers from each section of the Venn diagrams to your results in Activity 1. What conjecture can you make about the relationship between these numbers and your results in Activity 1?

The product of all the prime factors in the diagram is equal to the least common multiple of the numbers.

What Is Your Answer?

3. **IN YOUR OWN WORDS** How can you find the least common multiple of two numbers? Give examples to support your explanation.

See Additional Answers.

4. The Venn diagram shows the prime factors of two numbers. Use the diagram to do the following tasks.

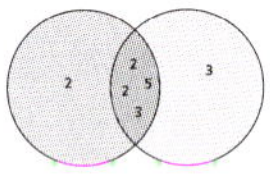

a. Identify the two numbers.

120, 180

b. Find the greatest common factor.

60

c. Find the least common multiple.

360

5. A student writes the prime factorizations of 8 and 12 in a table as shown. She claims she can use the table to find the greatest common factor and the least common multiple of 8 and 12. How is this possible?

8 =	2	2	2	
12 =	2	2		3

GCF: Multiply the prime factors that appear in the columns that have no empty spaces in the table; GCF $= 2 \cdot 2 = 4$

LCM: Multiply the prime factors that appear in any column of the table; LCM $= 2 \cdot 2 \cdot 2 \cdot 3 = 24$

6. Can you think of another way to find the least common multiple of two or more numbers? Explain.

***Sample answer:* Make a list of some common multiples of each number. Identify the common multiples and the least common multiple.**

Laurie's Notes

Activity 2

- Discuss the directions with students. Be sure they understand that they are using the prime factorization of each number as they did in the previous lesson so they should be comfortable with how to proceed.
- **MP7 Look for and Make Use of Structure:** Listen to the students' explanations to part (c). The least common multiple of two numbers can be found by finding the product of their prime factors listed in the Venn diagram.
- **Big Idea:** The Venn diagram is necessary in listing the prime factorization. The factors the two numbers have in common are written only once in the Venn diagram. This is an important connection for students to understand.

? "Think of two numbers that have no prime factors in common. What would the Venn diagram look like for their prime factorizations?" There would be no factors where the two circles intersect. "Does this method still work?" yes

What Is Your Answer?

- Question 4 summarizes greatest common factor (GCF) and least common multiple (LCM). Students are often confused by the two concepts. When students have finished, ask a volunteer to share his or her work.

Closure

- **Exit Ticket:** What is the least common multiple of 16 and 20? 80 Name two numbers whose least common multiple is 36. *Sample answer:* 9 and 12

2 ACTIVITY: Interpreting a Venn Diagram of Prime Factors

Work with a partner.

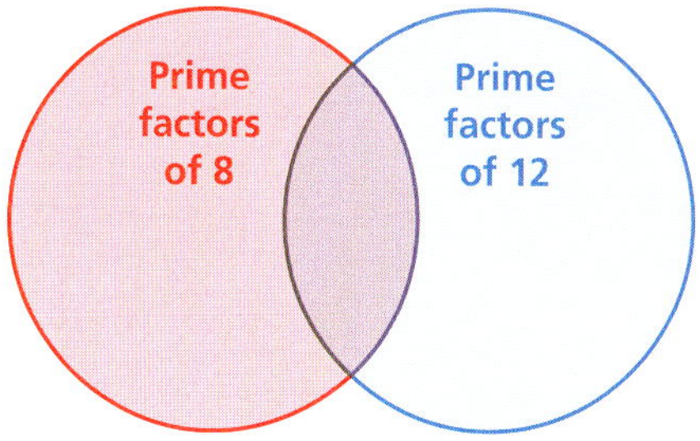

a. Write the prime factorizations of 8 and 12. Use the results to complete the Venn diagram.

b. Repeat part (a) for the remaining number pairs in Activity 1.

c. **STRUCTURE** Compare the numbers from each section of the Venn diagrams to your results in Activity 1. What conjecture can you make about the relationship between these numbers and your results in Activity 1?

What Is Your Answer?

Math Practice 3

Construct Arguments

How can you use diagrams to support your explanation?

3. **IN YOUR OWN WORDS** How can you find the least common multiple of two numbers? Give examples to support your explanation.

4. The Venn diagram shows the prime factors of two numbers.

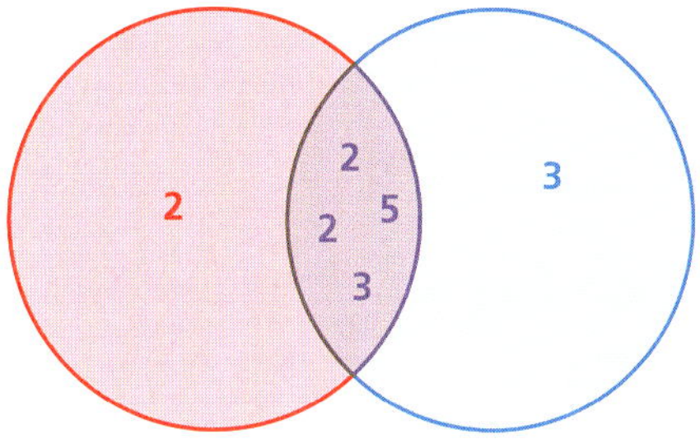

Use the diagram to do the following tasks.

a. Identify the two numbers.

b. Find the greatest common factor.

c. Find the least common multiple.

5. A student writes the prime factorizations of 8 and 12 in a table as shown. She claims she can use the table to find the greatest common factor and the least common multiple of 8 and 12. How is this possible?

8 =	2	2	2	
12 =	2	2		3

6. Can you think of another way to find the least common multiple of two or more numbers? Explain.

Practice Use what you learned about least common multiples to complete Exercises 3–5 on page 40.

1.6 Lesson

Check It Out
Lesson Tutorials
BigIdeasMath.com

Key Vocabulary

common multiples, *p. 38*

least common multiple, *p. 38*

Multiples that are shared by two or more numbers are called **common multiples**. The least of the common multiples is called the **least common multiple** (LCM). You can find the LCM of two or more numbers by listing multiples or using prime factors.

EXAMPLE 1 Finding the LCM Using Lists of Multiples

Find the LCM of 4 and 6.

List the multiples of each number.

Multiples of 4: 4, 8, (12), 16, 20, (24), 28, 32, (36), . . .

Multiples of 6: 6, (12), 18, (24), 30, (36), . . .

Circle the common multiples.

Some common multiples of 4 and 6 are 12, 24, and 36. The least of these common multiples is 12.

So, the LCM of 4 and 6 is 12.

On Your Own

Now You're Ready
Exercises 6–11

Find the LCM of the numbers using lists of multiples.

1. 3, 8 **2.** 9, 12 **3.** 6, 10

EXAMPLE 2 Finding the LCM Using Prime Factorizations

Find the LCM of 16 and 20.

Make a factor tree for each number.

Write the prime factorization of each number. Circle each different factor where it appears the greater number of times.

$16 = (2) \cdot (2) \cdot (2) \cdot (2)$ 2 appears more often here, so circle all 2s.

$20 = 2 \cdot 2 \cdot (5)$ 5 appears once. Do not circle the 2s again.

$2 \cdot 2 \cdot 2 \cdot 2 \cdot 5 = 80$ Find the product of the circled factors.

So, the LCM of 16 and 20 is 80.

On Your Own

Now You're Ready
Exercises 12–17

Find the LCM of the numbers using prime factorizations.

4. 14, 18 **5.** 28, 36 **6.** 24, 90

Multi-Language Glossary at BigIdeasMath.com

Laurie's Notes

Introduction

Connect

- **Yesterday:** Students used Venn diagrams to find common multiples of two numbers. (MP5, MP7)
- **Today:** Students will learn two methods for finding the least common multiple of two numbers.

Motivate

- **Puzzle Time:** A bell rings every 3 minutes, a dog barks every 4 minutes, and a person coughs every 5 minutes. If you just heard all 3 sounds, how long must you wait to hear
  - the bell ring and the dog bark? 12 minutes
  - the bell ring and the person cough? 15 minutes
  - the dog bark and the person cough? 20 minutes
  - all 3 sounds at the same time? 60 minutes
- Explain that there are different methods for answering these questions.

Lesson Notes

Example 1

? "What are the multiples of 4?" 4, 8, 12, 16, 20, 24, 26, 32, 36…

? "What are the multiples of 6?" 6, 12, 18, 24, 30, 36, 42…

? "What multiples appear in both lists?" 12, 24, and 36 "What do you notice about all of the common multiples?" They are the multiples of 12.

- **FYI:** Students should not just say, "The least common multiple is 12." The complete answer is, "The least common multiple *of 4 and 6* is 12."

On Your Own

- **Think-Pair-Share:** Students should read each question independently and then work in pairs to answer the questions. When they have answered the questions, the pair should compare their answers with another group and discuss any discrepancies.

Example 2

- Any factor that appears in either list is used in finding the LCM. This can be confusing. If needed, draw a Venn diagram to discuss this problem.
- The solution can be checked by using the first method of listing multiples.

? "Does it matter which method you use and why?" no; Each method will give you the correct answer.

? "How do you decide which method to use?" If the numbers are relatively small, use the listing method. If the numbers are relatively large, use the prime factorization method.

On Your Own

- Ask volunteers to share their work at the board or document camera to check solutions.

Goal Today's lesson is finding the **least common multiple (LCM)** of two or more numbers.

Technology for the Teacher

Dynamic Classroom

Lesson Tutorials
Lesson Plans
Answer Presentation Tool

Extra Example 1

Find the LCM of 4 and 9. 36

On Your Own

1. 24
2. 36
3. 30

Extra Example 2

Find the LCM of 16 and 24. 48

On Your Own

4. 126
5. 252
6. 360

Extra Example 3

Find the LCM of 3, 8, and 16. 48

On Your Own

7. 40 **8.** 60

9. *Sample answer:* 4, 10, 25

Extra Example 4

An advertising sign changes every 15 seconds. Another advertising sign changes every 25 seconds. Both signs just changed. After how many minutes will both signs change at the same time again? 75 seconds or $1\frac{1}{4}$ minutes

On Your Own

10. 90 seconds or $1\frac{1}{2}$ minutes

Differentiated Instruction

Visual

Show students the following method for finding the LCM of 300 and 45. This method uses exponents in the prime factorization.

$$300 = 2 \times 2 \times 3 \times 5 \times 5 = 2^2 \times 3 \times 5^2$$

$$45 = 3 \times 3 \times 5 = 3^2 \times 5$$

The greatest power of 2 is 2^2.
The greatest power of 3 is 3^2.
The greatest power of 5 is 5^2.
So the LCM is

$$2^2 \times 3^2 \times 5^2 = 900.$$

Laurie's Notes

Example 3

- ? "We want to find the LCM for 3 numbers. Which method do you think will be the most efficient and why?" prime factorization; Listing the multiples of 18 is more challenging than listing the multiples of 4 and 15. Writing the prime factorization of all 3 numbers is fairly quick.
- Work through the problem as shown.
- ? "Is it possible for the LCM of two numbers to be one of the numbers? Explain." yes; The greater of the two numbers could be the LCM if the greater number is a multiple of the lesser number.
- **Big Idea:** The LCM of two numbers will always be greater than or equal to the greater of the two original numbers.

On Your Own

- Ask students to explain which method they used to find the LCM and why they chose that method.
- **MP2 Reason Abstractly and Quantitatively** and **MP3a Construct Viable Arguments:** In Question 9, students are given the LCM and they need to find the two original numbers. This is not a trivial problem, so ask students to describe their reasoning process.

Example 4

- This problem is similar to the opening Motivate puzzle. Students should also see the connection between finding the LCM of 3 and 4 and finding the LCM of 30 and 40.
- Ask a volunteer to read the problem.
- ? "What do the multiples represent in this problem?" how many seconds pass before the light changes
- Although 120 seconds is the LCM, in the context of the problem, it is appropriate to convert 120 seconds to 2 minutes.

On Your Own

- **Neighbor Check:** Have students work independently and then have their neighbors check their work. Have students discuss any discrepancies.

Closure

- **Writing Prompt:** To find the LCM of 8 and 15, I would… *Sample answer:* Find the prime factorization of each number. Then circle each different factor where it appears the greatest number of times.
  $8 = ② \times ② \times ②$
  $15 = ③ \times ⑤$
  The least common multiple of 8 and 15 is $2 \times 2 \times 2 \times 3 \times 5 = 120$.

EXAMPLE 3 Finding the LCM of Three Numbers

Find the LCM of 4, 15, and 18.

Write the prime factorization of each number. Circle each different factor where it appears the greatest number of times.

$4 = (2) \cdot (2)$	2 appears most often here, so circle both 2s.
$15 = 3 \cdot (5)$	5 appears here only, so circle 5.
$18 = 2 \cdot (3) \cdot (3)$	3 appears most often here, so circle both 3s.
$2 \cdot 2 \cdot 5 \cdot 3 \cdot 3 = 180$	Find the product of the circled factors.

∴ So, the LCM of 4, 15, and 18 is 180.

On Your Own

Now You're Ready
Exercises 22–27

Find the LCM of the numbers.

7. 2, 5, 8

8. 6, 10, 12

9. Write a set of numbers whose least common multiple is 100.

EXAMPLE 4 Real-Life Application

A traffic light changes every 30 seconds. Another traffic light changes every 40 seconds. Both lights just changed. After how many minutes will both lights change at the same time again?

Find the LCM of 30 and 40 by listing multiples of each number. Circle the least common multiple.

Multiples of 30: 30, 60, 90, (120), . . .

Multiples of 40: 40, 80, (120), 160, . . .

The LCM is 120. So, both lights will change again after 120 seconds.

Because there are 60 seconds in 1 minute, there are $120 \div 60 = 2$ minutes in 120 seconds.

∴ Both lights will change at the same time again after 2 minutes.

On Your Own

10. **WHAT IF?** In Example 4, the traffic light that changes every 40 seconds is adjusted to change every 45 seconds. Both lights just changed. After how many minutes will both lights change at the same time again?

1.6 Exercises

Vocabulary and Concept Check

1. **VOCABULARY** What is the least common multiple (LCM) of two numbers?
2. **WRITING** Describe how to find the LCM of two numbers by using prime factorization.

Practice and Problem Solving

Use a Venn diagram to find the least common multiple of the numbers.

3. 3, 7
4. 6, 8
5. 12, 15

Find the LCM of the numbers using lists of multiples.

(1)
6. 2, 9
7. 3, 4
8. 8, 9
9. 5, 8
10. 15, 20
11. 12, 18

Find the LCM of the numbers using prime factorizations.

(2)
12. 9, 21
13. 12, 27
14. 18, 45
15. 22, 33
16. 36, 60
17. 35, 50

✗
$6 \times 9 = 54$
The LCM of 6 and 9 is 54.

18. **ERROR ANALYSIS** Describe and correct the error in finding the LCM.

19. **AQUATICS** You have diving lessons every fifth day and swimming lessons every third day. Today you have both lessons. In how many days will you have both lessons on the same day again?
20. **HOT DOGS** Hot dogs come in packs of 10, while buns come in packs of eight. What are the least numbers of packs you should buy in order to have the same numbers of hot dogs and buns?
21. **MODELING** Which model represents an LCM that is different from the other three? Explain your reasoning.

A.
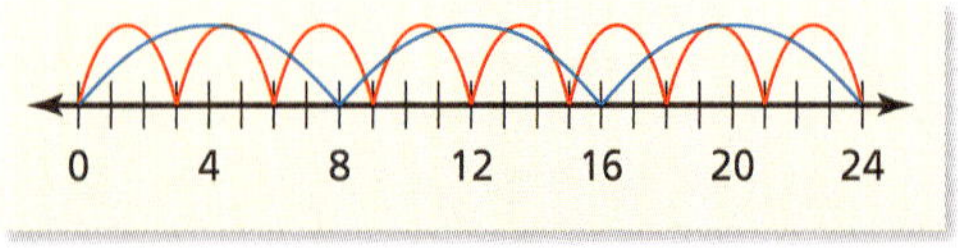

B.
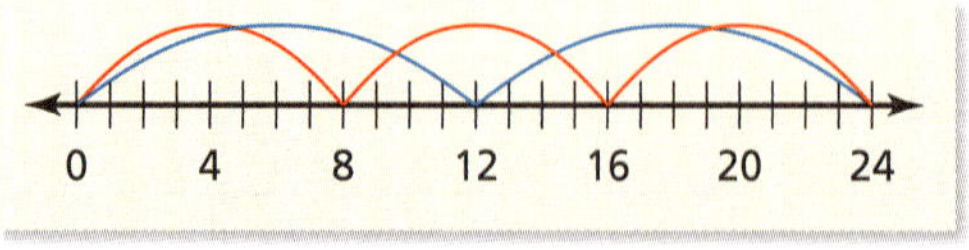

C.
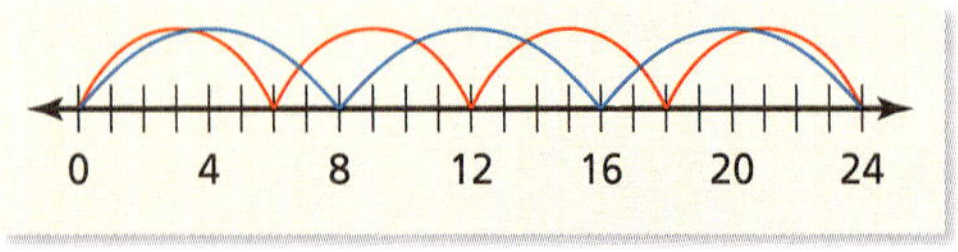

D.
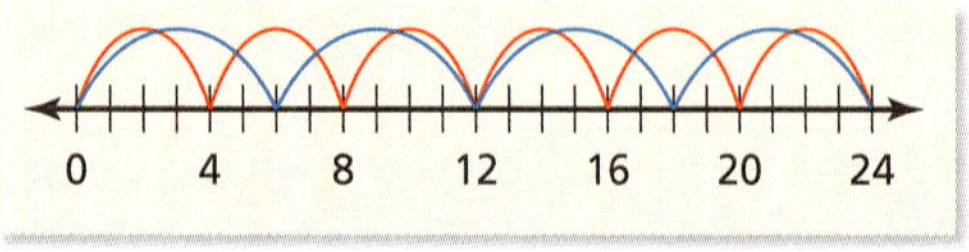

Assignment Guide and Homework Check

Level	Day 1 Activity Assignment	Day 2 Lesson Assignment	Homework Check
Basic	3–5, 37–40	1–2, 7–17 odd, 18, 19–23 odd	9, 15, 19, 23
Average	3–5, 37–40	1–2, 8–20 even, 25–35 odd	10, 14, 20, 25, 33
Advanced	3–5, 37–40	1–2, 10, 14–18 even, 22–32 even, 34–36	10, 16, 22, 26, 34

Common Errors

- **Exercises 3–5** Some students may struggle using a Venn diagram. Make sure they understand that the product of all prime factors in the circles is the LCM of the numbers represented by the circles.
- **Exercises 12–17** After writing the prime factorizations, students may struggle over which factors to use to find the LCM. If needed, students should draw a Venn diagram to clarify the problem.
- **Exercise 21** The number lines may confuse students. Explain to them that the loops of each number will intersect on common multiples.
- **Exercise 35** Some students may struggle if they use a Venn diagram for three sets. Tell them the same rules apply for sets of three as with a set of two. Make sure they understand the product of all prime factors in the circles is the LCM of the numbers represented by the circles.

1.6 Record and Practice Journal

Find the LCM of the numbers using lists of multiples.

1. 3, 8
   24
2. 8, 14
   56
3. 7, 21
   21
4. 5, 11
   55
5. 8, 20
   40
6. 14, 20
   140

Find the LCM of the numbers using prime factorizations.

7. 12, 36
   36
8. 5, 12
   60
9. 3, 17
   51
10. 10, 12
    60
11. 20, 30
    60
12. 32, 40
    160
13. A music store gives every 20th customer a $5 gift card. Every 50th customer gets a $10 gift card. Which customer will be the first to receive both types of gift cards?
    100th customer

Vocabulary and Concept Check

1. The LCM of two numbers is the least of the multiples shared by the two numbers.
2. First, find the prime factorization of both numbers. Next, circle each different factor where it appears the greatest number of times. Then, find the product of the circled factors.

Practice and Problem Solving

3. 21
4. 24
5. 60
6. 18
7. 12
8. 72
9. 40
10. 60
11. 36
12. 63
13. 108
14. 90
15. 66
16. 180
17. 350
18. The product of two numbers is not necessarily the LCM. Use prime factorization to see that the LCM is $2 \cdot 3 \cdot 3 = 18$.
19. 15 days
20. 4 packs of hot dogs and 5 packs of buns
21. D; This model represents multiples of 4 and 6 which have an LCM of 12. The other models represent multiples of 3 and 8, 8 and 12, and 6 and 8, which have an LCM of 24.

Practice and Problem Solving

22. 42 **23.** 165

24. 36 **25.** 120

26. 126 **27.** 1260

28. *Sample answer:* Prime factorization because it is tedious to list all of the multiples of two numbers that do not have any common factors.

29. always

30. sometimes

31. never

32. See *Taking Math Deeper.*

33. 300th caller

34. you: 7 mi; your friend: 6 mi

35. **a.** See Additional Answers.

b. 240

c. 80;120

36. The LCM of the two numbers is equal to their product when the two numbers have no common prime factors.

Fair Game Review

37. 3^2 **38.** 5^4

39. 17^5 **40.** B

Mini-Assessment

Find the LCM of the numbers.

1. 15, 18 90
2. 8, 14 56
3. 24, 32 96
4. 5, 8, 10 40
5. 6, 8, 12 48

Taking Math Deeper

Exercise 32

One way to solve this problem is to represent the arrival times visually using number lines.

1. Create three number lines that you can use to show the arrival times of the three subway lines. Make each number line large enough to show several arrivals for each subway line.

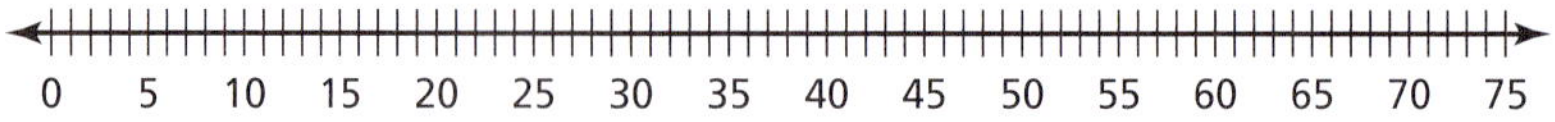

2. The three subway lines just arrived at the same time. Let zero represent this time on each number line. Show the arrival times of the subway lines on the number lines.

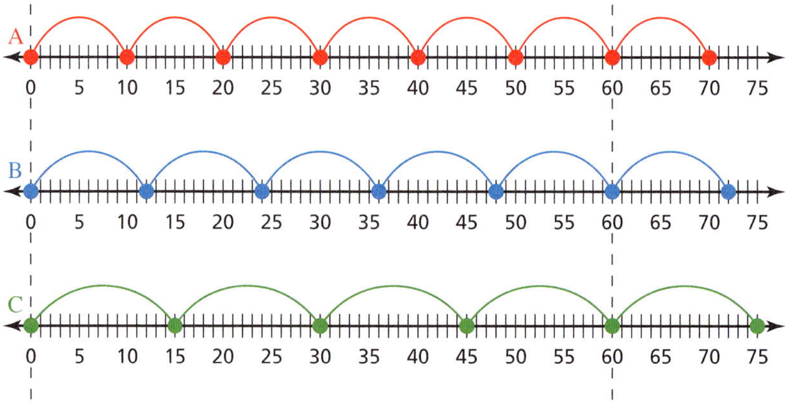

3. After 0, the next value plotted on all three number lines is 60. So, you must wait 60 minutes until all three subway lines arrive at the same time again.

Project

Research the subway system in New York City. Choose a major station and write a real-life problem involving arrival times of subway lines.

Reteaching and Enrichment Strategies

If students need help. . .	If students got it. . .
Resources by Chapter • Practice A and Practice B • Puzzle Time Record and Practice Journal Practice Differentiating the Lesson Lesson Tutorials Skills Review Handbook	Resources by Chapter • Enrichment and Extension • Technology Connection Start the next section

Find the LCM of the numbers.

3 **22.** 2, 3, 7 **23.** 3, 5, 11 **24.** 4, 9, 12

25. 6, 8, 15 **26.** 7, 18, 21 **27.** 9, 10, 28

28. **REASONING** You need to find the LCM of 13 and 14. Would you rather list their multiples or use their prime factorizations? Explain.

CRITICAL THINKING **Tell whether the statement is *always*, *sometimes*, or *never* true.**

29. The LCM of two different prime numbers is their product.

30. The LCM of a set of numbers is equal to one of the numbers in the set.

31. The GCF of two different numbers is the LCM of the numbers.

32. **SUBWAY** At Union Station, you notice that three subway lines just arrived at the same time. The table shows their arrival schedule. How long must you wait until all three lines arrive at Union Station at the same time again?

Subway Line	Arrival Time
A	every 10 min
B	every 12 min
C	every 15 min

33. **RADIO CONTEST** A radio station gives away $15 to every 15th caller, $25 to every 25th caller, and free concert tickets to every 100th caller. When will the station first give away *all* three prizes to one caller?

34. **TREADMILL** You and a friend are running on treadmills. You run 0.5 mile every 3 minutes, and your friend runs 2 miles every 14 minutes. You both start and stop running at the same time and run a whole number of miles. What is the least possible number of miles you and your friend can run?

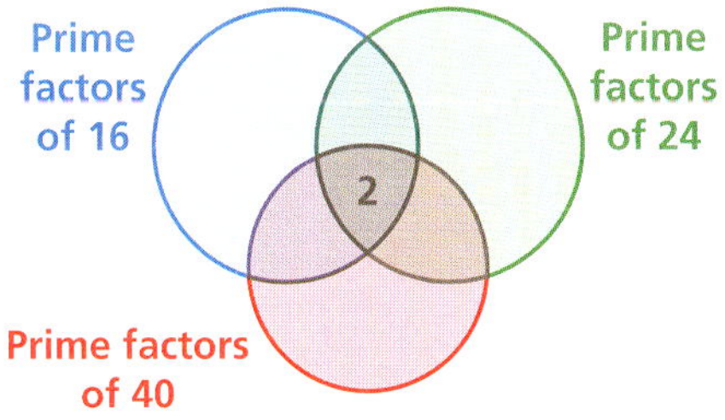

35. **VENN DIAGRAM** Refer to the Venn diagram.

a. Copy and complete the Venn diagram.

b. What is the LCM of 16, 24, and 40?

c. What is the LCM of 16 and 40? 24 and 40?

36. Number Sense When is the LCM of two numbers equal to their product?

Fair Game Review What you learned in previous grades & lessons

Write the product as a power. *(Section 1.2)*

37. 3×3 **38.** $5 \cdot 5 \cdot 5 \cdot 5$ **39.** $17 \times 17 \times 17 \times 17 \times 17$

40. **MULTIPLE CHOICE** Which two powers have the same value? *(Section 1.2)*

Ⓐ 1^3 and 3^1 Ⓑ 2^4 and 4^2 Ⓒ 3^2 and 2^3 Ⓓ 4^3 and 3^4

Extension 1.6 Adding and Subtracting Fractions

Check It Out
Lesson Tutorials
BigIdeasMath.com

Key Vocabulary
least common denominator, *p. 42*

Recall that you can add and subtract fractions with unlike denominators by writing equivalent fractions with a common denominator. One way to do this is by multiplying the numerator and the denominator of each fraction by the denominator of the other fraction.

EXAMPLE 1 Adding Fractions Using a Common Denominator

Find $\frac{5}{8} + \frac{1}{6}$.

Rewrite the fractions with a common denominator. Use the product of the denominators as the common denominator.

$\frac{5}{8} + \frac{1}{6} = \frac{5 \cdot 6}{8 \cdot 6} + \frac{1 \cdot 8}{6 \cdot 8}$ — Rewrite the fractions using a common denominator of 8 • 6 = 48.

$= \frac{30}{48} + \frac{8}{48}$ — Multiply.

$= \frac{38}{48}$ — Add the numerators.

$= \frac{\cancel{2}^{1} \cdot 19}{\cancel{2}_{1} \cdot 24}$ — Divide out the common factor 2.

$= \frac{19}{24}$ — Simplify.

Study Tip
A fraction is in *simplest form* when the numerator and the denominator have no common factors other than 1.

The **least common denominator** (LCD) of two or more fractions is the least common multiple (LCM) of the denominators. The LCD provides another method for adding and subtracting fractions with unlike denominators.

EXAMPLE 2 Adding Fractions Using the LCD

Find $\frac{5}{8} + \frac{1}{6}$.

Find the LCM of the denominators.

Multiples of 8: 8, 16, (24), 32, 40, (48), . . .

Multiples of 6: 6, 12, 18, (24), 30, 36, 42, (48), . . .

The LCM of 8 and 6 is 24. So, the LCD is 24.

$\frac{5}{8} + \frac{1}{6} = \frac{5 \cdot 3}{8 \cdot 3} + \frac{1 \cdot 4}{6 \cdot 4}$ — Rewrite the fractions using the LCD, 24.

$= \frac{15}{24} + \frac{4}{24}$ — Multiply.

$= \frac{19}{24}$ — Add the numerators.

Common Core
Common Multiples
In this extension, you will
- use least common multiples to add and subtract fractions.

Applying Standard 6.NS.4

Multi-Language Glossary at BigIdeasMath.com

Laurie's Notes

Introduction

Connect

- **Yesterday:** Students learned two methods for finding the least common multiple of two numbers. (MP2, MP3a)
- **Today:** Students will use the least common multiple as another possible method to add and subtract fractions with unlike denominators.

Motivate

? "Have you heard the phrase, 'You can't add apples and oranges'?"

- Ask them what they think the phrase means. They might talk about needing a common denominator.

Lesson Notes

Example 1

- Similar to what students have seen in prior grades, one way to get a common denominator is to use the product of the two denominators.
- **MP2 Reason Abstractly and Quantitatively:** Quantitative reasoning involves thinking about the units involved (eighths and sixths) and their meaning. Not just how to compute with them.

? "Can you add $\frac{5}{8} + \frac{1}{6}$ and get $\frac{6}{14}$? Explain." no; $\frac{6}{14}$ is less than $\frac{1}{2}$. Because $\frac{5}{8}$ is greater than $\frac{1}{2}$ and you are adding a positive amount to it, the sum must be greater than $\frac{1}{2}$.

? "What are equivalent fractions?" fractions that are equal but do not have the same denominator

- Explain that you want to rewrite $\frac{5}{8}$ and $\frac{1}{6}$ as equivalent fractions with a common denominator. The common denominator will be $8 \cdot 6$.
- In the first step of the problem, I use an alternate color to emphasize that when you multiply the numerator and denominator by the same number you're just multiplying by 1, so the value of the fraction is unchanged.

? "The sum is $\frac{38}{48}$. Can this fraction be simplified?" yes; by a factor of 2

Example 2

- Define least common denominator (LCD). Explain that using the LCD is another method to add and subtract fractions.
- This example uses the same two fractions as Example 1.

? "What is the least common multiple of 8 and 6?" 24

- Explain that to add the fractions, a common denominator of 24 is used.
- Work through the problem. The sum is the same as Example 1, and it is already in simplest form.
- Discuss the two strategies for finding equivalent fractions: the product of the denominators versus the least common multiple of the denominators.

? "Can you think of a problem where the product of the denominators and the least common multiple of the denominators would be the same?" Hopefully, students will understand that if the denominators have no common factors (relatively prime), the two methods are the same.

Common Core State Standards

6.NS.4 Find . . . the least common multiple of two whole numbers less than or equal to 12. . . .

Goal Today's lesson is using the least common multiple to add and subtract fractions with unlike denominators.

Technology for the Teacher

Dynamic Classroom

Lesson Tutorials
Lesson Plans
Answer Presentation Tool

Extra Example 1

Find $\frac{1}{3} + \frac{1}{4}$. $\frac{7}{12}$

Extra Example 2

Find $\frac{1}{6} + \frac{5}{9}$. $\frac{13}{18}$

Record and Practice Journal Extension 1.6 Practice

1. $\frac{25}{30}, \frac{9}{30}$
2. $\frac{20}{36}, \frac{33}{36}$
3. $>$
4. $<$
5. $=$
6. $<$
7. $1\frac{1}{12}$
8. $\frac{7}{8}$
9. $3\frac{27}{28}$
10. $6\frac{3}{10}$
11. $\frac{1}{4}$
12. $\frac{23}{60}$
13. $4\frac{17}{28}$
14. $\frac{4}{9}$
15. $\frac{1}{12}$ c
16. $2\frac{5}{12}$ lb

Extra Example 3

Find $5\frac{2}{3} + 2\frac{1}{4}$. $7\frac{11}{12}$

Practice

1. $\frac{4}{24}, \frac{9}{24}$
2. $\frac{40}{70}, \frac{21}{70}$
3. $\frac{15}{36}, \frac{8}{36}$
4. $\frac{30}{40}, \frac{25}{40}, \frac{4}{40}$
5. $<$
6. $<$
7. $=$
8. $>$
9. $1\frac{5}{12}$
10. $1\frac{5}{14}$
11. $\frac{17}{60}$
12. $\frac{7}{72}$
13. $5\frac{11}{18}$
14. $5\frac{23}{80}$
15. $1\frac{1}{12}$
16. $1\frac{10}{33}$
17. *Sample answer:* The LCD method uses numbers that are easier to work with, but there is extra work in finding the LCD. Using the other method, there are no preliminary steps for finding the LCD, but there may be more simplifying in the solution.

Mini-Assessment

1. Find $\frac{1}{4} + \frac{3}{8}$ using a common denominator. $\frac{5}{8}$
2. Find $\frac{3}{4} + \frac{1}{8}$ using the LCD. $\frac{7}{8}$
3. Find $5\frac{3}{5} - 1\frac{1}{7}$. $4\frac{16}{35}$

Laurie's Notes

Example 3

- Write the problem. Ask students to estimate an answer and explain their reasoning. A reasonable estimate is about $2\frac{1}{2}$.
- ? "How do you write $4\frac{3}{4}$ as an improper fraction?" Listen for an explanation that gives $\frac{19}{4}$ as an answer.
- ? "How do you write $2\frac{3}{10}$ as an improper fraction?" Listen for an explanation that gives $\frac{23}{10}$ as an answer.
- Explain that the first method uses the product of the two denominators as the common denominator. Work through the example as shown.
- Explain that the second method uses the least common multiple of the two denominators as the common denominator. Work through the example as shown.
- Take time to show students that instead of using improper fractions, the problem can be done by just renaming the fractional parts, and then subtracting the number parts and fraction parts.
- When asked which method they prefer, students will often say the second method because the numbers they work with are smaller. There is usually at least one student, however, who prefers the first method because they can skip the step of finding the LCD if it is not obvious.
- Discuss the Study Tip. Students should be familiar with $\frac{a}{b} \pm \frac{c}{d} = \frac{ad \pm bc}{bd}$ from Grade 5. (CCSS 5.NF.1) Explain the symbol $\pm$ if students are unfamiliar with it. The variables used to generalize the procedure from prior grades should be familiar to students.

Practice

- **Neighbor Check:** Have students work independently and then have their neighbors check their work. Have students discuss any discrepancies.

Closure

- **Exit Ticket:** Find $3\frac{1}{6} + 4\frac{2}{3}$ using each method. $7\frac{5}{6}$

To add or subtract mixed numbers, first rewrite the numbers as improper fractions. Then find the common denominator.

EXAMPLE 3 Subtracting Mixed Numbers

Find $4\frac{3}{4} - 2\frac{3}{10}$.

Write the difference using improper fractions.

$$4\frac{3}{4} - 2\frac{3}{10} = \frac{19}{4} - \frac{23}{10}$$

Study Tip

Notice that Method 1 uses the same procedure shown in Example 1. You can generalize the procedure using the rule

$\frac{a}{b} \pm \frac{c}{d} = \frac{ad \pm bc}{bd}$.

Method 1: Use the product of the denominators as the common denominator.

$$\frac{19}{4} - \frac{23}{10} = \frac{19 \cdot 10}{4 \cdot 10} - \frac{23 \cdot 4}{10 \cdot 4}$$ Rewrite the fractions using a common denominator of $4 \cdot 10 = 40$.

$$= \frac{190}{40} - \frac{92}{40}$$ Multiply.

$$= \frac{98}{40}$$ Subtract the numerators.

$$= \frac{49}{20}, \text{ or } 2\frac{9}{20}$$ Simplify.

Method 2: Use the LCD. The LCM of 4 and 10 is 20.

$$\frac{19}{4} - \frac{23}{10} = \frac{19 \cdot 5}{4 \cdot 5} - \frac{23 \cdot 2}{10 \cdot 2}$$ Rewrite the fractions using the LCD, 20.

$$= \frac{95}{20} - \frac{46}{20}$$ Multiply.

$$= \frac{49}{20}, \text{ or } 2\frac{9}{20}$$ Simplify.

Practice

Use the LCD to rewrite the fractions with the same denominator.

1. $\frac{1}{6}, \frac{3}{8}$
2. $\frac{4}{7}, \frac{3}{10}$
3. $\frac{5}{12}, \frac{2}{9}$
4. $\frac{3}{4}, \frac{5}{8}, \frac{1}{10}$

Copy and complete the statement using <, >, or =.

5. $\frac{4}{5}$ ☐ $\frac{5}{6}$
6. $\frac{5}{14}$ ☐ $\frac{3}{8}$
7. $2\frac{2}{5}$ ☐ $\frac{24}{10}$
8. $4\frac{9}{25}$ ☐ $4\frac{7}{20}$

Add or subtract. Write the answer in simplest form.

9. $\frac{2}{3} + \frac{3}{4}$
10. $\frac{6}{7} + \frac{1}{2}$
11. $\frac{7}{10} - \frac{5}{12}$
12. $\frac{13}{18} - \frac{5}{8}$
13. $2\frac{1}{6} + 3\frac{4}{9}$
14. $4\frac{3}{16} + 1\frac{1}{10}$
15. $1\frac{5}{6} - \frac{3}{4}$
16. $3\frac{2}{3} - 2\frac{4}{11}$

17. **COMPARING METHODS** List some advantages and disadvantages of each method shown in the examples. Which method do you prefer? Why?

1.4–1.6 Quiz

Check It Out
Progress Check
BigIdeasMath.com

List the factor pairs of the number. *(Section 1.4)*

1. 48

2. 56

Write the prime factorization of the number. *(Section 1.4)*

3. 60

4. 72

Find the GCF of the numbers using lists of factors. *(Section 1.5)*

5. 18, 42

6. 24, 44, 52

Find the GCF of the numbers using prime factorizations. *(Section 1.5)*

7. 38, 68

8. 68, 76, 92

Find the LCM of the numbers using lists of multiples. *(Section 1.6)*

9. 8, 14

10. 3, 6, 16

Find the LCM of the numbers using prime factorizations. *(Section 1.6)*

11. 18, 30

12. 6, 24, 32

Add or subtract. Write the answer in simplest form. *(Section 1.6)*

13. $\frac{3}{5} + \frac{2}{3}$

14. $\frac{7}{8} - \frac{3}{4}$

15. PICNIC BASKETS You are creating identical picnic baskets using 30 sandwiches and 42 apples. What is the greatest number of baskets that you can fill using all of the food? *(Section 1.5)*

16. RIBBON You have 52 inches of yellow ribbon and 64 inches of red ribbon. You want to cut the ribbons into pieces of equal length with no leftovers. What is the greatest length of the pieces that you can make? *(Section 1.5)*

17. MUSIC LESSONS You have piano lessons every fourth day and guitar lessons every sixth day. Today you have both lessons. In how many days will you have both lessons on the same day again? Explain. *(Section 1.6)*

18. HAMBURGERS Hamburgers come in packs of 20, while buns come in packs of 12. What is the least number of packs you should buy in order to have the same numbers of hamburgers and buns? *(Section 1.6)*

Alternative Assessment Options

Math Chat	Student Reflective Focus Question
Structured Interview	Writing Prompt

Math Chat

- Work in groups of four. Two members work together to discuss prime factorization and finding the greatest common factor. The other pair work together to discuss prime factorization and finding the least common multiple. When they are finished, they explain their findings to the other members of their group.
- The teacher should walk around the classroom listening to the groups and asking questions to ensure understanding.

Study Help Sample Answers

Remind students to complete Graphic Organizers for the rest of the chapter.

6.

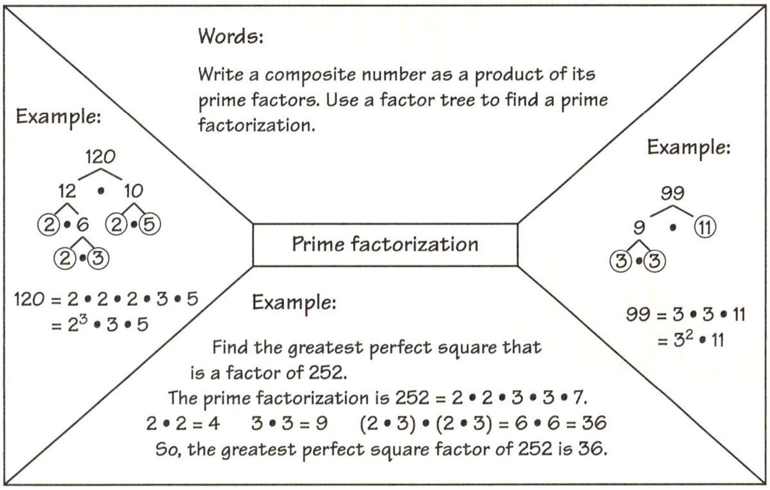

7–9. Available at *BigIdeasMath.com.*

Reteaching and Enrichment Strategies

If students need help. . .	If students got it. . .
Resources by Chapter • Practice A and Practice B • Puzzle Time Lesson Tutorials *BigIdeasMath.com*	Resources by Chapter • Enrichment and Extension • Technology Connection Game Closet at *BigIdeasMath.com* Start the Chapter Review

Answers

1. 1, 48; 2, 24; 3, 16; 4, 12; 6, 8

2. 1, 56; 2, 28; 4, 14; 7, 8

3. $2^2 \times 3 \times 5$

4. $2^3 \times 3^2$

5. 6

6. 4

7. 2

8. 4

9. 56

10. 48

11. 90

12. 96

13. $1\frac{4}{15}$

14. $\frac{1}{8}$

15. 6 baskets

16. 4 in.

17. 12 days; The LCM of 4 and 6 is 12. So, you will have both lessons on the same day in 12 days.

18. 3 packs of hamburgers and 5 packs of buns

Technology for the Teacher

Online Assessment
Assessment Book
ExamView® Assessment Suite

For the Teacher
Additional Review Options

- *BigIdeasMath.com*
- Online Assessment
- Game Closet at *BigIdeasMath.com*
- Vocabulary Help
- Resources by Chapter

Answers

1. 7281
2. 5156
3. 7296
4. $6\frac{30}{31}$
5. 343
6. 64
7. 256

Review of Common Errors

Exercises 1, 2

- Students may set up the problem incorrectly using the vertical method of adding or subtracting. Remind them to line up the numbers by place value, then add or subtract.

Exercise 3

- Students may incorrectly line up the multiplication of the tens digit. Remind them that the right-most digit should line up in the tens-digit place.

Exercises 5–7

- Students may make the mistake of writing the exponent as a factor. Again, remind them that the exponent is the number of times the base is used as a factor. You may want to demonstrate this point with a quick example. For instance, point out that $5^2 = 5 \times 5 = 25$ but $5 \times 2 = 10$.

Exercises 14–16

- Students may think that smaller numbers have fewer factors. Remind students to keep finding factors until all factors are prime numbers.

Exercises 26–28

- After writing the prime factorizations, students may struggle over which factors to use to find the LCM. If needed, students should draw a Venn diagram to clarify the problem.

1 Chapter Review

Check It Out
Vocabulary Help
BigIdeasMath.com

Review Key Vocabulary

power, *p. 12*
base, *p. 12*
exponent, *p. 12*
perfect square, *p. 13*
numerical expression, *p. 18*
evaluate, *p. 18*
order of operations, *p. 18*
factor pair, *p. 26*
prime factorization, *p. 26*
factor tree, *p. 26*
Venn diagram, *p. 30*
common factors, *p. 32*
greatest common factor (GCF), *p. 32*
common multiples, *p. 38*
least common multiple (LCM), *p. 38*
least common denominator (LCD), *p. 42*

Review Examples and Exercises

1.1 Whole Number Operations *(pp. 2–9)*

Use the tens place because 203 is less than 508.

$$\begin{array}{r} 2 \\ 203\overline{)\,5081} \\ -\,406 \\ \hline 102 \end{array}$$

Divide 508 by 203: There are two groups of 203 in 508.
Multiply 2 and 203.
Subtract 406 from 508.

Next, bring down the 1 and divide the ones.

$$\begin{array}{r} 25 \text{ R6} \\ 203\overline{)\,5081} \\ -\,406\downarrow \\ \hline 1021 \\ -\,1015 \\ \hline 6 \end{array}$$

Divide 1021 by 203: There are five groups of 203 in 1021.
Multiply 5 and 203.
Subtract 1015 from 1021.

The quotient of 5081 and 203 is $25\frac{6}{203}$.

Exercises

Find the value of the expression. Use estimation to check your answer.

1. $4382 + 2899$

2. $8724 - 3568$

3. 192×38

4. $216 \div 31$

1.2 Powers and Exponents *(pp. 10–15)*

Evaluate 6^2.

$6^2 = 6 \cdot 6 = 36$ Write as repeated multiplication and simplify.

Exercises

Find the value of the power.

5. 7^3

6. 2^6

7. 4^4

1.3 Order of Operations (pp. 16–21)

Evaluate $4^3 - 15 \div 5$.

$$\begin{aligned} 4^3 - 15 \div 5 &= 64 - 15 \div 5 && \text{Evaluate } 4^3. \\ &= 64 - 3 && \text{Divide 15 by 5.} \\ &= 61 && \text{Subtract 3 from 64.} \end{aligned}$$

Exercises

Evaluate the expression.

8. $3 \times 6 - 12 \div 6$ **9.** $20 \times (3^2 - 4) \div 50$ **10.** $5 + (4^2 + 2) \div 6$

1.4 Prime Factorization (pp. 24–29)

Write the prime factorization of 18.

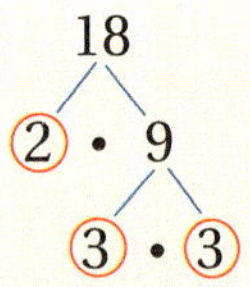

Find a factor pair and draw "branches."

Circle the prime factors as you find them.

Continue until each branch ends at a prime factor.

The prime factorization of 18 is $2 \cdot 3 \cdot 3$, or $2 \cdot 3^2$.

Exercises

List the factor pairs of the number.

11. 28 **12.** 44 **13.** 63

Write the prime factorization of the number.

14. 42 **15.** 50 **16.** 66

1.5 Greatest Common Factor (pp. 30–35)

a. Find the GCF of 32 and 76.

Factors of 32: ①, ②, ④, 8, 16, 32
Factors of 76: ①, ②, ④, 19, 38, 76

The greatest of the common factors is 4.

So, the GCF of 32 and 76 is 4.

b. Find the GCF of 45 and 63.

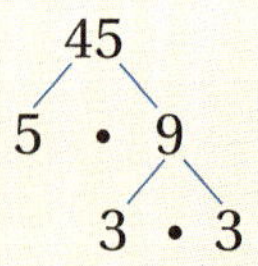

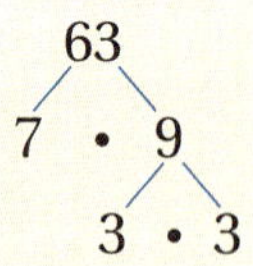

$45 = 3 \cdot 3 \cdot 5$

$63 = 3 \cdot 3 \cdot 7$

$3 \cdot 3 = 9$

So, the GCF of 45 and 63 is 9.

Review Game

Numerical Expressions and Factors

Materials

- copy of game
- pencil
- paper

Players: 2

Directions

- Play in pairs. There are five rounds, each with four parts.
- In each round, players take turns choosing a part and determining the answer. Each part may be used only once.
- Players add their answers in the round. The winner of the round is determined by the *Who Wins?* statement of the round.
- The winner of the round earns one point.

Who Wins? The player with the most points after the five rounds.

Round 1: Whole Number Operations
Direction Simplify the expressions.
Who Wins? The player with the greatest total wins.

A. $195 + 217$ **B.** $569 - 497$ **C.** 12×36 **D.** $583 \div 11$

Round 2: Powers and Exponents
Direction Simplify the expressions.
Who Wins? The player with the least total wins.

E. 3^5 **F.** 5^3 **G.** 4^4 **H.** 6^3

Round 3: Order of Operations
Direction Simplify the expressions.
Who Wins? The player with the greatest total wins.

I. $3 + (5 \cdot 4)$ **J.** $17 - 5 \cdot 6 \div 10$
K. $36 \div (7 - 1) \cdot (15 \div 3) + 4$ **L.** $7 - 4 + 5 \cdot 7$

Round 4: Greatest Common Factor
Direction Determine the greatest common factor of the numbers listed.
Who Wins? The player with the greatest total wins.

M. 16, 64 **N.** 42, 63 **P.** 18, 45 **Q.** 9, 39, 42

Round 5: Least Common Multiple
Direction Determine the least common multiple of the numbers listed.
Who Wins? The player with the least total wins.

R. 4, 6 **S.** 7, 4 **T.** 3, 8 **U.** 5, 6, 12

For the Student Additional Practice

- Lesson Tutorials
- Multi-Language Glossary
- Self-Grading Progress Check
- *BigIdeasMath.com*
  Dynamic Student Edition
  Student Resources

Answers

8. 16

9. 2

10. 8

11. 1, 28; 2, 14; 4, 7

12. 1, 44; 2, 22; 4, 11

13. 1, 63; 3, 21; 7, 9

14. $2 \times 3 \times 7$

15. 2×5^2

16. $2 \times 3 \times 11$

17. 9

18. 6

19. 4

20. 6

21. 4

22. 8

23. 28

24. 60

25. 84

26. 90

27. 60

28. 54

29. $\frac{15}{28}$

30. $\frac{67}{72}$

31. $1\frac{11}{30}$

32. $1\frac{8}{21}$ gal

My Thoughts on the Chapter

What worked. . .

> **Teacher Tip**
> Not allowed to write in your teaching edition? Use sticky notes to record your thoughts.

What did not work. . .

What I would do differently. . .

Exercises

Find the GCF of the numbers using lists of factors.

17. 27, 45 **18.** 30, 48 **19.** 28, 48, 64

Find the GCF of the numbers using prime factorizations.

20. 24, 90 **21.** 52, 68 **22.** 32, 56, 96

1.6 Least Common Multiple *(pp. 36–43)*

a. Find the LCM of 8 and 12.

Make a factor tree for each number.

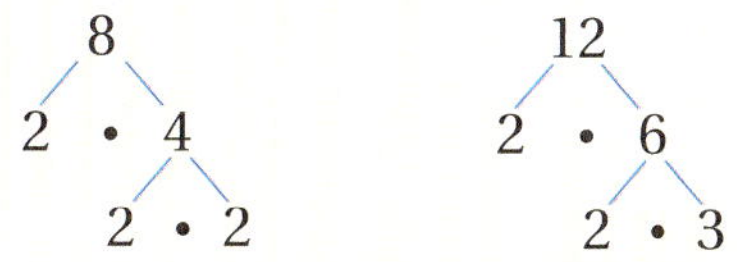

Write the prime factorization of each number. Circle each different factor where it appears the greater number of times.

$8 = 2 \cdot 2 \cdot 2$ — 2 appears more often here, so circle all 2s.

$12 = 2 \cdot 2 \cdot 3$ — 3 appears once. Do not circle the 2s again.

$2 \cdot 2 \cdot 2 \cdot 3 = 24$ — Find the product of the circled factors.

So, the LCM of 8 and 12 is 24.

b. Find $\frac{1}{2} + \frac{1}{3}$.

The LCM of 2 and 3 is 6. So, the LCD is 6.

$$\frac{1}{2} + \frac{1}{3} = \frac{1 \cdot 3}{2 \cdot 3} + \frac{1 \cdot 2}{3 \cdot 2} = \frac{3}{6} + \frac{2}{6} = \frac{5}{6}$$

Exercises

Find the LCM of the numbers using lists of multiples.

23. 4, 14 **24.** 6, 20 **25.** 12, 28

Find the LCM of the numbers using prime factorizations.

26. 6, 45 **27.** 10, 12 **28.** 18, 27

Add or subtract. Write the answer in simplest form.

29. $\frac{2}{7} + \frac{1}{4}$ **30.** $\frac{5}{9} + \frac{3}{8}$ **31.** $3\frac{5}{6} - 2\frac{7}{15}$

32. **WATER PITCHER** A water pitcher contains $\frac{2}{3}$ gallon of water. You add $\frac{5}{7}$ gallon of water to the pitcher. How much water does the pitcher contain?

1 Chapter Test

Check It Out
Test Practice
BigIdeasMath.com

Find the value of the expression. Use estimation to check your answer.

1. $3963 + 2379$

2. $6184 - 2348$

3. 184×26

4. $207 \div 23$

Find the value of the power.

5. 2^3

6. 15^2

7. 5^4

Evaluate the expression.

8. $11 \times 8 - 6 \div 2$

9. $5 + 2^3 \div 4 - 2$

10. $6 + 4(11 - 2) \div 3^2$

List the factor pairs of the number.

11. 52

12. 66

Write the prime factorization of the number.

13. 46

14. 28

Find the GCF of the numbers using lists of factors.

15. 24, 54

16. 16, 32, 72

Find the GCF of the numbers using prime factorizations.

17. 52, 65

18. 18, 45, 63

Find the LCM of the numbers using lists of multiples.

19. 14, 21

20. 9, 24

Find the LCM of the numbers using prime factorizations.

21. 26, 39

22. 6, 12, 14

23. **BRACELETS** You have 16 yellow beads, 20 red beads, and 24 orange beads to make identical bracelets. What is the greatest number of bracelets that you can make using all the beads?

24. **MARBLES** A bag contains equal numbers of green and blue marbles. You can divide all the green marbles into groups of 12 and all the blue marbles into groups of 16. What is the least number of each color of marble that can be in the bag?

25. **SCALE** You place a $3\frac{3}{8}$-pound weight on the left side of a balance scale and a $1\frac{1}{5}$-pound weight on the right side. How much weight do you need to add to the right side to balance the scale?

Test Item References

Chapter Test Questions	Section to Review	Common Core State Standards
1–4	1.1	6.NS.2
5–7	1.2	6.EE.1
8–10	1.3	6.EE.1
11–14	1.4	6.NS.4
15–18, 23	1.5	6.NS.4, 6.EE.2b
19–22, 24, 25	1.6	6.NS.4

Test-Taking Strategies

Remind students to quickly look over the entire test before they start so that they can budget their time. When they receive their tests, students should jot down simple examples of finding the greatest common factor and least common multiple on the back of the test. By doing this they will not become confused when they are under pressure. Have students use the **Stop** and **Think** strategy before they answer each question.

Common Errors

- **Exercises 5–7** Students may make the mistake of writing the exponent as a factor. Again, remind them that the exponent is the number of times the base is used as a factor. You may want to demonstrate this point with a quick example. For instance, point out that $5^2 = 5 \times 5 = 25$ but $5 \times 2 = 10$.
- **Exercises 13, 14, 17, 18, 21, 22** Students may think that smaller numbers have fewer factors. Remind students to keep finding factors until all factors are prime numbers.
- **Exercises 21, 22** After writing the prime factorizations, students may struggle over which factors to use to find the LCM. If needed, students should draw a Venn diagram to clarify the problem.

Reteaching and Enrichment Strategies

If students need help. . .	If students got it. . .
Resources by Chapter • Practice A and Practice B • Puzzle Time Record and Practice Journal Practice Differentiating the Lesson Lesson Tutorials *BigIdeasMath.com* Skills Review Handbook	Resources by Chapter • Enrichment and Extension • Technology Connection Game Closet at *BigIdeasMath.com* Start Standards Assessment

Answers

1. 6342
2. 3836
3. 4784
4. 9
5. 8
6. 225
7. 625
8. 85
9. 5
10. 10
11. 1, 52; 2, 26; 4, 13
12. 1, 66; 2, 33; 3, 22; 6, 11
13. 2×23
14. $2^2 \times 7$
15. 6
16. 8
17. 13
18. 9
19. 42
20. 72
21. 78
22. 84
23. 4 bracelets
24. 48 of each color of marble
25. $2\frac{7}{40}$ lb

Technology for the Teacher

Online Assessment
Assessment Book
ExamView® Assessment Suite

Test-Taking Strategies

Available at *BigIdeasMath.com*

After Answering Easy Questions, Relax
Answer Easy Questions First
Estimate the Answer
Read All Choices before Answering
Read Question before Answering
Solve Directly or Eliminate Choices
Solve Problem before Looking at Choices
Use Intelligent Guessing
Work Backwards

About this Strategy

When taking a multiple choice test, be sure to read each question carefully and thoroughly. Before answering a question, determine exactly what is being asked, then eliminate the wrong answers and select the best choice.

Answers

1. A
2. I
3. C
4. 98

Technology for the Teacher

Common Core State Standards Support Performance Tasks
Online Assessment
Assessment Book
ExamView® Assessment Suite

Item Analysis

1. **A.** Correct answer

   B. The student finds the GCF of 24 and 30 instead of the GCF of 24, 30, and 40.

   C. The student finds the GCF of 24 and 40 instead of the GCF of 24, 30, and 40.

   D. The student finds the GCF of 30 and 40 instead of the GCF of 24, 30, and 40.

2. **F.** The student thinks the area is equal to the side length.

   G. The student doubles the side length instead of squaring the side length.

   H. The student calculates the perimeter instead of the area.

   I. Correct answer

3. **A.** The student does not follow the correct order of operations and subtracts 8 from 2^3 first.

   B. The student does not follow the correct order of operations and subtracts 8 from $3 \cdot 2^3$ instead of dividing 8 by 4.

   C. Correct answer

   D. The student does not follow the correct order of operations and calculates $(3 \cdot 2)^3$ instead of $3 \cdot 2^3$.

4. **Gridded response:** Correct answer: 98

   Common error: The student calculates the GCF instead of the LCM.

Standards Assessment Icons

Gridded Response

Short Response (2-point rubric)

Extended Response (4-point rubric)

1 Standards Assessment

Test-Taking Strategy
Solve Directly or Eliminate Choices

"Which strategy would you use on this one: solve directly or eliminate choices?"

1. You are making identical bagel platters using 40 plain bagels, 30 raisin bagels, and 24 blueberry bagels. What is the greatest number of platters that you can make if there are no leftover bagels? *(6.NS.4)*

A. 2

B. 6

C. 8

D. 10

2. The top of an end table is a square with a side length of 16 inches. What is the area of the tabletop? *(6.EE.1)*

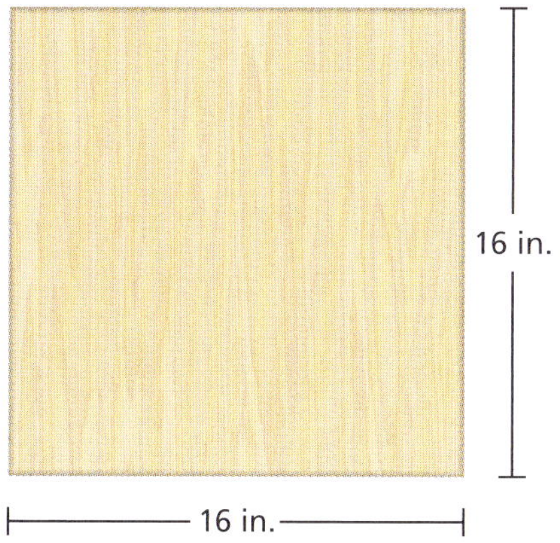

F. 16 in.^{2}

G. 32 in.^{2}

H. 64 in.^{2}

I. 256 in.^{2}

3. Which number is equivalent to the expression below? *(6.EE.1)*

$$3 \cdot 2^3 - 8 \div 4$$

A. 0

B. 4

C. 22

D. 214

4. What is the least common multiple of 14 and 49? *(6.NS.4)*

5. Which number is equivalent to the expression 7059 ÷ 301? *(6.NS.2)*

F. 23

G. $23\frac{136}{7059}$

H. $23\frac{136}{301}$

I. 136

6. You are building identical displays for the school fair using 65 blue boxes and 91 yellow boxes. What is the greatest number of displays you can build using all the boxes? *(6.NS.4)*

A. 13

B. 35

C. 91

D. 156

7. You hang the two strands of decorative lights shown below.

Strand 1: changes between red and blue every 15 seconds

Strand 2: changes between green and gold every 18 seconds

Both strands just changed color. After how many seconds will the strands change color at the same time again? *(6.NS.4)*

F. 3 seconds

G. 30 seconds

H. 90 seconds

I. 270 seconds

8. Which expression is equivalent to $\frac{29}{63}$? *(6.NS.4)*

A. $\frac{28}{60} + \frac{1}{3}$

B. $\frac{4}{27} + \frac{25}{36}$

C. $\frac{5}{21} + \frac{2}{9}$

D. $\frac{22}{47} + \frac{7}{16}$

9. Which expression is *not* equivalent to 32? *(6.EE.1)*

F. $6^2 - 8 \div 2$

G. $30 \div 2 + 5^2 - 8$

H. $30 + 4^2 \div (2 + 6)$

I. $8^2 \div 4 - 2$

10. Which number is equivalent to the expression 148 × 27? *(6.NS.2)*

A. 3696

B. 3896

C. 3946

D. 3996

Item Analysis (continued)

5. **F.** The student does not include the fractional part of the answer.
   G. The student incorrectly uses the dividend when writing the fractional part of the answer.
   H. Correct answer
   I. The student confuses the remainder with the quotient.

6. **A.** Correct answer
   B. The student multiplies 5 (a factor of 65) by 7 (a factor of 91) instead of finding the GCF of 65 and 91.
   C. The student chooses the greater of 65 and 91.
   D. The student adds 65 and 91.

7. **F.** The student finds the GCF instead of the LCM.
   G. The student incorrectly multiplies 2, 3, and 5 to find the LCM.
   H. Correct answer
   I. The student incorrectly multiplies 15 and 18 to find the LCM.

8. **A.** The student incorrectly adds the numerators and adds the denominators.
   B. The student incorrectly adds the numerators and adds the denominators.
   C. Correct answer
   D. The student incorrectly adds the numerators and adds the denominators.

9. **F.** The student does not apply the correct order of operations.
   G. The student does not apply the correct order of operations.
   H. The student does not apply the correct order of operations.
   I. Correct answer

10. **A.** The student does not carry the 3 to the hundreds place when multiplying 148 by 7.
    B. The student does not carry the 1 to the hundreds place when multiplying 148 by 20.
    C. The student does not carry the 5 to the tens place when multiplying 148 by 7.
    D. Correct answer

Answers

5. H
6. A
7. H
8. C
9. I
10. D

Answers

11. 6

12. H

13. 5, 1, 3; The GCF of the three numbers must be 1 because two of the numbers do not have any factors in common except 1.

14. C

15. I

16. B

Item Analysis (continued)

11. Gridded response: Correct answer: 6

Common error: The student does not find the GCF correctly.

12. **F.** The student does not apply the correct order of operations.

G. The student does not apply the correct order of operations.

H. Correct answer

I. The student does not apply the correct order of operations.

13. **2 points** The student demonstrates a thorough understanding of the GCF of the pairs of numbers and of the set of three numbers. GCF(10, 15) = 5, GCF(10, 21) = 1, GCF(15, 21) = 3, GCF(10, 15, 21) = 1; The GCF of the three numbers must be 1, because two of the numbers, 10 and 21, have only 1 as a common factor.

1 point The student demonstrates a partial understanding of GCF. For instance, the student does not find the GCF of the three numbers.

0 points The student demonstrates insufficient understanding of the GCF of a set of numbers.

14. **A.** The student performs a calculation incorrectly or does not understand the concept of a perfect square.

B. The student performs a calculation incorrectly or does not understand the concept of a perfect square.

C. Correct answer

D. The student performs a calculation incorrectly or does not understand the concept of a perfect square.

15. **F.** The student multiplies 4 by 12 instead of finding the LCM.

G. The student multiplies 6 by 8 instead of finding the LCM.

H. The student finds a common multiple of 8 and 24, but not the LCM.

I. Correct answer

16. **A.** The student does not follow the correct order of operations and uses the Distributive Property incorrectly, calculating $3 \cdot 6$ instead of 3 times the quantity $6 + 2^2$.

B. Correct answer

C. The student does not follow the correct order of operations and calculates, $3 \cdot 6 + (3 \cdot 2)^2$.

D. The student does not follow the correct order of operations and incorrectly adds $6 + 2$ before squaring.

11. You have 60 nickels, 48 dimes, and 42 quarters. You want to divide the coins into identical groups with no coins left over. What is the greatest number of groups that you can make? *(6.NS.4)*

12. Erica was evaluating the expression in the box below.

$$\begin{aligned} 56 \div (2^3 - 1) \times 4 &= 56 \div (8 - 1) \times 4 \\ &= 56 \div 7 \times 4 \\ &= 56 \div 28 \\ &= 2 \end{aligned}$$

What should Erica do to correct the error that she made? *(6.EE.1)*

F. Divide 56 by 8 because operations are performed left to right.

G. Multiply 1 by 4 because multiplication is done before subtraction.

H. Divide 56 by 7 because operations are performed left to right.

I. Divide 56 by 8 and multiply 1 by 4 because division and multiplication are performed before subtraction.

13. Find the greatest common factor for each pair of numbers.

Think
Solve
Explain

10 and 15   10 and 21   15 and 21

What can you conclude about the greatest common factor of 10, 15, and 21? Explain your reasoning. *(6.NS.4)*

14. Which number is *not* a perfect square? *(6.EE.1)*

A. 64   **C.** 96

B. 81   **D.** 100

15. Which number pair has a least common multiple of 48? *(6.NS.4)*

F. 4, 12   **H.** 8, 24

G. 6, 8   **I.** 16, 24

16. Which number is equivalent to the expression below? *(6.EE.1)*

$$\frac{3(6 + 2^2) + 2}{8}$$

A. 3   **C.** 7

B. 4   **D.** $24\frac{1}{4}$

2 Fractions and Decimals

"Dear Sir: You say that **MOST** humans use only a fraction of their brain power."

"But, $\frac{3}{2}$ is a fraction. So, does that mean that **SOME** humans use one and a half of their brain power?"

"One of my homework problems is 'How many halves are in five halves?'"

Common Core Progression

4th Grade
• Compare fractions. • Multiply fractions by whole numbers. • Compare decimals to the hundredths place.
5th Grade
• Multiply fractions or whole numbers by fractions. • Interpret a fraction as division; solve problems involving division of whole numbers leading to answers in the form of a mixed number. • Add, subtract, multiply, and divide decimals to the hundredths place.
6th Grade
• Divide fractions and mixed numbers. • Add, subtract, multiply, and divide decimals efficiently.

Chapter Summary

Section	Common Core State Standard	
2.1	Preparing for	6.NS.1
2.2	Learning	6.NS.1
2.3	Learning	6.NS.1 ★
2.4	Learning	6.NS.3
2.5	Learning	6.NS.3
2.6	Learning	6.NS.3 ★

★ Teaching is complete. Standard can be assessed.

Pacing Guide for Chapter 2

Chapter Opener	1 Day
Section 1 Activity Lesson	 1 Day 2 Days
Section 2 Activity Lesson	 1 Day 2 Days
Section 3 Activity Lesson	 1 Day 1 Day
Study Help / Quiz	1 Day
Section 4 Activity Lesson	 1 Day 1 Day
Section 5 Activity Lesson	 1 Day 2 Days
Section 6 Activity Lesson	 1 Day 2 Days
Chapter Review/ Chapter Tests	2 Days
Total Chapter 2	20 Days
Year-to-Date	38 Days

Technology for the Teacher

BigIdeasMath.com
Chapter at a Glance
Complete Materials List
Parent Letters: English and Spanish

Common Core State Standards

5.NBT.4 Use place value understanding to round decimals to any place.
5.NBT.5 Fluently multiply multi-digit whole numbers using the standard algorithm.
5.NBT.6 Find whole-number quotients of whole numbers up to four-digit dividends and two-digit divisors, ...

Additional Topics for Review

- Factors of Whole Numbers
- Simplifying Fractions
- Mixed Numbers and Improper Fractions
- Decimal Place Value

Try It Yourself

1.	200	**2.**	400
3.	1000	**4.**	2400
5.	4	**6.**	3
7.	6	**8.**	4
9.	3825	**10.**	12,978
11.	17,371	**12.**	6036
13.	40	**14.**	107
15.	38	**16.**	361

Record and Practice Journal Fair Game Review

1–9. Sample answers are given.

1.	1800	**2.**	1800
3.	4	**4.**	20
5.	5	**6.**	800
7.	1500	**8.**	13
9.	12 batches		
10.	5265	**11.**	5049
12.	37	**13.**	158
14.	95		

Math Background Notes

Vocabulary Review

- Product
- Quotient
- Estimating
- Evaluate

Estimating Whole Number Products and Quotients

- Students should be familiar with estimating whole number operations. They should also be familiar with multiplication and division of whole number multiples of 10, as well as with checking the result of a multiplication or division problem for reasonableness.
- You may want to review rounding with students. When estimating a quotient, students should use compatible numbers.
- **Common Error:** Students may multiply incorrectly after they round the factors in a product or divide incorrectly after they round the dividend and divisor in a quotient. For instance, in Example 1, students may end up with a product of 270. In Example 2, students may end up with a quotient of 30. As with adding and subtracting whole numbers, encourage students to get into the habit of checking their work.

Multiplying and Dividing Whole Numbers

- Students have been multiplying and dividing whole numbers.
- The review of multiplication and division skills with whole numbers will help students transition smoothly into the new material being presented in the chapter.
- You may want to review the multiplication and division algorithm with students.
- **Teaching Tip:** This section could be difficult for students who are still struggling to master their basic multiplication facts. You may consider allowing these students to use a multiplication chart to aid them through the problems, while encouraging them to master the facts.

Reteaching and Enrichment Strategies

If students need help. . .	If students got it. . .
Record and Practice Journal • Fair Game Review Skills Review Handbook Lesson Tutorials	Game Closet at *BigIdeasMath.com* Start the next section

What You Learned Before

"On a scale from 1 to 10, how do you like my painting?"

Estimating Whole Number Products and Quotients (5.NBT.4)

Example 1 Estimate 32 × 88.

32 is close to 30.

$32 \times 88 \approx 30 \times 90 = 2700$

88 is close to 90.

Example 2 Estimate 176 ÷ 57.

176 is close to 180.

$176 \div 57 \approx 180 \div 60 = 3$

57 is close to 60.

Try It Yourself

Estimate the product or the quotient.

1. 9×23 **2.** 19×22 **3.** 49×21 **4.** 38×61

5. $38 \div 9$ **6.** $63 \div 22$ **7.** $118 \div 19$ **8.** $245 \div 62$

Multiplying and Dividing Whole Numbers (5.NBT.5, 5.NBT.6)

Example 3 Find 356 × 21.

$$\begin{array}{r} {\scriptstyle 11} \\ 356 \\ \times \quad 21 \\ \hline 356 \\ +\ 7120 \\ \hline 7476 \end{array}$$

Example 4 Estimate 765 ÷ 3.

$$\begin{array}{r} 255 \\ 3\overline{)765} \\ -\ 6 \\ \hline 16 \\ -\ 15 \\ \hline 15 \\ -\ 15 \\ \hline 0 \end{array}$$

Try It Yourself

Find the product or the quotient.

9. $\begin{array}{r} 425 \\ \times \quad 9 \\ \hline \end{array}$ **10.** $\begin{array}{r} 721 \\ \times \quad 18 \\ \hline \end{array}$ **11.** $\begin{array}{r} 599 \\ \times \quad 29 \\ \hline \end{array}$ **12.** $\begin{array}{r} 503 \\ \times \quad 12 \\ \hline \end{array}$

13. $7\overline{)280}$ **14.** $4\overline{)428}$ **15.** $14\overline{)532}$ **16.** $23\overline{)8303}$

2.1 Multiplying Fractions

Essential Question What does it mean to multiply fractions?

1 ACTIVITY: Multiplying Fractions

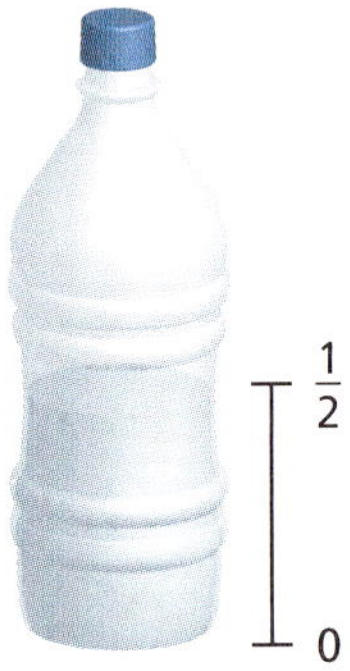

Work with a partner. A bottle of water is $\frac{1}{2}$ full. You drink $\frac{2}{3}$ of the water. How much of the bottle of water do you drink?

THINK ABOUT THE QUESTION: To help you think about this question, rewrite the question.

Words: What is $\frac{2}{3}$ of $\frac{1}{2}$? **Numbers:** $\frac{2}{3} \times \frac{1}{2} = ?$

Here is one way to get the answer.

- **Draw** a length of $\frac{1}{2}$.

Because you want to find $\frac{2}{3}$ of the length, divide it into 3 equal sections.

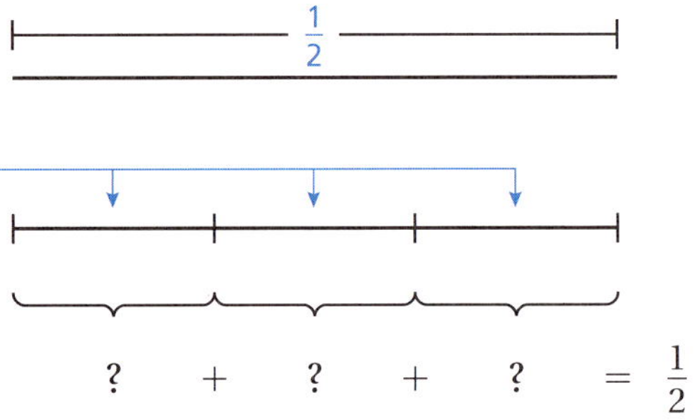

Now, you need to think of a way to divide $\frac{1}{2}$ into 3 equal parts.

- **Rewrite** $\frac{1}{2}$ as a fraction whose numerator is divisible by 3.

Because the length is divided into 3 equal sections, multiply the numerator and denominator by 3.

In this form, you see that $\frac{3}{6}$ can be divided into 3 equal parts of $\frac{1}{6}$.

- Each part is $\frac{1}{6}$ of the bottle of water, and you drank two of them. Written as multiplication, you have

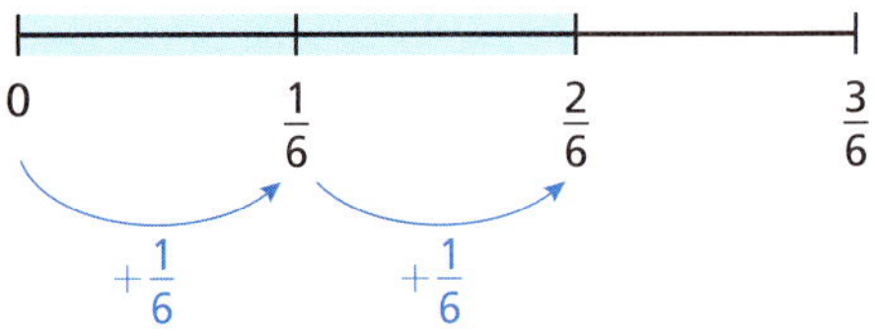

$$\frac{2}{3} \times \frac{1}{2} = \frac{2}{\square} = \frac{\square}{\square}.$$

So, you drank $\frac{\square}{\square}$ of the bottle of water.

COMMON CORE
Dividing Fractions
In this lesson, you will
- use models to multiply fractions.
- multiply fractions by fractions.

Preparing for Standard 6.NS.1

Laurie's Notes

Introduction

Standards for Mathematical Practice

- **MP1a Make Sense of Problems:** Visual and concrete models are useful tools to help students make sense of a problem. You want students to develop a habit of sketching models.

Motivate

- Have the students try the following problems with partners.

  What is $\frac{1}{2}$ of 20? What is $\frac{1}{2}$ of 6? What is $\frac{1}{2}$ of 2?

  What is $\frac{1}{2}$ of 1? What is $\frac{1}{2}$ of $\frac{1}{2}$? What is $\frac{1}{2}$ of $\frac{1}{4}$?

  What is $\frac{1}{2}$ of $\frac{1}{3}$? What is $\frac{1}{2}$ of $\frac{3}{4}$?
- Students may need to think a bit longer to get the last few answers. If so, give them a hint: "What would $\frac{1}{2}$ of $\frac{1}{4}$ of a sandwich look like?" "What would $\frac{1}{2}$ of $\frac{1}{3}$ of a pizza look like?" etc.
- **MP3a Construct Viable Arguments:** For younger students, drawing pictures is one way to construct a viable argument.

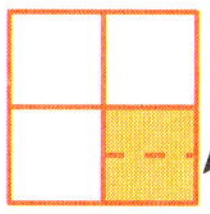

$\frac{1}{2}$ of $\frac{1}{4}$

$\frac{1}{2}$ of $\frac{1}{3}$

Activity Notes

Activity 1

- Divide the bottle into halves. Use a marker to label every $\frac{1}{2}$ interval.
- Fill the bottle with water until it is $\frac{1}{2}$ full. (The more cylindrical the bottle, the better. Also, a drop of food coloring may help the students in the back of the classroom to see the water.)
- ? "How much water is in the bottle?" $\frac{1}{2}$
- ? "I want to drink $\frac{2}{3}$ of this water. How much water will that be, and how do you know?" About $\frac{1}{4}$ $\left(\text{actually } \frac{1}{3}\right)$; Answers will vary.
- Discuss the various approaches students may have.
- Work through the linear model approach in the book.
- **Connection:** This method of multiplying fractions uses equivalent fractions. $\frac{2}{3}$ of $\frac{1}{2} = \frac{2}{3} \times \frac{1}{2} = \frac{2}{3} \times \frac{3}{6}$. If $\frac{1}{3}$ of $\frac{3}{6}$ is $\frac{1}{6}$, then $\frac{2}{3}$ of $\frac{3}{6}$ is $\frac{2}{6} = \frac{1}{3}$. By using number sense, students develop a sense of how fractions are multiplied (multiply the numerators and multiply the denominators).
- You want $\frac{2}{3}$ of a quantity that does not appear to be divisible by 3. By renaming the quantity $\left(\frac{1}{2} = \frac{3}{6}\right)$, then it is divisible by 3.

Common Core State Standards

6.NS.1 Interpret and compute quotients of fractions, and solve word problems involving division of fractions by fractions.

Previous Learning

Students should know how to multiply a fraction by a whole number.

Technology for the Teacher

Dynamic Classroom

Lesson Plans

Complete Materials List

2.1 Record and Practice Journal

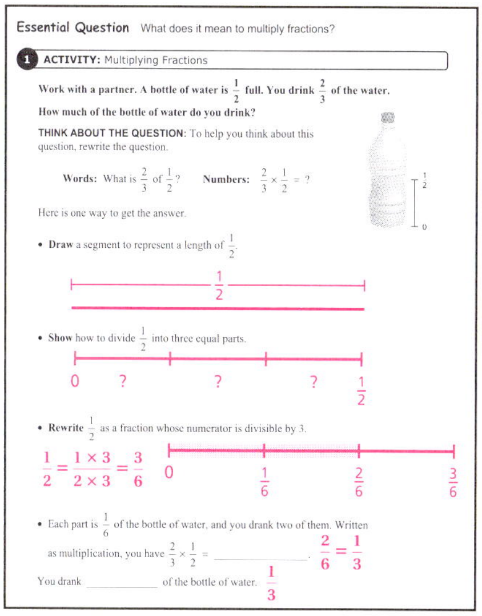

Essential Question What does it mean to multiply fractions?

1 ACTIVITY: Multiplying Fractions

Work with a partner. A bottle of water is $\frac{1}{2}$ full. You drink $\frac{2}{3}$ of the water. How much of the bottle of water do you drink?

THINK ABOUT THE QUESTION: To help you think about this question, rewrite the question.

Words: What is $\frac{2}{3}$ of $\frac{1}{2}$? **Numbers:** $\frac{2}{3} \times \frac{1}{2} = ?$

Here is one way to get the answer.

- **Draw** a segment to represent a length of $\frac{1}{2}$.
- **Show** how to divide $\frac{1}{2}$ into three equal parts.
- **Rewrite** $\frac{1}{2}$ as a fraction whose numerator is divisible by 3. $\frac{1}{2} = \frac{1 \times 3}{2 \times 3} = \frac{3}{6}$
- Each part is $\frac{1}{6}$ of the bottle of water, and you drank two of them. Written as multiplication, you have $\frac{2}{3} \times \frac{1}{2} =$ $\frac{2}{6} = \frac{1}{3}$.

You drank $\frac{1}{3}$ of the bottle of water.

English Language Learners

Analyzing Word Problems

Give students a copy of the word problems that are triple-spaced and have wide margins. The additional space gives students plenty of room to add notes. Demonstrate how to underline key words, phrases, and numbers; write down equivalent words in English (or their native language); and draw lines between elements to make the meaning clearer.

2.1 Record and Practice Journal

2 ACTIVITY: Multiplying Fractions

A park has a playground that is $\frac{3}{4}$ of its width and $\frac{4}{5}$ of its length. What fraction of the park is covered by the playground?

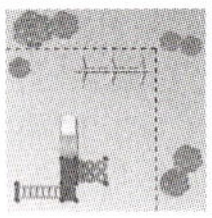

Fold a piece of paper horizontally into fourths and shade three of the fourths to represent $\frac{3}{4}$.

Fold the paper vertically into fifths and shade $\frac{4}{5}$ of the paper another color.

Count the total number of squares. This number is the denominator. The numerator is the number of squares shaded with both colors.

$\frac{3}{4} \times \frac{4}{5} = \frac{12}{20} = \frac{3}{5}$. So, $\frac{3}{5}$ of the park is covered by the playground.

Inductive Reasoning

Work with a partner. Complete the table by using a model or folding paper.

	Exercise	Verbal Expression	Answer
1	3. $\frac{2}{3} \times \frac{1}{2}$	$\frac{2}{3}$ of $\frac{1}{2}$	$\frac{1}{3}$
2	4. $\frac{3}{4} \times \frac{4}{5}$	$\frac{3}{4}$ of $\frac{4}{5}$	$\frac{3}{5}$
	5. $\frac{2}{3} \times \frac{5}{6}$	$\frac{2}{3}$ of $\frac{5}{6}$	$\frac{5}{9}$
	6. $\frac{1}{6} \times \frac{1}{4}$	$\frac{1}{6}$ of $\frac{1}{4}$	$\frac{1}{24}$
	7. $\frac{2}{5} \times \frac{1}{2}$	$\frac{2}{5}$ of $\frac{1}{2}$	$\frac{1}{5}$
	8. $\frac{5}{8} \times \frac{4}{5}$	$\frac{5}{8}$ of $\frac{4}{5}$	$\frac{1}{2}$

What Is Your Answer?

9. **IN YOUR OWN WORDS** What does it mean to multiply fractions?

It means to take a part of a fraction.

10. **STRUCTURE** Write a general rule for multiplying fractions.

Multiply the numerators and multiply the denominators.

Laurie's Notes

Activity 2

- In this activity, the students use an area model to multiply fractions. You are finding the area of a rectangle (length × width).
- Hand students a piece of paper and say, "This rectangle represents a park. $\frac{3}{4}$ of its width and $\frac{4}{5}$ of its length is a playground!"
- Have students fold the paper as shown in the text. This may be difficult, so it is a good idea to have students sketch the fifths before they fold the paper.
- **Discuss:** In order to name the amount of paper that is shaded (representing the playground), students should think of a name for each small piece (twentieths) and count how many are shaded (12).
- **MP1a Make Sense of Problems:** The concrete model will help students visualize and make sense of the problem, allowing them then to solve the problem.

Inductive Reasoning

- Students should work with partners to complete the remaining problems.
- Remind students that they should sketch a visual model (area or linear model) or paper fold to make sense of each problem.
- Have students share their work for the visual model at the board, including both the problem and the answer. For example: $\frac{2}{3} \times \frac{5}{6} = \frac{10}{18} = \frac{5}{9}$.
- **MP8 Look for and Express Regularity in Repeated Reasoning:** Mathematically proficient students notice if calculations are repeated. In this case the product, prior to simplifying the fraction, should be recognized as the product of the numerators over the product of the denominators.

? "Do you notice any patterns in your answers?" Listen for students to describe multiplying the numerators and multiplying the denominators.

What Is Your Answer?

- **Think-Pair-Share:** Students should read each question independently and then work in pairs to answer the questions. When they have answered the questions, the pair should compare their answers with another group and discuss any discrepancies.

Closure

- Draw a picture to show $\frac{1}{2}$ of $\frac{3}{4}$, then use your picture to find $\frac{1}{2} \times \frac{3}{4}$.

$\frac{3}{8}$; *Sample picture:*

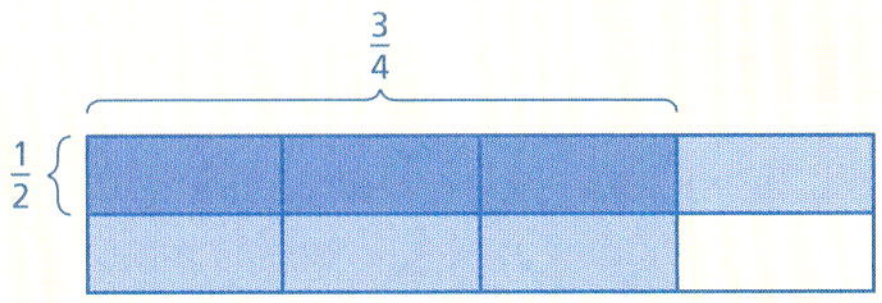

2 ACTIVITY: Multiplying Fractions

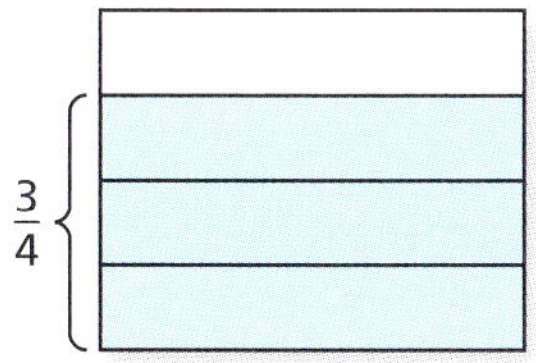

Work with a partner. A park has a playground that is $\frac{3}{4}$ of its width and $\frac{4}{5}$ of its length. What fraction of the park is covered by the playground?

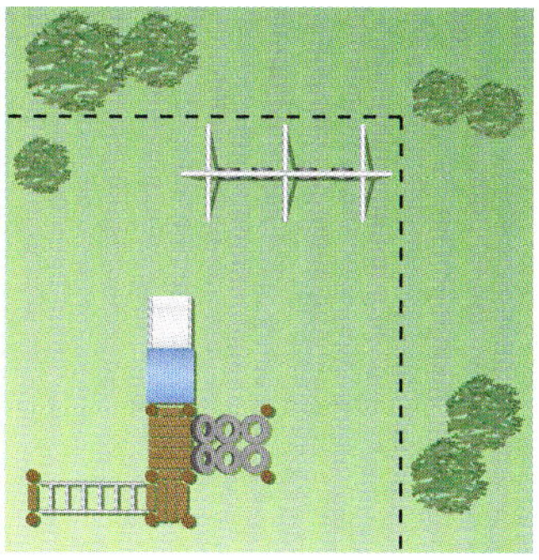

Fold a piece of paper horizontally into fourths and shade three of the fourths to represent $\frac{3}{4}$.

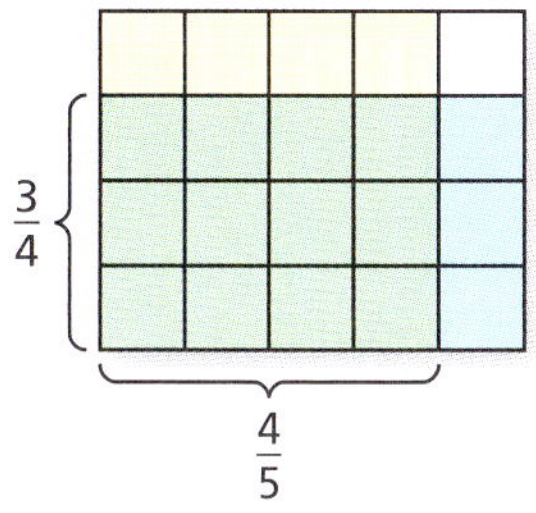

Fold the paper vertically into fifths and shade $\frac{4}{5}$ of the paper another color.

Count the total number of squares. This number is the denominator. The numerator is the number of squares shaded with both colors.

$\frac{3}{4} \times \frac{4}{5} = \frac{\square}{\square} = \frac{\square}{\square}$. So, $\frac{\square}{\square}$ of the park is covered by the playground.

Inductive Reasoning

Work with a partner. Complete the table by using a model or folding paper.

Exercise	Verbal Expression	Answer
1 **3.** $\frac{2}{3} \times \frac{1}{2}$	$\frac{2}{3}$ of $\frac{1}{2}$	$\frac{\square}{\square}$
2 **4.** $\frac{3}{4} \times \frac{4}{5}$	$\frac{3}{4}$ of $\frac{4}{5}$	$\frac{\square}{\square}$
5. $\frac{2}{3} \times \frac{5}{6}$		
6. $\frac{1}{6} \times \frac{1}{4}$		
7. $\frac{2}{5} \times \frac{1}{2}$		
8. $\frac{5}{8} \times \frac{4}{5}$		

Math Practice 1

Consider Similar Problems

What are the similarities in constructing the models for each problem? What are the differences?

What Is Your Answer?

9. **IN YOUR OWN WORDS** What does it mean to multiply fractions?
10. **STRUCTURE** Write a general rule for multiplying fractions.

Practice Use what you learned about multiplying fractions to complete Exercises 4–11 on page 59.

2.1 Lesson

Check It Out
Lesson Tutorials
BigIdeasMath.com

Key Idea

Multiplying Fractions

Words Multiply the numerators and multiply the denominators.

Numbers $\frac{3}{7} \times \frac{1}{2} = \frac{3 \times 1}{7 \times 2} = \frac{3}{14}$

Algebra $\frac{a}{b} \cdot \frac{c}{d} = \frac{a \cdot c}{b \cdot d}$, where $b, d \neq 0$

EXAMPLE 1 Multiplying Fractions

Find $\frac{1}{5} \times \frac{1}{3}$.

$\frac{1}{5} \times \frac{1}{3} = \frac{1 \times 1}{5 \times 3}$ ← Multiply the numerators. ← Multiply the denominators.

$= \frac{1}{15}$ Simplify.

EXAMPLE 2 Multiplying Fractions with Common Factors

Find $\frac{8}{9} \times \frac{3}{4}$. **Estimate** $1 \times \frac{3}{4} = \frac{3}{4}$

$\frac{8}{9} \times \frac{3}{4} = \frac{8 \times 3}{9 \times 4}$ ← Multiply the numerators. ← Multiply the denominators.

$= \frac{\cancel{8}^{2} \times \cancel{3}^{1}}{\cancel{9}_{3} \times \cancel{4}_{1}}$ Divide out common factors.

$= \frac{2}{3}$ Simplify.

∴ The product is $\frac{2}{3}$. **Reasonable?** $\frac{2}{3} \approx \frac{3}{4}$ ✓

Study Tip

When the numerator of one fraction is the same as the denominator of another fraction, you can use mental math to multiply. For example, $\frac{4}{5} \times \frac{5}{9} = \frac{4}{9}$ because you can divide out the common factor 5.

On Your Own

Now You're Ready
Exercises 4–19

Multiply. Write the answer in simplest form.

1. $\frac{1}{2} \times \frac{5}{6}$
2. $\frac{7}{8} \times \frac{1}{4}$
3. $\frac{3}{7} \times \frac{2}{3}$
4. $\frac{4}{9} \times \frac{3}{10}$

Laurie's Notes

Introduction

Connect

- **Yesterday:** Students used a visual model to multiply fractions. (MP1a, MP3a, MP8)
- **Today:** Students will formalize the process for multiplying fractions.

Motivate

- **FYI:** The New York City Public School District is the largest public school district in the nation. About $\frac{1}{20}$ of the New York population attends school in this district. The population of New York is about $\frac{1}{15}$ of the population of the United States.

? "What is $\frac{1}{20} \times \frac{1}{15}$, and what does the answer mean in the context of the problem given?" $\frac{1}{300}$; represents the fraction of the United States population that attends school in the New York City Public School District. This also means that for every 300 United States residents, 1 of them attends school in the New York City Public School District.

- Students should have a sense of how to multiply the fractions based upon yesterday's investigations.

Lesson Notes

Key Idea

- **Connections:** Previously, when a fraction was multiplied by a whole number, the whole number was written as a fraction with a denominator of 1. This rule still works!

Example 1

- **MP1 Make Sense of Problems:** Mathematically proficient students continually ask themselves, "Does this make sense?" The visual model helps them make sense of their answer and judge its reasonableness.
- Work through the example. It may help to show a visual model.
- Discuss the reasonableness of the answer: The product of $\frac{1}{5}$ and $\frac{1}{3}$ should be fairly small. Think of $\frac{1}{3}$ of the class. Now take $\frac{1}{5}$ of that group. It's not a lot!

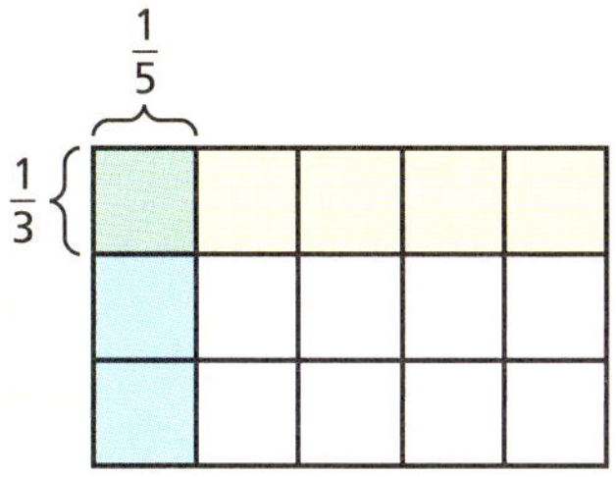

Example 2

- Work through the example. Note the Study Tip.
- **Connection:** If the common factors are not divided out, the product is $\frac{24}{36}$. Notice that the GCF of 24 and 36 is 12.

On Your Own

- **Neighbor Check:** Have students work independently and then have their neighbors check their work. Have students discuss any discrepancies.

Goal Today's lesson is multiplying fractions by fractions, fractions by mixed numbers, and mixed numbers by mixed numbers.

Technology for the Teacher

Dynamic Classroom

Lesson Tutorials
Lesson Plans
Answer Presentation Tool

Extra Example 1

Find $\frac{2}{7} \times \frac{3}{5}$. $\frac{6}{35}$

Extra Example 2

Find $\frac{2}{25} \times \frac{5}{6}$. $\frac{1}{15}$

On Your Own

1. $\frac{5}{12}$
2. $\frac{7}{32}$
3. $\frac{2}{7}$
4. $\frac{2}{15}$

Laurie's Notes

Extra Example 3

You have $\frac{5}{8}$ of a large bottle of shampoo. You pour $\frac{3}{10}$ of the shampoo into a smaller bottle. How much of the entire larger bottle did you pour into the smaller bottle?

$\frac{3}{16}$ of the entire larger bottle

On Your Own

5. $\frac{1}{6}$ of the entire bag

Extra Example 4

Find $6\frac{1}{3} \times \frac{3}{4}$. $4\frac{3}{4}$

English Language Learners

Vocabulary

Make sure that students understand the meanings of the phrases *mixed number* and *improper fraction*. Suggest that they add these phrases and examples to their notes.

Example 3

- **MP1a Make Sense of Problems:** The graphic model helps give an entry point into the problem.
- Read through the problem and refer to the graphic of the flour bag. Discuss the labels on the flour bag $\left(\text{i.e., } \frac{3}{4} \text{ of } \frac{2}{3}\right)$.
- **FYI:** An empanada is a Spanish word coming from *empanar* meaning *bread,* or *to bake pastry.* Although there are family-sized versions, it is basically an individual turnover with a pastry crust that is filled with chicken, meat, seafood, vegetables, or fruit.
- **Extension:** Some students find dividing out common factors to be pretty cool. If they want additional problems to try, here are two:

  $\frac{7}{12} \times \frac{3}{14}$ $\frac{10}{39} \times \frac{13}{15}$

? **Extension:** "Find the product of $\frac{2}{3} \times \frac{3}{4} \times \frac{4}{5} \times \frac{5}{6}$." $\frac{1}{3}$

On Your Own

- **Think-Pair-Share:** Students should read the question independently and then work in pairs to answer the question. When they have answered the question, the pair should compare their answer with another group and discuss any discrepancies.

Key Idea

- **Discuss:** When the mixed number is written as an improper fraction, the entire number is being multiplied by the simple fraction. Connect this idea to Example 2.

Example 4

? "Estimate $\frac{1}{2}$ of $2\frac{3}{4}$, and explain how you arrived at your estimate." $2\frac{3}{4}$ is close to 3, and $\frac{1}{2}$ of 3 is $1\frac{1}{2}$.

? "What is $2\frac{3}{4}$ as an improper fraction?" $\frac{11}{4}$

Work through the problem as shown.

EXAMPLE 3 Real-Life Application

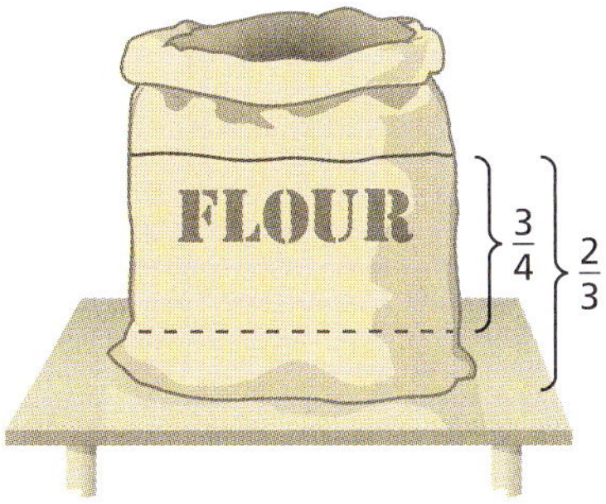

You have $\frac{2}{3}$ of a bag of flour. You use $\frac{3}{4}$ of the flour to make empanada dough. How much of the entire bag do you use to make the dough?

Method 1: Use a model. Six of the 12 squares have both types of shading.

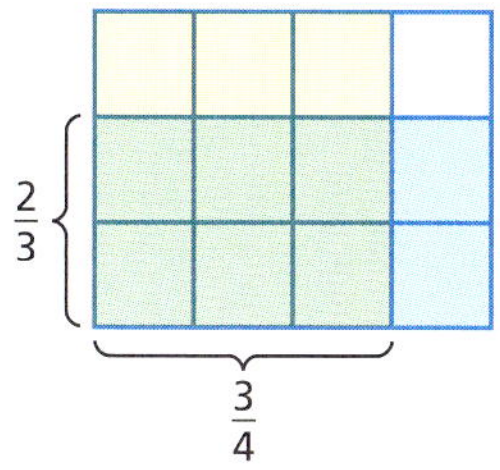

∴ So, you use $\frac{6}{12} = \frac{1}{2}$ of the entire bag.

Method 2: To find $\frac{3}{4}$ of $\frac{2}{3}$, multiply.

$$\frac{3}{4} \times \frac{2}{3} = \frac{\overset{1}{\cancel{3}} \times \overset{1}{\cancel{2}}}{\underset{2}{\cancel{4}} \times \underset{1}{\cancel{3}}}$$ Multiply the numerators and the denominators. Divide out common factors.

$$= \frac{1}{2}$$ Simplify.

∴ So, you use $\frac{1}{2}$ of the entire bag.

On Your Own

5. **WHAT IF?** In Example 3, you use $\frac{1}{4}$ of the flour to make the dough. How much of the entire bag do you use to make the dough?

Key Idea

Multiplying Mixed Numbers

Write each mixed number as an improper fraction. Then multiply as you would with fractions.

EXAMPLE 4 Multiplying a Fraction and a Mixed Number

Find $\frac{1}{2} \times 2\frac{3}{4}$.

Estimate $\frac{1}{2} \times 3 = 1\frac{1}{2}$

$$\frac{1}{2} \times 2\frac{3}{4} = \frac{1}{2} \times \frac{11}{4}$$ Write $2\frac{3}{4}$ as the improper fraction $\frac{11}{4}$.

$$= \frac{1 \times 11}{2 \times 4}$$ Multiply the numerators and the denominators.

$$= \frac{11}{8}, \text{ or } 1\frac{3}{8}$$ Simplify.

∴ The product is $1\frac{3}{8}$.

Reasonable? $1\frac{3}{8} \approx 1\frac{1}{2}$ ✓

EXAMPLE 5 Multiplying Mixed Numbers

Find $1\frac{4}{5} \times 3\frac{2}{3}$.

Estimate $2 \times 4 = 8$

$1\frac{4}{5} \times 3\frac{2}{3} = \frac{9}{5} \times \frac{11}{3}$ — Write $1\frac{4}{5}$ and $3\frac{2}{3}$ as improper fractions.

$= \frac{\cancel{9}^{3} \times 11}{5 \times \cancel{3}_{1}}$ — Multiply fractions. Divide out the common factor 3.

$= \frac{33}{5}$, or $6\frac{3}{5}$ — Simplify.

The product is $6\frac{3}{5}$.

Reasonable? $6\frac{3}{5} \approx 8$ ✓

On Your Own

Now You're Ready
Exercises 26–41

Multiply. Write the answer in simplest form.

6. $\frac{1}{3} \times 1\frac{1}{6}$ **7.** $3\frac{1}{2} \times \frac{4}{9}$ **8.** $1\frac{7}{8} \times 2\frac{2}{5}$ **9.** $5\frac{5}{7} \times 2\frac{1}{10}$

EXAMPLE 6 Real-Life Application

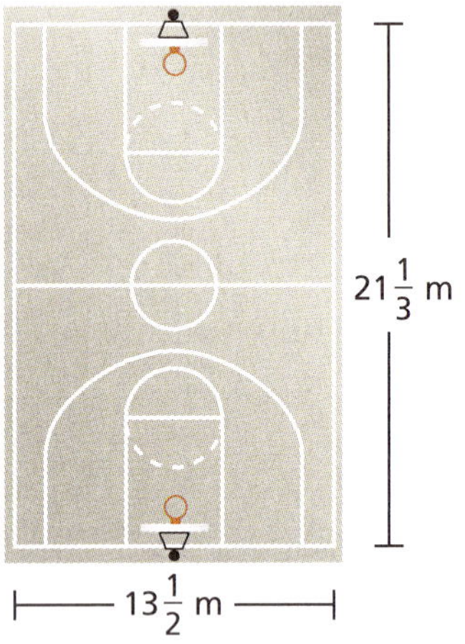

A city is resurfacing a basketball court. Find the area of the court.

Estimate $21 \times 14 = 294$

$A = \ell w$ — Write the formula for the area of a rectangle.

$= 21\frac{1}{3} \times 13\frac{1}{2}$ — Substitute $21\frac{1}{3}$ for ℓ and $13\frac{1}{2}$ for w.

$= \frac{64}{3} \times \frac{27}{2}$ — Write $21\frac{1}{3}$ and $13\frac{1}{2}$ as improper fractions.

$= \frac{\cancel{64}^{32} \times \cancel{27}^{9}}{\cancel{3}_{1} \times \cancel{2}_{1}}$ — Multiply fractions. Divide out common factors.

$= 288$ — Simplify.

So, the area of the court is 288 square meters.

Reasonable? $288 \approx 294$ ✓

On Your Own

10. Find the area of a rectangular air hockey table that is $8\frac{1}{4}$ feet by $4\frac{3}{8}$ feet.

Laurie's Notes

Example 5

- **Common Error:** Students may multiply the whole numbers and then multiply the fractions as shown: $1\frac{4}{5} \times 3\frac{2}{3} \neq 3\frac{8}{15}$.
- Write the problem. Estimate the product.
- ? "What is the first step in multiplying these mixed numbers?" Write the mixed numbers as improper fractions.
- As you work through this example, point out that by writing each of the mixed numbers as improper fractions, it ensures that all of the first number is being multiplied by all of the second number.
- Finish working through the problem. Note that the answer is more reasonable than $3\frac{8}{15}$.

On Your Own

- Have students work with partners on these problems. Encourage them to write an estimate first, reinforcing the concept of rounding. Finish with judging the reasonableness of the answer.

Example 6

- Students have been using an area model in this chapter, so the formula is familiar. Notice that an estimate is made.
- Work through the problem. Students should appreciate the technique of dividing out common factors in the problem.
- ? "Why is the answer labeled with square meters?" *Sample answer:* Area is measured in square units.
- ? "If the basketball court were the same size but measured in yards instead of meters, would the answer be *more than* or *less than* 288 square yards? Explain." more; Meters are greater than yards.

On Your Own

- **Neighbor Check:** Have students work independently and then have their neighbors check their work. Have students discuss any discrepancies.

Closure

- **Exit Ticket:** If you multiply two proper fractions, what do you know about the product? The product will be less than the greater fraction. If you multiply two mixed numbers, what do you know about the product? *Sample answer:* It will be greater than the product of the whole number portions of the mixed numbers.

Extra Example 5

Find $2\frac{3}{5} \times 3\frac{1}{3}$. $8\frac{2}{3}$

On Your Own

6. $\frac{7}{18}$
7. $1\frac{5}{9}$
8. $4\frac{1}{2}$
9. 12

Extra Example 6

A grass practice field is $80\frac{1}{2}$ yards by $30\frac{3}{4}$ yards. Find the area of the practice field. $2475\frac{3}{8}$ square yards

On Your Own

10. $36\frac{3}{32}$ ft^2

Vocabulary and Concept Check

1. Multiply numerators and multiply denominators, then simplify the fraction.
2. 4
3. *Sample answer:* $3\frac{1}{2} \times 3\frac{1}{7} = 11$

Practice and Problem Solving

4. $\frac{2}{21}$
5. $\frac{5}{16}$
6. $\frac{1}{10}$
7. $\frac{3}{28}$
8. $\frac{8}{21}$
9. $\frac{5}{8}$
10. $\frac{1}{24}$
11. $\frac{1}{3}$
12. $4\frac{1}{6}$
13. $5\frac{1}{4}$
14. $\frac{2}{5}$
15. $\frac{16}{45}$
16. $\frac{9}{49}$
17. $\frac{5}{27}$
18. $\frac{13}{21}$
19. $1\frac{19}{30}$
20. You did not multiply the denominators.
$$\frac{2}{5} \times \frac{3}{10} = \frac{2 \times 3}{5 \times 10} = \frac{\overset{1}{\cancel{2}} \times 3}{5 \times \underset{5}{\cancel{10}}} = \frac{3}{25}$$
21. $\frac{3}{10}$
22. $\frac{1}{4}$
23. $\frac{4}{7} > \left(\frac{9}{10} \times \frac{4}{7}\right)$; Because $\frac{9}{10} < 1$, the product will be less than $\frac{4}{7}$.
24. $\left(\frac{5}{8} \times \frac{22}{15}\right) > \frac{5}{8}$; Because $\frac{22}{15} > 1$, the product will be greater than $\frac{5}{8}$.
25. See Additional Answers.

Assignment Guide and Homework Check

Level	Day 1 Activity Assignment	Day 2 Lesson Assignment	Homework Check
Basic	4–11, 60–64	1–3, 13–19 odd, 20, 21–31 odd, 42	15, 17, 21, 29
Average	4–11, 60–64	1–3, 20–38 even, 42, 43, 45, 53, 55	26, 34, 38, 45, 53
Advanced	4–11, 60–64	1–3, 20, 22, 34–58 even	22, 34, 38, 44, 58

Common Errors

- **Exercises 4–19** Students may try to cross multiply instead of multiplying the numerators and denominators. Remind them that you cross multiply when an equal sign is between the fractions, but that you multiply the numerators and denominators when there is a multiplication symbol between the fractions.
- **Exercises 4–19** Students may try to get a common denominator and then multiply the fractions. Tell them this is not needed when multiplying, but is necessary for addition and subtraction. (Students will still end up with the same product, but the numbers will be greater before they simplify.)

2.1 Record and Practice Journal

Multiply. Write the answer in simplest form.

1. $\frac{1}{6} \times \frac{5}{8}$ $\frac{5}{48}$
2. $\frac{7}{9} \times 3$ $2\frac{1}{3}$
3. $\frac{8}{9} \times \frac{3}{5}$ $\frac{8}{15}$
4. $\frac{7}{8} \times 2\frac{1}{3}$ $2\frac{1}{24}$
5. $7 \times 3\frac{9}{14}$ $25\frac{1}{2}$
6. $5\frac{5}{9} \times 2\frac{7}{10}$ 15
7. You reserve $\frac{2}{5}$ of the seats on a tour bus. You are able to fill $\frac{5}{8}$ of the seats you reserve. What fraction of the seats on the bus are you able to fill? $\frac{1}{4}$
8. A triangle has a base of $5\frac{2}{3}$ inches and a height of 3 inches. What is the area of the triangle? $8\frac{1}{2}$ in.^{2}

2.1 Exercises

Check It Out
Help with Homework
BigIdeasMath.com

Vocabulary and Concept Check

1. **WRITING** Explain how to multiply two fractions.

2. **REASONING** Name the missing denominator.

$$\frac{3}{7} \times \frac{1}{\square} = \frac{3}{28}$$

3. **OPEN-ENDED** Write two mixed numbers between 3 and 4 that have a product between 9 and 12.

Practice and Problem Solving

Multiply. Write the answer in simplest form.

1 2

4. $\frac{1}{7} \times \frac{2}{3}$ 5. $\frac{5}{8} \times \frac{1}{2}$ 6. $\frac{1}{4} \times \frac{2}{5}$ 7. $\frac{3}{7} \times \frac{1}{4}$

8. $\frac{2}{3} \times \frac{4}{7}$ 9. $\frac{5}{7} \times \frac{7}{8}$ 10. $\frac{3}{8} \times \frac{1}{9}$ 11. $\frac{5}{6} \times \frac{2}{5}$

12. $\frac{5}{12} \times 10$ 13. $6 \times \frac{7}{8}$ 14. $\frac{3}{4} \times \frac{8}{15}$ 15. $\frac{4}{9} \times \frac{4}{5}$

16. $\frac{3}{7} \times \frac{3}{7}$ 17. $\frac{5}{6} \times \frac{2}{9}$ 18. $\frac{13}{18} \times \frac{6}{7}$ 19. $\frac{7}{9} \times \frac{21}{10}$

20. **ERROR ANALYSIS** Describe and correct the error in finding the product.

$$\frac{2}{5} \times \frac{3}{10} = \frac{4}{10} \times \frac{3}{10} = \frac{4 \times 3}{10} = \frac{12}{10} = 1\frac{1}{5}$$

21. **AQUARIUM** In an aquarium, $\frac{2}{5}$ of the fish are surgeonfish. Of these, $\frac{3}{4}$ are yellow tangs. What fraction of all fish in the aquarium are yellow tangs?

22. **JUMP ROPE** You exercise for $\frac{3}{4}$ of an hour. You jump rope for $\frac{1}{3}$ of that time. What fraction of the hour do you spend jumping rope?

Without finding the product, copy and complete the statement using <, >, or =. Explain your reasoning.

23. $\frac{4}{7}$ $\square$ $\left(\frac{9}{10} \times \frac{4}{7}\right)$ 24. $\left(\frac{5}{8} \times \frac{22}{15}\right)$ $\square$ $\frac{5}{8}$ 25. $\frac{5}{6}$ $\square$ $\left(\frac{5}{6} \times \frac{7}{7}\right)$

Multiply. Write the answer in simplest form.

4 5 26. $1\frac{1}{3} \times \frac{2}{3}$ 27. $6\frac{2}{3} \times \frac{3}{10}$ 28. $2\frac{1}{2} \times \frac{4}{5}$ 29. $\frac{3}{5} \times 3\frac{1}{3}$

30. $7\frac{1}{2} \times \frac{2}{3}$ 31. $\frac{5}{9} \times 3\frac{3}{5}$ 32. $\frac{3}{4} \times 1\frac{1}{3}$ 33. $3\frac{3}{4} \times \frac{2}{5}$

34. $4\frac{3}{8} \times \frac{4}{5}$ 35. $\frac{3}{7} \times 2\frac{5}{6}$ 36. $1\frac{3}{10} \times 18$ 37. $15 \times 2\frac{4}{9}$

38. $1\frac{1}{6} \times 6\frac{3}{4}$ 39. $2\frac{5}{12} \times 2\frac{2}{3}$ 40. $5\frac{5}{7} \times 3\frac{1}{8}$ 41. $2\frac{4}{5} \times 4\frac{1}{16}$

ERROR ANALYSIS **Describe and correct the error in finding the product.**

42. ✗ $4 \times 3\frac{7}{10} = 12\frac{7}{10}$

43. ✗ $2\frac{1}{2} \times 7\frac{4}{5} = (2 \times 7) + \left(\frac{1}{2} \times \frac{4}{5}\right)$
$= 14 + \frac{2}{5} = 14\frac{2}{5}$

44. **VITAMIN C** A vitamin C tablet contains $\frac{1}{40}$ of a gram of vitamin C. You take $1\frac{1}{2}$ tablets every day. How many grams of vitamin C do you take every day?

45. **SCHOOL BANNER** You make a banner for a football rally.

a. What is the area of the banner?

b. You add a $\frac{1}{4}$-foot border on each side. What is the new area of the banner?

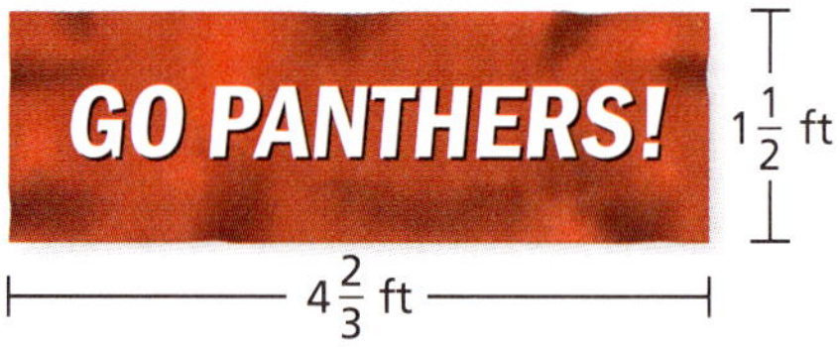

46. **NUMBER SENSE** Without calculating, is $1\frac{1}{6} \cdot \frac{4}{5}$ less than or greater than $1\frac{1}{6}$? Is the product less than or greater than $\frac{4}{5}$? Explain your reasoning.

Multiply. Write the answer in simplest form.

47. $\frac{1}{2} \times \frac{3}{5} \times \frac{4}{9}$ 48. $\frac{4}{7} \times 4\frac{3}{8} \times \frac{5}{6}$ 49. $1\frac{1}{15} \times 5\frac{2}{5} \times 4\frac{7}{12}$

50. $\left(\frac{3}{5}\right)^3$ 51. $\left(\frac{4}{5}\right)^2 \times \left(\frac{3}{4}\right)^2$ 52. $\left(\frac{5}{6}\right)^2 \times \left(1\frac{1}{10}\right)^2$

53. **PICTURES** Three pictures hang side by side on a wall. What is the total area of the wall that the pictures cover?

54. **OPEN-ENDED** Find a fraction that, when multiplied by $\frac{1}{2}$, is less than $\frac{1}{4}$.

Common Errors

- **Exercises 26–41** Students may forget to change the mixed number to an improper fraction. Show students visually that it is necessary to change the mixed number into an improper fraction.
- **Exercises 26–41** Students may write $6\frac{2}{3}$ as $\frac{3 \times 6 \times 2}{3}$ instead of as $\frac{3 \times 6 + 2}{3}$. Make sure they have a good understanding of how to convert mixed numbers to improper fractions and vice versa.
- **Exercises 26–41** Students may try to divide out common factors of the whole number with the numerator or denominator. Remind students that they should wait to divide out common factors until after converting mixed numbers to improper fractions.
- **Exercise 36** Students may forget how to write a whole number as an improper fraction. Remind them to express the whole number as an improper fraction by putting it over a denominator of 1.
- **Exercises 50–52** Students may not remember the definition of exponent and will multiply the fraction by the exponent. Remind them of the definition of exponent and also that exponents are evaluated before multiplication in the order of operations.

Differentiated Instruction

Inclusion

Some students may confuse the concepts of adding fractions and multiplying fractions. Remind students when adding fractions, you must have a common denominator.

Like denominators:

$$\frac{a}{b} + \frac{c}{b} = \frac{a+c}{b}, \text{ where } b \neq 0$$

Unlike denominators:

$$\frac{a}{b} + \frac{c}{d} = \frac{ad}{bd} + \frac{bc}{bd} = \frac{ad+bc}{bd}, \text{ where } b, d \neq 0$$

When multiplying fractions, you multiply the numerators and multiply the denominators.

$$\frac{a}{b} \cdot \frac{c}{d} = \frac{ac}{bd}, \text{ where } b, d \neq 0$$

Practice and Problem Solving

26. $\frac{8}{9}$ **27.** 2

28. 2 **29.** 2

30. 5 **31.** 2

32. 1 **33.** $1\frac{1}{2}$

34. $3\frac{1}{2}$ **35.** $1\frac{3}{14}$

36. $23\frac{2}{5}$ **37.** $36\frac{2}{3}$

38. $7\frac{7}{8}$ **39.** $6\frac{4}{9}$

40. $17\frac{6}{7}$ **41.** $11\frac{3}{8}$

42. You must first rewrite the mixed number as an improper fraction and then multiply.

$$4 \times 3\frac{7}{10} = 4 \times \frac{37}{10} = \frac{\overset{2}{\cancel{4}} \times 37}{\underset{5}{\cancel{10}}} = \frac{74}{5}, \text{ or } 14\frac{4}{5}$$

43–46. See Additional Answers.

Practice and Problem Solving

47. $\frac{2}{15}$ **48.** $2\frac{1}{12}$

49. $26\frac{2}{5}$ **50.** $\frac{27}{125}$

51. $\frac{9}{25}$ **52.** $\frac{121}{144}$

53. 462 in.^{2}

54. *Sample answer:* $\frac{1}{3}$

55. 4 mi

56. No; Positive mixed numbers are greater than 1 and the product of two numbers greater than 1 is always greater than 1.

57. See *Taking Math Deeper.*

58. **a.** $5\frac{1}{10}$ h

b. 11:39 A.M.

59. **a.** $\frac{3}{50}$ **b.** 45 people

Fair Game Review

60. $2^3 \cdot 3$

61. $3^2 \cdot 5$

62. 53 is prime

63. $2^2 \cdot 3 \cdot 5$

64. B

Mini-Assessment

Multiply. Write the answer in simplest form.

1. $\frac{1}{8} \times \frac{3}{5}$ $\frac{3}{40}$
2. $\frac{3}{8} \times \frac{2}{9}$ $\frac{1}{12}$
3. $\frac{1}{6} \times 4\frac{4}{5}$ $\frac{4}{5}$
4. $2\frac{1}{5} \times 3\frac{4}{7}$ $7\frac{6}{7}$

Taking Math Deeper

Exercise 57

This problem asks students to look for two ways to solve the problem. The purpose of this problem is to see that *changing units* can sometimes result in a lot less work with messy fractions.

1 **a.** Draw a diagram. Identify the area to be found.

I want to find the area of the green region.

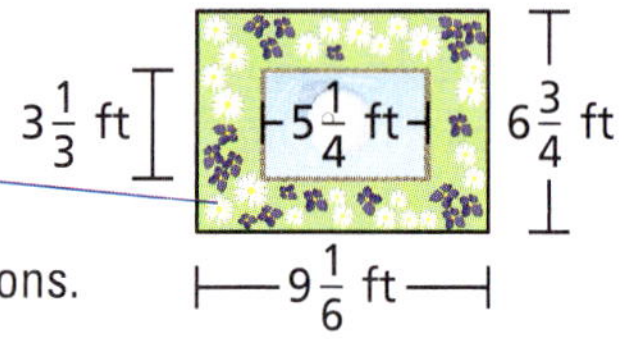

2 **b.** Use feet or inches for the dimensions.

Find the area using feet (involves fractions).

Area of Green = Area of Garden − Area of Fountain

$$= 6\frac{3}{4} \cdot 9\frac{1}{6} - 3\frac{1}{3} \cdot 5\frac{1}{4}$$

$$= \frac{27}{4} \cdot \frac{55}{6} - \frac{10}{3} \cdot \frac{21}{4}$$

$$= \frac{\overset{9}{\cancel{27}}}{4} \cdot \frac{55}{\underset{2}{\cancel{6}}} - \frac{\overset{5}{\cancel{10}}}{\underset{1}{\cancel{3}}} \cdot \frac{\overset{7}{\cancel{21}}}{\underset{2}{\cancel{4}}}$$

$$= \frac{495}{8} - \frac{35}{2}$$

$$= \frac{495}{8} - \frac{140}{8}$$

$$= \frac{355}{8} \text{ or } 44\frac{3}{8} \text{ ft}^2$$

3 Find the area using inches.

Area of Green = Area of Garden − Area of Fountain

$$= 81 \cdot 110 - 40 \cdot 63$$

$$= 8910 - 2520$$

$$= 6390 \text{ in.}^2$$

$$= 44\frac{3}{8} \text{ ft}^2$$ Divide by 144: 1 ft [1 ft] = 12 in. [12 in.]

c.

Project

You double the length and width of your garden. Now you can add paths through the garden, a flower bed and other features. Draw a diagram of your new garden with your added features. Be sure to include dimensions.

Reteaching and Enrichment Strategies

If students need help. . .	If students got it. . .
Resources by Chapter • Practice A and Practice B • Puzzle Time Record and Practice Journal Practice Differentiating the Lesson Lesson Tutorials Skills Review Handbook	Resources by Chapter • Enrichment and Extension • Technology Connection Start the next section

55. **DISTANCES** You are in a bike race. When you get to the first checkpoint, you are $\frac{2}{5}$ of the distance to the second checkpoint. When you get to the second checkpoint, you are $\frac{1}{4}$ of the distance to the finish. What is the distance from the start to the first checkpoint?

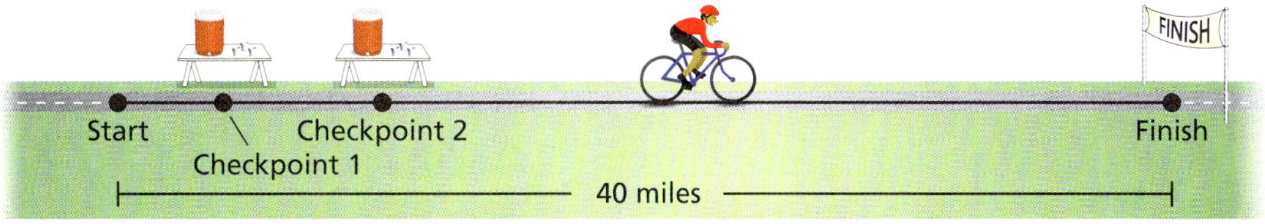

56. **NUMBER SENSE** Is the product of two positive mixed numbers ever less than 1? Explain.

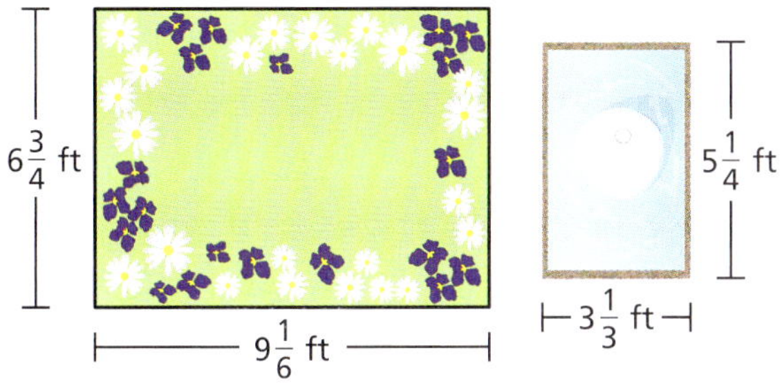

57. **MODELING** You plan to add a fountain to your garden.

a. Draw a diagram of the fountain in the garden. Label the dimensions.

b. Describe two methods for finding the area of the garden that surrounds the fountain.

c. Find the area. Which method did you use, and why?

58. **COOKING** The cooking time for a ham is $\frac{2}{5}$ of an hour for each pound.

a. How long should you cook a ham that weighs $12\frac{3}{4}$ pounds?

b. Dinner time is 4:45 P.M. What time should you start cooking the ham?

59. **PETS** You ask 150 people about their pets. The results show that $\frac{9}{25}$ of the people own a dog. Of the people who own a dog, $\frac{1}{6}$ of them also own a cat.

a. What fraction of the people own a dog and a cat?

b. Reasoning How many people own a dog but not a cat? Explain.

Fair Game Review What you learned in previous grades & lessons

Find the prime factorization of the number. *(Section 1.4)*

60. 24 61. 45 62. 53 63. 60

64. **MULTIPLE CHOICE** A science experiment calls for $\frac{3}{4}$ cup of baking powder. You have $\frac{1}{3}$ cup of baking powder. How much more baking powder do you need? *(Section 1.6)*

Ⓐ $\frac{1}{4}$ cup Ⓑ $\frac{5}{12}$ cup Ⓒ $\frac{4}{7}$ cup Ⓓ $1\frac{1}{12}$ cups

2.2 Dividing Fractions

Essential Question How can you divide by a fraction?

1 ACTIVITY: Dividing by a Fraction

Work with a partner. Write the division problem and solve it using a model.

a. How many two-thirds are in three?

The division problem is $\square \div \frac{\square}{\square}$.

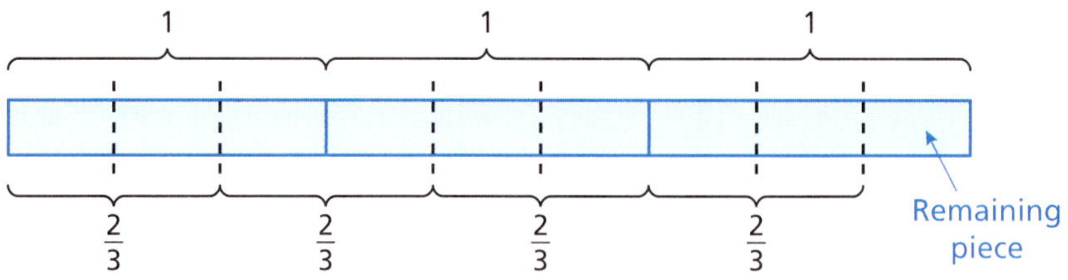

How many groups of $\frac{2}{3}$ are in 3? $\square$

The remaining piece represents $\frac{\square}{\square}$ of $\frac{2}{3}$.

So, there are $\square \frac{\square}{\square}$ groups of $\frac{2}{3}$ in 3.

So, $\square \div \frac{\square}{\square} = \square \frac{\square}{\square}$.

b. How many halves are in five halves?

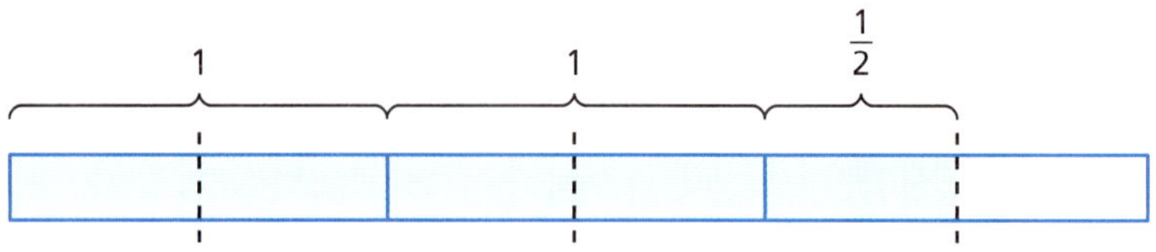

c. How many four-fifths are in eight?

d. How many one-thirds are in seven halves?

e. How many three-fourths are in five halves?

COMMON CORE

Dividing Fractions

In this lesson, you will

- write reciprocals of numbers.
- use models to divide fractions.
- divide fractions by fractions.
- solve real-life problems.

Learning Standard 6.NS.1

Laurie's Notes

Introduction

Standards for Mathematical Practice

- **MP7 Look for and Make Use of Structure:** You want students to develop a habit of looking closely to discern a pattern or structure. That is what students will do in the activities today.

Motivate

? "How many pieces of licorice would I need if I were to give each of you $\frac{1}{2}$ of a piece?" The answer is half of the number of students in your class. For 20 students, you need 10 pieces; for 21 students, you need $10\frac{1}{2}$ pieces.

- **Quick Model:** Show students six strips of paper. Say, "These are pieces of licorice. How many halves are there in 6?" You may want to cut the pieces of paper in half. Then write $6 \div \frac{1}{2} = 12$.

? "How many pieces of licorice would I need if I were to give you each $\frac{1}{3}$ of a piece?" The answer is one-third the number of students.

- **Quick Model:** Take 6 strips of paper. Say, "These are pieces of licorice. How many thirds are there in 6?" You may want to cut the pieces of paper into thirds. Then write: $6 \div \frac{1}{3} = 18$.
- **MP3 Construct Viable Arguments and Critique the Reasoning of Others:** For younger students, a physical model, such as the strip of paper, is one way to construct a viable argument.

Activity Notes

Activity 1

- You may want to give each student a strip of paper. Have them fold it to form 3 equal parts. Say, "This is a model of 3. How many $\frac{2}{3}$ are in 3?"
- **Explanation:** Have students use their pencils to divide each whole into thirds. Then count the number of $\frac{2}{3}$.

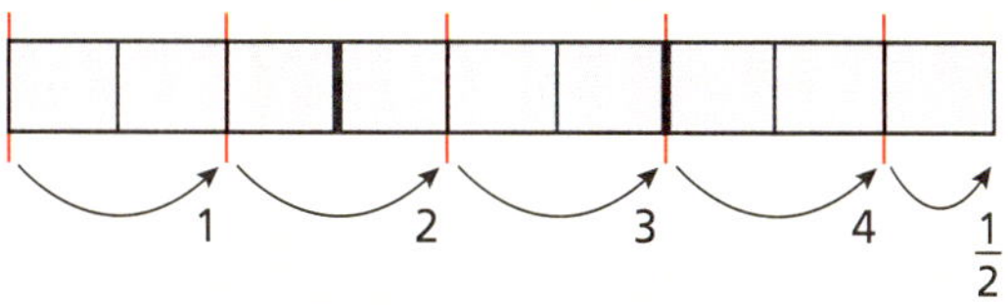

- There are 4 pieces of length $\frac{2}{3}$, and a remaining piece that is $\frac{1}{2}$ the length of $\frac{2}{3}$. So, there are a total of $4\frac{1}{2}$ pieces of length $\frac{2}{3}$ in 3.
- **MP3:** Take time to have a student talk through his or her approach to each part. Have the student display the model as the student explains his or her thinking. Other students should listen carefully to the language used and the explanation given. Create the culture in your class of students talking through a problem versus just giving an answer. You want students to be in the habit of communicating their thinking and having others critiquing their arguments.

Common Core State Standards

6.NS.1 Interpret and compute quotients of fractions, and solve word problems involving division of fractions by fractions.

Previous Learning

Students should know how to multiply a fraction by a fraction.

Technology for the Teacher

Dynamic Classroom

Lesson Plans
Complete Materials List

2.2 Record and Practice Journal

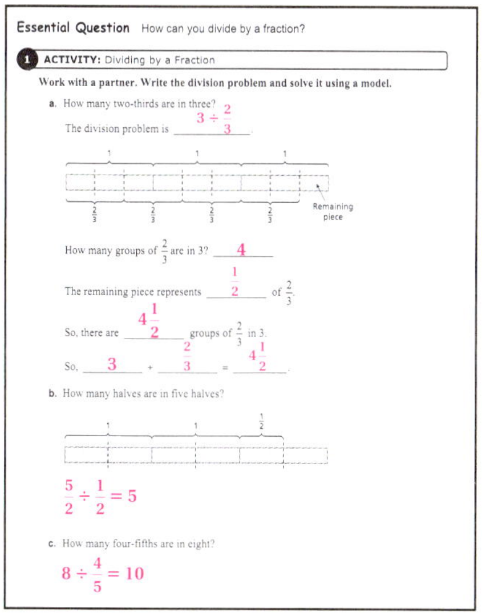
Essential Question How can you divide by a fraction?

1 ACTIVITY: Dividing by a Fraction

Work with a partner. Write the division problem and solve it using a model.

a. How many two-thirds are in three? The division problem is $3 \div \frac{2}{3}$

How many groups of $\frac{2}{3}$ are in 3? 4

The remaining piece represents $\frac{1}{2}$ of $\frac{2}{3}$.

So, there are $4\frac{1}{2}$ groups of $\frac{2}{3}$ in 3.

So, $3 \div \frac{2}{3} = 4\frac{1}{2}$.

b. How many halves are in five halves?

$\frac{5}{2} \div \frac{1}{2} = 5$

c. How many four-fifths are in eight?

$8 \div \frac{4}{5} = 10$

Differentiated Instruction

Inclusion

Some students may have difficulty dividing fractions when the divisor or the dividend is a whole number. When this occurs, suggest that the student write the whole number as a fraction for the first step.

Example 1

$$8 \div \frac{1}{2} = \frac{8}{1} \div \frac{1}{2} = \frac{8}{1} \times \frac{2}{1} = 16$$

Example 2

$$\frac{3}{4} \div 9 = \frac{3}{4} \div \frac{9}{1} = \frac{3}{4} \times \frac{1}{9} = \frac{1}{12}$$

2.2 Record and Practice Journal

d. How many one-thirds are in seven halves?

$\frac{7}{2} \div \frac{1}{3} = 10\frac{1}{2}$

e. How many three-fourths are in five halves?

$\frac{5}{2} \div \frac{3}{4} = 3\frac{1}{3}$

2 ACTIVITY: Using Tables to Recognize a Pattern

Work with a partner.

a. Complete each table.

Division Table

$8 \div 16$	$\frac{1}{2}$
$8 \div 8$	1
$8 \div 4$	2
$8 \div 2$	4
$8 \div 1$	8
$8 \div \frac{1}{2}$	16
$8 \div \frac{1}{4}$	32
$8 \div \frac{1}{8}$	64

Multiplication Table

$8 \times \frac{1}{16}$	$\frac{1}{2}$
$8 \times \frac{1}{8}$	1
$8 \times \frac{1}{4}$	2
$8 \times \frac{1}{2}$	4
8×1	8
8×2	16
8×4	32
8×8	64

b. Describe the relationship between the numbers in the right column of the division table and the numbers in the right column of the multiplication table.

They are the same.

c. Describe the relationship between the shaded numbers in the division table and the shaded numbers in the multiplication table.

The numerators and denominators are switched.

d. **STRUCTURE** Make a conjecture about how you can use multiplication to divide by a fraction.

Switch the numerator and denominator of the divisor; change division to multiplication.

e. Test your conjecture using the problems in Activity 1.

Check students' work.

What Is Your Answer?

3. **IN YOUR OWN WORDS** How can you divide by a fraction? Give an example.

Use a model or rewrite as multiplication.

4. How many halves are in a fourth? Explain how you found your answer.

$\frac{1}{4} \div \frac{1}{2} = \frac{1}{4} \times \frac{2}{1} = \frac{2}{4} = \frac{1}{2}$

Laurie's Notes

Activity 2

- ? Students may have a question about the first entry in the Division Table. Use language such as "if a team is composed of 16 people, how many teams could be made with 8 people?" none; You only have $\frac{1}{2}$ a team.
- The next four entries in the Division Table will not be a problem and the entries given in the Multiplication Table follow from the previous lesson.
- Listen to student conversations as they complete each table. It is likely they will rely on pattern recognition (answers are doubling) versus actually considering the written problems.
- ? When students have finished the activity ask, "What relationship was there between the red numbers in each table?" They were the same.
- ? "What was the relationship between the blue numbers in each table?" Listen for the relationship of reciprocals. Students may not have accurate language.
- Ask several pairs of students to share their conjectures for part (d).
- **MP3:** In listening to the conjectures, the mathematically proficient student will note that the conjectures are based upon the case of a whole number being divided by a fraction. They will want to consider the case of a fraction or mixed number being divided by a fraction.

What Is Your Answer?

- Invert the fraction and then multiply. The quotient (answer) is the number of fractional parts that are in the dividend.

Closure

- **Exit Ticket:** How many thirds are in 2? In other words, what is $2 \div \frac{1}{3}$? Draw a picture to explain your answer. 6

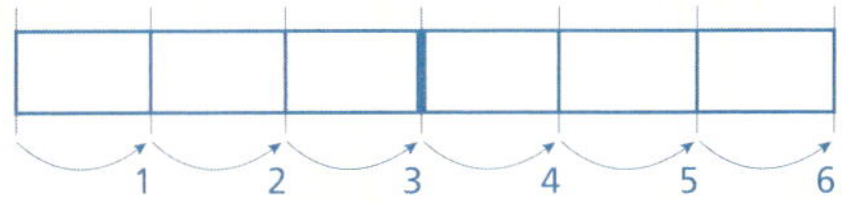

2 ACTIVITY: Using Tables to Recognize a Pattern

Work with a partner.

a. Complete each table.

Division Table

$8 \div 16$	$\frac{1}{2}$
$8 \div 8$	1
$8 \div 4$	2
$8 \div 2$	4
$8 \div 1$	8
$8 \div \frac{1}{2}$	
$8 \div \frac{1}{4}$	
$8 \div \frac{1}{8}$	

Multiplication Table

$8 \times \frac{1}{16}$	$\frac{1}{2}$
$8 \times \frac{1}{8}$	1
$8 \times \frac{1}{4}$	2
$8 \times \frac{1}{2}$	4
8×1	8
8×2	
8×4	
8×8	

Math Practice 4

Analyze Relationships

How is multiplying numbers similar to dividing numbers?

b. Describe the relationship between the red numbers in the division table and the red numbers in the multiplication table.

c. Describe the relationship between the blue numbers in the division table and the blue numbers in the multiplication table.

d. **STRUCTURE** Make a conjecture about how you can use multiplication to divide by a fraction.

e. Test your conjecture using the problems in Activity 1.

What Is Your Answer?

3. **IN YOUR OWN WORDS** How can you divide by a fraction? Give an example.

4. How many halves are in a fourth? Explain how you found your answer.

Practice Use what you learned about dividing fractions to complete Exercises 11–18 on page 67.

2.2 Lesson

Check It Out
Lesson Tutorials
BigIdeasMath.com

Key Vocabulary
reciprocals, *p. 64*

Two numbers whose product is 1 are **reciprocals**. To write the reciprocal of a number, write the number as a fraction. Then invert the fraction. So, the reciprocal of a fraction $\frac{a}{b}$ is $\frac{b}{a}$, where a and $b \neq 0$.

The Meaning of a Word ● Invert

When you **invert** a glass, you turn it over.

Study Tip

The product of a nonzero number and its reciprocal is 1.

$\frac{a}{b} \cdot \frac{b}{a} = 1$

This is called the *Multiplicative Inverse Property*. You will learn more about this property in Chapter 7.

EXAMPLE 1 Writing Reciprocals

	Original Number	*Fraction*	*Reciprocal*	*Check*
a.	$\frac{3}{5}$	$\frac{3}{5}$	$\frac{5}{3}$	$\frac{3}{5} \times \frac{5}{3} = 1$
b.	$\frac{9}{5}$	$\frac{9}{5}$	$\frac{5}{9}$	$\frac{9}{5} \times \frac{5}{9} = 1$
c.	2	$\frac{2}{1}$	$\frac{1}{2}$	$\frac{2}{1} \times \frac{1}{2} = 1$

Study Tip

When any number is multiplied by 0, the product is 0. So, the number 0 does not have a reciprocal.

On Your Own

Now You're Ready
Exercises 7–10

Write the reciprocal of the number.

1. $\frac{3}{4}$ **2.** 5 **3.** $\frac{7}{2}$ **4.** $\frac{4}{9}$

Key Idea

Dividing Fractions

Words To divide a number by a fraction, multiply the number by the reciprocal of the fraction.

Numbers $\frac{1}{5} \div \frac{3}{4} = \frac{1}{5} \times \frac{4}{3} = \frac{1 \times 4}{5 \times 3}$

Algebra $\frac{a}{b} \div \frac{c}{d} = \frac{a}{b} \cdot \frac{d}{c} = \frac{a \cdot d}{b \cdot c}$, where b, c, and $d \neq 0$

Multi-Language Glossary at BigIdeasMath.com

Laurie's Notes

Introduction

Connect

- **Yesterday:** Students used a visual model to divide by a fraction. (MP3, MP7)
- **Today:** Students will use a formal rule to divide by a fraction.

Motivate

- Hold up 3 quarters.
- ? "How many nickels are there in 3 quarters?" 15
- ? "How could I write that question as a division problem with fractions?"

  Think of what part of a dollar each coin represents → $\frac{3}{4} \div \frac{1}{20} = 15$
- ? "Does our technique from yesterday, *invert and multiply*, work for this problem?" yes: $\frac{3}{4} \div \frac{1}{20} = \frac{3}{4} \times \frac{20}{1} = 15$
- **Discuss:** This problem is a fraction divided by a fraction. Most of the problems yesterday were a whole number divided by a fraction. Today you will look at many types of division problems with fractions.
- Introduce reciprocal as shown in the text.
- **Teacher Note:** Today's lesson is fairly long. Do not rush through the lesson. It is better to spend two days than to rush through important concepts.

Lesson Notes

Example 1

- To write the reciprocal of a whole number, you begin by representing the whole number over one. Then switch the numerator and denominator.
- **Common Error:** When you ask students to write a number and its reciprocal, they frequently write: $\frac{3}{4} = \frac{4}{3}$. Reciprocals are not equal (except for 1). Students should write $\frac{3}{4}, \frac{4}{3}$ or "the reciprocal of $\frac{3}{4}$ is $\frac{4}{3}$."
- **Check:** The product of the **reciprocals** should equal 1.

Key Idea

- Write the Key Idea on the board.
- Note that the definition says "to divide a number *by a fraction*, ..." Students frequently ask, even two years later, which fraction to take the reciprocal of. Read this portion of the definition again. We are dividing a number (it could be a whole number or a fraction) *by a fraction* (Remind students that a whole number could also be written as a fraction).
- **Connection:** Notice that this definition can be used for dividing whole numbers, too: $8 \div 2 = 8 \div \frac{2}{1} = 8 \times \frac{1}{2} = \frac{8}{2} = 4$.
- The color-coding numbers helps all students, not just the visual learners.

Goal Today's lesson is dividing by a fraction.

Technology for the Teacher

Dynamic Classroom

Lesson Tutorials
Lesson Plans
Answer Presentation Tool

Extra Example 1

Write the reciprocal of the number.

a. $\frac{7}{8}$ $\frac{8}{7}$

b. $\frac{3}{2}$ $\frac{2}{3}$

c. 9 $\frac{1}{9}$

On Your Own

1. $\frac{4}{3}$
2. $\frac{1}{5}$
3. $\frac{2}{7}$
4. $\frac{9}{4}$

Extra Example 2

Find $\frac{3}{4} \div \frac{9}{10}$. $\frac{5}{6}$

Extra Example 3

A length of rope is 6 meters long. How many $\frac{3}{5}$-meter pieces can be cut from the length of rope? ten $\frac{3}{5}$-meter pieces

On Your Own

5. $\frac{6}{7}$
6. 4
7. $1\frac{1}{2}$
8. $1\frac{1}{3}$
9. 14

English Language Learners

Vocabulary

For English learners, discuss the meaning of the word *reciprocal* in nonmathematical and mathematical contexts. A *reciprocal agreement* between two states is an exchange of privileges. For example, a Washington driver's license is honored in Pennsylvania as long as Washington honors a Pennsylvania driver's license. In math, a reciprocal of a fraction exchanges the numerator and the denominator.

Laurie's Notes

Example 2

- Estimate the quotient first.
- This question is asking how many $\frac{2}{3}$ are in $\frac{1}{6}$. Because $\frac{1}{6}$ is much less than $\frac{2}{3}$, there aren't many! The answer must be very small, close to 0.
- Work through the example. Check the reasonableness of the answer.

Example 3

- Write the problem and sketch the model. Work through each method.
- A yardstick is a helpful physical model for this problem. Every 9 inches $\left(\frac{3}{4}\text{ of a foot}\right)$ can be marked. A rectangular model of 3 can also be drawn and marked as shown in the example.
- **MP6 Attend to Precision:** Units are an important aspect of the problem. The question is asking about a unit size of $\frac{3}{4}$ and how many of them are in 3 whole. Because the unit is less than 1 $\left(\frac{3}{4} < 1\right)$, 3 divided by something less than 1 must be greater than 3. Mathematically proficient students continually consider the units involved in a problem.
- Summarize with students what they have done so far: learned what a reciprocal is, divided two fractions, and divided a whole number by a fraction. Now they will look at dividing a fraction by a whole number.

Words of Wisdom

- It is important to have students think about the answer (estimate the answer) before they start working the problem. They should have a sense of the relative size of the answer before they begin.

On Your Own

- Before they begin, have students decide if the quotient is *less than* 1 or *greater than* 1. This is meant to tell if students can compare the fractions they are dividing.

EXAMPLE 2 Dividing a Fraction by a Fraction

Find $\frac{1}{6} \div \frac{2}{3}$.

$\frac{1}{6} \div \frac{2}{3} = \frac{1}{6} \times \frac{3}{2}$ — Multiply by the reciprocal of $\frac{2}{3}$, which is $\frac{3}{2}$.

$= \frac{1 \times \cancel{3}^{1}}{\cancel{6}_{2} \times 2}$ — Multiply fractions. Divide out the common factor 3.

$= \frac{1}{4}$ — Simplify.

EXAMPLE 3 Dividing a Whole Number by a Fraction

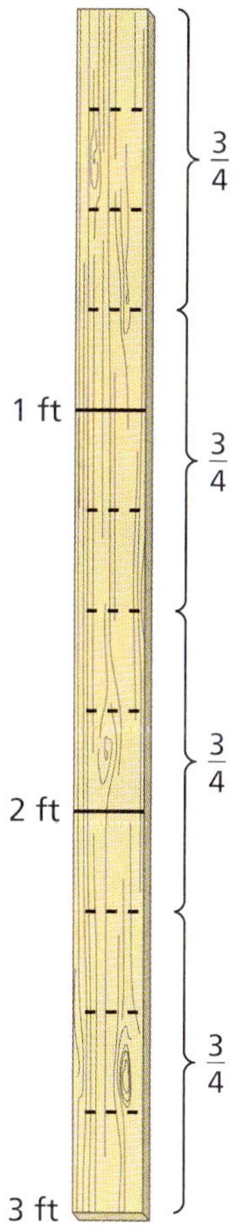

A piece of wood is 3 feet long. How many $\frac{3}{4}$-foot pieces can you cut from the piece of wood?

Method 1: Draw a diagram. Mark each foot on the diagram. Then divide each foot into $\frac{1}{4}$-foot sections.

Count the number of $\frac{3}{4}$-foot pieces of wood. There are four.

∴ So, you can cut four $\frac{3}{4}$-foot pieces from the piece of wood.

Method 2: Divide 3 by $\frac{3}{4}$ to find the number of $\frac{3}{4}$-foot pieces.

$3 \div \frac{3}{4} = 3 \times \frac{4}{3}$ — Multiply by the reciprocal of $\frac{3}{4}$, which is $\frac{4}{3}$.

$= \frac{\cancel{3}^{1} \times 4}{\cancel{3}_{1}}$ — Multiply. Divide out the common factor 3.

$= 4$ — Simplify.

∴ So, you can cut four $\frac{3}{4}$-foot pieces from the piece of wood.

On Your Own

Divide. Write the answer in simplest form.

5. $\frac{2}{7} \div \frac{1}{3}$

6. $\frac{1}{2} \div \frac{1}{8}$

7. $\frac{3}{8} \div \frac{1}{4}$

8. $\frac{2}{5} \div \frac{3}{10}$

9. How many $\frac{1}{2}$-foot pieces can you cut from a 7-foot piece of wood?

EXAMPLE 4 Dividing a Fraction by a Whole Number

Find $\frac{4}{5} \div 2$.

$\frac{4}{5} \div 2 = \frac{4}{5} \div \frac{2}{1}$ — Write 2 as an improper fraction.

$= \frac{4}{5} \times \frac{1}{2}$ — Multiply by the reciprocal of $\frac{2}{1}$, which is $\frac{1}{2}$.

$= \frac{\overset{2}{\cancel{4}} \times 1}{5 \times \underset{1}{\cancel{2}}}$ — Multiply fractions. Divide out the common factor 2.

$= \frac{2}{5}$ — Simplify.

On Your Own

Now You're Ready Exercises 11–26

Divide. Write the answer in simplest form.

10. $\frac{1}{2} \div 3$

11. $\frac{2}{3} \div 10$

12. $\frac{5}{8} \div 4$

13. $\frac{6}{7} \div 4$

EXAMPLE 5 Using Order of Operations

Evaluate $\frac{3}{8} + \frac{5}{6} \div 5$.

$\frac{3}{8} + \frac{5}{6} \div 5 = \frac{3}{8} + \frac{5}{6} \times \frac{1}{5}$ — Multiply by the reciprocal of 5, which is $\frac{1}{5}$.

$= \frac{3}{8} + \frac{\overset{1}{\cancel{5}} \times 1}{6 \times \underset{1}{\cancel{5}}}$ — Multiply $\frac{5}{6}$ and $\frac{1}{5}$. Divide out the common factor 5.

$= \frac{3}{8} + \frac{1}{6}$ — Simplify.

$= \frac{18}{48} + \frac{8}{48}$ — Rewrite fractions using a common denominator.

$= \frac{26}{48}$, or $\frac{13}{24}$ — Simplify.

Study Tip

You can use the LCD, 24, to add the fractions in Example 5.

$\frac{3}{8} + \frac{1}{6} = \frac{9}{24} + \frac{4}{24} = \frac{13}{24}$

On Your Own

Now You're Ready Exercises 43–51

Evaluate the expression. Write the answer in simplest form.

14. $\frac{4}{5} + \frac{2}{5} \div 4$

15. $\frac{3}{8} \div \frac{3}{4} - \frac{1}{6}$

16. $\frac{8}{9} \div 2 \div 8$

Laurie's Notes

Example 4

? "What do you know about the quotient when you divide a fraction by a whole number?" The quotient will be less than 1. Note: This would not necessarily be the case if the fraction was improper.

? "How would you estimate the answer first?" Listen for $\frac{4}{5}$ is close to 1 and 2 goes into 1 only a $\frac{1}{2}$ time. So, the answer is close to $\frac{1}{2}$.

- Continue to work through the problem, making note of dividing out the common factor of 2 before finding the product.

On Your Own

- Have pairs of students present their work at the board.

Example 5

- In this example, students must pull all of the operations together. After spending a good deal of time on multiplication and division, students start to forget about needing a common denominator when adding and subtracting fractions. They may also forget about the order of operations. It is good to continually review and integrate skills.

? Write the problem on the board, and ask a few questions.

- "What operations are involved in this problem?" division and addition
- "Which operation is performed first? Why?" division; order of operations
- "How do you divide by 5?" Multiply by $\frac{1}{5}$.

- **Common Error:** Students try to divide out what they think is a common factor of 3 because there is a 6 in the denominator. Remind students that the numbers must be *factors*.

? "What can be used as a common denominator?" Students may say 24 or 48. The problem uses the common denominator of 48, and the Study Tip uses the LCD of 24.

On Your Own

- If students are able to complete these problems correctly, they are ready for the assignment. Have pairs of students show their work at the board.

Closure

- Hold up 3 quarters. "How many dimes are in 3 quarters? Write and solve a division problem using fractions to represent this situation."
$7\frac{1}{2}$; $\frac{3}{4} \div \frac{1}{10} = \frac{3}{4} \cdot \frac{10}{1} = \frac{30}{4} = 7\frac{1}{2}$

Extra Example 4

Find $\frac{2}{3} \div 6$. $\frac{1}{9}$

On Your Own

10. $\frac{1}{6}$
11. $\frac{1}{15}$
12. $\frac{5}{32}$
13. $\frac{3}{14}$

Extra Example 5

Evaluate $\frac{5}{8} - \frac{7}{8} \div 5$. $\frac{9}{20}$

On Your Own

14. $\frac{9}{10}$
15. $\frac{1}{3}$
16. $\frac{1}{18}$

Differentiated Instruction

Kinesthetic

Give each student nine cards numbered 1 through 9. On a sheet of paper, have students draw a horizontal line halfway down the page. Ask one student to create a fraction using two of the cards. Have the rest of the class form the same fraction. Then ask the students how to find the reciprocal of the fraction. Have students make the reciprocal of the fraction with their cards and write a sentence naming the fraction and its reciprocal.

Vocabulary and Concept Check

1. *Sample answer:* $\frac{2}{5}, \frac{5}{2}$
2. $\frac{2}{9}$ does not belong because its reciprocal is not a whole number.
3. B
4. D
5. A
6. C

Practice and Problem Solving

7. $\frac{1}{8}$
8. $\frac{7}{6}$
9. $\frac{5}{2}$
10. $\frac{11}{8}$
11. $\frac{1}{2}$
12. $2\frac{11}{12}$
13. 16
14. 20
15. $\frac{1}{14}$
16. $\frac{3}{25}$
17. $\frac{1}{3}$
18. $\frac{2}{3}$
19. 3
20. $1\frac{13}{15}$
21. $\frac{2}{27}$
22. $\frac{1}{24}$
23. $\frac{27}{28}$
24. $\frac{26}{75}$
25. $20\frac{1}{4}$
26. 24
27. You need to invert the second fraction before you multiply. $\frac{4}{7} \div \frac{13}{28} = \frac{4}{7} \times \frac{28}{13} = \frac{4 \times \cancel{28}^{4}}{{}_{1}\cancel{7} \times 13} = \frac{16}{13}$, or $1\frac{3}{13}$
28. You invert the second fraction, not the first.

$$\frac{2}{5} \div \frac{8}{9} = \frac{2}{5} \times \frac{9}{8} = \frac{{}^{1}\cancel{2} \times 9}{5 \times \cancel{8}_{4}} = \frac{9}{20}$$

Assignment Guide and Homework Check

Level	Day 1 Activity Assignment	Day 2 Lesson Assignment	Homework Check
Basic	11–18, 58–62	1–6, 19–25 odd, 27–31, 43–51 odd	19, 25, 28, 30, 51
Average	11–18, 58–62	1–6, 19–25 odd, 28–30 even, 43–51 odd, 54, 55	19, 25, 28, 30, 47, 55
Advanced	11–18, 58–62	1–6, 24–28 even, 32–54 even, 56, 57	26, 28, 32, 48, 56

Common Errors

- **Exercise 7** Students may have difficulty rewriting a whole number as a fraction. Remind them that a whole number is a fraction with a denominator of 1.
- **Exercises 11–26** Students may find the reciprocal of the dividend instead of the divisor, or the reciprocal of both. Remind them that they only need to find the reciprocal of the divisor. Demonstrate this concept with a calculator. Enter the problem as a division problem, and then enter it as a multiplication problem with the reciprocal. Ask students to compare the answers. This comparison can also be made with the incorrect methods to demonstrate that they will not work.
- **Exercises 11–26** Students may find the reciprocal but forget to change the division to multiplication. Remind students that inverting the number is the *opposite*, so it requires the *opposite* operation.

2.2 Record and Practice Journal

Complete the statement.

1. $\frac{3}{8} \times \underline{\frac{8}{3}} = 1$
2. $7 \times \underline{\frac{1}{7}} = 1$
3. $3 \div \underline{\frac{1}{12}} = 36$
4. $\frac{4}{9} \div \underline{\frac{1}{27}} = 12$

Evaluate the expression.

5. $\frac{1}{3} \div \frac{1}{6}$ **2**
6. $\frac{3}{8} \div \frac{5}{8}$ **$\frac{3}{5}$**
7. $6 \div \frac{2}{5}$ **15**
8. $\frac{4}{9} \div \frac{2}{3} \div \frac{5}{6}$ **$\frac{4}{5}$**
9. $\frac{1}{3} \div \frac{4}{7} \div \frac{3}{10}$ **$2\frac{5}{21}$**
10. $\frac{7}{8} \bullet \frac{4}{5} \div \frac{7}{20}$ **2**
11. In a jewelry store, rings make up $\frac{5}{9}$ of the inventory. Earrings make up $\frac{4}{15}$ of the inventory. How many times greater is the ring inventory than the earring inventory?

$2\frac{1}{12}$

2.2 Exercises

Check It Out
Help with Homework
BigIdeasMath.com

Vocabulary and Concept Check

1. **OPEN-ENDED** Write a fraction and its reciprocal.

2. **WHICH ONE DOESN'T BELONG?** Which of the following does *not* belong with the other three? Explain your reasoning.

$\frac{1}{3}$ $\frac{1}{6}$ $\frac{2}{9}$ $\frac{1}{8}$

MATCHING **Match the expression with its value.**

3. $\frac{2}{5} \div \frac{8}{15}$ 4. $\frac{8}{15} \div \frac{2}{5}$ 5. $\frac{2}{15} \div \frac{8}{5}$ 6. $\frac{8}{5} \div \frac{2}{15}$

A. $\frac{1}{12}$ B. $\frac{3}{4}$ C. 12 D. $1\frac{1}{3}$

Practice and Problem Solving

Write the reciprocal of the number.

1 7. 8 8. $\frac{6}{7}$ 9. $\frac{2}{5}$ 10. $\frac{8}{11}$

Divide. Write the answer in simplest form.

2 3 4 11. $\frac{1}{8} \div \frac{1}{4}$ 12. $\frac{5}{6} \div \frac{2}{7}$ 13. $12 \div \frac{3}{4}$ 14. $8 \div \frac{2}{5}$

15. $\frac{3}{7} \div 6$ 16. $\frac{12}{25} \div 4$ 17. $\frac{2}{9} \div \frac{2}{3}$ 18. $\frac{8}{15} \div \frac{4}{5}$

19. $\frac{1}{3} \div \frac{1}{9}$ 20. $\frac{7}{10} \div \frac{3}{8}$ 21. $\frac{14}{27} \div 7$ 22. $\frac{5}{8} \div 15$

23. $\frac{27}{32} \div \frac{7}{8}$ 24. $\frac{4}{15} \div \frac{10}{13}$ 25. $9 \div \frac{4}{9}$ 26. $10 \div \frac{5}{12}$

ERROR ANALYSIS **Describe and correct the error in finding the quotient.**

27. ✗

$$\frac{4}{7} \div \frac{13}{28} = \frac{4}{7} \times \frac{13}{28} = \frac{\cancel{4}^{1} \times 13}{7 \times \cancel{28}_{7}} = \frac{13}{49}$$

28. ✗

$$\frac{2}{5} \div \frac{8}{9} = \frac{5}{2} \times \frac{8}{9} = \frac{5 \times \cancel{8}^{4}}{{}_{1}\cancel{2} \times 9} = \frac{20}{9}$$

29. **REASONING** How can you use estimation to show that the quotient in Exercise 28 is incorrect?

30. APPLE PIE You have $\frac{3}{5}$ of an apple pie. You divide the remaining pie into 5 equal slices. What fraction of the original pie is each slice?

31. ANIMALS How many times longer is the baby alligator than the baby gecko?

Determine whether the numbers are reciprocals. If not, write the reciprocal of each number.

32. $9, \frac{1}{9}$ **33.** $\frac{4}{5}, \frac{10}{8}$ **34.** $\frac{5}{6}, \frac{15}{18}$ **35.** $\frac{6}{5}, \frac{5}{6}$

Copy and complete the statement.

36. $\frac{5}{12} \times \square = 1$ **37.** $3 \times \square = 1$ **38.** $7 \div \square = 56$

Without finding the quotient, copy and complete the statement using <, >, or =. Explain your reasoning.

39. $5 \div \frac{7}{9} \; \square \; 5$ **40.** $\frac{3}{7} \div 1 \; \square \; \frac{3}{7}$ **41.** $8 \div \frac{3}{4} \; \square \; 8$ **42.** $\frac{5}{6} \div \frac{7}{8} \; \square \; \frac{5}{6}$

Evaluate the expression. Write the answer in simplest form.

5 **43.** $\frac{1}{6} \div 6 \div 6$ **44.** $\frac{7}{12} \div 14 \div 6$ **45.** $\frac{3}{5} \div \frac{4}{7} \div \frac{9}{10}$

46. $4 \div \frac{8}{9} - \frac{1}{2}$ **47.** $\frac{3}{4} + \frac{5}{6} \div \frac{2}{3}$ **48.** $\frac{7}{8} - \frac{3}{8} \div 9$

49. $\frac{9}{16} \div \frac{3}{4} \cdot \frac{2}{13}$ **50.** $\frac{3}{14} \cdot \frac{2}{5} \div \frac{6}{7}$ **51.** $\frac{10}{27} \cdot \left(\frac{3}{8} \div \frac{5}{24}\right)$

52. REASONING Use a model to evaluate the quotient $\frac{1}{2} \div \frac{1}{6}$. Explain.

53. VIDEO CHATTING You use $\frac{1}{8}$ of your battery for every $\frac{2}{5}$ of an hour that you video chat. You use $\frac{3}{4}$ of your battery video chatting. How long did you video chat?

Common Errors

- **Exercises 32–35** Unless the fraction is obviously the reciprocal, students may state that the fraction is not a reciprocal. Remind them that if the product of the two fractions is 1, then the fractions are reciprocals of one another.
- **Exercise 38** Students may read the problem quickly and think that it is a multiplication problem instead of a division problem and that the answer is 8. The answer is the reciprocal of 8. Tell them to think of the problem backwards. "56 multiplied by what number equals 7?" This should help them to see that the answer must be a fraction.
- **Exercises 43–51** Students may try to apply the Associative and Commutative Properties. Remind them that the Associative and Commutative Properties apply only to multiplication and addition. Also remind students to follow the order of operations for all the problems in this section.

Practice and Problem Solving

29. Round $\frac{2}{5}$ to $\frac{1}{2}$ and $\frac{8}{9}$ to 1. $\frac{1}{2} \div 1 = \frac{1}{2}$, which is not close to the incorrect answer of $\frac{20}{9}$.

30. $\frac{3}{25}$

31. $5\frac{5}{8}$ times

32. yes

33. yes

34. no; Both fractions are equal to $\frac{5}{6}$. The reciprocal of $\frac{5}{6}$ is $\frac{6}{5}$.

35. yes

36. $\frac{12}{5}$

37. $\frac{1}{3}$

38. $\frac{1}{8}$

39. >; When you divide a number by a fraction less than 1, the quotient is greater than the number.

40. =; When you divide a fraction by 1, the quotient is the fraction.

41. >; When you divide a number by a fraction less than 1, the quotient is greater than the number.

42. >; When you divide a fraction by a fraction less than 1, the quotient is greater than the fraction.

43. $\frac{1}{216}$

44. $\frac{1}{144}$

45. $1\frac{1}{6}$

46. 4

47. 2

48. $\frac{5}{6}$

49. $\frac{3}{26}$

50. $\frac{1}{10}$

51. $\frac{2}{3}$

52. 3

53. $2\frac{2}{5}$ h or 2 h 24 min

Practice and Problem Solving

54. when the fraction has, or can be simplified to have, a 1 in the numerator; The reciprocal will have, or can be simplified to have, a 1 in the denominator, so it is a whole number.

55. **a.** $3\frac{3}{4}$ times

b. $3\frac{1}{3}$ times

c. $\frac{1}{4}$

56. **a.** See *Taking Math Deeper.*

b. 2 plates; $\frac{1}{2}$ pt

c. See *Taking Math Deeper.*

57. **a.** $67\frac{1}{5}$ gal

b. $8\frac{2}{5}$ gal

c. $33\frac{3}{5}$ gal

Fair Game Review

58. 8 **59.** 6

60. 16 **61.** 5

62. D

Mini-Assessment

Divide. Write the answer in simplest form.

1. $\frac{1}{3} \div \frac{1}{9}$ 3

2. $\frac{4}{5} \div \frac{2}{3}$ $1\frac{1}{5}$

3. $\frac{6}{10} \div \frac{1}{2}$ $1\frac{1}{5}$

4. $\frac{12}{15} \div 6$ $\frac{2}{15}$

5. You have $\frac{3}{4}$ of a pumpkin pie remaining. You divide the remaining pie into six equal slices. What fraction of the original pie is each slice? $\frac{1}{8}$

Taking Math Deeper

Exercise 56, parts (a) and (c)

Part (c) of this problem is a nice opportunity to show students how "making a table" can be a good problem-solving strategy.

1 Organize the given information.

$\frac{7}{8}$ pint to glaze a bowl

$\frac{9}{16}$ pint to glaze a plate

6 pints of glaze

2 **a.** I can glaze $6 \div \frac{7}{8}$ bowls.

$$6 \div \frac{7}{8} = 6 \cdot \frac{8}{7} = \frac{48}{7} = 6\frac{6}{7} \text{ or about 6 bowls.}$$

I can glaze $6 \div \frac{9}{16}$ plates.

$$6 \div \frac{9}{16} = 6 \cdot \frac{16}{9} = \frac{32}{3} = 10\frac{2}{3} \text{ or about 10 plates.}$$

3 Make a table for bowls and for plates.

Bowls	1	2	3	4	5	6
Glaze (pints)	$\frac{7}{8}$	$1\frac{3}{4}$	$2\frac{5}{8}$	$3\frac{1}{2}$	$4\frac{3}{8}$	$5\frac{1}{4}$

Plates	1	2	3	4	5	6
Glaze (pints)	$\frac{9}{16}$	$1\frac{1}{8}$	$1\frac{11}{16}$	$2\frac{1}{4}$	$2\frac{13}{16}$	$3\frac{3}{8}$

Use *Guess, Check, and Revise* to find two glaze amounts that add to 6 pints.

c. $2\frac{5}{8} + 3\frac{3}{8} = 6$ pints

3 bowls 6 plates

Reteaching and Enrichment Strategies

If students need help. . .	If students got it. . .
Resources by Chapter • Practice A and Practice B • Puzzle Time Record and Practice Journal Practice Differentiating the Lesson Lesson Tutorials Skills Review Handbook	Resources by Chapter • Enrichment and Extension • Technology Connection Start the next section

54. **NUMBER SENSE** When is the reciprocal of a fraction a whole number? Explain.

55. **BUDGETS** The table shows the portions of a family budget that are spent on several expenses.

Expense	Portion of Budget
Housing	$\frac{1}{4}$
Food	$\frac{1}{12}$
Automobiles	$\frac{1}{15}$
Recreation	$\frac{1}{40}$

  a. How many times more is the expense for housing than for automobiles?

  b. How many times more is the expense for food than for recreation?

  c. The expense for automobile fuel is $\frac{1}{60}$ of the total expenses. What fraction of the automobile expense is spent on fuel?

56. **PROBLEM SOLVING** You have 6 pints of glaze. It takes $\frac{7}{8}$ of a pint to glaze a bowl and $\frac{9}{16}$ of a pint to glaze a plate.

  a. How many bowls could you glaze? How many plates could you glaze?

  b. You want to glaze 5 bowls, and then use the rest for plates. How many plates can you glaze? How much glaze will be left over?

  c. How many of each object could you glaze so that there is no glaze left over? Explain how you found your answer.

57. **Reasoning** A water tank is $\frac{1}{8}$ full. The tank is $\frac{3}{4}$ full when 42 gallons of water are added to the tank.

  a. How much water can the tank hold?

  b. How much water was originally in the tank?

  c. How much water is in the tank when it is $\frac{1}{2}$ full?

Fair Game Review What you learned in previous grades & lessons

Find the GCF of the numbers. *(Section 1.5)*

58. 8, 16 59. 24, 66 60. 48, 80 61. 15, 45, 100

62. **MULTIPLE CHOICE** How many inches are in $5\frac{1}{2}$ yards? *(Skills Review Handbook)*

Ⓐ $15\frac{1}{2}$ Ⓑ $16\frac{1}{2}$ Ⓒ 66 Ⓓ 198

2.3 Dividing Mixed Numbers

Essential Question How can you model division by a mixed number?

1 ACTIVITY: Writing a Story

Work with a partner. Think of a story that uses division by a mixed number.

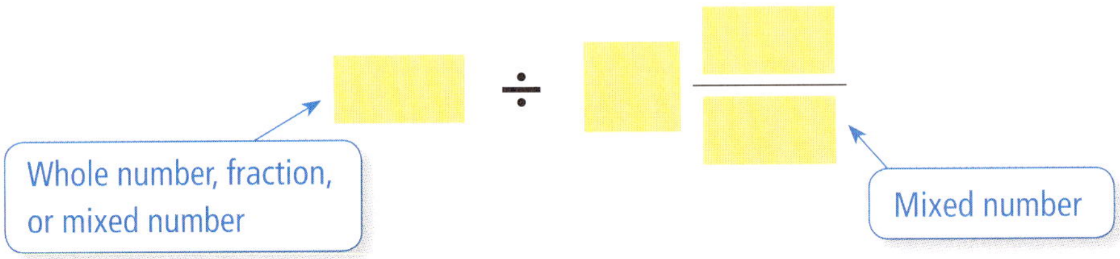

a. Write your story. Then draw pictures for your story.

b. Solve the division problem and use the answer in your story. Include a diagram of the division problem.

There are many possible stories. Here is one that uses $6 \div 1\frac{1}{2}$.

Joe goes on a camping trip with his aunt, his uncle, and three cousins. They leave at 5:00 P.M. and drive 2 hours to the campground.

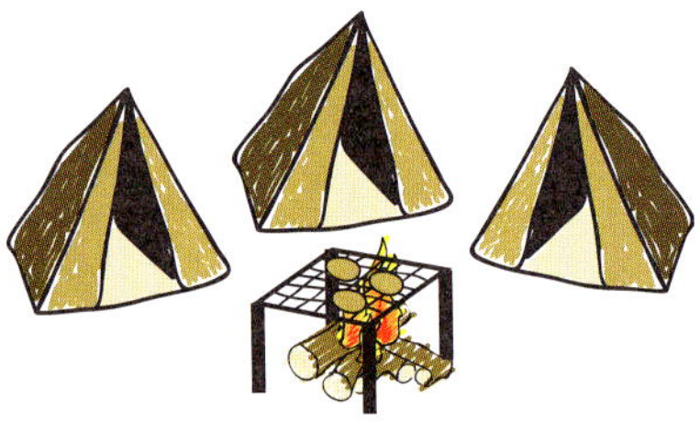

Joe helps his uncle put up three tents. His aunt cooks hamburgers on a grill that is over a fire.

In the morning, Joe tells his aunt that he is making pancakes. He decides to triple the recipe so there will be plenty of pancakes for everyone. A single recipe uses 2 cups of water, so he needs a total of 6 cups.

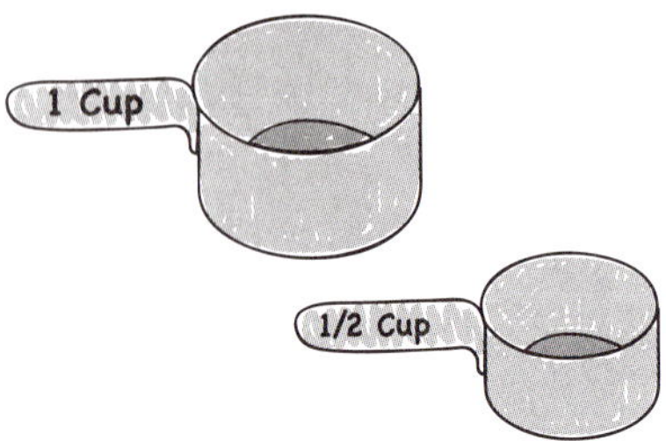

Joe's aunt has a 1-cup measuring cup and a ½-cup measuring cup. The water faucet is about 50 yards from the campsite. Joe tells his cousins that he can get 6 cups of water in only 4 trips.

When his cousins ask him how he knows that, he uses a stick to draw a diagram in the dirt. Joe says, "This diagram shows that there are four 1½s in 6." In other words, $6 \div 1\frac{1}{2} = 4$.

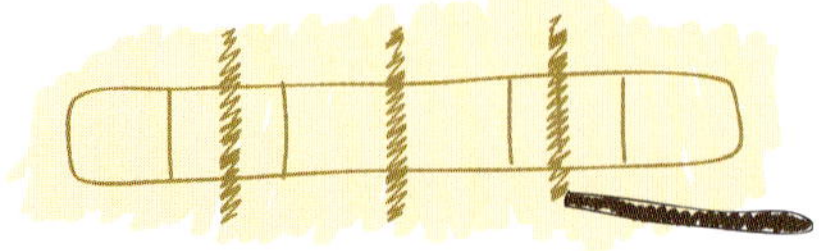

Common Core

Dividing Fractions

In this lesson, you will

- use models to divide mixed numbers.
- divide mixed numbers.
- solve real-life problems.

Learning Standard 6.NS.1

Laurie's Notes

Introduction

Standards for Mathematical Practice

- **MP6 Attend to Precision:** Mathematically proficient students continually consider the units involved in a problem. As part of writing the story problems, students need to pay attention to the units associated with each number. This is a subtle introduction to dimensional analysis, something the science teachers will be happy to see.
- This is a very different activity, one which we use in each book of the series. The goal is to have students use their literacy skills, putting mathematical skills in a context. Students need practice with writing stories, and many will find the first attempt challenging.

Motivate

- Share a colorful story while holding a travel coffee mug. The essence of the story should result in the following facts: a coffee pot holds 12 cups and the mug holds about $1\frac{1}{2}$ cups. How many times can the travel mug be filled?

? "What expression can you use to answer this question? What is the answer?" $12 \div 1\frac{1}{2}$; Listen for methods to find the solution.

Activity Notes

Activity 1

- Ask a student to read the sample story to the class. Discuss the story.

? "Have any of you gone on a camping trip?"

- Discuss the pictures.

? "Would you prefer to read just the text, or do you enjoy having pictures that relate to the text?"

- Draw the diagram at the bottom of the page. Ask a student to explain the solution.
- **Big Idea:** When writing the final problem, 6 ÷ 1½ = 4, identify the units associated with each number. The 6 is *cups*, the $1\frac{1}{2}$ is *cups per trip,* and the 4 is *trips.* If students have not yet written a story, suggest that they consider a context that involves a rate (e.g., miles per hour, songs per minute, cookies per team, tons per truckload). The rate helps define the context.
- **Common Error:** Students might decide on a rate where a mixed number does not make sense in terms of trying to write a story. For instance, they may want to work with "cars per group." You can have 6 cars and 4 groups, but you would not have $1\frac{1}{2}$ cars per group.
- Once students have decided on the context, allow 20–30 minutes of writing time.
- Allow a few students to share their short stories.

Common Core State Standards

6.NS.1 Interpret and compute quotients of fractions, and solve word problems involving division of fractions by fractions.

Previous Learning

Students should know how to divide a number by a fraction.

Technology for the Teacher

Dynamic Classroom

Lesson Plans
Complete Materials List

2.3 Record and Practice Journal

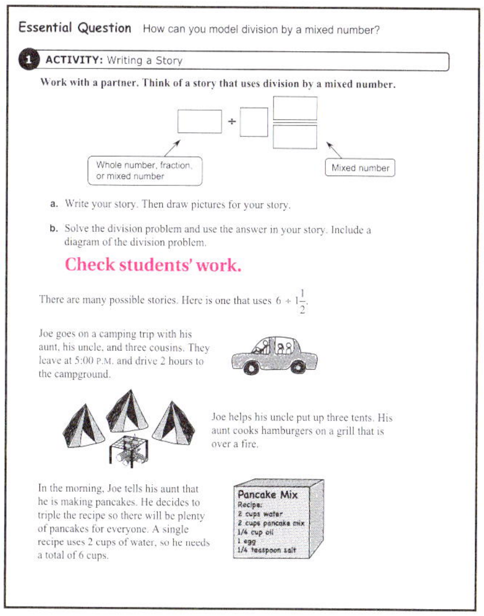
Essential Question How can you model division by a mixed number?

1 ACTIVITY: Writing a Story

Work with a partner. Think of a story that uses division by a mixed number.

☐ ÷ ☐ $\frac{☐}{☐}$

Whole number, fraction, or mixed number — Mixed number

a. Write your story. Then draw pictures for your story.

b. Solve the division problem and use the answer in your story. Include a diagram of the division problem.

Check students' work.

There are many possible stories. Here is one that uses $6 \div 1\frac{1}{2}$.

Joe goes on a camping trip with his aunt, his uncle, and three cousins. They leave at 5:00 P.M. and drive 2 hours to the campground.

Joe helps his uncle put up three tents. His aunt cooks hamburgers on a grill that is over a fire.

In the morning, Joe tells his aunt that he is making pancakes. He decides to triple the recipe so there will be plenty of pancakes for everyone. A single recipe uses 2 cups of water, so he needs a total of 6 cups.

Pancake Mix
Recipe:
2 cups water
2 cups pancake mix
1/4 cup oil
1 egg
1/4 teaspoon salt

English Language Learners

Pair Activity

For Activity 3, pair English learners with English speakers to write stories based on the English learners' cultures. Encourage students to share how to incorporate math into the story.

2.3 Record and Practice Journal

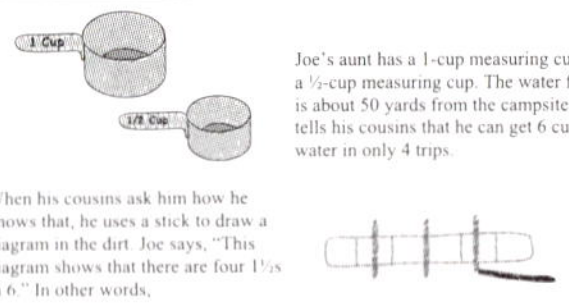

Joe's aunt has a 1-cup measuring cup and a ½-cup measuring cup. The water faucet is about 50 yards from the campsite. Joe tells his cousins that he can get 6 cups of water in only 4 trips.

When his cousins ask him how he knows that, he uses a stick to draw a diagram in the dirt. Joe says, "This diagram shows that there are four 1½s in 6." In other words,

$$6 \div 1\frac{1}{2} = 4.$$

2 ACTIVITY: Dividing Mixed Numbers

Work with a partner. Write the division problem and solve it using a model.

a. How many three-fourths are in four and one-half? $4\frac{1}{2} \div \frac{3}{4} = 6$

b. How many five-sixths are in three and one-third? $3\frac{1}{3} \div \frac{5}{6} = 4$

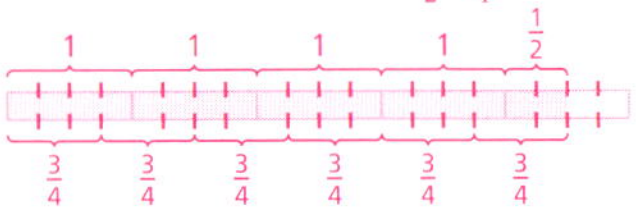

c. How many three-eighths are in three and three-fourths? $3\frac{3}{4} \div \frac{3}{8} = 10$

d. How many one and one-halves are in six? $6 \div 1\frac{1}{2} = 4$

e. How many one and one-fifths are in five? $5 \div 1\frac{1}{5} = 4\frac{1}{6}$

f. How many one and one-fourths are in four and one-half? $4\frac{1}{2} \div 1\frac{1}{4} = 3\frac{3}{5}$

g. How many two and one-thirds are in five and five-sixths? $5\frac{5}{6} \div 2\frac{1}{3} = 2\frac{1}{2}$

What Is Your Answer?

3. IN YOUR OWN WORDS How can you model division by a mixed number?

Check students' work.

4. Can you think of another method you can use to obtain your answers in Activity 2?

Write mixed numbers as improper fractions and divide as you would with proper fractions.

Laurie's Notes

Activity 2

- A visual model is drawn to represent the division problem. The labels above the diagram represent the quantity you have (dividend). Below the diagram students mark the size of the unit (divisor) they want to fit into the quantity. In part (a), they are counting the number of $\frac{3}{4}$-units in $4\frac{1}{2}$.

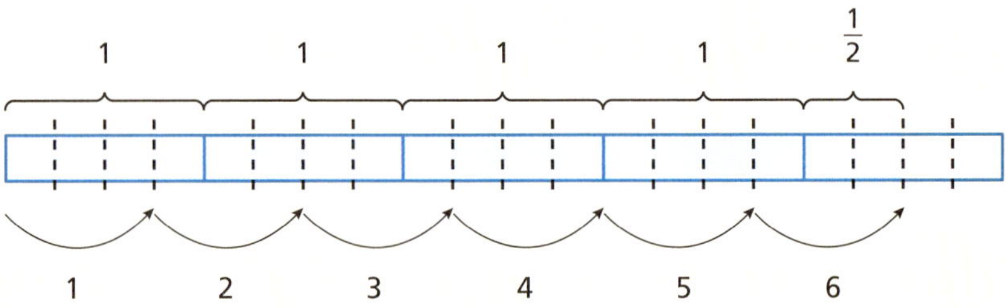

- Students will have more difficulty with parts (e), (f), and (g) because the quotient is not a whole number. Help students by asking them to think about what portion of the unit they are counting is represented by the amount that is left. For example, in part (e), the remaining piece is $\frac{1}{6}$ of the $1\frac{1}{5}$ unit they are counting.
- **MP3 Construct Viable Arguments and Critique the Reasoning of Others:** Take time to have a student talk through his or her approach to each problem in the activity. The student's rectangular model could be displayed or drawn as the student explains his or her thinking. Other students should listen carefully to the language used and the explanation given. You want students to be in the habit of communicating their thinking and having others critiquing their arguments.

What Is Your Answer?

- Students should describe the method of drawing a rectangular model of the dividend and counting the number of divisor units.
- For Question 4, there may be students that test out their conjectures from the previous lesson of inverting and multiplying. If successful, they would have written each number as an improper fraction.

Closure

- **Writing Prompt:** Dividing by a mixed number is like dividing by a fraction because . . . You can write a mixed number as an improper fraction.

2 ACTIVITY: Dividing Mixed Numbers

Work with a partner. Write the division problem and solve it using a model.

Math Practice 2

Make Sense of Quantities

What values do the parts of the model represent?

a. How many three-fourths are in four and one-half?

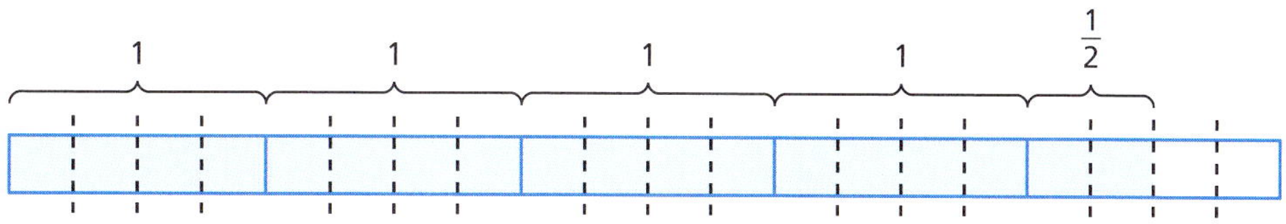

b. How many five-sixths are in three and one-third?

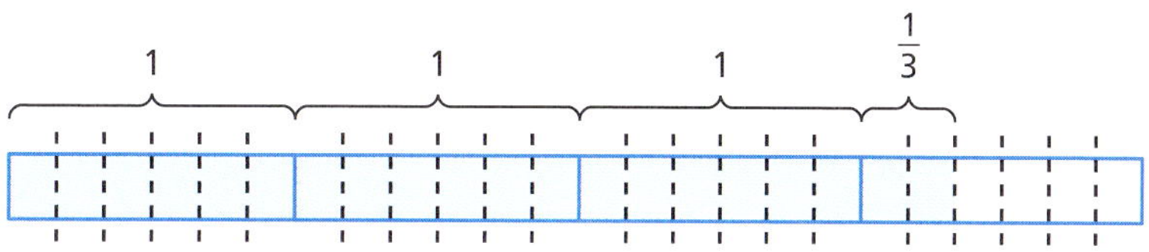

c. How many three-eighths are in three and three-fourths?

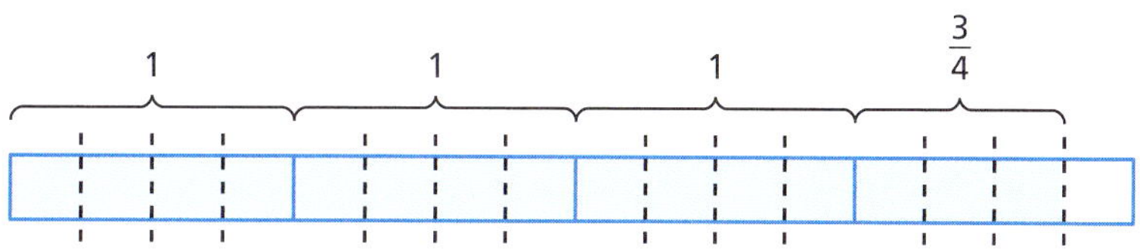

d. How many one and one-halves are in six?

e. How many one and one-fifths are in five?

f. How many one and one-fourths are in four and one-half?

g. How many two and one-thirds are in five and five-sixths?

What Is Your Answer?

3. **IN YOUR OWN WORDS** How can you model division by a mixed number?

4. Can you think of another method you can use to obtain your answers in Activity 2?

Practice Use what you learned about dividing mixed numbers to complete Exercises 5–12 on page 74.

2.3 Lesson

Check It Out
Lesson Tutorials
BigIdeasMath.com

Key Idea

Dividing Mixed Numbers

Write each mixed number as an improper fraction. Then divide as you would with proper fractions.

EXAMPLE 1 Dividing a Mixed Number by a Fraction

Find $2\frac{1}{4} \div \frac{3}{8}$.

$2\frac{1}{4} \div \frac{3}{8} = \frac{9}{4} \div \frac{3}{8}$ Write $2\frac{1}{4}$ as the improper fraction $\frac{9}{4}$.

$= \frac{9}{4} \times \frac{8}{3}$ Multiply by the reciprocal of $\frac{3}{8}$, which is $\frac{8}{3}$.

$= \frac{\overset{3}{\cancel{9}} \times \overset{2}{\cancel{8}}}{\underset{1}{\cancel{4}} \times \underset{1}{\cancel{3}}}$ Multiply fractions. Divide out common factors.

$= 6$ Simplify.

Check

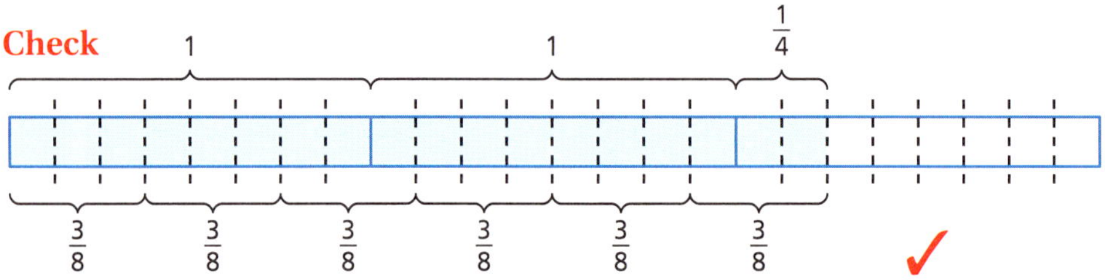

EXAMPLE 2 Dividing Mixed Numbers

Find $3\frac{5}{6} \div 1\frac{2}{3}$.

Estimate $4 \div 2 = 2$

$3\frac{5}{6} \div 1\frac{2}{3} = \frac{23}{6} \div \frac{5}{3}$ Write each mixed number as an improper fraction.

$= \frac{23}{6} \times \frac{3}{5}$ Multiply by the reciprocal of $\frac{5}{3}$, which is $\frac{3}{5}$.

$= \frac{23 \times \overset{1}{\cancel{3}}}{\underset{2}{\cancel{6}} \times 5}$ Multiply fractions. Divide out common factors.

$= \frac{23}{10}$, or $2\frac{3}{10}$ Simplify.

So, the quotient is $2\frac{3}{10}$.

Reasonable? $2\frac{3}{10} \approx 2$ ✓

On Your Own

Now You're Ready
Exercises 5–20

Divide. Write the answer in simplest form.

1. $1\frac{3}{7} \div \frac{2}{3}$ **2.** $2\frac{1}{6} \div \frac{3}{4}$ **3.** $8\frac{1}{4} \div 1\frac{1}{2}$ **4.** $6\frac{4}{5} \div 2\frac{1}{8}$

Laurie's Notes

Introduction

Connect

- **Yesterday:** Students wrote story problems related to dividing with a mixed number. (MP3, MP6)
- **Today:** Students will use a formal rule to divide with mixed numbers.

Motivate

- Do a warm-up with students where you ask them to give examples of different types of numbers. The goal is to review vocabulary and check on number sense.
- Examples: a fraction less than $\frac{1}{2}$, an improper fraction, a mixed number between 2 and 3, an improper fraction with an odd denominator, two fractions whose product is 1, a division problem involving fractions whose quotient is greater than 1, and a division problem involving fractions whose quotient is less than 1.

Lesson Notes

Key Idea

? "How do you write a mixed number as an improper fraction? Give an example." *Sample answer:* Multiply the denominator of the fraction by the whole number part and then add the numerator. Put this quantity over the denominator of the fraction. $2\frac{3}{5} = \frac{5 \cdot 2 + 3}{5} = \frac{13}{5}$

? "How do you write a whole number as a fraction? Give an example." Write the number as the numerator of the fraction with a denominator of 1.

Example 1

- Estimate the quotient first.

? "What is the first step?" Write $2\frac{1}{4}$ as an improper fraction.

- **MP4 Model with Mathematics:** You want students to develop the habit of checking their work. In this example, the visual model provided helps to connect the abstract problem with a visual model of the solution.
- **Check**: Ask a student to explain why the model verifies the quotient found.

Example 2

- Check the problem with a model. In words, there are 2 groups of $1\frac{2}{3}$ in $3\frac{5}{6}$, with 3 left over out of the 10 needed for another group. So, it is $2\frac{3}{10}$.

On Your Own

- Before they begin, have students decide if the quotient is *greater than 1* or *less than 1.*
- Students should work with partners on these problems. Have pairs of students share their work at the board. Check students' confidence with dividing out common factors.

Goal Today's lesson is dividing mixed numbers.

Technology for the Teacher

Dynamic Classroom

Lesson Tutorials
Lesson Plans
Answer Presentation Tool

Extra Example 1

Find $5\frac{1}{5} \div \frac{1}{2}$. $10\frac{2}{5}$

Extra Example 2

Find $7\frac{3}{4} \div 2\frac{2}{3}$. $2\frac{29}{32}$

On Your Own

1. $2\frac{1}{7}$
2. $2\frac{8}{9}$
3. $5\frac{1}{2}$
4. $3\frac{1}{5}$

Extra Example 3

Evaluate $\frac{3}{4} - 3\frac{1}{2} \div 8\frac{1}{6}$. $\frac{9}{28}$

Extra Example 4

One serving of French onion soup is $1\frac{3}{4}$ cups. A restaurant cook makes 75 cups of soup. Is there enough to serve 40 people? Yes, because $42\frac{6}{7}$ is the number of $1\frac{3}{4}$-cup servings you have in 75 cups of soup. Because $42\frac{6}{7}$ is greater than 40, there is enough soup to serve 40 people.

On Your Own

5. $8\frac{1}{8}$
6. $4\frac{8}{9}$
7. 2
8. $\frac{1}{3}$
9. yes; There are 18 servings of soup, so you can serve 15 people.

Differentiated Instruction

Visual

Students who have difficulty dividing out common factors may benefit from a color-coding system. Have students circle pairs of numbers that share a common factor with the same color. For instance, in Example 3, 21 and 9 could be circled in blue because they share a common factor of three. The factors 8 and 4 could be circled in red, because they share a common factor of 4. To divide out common factors, the students only need to look at one color at a time.

Laurie's Notes

Example 3

- Write the problem on the board.
- ? "How many operations are involved in this problem? What operation is performed first?" division and subtraction; division
- **FYI:** Students do not always see the common factors in this problem. They sometimes believe that you need to divide out the entire number (written in the numerator or denominator), instead of a factor of the number. In this case, 9 and 21 have the common factor of 3, which is divided out.

Example 4

- Do the problem together as a class.
- ? "What does the quotient represent when you divide 50 by $1\frac{2}{3}$?" the number of servings from 50 cups of soup
- It is important to ask this question so that students will know how to interpret the answers.
- ? **Alternative Method:** Ask questions about solving the problem differently.
  - "What does the product of 35 and $1\frac{2}{3}$ represent?" the amount of soup needed to serve 35 people
  - "Is either approach $\left(50 \div 1\frac{2}{3} \text{ or } 35 \times 1\frac{2}{3}\right)$ sufficient to answer the question?" Yes, both approaches work.
- **MP2 Reason Abstractly and Quantitatively:** In thinking about this problem, mathematically proficient students have the ability to reason quantitatively from the context given.

On Your Own

- Questions 5–8 require a bit more time. Have students work the problems individually first, and then check with neighbors. There are many skills involved with each. Although all four problems involve the order of operations, Questions 7 and 8 are not solved from left to right.

Closure

- **Connection:** How is division of mixed numbers like multiplication of mixed numbers? In each case, you must first rewrite the mixed numbers as improper fractions.

EXAMPLE 3 Using Order of Operations

Evaluate $5\frac{1}{4} \div 1\frac{1}{8} - \frac{2}{3}$.

$$5\frac{1}{4} \div 1\frac{1}{8} - \frac{2}{3} = \frac{21}{4} \div \frac{9}{8} - \frac{2}{3}$$ Write each mixed number as an improper fraction.

$$= \frac{21}{4} \times \frac{8}{9} - \frac{2}{3}$$ Multiply by the reciprocal of $\frac{9}{8}$, which is $\frac{8}{9}$.

$$= \frac{\overset{7}{\cancel{21}} \times \overset{2}{\cancel{8}}}{\underset{1}{\cancel{4}} \times \underset{3}{\cancel{9}}} - \frac{2}{3}$$ Multiply $\frac{21}{4}$ and $\frac{8}{9}$. Divide out common factors.

$$= \frac{14}{3} - \frac{2}{3}$$ Simplify.

$$= \frac{12}{3}, \text{ or } 4$$ Subtract.

Remember

Be sure to check your answers whenever possible. In Example 3, you can use estimation to check that your answer is reasonable.

$5\frac{1}{4} \div 1\frac{1}{8} - \frac{2}{3}$

$\approx 5 \div 1 - 1$

$= 5 - 1$

$= 4$ ✓

EXAMPLE 4 Real-Life Application

One serving of tortilla soup is $1\frac{2}{3}$ cups. A restaurant cook makes 50 cups of soup. Is there enough to serve 35 people? Explain.

Divide 50 by $1\frac{2}{3}$ to find the number of available servings.

$$50 \div 1\frac{2}{3} = \frac{50}{1} \div \frac{5}{3}$$ Rewrite each number as an improper fraction.

$$= \frac{50}{1} \cdot \frac{3}{5}$$ Multiply by the reciprocal of $\frac{5}{3}$, which is $\frac{3}{5}$.

$$= \frac{\overset{10}{\cancel{50}} \cdot 3}{1 \cdot \underset{1}{\cancel{5}}}$$ Multiply fractions. Divide out common factors.

$$= 30$$ Simplify.

No. Because 30 is less than 35, there is not enough soup to serve 35 people.

On Your Own

Now You're Ready
Exercises 26–37

Evaluate the expression. Write the answer in simplest form.

5. $1\frac{1}{2} \div \frac{1}{6} - \frac{7}{8}$
6. $3\frac{1}{3} \div \frac{5}{6} + \frac{8}{9}$
7. $\frac{2}{5} + 2\frac{4}{5} \div 1\frac{3}{4}$
8. $\frac{2}{3} - 1\frac{4}{7} \div 4\frac{5}{7}$
9. In Example 4, can 30 cups of tortilla soup serve 15 people? Explain.

2.3 Exercises

Check It Out
Help with Homework
BigIdeasMath.com

Vocabulary and Concept Check

1. **VOCABULARY** What is the reciprocal of $7\frac{1}{3}$?

2. **NUMBER SENSE** Is $5\frac{1}{4} \div 3\frac{1}{2}$ the same as $3\frac{1}{2} \div 5\frac{1}{4}$? Explain.

3. **NUMBER SENSE** Is the reciprocal of an improper fraction *sometimes, always,* or *never* a proper fraction? Explain.

4. **DIFFERENT WORDS, SAME QUESTION** Which is different? Find "both" answers.

What is $5\frac{1}{2}$ divided by $\frac{1}{8}$?	Find the quotient of $5\frac{1}{2}$ and $\frac{1}{8}$.
What is $5\frac{1}{2}$ times 8?	Find the product of $5\frac{1}{2}$ and $\frac{1}{8}$.

Practice and Problem Solving

Divide. Write the answer in simplest form.

1 2

5. $2\frac{1}{4} \div \frac{3}{4}$
6. $3\frac{4}{5} \div \frac{2}{5}$
7. $8\frac{1}{8} \div \frac{5}{6}$
8. $7\frac{5}{9} \div \frac{4}{7}$
9. $7\frac{1}{2} \div 1\frac{9}{10}$
10. $3\frac{3}{4} \div 2\frac{1}{12}$
11. $7\frac{1}{5} \div 8$
12. $8\frac{4}{7} \div 15$
13. $8\frac{1}{3} \div \frac{2}{3}$
14. $9\frac{1}{6} \div \frac{5}{6}$
15. $13 \div 10\frac{5}{6}$
16. $12 \div 5\frac{9}{11}$
17. $\frac{7}{8} \div 3\frac{1}{16}$
18. $\frac{4}{9} \div 1\frac{7}{15}$
19. $4\frac{5}{16} \div 3\frac{3}{8}$
20. $6\frac{2}{9} \div 5\frac{5}{6}$

21. **ERROR ANALYSIS** Describe and correct the error in finding the quotient.

$$3\frac{1}{2} \div 1\frac{2}{3} = 3\frac{1}{2} \times 1\frac{3}{2} = \frac{7}{2} \times \frac{5}{2} = \frac{35}{4} = 8\frac{3}{4}$$

22. **DOG FOOD** A bag contains 42 cups of dog food. Your dog eats $2\frac{1}{3}$ cups of dog food each day. How many days does the bag of dog food last?

23. **HAMBURGERS** How many $\frac{1}{4}$-pound hamburgers can you make from $3\frac{1}{2}$ pounds of ground beef?

24. **BOOKS** How many $1\frac{3}{5}$-inch-thick books can fit on a $14\frac{1}{2}$-inch-long bookshelf?

Assignment Guide and Homework Check

Level	Day 1 Activity Assignment	Day 2 Lesson Assignment	Homework Check
Basic	5–12, 41–45	1–4, 13–21 odd, 22–24, 27–33 odd	2, 15, 19, 24, 33
Average	5–12, 41–45	1–4, 13–23 odd, 26–38 even	15, 19, 30, 36, 38
Advanced	5–12, 41–45	1–4, 21–25 odd, 26–38 even, 39, 40	27, 30, 36, 39

Common Errors

- **Exercises 5–20** Students may try to divide the whole numbers and the fractions separately. Remind them to rewrite the mixed numbers as improper fractions and to think of each mixed number as one number.
- **Exercises 5–20** Students may find the reciprocal before writing the mixed number as an improper fraction. Remind them that the order of the steps in solving the problem does matter. Tell students to write out each step so that they can keep track of the correct order.
- **Exercises 26–37** Students may forget to use the order of operations. Remind them about the order of operations. Have them use parentheses to help remember which parts to evaluate first.

2.3 Record and Practice Journal

Divide. Write the answer in simplest form.

1. $4\frac{1}{6} \div 5$ **$\frac{5}{6}$**
2. $\frac{5}{8} \div 5\frac{3}{4}$ **$\frac{5}{46}$**
3. $8\frac{1}{6} \div 2\frac{1}{24}$ **4**
4. $2\frac{3}{10} \div 3\frac{3}{5}$ **$\frac{23}{36}$**
5. $6\frac{6}{7} \div 3\frac{3}{5}$ **$1\frac{19}{21}$**
6. $3\frac{3}{5} \div 6\frac{6}{7}$ **$\frac{21}{40}$**

Evaluate the expression.

7. $4\frac{7}{12} \div \frac{3}{4} \times \frac{3}{11}$ **$1\frac{2}{3}$**
8. $9 \div 8\frac{1}{10} - \frac{5}{9}$ **$\frac{5}{9}$**
9. $5\frac{7}{8} \times \left(2\frac{4}{5} \div 7\right)$ **$2\frac{7}{20}$**

10. At a road race, you have $60\frac{3}{4}$ feet available for a water station. Your tables are $6\frac{3}{4}$ feet long. How many tables can you line up for the water station? **9 tables**

Vocabulary and Concept Check

1. $\frac{3}{22}$
2. no; Division is not commutative.
3. sometimes; The reciprocal of $\frac{2}{2}$ is $\frac{2}{2}$, which is improper.
4. Find the product of $5\frac{1}{2}$ and $\frac{1}{8}$.; $\frac{11}{16}$; 44

Practice and Problem Solving

5. 3
6. $9\frac{1}{2}$
7. $9\frac{3}{4}$
8. $13\frac{2}{9}$
9. $3\frac{18}{19}$
10. $1\frac{4}{5}$
11. $\frac{9}{10}$
12. $\frac{4}{7}$
13. $12\frac{1}{2}$
14. 11
15. $1\frac{1}{5}$
16. $2\frac{1}{16}$
17. $\frac{2}{7}$
18. $\frac{10}{33}$
19. $1\frac{5}{18}$
20. $1\frac{1}{15}$
21. The mixed number $1\frac{2}{3}$ was not written as an improper fraction before inverting.

$$3\frac{1}{2} \div 1\frac{2}{3} = \frac{7}{2} \div \frac{5}{3} = \frac{7}{2} \times \frac{3}{5} = \frac{7 \times 3}{2 \times 5} = \frac{21}{10}, \text{ or } 2\frac{1}{10}$$

22. 18 days
23. 14 hamburgers
24. 9 books

Practice and Problem Solving

25. no; The model shows $2\frac{1}{2} \div 1\frac{1}{6} = 2\frac{1}{7}$. There are 2 full groups of $1\frac{1}{6}$ plus one piece remaining, which represents $\frac{1}{7}$ of $1\frac{1}{6}$.

26. 3 **27.** 4

28. 3 **29.** $1\frac{1}{3}$

30. $6\frac{9}{26}$ **31.** $5\frac{1}{6}$

32. $13\frac{1}{5}$ **33.** $\frac{7}{54}$

34. $3\frac{1}{3}$ **35.** $12\frac{1}{2}$

36. $\frac{7}{108}$ **37.** $\frac{22}{35}$

38. See *Taking Math Deeper*.

39. **a.** 6 ramps; *Sample answer:* The estimate is reasonable because $12\frac{1}{2}$ was rounded down.

b. 6 ramps; $1\frac{1}{4}$ feet left over

40. $1\frac{1}{3}$ times greater

Fair Game Review

41. 0.43 **42.** 0.013

43. 3.8 **44.** 7.009

45. C

Mini-Assessment

Divide. Write the answer in simplest form.

1. $6\frac{1}{2} \div 2\frac{3}{4}$ $2\frac{4}{11}$ **2.** $4\frac{2}{3} \div 1\frac{1}{6}$ 4

3. $8\frac{4}{5} \div \frac{4}{10}$ 22 **4.** $6 \div 2\frac{3}{5}$ $2\frac{4}{13}$

5. A bag contains 13 cups of cat food. Your cats eat $1\frac{1}{2}$ cups of food each day. How many full days does the bag of cat food last? 8 days

Taking Math Deeper

Exercise 38

This problem is a nice opportunity to show how "making a table" can be a good problem-solving strategy. In this case, notice how it is convenient to make a table with 3 rows.

1 Organize the given information.

12 cups of granola

$8\frac{1}{2}$ cups of peanuts

Trail Mix

$2\frac{3}{4}$ cups granola

$1\frac{1}{3}$ cups peanuts

2 Make a table.

Batches	1	2	3	4	5
Granola	$2\frac{3}{4}$	$5\frac{1}{2}$	$8\frac{1}{4}$	11	$13\frac{3}{4}$
Peanuts	$1\frac{1}{3}$	$2\frac{2}{3}$	4	$5\frac{1}{3}$	$6\frac{2}{3}$

3 Use the table to make a conclusion.

Batches	1	2	3	4	5
Granola	$2\frac{3}{4}$	$5\frac{1}{2}$	$8\frac{1}{4}$	11	$13\frac{3}{4}$
Peanuts	$1\frac{1}{3}$	$2\frac{2}{3}$	4	$5\frac{1}{3}$	$6\frac{2}{3}$

Five batches use more than 12 cups of granola.

So, you can make 4 full batches of trail mix.

Project

Write your own recipe for trail mix. Be sure to include the amount of each ingredient, a serving size, and how many people your recipe serves. Give your recipe a name that will make people want to try it!

Reteaching and Enrichment Strategies

If students need help. . .	If students got it. . .
Resources by Chapter • Practice A and Practice B • Puzzle Time Record and Practice Journal Practice Differentiating the Lesson Lesson Tutorials Skills Review Handbook	Resources by Chapter • Enrichment and Extension • Technology Connection Start the next section

25. **LOGIC** Alexei uses the model shown to state that $2\frac{1}{2} \div 1\frac{1}{6} = 2\frac{1}{6}$. Is Alexei correct? Justify your answer using the model.

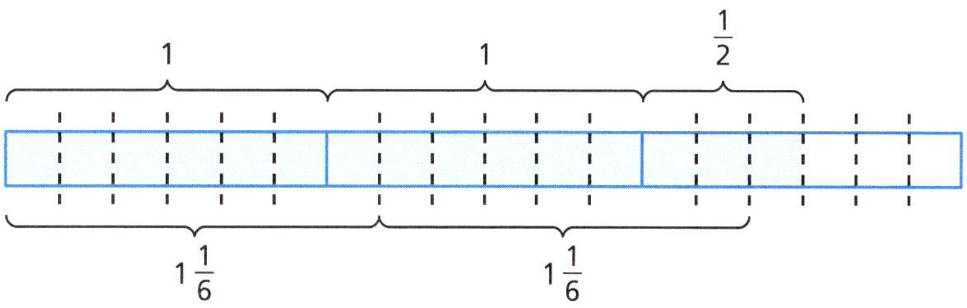

Evaluate the expression. Write the answer in simplest form.

3 26. $3 \div 1\frac{1}{5} + \frac{1}{2}$

27. $4\frac{2}{3} - 1\frac{1}{3} \div 2$

28. $\frac{2}{5} + 2\frac{1}{6} \div \frac{5}{6}$

29. $5\frac{5}{6} \div 3\frac{3}{4} - \frac{2}{9}$

30. $6\frac{1}{2} - \frac{7}{8} \div 5\frac{11}{16}$

31. $9\frac{1}{6} \div 5 + 3\frac{1}{3}$

32. $3\frac{3}{5} + 4\frac{4}{15} \div \frac{4}{9}$

33. $\frac{3}{5} \times \frac{7}{12} \div 2\frac{7}{10}$

34. $4\frac{3}{8} \div \frac{3}{4} \times \frac{4}{7}$

35. $1\frac{9}{11} \times 4\frac{7}{12} \div \frac{2}{3}$

36. $3\frac{4}{15} \div \left(8 \times 6\frac{3}{10}\right)$

37. $2\frac{5}{14} \div \left(2\frac{5}{8} \times 1\frac{3}{7}\right)$

38. **TRAIL MIX** You have 12 cups of granola and $8\frac{1}{2}$ cups of peanuts to make trail mix. What is the greatest number of full batches of trail mix you can make? Explain how you found your answer.

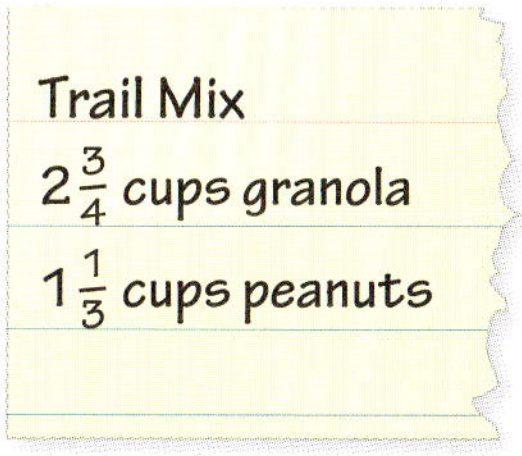

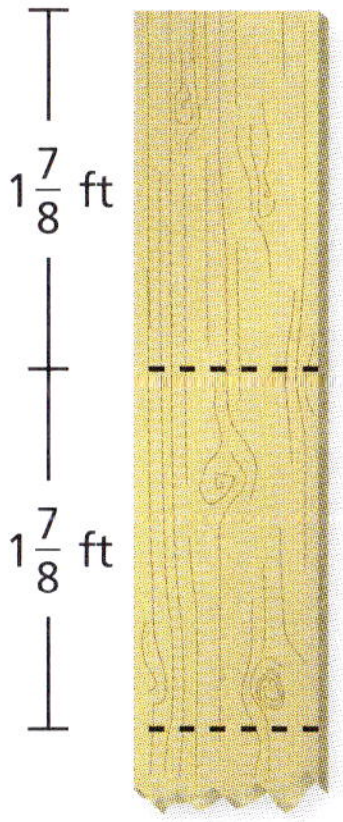

39. **RAMPS** You make skateboard ramps by cutting pieces from a board that is $12\frac{1}{2}$ feet long.

a. Estimate how many ramps you can cut from the board. Is your estimate reasonable? Explain.

b. How many ramps can you cut from the board? How much wood is left over?

40. **Reasoning** At a track meet, the longest shot-put throw by a boy is 25 feet 8 inches. The longest shot-put throw by a girl is 19 feet 3 inches. How many times greater is the longest shot-put throw by the boy than by the girl?

Fair Game Review What you learned in previous grades & lessons

Write the number as a decimal. *(Skills Review Handbook)*

41. forty-three hundredths

42. thirteen thousandths

43. three and eight tenths

44. seven and nine thousandths

45. **MULTIPLE CHOICE** The winner in a vote for class president received $\frac{3}{4}$ of the 240 votes. How many votes did the winner receive? *(Skills Review Handbook)*

Ⓐ 60 Ⓑ 150 Ⓒ 180 Ⓓ 320

2 Study Help

Check It Out
Graphic Organizer
BigIdeasMath.com

You can use a **notetaking organizer** to write notes, vocabulary, and questions about a topic. Here is an example of a notetaking organizer for dividing fractions.

Write important vocabulary or formulas in this space. →

$\frac{a}{b} \div \frac{c}{d} = \frac{a}{b} \cdot \frac{d}{c}$

$= \frac{a \cdot d}{b \cdot c}$

(where b, c, and $d \neq 0$)

Dividing fractions

To divide a number by a fraction, multiply the number by the reciprocal of the fraction.

Example:

$\frac{1}{5} \div \frac{3}{4} = \frac{1}{5} \times \frac{4}{3} = \frac{1 \times 4}{5 \times 3} = \frac{4}{15}$

← Write your notes about the topic in this space.

How do you divide a mixed number by a fraction?

← Write your questions about the topic in this space.

On Your Own

Make notetaking organizers to help you study these topics.

1. multiplying fractions
2. multiplying mixed numbers
3. dividing mixed numbers

After you complete this chapter, make notetaking organizers for the following topics.

4. adding and subtracting decimals
5. multiplying decimals by whole numbers
6. multiplying decimals by decimals
7. dividing decimals by whole numbers
8. dividing decimals by decimals

"The notetaking organizer in my math class gave me an idea of how to organize my doggy biscuits."

Sample Answers

1.

$\frac{a}{b} \cdot \frac{c}{d} = \frac{a \cdot c}{b \cdot d}$ ($b \neq 0$ and $d \neq 0$)	Multiplying fractions Multiply the numerators and multiply the denominators. The denominators cannot be 0. Example: $\frac{3}{4} \times \frac{5}{8} = \frac{3 \times 5}{4 \times 8} = \frac{15}{32}$
How do you multiply two mixed numbers?	

2.

$\frac{a}{b} \cdot \frac{c}{d} = \frac{a \cdot c}{b \cdot d}$ ($b \neq 0$ and $d \neq 0$)	Multiplying mixed numbers 1. Write each mixed number as an improper fraction. 2. Multiply as you would with fractions. Example: $1\frac{2}{3} \times 2\frac{3}{4} = \frac{5}{3} \times \frac{11}{4}$ $= \frac{5 \times 11}{3 \times 4}$ $= \frac{55}{12}$, or $4\frac{7}{12}$
Why can't b and d equal 0? How do you divide mixed numbers?	

3.

$\frac{a}{b} \div \frac{c}{d} = \frac{a}{b} \times \frac{d}{c}$ (b, c, and $d \neq 0$)	Dividing mixed numbers 1. Write each mixed number as an improper fraction. 2. Divide as you would with fractions. Example: $4\frac{3}{4} \div 2\frac{1}{2} = \frac{19}{4} \div \frac{5}{2}$ $= \frac{19}{4} \times \frac{2}{5}$ $= \frac{19 \times \cancel{2}^{1}}{{}_{2}\cancel{4} \times 5}$ $= \frac{19}{10}$, or $1\frac{9}{10}$
How do you divide decimals?	

List of Organizers
Available at *BigIdeasMath.com*

Comparison Chart
Concept Circle
Definition (Idea) and Example Chart
Example and Non-Example Chart
Formula Triangle
Four Square
Information Frame
Information Wheel
Notetaking Organizer
Process Diagram
Summary Triangle
Word Magnet
Y Chart

About this Organizer
A **Notetaking Organizer** can be used to write notes, vocabulary, and questions about a topic. In the space on the left, students write important vocabulary or formulas. In the space on the right, students write their notes about the topic. In the space at the bottom, students write their questions about the topic. A notetaking organizer can also be used as an assessment tool, in which blanks are left for students to complete.

Technology for the Teacher
Editable Graphic Organizer

Answers

1. $\frac{3}{28}$
2. $\frac{3}{5}$
3. $\frac{7}{15}$
4. $19\frac{19}{20}$
5. $\frac{1}{3}$
6. $11\frac{1}{5}$
7. 39
8. $6\frac{9}{10}$
9. $9\frac{1}{2}$
10. $1\frac{1}{2}$
11. 15
12. $1\frac{1}{3}$
13. 12 stores
14. $7606\frac{5}{8}$ ft^2
15. $3\frac{1}{3}$ times more profit
16. 4 skateboards

Technology for the Teacher

Online Assessment
Assessment Book
ExamView® Assessment Suite

Alternative Quiz Ideas

100% Quiz	Math Log
Error Notebook	Notebook Quiz
Group Quiz	Partner Quiz
Homework Quiz	**Pass the Paper**

Pass the Paper

- Work in groups of four. The first student copies the problem and completes the first step, explaining his or her work.
- The paper is passed and the second student works through the next step, also explaining his or her work.
- This process continues until the problem is completed.
- The second member of the group starts the next problem. Students should be allowed to question and debate as they are working through the quiz.
- Student groups can be selected by the teacher, by students, through a random process, or any way that works for your class.
- The teacher walks around the classroom listening to the groups and asks questions to ensure understanding.

Reteaching and Enrichment Strategies

If students need help. . .	If students got it. . .
Resources by Chapter • Practice A and Practice B • Puzzle Time Lesson Tutorials *BigIdeasMath.com*	Resources by Chapter • Enrichment and Extension • Technology Connection Game Closet at *BigIdeasMath.com* Start the next section

2.1–2.3 Quiz

Check It Out
Progress Check
BigIdeasMath.com

Multiply. Write the answer in simplest form. *(Section 2.1)*

1. $\frac{3}{7} \times \frac{1}{4}$

2. $\frac{9}{10} \times \frac{2}{3}$

3. $1\frac{1}{6} \times \frac{2}{5}$

4. $3\frac{1}{2} \times 5\frac{7}{10}$

Divide. Write the answer in simplest form. *(Section 2.2 and Section 2.3)*

5. $\frac{1}{9} \div \frac{1}{3}$

6. $7 \div \frac{5}{8}$

7. $4\frac{7}{8} \div \frac{1}{8}$

8. $7\frac{2}{3} \div 1\frac{1}{9}$

Evaluate the expression. Write the answer in simplest form. *(Section 2.2 and Section 2.3)*

9. $6 \div \frac{2}{3} + \frac{1}{2}$

10. $\frac{7}{12} \div \frac{1}{4} \times \frac{9}{14}$

11. $3\frac{1}{3} \times 3\frac{3}{4} \div \frac{5}{6}$

12. $6\frac{2}{9} \div \left(4 \times 1\frac{1}{6}\right)$

13. MALL In a mall, $\frac{1}{15}$ of the stores sell shoes. There are 180 stores in the mall. How many of the stores sell shoes? *(Section 2.1)*

14. CONCERT FLOOR The floor of a concert venue is $100\frac{3}{4}$ feet by $75\frac{1}{2}$ feet. What is the area of the floor? *(Section 2.1)*

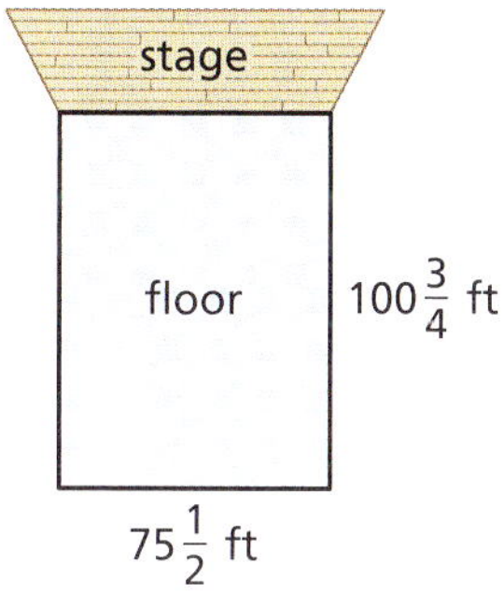

15. BAND Band members make $\frac{2}{3}$ of their profit from selling concert tickets. They make $\frac{1}{5}$ of their profit from selling band merchandise at the concerts. How many times more profit do they make from ticket sales than from merchandise sales? *(Section 2.2)*

16. SKATEBOARDS You are cutting as many $32\frac{1}{4}$-inch sections as you can out of the board to make skateboards. How many skateboards can you make? *(Section 2.3)*

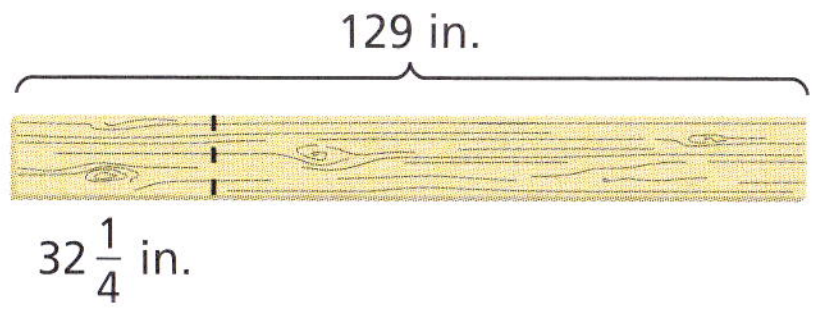

2.4 Adding and Subtracting Decimals

Essential Question How can you add and subtract decimals?

Base ten blocks can be used to model numbers.

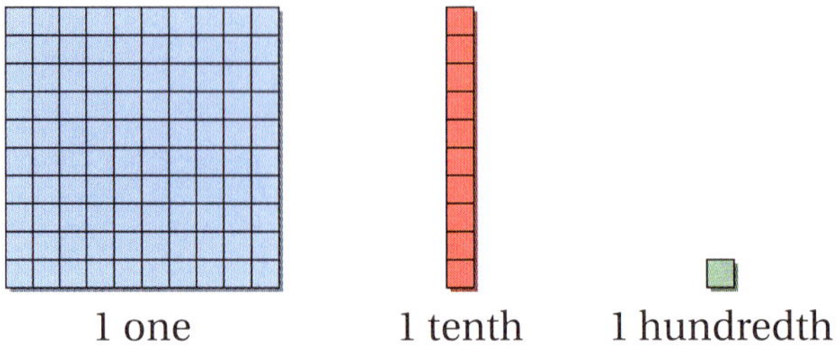

1 ACTIVITY: Modeling a Sum

Work with a partner. Use base ten blocks to find the sum.

a. 1.23 + 0.87

Which base ten blocks do you need to model the numbers in the sum? How many of each do you need?

+

1.23 + 0.87

How many of each base ten block do you have when you combine the blocks?

___ ones ___ tenths ___ hundredths

How many of each base ten block do you have when you trade the blocks?

___ ones ___ tenths ___ hundredths

So, 1.23 + 0.87 = ___.

b. 1.25 + 1.35 **c.** 2.14 + 0.92 **d.** 0.73 + 0.86

Common Core

Adding and Subtracting Decimals

In this lesson, you will

- use models to add and subtract decimals.
- add and subtract decimals.

Learning Standard
6.NS.3

2 ACTIVITY: Modeling a Difference

Work with a partner. Use base ten blocks to find the difference.

a. 2.43 − 0.73

Which number is shown by the model? ___

Circle the portion of the model that represents 0.73.

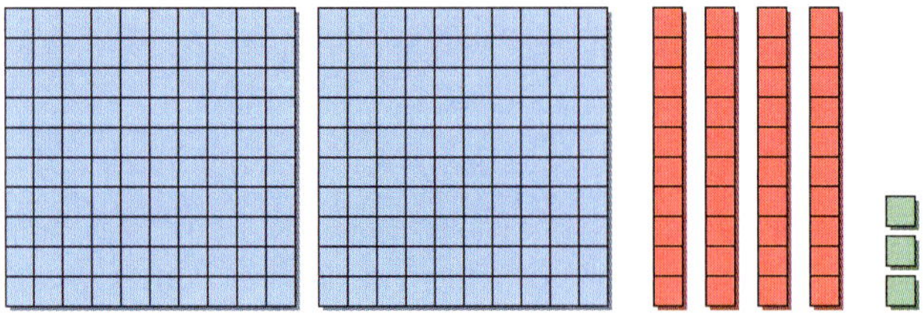

So, 2.43 − 0.73 = ___.

b. 1.86 − 1.26 **c.** 3.72 − 0.5 **d.** 1.58 − 0.09

Laurie's Notes

Introduction

Standards for Mathematical Practice

- **MP7 Look for and Make Use of Structure:** Decimals are another form, or representation of numbers. The rules governing operations with decimals are connected to operations with fractions. The goal is for students to see and make sense of these connections.

Motivate

- Display the following base-ten pieces and ask what number they represent.

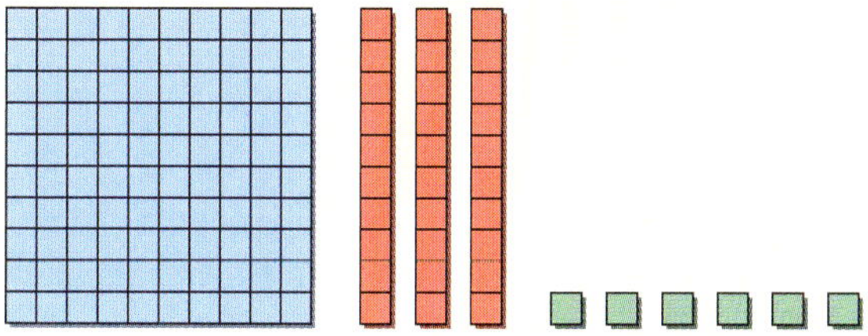

- If students recall working with the pieces, they will say they represent 136.
- ? Point to the blue piece. "If this piece were equal to 1, what would each of the other pieces represent?" The red piece would be one-tenth, and the green would be one-hundredth.
- Write the name for each piece next to it.
  blue = 1; red = 0.1; green = 0.01
- ? "So what number would all of the original pieces represent?" 1.36
- Explain that today they will be adding and subtracting decimals.

Activity Notes

Activity 1

- Encourage students to use the base-ten pieces even if they think they know the answer.
- **FYI:** Students should correctly read 2.1 as "two and one-tenth" and not "two point one."
- **MP6 Attend to Precision:** Students should be in the habit of paying attention to the units, which "two point one" does not.
- Note that all of these problems have addends with the same place values.

Activity 2

- Students will need to circle 3 strips (0.3) of the unit square to represent 0.73. The connection you want students to make is that renaming will need to be done in the standard algorithm.
- Encourage students to use the base-ten pieces even if they think they know the answer. In completing part (d), trading will need to be done.
- When students have finished, ask a volunteer to share his or her process and answer. Listen to the explanation for part (c). The tenths (strips) were removed with the subtraction. Without "lining up the decimal point," students have correctly subtracted like place values.

Common Core State Standards

6.NS.3 Fluently add, subtract, multiply, and divide multi-digit decimals using the standard algorithm for each operation.

Previous Learning

Students should be able to add and subtract decimals to the hundredths place.

Technology for the Teacher

Dynamic Classroom

Lesson Plans
Complete Materials List

2.4 Record and Practice Journal

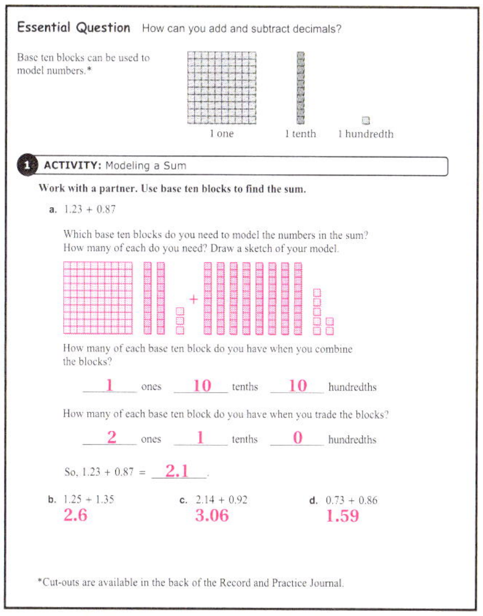
Essential Question How can you add and subtract decimals?

Base ten blocks can be used to model numbers.*

1 one 1 tenth 1 hundredth

1 ACTIVITY: Modeling a Sum

Work with a partner. Use base ten blocks to find the sum.

a. 1.23 + 0.87

Which base ten blocks do you need to model the numbers in the sum? How many of each do you need? Draw a sketch of your model.

How many of each base ten block do you have when you combine the blocks?

1 ones 10 tenths 10 hundredths

How many of each base ten block do you have when you trade the blocks?

2 ones 1 tenths 0 hundredths

So, 1.23 + 0.87 = 2.1.

b. 1.25 + 1.35 2.6 c. 2.14 + 0.92 3.06 d. 0.73 + 0.86 1.59

*Cut-outs are available in the back of the Record and Practice Journal.

Laurie's Notes

English Language Learners

English language learners may struggle with the base-ten place value system. Use the place value chart to help organize and connect the different forms of a number.

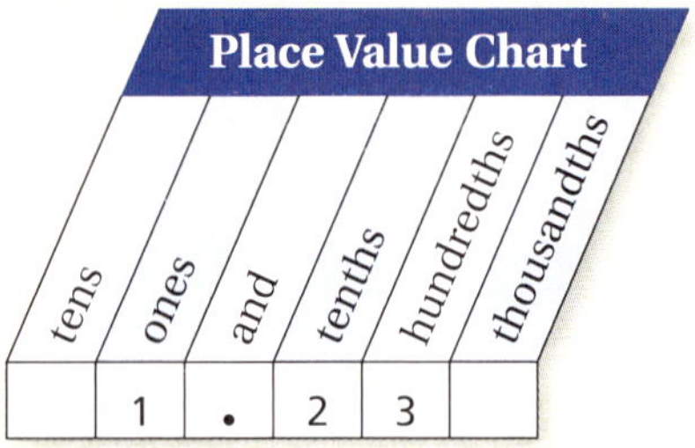

1.23

one and twenty-three hundredths

$1\frac{23}{100}$

2.4 Record and Practice Journal

2 ACTIVITY: Modeling a Difference

Work with a partner. Use base ten blocks to find the difference.

a. 2.43 − 0.73

Which number is shown by the model?

2.43

Circle the portion of the model that represents 0.73.

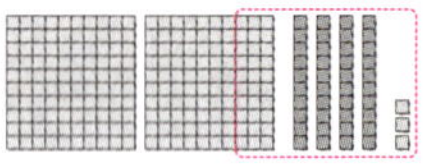

So, 2.43 − 0.73 = 1.7.

b. 1.86 − 1.26 0.6

c. 3.72 − 0.5 3.22

d. 1.58 − 0.09 1.49

3 ACTIVITY: Making a Conjecture

Work with a partner.

a. Find each sum or difference.

123 + 87 210 125 + 135 260 214 + 92 306 73 + 86 159

243 − 73 170 186 − 126 60 372 − 50 322 158 − 9 149

b. How are the numerical expressions in part (a) related to the numerical expressions in Activities 1 and 2? How are the sums and differences related?

Sequence of digits and operations are the same; Activities 1 and 2 answers have a decimal point.

c. **STRUCTURE** There is a relationship between adding and subtracting decimals and adding and subtracting whole numbers. What conjecture can you make about this relationship?

Check students' work.

4 ACTIVITY: Using a Place Value Chart

Work with a partner. Use the place value chart to find the sum or difference.

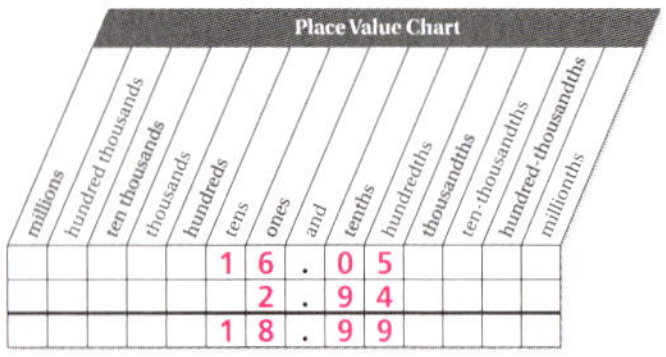

a. 16.05 + 2.94 18.99

b. 7.421 + 92.55 99.971

c. 38.72 − 8.61 30.11

d. 64.968 − 51.167 13.801

What Is Your Answer?

5. **MODELING** Describe two real-life examples of when you would need to add and subtract decimals.

Check students' work.

6. **IN YOUR OWN WORDS** How can you add and subtract decimals?

align decimal points, add and subtract as you would with whole numbers

Activity 3

- **MP7:** Students will quickly realize that the problems in Activity 3 are the whole number equivalent of the problems in Activities 1 and 2. When you need to regroup or rename with the whole-number problem, you regroup or rename in the same way with the related decimal problem. Mathematically proficient students can shift perspective. The base-ten pieces are the same, only the naming of the unit changes.

Activity 4

- This activity begins to focus students' attention on the role of place value when adding or subtracting decimals.
- Ask students to share their work.
- In part (b), the two numbers each contain four digits. Check to see that students place the numbers in the chart correctly.
- Discuss the usefulness of the place value chart when adding and subtracting decimals. Students' comments should include keeping digits and decimal points aligned, placing the digits of each number in the correct place-value position, and so on.

What Is Your Answer?

- Students should describe the process of adding and subtracting like place values.

Closure

- Give an example of a decimal addition problem whose sum is greater than 3.05 *Sample answer:* 1.783 + 2.045
- Give an example of a decimal subtraction problem whose difference is less than 1.05. *Sample answer:* 2.994 − 2.045

3 ACTIVITY: Making a Conjecture

Work with a partner.

a. Find each sum or difference.

123 + 87	125 + 135	214 + 92	73 + 86
243 − 73	186 − 126	372 − 50	158 − 9

b. How are the numerical expressions in part (a) related to the numerical expressions in Activities 1 and 2? How are the sums and differences related?

c. **STRUCTURE** There is a relationship between adding and subtracting decimals and adding and subtracting whole numbers. What conjecture can you make about this relationship?

4 ACTIVITY: Using a Place Value Chart

Work with a partner. Use the place value chart to find the sum or difference.

Math Practice 3

Analyze Conjectures

How can the conjecture you wrote in Activity 3 help you to solve these problems?

Place Value Chart

millions	hundred thousands	ten thousands	thousands	hundreds	tens	ones	and	tenths	hundredths	thousandths	ten-thousandths	hundred-thousandths	millionths
							.						
							.						
							.						

a. 16.05 + 2.94 **b.** 7.421 + 92.55

c. 38.72 − 8.61 **d.** 64.968 − 51.167

What Is Your Answer?

5. **MODELING** Describe two real-life examples of when you would need to add and subtract decimals.

6. **IN YOUR OWN WORDS** How can you add and subtract decimals?

Practice Use what you learned about adding and subtracting decimals to complete Exercises 3–4 on page 82.

2.4 Lesson

Check It Out
Lesson Tutorials
BigIdeasMath.com

Key Idea

Adding and Subtracting Decimals

To add or subtract decimals, write the numbers vertically and line up the decimal points. Then bring down the decimal point and add or subtract as you would with whole numbers.

EXAMPLE 1 Adding Decimals

a. Add 8.13 + 2.76. **Estimate** $8.13 + 2.76 \approx 8 + 3 = 11$

Line up the decimal points.

$$\begin{array}{r} 8.13 \\ +\ 2.76 \\ \hline 10.89 \end{array}$$

Add as you would with whole numbers.

Reasonable? $10.89 \approx 11$ ✓

Study Tip

Be sure to add or subtract only digits that have the same place value.

b. Add 1.459 + 23.7.

$$\begin{array}{r} {}^{1}\ \ \ \ \ \ \\ 1.459 \\ +\ 23.700 \\ \hline 25.159 \end{array}$$

Insert zeros so that both numbers have the same number of decimal places.

EXAMPLE 2 Subtracting Decimals

a. Subtract 5.508 − 3.174. **Estimate** $5.508 - 3.174 \approx 6 - 3 = 3$

Line up the decimal points.

$$\begin{array}{r} 4\ 10\ \ \\ 5.\not{5}\,\not{0}\,8 \\ -\ 3.1\,7\,4 \\ \hline 2.3\,3\,4 \end{array}$$

Subtract as you would with whole numbers.

Reasonable? $2.334 \approx 3$ ✓

b. Subtract 21.9 − 1.605.

$$\begin{array}{r} 9\ \ \ \\ 8\ 10\ 10 \\ 21.\not{9}\,\not{0}\,\not{0} \\ -\ 1.6\,0\,5 \\ \hline 20.2\,9\,5 \end{array}$$

Insert zeros so that both numbers have the same number of decimal places.

On Your Own

Now You're Ready
Exercises 5–16

Add or subtract.

1. 4.206 + 10.85
2. 15.5 + 8.229
3. 78.41 + 90.99
4. 6.34 − 5.33
5. 27.9 − 0.905
6. 18.626 − 13.88

Laurie's Notes

Introduction

Connect

- **Yesterday:** Students explored adding and subtracting decimals with base-ten pieces. (MP6, MP7)
- **Today:** Students will use a formal rule to add and subtract decimals.

Motivate

- **Acting time!** Write the addition problem as shown on the board before students arrive.

$$\begin{array}{r} 317 \\ +\ 24 \\ \hline 557 \end{array}$$

- When a student comments on the error, respond with, "It's easier this way. There's no regrouping. It also works when subtracting."
- Write the subtraction problem as shown on the board. "See! It's easier, and you don't have to do any of the renaming."

$$\begin{array}{r} 960 \\ -\ 64 \\ \hline 320 \end{array}$$

- This should lead to a discussion of why we right-justify whole numbers when we add and subtract. Place value is the key.
- **MP3a Construct Viable Arguments:** It is important to keep referring to place value and like units being added or subtracted. Students should do more than state the rule of "line up the decimal point." There needs to be reasoning that students could communicate to others.

Lesson Notes

Key Idea

- **Connection:** To add and subtract fractions, you need a common denominator. To add and subtract decimals, you need common place values, adding tenths to tenths, hundredths to hundredths, and so on.

Example 1

- Ask a volunteer to read the problems as you write them.
- As you work through the problem, ask about the value of the digits being added. For instance, the 3 and 6 are both hundredths.
- ? For part (b), "What is a reasonable estimate? Explain." 25; 1 + 24
- Discuss with students why 23.7 = 23.700.
- Note in this problem the need to regroup from the tenths to the ones. Check for student understanding of this process and notation.

Example 2

- **Common Error:** In part (b), students will want to subtract 21.9 from 1.605 because there are fewer digits. Students making this mistake have not understood the need to subtract like units. It would be helpful to return to the visual model—base-ten pieces to do additional problems.
- Renaming is necessary. Check for student understanding.

On Your Own

- Check on Question 5, which requires 0s to be added, and renaming.

Goal Today's lesson is adding and subtracting decimals.

Technology for the Teacher

Dynamic Classroom

Lesson Tutorials
Lesson Plans
Answer Presentation Tool

Extra Example 1

a. Add 5.44 + 6.31. 11.75

b. Add 2.8 + 5.547. 8.347

Extra Example 2

a. Subtract 6.124 − 2.409. 3.715

b. Subtract 15.7 − 2.375. 13.325

On Your Own

1. 15.056 **2.** 23.729

3. 169.4 **4.** 1.01

5. 26.995 **6.** 4.746

Extra Example 3

A binder at the store costs $4.87. A pack of pens at the same store costs $2.94. You pay for both the binder and pack of pens with a $20 bill. How much change do you receive? $12.19

Extra Example 4

Find the perimeter of the trapezoid.

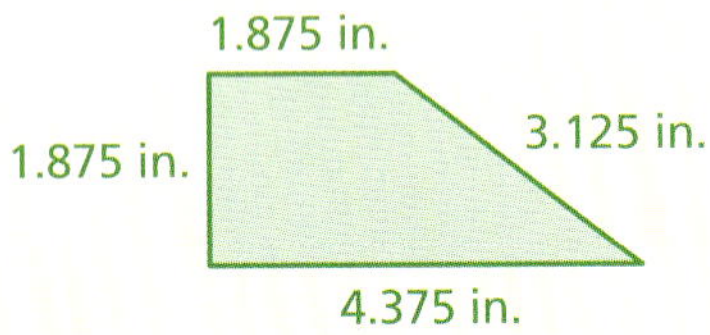

11.25 inches

On Your Own

7. $12.25

8. 152.4 cm

Differentiated Instruction

Kinesthetic

Have students model Example 3 using base-ten blocks. Then have them calculate the expression by hand and using a calculator. Discuss the similarities and differences of the three methods.

Laurie's Notes

Example 3

- Ask a volunteer to read the problem.
- Ask a student to describe a method for solving the problem. This should lead to the verbal model shown.
- Mental math could be used to find the sum of the two meals, while some students may need to see the vertical layout.
- ? "Are there other methods for solving the problem?" yes; Each value could be subtracted from the $10, meaning $10 − $3.45 − $3.90.

Example 4

- **FYI:** The Lincoln Memorial Reflecting Pool was constructed in 1922–23, has a depth of about 18 inches on the sides and 30 inches in the center, and holds approximately 6.75 million gallons of water. The pool was renovated in 2011–12 at a cost of $30.7 million.
- ? Ask if any of your students have visited Washington, D.C., and this famous reflecting pool. If they have, they should have a sense as to the size.
- Read the problem and ask what the phrase "walk the perimeter" means. It is a common phrase, but not all students will understand that it literally means to walk around the pool.
- Draw and label a sketch of the pool.
- Work through the problem as shown. Ask if there are other methods that could be used. Some students might suggest adding the length and width and then doubling, meaning adding that sum to itself.
- **Extension:** Have students estimate the distance in feet. A meter is a bit longer than a yard, or 3 feet. If the perimeter is multiplied by 3, it would be approximately 4017 feet, or about $\frac{4}{5}$ mile.

On Your Own

- ? In Question 8 ask, "What does the red symbol inside the triangle denote?" a right angle

Closure

- **Writing:** Write a note to a friend that explains how to add and subtract decimals, and why the rule is necessary.

EXAMPLE 3 Real-Life Application

Your meal at the school cafeteria costs $3.45. Your friend's meal costs $3.90. You pay for both meals with a $10 bill. How much change do you receive?

Use a verbal model to solve the problem.

$$\text{amount of change} = \text{amount given} - \left(\text{cost of your meal} + \text{cost of friend's meal}\right)$$

$= 10.00 - (3.45 + 3.90)$ Substitute values.

$= 10.00 - 7.35$ Add inside parentheses.

$= 2.65$ Subtract.

So, you receive $2.65.

EXAMPLE 4 Real-Life Application

The Lincoln Memorial Reflecting Pool is approximately rectangular. Its width is 50.9 meters, and its length is 618.44 meters. You walk the perimeter of the pool. About how many meters do you walk?

Draw a diagram and label the dimensions.

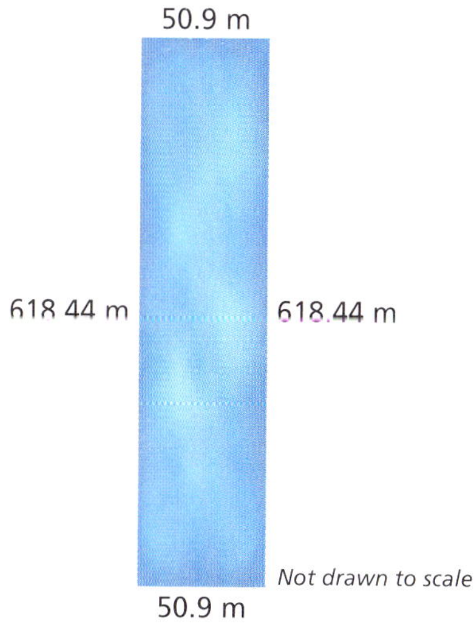

Find the sum of the side lengths.

$$\begin{array}{r} {}^{1\,1\,2} \\ 618.44 \\ 50.90 \\ 618.44 \\ +\ \ 50.90 \\ \hline 1338.68 \end{array}$$

So, you walk about 1339 meters.

On Your Own

Now You're Ready
Exercises 21–26

7. **WHAT IF?** In Example 3, your meal costs $4.10 and your friend's meal costs $3.65. You pay for both meals with a $20 bill. How much change do you receive?

8. Find the perimeter of the triangle.

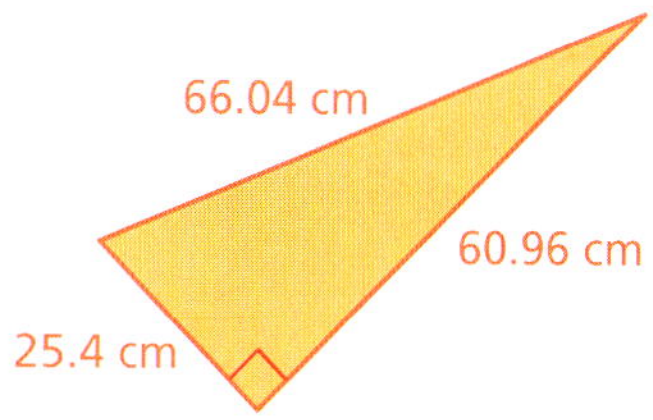

2.4 Exercises

Check It Out
Help with Homework
BigIdeasMath.com

Vocabulary and Concept Check

1. **CHOOSE TOOLS** Why is it helpful to estimate the answer before adding or subtracting decimals?

2. **WRITING** When adding or subtracting decimals, how can you be sure to add or subtract only digits that have the same place value?

Practice and Problem Solving

Write and evaluate the numerical expression modeled by the base ten blocks.

3. 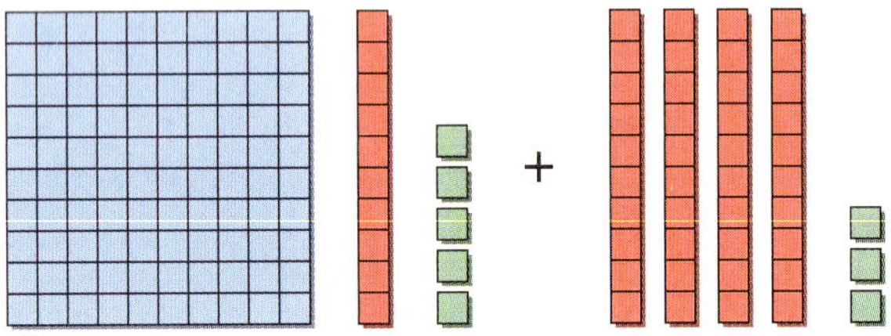

4. 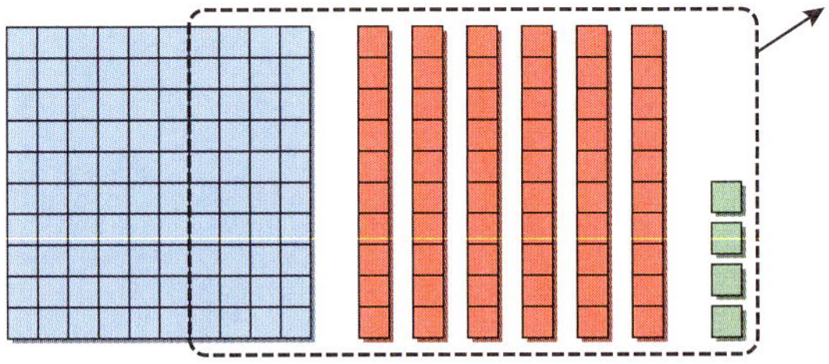

Add.

(1) 5. 7.82 + 3.209   6. 3.7 + 2.774   7. 12.829 + 10.07

8. 20.35 + 13.748   9. 17.440 + 12.497   10. 15.255 + 19.058

Subtract.

(2) 11. 4.58 − 3.12   12. 8.629 − 5.309   13. 6.98 − 2.614

14. 15.131 − 11.57   15. 13.5 − 10.856   16. 25.82 − 22.936

ERROR ANALYSIS **Describe and correct the error in the solution.**

17. ✗

$$\begin{array}{r} 6.058 \\ +\ 3.95 \\ \hline 6.453 \end{array}$$

18. ✗

$$\begin{array}{r} 9.5 \\ -\ 7.18 \\ \hline 2.48 \end{array}$$

Breakfast Menu
7:30 A.M. to 11:00 A.M.

Item	Price	Item	Price
2 Eggs (any style)	\$2.95	Bacon & Eggs	\$3.95
Steak & Eggs	\$6.25	Cheese Omelet	\$3.55
Ham & Eggs	\$3.95	Ham Omelet	\$4.35
Sausage & Eggs	\$3.95	Ham & Cheese Omelet	\$4.95
Salami & Eggs	\$3.95		

19. **BREAKFAST** You order the sausage and eggs breakfast, and your friend orders the ham omelet. How much is the bill before taxes and tip?

20. **HAM & CHEESE** How much more does the ham and cheese omelet cost than the cheese omelet?

Assignment Guide and Homework Check

Level	Day 1 Activity Assignment	Day 2 Lesson Assignment	Homework Check
Basic	3, 4, 36–40	1, 2, 5–29 odd	7, 9, 13, 19, 23
Average	3, 4, 36–40	1, 2, 8–20 even, 21–25 odd, 31–35 odd	8, 10, 14, 20, 25
Advanced	3, 4, 36–40	1, 2, 8, 10, 14–34 even, 35	8, 14, 24, 26, 30

Common Errors

- **Exercises 5–16 and 21–26** Students may forget to insert zeros as place holders. Remind them that each number added or subtracted needs to have the same number of decimal places.
- **Exercises 21–26** Students may get confused with the order of operations (they may want to use addition before subtraction). Remind them that when an expression has both addition and subtraction, they need to work left to right.

2.4 Record and Practice Journal

Add.

1. 3.02 + 1.67
   4.69
2. 1.4 + 8.68
   10.08
3. 11.514 + 4.29
   15.804
4. 15.71 + 12.643
   28.353
5. 9.562 + 21.764
   31.326
6. 15.602 + 2.47
   18.072

Subtract.

7. 2.64 − 1.52
   1.12
8. 4.023 − 3.146
   0.877
9. 7.87 − 5.152
   2.718
10. 16.045 − 12.63
    3.415
11. 17.1 − 11.457
    5.643
12. 5.18 − 2.487
    2.693
13. You buy a movie for $19.99 and a set of earphones for $12.49. How much is the bill before taxes?
    $32.48

Vocabulary and Concept Check

1. Estimating allows you to check that your answer is reasonable.
2. Line up the decimal points and insert zeros if needed.

Practice and Problem Solving

3. $1.15 + 0.43 = 1.58$
4. $1.64 - 1.04 = 0.6$
5. 11.029
6. 6.474
7. 22.899
8. 34.098
9. 29.937
10. 34.313
11. 1.46
12. 3.32
13. 4.366
14. 3.561
15. 2.644
16. 2.884
17. Line up the decimal points before adding. Insert a 0 at the end of the second number so that both numbers have the same number of decimal places. $6.058 + 3.95 = 10.008$.
18. You need to regroup and subtract 8 from 10. $9.5 - 7.18 = 2.32$.
19. $8.30
20. $1.40

Practice and Problem Solving

21. 19.58 **22.** 6.772

23. 10 **24.** 10.343

25. 15.606 **26.** 21.582

27. the decimal parts in the sum total 1; the decimal parts in the difference are exactly the same

28. *Sample answer:* 9.55, 10.6, 7.755

29. 34.995 m

30. 3.6 h

31. 4.816 AU

32. 28.546 AU

33. 20.189 AU

34. See *Taking Math Deeper.*

35. 6.85 units

Fair Game Review

36. $\frac{1}{2}$ **37.** $\frac{1}{4}$

38. $\frac{1}{6}$ **39.** $\frac{1}{20}$

40. C

Mini-Assessment

Evaluate the expression.

1. $2.48 + 5.15$ 7.63
2. $4.19 + 5.8$ 9.99
3. $6.16 - 4.08$ 2.08
4. $5.8 - 2.382$ 3.418
5. At a home improvement store, you decide to purchase a hammer and a pack of nails. The hammer costs \$8.53. The pack of nails costs \$0.87. What is the total cost? How much change will you receive if you pay with a \$10 bill? a \$20 bill? \$9.40; \$0.60; \$10.60

Taking Math Deeper

Exercise 34

This is a short, but nice problem. Students need to choose the correct operation. When a student is unsure which operation to choose, drawing a diagram may help.

① Draw and label a diagram of our solar system.

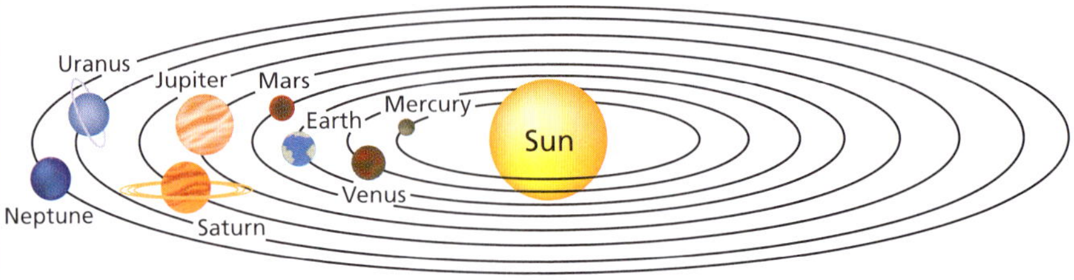

② The orbits of the planets are approximately coplanar, meaning they lie on the same plane. so, you can conclude that the greatest distance between Venus and Saturn occurs when the two planets are on opposite sides of the Sun, as shown.

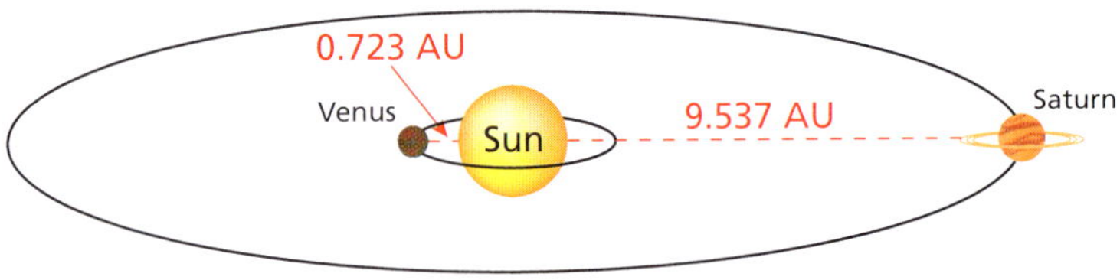

From the diagram, you can see that the sum of the average distances will give an estimate of the greatest distance.

③ Answer the question. Add the average distances to get $0.723 + 9.537 = 10.26$. So, the greatest distance between Venus and Saturn is about 10.26 AU.

Project

Research how our solar system was formed and why the orbits of the planets are approximately coplanar. Also research the terms *perihelion* and *aphelion*. Can you use these concepts to find better estimates in Exercises 33 and 34?

Reteaching and Enrichment Strategies

If students need help...	If students got it...
Resources by Chapter • Practice A and Practice B • Puzzle Time Record and Practice Journal Practice Differentiating the Lesson Lesson Tutorials Skills Review Handbook	Resources by Chapter • Enrichment and Extension • Technology Connection Start the next section

Evaluate the expression.

3 **21.** 6.105 + 10.4 + 3.075

22. 22.6 − 12.286 − 3.542

23. 15.35 + 7.604 − 12.954

24. 16.5 − 13.45 + 7.293

25. 25.92 − 18.478 + 8.164

26. 23.45 + 17.75 − 19.618

27. **STRUCTURE** When is the sum of two decimals equal to a whole number? When is the difference of two decimals equal to a whole number?

28. **OPEN-ENDED** Write three decimals that have a sum of 27.905.

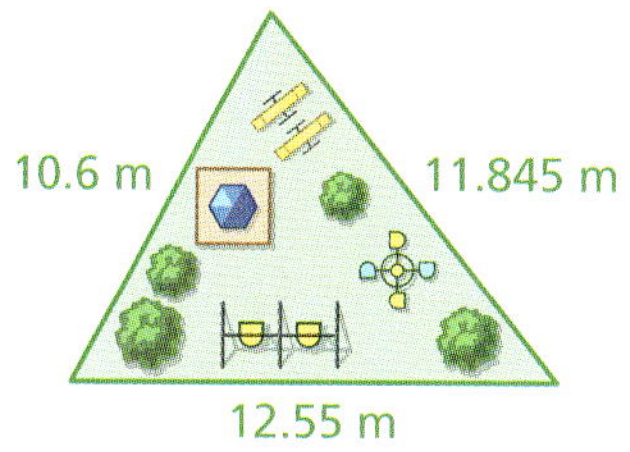

29. **DAY CARE** A day-care center is building a new outdoor play area. The diagram shows the dimensions in meters. How much fencing is needed to enclose the play area?

30. **HOMEWORK** You work 1.15 hours on English homework and 1.75 hours on math homework. Your science homework takes 1.05 hours less than your math homework. How many hours do you work on homework?

ASTRONOMY **An astronomical unit (AU) is the average distance of Earth from the Sun. In Exercises 31–34, use the table that shows the average distance of each planet in our solar system from the Sun.**

31. How much farther is Jupiter from the Sun than Mercury?

32. How much farther is Neptune from the Sun than Mars?

33. Estimate the greatest distance between Earth and Uranus.

34. Estimate the greatest distance between Venus and Saturn.

Planet	Average Distance from the Sun (AU)
Mercury	0.387
Venus	0.723
Earth	1.000
Mars	1.524
Jupiter	5.203
Saturn	9.537
Uranus	19.189
Neptune	30.07

35. Critical Thinking The length of a rectangle is twice the width. The perimeter of the rectangle can be expressed as 3 • 13.7. What is the width?

Fair Game Review What you learned in previous grades & lessons

Multiply. Write the answer in simplest form. *(Section 2.1)*

36. $\frac{7}{10} \times \frac{5}{7}$

37. $\frac{5}{6} \times \frac{3}{10}$

38. $\frac{3}{4} \times \frac{2}{9}$

39. $\frac{2}{5} \times \frac{1}{8}$

40. **MULTIPLE CHOICE** What is the LCM of 6, 12, and 18? *(Section 1.6)*

Ⓐ 6 Ⓑ 18 Ⓒ 36 Ⓓ 72

2.5 Multiplying Decimals

Essential Question How can you multiply decimals?

1 ACTIVITY: Multiplying Decimals Using a Rectangle

Work with a partner. Use a rectangle to find the product.

a. 2.7 • 1.3

Arrange base ten blocks to form a rectangle of length 2.7 units and width 1.3 units.

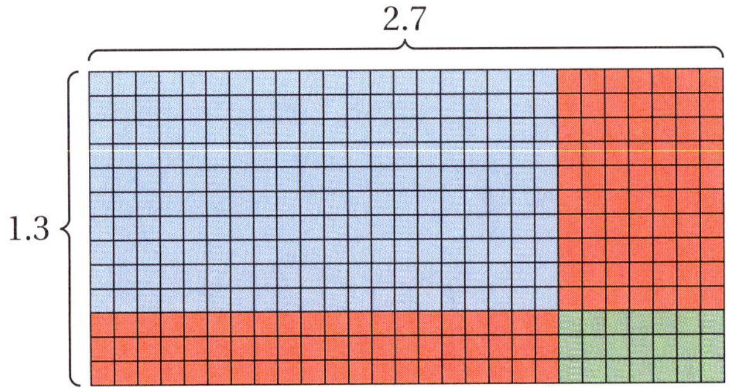

The area of the rectangle represents the product.

Find the total area represented by each grouping of base ten blocks.

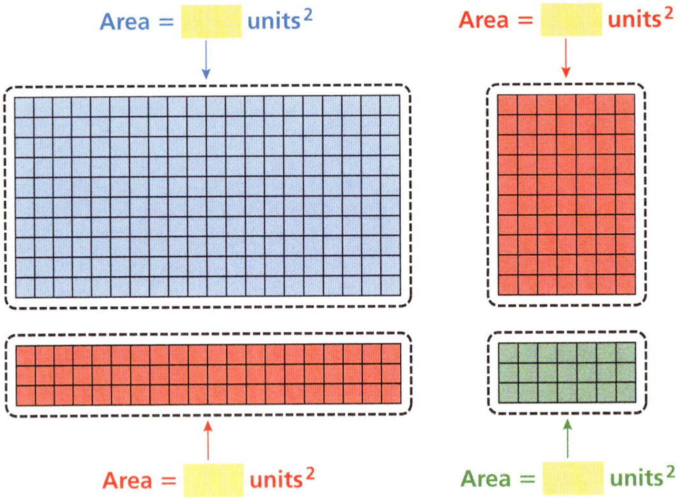

The area of the rectangle is:

+

units^2

So, 2.7 • 1.3 = ___.

b. 1.8 • 1.1 **c.** 4.6 • 1.2 **d.** 3.2 • 2.4

Common Core

Multiplying Decimals

In this lesson, you will
- use models to multiply decimals.
- multiply decimals.

Learning Standard 6.NS.3

Laurie's Notes

Introduction

Standards for Mathematical Practice

- **MP1 Make Sense of Problems and Persevere in Solving Them:** The area model for multiplication is the visual tool used to help students make sense of decimal multiplication. The dimensions of the rectangle represent the factors and the area is the product. Making connections within mathematics is a habit you want all students to develop.

Motivate

- ? Show the class a quarter. Ask students to tell you the value of x quarters, where x is a multiple or power of 10. Record the value of the quarters.
  "What is the value of 10 quarters?" Write $10 \times \$0.25 = \$2.50$ on the board.
  "What is the value of 20 quarters?" Write $20 \times \$0.25 = \$5.00$ on the board.
  "What is the value of 100 quarters?" Write $100 \times \$0.25 = \$25.00$ on the board.
- Tell students that today's lesson is about decimal multiplication. Money is represented as a decimal. Without the context of money, they will develop an understanding of decimal multiplication.

Words of Wisdom

- You may prefer base-ten paper for students to draw sketches. The advantage of base-ten blocks is that they can be manipulated, grouped, and traded. The disadvantage is the time it takes some students to form the rectangles. You need to judge what the correct tool is for your students.

Activity Notes

Activity 1

- Students should have prior knowledge of the area model for multiplication. If this is not the case, do a quick problem (12×14) with base-ten blocks, being VERY careful with naming the pieces correctly. The small square would now be 1, not 0.01.
- ? "Why is there a *raised 2* after the word units in the problem?" The correct units for area are square units, written units^2.
- Have students work with their partners on part (a) as you circulate to check for understanding. You might ask about an estimate for the problem.
- ? "Multiplying a number close to 3 by a number a bit more than 1 should give a product of about what?" about 3
- **MP8 Look for and Express Regularity in Repeated Reasoning:** As students work through the problems, they should consider the reasonableness of their intermediate results and their final answer.
- **MP3 Construct Viable Arguments and Critique the Reasoning of Others:** Take time for students to share their work and think about each problem. Students need to practice communicating their thinking. You want it to become the norm in your classroom that students regularly present their work and that classmates listen carefully. If the work is not correct, it should be pointed out in a respectful manner.

Common Core State Standards

6.NS.3 Fluently add, subtract, multiply, and divide multi-digit decimals using the standard algorithm for each operation.

Previous Learning

Students should know powers and multiples of 10, and how to multiply whole numbers.

Technology for the Teacher

Dynamic Classroom

Lesson Plans
Complete Materials List

2.5 Record and Practice Journal

Essential Question How can you multiply decimals?

1 ACTIVITY: Multiplying Decimals Using a Rectangle

Work with a partner. Use a rectangle to find the product.

a. $2.7 \bullet 1.3$

Arrange base ten blocks to form a rectangle of length 2.7 units and width 1.3 units. Sketch your model.

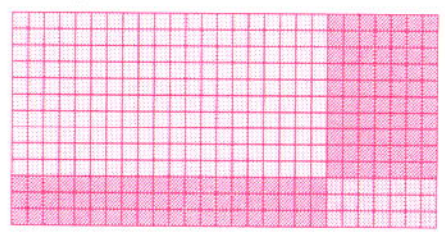

The area of the rectangle represents the product.

Find the total area represented by each grouping of base ten blocks.

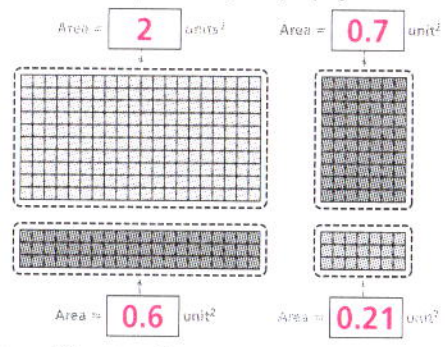

The area of the rectangle is

$2.0 + 0.7 + 0.6 + 0.21 = 3.51$ units^2.

So, $2.7 \bullet 1.3 = 3.51$.

Differentiated Instruction

Kinesthetic

Have students use base ten blocks to model decimal and whole number multiplication. Assign the products below and ask students to describe any patterns they see.

2×3	2×0.3	2×0.03
2×4	2×0.4	2×0.04
2×5	2×0.5	2×0.05
2×6	2×0.6	2×0.06

2.5 Record and Practice Journal

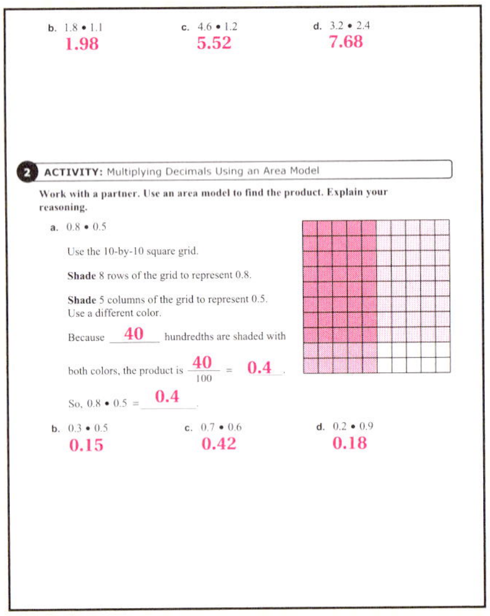

b. 1.8 • 1.1 **1.98** c. 4.6 • 1.2 **5.52** d. 3.2 • 2.4 **7.68**

2 ACTIVITY: Multiplying Decimals Using an Area Model

Work with a partner. Use an area model to find the product. Explain your reasoning.

a. 0.8 • 0.5

Use the 10-by-10 square grid.

Shade 8 rows of the grid to represent 0.8.

Shade 5 columns of the grid to represent 0.5. Use a different color.

Because **40** hundredths are shaded with both colors, the product is $\frac{40}{100}$ = **0.4**.

So, 0.8 • 0.5 = **0.4**.

b. 0.3 • 0.5 **0.15** c. 0.7 • 0.6 **0.42** d. 0.2 • 0.9 **0.18**

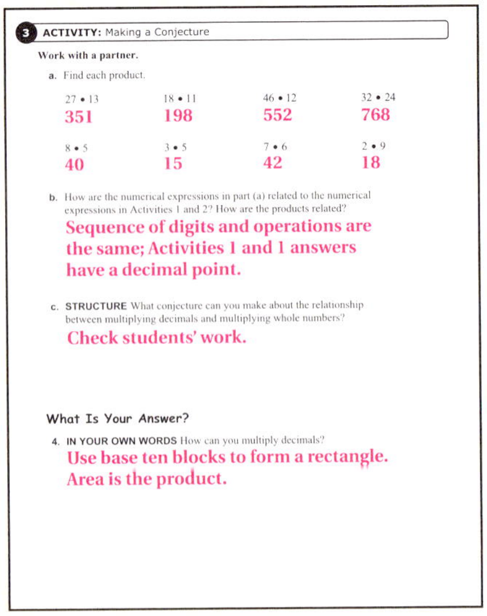

3 ACTIVITY: Making a Conjecture

Work with a partner.

a. Find each product.

27 • 13 **351**	18 • 11 **198**	46 • 12 **552**	32 • 24 **768**
8 • 5 **40**	3 • 5 **15**	7 • 6 **42**	2 • 9 **18**

b. How are the numerical expressions in part (a) related to the numerical expressions in Activities 1 and 2? How are the products related?
Sequence of digits and operations are the same; Activities 1 and 1 answers have a decimal point.

c. **STRUCTURE** What conjecture can you make about the relationship between multiplying decimals and multiplying whole numbers?
Check students' work.

What Is Your Answer?

4. **IN YOUR OWN WORDS** How can you multiply decimals?
Use base ten blocks to form a rectangle. Area is the product.

Laurie's Notes

Activity 2

- Discuss the model. The large square is the unit square, representing the whole number 1. It might be helpful to do the sketch in stages.

- Help students to understand that the area being considered is the rectangle with dimensions 0.5 by 0.8. Because the large square represents 1, the area must be less than 1.
- If available, offer each pair of students colored pencils and base-10 paper.
- When students have finished ask a volunteer to share his or her process and answer.
- ? What is true about all of the factors in the four problems?" All factors are a decimal less than 1.
- ? "What is true about all of the products in the four problems? All products are a decimal less than 1.
- You may need to probe more at this point, asking students to compare the product in each problem to the factors in the same problem.
- **Big Idea:** Multiplying two numbers, each less than 1, will result in a product less than either number. A tenth times a tenth is a hundredth.

Activity 3

- **MP7 Look for and Make Use of Structure:** Students will quickly realize that the problems in Activity 3 are the whole number equivalent of the problems in Activities 1 and 2. The digits are the same, only the location of the decimal point is different. Mathematically proficient students can shift perspective. The decimal places in the factors influence the location of the decimal point in the product, and hence the number of decimal places in the product.
- As time permits, ask students to consider the decimal places in the factors and the decimal places in the product. The cases thus far are limited, so their conjectures may or may not be correct.

What Is Your Answer?

- Students should describe the process of multiplying decimals as the same as multiplying whole numbers. The connection between the number of place values in the factors summing to the number of place values in the product may not be secure with all students.

Closure

- ? "What problem is suggested by this sketch?" 1.3 • 0.6 "What is the product?" 0.78

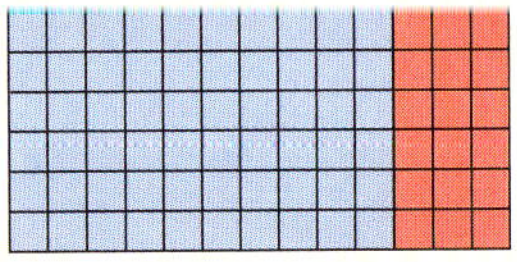

2 ACTIVITY: Multiplying Decimals Using an Area Model

Work with a partner. Use an area model to find the product. Explain your reasoning.

a. 0.8 • 0.5

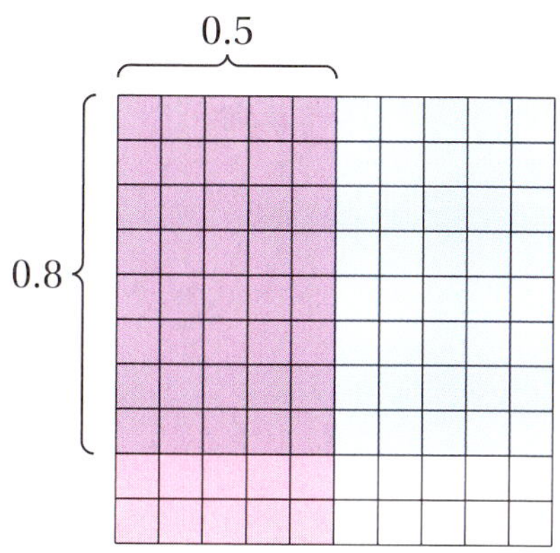

Because ▢ hundredths are shaded with both colors, the product is

$\frac{\square}{100} = \square$.

So, 0.8 • 0.5 = ▢.

b. 0.3 • 0.5 **c.** 0.7 • 0.6 **d.** 0.2 • 0.9

Math Practice 7

View as Components

How can you use an area model to find the product?

3 ACTIVITY: Making a Conjecture

Work with a partner.

a. Find each product.

27 • 13	18 • 11	46 • 12	32 • 24
8 • 5	3 • 5	7 • 6	2 • 9

b. How are the numerical expressions in part (a) related to the numerical expressions in Activities 1 and 2? How are the products related?

c. **STRUCTURE** What conjecture can you make about the relationship between multiplying decimals and multiplying whole numbers?

What Is Your Answer?

4. **IN YOUR OWN WORDS** How can you multiply decimals?

Practice Use what you learned about multiplying decimals to complete Exercises 9–12 on page 89.

2.5 Lesson

Check It Out
Lesson Tutorials
BigIdeasMath.com

Key Idea

Multiplying Decimals by Whole Numbers

Words Multiply as you would with whole numbers. Then count the number of decimal places in the decimal factor. The product has the same number of decimal places.

Numbers

$$\begin{array}{r} 13.91 \\ \times \quad 7 \\ \hline 97.37 \end{array}$$ 2 decimal places

$$\begin{array}{r} 6.218 \\ \times \quad 4 \\ \hline 24.872 \end{array}$$ 3 decimal places

EXAMPLE 1 Multiplying Decimals and Whole Numbers

a. Find 6 × 3.91.

Estimate $6 \times 4 = 24$

$$\begin{array}{r} \overset{5}{3.91} \\ \times \quad 6 \\ \hline 23.46 \end{array}$$

3.91 ← 2 decimal places

23.46 ← Count 2 decimal places from right to left.

So, $6 \times 3.91 = 23.46$.

Reasonable? $23.46 \approx 24$ ✓

b. Find 3 × 0.016.

Estimate $3 \times 0 = 0$

$$\begin{array}{r} 0.0\overset{1}{1}6 \\ \times \quad 3 \\ \hline 0.048 \end{array}$$

0.016 ← 3 decimal places

0.048 ← To have 3 decimal places, insert zeros to the left of 48.

So, $3 \times 0.016 = 0.048$.

Reasonable? $0.048 \approx 0$ ✓

EXAMPLE 2 Use Mental Math

How high is a stack of 100 dimes?

Method 1: Multiply 1.35 by 100.

$$\begin{array}{r} 1.35 \\ \times 1\,00 \\ \hline 0\,00 \\ 00\,0 \\ 135 \\ \hline 135.00 \end{array}$$

2 decimal places

Method 2: You are multiplying by a power of 10. Use mental math.

There are two zeros in 100. So, move the decimal point in 1.35 two places to the right.

$1.35 \times 100 = 135. = 135$

So, a stack of 100 dimes is 135 millimeters high.

On Your Own

Now You're Ready
Exercises 13–24

Multiply. Use estimation to check your answer.

1. 12.3×8
2. 5×14.51
3. 0.88×9
4. 0.003×10
5. A quarter is 1.75 millimeters thick. How high is a stack of 1000 quarters? Solve using both methods.

Laurie's Notes

Introduction

Connect

- **Yesterday:** Students explored multiplying decimals with base-ten blocks and base-ten paper. (MP1, MP3, MP7, MP8)
- **Today:** Students will use a formal rule to multiply decimals.

Motivate

- ? "Is the product of two numbers always greater than either of the two factors?" No, only if both factors are greater than 1.
- ? "Can you give an example of a multiplication problem where the product is greater than one factor, but less than the other factor?" $4 \times 0.5 = 2$
- ? "Can you give an example of a multiplication problem where the product is less than both factors?" $0.4 \times 0.2 = 0.08$
- **Teacher Note:** Today's lesson is fairly long. Do not rush through the lesson. It is better to spend two days than to rush through important concepts.

Lesson Notes

Example 1

- Encourage students to estimate the product before beginning.
- Work through the two examples. In order to place the decimal point correctly, students should count from the right in the product.
- **Common Error:** A common error is for students to count the number of decimal places from left to right when placing the decimal point in the product.
- **Connection:** When the decimal is very small, it may be helpful to have students write the number as a fraction and then multiply.

$$3 \times \frac{16}{1000} = \frac{48}{1000} = 0.048$$

- Remind students that this is another strategy they can use, like estimating, to check their answer when they multiply decimals.

Example 2

- Work through this example together as a class. Have students estimate the answer to 1.35×100. Then do Method 1.
- ? After completing Method 1, ask students if they can think of how to do the problem using mental math. See if anyone describes the process in Method 2. If not, review Method 2.
- ? Ask students which method they prefer. Some may prefer the accuracy of Method 1. Most will probably like to use mental math, as in Method 2.

On Your Own

- Students should be able to use mental math to find the estimate in each problem.
- **Neighbor Check:** Have students work independently and then have their neighbors check their work. Have students discuss any discrepancies.

Goal Today's lesson is finding the product of a whole number and a decimal and finding the product of two decimals.

Technology for the Teacher

Dynamic Classroom

Lesson Tutorials
Lesson Plans
Answer Presentation Tool

Extra Example 1

Multiply.

a. 4.8×3 14.4

b. 0.0045×5 0.0225

Extra Example 2

A half-dollar is 2.15 millimeters thick. How high is a stack of 100 half-dollars? 215 mm

On Your Own

1. 98.4
2. 72.55
3. 7.92
4. 0.03
5. 1750 mm

Laurie's Notes

Key Idea

- Students generally have little difficulty at this stage with the rule for multiplying decimals. You want to be sure that they have an understanding of why the rule works.
- Connect place value and fraction multiplication.
- **FYI:** If no multiplication error is made, locating the decimal point in the product is usually not difficult for students.
- The concepts students find most difficult are often related to trailing zeros in the product or inserting zeros in the product.

Example 3

- Work through both parts of the example.
- Begin each problem by estimating the product. The last step should always be to judge the reasonableness of the solution compared to your estimate. This is especially important in part (b), where the product is much less than the first factor.
- In part (b), it is more difficult to estimate the size of the answer. You may want to remind students they can also use fractions to estimate the answer.

Words of Wisdom

- Ask students to record the number of decimal places in each factor. They should do this in all of the problems at this point.

Example:

$$\begin{array}{rl} 3.1 & (1) \leftarrow \text{Number of decimal places} \\ \times\ 0.05 & (2) \\ \hline 0.155 & (3) \end{array}$$

On Your Own

- **Neighbor Check:** Have students work independently and then have their neighbors check their work. Have students discuss any discrepancies.
- Watch for multiplication errors in Question 7 due to the 0 in 9.04.

Extra Example 3

Multiply.

a. 1.3×3.5 4.55

b. 5.8×0.02 0.116

On Your Own

6. 45.36 **7.** 24.408

8. 0.5688 **9.** 0.3570

English Language Learners

Numbers and Words

English learners are more comfortable with numbers than with words. Provide students with the opportunity to improve their English. On a piece of paper, give students decimal numbers written in numbers and words. Ask the students to match the number with the words.

Numbers	Words
4.35	twelve and five-thousandths
12.005	six and seven-tenths
0.83	four and thirty-five hundredths
6.7	eighty-three hundredths

Have students copy the number and then write the corresponding words next to the number. Encourage students to say the number out loud.

The rule for multiplying two decimals is similar to the rule for multiplying a decimal by a whole number.

Key Idea

Multiplying Decimals by Decimals

Words Multiply as you would with whole numbers. Then add the number of decimal places in the factors. The sum is the number of decimal places in the product.

Numbers

$$\begin{array}{r} 4.716 \\ \times \quad 0.2 \\ \hline 0.9432 \end{array} \quad \begin{array}{l} \leftarrow \text{3 decimal places} \\ \leftarrow \text{+ 1 decimal place} \\ \leftarrow \text{4 decimal places} \end{array}$$

EXAMPLE 3 Multiplying Decimals

a. Multiply 4.8 × 7.2. **Estimate** $5 \times 7 = 35$

$$\begin{array}{r} 4.8 \\ \times 7.2 \\ \hline 96 \\ 336 \\ \hline 34.56 \end{array} \quad \begin{array}{l} \leftarrow \text{1 decimal place} \\ \leftarrow \text{+ 1 decimal place} \\ \\ \\ \leftarrow \text{2 decimal places} \end{array}$$

So, $4.8 \times 7.2 = 34.56$. **Reasonable?** $34.56 \approx 35$ ✓

b. Multiply 3.1 × 0.05. **Estimate** $3 \times 0 = 0$

$$\begin{array}{r} 3.1 \\ \times 0.05 \\ \hline 0.155 \end{array} \quad \begin{array}{l} \leftarrow \text{1 decimal place} \\ \leftarrow \text{+ 2 decimal places} \\ \leftarrow \text{3 decimal places} \end{array}$$

So, $3.1 \times 0.05 = 0.155$. **Reasonable?** $0.155 \approx 0$ ✓

On Your Own

Now You're Ready
Exercises 30–45

Multiply. Use estimation to check your answer.

6. 8.1×5.6

7. 2.7×9.04

8. 6.32×0.09

9. 1.785×0.2

EXAMPLE 4 Evaluating an Expression

What is the value of 2.44(4.5 − 3.175)?

Ⓐ 3.233 Ⓑ 3.599 Ⓒ 7.805 Ⓓ 32.33

Step 1: Subtract first because the minus sign is in parentheses.

$$\begin{array}{r} 4.500 \\ -\,3.175 \\ \hline 1.325 \end{array}$$

So, 2.44(4.5 − 3.175) = 2.44(1.325).

Step 2: Multiply the result from Step 1 by 2.44.

$$\begin{array}{r} 1.325 \\ \times\ 2.44 \\ \hline 5300 \\ 5300 \\ 2650 \\ \hline 3.23300 \end{array}$$

∴ The correct answer is Ⓐ.

On Your Own

Now You're Ready
Exercises 52–60

Evaluate the expression.

10. 12.67 + 8.2 • 1.9

11. 6.4(1.8 • 7.5)

EXAMPLE 5 Real-Life Application

You buy 2.75 pounds of tomatoes. You hand the cashier a $10 bill. How much change will you receive?

Step 1: Find the cost of the tomatoes. Multiply 1.89 by 2.75.

$$\begin{array}{rl} 1.89 & \leftarrow \text{2 decimal places} \\ \times\ 2.75 & \leftarrow +\ \text{2 decimal places} \\ \hline 945 & \\ 1323 & \\ 378 & \\ \hline 5.1975 & \leftarrow \text{4 decimal places} \end{array}$$

The cost of 2.75 pounds of tomatoes is $5.20.

Step 2: Subtract the cost of the tomatoes from the amount of money you hand the cashier.

10.00 − 5.20 = $4.80

∴ So, you will receive $4.80 in change.

On Your Own

12. WHAT IF? You buy 2.25 pounds of grapes. You hand the cashier a $5 bill. How much change will you receive?

Laurie's Notes

Example 4

- ? "What operations are involved in this problem?" multiplication and subtraction
- You may need to remind students that a number written next to the parentheses in this form implies multiplication.
- ? "Where should we start and why?" subtracting inside the parentheses, because of the order of operations
- **Common Error:** Students may write 4.5 under 3.175 due to the number of digits in the number. Regardless of the number of digits, 3.175 is being subtracted *from* 4.5.
- ? In Step 2 ask, "How many decimal places are in the first factor?" 3 "How many decimal places are in the second factor?" 2 "How many decimal places are needed in the product?" 5
- Remind students that when counting the 5 decimal places, they will count from right to left.
- **MP6 Attend to Precision:** Once the decimal point is located, trailing zeros generally are not written. Mathematically proficient students understand that the zeros are necessary for determining the location of the decimal point, but not in representing the decimal product.

On Your Own

- Students must again think about order of operations for Question 10.
- **Neighbor Check:** Have students work independently and then have their neighbors check their work. Have students discuss any discrepancies.

Example 5

- ? It may have been some time since students multiplied two 3-digit numbers. This problem may prompt questions about multiplication in general.
  - "What is a reasonable estimate for the cost of 2.75 pounds at $1.89 a pound?" $3 \times \$2 = \$6$
  - "What is 2.75 written as a fraction?" $2\frac{3}{4}$; This question is simply to review this benchmark fraction. It should be automatic at this point.
  - "Why is the exact answer (5.1975) written as $5.20?" You need to round the answer to the nearest hundredth, because money is only written with two decimal places to represent the cents.
- The operations in Examples 4 and 5 are the same (subtraction and multiplication). Discuss with students why the operations were performed in different orders in the two problems.

Closure

- If $4321 \times 56 = 241{,}976$ then $4.321 \times ? = 2.41976$. Explain your reasoning. 0.56; The number of decimal places in the product is the sum of the number of decimal places in the factors.

Extra Example 4

What is the value of $5.7(5.1 + 4.86)$? 56.772

On Your Own

10. 28.25 **11.** 86.4

Extra Example 5

You buy 4.5 pounds of bananas at $0.49 a pound. You hand the cashier a $20 bill. How much change will you get back? $17.79

On Your Own

12. $0.52

Vocabulary and Concept Check

1. Place the decimal point so that there are two decimal places. $1.2 \times 2.4 = 2.88$
2. Greater than 8; Because 8 is being multiplied by a number greater than 1.
3. 8.722
4. 6.2832
5. 19.5750
6. 3
7. 4
8. 5

Practice and Problem Solving

9. 3.15
10. 0.24
11. 0.21
12. 6.21
13. 33.6
14. 31.5
15. 115.04
16. 18.27
17. 21.45
18. 29.45
19. 13.888
20. 98.256
21. 2.4
22. 0.076
23. 0.0342
24. 0.093
25. The decimal is in the wrong place. $0.0045 \times 9 = 0.0405$
26. The decimal is in the wrong place. $0.32 \times 5 = 1.60$
27. 30.06 lb
28. 8.75 ft
29. 3 mm; 9 mm

Assignment Guide and Homework Check

Level	Day 1 Activity Assignment	Day 2 Lesson Assignment	Homework Check
Basic	9–12, 70–74	1–8, 13, 15, 21, 25, 27, 31, 35, 53, 55	15, 21, 27, 31, 55
Average	9–12, 70–74	1–8, 15, 21, 26, 28, 42–48 even, 54, 56	15, 21, 28, 42, 44
Advanced	9–12, 70–74	1–8, 20, 26, 28, 36, 44, 48, 50, 54–58 even, 68	20, 28, 36, 44, 54

Common Errors

- **Exercises 13–24** Students might count the decimal places to the left of the decimal point instead of the right. Remind them that they are looking to the right of the decimal point, just like they did when writing decimals as fractions. Stress that estimating the product first will help.
- **Exercises 13–24** When determining where to place the decimal point, students might count left to right instead of right to left. Tell students to count from the least to greatest place value, so that means they must start counting decimal places from the right.
- **Exercises 21–24** Students might ignore the zeros when counting the number of decimal places, or forget to add them when finding the answer. Remind them that zeros are placeholders and are important for determining the correct answer. For example, 0.024 is not the same number as 0.24. Students may be more successful if they write these decimals as fractions and then multiply.

2.5 Record and Practice Journal

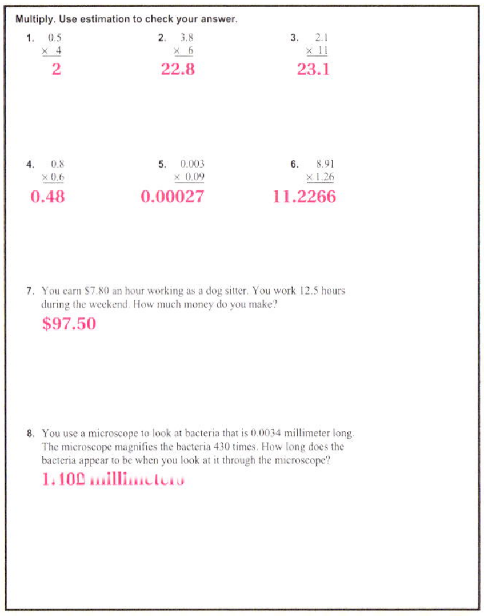

Multiply. Use estimation to check your answer.

1. 0.5×4 — **2**
2. 3.8×6 — **22.8**
3. 2.1×11 — **23.1**
4. 0.8×0.6 — **0.48**
5. 0.003×0.09 — **0.00027**
6. 8.91×1.26 — **11.2266**
7. You earn $7.80 an hour working as a dog sitter. You work 12.5 hours during the weekend. How much money do you make?
   $97.50
8. You use a microscope to look at bacteria that is 0.0034 millimeter long. The microscope magnifies the bacteria 430 times. How long does the bacteria appear to be when you look at it through the microscope?
   [illegible] millimeters

2.5 Exercises

Check It Out
Help with Homework
BigIdeasMath.com

Vocabulary and Concept Check

1. **NUMBER SENSE** If you know $12 \times 24 = 288$, how can you find 1.2×2.4?
2. **NUMBER SENSE** Is the product 1.23×8 greater than or less than 8? Explain.

Copy the problem and place the decimal point in the product.

3. $\begin{array}{r} 1.78 \\ \times\ \ 4.9 \\ \hline 8722 \end{array}$

4. $\begin{array}{r} 9.24 \\ \times\ \ 0.68 \\ \hline 62832 \end{array}$

5. $\begin{array}{r} 3.75 \\ \times\ \ 5.22 \\ \hline 195750 \end{array}$

How many decimal places are in the product?

6. 6.17×8.2
7. 1.684×10.2
8. 0.053×2.78

Practice and Problem Solving

Use base ten blocks or an area model to find the product.

9. $\begin{array}{r} 2.1 \\ \times\ 1.5 \\ \hline \end{array}$

10. $\begin{array}{r} 0.6 \\ \times\ 0.4 \\ \hline \end{array}$

11. $\begin{array}{r} 0.7 \\ \times\ 0.3 \\ \hline \end{array}$

12. $\begin{array}{r} 2.7 \\ \times\ 2.3 \\ \hline \end{array}$

Multiply. Use estimation to check your answer.

1 2 13. $\begin{array}{r} 4.8 \\ \times\ \ 7 \\ \hline \end{array}$

14. $\begin{array}{r} 6.3 \\ \times\ \ 5 \\ \hline \end{array}$

15. $\begin{array}{r} 7.19 \\ \times\ \ 16 \\ \hline \end{array}$

16. $\begin{array}{r} 0.87 \\ \times\ \ 21 \\ \hline \end{array}$

17. $\begin{array}{r} 1.95 \\ \times\ \ 11 \\ \hline \end{array}$

18. $\begin{array}{r} 5.89 \\ \times\ \ 5 \\ \hline \end{array}$

19. $\begin{array}{r} 3.472 \\ \times\ \ 4 \\ \hline \end{array}$

20. $\begin{array}{r} 8.188 \\ \times\ \ 12 \\ \hline \end{array}$

21. 100×0.024
22. 19×0.004
23. 0.0038×9
24. 10×0.0093

ERROR ANALYSIS **Describe and correct the error in the solution.**

25. ✗ $\begin{array}{r} 0.0045 \\ \times\ \ \ 9 \\ \hline 4.05 \end{array}$

26. ✗ $\begin{array}{r} 0.32 \\ \times\ \ 5 \\ \hline 0.160 \end{array}$

27. **MOON** The weight of an object on the Moon is about 0.167 of its weight on Earth. How much does a 180-pound astronaut weigh on the Moon?
28. **BAMBOO** A bamboo plant grows about 1.25 feet each day. Find the growth in one week.
29. **NAILS** A fingernail grows about 0.1 millimeter each day. How much does a fingernail grow in 30 days? 90 days?

Multiply.

③ **30.** $\begin{array}{r} 0.7 \\ \times\ 0.2 \\ \hline \end{array}$ **31.** $\begin{array}{r} 0.08 \\ \times\ 0.3 \\ \hline \end{array}$ **32.** $\begin{array}{r} 0.007 \\ \times\ 0.03 \\ \hline \end{array}$ **33.** $\begin{array}{r} 0.0008 \\ \times\ 0.09 \\ \hline \end{array}$

34. $\begin{array}{r} 0.004 \\ \times\ 0.9 \\ \hline \end{array}$ **35.** $\begin{array}{r} 0.06 \\ \times\ 0.5 \\ \hline \end{array}$ **36.** $\begin{array}{r} 0.0008 \\ \times\ 0.004 \\ \hline \end{array}$ **37.** $\begin{array}{r} 0.0002 \\ \times\ 0.06 \\ \hline \end{array}$

38. 12.4×0.2 **39.** 18.6×5.9 **40.** 7.91×0.72 **41.** 1.16×3.35

42. 6.478×18.21 **43.** 1.9×7.216 **44.** 0.0021×18.2 **45.** 6.109×8.4

46. **ERROR ANALYSIS** Describe and correct the error in the solution.

✗ $\begin{array}{r} 4.9 \\ \times\ 3.8 \\ \hline 186.2 \end{array}$

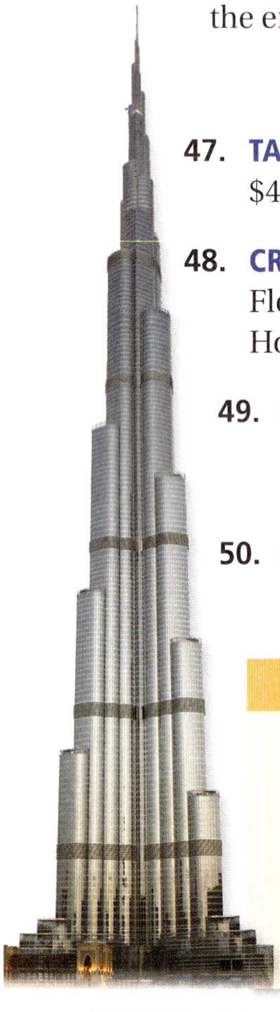

47. **TAKEOUT** A Chinese restaurant offers buffet takeout for \$4.99 per pound. How much does your takeout meal cost?

48. **CROPLAND** Alabama has about 2.51 million acres of cropland. Florida has about 1.15 times as much cropland as Alabama. How much cropland does Florida have?

49. **GOLD** On a tour of an old gold mine, you find a nugget containing 0.82 ounce of gold. Gold is worth \$1566.80 per ounce. How much is your nugget worth?

50. **BUILDING HEIGHTS** One meter is approximately 3.28 feet. Find the height of each building in feet by multiplying its height in meters by 3.28.

Continent	Tallest Building	Height (meters)
Africa	Carlton Centre Office Tower	223
Asia	Burj Khalifa	828
Australia	Q1 Tower	323
Europe	The Shard	310
North America	Willis Tower	442
South America	Gran Torre	300

51. **REASONING** Show how to evaluate $7.12 \times 8.22 \times 100$ without multiplying the two decimals.

ORDER OF OPERATIONS **Evaluate the expression.**

④ **52.** $2.4 \times 16 + 7$ **53.** $6.85 \times 2 \times 10$ **54.** $1.047 \times 5 - 0.88$

55. $4.32(3.7 + 1.65)$ **56.** $23.98 - 1.7^2 \cdot 7.6$ **57.** $12 \cdot 5.16 + 10.064$

58. $0.9(8.2 \cdot 20.35)$ **59.** $7.5^2(6.084 - 5.44)$ **60.** $6.8 \cdot 2.18 \cdot 3.95$

61. **REASONING** Without multiplying, how many decimal places does 3.4^2 have? 3.4^3? 3.4^4? Explain your reasoning.

Common Errors

- **Exercises 30–45** Students may only count the number of decimal places in one number instead of two. Remind them that they need to count the number of decimal places in both numbers.
- **Exercises 30–45** Suggest to students who are struggling with the concept to write the decimals as fractions. They can also write the number of decimal places to the right of the problem, as shown in the example, so that they remember where to place the decimal point.

Practice and Problem Solving

30. 0.14 **31.** 0.024

32. 0.00021 **33.** 0.000072

34. 0.0036 **35.** 0.03

36. 0.0000032 **37.** 0.000012

38. 2.48 **39.** 109.74

40. 5.6952 **41.** 3.886

42. 117.96438 **43.** 13.7104

44. 0.03822 **45.** 51.3156

46. The product should have two decimal places.

$$\begin{array}{r} 4.9 \\ \times\ 3.8 \\ \hline 18.62 \end{array}$$

47. \$3.24

48. 2.8865 million acres

49. \$1284.78

50. Carlton Centre Office Tower: 731.44 ft;
Burj Khalifa: 2715.84 ft;
Q1 Tower: 1059.44 ft;
The Shard: 1016.8 ft;
Willis Tower: 1449.76 ft;
Gran Torre: 984 ft

51. $7.12 \times 8.22 \times 100$
$= 7.12 \times 822 = 5852.64$

52. 45.4 **53.** 137

54. 4.355 **55.** 23.112

56. 2.016 **57.** 71.984

58. 150.183 **59.** 36.225

60. 58.5548 **61.** 2; 3; 4

English Language Learners

Vocabulary

Students should be warned not to drop the *trailing zeros* before placing the decimal point. Students need to include the zeros in the number of decimal places to be counted. Trailing zeros may be dropped after the decimal point has been placed.

$$\begin{array}{r} 3.25 \\ \times\ 0.8 \\ \hline 2.600 \end{array}$$

$3.25 \times 0.8 = 2.6$

Practice and Problem Solving

62. Each number is 0.6 times the previous number; 0.1296, 0.07776, 0.046656
63. Each number is 0.1 times the previous number; 0.0015, 0.00015, 0.000015
64. Each number is 0.5 times the previous number; 0.0025, 0.00125, 0.000625
65. Each number is 1.5 times the previous number; 25.3125, 37.96875, 56.953125
66. \$14.03
67. **a.** 190.06 mi
    b. 91.29 mi
68. See Additional Answers.
69. See *Taking Math Deeper*.

Fair Game Review

70. 26 **71.** 5
72. 3 **73.** 7
74. D

Mini-Assessment

Multiply.

1. 9.6×5 48
2. 4.321×2 8.642
3. 20.25×0.6 12.15
4. 2.18×3.24 7.0632
5. Apples are on sale for \$1.29 per pound. You buy 4.25 pounds of apples. How much will it cost to buy the apples? \$5.48

Taking Math Deeper

Exercise 69

This problem gives students a chance to factor with decimals. The key to part (a) is to write 952 as the product of two whole number factors. Then insert a decimal point to get the factors for 9.52.

1 **a.** Factor 9.52.

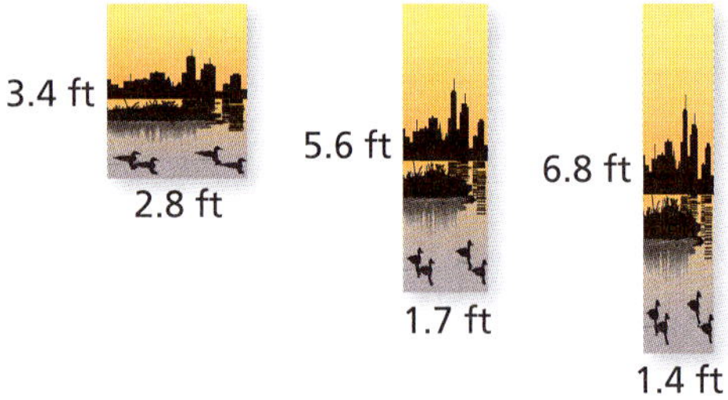

2 Find the perimeter of each painting. The painting with the shortest perimeter is the least expensive to frame.

b. $P = 2(3.4) + 2(2.8) = 12.4$ ft
$P = 2(5.6) + 2(1.7) = 14.6$ ft
$P = 2(6.8) + 2(1.4) = 16.4$ ft

Regardless of the frame, the perimeter of the painting does not change. \$5(12.4) is five times more than \$1(12.4).

c. Yes; the fancy framing will cost 5 times more than the thin black framing.

3 Draw a diagram. Remember to include the mitered corners. When finding the outside perimeter, use estimation.

$P = 20.4$ ft
$C = 5(20.4) = \$102$

$P = 14.0$ ft
$C = 1(14) = \$14$

$5(14) = \$70; \$102 > \$70$

d. Yes; the fancy framing costs *more than* 5 times as much as the thin black framing.

Reteaching and Enrichment Strategies

If students need help. . .	If students got it. . .
Resources by Chapter • Practice A and Practice B • Puzzle Time Record and Practice Journal Practice Differentiating the Lesson Lesson Tutorials Skills Review Handbook	Resources by Chapter • Enrichment and Extension • Technology Connection Start the next section

REPEATED REASONING **Describe the pattern. Find the next three numbers.**

62. 1, 0.6, 0.36, 0.216, . . .

63. 15, 1.5, 0.15, 0.015, . . .

64. 0.04, 0.02, 0.01, 0.005, . . .

65. 5, 7.5, 11.25, 16.875, . . .

66. **FOOD** You buy 2.6 pounds of apples and 1.475 pounds of peaches. You hand the cashier a $20 bill. How much change will you receive?

67. **MILEAGE** A car can travel 22.36 miles on one gallon of gasoline.

a. How far can the car travel on 8.5 gallons of gasoline?

b. A hybrid car can travel 33.1 miles on one gallon of gasoline. How much farther can the hybrid car travel on 8.5 gallons of gasoline?

68. **OPEN-ENDED** You and four friends have dinner at a restaurant.

a. Draw a restaurant menu that has main items, desserts, and beverages, with their prices.

b. Write a guest check that shows what each of you ate. Find the subtotal.

c. Multiply by 0.07 to find the tax. Then find the total.

d. Round the total to the nearest whole number. Multiply by 0.20 to estimate a tip. Including the tip, how much did you spend?

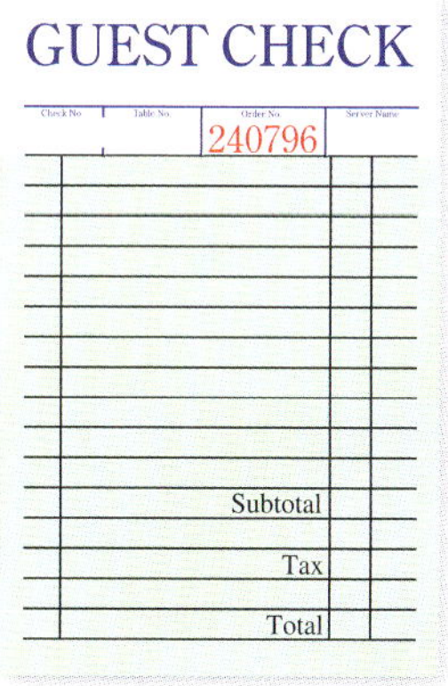

69. **Geometry** A rectangular painting has an area of 9.52 square feet.

a. Draw three different ways in which this can happen.

b. The cost of a frame depends on the perimeter of the painting. Which of your drawings from part (a) is the least expensive to frame? Explain your reasoning.

c. The thin, black framing costs $1 per foot. The fancy framing costs $5 per foot. Will the fancy framing cost five times as much as the black framing? Explain why or why not.

d. Suppose the cost of a frame depends on the outside perimeter of the frame. Does this change your answer to part (c)? Explain why or why not.

Fair Game Review What you learned in previous grades & lessons

Divide. *(Skills Review Handbook)*

70. $78 \div 3$

71. $65 \div 13$

72. $57 \div 19$

73. $84 \div 12$

74. **MULTIPLE CHOICE** How many edges does the rectangular prism at the right have? *(Skills Review Handbook)*

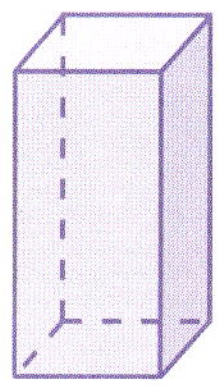

Ⓐ 4　　Ⓑ 6

Ⓒ 8　　Ⓓ 12

2.6 Dividing Decimals

Essential Question How can you use base ten blocks to model decimal division?

1 ACTIVITY: Dividing Decimals

Work with a partner. Use base ten blocks to model the division. Then find the quotient.

a. 2.4 ÷ 0.6

Begin by modeling 2.4.

2.4

How many of each base ten block did you use?

____ ones

____ tenths

____ hundredths

Next, think of the division problem 2.4 ÷ 0.6 as the question,

"How can you divide 2.4 into groups of 0.6?"

Rearrange the model for 2.4 into groups of 0.6. There are ____ groups of 0.6.

So, 2.4 ÷ 0.6 = ____.

b. 1.8 ÷ 2 **c.** 3.9 ÷ 3 **d.** 2.8 ÷ 0.7 **e.** 3.2 ÷ 0.4

f. Write and solve the division problem represented by the model.

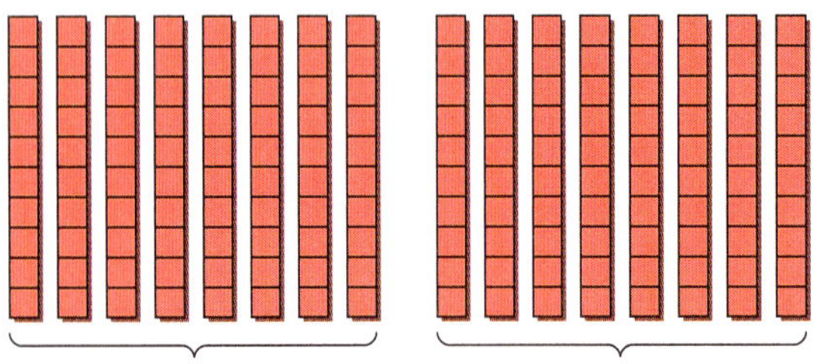

0.8 0.8

Common Core

Dividing Decimals

In this lesson, you will

- use models to divide decimals.
- divide decimals.

Learning Standard
6.NS.3

Laurie's Notes

Introduction

Standards for Mathematical Practice

- **MP1 Make Sense of Problems** and **MP5 Use Appropriate Tools Strategically:** Base ten blocks are a useful tool to help students make sense of a problem. Students should develop a habit of sketching models.

Motivate

- Ask for 12 volunteers at the front of the room. When all of the students are standing, ask them to arrange themselves in six groups of two.
- ? "What division problem could you write to describe what the students just did?" Some will say $12 \div 6 = 2$, and some will say $12 \div 2 = 6$.
- Write the problem $12 \div 4 = 3$ on the board, and ask students to describe what the 12 people should do. They are just as likely to say make 3 groups of 4 as they are to say make 4 groups of 3.
- Without a context stated, either would be correct. Interpreting a division equation is needed in the first Activity.
- **Standards Progression:** In Grade 5, students have found whole number quotients and divided decimals related to addition and subtraction. In Grade 6, students are expected to become fluent with division, showing remainders. In Grade 7, students will convert a rational number to a decimal using long division.

Activity Notes

Words of Wisdom

- **Connections:** In Activity 1, there are two possible ways to model $2.4 \div 0.6$.
  - Think of 2.4 as the improper fraction, 24-tenths.
  - Think of 2.4 as the mixed number, 2 and 4-tenths.

Activity 1

- Have students model 2.4 with their base ten blocks. You want to divide this amount by 0.6. Read it as "sixth-tenths," not "point six."
- ? "How do you model 2.4 so that only tenths are used?" Exchange 2 whole units for 20-tenths.
- ? "What does it mean to divide 24-tenths by 6-tenths?" It means "How many groups of 6-tenths can you make from 24-tenths?"
- Ask a student to show how 24-tenths can be divided into groups of 6-tenths. Record the problem and the quotient on the board: $2.4 \div 0.6 = 4$.
- **MP2 Reason Abstractly and Quantitatively:** Mathematically proficient students are able to create equivalent representations of a problem.
- **Connection:** The rule for dividing decimals is to multiply the divisor and dividend by a power of 10. This is accomplished using base ten blocks, when 2 wholes are exchanged for 20-tenths. You now have 24-tenths $\div$ 6-tenths, which is equivalent to $24 \div 6$.
- Have students work in pairs on parts (b)–(e). Note that in parts (b) and (c) the divisor is a whole number. The divisor can be interpreted as the number of groups versus the size of the group as was done in the Motivate.

Common Core State Standards

6.NS.3 Fluently add, subtract, multiply, and divide multi-digit decimals using the standard algorithm for each operation.

Previous Learning

Students have used base ten blocks to model operations with decimals.

Technology for the Teacher

Dynamic Classroom

Lesson Plans
Complete Materials List

2.6 Record and Practice Journal

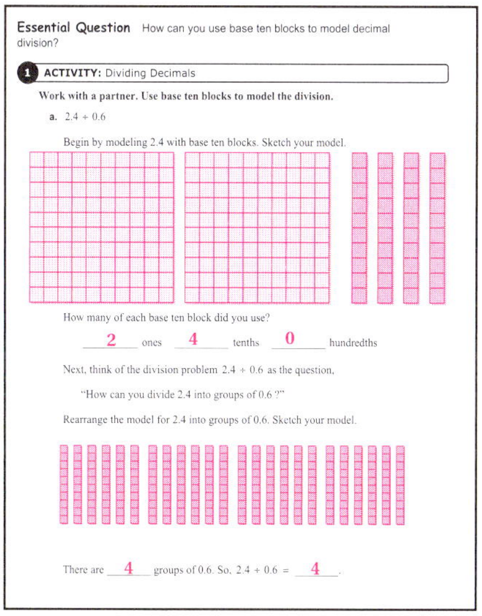

Essential Question How can you use base ten blocks to model decimal division?

1 ACTIVITY: Dividing Decimals

Work with a partner. Use base ten blocks to model the division.

a. $2.4 \div 0.6$

Begin by modeling 2.4 with base ten blocks. Sketch your model.

How many of each base ten block did you use?

2 ones 4 tenths 0 hundredths

Next, think of the division problem $2.4 \div 0.6$ as the question,

"How can you divide 2.4 into groups of 0.6 ?"

Rearrange the model for 2.4 into groups of 0.6. Sketch your model.

There are 4 groups of 0.6. So, $2.4 \div 0.6 = 4$.

English Language Learners

Graphic Organizer

Have students copy the table in their notes as a reference. Encourage them to add their own examples using other powers of ten.

Multiplying by Whole Number Powers of 10	
Move decimal point right.	$10 \times 2.5 = 25$
Multiplying by Decimal Powers of 10	
Move decimal point left.	$0.1 \times 2.5 = 0.25$
Dividing by Whole Number Powers of 10	
Move decimal point left.	$2.5 \div 10 = 0.25$
Dividing by Decimal Powers of 10	
Move decimal point right.	$2.5 \div 0.1 = 25$

2.6 Record and Practice Journal

b. $1.8 \div 2$
0.9

c. $3.9 \div 3$
1.3

d. $2.8 \div 0.7$
4

e. $3.2 \div 0.4$
8

f. Write and solve the division problem represented by the model.
$1.6 \div 0.8 = 2$
or $1.6 \div 2 = 0.8$

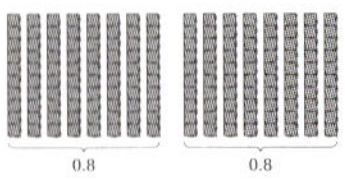

2 ACTIVITY: Dividing Decimals

Work with a partner. Use base ten blocks to model the division.

a. $0.3 \div 0.06$

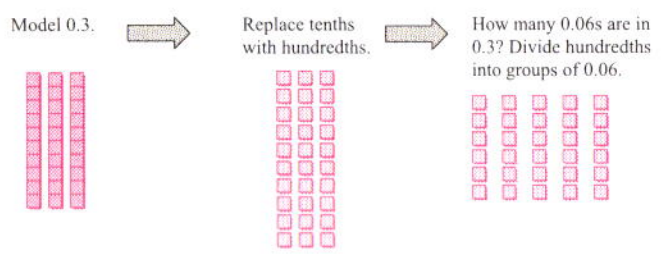

There are **5** groups of 0.06. So, $0.3 \div 0.06 =$ **5**.

b. $0.2 \div 0.04$
5

c. $0.6 \div 0.01$
60

d. $0.16 \div 0.08$
2

e. $0.28 \div 0.07$
4

What Is Your Answer?

3. **IN YOUR OWN WORDS** How can you use base ten blocks to model decimal division? Use examples from Activity 1 and Activity 2 as part of your answer.
Check students' work.

4. **WRITING** Newton's poem is about dividing fractions. Write a poem about dividing decimals.

"When you must divide a fraction, do this very simple action: Flip what you're dividing BY, and then it's easy—multiply!"

Check students' work.

5. Think of your own cartoon about dividing decimals. Draw your cartoon.
Check students' work.

Laurie's Notes

Activity 2

? "Is $0.3 \div 0.06$ *greater than* 1 or *less than* 1? How do you know?" greater than 1; The dividend (0.3) is greater than the divisor (0.06).

- Work through part (a) with students as they work in groups.
- **MP1:** The concrete model will help students visualize and make sense of the problem, allowing them then to solve the problem.
- When students exchange tenths for hundredths, it would be helpful for them to say or write "two-tenths is equivalent to twenty-hundredths."
- Note: Parts (d) and (e) already have like units (16-hundredths divided by 8-hundredths, and 28-hundredths divided by 7-hundredths).

Words of Wisdom

- The base ten blocks are used in the activities so that students can develop a visual model for division.
- The base ten pieces work easily because the numbers are compatible *and* because the quotients are all whole numbers. Notice that none of the quotients are decimals.
- To help students develop an understanding of decimal division, ask them to look at the following.

  $0.16 \div 0.08 = 2$ (16-hundredths divided by 8-hundredths)
  $1.6 \div 0.8 = 2$ (16-tenths divided by 8-tenths)
  $16 \div 8 = 2$ (16 ones divided by 8 ones)
  $160 \div 80 = 2$ (16 tens divided by 8 tens)

  This concept is a big idea in the lesson that follows. Notice that when the dividend and divisor are both multiplied by 10, the quotient stays the same.

What Is Your Answer?

- When students explain how base ten blocks are used to model decimal division, listen for references to like-sized pieces (the same units) being divided into groups of the same size.

? "You can use base ten blocks to model $0.8 \div 0.4$. Can you model $0.4 \div 0.8$?" You can't model $0.4 \div 0.8$ using base ten blocks, but you can model $\frac{1}{2}$ of the group that you need.

In tomorrow's lesson, problems of this type will be evaluated.

? "Could you use fractions to find $0.4 \div 0.8$?" yes; $\frac{4}{10} \div \frac{8}{10} = \frac{4}{8} = \frac{1}{2} = 0.5$

Closure

- Write each problem as a decimal division problem and solve.

  a. 54-hundredths divided by 9-hundredths $0.54 \div 0.09 = 6$
  b. 8-tenths divided by 5-hundredths $0.8 \div 0.05 = 16$

2 ACTIVITY: Dividing Decimals

Work with a partner. Use base ten blocks to model the division. Then find the quotient.

Math Practice 8

Evaluate Results

What can you do to check the reasonableness of your answer?

a. 0.3 ÷ 0.06

Model 0.3. | Replace tenths with hundredths. | How many 0.06s are in 0.3? Divide hundredths into groups of 0.06.

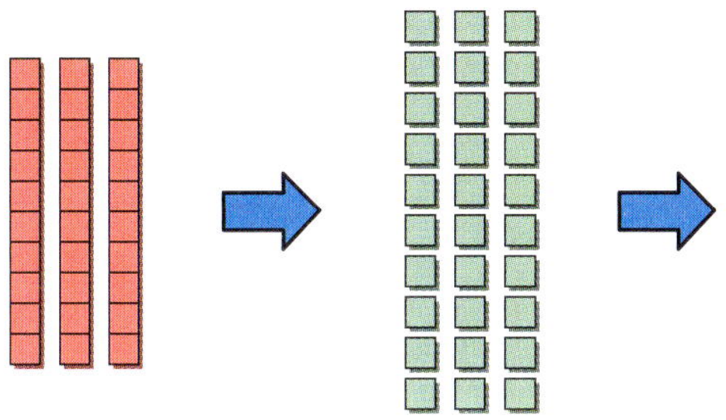

There are ___ groups of 0.06. So, 0.3 ÷ 0.06 = ___.

b. 0.2 ÷ 0.04

c. 0.6 ÷ 0.01

d. 0.16 ÷ 0.08

e. 0.28 ÷ 0.07

What Is Your Answer?

3. **IN YOUR OWN WORDS** How can you use base ten blocks to model decimal division? Use examples from Activity 1 and Activity 2 as part of your answer.

4. **WRITING** Newton's poem is about dividing fractions. Write a poem about dividing decimals.

"When you must divide a fraction, do this very simple action: Flip what you're dividing BY, and then it's easy—multiply!"

5. Think of your own cartoon about dividing decimals. Draw your cartoon.

Practice Use what you learned about dividing decimals to complete Exercises 8–11 on page 97.

2.6 Lesson

Check It Out
Lesson Tutorials
BigIdeasMath.com

Key Idea

Dividing Decimals by Whole Numbers

Words Place the decimal point in the quotient above the decimal point in the dividend. Then divide as you would with whole numbers. Continue until there is no remainder.

Numbers

$$\begin{array}{r} 1.83 \\ 4\overline{)7.32} \end{array}$$

Place the decimal point in the quotient above the decimal point in the dividend.

EXAMPLE 1 Dividing Decimals by Whole Numbers

a. Find 7.6 ÷ 4. **Estimate** $8 \div 4 = 2$

$$\begin{array}{r} 1.9 \\ 4\overline{)7.6} \\ -4 \\ \hline 3\,6 \\ -3\,6 \\ \hline 0 \end{array}$$

Place the decimal point in the quotient above the decimal point in the dividend.

So, 7.6 ÷ 4 = 1.9. **Reasonable?** $1.9 \approx 2$ ✓

b. Find 4.38 ÷ 12.

$$\begin{array}{r} 0.365 \\ 12\overline{)4.380} \\ -3\,6 \\ \hline 78 \\ -72 \\ \hline 60 \\ -60 \\ \hline 0 \end{array}$$

Place the decimal point in the quotient above the decimal point in the dividend.

Insert a zero and continue to divide.

So, 4.38 ÷ 12 = 0.365. **Check** $0.365 \times 12 = 4.38$ ✓

On Your Own

Now You're Ready
Exercises 12–23

Divide. Use estimation to check your answer.

1. 36.4 ÷ 2 **2.** 22.2 ÷ 6 **3.** 59.64 ÷ 7

4. 43.26 ÷ 14 **5.** 6.2 ÷ 4 **6.** 3.12 ÷ 16

Laurie's Notes

Introduction

Connect

- **Yesterday:** Students used base ten blocks to explore decimal division. (MP1, MP2, MP5)
- **Today:** Students will use a formal rule to divide decimals.

Motivate

- Give students some contexts for dividing a dollar amount among three or four people. Have students estimate the whole dollar amount of the quotient using compatible numbers.
- ? "Three siblings purchase a gift for their parents that costs $22.98. About how much does each sibling pay?" round $22.98 to $24; $24 ÷ 3 = $8
- ? "Four friends share the cost of a pizza and drinks that total $18.75. About how much does each friend owe?" round $18.75 to $20; $20 ÷ 4 = $5
- ? "You pay $4.32 for three pens. Were the pens more than $2 each?" No, the cost for each pen is somewhere between $1 and $2.

Lesson Notes

Words of Wisdom

- Placing the decimal point in the quotient above the decimal point in the dividend makes sense to the students.
- **FYI:** Some students may have difficulty lining up the columns when they do long division. Have them do their work on $\frac{1}{4}$-inch graph paper.

Example 1

- Work through both examples. Begin by estimating the quotient and end with judging the reasonableness of your answer.
- Students often ask where the decimal point is written in the work below the dividend. "Why don't you bring down the decimal point?" In theory, you could bring the decimal point down. But, once the decimal point is placed in the quotient, you treat the problem as if you were dividing whole numbers.
- ? "If the divisor is greater than the dividend, what do you know about the quotient?" It will be less than 1.
- You may want to share whole number examples to illustrate the above question. For example, use 8 ÷ 4 and 4 ÷ 8.
- ? When you start part (b), place the decimal point in the quotient. Then ask "Does 12 go into 4?" no "Does 12 go into 43?" yes
- ? In this problem, it is necessary to add a zero to the dividend in order to continue the division. Ask, "Is 4.38 equivalent to 4.380? 4.3800?" yes; yes
- Students want to know how many zeros to add. Tell students that you need to continue to add zeros until the division ends with a zero remainder, or until a pattern occurs, as it did when they studied repeating decimals.

On Your Own

- Question 4 is a little more challenging due to the two-digit divisor and the zero in the quotient.

Goal Today's lesson is dividing decimals.

Technology for the Teacher

Dynamic Classroom

Lesson Tutorials
Lesson Plans
Answer Presentation Tool

Extra Example 1

Divide.

a. 3.5 ÷ 7 0.5

b. 6.92 ÷ 8 0.865

On Your Own

1.	18.2	**2.**	3.7
3.	8.52	**4.**	3.09
5.	1.55	**6.**	0.195

Laurie's Notes

Key Idea

- **MP2 Reason Abstractly and Quantitatively** and **MP3 Construct Viable Arguments and Critique the Reasoning of Others:** Throughout the examples, students should be encouraged to explain their thinking about the process. Multiplying the dividend and divisor by the same power of ten is equivalent to multiplying the numerator and denominator of a fraction by the same number. Make sure students make this connection.
- Discuss why the rule works: When you multiply both the dividend and the divisor by a power of ten, the quotient is not changed. You may want to remind students of the activity from the motivation of the lesson.
- Remind students that they need to multiply by a large enough power of ten to make the divisor a whole number. Only when you divide by a whole number can you place the decimal point in the quotient.

Example 2

- Work through parts (a) and (b). The original problem in part (a) can be read "182-tenths ÷ 14-tenths." Multiply the divisor and the dividend by 10 to get the equivalent problem 182 ÷ 14.
- Do the check as part of each example to model this good habit for students.

On Your Own

- Note the two representations of division in these four problems.
- Ask students to estimate an answer before they begin. Do this as a class.

Extra Example 2

Divide.

a. 68.4 ÷ 5.7 12

b. 0.336 ÷ 0.42 0.8

On Your Own

7. 8 **8.** 17

9. 9.41 **10.** 0.9

Differentiated Instruction

Visual

When dividing decimals, students may forget which number to make the whole number (the *divisor*) or they may forget to multiply the dividend by the same power of 10. Students may be careless in placing the decimal point in the quotient. The decimal point should be placed above the new decimal point in the dividend. Create a bulletin board display as a visual reminder.

whole number $0.05\overline{)2.655}$

Key Idea

Dividing Decimals by Decimals

Words Multiply the divisor *and* the dividend by a power of 10 to make the divisor a whole number. Then place the decimal point in the quotient and divide as you would with whole numbers. Continue until there is no remainder.

Numbers $1.2\overline{)4.56}$ → $12\overline{)45.6}$ with quotient 3.8

Multiply each number by 10.

Place the decimal point above the decimal point in the dividend 45.6.

EXAMPLE 2 Dividing Decimals

a. Find 18.2 ÷ 1.4.

$1.4\overline{)18.2}$ → $14\overline{)182.}$

$$\begin{array}{r} 13. \\ 14\overline{)182.} \\ -14 \\ \hline 42 \\ -42 \\ \hline 0 \end{array}$$

Multiply each number by 10.

Place the decimal point above the decimal point in the dividend 182.

So, 18.2 ÷ 1.4 = 13. **Check** $13 \times 1.4 = 18.2$ ✓

Study Tip

Multiplying the divisor and the dividend by a power of 10 does not change the quotient.

For example:

18.2 ÷ 1.4 = 13

182 ÷ 14 = 13

1820 ÷ 140 = 13

b. Find 0.273 ÷ 0.39.

$0.39\overline{)0.273}$ → $39\overline{)27.3}$

$$\begin{array}{r} 0.7 \\ 39\overline{)27.3} \\ -27.3 \\ \hline 0 \end{array}$$

Multiply each number by 100.

So, 0.273 ÷ 0.39 = 0.7. **Check** $0.7 \times 0.39 = 0.273$ ✓

On Your Own

Now You're Ready
Exercises 36–39

Divide. Check your answer.

7. $1.2\overline{)9.6}$

8. $3.4\overline{)57.8}$

9. 21.643 ÷ 2.3

10. 0.459 ÷ 0.51

EXAMPLE 3 Inserting Zeros in the Dividend and the Quotient

Divide 2.45 ÷ 0.007.

$0.007\overline{)2.450} \rightarrow 7\overline{)2450}$

Multiply each number by 1000. Insert a zero in the dividend.

```
    350
 7)2450
  -21
    35
   -35
    00
```

Because 0 ÷ 7 = 0, insert a zero in the quotient.

So, 2.45 ÷ 0.007 = 350.

Study Tip

Remember to check your answer by multiplying the quotient by the divisor.

On Your Own

Now You're Ready Exercises 40–43

Divide. Check your answer.

11. 3.8 ÷ 0.16

12. 15.6 ÷ 0.78

13. 7.2 ÷ 0.048

14. 0.18 ÷ 0.003

EXAMPLE 4 Real-Life Application

How many times more cellular phone subscribers were there in 2011 than in 1991? Round to the nearest whole number.

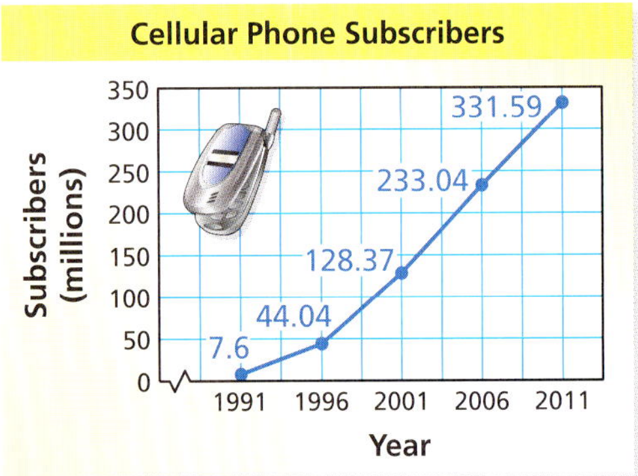

From the graph, there were 331.59 million subscribers in 2011 and 7.6 million in 1991. So, divide 331.59 by 7.6.

Estimate 320 ÷ 8 = 40

$7.6\overline{)331.59} \rightarrow 76\overline{)3315.9}$

```
      43.6   ← Rounds to 44.
 76)3315.9
   -304
     275
    -228
     47 9
    -45 6
      2 3
```

So, there were about 44 times more subscribers in 2011 than in 1991.

Reasonable? 44 ≈ 40 ✓

On Your Own

15. How many times more subscribers were there in 2006 than in 1996? Round to the nearest whole number.

Laurie's Notes

Example 3

- ? "How can you estimate the answer to 2.45 ÷ 0.007?" Listen for students to explain that the quotient has to be greater than 2.45.
- Sometimes students focus on the process and forget about the mathematics behind the process. For example, they may talk about moving decimal points but lose sight of why doing so doesn't change the problem.
- To help students understand (and remind them of) the mathematics, write 2.4 × 1000 and 0.007 × 1000 off to the side.
- Remind them that 2.45 ÷ 0.007 and 2450 ÷ 7 will have the same quotient.
- ? "Is 0.03 ÷ 0.004 *greater than one* or *less than one*? Why?" greater than one; The dividend is greater than the divisor.
- ? "What number do you need to multiply by to make the divisor a whole number?" 1000

On Your Own

- **Neighbor Check:** Have students work independently and then compare their answers with their neighbors. Have students discuss any differences.

Example 4

- ? **Review:** Ask questions about the graph: How many cell phone subscribers were there in 1996? In what year were there about 128 million cell phone subscribers? Describe the general trend of the graph. 44.04 million; 2001; The graph is increasing.
- **Common Error:** Students often read this type of question as, "How many more?" versus, "How many *times* more?" Relate this to a question about themselves that they might understand, such as, "How much older is your mom than you?" versus, "How many times older is your mom than you?"
- ? "If you want to round to the nearest whole number, what place value do you need to look at?" the tenths
- Rounding to the nearest whole number may be difficult for students due to the decimal point. Remind students that rounding to the nearest whole number is still done in the same way as any other place value.
- **Extension:** Suggest extending the example by having students continue to divide and see what happens. In this case, it does not terminate. This is a concept students will study in Grade 7.

On Your Own

- Have students estimate first. Then work through the solution to the question with the class.

Closure

- **Writing:** Use 8.4 and 0.42 to write an example whose quotient is greater than 1 and an example whose quotient is less than 1. Solve each problem. 8.4 ÷ 0.42 = 20; 0.42 ÷ 8.4 = 0.05

Extra Example 3

Find 1.76 ÷ 0.004. 440

On Your Own

11. 23.75 **12.** 20
13. 150 **14.** 60

Extra Example 4

In 2011, there were 6.5 thousand commercial FM radio stations. In 1990, there were 4.4 thousand. How many times more commercial FM radio stations were there in 2011 than in 1990? Round your answer to the nearest tenth. about 1.5 times more

On Your Own

15. about 5 times more

Vocabulary and Concept Check

1. $4\overline{)2.44}$ = 0.61
2. $18.6 \div 4 = 4.65$
3. $6.38 \div 11 = 0.58$
4. $88.27 \div 7 = 12.61$
5. $47\overline{)136}$
6. $21\overline{)1766}$
7. $216\overline{)1850}$

Practice and Problem Solving

8. 12 **9.** 13

10. 12 **11.** 9

12. 4.2 **13.** 6.7

14. 0.5 **15.** 1.3

16. 4.3 **17.** 9.4

18. 6.2 **19.** 3.2

20. 5.58 **21.** 0.098

22. 0.15 **23.** 3.57

24. They brought down 2 zeros instead of 1.

$$
\begin{array}{r}
3.112 \\
9\overline{)28.008} \\
\underline{27} \\
10 \\
\underline{9} \\
10 \\
\underline{9} \\
18 \\
\underline{18} \\
0
\end{array}
$$

25. A zero should be placed before the 8 in the quotient.

$$
\begin{array}{r}
0.086 \\
6\overline{)0.516} \\
\underline{48} \\
36 \\
\underline{36} \\
0
\end{array}
$$

26. $0.12

27. the 5-ounce bottle; The price per ounce is $2.06 for the 5-ounce bottle and $2.12 for the 4-ounce bottle.

Assignment Guide and Homework Check

Level	Day 1 Activity Assignment	Day 2 Lesson Assignment	Homework Check
Basic	8–11, 61–65	1–7, 13, 15, 25–29 odd, 35–39 odd	13, 15, 27, 29
Average	8–11, 61–65	1–7, 18, 20, 24, 26, 30–34 even, 39, 41, 45, 51	20, 26, 30, 39, 45
Advanced	8–11, 61–65	1–7, 18, 20, 24, 26, 32, 34, 40, 44, 46, 52, 54, 58	20, 26, 32, 40, 46

Common Errors

- **Exercises 12–23** Students may divide without putting the decimal point in the quotient. Tell them to place the decimal point immediately after they put the numbers in long division form. Remind students to use estimation to check the reasonableness of their answers.
- **Exercises 12–23** Students may have difficulty keeping their columns organized and will forget to include placeholder zeros or put in extra placeholders. Have them use graph paper as an organizational tool to help line up the numbers. Remind students to use estimation to check the reasonableness of their answers.

2.6 Record and Practice Journal

Divide. Check your answer.

1. $3\overline{)18.6}$ **6.2**
2. $6\overline{)46.8}$ **7.8**
3. $4\overline{)7.6}$ **1.9**
4. $24.5 \div 7$ **3.5**
5. $0.096 \div 8$ **0.012**
6. $15.65 \div 5$ **3.13**
7. $3.1\overline{)17.36}$ **5.6**
8. $6.4\overline{)43.52}$ **6.8**
9. $7.05\overline{)8.46}$ **1.2**
10. $9.24 \div 15.4$ **0.6**
11. $7.06 \div 0.353$ **20**
12. $0.015 \div 0.003$ **5**

13. It costs $859.32 to have a school dance.

School Dance
October 28th
Tickets $8

a. How many tickets must be sold to cover the cost? **108 tickets**

b. How many tickets must be sold to make a $980.68 profit? **230 tickets**

2.6 Exercises

Check It Out
Help with Homework
BigIdeasMath.com

Vocabulary and Concept Check

1. **NUMBER SENSE** Fix the one that is not correct.

$$\begin{array}{r} 6.1 \\ 4\overline{)24.4} \end{array} \qquad \begin{array}{r} 61 \\ 4\overline{)244} \end{array} \qquad \begin{array}{r} 6.1 \\ 4\overline{)2.44} \end{array}$$

Copy the problem and place the decimal point in the correct location.

2. $18.6 \div 4 = 465$
3. $6.38 \div 11 = 58$
4. $88.27 \div 7 = 1261$

Rewrite the problem so that the divisor is a whole number.

5. $4.7\overline{)13.6}$
6. $0.21\overline{)17.66}$
7. $2.16\overline{)18.5}$

Practice and Problem Solving

Use base ten blocks to find the quotient.

8. $3.6 \div 0.3$
9. $2.6 \div 0.2$
10. $0.72 \div 0.06$
11. $0.36 \div 0.04$

Divide. Use estimation to check your answer.

1 12. $6\overline{)25.2}$
13. $5\overline{)33.5}$
14. $7\overline{)3.5}$
15. $8\overline{)10.4}$
16. $38.7 \div 9$
17. $37.6 \div 4$
18. $43.4 \div 7$
19. $25.6 \div 8$
20. $44.64 \div 8$
21. $0.294 \div 3$
22. $3.6 \div 24$
23. $64.26 \div 18$

ERROR ANALYSIS **Describe and correct the error in finding the quotient.**

24. ✗

$$\begin{array}{r} 3.922 \\ 9\overline{)28.008} \\ \underline{27} \\ 1\,00 \\ \underline{81} \\ 198 \\ \underline{198} \\ 0 \end{array}$$

25. ✗

$$\begin{array}{r} 0.86 \\ 6\overline{)0.516} \\ \underline{48} \\ 36 \\ \underline{36} \\ 0 \end{array}$$

26. **TEXT MESSAGING** You send 40 text messages in one month. The total cost is $4.80. How much does each text message cost?

27. **SUNBLOCK** Of the two bottles of sunblock shown, which is the better buy? Explain.

4-ounce bottle $8.49 | 5-ounce bottle $10.29

ORDER OF OPERATIONS Evaluate the expression.

28. $7.68 + 3.18 \div 12$ **29.** $10.56 \div 3 - 1.9$ **30.** $19.6 \div 7 \times 9$

31. $5.5 \times 16.56 \div 9$ **32.** $35.25 \div 5 \div 3$ **33.** $13.41 \times (5.4 \div 9)$

34. **FRUIT PUNCH** Which pack of fruit punch is the best buy? Explain.

35. **SALE** You buy 3 pairs of jeans for \$35.95 each and get a fourth pair for free. What is your cost per pair of jeans?

Divide. Check your answer.

2 **36.** $2.1\overline{)25.2}$ **37.** $3.8\overline{)34.2}$ **38.** $36.47 \div 0.7$ **39.** $0.984 \div 12.3$

3 **40.** $4.23 \div 0.012$ **41.** $0.52 \div 0.0013$ **42.** $95.04 \div 0.0132$ **43.** $32.2 \div 0.07$

Divide. Round to the nearest hundredth if necessary.

44. $80.88 \div 8.425$ **45.** $0.8 \div 0.6$ **46.** $38.9 \div 6.44$ **47.** $11.6 \div 0.95$

48. **ERROR ANALYSIS** Describe and correct the error in rewriting the problem.

✗ $0.32\overline{)146.4} \rightarrow 32\overline{)1.464}$

49. **TICKETS** Tickets to the school musical cost \$6.25. The amount received from ticket sales is \$706.25. How many tickets were sold?

50. **HEIGHT** A person's running stride is about 1.14 times the person's height. Your friend's stride is 5.472 feet. How tall is your friend?

51. **MP3 PLAYER** You have 3.4 gigabytes available on your MP3 player. Each song is about 0.004 gigabyte. How many more songs can you download onto your MP3 player?

52. **SWIMMING** The table shows the top three times in a swimming event at the Summer Olympics. The event consists of a team of four women swimming 100 meters each.

Women's 4 × 100 Freestyle Relay		
Medal	**Country**	**Time (seconds)**
Gold	Australia	215.94
Silver	United States	216.39
Bronze	Netherlands	217.59

a. Suppose the times of all four swimmers on each team were the same. For each team, how much time does it take a swimmer to swim 100 meters?

b. Suppose each U.S. swimmer completed 100 meters a quarter second faster. Would the U.S. team have won the gold medal? Explain your reasoning.

Common Errors

- **Exercise 35** Students may automatically divide $35.95 by four without taking into account that $35.95 is the cost for *each* pair of jeans (except the free pair!). Remind them to read the problem carefully. Have students write out the cost for each of the four pairs of jeans and find the total. Then they can find the cost per pair. This type of question is common on state assessments, so take time to go over this question as a class.

Exercises 36–47

- Students may try to multiply by a power of 10 to make the dividend a whole number, instead of the divisor. While this method will still provide the same answer, it is easier for students to divide when the divisor is the smallest whole number possible. Remind them to multiply by whatever power of 10 will make the divisor a whole number.
- Some students may forget to multiply the dividend by the same multiple of 10 that they used to multiply the divisor. Remind them that they must multiply every number in the problem by the same multiple of 10.
- Students may place the decimal point in the wrong place as a result of sloppiness. Encourage them to use graph paper to keep their work organized.
- Remind students that they should check their answers by multiplying the quotient by the divisor and confirm that the product is equal to the dividend.

Practice and Problem Solving

28. 7.945 **29.** 1.62

30. 25.2 **31.** 10.12

32. 2.35 **33.** 8.046

34. the 12-pack; The price per unit is $0.74 for the 4-pack, $0.72 for the 12-pack, and $0.73 for the 24-pack. So, the 12-pack is the best buy.

35. $26.96

36. 12 **37.** 9

38. 52.1 **39.** 0.08

40. 352.5 **41.** 400

42. 7200 **43.** 460

44. 9.6 **45.** about 1.33

46. about 6.04

47. about 12.21

48. The decimal point in the dividend should be moved to the right instead of the left;

$0.32\overline{)146.4} \longrightarrow 32\overline{)14{,}640}$

49. 113 tickets

50. 4.8 ft

51. 850 songs

52. **a.** Australia: 53.985 sec
United States: 54.0975 sec
Netherlands: 54.3975 sec

b. yes; The team total would have been 215.39 seconds, which is faster than 215.94 seconds.

Differentiated Instruction

Inclusion

Remind students that one difference between multiplying decimals and dividing decimals is the placement of the decimal point. In multiplication, the decimal point is placed after the decimals are multiplied. In division, the placement of the decimal point is determined before dividing.

Practice and Problem Solving

53. = **54.** >

55. < **56.** <

57. about 5357 bees

58. a. 9 weeks

b. 3 more weeks; *Sample answer:* Use the same method as part (a) to find the total number of weeks needed, then subtract the number of weeks in part (a).

59. See *Taking Math Deeper.*

60. a. 3 times

b. 9 times

c. no; The answers are different because perimeter is measured in units and area is measured in square units.

d. The perimeter of the large rectangle is two times greater than the perimeter of the small rectangle. The area of the large rectangle is four times greater than the area of the small rectangle.

Fair Game Review

61. $1\frac{1}{6}$ **62.** $1\frac{3}{20}$

63. $\frac{1}{20}$ **64.** $\frac{1}{24}$

65. B

Mini-Assessment

Divide.

1. $6.8 \div 2$ 3.4 **2.** $0.963 \div 3$ 0.321

3. $7.6 \div 3.2$ 2.375 **4.** $15.47 \div 2.6$ 5.95

5. Cookies for a school fundraiser cost $3.75 per box. The total amount received from the cookie sale is $2718.75. How many boxes of cookies were sold? 725

Taking Math Deeper

Exercise 59

This is a good opportunity to help students practice unit analysis. With addition, subtraction, and multiplication, unit analysis is straightforward. With division, you need to "invert and multiply" just as you do when dividing fractions so that units divide out.

1 Organize the given information.

24 oz
$2.63

3.9 oz
$0.52

2 How many bowls of applesauce fit into the jar?

$$24\,\frac{\text{oz}}{\text{jar}} \div 3.9\,\frac{\text{oz}}{\text{bowl}}$$

$$= 24\,\frac{\cancel{\text{oz}}}{\text{jar}} \cdot \frac{1}{3.9}\,\frac{\text{bowls}}{\cancel{\text{oz}}}$$

a. $\approx$ 6.15 bowls per jar "Bowls per jar" is a check that I did it right.

3 **b. and c.** Which is the better buy?

- One way: Multiply. At the bowl price, you pay more for a jar.

$$0.52\,\frac{\$}{\cancel{\text{bowl}}} \cdot 6.15\,\frac{\cancel{\text{bowls}}}{\text{jar}} = \$3.20 \text{ per jar}$$

- Another way: Find the cost per ounce.

$$\frac{2.63}{24}\,\frac{\$}{\text{oz}} \approx \$0.11 \text{ per oz} \quad \text{Jar}$$

$$\frac{0.52}{3.9}\,\frac{\$}{\text{oz}} \approx \$0.13 \text{ per oz} \quad \text{Bowl}$$

Project

Find a recipe for applesauce. How many apples would it take to make a 24-ounce jar of applesauce? How many apples are in a 3.9-ounce bowl?

Reteaching and Enrichment Strategies

If students need help. . .	If students got it. . .
Resources by Chapter • Practice A and Practice B • Puzzle Time Record and Practice Journal Practice Differentiating the Lesson Lesson Tutorials Skills Review Handbook	Resources by Chapter • Enrichment and Extension • Technology Connection Start the next section

Without finding the quotient, copy and complete the statement using <, >, or =.

53. 6.66 ÷ 0.74 ____ 66.6 ÷ 7.4

54. 32.2 ÷ 0.7 ____ 3.22 ÷ 7

55. 160.72 ÷ 16.4 ____ 160.72 ÷ 1.64

56. 75.6 ÷ 63 ____ 7.56 ÷ 0.63

57. **BEES** To approximate the number of bees in a hive, multiply the number of bees that leave the hive in one minute by 3 and divide by 0.014. You count 25 bees leaving a hive in one minute. How many bees are in the hive?

58. **PROBLEM SOLVING** You are saving money to buy a new bicycle that costs $155.75. You have $30 and plan to save $5 each week. Your aunt decides to give you an additional $10 each week.

a. How many weeks will you have to save until you have enough money to buy the bicycle?

b. How many more weeks would you have to save to buy a new bicycle that costs $203.89? Explain how you found your answer.

Applesauce	
3.9-ounce bowl	$0.52
24-ounce jar	$2.63

59. **PRECISION** A store sells applesauce in two sizes.

a. How many *bowls* of applesauce fit in a *jar*? Round your answer to the nearest hundredth.

b. Explain two ways to find the better buy.

c. What is the better buy?

60. **Geometry** The large rectangle's dimensions are three times the dimensions of the small rectangle.

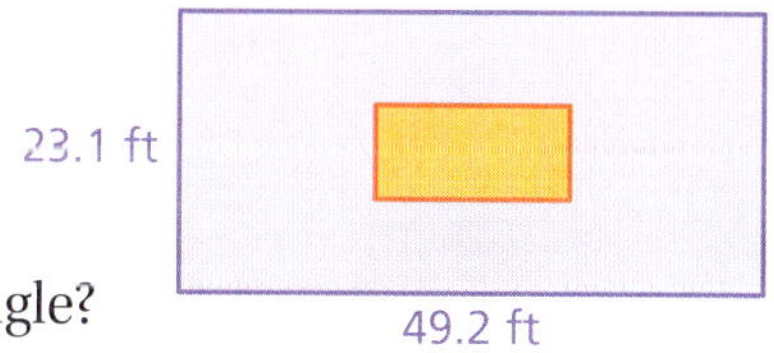

a. How many times greater is the perimeter of the large rectangle compared to the perimeter of the small rectangle?

b. How many times greater is the area of the large rectangle compared to the area of the small rectangle?

c. Are the answers to parts (a) and (b) the same? *Explain* why or why not.

d. What happens in parts (a) and (b) if the dimensions of the large rectangle are two times the dimensions of the small rectangle?

Fair Game Review What you learned in previous grades & lessons

Add or subtract. Write your answer in simplest form. *(Section 1.6)*

61. $\frac{1}{2} + \frac{2}{3}$

62. $\frac{2}{5} + \frac{3}{4}$

63. $\frac{3}{10} - \frac{1}{4}$

64. $\frac{11}{12} - \frac{7}{8}$

65. **MULTIPLE CHOICE** Melissa earns $7.40 an hour working at a grocery store. She works 14.25 hours this week. How much does she earn? *(Section 2.5)*

Ⓐ $83.13 Ⓑ $105.45 Ⓒ $156.75 Ⓓ $1054.50

2.4–2.6 Quiz

Check It Out
Progress Check
BigIdeasMath.com

Add or subtract. *(Section 2.4)*

1. $6.329 + 14.38$

2. $43.56 + 41.82$

3. $85.8 - 2.354$

4. $26.782 - 14.96$

Multiply. Use estimation to check your answer. *(Section 2.5)*

5. $\begin{array}{r} 7.6 \\ \times\ 5 \\ \hline \end{array}$

6. $\begin{array}{r} 0.62 \\ \times\ 17 \\ \hline \end{array}$

7. $\begin{array}{r} 0.54 \\ \times\ 0.9 \\ \hline \end{array}$

8. $\begin{array}{r} 4.16 \\ \times\ 0.7 \\ \hline \end{array}$

Divide. Use estimation to check your answer. *(Section 2.6)*

9. $5\overline{)8.4}$

10. $6\overline{)6.48}$

11. $5.6 \div 0.7$

12. $1.8 \div 0.03$

13. FIELD HOCKEY A field hockey field is rectangular. Its width is 54.88 meters, and its length is 91.46 meters. Find the perimeter of the field. *(Section 2.4)*

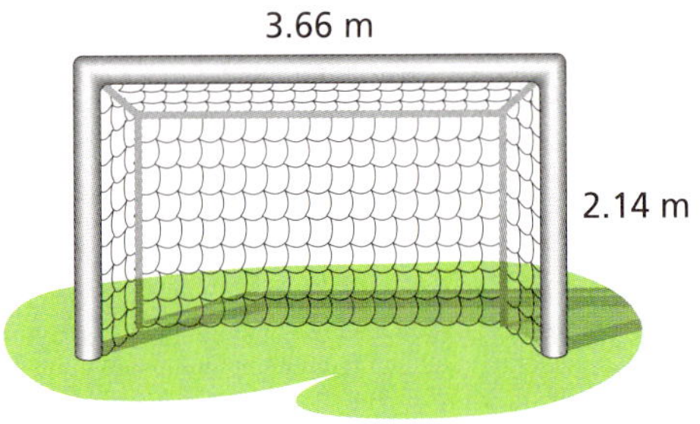

14. GEOMETRY Find the area of the mouth of the field hockey goal. *(Section 2.5)*

15. BROADWAY The bar graph shows the yearly attendance at traveling Broadway shows. *(Section 2.6)*

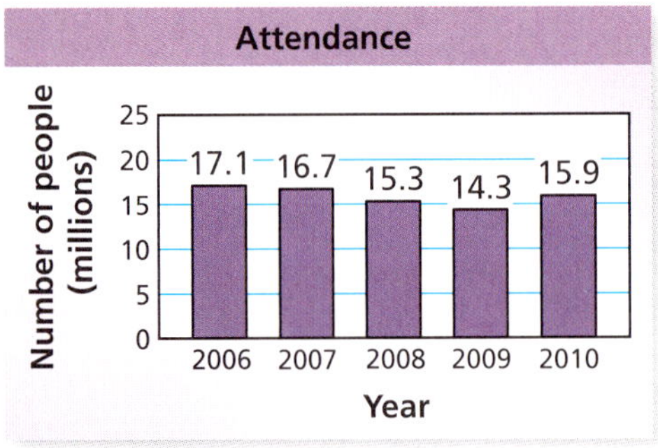

a. Suppose the attendance was the same each month in 2008. How many people attended each month?

b. How many times more people attended shows in 2006 than in 2009? Round your answer to the nearest tenth.

Alternative Assessment Options

Math Chat	Student Reflective Focus Question
Structured Interview	Writing Prompt

Math Chat

- Work in groups of four. Two members work together to discuss the similarities and differences of multiplying two fractions compared to multiplying two decimals. The other pair does the same for division. When they are finished, they explain their findings to the other members of their group.
- The teacher should walk around the classroom listening to the groups and asking questions to ensure understanding.

Study Help Sample Answers

Remind students to complete Graphic Organizers for the rest of the chapter.

4.

Be sure to line up the decimal points so that you add or subtract only the digits that have the same place value. You may have to insert zeros. addend + addend sum	Adding and subtracting decimals 1. Line up the decimal points. 2. Bring down the decimal point. 3. Add or subtract as you would with whole numbers. Example: 17.625 + 108.3 1 17.625 + 108.300 ← Insert zeros. 125.925
How do you add or subtract a decimal and a fraction?	

5–8. Available at *BigIdeasMath.com*

Reteaching and Enrichment Strategies

If students need help. . .	If students got it. . .
Resources by Chapter • Practice A and Practice B • Puzzle Time Lesson Tutorials *BigIdeasMath.com*	Resources by Chapter • Enrichment and Extension • Technology Connection Game Closet at *BigIdeasMath.com* Start the Chapter Review

Answers

1. 20.709
2. 85.38
3. 83.446
4. 11.822
5. 38
6. 10.54
7. 0.486
8. 2.912
9. 1.68
10. 1.08
11. 8
12. 60
13. 292.68 m
14. 7.8324 m^2
15. **a.** 1.275 million

    b. about 1.2 times as many

Technology for the Teacher

Online Assessment
Assessment Book
ExamView® Assessment Suite

For the Teacher
Additional Review Options

- *BigIdeasMath.com*
- Online Assessment
- Game Closet at *BigIdeasMath.com*
- Vocabulary Help
- Resources by Chapter

Answers

1. $\frac{5}{56}$
2. $\frac{3}{10}$
3. $\frac{1}{6}$
4. $\frac{6}{25}$
5. $2\frac{2}{15}$
6. $1\frac{17}{63}$
7. $4\frac{17}{48}$
8. $12\frac{4}{15}$
9. $\frac{5}{18}$
10. $\frac{9}{10}$
11. 15
12. $2\frac{26}{27}$

Review of Common Errors

Exercises 1–4

- Students may cross multiply instead of multiplying the numerators and denominators, or try to find a common denominator before multiplying.

Exercises 5–8

- Students may not change the mixed number into an improper fraction.

Exercises 9–12

- Students may find the reciprocal of the dividend instead of the divisor, the reciprocal of both, or forget to switch from division to multiplication.

Exercises 13–17

- Students may not change the mixed number into an improper fraction.

Exercises 18–21

- Students may forget to insert zeros as place holders. Remind them that each number added or subtracted needs to have the same number of decimal places.

Exercises 22–24

- Students may count the decimal places to the left of the decimal point instead of the right.
- Students may ignore the zeros when counting the number of decimal places, or forget to add them when finding the answer.

Exercises 25–28

- Students may not count the decimal places of both decimals factors.

Exercises 29–31

- Students may divide without putting the decimal point in the quotient.
- Students may have difficulty keeping their numbers organized and will forget to include placeholder zeros or put in extra placeholders.

Exercises 32–34

- Some students may struggle with determining the multiple of 10 to multiply by to get a whole number divisor.
- Students may try to multiply by the number of decimal places in the dividend instead of the divisor.

2 Chapter Review

Check It Out
Vocabulary Help
BigIdeasMath.com

Review Key Vocabulary

reciprocals, *p. 64*

Review Examples and Exercises

2.1 Multiplying Fractions *(pp. 54–61)*

a. Find $\frac{1}{4} \times \frac{3}{5}$.

$\frac{1}{4} \times \frac{3}{5} = \frac{1 \times 3}{4 \times 5} = \frac{3}{20}$ — Multiply the numerators and the denominators.

b. Find $\frac{3}{5} \times 1\frac{1}{8}$.

$\frac{3}{5} \times 1\frac{1}{8} = \frac{3}{5} \times \frac{9}{8}$ — Write $1\frac{1}{8}$ as the improper fraction $\frac{9}{8}$.

$= \frac{3 \times 9}{5 \times 8} = \frac{27}{40}$ — Multiply the numerators and the denominators.

Exercises

Multiply. Write the answer in simplest form.

1. $\frac{1}{8} \times \frac{5}{7}$
2. $\frac{3}{5} \times \frac{1}{2}$
3. $\frac{2}{9} \times \frac{3}{4}$
4. $\frac{3}{10} \times \frac{4}{5}$
5. $2\frac{2}{3} \times \frac{4}{5}$
6. $\frac{2}{7} \times 4\frac{4}{9}$
7. $1\frac{5}{6} \times 2\frac{3}{8}$
8. $2\frac{3}{10} \times 5\frac{1}{3}$

2.2 Dividing Fractions *(pp. 62–69)*

Find $\frac{3}{7} \div \frac{5}{8}$.

$\frac{3}{7} \div \frac{5}{8} = \frac{3}{7} \times \frac{8}{5}$ — Multiply by the reciprocal of $\frac{5}{8}$, which is $\frac{8}{5}$.

$= \frac{3 \times 8}{7 \times 5} = \frac{24}{35}$ — Multiply fractions and simplify.

Exercises

Divide. Write the answer in simplest form.

9. $\frac{1}{9} \div \frac{2}{5}$
10. $\frac{3}{4} \div \frac{5}{6}$
11. $5 \div \frac{1}{3}$
12. $\frac{8}{9} \div \frac{3}{10}$

2.3 Dividing Mixed Numbers (pp. 70–75)

Find $3\frac{3}{4} \div 1\frac{1}{2}$.

$3\frac{3}{4} \div 1\frac{1}{2} = \frac{15}{4} \div \frac{3}{2}$ — Write each mixed number as an improper fraction.

$= \frac{15}{4} \times \frac{2}{3}$ — Multiply by the reciprocal of $\frac{3}{2}$, which is $\frac{2}{3}$.

$= \frac{\overset{5}{\cancel{15}} \times \overset{1}{\cancel{2}}}{\underset{2}{\cancel{4}} \times \underset{1}{\cancel{3}}}$ — Multiply fractions. Divide out common factors.

$= \frac{5}{2}$, or $2\frac{1}{2}$ — Simplify.

Exercises

Divide. Write the answer in simplest form.

13. $1\frac{2}{5} \div \frac{4}{7}$ **14.** $2\frac{3}{8} \div \frac{3}{5}$ **15.** $4\frac{1}{8} \div 2\frac{1}{4}$ **16.** $5\frac{5}{8} \div 1\frac{2}{9}$

17. **PANCAKES** A box contains 10 cups of pancake mix. You use $\frac{2}{3}$ cup each time you make pancakes. How many times can you make pancakes?

2.4 Adding and Subtracting Decimals (pp. 78–83)

a. Add 7.36 + 2.22.

$$\begin{array}{r} 7.36 \\ +\ 2.22 \\ \hline 9.58 \end{array}$$

Line up the decimal points.

Add as you would with whole numbers.

b. Subtract 5.467 − 2.736.

$$\begin{array}{r} \overset{4}{\cancel{5}}.\overset{14}{\cancel{4}}67 \\ -\ 2.736 \\ \hline 2.731 \end{array}$$

Line up the decimal points.

Subtract as you would with whole numbers.

Exercises

Add or subtract.

18. 3.78 + 8.94 **19.** 19.89 + 4.372

20. 7.638 − 2.365 **21.** 14.21 − 4.103

Review Game

Multiplying and Dividing Decimals

Materials per Group

- 2 number cubes
- 2 coins (or integer counters)
- paper
- pencil

Players: groups of 4

Directions

- Each group of four divides into teams of two. Each team competes with the other team.
- Each team tosses a coin. One coin determines whether multiplication (heads) or division (tails) will be used by the group. The other coin determines whether estimation (heads) or exact calculation (tails) will be used by the group.
- Each team rolls a number cube twice to determine the numbers to be used. For example, if one team rolls a 5 and then a 3, the first number is 5.3. If the other team rolls a 4 and then a 2, the other number is 4.2.

Now the race is on!

Using the determined numbers and operations, each team estimates or calculates the answer. The team that finishes first has their work checked by the challenging team. If the challenging team accepts the answer, the team that finishes first wins the round. If the challenging team finds an error, the challenging team wins the round.

If the operation is carried out with estimation, the winning team receives one point. If the operation is carried out with exact calculation, the winning team receives two points.

Who Wins?

After a set amount of time, the team with the most points wins.

For the Student Additional Practice

- Lesson Tutorials
- Multi-Language Glossary
- Self-Grading Progress Check
- *BigIdeasMath.com*
  Dynamic Student Edition
  Student Resources

Answers

13. $2\frac{9}{20}$

14. $3\frac{23}{24}$

15. $1\frac{5}{6}$

16. $4\frac{53}{88}$

17. 15

18. 12.72

19. 24.262

20. 5.273

21. 10.107

22. 42.4

23. 42.7

24. 14.04

25. 76.08425

26. 1.1934

27. 0.012444

28. 143.52 in.^{2}

29. 1.7

30. 6.2

31. 7.1

32. 30

33. 0.05

34. 5.7

My Thoughts on the Chapter

Teacher Tip

Not allowed to write in your teaching edition? Use sticky notes to record your thoughts.

What worked. . .

What did not work. . .

What I would do differently. . .

2.5 Multiplying Decimals (pp. 84–91)

Find 7.5 × 5.3.

$$\begin{array}{r} 7.5 \\ \times\ 5.3 \\ \hline 2\,2\,5 \\ +\ 3\,7\,5\ \ \\ \hline 3\,9.7\,5 \end{array}$$

7.5 ← 1 decimal place
× 5.3 ← + 1 decimal place
39.75 ← 2 decimal places

So, 7.5 × 5.3 = 39.75.

Exercises

Multiply. Use estimation to check your answer.

22. 5.3 × 8 **23.** 6.1 × 7 **24.** 4.68 × 3

25. 9.475 × 8.03 **26.** 0.27 × 4.42 **27.** 0.051 × 0.244

28. **AREA** Find the area of the computer screen.

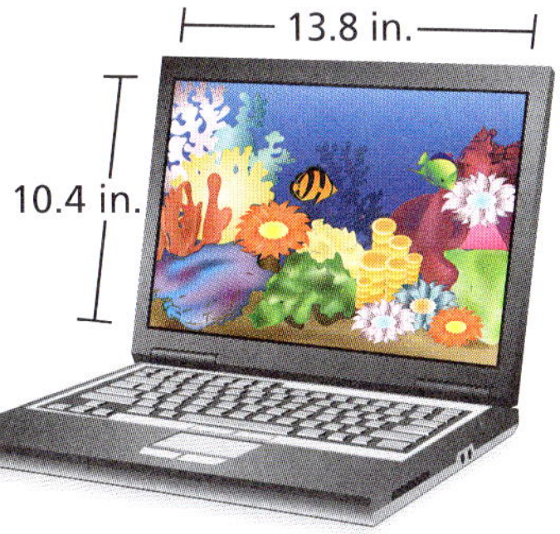

2.6 Dividing Decimals (pp. 92–99)

Find 22.8 ÷ 1.2.

Multiply 1.2 by 10.

$1.2\overline{)22.8}$ → $12\overline{)228.}$

Multiply 22.8 by 10.

Place the decimal point above the decimal point in the dividend 228.

$$\begin{array}{r} 19. \\ 12\overline{)228.} \\ -\ 12\ \ \\ \hline 108\ \\ -\ 108\ \\ \hline 0\ \end{array}$$

So, 22.8 ÷ 1.2 = 19.

Exercises

Divide. Use estimation to check your answer.

29. 6.8 ÷ 4 **30.** 13.2 ÷ 6 + 4 **31.** 49.7 ÷ 7

32. $0.12\overline{)3.6}$ **33.** $2.5\overline{)0.125}$ **34.** $3.9\overline{)22.23}$

2 Chapter Test

Check It Out
Test Practice
BigIdeasMath.com

Multiply. Write the answer in simplest form.

1. $\frac{9}{16} \times \frac{2}{3}$
2. $\frac{1}{10} \times \frac{5}{6}$
3. $1\frac{3}{7} \times 6\frac{7}{10}$

Divide. Write the answer in simplest form.

4. $\frac{1}{6} \div \frac{1}{3}$
5. $10 \div \frac{2}{5}$
6. $8\frac{3}{4} \div 2\frac{7}{8}$

Add or subtract.

7. $4.92 + 3.79$
8. $5.138 + 2.624$
9. $5.316 - 1.942$

Multiply. Use estimation to check your answer.

10. 6.7×8
11. 0.4×0.7
12. 4.87×7.23

Divide. Use estimation to check your answer.

13. $5.6 \div 7$
14. $2.6 \div 0.02$
15. $4\overline{)9.32}$
16. $0.25\overline{)5.46}$

17. **DVD SALE** Which deal is the better buy?

18. **BLOG** You spend $2\frac{1}{2}$ hours online. You spend $\frac{1}{5}$ of that time writing a blog. How long do you spend writing your blog?

19. **GRAPES** A grocery store sells grapes for \$1.99 per pound. You buy 2.34 pounds of the grapes. How much do you pay?

20. **PHOTOGRAPHY** A motocross rider is in the air for 2.5 seconds. Your camera can take a picture every 0.125 second. Your friend's camera can take a picture every 0.15 second.

   a. How many times faster is your camera than your friend's camera?

   b. How many more pictures can you take while the rider is in the air?

Test Item References

Chapter Test Questions	Section to Review	Common Core State Standards
1–3, 18	2.1	6.NS.1
4, 5	2.2	6.NS.1
6	2.3	6.NS.1
7–9	2.4	6.NS.3
10–12, 19, 20	2.5	6.NS.3
13–17, 20	2.6	6.NS.3

Test-Taking Strategies

Remind students to quickly look over the entire test before they start so that they can budget their time. When students hurry they may make unintentional mistakes, such as writing a mixed number as an improper fraction incorrectly or inverting the wrong fraction in a division problem. Remind students to **Stop** and **Think** before they write their answers.

Common Errors

- **Exercises 1–3** Students may try to get a common denominator when multiplying fractions.
- **Exercises 4–6** Students might find the reciprocal of the dividend instead of the divisor.
- **Exercises 10–12** Students may only count the number of decimal places in one number instead of two. Remind them that they need to count *both* sets of decimal places. Stress estimation as a check for reasonableness.
- **Exercises 13–16** Students may place the decimal point in the wrong place. Encourage them to use grid paper to keep their work organized. Remind students to check their answers by multiplying.
- **Exercises 14 and 16** Students may forget to multiply the dividend by the same multiple of 10 that they used to multiply the divisor. Remind them that they must multiply *both* numbers by the same multiple of 10.

Reteaching and Enrichment Strategies

If students need help. . .	If students got it. . .
Resources by Chapter • Practice A and Practice B • Puzzle Time Record and Practice Journal Practice Differentiating the Lesson Lesson Tutorials *BigIdeasMath.com* Skills Review Handbook	Resources by Chapter • Enrichment and Extension • Technology Connection Game Closet at *BigIdeasMath.com* Start Standards Assessment

Answers

1. $\frac{3}{8}$
2. $\frac{1}{12}$
3. $9\frac{4}{7}$
4. $\frac{1}{2}$
5. 25
6. $3\frac{1}{23}$
7. 8.71
8. 7.762
9. 3.374
10. 53.6
11. 0.28
12. 35.2101
13. 0.8
14. 130
15. 2.33
16. 21.84
17. 5 for $58.99
18. $\frac{1}{2}$ h
19. $4.66
20. **a.** 1.2 times faster
    b. 4 more pictures

Technology for the Teacher

Online Assessment
Assessment Book
ExamView® Assessment Suite

Test-Taking Strategies

Available at *BigIdeasMath.com*

After Answering Easy Questions, Relax
Answer Easy Questions First
Estimate the Answer
Read All Choices before Answering
Read Question before Answering
Solve Directly or Eliminate Choices
Solve Problem before Looking at Choices
Use Intelligent Guessing
Work Backwards

About this Strategy

When taking a multiple choice test, be sure to read each question carefully and thoroughly. After reading the question, estimate the answer before trying to solve it.

Answers

1. B
2. 9.393
3. G
4. D
5. \$8.06

Technology for the Teacher

Common Core State Standards Support
Performance Tasks
Online Assessment
Assessment Book
ExamView® Assessment Suite

Item Analysis

1. **A.** The student ignores the $\frac{1}{2}$ and finds the quotient of 2 and 10.

   B. Correct answer

   C. The student recognizes the quotient of $2\frac{1}{2}$ and 10 as 0.25, and then uses the digits of the decimal part of the number as the numerator and denominator of the fraction.

   D. The student finds the quotient of $2\frac{1}{2}$ and 10, and then subtracts this quotient from 1.

2. **Gridded response:** Correct answer: 9.393

   Common errors: The student places the decimal point incorrectly. The student does not perform the division correctly. The student does not follow the correct order of operations.

3. **F.** The student does not follow the correct order of operations and multiplies 3×6 instead of dividing 6 by 2.

   G. Correct answer

   H. The student does not follow the correct order of operations and calculates $(2 \cdot 4)^2$ instead of $2 \cdot 4^2$.

   I. The student does not follow the correct order of operations and adds the 3 before division and multiplication are completed, i.e., $(2 \cdot 4^2 + 3)(6 \div 2)$.

4. **A.** The student moves the decimal point in the dividend two places, but in the wrong direction, and does not move the decimal point in the divisor.

   B. The student moves the decimal point in the dividend correctly but does not move the decimal point in the divisor.

   C. The student moves the decimal point in the divisor correctly but moves the decimal point two places in the wrong direction.

   D. Correct answer

5. **Gridded response:** Correct answer: \$8.06

   Common error: The student places the decimal point incorrectly. The student does not perform the multiplication correctly.

2 Standards Assessment

1. At a party, 10 people equally shared $2\frac{1}{2}$ gallons of ice cream. How much ice cream did each person eat? *(6.NS.1)*

   A. $\frac{1}{5}$ gal **C.** $\frac{2}{5}$ gal

   B. $\frac{1}{4}$ gal **D.** $\frac{3}{4}$ gal

Test-Taking Strategy

Estimate the Answer

"Using **estimation** you can see that the answer is about 3. So, you should choose B."

2. What is the value of the expression below? *(6.NS.3)*

$$4.643 + 11.02 \div 2.32$$

3. Which number is equivalent to the expression below? *(6.EE.1)*

$$2 \cdot 4^2 + 3(6 \div 2)$$

   F. 25 **H.** 73

   G. 41 **I.** 105

4. Your friend divided two decimal numbers. Her work is shown in the box below. What should your friend change in order to divide the two decimal numbers correctly? *(6.NS.3)*

$$0.07\overline{)14.56} \rightarrow 7\overline{)14.56}\quad \text{quotient } 2.08$$

   A. Rewrite the problem as $0.07\overline{)0.1456}$. **C.** Rewrite the problem as $7\overline{)0.1456}$.

   B. Rewrite the problem as $0.07\overline{)1456}$. **D.** Rewrite the problem as $7\overline{)1456}$.

5. You bought some grapes at a farm stand. You paid \$2.48 per pound.

   What was the total amount that you paid for the grapes? *(6.NS.3)*

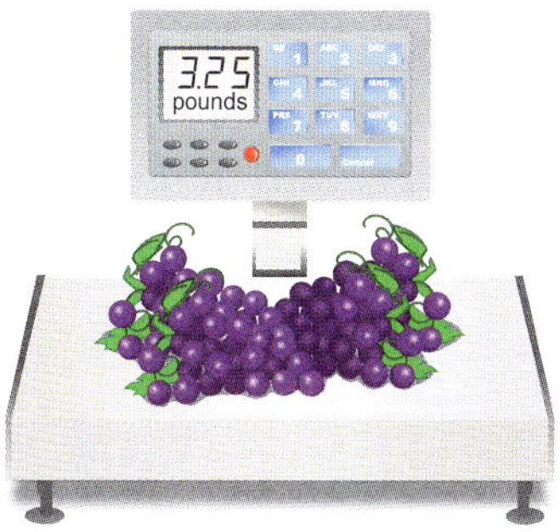

6. The steps your friend took to divide two mixed numbers are shown below.

$$4\frac{2}{3} \div 2\frac{1}{4} = \frac{14}{3} \times \frac{9}{4}$$
$$= \frac{21}{2}$$
$$= 10\frac{1}{2}$$

What should your friend change in order to divide the two mixed numbers correctly? *(6.NS.1)*

F. Find a common denominator of 3 and 4.

G. Multiply by the reciprocal of $\frac{14}{3}$.

H. Multiply by the reciprocal of $\frac{9}{4}$.

I. Rename $4\frac{2}{3}$ as $3\frac{5}{3}$.

7. Which pair of numbers does *not* have a least common multiple less than 100? *(6.NS.4)*

A. 10, 15

B. 12, 16

C. 16, 18

D. 18, 24

8. You are making identical snack bags. You have 18 fruit-chew snacks and 24 granola snacks. What is the greatest number of snack bags that you can make with no snacks left over? *(6.NS.4)*

F. 1

G. 2

H. 3

I. 6

9. Which expression is *not* equivalent to $\frac{2}{3}$? *(6.NS.1)*

A. $\frac{1}{4} + \frac{1}{3} \div \frac{4}{5}$

B. $\frac{13}{30} + \frac{1}{5} \div \frac{6}{7}$

C. $\frac{5}{6} - \frac{1}{8} \div \frac{1}{2}$

D. $\frac{13}{18} - \frac{1}{26} \div \frac{9}{13}$

Item Analysis (continued)

6. **F.** The student confuses the intended division algorithm with an addition/subtraction algorithm.

   G. The student finds the reciprocal of the dividend rather than the reciprocal of the divisor.

   H. Correct answer

   I. The student renames the first number of the expression as if the student were going to follow a subtraction algorithm.

7. **A.** The student does not find the LCM correctly.

   B. The student does not find the LCM correctly.

   C. Correct answer

   D. The student does not find the LCM correctly.

8. **F.** The student finds the least common factor of 18 and 24.

   G. The student finds a common factor of 18 and 24, but not the greatest common factor.

   H. The student finds a common factor of 18 and 24, but not the greatest common factor.

   I. Correct answer

9. **A.** The student does not follow the correct order of operations, adds incorrectly, or divides incorrectly.

   B. The student does not follow the correct order of operations, adds incorrectly, or divides incorrectly.

   C. Correct answer

   D. The student does not follow the correct order of operations, subtracts incorrectly, or divides incorrectly.

Answers

6. H
7. C
8. I
9. C

Answers

10. F

11. B

12. G

13. D

14. *Part A* $112\frac{3}{4}$ inches

Part B $5\frac{5}{9}$ times as long

Item Analysis (continued)

10. **F.** Correct answer

G. The student does not subtract correctly in the tenths place.

H. The student subtracts the 3 from the 4 in the hundredths place, instead of subtracting 4 from 13, and does not borrow from the ones place.

I. The student does not subtract correctly in the tenths place and ones place.

11. **A.** The student incorrectly calculates the expression by multiplying the numerators and multiplying the denominators.

B. Correct answer

C. The student incorrectly calculates the expression by inverting the dividend instead of the divisor.

D. The student incorrectly calculates the expression by inverting both the divisor and the dividend.

12. **F.** The student does not use the order of operations correctly.

G. Correct answer

H. The student does not use the order of operations correctly.

I. The student does not use the order of operations correctly.

13. **A.** The student does not find the greatest common factor.

B. The student finds the product of the two common prime factors of the three numbers.

C. The student does not find the greatest common factor.

D. Correct answer

14. **4 points** The student demonstrates a thorough understanding of multiplying and dividing mixed numbers and whole numbers. The student calculates the length and width of the walkway correctly and gets answers of $112\frac{3}{4}$ inches for Part A and $5\frac{5}{9}$ times as long for Part B. The student shows accurate, complete work for both parts and provides clear and complete explanations.

3 points The student demonstrates an understanding of multiplying and dividing mixed numbers and whole numbers, but the student's work and explanations demonstrate an essential but less than thorough understanding.

2 points The student demonstrates a partial understanding of multiplying and dividing mixed numbers and whole numbers. The student's work and explanations demonstrate a lack of essential understanding.

1 point The student demonstrates very limited understanding of multiplying and dividing mixed numbers and whole numbers. The student's response is incomplete and exhibits many flaws.

0 points The student provided no response, a completely incorrect or incomprehensible response, or a response that demonstrates insufficient understanding of multiplying and dividing mixed numbers and whole numbers.

10. Which number is equivalent to $5.139 - 2.64$? *(6.NS.3)*

F. 2.499

G. 2.599

H. 3.519

I. 3.599

11. Which expression is equivalent to $\frac{4}{9} \div \frac{5}{7}$? *(6.NS.1)*

A. $\frac{20}{63}$

B. $\frac{28}{45}$

C. $\frac{45}{28}$

D. $\frac{63}{20}$

12. Which of the following expressions is equivalent to a perfect square? *(6.EE.1)*

F. $3 + 2^2 \times 7$

G. $34 + 18 \div 3^2$

H. $(80 + 4) \div 4$

I. $3^2 + 6 \times 5 \div 3$

13. You are filling baskets using 18 green eggs, 36 red eggs, and 54 blue eggs. What is the greatest number of baskets that you can fill so that the baskets are identical and there are no eggs left over? *(6.NS.4)*

A. 3

B. 6

C. 9

D. 18

14. A walkway was built using identical concrete blocks. *(6.NS.1)*

Think
Solve
Explain

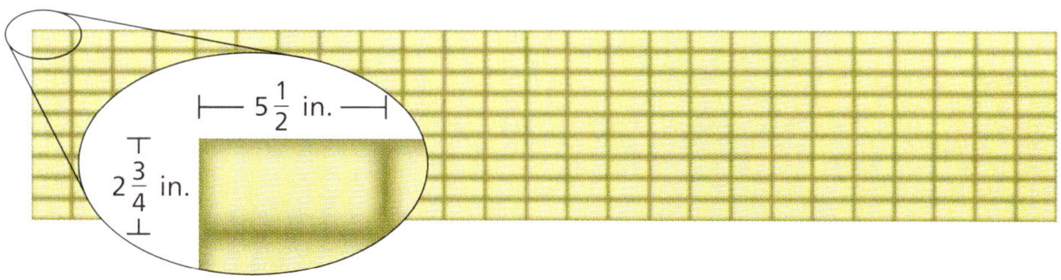

Part A How much longer, in inches, is the length of the walkway than the width of the walkway? Show your work and explain your reasoning.

Part B How many times longer is the length of the walkway than the width of the walkway? Show your work and explain your reasoning.

3 Algebraic Expressions and Properties

3.1 Algebraic Expressions
3.2 Writing Expressions
3.3 Properties of Addition and Multiplication
3.4 The Distributive Property

"Did you know that $5 \times 6 = 6 \times 5$, but $5 \div 6 \neq 6 \div 5$?"

"Only certain operations like addition and multiplication preserve equality when you switch the numbers around."

"Descartes, evaluate this expression when $x = 2$ to determine the number of cat treats you are going to eat today."

"Remember that you evaluate an algebraic expression by substituting the value of x into the expression."

Common Core Progression

4th Grade
• Add, subtract, multiply, and divide with whole numbers. • Find factors and multiples from 1–100.
5th Grade
• Write and interpret numerical expressions. • Use parentheses, brackets, or braces in numerical expressions.
6th Grade
• Write and evaluate algebraic expressions. • Find the GCF in algebraic expressions. • Apply the Commutative, Associative, and Distributive Properties to show expressions are equivalent.

Chapter Summary

Section	Common Core State Standard	
3.1	Learning	6.EE.2c
3.2	Learning	6.EE.2a ★
3.3	Learning	6.EE.3, 6.EE.4
3.4	Learning	6.NS.4 ★, 6.EE.3 ★, 6.EE.4 ★

★ Teaching is complete. Standard can be assessed.

Pacing Guide for Chapter 3

Chapter Opener	1 Day
Section 1	
Activity	1 Day
Lesson	1 Day
Section 2	
Activity	1 Day
Lesson	1 Day
Study Help / Quiz	1 Day
Section 3	
Activity	1 Day
Lesson	1 Day
Section 4	
Activity	1 Day
Lesson	1 Day
Extension	1 Day
Chapter Review/ Chapter Tests	2 Days
Total Chapter 3	13 Days
Year-to-Date	51 Days

Technology for the Teacher

BigIdeasMath.com
Chapter at a Glance
Complete Materials List
Parent Letters: English and Spanish

Common Core State Standards

5.OA.1 Use parentheses, brackets, or braces in numerical expressions, and evaluate expressions with these symbols.
5.OA.2 Write simple expressions that record calculations with numbers, and interpret numerical expressions without evaluating them.
6.EE.1 Write and evaluate numerical expressions involving whole-number exponents.

Additional Topics for Review

- Greatest Common Factor
- Multiples of Whole Numbers
- Factors of Whole Numbers

Try It Yourself

1. $3 \times (372 + 20{,}967)$ is 3 times as large as $372 + 20{,}967$.
2. $2 \times (432 + 346{,}322)$ is 2 times as large as $432 + 346{,}322$.
3. $4 \times (6722 + 4086)$ is 4 times as large as $6722 + 4086$.
4. $(115 + 36{,}372) + 6$ is 6 more than $(115 + 36{,}372)$.
5. $(392 + 75{,}325) + 78$ is 78 more than $392 + 75{,}325$.
6. $(352 + 46{,}795) + 100$ is 100 more than $352 + 46{,}795$.
7. $(30{,}929 + 425) \div 2$ is half as large as $30{,}929 + 425$.
8. $(58{,}742 - 721) \div 2$ is half as large as $58{,}742 - 721$.
9. $(96{,}792 + 564) \div 3$ is one-third as large as $96{,}792 + 564$.
10. 19
11. 5
12. 6
13. 38
14. 16
15. 9

Record and Practice Journal Fair Game Review

1–16. See Additional Answers.

Math Background Notes

Vocabulary Review

- Numerical Expression
- Order of Operations
- Exponent

Interpreting Numerical Expressions

- Students may use different words when interpreting the numerical expression.

? "What is another way to interpret Example 2?" (316 + 43,449) is increased by 5.

- Have students create a table with the headings *Addition, Subtraction, Multiplication,* and *Division.* As they work through the exercises, they can list key words and phrases that imply the different math operations.

Using Order of Operations

- Students should know the order of operations.
- You may want to review the correct order of operations with students. Many students probably learned the pneumonic device *Please Excuse My Dear Aunt Sally.* Ask a volunteer to explain why this phrase is helpful.
- You may choose to review exponents with students. Remind students that the exponent tells you how many times the base acts as a factor.
- **Common Error:** Students may misinterpret the exponent in Example 4. Watch for students who incorrectly compute $4^2 = 4 \cdot 2 = 8$.

Reteaching and Enrichment Strategies

If students need help...	If students got it...
Record and Practice Journal • Fair Game Review Skills Review Handbook Lesson Tutorials	Game Closet at *BigIdeasMath.com* Start the next section

What You Learned Before

"Great! You're up to $x = 2$. Let's keep going."

Interpreting Numerical Expressions (5.OA.2)

Example 1 Write a sentence interpreting the expression $3 \times (19{,}762 + 418)$.

$3 \times (19{,}762 + 418)$ is 3 times as large as $19{,}762 + 418$.

Example 2 Write a sentence interpreting the expression $(316 + 43{,}449) + 5$.

$(316 + 43{,}449) + 5$ is 5 more than $316 + 43{,}449$.

Example 3 Write a sentence interpreting the expression $(20{,}008 - 752) \div 2$.

$(20{,}008 - 752) \div 2$ is half as large as $20{,}008 - 752$.

Try It Yourself

Write a sentence interpreting the expression.

1. $3 \times (372 + 20{,}967)$ **2.** $2 \times (432 + 346{,}322)$ **3.** $4 \times (6722 + 4086)$

4. $(115 + 36{,}372) + 6$ **5.** $(392 + 75{,}325) + 78$ **6.** $(352 + 46{,}795) + 100$

7. $(30{,}929 + 425) \div 2$ **8.** $(58{,}742 - 721) \div 2$ **9.** $(96{,}792 + 564) \div 3$

Using Order of Operations (5.OA.1, 6.EE.1)

Example 4 Simplify $4^2 \div 2 + 3(9 - 5)$.

First:	Parentheses	$4^2 \div 2 + 3(9 - 5) = 4^2 \div 2 + 3 \cdot 4$
Second:	Exponents	$= 16 \div 2 + 3 \cdot 4$
Third:	Multiplication and Division (from left to right)	$= 8 + 12$
Fourth:	Addition and Subtraction (from left to right)	$= 20$

Try It Yourself

Simplify the expression.

10. $3^2 + 5(4 - 2)$ **11.** $3 + 4 \div 2$ **12.** $10 \div 5 \cdot 3$

13. $4(3^3 - 8) \div 2$ **14.** $3 \cdot 6 - 4 \div 2$ **15.** $12 + 7 \cdot 3 - 24$

3.1 Algebraic Expressions

Essential Question How can you write and evaluate an expression that represents a real-life problem?

1 ACTIVITY: Reading and Re-Reading

Work with a partner.

a. You babysit for 3 hours. You receive \$12. What is your hourly wage?

- Write the problem. Underline the important numbers and units you need to solve the problem.
- Read the problem carefully a second time. Circle the key word for the question.

- Write each important number or word, with its units, on a piece of paper. Write $+$, $-$, $\times$, $\div$, and $=$ on five other pieces of paper.

- Arrange the pieces of paper to answer the key word question, "What is your hourly wage?"
- Evaluate the expression that represents the hourly wage.

$$\text{hourly wage} = \square \div \square \qquad \text{Write.}$$

$$= \square \qquad \text{Evaluate.}$$

So, your hourly wage is \$ ☐ per hour.

b. How can you use your hourly wage to find how much you will receive for any number of hours worked?

Common Core

Algebraic Expressions

In this lesson, you will

- use order of operations to evaluate algebraic expressions.
- solve real-life problems.

Learning Standard
6.EE.2c

Laurie's Notes

Introduction

Standards for Mathematical Practice

- **MP6 Attend to Precision:** Students need the opportunity to communicate using mathematical language, and in this activity, their language is something you will want to pay attention to. Listen to how students refer to the quantities in the activities, particularly their reference to units.

Motivate and Discuss

- **Background:** Many students struggle with solving word problems, but not because the mathematics is too difficult. They struggle because they do not know how to approach word problems. In this first lesson, you want to help students hone their reading skills in order to become better problem solvers.

? Ask students to name jobs they have had where they earned money. Students may name babysitting, delivering newspapers, doing odd jobs for the neighbors, grocery shopping for an elderly person, and walking dogs. Discuss the difference between paid by the hour and paid by the job performed. What are the advantages and disadvantages of each method of payment?

Activity Notes

Activity 1

- Have small pieces of paper ready for students to use in Example 1. Each pair of students will need eight pieces of paper. You may want to have eight transparency pieces ready for the overhead projector, or a prepared file for an interactive white board.
- Manipulating the pieces allows students to make sense of writing a mathematical model. They are writing a sentence as they do in English class.
- **Common Question:** Students may ask, "Why go through all this work when we already knew the answer?" The focus needs to be on modeling a process: read and re-read, underline important numbers *and* units, underline what you are trying to solve for, and then write an expression.
- **Another Way:** You can also set up the problem as: \$12 = hourly wage (\$/hour) × 3 hours. Although this is a true statement, it is helpful to have all of the computations on one side of the equation. Let students know that they will learn how to solve equations similar to this at a later time, when the answer may not be so obvious.

Common Core State Standards

6.EE.2c Evaluate expressions at specific values of their variables. Include expressions that arise from formulas used in real-world problems. Perform arithmetic operations, including those involving whole number exponents, in the conventional order when there are no parentheses to specify a particular order (Order of Operations).

Previous Learning

Students need to be familiar with variables and how they are used.

Technology for the Teacher

Dynamic Classroom

Lesson Plans
Complete Materials List

3.1 Record and Practice Journal

Essential Question How can you write and evaluate an expression that represents a real-life problem?

1 ACTIVITY: Reading and Re-Reading

a. You babysit for 3 hours. You receive \$12.
What is your hourly wage?

- Underline the important numbers and units you need to solve the problem.
- Read the problem carefully a second time. Circle the key word for the question.
- Write each important number or word, with its units, on a piece of paper. Write +, −, ×, ÷, and = on five other pieces of paper.
- Arrange the pieces of paper to answer the key word question, "What is your hourly wage?"
- Evaluate the expression that represents the hourly wage.

hourly wage = 12 ÷ 3
= 4

Your hourly wage is **\$4 per hour**.

b. How can you use your hourly wage to find how much you will receive for any number of hours worked?

Sample answer: **Multiply the hourly wage by the number of hours worked.**

Differentiated Instruction

Auditory/Visual

Stress the importance of reading a real-life problem and determining what question is being asked. Have students use colored pencils or markers to circle or underline key words, numbers, and units that are needed to solve the problem. Ask why it is important to identify these words. Show students how to check their work when they write an expression for a real-life problem.

3.1 Record and Practice Journal

2 ACTIVITY: Reading and Re-Reading

Work with a partner. Use the strategy shown in Activity 1 to write an expression for each problem. After you have written the expression, evaluate it using mental math or some other method.

a. You wash cars for 2 hours. You receive $6. How much do you earn per hour?

Expression: $6 \div 2$

Amount you earn per hour: $3

b. You have $60. You buy a pair of jeans and a shirt. The pair of jeans costs $27. You come home with $15. How much did you spend on the shirt?

Expression: $60 - 27 - 15$

Amount you spend on shirt: $18

c. For lunch, you buy 5 sandwiches that cost $3 each. How much do you spend?

Expression: 5×3

Amount you spend on sandwiches: $15

d. You are running a 4500-foot race. How much farther do you have to go after running 2000 feet?

Expression: $4500 - 2000$

Amount left to go: 2500 ft

e. A young rattlesnake grows at a rate of about 20 centimeters per year. How much does a young rattlesnake grow in 2 years?

Expression: 2×20

Amount rattlesnake grows in 2 years: 40 cm

What Is Your Answer?

3. **IN YOUR OWN WORDS** How can you write and evaluate an expression that represents a real-life problem? Give one example with addition, one with subtraction, one with multiplication, and one with division.

Write the problem.
Underline key words.
Write an expression.
Evaluate.

Laurie's Notes

Activity 2

- Have additional slips of paper available for students to use to write the numbers and units in the activity. You can reuse the operation symbols from Activity 1 but will need about 20 additional slips of paper.
- Remind students that the focus is on process and recording the operations needed to solve each problem. In writing the problem, each number should be labeled with units. This habit will pay dividends throughout the year.
- **Extension:** Ask students to think about more than one way to solve the problem. For example, part (a) could be thought of two ways.

  hours worked $\times$ earnings per hour $=$ earnings

  amount earned $\div$ hours worked $=$ earnings per hour
- **MP6:** Listen for accurate language when they explain their method(s).
- Ask for volunteers to share their equations and their solutions.

What Is Your Answer?

- Have students share their examples with the class.

Closure

- Evaluate the expression: 4 egg yolks per serving $\times$ 12 servings. What question does it answer? Be sure to include units.
  How many egg yolks are needed? 48 egg yolks

2 ACTIVITY: Reading and Re-Reading

Math Practice 2

Make Sense of Quantities

What are the units in the problem? How does this help you write an expression?

Work with a partner. Use the strategy shown in Activity 1 to write an expression for each problem. After you have written the expression, evaluate it using mental math or some other method.

a. You wash cars for 2 hours. You receive \$6. How much do you earn per hour?

b. You have \$60. You buy a pair of jeans and a shirt. The pair of jeans costs \$27. You come home with \$15. How much did you spend on the shirt?

c. For lunch, you buy 5 sandwiches that cost \$3 each. How much do you spend?

d. You are running a 4500-foot race. How much farther do you have to go after running 2000 feet?

e. A young rattlesnake grows at a rate of about 20 centimeters per year. How much does a young rattlesnake grow in 2 years?

What Is Your Answer?

3. **IN YOUR OWN WORDS** How can you write and evaluate an expression that represents a real-life problem? Give one example with addition, one with subtraction, one with multiplication, and one with division.

Practice

Use what you learned about evaluating expressions to complete Exercises 4–7 on page 115.

3.1 Lesson

Check It Out
Lesson Tutorials
BigIdeasMath.com

Key Vocabulary
algebraic expression, *p. 112*
terms, *p. 112*
variable, *p. 112*
coefficient, *p. 112*
constant, *p. 112*

An **algebraic expression** is an expression that may contain numbers, operations, and one or more symbols. Parts of an algebraic expression are called **terms**.

A symbol that represents one or more numbers is called a **variable**.

$$5p + 4$$

The numerical factor of a term that contains a variable is a **coefficient**.

A term without a variable is called a **constant**.

EXAMPLE 1 Identifying Parts of an Algebraic Expression

Identify the terms, coefficients, and constants in each expression.

a. $5x + 13$

$5x + 13$

Terms: $5x$, 13

Coefficient: 5

Constant: 13

b. $2z^2 + y + 3$

$2z^2 + y + 3$

Terms: $2z^2$, $1y$, 3

Coefficients: 2, 1

Constant: 3

Study Tip
A variable by itself has a coefficient of 1. So, the term y in Example 1(b) has a coefficient of 1.

On Your Own

Now You're Ready
Exercises 8–13

Identify the terms, coefficients, and constants in the expression.

1. $12 + 10c$ **2.** $15 + 3w + \frac{1}{2}$ **3.** $z^2 + 9z$

EXAMPLE 2 Writing Algebraic Expressions Using Exponents

Write each expression using exponents.

a. $d \cdot d \cdot d \cdot d$

Because d is used as a factor 4 times, its exponent is 4.

So, $d \cdot d \cdot d \cdot d = d^4$.

b. $1.5 \cdot h \cdot h \cdot h$

Because h is used as a factor 3 times, its exponent is 3.

So, $1.5 \cdot h \cdot h \cdot h = 1.5h^3$.

Laurie's Notes

Introduction

Connect

- **Yesterday:** Students explored how to write and evaluate an expression written in words, focusing on units.
- **Today:** The lesson explains how to evaluate an algebraic expression containing a variable.

Motivate

- **Whole Class Activity:** Ask five students to stand at the front of the room. Give each student an index card with a number written on the card (possible numbers: 1, 3, 4, 5, 8). Write simple expressions on the board, drawing a box where the variable would be. Students should walk in front of the expression, place their index card in the box, and announce the value of the expression for their number. Examples:

  $\square + 14 \qquad 27 - \square \qquad \square^2$

- You can vary the expressions and numbers to fit the level of your class. The goal is for students to recognize that the value changes (or varies) for each number substituted.
- ? After several examples, have the students sit down. Erase the boxes and write variables to replace them. "What do the letters mean? Does anyone know what the expressions are called?" The variables have the same meaning as the boxes (an unknown value that represents a number). The expressions are called algebraic expressions.

Lesson Notes

Example 1

- Write $5x + 13$. Discuss the vocabulary: term, variable, coefficient, and constant. Reassure students the vocabulary will become familiar with use.
- ? "How many terms are there in $5x + 13$?" 2 "Name the terms." $5x$, 13
- In working through part (b), ask students to identify the variables first.
- ? "What is the coefficient of z^2?" 2 "What is the coefficient of y?" 1; It is common for students to say there isn't one.
- Explain that $1y = y$. We typically do not write the coefficient 1.

On Your Own

- Have students work together on questions. They should share their answers orally or on the board. Check students' work.

Example 2

- ? Write 3^2 and ask, "How is this read and what does the exponent mean?" 3 squared or 3 raised to the second power; how many times 3 is a factor
- This example connects to Lesson 1.2 on powers and exponents, now applied to variables.
- In part (b), discuss that only the variable h is being raised to a power.

Goal Today's lesson is evaluating **algebraic expressions**.

Technology for the Teacher

Dynamic Classroom

Lesson Tutorials
Lesson Plans
Answer Presentation Tool

Extra Example 1

Identify the terms, coefficients, and constants in each expression.

a. $3a + 17$
Terms: $3a$, 17
Coefficient: 3
Constant: 17

b. $4x^2 + 5x + 7$
Terms: $4x^2$, $5x$, 7
Coefficients: 4, 5
Constant: 7

On Your Own

1. Terms: 12, $10c$; Coefficient: 10; Constant: 12
2. Terms: 15, $3w$, $\frac{1}{2}$; Coefficient: 3; Constants: 15, $\frac{1}{2}$
3. Terms: z^2, $9z$; Coefficients: 1, 9; Constant: none

Extra Example 2

Write each expression using exponents.

a. $x \cdot x \cdot x \cdot x \cdot x$ x^5
b. $7 \cdot d \cdot d \cdot d$ $7d^3$

On Your Own

4. j^6 **5.** $9k^5$

English Language Learners

Vocabulary

The word *variable* is often used in algebra and is represented by a letter. The letter stands for a number that changes, or *varies*. Students find it helpful when the letter is meaningful to the problem. For instance, use the letter *t* to represent a unit of time or use the letter *d* to represent distance.

Extra Example 3

a. Evaluate $15 - y$ when $y = 3$. 12

b. Evaluate $24 \div w$ when $w = 4$. 6

On Your Own

6. 33 **7.** 13

Extra Example 4

Evaluate $f \cdot g$ when $f = 7$ and $g = 11$. 77

On Your Own

8. 3 **9.** 32

10. 16 **11.** 192

Laurie's Notes

On Your Own

- **Neighbor Check:** Have students work independently and then have their neighbors check their work. Have students discuss any discrepancies.

Discuss

- **MP6 Attend to Precision:** Explain what it means to evaluate an algebraic expression. A value is substituted for the variable, then the order of operations is used to find the value of the expression. Students should not use the phrase "plug in for the variable." Students should communicate with precise language.

Example 3

- Work through the written examples in part (a) and part (b) as a class. Use color-coding to help all students recognize that substitution has occurred.
- There is only one operation, therefore the order of operations does not come into play.
- Discuss the Study Tip. Students should be comfortable seeing multiplication being represented in each of the three forms.

? "Could you use all three forms to represent multiplication of two numbers?" No. If the numbers were 3 and 4, $3 \cdot 4$ and 3(4) are okay but writing 34 is not.

- Note: Now that variables have been introduced, multiplication will not be represented with $\times$ in order to avoid confusion with the variable *x*.

Example 4

- Encourage students to use color-coding to differentiate between the two values to be substituted.

? "How do you divide a whole number by a fraction? Multiply by $\frac{3}{2}$, the reciprocal of $\frac{2}{3}$.

- **MP2 Reason Abstractly and Quantitatively:** Giving students the opportunity to share different strategies to perform a computation helps them to develop computational fluency and the ability to reason quantitatively.

? "What strategies can you use to multiply 16 by $\frac{3}{2}$?" Listen for a variety of strategies such as (a) dividing out the common factor of 2 and then multiplying 8 times 3, (b) multiplying 16 times 3 and dividing by 2, and (c) multiplying 16 by $1\frac{1}{2}$ means take half of 16 (8) and add it to 16.

On Your Own

- Have students work together on the questions. They should share their answers orally or on the board. Check students' work.

On Your Own

Now You're Ready
Exercises 16–21

Write the expression using exponents.

4. $j \cdot j \cdot j \cdot j \cdot j \cdot j$

5. $9 \cdot k \cdot k \cdot k \cdot k \cdot k$

To evaluate an algebraic expression, substitute a number for each variable. Then use the order of operations to find the value of the numerical expression.

EXAMPLE 3 Evaluating Algebraic Expressions

a. Evaluate $k + 10$ when $k = 25$.

$k + 10 = 25 + 10$ — Substitute 25 for k.

$= 35$ — Add 25 and 10.

b. Evaluate $4 \cdot n$ when $n = 12$.

$4 \cdot n = 4 \cdot 12$ — Substitute 12 for n.

$= 48$ — Multiply 4 and 12.

Study Tip

You can write the product of 4 and n in several ways.

$4 \cdot n$

$4n$

$4(n)$

On Your Own

Now You're Ready
Exercises 25–32

6. Evaluate $24 + c$ when $c = 9$.

7. Evaluate $d - 17$ when $d = 30$.

EXAMPLE 4 Evaluating an Expression with Two Variables

Evaluate $a \div b$ when $a = 16$ and $b = \frac{2}{3}$.

$a \div b = 16 \div \frac{2}{3}$ — Substitute 16 for a and $\frac{2}{3}$ for b.

$= 16 \cdot \frac{3}{2}$ — Multiply by the reciprocal of $\frac{2}{3}$, which is $\frac{3}{2}$.

$= 24$ — Multiply.

On Your Own

Now You're Ready
Exercises 33–36

Evaluate the expression when $p = 24$ and $q = 8$.

8. $p \div q$

9. $q + p$

10. $p - q$

11. pq

EXAMPLE 5 Evaluating Expressions with Two Operations

a. Evaluate $3x - 14$ when $x = 5$.

$3x - 14 = 3(5) - 14$ Substitute 5 for x.

$= 15 - 14$ Using order of operations, multiply 3 and 5.

$= 1$ Subtract 14 from 15.

b. Evaluate $z^2 + 8.5$ when $z = 2$.

$z^2 + 8.5 = 2^2 + 8.5$ Substitute 2 for z.

$= 4 + 8.5$ Using order of operations, evaluate 2^2.

$= 12.5$ Add 4 and 8.5.

On Your Own

Now You're Ready
Exercises 43–51

Evaluate the expression when $y = 6$.

12. $5y + 1$ **13.** $30 - 24 \div y$ **14.** $y^2 - 7$ **15.** $1.5 + y^2$

EXAMPLE 6 Real-Life Application

You are saving money to buy a skateboard. You begin with \$45 and you save \$3 each week. The expression $45 + 3w$ gives the amount of money you save after w weeks.

a. How much will you have after 4 weeks, 10 weeks, and 20 weeks?

b. After 20 weeks, can you buy the skateboard? Explain.

Substitute the given number of weeks for w.

a.

Number of Weeks, w	$45 + 3w$	Amount Saved
4	$45 + 3(4)$	$45 + 12 = \$57$
10	$45 + 3(10)$	$45 + 30 = \$75$
20	$45 + 3(20)$	$45 + 60 = \$105$

b. After 20 weeks, you have \$105. So, you cannot buy the \$125 skateboard.

On Your Own

16. WHAT IF? In Example 6, the expression for how much money you have after w weeks is $45 + 4w$. Can you buy the skateboard after 20 weeks? Explain.

Laurie's Notes

Example 5

- Students are asked to evaluate expressions involving more than one operation. They will need to remember the order of operations.
- Write the expression from the example and ask how many operations will be performed.

? "What operation should you do first?" Part a. multiplication Part b. exponent

- This question usually elicits a comment about the order of operations, at which time you can probe for understanding.

On Your Own

- Students should work the questions and then share their answers either orally, on individual white boards, or at the board. Check students' work looking for errors with the order of operations.

Example 6

- Work through the example. Model the strategy from yesterday's class. Money saved (\$) = amount at the beginning (\$) + \$3 per week × # of weeks
- This example is good practice for students to set up a table to help to organize their work.

On Your Own

? "Describe what changes in the example to produce the expression $45 + 4w$." saving \$4 per week instead of \$3

Words of Wisdom

- If students are having difficulty, determine if the problem is a matter of computation, reading comprehension, order of operations, or some other difficulty.

Closure

- **Exit Ticket:** In Question 16, you found that by saving \$4 per week for 20 weeks you would have enough to purchase a skateboard. Will you have enough after 40 weeks to purchase two skateboards? Justify your answer. No. You had \$45 at the beginning, and you were adding 4 dollars a week to it. Unless you had another \$45 to add in, you would not be able to afford 2 skateboards.

Extra Example 5

a. Evaluate $\frac{t}{5} + 6$ when $t = 45$. 15

b. Evaluate $v^2 - 4$ when $v = 9$. 77

On Your Own

12. 31 **13.** 26

14. 29 **15.** 37.5

Extra Example 6

Your uncle gives you \$10 and you save \$2 a week. The expression $10 + 2w$ gives the amount you save after w weeks.

a. How much money will you have saved after 2 weeks, 5 weeks, and 10 weeks? \$14, \$20, \$30

b. After 10 weeks, do you have enough money to buy a video game that costs \$25? Explain. Yes, after 10 weeks you have \$30 and the video game costs \$25.

On Your Own

16. yes; Because $45 + 4(20) = 125$, you will be able to buy the skateboard.

Vocabulary and Concept Check

1. 3(4) + 5 does not belong because it is a numerical expression and the other three are algebraic expressions.
2. First: Perform operations in parentheses. Second: Evaluate terms with exponents. Third: Multiply or divide from left to right. Fourth: Add or subtract from left to right.
3. decrease; When you subtract greater and greater values from 20, you will have less and less left.

Practice and Problem Solving

4. \$4 per hour
5. \$120
6. 13 points
7. \$8
8. Terms: $7h$, 3
Coefficient: 7
Constant: 3
9. Terms: g, 12, $9g$
Coefficients: 1, 9
Constant: 12
10. Terms: $5c^2$, $7d$
Coefficients: 5, 7
Constant: none
11. Terms: $2m^2$, 15, $2p^2$
Coefficients: 2, 2
Constant: 15
12. Terms: 6, n^2, $\frac{1}{2}d$
Coefficients: 1, $\frac{1}{2}$
Constant: 6
13. Terms: $8x$, $\frac{x^2}{3}$
Coefficients: 8, $\frac{1}{3}$
Constant: none

14–15. See Additional Answers.

Assignment Guide and Homework Check

Level	Day 1 Activity Assignment	Day 2 Lesson Assignment	Homework Check
Basic	4–7, 56–60	1–3, 9–13 odd, 14, 15–21 odd, 22, 25–29 odd	11, 15, 17, 25, 29
Average	4–7, 56–60	1–3, 13–15, 16–24 even, 25–31 odd, 37, 43, 45	15, 18, 25, 29, 43
Advanced	4–7, 56–60	1–3, 14, 22, 24, 38–50 even, 53, 55	38, 46, 48, 50, 53

Common Errors

- **Exercises 8–13** Students may not list all of the coefficients of the variable terms. Remind them that variables such as g and d have a coefficient of 1 and that coefficients may also be fractions as in Exercises 12 and 13.

3.1 Record and Practice Journal

Evaluate the expression when $a = 4$, $b = 5$, and $c = 10$.

1. $a + 7$
11
2. $b - 3$
2
3. $9c$
90
4. $25 \div b$
5
5. $a \bullet c$
40
6. $b - a$
1
7. $a + b + c$
19
8. $\frac{c}{b}$
2
9. $4a - 7$
9
10. You need $2b$ cups of flour for making b loaves of bread. You have 8 cups of flour. Do you have enough for 5 loaves of bread? Explain.
no; five loaves of bread requires 10 cups of flour. You only have 8 cups of flour.
11. The expression $9a + 6s$ is the cost for a adults and s students to see a musical performance.
    a. Find the total cost for three adults and five students.
    \$57
    b. Find the total cost for four adults and four students.
    \$60

3.1 Exercises

Check It Out
Help with Homework
BigIdeasMath.com

Vocabulary and Concept Check

1. **WHICH ONE DOESN'T BELONG?** Which expression does *not* belong with the other three? Explain your reasoning.

$2x + 1$	$5w \cdot c$	$3(4) + 5$	$y \div z$

2. **NUMBER SENSE** Which step in the order of operations is first? second? third? fourth?

Add or subtract from left to right.	Multiply or divide from left to right.
Evaluate terms with exponents.	Perform operations in parentheses.

3. **NUMBER SENSE** Will the value of the expression $20 - x$ *increase*, *decrease*, or *stay the same* as x increases? Explain.

Practice and Problem Solving

Write and evaluate an expression for the problem.

4. You receive \$8 for raking leaves for 2 hours. What is your hourly wage?
5. Music lessons cost \$20 per week. How much do 6 weeks of lessons cost?
6. The scores on your first two history tests were 82 and 95. By how many points did you improve on your second test?
7. You buy a hat for \$12 and give the cashier a \$20 bill. How much change do you receive?

Identify the terms, coefficients, and constants in the expression.

1 8. $7h + 3$

9. $g + 12 + 9g$

10. $5c^2 + 7d$

11. $2m^2 + 15 + 2p^2$

12. $6 + n^2 + \frac{1}{2}d$

13. $8x + \frac{x^2}{3}$

✗
Terms: $2, x^2, y$
Coefficient: 2
Constant: none

14. **ERROR ANALYSIS** Describe and correct the error in identifying the terms, coefficients, and constants in the algebraic expression $2x^2y$.

15. **PERIMETER** You can use the expression $2\ell + 2w$ to find the perimeter of a rectangle where ℓ is the length and w is the width.

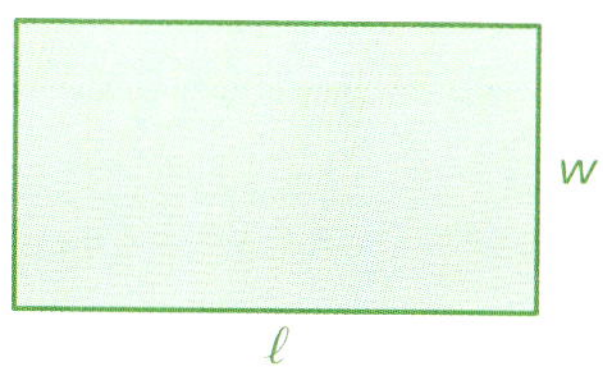

a. Identify the terms, coefficients, and constants in the expression.

b. Interpret the coefficients of the terms.

Write each expression using exponents.

2 **16.** $b \cdot b \cdot b$ **17.** $g \cdot g \cdot g \cdot g \cdot g$ **18.** $8 \cdot w \cdot w \cdot w \cdot w$

19. $5.2 \cdot y \cdot y \cdot y$ **20.** $a \cdot a \cdot c \cdot c$ **21.** $2.1 \cdot x \cdot z \cdot z \cdot z \cdot z$

✗ $3 \cdot n \cdot n \cdot n \cdot n = 4n^3$

22. ERROR ANALYSIS Describe and correct the error in writing the product using exponents.

23. AREA Write an expression using exponents that represents the area of the square.

As I was going to St. Ives
I met a man with seven wives
Each wife had seven sacks
Each sack had seven cats
Each cat had seven kits
Kits, cats, sacks, wives
How many were going to St. Ives?

24. ST. IVES Suppose the man in the St. Ives poem has x wives, each wife has x sacks, each sack has x cats, and each cat has x kits. Write an expression using exponents that represents the total number of kits, cats, sacks, and wives going to St. Ives.

ALGEBRA Evaluate the expression when $a = 3$, $b = 2$, and $c = 12$.

3 **25.** $6 + a$ **26.** $b \cdot 5$ **27.** $c - 1$ **28.** $27 \div a$

29. $12 - b$ **30.** $c + 5$ **31.** $2a$ **32.** $c \div 6$

4 **33.** $a + b$ **34.** $c - a$ **35.** $\frac{c}{a}$ **36.** $b \cdot c$

37. ERROR ANALYSIS Describe and correct the error in evaluating the expression when $m = 8$.

✗
$$5m + 3 = 5 \cdot 8 + 3$$
$$= 5 \cdot 11$$
$$= 55$$

38. LAWNS You earn $15n$ dollars for mowing n lawns. How much do you earn for mowing one lawn? seven lawns?

39. PLANT After m months, the height of a plant is $10 + 3m$ millimeters. How tall is the plant after eight months? three years?

Copy and complete the table.

40.

x	3	6	9
$x \cdot 8$			

41.

x	2	4	8
$64 \div x$			

42. FALLING OBJECT An object falls $16t^2$ feet in t seconds. You drop a rock from a bridge that is 75 feet above the water. Will the rock hit the water in 2 seconds? Explain.

Common Errors

- **Exercises 18–21** Students may raise all the factors to the same power. Discuss with them the difference between $8w^4 = 8 \cdot w \cdot w \cdot w \cdot w$ and $(8w)^4 = 8w \cdot 8w \cdot 8w \cdot 8w$.
- **Exercises 25–41** Students may substitute the wrong value for the variable. Tell students to write out the expression, and then write the value of the variable underneath the variable before substituting the value into the expression.
- **Exercises 43–51** Students may forget about the order of operations or they may substitute the wrong value for the variable. Review the order of operations. Before substituting, have students write or identify what operations are involved in the problem and which will be evaluated first.
- **Exercise 52** Students may hurry through the problem and choose the first expression. Remind them to read the problem carefully.

Practice and Problem Solving

16. b^3 **17.** g^5

18. $8w^4$ **19.** $5.2y^3$

20. a^2c^2 **21.** $2.1xz^4$

22. The coefficient and exponent are reversed.
$3 \cdot n \cdot n \cdot n \cdot n = 3n^4$

23. $25d^2$ **24.** x^4

25. 9 **26.** 10

27. 11 **28.** 9

29. 10 **30.** 17

31. 6 **32.** 2

33. 5 **34.** 9

35. 4 **36.** 24

37. Multiplication should be done first, then addition.

$$\begin{aligned} 5m + 3 &= 5 \cdot 8 + 3 \\ &= 40 + 3 \\ &= 43 \end{aligned}$$

38. \$15; \$105

39. 34 mm; 118 mm

40.

x	3	6	9
$x \cdot 8$	24	48	72

41.

x	2	4	8
$64 \div x$	32	16	8

42. no; In 2 seconds, the rock has only fallen 64 feet.

Differentiated Instruction

Visual

Some students may encounter difficulty when substituting a value for a variable. Have them substitute parentheses for the variable and then insert the value. For example, using Exercise 25:

$$\begin{aligned} 6 + a &= 6 + (\ ) \\ &= 6 + (2) \\ &= 8 \end{aligned}$$

For exercises with more than one variable, suggest they use colored pencils for the different variables, as shown in Example 4.

Practice and Problem Solving

43. 23 **44.** 8.2

45. $2\frac{5}{6}$ **46.** $10\frac{2}{3}$

47. 22 **48.** 82

49. 46 **50.** 48.3

51. 24 **52.** $4x + 3y$

53. See *Taking Math Deeper.*

54. See Additional Answers.

55. 64 in.^{3}

Fair Game Review

56. 243 **57.** 512

58. 2401 **59.** 256

60. C

Mini-Assessment

Evaluate the expression when $x = 2$, $y = 5$, and $z = 10$.

1. $5 + x$ 7
2. $3y$ 15
3. $z - y$ 5
4. $x \cdot y$ 10
5. Your friend earns $7x$ dollars for working x hours. How much does your friend earn for working 16 hours? 40 hours? $112; $280

Taking Math Deeper

Exercise 53

This exercise can be used to help build student intuition for distances and for times.

1. How fast would you have to move to travel 2000 feet in less than 10 minutes?

   It helps for students to draw a visual image of the problem.

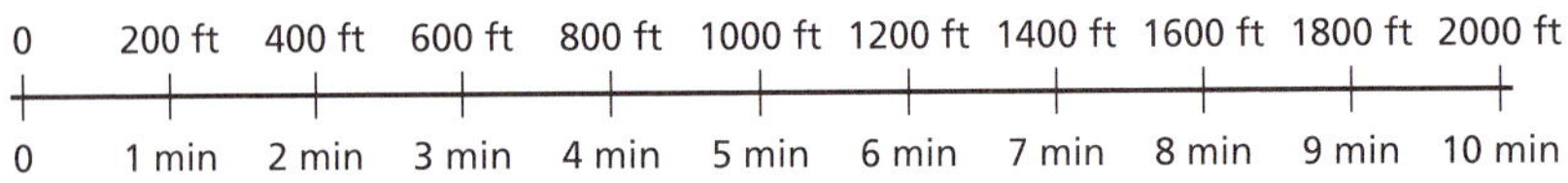

   To float 2000 feet in exactly 10 minutes, you would need to move 200 feet every minute. So, to travel this distance in less than 10 minutes, you would need to move faster . . . say 210 feet every minute or 220 feet every minute.

2. To help students get a sense of how fast this is, ask them to measure the width of your classroom and to use a watch with a second hand to walk at a speed of 200 feet every minute. For instance, if your classroom is 40 feet wide, a student would need to walk this distance in one-fifth of 1 minute, or 12 seconds.

3.

2012 SeaWorld Parks and Entertainment. Reproduced by permission of Busch Gardens/WaterCountry USA. All rights reserved.

Project

Ask students to use the Internet or some other reference source to find a map of an actual water theme park with the "river ride." Ask students to estimate the dimensions of the ride, including its total length. The one shown above is the Rambling River Ride in Virginia.

Reteaching and Enrichment Strategies

If students need help. . .	If students got it. . .
Resources by Chapter • Practice A and Practice B • Puzzle Time Record and Practice Journal Practice Differentiating the Lesson Lesson Tutorials Skills Review Handbook	Resources by Chapter • Enrichment and Extension • Technology Connection Start the next section

ALGEBRA Evaluate the expression when $a = 10$, $b = 9$, and $c = 4$.

5 **43.** $2a + 3$ **44.** $4c - 7.8$ **45.** $\frac{a}{4} + \frac{1}{3}$

46. $\frac{24}{b} + 8$ **47.** $c^2 + 6$ **48.** $a^2 - 18$

49. $a + 9c$ **50.** $bc + 12.3$ **51.** $3a + 2b - 6c$

Standard Rentals
$3

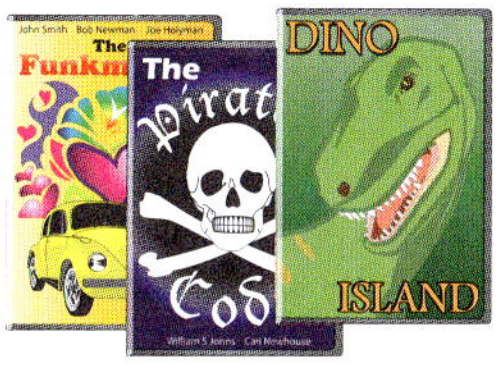

New Releases
$4

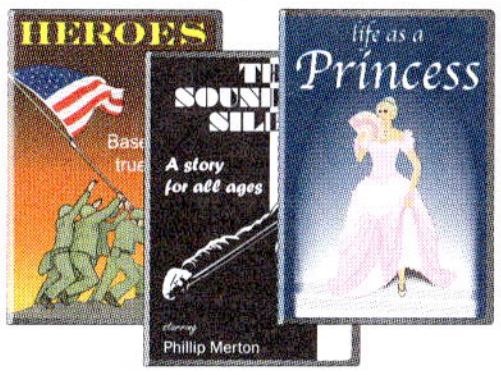

52. MOVIES You rent x new releases and y standard rentals. Which expression tells you how much money you will need?

$3x + 4y$ $4x + 3y$ $7(x + y)$

53. WATER PARK You float 2000 feet along a "Lazy River" water ride. The ride takes less than 10 minutes. Give two examples of possible times and speeds. Illustrate the water ride with a drawing.

54. SCIENCE CENTER The expression $20a + 13c$ is the cost (in dollars) for a adults and c students to enter a science center.

a. How much does it cost for an adult? a student? Explain your reasoning.

b. Find the total cost for 4 adults and 24 students.

c. You find the cost for a group. Then the numbers of adults and students in the group both double. Does the cost double? Explain your answer using an example.

d. In part (b), the number of adults is cut in half, but the number of students doubles. Is the cost the same? Explain your answer.

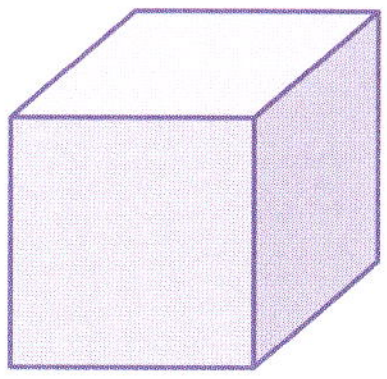

x in.

55. Reasoning The volume of the cube is equal to four times the area of one of its faces. What is the volume of the cube?

Fair Game Review What you learned in previous grades & lessons

Find the value of the power. *(Section 1.2)*

56. 3^5 **57.** 8^3 **58.** 7^4 **59.** 2^8

60. MULTIPLE CHOICE Which numbers have a least common multiple of 24? *(Section 1.6)*

Ⓐ 4, 6 Ⓑ 2, 22 Ⓒ 3, 8 Ⓓ 6, 12

3.2 Writing Expressions

Essential Question How can you write an expression that represents an unknown quantity?

1 ACTIVITY: Ordering Lunch

Work with a partner. You use a $20 bill to buy lunch at a café. You order a sandwich from the menu board shown.

a. Complete the table. In the last column, write a numerical expression for the amount of change received.

Sandwich	Price (dollars)	Change Received (dollars)
Reuben		
BLT		
Egg salad		
Roast beef		

b. **REPEATED REASONING** Write an expression for the amount of change you receive when you order any sandwich from the menu board.

c. Compare the expression you wrote in part (b) with the expressions in the last column of the table in part (a).

d. The café offers several side dishes, each at the same price. You order a chicken salad sandwich and two side dishes. Write an expression for the total amount of money you spend. Explain how you wrote your expression.

e. The expression $20 - 4.65s$ represents the amount of change one customer receives after ordering from the menu board. Explain what each part of the expression represents. Do you know what the customer ordered? Explain your reasoning.

Common Core

Algebraic Expressions

In this lesson, you will

- use variables to represent numbers in algebraic expressions.
- write algebraic expressions.

Learning Standard
6.EE.2a

Laurie's Notes

Introduction

Standards for Mathematical Practice

- **MP8 Look for and Express Regularity in Repeated Reasoning:** When students write and then evaluate algebraic expressions for different values, they develop a general method, or shortcut. This occurs as a result of the calculations being repeated.

Motivate

- A homonym is one of two or more words that have the same sound and often the same spelling, but differ in meaning. Examples: bank (embankment) and bank (where money is kept), and their, there, and they're.
- As a fun warm up, have groups of students make lists of homonyms (both with the same spelling and different spellings). When students are done, share each group's list with the class.

Discuss

- In math class you often hear words or phrases that suggest a particular operation. In fact, there are several words that suggest the same operation. Today's lesson is about translating phrases into algebraic expressions.

Activity Notes

Activity 1

- **MP8:** One goal of this activity is to get students to see the difference between numerical and algebraic expressions and when writing an algebraic expression, seeing that the operations can be performed repeatedly with different numbers.
- Introduce the activity by asking about favorite sandwiches and which sandwich they would order from those recommended by the chef.
- Check as students are working on part (a) to see that they are writing a numerical expression versus writing the result of the numerical expression. You want students to see the repetition in the last column.
- After students complete part (c), ask them to share their observations.
- ? After students have finished the activity, ask several students to share their explanations of part (d). "Is there more than one correct expression? Explain." yes; $4.95 + $2n$ and $2n$ + $4.95 "What does n represent?" the cost of a side dish
- **Common Misconception:** Some students will write an expression in part (d) that involves words, such as $4.95 and 2 times the cost of a side dish. They believe you cannot write the expression because the cost of a side dish is unknown. This is the concept of a variable.
- **MP3 Construct Viable Arguments and Critique the Reasoning of Others:** Ask one student to give his or her explanation of the expression $20 - 4.65s$. Ask if there are other interpretations. Students should feel comfortable justifying their statements.

Common Core State Standards

6.EE.2a Write, read, and evaluate expressions in which letters stand for numbers. Write expressions that record operations with numbers and with letters standing for numbers.

Previous Learning

Students need to be familiar with variables and how they are used.

Technology for the Teacher

Dynamic Classroom

Lesson Plans
Complete Materials List

3.2 Record and Practice Journal

Essential Question How can you write an expression that represents an unknown quantity?

1 ACTIVITY: Ordering Lunch

Work with a partner. You use a $20 bill to buy lunch at a café. You order a sandwich from the menu board.

a. Complete the table. In the last column, write a numerical expression for the amount of change received.

Sandwich	Price (dollars)	Change Received (dollars)
Reuben	6.45	20 − 6.45
BLT	5.25	20 − 5.25
Egg salad	4.65	20 − 4.65
Roast beef	6.75	20 − 6.75

b. **REPEATED REASONING** Write an expression for the amount of change you receive when you order any sandwich from the menu board.

Sample answer: $20 - s$

c. Compare the expression you wrote in part (b) with the expression in the last column of the table in part (a).

Part (a) are numerical expressions.
Part (b) are algebraic expressions.

English Language Learners

Organization

Create a table so that students can practice translating variable expressions. Add blank rows for students to add additional expressions.

Variable Expression	Meaning	Operation
$x + 14$	x plus 14	Addition
$y - 75$	y minus 75	Subtraction
$3 \div z$, $\frac{3}{z}$	3 divided by z	Division
$6n$, $6 \cdot n$, $(6)(n)$	6 times n	Multiplication

Use the table at the beginning of the lesson to review the key words and phrases.

3.2 Record and Practice Journal

d. The café offers several side dishes, each at the same price. You order a chicken salad sandwich and two side dishes. Write an expression for the total amount of money you spend. Explain how you wrote your expression.

Sample answer: $4.95 + 2d$

e. The expression $20 - 4.65s$ represents the amount of change one customer receives after ordering from the menu board. Explain what each part of the expression represents. Do you know what the customer ordered? Explain your reasoning.

20: customer paid using $20
s: number of items ordered
4.65: price per item
You do not know what the customer ordered.

2 ACTIVITY: Words That Imply Addition or Subtraction

Work with a partner.

a. Complete the table.

Variable	Phrase	Expression
n	4 more than a number	$n + 4$ or $4 + n$
m	the difference of a number and 3	$m - 3$
x	the sum of a number and 8	$x + 8$ or $8 + x$
p	10 less than a number	$p - 10$
n	7 units farther away	$n + 7$ or $7 + n$
t	8 minutes sooner	$t - 8$
w	12 minutes later	$w + 12$ or $12 + w$
y	a number increased by 9	$y + 9$ or $9 + y$

b. Here is a word problem that uses one of the expressions in the table.

You arrive at the café 8 minutes sooner than your friend. Your friend arrives at 6:42 P.M. When did you arrive?

Which expression from the table on the previous page can you use to solve the problem?

$t - 8$

c. Write a problem that uses a different expression from the table.

Sample answer: You have 4 more DVDs than your friend. Your friend has 16 DVDs. How many DVDs do you have?

3 ACTIVITY: Words That Imply Multiplication or Division

Work with a partner. Match each phrase with an expression.

- the product of a number and 3
- the quotient of 3 and a number
- 4 times a number
- a number divided by 3
- twice a number

- $n \div 3$
- $4p$
- $n \cdot 3$
- $2m$
- $3 \div n$

What Is Your Answer?

4. IN YOUR OWN WORDS How can you write an expression that represents an unknown quantity? Give examples to support your explanation.

Sample answer: Use variables to represent any unknown amounts. To determine the math operations for the expression, look for key words or phrases.

Laurie's Notes

Activity 2

- Introduce the activity by telling students that they are going to write some expressions using the variable given.
- **Common Error:** Subtraction is not commutative, and therefore students must be careful with this operation. Make a point of explaining that the difference of a number and 3, $m - 3$, is not the same as the difference of 3 and a number, $3 - m$.
- **Struggling Students:** Some students may find Activity 3 easier than Activity 2. You may want to do Activity 3 first, or do Activity 2 together as a class.

Activity 3

- **Common Error:** Give students the same reminder about division as was given about subtraction. Explain that the quotient of a number and 3, $n \div 3$, is not the same as the quotient of 3 and a number, $3 \div n$.

What Is Your Answer?

- *Sample answer:* Use variables to represent any unknown quantities. To determine the math operations for the expressions, look for key words or phrases.

Closure

Classroom Activity: This is a nice summary activity that can be done with the whole class.

- Write words or phrases on index cards for each of the four operations so that each student can have an index card. Distribute the cards. Some of the phrases can be the same.
- Designate a corner of the room for each of the four operations. Have students move to the corner of the room associated with the operation suggested by the word or phrase on their index card.
- Each of the four clusters of students should verify that all of the group's index cards match the correct operation.
- Have each cluster of students read the words on their index cards aloud to the class.
- Collect the cards to use again.

2 ACTIVITY: Words That Imply Addition or Subtraction

Math Practice 2

Use Expressions

How do the key words in the phrase help you write the given relationship as an expression?

Work with a partner.

a. Complete the table.

Variable	Phrase	Expression
n	4 **more than** a number	
m	the **difference** of a number and 3	
x	the **sum** of a number and 8	
p	10 **less than** a number	
n	7 units **farther** away	
t	8 minutes **sooner**	
w	12 minutes **later**	
y	a number **increased** by 9	

b. Here is a word problem that uses one of the expressions in the table.

You arrive at the café 8 minutes sooner than your friend. Your friend arrives at 6:42 P.M. When did you arrive?

Which expression from the table can you use to solve the problem?

c. Write a problem that uses a different expression from the table.

3 ACTIVITY: Words That Imply Multiplication or Division

Work with a partner. Match each phrase with an expression.

the product of a number and 3	$n \div 3$
the quotient of 3 and a number	$4p$
4 times a number	$n \cdot 3$
a number divided by 3	$2m$
twice a number	$3 \div n$

What Is Your Answer?

4. **IN YOUR OWN WORDS** How can you write an expression that represents an unknown quantity? Give examples to support your explanation.

Practice Use what you learned about writing expressions to complete Exercises 9–12 on page 122.

3.2 Lesson

Check It Out
Lesson Tutorials
BigIdeasMath.com

Some words imply math operations.

Operation	Addition	Subtraction	Multiplication	Division
Key Words and Phrases	added to plus sum of more than increased by total of and	subtracted from minus difference of less than decreased by fewer than take away	multiplied by times product of twice of	divided by quotient of

EXAMPLE 1 Writing Numerical Expressions

Write the phrase as an expression.

a. 8 fewer than 21

$21 - 8$ — The phrase *fewer than* means *subtraction.*

b. the product of 30 and 9

30×9, or $30 \cdot 9$ — The phrase *product of* means *multiplication.*

EXAMPLE 2 Writing Algebraic Expressions

Write the phrase as an expression.

a. 14 more than a number x

$x + 14$ — The phrase *more than* means *addition.*

b. a number y minus 75

$y - 75$ — The word *minus* means *subtraction.*

c. the quotient of 3 and a number z

$3 \div z$, or $\frac{3}{z}$ — The phrase *quotient of* means *division.*

Common Error

When writing expressions involving subtraction or division, order is important. For example, the quotient of a number x and 2 means

$x \div 2$, not $2 \div x$.

On Your Own

Now You're Ready
Exercises 3–18

Write the phrase as an expression.

1. the sum of 18 and 35
2. 6 times 50
3. 25 less than a number b
4. a number x divided by 4
5. the total of a number t and 11
6. 100 decreased by a number k

Laurie's Notes

Introduction

Connect

- **Yesterday:** Students explored how to write an algebraic expression that represented a verbal phrase. (MP3, MP8)
- **Today:** Today's lesson presents the content more formally.

Motivate

- Write "price + \$4" on the board and ask how this phrase might be expressed in words, for example: "the price increased by \$4" or "\$4 added to the price." Students should be able to think of several more interpretations.
- Below "price + \$4," write "$p + 4$." Explain that you can use the variable p to represent the price, instead of writing it out.
- Additional examples:

Age − 2	\$2.50 × number	Quarts ÷ 4
$a - 2$	$2.5n$	$q \div 4$ or $\frac{q}{4}$

- **Whole Class Activity:** Recall the sorting activity that closed the class yesterday. Make a chart on the board similar to the start of Lesson 3.2 and fill it in as a class. Students should *not* have their books open. Use the sorted index cards from yesterday. Ask for 4 volunteers to simply read any versions not already listed.

Lesson Notes

Example 1

- Review the table of words that imply math operations. Ask students to add to the list.

Example 2

- Work through the example. Again note that subtraction and division are not commutative. For example in 2(c), note that the solution is *not* $z \div 3$.
- Discuss the Common Error box in the left margin.

? "Is order important in addition and multiplication? Explain." No; you can add in any order and multiply in any order.

Words of Wisdom

- No example is shown for multiplication; however, you may wish to include one in order to point out the conventional way in which multiplication with a variable is represented.

? "Are $n3$ and $3n$ equivalent? Why?" Multiplication is commutative, so they are equivalent.

On Your Own

- **Neighbor Check:** Have students work independently and then have their neighbors check their work. Have students discuss any discrepancies.

Goal Today's lesson is writing mathematical expressions that correspond to given situations.

Technology for the Teacher

Dynamic Classroom

Lesson Tutorials
Lesson Plans
Answer Presentation Tool

Differentiated Instruction

Auditory

On the board, write a few expressions containing a variable and one operation. Have students practice reading them aloud in various ways. For example, $x - 4$ can be read as "x minus four," "the difference of x and four," or "4 less than x."

Extra Example 1

Write the phrase as an expression.

a. 34 divided by 2 $34 \div 2$

b. 35 increased by 43 $35 + 43$

Extra Example 2

Write the phrase as an expression.

a. the difference of a number t and 95 $t - 95$

b. twice a number w $2w$

On Your Own

1. $18 + 35$ or $35 + 18$
2. $6 \cdot 50$ or $50 \cdot 6$
3. $b - 25$
4. $x \div 4$
5. $t + 11$ or $11 + t$
6. $100 - k$

Extra Example 3

The number of students in the science club is 4 more than twice the number of students in the math club. Let m be the number of students in the math club. Write an expression to find the number of students in the science club.

$2m + 4$

Extra Example 4

A bamboo shoot is 4 inches tall. During its growing season it grows 3 inches a day.

a. Make a table that shows the height of the bamboo for 4 days. Then write an expression for the height after d days.

Day, d	Height (inches)
0	4
1	$4 + 3(1) = 7$
2	$4 + 3(2) = 10$
3	$4 + 3(3) = 13$
4	$4 + 3(4) = 16$

$4 + 3d$

b. What is the height after 14 days?

46 inches

On Your Own

7. $2t + 5$ **8.** 250 in.

Laurie's Notes

Example 3

- **MP1 Make Sense of Problems and Persevere in Solving Them:** Ask a volunteer to read the problem. Given the amount of words, some students may have difficulty getting started. Refer to yesterday's activity and how it was necessary to reread. Underlining key words may be another way in which students make sense of the problem.
- Students need a context for some of the words in the problem. Ask if they have been on these interstates, if they are close to home, in their state, etc. If not, what interstates are close? Why do we have interstates?
- **Common Error:** Choice C is a very common *wrong* answer. Ask students to think about what this expression would equal when $m = 100$. (Something negative is the common response!)

Example 4

- Say, "We want to organize some information about this problem, so let's set up a table to keep track of the height each year."
- Make a table with two columns (year, height). It is not natural to start with the year 0, so be sure to discuss this. Each year students should add 15 to the last number.
- ? "Can you find the height after 9 years without computing all of the years between 4 and 9?" Write an expression.
- Work with students to develop the general expression $10 + 15t$ and verify that it indeed works for years 0 through 4.

On Your Own

- Note that for Question 7, some students might write $5 + 2t$ and others might write $2t + 5$. Addition is commutative, so this is fine. You may want to caution students that subtraction is not commutative, so there is only one way to correctly write "five less than twice as many tokens."

Words of Wisdom

- English Language Learners will likely need additional vocabulary support.

Closure

- **Matching:**

$3x - 2$ A. The difference of 2 and 3 times a number

$2 - 3x$ B. Three less than twice a number

$2x - 3$ C. The difference of three times a number and 2

Answers in order: C, A, B

EXAMPLE 3 Writing an Algebraic Expression

The length of Interstate 90 from the West Coast to the East Coast is 153.5 miles more than 2 times the length of Interstate 15 from southern California to northern Montana. Let *m* be the length of Interstate 15. Which expression can you use to represent the length of Interstate 90?

Ⓐ $2m + 153.5$ Ⓑ $2m - 153.5$ Ⓒ $153.5 - 2m$ Ⓓ $153.5m + 2$

The word *times* means *multiplication*. So, multiply 2 and *m*.

The phrase *more than* means *addition*. So, add 2*m* and 153.5.

$2m + 153.5$

∴ The correct answer is Ⓐ.

EXAMPLE 4 Real-Life Application

You plant a cypress tree that is 10 inches tall. Each year, its height increases by 15 inches.

a. Make a table that shows the height of the tree for 4 years. Then write an expression for the height after *t* years.

b. What is the height after 9 years?

a. The height is *increasing*, so *add* 15 each year as shown in the table.

Year, *t*	Height (inches)
0	10
1	$10 + 15(1) = 25$
2	$10 + 15(2) = 40$
3	$10 + 15(3) = 55$
4	$10 + 15(4) = 70$

When *t* is 0, the height is 10 inches.

You can see that an expression is $10 + 15t$.

∴ So, the height after year *t* is $10 + 15t$.

b. Evaluate $10 + 15t$ when $t = 9$.

$$10 + 15t = 10 + 15(9) = 145$$

∴ After 9 years, the height of the tree is 145 inches.

Study Tip

Sometimes, like in Example 3, a variable represents a single value. Other times, like in Example 4, a variable can represent more than one value.

On Your Own

Now You're Ready Exercises 27–30

7. Your friend has 5 more than twice as many game tokens as your sister. Let *t* be the number of game tokens your sister has. Write an expression for the number of game tokens your friend has.

8. **WHAT IF?** In Example 4, what is the height of the cypress tree after 16 years?

3.2 Exercises

Check It Out
Help with Homework
BigIdeasMath.com

Vocabulary and Concept Check

1. **DIFFERENT WORDS, SAME QUESTION** Which is different? Write "both" expressions.

12 more than x	x increased by 12	x take away 12	the sum of x and 12

2. **REASONING** You pay $0.25p$ dollars to print p photos. What does the coefficient represent?

Practice and Problem Solving

Write the phrase as an expression.

1 2

3. 5 less than 8
4. the product of 3 and 12
5. 28 divided by 7
6. the total of 6 and 10
7. 3 fewer than 18
8. 17 added to 15
9. 13 subtracted from a number x
10. 5 times a number d
11. the quotient of 18 and a number a
12. the difference of a number s and 6
13. 7 increased by a number w
14. a number b squared
15. the sum of a number y and 4
16. the difference of 12 and a number x
17. twice a number z
18. a number t cubed

ERROR ANALYSIS **Describe and correct the error in writing the phrase as an expression.**

19. the quotient of 8 and a number y

✗ $\frac{y}{8}$

20. 16 decreased by a number x

✗ $x - 16$

21. **DINNER** Five friends share the cost of a dinner equally.
    a. Write an expression for the cost per person.
    b. Make up a total cost and test your expression. Is the result reasonable?

22. **TV SHOW** A television show has 19 episodes per season.
    a. Copy and complete the table.
    b. Write an expression for the number of episodes in n seasons.

Seasons	1	2	3	4	5
Episodes					

Give two ways to write the expression as a phrase.

23. $n + 6$
24. $4w$
25. $15 - b$
26. $14 - 3z$

Assignment Guide and Homework Check

Level	Day 1 Activity Assignment	Day 2 Lesson Assignment	Homework Check
Basic	9–12, 36–40	1, 2, 3, 5, 7, 13–21 odd, 22, 27, 28	13, 17, 22, 27
Average	9–12, 36–40	1, 2, 13–19 odd, 20–26 even, 27–33 odd	17, 26, 27, 33
Advanced	9–12, 36–40	1, 2, 16–22 even, 24–30 even, 31–35	26, 30, 33, 34

Common Errors

- **Exercises 3–18** Students may write the subtraction and/or division problems in the wrong order. For example, they may write "5 less than 8" as 5 − 8 instead of 8 − 5.
- **Exercises 3–18** For subtraction problems, have students look for the key words "from" and "than" to know that the order of the numbers must be switched from the way it is written. For example, "9 subtracted *from* 14" means 14 − 9; "9 fewer *than* 14" means 14 − 9.
- **Exercises 27–30** Students may write the problem in the wrong order such as $\frac{y-3}{4}$ instead of $\frac{y}{4} - 3$. Tell students to write out the phrase and put parentheses around the different parts of the expression. Then evaluate the expression using order of operations. For example, "3 less than (the quotient of a number *y* and 4)."

3.2 Record and Practice Journal

Write the phrase as an expression.

1. the total of 8 and 13
   $8 + 13$
2. 42 divided by 7
   $42 \div 7$

Give two ways to write the expression as a phrase.

3. $6 + p$
   Sample answer: The sum of 6 and a number *p*. 6 plus a number *p*.
4. $9m$
   Sample answer: The product of 9 and a number *m*. 9 multiplied by a number *m*.

Write the phrase as an expression. Then evaluate when $x = 3$ and $y = 15$.

5. 7 more than the quotient of a number *y* and 5
   $\frac{y}{5} + 7$; 10
6. twice the sum of a number *x* and 8
   $2(x + 8)$; 22

7. You earn $7 for every hour that you babysit.
   a. Complete the table.

Hours	1	2	3	4	5	6	7	8
Earnings	$7	$14	$21	$28	$35	$42	$49	$56

   b. Write an expression for the amount you earn after *h* hours.
   $7h$

Vocabulary and Concept Check

1. x take away 12; $x - 12$; $x + 12$
2. The coefficient represents the cost to print each photo, 25 cents.

Practice and Problem Solving

3. $8 - 5$
4. $3 \cdot 12$ or $12 \cdot 3$
5. $28 \div 7$
6. $6 + 10$ or $10 + 6$
7. $18 - 3$
8. $15 + 17$ or $17 + 15$
9. $x - 13$
10. $5 \cdot d$ or $d \cdot 5$
11. $18 \div a$
12. $s - 6$
13. $7 + w$ or $w + 7$
14. b^2
15. $y + 4$ or $4 + y$
16. $12 - x$
17. $2 \cdot z$ or $z \cdot 2$
18. t^3
19. The expression is not written in the correct order; $\frac{8}{y}$
20. The expression is not written in the correct order; $16 - x$
21. a. $x \div 5$
    b. *Sample answer:* If the total cost is $30, then the cost per person is $x \div 5 = 30 \div 5 = \$6$. The result is reasonable.
22. a.

Seasons	1	2	3	4	5
Episodes	19	38	57	76	95

    b. $19n$

Practice and Problem Solving

23–26. Sample answers are given.

23. The sum of n and 6; 6 more than a number n

24. 4 times a number w; The product of 4 and a number w

25. A number b less than 15; 15 take away a number b

26. 3 times a number z less than 14; 14 minus 3 times a number z

27. $\frac{y}{4} - 3$; 2 **28.** $\frac{x+4}{3}$; 3

29. $8x + 6$; 46 **30.** $\frac{40}{y-16}$; 10

31. See Additional Answers.

32. **a.** $5a - 8$

b. $2f + 25$

c. Florida has 67 counties. Georgia has 159 counties.

33. See *Taking Math Deeper*.

34. $a + 8$

35. $\frac{x}{4}$

Fair Game Review

36. 45 **37.** 59

38. 390 **39.** 140

40. D

Mini-Assessment

Write the phrase as an expression.

1. the product of 5 and 7 $5 \cdot 7$
2. the sum of a number b and 3 $b + 3$
3. the quotient of 9 and a number y $\frac{9}{y}$
4. 8 less than a number x $x - 8$
5. It costs \$25 per day to rent a surfboard and \$15 to rent a wetsuit. Write an expression for the cost of renting the surfboard for y days. $25y + 15$

Taking Math Deeper

Exercise 33

Tables are wonderful ways to organize information. This problem doesn't ask students to make a table. That is something they can do *just because it helps*.

1 Make a table. Find a pattern.

Round	0	1	2	3	4	5
Singers	140	125	110	95	80	65

15 fewer for each round

2 **a.** Use the pattern to write an expression.

Number of Singers after Round n $= 140 - 15n$

Test the expression.

When $n = 1$, $140 - 15(1) = 125$. ✓

When $n = 2$, $140 - 15(2) = 110$. ✓

3 **b.** Use the expression to evaluate.

When $n = 8$, $140 - 15(8) = 20$.

There are 20 people left after the eighth round. So, 20 people compete in the ninth round.

Check by extending the table.

Round	5	6	7	8
Singers	65	50	35	20

Subtract 15 for each round.

Project

Investigate and write a report on singing competitions for middle school choruses.

Reteaching and Enrichment Strategies

If students need help. . .	If students got it. . .
Resources by Chapter • Practice A and Practice B • Puzzle Time Record and Practice Journal Practice Differentiating the Lesson Lesson Tutorials Skills Review Handbook	Resources by Chapter • Enrichment and Extension • Technology Connection Start the next section

3 4 Write the phrase as an expression. Then evaluate when $x = 5$ and $y = 20$.

27. 3 less than the quotient of a number y and 4

28. the sum of a number x and 4, all divided by 3

29. 6 more than the product of 8 and a number x

30. the quotient of 40 and the difference of a number y and 16

31. MODELING It costs \$3 to bowl a game and \$2 for shoe rental.

a. Make a table for the cost of up to 5 games.

b. Write an expression for the cost of g games.

c. Use your expression to find the cost of 8 games.

32. PUZZLE Florida has 8 less than 5 times the number of counties in Arizona. Georgia has 25 more than twice the number of counties in Florida.

a. Write an expression for the number of counties in Florida.

b. Write an expression for the number of counties in Georgia.

c. Arizona has 15 counties. How many do Florida and Georgia have?

33. PATTERNS There are 140 people in a singing competition. The graph shows the results for the first five rounds.

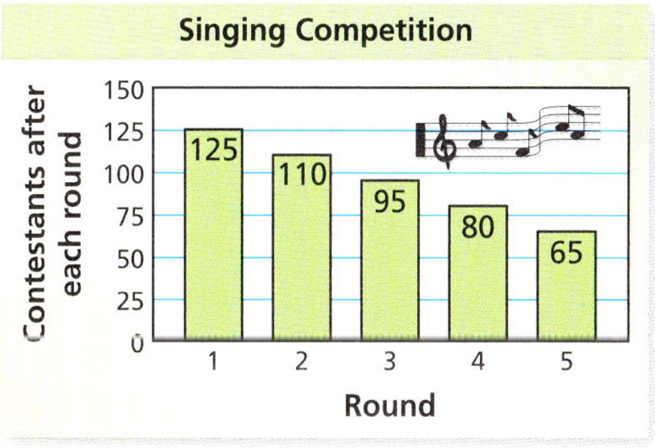

a. Write an expression for the number of people after each round.

b. How many people compete in the ninth round? Explain your reasoning.

34. NUMBER SENSE The difference between two numbers is 8. The lesser number is a. Write an expression for the greater number.

35. Reasoning One number is four times another. The greater number is x. Write an expression for the lesser number.

Fair Game Review What you learned in previous grades & lessons

Evaluate the expression. *(Skills Review Handbook)*

36. $8 + (22 + 15)$ **37.** $(13 + 9) + 37$ **38.** $(13 \times 6) \times 5$ **39.** $4 \times (7 \times 5)$

40. MULTIPLE CHOICE A grocery store is making fruit baskets using 144 apples, 108 oranges, and 90 pears. Each basket will be identical. What is the greatest number of fruit baskets the store can make using all the fruit? *(Section 1.5)*

Ⓐ 6 Ⓑ 9 Ⓒ 16 Ⓓ 18

3 Study Help

Check It Out
Graphic Organizer
BigIdeasMath.com

You can use an **information wheel** to organize information about a topic. Here is an example of an information wheel for identifying parts of an algebraic expression.

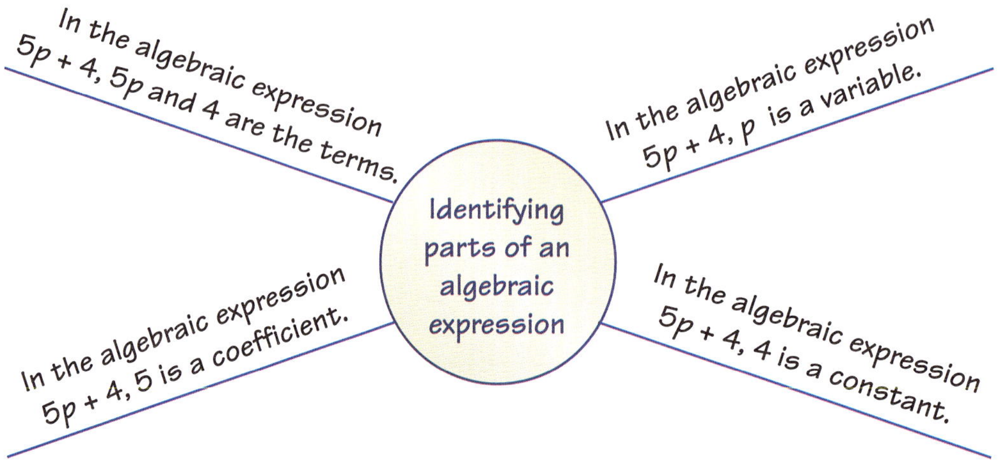

On Your Own

Make information wheels to help you study these topics.

1. evaluating algebraic expressions
2. writing algebraic expressions

After you complete this chapter, make information wheels for the following topics.

3. Commutative Properties of Addition and Multiplication
4. Associative Properties of Addition and Multiplication
5. Addition Property of Zero
6. Multiplication Properties of Zero and One
7. Distributive Property
8. factoring expressions

"My **information wheel** for Fluffy has matching adjectives and nouns."

Sample Answers

1.

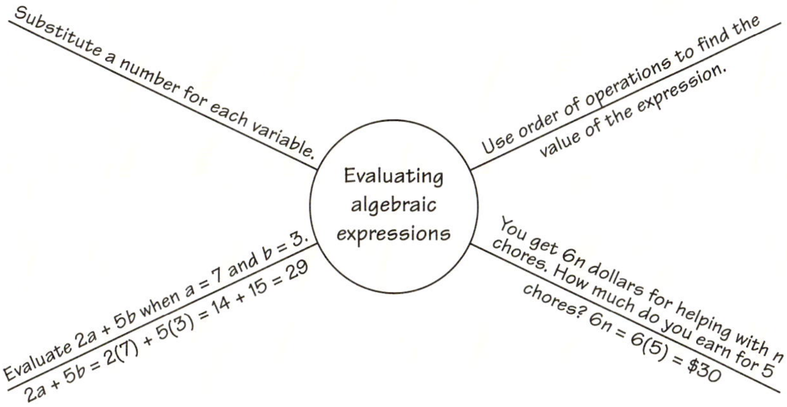

2.

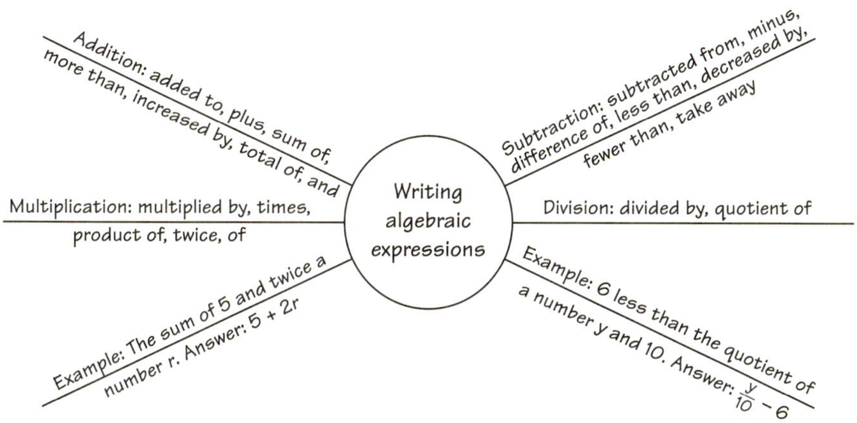

List of Organizers

Available at *BigIdeasMath.com*

Comparison Chart
Concept Circle
Definition (Idea) and Example Chart
Example and Non-Example Chart
Formula Triangle
Four Square
Information Frame
Information Wheel
Notetaking Organizer
Process Diagram
Summary Triangle
Word Magnet
Y Chart

About this Organizer

An **Information Wheel** can be used to organize information about a concept. Students write the concept in the middle of the "wheel." Then students write information related to the concept on the "spokes" of the wheel. Related information can include, but is not limited to: vocabulary words or terms, definitions, formulas, procedures, examples, and visuals. This type of organizer serves as a good summary tool because any information related to a concept can be included.

Technology for the Teacher

Editable Graphic Organizer

Answers

1. Terms: $6q$, 1;
   Coefficient: 6;
   Constant: 1

2. Terms: $3r^2$, $4r$, 8;
   Coefficients: 3, 4;
   Constant: 8

3. s^4

4. $2t^5$

5. 13

6. 16

7. 58

8.

x	$x + 6$
1	7
2	8
3	9

9.

x	$3x - 5$
3	4
6	13
9	22

10. $28 + 35$ or $35 + 28$

11. $x \div 2$

12. $m \cdot 23$ or $23m$

13. $a - 10$

14. $68

15. $195

16. **a.** $33 + m$

    b. $333

Technology for the Teacher

Online Assessment
Assessment Book
ExamView® Assessment Suite

Alternative Quiz Ideas

100% Quiz	Math Log
Error Notebook	Notebook Quiz
Group Quiz	Partner Quiz
Homework Quiz	Pass the Paper

Error Notebook

- An error notebook provides an opportunity for students to analyze and learn from their errors. Have students make an error notebook for this chapter. They should work in their notebook a little each day. Give students the following directions.
- Use a notebook and divide the page into three columns.
- Label the first column *problem,* second column *error,* and third column *correction.*
- In the first column, write down the problem in which errors were made. Record the source of the problem (homework, quiz, in-class assignment).
- The second column should show the exact error that was made. Include a statement of why you think the error was made. Statements might include, "I didn't write the subtraction in the correct order." This is where the learning takes place, so it is helpful to use a different color ink for the work in this column.
- The last column contains the corrected problems and comments that will help with future work.
- Separate each problem with horizontal lines.

Reteaching and Enrichment Strategies

If students need help. . .	If students got it. . .
Resources by Chapter • Practice A and Practice B • Puzzle Time Lesson Tutorials *BigIdeasMath.com*	Resources by Chapter • Enrichment and Extension • Technology Connection Game Closet at *BigIdeasMath.com* Start the next section

3.1–3.2 Quiz

Check It Out
Progress Check
BigIdeasMath.com

Identify the terms, coefficients, and constants of the expression. *(Section 3.1)*

1. $6q + 1$

2. $3r^2 + 4r + 8$

Write the expression using exponents. *(Section 3.1)*

3. $s \cdot s \cdot s \cdot s$

4. $2 \cdot t \cdot t \cdot t \cdot t \cdot t$

Evaluate the expression when $a = 8$ and $b = 2$. *(Section 3.1)*

5. $a + 5$

6. ab

7. $a^2 - 6$

Copy and complete the table. *(Section 3.1)*

8.

x	$x + 6$
1	
2	
3	

9.

x	$3x - 5$
3	
6	
9	

Write the phrase as an expression. *(Section 3.2)*

10. the sum of 28 and 35

11. a number x divided by 2

12. the product of a number m and 23

13. 10 less than a number a

14. **COUPON** The expression $p - 15$ is the amount you pay after using the coupon on a purchase of p dollars. How much do you pay for a purchase of \$83? *(Section 3.1)*

Coupon

Good for \$15 off any purchase of \$75 or more

15. **AMUSEMENT PARK** The expression $15a + 12c$ is the cost (in dollars) of admission at an amusement park for a adults and c children. Find the total cost for 5 adults and 10 children. *(Section 3.1)*

16. **MOVING TRUCK** To rent a moving truck for the day, it costs \$33 plus \$1 for each mile driven. *(Section 3.2)*

a. Write an expression for the cost to rent the truck.

b. You drive the truck 300 miles. How much do you pay?

3.3 Properties of Addition and Multiplication

Essential Question Does the order in which you perform an operation matter?

1 ACTIVITY: Does Order Matter?

Work with a partner. Place each statement in the correct oval.

a. Fasten 5 shirt buttons.
b. Put on a shirt and tie.
c. Fill and seal an envelope.
d. Floss your teeth.
e. Put on your shoes.
f. Chew and swallow.

Order Matters	**Order Doesn't Matter**

Think of three math problems using the four operations where order matters and three where order doesn't matter.

The Meaning of a Word ● Commute

When you **commute** the positions of two stuffed animals on a shelf,

you switch their positions.

Common Core

Equivalent Expressions

In this lesson, you will
- use properties of operations to generate equivalent expressions.

Learning Standards
6.EE.3
6.EE.4

2 ACTIVITY: Commutative Properties

Work with a partner.

a. Which of the following are true?

$3 + 5 \stackrel{?}{=} 5 + 3$ $\qquad$ $3 - 5 \stackrel{?}{=} 5 - 3$

$9 \times 3 \stackrel{?}{=} 3 \times 9$ $\qquad$ $9 \div 3 \stackrel{?}{=} 3 \div 9$

b. The true equations show the Commutative Properties of Addition and Multiplication. Why do you think they are called *commutative*?

Laurie's Notes

Introduction

Standards for Mathematical Practice

- **MP3 Construct Viable Arguments and Critique the Reasoning of Others:** Mathematically proficient students justify their conclusions, communicate them to others, and respond to the arguments of others. In this lesson, students will make conjectures about which operations are commutative and associative. Expect clear communication of their thinking.

Motivate

- **Act It Out:** If there is a bit of an actor in you, start the class by pretending that you're brushing your teeth (about 10 seconds) and then take the tube of toothpaste out and apply some to the toothbrush. This will clearly evoke a few comments about the order in which you performed the two tasks.

? "Oh, does the order matter? Hmmm…"

- If brushing your teeth doesn't work for you, wear a sock over a shoe, or push the number on the intercom phone and then pick up the receiver (anything obviously in the wrong order). Do something that students will remember. Catch their attention!

Activity Notes

Activity 1

- Students have seen the Commutative Property before, so they may know which operations are commutative and which are not.
- Make a two-column list on the board: Order Matters and Order Doesn't Matter. While you might expect subtraction and division problems under the first heading, and addition and multiplication problems under the second heading, I had a student who wrote $1\frac{1}{2} - 1\frac{2}{4}$ under the second heading.
- I had heard partner conversation. "Only adding and multiplying don't matter." The partner responded, "I bet I can do something else." This was the student who wrote the problem.
- **MP2 Reason Abstractly and Quantitatively, MP3, MP6 Attend to Precision:** It became a teachable moment. The class discussed what it meant to write an operation where the order matters. They concluded this was a special case and summarized, $m - n \neq n - m$, unless $m = n$.

Meaning of the Word

- Discuss the meaning of the word *commute*. An example is shown. Talk about *commuters* who go to work and back home, switching the direction.

Activity 2

- This activity is *not* asking students to evaluate each side of the equation, only which statements are true.
- Point out that the question mark above the equal sign is used when *questioning* or checking whether two sides are equal.

? "Can anyone write the property using variables (like *a* and *b*)?"

Common Core State Standards

6.EE.3 Apply the properties of operations to generate equivalent expressions.

6.EE.4 Identify when two expressions are equivalent.

Previous Learning

Students need to be familiar with basic computation facts. Applying mental math strategies will be helpful.

Technology for the Teacher

Dynamic Classroom

Lesson Plans
Complete Materials List

3.3 Record and Practice Journal

Essential Question Does the order in which you perform an operation matter?

1 ACTIVITY: Does Order Matter?

Work with a partner. Place each statement in the correct oval.

a. Fasten 5 shirt buttons. b. Put on a shirt and tie.

c. Fill and seal an envelope. d. Floss your teeth.

e. Put on your shoes. f. Chew and swallow.

Order Matters	Order Doesn't Matter
b, c, f	a, d, e

Write three math problems using the four operations where order matters and three where order doesn't matter.

Sample answer:
Order doesn't matter: $3 + 4 - 2$
Order matters: $3 + 4 \times 2$, $4 \div 2 + 1$

Laurie's Notes

The Meaning of a Word

- Discuss the meaning of the word *associate*, putting two things together. You *associate* dogs with barking, Arizona with the Grand Canyon, and sports cars with speed.
- For each of these examples, it is possible to *associate* a third word, like Arizona, Grand Canyon, and national parks. Numbers are different because you can only operate on two numbers at a time, even though there might be three numbers. So, you use parentheses to group two numbers at a time.
- Remind students that grouping symbols, such as parentheses, tell us to perform the operations inside them first.

Activity 3

- This activity is asking students to decide if the statement is true. In other words, does each side of the equation have the same value?
- Discuss class answers to part (a). Ask for a volunteer to share the answer to part (b). Again, encourage the use of variables.

What Is Your Answer?

- For Question 5, students should recognize the pairs that sum to 20, although some students might think there are easier combinations.
- For Question 6, encourage students to share their reasoning. The last two problems are key. Have students share their own codes, if time permits.

Closure

- **Share This Story:** Years ago it was important for shopkeepers to add quickly and efficiently in their heads. Or, if there were many items being purchased, they would write the prices down on the brown paper used to wrap the items (or a brown paper bag) and quickly add the prices together. There were no adding machines or cash registers. Here is a list of items being purchased.

? "How would you add them if you weren't allowed to use a calculator, adding machine, or cash register?"

$4.40
$1.22
$3.30
$1.08
$2.30

English Language Learners

Vocabulary

Write the word *association* on the board. Emphasize the meaning as *a group of people*. Demonstrate that the Associative Properties of Addition and Multiplication state that you can *group numbers so they are associated in different ways* and the answer will be the same.

3.3 Record and Practice Journal

2 ACTIVITY: Commutative Properties

Work with a partner.

a. Circle the statements that are true.

$3 + 5 \stackrel{?}{=} 5 + 3$ (circled) $\qquad 3 - 5 \stackrel{?}{=} 5 - 3$

$9 \times 3 \stackrel{?}{=} 3 \times 9$ (circled) $\qquad 9 \div 3 \stackrel{?}{=} 3 \div 9$

b. The true equations show the Commutative Properties of Addition and Multiplication. Why do you think they are called *commutative*?

Order does not matter.

3 ACTIVITY: Associative Properties

Work with a partner.

a. Circle the statements that are true.

$8 + (3 + 1) \stackrel{?}{=} (8 + 3) + 1$ (circled) $\qquad 8 - (3 - 1) \stackrel{?}{=} (8 - 3) - 1$

$12 \times (6 \times 2) \stackrel{?}{=} (12 \times 6) \times 2$ (circled) $\qquad 12 \div (6 \div 2) \stackrel{?}{=} (12 \div 6) \div 2$

b. The true equations show the Associative Properties of Addition and Multiplication. Why do you think they are called *associative*?

Location of grouping symbols does not matter.

What Is Your Answer?

4. **IN YOUR OWN WORDS** Does the order in which you perform an operation matter? Give examples to support your explanation.

yes; Check students' work.

5. **MENTAL MATH** Explain how you can add the sum in your head.

$11 + 7 + 12 + 13 + 8 + 9$

See Additional Answers.

6. **SECRET CODE** The creatures on a distant planet use the symbols ■, ◆, ★, and ● for the four operations.

a. Use the codes to decide which symbol represents addition and which symbol represents multiplication. Explain your reasoning.

3 ● 4 = 4 ● 3
3 ★ 4 = 4 ★ 3
2 ● (5 ● 3) = (2 ● 5) ● 3
2 ★ (5 ★ 3) = (2 ★ 5) ★ 3
0 ● 4 = 0
0 ★ 4 = 4

• is multiplication; ★ is addition

b. Make up your own symbols for addition and multiplication. Write codes using your symbols. Trade codes with a classmate. Decide which symbol represents addition and which symbol represents multiplication.

Check students' work.

The Meaning of a Word ● Associate

You have two best friends. Sometimes you **associate** with one of them.

And sometimes you **associate** with the other.

3 ACTIVITY: Associative Properties

Math Practice 3

Use Counterexamples

What do the false equations tell you about the Associative Properties?

Work with a partner.

a. Which of the following are true?

$8 + (3 + 1) \stackrel{?}{=} (8 + 3) + 1$ $\qquad$ $8 - (3 - 1) \stackrel{?}{=} (8 - 3) - 1$

$12 \times (6 \times 2) \stackrel{?}{=} (12 \times 6) \times 2$ $\qquad$ $12 \div (6 \div 2) \stackrel{?}{=} (12 \div 6) \div 2$

b. The true equations show the Associative Properties of Addition and Multiplication. Why do you think they are called *associative*?

What Is Your Answer?

4. IN YOUR OWN WORDS Does the order in which you perform an operation matter? Give examples to support your explanation.

5. MENTAL MATH Explain how you can add the sum in your head.

$$11 + 7 + 12 + 13 + 8 + 9$$

6. SECRET CODE The creatures on a distant planet use the symbols ■, ◆, ★, and ● for the four operations.

a. Use the codes to decide which symbol represents addition and which symbol represents multiplication. Explain your reasoning.

3 ● 4 = 4 ● 3

3 ★ 4 = 4 ★ 3

2 ● (5 ● 3) = (2 ● 5) ● 3

2 ★ (5 ★ 3) = (2 ★ 5) ★ 3

0 ● 4 = 0

0 ★ 4 = 4

b. Make up your own symbols for addition and multiplication. Write codes using your symbols. Trade codes with a classmate. Decide which symbol represents addition and which symbol represents multiplication.

Practice Use what you learned about the properties of addition and multiplication to complete Exercises 5–8 on page 130.

3.3 Lesson

Check It Out
Lesson Tutorials
BigIdeasMath.com

Key Vocabulary
equivalent expressions, *p. 128*

Expressions with the same value, like $12 + 7$ and $7 + 12$, are **equivalent expressions**. You can use the Commutative and Associative Properties to write equivalent expressions.

Key Ideas

Commutative Properties

Words Changing the order of addends or factors does not change the sum or product.

Numbers $5 + 8 = 8 + 5$
$5 \cdot 8 = 8 \cdot 5$

Algebra $a + b = b + a$
$a \cdot b = b \cdot a$

Associative Properties

Words Changing the grouping of addends or factors does not change the sum or product.

Numbers $(7 + 4) + 2 = 7 + (4 + 2)$
$(7 \cdot 4) \cdot 2 = 7 \cdot (4 \cdot 2)$

Algebra $(a + b) + c = a + (b + c)$
$(a \cdot b) \cdot c = a \cdot (b \cdot c)$

EXAMPLE 1 Using Properties to Write Equivalent Expressions

a. Simplify the expression $7 + (12 + x)$.

$7 + (12 + x) = (7 + 12) + x$ Associative Property of Addition
$= 19 + x$ Add 7 and 12.

b. Simplify the expression $(6.1 + x) + 8.4$.

$(6.1 + x) + 8.4 = (x + 6.1) + 8.4$ Commutative Property of Addition
$= x + (6.1 + 8.4)$ Associative Property of Addition
$= x + 14.5$ Add 6.1 and 8.4.

c. Simplify the expression $5(11y)$.

$5(11y) = (5 \cdot 11)y$ Associative Property of Multiplication
$= 55y$ Multiply 5 and 11.

Study Tip

One way to check whether expressions are equivalent is to evaluate each expression for any value of the variable. In Example 1(a), use $x = 2$.

$7 + (12 + x) = 19 + x$

$7 + (12 + 2) \stackrel{?}{=} 19 + 2$

$21 = 21$ ✓

On Your Own

Now You're Ready
Exercises 5–8

Simplify the expression. Explain each step.

1. $10 + (a + 9)$ **2.** $\left(c + \frac{2}{3}\right) + \frac{1}{2}$ **3.** $5(4n)$

Multi-Language Glossary at BigIdeasMath.com

Laurie's Notes

Introduction

Connect

- **Yesterday:** Students developed an understanding of the Commutative (order) and Associative (grouping) Properties. (MP2, MP3, MP6)
- **Today:** Students will use those two properties, and three other properties, to show that expressions are equivalent.

Discuss

- Begin with a discussion of what the word *equivalent* means. Examples: a dozen and 12 items, cell phone and mobile phone, etc.
- In mathematics you have equivalent fractions $\left(\text{like } \frac{1}{2} \text{ and } \frac{2}{4}\right)$ and equivalent measures (like 12 inches and 1 foot).
- Define **equivalent expressions**. Relate this to the *equivalent* things previously discussed.
- Previoiusly, students learned these properties using numerical expressions. Now, they will apply the properties to algebraic expressions.

Lesson Notes

Key Ideas

- Write the Commutative and Associative Properties. The key words to remember are *order* for the Commutative Property and *grouping* for the Associative Property.
- For the Commutative Property, multiplication can be represented in different ways: $ab = ba$ or $a \cdot b = b \cdot a$. This is also true for the Associative Property: $(ab)c = a(bc)$ versus $(a \cdot b) \cdot c = a \cdot (b \cdot c)$.
- Discuss that the variables represent all numbers that they have studied: whole numbers, fractions, and decimals. Share that they will want to verify that the properties hold for other types of numbers (integers) as well.

Example 1

- **MP6 Attend to Precision:** The challenges students have with these problems are that the solutions are obvious! Students will say that it seems okay to simply combine the numbers and then tag on the variable. Remind students that it is okay to *just combine numbers* for only the operations of addition and multiplication. If the first problem was $7 - (12 - x)$, combining the numbers would result in the wrong answer.
- **Connection:** Part (b) is a good place to check previous skills. Ask students how they would simplify $(6 + x) + 8.4$. You are trying to determine if students remember that when adding a whole number and a decimal, you must locate the decimal point to the right of the whole number and line up the decimal points. Check for understanding of adding like place values.

On Your Own

- **Neighbor Check:** Have students work independently and then have their neighbors check their work. Have students discuss any discrepancies.
- Note how Question 3 is represented without the multiplication dot.

Goal Today's lesson is using properties to show that two expressions are equivalent.

Technology for the Teacher

Dynamic Classroom

Lesson Tutorials
Lesson Plans
Answer Presentation Tool

Differentiated Instruction

Visual

Write numbers, addition operators, and multiplication operators on index cards. Model expressions and let students rearrange the cards to model the Commutative and Associative Properties. Create expressions so students see the benefits of using the properties.

Sum to multiples of 10:

$$17 + 6 + 3 + 4 = (17 + 3) + (6 + 4)$$

Multiply to get a product of 100:

$$\begin{aligned} 25 \times 15 \times 4 &= (25 \times 4) \times 15 \\ &= 100 \times 15 \\ &= 1500 \end{aligned}$$

Extra Example 1

a. Simplify the expression $(r + 3) + 6$.
$r + 9$

b. Simplify the expression $4(3p)$.
$12p$

c. Simplify the expression $(6 \cdot b) \cdot 7$.
$42b$

On Your Own

1. $10 + (a + 9) =$
$10 + (9 + a)$
Comm. Prop. of Add.
$= (10 + 9) + a$
Assoc. Prop. of Add.
$= 19 + a$
Add 10 and 9.

On Your Own

2. $\left(c+\frac{2}{3}\right)+\frac{1}{2}=c+\left(\frac{2}{3}+\frac{1}{2}\right)$
Assoc. Prop. of Add.
$=c+1\frac{1}{6}$
Add $\frac{2}{3}$ and $\frac{1}{2}$.

3. $5(4n)=(5 \cdot 4)n$
Assoc. Prop. of Mult.
$=20n$
Multiply 5 and 4.

Extra Example 2

a. Simplify the expression $0+(s+4)$. $s+4$

b. Simplify the expression $(h \cdot 1) \cdot 8$. $8h$

Extra Example 3

You hand out 425 programs on the first night of your school's variety show, p programs on the second night, and 520 programs on the third night. Write an expression for the total number of programs you handed out. Then simplify the expression. $425+p+520$; $p+945$

On Your Own

4. $12 \cdot b \cdot 0=12 \cdot (b \cdot 0)$
Assoc. Prop. of Mult.
$=12 \cdot 0$
Mult. Prop. of Zero
$=0$
Mult. Prop. of Zero

5. $1 \cdot m \cdot 24=(1 \cdot m) \cdot 24$
Assoc. Prop. of Mult.
$=m \cdot 24$
Mult. Prop. of One
$=24m$
Comm. Prop. of Mult.

6. $(t+15)+0=t+(15+0)$
Assoc. Prop. of Add.
$=t+15$
Add. Prop. of Zero

7. $7x+154.75$

Laurie's Notes

Key Ideas

- Write and discuss the three new properties. Students may say (or at least think) that these properties are obvious.
- Multiplying by zero, in any form, produces a product of 0.
- Multiplying by 1, in any form, does not change the value of the quantity.

  Remind students that 1 can be represented in different ways: 1, $\left(\frac{1}{2}+\frac{1}{2}\right)$, or $\frac{3}{3}$. This is also called the Multiplicative Identity.

Example 2

- Work through the two parts and point out the properties being used. Mention to students that the first step in both parts involves the Associative Property. In part (a) you would get the same result if $(0 \cdot p)$ had been grouped together first.

Example 3

- Ask a student to read the problem aloud. Ask how many T-shirts were purchased (note the common error in the margin next to the example). Explain that the total amount spent on T-shirts was 7 times the cost of one shirt.
- **Connection:** Here is an example of a whole number being added to a decimal. Did your students compute correctly?
- ? **Extension:** "What if the sponsor bought 3 extra T-shirts, for a total of 10, and the shirts were $12.45 each? What would be the cost of the T-shirts?" $124.50
- This is a quick review of how to multiply by 10.

On Your Own

- **Think-Pair-Share:** Students should read each question independently and then work in pairs to answer the questions. When they have answered the questions, the pair should compare their answers with another group and discuss any discrepancies.

Closure

- On the board in your class, ask student volunteers to share examples of:
  Commutative Property using a variable and a number
  Commutative Property using two variables
  Associative Property using 3 numbers

Key Ideas

Addition Property of Zero

Words The sum of any number and 0 is that number.

Numbers $7 + 0 = 7$ **Algebra** $a + 0 = a$

Multiplication Properties of Zero and One

Words The product of any number and 0 is 0.

The product of any number and 1 is that number.

Numbers $9 \cdot 0 = 0$ **Algebra** $a \cdot 0 = 0$

$4 \cdot 1 = 4$ $a \cdot 1 = a$

EXAMPLE 2 Using Properties to Write Equivalent Expressions

a. Simplify the expression $9 \cdot 0 \cdot p$.

$9 \cdot 0 \cdot p = (9 \cdot 0) \cdot p$ Associative Property of Multiplication

$= 0 \cdot p = 0$ Multiplication Property of Zero

b. Simplify the expression $4.5 \cdot r \cdot 1$.

$4.5 \cdot r \cdot 1 = 4.5 \cdot (r \cdot 1)$ Associative Property of Multiplication

$= 4.5 \cdot r$ Multiplication Property of One

$= 4.5r$

EXAMPLE 3 Real-Life Application

Common Error

You **and** six friends are on the team, so use the expression $7x$, not $6x$, to represent the cost of the T-shirts.

You and six friends play on a basketball team. A sponsor paid \$100 for the league fee, x dollars for each player's T-shirt, and \$68.25 for trophies. Write an expression for the total amount the sponsor paid.

Add the league fee, the cost of the T-shirts, and the cost of the trophies.

$100 + 7x + 68.25 = 7x + 100 + 68.25$ Commutative Property of Addition

$= 7x + 168.25$ Add 100 and 68.25.

An expression for the total amount is $7x + 168.25$.

On Your Own

Now You're Ready
Exercises 9–23

Simplify the expression. Explain each step.

4. $12 \cdot b \cdot 0$ **5.** $1 \cdot m \cdot 24$ **6.** $(t + 15) + 0$

7. **WHAT IF?** In Example 3, your sponsor paid \$54.75 for trophies. Write an expression for the total amount the sponsor paid.

3.3 Exercises

Check It Out
Help with Homework
BigIdeasMath.com

Vocabulary and Concept Check

1. **NUMBER SENSE** Write an example of a sum of fractions. Show that the Commutative Property of Addition is true for the sum.

2. **OPEN-ENDED** Write an algebraic expression that can be simplified using the Associative Property of Addition.

3. **OPEN-ENDED** Write an algebraic expression that can be simplified using the Associative Property of Multiplication and the Multiplication Property of One.

4. **WHICH ONE DOESN'T BELONG?** Which statement does *not* belong with the other three? Explain your reasoning.

$7 + (x + 4) = 7 + (4 + x)$	$(3 + b) + 2 = (b + 3) + 2$
$9 + (7 + w) = (9 + 7) + w$	$(4 + n) + 6 = (n + 4) + 6$

Practice and Problem Solving

Tell which property the statement illustrates.

1 5. $5 \cdot p = p \cdot 5$

6. $2 + (12 + r) = (2 + 12) + r$

7. $4 \cdot (x \cdot 10) = (4 \cdot x) \cdot 10$

8. $x + 7.5 = 7.5 + x$

2 9. $(c + 2) + 0 = c + 2$

10. $a \cdot 1 = a$

11. **ERROR ANALYSIS** Describe and correct the error in stating the property that the statement illustrates.

✗ $(7 + x) + 3 = (x + 7) + 3$
Associative Property of Addition

Simplify the expression. Explain each step.

12. $6 + (5 + x)$

13. $(14 + y) + 3$

14. $6(2b)$

15. $7(9w)$

16. $3.2 + (x + 5.1)$

17. $(0 + a) + 8$

18. $9 \cdot c \cdot 4$

19. $(18.6 \cdot d) \cdot 1$

20. $\left(3k + 4\frac{1}{5}\right) + 8\frac{3}{5}$

21. $(2.4 + 4n) + 9$

22. $(3s) \cdot 8$

23. $z \cdot 0 \cdot 12$

24. **GEOMETRY** The expression $12 + x + 4$ represents the perimeter of a triangle. Simplify the expression.

25. **RAISIN COOKIES** A case of raisin cookies has 10 cartons. A carton has 12 boxes. The amount you earn on a whole case is $10(12x)$ dollars.

   a. What does x represent?

   b. Simplify the expression.

Assignment Guide and Homework Check

Level	Day 1 Activity Assignment	Day 2 Lesson Assignment	Homework Check
Basic	5–8, 35–43	1–4, 9–23 odd, 24, 25, 27	7, 17, 21, 24, 27
Average	5–8, 35–43	1–4, 11–23 odd, 24–32 even	17, 21, 24, 26, 30
Advanced	5–8, 35–43	1–4, 11, 12–34 even	16, 22, 32, 34

Common Errors

- **Exercises 12–23** Students are often confused by the differences between the Commutative and Associative Properties and will incorrectly label steps. Show students that the Associative Property moves the parentheses but does not change the position of the numbers. The Commutative Property changes the positions of the numbers.
- **Exercises 29–33** Some students may still be struggling with the differences between the two properties. Ask students to think about how they would complete the statement without identifying a property. Once they know how to simplify it, ask them to look at the definitions of the properties and identify which property they used.
- **Exercise 34** Students may think that there is only one correct expression. Ask students to identify which property leads to the other expression.

3.3 Record and Practice Journal

Tell which property illustrates the statement.

1. $x \cdot 1 = x$
   Multiplication Property of One
2. $4.8 + k = k + 4.8$
   Commutative Property of Addition

Simplify the expression. Explain each step.

3. $8 + (7 + x)$
   $15 + x$
4. $10(11a)$
   $110a$

Complete the statement using the specified property.

	Property	Statement
5.	Addition Property of Zero	$(b + 0) + 6 = b + 6$
6.	Commutative Property of Multiplication	$3 \cdot (n \cdot 5) = 3 \cdot (5 \cdot n)$

7. You earn 10 points for every coin you collect in a video game. Then you find a star that triples your score.
   a. Write an expression for the number of points you earn from the coins.
   $10c$
   b. Write and simplify an expression for the total number of points you earn.
   $3(10c) = 30c$

Vocabulary and Concept Check

1. *Sample answer:*
   $$\frac{1}{5} + \frac{3}{5} = \frac{3}{5} + \frac{1}{5}$$
   $$\frac{4}{5} = \frac{4}{5}$$
2. *Sample answer:*
   $$(x + 8) + 4 = x + (8 + 4)$$
   $$= x + 12$$
3. *Sample answer:*
   $$(5 \cdot x) \cdot 1 = 5 \cdot (x \cdot 1)$$
   $$= 5x$$
4. $9 + (7 + w) = (9 + 7) + w$ does not belong because it demonstrates the Associative Property of Addition whereas the other statements demonstrate the Commutative Property of Addition.

Practice and Problem Solving

5. Comm. Prop. of Mult.
6. Assoc. Prop. of Add.
7. Assoc. Prop. of Mult.
8. Comm. Prop. of Add.
9. Add. Prop. of Zero
10. Mult. Prop. of One
11. The grouping of the numbers did not change. The statement illustrates the Commutative Property of Addition because the order of the addends changed.
12. $6 + (5 + x) = (6 + 5) + x$
    Assoc. Prop. of Add.
    $= 11 + x$
    Add 6 and 5.
13. $(14 + y) + 3 = (y + 14) + 3$
    Comm. Prop. of Add.
    $= y + (14 + 3)$
    Assoc. Prop. of Add.
    $= y + 17$
    Add 14 and 3.

Practice and Problem Solving

14. $6(2b) = (6 \cdot 2)b$
Assoc. Prop. of Mult.
$= 12b$
Multiply 6 and 2.

15. $7(9w) = (7 \cdot 9)w$
Assoc. Prop. of Mult.
$= 63w$
Multiply 7 and 9.

16. $3.2 + (x + 5.1) =$
$3.2 + (5.1 + x)$
Comm. Prop. of Add.
$= (3.2 + 5.1) + x$
Assoc. Prop. of Add.
$= 8.3 + x$
Add 3.2 and 5.1.

17–25. See Additional Answers.

26. See *Taking Math Deeper*.

27–34. See Additional Answers.

Fair Game Review

35. 98 **36.** 108

37. 90 **38.** 200

39. 37 is already prime.

40. $2^4 \times 3^2$

41. 3×7^2

42. 5×41

43. B

Mini-Assessment

Simplify the expression. Explain each step.

1. $7 + (3 + y)$ $10 + y$; Assoc. Prop. of Add.
2. $4(3x)$ $12x$; Assoc. Prop. of Mult.
3. $6 + (b + 4)$ $10 + b$; Comm. Prop. of Add.; Assoc. Prop. of Add.
4. $(7 \cdot d) \cdot 1$ $7d$; Assoc. Prop. of Mult.; Mult. Prop. of One
5. The expression $5 + x + 5$ represents the perimeter of a triangle. Simplify the expression. $10 + x$

Taking Math Deeper

Exercise 26

This problem emphasizes the power of visualization. In the drawing, x appears to be about half of the length, or about 6. Without redrawing, it is difficult for most people to "see" a different shape. So, encourage students to draw the prism for each of the values of x: 0.25, 12.5, and 144.

1 Given drawing:

a. Volume:
$V = 12.5 \cdot x \cdot 1$
$= 12.5x$

2 1 by 12.5 by 0.25

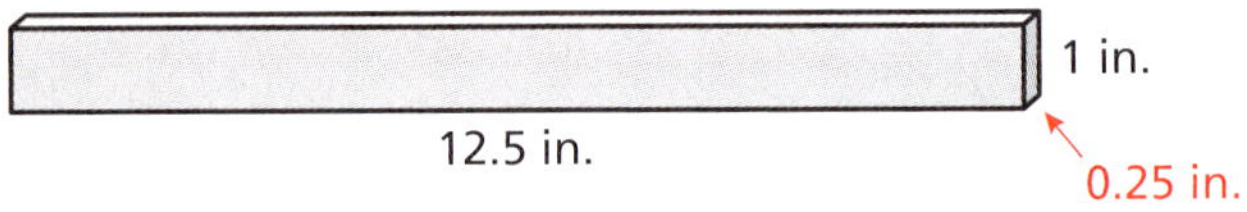

Looks like a ruler.

1 by 12.5 by 12.5

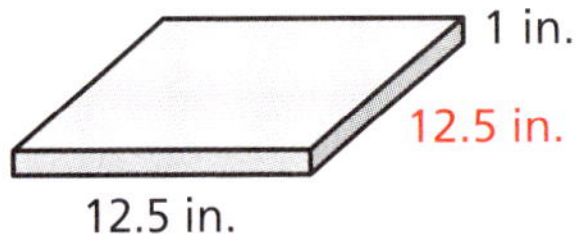

Looks like a floor tile.

1 by 12.5 by 144

Looks like siding.

3 **b.** So when $x = 0.25$, the object is B, the ruler. When $x = 12.5$, the object is C, the square floor tiles. When $x = 144$, the object is A, siding for a house.

Project

Think of three different objects you could use for the items in Exercise 26. Draw the objects and give reasonable values for x.

Reteaching and Enrichment Strategies

If students need help. . .	If students got it. . .
Resources by Chapter • Practice A and Practice B • Puzzle Time Record and Practice Journal Practice Differentiating the Lesson Lesson Tutorials Skills Review Handbook	Resources by Chapter • Enrichment and Extension • Technology Connection Start the next section

26. **STRUCTURE** The volume of the rectangular prism is $12.5 \cdot x \cdot 1$.

   a. Simplify the expression.

   b. Match $x = 0.25$, 12.5, and 144 with the object. Explain.

   A. siding for a house B. ruler C. square floor tile

Write the phrase as an expression. Then simplify the expression.

27. 7 plus the sum of a number x and 5

28. the product of 8 and a number y multiplied by 9

Copy and complete the statement using the specified property.

	Property	Statement
29.	Associative Property of Multiplication	$7(2y) =$ ▬
30.	Commutative Property of Multiplication	$13.2 \cdot (x \cdot 1) =$ ▬
31.	Associative Property of Addition	$17 + (6 + 2x) =$ ▬
32.	Addition Property of Zero	$2 + (c + 0) =$ ▬
33.	Multiplication Property of One	$1 \cdot w \cdot 16 =$ ▬

34. **HATS** You and a friend sell hats at a fair booth. You sell 16 hats on the first shift and 21 hats on the third shift. Your friend sells x hats on the second shift.

   a. Write an expression for the number of hats sold.

   b. The expression $37(14) + 10x$ represents the amount that you both earned. How can you tell that your friend was selling the hats for a discounted price?

   c. **Reasoning** You earned more money than your friend. What can you say about the value of x?

Fair Game Review What you learned in previous grades & lessons

Evaluate the expression. *(Section 1.3)*

35. $7(10 + 4)$ 36. $12(10 - 1)$ 37. $6(5 + 10)$ 38. $8(30 - 5)$

Find the prime factorization of the number. *(Section 1.4)*

39. 37 40. 144 41. 147 42. 205

43. **MULTIPLE CHOICE** A bag has 16 blue, 20 red, and 24 green marbles. What fraction of the marbles in the bag are blue? *(Skills Review Handbook)*

   (A) $\frac{1}{5}$ (B) $\frac{4}{15}$ (C) $\frac{4}{11}$ (D) $\frac{11}{15}$

3.4 The Distributive Property

Essential Question How do you use mental math to multiply two numbers?

The Meaning of a Word ● Distribute

When you **distribute** something to each person in a group,

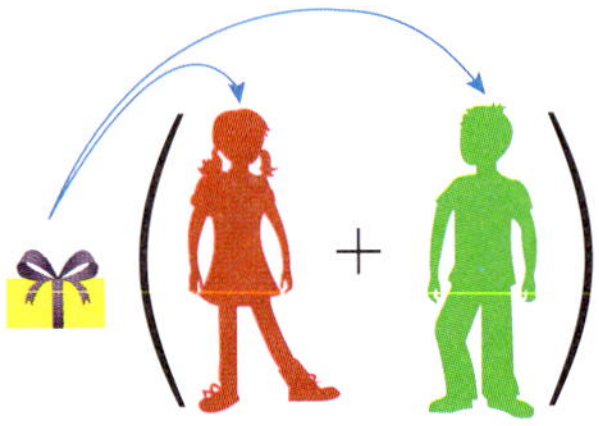

you give that thing to each person in the group.

1 ACTIVITY: Modeling a Property

Work with a partner.

a. **MODELING** Draw two rectangles of the same width but with different lengths on a piece of grid paper. Label the dimensions.

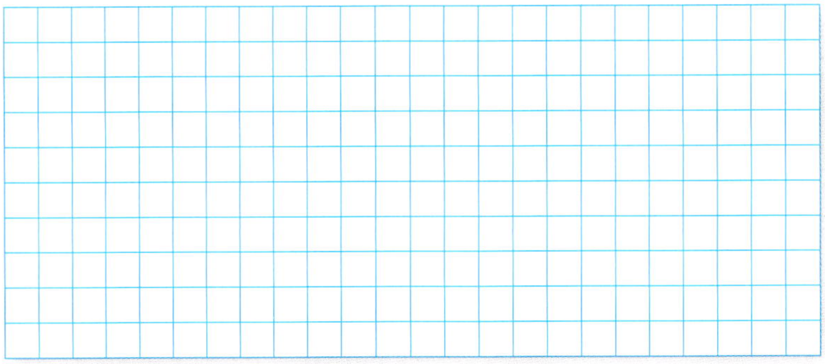

b. Write an expression for the total area of the rectangles.

$$(\square \times \square) + (\square \times \square)$$

c. Rearrange the rectangles by aligning the shortest sides to form one rectangle. Label the dimensions. Write an expression for the area.

$$\square \times (\square + \square)$$

d. Can the expressions from parts (b) and (c) be set equal to each other? Explain.

e. **REPEATED REASONING** Repeat this activity using different rectangles. Explain how this illustrates the Distributive Property. Write a rule for the Distributive Property.

Common Core

Equivalent Expressions

In this lesson, you will
- use the Distributive Property to find products.
- use the Distributive Property to simplify algebraic expressions.

Learning Standards
6.NS.4
6.EE.2b
6.EE.3
6.EE.4

Laurie's Notes

Introduction

Standards for Mathematical Practice

- **MP4 Model with Mathematics:** Mathematically proficient students are able to use models to represent relationships in a problem. In this lesson, students use an area model to represent multiplication.

Motivate

- Play *Equal Corners.* Write four problems in the corners of a rectangle on the board. Two of the four problems should have the same product. Students then circle the two equal products. Try several rounds. *Sample:*

(20 × 8)	(4 × 40)
60 × 10	20 × 40

Discuss

? "How did you determine which products were equal?"

- **Listen:** Pay attention to how students explain multiplying by multiples of 10. Students should multiply the nonzero numbers, then insert zeros equal to the total number in the two factors.

Activity Notes

Activity 1

- This activity can be modeled using square tiles or with an interactive whiteboard. The area model, a visual representation, helps students make sense of this numeric property.
- If initial rectangles in part (a) are not too large, there should be sufficient room to complete part (c). If necessary, have additional grid paper available.

? When students have finished ask, "What expression did you write in part (b) for the total area of the rectangles?" Answers will vary.

? "What expression did you write in part (c) for the single, larger rectangle?" Answers will vary.

- **MP3 Construct Viable Arguments and Critique the Reasoning of Others:** Ask a volunteer to share his or her explanation for part (d).
- **MP8 Look for and Express Regularity in Repeated Reasoning:** Students should work through several trials, trying different-sized rectangles. Generalizations should be evident from the samples students have drawn.

Common Core State Standards

6.NS.4 . . . Use the distributive property to express a sum of two whole numbers 1–100 with a common factor as a multiple of a sum of two whole numbers with no common factor.

6.EE.2b Identify parts of an expression. . .

6.EE.3 Apply the properties of operations to generate equivalent expressions.

6.EE.4 Identify when two expressions are equivalent.

Previous Learning

Students need to be familiar with basic computation facts. Applying mental math strategies will be helpful.

Technology for the Teacher

Dynamic Classroom

Lesson Plans

Complete Materials List

3.4 Record and Practice Journal

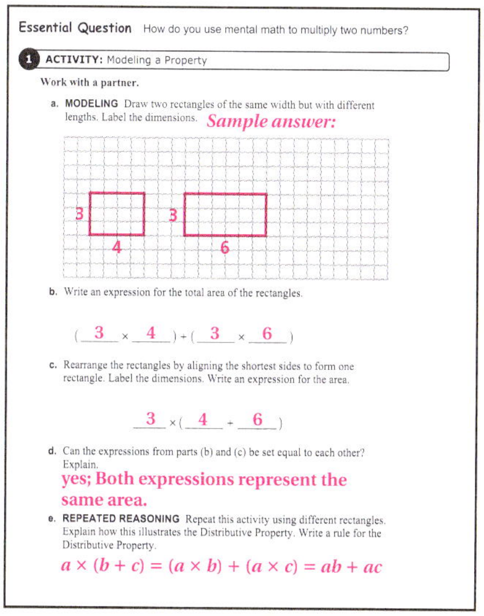

Essential Question How do you use mental math to multiply two numbers?

1 ACTIVITY: Modeling a Property

Work with a partner.

a. MODELING Draw two rectangles of the same width but with different lengths. Label the dimensions. *Sample answer:*

b. Write an expression for the total area of the rectangles.

(3 × 4) + (3 × 6)

c. Rearrange the rectangles by aligning the shortest sides to form one rectangle. Label the dimensions. Write an expression for the area.

3 × (4 + 6)

d. Can the expressions from parts (b) and (c) be set equal to each other? Explain.

yes; Both expressions represent the same area.

e. REPEATED REASONING Repeat this activity using different rectangles. Explain how this illustrates the Distributive Property. Write a rule for the Distributive Property.

$a \times (b + c) = (a \times b) + (a \times c) = ab + ac$

Differentiated Instruction

Auditory

Some students may prefer to learn a verbal statement of the Distributive Property: "*a* times the quantity *b* plus *c* equals *a* times *b* plus *a* times *c*" or "*a* times the quantity *b* minus *c* equals *a* times *b* minus *a* times *c*." Remind students that the quantity *b* plus *c* means that *b* and *c* are grouped together.

3.4 Record and Practice Journal

2 ACTIVITY: Using Mental Math

Work with a partner. Use mental math to find the product.

a. 23×6 **138**

b. 33×7 **231**

c. 47×9 **423**

d. 28×5 **140**

e. 17×4 **68**

3 ACTIVITY: Using Mental Math

Work with a partner. Use the Distributive Property and mental math to find the product.

a. 6×23 **138**

b. 5×17 **85**

c. 8×26 **208**

d. 20×19 **380**

e. 40×29 **1160**

f. 25×39 **975**

g. 15×47 **705**

What Is Your Answer?

4. Compare the methods in Activities 2 and 3.
Check students' work.

5. IN YOUR OWN WORDS How do you use mental math to multiply two numbers? Give examples to support your explanation.
Rewrite a multiplication problem as the sum or difference of two simpler products.

Laurie's Notes

Activity 2

- **Connection:** When students learned multiplication of a single-digit number times a two-digit number in grade 3 (3.NBT.3), they likely used base-ten pieces to make an area model. In multiplying 6×23, they would have made a rectangle similar to the following.

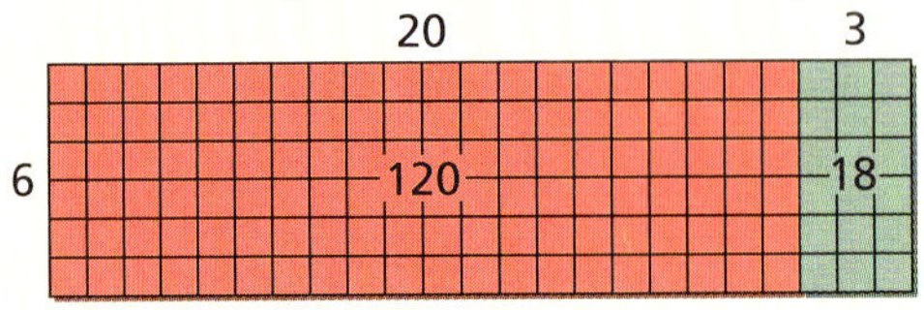

- Take time to contrast and compare: the model shown above, the model shown in Activity 1, and the traditional multiplication algorithm.
- As a true mental math activity, students must hold the first product in their heads while they perform the second multiplication. When you ask students to share their answers, ask them to explain their methods.

Activity 3

- This activity should help students recognize that any 2-digit number can be represented as a sum or difference. Example : $49 = 40 + 9 = 50 - 1$
- The last two parts involve *nice numbers*. Ask students how they did these problems.
  Example: $25 \times 39 = 25 \times (40 - 1) = (25 \times 40) - (25 \times 1) = 1000 - 25 = 975$
- If time permits, discuss how order of operations is involved in this problem. The parentheses are not needed around the two products (25×40) and (25×1).

What Is Your Answer?

- For Question 5, you are listening for the idea that a 2-digit number can be written as a sum or a difference.
- Have students share their answers.

Closure

- Discuss with students that even though 8×36 is represented with one multiplication symbol, two products are found. Multiplying 8×36 involves distributing the 8 over the 36 by multiplying 8×30 and 8×6.

2 ACTIVITY: Using Mental Math

Math Practice 1

Find Entry Points

How can you rewrite the larger number as the sum of two numbers so that you can use mental math?

Work with a partner. Use the method shown to find the product.

a. Sample: 23×6

$$\begin{array}{r} 23 \\ \times\ 6 \\ \hline 120 \\ +\ 18 \\ \hline 138 \end{array}$$

23 is 20 + 3.

Multiply 20 and 6.

Multiply 3 and 6.

Add.

So, $23 \times 6 = 138$.

b. 33×7 **c.** 47×9

d. 28×5 **e.** 17×4

3 ACTIVITY: Using Mental Math

Work with a partner. Use the Distributive Property and mental math to find the product.

a. Sample: 6×23

$6 \times 23 = 6 \times (20 + 3)$	Write 23 as the sum of 20 and 3.
$= (6 \times 20) + (6 \times 3)$	Distribute the 6 over the sum.
$= 120 + 18$	Find the products.
$= 138$	Add.

So, $6 \times 23 = 138$.

b. 5×17 **c.** 8×26

d. 20×19 **e.** 40×29

f. 25×39 **g.** 15×47

What Is Your Answer?

4. Compare the methods in Activities 2 and 3.

5. **IN YOUR OWN WORDS** How do you use mental math to multiply two numbers? Give examples to support your explanation.

Practice Use what you learned about the Distributive Property to complete Exercises 5–8 on page 137.

3.4 Lesson

Check It Out
Lesson Tutorials
BigIdeasMath.com

Key Vocabulary
like terms, *p. 136*

Key Idea

Distributive Property

Words To multiply a sum or difference by a number, multiply each number in the sum or difference by the number outside the parentheses. Then evaluate.

Numbers $3(7 + 2) = 3 \times 7 + 3 \times 2$

$3(7 - 2) = 3 \times 7 - 3 \times 2$

Algebra $a(b + c) = ab + ac$

$a(b - c) = ab - ac$

EXAMPLE 1 Using Mental Math

Use the Distributive Property and mental math to find 8×53.

$8 \times 53 = 8(50 + 3)$	Write 53 as 50 + 3.
$= 8(50) + 8(3)$	Distributive Property
$= 400 + 24$	Multiply.
$= 424$	Add.

EXAMPLE 2 Using the Distributive Property

Use the Distributive Property to find $\frac{1}{2} \times 2\frac{3}{4}$.

$\frac{1}{2} \times 2\frac{3}{4} = \frac{1}{2} \times \left(2 + \frac{3}{4}\right)$	Rewrite $2\frac{3}{4}$ as the sum $2 + \frac{3}{4}$.
$= \left(\frac{1}{2} \times 2\right) + \left(\frac{1}{2} \times \frac{3}{4}\right)$	Distributive Property
$= 1 + \frac{3}{8}$	Multiply.
$= 1\frac{3}{8}$	Add.

On Your Own

Now You're Ready
Exercises 5–16

Use the Distributive Property to find the product.

1. 5×41
2. 9×19
3. $6(37)$
4. $\frac{2}{3} \times 1\frac{1}{2}$
5. $\frac{1}{4} \times 4\frac{1}{5}$
6. $\frac{2}{7} \times 3\frac{3}{4}$

Laurie's Notes

Introduction

Connect

- **Yesterday:** Students developed an understanding that multiplying numbers with more than one digit involves partial products: the one digit is *distributed over* the digits of the 2-digit number. (MP3, MP4, MP8)
- **Today:** Students will represent this process with algebra and refer to it as the Distributive Property.

Discuss

- Begin with a discussion of how multiplication can be represented.
  8×53 $8 \cdot 53$ $8(53)$
- There is also a fourth representation for variables.
  $8 \times n$ $8 \cdot n$ $8(n)$ $8n$

Lesson Notes

Key Idea

- **Representation:** Draw attention to how the multiplication is being shown.
- Students may ask why you would write 3×9 as $3(7 + 2)$ and 3×5 as $3(7 - 2)$. The answer is that you wouldn't! Simple values are being used to demonstrate the property instead of using a problem such as 8×67.
- Spend more time discussing the algebraic representation. In the problem $a(b + c)$, ask what *operations* students see. Not all students will see two. Some may only see addition.
- Explain that the blue arrows are used to remind us that we are distributing one number over another. The arrows can also be written above the problem.

Example 1

- Work through the problem. Draw a connection to what students did yesterday. The only thing that has changed is how multiplication is being represented, which is easier for math teachers to see than for students!

Example 2

- **MP7 Look for and Make Use of Structure:** Students might think of $2\frac{3}{4}$ as $\frac{11}{4}$ or as $2 + \frac{3}{4}$. Discuss the advantage of thinking of it as a sum versus an improper fraction.

? "How can the Distributive Property be used to help you multiply $\frac{1}{2} \times 2\frac{3}{4}$?" Think of $2\frac{3}{4}$ as $2 + \frac{3}{4}$.

On Your Own

- **Neighbor Check:** Have students work independently and then have their neighbors check their work. Have students discuss any discrepancies.

Goal Today's lesson is applying the Distributive Property to show that two expressions are equivalent.

Technology for the Teacher

Dynamic Classroom

Lesson Tutorials
Lesson Plans
Answer Presentation Tool

Extra Example 1

Use the Distributive Property and mental math to find 6×82. 492

Extra Example 2

Use the Distributive Property to find $\frac{5}{6} \times 3\frac{1}{10}$. $2\frac{7}{12}$

On Your Own

1. 205
2. 171
3. 222
4. 1
5. $1\frac{1}{20}$
6. $1\frac{1}{14}$

Extra Example 3

Use the Distributive Property to simplify the expression.

a. $11(x + 7)$ $11x + 77$

b. $8(c - 4)$ $8c - 32$

c. $3(4 + r + 3)$ $3r + 21$

On Your Own

7. $7a + 14$
8. $3d - 33$
9. $56 - 28d$

Extra Example 4

Maria is x years old. Her father is 25 years older than Maria. Maria's grandfather is two times older than Maria's father. Write and simplify an expression that represents Maria's grandfather's age in years.

$2(x + 25) = 2x + 50$

English Language Learners

Vocabulary

When you give examples of the Distributive Property, you may want to refer to the number outside the parentheses as the *outside number* and the numbers inside the parentheses as the *inside numbers*.

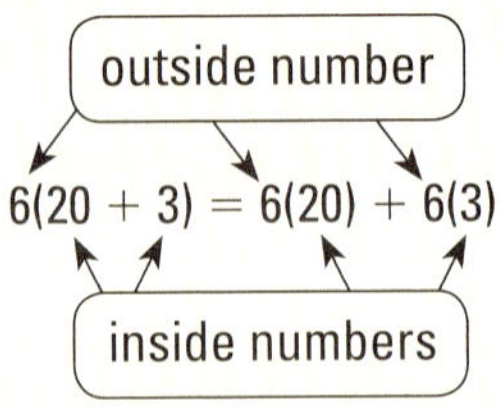

Laurie's Notes

Example 3

- Slowly work through the three parts. Introducing a variable expression is *not* a small change! The use of arrows to show that the 4 is being distributed over the n and the 5 is helpful.

On Your Own

- **Think-Pair-Share:** Students should read each question independently and then work in pairs to answer the questions. When they have answered the questions, the pair should compare their answers with another group and discuss any discrepancies.
- Some students may recognize that $7(2 + 6 - 4d)$ is equivalent to $7(8 - 4d)$.

Example 4

- Write the table representing each person's age in words and algebraically. Connect this problem back to Section 3.1.

? **Extension:** "If José is 12, find each person's age." Felipe is 14 and Maria is 42.

EXAMPLE 3 Simplifying Algebraic Expressions

Use the Distributive Property to simplify the expression.

a. $4(n+5)$

$4(n+5) = 4(n) + 4(5)$	Distributive Property
$= 4n + 20$	Multiply.

b. $12(2y-3)$

$12(2y-3) = 12(2y) - 12(3)$	Distributive Property
$= 24y - 36$	Multiply.

c. $9(6+x+2)$

$9(6+x+2) = 9(6) + 9(x) + 9(2)$	Distributive Property
$= 54 + 9x + 18$	Multiply.
$= 9x + 54 + 18$	Commutative Property of Addition
$= 9x + 72$	Add 54 and 18.

Study Tip

You can use the Distributive Property when there are more than two terms in the sum or difference.

On Your Own

Now You're Ready
Exercises 17–32

Use the Distributive Property to simplify the expression.

7. $7(a+2)$ **8.** $3(d-11)$ **9.** $7(2+6-4d)$

EXAMPLE 4 Real-Life Application

José is x years old. His brother, Felipe, is 2 years older than José. Their aunt, Maria, is three times as old as Felipe. Write and simplify an expression that represents Maria's age in years.

Name	Description	Expression
José	He is x years old.	x
Felipe	He is 2 years *older* than José. So, *add* 2 to x.	$x + 2$
Maria	She is three *times* as old as Felipe. So, *multiply* 3 and $(x + 2)$.	$3(x+2)$

$3(x+2) = 3(x) + 3(2)$	Distributive Property
$= 3x + 6$	Multiply.

∴ Maria's age in years is represented by the expression $3x + 6$.

On Your Own

10. Alexis is x years old. Her sister, Gloria, is 7 years older than Alexis. Their grandfather is five times as old as Gloria. Write and simplify an expression that represents their grandfather's age in years.

In an algebraic expression, **like terms** are terms that have the same variables raised to the same exponents. Constant terms are also like terms.

Like terms ($5x$ and $2x$)

$$5x + 19 + 2x + 2$$

Like terms (19 and 2)

Use the Distributive Property to *combine* like terms.

EXAMPLE 5 Combining Like Terms

Simplify each expression.

a. $3x + 9 + 2x - 5$

$3x + 9 + 2x - 5 = 3x + 2x + 9 - 5$	Commutative Property of Addition
$= (3 + 2)x + 9 - 5$	Distributive Property
$= 5x + 4$	Simplify.

b. $y + y + y$

$y + y + y = 1y + 1y + 1y$	Multiplication Property of One
$= (1 + 1 + 1)y$	Distributive Property
$= 3y$	Add coefficients.

c. $7z + 2(z - 5y)$

$7z + 2(z - 5y) = 7z + 2(z) - 2(5y)$	Distributive Property
$= 7z + 2z - 10y$	Multiply.
$= (7 + 2)z - 10y$	Distributive Property
$= 9z - 10y$	Add coefficients.

On Your Own

Now You're Ready
Exercises 39–53

Simplify the expression.

11. $8 + 3z - z$

12. $3(b + 5) + b + 2$

Laurie's Notes

On Your Own

- **Think-Pair-Share:** Students should read each question independently and then work in pairs to answer the questions. When they have answered the questions, the pair should compare their answers with another group and discuss any discrepancies.

Discuss

- Write the expression $3x + 2x + 12 - 6$ on the board and ask a student to read it.

? "How many terms are there?" 4 "What are the constant terms?" 12 and 6 Note that technically the constant terms are 12 and –6, although integers have not been introduced.

- Discuss that the terms $3x$ and $2x$ are called like terms because they have the same variable being raised to the same exponent. The Distributive Property is used to simplify like terms.

Example 5

- Work through the first two problems as shown. These problems make sense to students. The third problem is more involved.

? Write the third expression and ask, "How many operations do you see? What are they?" 5 operations; multiply, add, multiply, subtract, multiply

- **MP7:** Help students recognize the structure of this expression. I like to read it as "7 times the variable z plus 2 times the quantity z subtract $5y$." Before working with the $7z$, the Distributive Property must be used.
- Use arrows to help students see the 2 being distributed over the z and the $5y$.

On Your Own

- **Neighbor Check:** Have students work independently and then have their neighbors check their work. Have students discuss any discrepancies.

Closure

- **Writing:** Ask students to explain the Distributive Property in words. As a writing activity, have students write a paragraph explaining how to do the problem $5(n + 8)$.

On Your Own

10. $5(x + 7) = 5x + 35$

Extra Example 5

Simplify each expression.

a. $7w - 4 + 3w + 11$ $10w + 7$

b. $b + b + b + b$ $4b$

c. $2c + 3(f + 5c)$ $17c + 3f$

On Your Own

11. $8 + 2z$

12. $4b + 17$

Vocabulary and Concept Check

1. *Sample answer:* You must distribute or give the number outside the parentheses to the numbers inside the parentheses.
2. *Sample answer:*
   $6 + [2(2 + x)]$
   $= 6 + [2(2) + 2(x)]$
   $= 6 + (4 + 2x)$
   $= (6 + 4) + 2x$
   $= 10 + 2x$
3. $4 + (x \cdot 4)$ does not belong because it does not represent the Distributive Property.
4. $8x$ and $7x$, 1 and 4

Practice and Problem Solving

5. 63 **6.** 684

7. 516 **8.** 440

9. 936 **10.** 216

11. 504 **12.** 196

13. $\frac{4}{7}$ **14.** 2

15. $2\frac{1}{2}$ **16.** $\frac{3}{4}$

17. $3x + 12$ **18.** $10b - 60$

19. $6s - 54$ **20.** $56 + 7y$

21. $96 + 8a$ **22.** $18n + 9$

23. $72 - 12k$ **24.** $90 - 54w$

25. $63 + 9c$ **26.** $70 + 7x$

27. $40g + 24$ **28.** $78 + 6z$

29. $4x + 4y$ **30.** $25x - 25y$

31. $7p + 7q + 63$

32. $13n + 52 + 91m$

33. The 6 was not distributed to the 8 inside the parentheses. $6(y + 8) = 6y + 48$

Assignment Guide and Homework Check

Level	Day 1 Activity Assignment	Day 2 Lesson Assignment	Homework Check
Basic	5–8, 66–70	1–4, 17, 19, 25, 27, 33, 35, 39–43 odd, 54	17, 19, 27, 39, 43
Average	5–8, 66–70	1–4, 22–32 even, 33, 34, 39–49 odd, 54	22, 28, 30, 41, 43
Advanced	5–8, 66–70	1–4, 22–32 even, 33, 34, 42–54 even, 58	28, 30, 44, 46, 52

Common Errors

- **Exercises 13–16** Students may multiply the fractions and then *add on* the whole number. For example, $\frac{1}{2} \times 2\frac{3}{4} \neq 2\frac{3}{8}$. Remind them to multiply both the whole part and the fraction part by the fraction and then add.
- **Exercises 17–32** Students might forget to distribute to each part of the expression especially when there are more than two parts within the parentheses. Tell students to write the expression on their paper and draw arrows from the number being distributed to each part.

3.4 Record and Practice Journal

Use the Distributive Property and mental math to find the product.

1. 4×31 — **124**
2. 7×49 — **343**
3. $16(38)$ — **608**

Use the Distributive Property to simplify the expression.

4. $8(5 + w)$ — $40 + 8w$
5. $11(9 + d)$ — $99 + 11d$
6. $15(p - 4 + 2)$ — $15p - 30$

Simplify the expression.

7. $2x - 4 + 3x$ — $5x - 4$
8. $4y - 1 - 3y + 2$ — $y + 1$
9. $x + 2(x - 4)$ — $3x - 8$

10. A jazz band sells 31 large boxes of fruit and 74 small boxes of fruit for a fundraiser.

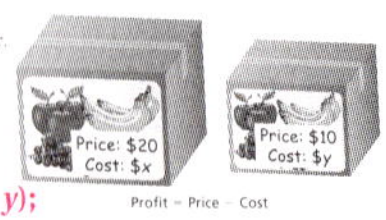

Profit = Price − Cost

   a. Use the Distributive Property to write and simplify an expression for the profit.
   $31(20 - x) + 74(10 - y)$; $1360 - 31x - 74y$
   b. A large box of fruit costs $9 and a small box of fruit costs $4. What is the jazz band's profit?
   $785

3.4 Exercises

Check It Out
Help with Homework
BigIdeasMath.com

Vocabulary and Concept Check

1. **WRITING** One meaning of the word *distribute* is "to give something to each member of a group." How can this help you remember the Distributive Property?

2. **OPEN-ENDED** Write an algebraic expression in which you use the Distributive Property and then the Associative Property of Addition to simplify.

3. **WHICH ONE DOESN'T BELONG?** Which expression does *not* belong with the other three? Explain your reasoning.

   $2(x + 2)$ $5(x - 8)$ $4 + (x \cdot 4)$ $8(9 - x)$

4. Identify the like terms in the expression $8x + 1 + 7x + 4$.

Practice and Problem Solving

Use the Distributive Property and mental math to find the product.

1 **5.** 3×21 **6.** 9×76 **7.** $12(43)$ **8.** $5(88)$

9. 18×52 **10.** 8×27 **11.** $8(63)$ **12.** $7(28)$

Use the Distributive Property to find the product.

2 **13.** $\frac{1}{4} \times 2\frac{2}{7}$ **14.** $\frac{5}{6} \times 2\frac{2}{5}$ **15.** $\frac{5}{9} \times 4\frac{1}{2}$ **16.** $\frac{2}{15} \times 5\frac{5}{8}$

Use the Distributive Property to simplify the expression.

3 **17.** $3(x + 4)$ **18.** $10(b - 6)$ **19.** $6(s - 9)$ **20.** $7(8 + y)$

21. $8(12 + a)$ **22.** $9(2n + 1)$ **23.** $12(6 - k)$ **24.** $18(5 - 3w)$

25. $9(3 + c + 4)$ **26.** $7(8 + x + 2)$ **27.** $8(5g + 5 - 2)$ **28.** $6(10 + z + 3)$

29. $4(x + y)$ **30.** $25(x - y)$ **31.** $7(p + q + 9)$ **32.** $13(n + 4 + 7m)$

33. **ERROR ANALYSIS** Describe and correct the error in rewriting the expression.

✗ $6(y + 8) = 6y + 8$

34. **ART MUSEUM** A class of 30 students visits an art museum and a special exhibit while there.

a. Use the Distributive Property to write and simplify an expression for the cost.

b. Estimate a reasonable value for x. Explain.

c. Use your estimate for x to evaluate the original expression and the simplified expression in part (a). Are the values the same?

PRICES

	Museum	Exhibit
Child (under 5)	Free	Free
Student	\$8	\$x
Regular	\$12	\$4
Senior	\$10	\$3

35. **FITNESS** Each day, you run on a treadmill for r minutes and lift weights for 15 minutes. Which expressions can you use to find how many minutes of exercise you do in 5 days? Explain your reasoning.

$5(r + 15)$ $5r + 5 \cdot 15$ $5r + 15$ $r(5 + 15)$

36. **SPEED** A cheetah can run 103 feet per second. A zebra can run x feet per second. Use the Distributive Property to write and simplify an expression for how much farther the cheetah can run in 10 seconds.

UNIFORMS **Your baseball team has 16 players. Use the Distributive Property to write and simplify an expression for the total cost of buying the items shown for all the players.**

37.

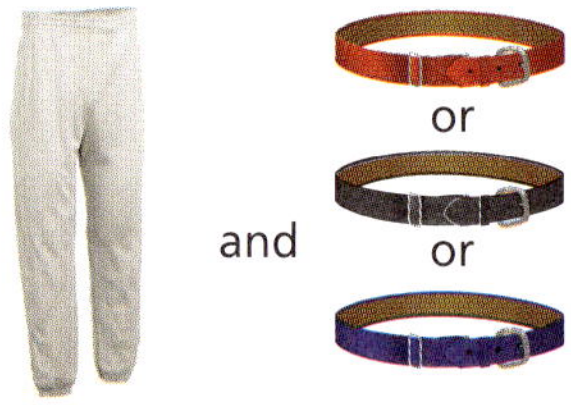

38.

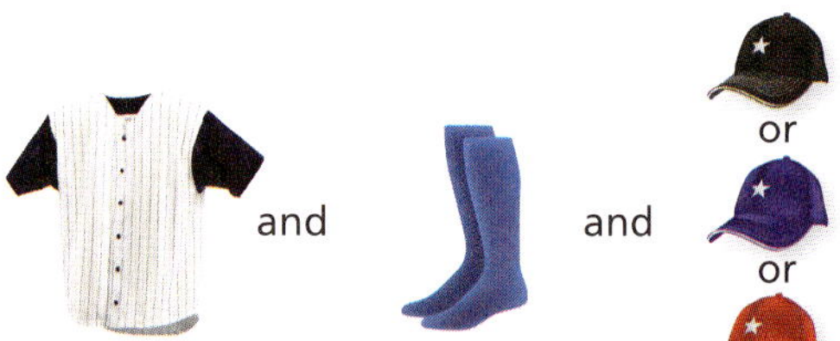

5 **Simplify the expression.**

39. $6(x + 4) + 1$

40. $5 + 8(3 + x)$

41. $7(8 + 4k) + 12$

42. $x + 3 + 5x$

43. $7y + 6 - 1 + 12y$

44. $w + w + 5w$

45. $4d + 9 - d - 8$

46. $n + 3(n - 1)$

47. $2v + 8v - 5v$

48. $5(z + 4) + 5(2 - z)$

49. $2.7(w - 5.2)$

50. $\frac{2}{3}y + \frac{1}{6}y + y$

51. $\frac{3}{4}\left(z + \frac{2}{5}\right) + 2z$

52. $7(x + y) - 7x$

53. $4x + 9y + 3(x + y)$

54. **ERROR ANALYSIS** Describe and correct the error in simplifying the expression.

✗
$$\begin{aligned} 8x - 2x + 5x &= 8x - 7x \\ &= (8 - 7)x \\ &= x \end{aligned}$$

ALGEBRA **Find the value of x that makes the expressions equivalent.**

55. $4(x - 5)$; $32 - 20$

56. $2(x + 9)$; $30 + 18$

57. $7(8 - x)$; $56 - 21$

58. **REASONING** Simplify the expressions and compare. What do you notice? Explain.

$4(x + 6)$ $(x + 6) + (x + 6) + (x + 6) + (x + 6)$

Common Errors

- **Exercises 39–53** Students may struggle using the Distributive Property to combine like terms. Remind them that the factor common to both terms is outside of the parentheses leaving the terms to be added or subtracted inside the parentheses.
- **Exercises 55–57** Students might only use *Guess, Check, and Revise* to solve. Encourage students to use the Distributive Property and then compare the expressions to find the value of x.

Practice and Problem Solving

34. **a.** $30(8 + x) = 240 + 30x$

b. *Sample answer:* \$2; It is less than the regular price to the exhibit.

c. *Sample answer:* \$300; yes

35. $5(r + 15)$ and $5r + 5 \cdot 15$, because they are equivalent expressions.

36. $10(103 - x) = 1030 - 10x$

37. $16(10 + x) = 160 + 16x$

38. $16(12 + 4 + x) = 256 + 16x$

39. $6x + 25$ **40.** $29 + 8x$

41. $68 + 28k$ **42.** $6x + 3$

43. $19y + 5$ **44.** $7w$

45. $3d + 1$ **46.** $4n - 3$

47. $5v$ **48.** 30

49. $2.7w - 14.04$

50. $\frac{11}{6}y$ **51.** $\frac{11}{4}z + \frac{3}{10}$

52. $7y$ **53.** $7x + 12y$

54. Adding before subtracting.

$$\begin{aligned} 8x - 2x + 5x &= (8 - 2)x + 5x \\ &= 6x + 5x \\ &= (6 + 5)x \\ &= 11x \end{aligned}$$

55. $x = 8$

56. $x = 15$ **57.** $x = 3$

English Language Learners

Pair Activity

Have students work in pairs and assign a group of problems. One student completes the first step in simplifying the expression. The second student completes the next step and returns it to the first student. Students continue to alternate the writing of steps until the expression is written in simplest form. This interaction allows students to develop an understanding of the concept and allows the teacher the opportunity to provide one-on-one assistance.

Practice and Problem Solving

58. $4x + 24$; $4x + 24$; The results are the same. Multiplying $(x + 6)$ by 4 and adding 4 groups of $(x + 6)$ are equivalent operations.

59. Area: $8x + 64$
Perimeter: $2x + 32$

60. Area: $12x + 66$
Perimeter: $2x + 35$

61. Area: $9x + 108$
Perimeter: $2x + 42$

62. See *Taking Math Deeper*.

63–65. See Additional Answers.

Fair Game Review

66. 10.641 **67.** 34.006

68. 135.213 **69.** 0.387

70. B

Mini-Assessment

Use the Distributive Property to simplify the expression.

1. $6(x + 4)$ $6x + 24$
2. $14(g - 2)$ $14g - 28$
3. $3(a + 10)$ $3a + 30$
4. $8(n + 3 + 4)$ $8n + 56$
5. Mike is y years old. His cousin, Jeremy, is 3 years older than Mike. Their friend, Ryan, is two times as old as Jeremy. Write and simplify an expression that represents Ryan's age in years. $2(y + 3)$; $2y + 6$

Taking Math Deeper

Exercise 62

The key to this problem is to read all of the information *carefully!* There are two types of dollar amounts in the problem. One is *cost* (what a business pays). The other is *price* (what a customer pays). The difference between price and cost is the profit for the company.

1 Organize information.

2 Write expressions for the profit.

Profit for Large Candle: $42(10 - x)$ — Profit = Price − Cost

Profit for Small Candle: $56(5 - y)$ — Profit = Price − Cost

a. Total Profit $= 42(10 - x) + 56(5 - y)$ Add expressions.
$= 420 - 42x + 280 - 56y$ Distributive Property
$= 700 - 42x - 56y$ Simplify.

3 Find the profit when $x = 5$ and $y = 3$.

b. Total Profit $= 700 - 42x - 56y$
$= 700 - 42(5) - 56(3)$
$= 700 - 210 - 168$
$= \$322$

Project

Make a poster to market and sell your candles. Include mathematical information in your poster. Do you think that advertising will help your sales?

Reteaching and Enrichment Strategies

If students need help. . .	If students got it. . .
Resources by Chapter • Practice A and Practice B • Puzzle Time Record and Practice Journal Practice Differentiating the Lesson Lesson Tutorials Skills Review Handbook	Resources by Chapter • Enrichment and Extension • Technology Connection Start the next section

GEOMETRY **Write and simplify expressions for the area and perimeter of the rectangle.**

59.

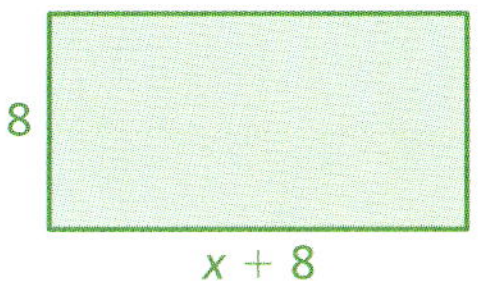

60.

61.

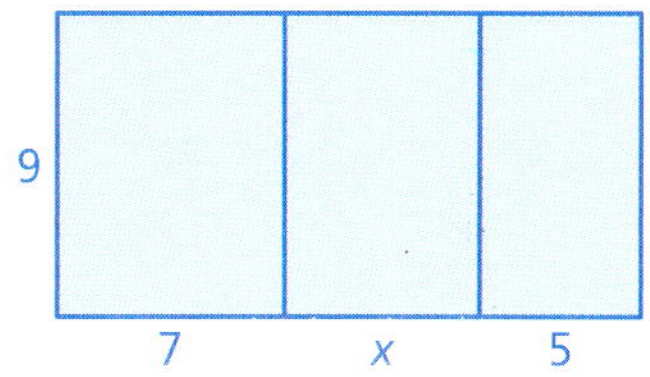

62. **FUNDRAISER** An art club sells 42 large candles and 56 small candles.

Profit = Price − Cost

a. Use the Distributive Property to write and simplify an expression for the profit.

b. A large candle costs \$5, and a small candle costs \$3. What is the club's profit?

63. **REASONING** Evaluate each expression by (1) using the Distributive Property and (2) evaluating inside the parentheses first. Which method do you prefer? Is your preference the same for both expressions? Explain your reasoning.

a. $2(3.22 - 0.12)$ **b.** $12\left(\frac{1}{2} + \frac{2}{3}\right)$

64. **REASONING** Write and simplify an expression for the difference between the perimeters of the rectangle and the hexagon. Interpret your answer.

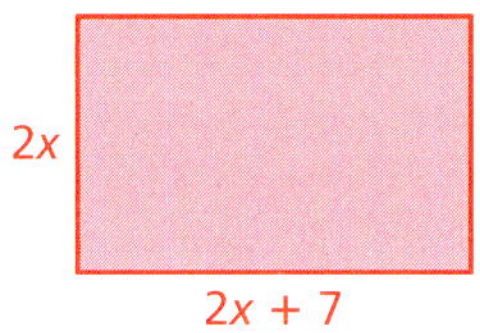

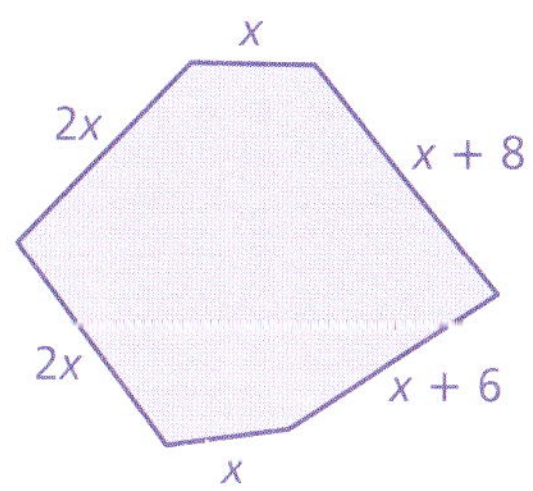

65. Puzzle Add one set of parentheses to the expression $7 \cdot x + 3 + 8 \cdot x + 3 \cdot x + 8 - 9$ so that it is equivalent to $2(9x + 10)$.

Fair Game Review What you learned in previous grades & lessons

Evaluate the expression. *(Section 2.4, Section 2.5, and Section 2.6)*

66. $4.871 + 7.4 - 1.63$

67. $25.06 - 0.049 + 8.995$

68. $15.3 \cdot 9.1 - 4.017$

69. $29.24 \div 3.4 \cdot 0.045$

70. **MULTIPLE CHOICE** What is the GCF of 48, 80, and 96? *(Section 1.5)*

Ⓐ 12 Ⓑ 16

Ⓒ 24 Ⓓ 480

Extension 3.4 Factoring Expressions

Check It Out
Lesson Tutorials
BigIdeasMath.com

Key Idea

Key Vocabulary
factoring an expression, p. 140

Factoring an Expression

Words Writing a numerical expression or algebraic expression as a product of factors is called **factoring the expression**. You can use the Distributive Property to factor expressions.

Numbers $3 \cdot 7 + 3 \cdot 2 = 3(7 + 2)$

$3 \cdot 7 - 3 \cdot 2 = 3(7 - 2)$

Algebra $ab + ac = a(b + c)$

$ab - ac = a(b - c)$

EXAMPLE 1 Factoring a Numerical Expression

Factor 20 − 12 using the GCF.

Find the GCF of 20 and 12 by listing their factors.

Factors of 20: (1), (2), (4), 5, 10, 20

Factors of 12: (1), (2), 3, (4), 6, 12

Circle the common factors.

The GCF of 20 and 12 is 4.

Write each term of the expression as a product of the GCF and the remaining factor. Then use the Distributive Property to factor the expression.

$20 - 12 = 4(5) - 4(3)$ Rewrite using GCF.

$= 4(5 - 3)$ Distributive Property

Study Tip
When you factor an expression, you can *factor out* any common factor.

EXAMPLE 2 Identifying Equivalent Expressions

Which expression is not equivalent to $16x + 24$?

(A) $2(8x + 12)$ (B) $4(4x + 6)$ (C) $6(3x + 4)$ (D) $(2x + 3)8$

Each choice is a product of two factors in which one is a whole number and the other is the sum of two terms. For an expression to be equivalent to $16x + 24$, its whole number factor must be a common factor of 16 and 24.

Factors of 16: (1), (2), (4), (8), 16

Factors of 24: (1), (2), 3, (4), 6, (8), 12, 24

Circle the common factors.

The common factors of 16 and 24 are 1, 2, 4, and 8. Because 6 is not a common factor of 16 and 24, Choice C cannot be equivalent to $16x + 24$.

Check: $6(3x + 4) = 6(3x) + 6(4) = 18x + 24 \neq 16x + 24$ ✗

∴ So, the correct answer is (C).

Common Core
Equivalent Expressions
In this extension, you will
- use the Distributive Property to produce equivalent expressions.

Learning Standards
6.NS.4
6.EE.3
6.EE.4

Laurie's Notes

Introduction

Connect

- **Yesterday:** Students used the Distributive Property to write and simplify expressions. (MP7)
- **Today:** Students will factor numerical and algebraic expressions.

Motivate

- Write this challenge problem on the board: $\frac{2}{3} \cdot \frac{4}{5} \cdot \frac{5}{8} \cdot \frac{3}{11} = ?$
- Students should quickly see that there are many common factors that can be divided out leaving a product of $\frac{1}{11}$.
- Remind students that in simplifying a fraction, they should look for common factors between the numerator and denominator.

Lesson Notes

Key Idea

? "What does the Distributive Property state?" $a(b + c) = ab + ac$

- **MP7 Look for and Make Use of Structure:** Discuss equality. Students often view the equal sign as computation on the left side that results in *the answer* on the right. Equality means that the quantities on each side of the equal sign have the same value. The Distributive Property can be written as (the numerical or algebraic expression) = (the factored form of the expression).
- Students will think of this as the reverse of the Distributive Property.
- **MP6 Attend to Precision:** Be precise with language. When an expression is rewritten as the product of factors, it is called *factoring the expression.*
- Explain that when they combined like terms in the last lesson, they were essentially factoring out the variable.

Example 1

? "What is one way to find the greatest common factor of two numbers?" List the factors of each and select the greatest number common to both.

- Discuss that you could factor out a 2. The result will have factors of 2 and a difference of two even numbers, which they could again factor out a 2.
- **MP7:** While 20 − 12 = 8 is a true statement, it does not answer the question. The GCF of 20 and 12 is 4, so there will be a factor of 4 in the expression.

Example 2

- Students generally have less difficulty factoring algebraic expressions.

? "What are the common factors for 16 and 24?" 1, 2, 4, and 8

- **Test-Taking Strategy:** Students could use the Distributive Property to multiply the answers. Answer C is the only answer that does not result in $16x + 24$.

Common Core State Standards

6.NS.4 . . . Use the distributive property to express a sum of two whole numbers 1–100 with a common factor as a multiple of a sum of two whole numbers with no common factor.
6.EE.3 Apply the properties of operations to generate equivalent expressions.
6.EE.4 Identify when two expressions are equivalent.

Goal Today's lesson is factoring numerical and algebraic expressions.

Technology for the Teacher

Dynamic Classroom

Lesson Tutorials
Lesson Plans
Answer Presentation Tool

Extra Example 1

Factor 20 + 8 using the GCF. 4(5 + 2)

Extra Example 2

Is $15x + 35$ equivalent to $5(3x + 7)$? Explain.
Yes. *Sample answer:* When you use the Distributive Property on $5(3x + 7)$, you get $15x + 35$, which is equivalent.

Record and Practice Journal Extension 3.4 Practice

1. $2(1 + 4)$
2. $3(3 - 1)$
3. $5(6 + 5)$
4. $7(5 - 2)$
5. $9(9 - 2)$
6. $20(3 + 5)$
7. $4(7 - 5)$
8. $24(3 + 2)$
9. $6(2x + 3)$
10. $2(2y + 5)$
11. $16(2y - 3)$
12. $5(3y + 8)$
13. $8(2x + 3)$
14. $11(x + 3)$
15. $13(x + 3y)$
16. $21(x - 2y)$
17. $(3x + 4)$ in.

Laurie's Notes

Extra Example 3

You receive a discount on each song you purchase for your MP3 player. The original price of each song is x dollars. You purchase 3 songs for a total of $(3x - 6)$ dollars. Factor the expression. What can you conclude about the discount?

Sample answer: The factored expression shows a $2 discount for every song you purchase. The original expression shows a total savings of $6.

Practice

1. $7(1 + 2)$
2. $11(4 - 1)$
3. $6(3 - 2)$
4. $5(14 + 19)$
5. $12(5 - 3)$
6. $20(5 - 4)$
7. $28(3 + 1)$
8. $16(3 + 5)$
9. $2(x + 5)$
10. $3(5x + 2)$
11. $13(2x - 1)$
12. $10(5x - 6)$
13. $9(4x + 1)$
14. $14(x - 7)$
15. $5(2x - 5y)$
16. $8(3y + 11x)$
17. yes; yes; Because a and b are divisible by c, you can factor c out of each expression. Because c is a factor of the sum and the difference, each expression is divisible by c.
18. *Sample answer:* $16 + 8x$, $8(x + 2)$, $4(2x + 4)$, $2(4x + 8)$, $(4x + 8)2$
19. $(x + 4)$ ft
20. See Additional Answers.

Mini-Assessment

Factor the expression using the GCF.

1. $24 - 18$ $6(4 - 3)$
2. $32 + 16$ $16(2 + 1)$
3. $6x + 42$ $6(x + 7)$

Example 3

- Ask a volunteer to read the problem.
- ? "How many books did you purchase and what was the cost of each?" You bought 5 books and each cost x dollars.
- ? "Do $5x$ and 15 have any common factors?" yes; 5
- Write the expression $5x - 15$ in factored form by factoring out the common factor of 5.
- **MP1 Make Sense of Problems and Persevere in Solving Them:** It helps students make sense of the problem when you look at the factored form (the answer) and ask what it would be equivalent to if you distributed the 5. In other words, you are checking that the answer is correct.
- ? "Because $5x - 15 = 5(x - 3)$, what does the factored form mean in the context of the problem?" There is a $3 discount on each book purchased.

Practice

- **Think-Pair-Share:** Students should read each question independently and then work in pairs to answer the questions. When they have answered the questions, the pair should compare their answers with another group and discuss any discrepancies.

Closure

- Factor the expression using the GCF.
  a. $12x - 4$ $4(3x - 1)$
  b. $16 + 24x$ $8(2 + 3x)$

EXAMPLE 3 Factoring an Algebraic Expression

You receive a discount on each book you buy for your electronic reader. The original price of each book is x dollars. You buy 5 books for a total of $(5x - 15)$ dollars. Factor the expression. What can you conclude about the discount?

Find the GCF of $5x$ and 15 by writing their prime factorizations.

$$5x = 5 \cdot x$$
$$15 = 5 \cdot 3$$

Circle the common prime factor.

So, the GCF of $5x$ and 15 is 5. Use the GCF to factor the expression.

$$5x - 15 = 5(x) - 5(3) \quad \text{Rewrite using GCF.}$$
$$= 5(x - 3) \quad \text{Distributive Property}$$

The factor 5 represents the number of books purchased. The factor $(x - 3)$ represents the price of each book. This factor is a difference of two terms, showing that the price x of each book is decreased by \$3.

So, the factored expression shows a \$3 discount for every book you buy. The original expression shows a total savings of \$15.

Practice

Factor the expression using the GCF.

1. $7 + 14$
2. $44 - 11$
3. $18 - 12$
4. $70 + 95$
5. $60 - 36$
6. $100 - 80$
7. $84 + 28$
8. $48 + 80$
9. $2x + 10$
10. $15x + 6$
11. $26x - 13$
12. $50x - 60$
13. $36x + 9$
14. $14x - 98$
15. $10x - 25y$
16. $24y + 88x$

17. **REASONING** The whole numbers a and b are divisible by c. Is $a + b$ divisible by c? Is $b - a$ divisible by c? Explain your reasoning.

18. **OPEN-ENDED** Write five expressions that are equivalent to $8x + 16$.

19. **GEOMETRY** The area of the parallelogram is $(4x + 16)$ square feet. Write an expression for the base.

20. **STRUCTURE** You buy 37 concert tickets for \$8 each, and then sell all 37 tickets for \$11 each. The work below shows two ways you can determine your profit. Describe each solution method. Which do you prefer? Explain your reasoning.

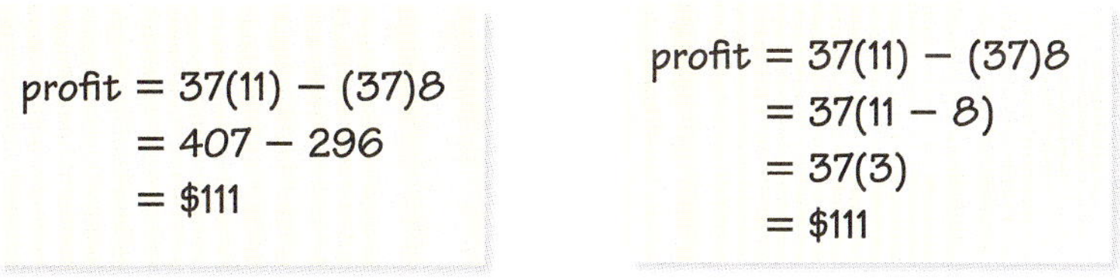

3.3–3.4 Quiz

Check It Out
Progress Check
BigIdeasMath.com

Tell which property the statement illustrates. *(Section 3.3)*

1. $3.5 \cdot z = z \cdot 3.5$

2. $14 + (35 + w) = (14 + 35) + w$

Simplify the expression. Explain each step. *(Section 3.3)*

3. $3.2 + (b + 5.7)$

4. $6 \cdot (10 \cdot k)$

Use the Distributive Property and mental math to find the product. *(Section 3.4)*

5. 6×49

6. 7×86

Use the Distributive Property to simplify the expression. *(Section 3.4)*

7. $5(x - 8)$

8. $7(y + 3)$

Simplify the expression. *(Section 3.4)*

9. $6q + 2 + 3q + 5$

10. $4r + 3(r - 2)$

Factor the expression using the GCF. *(Section 3.4)*

11. $12 + 21$

12. $16x - 36$

13. **GEOMETRY** The expression $18 + 7 + (18 + 2x) + 7$ represents the perimeter of the trapezoid. Simplify the expression. *(Section 3.3)*

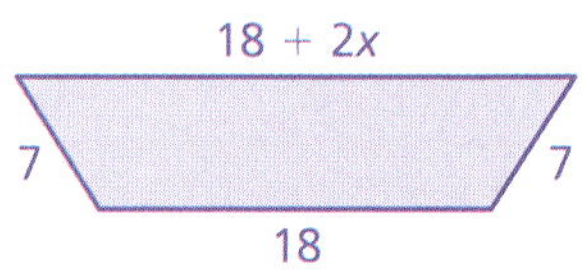

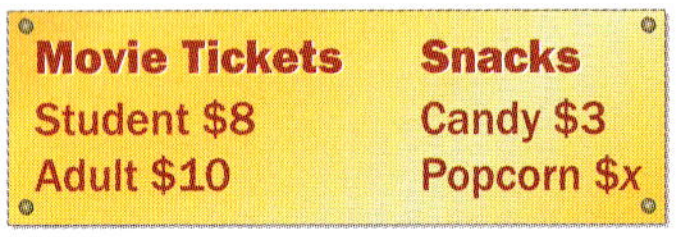

14. **MOVIES** You and four of your friends go to a movie and each buy popcorn. *(Section 3.4)*

a. Use the Distributive Property to write an expression for the total cost to buy movie tickets and popcorn. Simplify the expression.

b. Choose a reasonable value for x. Evaluate the expression.

15. **GEOMETRY** The length of a rectangle is 16 inches, and its area is $(32x + 48)$ square inches. Factor the expression for the area. Write an expression for the width. *(Section 3.4)*

Alternative Assessment Options

Math Chat
Structured Interview
Student Reflective Focus Question
Writing Prompt

Writing Prompt

Ask students to write expressions with variables using at least three of the properties from the chapter. Then have students simplify their expressions and identify the properties at each step.

Study Help Sample Answers

Remind students to complete Graphic Organizers for the rest of the chapter.

3.

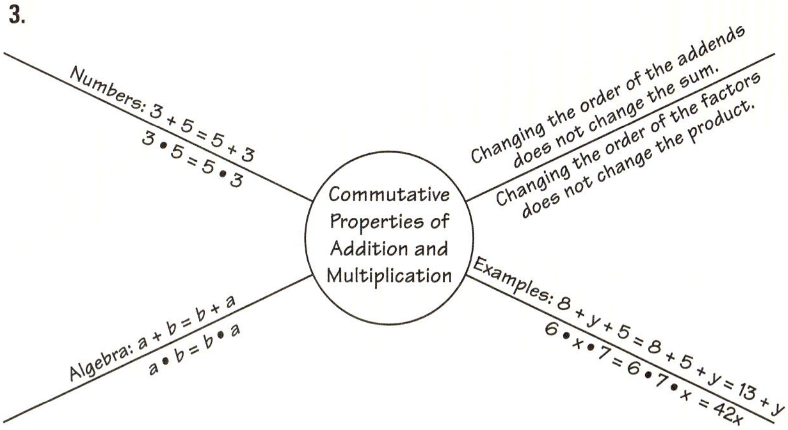

4.

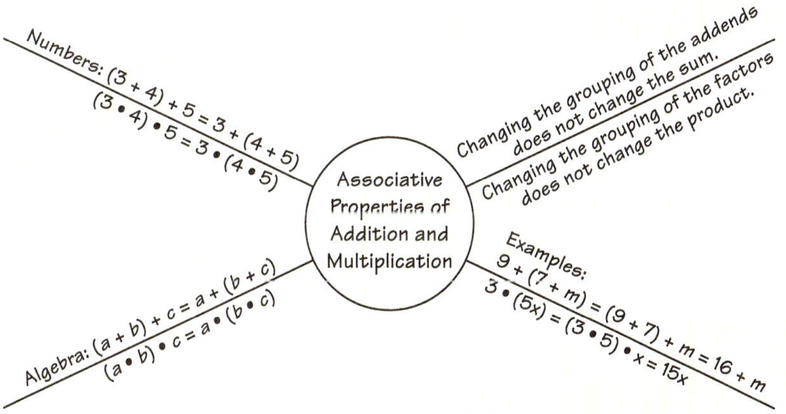

5–8. Available at *BigIdeasMath.com*.

Reteaching and Enrichment Strategies

If students need help...	If students got it...
Resources by Chapter • Practice A and Practice B • Puzzle Time Lesson Tutorials *BigIdeasMath.com*	Resources by Chapter • Enrichment and Extension • Technology Connection Game Closet at *BigIdeasMath.com* Start the Chapter Review

Answers

1. Comm. Prop. of Mult.
2. Assoc. Prop. of Add.
3. $3.2 + (b + 5.7) =$
   $3.2 + (5.7 + b)$
   Comm. Prop. of Add.
   $= (3.2 + 5.7) + b$
   Assoc. Prop. of Add.
   $= 8.9 + b$
   Add 3.2 and 5.7.
4. $6 \cdot (10 \cdot k) = (6 \cdot 10) \cdot k$
   Assoc. Prop. of Mult.
   $= 60k$
   Multiply 6 and 10.
5. 294
6. 602
7. $5x - 40$
8. $7y + 21$
9. $9q + 7$
10. $7r - 6$
11. $3(4 + 7)$
12. $4(4x - 9)$
13. $2x + 50$
14. **a.** $5(x + 8) = 5x + 40$

    b. *Sample answer:* $x = 4$; \$60
15. $16(2x + 3)$; $(2x + 3)$ in.

Technology for the Teacher

Online Assessment
Assessment Book
ExamView® Assessment Suite

For the Teacher
Additional Review Options

- *BigIdeasMath.com*
- Online Assessment
- Game Closet at *BigIdeasMath.com*
- Vocabulary Help
- Resources by Chapter

Answers

1. 4
2. 24
3. 12
4. 305
5. $b - 11$
6. $d \cdot 32$ or $32d$
7. $n + 18$ or $18 + n$
8. $t - 17$
9. **a.** $2p - 4$

   b. 44 points
10. $10 + (2 + y) =$
    $(10 + 2) + y$
    Assoc. Prop. of Add.
    $= 12 + y$
    Add 10 and 2.
11. $(21 + b) + 1 =$
    $(b + 21) + 1$
    Comm. Prop. of Add.
    $= b + (21 + 1)$
    Assoc. Prop. of Add.
    $= b + 22$
    Add 21 and 1.
12. $3(7x) =$
    $(3 \cdot 7)x$
    Assoc. Prop. of Mult.
    $= 21x$
    Multiply 3 and 7.
13. $1(3.2w) =$
    $(1 \cdot 3.2)w$
    Assoc. Prop. of Mult.
    $= 3.2w$
    Mult. Prop. of One

Review of Common Errors

Exercises 1–4

- Students may substitute the wrong value for the variable.
- Tell students to write out the expression and then write the value of the variable needed before substituting the value into the expression.

Exercises 5–9

- Students may write the subtraction and/or division problems in the wrong order.
- For subtraction problems, have students look for the key words "from" and "than" to know that the order of the numbers must be switched from the way it is written.

Exercises 10–15

- Students often get confused about the difference between the Commutative and Associative Properties and will incorrectly label steps.
- Show students that the Associative Property moves the parentheses (but does not change the positions of the numbers) so that there are different numbers or variables inside, while the Commutative Property changes the positions.

Exercise 16

- Students may try to add $3x$ to 7 and/or 4.
- Although students are not yet familiar with *like terms*, you can point out to them that because $3x$ has a variable and you do not know the value of that variable, you cannot add $3x$ to 7 or 4 and get a meaningful result.

Exercises 17–23

- Students may forget to distribute to each term, especially when there are more than two terms within the parentheses.
- Tell students to write the expression on their paper and draw arrows to each term in the parentheses that needs to be distributed to, and then distribute.

3 Chapter Review

Check It Out
Vocabulary Help
BigIdeasMath.com

Review Key Vocabulary

algebraic expression, *p. 112*
terms, *p. 112*
variable, *p. 112*
coefficient, *p. 112*
constant, *p. 112*
equivalent expressions, *p. 128*
like terms, *p. 136*
factoring an expression, *p. 140*

Review Examples and Exercises

3.1 Algebraic Expressions *(pp. 110–117)*

a. Evaluate $a \div b$ when $a = 48$ and $b = 8$.

$a \div b = 48 \div 8$ — Substitute 48 for *a* and 8 for *b*.

$= 6$ — Divide 48 by 8.

b. Evaluate $y^2 - 14$ when $y = 5$.

$y^2 - 14 = 5^2 - 14$ — Substitute 5 for *y*.

$= 25 - 14$ — Using order of operations, evaluate 5^2.

$= 11$ — Subtract 14 from 25.

Exercises

Evaluate the expression when $x = 20$ and $y = 4$.

1. $x \div 5$ **2.** $y + x$ **3.** $8y - x$

4. **GAMING** In a video game, you score p game points and b triple bonus points. An expression for your score is $p + 3b$. What is your score when you earn 245 game points and 20 triple bonus points?

3.2 Writing Expressions *(pp. 118–123)*

Write the phrase as an expression.

a. a number z decreased by 18

$z - 18$ — The phrase *decreased by* means *subtraction*.

b. the sum of 7 and the product of a number x and 12

$7 + 12x$ — The phrase *sum of* means *addition*. The phrase *product of* means *multiplication*.

Exercises

Write the phrase as an expression.

5. 11 fewer than a number b

6. the product of a number d and 32

7. 18 added to a number n

8. a number t decreased by 17

9. **BASKETBALL** Your basketball team scored 4 fewer than twice as many points as the other team.

a. Write an expression for the number of points your team scored.

b. The other team scored 24 points. How many points did your team score?

3.3 Properties of Addition and Multiplication *(pp. 126–131)*

a. Simplify the expression $(x + 18) + 4$.

$(x + 18) + 4 = x + (18 + 4)$ — Associative Property of Addition

$= x + 22$ — Add 18 and 4.

b. Simplify the expression $(5.2 + a) + 0$.

$(5.2 + a) + 0 = 5.2 + (a + 0)$ — Associative Property of Addition

$= 5.2 + a$ — Addition Property of Zero

c. Simplify the expression $36 \cdot r \cdot 1$.

$36 \cdot r \cdot 1 = 36 \cdot (r \cdot 1)$ — Associative Property of Multiplication

$= 36 \cdot r$ — Multiplication Property of One

$= 36r$

Exercises

Simplify the expression. Explain each step.

10. $10 + (2 + y)$

11. $(21 + b) + 1$

12. $3(7x)$

13. $1(3.2w)$

14. $5.3 + (w + 1.2)$

15. $(0 + t) + 9$

16. **GEOMETRY** The expression $7 + 3x + 4$ represents the perimeter of the triangle. Simplify the expression.

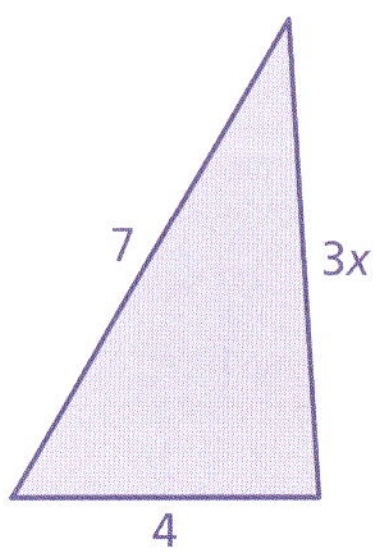

Review Game

Evaluating Expressions

Materials

- blank bingo cards (5 rows by 5 columns)
- 51 index cards
- blank paper
- bingo chips or small pieces of paper

Players: whole class

Directions

Create 51 index cards with problems involving exponents, order of operations, evaluating expressions, writing expressions, properties of addition and multiplication, and the Distributive Property. Each problem should have a whole number answer between 0 and 50, and each answer should only be used once.

The student fills in the empty spaces of a blank bingo card with whole numbers between 0 and 50. Students can choose the numbers to fill in the spaces, but should not use the same number twice.

The teacher shuffles the index cards, selects one card, and writes the problem from that card on the board.

The student uses the blank paper to solve the problem. If the student has the answer on his or her bingo card, the student places a chip or small piece of paper over the answer.

The class continues to solve problems until a student raises his or her hand and says, "BINGO!"

Who Wins?

The student who has five chips in a row horizontally, vertically, or diagonally wins.

For the Student Additional Practice

- Lesson Tutorials
- Multi-Language Glossary
- Self-Grading Progress Check
- *BigIdeasMath.com*
  Dynamic Student Edition
  Student Resources

Answers

14. $5.3 + (w + 1.2)$
$= 5.3 + (1.2 + w)$
Comm. Prop. of Add.
$= (5.3 + 1.2) + w$
Assoc. Prop. of Add.
$= 6.5 + w$
Add 5.3 and 1.2.

15. $(0 + t) + 9 = t + 9$
Add. Prop. of Zero

16. $3x + 11$

17. $1\frac{3}{4}$

18. $2\frac{9}{14}$

19. $1\frac{2}{11}$

20. $2x + 24$

21. $11b - 33$

22. $8s - 8$

23. $36 + 6y$

24. $25z - 100$

25. $35w - 70$

26. $52

27. $9n + 15$

28. $7t + 2$

29. $8z - 5$

30. $5(3 + 7)$

31. $4(9x - 7)$

32. $8(2x + 7y)$

My Thoughts on the Chapter

What worked. . .

Teacher Tip

Not allowed to write in your teaching edition? Use sticky notes to record your thoughts.

What did not work. . .

What I would do differently. . .

3.4 The Distributive Property (pp. 132–141)

a. Use the Distributive Property to simplify $3(n + 9)$.

$3(n + 9) = 3(n) + 3(9)$ — Distributive Property

$= 3n + 27$ — Multiply.

b. Simplify $5x + 7 + 3x - 2$.

$5x + 7 + 3x - 2 = 5x + 3x + 7 - 2$ — Commutative Property of Addition

$= (5 + 3)x + 7 - 2$ — Distributive Property

$= 8x + 5$ — Simplify.

c. Factor $14x - 49$ using the GCF.

Find the GCF of $14x$ and 49 by writing their prime factorizations.

$14x = 2 \cdot 7 \cdot x$ — Circle the common prime factor.

$49 = 7 \cdot 7$

So, the GCF of $14x$ and 49 is 7. Use the GCF to factor the expression.

$14x - 49 = 7(2x) - 7(7)$ — Rewrite using GCF.

$= 7(2x - 7)$ — Distributive Property

Exercises

Use the Distributive Property to find the product.

17. $\frac{3}{4} \times 2\frac{1}{3}$ **18.** $\frac{4}{7} \times 4\frac{5}{8}$ **19.** $\frac{1}{5} \times 5\frac{10}{11}$

Use the Distributive Property to simplify the expression.

20. $2(x + 12)$ **21.** $11(b - 3)$ **22.** $8(s - 1)$

23. $6(6 + y)$ **24.** $25(z - 4)$ **25.** $35(w - 2)$

26. **HAIRCUT** A family of four goes to a salon for haircuts. The cost of each haircut is \$13. Use the Distributive Property and mental math to find the product 4×13 for the total cost.

Simplify the expression.

27. $5(n + 3) + 4n$ **28.** $t + 2 + 6t$ **29.** $3z + 4 + 5z - 9$

Factor the expression using the GCF.

30. $15 + 35$ **31.** $36x - 28$ **32.** $16x + 56y$

3 Chapter Test

Check It Out
Test Practice
BigIdeasMath.com

Evaluate the expression when $a = 6$ and $b = 8$.

1. $4 + a$
2. $a - 6$
3. ab

Write the phrase as an expression.

4. twice a number x
5. 25 more than 50
6. 40 divided by 5

Simplify the expression. Explain each step.

7. $3.1 + (8.6 + m)$
8. $(10 \cdot n) \cdot 7$
9. $3(15w)$

Use the Distributive Property to simplify the expression.

10. $4(x + 8)$
11. $12(y - 5)$

Simplify the expression.

12. $4(q + 2) - 6$
13. $3(2 + 5r) + 11$
14. $s + 3s + 4s$
15. $4t - 2 - 2t + 7$

Factor the expression using the GCF.

16. $18 + 24$
17. $40 - 16$
18. $15x + 20$
19. $32x - 40y$

20. **SOCCER GAME** Playing time is added at the end of a soccer game to make up for stoppages. An expression for the length of a 90-minute soccer game with x minutes of stoppage time is $90 + x$. How long is a game with 4 minutes of stoppage time?

21. **GEOMETRY** The expression $15 \cdot x \cdot 6$ represents the volume of a rectangular prism with a length of 15, a width of x, and a height of 6. Simplify the expression.

22. **PARTY FAVORS** You make party favors for an event. You tie 9 inches of ribbon around each party favor. Write an expression for the amount of ribbon you need for n party favors. The ribbon costs \$3 for each *yard*. Write an expression for the total cost of the ribbon.

Test Item References

Chapter Test Questions	Section to Review	Common Core State Standards
1–3, 20	3.1	6.EE.2c
4–6, 22	3.2	6.EE.2a
7–9, 21	3.3	6.EE.3, 6.EE.4
10–19	3.4	6.NS.4, 6.EE.3, 6.EE.4

Test-Taking Strategies

Remind students to quickly look over the entire test before they start so that they can budget their time. They should not spend too much time on any single problem. Urge students to try to work on a part of each problem because partial credit is better than no credit. Teach students to use the Stop and Think strategy before answering. **Stop** and carefully read the question, and **Think** about what the answer should look like.

For Your Information

- **Exercise 20** Students may not know the meaning of the word *stoppage*. Stoppage is discretionary time that is added to make up for time lost due to major fouls or injuries.

Common Errors

- **Exercises 1–3** Students may not replace the variable correctly. Sometimes it helps to have students rewrite the expression with parentheses and put the value of the variable in the parentheses.
- **Exercise 7** Students miss this exercise because of the decimal addition. Remind them to line up the decimal points before adding.
- **Exercises 10 and 11** The Distributive Property could be the reason why students miss these exercises. Have them review the property.

Reteaching and Enrichment Strategies

If students need help. . .	If students got it. . .
Resources by Chapter • Practice A and Practice B • Puzzle Time Record and Practice Journal Practice Differentiating the Lesson Lesson Tutorials *BigIdeasMath.com* Skills Review Handbook	Resources by Chapter • Enrichment and Extension • Technology Connection Game Closet at *BigIdeasMath.com* Start Standards Assessment

Answers

1. 10
2. 0
3. 48
4. $2x$ or $x \cdot 2$
5. $25 + 50$ or $50 + 25$
6. $40 \div 5$
7. $3.1 + (8.6 + m) = (3.1 + 8.6) + m$
   Assoc. Prop. of Add.
   $= 11.7 + m$
   Add 3.1 and 8.6.
8. $(10 \cdot n) \cdot 7 = (n \cdot 10) \cdot 7$
   Comm. Prop. of Mult.
   $= n \cdot (10 \cdot 7)$
   Assoc. Prop. of Mult.
   $= n \cdot 70$
   Multiply 10 and 7.
   $= 70n$
   Comm. Prop. of Mult.
9. $3(15w) = (3 \cdot 15)w$
   Assoc. Prop. of Mult.
   $= 45w$
   Multiply 3 and 15.
10. $4x + 32$
11. $12y - 60$
12. $4q + 2$
13. $15r + 17$
14. $8s$
15. $2t + 5$
16. $6(3 + 4)$
17. $8(5 - 2)$
18. $5(3x + 4)$
19. $8(4x - 5y)$
20. 94 min
21. $90x$
22. $9n$; $3\left(\frac{1}{4}n\right)$

Technology for the Teacher

Online Assessment
Assessment Book
ExamView® Assessment Suite

Test-Taking Strategies

Available at *BigIdeasMath.com*

After Answering Easy Questions, Relax

Answer Easy Questions First

Estimate the Answer

Read All Choices before Answering

Read Question before Answering

Solve Directly or Eliminate Choices

Solve Problem before Looking at Choices

Use Intelligent Guessing

Work Backwards

About this Strategy

When taking a multiple choice test, be sure to read each question carefully and thoroughly. After skimming the test and answering the easy questions, stop for a few seconds, take a deep breath, and relax. Work through the remaining questions carefully, using your knowledge and test-taking strategies. Remember, you already completed many of the questions on the test!

Answers

1. C
2. F
3. $26
4. B

Technology for the Teacher

Common Core State Standards Support Performance Tasks

Online Assessment

Assessment Book

ExamView® Assessment Suite

Item Analysis

1. **A.** The student does not consider that there is a cost for *each* game. The student adds $2.00 and $0.25 and does not use a variable to represent the number of games played.

   B. The student multiplies the number of games played by the sum of the admission price and the games price. This expression multiplies the admission price by the number of games played, not just the price for each game.

   C. Correct answer

   D. The student uses x to represent the number of admissions, not the number of games played. This expression assumes that there is a flat rate for the games, rather than for the admission.

2. **F.** Correct answer

   G. The student misidentifies the property being used as the Multiplication Property of One.

   H. The student misidentifies the property being used as the Associative Property of Multiplication.

   I. The student misidentifies the property being used as the Commutative Property of Multiplication.

3. **Gridded Response:** Correct answer: $26

   Common Error: The student multiplies the cost of a hardcover book by the number of paperback books and the cost of a paperback book by the number of hardcover books to get an answer of $24.

4. **A.** The student counts the number of decimal places incorrectly.

   B. Correct answer

   C. The student counts only the decimal places in the second factor.

   D. The student subtracts the number of decimal places in the two factors instead of adding them.

3 Standards Assessment

Test-Taking Strategy
After Answering Easy Questions, Relax

"After answering easy questions, relax and try the harder ones. For this, 3(4) = 12. So, it is B."

1. The student council is organizing a school fair. Council members are making signs to show the prices for admission and for each game a person can play.

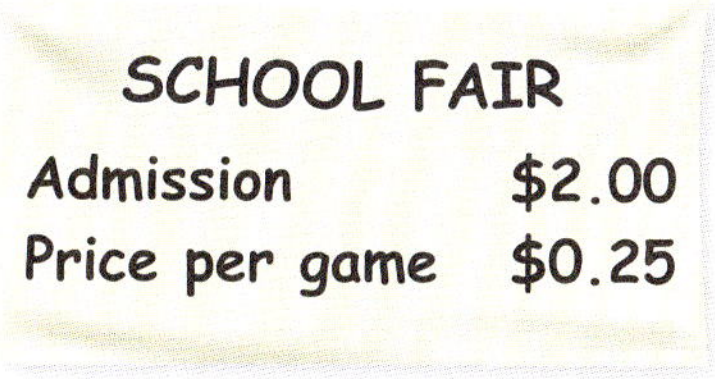

Let x represent the number of games. Which expression can you use to determine the total amount, in dollars, a person pays for admission and playing x games? *(6.EE.2a)*

A. 2.25 **C.** $2 + 0.25x$

B. $2.25x$ **D.** $2x + 0.25$

2. Which property does the equation below represent? *(6.EE.3)*

$$17 \cdot 44 + 17 \cdot 56 = 17 \cdot 100$$

F. Distributive Property **H.** Associative Property of Multiplication

G. Multiplication Property of One **I.** Commutative Property of Multiplication

3. At a used book store, you can purchase two types of books.

You can use the expression $3h + 2p$ to find the total cost for h hardcover books and p paperback books. What is the total cost, in dollars, for 6 hardcover books and 4 paperback books? *(6.EE.2c)*

4. What is the value of 9.6×12.643? *(6.NS.3)*

A. 12.13728 **C.** 1213.728

B. 121.3728 **D.** 12,137.28

5. What is the value of 4.391 + 5.954? *(6.NS.3)*

F. 9.12145

G. 9.245

H. 9.345

I. 10.345

6. Which number pair has a greatest common factor of 6? *(6.NS.4)*

A. 18, 54

B. 30, 42

C. 30, 60

D. 36, 60

7. Properties of Addition and Multiplication are used to simplify an expression.

$$\begin{aligned} 36 \cdot 23 + 33 \cdot 64 &= 36 \cdot 23 + 64 \cdot 33 \\ &= 36 \cdot 23 + 64 \cdot (23 + 10) \\ &= 36 \cdot 23 + 64 \cdot 23 + 64 \cdot 10 \\ &= x \cdot 23 + 64 \cdot 10 \\ &= 2300 + 640 \\ &= 2940 \end{aligned}$$

What number belongs in place of the x? *(6.EE.3)*

8. Which property was used to simplify the expression? *(6.EE.3)*

$$\begin{aligned} (47 \times 125) \times 8 &= 47 \times (125 \times 8) \\ &= 47 \times 1000 \\ &= 47{,}000 \end{aligned}$$

F. Distributive Property

G. Multiplication Property of One

H. Associative Property of Multiplication

I. Commutative Property of Multiplication

9. What is the value of the expression below when $a = 5$, $b = 7$, and $c = 6$? *(6.EE.2c)*

$$9b - 4a + 2c$$

A. 29

B. 31

C. 55

D. 78

Item Analysis (continued)

5. **F.** The student adds each place value and inserts the 1s instead of carrying them.

   G. The student does not carry any 1s.

   H. The student does not carry the 1 from the tenths place to the units place.

   I. Correct answer

6. **A.** The student finds a common factor, but not the GCF.

   B. Correct answer

   C. The student finds a common factor, but not the GCF.

   D. The student finds a common factor, but not the GCF.

7. **Gridded Response:** Correct answer: 100

   Common Error: The student does not recognize the application of the Distributive Property and instead tries to work backwards from 2300. The student makes a place value error while dividing 2300 by 23 to get 1000.

8. **F.** The student misidentifies the property being used as the Distributive Property.

   G. The student misidentifies the property being used as the Multiplication Property of One.

   H. Correct answer

   I. The student misidentifies the property being used as the Commutative Property of Multiplication.

9. **A.** The student substitutes 5 for b and 7 for a.

   B. The student makes an order of operations error, adding $4a$ to $2c$ and then subtracting this sum from $9b$.

   C. Correct answer

   D. The student makes two-digit numbers out of the coefficient and the variable, such as 97 instead of 9×7.

Answers

5. I
6. B
7. 100
8. H
9. C

Answers

10. G

11. B

12. F

13. 1800

14. D

15. I

Item Analysis (continued)

10. **F.** The student does not distribute the multiplication by *a* to the second addend, *c*.

G. Correct answer

H. The student incorrectly distributes the addition of *a* instead of multiplication by *a*.

I. The student incorrectly distributes the addition of *a* instead of multiplication by *a* and then multiplies the resulting sums.

11. **A.** The student incorrectly inverts both the dividend and the divisor instead of only inverting the divisor.

B. Correct answer

C. The student does not rewrite the mixed numbers as improper fractions correctly.

D. The student incorrectly inverts the dividend instead of the divisor.

12. **F.** Correct answer

G. The student does not find the LCM correctly.

H. The student does not find the LCM correctly.

I. The student does not find the LCM correctly.

13. **2 points** The student demonstrates a thorough understanding of the Associative and Commutative Properties of Multiplication. For example, the student explains that the Commutative Property can be used to rewrite the expression as $(18 \times 25) \times 4$. The student then explains that the Associative Property can be used to rewrite the expression as $18 \times (25 \times 4)$. The student then uses compatible numbers to get 18×100 and a final answer of 1800.

1 point The student demonstrates a partial understanding of the Associative and Commutative Properties of Multiplication. For example, the student shows correct, efficient steps but provides no explanation.

0 points The student demonstrates insufficient understanding of the Associative and Commutative Properties of Multiplication. The student either simplifies the expression inefficiently (such as left-to-right) or simplifies the expression incorrectly.

14. **A.** The student makes two-digit numbers out of the coefficient and the variable, such as 36 instead of 3×6.

B. The student substitutes 9 for *x* and 6 for *y*.

C. The student uses the values of the variables as coefficients.

D. Correct answer

15. **F.** The student subtracts $12 - 8$ before finding the power, multiplying, and dividing.

G. The student subtracts 8 from the product of 2×12 before finding the power and dividing.

H. The student finds the power of $8 \div 2$ instead of finding the power 2^2 first.

I. Correct answer

10. Which equation correctly demonstrates the Distributive Property? *(6.EE.4)*

F. $a(b + c) = ab + c$

G. $a(b + c) = ab + ac$

H. $a + (b + c) = (a + b) + (a + c)$

I. $a + (b + c) = (a + b) \cdot (a + c)$

11. Which expression is equivalent to $3\frac{3}{5} \div 6\frac{1}{2}$? *(6.NS.1)*

A. $\frac{5}{18} \times \frac{13}{2}$

B. $\frac{18}{5} \times \frac{2}{13}$

C. $\frac{9}{5} \div \frac{6}{2}$

D. $\frac{18}{5} \div \frac{2}{13}$

12. Which number pair does *not* have a least common multiple of 24? *(6.NS.4)*

F. 2, 12

G. 3, 8

H. 6, 8

I. 12, 24

13. Use the Properties of Multiplication to simplify the expression in an efficient way. Show your work and explain how you used the Properties of Multiplication. *(6.EE.3)*

Think
Solve
Explain

$$(25 \times 18) \times 4$$

14. You evaluated an expression using $x = 6$ and $y = 9$. You correctly got an answer of 105. Which expression did you evaluate? *(6.EE.2c)*

A. $3x + 6y$

B. $5x + 10y$

C. $6x + 9y$

D. $10x + 5y$

15. Which number is equivalent to the expression below? *(6.EE.1)*

$$2 \times 12 - 8 \div 2^2$$

F. 2

G. 4

H. 8

I. 22

4 Areas of Polygons

4.1 Areas of Parallelograms

4.2 Areas of Triangles

4.3 Areas of Trapezoids

4.4 Polygons in the Coordinate Plane

"Remember, Descartes, you don't have to measure area in standard units like square inches or square centimeters."

"You can also use nonstandard units ... like the length of your paw squared."

"Yummy. I smell cheese. I LOVE cheese."

Common Core Progression

4th Grade

- Find the area and perimeter of a rectangle using formulas.
- Draw points, line segments, parallel lines, and perpendicular lines.
- Classify two-dimensional figures based on angles, parallel lines, and perpendicular lines.

5th Grade

- Find the area of a rectangle with fractional side lengths.
- Classify two-dimensional figures into categories based on properties.
- Generate numerical patterns given rule, identify the relationship, and form ordered pairs.
- Plot points in the first quadrant of the coordinate plane.

6th Grade

- Find areas of triangles, special quadrilaterals, and polygons.
- Find the distance between points with the same *x*- or *y*-coordinate.
- Draw polygons in the coordinate plane given vertices and find lengths of sides.

Pacing Guide for Chapter 4

Chapter Opener	1 Day
Section 1 Activity Lesson	 1 Day 1 Day
Section 2 Activity Lesson	 1 Day 1 Day
Study Help / Quiz	1 Day
Section 3 Activity Lesson Extension	 1 Day 1 Day 1 Day
Section 4 Activity Lesson	 1 Day 1 Day
Chapter Review/ Chapter Tests	2 Days
Total Chapter 4	13 Days
Year-to-Date	64 Days

Chapter Summary

Section	Common Core State Standard	
4.1	Learning	6.G.1
4.2	Learning	6.G.1
4.3	Learning	6.G.1 ★
4.4	Learning	6.G.3 ★

★ Teaching is complete. Standard can be assessed.

Technology for the Teacher

BigIdeasMath.com
Chapter at a Glance
Complete Materials List
Parent Letters: English and Spanish

Common Core State Standards

4.MD.3 Apply the area and perimeter formulas for rectangles in real world and mathematical problems.

5.G.1 Use a pair of perpendicular number lines, called axes, to define a coordinate system, with the intersection of the lines (the origin) arranged to coincide with the 0 on each line and a given point in the plane located by using an ordered pair of numbers, called its coordinates. Understand that the first number indicates how far to travel from the origin in the direction of one axis, and the second number indicates how far to travel in the direction of the second axis, with the convention that the names of the two axes and the coordinates correspond (e.g., x-axis and x-coordinate, y-axis and y-coordinate).

Additional Topics for Review

- Parallel and Perpendicular Lines
- Classify Two-Dimensional Figures

Try It Yourself

1. 49 m^2 **2.** 180 yd^2

3. 5850 mm^2

4–6.

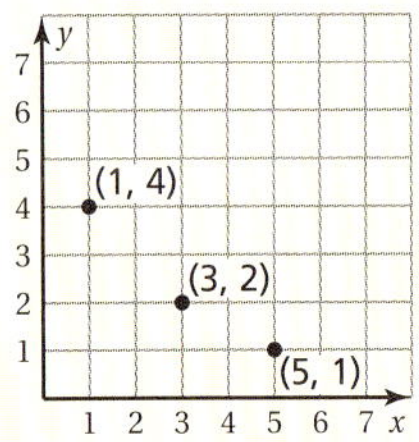

Record and Practice Journal Fair Game Review

1. 50 ft^2 **2.** 9 cm^2

3. 16 in.^2 **4.** 14 yd^2

5. 80 ft^2

6.

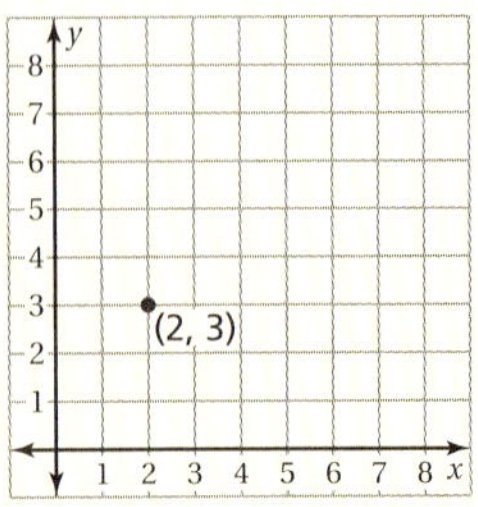

7–11. See Additional Answers.

Math Background Notes

Vocabulary Review

- Area
- Coordinate Plane
- x-axis
- y-axis
- Origin
- Ordered Pair
- x-coordinate
- y-coordinate

Finding Areas of Squares and Rectangles

- Remind students that the exponent means to multiply the base by itself. 15^2 is 15×15 and not 15×2.
- ? "What type of units do you use to measure area?" square units
- ? "Why is area measured using square units?" There are two dimensions with multiplication in the formula. $\text{cm} \times \text{cm} = \text{cm}^2$
- Review how square units are written: cm^2, sq cm, or square centimeters.

Plotting Ordered Pairs

- Use a model of a coordinate grid to identify key vocabulary: coordinate plane, x-axis, y-axis, origin, ordered pair, x-coordinate, y-coordinate.
- **Connections:** Hold your arms out horizontally and relate them to the horizon to help students remember which way is horizontal. Origin means "where something starts."
- Students should understand that each point in the coordinate plane has a unique name. The x- and y-coordinates specify an exact spot, similar to latitude and longitude on a map.
- Stress that the order in which you plot the points is important. The ordered pairs (2, 4) and (4, 2) are not the same.
  - The x-coordinate is always first. It tells how far to go horizontally.
  - The y-coordinate is always second. It tells how far to go vertically.
- The ordered pairs are always (x, y). If it helps, tell students that the order of the coordinates is in alphabetical order, x and then y.

Reteaching and Enrichment Strategies

If students need help...	If students got it...
Record and Practice Journal • Fair Game Review Skills Review Handbook Lesson Tutorials	Game Closet at *BigIdeasMath.com* Start the next section

What You Learned Before

Finding Areas of Squares and Rectangles (4.MD.3)

Example 1 Find the area of the square or rectangle.

a.

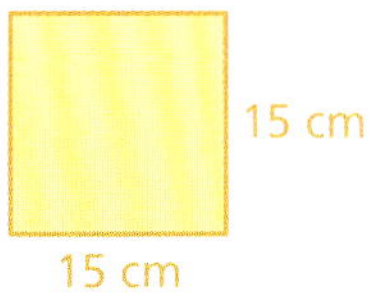

b.

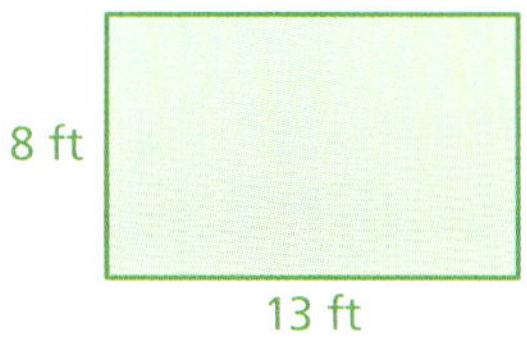

a.

$A = s^2$ Write formula.

$= 15^2$ Substitute.

$= 225$ Simplify.

The area of the square is 225 square centimeters.

b.

$A = \ell w$

$= 13(8)$

$= 104$

The area of the rectangle is 104 square feet.

Try It Yourself

Find the area of the square or rectangle.

1.

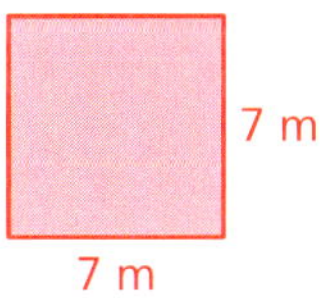

2.

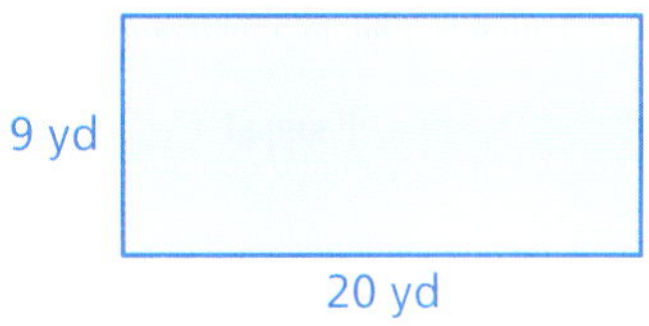

3.

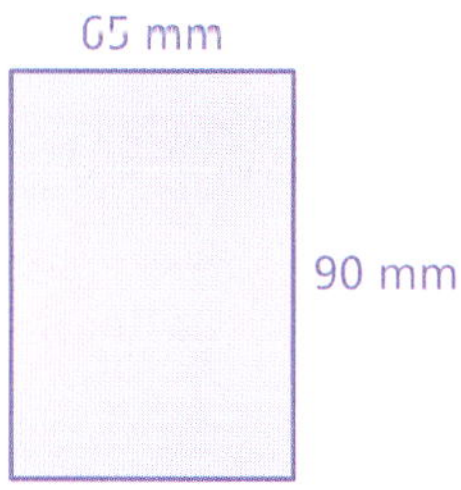

Plotting Ordered Pairs (5.G.1)

Example 2 Plot (2, 3) in a coordinate plane.

Start at the origin. Move 2 units right and 3 units up. Then plot the point.

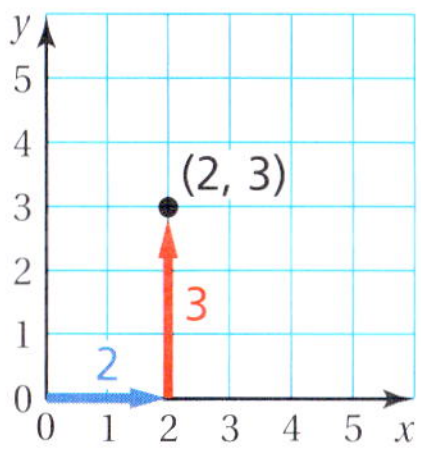

Try It Yourself

Plot the ordered pair in a coordinate plane.

4. (1, 4) **5.** (3, 2) **6.** (5, 1)

4.1 Areas of Parallelograms

Essential Question How can you derive a formula for the area of a parallelogram?

A **polygon** is a closed figure in a plane that is made up of three or more line segments that intersect only at their endpoints. Several examples of polygons are parallelograms, triangles, and trapezoids.

The formulas for the areas of polygons can be derived from one area formula, the area of a rectangle. Recall that the area of a rectangle is the product of its length ℓ and its width w. The process you use to derive these other formulas is called *deductive reasoning*.

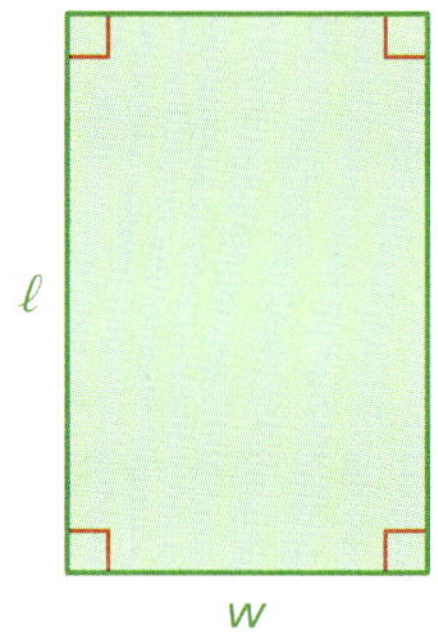

Area $= \ell w$

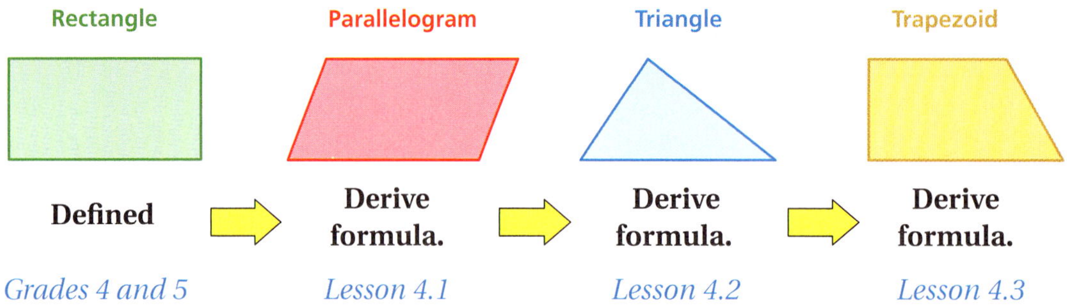

1 ACTIVITY: Deriving the Area Formula of a Parallelogram

Work with a partner.

a. Draw *any* rectangle on a piece of grid paper. An example is shown below. Label the length and width. Then find the area of your rectangle.

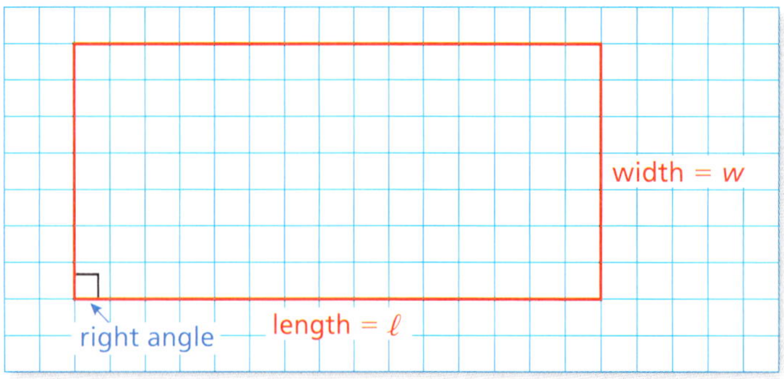

b. Cut your rectangle into two pieces to form a parallelogram. Compare the area of the rectangle with the area of the parallelogram. What do you notice? Use your results to write a formula for the area A of a parallelogram.

$A =$ ______ Formula

COMMON CORE

Geometry

In this lesson, you will
- find areas of parallelograms.
- solve real-life problems.

Learning Standard
6.G.1

Laurie's Notes

Introduction

Standards for Mathematical Practice

- **MP3 Construct Viable Arguments and Critique the Reasoning of Others:** In this chapter students will derive area formulas from a known formula using deductive reasoning. Students should listen carefully to the statements made by classmates and critique their reasoning.

Motivate

- **Story Time:** Tell students you worked as a short-order cook at a diner while in college. Combining your love of math with cooking, you cut a piece of toast on a diagonal. Then you arranged the pieces so they were no longer square.
- Distribute two right triangle pieces, or a square piece of paper that students can fold on a diagonal and tear apart.

? "How many different shapes can you make with the toast?" three

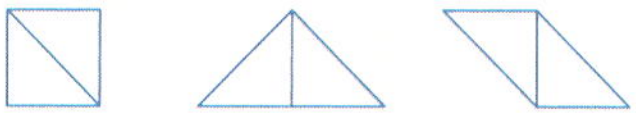

? "Did the amount of toast change?" no "What did change?" the shape

- Use student responses to assess knowledge of simple geometric shapes: rectangle, parallelogram, triangle, and trapezoid.
- Explain that in the next few lessons students will start with a shape, cut and rearrange it to make a shape for which they already know the formula.

Activity Notes

Discuss

- Discuss the definition of a polygon. Draw the following on the board and then ask students, "So what's a polygon?"

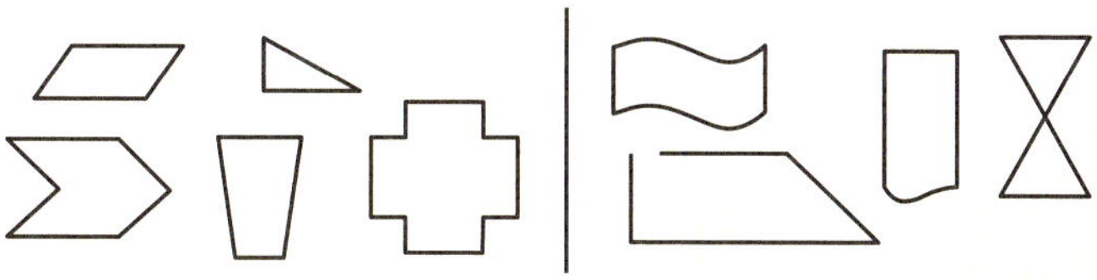

- Explain that in this chapter they will use deductive reasoning to derive area formulas for other polygons from the area formula for a rectangle.

Activity 1

- Suggest that students draw a rectangle large enough to manipulate. They may count squares to find the area or find the product of the dimensions.
- **Extension:** Students may want to explore several rectangles and several types of cuts to make a parallelogram.
- **MP3:** Ask a volunteer to share his or her rectangle and parallelogram. Did all students try essentially the same cut or are there multiple versions to present? Are there any cuts that did not allow a parallelogram to be made?
- Students should clearly state the formula. It is not width × length. If they are stuck, ask them where the width and length of the rectangle are found in the parallelogram.

Common Core State Standards

6.G.1 Find the area of . . . special quadrilaterals . . . by composing into rectangles . . .; apply these techniques in the context of solving real-world and mathematical problems.

Previous Learning

In earlier grades, students have used an area model to represent multiplication. Students have found the area of a rectangle.

Technology for the Teacher

Dynamic Classroom

Lesson Plans
Complete Materials List

4.1 Record and Practice Journal

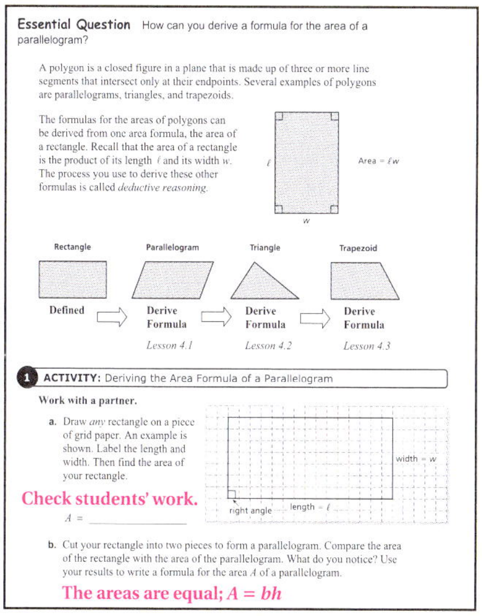

Essential Question How can you derive a formula for the area of a parallelogram?

A polygon is a closed figure in a plane that is made up of three or more line segments that intersect only at their endpoints. Several examples of polygons are parallelograms, triangles, and trapezoids.

The formulas for the areas of polygons can be derived from one area formula, the area of a rectangle. Recall that the area of a rectangle is the product of its length ℓ and its width w. The process you use to derive these other formulas is called *deductive reasoning*.

1 ACTIVITY: Deriving the Area Formula of a Parallelogram

Work with a partner.

a. Draw *any* rectangle on a piece of grid paper. An example is shown. Label the length and width. Then find the area of your rectangle.

Check students' work.

A = ______

b. Cut your rectangle into two pieces to form a parallelogram. Compare the area of the rectangle with the area of the parallelogram. What do you notice? Use your results to write a formula for the area A of a parallelogram.

The areas are equal; $A = bh$

Differentiated Instruction

Auditory

Students may have difficulty remembering which measurement (units or *square* units) should be used for perimeter. Remind students that a *meter* is a unit of length, so the perimeter of a figure has units of length (*not* square units of area).

4.1 Record and Practice Journal

2 ACTIVITY: Finding the Areas of Parallelograms

Work with a partner.

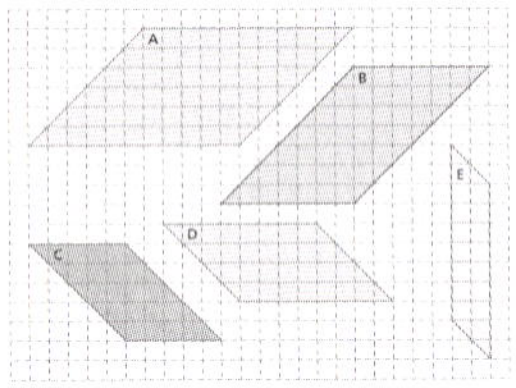

a. Find the area of each parallelogram by cutting it into two pieces to form a rectangle.*

A: 66 $units^2$ D: 32 $units^2$
B: 49 $units^2$ E: 18 $units^2$
C: 25 $units^2$

b. Use the formula you wrote in Activity 1 to find the area of each parallelogram. Compare your answers to those in part (a).

Check students' work.

c. Count unit squares for each parallelogram to check your results.

Check students' work.

*Cut-outs are available in the back of the Record and Practice Journal.

What Is Your Answer?

3. **IN YOUR OWN WORDS** How can you derive a formula for the area of a parallelogram?

Sample answer: Draw a parallelogram and cut it along the height. Slide the cut off piece to the other end to form a rectangle. Then use the area formula for a rectangle to find the area of the parallelogram.

4. **REASONING** The areas of a rectangle and a parallelogram are equal. The length of the rectangle is equal to the base of the parallelogram. What can you say about the width of the rectangle and the height of the parallelogram? Draw a diagram to support your answer.

They are equal.

5. What is the height of the parallelogram shown? How do you know?

4 units

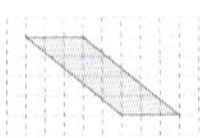

Laurie's Notes

Activity 2

- This activity is the converse of the previous activity. Knowing how they cut the rectangle to make the parallelogram should give students a sense of how to start with the parallelogram and cut to make a rectangle.
- Students should try to visualize the cut and how the pieces would be rearranged to form the rectangle before they actually make the cut.
- Display the parallelograms as you ask for a volunteer to count the unit squares and half-unit squares. The volunteer should count aloud for the class.
- Check to see that students understand how to find and label the height of the parallelogram.

What Is Your Answer?

- **Neighbor Check:** Have students work independently and then have their neighbors check their work. Have students discuss any discrepancies.

Closure

- **Exit Ticket:** Find the area and perimeter of the parallelogram shown.

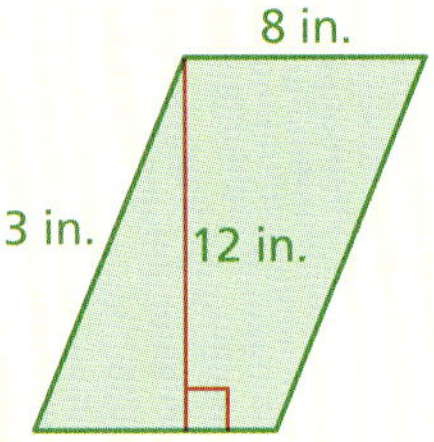

96 in.^{2}; 42 in.

2 ACTIVITY: Finding Areas of Parallelograms

Math Practice 3

Use Assumptions

How are rectangles and parallelograms similar? How can you use this information to solve the problem?

Work with a partner.

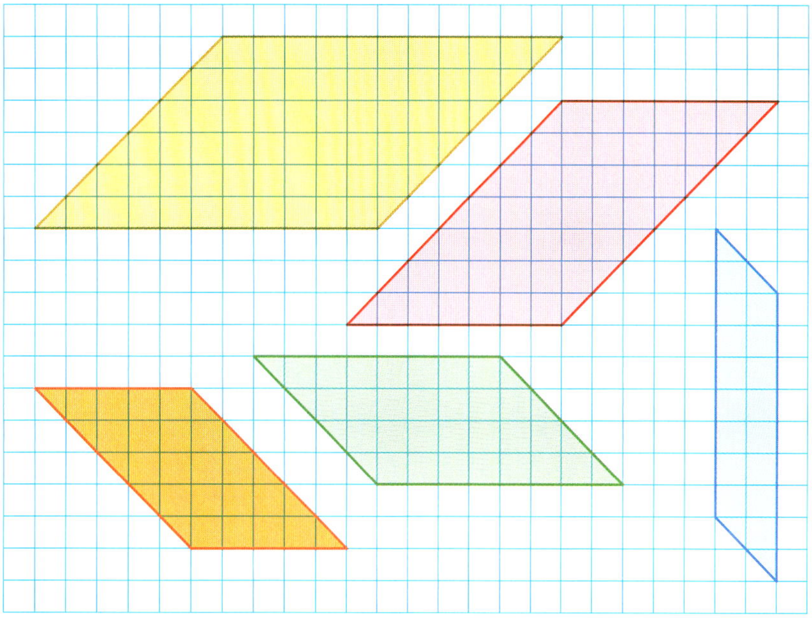

a. Find the area of each parallelogram by cutting it into two pieces to form a rectangle.

b. Use the formula you wrote in Activity 1 to find the area of each parallelogram. Compare your answers to those in part (a).

c. Count unit squares for each parallelogram to check your results.

What Is Your Answer?

3. **IN YOUR OWN WORDS** How can you derive a formula for the area of a parallelogram?

4. **REASONING** The areas of a rectangle and a parallelogram are equal. The length of a rectangle is equal to the base of the parallelogram. What can you say about the width of the rectangle and the height of the parallelogram? Draw a diagram to support your answer.

5. What is the height of the parallelogram shown? How do you know?

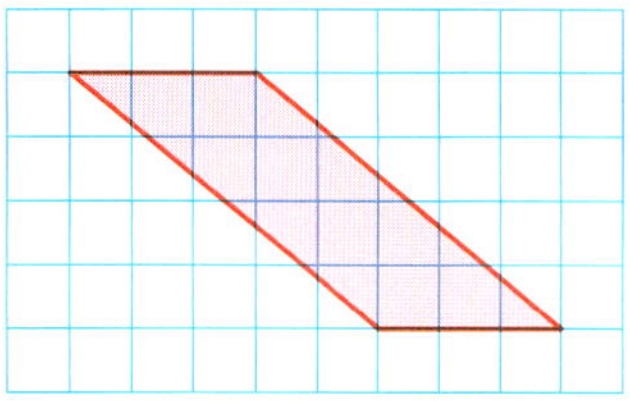

Practice Use what you learned about the areas of parallelograms to complete Exercises 3–5 on page 156.

4.1 Lesson

Check It Out Lesson Tutorials BigIdeasMath.com

Key Vocabulary
polygon, *p. 152*

The *area* of a polygon is the amount of surface it covers. You can find the area of a parallelogram in much the same way as you can find the area of a rectangle.

Key Idea

Area of a Parallelogram

Words The area A of a parallelogram is the product of its base b and its height h.

Algebra $A = bh$

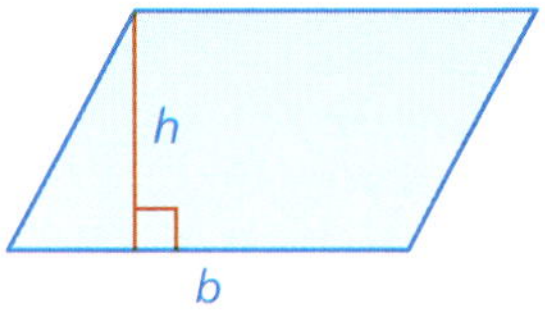

EXAMPLE 1 Finding Areas of Parallelograms

Find the area of each parallelogram.

a.

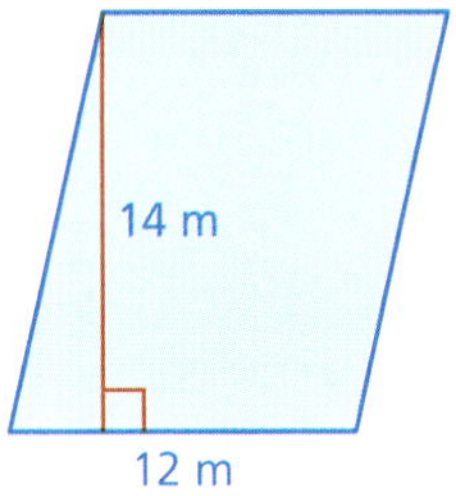

$A = bh$ — Write formula.

$= 12(14)$ — Substitute values.

$= 168$ — Multiply.

The area of the parallelogram is 168 square meters.

b.

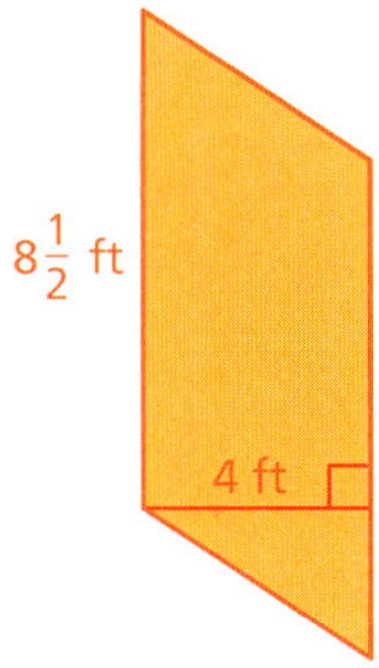

$A = bh$

$= 8\frac{1}{2}(4)$

$= 34$

The area of the parallelogram is 34 square feet.

Remember
Area is measured in square units.

On Your Own

Now You're Ready Exercises 3–8

Find the area of the parallelogram.

1.

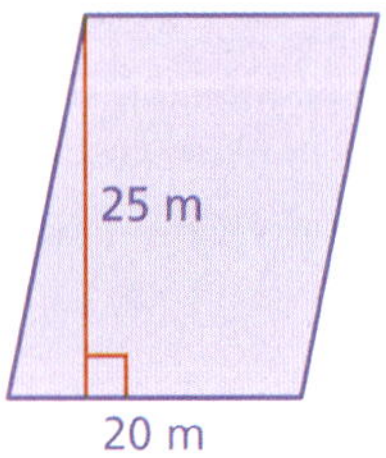

2.

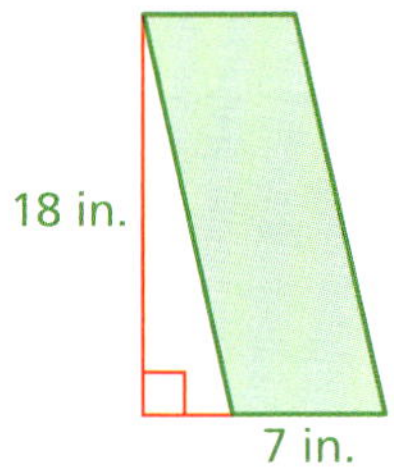

3.

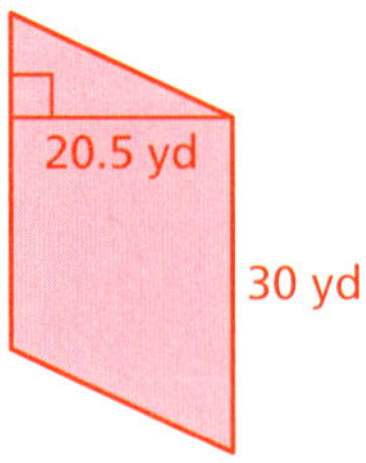

Multi-Language Glossary at BigIdeasMath.com

Laurie's Notes

Introduction

Connect

- **Yesterday:** Students derived the formula for the area of a parallelogram. (MP3)
- **Today:** Students will use the formula to find areas of parallelograms.

Motivate

- Draw a sketch similar to the one below. Ask each student to answer:

? "Which of the shaded figures has the greatest area?" The areas are all equal. How students answer this will give insight into how well they have internalized yesterday's activity.

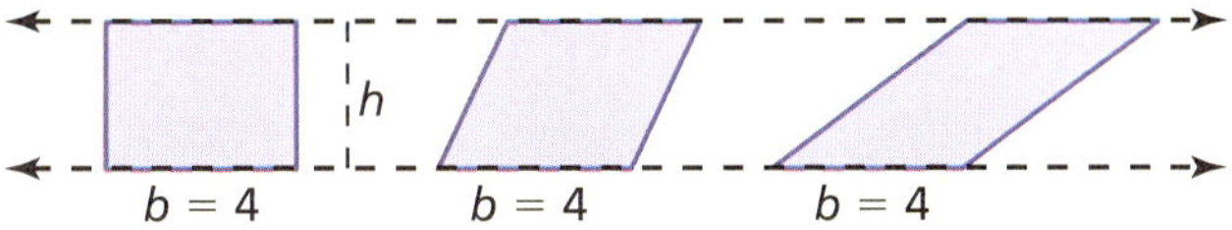

- Ask a student to explain this result. Hopefully the student will talk about cutting off a piece of the parallelogram and moving the piece to form a rectangle that is congruent (the same size and same shape) to the first quadrilateral.

Lesson Notes

Key Idea

- The *height* of a parallelogram is the perpendicular distance from a base to the opposite side.
- It is very important that students recognize that the height of a parallelogram is NOT a side of the parallelogram, unless the figure is a rectangle or square.
- Try to connect the concept of the height of a parallelogram to the height (altitude) of a kite. The length of the string is *not* the height of the kite.

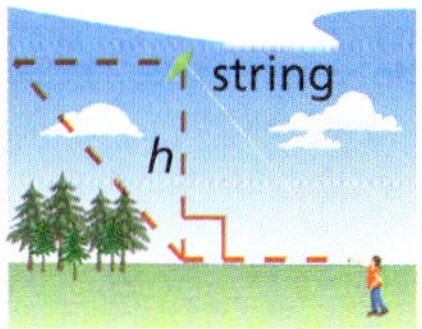

- Ask students to think about b and h in the formula as variables, asking them what value(s) they may stand for, getting them to realize that they are ranges of values (> 0).

Example 1

? "What type of units do you use to measure area?" square units

? "Why is area measured using square units?" There are two dimensions with multiplication in the formula. $\text{cm} \times \text{cm} = \text{cm}^2$

- Review how square units are written: cm^2, sq cm, or square centimeters.
- **MP6 Attend to Precision:** There is a difference between 168 meters and 168 square meters. Be sure that students are labeling results correctly.
- Begin good habits now. Be explicit in writing the formula, substituting the known values, and computing the answer.

On Your Own

- Students should note that in Question 2 the height is shown outside the parallelogram. The same could have been done for the other two questions.

Goal Today's lesson is finding the areas of parallelograms.

Technology for the Teacher

Dynamic Classroom

Lesson Tutorials
Lesson Plans
Answer Presentation Tool

Extra Example 1

Find the area of each parallelogram.

a.

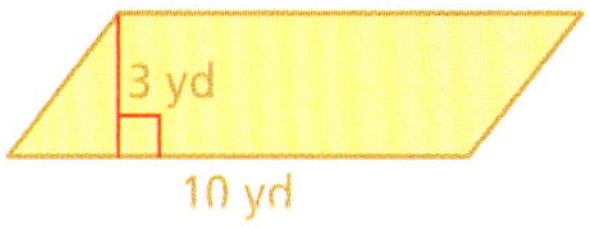

30 yd^2

b.

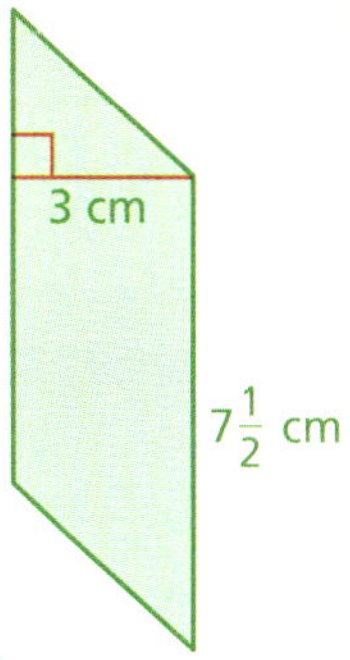

$22\frac{1}{2} \text{ cm}^2$

On Your Own

1. 500 m^2
2. 126 in.^2
3. 615 yd^2

Extra Example 2

Find the area of the parallelogram.

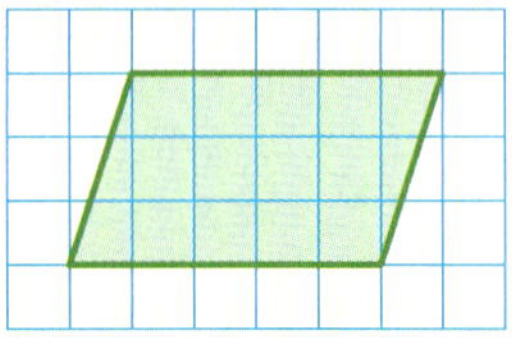

15 square units

On Your Own

4. 12 square units

Extra Example 3

In Example 3, you cut an 8-inch square out of the piece of wood. What is the area of the photo prop? 4544 $in.^2$

On Your Own

5. 32 ft^2

6. 4464 $in.^2$

English Language Learners

Vocabulary

English language learners may be confused by the abbreviations used for customary and metric measures. Have students write the measure and its abbreviation in their notebooks. List the customary measures separate from the metric measures.

Customary

inches	in.
feet	ft
yard	yd
miles	mi

Metric

millimeter	mm
centimeter	cm
meter	m
kilometer	km

Laurie's Notes

Example 2

- This example follows from yesterday's activity.
- Students may need help finding the height. This is an example of the height being outside of the polygon. This example is a preview of sorts to working in the coordinate plane in Lesson 4.4 and Lesson 6.5.

? To assist students in seeing the height, you could extend the lines through the bases and ask, "What is the vertical distance between the two bases?"

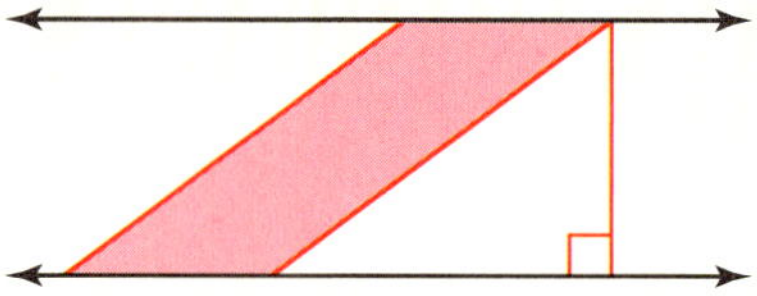

Example 3

- Discuss each shape and the formulas used.
- The piece of wood is measured in feet and the square is measured in inches. The question asks what is left after the square is removed. So, the first step is to convert feet to inches.
- Students may ask why they can't just multiply by 12. The dimensional analysis is a strategy that students will use throughout mathematics and science.
- Take the time to write the verbal model. Again, this develops good habits in problem solving.

On Your Own

- **Think-Pair-Share:** Students should read each question independently and then work in pairs to answer the questions. When they have answered the questions, the pair should compare their answers with another group and discuss any discrepancies.

Closure

? Draw all three quadrilaterals and ask, "Are all of these polygons parallelograms?" yes

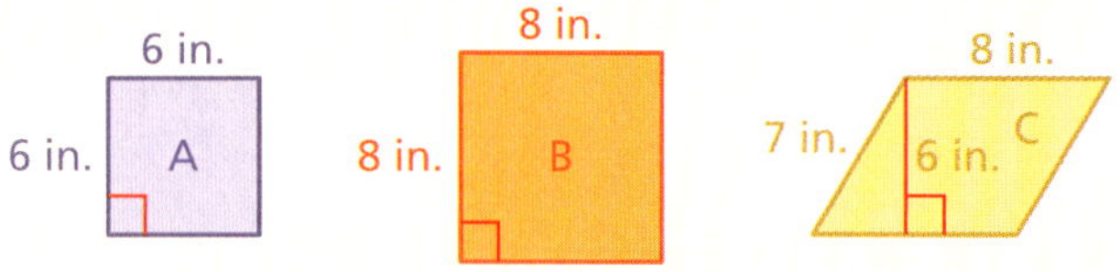

- Determine the area and perimeter of each figure. Then order the quadrilaterals from least to greatest area and least to greatest perimeter. A: 36 $in.^2$, 24 in.; B: 64 $in.^2$, 32 in.; C: 48 $in.^2$, 30 in.; A, C, B; A, C, B

EXAMPLE 2 Finding the Area of a Parallelogram on a Grid

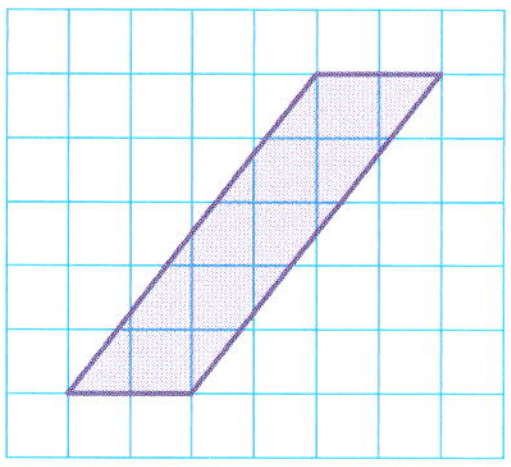

Find the area of the parallelogram.

Count grid lines to find the dimensions.
The base b is 2 units, and the height h is 5 units.

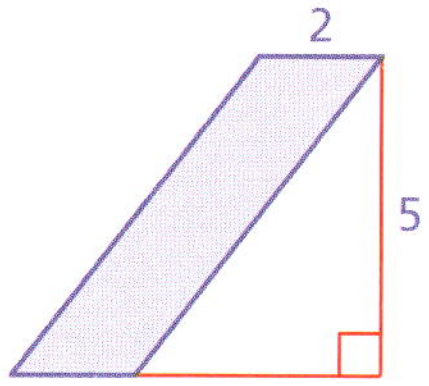

$A = bh$	Write formula.
$= 2(5)$	Substitute values.
$= 10$	Multiply.

∴ The area of the parallelogram is 10 square units.

On Your Own

Now You're Ready
Exercises 11–13

4. Find the area of the parallelogram.

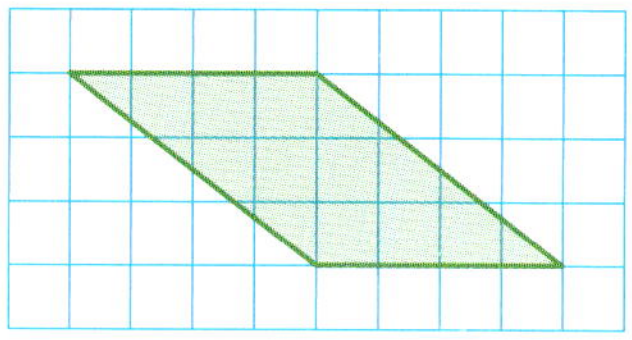

EXAMPLE 3 Real-Life Application

You make a photo prop for a school fair. You cut a 10-inch square out of a parallelogram-shaped piece of wood. What is the area of the photo prop?

Convert the dimensions of the piece of wood to inches.

There are 12 inches in 1 foot, so the base is $4 \cdot 12 = 48$ inches and the height is $8 \cdot 12 = 96$ inches.

Use a verbal model to solve the problem.

area of photo prop	=	area of wood	−	area of square	
	=	96(48)	−	10^2	Substitute.
	=	96(48)	−	100	Evaluate 10^2.
	=	4608	−	100	Multiply 96 and 48.
	=	4508			Subtract 100 from 4608.

∴ The area of the photo prop is 4508 square inches.

On Your Own

Now You're Ready
Exercises 14–16

5. Find the area of the shaded region.

6. **WHAT IF?** In Example 3, you cut a 12-inch square out of the piece of wood. What is the area of the photo prop?

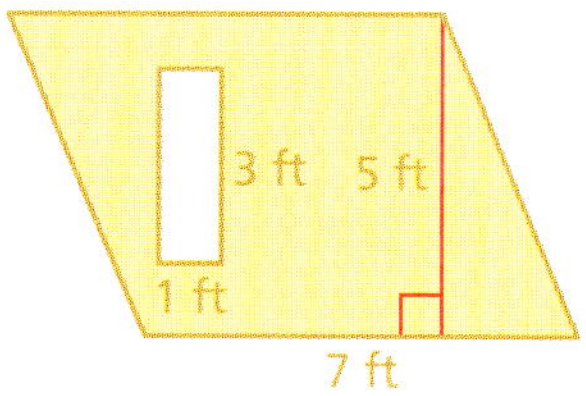

4.1 Exercises

Check It Out
Help with Homework
BigIdeasMath.com

Vocabulary and Concept Check

1. **WRITING** What is the area of a polygon? Explain how the perimeter and the area of the polygon are different.

2. **CHOOSE TOOLS** Construct a parallelogram that has an area of 24 square inches. Explain your method.

Practice and Problem Solving

Find the area of the parallelogram.

1 3.

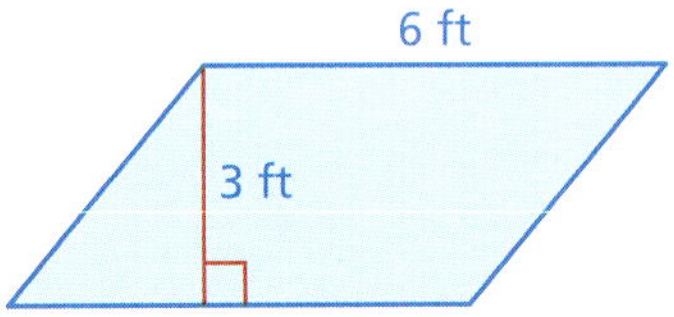

4.

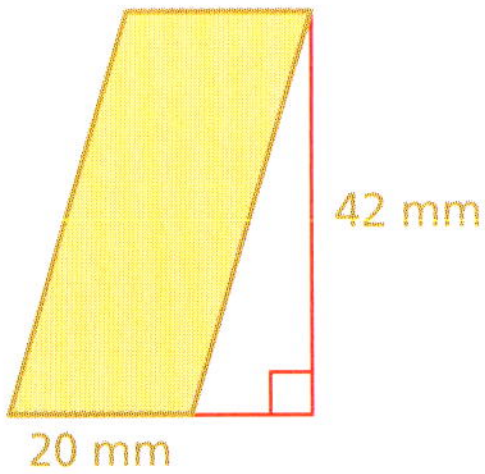

5.

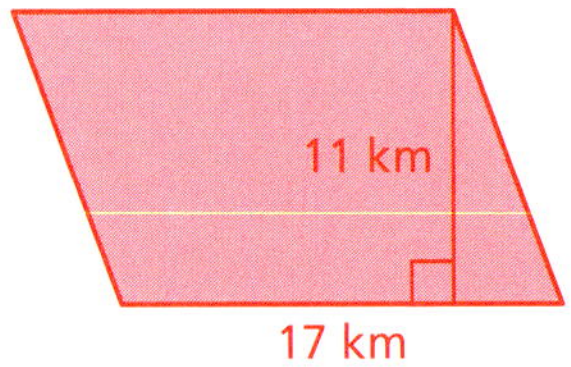

6.

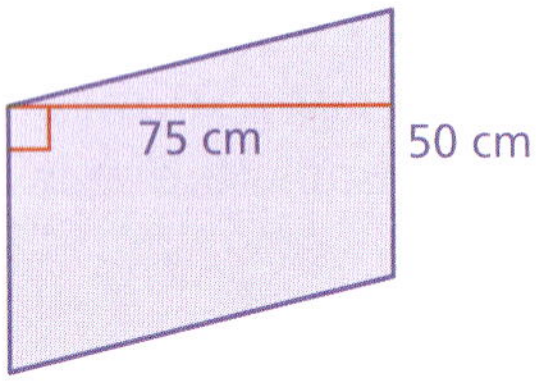

7.

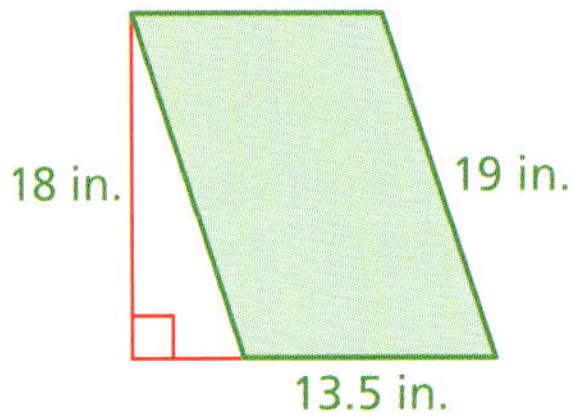

8.

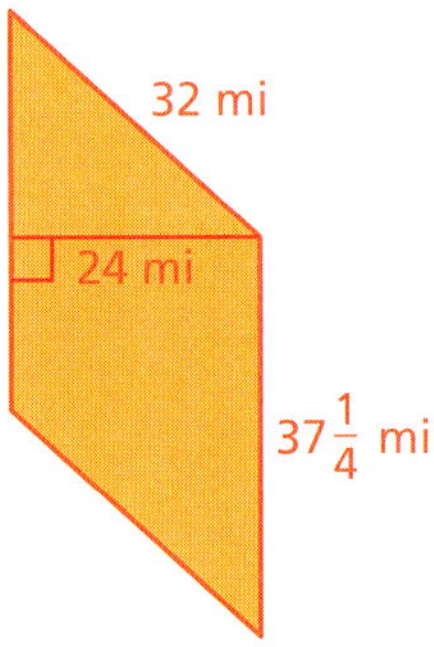

9. **ERROR ANALYSIS** Describe and correct the error in finding the area of the parallelogram.

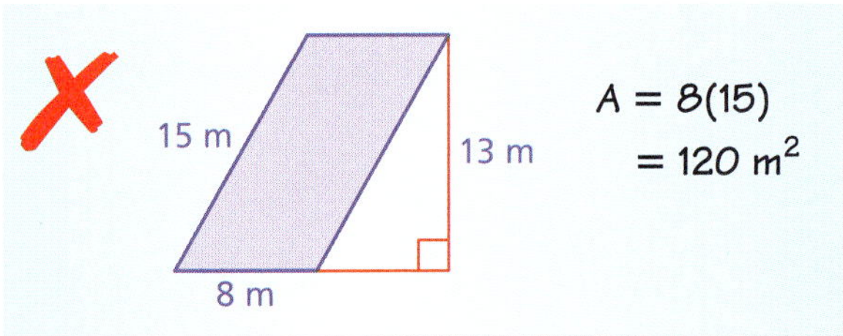

10. **CERAMIC TILE** A ceramic tile in the shape of a parallelogram has a base of 4 inches and a height of 1.5 inches. What is the area of the tile?

Find the area of the parallelogram.

2 11.

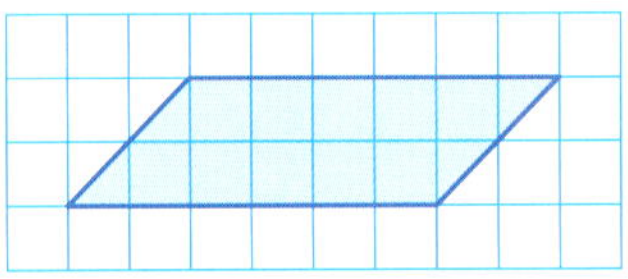

12.

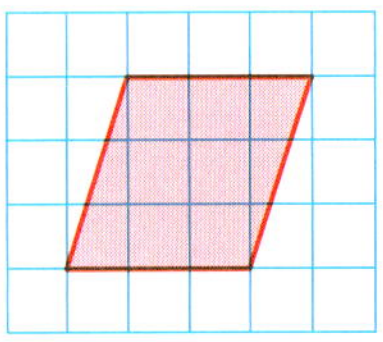

13.

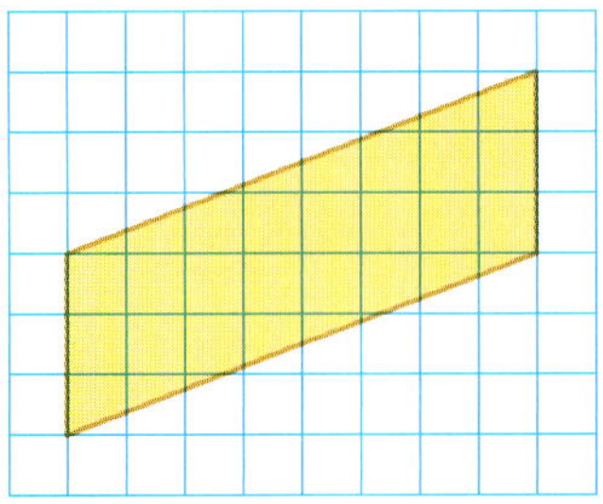

Assignment Guide and Homework Check

Level	Day 1 Activity Assignment	Day 2 Lesson Assignment	Homework Check
Basic	3–5, 22–26	1, 2, 7–11 odd, 12, 13–21 odd	7, 12, 15, 17, 21
Average	3–5, 22–26	1, 2, 6–10 even, 9, 11–14, 16, 17, 19–21	10, 12, 14, 17, 21
Advanced	3–5, 22–26	1, 2, 6–16 even, 9, 17–21	10, 12, 14, 18, 21

Common Errors

- **Exercises 3–13** Students may use the formula for perimeter instead of the formula for area. Tell students to write the formula and then identify the value of each variable before substituting.
- **Exercises 14–16** Students may forget to subtract the area of the white rectangle or parallelogram when finding the area of the shaded region. Draw the parallelogram on a piece of paper and cut out the part that the students must subtract to find the area.
- **Exercises 14–16** Some students may try to break the shaded region into smaller regions and then find the area of each smaller region. This is much more difficult.

4.1 Record and Practice Journal

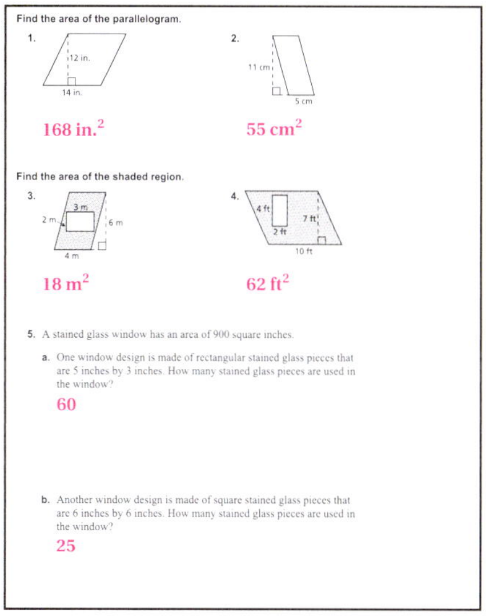
Find the area of the parallelogram.

1. 12 in., 14 in.

168 in.^{2}

2. 11 cm, 5 cm

55 cm^2

Find the area of the shaded region.

3. 3 m, 2 m, 6 m, 4 m

18 m^2

4. 4 ft, 7 ft, 2 ft, 10 ft

62 ft^2

5. A stained glass window has an area of 900 square inches.

a. One window design is made of rectangular stained glass pieces that are 5 inches by 3 inches. How many stained glass pieces are used in the window?

60

b. Another window design is made of square stained glass pieces that are 6 inches by 6 inches. How many stained glass pieces are used in the window?

25

Vocabulary and Concept Check

1. The area of a polygon is the amount of surface it covers. The perimeter of a polygon is the distance around the polygon.
2. *Sample answer:* Any parallelogram where the product of the base and height is 24 in.^{2}.

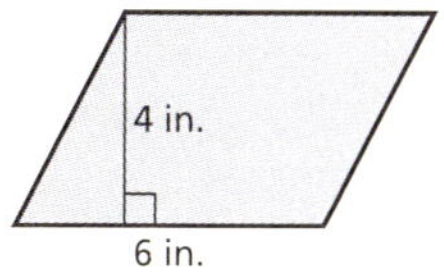

Practice and Problem Solving

3. 18 ft^2
4. 840 mm^2
5. 187 km^2
6. 3750 cm^2
7. 243 in.^{2}
8. 894 mi^2
9. 15 meters was used for the height instead of 13 meters. $A = 8(13) = 104$ m^2
10. 6 in.^{2}
11. 12 units^2
12. 9 units^2
13. 24 units^2

Practice and Problem Solving

14. 64 cm^2

15 72 m^2

16. 96 ft^2

17. See *Taking Math Deeper.*

18. 22 times

19. 287 in.^2

20–21. See Additional Answers.

Fair Game Review

22. 13	**23.** 1640
24. 480	**25.** 118
26. B	

Mini-Assessment

Find the area of the square, rectangle, or parallelogram.

1.

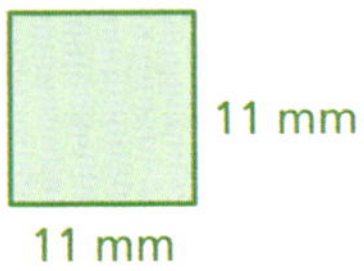

121 mm^2

2.

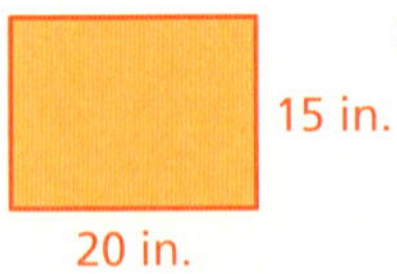

300 in.^2

3.

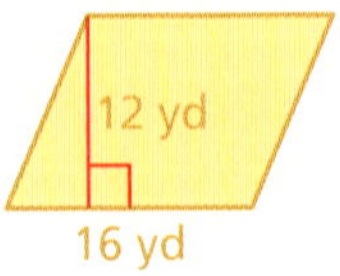

192 yd^2

4. Find the area of the shaded region.

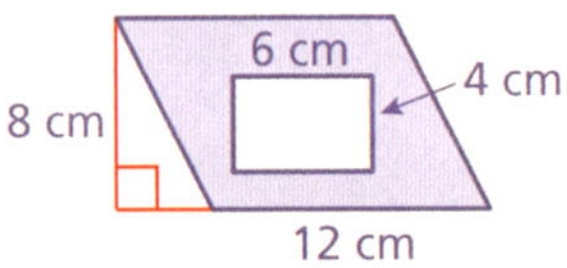

72 cm^2

Taking Math Deeper

Exercise 17

Remind students that they don't have to see how to start the solution before they put their pencil to paper.

- Doodle some notes.
- Draw a diagram.
- Jot down some facts.
- Rewrite the problem.
- Underline the important words.

These things kick students' brains into gear. Soon they see a way to the solution.

1 How can an area be represented by $s^2 + 128$?
128 could be a rectangle. Draw a couple.

It could be 4 by 32.

It could be 8 by 16. — This one looks like a good deck.

2 Here is a straightforward solution.

- You have the 128. How about the s^2?
- It could be the area of a square. Add this to your diagram.

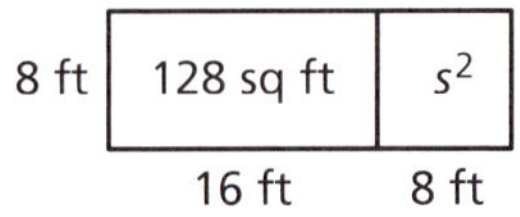

3 Here are more creative solutions.

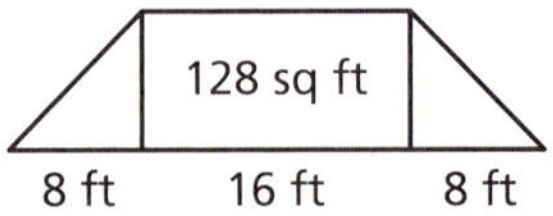

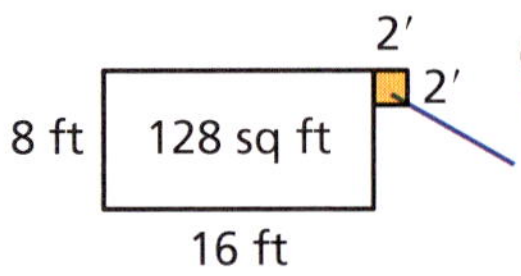

Project

A builder needs to present four different decks to a customer using the information in Exercise 17. Design four different decks for him to use in his sales presentation. Explain how they meet the requirements of Exercise 17.

Reteaching and Enrichment Strategies

If students need help. . .	If students got it. . .
Resources by Chapter • Practice A and Practice B • Puzzle Time Record and Practice Journal Practice Differentiating the Lesson Lesson Tutorials Skills Review Handbook	Resources by Chapter • Enrichment and Extension • Technology Connection Start the next section

Find the area of the shaded region.

3 **14.**

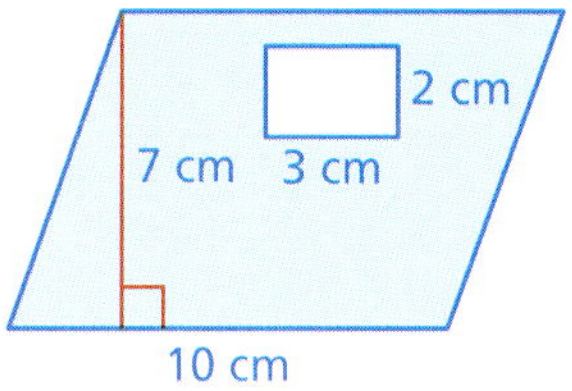

15.

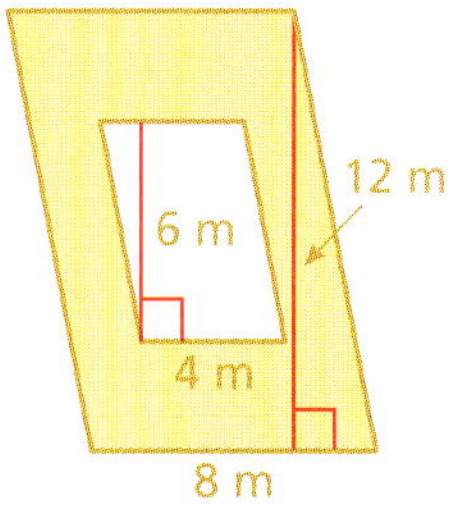

16.

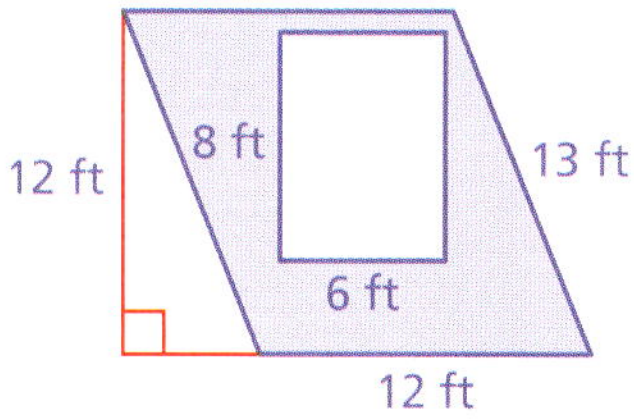

17. **DECK** Your deck has an area of 128 square feet. After adding a section, the area will be $s^2 + 128$ square feet. Draw a diagram of how this can happen.

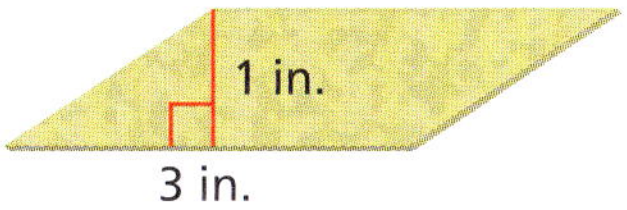

18. **T-SHIRT DESIGN** You use the parallelogram-shaped sponge to create the T-shirt design. The area of the design is 66 square inches. How many times do you use the sponge to create the design? Draw a diagram to support your answer.

19. **STAIRCASE** The staircase has three parallelogram-shaped panels that are the same size. The horizontal distance between each panel is 4.25 inches. What is the area of one panel?

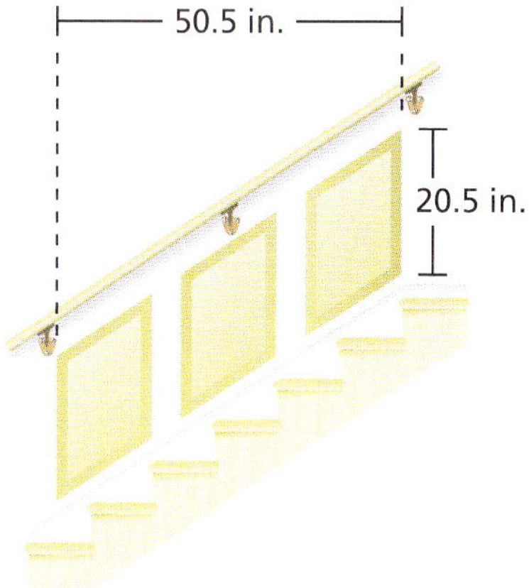

20. **REASONING** Find the missing dimensions in the table.

Parallelogram	Base	Height	Area
A	$x + 4$		$5x + 20$
B		8	$8x - 24$
C	6		$12x + 6y$

21. **Logic** Each dimension of a parallelogram is multiplied by a positive number n. Write an expression for the area of the new parallelogram.

Fair Game Review What you learned in previous grades & lessons

Use mental math to multiply. *(Skills Review Handbook)*

22. $\frac{1}{2} \times 26$ **23.** 82×20 **24.** 16×30 **25.** $\frac{1}{2} \times 236$

26. **MULTIPLE CHOICE** Which of the following describes angle B? *(Skills Review Handbook)*

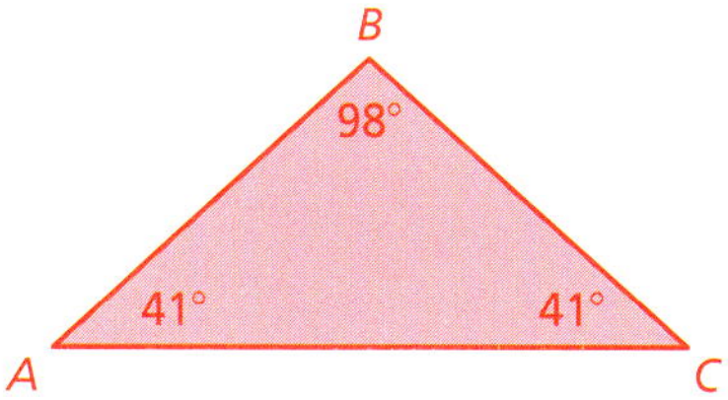

Ⓐ acute Ⓑ obtuse

Ⓒ right Ⓓ isosceles

4.2 Areas of Triangles

Essential Question How can you derive a formula for the area of a triangle?

1 ACTIVITY: Deriving the Area Formula of a Triangle

Work with a partner.

a. Draw *any* rectangle on a piece of grid paper. An example is shown below. Label the length and width. Then find the area of your rectangle.

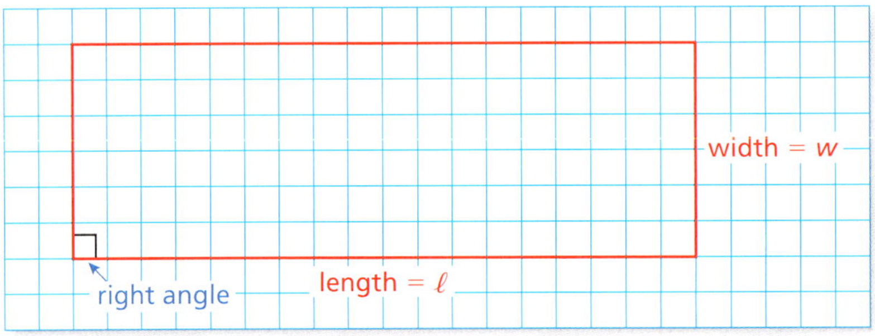

b. Draw a diagonal from one corner of your rectangle to the opposite corner. Cut along the diagonal. Compare the area of the rectangle with the area of the two pieces you cut. What do you notice? Use your results to write a formula for the area A of a triangle.

$A =$ ______ Formula

2 ACTIVITY: Deriving the Area Formula of a Triangle

Work with a partner.

a. Fold a piece of grid paper in half. Draw a triangle so that its base lies on one of the horizontal lines of the paper. Do not use a right triangle. Label the height and the base *inside* the triangle.

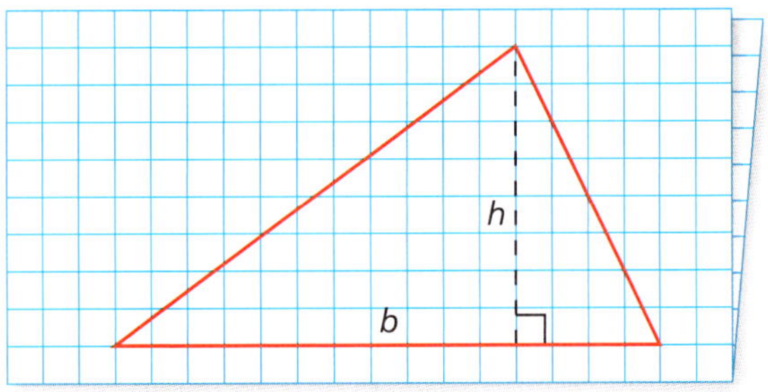

fold

b. Estimate the area of your triangle by counting unit squares.

Area ≈ ______ Estimate

c. Cut out the triangle so that you end up with two identical triangles. Form a quadrilateral whose area you know. What type of quadrilateral is it? Explain how you *know* it is this type.

d. Use your results to write a formula for the area of a triangle. Then use your formula to find the exact area of your triangle. Compare this area with your estimate in part (b).

COMMON CORE

Geometry

In this lesson, you will
- find areas of triangles.
- solve real-life problems.

Learning Standard
6.G.1

Laurie's Notes

Introduction

Standards for Mathematical Practice

- **MP3 Construct Viable Arguments and Critique the Reasoning of Others:** Deductive reasoning is again used to make a conjecture about a new formula, the area of a triangle. Students should be able to analyze the two approaches used to derive the formula for the area of a triangle.

Motivate

- The Flatiron (or Fuller) Building in New York City was built in 1902 on a triangular block formed by Fifth Avenue, Broadway and East 22nd Street.

? "Why do you think it is called the Flatiron Building?" Its shape is similar to a flatiron (for pressing clothes).

? "If you had an office in that building, where would you want it to be and why?" I personally would want the top floor at the most acute vertex!

- Triangles are a fairly uncommon shape for a building.

Activity Notes

Activity 1

- **MP3:** Listen for a logical argument: because there are two triangles of equal area, each must be half of the area of the rectangle. Students will likely write $A = \ell w \div 2$

? "Is there another way to write the formula?" yes; $A = \frac{1}{2}\ell w$

- Remind students that when two variables, or a variable and number, are written adjacent to one another it implies multiplication.
- **Extension:** Ask, "Can you can start with a non-right parallelogram and cut along the diagonal?" yes; Take time to explore this question.

Activity 2

- **Common Error:** In part (c), students may arrange their triangles with the bases together. The labels for the base and height have no meaning for this quadrilateral. Students need to arrange the triangles so that the base of the triangle is also the base of the quadrilateral.

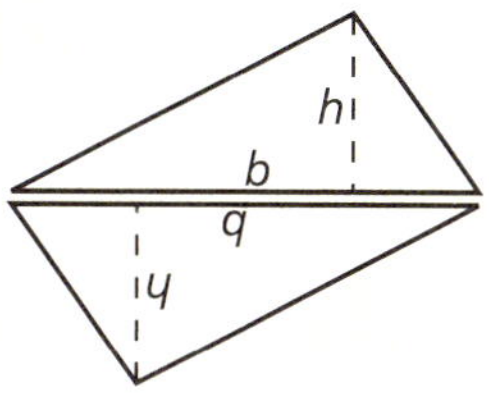

- When students have finished, ask how they estimated the area of their triangle. You want to hear students describe how they dealt with nonsquare pieces of the grid paper.

? "Did anyone end up with a rectangle when they arranged their triangles? How would this be possible?" Their triangle was a right triangle.

? "Did anyone end up with a square when they arranged their triangles? How would this be possible?" Their triangle was a right isosceles triangle.

? "How are the first two activities the same?" Both activities are deriving the formula for area of a triangle, but using different methods.

Common Core State Standards

6.G.1 Find the area of right triangles, other triangles, . . . by composing into rectangles . . . ; apply these techniques in the context of solving real-world and mathematical problems.

Previous Learning

Students have found the area of a square, rectangle, and parallelogram.

Technology for the Teacher

Dynamic Classroom

Lesson Plans
Complete Materials List

4.2 Record and Practice Journal

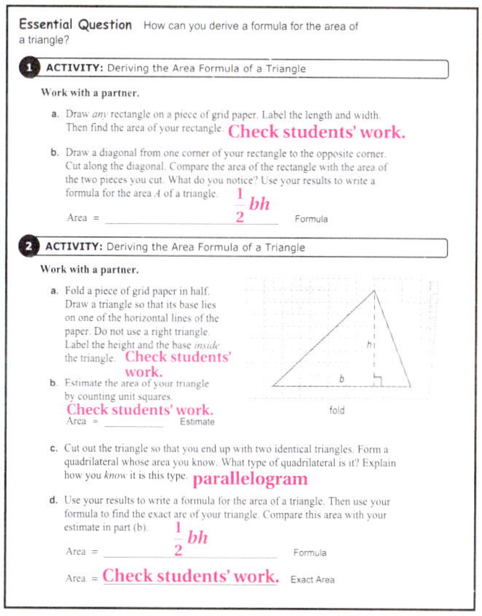

Essential Question How can you derive a formula for the area of a triangle?

1 ACTIVITY: Deriving the Area Formula of a Triangle

Work with a partner.

a. Draw *any* rectangle on a piece of grid paper. Label the length and width. Then find the area of your rectangle. Check students' work.

b. Draw a diagonal from one corner of your rectangle to the opposite corner. Cut along the diagonal. Compare the area of the rectangle with the area of the two pieces you cut. What do you notice? Use your results to write a formula for the area A of a triangle.

Area = $\frac{1}{2}bh$ Formula

2 ACTIVITY: Deriving the Area Formula of a Triangle

Work with a partner.

a. Fold a piece of grid paper in half. Draw a triangle so that its base lies on one of the horizontal lines of the paper. Do not use a right triangle. Label the height and the base *inside* the triangle. Check students' work.

b. Estimate the area of your triangle by counting unit squares.

Area = Check students' work. Estimate

fold

c. Cut out the triangle so that you end up with two identical triangles. Form a quadrilateral whose area you know. What type of quadrilateral is it? Explain how you *know* it is this type. parallelogram

d. Use your results to write a formula for the area of a triangle. Then use your formula to find the exact are of your triangle. Compare this area with your estimate in part (b).

Area = $\frac{1}{2}bh$ Formula

Area = Check students' work. Exact Area

English Language Learners

Vocabulary

Review the definition of perpendicular. Two lines are perpendicular if the lines form a right angle. The corner of a sheet of paper is a right angle. For obtuse triangles, the height (the line perpendicular to the base) may lie outside of the triangle.

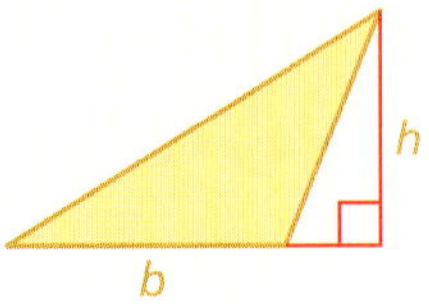

4.2 Record and Practice Journal

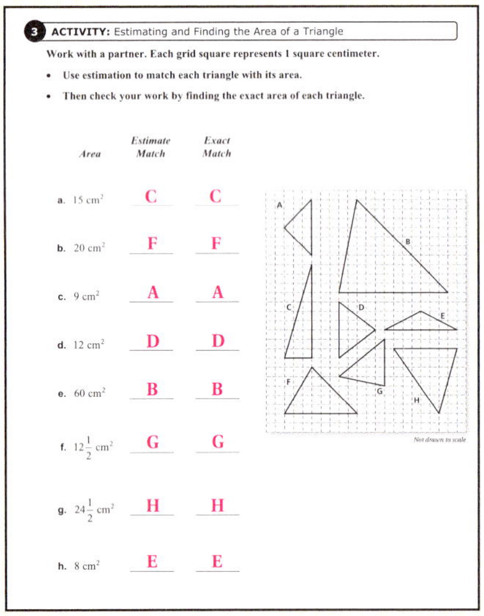

3 ACTIVITY: Estimating and Finding the Area of a Triangle

Work with a partner. Each grid square represents 1 square centimeter.

- **Use estimation to match each triangle with its area.**
- **Then check your work by finding the exact area of each triangle.**

	Area	*Estimate Match*	*Exact Match*
a.	15 cm²	C	C
b.	20 cm²	F	F
c.	9 cm²	A	A
d.	12 cm²	D	D
e.	60 cm²	B	B
f.	$12\frac{1}{2}$ cm²	G	G
g.	$24\frac{1}{2}$ cm²	H	H
h.	8 cm²	E	E

Not drawn to scale

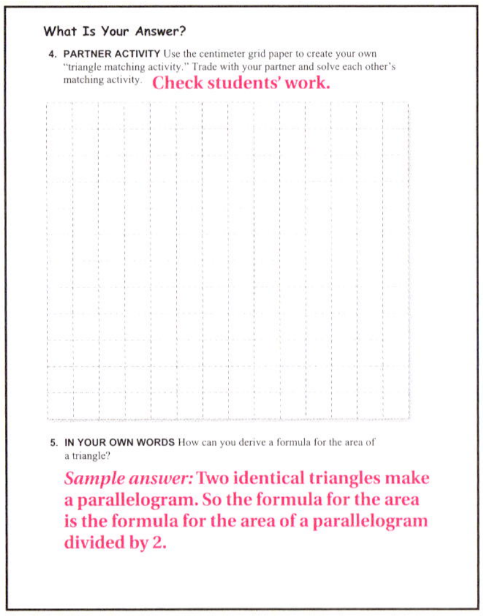

What Is Your Answer?

4. **PARTNER ACTIVITY** Use the centimeter grid paper to create your own "triangle matching activity." Trade with your partner and solve each other's matching activity. **Check students' work.**

5. **IN YOUR OWN WORDS** How can you derive a formula for the area of a triangle?

   Sample answer: **Two identical triangles make a parallelogram. So the formula for the area is the formula for the area of a parallelogram divided by 2.**

Laurie's Notes

Activity 3

- Students will first estimate the area of each triangle. Again, they will have to take into account nonsquare pieces of the grid paper.
- Given the different orientations of the triangles, students will need to decide which side of the triangle should be the base. If they need assistance, suggest that they use the horizontal or vertical side of the triangle as the base.

? "How well did you do with your estimate?" Answers will vary.

- For each triangle, ask a student to explain how they found the exact area.

What Is Your Answer?

- **FYI:** If one of the sides of the triangle is not horizontal or vertical, it will be difficult to find the exact area.

Closure

- Another famous triangle is in the Atlantic Ocean. It is the Bermuda Triangle. Draw a sketch of the triangle and ask students to compute the area of this triangular region. about 450,000 mi^2

Math Practice 6

Calculate Accurately

How can you estimate the area of each triangle so that the answer is close to the exact area?

3 ACTIVITY: Estimating and Finding the Area of a Triangle

Work with a partner. Each grid square represents 1 square centimeter.

- **Use estimation to match each triangle with its area.**
- **Then check your work by finding the exact area of each triangle.**

	Area	*Estimate Match*	*Exact Match*
a.	15 cm^2		
b.	20 cm^2		
c.	9 cm^2		
d.	12 cm^2		
e.	60 cm^2		
f.	$12\frac{1}{2}$ cm^2		
g.	$24\frac{1}{2}$ cm^2		
h.	8 cm^2		

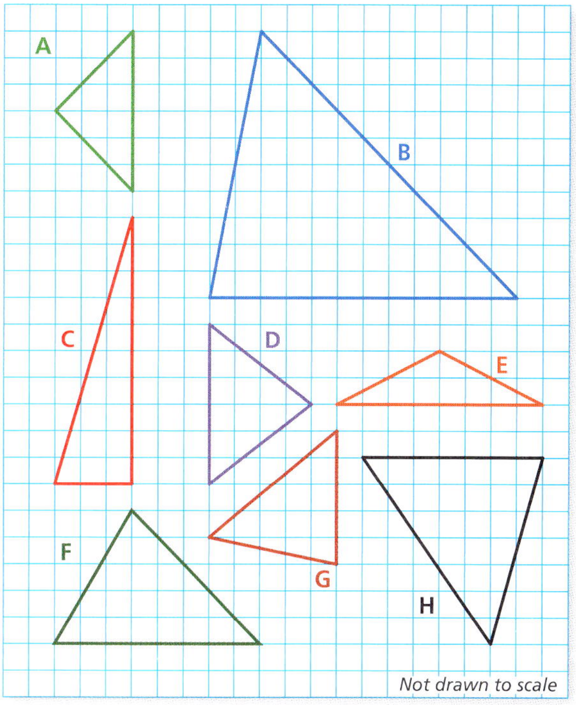

What Is Your Answer?

4. **PARTNER ACTIVITY** Use a piece of centimeter grid paper to create your own "triangle matching activity." Trade with your partner and solve each other's matching activity.

5. **IN YOUR OWN WORDS** How can you derive a formula for the area of a triangle?

Practice Use what you learned about the areas of triangles to complete Exercises 3–5 on page 162.

4.2 Lesson

Check It Out
Lesson Tutorials
BigIdeasMath.com

Key Idea

Area of a Triangle

Words The area A of a triangle is one-half the product of its base b and its height h.

Algebra $A = \frac{1}{2}bh$

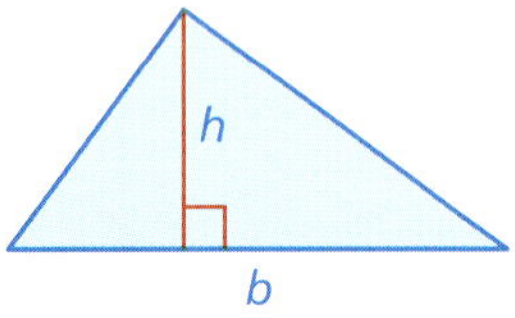

EXAMPLE 1 Finding the Area of a Triangle

Find the area of the triangle.

$A = \frac{1}{2}bh$	Write formula.
$= \frac{1}{2}(5)(8)$	Substitute 5 for b and 8 for h.
$= \frac{1}{2}(40)$	Multiply 5 and 8.
$= 20$	Multiply $\frac{1}{2}$ and 40.

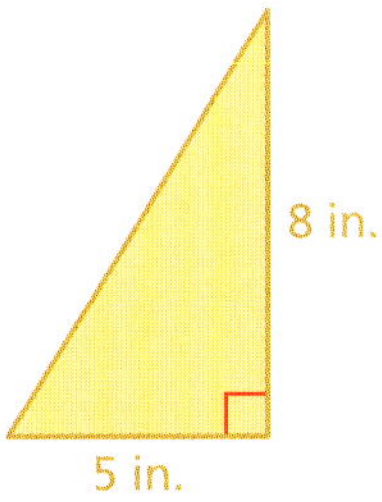

Remember

In Example 1, use the Associative Property of Multiplication to multiply 5 and 8 first.

∴ The area of the triangle is 20 square inches.

Reasonable? Draw the triangle on grid paper and count unit squares. Each square in the grid represents 1 square inch.

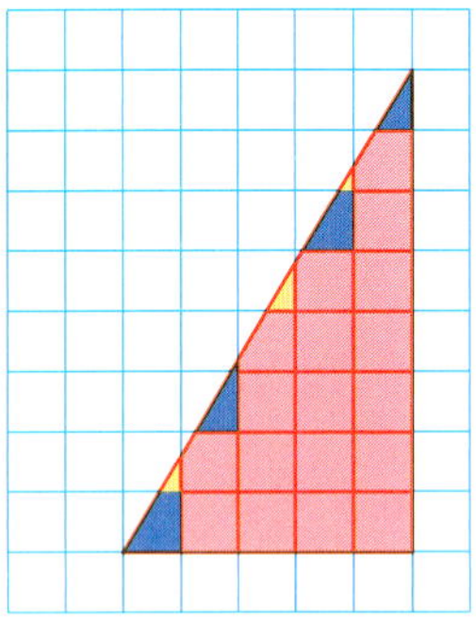

Squares full or nearly full: 18

Squares about half full: 4

The area is $18(1) + 4\left(\frac{1}{2}\right) = 20$ square inches.

So, the answer is reasonable. ✓

On Your Own

Now You're Ready
Exercises 3–8

Find the area of the triangle.

1.

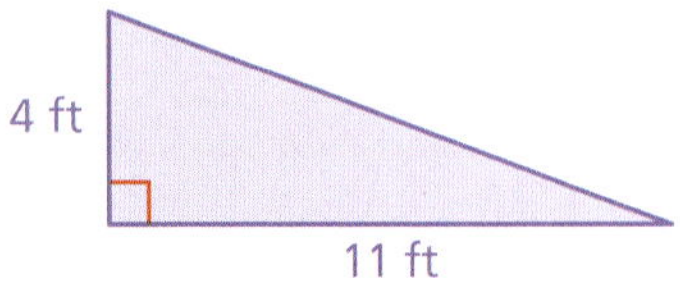

2.

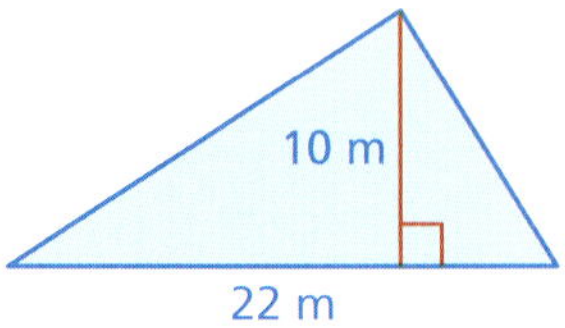

Laurie's Notes

Introduction

Connect

- **Yesterday:** Students derived the formula for the area of a triangle. (MP3)
- **Today:** Students will use the formula to find the areas of several triangles.

Motivate

- **Preparation:** Before class, make a triangular hat by folding a newspaper. Alternatively, you could search the Internet for a video of the hat (newspaper sailor hat) or a napkin (pyramid napkin folding) being folded. You want students to see a common folded triangular shape.
- The goal is to have students think about how much area of the original newspaper (or napkin) is showing.

Lesson Notes

Key Idea

- The *base* of a triangle can be any of its sides.
- The *height* of a triangle is the perpendicular distance from a base to the opposite vertex.
- **FYI:** The formula $A = \frac{1}{2}bh$ is sometimes written $A = \frac{bh}{2}$. Students may make computational mistakes due to the fraction in this formula.

Example 1

- The idea of dividing by 2 as another way of taking $\frac{1}{2}$ of a number makes sense to students. For example: $\frac{1}{2}$ of 10 is the same as $10 \div 2$.
- **Common Error:** Students may want to take $\frac{1}{2}$ of 5 *and* $\frac{1}{2}$ of 8 because they think it is the Distributive Property.
- Note that the product of 5 and 8 was found (Associative Property), and then $\frac{1}{2}$ of that product was computed. Because multiplication is commutative, it would also be correct to find the product of $\frac{1}{2}$ and 8 (which is 4), and then multiply 4 by 5.
- The visual check for reasonableness is the approach used.

On Your Own

- **Extension:** Draw the following on the board and ask students how the area of the rectangle is related to the answers they found for Questions 1 and 2.

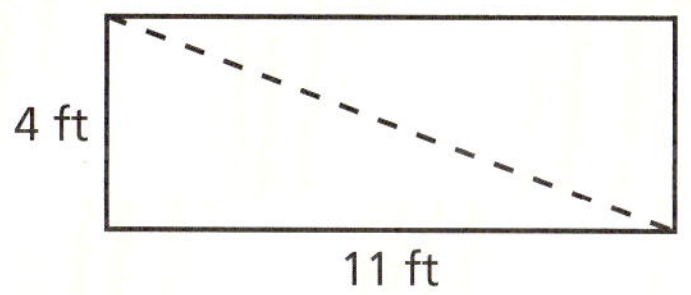

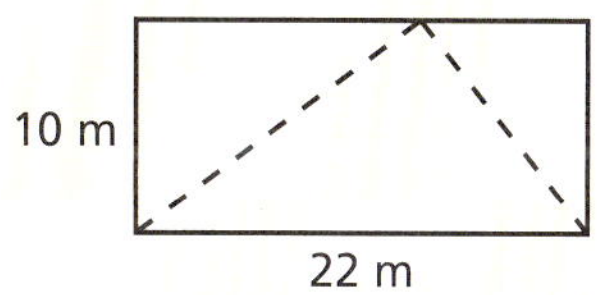

Goal Today's lesson is finding the area of a triangle.

Technology for the Teacher

Dynamic Classroom

Lesson Tutorials
Lesson Plans
Answer Presentation Tool

Differentiated Instruction

Inclusion

When using the formula for the area of a triangle, $A = \frac{1}{2}bh$, tell students to simplify the calculation by dividing b or h by 2, whichever is even.

Extra Example 1

Find the area of the triangle.

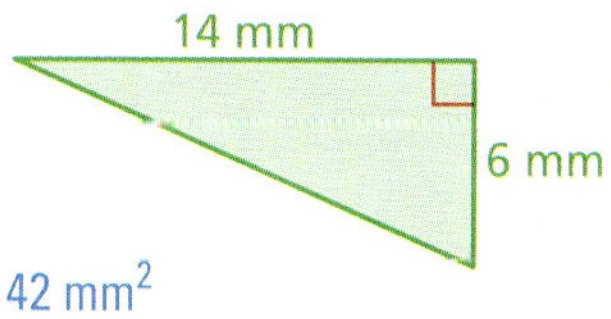

42 mm^2

On Your Own

1. 22 ft^2
2. 110 m^2

Laurie's Notes

Extra Example 2

Find the area of the triangle.

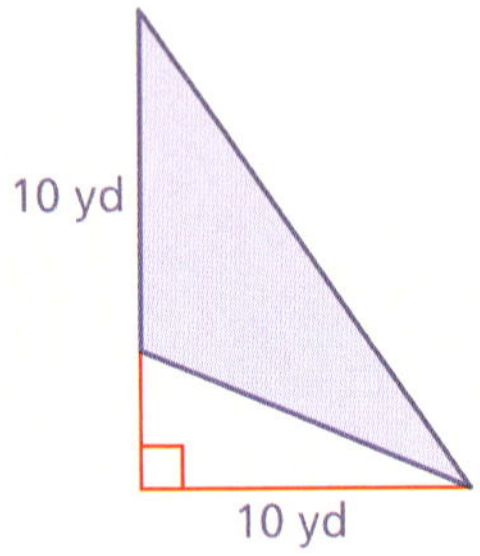

50 yd^2

Extra Example 3

A triangle has a base of 2 inches and a height of 3 inches. The base and height of a second triangle are four times greater than the base and height of the first triangle. How many times greater is the area of the second triangle than the area of the first triangle? 16 times greater

On Your Own

3. 60 cm^2

4. 9 times greater

Example 2

- **MP1 Make Sense of Problems and Persevere in Solving Them:** Mathematically proficient students will check their answers to problems using different methods. Finding $\frac{1}{2}$ of 12 and then multiplying by 9 is different from finding the product of 12 and 9 and then taking $\frac{1}{2}$. Ask students if they can think of examples where one of the two approaches is easier than the other. Example: $\frac{1}{2}(2.5)(4)$

? "How do you find $\frac{1}{2} \times 12 \times 9$?" Listen for the strategies of taking $\frac{1}{2}$ of 12 first, and then multiplying by 9; or finding the product of 12 and 9, and then taking $\frac{1}{2}$ of the product.

- In some cases, the base of a triangle needs to be extended in order to find the height.

Example 3

- Take time to write the formula for the area of a triangle. It is good practice for students to go through the substitution process.
- **Common Error:** The question asks how many times greater, not how much greater. Students may give an answer of 3 square centimeters because they subtracted the two areas.

On Your Own

- A diagram of this type often confuses students. The red segments are not part of the lengths of the sides of the triangle.

Closure

- Have students find the area of the newspaper hat (or triangular napkin) by measuring the necessary dimensions.
- If you did not make the hat or napkin, you could ask students to finish the following writing prompt.
  The area of a triangle is similar to the area of a parallelogram because. . .

EXAMPLE 2 Finding the Area of a Triangle

Find the area of the triangle.

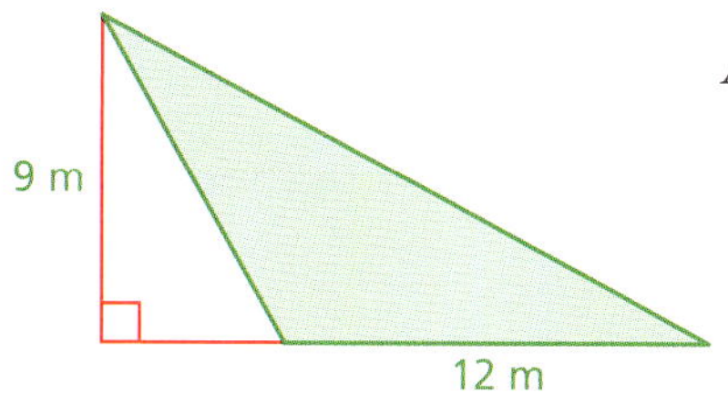

$$A = \frac{1}{2}bh \quad \text{Write formula.}$$

$$= \frac{1}{2}(12)(9) \quad \text{Substitute 12 for } b \text{ and 9 for } h.$$

$$= 54 \quad \text{Multiply.}$$

∴ The area of the triangle is 54 square meters.

EXAMPLE 3 Real-Life Application

The base and height of the red butterfly wing are two times greater than the base and height of the blue butterfly wing. How many times greater is the area of the red wing than the area of the blue wing?

Find the area of the blue wing.

$$A = \frac{1}{2}bh \quad \text{Write formula.}$$

$$= \frac{1}{2}(2)(1) \quad \text{Substitute 2 for } b \text{ and 1 for } h.$$

$$= 1 \text{ cm}^2 \quad \text{Multiply.}$$

The red wing dimensions are 2 times greater, so the base is $2 \times 2 = 4$ cm and the height is $2 \times 1 = 2$ cm. Find the area of the red wing.

$$A = \frac{1}{2}bh \quad \text{Write formula.}$$

$$= \frac{1}{2}(4)(2) \quad \text{Substitute 4 for } b \text{ and 2 for } h.$$

$$= 4 \text{ cm}^2 \quad \text{Multiply.}$$

∴ Because $\frac{4 \text{ cm}^2}{1 \text{ cm}^2} = 4$, the area of the red wing is 4 times greater.

On Your Own

Now You're Ready
Exercises 12–14

3. Find the area of the triangle.

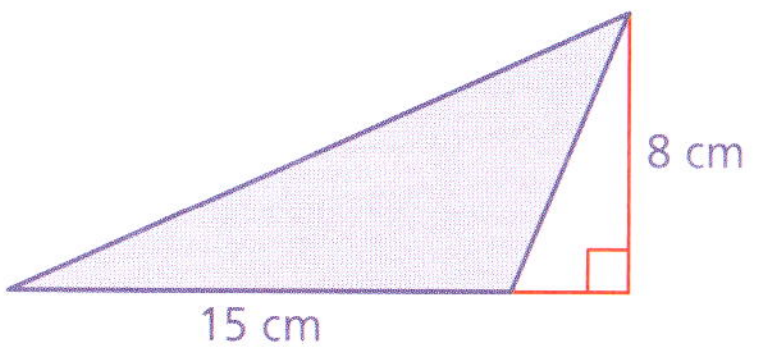

4. **WHAT IF?** In Example 3, the base and the height of the red butterfly wing are three times greater than those of the blue wing. How many times greater is the area of the red wing?

4.2 Exercises

Check It Out
Help with Homework
BigIdeasMath.com

Vocabulary and Concept Check

1. **CRITICAL THINKING** Can *any* side of a triangle be labeled as its base? Explain.

2. **DIFFERENT WORDS, SAME QUESTION** Which is different? Find "both" answers.

What is the area of the triangle?	What is the distance around the triangle?
How many unit squares fit in the triangle?	What is one-half the product of the base and the height?

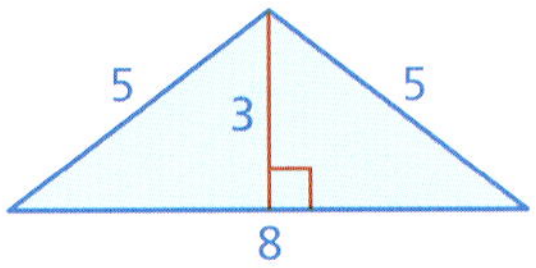

Practice and Problem Solving

Find the area of the triangle.

1 3. 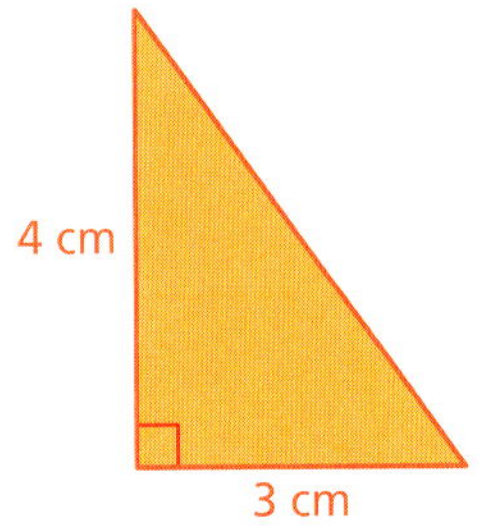

4. 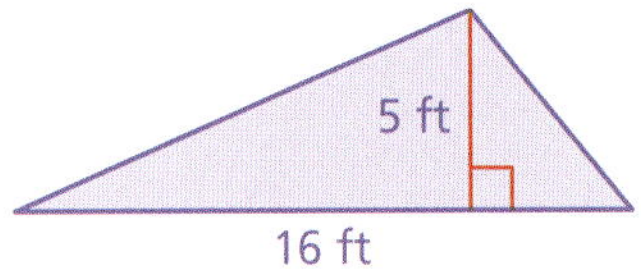

5. 

6. 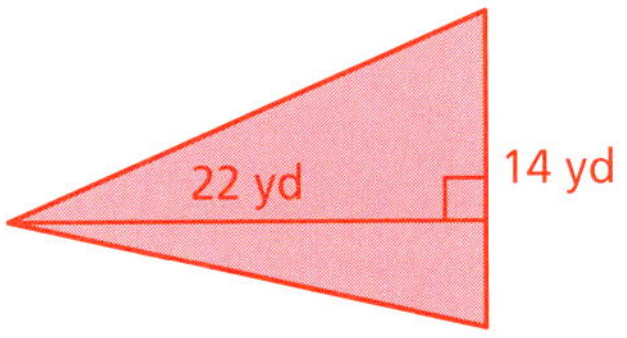

7. 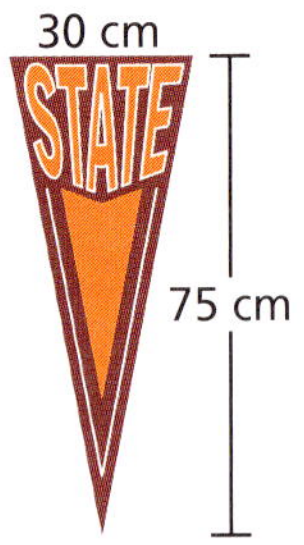

8. 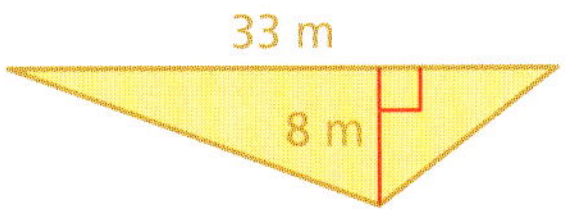

9. **ERROR ANALYSIS** Describe and correct the error in finding the area of the triangle.

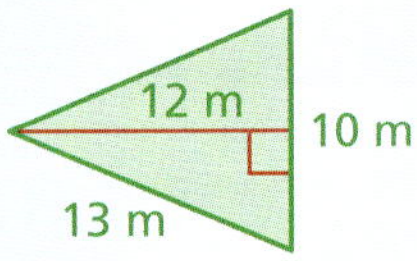

$A = \frac{1}{2}(10)(13)$

$= 65 \text{ m}^2$

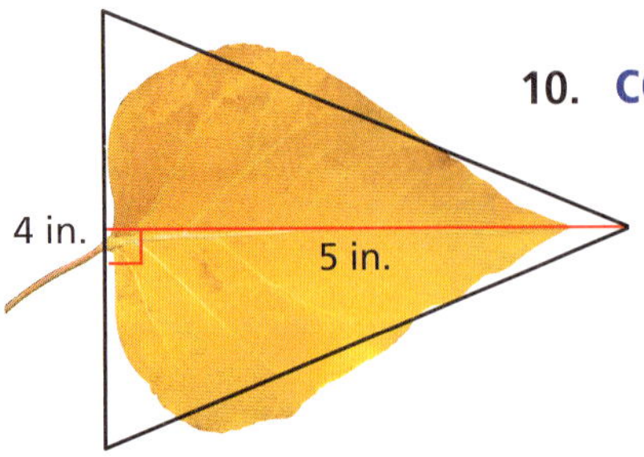

10. **COTTONWOOD LEAF** Estimate the area of the cottonwood leaf.

11. **CORNER SHELF** A shelf has the shape of a triangle. The base of the shelf is 36 centimeters, and the height is 18 centimeters. Find the area of the shelf.

Assignment Guide and Homework Check

Level	Day 1 Activity Assignment	Day 2 Lesson Assignment	Homework Check
Basic	3–5, 21–24	1, 2, 6–13, 15	8, 11, 13, 15
Average	3–5, 21–24	1, 2, 6–14, 16, 18, 19	8, 11, 14, 16
Advanced	3–5, 21–24	1, 2, 6–9, 12–14, 16, 17, 19, 20	8, 12, 14, 19

Common Errors

- **Exercises 3–8** Students may multiply the base and the height and forget to multiply by $\frac{1}{2}$. Tell students to write the formula, and then identify the value of each variable before substituting.
- **Exercises 3–8** Students may mentally find the area by multiplying one-half the base and one-half the height. The answer will be one-half of the value of the correct answer. Tell students to write the formula, and then identify the value of each variable before substituting.
- **Exercises 12–14** Students may be confused by the drawing and labeling of obtuse triangles. Remind students to identify the base of the triangle. Then the height of the triangle is the perpendicular distance from the base to the opposite vertex. For obtuse triangles, the height may be labeled outside of the triangle.

4.2 Record and Practice Journal

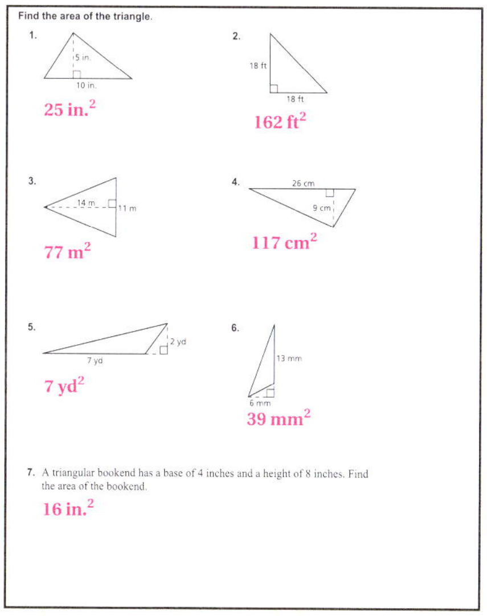
Find the area of the triangle.

1. 25 in.^{2}

2. 162 ft^2

3. 77 m^2

4. 117 cm^2

5. 7 yd^2

6. 39 mm^2

7. A triangular bookend has a base of 4 inches and a height of 8 inches. Find the area of the bookend.
16 in.^{2}

Vocabulary and Concept Check

1. yes; To find the area of the triangle, you must also know the height of the triangle. That is, the perpendicular distance from the base to the opposite vertex.
2. What is the distance around the triangle?; 18 units; 12 units^2

Practice and Problem Solving

3. 6 cm^2
4. 40 ft^2
5. 1620 in.^{2}
6. 154 yd^2
7. 1125 cm^2
8. 132 m^2
9. The side length of 13 meters was used instead of the height.
   $A = \frac{1}{2}(10)(12) = 60 \text{ m}^2$
10. about 10 in.^{2}
11. 324 cm^2
12. 68 m^2
13. 90 mi^2
14. 189 mm^2
15. *Sample answer:*

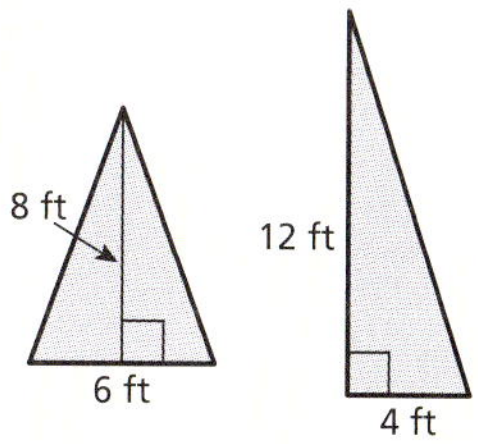

Practice and Problem Solving

16. See *Taking Math Deeper.*

17. x^2 times greater

18. Find the length of the base by dividing the perimeter by 3. Then multiply one-half of the base by the height.

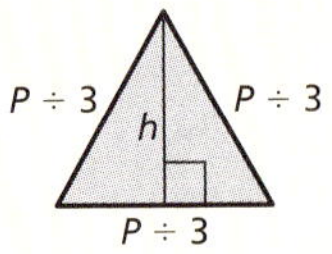

19. 4 times greater

20. 6 ft

Fair Game Review

21. Mult. Prop. of One

22. Comm. Prop. of Mult.

23. Assoc. Prop. of Add.

24. C

Mini-Assessment

Find the area of the triangle.

1.

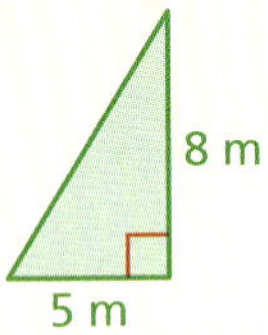

$20\ m^2$

2.

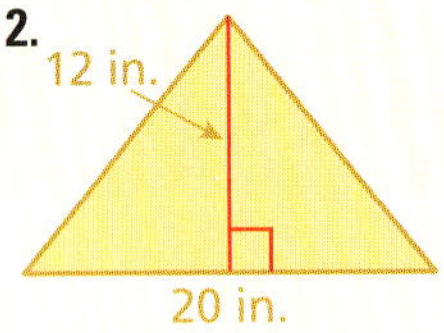

$120\ in.^2$

3. $21\ ft^2$

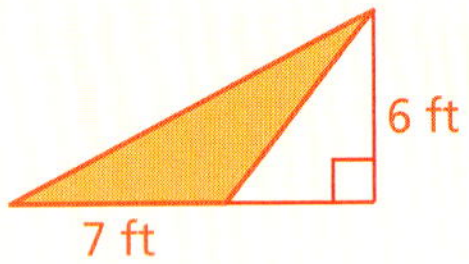

4. $108\ mm^2$

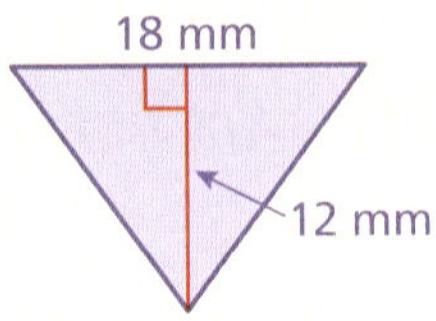

5. A right triangle has a base of 4 inches and a height of 2 inches. The base and height of a second right triangle are two times greater than the base and the height of the first right triangle. How many times greater is the area of the second triangle than the area of the first triangle? 4 times greater

Taking Math Deeper

Exercise 16

Some questions involving mathematics have a straightforward "math" answer. But when you think about the real-life situation, you realize that the answer is more complicated. This is an example of one of those types of questions.

1 Finding the area of the sail of the hang glider is straightforward.

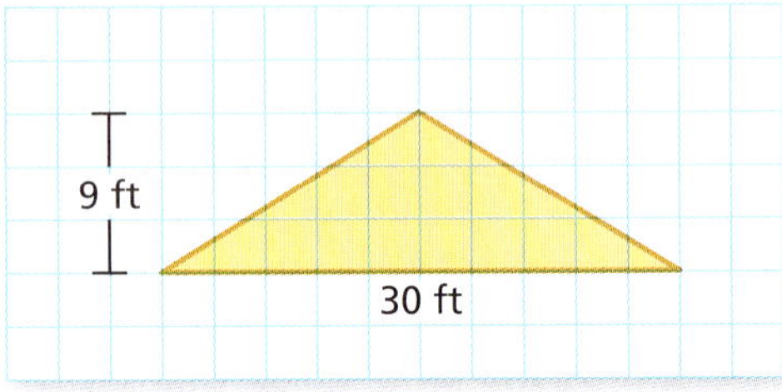

a. $A = \frac{1}{2}bh$

$= \frac{1}{2} \cdot 30 \cdot 9$

$= 135\ ft^2$

2 We have found the area of the fabric for the completed hang glider. You will most likely choose to give full credit for this answer. However, you might also choose to take the question more literally. Anyone who sews knows that you need extra fabric, to account for seams and losses in cutting.

From the photograph, you can see the seams in the hang glider. Suppose the fabric comes in a width of 3 feet. How could you layout the fabric to produce an isosceles triangle with a base of 30 feet and a height of 9 feet? This is a challenging question.

3 **b.** In general, the greater the weight of the pilot, the greater the area of the hang glider's sail.

For example:

Sail Area	*Pilot Weight*
$126\ ft^2$	110–176 lbs
$148\ ft^2$	150–215 lbs
$160\ ft^2$	185–275 lbs

Reteaching and Enrichment Strategies

If students need help. . .	If students got it. . .
Resources by Chapter • Practice A and Practice B • Puzzle Time Record and Practice Journal Practice Differentiating the Lesson Lesson Tutorials Skills Review Handbook	Resources by Chapter • Enrichment and Extension • Technology Connection Start the next section

Find the area of the triangle.

2 **12.**

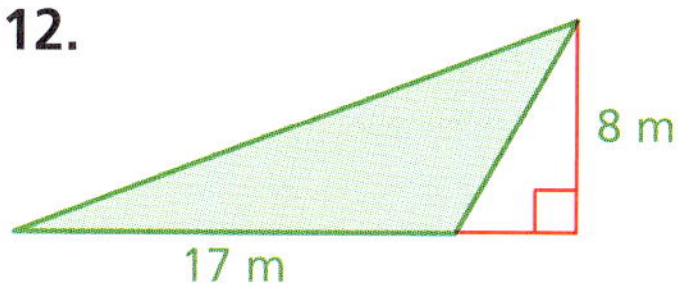

13.

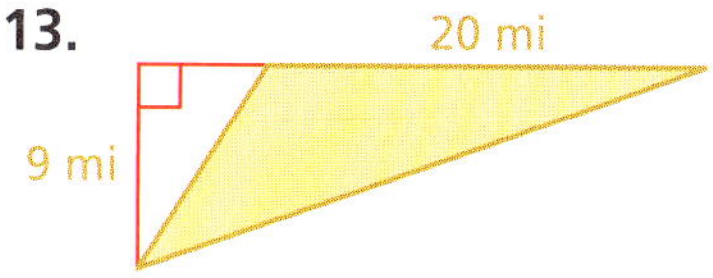

14.

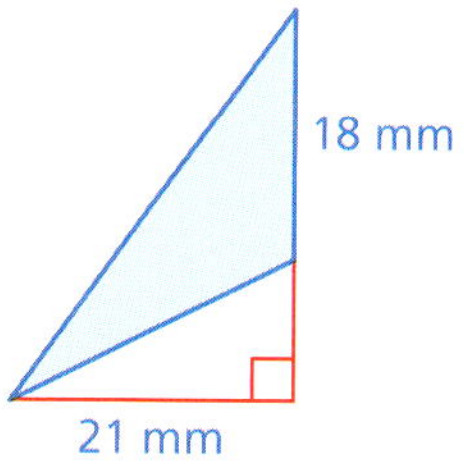

15. OPEN-ENDED Draw and label two triangles that each have an area of 24 square feet.

16. HANG GLIDING The wingspan of the triangular hang glider is 30 feet.

a. How much fabric is needed to make the sail?

b. RESEARCH Use the Internet or some other source to find how the area of the sail is related to the weight limit of the pilot.

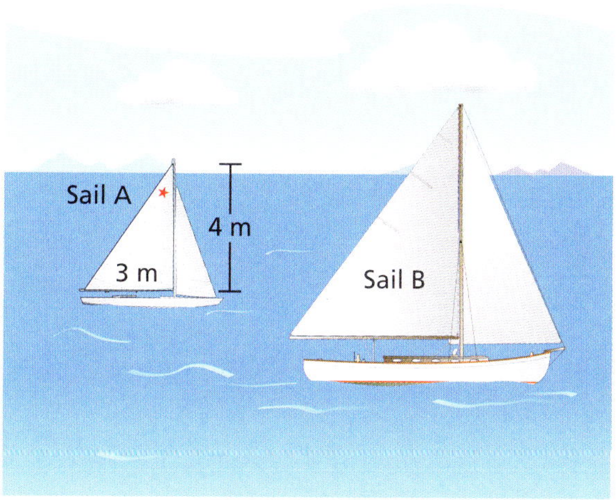

17. SAILBOATS The base and the height of Sail B are x times greater than the base and the height of Sail A. How many times greater is the area of Sail B? Write your answer as a power.

18. WRITING You know the height and the perimeter of an equilateral triangle. Explain how to find the area of the triangle. Draw a diagram to support your reasoning.

19. REASONING The base and the height of Triangle A are half the base and the height of Triangle B. How many times greater is the area of Triangle B?

20. Critical Thinking The total area of the polygon is 176 square feet. Find the value of x.

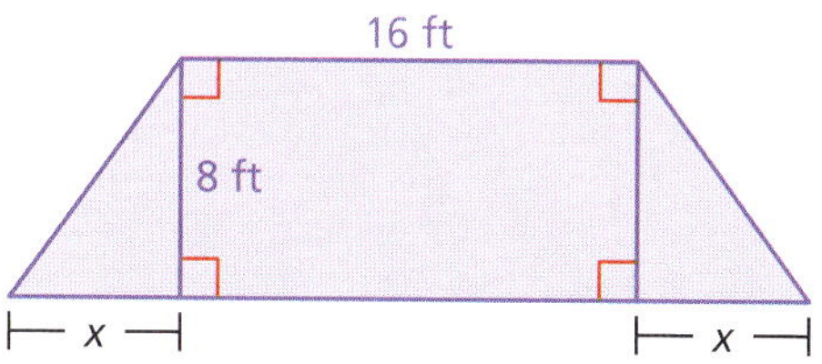

Fair Game Review What you learned in previous grades & lessons

Tell which property is illustrated by the statement. *(Section 3.3)*

21. $n \cdot 1 = n$

22. $4 \cdot m = m \cdot 4$

23. $(x + 2) + 5 = x + (2 + 5)$

24. MULTIPLE CHOICE What is the first step when using order of operations? *(Section 1.3)*

Ⓐ Multiply or divide from left to right.

Ⓑ Add or subtract from left to right.

Ⓒ Perform operations in parentheses.

Ⓓ Evaluate numbers with exponents.

4 Study Help

Check It Out
Graphic Organizer
BigIdeasMath.com

You can use a **four square** to organize information about a topic. Each of the four squares can be a category, such as *definition, vocabulary, example, non-example, words, algebra, table, numbers, visual, graph,* or *equation*. Here is an example of a four square for the area of a parallelogram.

Area of a parallelogram

Words
The area A of a parallelogram is the product of its base b and its height h.

Algebra
$A = bh$

Example
$A = bh$
$= 8(5)$
$= 40$

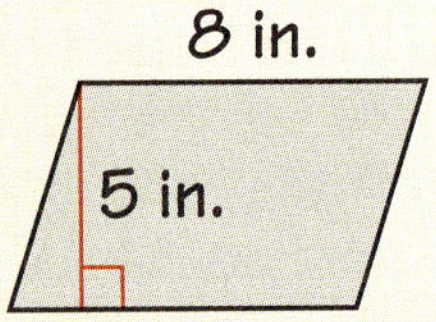

The area of the parallelogram is 40 square inches.

Example
$A = bh$
$= 6(10)$
$= 60$

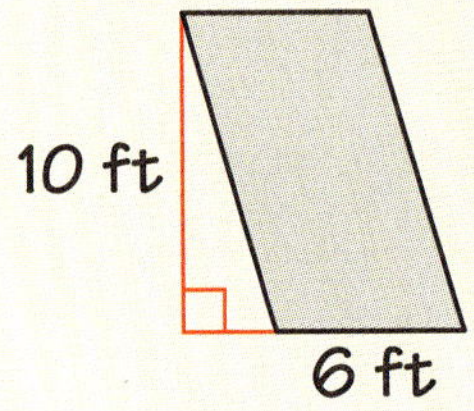

The area of the parallelogram is 60 square feet.

On Your Own

Make a four square to help you study the topic.

1. area of a triangle

After you complete this chapter, make four squares for the following topics.

2. area of a trapezoid
3. area of a composite figure
4. drawing a polygon in a coordinate plane
5. finding distances in the first quadrant

"Sorry, but I have limited space in my **four square**. I needed pet names with only three letters."

Sample Answer

1.

Words

The area A of a triangle is one-half the product of its base b and its height h.

Algebra

$$A = \frac{1}{2}bh$$

Area of a triangle

Example

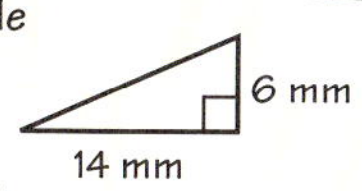

$$A = \frac{1}{2}bh$$
$$= \frac{1}{2}(14)(6)$$
$$= 42$$

The area of the triangle is 42 square millimeters.

Example

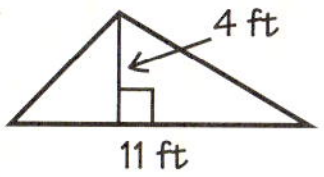

$$A = \frac{1}{2}bh$$
$$= \frac{1}{2}(11)(4)$$
$$= 22$$

The area of the triangle is 22 square feet.

List of Organizers

Available at *BigIdeasMath.com*

Comparison Chart
Concept Circle
Definition (Idea) and Example Chart
Example and Non-Example Chart
Formula Triangle
Four Square
Information Frame
Information Wheel
Notetaking Organizer
Process Diagram
Summary Triangle
Word Magnet
Y Chart

About this Organizer

A **Four Square** can be used to organize information about a topic. Students write the topic in the "bubble" in the middle of the four square. Then students write concepts related to the topic in the four squares surrounding the bubble. Any concept related to the topic can be used. Encourage students to include concepts that will help them learn the topic. Students can place their four squares on note cards to use as a quick study reference.

Technology for the Teacher

Editable Graphic Organizer

Answers

1. 14 cm^2
2. 5950 $in.^2$
3. 384 yd^2
4. 882 mi^2
5. 28 cm^2
6. 78 m^2
7. 9 mi^2
8. 3 sheets

Technology for the Teacher

Online Assessment
Assessment Book
ExamView® Assessment Suite

Alternative Quiz Ideas

100% Quiz
Error Notebook
Group Quiz
Homework Quiz
Math Log
Notebook Quiz
Partner Quiz
Pass the Paper

Group Quiz

Students work in groups. Give each group a large index card. Each group writes five questions that they feel evaluate the material they have been studying. On a separate piece of paper, students solve the problems. When they are finished, they exchange cards with another group. The new groups work through the questions on the card.

Reteaching and Enrichment Strategies

If students need help...	If students got it...
Resources by Chapter • Practice A and Practice B • Puzzle Time Lesson Tutorials *BigIdeasMath.com*	Resources by Chapter • Enrichment and Extension • Technology Connection Game Closet at *BigIdeasMath.com* Start the next section

4.1–4.2 Quiz

Check It Out
Progress Check
BigIdeasMath.com

Find the area of the parallelogram. *(Section 4.1)*

1.

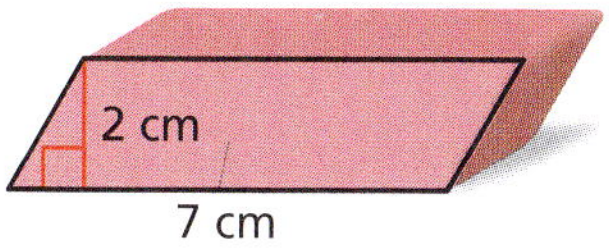

2.

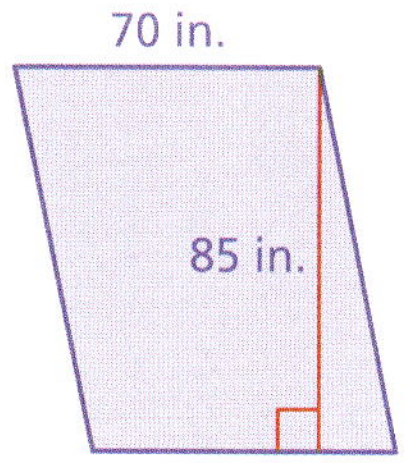

3.

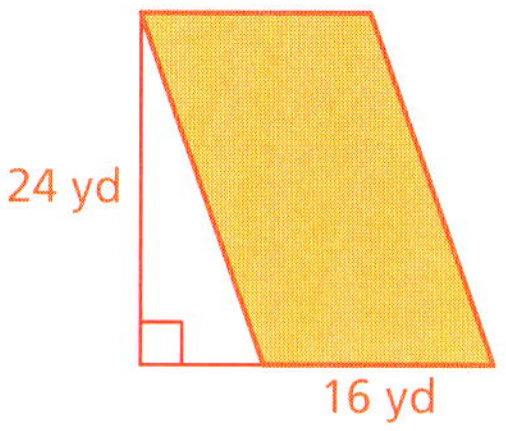

4.

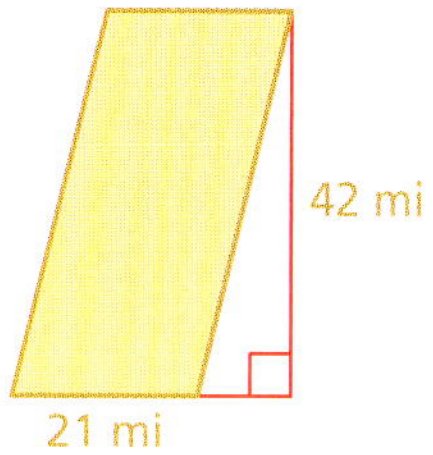

Find the area of the triangle. *(Section 4.2)*

5.

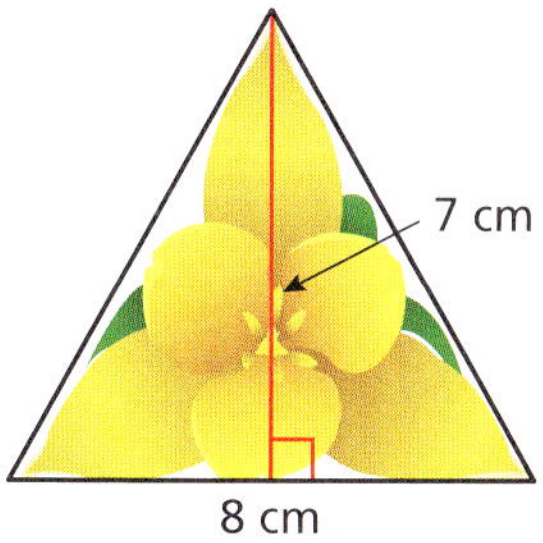

6.

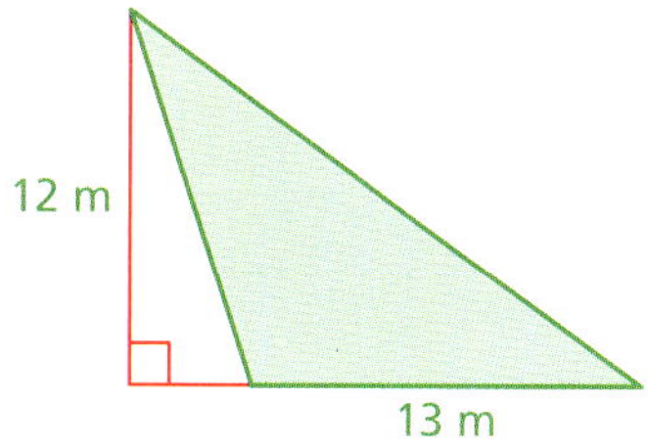

7. LAND A wildlife conservation group buys a plot of land. How much land does it buy? *(Section 4.2)*

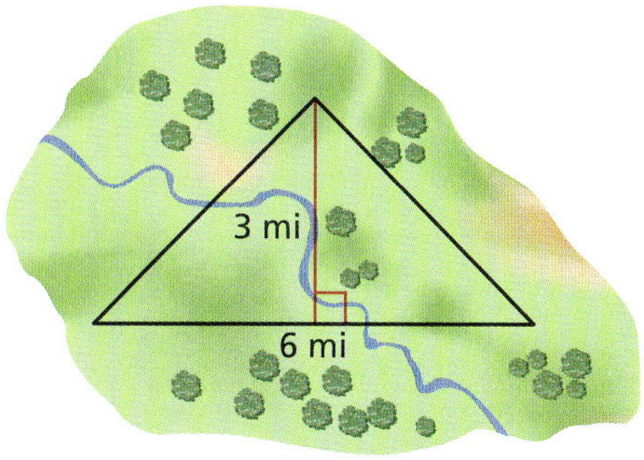

8. FRAMING A sheet of plywood is 4 feet wide by 8 feet long. What is the minimum number of sheets of plywood needed to cover the frame? Justify your answer. *(Section 4.2)*

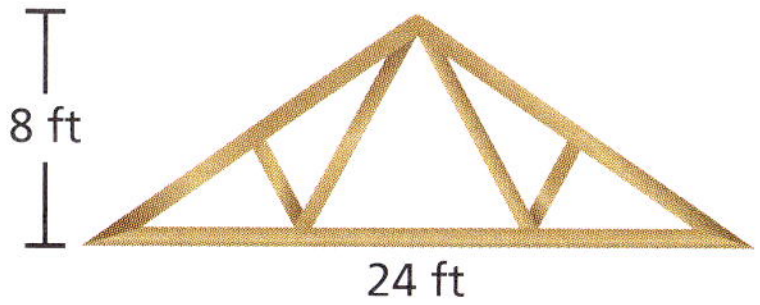

4.3 Areas of Trapezoids

Essential Question How can you derive a formula for the area of a trapezoid?

1 ACTIVITY: Deriving the Area Formula of a Trapezoid

Work with a partner. Use a piece of centimeter grid paper.

a. Draw *any* trapezoid so that its base lies on one of the horizontal lines of the paper.

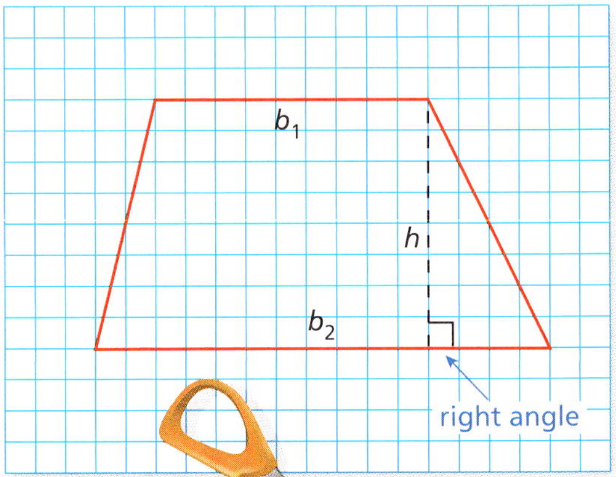

b. Estimate the area of your trapezoid (in square centimeters) by counting unit squares.

Area ≈ ________ Estimate

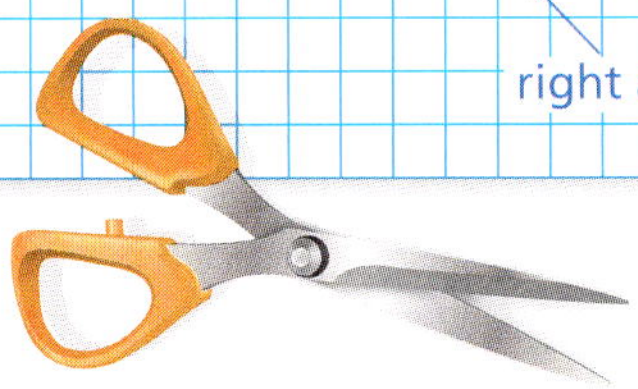

c. Label the height and the bases *inside* the trapezoid.

d. Cut out the trapezoid. Mark the midpoint of the side opposite the height. Draw a line from the midpoint to the opposite upper vertex.

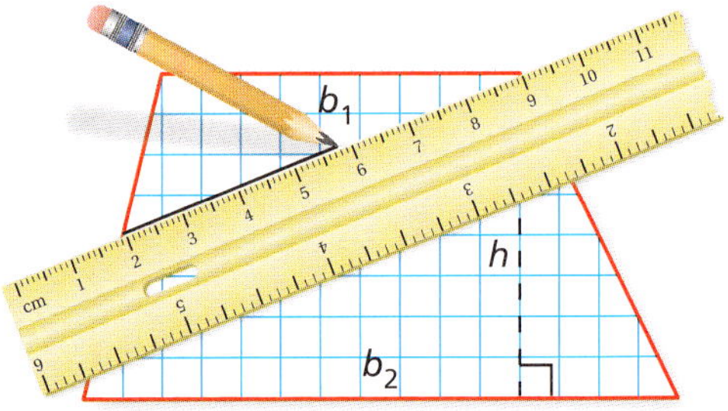

e. Cut along the line. You will end up with a triangle and a quadrilateral. Arrange these two figures to form a figure whose area you know.

f. Use your result to write a *formula* for the area of a trapezoid.

Area = ________ Formula

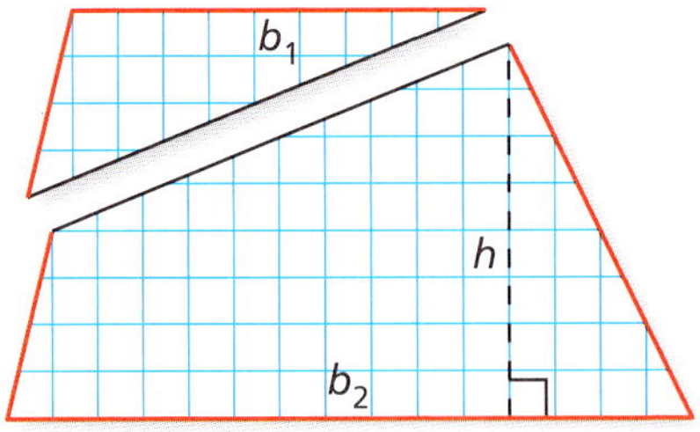

g. Use your formula to find the area of your trapezoid (in square centimeters).

Area = ________ Exact Area

h. Compare this area with your estimate in part (b).

COMMON CORE

Geometry

In this lesson, you will
- find areas of trapezoids.
- solve real-life problems.

Learning Standard
6.G.1

Laurie's Notes

Introduction

Standards for Mathematical Practice

- **MP3 Construct Viable Arguments and Critique the Reasoning of Others:** Deductive reasoning is again used to make a conjecture about a new formula, the area of a trapezoid. Students should be able to construct an argument for deriving this formula.

Motivate

- Display this common optical design. Identify the concave hexagons that fit together to make this tessellation.
- ? "Can you draw a segment to change each hexagon into two congruent quadrilaterals? yes What type of quadrilateral will you have?" trapezoid

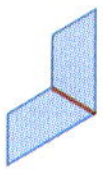

- Explain that today's activity involves exploring the area of a trapezoid.

Activity Notes

Activity 1

- ? "How are trapezoids and parallelograms alike? different?" Both are 4-sided. A parallelogram has two pairs of opposite sides that are parallel and congruent. A trapezoid has only one pair of opposite sides that are parallel.
- Review the dimensions of a trapezoid: the two parallel bases are often called b_1 and b_2. The dimensions are read "b sub 1" and "b sub 2." The two bases are not the same length, so to distinguish between them the same variable is used with different subscripts.
- Students may need assistance with part (d). If the height is an even number, students can locate the midpoint by moving along the vertical grid line.

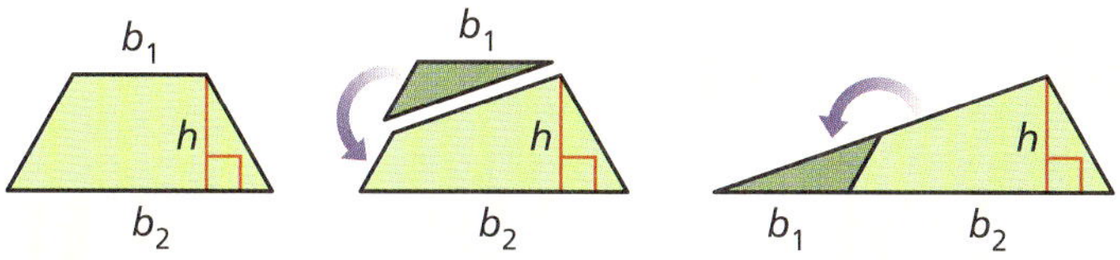

- Students will need to experiment to see how to form a new shape from the triangle and quadrilateral. The new triangle will have the same height as the original trapezoid. The base of the new triangle is the sum of the two bases of the trapezoid. Students may need help in writing the formula for the area of a trapezoid because parentheses are needed.
- Discuss the formula students wrote for the area of a trapezoid. Ask them to explain how they derived the formula.
- **Alternate Approach:** Have students cut out 2 congruent trapezoids and label the bases and height. Arrange the trapezoids to form a parallelogram.
  - **MP3:** Students should be able to construct an argument to find the area of one trapezoid. The constructed parallelogram has an area of $(b_1 + b_2)h$. Half of the area of this parallelogram is the area of the trapezoid.

Common Core State Standards

6.G.1 Find the area of . . . special quadrilaterals, . . . by . . . decomposing into triangles . . . ; apply these techniques in the context of solving real-world and mathematical problems.

Previous Learning

Students have found the area of a square, rectangle, parallelogram, and triangle.

Technology for the Teacher

Dynamic Classroom

Lesson Plans

Complete Materials List

4.3 Record and Practice Journal

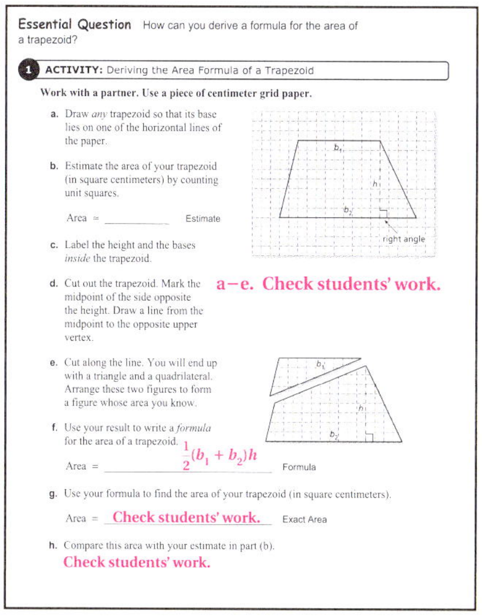

Essential Question How can you derive a formula for the area of a trapezoid?

1 ACTIVITY: Deriving the Area Formula of a Trapezoid

Work with a partner. Use a piece of centimeter grid paper.

a. Draw *any* trapezoid so that its base lies on one of the horizontal lines of the paper.

b. Estimate the area of your trapezoid (in square centimeters) by counting unit squares.

Area ≈ ________ Estimate

c. Label the height and the bases *inside* the trapezoid.

d. Cut out the trapezoid. Mark the midpoint of the side opposite the height. Draw a line from the midpoint to the opposite upper vertex.

a–e. Check students' work.

e. Cut along the line. You will end up with a triangle and a quadrilateral. Arrange these two figures to form a figure whose area you know.

f. Use your result to write a *formula* for the area of a trapezoid.

Area = $\frac{1}{2}(b_1 + b_2)h$ Formula

g. Use your formula to find the area of your trapezoid (in square centimeters).

Area = Check students' work. Exact Area

h. Compare this area with your estimate in part (b).

Check students' work.

Differentiated Instruction

Kinesthetic

Have students draw a trapezoid on grid paper. Then draw a diagonal to form two triangles that are the same height as the trapezoid.

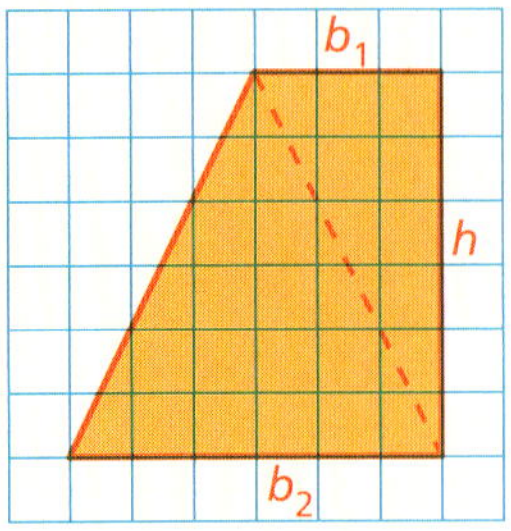

The bases of the trapezoid become the bases of the triangles. Find the area of the trapezoid by summing the areas of the two triangles.

4.3 Record and Practice Journal

2 ACTIVITY: Writing a Math Lesson

Work with a partner. Use your results from Activity 1 to write a lesson on finding the area of a trapezoid. Check students' work.

Area of a Trapezoid

Key Idea Use the following steps to find the area of a trapezoid.

1.

2.

3.

Describe steps you can use to find the area of a trapezoid.

Examples ← Write 2 examples for finding the area of a trapezoid. Include a drawing for each.

a. b.

Exercises ← Write 2 exercises for finding the area of a trapezoid. Include an answer sheet.

Find the area.

1. 2.

What Is Your Answer?

3. **IN YOUR OWN WORDS** How can you derive a formula for the area of a trapezoid?

Sample answer: A trapezoid can be cut and rearranged to form a triangle where the base is the sum of the bases of the trapezoid. So, the area of a trapezoid is the area of a triangle whose base is the sum of the bases of the trapezoid.

4. In this chapter, you used deductive reasoning to derive new area formulas from area formulas you have already learned. Describe a real-life career in which deductive reasoning is important.

Check students' work.

Laurie's Notes

Activity 2

- This activity will give you insight into how students think about finding the area of a trapezoid.
- When partners have finished, have them exchange their lessons with another pair of students. Everyone should read the lesson they receive and write a critique to the authors of the lesson. They should provide feedback on what is clear in the lesson and what can be made clearer.
- Have students complete the exercises of the lesson they received.

What Is Your Answer?

- For Question 4, there are several popular television shows which feature deductive reasoning. Hopefully students will be able to list several careers, such as crime investigation and medical research.

Closure

- Draw the following figure for students. Explain that a diagonal of the trapezoid has been drawn.

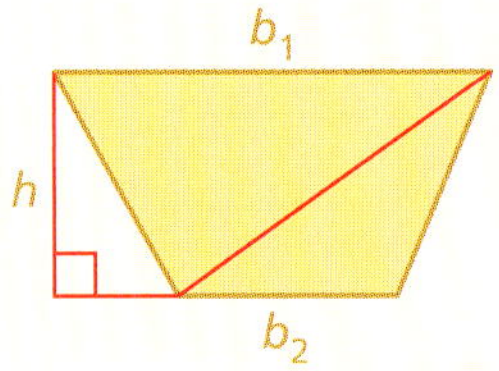

- Have students explain to their partners how they might derive the area of a trapezoid from this figure.

2 ACTIVITY: Writing a Math Lesson

Work with a partner. Use your results from Activity 1 to write a lesson on finding the area of a trapezoid.

Math Practice 6

Use Clear Definitions

Do your steps for the *Key Idea* help another person understand how to solve the problem? Do the examples follow your steps?

Area of a Trapezoid

Describe steps you can use to find the area of a trapezoid.

Key Idea Use the following steps to find the area of a trapezoid.

1.
2.
3.

Write two examples for finding the area of a trapezoid. Include a drawing for each.

Examples

a. b.

Write two exercises for finding the area of a trapezoid. Include an answer sheet.

Exercises

Find the area.

1. 2.

What Is Your Answer?

3. **IN YOUR OWN WORDS** How can you derive a formula for the area of a trapezoid?

4. In this chapter, you used deductive reasoning to derive new area formulas from area formulas you have already learned. Describe a real-life career in which deductive reasoning is important.

Practice Use what you learned about the areas of trapezoids to complete Exercises 4–6 on page 170.

4.3 Lesson

Check It Out
Lesson Tutorials
BigIdeasMath.com

Key Idea

Area of a Trapezoid

Words The area A of a trapezoid is one-half the product of its height h and the sum of its bases b_1 and b_2.

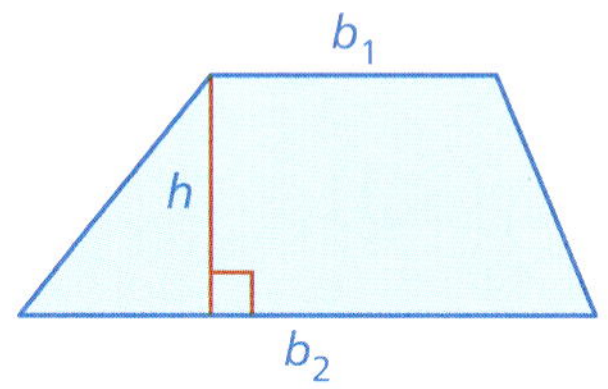

Algebra $A = \frac{1}{2}h(b_1 + b_2)$

EXAMPLE 1 Finding Areas of Trapezoids

Find the area of each trapezoid.

a.

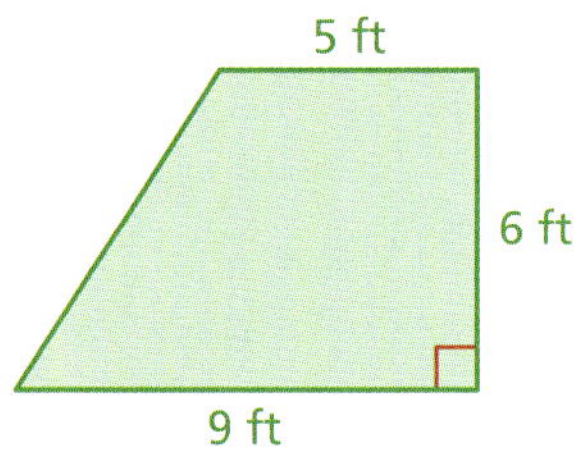

$A = \frac{1}{2}h(b_1 + b_2)$ Write formula.

$= \frac{1}{2}(6)(5 + 9)$ Substitute.

$= \frac{1}{2}(6)(14)$ Add.

$= 42$ Multiply.

The area of the trapezoid is 42 square feet.

b.

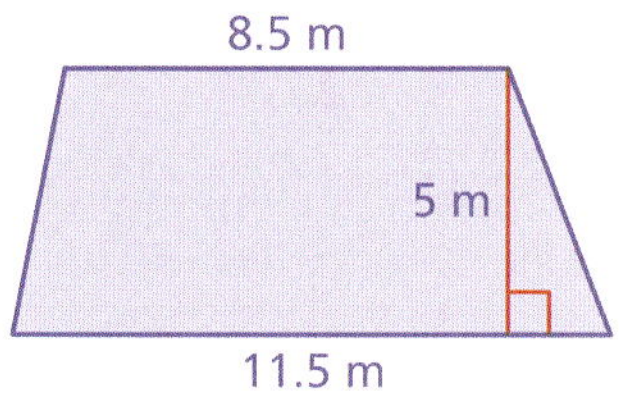

$A = \frac{1}{2}h(b_1 + b_2)$

$= \frac{1}{2}(5)(8.5 + 11.5)$

$= \frac{1}{2}(5)(20)$

$= 50$

The area of the trapezoid is 50 square meters.

On Your Own

Now You're Ready
Exercises 7–9

Find the area of the trapezoid.

1.

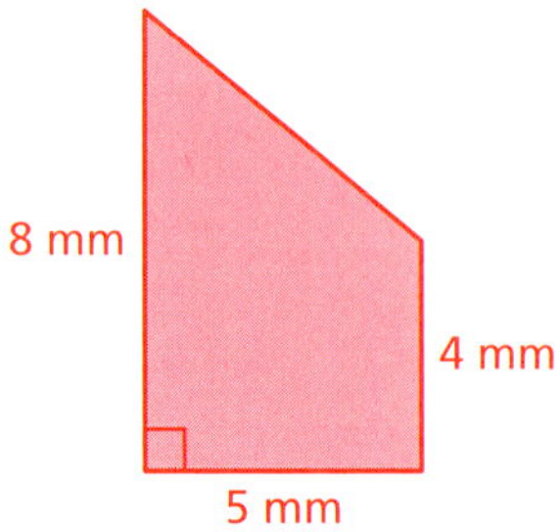

2.

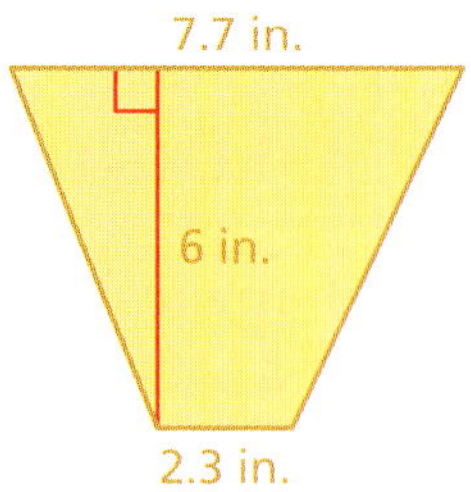

Laurie's Notes

Introduction

Connect

- **Yesterday:** Students derived the formula for the area of a trapezoid. (MP3)
- **Today:** Students will use the formula to find the areas of several trapezoids.

Motivate

- Historically, the word *trapezoid* refers to a quadrilateral with no parallel sides and a *trapezium* refers to a quadrilateral with only one pair of opposite sides that are parallel. In the 1700s a mathematical dictionary published in the United States reversed the meanings of the words and the definition of a trapezoid as a quadrilateral with only one pair of opposite sides that are parallel became accepted. Some countries today, such as England, still use the historical definitions.

Lesson Notes

Key Idea

- The *height* of a trapezoid is the perpendicular distance between the parallel bases.
- Write and discuss the area of a trapezoid formula.
- Again, remind students that to find $\frac{1}{2}$ of a number you can divide the number by 2.

Example 1

- Work through both examples.
- Write the formula first. Show how the values are substituted for the variables. In performing the operations, find the sum first. Then find the product of the sum and the height of the trapezoid. Finally, divide the product by 2.
- Discuss the factors and terms involved in the formula: $\frac{1}{2}$, h, and $(b_1 + b_2)$ and the order of operations.

? "How else might the computations be done in the first example?" Listen for: First multiply $\frac{1}{2}$ and 6. Then use the Distributive Property, or add $5 + 9$ and then multiply by 3.

- **MP7 Look for and Make Use of Structure:** You want students to be able to recognize that while there are three variables, $(b_1 + b_2)$ is one factor in the formula. There are three factors, $\frac{1}{2}$, h, and $(b_1 + b_2)$, and there are different ways in which the factors can be multiplied. If the height is an odd number, you might not want to take $\frac{1}{2}$ of it as the first step.

On Your Own

- In Question 1, note that the height is a side of the trapezoid, and that it is not a vertical segment.

Goal Today's lesson is finding the area of a trapezoid.

Technology for the Teacher

Dynamic Classroom

Lesson Tutorials
Lesson Plans
Answer Presentation Tool

English Language Learners

Illustrate

Have students look around the classroom, school building, and school grounds for shapes they have studied in this chapter. Students can find the perimeter or the area of a shape and can identify what types of workers would need to know this information for their jobs.

Extra Example 1

Find the area of each trapezoid.

a.

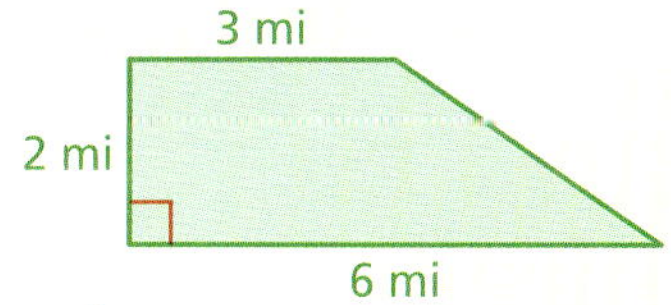

9 mi^2

b.

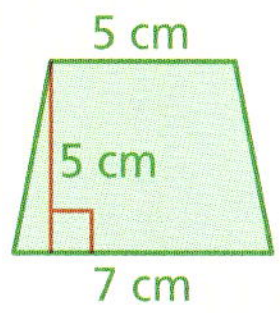

30 cm^2

On Your Own

1. 30 mm^2
2. 30 in.^2

Extra Example 2

Find the area of the trapezoid.

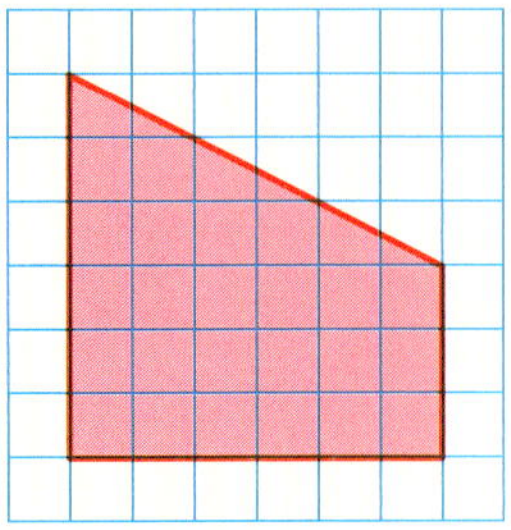

27 units^2

Extra Example 3

A trapezoid can be used to approximate the shape of Smyth County, Virginia. The population is about 32,000. About how many people are there per square mile?

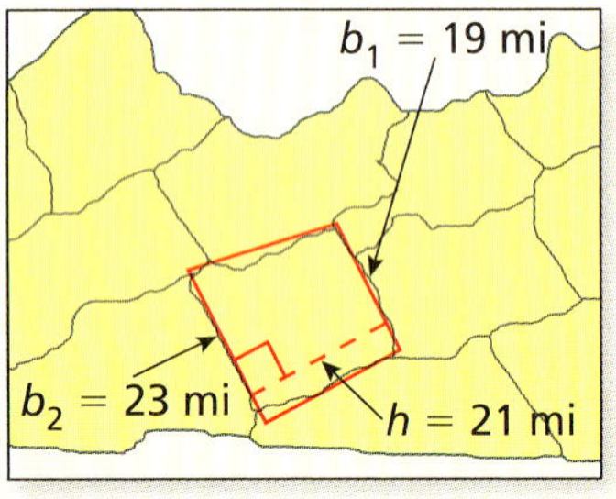

about 73 people per square mile

On Your Own

3. $10\frac{1}{2}$ units^2

4. about 1 person per square mile; $\frac{22{,}650 \text{ people}}{530 \text{ mi}^2} \approx 43$ people per square mile

Laurie's Notes

Example 2

? "How can you find the dimensions of the trapezoid?" Count the grid lines.

- **MP1 Make Sense of Problems and Persevere in Solving Them:** Mathematically proficient students will check their answers to problems using different methods. Finding one-half of 6 and then multiplying by 3 is different from finding the product of 6 and 3 and then taking one-half. Ask students if they can think of examples where one of the two approaches is easier than the other. Example: $\frac{1}{2}(2.5)(4)$
- **Common Error:** Be sure that students count grid lines correctly. Often they will count points versus the segments between the points. Students who do this will have bases of 2 and 3, and a height of 7.

Example 3

- Students will be dividing whole numbers where the remainder is not zero. From the context of the problem, it is clear that the nearest whole number is an appropriate answer. In Grade 7, students will study decimal remainders (7.NS.2d).
- This is an example of finding population density. Help students think about the size of 1 square mile. Relate it to a physical location in your area that students would be familiar with. In Scott County, there are approximately 44 people living in every square mile.
- **Extension:** Although the county you live and/or teach in may not be similar to a common geometric figure, you can find the land areas and population estimates at *quickfacts.census.gov/qfd/index.html.*

On Your Own

- It is good for students to practice *reading* the dimensions of a figure drawn on grid paper.
- Encourage students to estimate their answers first by counting.
- This is an example where both the height and the sum of the bases are odd. Ask a student to explain how he or she found the answer. A common approach will be $3 \times (3 + 4)$ is 21 and one-half of 21 is 10.5, so the area is 10.5 units^2.
- For Question 4, guide students to make an approximation without calculating. The decrease is close to the number of square miles in the country, so there should be about 1 person per square mile decrease.

Closure

- Find the area of the trapezoid in two different ways. Which method do you prefer? Explain. 36 in.^{2}; Method 1: Area of a Trapezoid Formula, Method 2: sum the areas of the two triangles and the rectangle

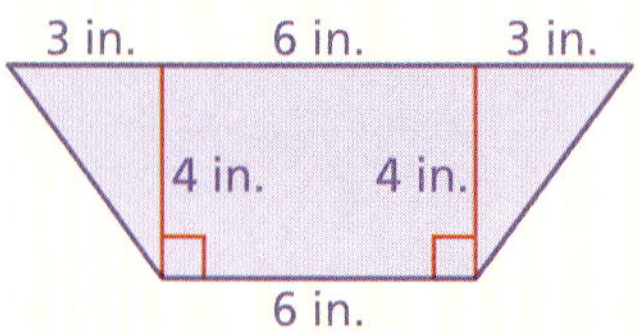

EXAMPLE 2 Finding the Area of a Trapezoid on a Grid

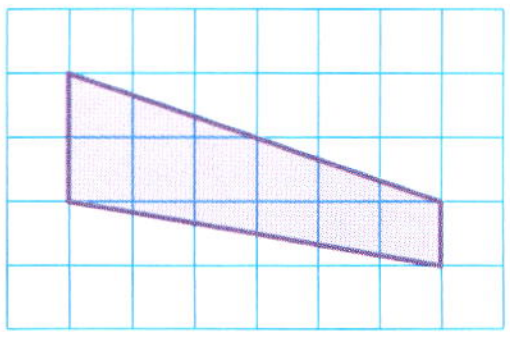

What is the area of the trapezoid?

Ⓐ 6 $units^2$ Ⓑ 7 $units^2$ Ⓒ 9 $units^2$ Ⓓ 12 $units^2$

Count grid lines to find the dimensions. The height h is 6 units, base b_1 is 1 unit, and base b_2 is 2 units.

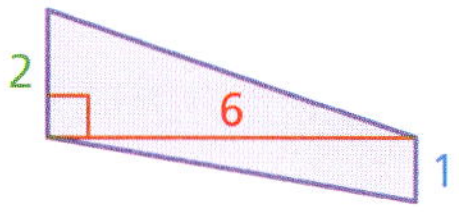

$$A = \frac{1}{2}h(b_1 + b_2)$$ Write formula.

$$= \frac{1}{2}(6)(1 + 2)$$ Substitute values.

$$= \frac{1}{2}(6)(3)$$ Add.

$$= 9$$ Multiply.

The area of the trapezoid is 9 square units.
The correct answer is Ⓒ.

EXAMPLE 3 Real-Life Application

You can use a trapezoid to approximate the shape of Scott County, Virginia. The population is about 23,200. About how many people are there per square mile?

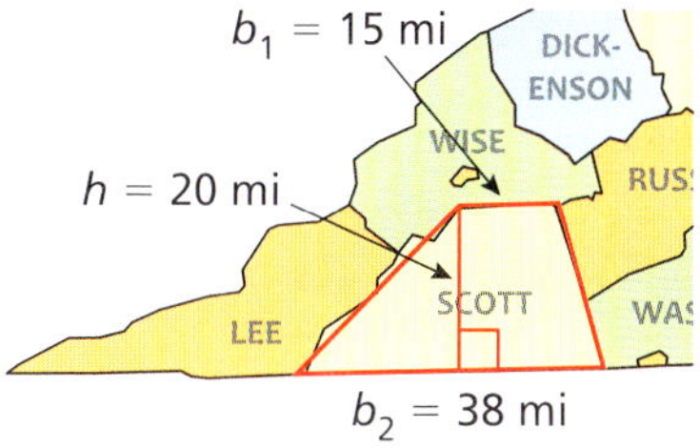

Find the area of Scott County.

$$A = \frac{1}{2}h(b_1 + b_2)$$ Write formula for area of a trapezoid.

$$= \frac{1}{2}(20)(15 + 38)$$ Substitute 20 for h, 15 for b_1, and 38 for b_2.

$$= \frac{1}{2}(20)(53) = 530$$ Simplify.

The area of Scott County is about 530 square miles. Divide the population by the area to find the number of people per square mile.

So, there are about $\frac{23{,}200 \text{ people}}{530 \text{ mi}^2} \approx 44$ people per square mile.

On Your Own

Now You're Ready
Exercises 11–13

3. Find the area of the trapezoid.

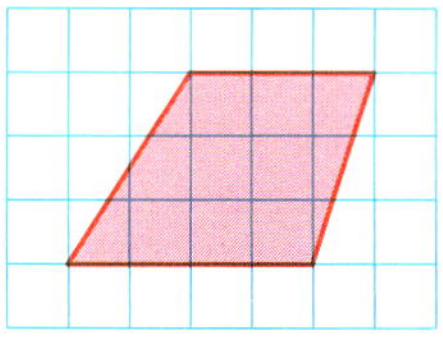

4. **WHAT IF?** In Example 3, the population of Scott County decreases by 550. By how much does the number of people per square mile change? Explain.

4.3 Exercises

Vocabulary and Concept Check

1. **VOCABULARY** Identify the bases and the height of the trapezoid.

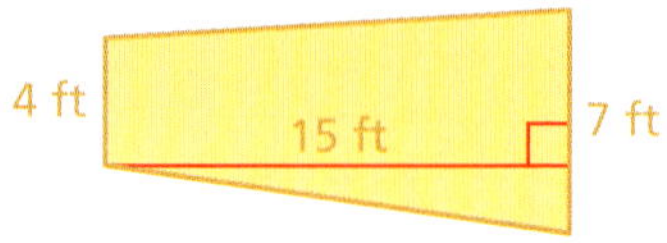

2. **REASONING** What measures do you need to find the area of a trapezoid?

3. **WHICH ONE DOESN'T BELONG?** Which one does *not* belong with the other three? Explain your reasoning.

$\frac{1}{2}bh$ $\qquad$ ℓw $\qquad$ $2\ell + 2w$ $\qquad$ $\frac{1}{2}h(b_1 + b_2)$

Practice and Problem Solving

Find the area of the trapezoid.

4. $b_1 = 4, b_2 = 8, h = 2$

5. $b_1 = 5, b_2 = 7, h = 4$

6. $b_1 = 12, b_2 = 6, h = 3$

1 7.

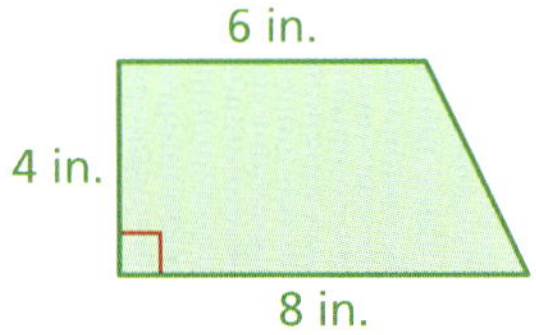

8.

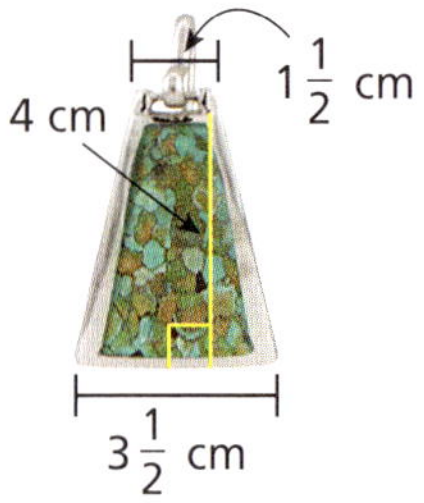

9.

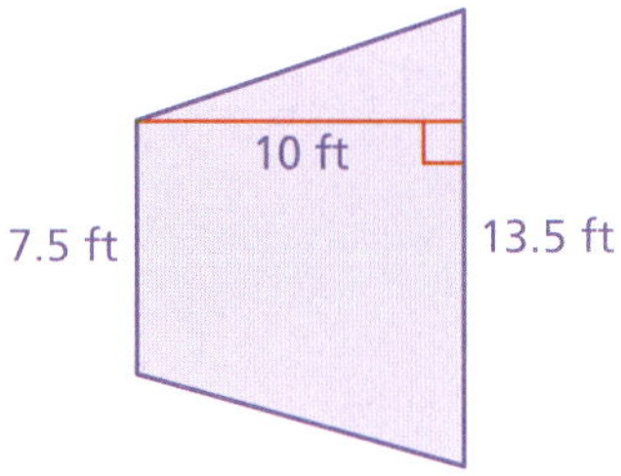

10. **ERROR ANALYSIS** Describe and correct the error in finding the area of the trapezoid.

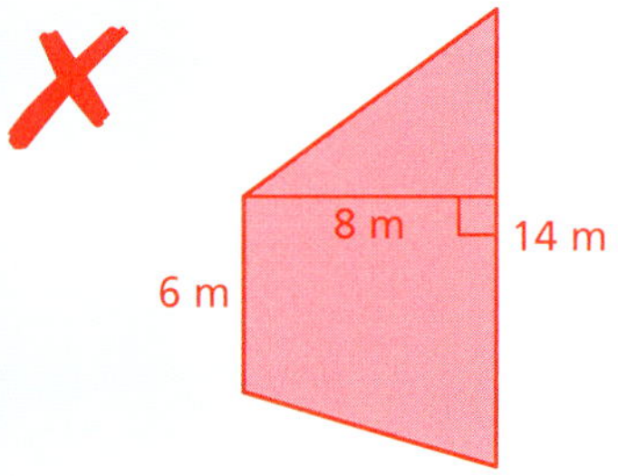

$$\text{Area} = \frac{1}{2}(6 + 14)$$
$$= 10 \text{ m}^2$$

Find the area of the trapezoid.

2 11.

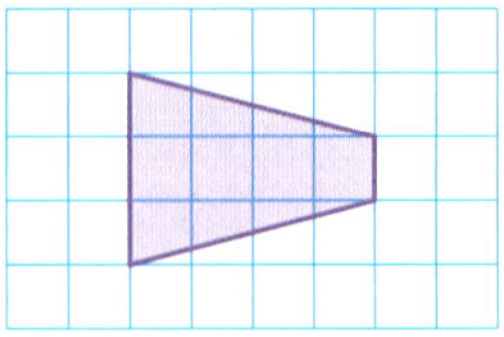

12.

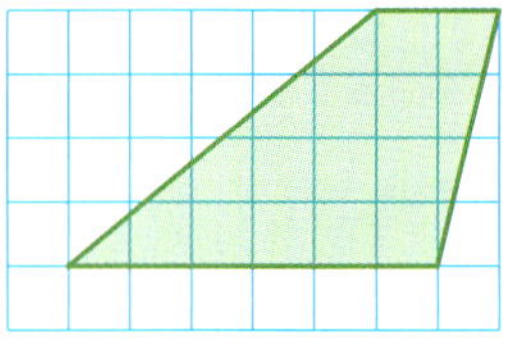

13.

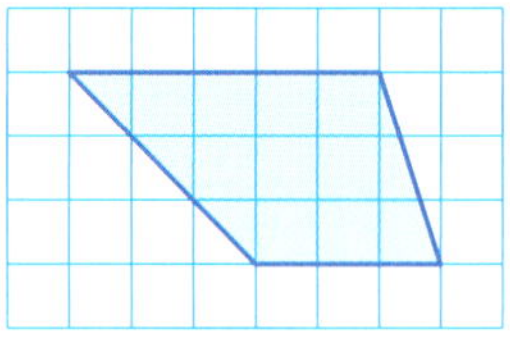

14. **LIGHT** Light shines through a window. What is the area of the trapezoid-shaped region created by the light?

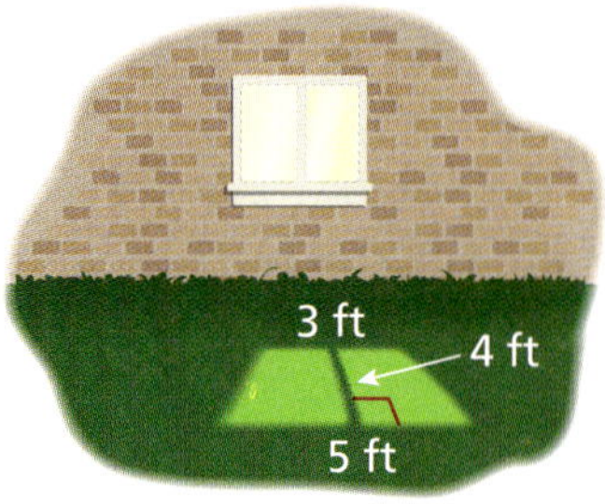

Assignment Guide and Homework Check

Level	Day 1 Activity Assignment	Day 2 Lesson Assignment	Homework Check
Basic	4–6, 23–27	1–3, 7–15, 17, 19	9, 13, 14, 15
Average	4–6, 23–27	1–3, 7–16, 19–21 odd	9, 13, 16, 21
Advanced	4–6, 23–27	1–3, 8–16, 18–22	8, 12, 18, 21

Common Errors

- **Exercises 7–9** Students may forget to multiply by $\frac{1}{2}$. Tell students to write out the formula for the area of a trapezoid, and then identify the value of each variable before substituting.
- **Exercise 9** Students may forget to align the decimal points when adding decimals.
- **Exercises 11–13** Students may not count grid lines correctly. Often they will count points versus the segments between the points. When this happens the bases and the heights will be one greater than they should be.
- **Exercises 15–18** Students may substitute the height for one of the bases. Tell students to write out the formula for the area of a trapezoid, and then identify the value of each variable before substituting.

4.3 Record and Practice Journal

Find the area of the trapezoid.

1. 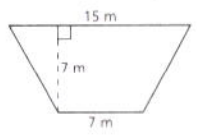

77 m^2

2. 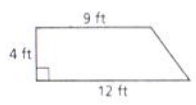

42 ft^2

Find the area of the trapezoid with height h and bases b_1 and b_2.

3. $h = 10$ yd
$b_1 = 17$ yd
$b_2 = 21$ yd
190 yd^2

4. $h = 9$ cm
$b_1 = 4.5$ cm
$b_2 = 5.5$ cm
45 cm^2

5. The triangle and the trapezoid have the same area. Base b_2 is twice the length of base b_1. What are the lengths of the bases of the trapezoid?

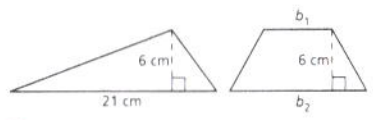

$b_1 = 7$ cm
$b_2 = 14$ cm

Vocabulary and Concept Check

1. bases: 4 ft and 7 ft; height: 15 ft
2. height h and bases b_1 and b_2
3. $2\ell + 2w$; This is an expression for the perimeter of a rectangle. The other three are expressions for area (triangle, rectangle, and trapezoid).

Practice and Problem Solving

4. 12 units^2
5. 24 units^2
6. 27 units^2
7. 28 in.^{2}
8. 10 cm^2
9. 105 ft^2
10. The height was not included in the formula.
$A = \frac{1}{2}(8)(6 + 14) = 80\text{ m}^2$
11. 8 units^2
12. 16 units^2
13. 12 units^2
14. 16 ft^2
15. 60 in.^{2}
16. 253 cm^2
17. 78 mi^2
18. 301 m^2
19. 18 ft

Practice and Problem Solving

20. *Sample answers:*
$b_1 = 2$ ft, $b_2 = 3$ ft;
$b_1 = 1.5$ ft, $b_2 = 3.5$ ft

21. See *Taking Math Deeper*.

22. **a.** $x > 0$ and $x < 15$ inches; For $x = 15$ inches the area of the trapezoid is twice the area of the triangle, so x must be less than 15 inches.

b. no; When $x = 15$ inches the quadrilateral is a rectangle, not a trapezoid.

Fair Game Review

23–26. See Additional Answers.

27. C

Mini-Assessment

Find the area of the trapezoid.

1. 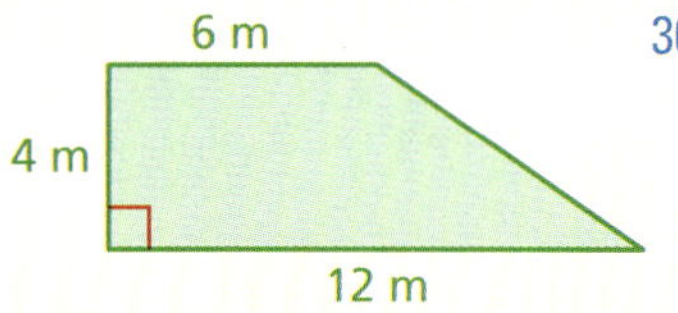

36 m^2

2. 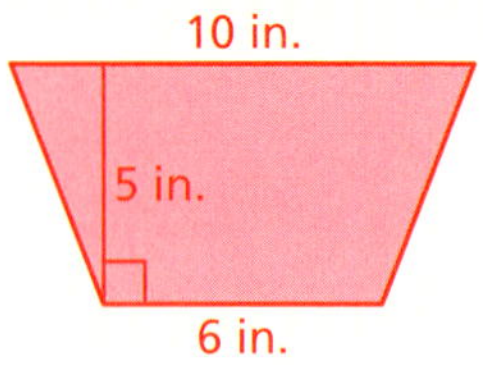

40 in.^{2}

3. 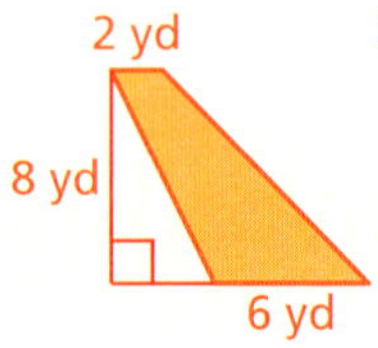

32 yd^2

4. $h = 10$ cm, $b_1 = 3$ cm, $b_2 = 5$ cm
40 cm^2

Taking Math Deeper

Exercise 21

This problem gives students several options for problem-solving strategies. One is to assign values for h, b_1, and b_2 and then *Solve a Simpler Problem.*

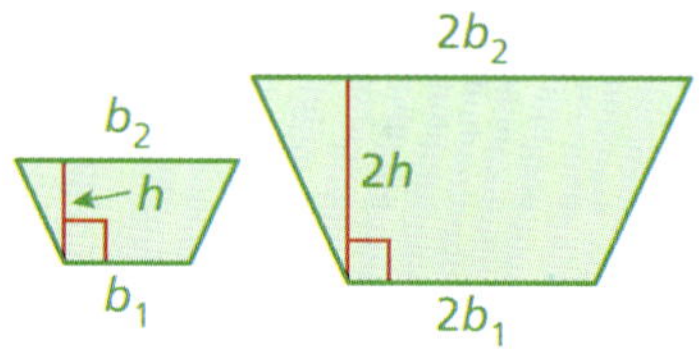

1 *Solve a Simpler Problem:* Let $h = 2$, $b_1 = 4$, and $b_2 = 6$.

Area of Small Trapezoid	Area of Large Trapezoid
$= \frac{1}{2}h(b_1 + b_2)$	$= \frac{1}{2}(2h)(2b_1 + 2b_2)$
$= \frac{1}{2}(2)(4 + 6)$	$= \frac{1}{2}(4)(8 + 12)$
$= 10 \text{ units}^2$	$= 40 \text{ units}^2$

So, the area of the floor covered by the larger speaker is 4 times greater than the area of the floor covered by the smaller speaker.

2 At this stage in their study of algebra, most students do not know enough about simplifying algebraic expressions to solve the problem abstractly.

$$\frac{\text{Area of Large Trapezoid}}{\text{Area of Small Trapezoid}} = \frac{\frac{1}{2}(2h)(2b_1 + 2b_2)}{\frac{1}{2}h(b_1 + b_2)} = \frac{4h(b_1 + b_2)}{h(b_1 + b_2)} = 4$$

So, the best approach is to try a few other values for h, b_1, and b_2, and then use inductive reasoning to conclude that the area of the large trapezoid is always 4 times greater than the area of the small trapezoid.

3 **Extension:** Have students review Example 3 on page 161. In that example, they can see that doubling the dimensions has the effect of making the area 4 times greater.

Ask students to experiment with all the shapes they have studied in this chapter. In each case, they can conclude that doubling the dimensions of the figure increases the area by a factor of 4.

Reteaching and Enrichment Strategies

If students need help. . .	If students got it. . .
Resources by Chapter • Practice A and Practice B • Puzzle Time Record and Practice Journal Practice Differentiating the Lesson Lesson Tutorials Skills Review Handbook	Resources by Chapter • Enrichment and Extension • Technology Connection Start the next section

Find the area of a trapezoid with height h and bases b_1 and b_2.

15. $h = 6$ in.
$b_1 = 9$ in.
$b_2 = 11$ in.

16. $h = 22$ cm
$b_1 = 10.5$ cm
$b_2 = 12.5$ cm

17. $h = 12$ mi
$b_1 = 5.6$ mi
$b_2 = 7.4$ mi

18. $h = 14$ m
$b_1 = 21$ m
$b_2 = 22$ m

19. **REASONING** The rectangle and the trapezoid have the same area. What is the length ℓ of the rectangle?

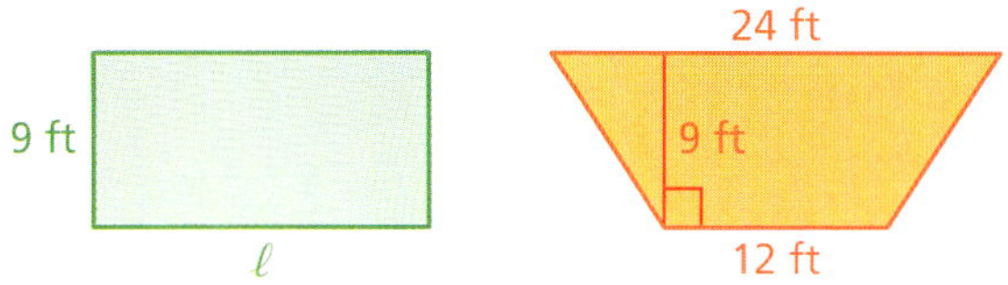

20. **OPEN-ENDED** The area of the trapezoidal student election sign is 5 square feet. Find two possible values for each base length.

21. **AUDIO** How many times greater is the area of the floor covered by the larger speaker than by the smaller speaker?

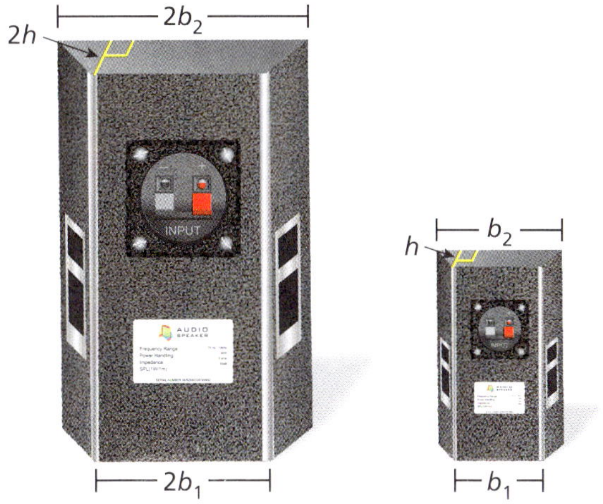

22. Critical Thinking The triangle and the trapezoid share a 15-inch base and a height of 10 inches.

a. The area of the trapezoid is less than twice the area of the triangle. Find the values of x. Explain your reasoning.

b. Can the area of the *trapezoid* be exactly twice the area of the triangle? Explain your reasoning.

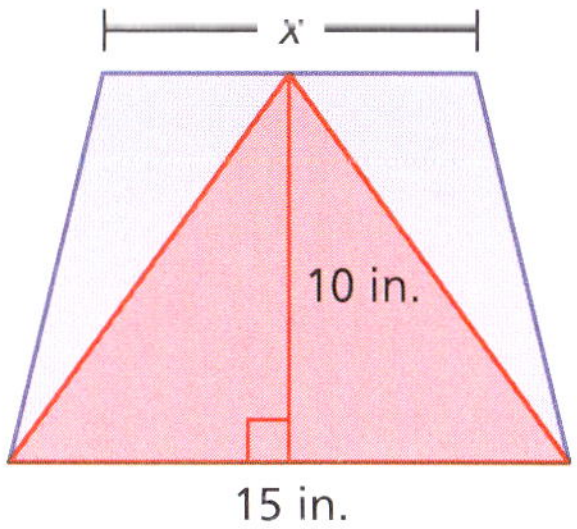

Fair Game Review What you learned in previous grades & lessons

Plot the ordered pair in a coordinate plane. *(Skills Review Handbook)*

23. (5, 0) **24.** (2, 4) **25.** (0, 3) **26.** (6, 1)

27. **MULTIPLE CHOICE** Which expression represents "6 more than x"? *(Section 3.2)*

(A) $6 - x$ (B) $6x$ (C) $x + 6$ (D) $\frac{6}{x}$

Extension 4.3 Areas of Composite Figures

Check It Out
Lesson Tutorials
BigIdeasMath.com

Key Vocabulary
composite figure, p. 172

A **composite figure** is made up of triangles, squares, rectangles, and other two-dimensional figures. Here are two examples.

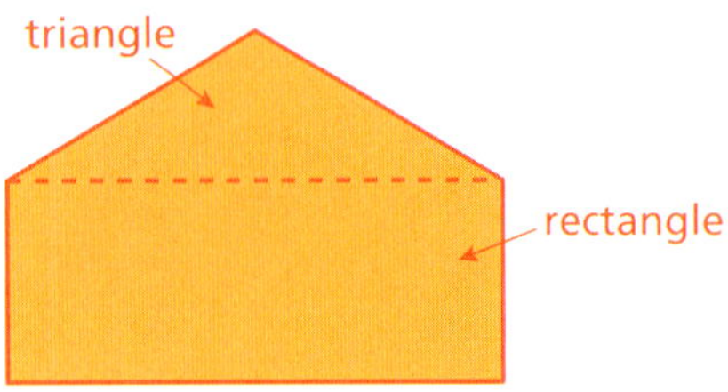

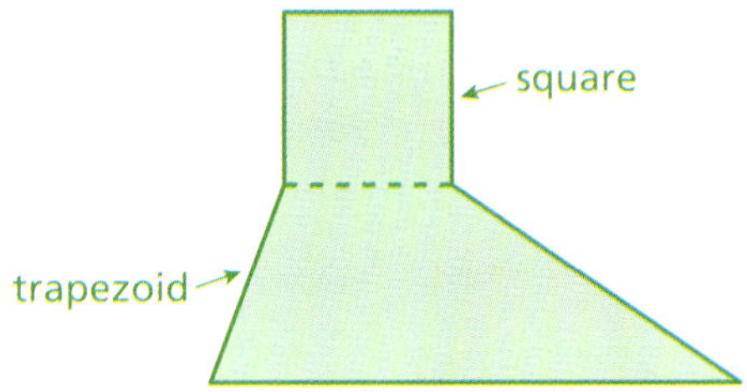

To find the area of a composite figure, separate it into figures with areas you know how to find. Then find the sum of the areas of those figures.

EXAMPLE 1 Finding the Area of a Composite Figure

Find the area of the purple figure.

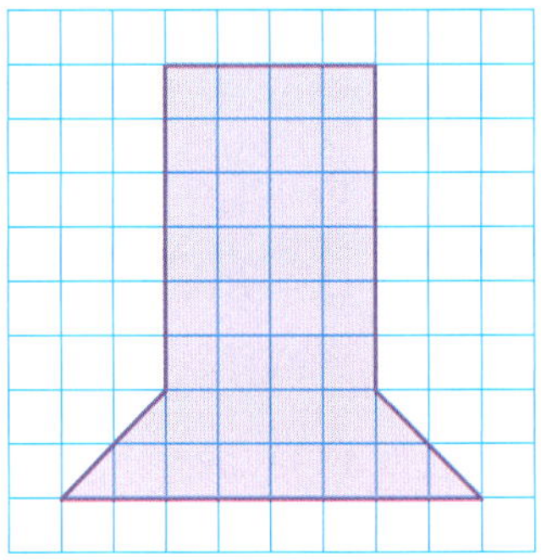

You can separate the figure into a rectangle and a trapezoid. Count grid lines to find the dimensions of each figure. Then find the area of each figure.

Study Tip
There is often more than one way to separate composite figures. In Example 1, you can separate the figure into one rectangle and two triangles.

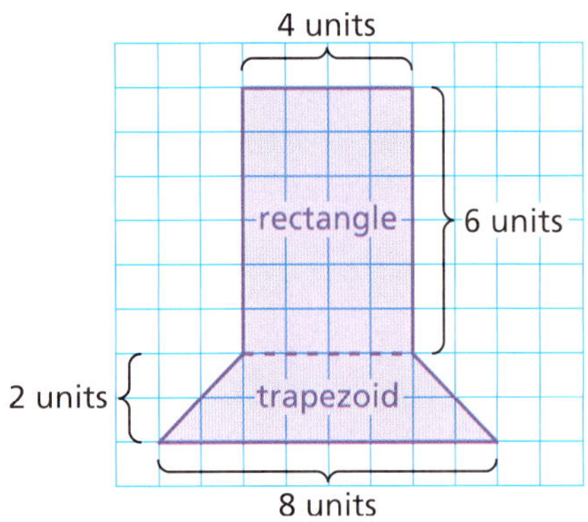

Area of Rectangle

$A = \ell w$

$= 6(4)$

$= 24$

Area of Trapezoid

$A = \frac{1}{2}h(b_1 + b_2)$

$= \frac{1}{2}(2)(4 + 8)$

$= 12$

So, the area of the purple figure is $24 + 12 = 36$ square units.

Reasonable? You can check your result by counting unit squares.

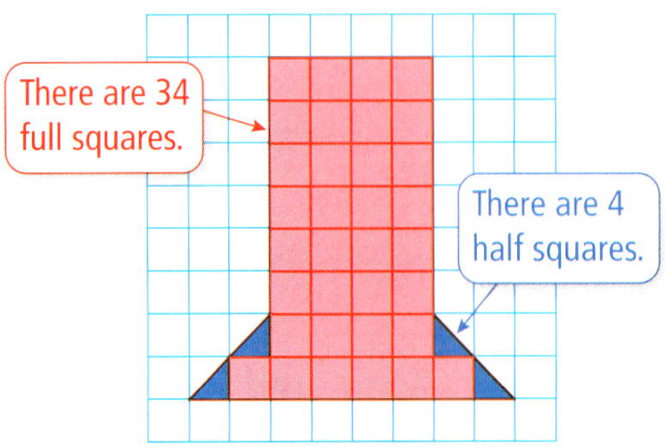

Full squares: 34

Half squares: 4

The area is

$$34(1) + 4\left(\frac{1}{2}\right) = 36 \text{ square units.}$$

So, the answer is reasonable. ✓

Common Core
Geometry
In this extension, you will
- find areas of composite figures.
- solve real-life problems.

Applying Standard 6.G.1

Laurie's Notes

Introduction

Connect

- **Yesterday:** Students have learned to find the areas of several polygons. (MP1)
- **Today:** Students will use different area formulas to find the area of a composite figure.

FYI

- A trapezoid can be decomposed into different polygons. Students have already worked with composite figures. Thinking of a trapezoid as a composite figure was actually how its area formula was derived.

Motivate

- Give students a piece of paper with four copies of a trapezoid drawn on it. Ask them to "decompose" the trapezoid into polygons that they would be able to find the area of. Sample answers are shown.

- Give time for students to share their thinking with the class. In each case, they should identify the names of the shapes they formed.
- Discuss that the area of the trapezoid is the sum of the areas of the figures of which it is composed.

Discuss

- ? Draw a couple of samples of composite figures and ask, "These are examples of a composite figure. What do you think a composite figure is?" Listen for a shape that is made up of other polygons like triangles and rectangles.
- **FYI:** Composite figures do not have to be composed of only polygons. The oval track shown is two semi-circles and a rectangle.

Lesson Notes

Example 1

- ? "What are the dimensions of this shape?" The rectangle is 6 by 4 units. The trapezoid has bases of 4 and 8 and a height of 2.
- Write the formula for the area of each polygon, substitute the known values, and perform the computations.
- Check the reasonableness of the answer as shown.
- ? "Could the figure have been decomposed a different way? Explain." Yes; have students show different methods.
- If time allows, work through different methods so that students understand that the total area will not change.

Common Core State Standards

6.G.1 Find the area of right triangles, other triangles, special quadrilaterals, and polygons by composing into rectangles or decomposing into triangles and other shapes; apply these techniques in the context of solving real-world and mathematical problems.

Goal Today's lesson is using different area formulas to find the area of a **composite figure**.

Technology for the Teacher

Dynamic Classroom

Lesson Tutorials
Lesson Plans
Answer Presentation Tool

Extra Example 1

Find the area of the red figure.

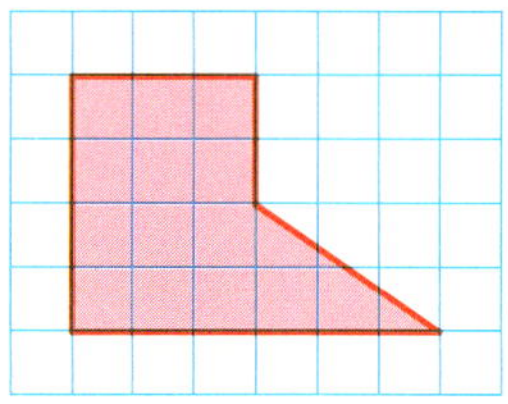

15 square units

Record and Practice Journal Extension 4.3 Practice

1.	32 units^2	**2.**	22 units^2
3.	26 units^2	**4.**	12 units^2
5.	24 cm^2	**6.**	28 m^2
7.	64 ft^2	**8.**	36 in.^2
9.	64 ft^2		

Extra Example 2

Find the area of the swimming pool.

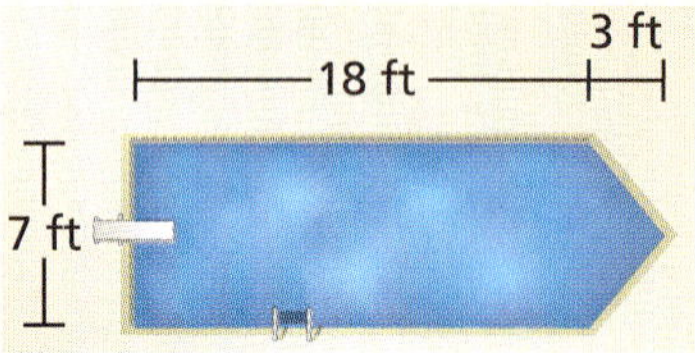

136.5 square feet

Practice

1. 36 units^2
2. 32 units^2
3. 20 units^2
4. 120 ft^2
5. $126\frac{1}{2}$ cm^2
6. 132 in.^{2}
7. See Additional Answers.

Mini-Assessment

1. Find the area of the figure.

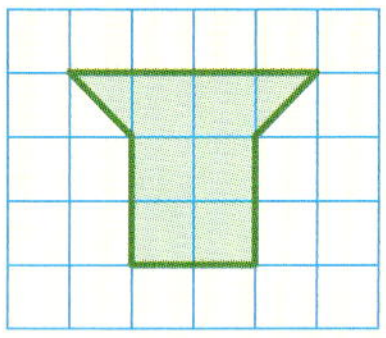

   7 square units

2. Find the area of the softball field.

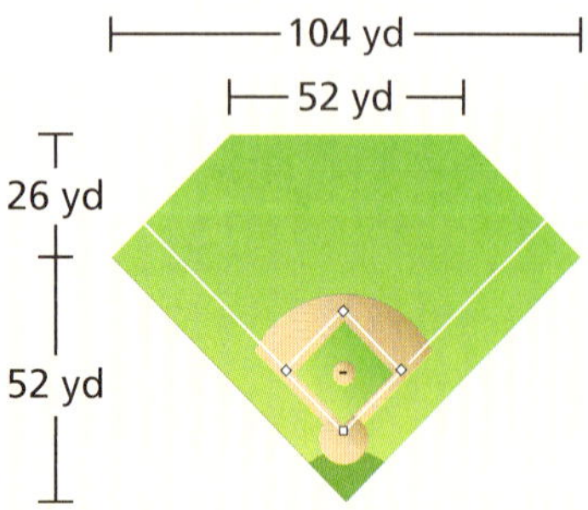

   4732 square feet

Laurie's Notes

Example 2

- Draw the composite figure shown and label all of the given dimensions.
- Ask students to talk with their partners about different ways in which the figure could be decomposed. Suggest to them that their methods need to result in figures where the dimensions are known or could be found.
- Give students a few minutes and then share as a class.
- **MP1 Make Sense of Problems and Persevere in Solving Them:** Instead of modeling the problem for them, have students work the problem with their partners. Students should have a pathway into the problem so that they can be successful.
- Ask different students to present their solutions, looking for a selection of methods.

Practice

- **Neighbor Check:** Have students work independently and then have their neighbors check their work. Have students discuss any discrepancies.

Closure

- Describe how to find the area of the following.

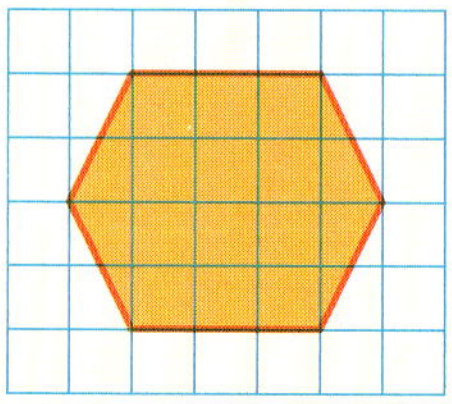

Sample answer:

Separate the hexagon with a horizontal line to create two trapezoids.

$$A = \frac{1}{2}(2)(3 + 5) + \frac{1}{2}(2)(5 + 3)$$
$$= 8 + 8$$
$$= 16 \text{ square units}$$

EXAMPLE 2 Real-Life Application

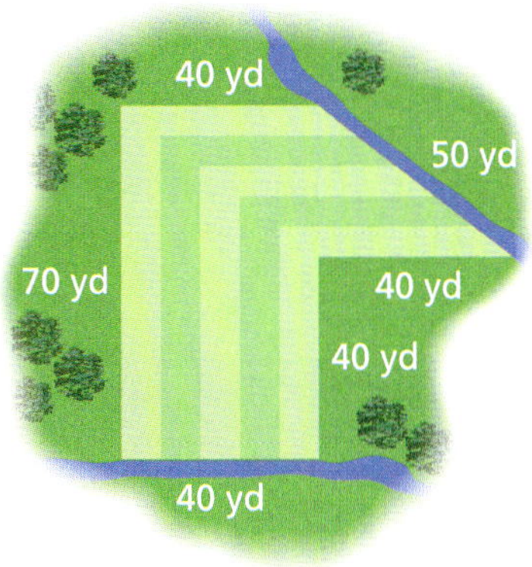

Find the area of the fairway between two streams on a golf course.

There are several ways to separate the fairway into figures whose areas you can find using formulas. It appears that one way is to separate it into a right triangle and a rectangle.

Identify each shape and find any missing dimensions.

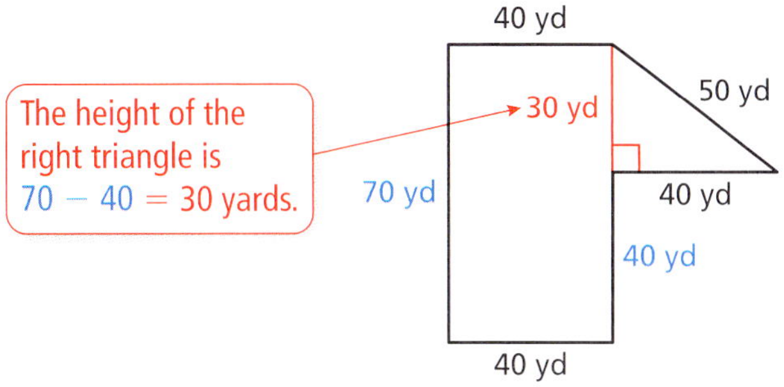

Area of Rectangle	*Area of Right Triangle*
$A = \ell w$	$A = \frac{1}{2}bh$
$= 70(40)$	$= \frac{1}{2}(40)(30)$
$= 2800$	$= 600$

So, the area of the fairway is 2800 + 600 = 3400 square yards.

Practice

Find the area of the shaded figure.

1. 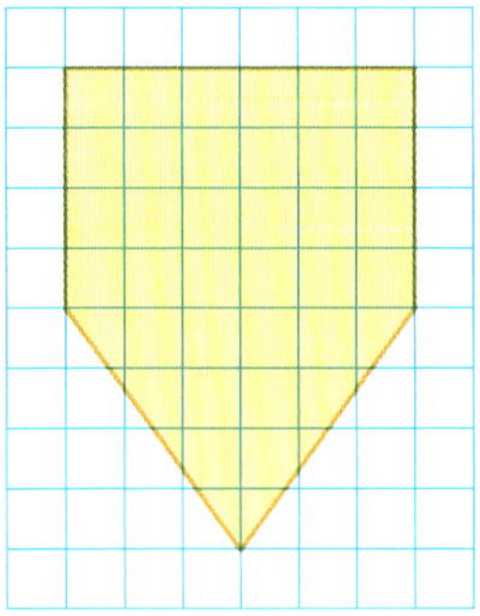

2. 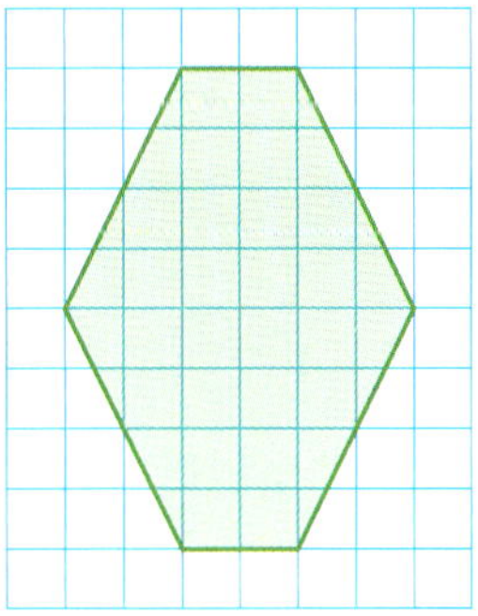

3. 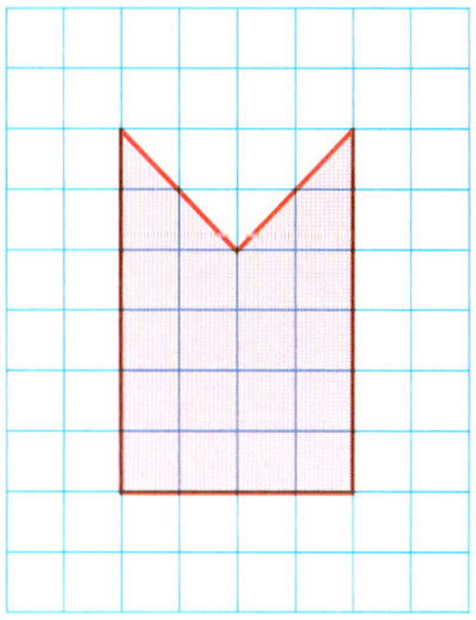

Find the area of the figure.

4. 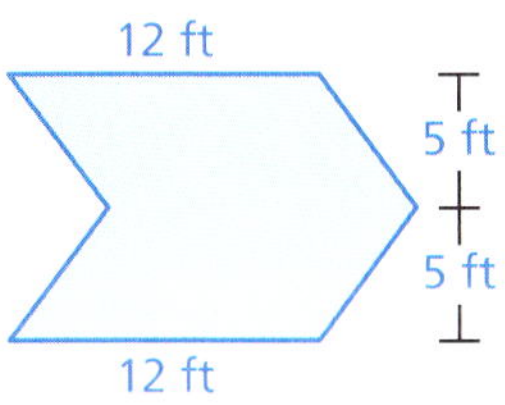

5. 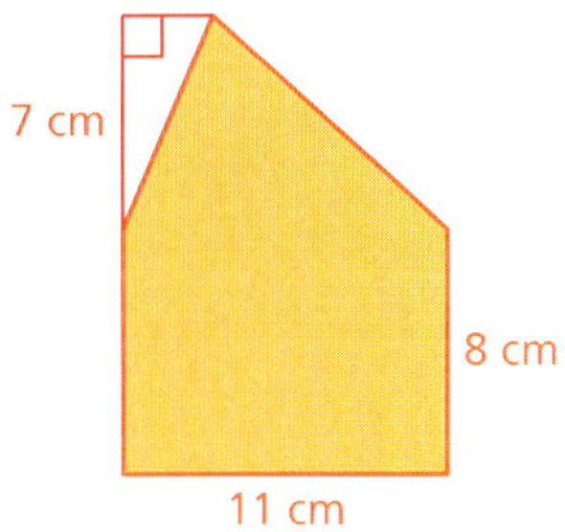

6. 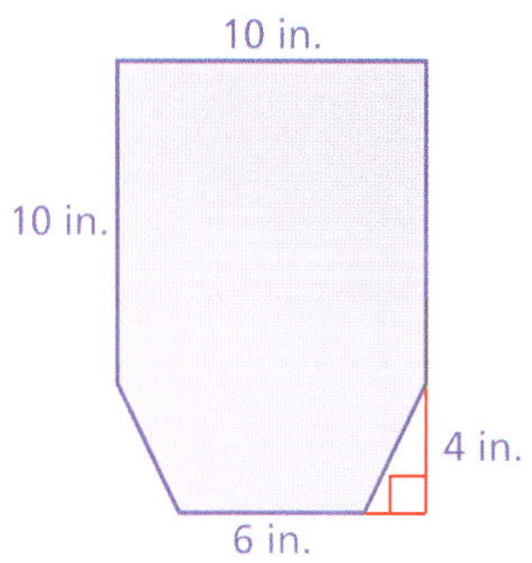

7. **ANOTHER METHOD** Find the area in Example 2 using a different method.

4.4 Polygons in the Coordinate Plane

Essential Question How can you find the lengths of line segments in a coordinate plane?

1 ACTIVITY: Finding Distances on a Map

Work with a partner. The coordinate grid shows a portion of a city. Each square on the grid represents one square mile.

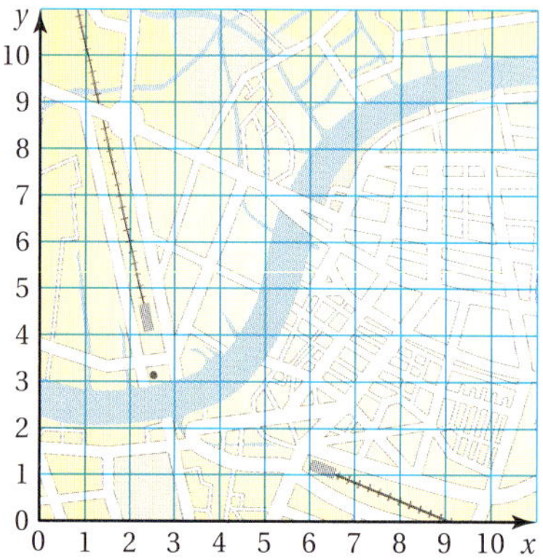

a. A public library is located at (4, 5). City Hall is located at (7, 5). Plot and label these points.

b. How far is the public library from City Hall?

c. A stadium is located 4 miles from the public library. Give the coordinates of several possible locations of the stadium. Justify your answers by graphing.

d. Connect the three locations of the public library, City Hall, and the stadium using your answers in part (c). What shapes are formed?

2 ACTIVITY: Graphing Polygons

Work with a partner. Plot and label each set of points in the coordinate plane. Then connect each set of points to form a polygon.

Rectangle: $A(2, 3)$, $B(2, 10)$, $C(6, 10)$, $D(6, 3)$

Triangle: $E(8, 3)$, $F(14, 8)$, $G(14, 3)$

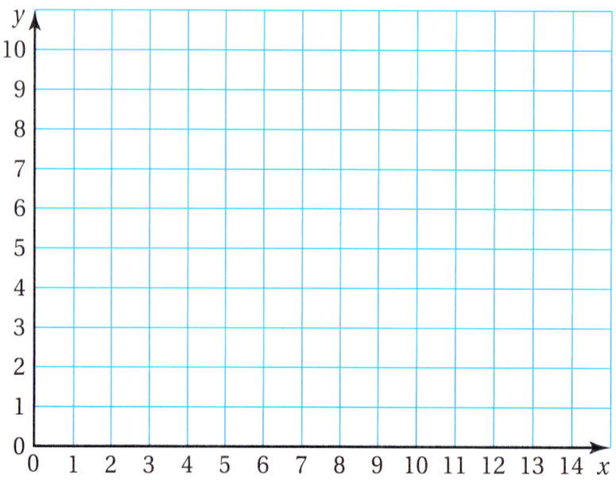

COMMON CORE

Geometry

In this lesson, you will
- draw polygons in the coordinate plane.
- find distances in the coordinate plane.
- solve real-life problems.

Learning Standard
6.G.3

Laurie's Notes

Introduction

Standards for Mathematical Practice

- **MP8 Look for and Express Regularity in Repeated Reasoning:** Mathematically proficient students recognize patterns and make connections among mathematical ideas.

Motivate

? "Who has played the game of *Battleship* before?" Chances are most students have or will have a familiarity with it.

- Tell them that you have hidden something in the classroom. It's their job to guess "coordinates" of the location.
- Identify the corner of the room that will be the origin (0, 0). Give some sense of size by labeling the opposite corner of your room. A sketch of my classroom is shown.

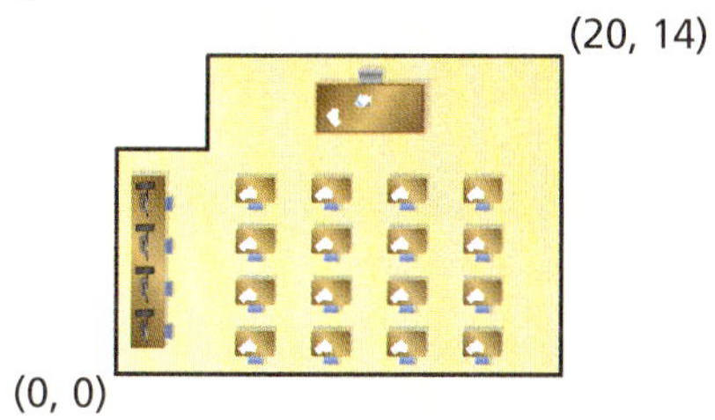

- Have students guess coordinates of the location of your "hide." This will allow you to review concepts necessary for this lesson, such as how to start at the origin, (0, 0) and locate an ordered pair.
- **FYI:** Graphing in the first quadrant is a Grade 5 standard. (5.G.2)

Activity Notes

Activity 1

- Students should be able to plot the ordered pairs without difficulty if you have done the motivation described above.
- **Connection:** Students have found the dimensions of a polygon on a grid. Finding distances of segments is an equivalent skill.
- Ask students to share their answers to parts (c) and (d).
- **Extension:** Ask if there are other points that would also be 4 miles from the library. The solution is a circle with radius of 4 centered at the library.

Activity 2

- When students have finished, have students share their results at the overhead, document camera, or interactive board. This will ensure that students have successfully plotted the points prior to beginning Activity 3.

Common Core State Standards

6.G.3 Draw polygons in the coordinate plane given coordinates for the vertices; use coordinates to find the length of a side joining points with the same first coordinate or the same second coordinate. Apply these techniques in the context of solving real-world and mathematical problems.

Previous Learning

Students need to be familiar with plotting points in the first quadrant of the coordinate plane.

Technology for the Teacher

Dynamic Classroom

Lesson Plans
Complete Materials List

4.4 Record and Practice Journal

Essential Question How can you find the lengths of line segments in a coordinate plane?

1 ACTIVITY: Finding Distances on a Map

Work with a partner. The coordinate grid shows a portion of a city. Each square on the grid represents one square mile.

a. A public library is located at (4, 5). City Hall is located at (7, 5). Plot and label these points.

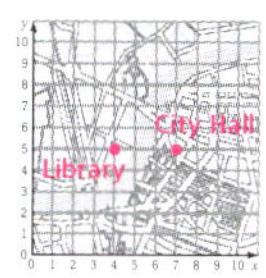

b. How far is the public library from City Hall?
3 miles

c. A stadium is located 4 miles from the public library. Give the coordinates of several possible locations of the stadium. Justify your answers by graphing.
Sample answers: (4, 9), (4, 1), (0, 5), (8. 5)

d. Connect the three locations of the public library, City Hall, and the stadium using your answers in part (c). What shapes are formed?
triangle, triangle, line segment, line segment

2 ACTIVITY: Graphing Polygons

Work with a partner. Plot and label each set of points in the coordinate plane. Then connect each set of points to form a polygon.

Rectangle: $A(2, 3)$, $B(2, 10)$, $C(6, 10)$, $D(6, 3)$

Triangle: $E(8, 3)$, $F(14, 8)$, $G(14, 3)$

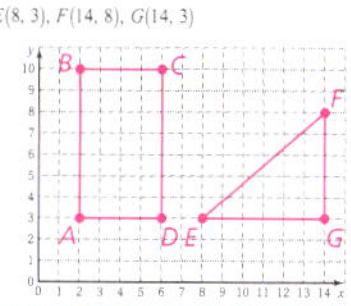

Differentiated Instruction

Organization

Point out to students that the coordinate plane is made up of two number lines. The horizontal line is similar to the number line with which they are familiar. This is the *x*-axis and the *x*-coordinate corresponds to the first number of a coordinate pair. The second number line is perpendicular to the first. This is the *y*-axis and the *y*-coordinate corresponds to the second number of a coordinate pair.

4.4 Record and Practice Journal

3 ACTIVITY: Finding Distances in a Coordinate Plane

Work with a partner.

a. Find the length of each horizontal line segment in Activity 2.

$AD = 4$ units, $BC = 4$ units, $EG = 6$ units

b. **STRUCTURE** What relationship do you notice between the lengths of the line segments in part (a) and the coordinates of their endpoints? Explain.

The lengths are equal to the differences of the *x*-coordinates.

c. Find the length of each vertical line segment in Activity 2.

$AB = 7$ units, $CD = 7$ units, $FG = 5$ units

d. **STRUCTURE** What relationship do you notice between the lengths of the line segments in part (c) and the coordinates of their endpoints? Explain.

The lengths are equal to the differences of the *y*-coordinates.

e. Plot and label the points below in the coordinate plane. Then connect each pair of points with a line segment. Use the relationships you discovered in parts (b) and (d) above to find the length of each line segment. Show your work.

$S(3, 1)$ and $T(14, 1)$ **11 units**

$U(9, 8)$ and $V(9, 0)$ **8 units**

$W(0, 7)$ and $X(0, 10)$ **3 units**

$Y(1, 9)$ and $Z(7, 9)$ **6 units**

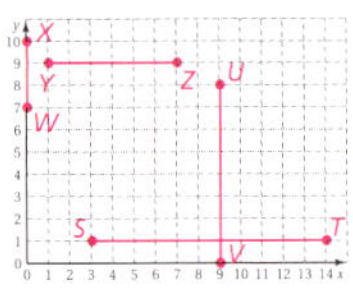

f. Check your answers in part (e) by counting grid lines.

Check students' work.

What Is Your Answer?

4. **IN YOUR OWN WORDS** How can you find the lengths of line segments in a coordinate plane? Give examples to support your explanation.

For vertical and horizontal line segments
- **count grid lines**
- **find difference of *y*-coordinates or *x*-coordinates**

5. Do the methods you used in Activity 3 work for diagonal line segments? Explain why or why not.

no

6. Use the Internet or some other reference to find an example of how "finding distances in a coordinate plane" is helpful in each of the following careers. **Check students' work.**

a. Archaeologist

b. Surveyor

c. Pilot

Laurie's Notes

Activity 3

- Give students sufficient time to work through the activity. As you circulate, make sure that students are paying attention to the *x*- and *y*-coordinates of the ordered pairs.
- **MP7 Look for and Make Use of Structure:** Mathematically proficient students will observe the same *x*-coordinate of vertical segments and the same *y*-coordinate of horizontal segments.
- Sometimes students will simply note that the points are "up and down" or "straight across" from one another. Focus their attention on the actual ordered pairs versus the location of the points in the coordinate grid.
- Students have no formal conception of absolute value so they will describe the length of the segments as the difference between the *x*- or *y*-coordinate. At this point, it is fine that they notice you simply subtract the lesser coordinate from the greater coordinate to find the length of the line segment.

What Is Your Answer?

- In Question 4, students will only reference horizontal and vertical segments.
- Depending upon your students, you may want to plant the seed that in later mathematics classes they will not be restricted to horizontal and vertical segments. Segments drawn on a diagonal in the coordinate plane also have a length, and there will be a method to determine that length. This is the question posed in Question 5.
- Ask if there are parents whose jobs involve finding distances. Ask students to share information.

Closure

- Plot a rectangle with a perimeter of 20. What are the coordinates? *Sample answer:* (1, 2), (7, 2), (7, 6), (1, 6)

3 ACTIVITY: Finding Distances in a Coordinate Plane

Work with a partner.

a. Find the length of each horizontal line segment in Activity 2.

b. **STRUCTURE** What relationship do you notice between the lengths of the line segments in part (a) and the coordinates of their endpoints? Explain.

c. Find the length of each vertical line segment in Activity 2.

d. **STRUCTURE** What relationship do you notice between the lengths of the line segments in part (c) and the coordinates of their endpoints? Explain.

e. Plot and label the points below in the coordinate plane. Then connect each pair of points with a line segment. Use the relationships you discovered in parts (b) and (d) above to find the length of each line segment. Show your work.

$S(3, 1)$ and $T(14, 1)$ $U(9, 8)$ and $V(9, 0)$

$W(0, 7)$ and $X(0, 10)$ $Y(1, 9)$ and $Z(7, 9)$

f. Check your answers in part (e) by counting grid lines.

Math Practice 8

Repeat Calculations

What calculations are repeated? How can you use this information to write a rule about the length of a line segment?

What Is Your Answer?

4. **IN YOUR OWN WORDS** How can you find the lengths of line segments in a coordinate plane? Give examples to support your explanation.

5. Do the methods you used in Activity 3 work for diagonal line segments? Explain why or why not.

6. Use the Internet or some other reference to find an example of how "finding distances in a coordinate plane" is helpful in each of the following careers.

a.

Archaeologist

b.

Surveyor

c.

Pilot

Practice Use what you learned about finding the lengths of line segments to complete Exercises 3–5 on page 178.

4.4 Lesson

Check It Out
Lesson Tutorials
BigIdeasMath.com

You can use ordered pairs to represent vertices of polygons. To draw a polygon in a coordinate plane, plot and connect the ordered pairs.

EXAMPLE 1 Drawing a Polygon in a Coordinate Plane

The vertices of a quadrilateral are $A(2, 4)$, $B(3, 9)$, $C(7, 8)$, and $D(8, 1)$. Draw the quadrilateral in a coordinate plane.

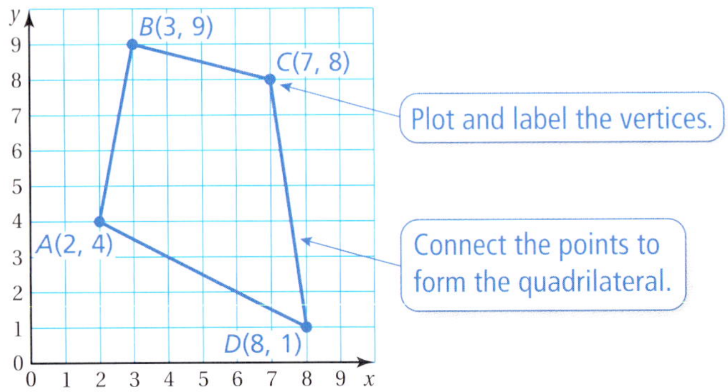

Study Tip

After you plot the vertices, connect them *in order* to draw the polygon.

On Your Own

Now You're Ready
Exercises 6–11

Draw the polygon with the given vertices in a coordinate plane.

1. $A(0, 0)$, $B(5, 7)$, $C(7, 4)$
2. $W(4, 4)$, $X(7, 4)$, $Y(7, 1)$, $Z(4, 1)$
3. $F(1, 3)$, $G(3, 6)$, $H(5, 6)$, $J(3, 3)$
4. $P(1, 4)$, $Q(3, 5)$, $R(7, 3)$, $S\left(6, \frac{1}{2}\right)$, $T\left(2, \frac{1}{2}\right)$

Key Idea

Finding Distances in the First Quadrant

You can find the length of a horizontal or vertical line segment in a coordinate plane by using the coordinates of the endpoints.

- When the x-coordinates are the same, the vertical distance between the points is the difference of the y-coordinates.
- When the y-coordinates are the same, the horizontal distance between the points is the difference of the x-coordinates.

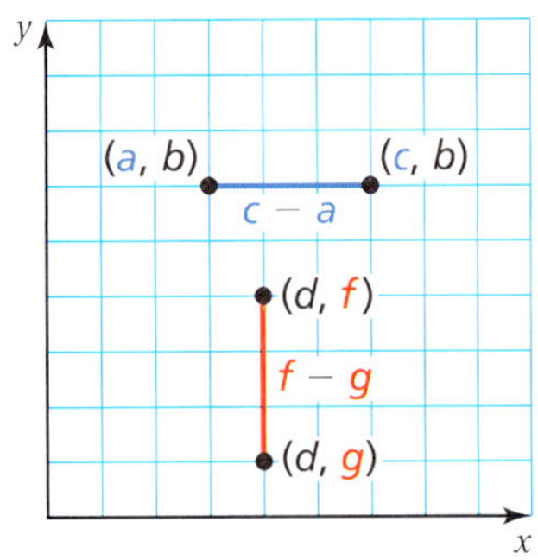

Be sure to subtract the lesser coordinate from the greater coordinate.

Laurie's Notes

Introduction

Connect

- **Yesterday:** Students plotted ordered pairs and explored lengths of line segments in a coordinate plane. (MP7, MP8)
- **Today:** Students will draw polygons in the coordinate plane and find the length of their sides.

Motivate

- Draw two axes without any scale, and a rectangle, as shown.
- Tell students to use the numbers 2, 5, 6, and 8 to fill in all 8 blanks so that the ordered pairs represent the vertices of a rectangle.
- Yes, each number is used twice, but that is not something that needs to be stated at the beginning.

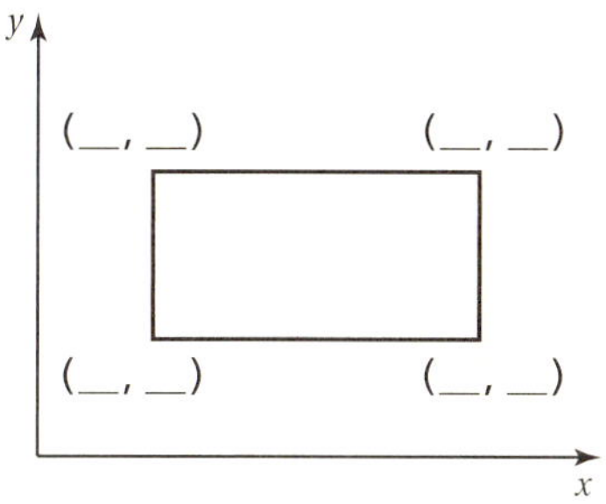

- *Sample answer:* (2, 5), (8, 5), (8, 6), and (2, 6)
- Ask students to discuss their reasoning and process. The goal is for students to think about the *x*-coordinates of vertical segments and the *y*-coordinates of horizontal segments.

Lesson Notes

Example 1

- Write the ordered pairs and ask students to plot them. Note that the labels: *A*, *B*, *C*, and D are also written next to the ordered pair in the coordinate plane. In this book, the vertices will always be in alphabetical order.
- ? "When the ordered pairs are connected in order, which polygon is formed?" quadrilateral
- ? "Is it a special quadrilateral with a more specific name?" No

On Your Own

- Students should plot each problem on a different coordinate plane.
- Have volunteers share their polygons with the class. Have students identify the type of polygon they have plotted.

Key Idea

- Write the Key Idea. Sketch a horizontal segment and a vertical segment that can be referenced as you discuss how to find the length of the segment. Be sure to identify what the term *endpoint* means.
- There are many words used to describe how to find distances in the first quadrant. Having sample segments to reference will help students see that it is a simple subtraction problem.
- All of the ordered pairs will be in the first quadrant. For clarity, the Key Idea tells students to subtract the lesser number from the greater number. In Section 6.5, students will plot ordered pairs in all four quadrants.

Goal Today's lesson is drawing polygons in the coordinate plane and finding the lengths of the sides of the polygons.

Technology for the Teacher

Dynamic Classroom

Lesson Tutorials
Lesson Plans
Answer Presentation Tool

Extra Example 1

The vertices of a quadrilateral are *A*(2, 8), *B*(5, 7), *C*(7, 4), and *D*(1, 1). Draw the quadrilateral in a coordinate plane.

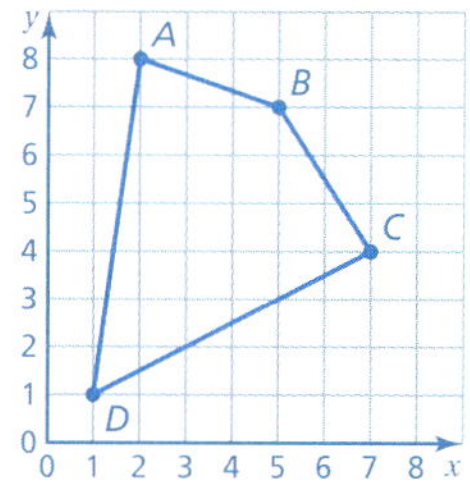

On Your Own

1. 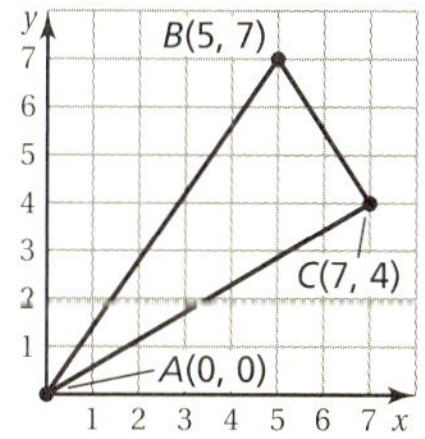

2. 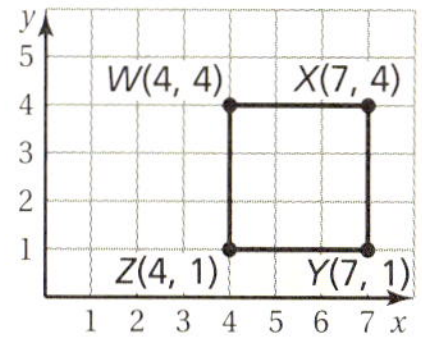

3. 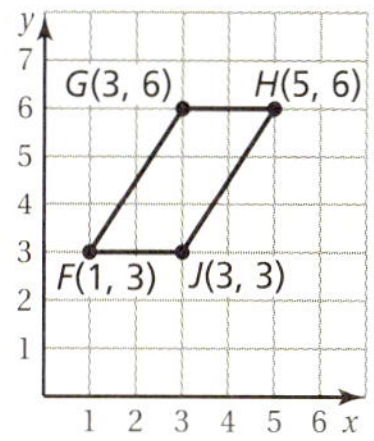

4. 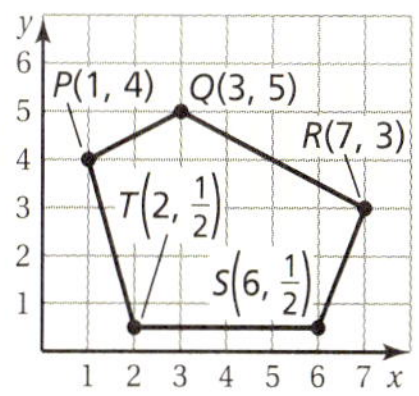

English Language Learners

Vocabulary

Encourage English language learners to answer in complete sentences. For instance, in Example 3, a student could respond, "The area of the giraffe exhibit is 4800 square feet." This provides language support and reinforces the units of measure in the answer.

Extra Example 2

The vertices of a rectangle are $R(2, 8)$, $S(5, 8)$, $T(5, 1)$, and $U(2, 1)$. Draw the rectangle in a coordinate plane, and find its perimeter.

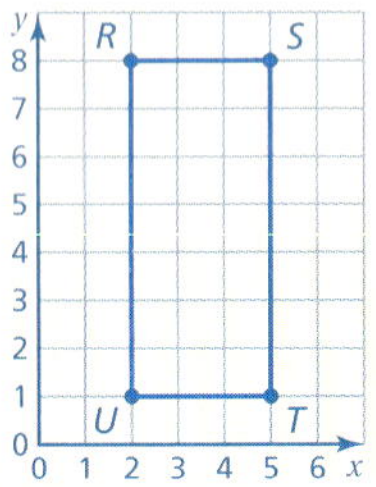

20 units

Extra Example 3

In a grid of a housing development, the vertices of a children's play area are $J(20, 60)$, $K(80, 70)$, $L(80, 0)$, and $M(20, 0)$. The coordinates are measured in feet. What is the area of the children's play area? 3900 square feet

On Your Own

5. 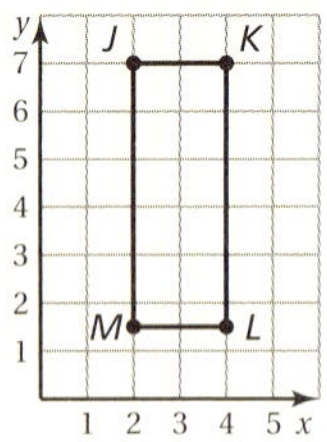

15 units; 11 square units

6. Increases it by 600 square feet for a total of 5400 square feet.

Laurie's Notes

Example 2

- Plot the ordered pairs as shown.
- ? "What information is needed to find the perimeter of the rectangle?" the length of each side
- ? "What is the length of the base? Explain." 6; $7 - 1$
- ? "What is the height of the rectangle? Explain." 4; $6 - 2$
- ? "What is the perimeter? Explain." 20; $6 \times 2 + 4 \times 2$
- Note that some students may find the perimeter by adding, $6 + 6 + 4 + 4$.
- Because no units have been specified, write the answer as 20 units.

Example 3

- **MP6 Attend to Precision:** The last example asked students to find a perimeter. This example asks students to find area. Students should pay attention to units in labeling their answers.
- Ask a student to read the problem.
- ? "What information is needed to find the area of the trapezoid?" the lengths of both bases and the height
- ? "How do you find the length of each base?" The longer base has a length of $100 - 0 = 100$. The shorter base has a length of $90 - 0 = 90$.
- ? "What ordered pairs can you use to find the height of the trapezoid?" $H(0, 30)$ and $E(0, 90)$
- ? "What is the height?" $90 - 30 = 60$
- Write the formula for the area of a trapezoid and substitute the known values.
- Finish the problem as shown. Be sure that the answer is labeled with the correct units.

On Your Own

- **Think-Pair-Share:** Students should read each question independently and then work in pairs to answer the questions. When they have answered the questions, the pair should compare their answers with another group and discuss any discrepancies.

Closure

- Give the endpoints of a horizontal segment with a length of 6. *Sample answer:* (1, 3) and (12, 3)
- Give the endpoints of a vertical segment with a length of 11. *Sample answer:* (3, 2) and (3, 8)

EXAMPLE 2 Finding a Perimeter

The vertices of a rectangle are $F(1, 6)$, $G(7, 6)$, $H(7, 2)$, and $J(1, 2)$. Draw the rectangle in a coordinate plane and find its perimeter.

Draw the rectangle and use the vertices to find its dimensions.

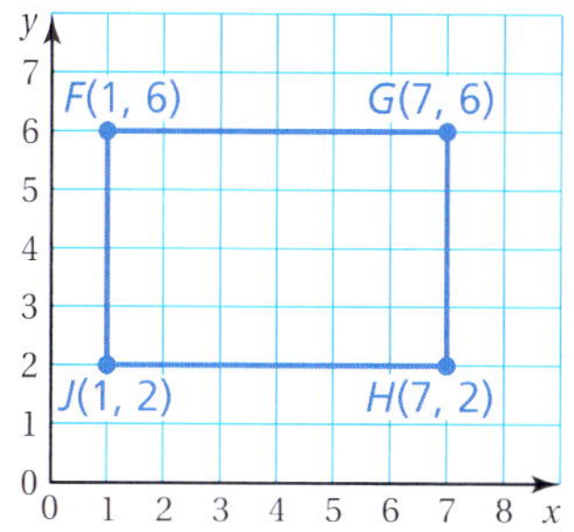

The length is the horizontal distance between $F(1, 6)$ and $G(7, 6)$, which is the difference of the x-coordinates.

$$\text{length} = 7 - 1 = 6 \text{ units}$$

The width is the vertical distance between $G(7, 6)$ and $H(7, 2)$, which is the difference of the y-coordinates.

$$\text{width} = 6 - 2 = 4 \text{ units}$$

So, the perimeter of the rectangle is $2(6) + 2(4) = 20$ units.

Study Tip

You can also find the length using vertices H and J. You can find the width using vertices F and J.

EXAMPLE 3 Real-Life Application

In a grid of the exhibits at a zoo, the vertices of the giraffe exhibit are $E(0, 90)$, $F(60, 90)$, $G(100, 30)$, and $H(0, 30)$. The coordinates are measured in feet. What is the area of the giraffe exhibit?

Plot and connect the vertices using a coordinate grid to form a trapezoid. Use the coordinates to find the lengths of the bases and the height.

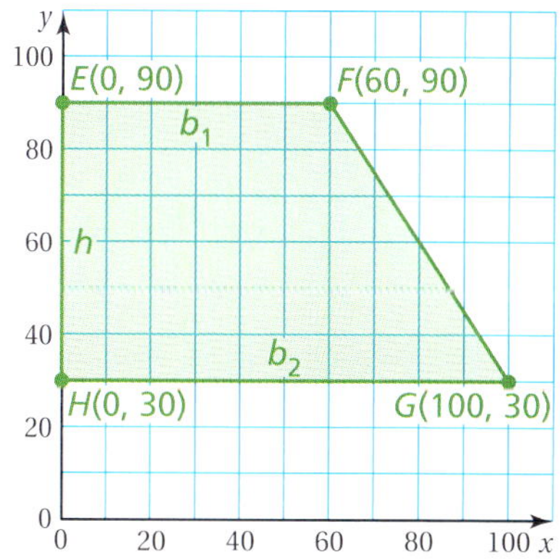

$$b_1 = 60 - 0 = 60$$

$$b_2 = 100 - 0 = 100$$

$$h = 90 - 30 = 60$$

Use the formula for the area of a trapezoid.

$$A = \frac{1}{2}(60)(60 + 100)$$

$$= \frac{1}{2}(60)(160) = 4800$$

The area of the giraffe exhibit is 4800 square feet.

Common Error

You can count grid lines to find the dimensions, but make sure you consider the scale of the axes.

On Your Own

Now You're Ready
Exercises 12–15

5. The vertices of a rectangle are $J(2, 7)$, $K(4, 7)$, $L(4, 1.5)$, and $M(2, 1.5)$. Find the perimeter and the area of the rectangle.
6. **WHAT IF?** In Example 3, the giraffe exhibit is enlarged by moving vertex F to $(80, 90)$. How does this affect the area? Explain.

4.4 Exercises

Check It Out
Help with Homework
BigIdeasMath.com

Vocabulary and Concept Check

1. **WRITING** How can you use a coordinate plane to draw a polygon?
2. **WRITING** How can you find the perimeter of a rectangle in a coordinate plane?

Practice and Problem Solving

Plot and label each pair of points in a coordinate plane. Find the length of the line segment connecting the points.

3. $C(0, 1)$, $D(8, 1)$
4. $K(5, 2)$, $L(5, 6)$
5. $Q(3, 4)$, $R(3, 9)$

Draw the polygon with the given vertices in a coordinate plane.

1 6. $A(4, 7)$, $B(6, 2)$, $C(0, 0)$

7. $D\left(\frac{1}{2}, 2\right)$, $E(5, 5)$, $F(4, 1)$

8. $G\left(1\frac{1}{2}, 4\right)$, $H\left(1\frac{1}{2}, 8\right)$, $J(5, 8)$, $K(5, 4)$

9. $L(3, 2)$, $M(3, 5)$, $N(9, 5)$, $P(9, 2)$

10. $Q(0, 4)$, $R(10, 8)$, $S(7, 4)$, $T(10, 2)$, $U(5, 0)$

11. $V(2, 2)$, $W\left(3, 7\frac{1}{2}\right)$, $X\left(8, 7\frac{1}{2}\right)$, $Y(10, 4)$, $Z(7, 0)$

Find the perimeter and the area of the polygon with the given vertices.

2 12. $C(1, 1)$, $D(1, 4)$, $E(4, 4)$, $F(4, 1)$

13. $J(1, 2)$, $K(7, 2)$, $L(7, 8)$, $M(1, 8)$

14. $N(0, 2)$, $P(5, 2)$, $Q(5, 5)$, $R(0, 5)$

15. $S(3, 0)$, $T(3, 9)$, $U(8, 9)$, $V(8, 0)$

16. **ERROR ANALYSIS** Describe and correct the error in drawing a triangle with vertices $A(5, 1)$, $B(7, 6)$, and $C(1, 3)$.

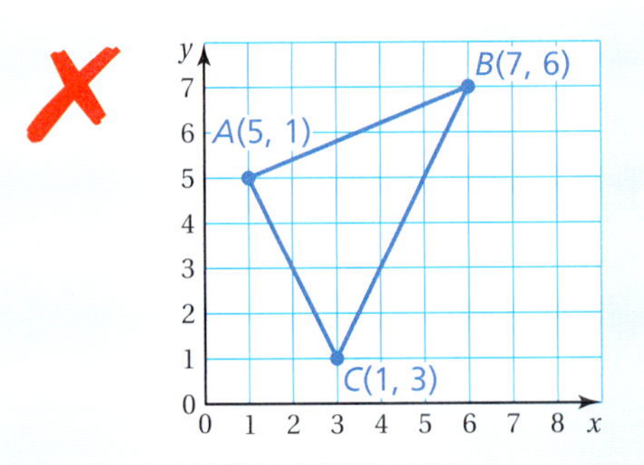

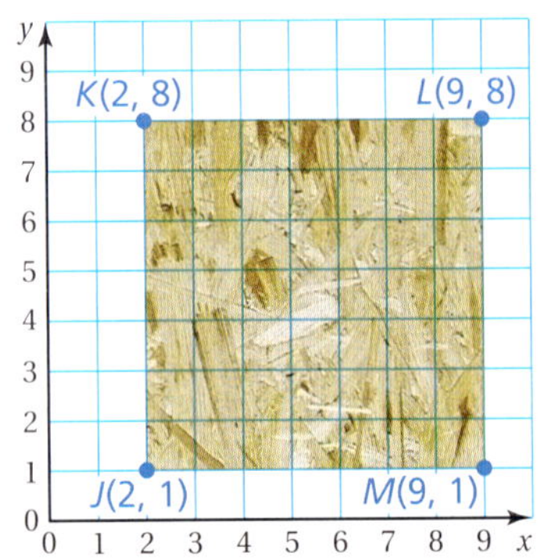

17. **TREE HOUSE** You design a tree house using a coordinate plane. You plot the vertices of the floor at $J(2, 1)$, $K(2, 8)$, $L(9, 8)$, and $M(9, 1)$. The coordinates are measured in feet.

  a. What is the shape of the floor?

  b. What are the perimeter and the area of the floor?

Assignment Guide and Homework Check

Level	Day 1 Activity Assignment	Day 2 Lesson Assignment	Homework Check
Basic	3–5, 27–31	1, 2, 7–15 odd, 16, 17, 19	7, 9, 13, 17
Average	3–5, 27–31	1, 2, 6–16 even, 17, 19, 22–24	6, 10, 14, 24
Advanced	3–5, 27–31	1, 2, 8–24 even, 25, 26	10, 12, 14, 24

For Your Information

- **Exercise 23** Students may be confused by the statement, "Each grid square represents 9 square miles." Students need to realize that if 1 grid square is 9 square miles, then 1 unit length is equal to 3 miles.

Common Errors

- **Exercises 7, 8, and 11** The fractions in the coordinates may confuse students. Refresh the concept of graphing a coordinate involving a fraction.
- **Exercises 12–15** Students may confuse perimeter and area. Have them write the formulas and then substitute the values. Students should also include "units" or "square units" as part of the answer.

4.4 Record and Practice Journal

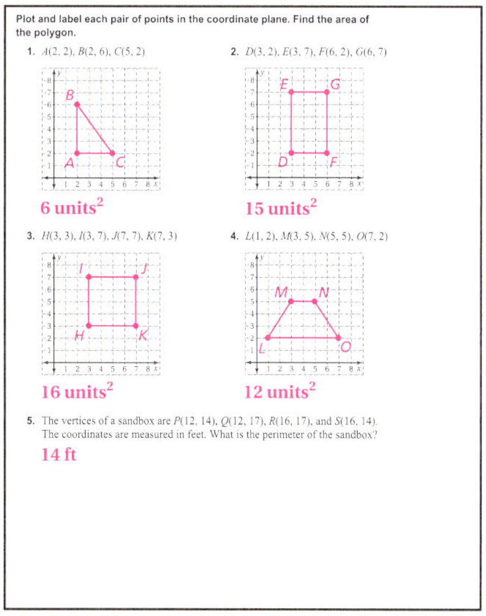
Plot and label each pair of points in the coordinate plane. Find the area of the polygon.

1. A(2, 2), B(2, 6), C(5, 2) — 6 units²
2. D(3, 2), E(3, 7), F(6, 2), G(6, 7) — 15 units²
3. H(3, 3), I(3, 7), J(7, 7), K(7, 3) — 16 units²
4. L(1, 2), M(3, 5), N(5, 5), O(7, 2) — 12 units²
5. The vertices of a sandbox are P(12, 14), Q(12, 17), R(16, 17), and S(16, 14). The coordinates are measured in feet. What is the perimeter of the sandbox? 14 ft

Vocabulary and Concept Check

1. Plot the points that represent the vertices of the polygon and connect the points in order.
2. Assume that all of the sides are vertical or horizontal. Find the points with the same y-coordinates. Subtract the x-coordinates of those points to find the length of the horizontal sides. Find the points with the same x-coordinates. Subtract the y-coordinates of those points to find the length of the vertical sides. Last, add the lengths of all four sides to find the perimeter.

Practice and Problem Solving

3. 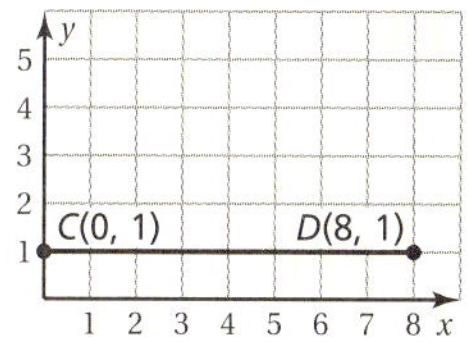

Length of CD is 8 units.

4. 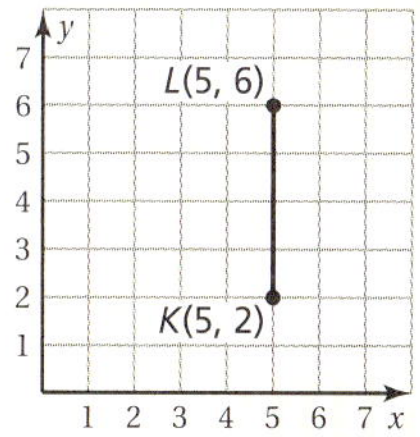

Length of KL is 4 units.

5. 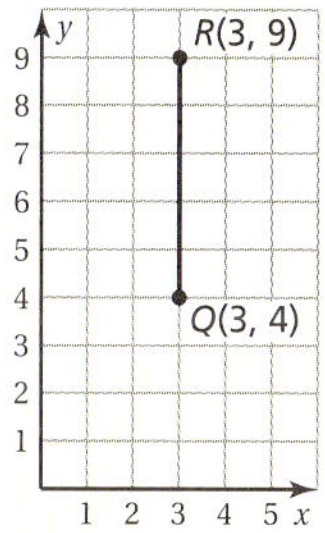

Length of QR is 5 units.

Practice and Problem Solving

6.

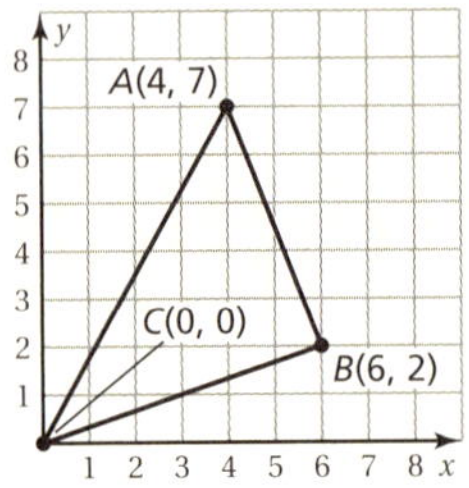

7–25. See Additional Answers.

26. See *Taking Math Deeper*.

Fair Game Review

27. 2   **28.** $8\frac{4}{5}$

29. $\frac{5}{16}$ **30.** $3\frac{19}{27}$

31. D

Mini-Assessment

1. The vertices of a quadrilateral are $A(1, 3)$, $B(4, 8)$, $C(8, 6)$, and $D(9, 2)$. Draw the quadrilateral in a coordinate plane.

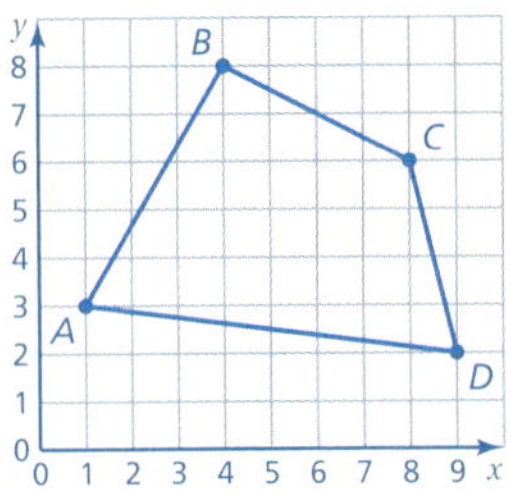

2. The vertices of a rectangle are $F(2, 5)$, $G(7, 5)$, $H(7, 1)$, and $J(2, 1)$. Draw the rectangle in a coordinate plane and find its perimeter.

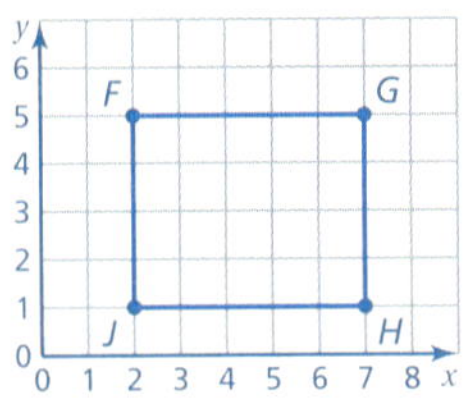

18 units

3. In a grid of the swimming area at a beach, the vertices are $E(1, 0)$, $F(1, 5)$, $G(4, 7)$, and $H(4, 0)$. The coordinates are measured in meters. What is the area of the swimming area? 18 square meters

Taking Math Deeper

Exercise 26

Note that in this exercise, we are assuming a and b are positive because students have not yet plotted ordered pairs outside of Quadrant I.

1 Draw the figures. Place a tick mark at some value a on the vertical axis. Then use the given vertices to draw a rectangle.

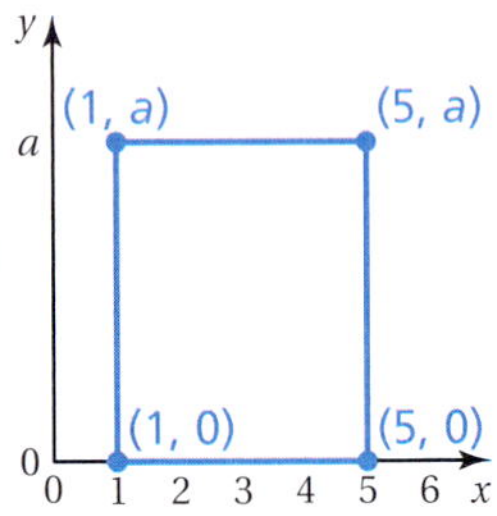

Place a tick mark at some value b on the vertical axis. Because the value of a is greater than the value of b, b must be between 0 and a on the vertical axis. Use the given vertices to draw a parallelogram. Here are two possible examples.

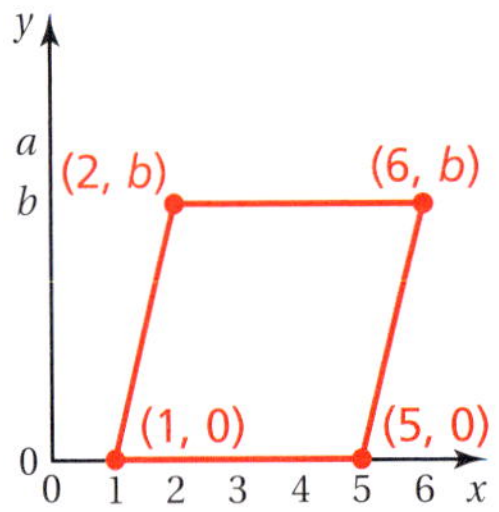

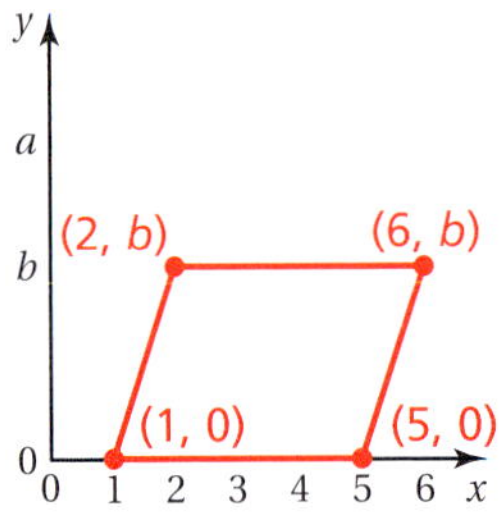

2 Look at the dimensions. The base of each figure can be represented by the side length with endpoints (1, 0) and (5, 0). So, the bases have the same length. The heights of the quadrilaterals are different. The height a of the rectangle is greater than the height b of the parallelogram.

3 Answer the question. The area of each figure can be found by multiplying the base and height. The height of the rectangle is always greater, so you are multiplying the base by a greater number compared to the height of the parallelogram. So, the rectangle has the greater area.

Project

Research the discovery of the coordinate plane. Who invented it? When was it invented? Explain why you think the coordinate plane is helpful in mathematics.

Reteaching and Enrichment Strategies

If students need help. . .	If students got it. . .
Resources by Chapter • Practice A and Practice B • Puzzle Time Record and Practice Journal Practice Differentiating the Lesson Lesson Tutorials Skills Review Handbook	Resources by Chapter • Enrichment and Extension • Technology Connection Start the next section

OPEN-ENDED **Draw a polygon with the given conditions in a coordinate plane.**

18. a square with a perimeter of 20 units

19. a rectangle with a perimeter of 18 units

20. a rectangle with an area of 24 units^2

21. a triangle with an area of 15 units^2

22. **STRUCTURE** The coordinate plane shows three vertices of a parallelogram. Find two possible points that could represent the fourth vertex.

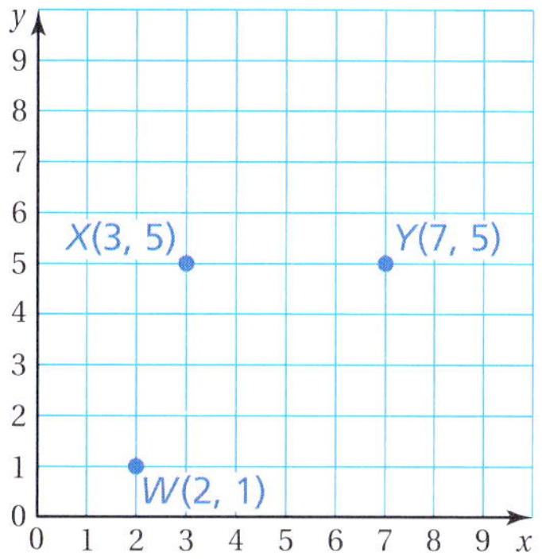

23. **BUS ROUTE** Polygon *JKLMNP* represents a bus route. Each grid square represents 9 square miles. What is the shortest distance, in miles, from station *P* to station *L* using the bus route? Explain your reasoning.

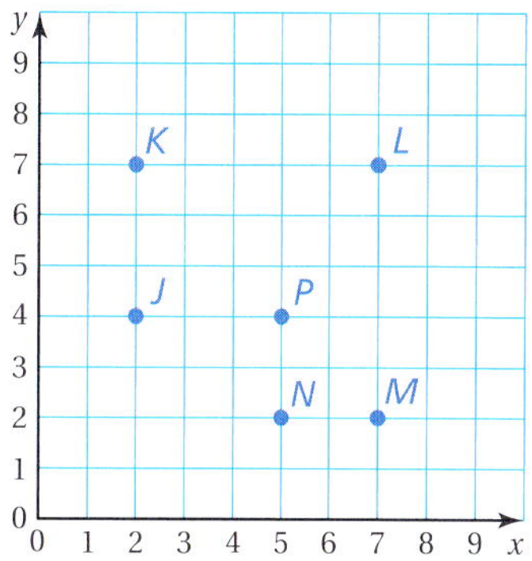

24. **CITY LIMITS** In a topographical map of a city, the vertices of the city limits are $A(10, 9)$, $B(18, 9)$, $C(18, 2)$, $D(14, 4.5)$, and $E(10, 4.5)$. The coordinates are measured in miles. What is the area of the city?

25. **BACKYARD** The vertices of a backyard are $W(10, 30)$, $X(10, 100)$, $Y(110, 100)$, and $Z(50, 30)$. The coordinates are measured in feet. The line segment XZ separates the backyard into a lawn and a garden. The area of the lawn is greater than the area of the garden. How many times larger is the lawn than the garden?

26. **Precision** The vertices of a rectangle are $(1, 0)$, $(1, a)$, $(5, a)$, and $(5, 0)$. The vertices of a parallelogram are $(1, 0)$, $(2, b)$, $(6, b)$, and $(5, 0)$. The value of a is greater than the value of b. Which polygon has a greater area? Explain your reasoning.

Fair Game Review What you learned in previous grades & lessons

Divide. Write the answer in simplest form. *(Section 2.3)*

27. $1\frac{1}{3} \div \frac{2}{3}$

28. $6\frac{3}{5} \div \frac{3}{4}$

29. $2\frac{1}{2} \div 8$

30. $4\frac{1}{6} \div 1\frac{1}{8}$

31. **MULTIPLE CHOICE** You are filling bottles from 5 gallons of lemonade. How many bottles can you fill when each bottle is $\frac{3}{8}$ of a gallon? *(Section 2.2)*

Ⓐ $1\frac{7}{8}$ Ⓑ 3 Ⓒ 8 Ⓓ $13\frac{1}{3}$

4.3–4.4 Quiz

Check It Out
Progress Check
BigIdeasMath.com

Find the area of the trapezoid. *(Section 4.3)*

1.

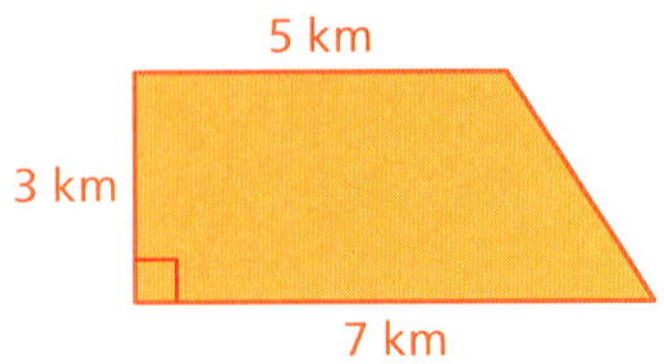

2.

3.

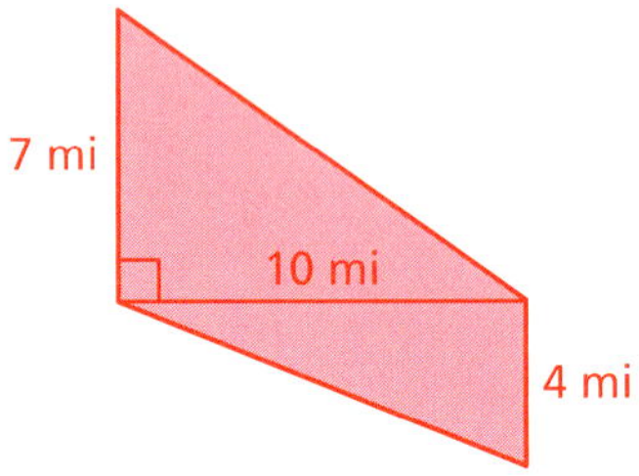

Find the area of the figure. *(Section 4.3)*

4.

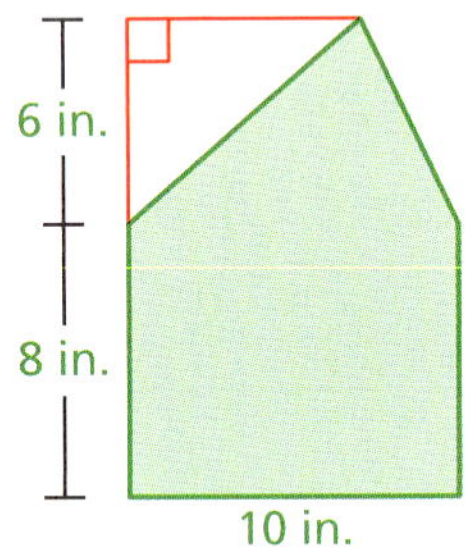

5.

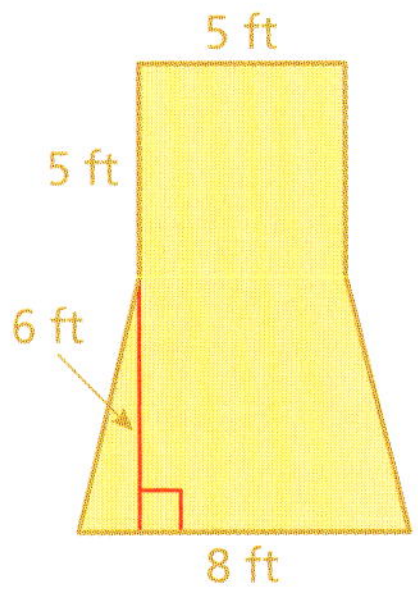

6.

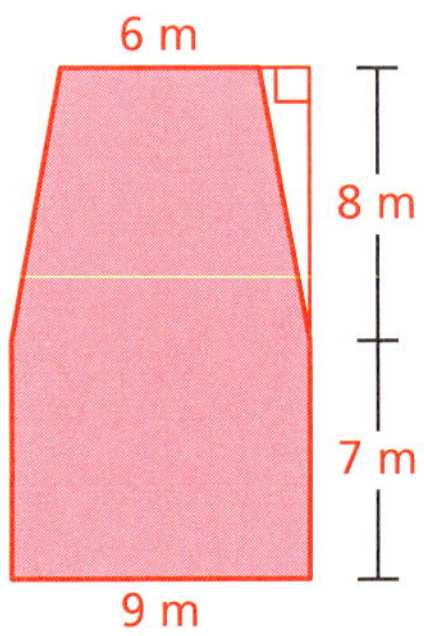

Draw the polygon with the given vertices in a coordinate plane. *(Section 4.4)*

7. $A(1, 2)$, $B(3, 5)$, $C(6, 1)$

8. $E(1, 2)$, $F(3, 6)$, $G(8, 6)$, $H(6, 2)$

Find the perimeter and the area of the polygon with the given vertices. *(Section 4.4)*

9. $J(1, 3)$, $K(1, 8)$, $L(5, 8)$, $M(5, 3)$

10. $P(1, 2)$, $Q(1, 7)$, $R(7, 7)$, $S(7, 2)$

11. BACK POCKET How much material do you need to make two back pockets? *(Section 4.3)*

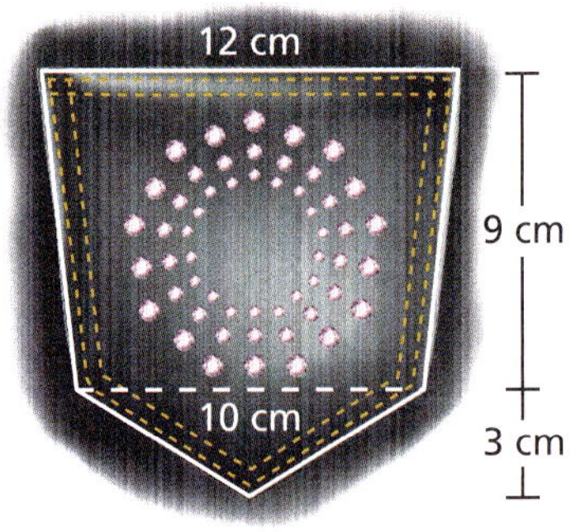

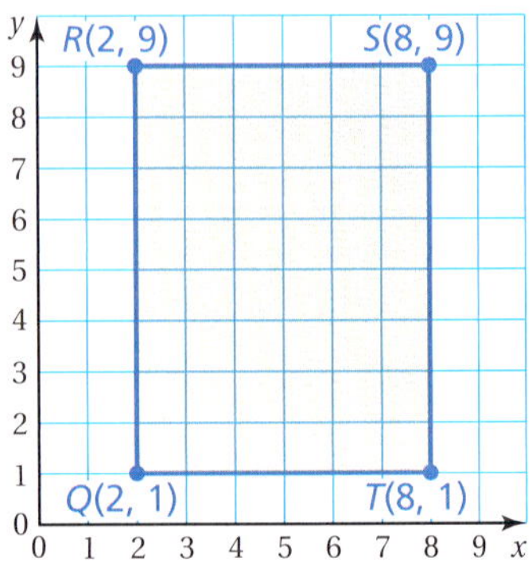

12. PATIO Plans for a patio are shown in the coordinate plane at the left. The coordinates are measured in feet. Find the perimeter and the area of the patio. *(Section 4.4)*

Alternative Assessment Options

Math Chat	**Student Reflective Focus Question**
Structured Interview	Writing Prompt

Student Reflective Focus Question

Ask students to use their own words to summarize the similarities and differences among the area formulas for a rectangle, parallelogram, triangle, and trapezoid. Be sure that they include examples. Select students at random to present their summaries to the class.

Study Help Sample Answers

Remind students to complete Graphic Organizers for the rest of the chapter.

2.

Area of a trapezoid

Words

The area A of a trapezoid is one-half the product of its height h and the sum of its bases b_1 and b_2.

Algebra

$$A = \frac{1}{2}h(b_1 + b_2)$$

Example

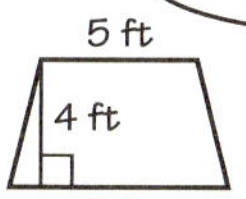

$A = \frac{1}{2}h(b_1 + b_2)$

$= \frac{1}{2}(4)(5 + 7)$

$= 24$

The area of the trapezoid is 24 square feet.

Example

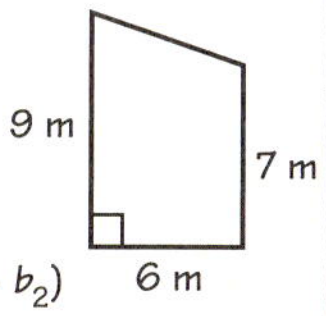

$A = \frac{1}{2}h(b_1 + b_2)$

$= \frac{1}{2}(6)(7 + 9)$

$= 48$

The area of the trapezoid is 48 square meters.

3–5. Available at *BigIdeasMath.com.*

Reteaching and Enrichment Strategies

If students need help. . .	If students got it. . .
Resources by Chapter • Practice A and Practice B • Puzzle Time Lesson Tutorials *BigIdeasMath.com*	Resources by Chapter • Enrichment and Extension • Technology Connection Game Closet at *BigIdeasMath.com* Start the Chapter Review

Answers

1. 18 km^2
2. 100 in.^{2}
3. 55 mi^2
4. 110 in.^{2}
5. 64 ft^2
6. 123 m^2
7. 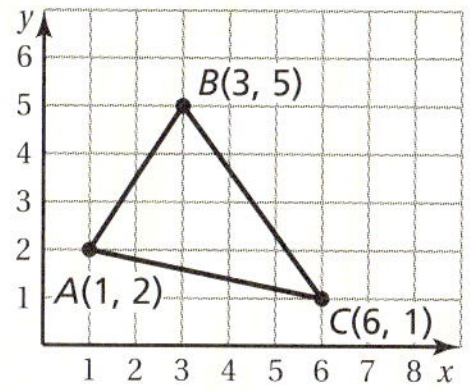

8. 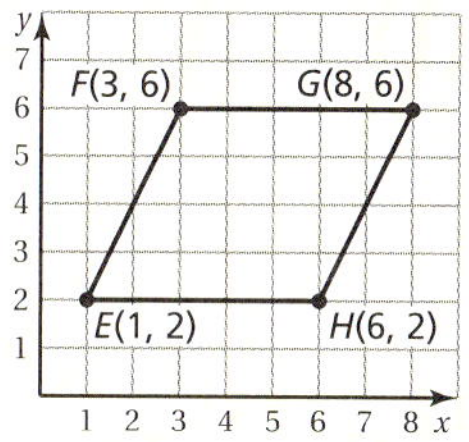

9. 18 units; 20 units^2
10. 22 units; 30 units^2
11. 228 cm^2
12. 28 ft; 48 ft^2

Technology for the Teacher

Online Assessment
Assessment Book
ExamView® Assessment Suite

For the Teacher
Additional Review Options

- *BigIdeasMath.com*
- Online Assessment
- Game Closet at *BigIdeasMath.com*
- Vocabulary Help
- Resources by Chapter

Answers

1. 500 yd^2
2. 242 mm^2
3. 80 km^2
4. 175 cm^2
5. 105 m^2
6. 6 in.^2
7. 49 mi^2
8. 76 ft^2
9. 120 cm^2
10. 90 in.^2

Review of Common Errors

Exercises 1 and 2

- Students may use the formula for perimeter instead of the formula for area. Tell students to write the formula, and then identify the value of each variable before substituting.

Exercises 3 and 4

- Students may multiply the base and the height and forget to multiply by $\frac{1}{2}$. Tell students to write the formula, and then identify the value of each variable before substituting.

Exercises 3 and 4

- Students may mentally find the area by multiplying one-half the base and one-half the height. The answer will be one-half of the value of the correct answer. Tell students to write the formula, and then identify the value of each variable before substituting.

Exercise 4

- Students may be confused by the drawing and labeling of obtuse triangles. Remind students to identify the base of the triangle. Then the height of the triangle is the perpendicular distance from the base to the opposite vertex. For obtuse triangles, the height may be labeled outside of the triangle.

Exercises 5–7

- Students may forget to multiply by $\frac{1}{2}$. Tell students to write the formula for the area of a trapezoid, and then identify the value of each variable before substituting.

Exercises 5–7

- Students may substitute the height for one of the bases. Tell students to write the formula for the area of a trapezoid, and then identify the value of each variable before substituting.

Exercise 14

- The fractions in the coordinates may confuse students. Refresh the concept of graphing a coordinate involving a fraction.

Exercises 15–18

- Students may confuse perimeter and area. Have them write the formulas and then substitute the values. Students should also include "units" or "square units" as part of the answer.

4 Chapter Review

Check It Out
Vocabulary Help
BigIdeasMath.com

Review Key Vocabulary

polygon, *p. 152*

composite figure, *p. 172*

Review Examples and Exercises

4.1 Areas of Parallelograms *(pp. 152–157)*

Find the area of the parallelogram.

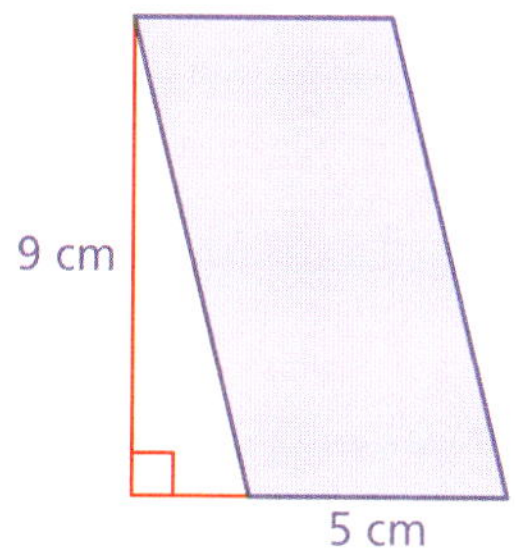

$A = bh$	Write formula.
$= 5(9)$	Substitute 5 for b and 9 for h.
$= 45$	Multiply.

∴ The area of the parallelogram is 45 square centimeters.

Exercises

Find the area of the parallelogram.

1.

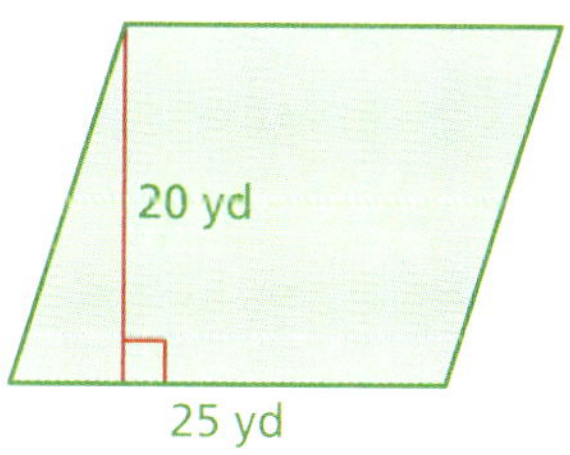

2.

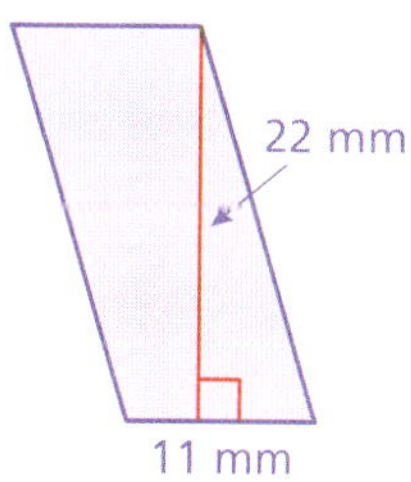

4.2 Areas of Triangles *(pp. 158–163)*

Find the area of the triangle.

$A = \frac{1}{2}bh$	Write formula.
$= \frac{1}{2}(10)(7)$	Substitute.
$= 35$	Multiply.

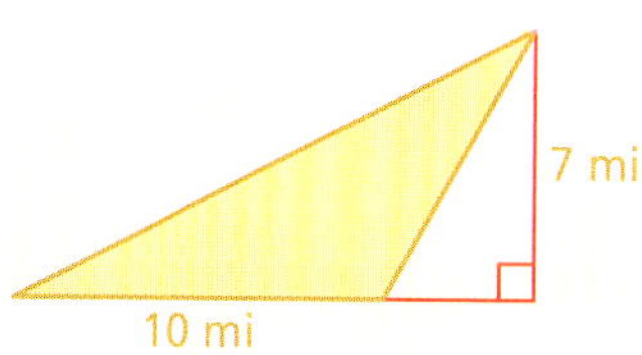

∴ The area of the triangle is 35 square miles.

Exercises

Find the area of the triangle.

3. 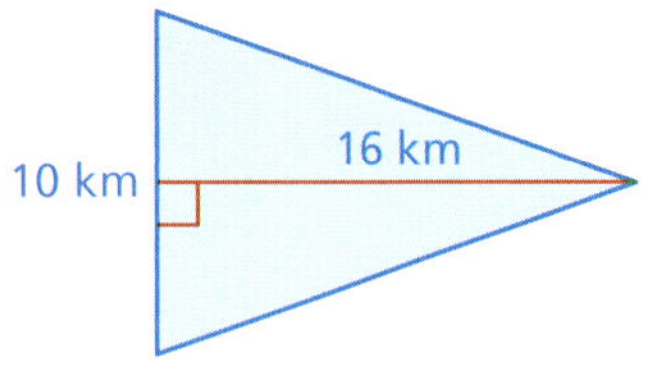

4. 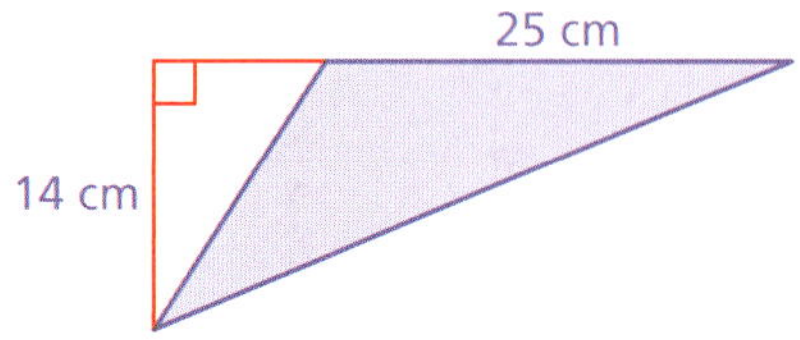

4.3 Areas of Trapezoids *(pp. 166–173)*

Find the area of the trapezoid.

$$A = \frac{1}{2}h(b_1 + b_2) \quad \text{Write formula.}$$

$$= \frac{1}{2}(10)(8 + 18) \quad \text{Substitute.}$$

$$= \frac{1}{2}(10)(26) = 130 \quad \text{Multiply.}$$

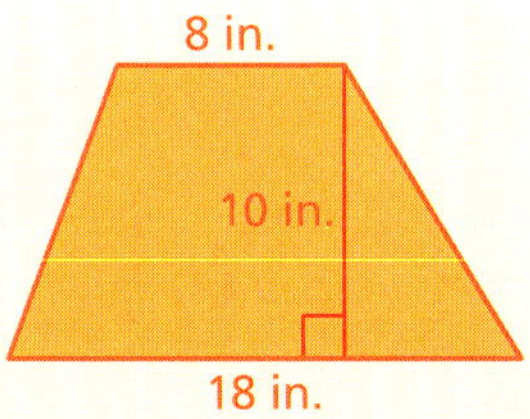

∴ The area of the trapezoid is 130 square inches.

Exercises

Find the area of the trapezoid.

5. 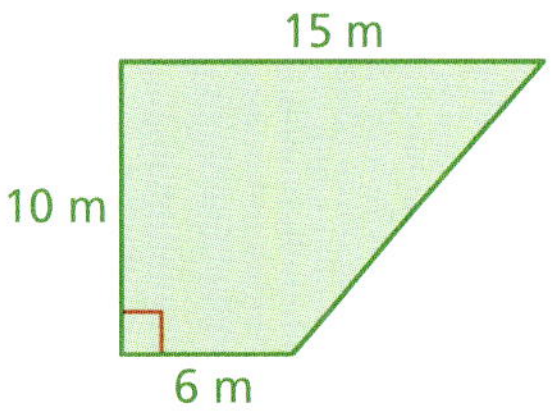

6. 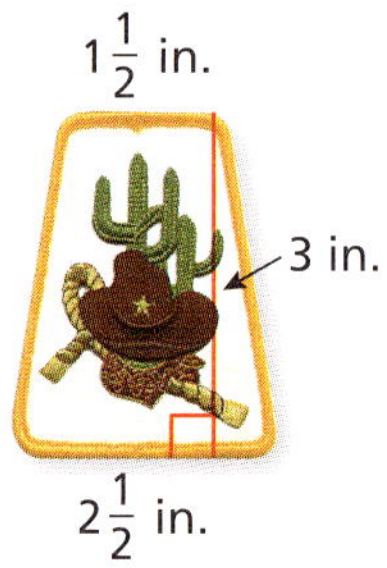

7. 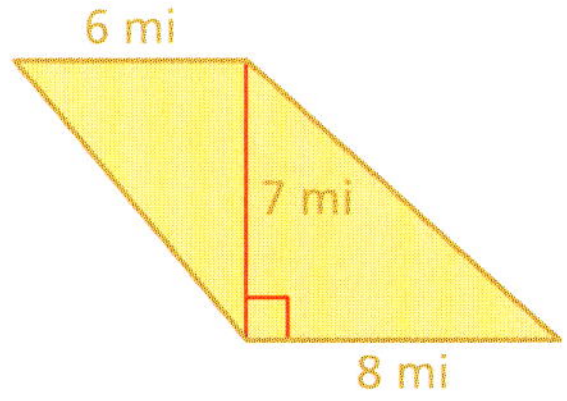

Find the area of the figure.

8. 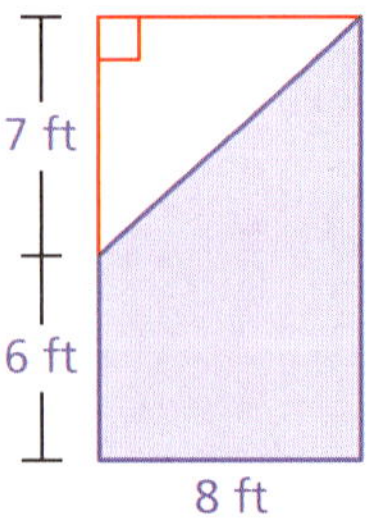

9. 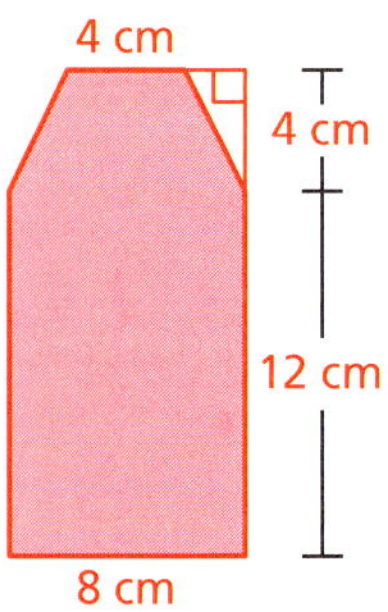

10. 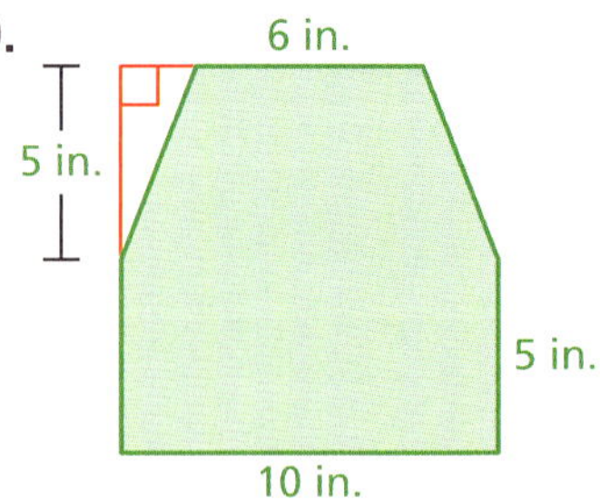

Review Game

Area and Perimeter Tic-Tac-Toe

Materials

- Tic-Tac-Toe Game Boards (Each board will have three different parallelograms, three different triangles, and three different trapezoids with the lengths, widths, and heights labeled.)
- Coin
- 5 red and 5 blue chips

Players: 2

Directions

Each student has 5 chips of the same color.

The first player selects a shape and then tosses a coin. If the coin is heads, the student finds the area of the shape. If the coin is tails, the student finds the perimeter of the shape.

If the answer is correct, the student covers the shape with a chip. If the answer is incorrect, the student loses a turn and the second player has an opportunity to solve the problem and cover the shape with a chip.

Students continue taking turns until one player gets three chips in a row, vertically, horizontally, or diagonally.

Students can exchange boards with another group and continue to play another round.

Who Wins?

A player earns one point for getting Tic-Tac-Toe. Continue to play until one student gets three points or until the teacher calls time.

For the Student
Additional Practice

- Lesson Tutorials
- Multi-Language Glossary
- Self-Grading Progress Check
- *BigIdeasMath.com*
  Dynamic Student Edition
  Student Resources

Answers

11.

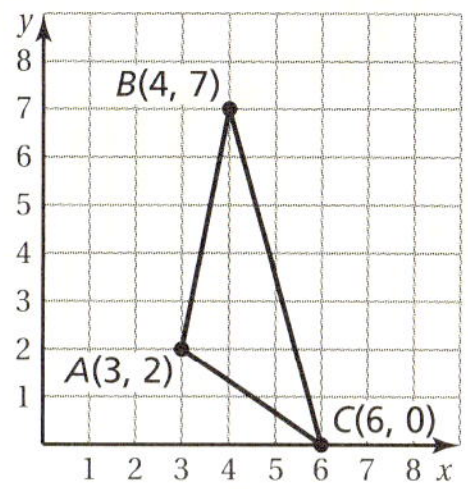

12.

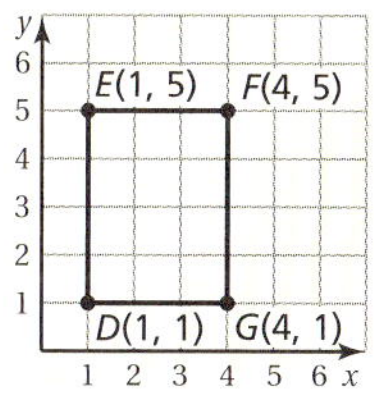

13.

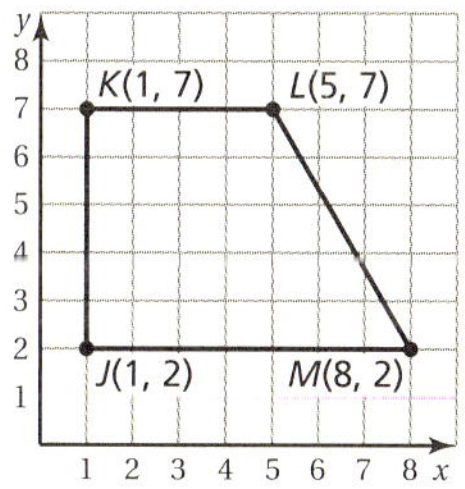

14.

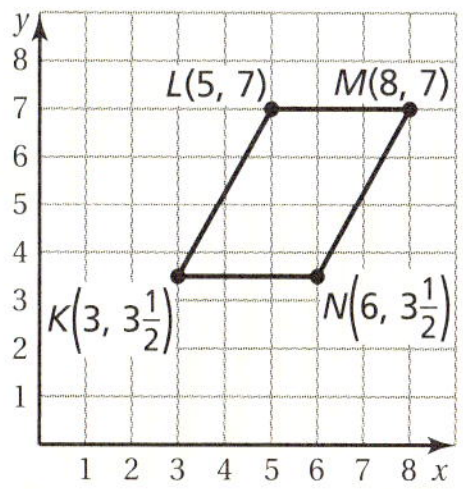

15. 18 units; 20 units^2

16. 10 units; 6 units^2

17. 18 units; 18 units^2

18. 28 units; 33 units^2

My Thoughts on the Chapter

What worked. . .

What did not work. . .

What I would do differently. . .

Teacher Tip

Not allowed to write in your teaching edition? Use sticky notes to record your thoughts.

4.4 Polygons in the Coordinate Plane (pp. 174–179)

a. The vertices of a triangle are $A(1, 3)$, $B(5, 9)$, and $C(8, 2)$. Draw the triangle in a coordinate plane.

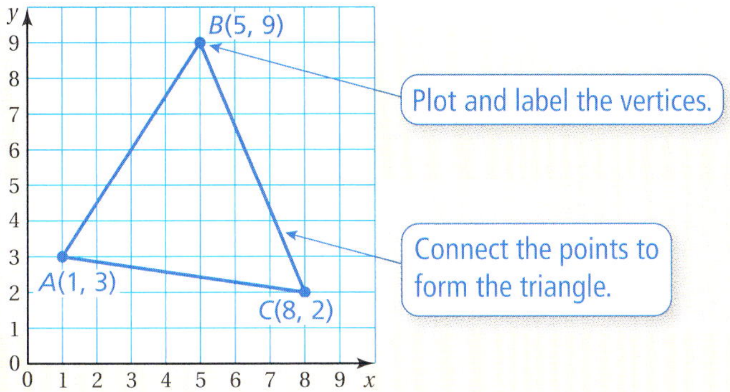

b. The vertices of a rectangle are $F(2, 6)$, $G(8, 6)$, $H(8, 1)$, and $J(2, 1)$. Draw the rectangle in a coordinate plane and find its perimeter.

Draw the rectangle and use the vertices to find its dimensions.

The length is the horizontal distance between $F(2, 6)$ and $G(8, 6)$, which is the difference of the x-coordinates.

length $= 8 - 2 = 6$ units

The width is the vertical distance between $G(8, 6)$ and $H(8, 1)$, which is the difference of the y-coordinates.

width $= 6 - 1 = 5$ units

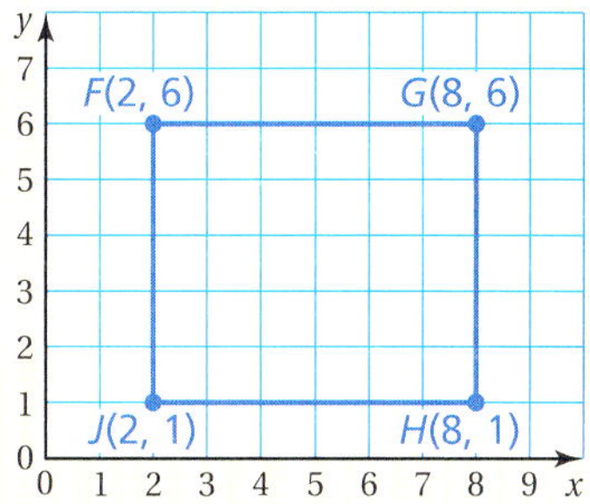

So, the perimeter of the rectangle is $2(6) + 2(5) = 22$ units.

Exercises

Draw the polygon with the given vertices in a coordinate plane.

11. $A(3, 2)$, $B(4, 7)$, $C(6, 0)$

12. $D(1, 1)$, $E(1, 5)$, $F(4, 5)$, $G(4, 1)$

13. $J(1, 2)$, $K(1, 7)$, $L(5, 7)$, $M(8, 2)$

14. $K\left(3, 3\frac{1}{2}\right)$, $L(5, 7)$, $M(8, 7)$, $N\left(6, 3\frac{1}{2}\right)$

Find the perimeter and the area of the polygon with the given vertices.

15. $P(4, 3)$, $Q(4, 7)$, $R(9, 7)$, $S(9, 3)$

16. $T(2, 7)$, $U(2, 9)$, $V(5, 9)$, $W(5, 7)$

17. $W(11, 2)$, $X(11, 8)$, $Y(14, 8)$, $Z(14, 2)$

18. $A(12, 2)$, $B(12, 13)$, $C(15, 13)$, $D(15, 2)$

4 Chapter Test

Check It Out
Test Practice
BigIdeasMath.com

Find the area of the parallelogram, triangle, or trapezoid.

1.

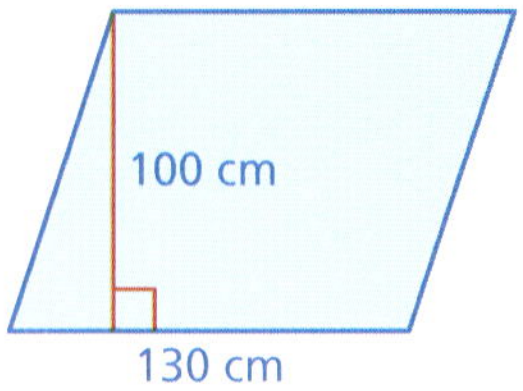

2.

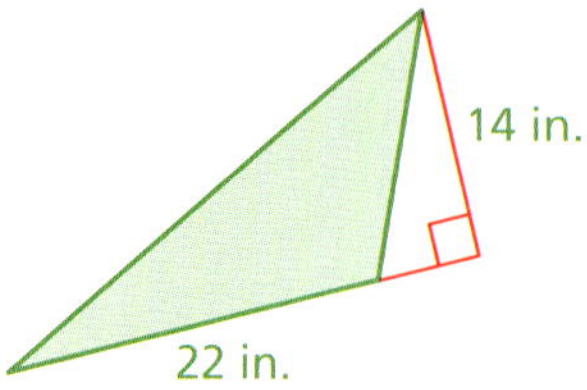

3.

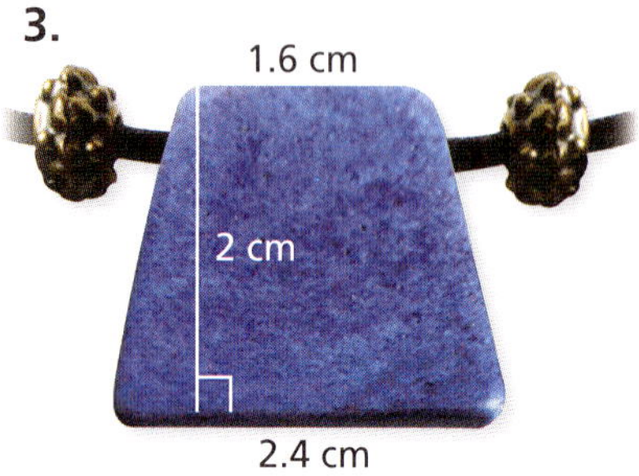

Find the area of the figure.

4.

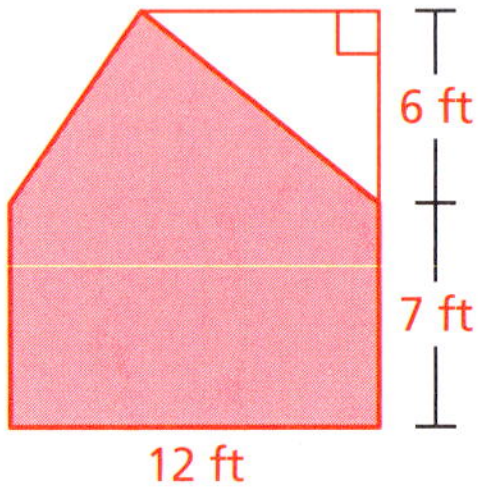

5.

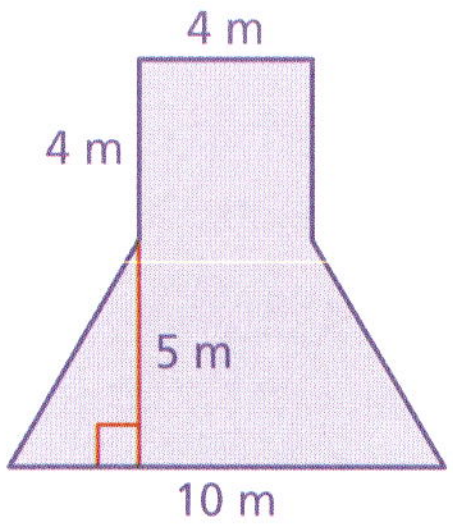

6.

Draw the polygon with the given vertices in a coordinate plane.

7. $A(4, 2)$, $B(5, 6)$, $C(7, 4)$

8. $D(3, 4)$, $E(5, 8)$, $F(8, 8)$, $G(6, 4)$

Find the perimeter and the area of the polygon with the given vertices.

9. $Q(5, 6)$, $R(5, 10)$, $S(9, 10)$, $T(9, 6)$

10. $W(2, 8)$, $X(2, 16)$, $Y(8, 16)$, $Z(8, 8)$

11. TABLETOP The base lengths of a trapezoidal tabletop are 6 feet and 8 feet. The height is 5 feet. What is the area of the tabletop?

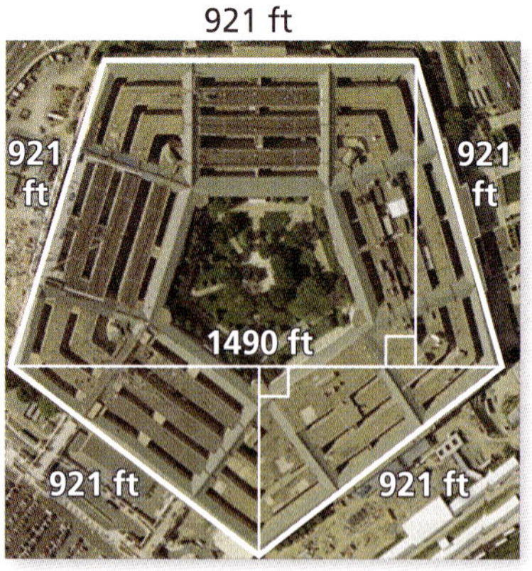

12. PENTAGON The Pentagon in Arlington, Virginia, is the headquarters of the U.S. Department of Defense.

a. Find the perimeter of the Pentagon.

b. A pentagon is made of a triangle and a trapezoid. The height of the triangle shown is about 541 feet, and the height of the trapezoid shown is about 876 feet. Estimate the land area of the Pentagon.

13. CAMPING The vertices of a campsite are (25, 15), (25, 30), (55, 30), and (55, 15). The vertices of your tent are (30, 20), (30, 25), (40, 25), and (40, 20). The coordinates are measured in feet. What is the area of the campsite not covered by your tent?

Test Item References

Chapter Test Questions	Section to Review	Common Core State Standards
1	4.1	6.G.1
2	4.2	6.G.1
3–6, 11, 12	4.3	6.G.1
7–10, 13	4.4	6.G.3

Test-Taking Strategies

Remind students to quickly look over the entire test before they start so that they can budget their time. On this test, it is very important for students to **Stop** and **Think**. When students hurry on a test dealing with perimeter and area formulas, they may use the wrong formula. Encourage students to write the formula first, and then substitute the values for the variables.

Common Errors

- **Exercises 1–3** Students may use the wrong formula. Tell students to write the formula, and then identify the value of each variable before substituting.
- **Exercise 2** Students may multiply the base and the height and forget to multiply by $\frac{1}{2}$. Tell students to write the formula, and then identify the value of each variable before substituting.
- **Exercise 2** Students may mentally find the area by multiplying one-half the base and one-half the height. The answer will be one-half of the value of the correct answer. Tell students to write the formula, and then identify the value of each variable before substituting.
- **Exercise 2** Students may be confused by the drawing and labeling of obtuse triangles. Remind students to identify the base of the triangle. Then the height of the triangle is the perpendicular distance from the base to the opposite vertex. For obtuse triangles, the height may be labeled outside of the triangle.
- **Exercises 9 and 10** Students may confuse perimeter and area. Have them write the formulas and then substitute the values. Students should also include "units" or "square units" as part of the answer.
- **Exercises 3, 11, and 12** Students may substitute the height for one of the bases. Tell students to write the formula for the area of a trapezoid, and then identify the value of each variable before substituting.

Reteaching and Enrichment Strategies

If students need help. . .	If students got it. . .
Resources by Chapter • Practice A and Practice B • Puzzle Time Record and Practice Journal Practice Differentiating the Lesson Lesson Tutorials *BigIdeasMath.com* Skills Review Handbook	Resources by Chapter • Enrichment and Extension • Technology Connection Game Closet at *BigIdeasMath.com* Start Standards Assessment

Answers

1. 13,000 cm^2
2. 154 $in.^2$
3. 4 cm^2
4. 120 ft^2
5. 51 m^2
6. 128 $in.^2$
7. 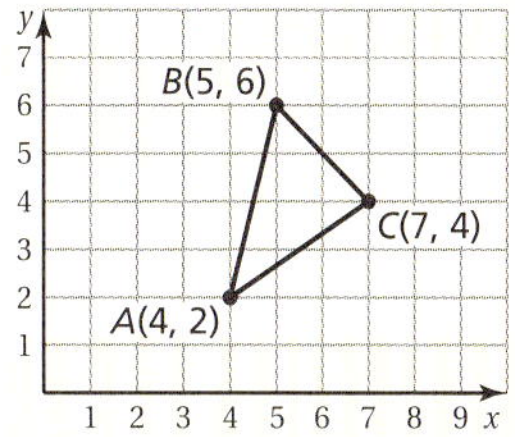

8. 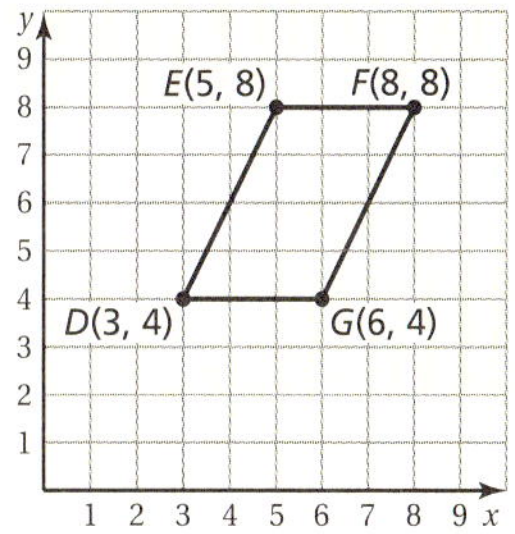

9. 16 units; 16 $units^2$
10. 28 units; 48 $units^2$
11. 35 ft^2
12. a. 4605 ft
    b. 1,459,063 ft^2
13. 400 ft^2

Technology for the Teacher

Online Assessment
Assessment Book
ExamView® Assessment Suite

Test-Taking Strategies
Available at *BigIdeasMath.com*

After Answering Easy Questions, Relax
Answer Easy Questions First
Estimate the Answer
Read All Choices before Answering
Read Question before Answering
Solve Directly or Eliminate Choices
Solve Problem before Looking at Choices
Use Intelligent Guessing
Work Backwards

About this Strategy
When taking a multiple choice test, be sure to read each question carefully and thoroughly. When taking a timed test, it is often best to skim the test and answer the easy questions first. Be careful that you record your answer in the correct position on the answer sheet.

Answers
1. B
2. 5832
3. F
4. C

Technology for the Teacher

Common Core State Standards Support Performance Tasks
Online Assessment
Assessment Book
ExamView® Assessment Suite

Item Analysis

1. **A.** The student finds the area of a trapezoid with bases of 2 and 6 and a height of 8.
   B. Correct answer
   C. The student counts all squares with any shading.
   D. The student uses $A = h(b_1 + b_2)$ to find the area of a trapezoid.
2. **Gridded Response:** Correct answer: 5832
   Common Error: The student adds 18 three times and gets 54.
3. **F.** Correct answer
   G. The student found a factor of 36 but not a factor of 42.
   H. The student found a factor of 36 but not a factor of 42.
   I. The student found a factor of 36 but not a factor of 42.
4. **A.** The student only finds the length.
   B. The student adds the length and width.
   C. Correct answer
   D. The student adds all of the *x*- and *y*-coordinates of the vertices.

4 Standards Assessment

Test-Taking Strategy
Answer Easy Questions First

"Scan the test and answer the easy questions first. You know that the area of a triangle is one-half times the product of its base and its height."

1. What is the area of the shaded figure shown below? *(6.G.1)*

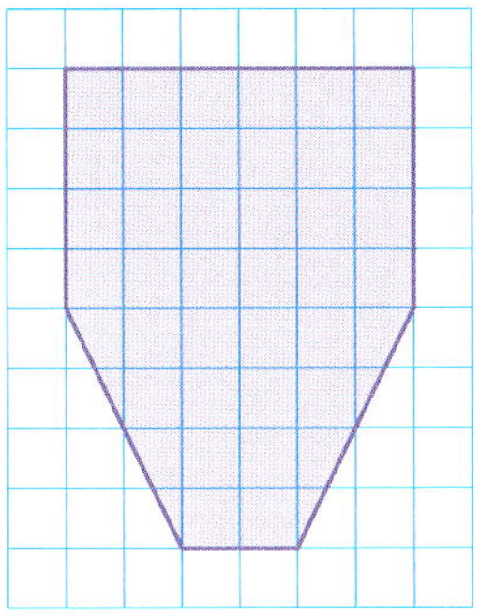

A. 32 units^2

B. 40 units^2

C. 44 units^2

D. 56 units^2

2. What is the value of the expression below? *(6.EE.1)*

$$18^3$$

3. You have 36 red apples and 42 green apples. What is the greatest number of identical fruit baskets you can make with no apples left over? *(6.NS.4)*

F. 6

G. 9

H. 12

I. 18

4. What is the perimeter of the rectangle with the vertices shown below? *(6.G.3)*

$$A(4, 7), B(4, 15), C(9, 15), D(9,7)$$

A. 8 units

B. 13 units

C. 26 units

D. 70 units

5. What property was used to simplify the expression? *(6.EE.3)*

$$5 \times 78 = 5(70 + 8)$$
$$= 5(70) + 5(8)$$
$$= 350 + 40$$
$$= 390$$

F. Associative Property of Multiplication

G. Commutative Property of Addition

H. Distributive Property

I. Multiplication Property of One

6. What is the area, in square yards, of the triangle below? *(6.G.1)*

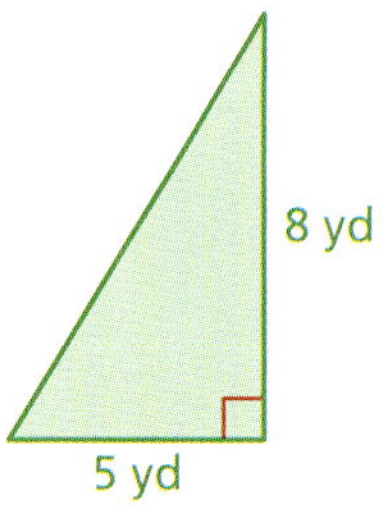

7. Which of the following is equivalent to $\frac{12}{35}$? *(6.NS.1)*

A. $\frac{5}{6} \div \frac{2}{7}$

B. $\frac{2}{7} \div \frac{6}{5}$

C. $\frac{2}{7} \div \frac{5}{6}$

D. $\frac{5}{6} \div \frac{7}{2}$

8. The description below represents the area of which polygon? *(6.G.1)*

"one-half the product of its height and the sum of the lengths of its bases"

F. rectangle

G. square

H. trapezoid

I. triangle

Item Analysis (continued)

5. **F.** The student incorrectly identifies the property being used as the Associative Property of Multiplication.

   G. The student incorrectly identifies the property being used as the Commutative Property of Addition.

   H. Correct answer

   I. The student incorrectly identifies the property being used as the Multiplication Property of One.

6. **Gridded Response:** Correct answer: 20

   Common Error: The student multiplies 5 by 8 to get 40 and forgets to multiply by one-half.

7. **A.** The student incorrectly inverts the dividend and multiplies instead of inverting the divisor and multiplying.

   B. The student incorrectly multiplies the fractions instead of inverting the divisor and multiplying.

   C. Correct answer

   D. The student incorrectly inverts both the dividend and the divisor instead of inverting only the divisor before multiplying.

8. **F.** The student incorrectly identifies the description as the formula for the area of a rectangle.

   G. The student incorrectly identifies the description as the formula for the area of a square.

   H. Correct answer

   I. The student incorrectly identifies the description as the formula for the area of a triangle.

Answers

5. H
6. 20
7. C
8. H

Answers

9. B

10. F

11. *Part A* 435 ft^2

Part B 435 ft^2

Part C The trapezoid can be decomposed into two triangles and a rectangle. The formulas in Part A are simpler. Part B requires only one formula.

Item Analysis (continued)

9. **A.** The student incorrectly performs the order of operations.

B. Correct answer

C. The student incorrectly performs the order of operations.

D. The student incorrectly performs the order of operations.

10. **F.** Correct answer

G. The student writes an expression that represents addition instead of multiplication.

H. The student writes an expression that represents subtraction instead of multiplication.

I. The student writes an expression that represents division instead of multiplication.

11. **4 points** The student demonstrates a thorough understanding of finding areas of polygons. The student calculates the area of the yard correctly and gets an answer of 435 square feet. The student is able to explain why the two methods give the same result and provides an advantage for using each method. The student shows accurate, complete work for all parts and provides clear and complete explanations.

3 points The student demonstrates an understanding of finding areas of polygons, but the student's work and explanations demonstrate an essential but less than thorough understanding.

2 points The student demonstrates a partial understanding of finding areas of polygons. The student's work and explanations demonstrate a lack of essential understanding.

1 point The student demonstrates very limited understanding of finding areas of polygons. The student's response is incomplete and exhibits many flaws.

0 points The student provided no response, a completely incorrect or incomprehensible response, or a response that demonstrates insufficient understanding of finding areas of polygons.

9. Edward was evaluating the expression in the box.

$$
\begin{aligned}
180 \div 9 + 3^4 - 1 &= 180 \div 9 + 81 - 1 \\
&= 180 \div 90 - 1 \\
&= 2 - 1 \\
&= 1
\end{aligned}
$$

What should Edward do to correct the error that he made? *(6.EE.1)*

A. Add 9 and 81 then subtract 1 before dividing.

B. Divide 180 by 9 before adding or subtracting.

C. Divide 180 by 9 then subtract 1 before adding 3^4.

D. Subtract 1 from 90 before dividing.

10. You have 3 times as many guitar picks as your cousin. Let v be the number of guitar picks that your cousin has. Which expression represents the number of guitar picks you have? *(6.EE.2a)*

F. $3v$

G. $v + 3$

H. $3 - v$

I. $\frac{v}{3}$

11. Your family hires a company to install invisible fencing around your yard. *(6.G.1)*

Think Solve Explain

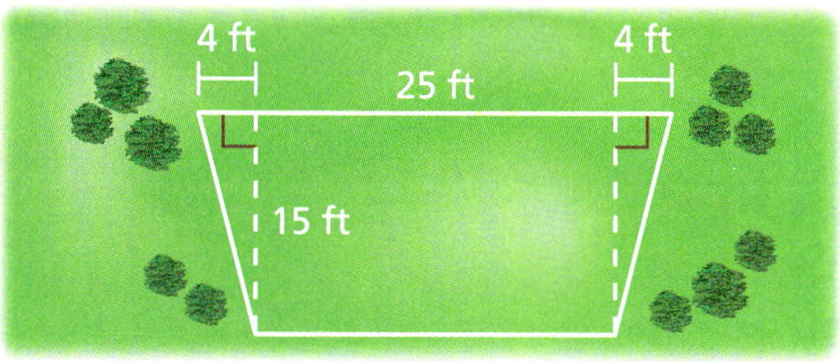

Part A Find the area of the yard using only the area formulas for rectangles and triangles. Show your work.

Part B Find the area of the yard using the area formula for trapezoids.

Part C Explain why the two methods of finding the area of the yard give the same result. Describe the advantages of each method.

5 Ratios and Rates

Common Core Progression

4th Grade

- Equate fractions and decimals with denominators of 10 and 100.
- Generate equivalent fractions.
- Convert within a measurement system from a larger unit to a smaller unit.

5th Grade

- Multiply and divide decimals.
- Convert standard measurement units within a measurement system.

6th Grade

- Understand ratios, rates, and unit rates.
- Compare ratios using tables.
- Find percent as a rate per 100.
- Solve problems involving finding the whole, given a part and the percent.
- Use ratio reasoning to convert measurement units.

Pacing Guide for Chapter 5

Chapter Opener	1 Day
Section 1 Activity Lesson	 1 Day 1 Day
Section 2 Activity Lesson	 1 Day 1 Day
Section 3 Activity Lesson	 1 Day 1 Day
Section 4 Activity Lesson	 1 Day 1 Day
Study Help / Quiz	1 Day
Section 5 Activity Lesson	 1 Day 1 Day
Section 6 Activity Lesson	 1 Day 1 Day
Section 7 Activity Lesson	 1 Day 1 Day
Chapter Review/ Chapter Tests	2 Days
Total Chapter 5	18 Days
Year-to-Date	82 Days

Chapter Summary

Section	Common Core State Standard	
5.1	Learning	6.RP.1
5.2	Learning	6.RP.1 ★, 6.RP.3a
5.3	Learning	6.RP.2, 6.RP.3a, 6.RP.3b
5.4	Learning	6.RP.2 ★, 6.RP.3a
5.5	Learning	6.RP.3c
5.6	Learning	6.RP.3c
5.7	Learning	6.RP.3d ★

★ Teaching is complete. Standard can be assessed.

Technology for the Teacher

BigIdeasMath.com
Chapter at a Glance
Complete Materials List
Parent Letters: English and Spanish

Common Core State Standards

5.OA.3 . . . Identify apparent relationships between corresponding terms . . ..
5.NF.4a Interpret the product $(a/b) \times q$ as a parts of a partition of q into b equal parts; equivalently, as the result of a sequence of operations $a \times q \div b$.
6.NS.1 Interpret and compute quotients of fractions, ...

Additional Topics for Review

- Equivalent Fractions
- Multiply and Divide Decimals
- Convert Measures within Customary or Metric System

Try It Yourself

1. The x-value times 2 equals the y-value; 14
2. The x-value times 4 equals the y-value; 32
3. The x-value times 5 equals the y-value; 20
4. $\frac{13}{100}$
5. $\frac{39}{100}$
6. $\frac{70}{9}$
7. $\frac{17}{4}$

Record and Practice Journal Fair Game Review

1. The In value plus three equals the Out value.
2. The In value times three equals the Out value.
3. The In value minus 10 equals the Out value.
4. The In value divided by two equals the Out value.
5.

In	Out
0.65	0
0.70	0.05
0.75	0.10
1.00	0.35

The In value minus 0.65 equals the Out value.

6–12. See Additional Answers.

Math Background Notes

Vocabulary Review

- Divisor
- Reciprocal

Identifying Patterns

- Students should be familiar with recognizing and describing patterns.
- The skill of identifying patterns using a table is a precursor to working with functions. Try to foster the upcoming skill by using words such as "relationship" and "rule" when describing how the x-value relates to the y-value.
- **Common Error:** Some students may search for a pattern among the x-values or among the y-values. Remind students that they are searching for the relationship between x- and y-values. Encourage them to relate the rows of the table, not the columns.

Multiplying and Dividing by Fractions

- Students should know how to multiply and divide by fractions.
- Remind students that the rules for multiplying and dividing by fractions are different from the rules for adding and subtracting fractions. Multiplying and dividing by fractions does not require a common denominator.
- **Teaching Tip:** Most students will remember the process to divide by fractions. If your students are comfortable with the process, encourage them to describe it using math vocabulary. Instead of "change the sign and flip the fraction," encourage "multiply by the reciprocal of the divisor."

Reteaching and Enrichment Strategies

If students need help. . .	If students got it. . .
Record and Practice Journal • Fair Game Review Skills Review Handbook Lesson Tutorials	Game Closet at *BigIdeasMath.com* Start the next section

What You Learned Before

"I like seventy-five percent of them."

Identifying Patterns (5.OA.3)

Example 1 Using the numbers from the table, find and state the rule in words. Then find the missing value.

x	*y*
1	6
2	12
3	18
4	

Each y-value is 6 times the x-value.

The x-value times 6 equals the y-value.
The missing value is $6(4) = 24$.

Try It Yourself

Using the numbers from the table, find and state the rule in words. Then find the missing value.

1.

x	*y*
1	2
3	6
5	10
7	

2.

x	*y*
2	8
4	16
6	24
8	

3.

x	*y*
1	5
2	10
3	15
4	

Multiplying and Dividing by Fractions (5.NF.4a, 6.NS.1)

Example 2 Find $\frac{5}{6} \cdot \frac{3}{4}$.

$$\frac{5}{6} \cdot \frac{3}{4} = \frac{5 \cdot \overset{1}{\cancel{3}}}{\underset{2}{\cancel{6}} \cdot 4}$$

$$= \frac{5}{8}$$

Example 3 Find $2 \div \frac{9}{10}$.

$$2 \div \frac{9}{10} = 2 \cdot \frac{10}{9}$$

Multiply by the reciprocal of the divisor.

$$= \frac{2 \cdot 10}{9}$$

$$= \frac{20}{9}$$

Try It Yourself

Evaluate the expression. Write the answer in simplest form.

4. $\frac{1}{5} \cdot \frac{13}{20}$

5. $\frac{3}{4} \cdot \frac{13}{25}$

6. $7 \div \frac{9}{10}$

7. $4 \div \frac{16}{17}$

5.1 Ratios

Essential Question How can you represent a relationship between two quantities?

1 ACTIVITY: Comparing Quantities

Work with a partner. Use the collection of objects to complete each statement.

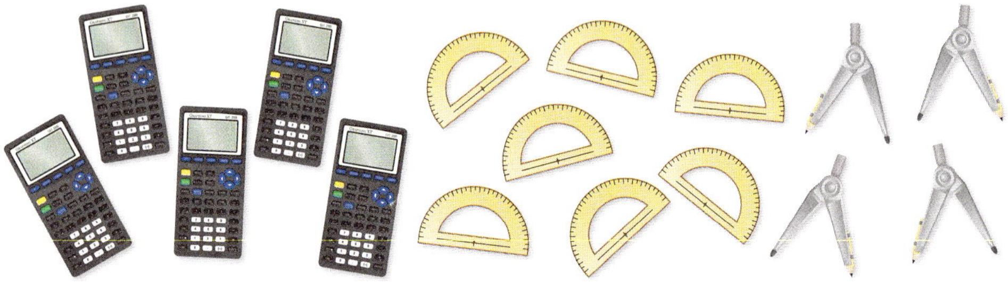

There are ___ graphing calculators to ___ protractors.

There are ___ protractors to ___ graphing calculators.

There are ___ compasses to ___ protractors.

There are ___ graphing calculators to ___ compasses.

There are ___ protractors to ___ total objects.

The number of graphing calculators is $\frac{\square}{\square}$ of the total number of objects.

2 ACTIVITY: Playing Garbage Basketball

Work with a partner.

- **Take turns shooting a ball or other object into a wastebasket from a reasonable distance.**
- **Organize the numbers of shots you made and shots you missed in a chart.**

a. Write a statement similar to those in Activity 1 that describes the relationship between the number of shots you made and the number of shots you missed.

b. Write a statement similar to those in Activity 1 that describes the relationship between the number of shots you made and the total number of shots.

c. What fraction of your shots did you make? What fraction did you miss?

Common Core

Ratios

In this lesson, you will
- understand the concept of a ratio.
- use ratios to describe the relationship between two quantities.

Learning Standard
6.RP.1

Laurie's Notes

Introduction

Standards for Mathematical Practice

- **MP3 Construct Viable Arguments and Critique the Reasoning of Others:** In these activities, students will have the chance to justify their conclusions and communicate them to others.

Motivate

- Use a portion of a page from a telephone book to find the numbers of telephone numbers that end in 0, 1, 2, . . . , 9.
- ? "Which number appears most often? Least often?" Answers will vary.
- Use the responses to write a comparison similar to Activity 1. Say, "Comparing the least to the greatest would be _____ to _____."
- Ask additional comparison questions such as the number of evens compared to the number of odds and the number of primes compared to the number of composites.

Activity Notes

Activity 1

- When students have finished, discuss the difference between comparing two objects to one another and comparing one object to the whole collection. Ratios often compare *a part* to *a part*. Fractions compare *a part* to *a whole*. *Ratio* is not defined until tomorrow's lesson. Today, students become comfortable with representing a relationship between two quantities.
- ? "Is the number of calculators to protractors the same as the number of protractors to calculators? Explain." no; 7 to 5 is different than 5 to 7.
- ? "How many total objects are there?" 16 "What fraction of the objects are protractors?" $\frac{7}{16}$ "What fraction are compasses?" $\frac{4}{16}$, or $\frac{1}{4}$

Activity 2

- This activity may be difficult to manage in a large class of students. It can be simulated in a hallway with students using paper clips. It could also be done with one student volunteering as the "shooter." It is more enjoyable for students when they can collect information themselves.
- Students may organize their results differently. Most will probably organize the makes and misses but some may keep track of the makes, misses, and total shots attempted.
- This is a good opportunity to review making tally marks.
- For part (a), listen for a statement such as, "There were 14 shots made to 6 shots missed. "For part (b), listen for a statement such as, "There were 14 shots made to 20 shots attempted."
- Discuss the difference between parts (a) and (b): part to part and part to whole.
- ? "In part (c), what is the relationship between the two fractions?" They sum to 1.

Common Core State Standards

6.RP.1 Understand the concept of a ratio and use ratio language to describe a ratio relationship between two quantities.

Explaining the Standards

Per the Common Core standards and the progression documents, and to avoid confusion, fraction notation is not used in this or the next section to represent ratios. This may not be a traditional approach for many teachers, but applying fractions here can be a distraction. Students will use fractions in Section 5.7 for unit analysis, in Grade 7 to write proportions, and in high school.

Previous Learning

Students should know how to recognize and generate equivalent fractions.

Technology for the Teacher

Dynamic Classroom

Lesson Plans
Complete Materials List

5.1 Record and Practice Journal

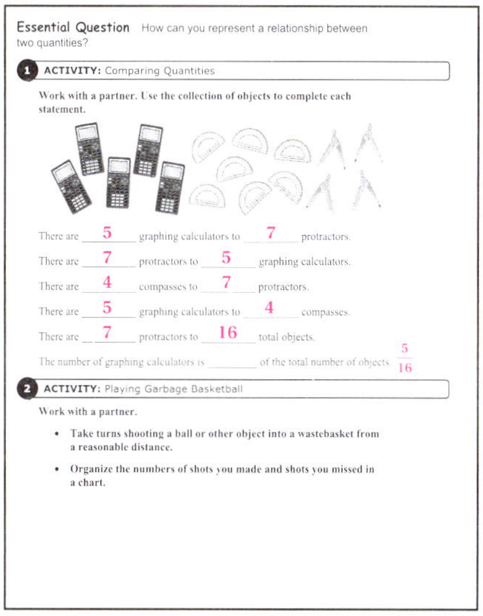

Essential Question How can you represent a relationship between two quantities?

1 ACTIVITY: Comparing Quantities

Work with a partner. Use the collection of objects to complete each statement.

There are 5 graphing calculators to 7 protractors.
There are 7 protractors to 5 graphing calculators.
There are 4 compasses to 7 protractors.
There are 5 graphing calculators to 4 compasses.
There are 7 protractors to 16 total objects.
The number of graphing calculators is $\frac{5}{16}$ of the total number of objects.

2 ACTIVITY: Playing Garbage Basketball

Work with a partner.

- Take turns shooting a ball or other object into a wastebasket from a reasonable distance.
- Organize the numbers of shots you made and shots you missed in a chart.

Differentiated Instruction

Visual

Ratios can be part-to-part, part-to-whole or whole-to-part comparisons. Students may benefit from seeing a visual display of the different ways that ratios can be expressed.

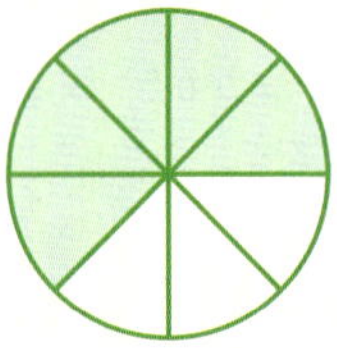

Part-to-part
shaded parts to unshaded parts = 5 : 3

Part-to-whole
shaded parts to total parts = 5 : 8

Whole-to-part
total parts to unshaded parts = 8 : 3

5.1 Record and Practice Journal

a. Write a statement similar to those in Activity 1 that describes the relationship between the number of shots you made and the number of shots you missed.

(a)–(c) Check students' work.

b. Write a statement similar to those in Activity 1 that describes the relationship between the number of shots you made and the total number of shots.

c. What fraction of your shots did you make? What fraction did you miss?

3 ACTIVITY: Reading a Diagram

Work with a partner. You mix different amounts of paint to create new colors. Write a statement that describes the relationship between the amounts of paint shown in each diagram.

a. Blue / Green

There are 4 parts blue for every 3 parts green.

b. Orange / Yellow

There are 2 parts orange for every 3 parts yellow.

c. Red / Blue

There are 4 parts red for every 2 parts blue.

d. White / Purple

There is 1 part white for every 5 parts purple.

4 ACTIVITY: Describing Relationships

Work with a partner. Use a table or a diagram to represent the relationship between the two quantities.

a. For every 3 boys standing in a line, there are 4 girls.

Sample answer: Boys (3 parts), Girls (4 parts)

b. For each vote Brian received, Sasha received 6 votes.

Sample answer: Brian (1 part), Sasha (6 parts)

c. A class counts the number of vehicles that pass by its school from 1:00 to 2:00 P.M. There are 3 times as many cars as trucks.

Sample answer: Cars (3 parts), Trucks (1 part)

d. A hand sanitizer contains 5 parts aloe for every 2 parts distilled water.

Sample answer: Aloe (5 parts), Distilled water (2 parts)

What Is Your Answer?

5. **IN YOUR OWN WORDS** How can you represent a relationship between two quantities? Give examples to support your explanation.

use statements such as "for every," "to," "for each" or concrete models or tape diagrams

6. **MODELING** You make 48 pints of pink paint by using 5 pints of red paint for every 3 pints of white paint. Use a diagram to find the number of pints of red paint and white paint in your mixture. Explain.

30 pints of red paint
18 pints of white paint

Laurie's Notes

Activity 3

- Students have likely heard the phrase *2 parts ____ to every 3 parts ____*. This activity will make sense to students.
- Discuss the model used. It is a tape diagram, similar to the bar models and fraction strips that students used for modeling fractions in Chapter 2.
- Check the statements written for parts (c) and (d).
- **Extension:** What color do you produce from the mixtures in each part? (a) teal or turquoise; (b) yellow-orange; (c) reddish purple; (d) light purple

Activity 4

- In representing these relationships, students may be creative with pictures or diagrams. Be sure to give students a chance to share their work.
- If students make a table, then they will likely use tally marks or simply write the numbers.
- **MP3:** Note that some students may represent an equivalent ratio without using that language. For instance, in part (b), they may have two votes represented for Brian and 12 votes represented for Sasha. If this occurs, then ask the student who does this to explain why it is correct. In part (c), different answers are likely, given the way the relationship is stated.

What Is Your Answer?

- Students should consider their work in Activity 3 to answer Question 5.

Closure

- **Exit Ticket:** Conduct a survey in class and record how many students have brown, green, blue, and hazel eyes. What is the ratio of students with blue eyes to the total number of students? What is the ratio of students with brown eyes to the total number of students? What conclusions can you draw from these ratios? Answers will vary. Typically, more students will have brown eyes than blue eyes. Because the total number of students is the same for each ratio, students should be able to determine that the ratio of brown eyes to total students is greater than the ratio of blue eyes to total students.

3 ACTIVITY: Reading a Diagram

Work with a partner. You mix different amounts of paint to create new colors. Write a statement that describes the relationship between the amounts of paint shown in each diagram.

a. Blue

Green

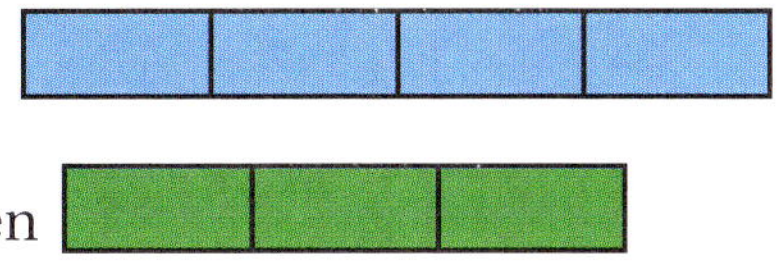

There are ___ parts blue for every ___ parts green.

b. Orange

Yellow

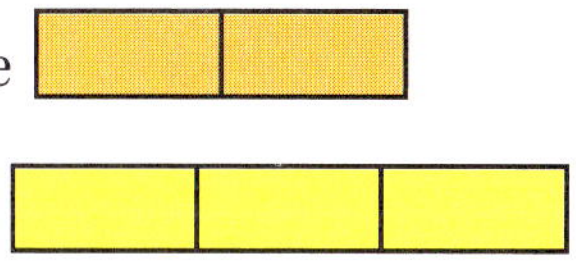

There are ___ for every ___.

c. Red

Blue

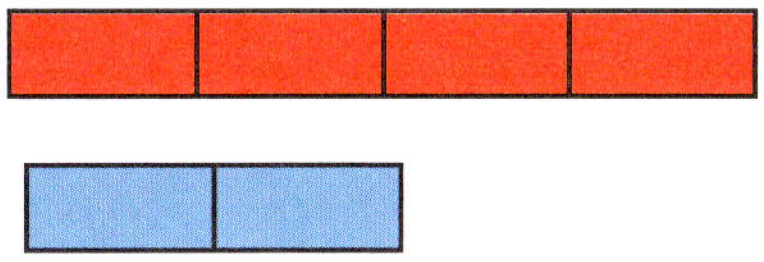

___.

d. White

Purple

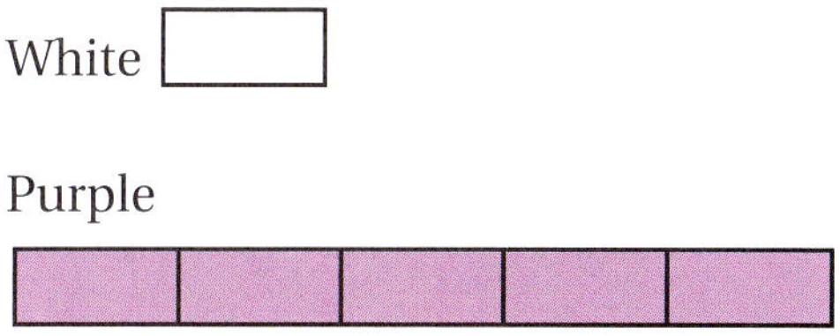

___.

4 ACTIVITY: Describing Relationships

Math Practice 4

Use a Table or Diagram

What are the quantities in this problem? How does a table or diagram represent the relationship between the quantities?

Work with a partner. Use a table or a diagram to represent the relationship between the two quantities.

a. For every 3 boys standing in a line, there are 4 girls.

b. For each vote Brian received, Sasha received 6 votes.

c. A class counts the number of vehicles that pass by its school from 1:00 to 2:00 P.M. There are 3 times as many cars as trucks.

d. A hand sanitizer contains 5 parts aloe for every 2 parts distilled water.

What Is Your Answer?

5. **IN YOUR OWN WORDS** How can you represent a relationship between two quantities? Give examples to support your explanation.

6. **MODELING** You make 48 pints of pink paint by using 5 pints of red paint for every 3 pints of white paint. Use a diagram to find the number of pints of red paint and white paint in your mixture. Explain.

Practice Use what you learned about comparing two quantities to complete Exercises 4 and 5 on page 194.

5.1 Lesson

Check It Out
Lesson Tutorials
BigIdeasMath.com

Key Vocabulary
ratio, *p. 192*

Key Idea

Ratio

Words A **ratio** is a comparison of two quantities. Ratios can be part-to-part, part-to-whole, or whole-to-part comparisons.

Examples
2 red crayons *to* 6 blue crayons
1 red crayon *for every* 3 blue crayons
3 blue crayons *per* 1 red crayon
3 blue crayons *for each* red crayon
3 blue crayons *out of every* 4 crayons
2 red crayons *out of* 8 crayons

Algebra The ratio of a to b can be written as $a : b$.

EXAMPLE 1 Writing Ratios

You have the coins shown.

a. Write the ratio of pennies to quarters.

6 pennies → 6 to 7 ← 7 quarters

So, the ratio of pennies to quarters is 6 to 7, or 6 : 7.

b. Write the ratio of quarters to dimes.

7 quarters → 7 to 3 ← 3 dimes

So, the ratio of quarters to dimes is 7 to 3, or 7 : 3

c. Write the ratio of dimes to the total number of coins.

3 dimes → 3 to 16 ← 16 coins

So, the ratio of dimes to the total number of coins is 3 to 16, or 3 : 16.

Remember

Part-to-whole relationships compare a part of a whole to the whole. Fractions represent part-to-whole relationships. Part-to-part relationships compare a part of a whole to another part of the whole.

On Your Own

Now You're Ready
Exercises 6–13

1. In Example 1, write the ratio of dimes to pennies.
2. The circle graph shows the favorite ice-cream toppings of several students. Use ratio language to compare the number of students who favor peanuts to the total number of students.

Favorite Toppings

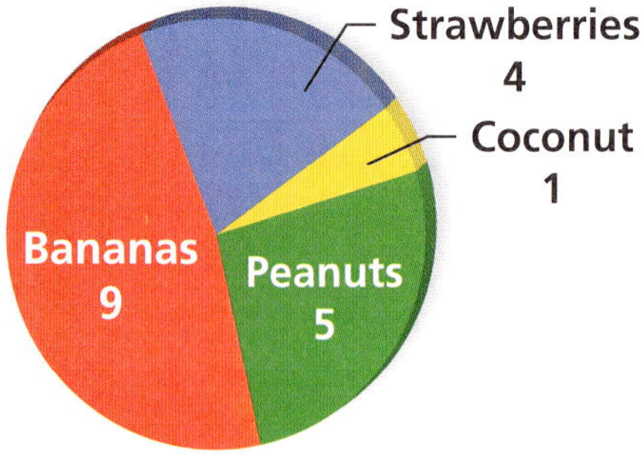

Laurie's Notes

Introduction

Connect

- **Yesterday:** Students explored how to express the relationship between two quantities (MP3).
- **Today:** Students will write ratios.

Motivate

- Have 2 boys and 4 girls stand at the front of the room. Then ask students to complete each statement.
  (a) There are _____ boy(s) to every _____ girl(s). 2, 4; or 1, 2
  (b) There are _____ girl(s) to every _____ boy(s). 4, 2; or 2, 1; This helps to point out that order matters in a ratio.
  (c) There are _____ girl(s) to _____ people standing. 4, 6; or 2, 3
  (d) _____ boys need to stand so there are 3 boys to every 2 girls. 6 total

Lesson Notes

Key Idea

- Write the Key Idea. Connect to yesterday's investigation as you explicitly state that ratios can be part-to-part, part-to-whole, or whole-to-part comparisons.
- Refer to the red and blue crayons as you discuss each example. It would be helpful to have physical models, such as 6 blue crayons and 2 red crayons, 6 pencils and 2 pens, or 6 yellow tiles and 2 green tiles. Manipulate, or ask a student to manipulate the objects as you state each ratio in the examples. Make sure students understand the subtle differences in the wording.
- When discussing the second and third examples, be sure to note that the ratio $a : b$ is not the same as the ratio $b : a$. In comparing two unequal quantities, there are two ways of writing the ratio.
- Note that although fractional representation is not introduced at this time, students may see the connection.

Example 1

- Work through each part of the example.
- Discuss the *Remember* note in the margin. Ask students to identify each of the examples as being part-to-part, part-to-whole, or whole-to-part comparisons.
- **Extension:** "What is the ratio of the value of the dimes to the value of the quarters?" Three dimes equals $0.30. Six quarters equals $1.50. So, the ratio is $0.30 to $1.50.

On Your Own

- **Neighbor Check:** Have students work independently and then have their neighbors check their work. Have students discuss any discrepancies.

Goal Today's lesson is writing **ratios**.

Technology for the Teacher

Dynamic Classroom

Lesson Tutorials
Lesson Plans
Answer Presentation Tool

Extra Example 1

You have the marbles shown.

a. Write the ratio of green marbles to red marbles. 4 to 7, or 4 : 7
b. Write the ratio of blue marbles to green marbles. 9 to 4, or 9 : 4
c. Write the ratio of red marbles to the total number of marbles. 7 to 20, or 7 : 20

On Your Own

1. 3 to 6, or 3 : 6
2. 5 out of 19 students favor peanuts

Laurie's Notes

Discuss

- Define *tape diagram*. This model was used in yesterday's investigation.
- **MP1a Make Sense of Problems** and **MP5 Use Appropriate Tools Strategically:** You may find it helpful to use colored square tiles, placed end-to-end, to model tape diagrams.

Example 2

- Ask a volunteer to read the example.
- ? "Do you know either allowance?" no
- ? "What does \$40 represent?" the total of both allowances
- Draw and label the tape diagram as shown.
- ? "How many total parts are there?" 8
- ? "What does each part represent? Explain." \$5; Eight parts represent \$40, and \$40 ÷ 8 = \$5.
- Finish the problem as shown. As a check, note that the allowances sum to \$40. It may help to fill in each part of the tape diagram with the value it represents (\$5).

Example 3

- Ask a volunteer to read the problem. Discuss and make a list of the known information.
- Draw and label the tape diagram as shown.
- ? "What do the parts of the tape diagram represent?" garlic bulbs
- **Common Error:** Each part does *not* represent 1 garlic bulb. The 3 blue parts represent the total number of bulbs you will plant and the 4 orange parts represent the total number of bulbs you will use for cooking.
- ? "How many total parts are there?" 7
- ? "What does each part represent? Explain." 6 bulbs; Seven parts represent 42 bulbs, and 42 ÷ 7 = 6.
- Finish the problem as shown.

On Your Own

- Ask a volunteer to explain his or her work for Question 4.

Closure

- **Exit Ticket:** The following tape diagram represents the ratio of shirts to sweaters worn by students in a class.

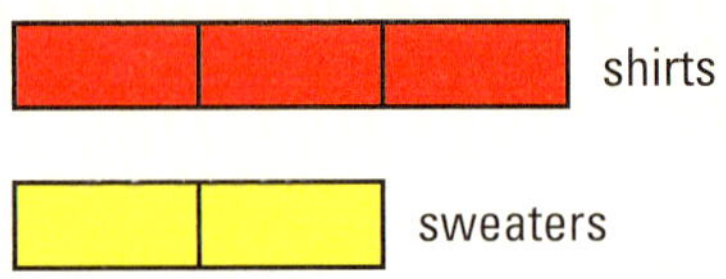

a. Write the ratio of shirts to sweaters. 3 to 2, or 3 : 2

b. There are 10 sweaters. What does each part represent? Explain.
5 articles of clothing; Two parts represent 10 sweaters, and 10 ÷ 2 = 5.

c. How many shirts are there? 15

Extra Example 2

The ratio of time a student spends on science homework to time the student spends on math homework is 2 : 5. The total amount of time the student spends on homework for these classes is 63 minutes. How much time does the student spend on homework for each subject? Science: 18 minutes; math: 45 minutes

Extra Example 3

You separate 72 walnuts into two groups: one for snacks and one for baking. You will keep 3 walnuts as snacks for every 5 walnuts that you will use for baking. Each walnut has 2 pieces. How many walnut pieces will you keep for snacks? 54 pieces

On Your Own

3. Yours: \$16; Your friend's: \$24

4. Fewer; Using this ratio, you will plant 14 bulbs or 112 cloves, which is 32 fewer than originally planned.

English Language Learners

Vocabulary
Have students add a glossary to their math notebooks. Key vocabulary words should be added as they are introduced. Illustrations next to vocabulary words will help in understanding and reinforcing the concept.

A *tape diagram* is a diagram that looks like a segment of tape. It shows the relationship between two quantities.

EXAMPLE 2 Using a Tape Diagram

The ratio of your monthly allowance to your friend's monthly allowance is 5 : 3. The monthly allowances total \$40. How much is each allowance?

To help visualize the problem, express the ratio 5 : 3 using a tape diagram.

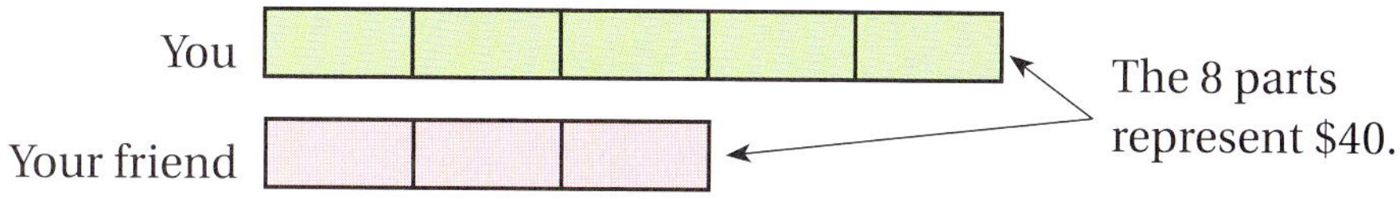

Because there are 8 parts, you know that 1 part represents $\$40 \div 8 = \$5$.

5 parts represent $\$5 \cdot 5 = \$25$.

3 parts represent $\$5 \cdot 3 = \$15$.

So, your monthly allowance is \$25, and your friend's monthly allowance is \$15.

EXAMPLE 3 Using a Tape Diagram

You separate 42 bulbs of garlic into two groups: one for planting and one for cooking. You will plant 3 bulbs for every 4 bulbs that you will use for cooking. Each bulb has about 8 cloves. About how many cloves will you plant?

To help visualize the problem, express the ratio *3 for every 4* using a tape diagram.

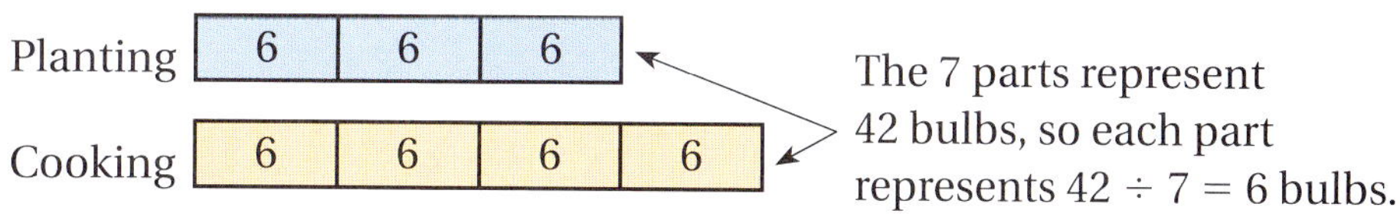

There are $3 \cdot 6 = 18$ bulbs for planting and $4 \cdot 6 = 24$ bulbs for cooking. The group of 18 bulbs has about $18 \cdot 8 = 144$ cloves.

So, you will plant about 144 cloves.

On Your Own

Now You're Ready
Exercises 15 and 16

3. **WHAT IF?** In Example 2, the ratio is 2 to 3. How much is each allowance?

4. **WHAT IF?** In Example 3, you will plant 1 bulb for every 2 bulbs that you will use for cooking. Will you plant more or fewer cloves than originally planned? Explain your reasoning.

5.1 Exercises

Check It Out
Help with Homework
BigIdeasMath.com

Vocabulary and Concept Check

1. **VOCABULARY** The ratio of vowels to consonants in a word is 5 to 7. Are there more vowels or consonants in the word? Explain.

2. **NUMBER SENSE** You are comparing apples to oranges in a fruit bowl. Is the ratio 2 : 3 the same as the ratio 3 : 2? Explain.

3. **WHICH ONE DOESN'T BELONG?** Which ratio does *not* belong with the other three? Explain your reasoning.

2 parts to 5 parts	2 out of every 5	2 for each 5	2 for every 5

Practice and Problem Solving

Use a table or a diagram to represent the relationship between the two quantities.

4. For each lion, there are 7 giraffes.

5. For every 5 seats, there are 4 fans.

Write the ratio. Explain what the ratio means.

1 6. frogs to turtles

7. basketballs to soccer balls

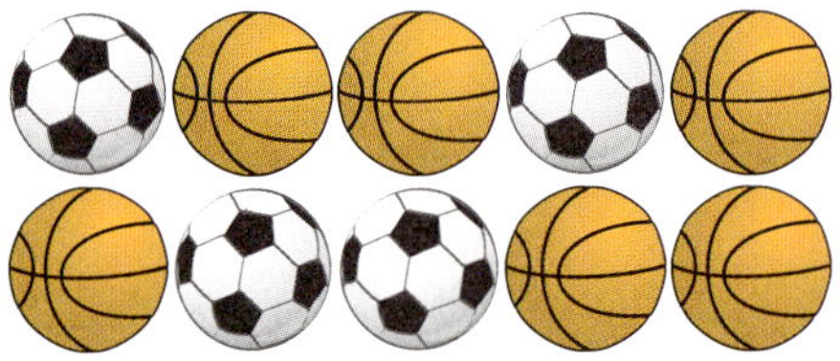

8. calculators : pencils

9. shirts : pants

Use the table to write the ratio. Explain what the ratio means.

10. dramas to movies

11. comedies to movies

12. movies : action

13. movies : dramas

Movie	Number
Drama	3
Comedy	8
Action	4

Topic	Stamps
Birds	7
Celebrity	14
Horses	5
Ships	9

14. **STAMP COLLECTING** The table shows the numbers of stamps in a new stamp collection. Use ratio language to compare the number of celebrity stamps to the total number of stamps.

Assignment Guide and Homework Check

Level	Day 1 Activity Assignment	Day 2 Lesson Assignment	Homework Check
Basic	4, 5, 25–29	1–3, 7–23 odd	7, 11, 13, 14, 15
Average	4, 5, 25–29	1–3, 8–21	8, 13, 14, 16, 18
Advanced	4, 5, 25–29	1–3, 8–24	8, 12, 16, 18, 20

Common Errors

- **Exercise 14** Students may come up with another correct answer, 2 celebrity stamps for every 5 total stamps. Although students are not expected to simplify ratios yet, encourage those who are already simplifying.
- **Exercises 22 and 23** Students may have trouble with compound ratios. Let them know that a ratio with 3 parts compares 3 quantities instead of 2 quantities.
- **Exercise 23** Students could convert gallons to pints before solving the problem to avoid having answers that are incorrect by a factor of 4. Remind them that 1 gallon = 8 pints.

5.1 Record and Practice Journal

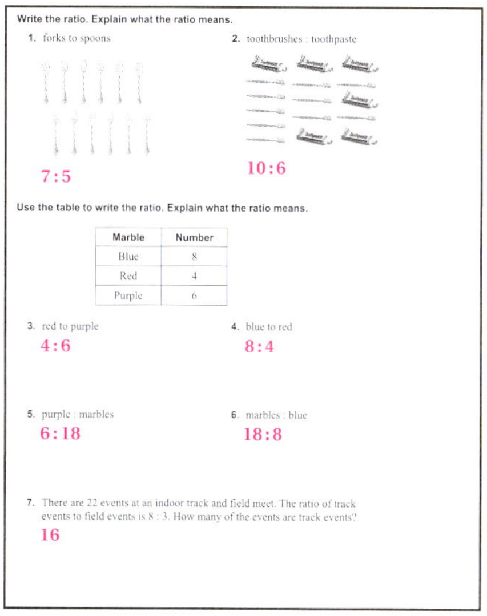

Write the ratio. Explain what the ratio means.

1. forks to spoons

7:5

2. toothbrushes : toothpaste

10:6

Use the table to write the ratio. Explain what the ratio means.

Marble	Number
Blue	8
Red	4
Purple	6

3. red to purple

4:6

4. blue to red

8:4

5. purple : marbles

6:18

6. marbles : blue

18:8

7. There are 22 events at an indoor track and field meet. The ratio of track events to field events is 8 : 3. How many of the events are track events?

16

Vocabulary and Concept Check

1. consonants; For every 5 vowels there are 7 consonants, so consonants outnumber the vowels.
2. no; The order of the quantities is important. Two apples for every 3 oranges is *not* the same as 3 apples for every 2 oranges.
3. 2 out of every 5; This ratio is 2 : 3, all other ratios are 2 : 5.

Practice and Problem Solving

4–5. See Additional Answers.

6. 2 to 5, or 2 : 5; For every 2 frogs, there are 5 turtles.
7. 6 to 4, or 6 : 4; For every 6 basketballs, there are 4 soccer balls.
8. 2 to 6, or 2 : 6; For every 2 calculators, there are 6 pencils.
9. 3 to 7, or 3 : 7; For every 3 shirts, there are 7 pants.
10. 3 to 15, or 3 : 15; 3 out of 15 movies are dramas.
11. 8 to 15, or 8 : 15; 8 out of 15 movies are comedies.
12. 15 to 4, or 15 : 4; Out of 15 movies, 4 are action.
13. 15 to 3, or 15 : 3; Out of 15 movies, 3 are dramas.
14. 14 out of 35 stamps have celebrities on them.

Practice and Problem Solving

15. 9 h **16.** 4 h

17. 12 : 16 **18.** 21 states

19. 6 black pieces; The ratio of black to red is 3 : 5, so each part is 16 ÷ 8 = 2. So, there are 3 • 2 = 6 black pieces and 5 • 2 = 10 red pieces.

20. 8; The ratio of boys to girls is 5 : 7, so each part is 48 ÷ 12 = 4. So, there are 5 • 4 = 20 boys and 7 • 4 = 28 girls.

21. See *Taking Math Deeper*.

22. 67.5 in.; *Sample answer:* Using a tape diagram, 2 parts represents 15 inches, so each part is 7.5 inches. There are 9 total parts, which represents 67.5 inches.

23–24. See Additional Answers.

Fair Game Review

25. 4.6 **26.** 3.29

27. 2.53 **28.** 1.478

29. B

Mini-Assessment

Use the table to write the ratio. Explain what the ratio means.

Pet Type	Number Sold
Dog	12
Cat	9
Fish	17
Bird	5

1. cat to dog 9 to 12, or 9 : 12; 9 cats are sold for every 12 dogs sold.
2. fish to dog 17 to 12, or 17 : 12; 17 fish are sold for every 12 dogs sold.
3. bird to fish 5 to 17, or 5 : 17; 5 birds are sold for every 17 fish sold.
4. cat to all animals sold 9 to 43, or 9 : 43; 9 cats are sold for every 43 animals sold.

Taking Math Deeper

Exercise 21

This problem introduces students to a fundamental concept regarding perimeter and area. As the dimensions of a figure are increased by a factor of k, the perimeter increases by a factor of k, but the area increases by a factor of k^2.

1 Make a table.

Figure	2 ft × 3 ft rectangle	4 ft × 6 ft rectangle
Length	3 ft	6 ft
Width	2 ft	4 ft
Perimeter	10 ft	20 ft
Area	6 sq ft	24 sq ft

2 Find ratios.

		a.		**b.**
Blue Length : Green Length	=	3 ft : 6 ft	=	1 : 2
Blue Width : Green Width	=	2 ft : 4 ft	=	1 : 2
Blue Perimeter : Green Perimeter	=	10 ft : 20 ft	=	1 : 2
Blue Area : Green Area	=	6 sq ft : 24 sq ft	=	1 : 4

1 to 4

3 Cut it out to help me see it.

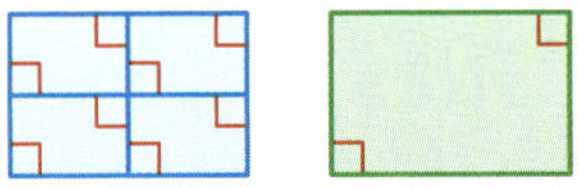

4 of these = 1 of these

Reteaching and Enrichment Strategies

If students need help. . .	If students got it. . .
Resources by Chapter • Practice A and Practice B • Puzzle Time Record and Practice Journal Practice Differentiating the Lesson Lesson Tutorials Skills Review Handbook	Resources by Chapter • Enrichment and Extension • Technology Connection Start the next section

You and a friend tutor for a total of 12 hours. Use the tape diagram to find how many hours you tutor.

2 **15.** 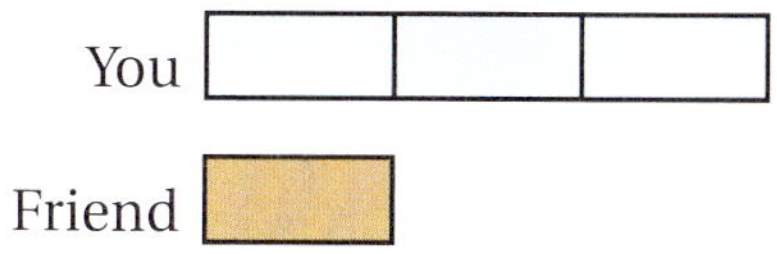

16. 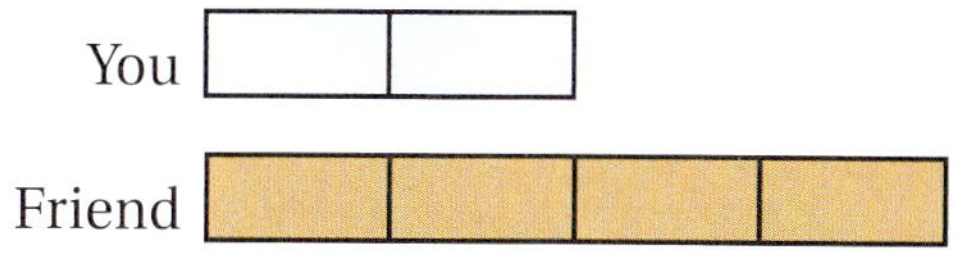

17. **REASONING** Twelve of the 28 students in a class have a dog. What is the ratio of students who have a dog to students who do not?

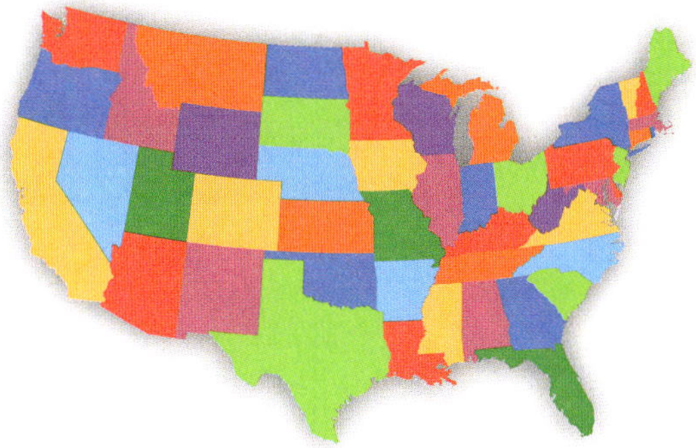

18. **GEOGRAPHY** In the continental United States, the ratio of states that border an ocean to states that do not border an ocean is 7 : 9. How many of the states border an ocean?

19. **CHECKERS** During a checkers game, there are 16 pieces left. The ratio of black to red is 3 : 5. How many black pieces are on the board? Explain how you found your answer.

20. **SCHOOL PLAY** There are 48 students in a school play. The ratio of boys to girls is 5 : 7. How many more girls than boys are in the play? Explain how you found your answer.

21. **GEOMETRY** Use the blue and green rectangles.

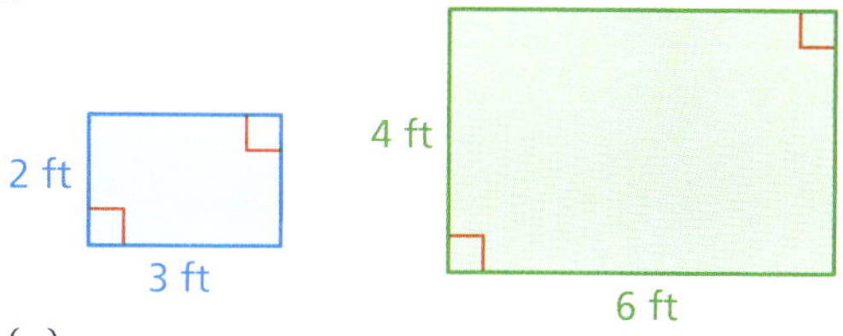

a. Find the ratio of the length of the blue rectangle to the length of the green rectangle. Repeat this for width, perimeter, and area.

b. Compare and contrast your ratios in part (a).

22. **PERIMETER** The ratio of the side lengths of a triangle is 2 : 3 : 4. The shortest side is 15 inches. What is the perimeter? Explain.

23. **PRECISION** You mix soda water, fruit punch concentrate, and ginger ale in the ratio of 1 : 2 : 5 to make fruit punch. How many pints of each ingredient should you use to make 4 gallons of fruit punch? Is your answer reasonable? Explain.

24. **Reasoning** There are 12 boys and 10 girls in your gym class. If 6 boys joined the class, how many girls would need to join for the ratio of boys to girls to remain the same? Justify your answer.

Fair Game Review What you learned in previous grades & lessons

Divide. *(Section 2.6)*

25. $13.8 \div 3$ **26.** $16.45 \div 5$ **27.** $53.13 \div 21$ **28.** $19.214 \div 13$

29. **MULTIPLE CHOICE** What is the value of the expression $x \div y$ when $x = 30$ and $y = 18$? *(Section 3.1)*

(A) $\frac{3}{5}$ (B) $1\frac{2}{3}$ (C) 12 (D) 48

5.2 Ratio Tables

Essential Question How can you find two ratios that describe the same relationship?

1 ACTIVITY: Making a Mixture

Work with a partner. A mixture calls for 1 cup of lemonade and 3 cups of iced tea.

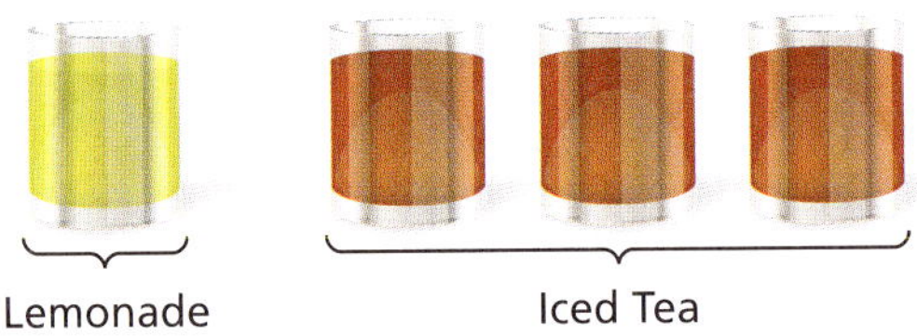

a. How many total cups does the mixture contain? ___ cups

For every ___ cup of lemonade, there are ___ cups of iced tea.

b. How do you make a larger batch of this mixture? Describe your procedure and use the table below to organize your results. Add more columns to the table if needed.

Cups of Lemonade						
Cups of Iced Tea						
Total Cups						

c. Which operations did you use to complete your table? Do you think there is more than one way to complete the table? Explain.

d. How many total cups are in your final mixture? How many of those cups are lemonade? How many are iced tea? Compare your results with those of other groups in your class.

e. Suppose you take a sip from every group's final mixture. Do you think all the mixtures should taste the same? Do you think the color of all the mixtures should be the same? Explain your reasoning.

f. Why do you think it is useful to use a table when organizing your results in this activity? Explain.

Common Core

Ratios

In this lesson, you will
- use ratio tables to find equivalent ratios.
- solve real-life problems.

Learning Standards
6.RP.1
6.RP.3a

Laurie's Notes

Introduction

Standards for Mathematical Practice

- **MP2 Reason Abstractly and Quantitatively:** In this lesson, students make sense about quantities and their relationships in problem situations. The context helps students to make sense of the problem, and by decontextualizing the problem, the students are able to think about additive and multiplicative relationships.

Motivate

- **Preparation:** Before class, make mixtures of blue water and red water in two clear glasses or science beakers. Use 6 to 8 drops of food coloring to make each colored water mixture. You will also need a third clear glass (or beaker) and an eye dropper.

? "What color will you get mixing red and blue?" purple

- Make a shade of purple by adding 3 parts blue and 2 parts red (or whatever ratio you wish). The parts could be a full eye dropper or some other convenient amount.
- Have the students decide on a name for this shade of purple.

? "I would like to make more of this color. How can I do this?" Answers will vary. Look for correct ratios.

- Record different responses in a table similar to what they will use in Activity 1.
- Offer an incorrect response (such as 4 parts blue and 4 parts red). Students should be able to reason that the shade of purple will change.

Activity Notes

Activity 1

- This can be a hands-on activity if you wish to provide lemonade and iced tea. It can be done as a whole class or with partners, which would require more supplies.
- Have students complete part (a) and then stop. Point out that the quantities have the same units–the amounts used are cups. Explain that you want to make a larger batch, and they get to decide how to mix it.
- Have students complete the rest of the activity. After they have finished, take time to discuss each part.
- **MP2:** Ask students to share methods used to complete the table in part (c). Students have the opportunity to reason quantitatively by using an additive or a multiplicative structure. As an extension, add a column to the table and give students the number of cups of lemonade. Ask them to find the other two amounts and explain their methods.
- **MP3 Construct Viable Arguments and Critique the Reasoning of Others:** For part (e), students should construct a viable argument, and then state it aloud to classmates. Listen for valid reasoning.

Common Core State Standards

6.RP.1 Understand the concept of a ratio and use ratio language to describe a ratio relationship between two quantities.

6.RP.3a Make tables of equivalent ratios relating quantities with whole-number measurements, find missing values in the tables . . . . Use tables to compare ratios.

Previous Learning

Students should know how to write ratios.

Technology for the Teacher

Dynamic Classroom

Lesson Plans
Complete Materials List

5.2 Record and Practice Journal

Essential Question How can you find two ratios that describe the same relationship?

1 ACTIVITY: Making a Mixture

Work with a partner. A mixture calls for 1 cup of lemonade and 3 cups of iced tea.

a. How many total cups does the mixture contain? 4 cups

For every 1 cup of lemonade, there are 3 cups of iced tea.

b. How do you make a larger batch of this mixture? Describe your procedure and use the table below to organize your results. Add more columns to the table if needed.

Cups of Lemonade	1	2	3	4	5	6	7	8
Cups of Iced Tea	3	6	9	12	15	18	21	24
Total Cups	4	8	12	16	20	24	28	32

c. Which operations did you use to complete your table? Do you think there is more than one way to complete the table? Explain.

repeated addition or multiplication

d. How many total cups are in your final mixture? How many of those cups are lemonade? How many are iced tea? Compare your results with those of other groups in your class.

32 cups;
8 cups lemonade and 24 cups iced tea

Differentiated Instruction

Kinesthetic

Use manipulatives to reinforce part-to-part, part-to-whole, and whole-to-part ratios. Give each student 4 yellow chips and 6 red chips. Ask students to find the following ratios.

Part-to-part	4 yellow to 6 red 6 red to 4 yellow
Part-to-whole	4 yellow to 10 total 6 red to 10 total
Whole-to-part	10 total to 4 yellow 10 total to 6 red

5.2 Record and Practice Journal

e. Suppose you take a sip from every group's final mixture. Do you think all the mixtures should taste the same? Do you think the color of all the mixtures should be the same? Explain your reasoning.

yes; yes

f. Why do you think it is useful to use a table when organizing your results in this activity? Explain.

Check students' work.

2 ACTIVITY: Using a Multiplication Table

Work with a partner. Use the information in Activity 1 and the multiplication table below.

	1	2	3	4	5	6	7	8	9	10	11	12
1	1	2	3	4	5	6	7	8	9	10	11	12
2	2	4	6	8	10	12	14	16	18	20	22	24
3	3	6	9	12	15	18	21	24	27	30	33	36
4	4	8	12	16	20	24	28	32	26	40	44	48

a. A mixture contains 8 cups of lemonade. How many cups of iced tea are in the mixture?

24 cups iced tea

b. A mixture contains 21 cups of iced tea. How many cups of lemonade are in the mixture?

7 cups lemonade

c. A mixture has a total of 40 cups. How many cups are lemonade? How many cups are iced tea?

10 cups lemonade, 30 cups iced tea

d. Explain how a multiplication table may have helped you in Activity 1.

Check students' work.

3 ACTIVITY: Using More than One Ratio to Describe a Quantity

Work with a partner.

a. Find the ratio of pitchers of lemonade to pitchers of iced tea.

9 : 6

b. How can you divide the pitchers into equal groups? Is there more than one way? Use your results to describe the entire collection of pitchers.

3 pitchers of lemonade for every 2 pitchers of iced tea

c. Three more pitchers of lemonade are added. Is there more than one way to divide the pitchers into equal groups? Explain.

yes

d. The number of pitchers of lemonade and iced tea are doubled. Can you use the ratio in part (b) to describe the entire collection of pitchers? Explain.

yes

What Is Your Answer?

4. **IN YOUR OWN WORDS** How can you find two ratios that describe the same relationship? Give examples to support your explanation.

Make a table and use repeated addition or multiplication.

Laurie's Notes

Activity 2

- This activity is a preview of ratio tables.
- Remind students that the ratio of lemonade to iced tea is 1 to 3.
- Students are likely to answer each question without using the multiplication table. Remind them to think about how the table could be used to find the answer. They will most likely recognize that row 3 (iced tea) is involved, but it may not be obvious that row 1 (lemonade) is also involved. When discussing answers, this would be the time to make such a connection.
- Encourage students to think about the relationships between the numbers in the multiplication table, both additive and multiplicative.
- Have students discuss with others how they found their answers.
- **Connection:** The multiplication table can be used to show both the additive and multiplicative methods that students may have used in Activity 1.

Activity 3

- Before students begin, be sure they understand that this is a new activity, not a continuation of Activities 1 and 2.
- ? "What is the ratio of pitchers of lemonade to pitchers of iced tea?" 9 : 6
- If students are stuck on part (b), use physical models such as 9 yellow tiles and 6 red tiles. Tell students that each group must be the same.
- ? "Is there a way to use the multiplication table to answer this question?" Students may need guidance to think about how to use the table.
- In part (c), use tiles to represent the pitchers. A physical model will help students visualize the problem. They will then be able to make the connection to the multiplication table.
- **MP3:** For part (d), students should construct a viable argument, and then state it aloud to classmates. Listen for valid reasoning.

What Is Your Answer?

- Students should describe both additive and multiplicative approaches.

Closure

- Refer to the colored water in the opener and ask, "If you use 2 parts red and 5 parts blue to form a shade of purple, then what other quantities could you also use to produce the same shade? Explain."

2 ACTIVITY: Using a Multiplication Table

Math Practice 2

Use Operations

For each part of this problem, how do you know which operation to use?

Work with a partner. Use the information in Activity 1 and the multiplication table below.

	1	2	3	4	5	6	7	8	9	10	11	12
1	1	2	3	4	5	6	7	8	9	10	11	12
2	2	4	6	8	10	12	14	16	18	20	22	24
3	3	6	9	12	15	18	21	24	27	30	33	36
4	4	8	12	16	20	24	28	32	36	40	44	48

a. A mixture contains 8 cups of lemonade. How many cups of iced tea are in the mixture?

b. A mixture contains 21 cups of iced tea. How many cups of lemonade are in the mixture?

c. A mixture has a total of 40 cups. How many cups are lemonade? How many are iced tea?

d. **REPEATED REASONING** Explain how a multiplication table may have helped you in Activity 1.

3 ACTIVITY: Using More than One Ratio to Describe a Quantity

Work with a partner.

a. Find the ratio of pitchers of lemonade to pitchers of iced tea.

b. How can you divide the pitchers into equal groups? Is there more than one way? Use your results to describe the entire collection of pitchers.

c. Three more pitchers of lemonade are added. Is there more than one way to divide the pitchers into equal groups? Explain.

d. The number of pitchers of lemonade and iced tea are doubled. Can you use the ratio in part (b) to describe the entire collection of pitchers? Explain.

What Is Your Answer?

4. **IN YOUR OWN WORDS** How can you find two ratios that describe the same relationship? Give examples to support your explanation.

Practice Use what you learned about ratios to complete Exercises 4 and 5 on page 201.

5.2 Lesson

Check It Out Lesson Tutorials BigIdeasMath.com

Key Vocabulary
equivalent ratios, *p. 198*
ratio table, *p. 198*

Two ratios that describe the same relationship are **equivalent ratios**. You can find equivalent ratios by:

- adding or subtracting quantities in equivalent ratios.
- multiplying or dividing each quantity in a ratio by the same number.

You can find and organize equivalent ratios in a **ratio table**.

EXAMPLE 1 Completing Ratio Tables

Find the missing value(s) in each ratio table. Then write the equivalent ratios.

a.

Pens	1	2	
Pencils	3		9

b.

Dogs	4		24
Cats	6	12	

a. You can use repeated addition with the original ratio to find the missing values.

+1 +1

Pens	1	2	3
Pencils	3	6	9

+3 +3

The equivalent ratios are 1 : 3, 2 : 6, and 3 : 9.

b. You can use multiplication to find the missing values.

× 2 × 3

Dogs	4	8	24
Cats	6	12	36

× 2 × 3

The equivalent ratios are 4 : 6, 8 : 12, and 24 : 36.

On Your Own

Now You're Ready Exercises 6–11

Find the missing value(s) in the ratio table. Then write the equivalent ratios.

1.

Plantains	4		12
Bananas	3	6	

2.

Euros	5	10	
Dollars	4		32

Multi-Language Glossary at BigIdeasMath.com

Laurie's Notes

Introduction

Connect

- **Yesterday:** Students explored how to find two or more ratios that describe the same relationship (MP2, MP3).
- **Today:** Students will make ratio tables and use them to solve problems.

Motivate

- **Story Time:** Tell students that you took a 6-day float trip down a river.
- Make a table as shown and ask students to fill in the missing values.

Number of miles	1	2	3	5
Drop in elevation (feet)	40	80	120	200

- Ask students to explain how they found the missing values.

Lesson Notes

Discuss

- Define **equivalent ratios**. Give examples such as 2 : 3 and 4 : 6.
- Discuss the two methods for finding equivalent ratios.
- **Common Error:** Students may add or subtract the same quantity to each number in the ratio, saying, for example, that 2 : 3 and 4 : 5 are equivalent. Make it very clear that the numbers you add or subtract must be in the same ratio. You cannot add or subtract the same number, as when multiplying or dividing.
- Discuss the tables they have used and name them as **ratio tables**.

Example 1

? "In part (a), what is the ratio of pens to pencils?" 1 : 3

- Use repeated addition to complete the table in part (a).
- Point out that the equivalent ratios in part (a) are 1 : 3, 2 : 6, and 3 : 9.
- Use multiplication to complete the table in part (b). Make sure students understand that you do not have to use the same factor between consecutive columns.
- You could follow up this example by asking how students could use multiplication in part (a) and addition in part (b). For instance, in part (b), students could add columns to the table and use repeated addition, as follows. You could also use the following to show that the repeated addition justifies the multiplication and vice versa.

Dogs	**4**	**8**	12	16	20	**24**
Cats	**6**	**12**	18	24	30	**36**

On Your Own

- **MP3 Construct Viable Arguments and Critique the Reasoning of Others:** Ask students to share the method(s) they used to answer these questions to explain their reasoning for choosing the method(s).

Goal Today's lesson is making and using **ratio tables**.

Technology for the Teacher

Dynamic Classroom

Lesson Tutorials
Lesson Plans
Answer Presentation Tool

Extra Example 1

Find the missing values in each ratio table. Then write the equivalent ratios.

a.

Cars	2		6
Trucks	5	10	

4, 15; 2 : 5, 4 : 10, 6 : 15

b.

Markers	3	9	
Crayons	4		48

12, 36; 3 : 4, 9 : 12, 36 : 48

On Your Own

1. 8, 9; 4 : 3, 8 : 6, 12 : 9
2. 8, 40; 5 : 4, 10 : 8, 40 : 32

Extra Example 2

You are making a juice drink. The directions on the can of juice concentrate indicate to use 6 parts of water for every 1 part of juice concentrate. You use 24 cups of water. How much juice concentrate do you need? 4 cups

English Language Learners

Vocabulary

Students may benefit by seeing word problems in picture format.

For every 2 tickets you sell, your friend sells 3. You sell a total of 6 tickets. How many tickets does your friend sell?

When students are not yet fluent in English, this can help them focus on the math concepts.

On Your Own

3. 6 cups
4. 3 quarts

Laurie's Notes

Example 2

- Ask a volunteer to read the problem.
- ? "Does anyone have a hummingbird feeder at home?" Answers will vary.
- **FYI:** The ideal mixture during hot or dry weather is 4 parts of water for every 1 part of sugar. During rainy, foggy, or cold conditions when water is plentiful but when birds need more energy, the mixture can be 3 parts of water for every 1 part of sugar. In natural nectars, concentrations of sugars can vary within that range.
- ? "Is there more than one way to make a ratio table to solve this problem?" yes
- Draw a ratio table that has more columns than necessary and use addition to complete the table.
- ? "What should you add in the first row?" 4 "What should you add in the second row?" 1
- Interpret the results given in the ratio table. There are five equivalent ratios, and the equivalent ratio given in the last column answers the question. That is, you need 5 cups of sugar for 20 cups of water.
- Discuss the Study Tip.
- **Big Idea:** The units, cups, are the same for each item. The units could also have been teaspoons, tablespoons, etc.
- Draw a ratio table that has more columns than necessary and use multiplication to complete the table.
- ? "What should you multiply each quantity (water and sugar) by to find the equivalent ratio with 20 cups of water? Explain." 5; Because $20 \div 4 = 5$, you multiply each quantity by 5.
- **Extension:** Students could also use a mixed approach:

$\times 4$ $\quad$ $+ 4$

Water (cups)	4	16	20
Sugar (cups)	1	4	5

$\times 4$ $\quad$ $+ 1$

The above illustrates that there are often countless ways to arrive at an answer using ratio tables. You could ask the class if anyone can come up with a different way. Many students are eager to share their methods, knowing that they resulted in the correct answer.

On Your Own

- **Neighbor Check:** Have students work independently and then have their neighbors check their work. Have students discuss any discrepancies.

EXAMPLE 2 Making a Ratio Table

You are making sugar water for your hummingbird feeder. A website indicates to use 4 parts of water for every 1 part of sugar. You use 20 cups of water. How much sugar do you need?

You can solve this problem by using equivalent ratios. The ratio of water to sugar is 4 parts to 1 part. So, for every 4 cups of water, you need 1 cup of sugar. Find an equivalent ratio with 20 parts water.

Method 1: Use a ratio table and addition.

You can think of making a larger batch of sugar water as combining several batches of 4 to 1 mixtures. Use addition to obtain 20 in the water column.

+4 +4 +4 +4

Water (cups)	4	8	12	16	20
Sugar (cups)	1	2	3	4	5

+1 +1 +1 +1

The ratio 20 to 5 is equivalent to 4 to 1.

So, you need 5 cups of sugar.

Method 2: Use a ratio table and multiplication.

You multiplied the amount of water in the recipe by 5 because $20 \div 4 = 5$. So, you need to multiply the amount of sugar by 5. Multiply each part of the ratio in the original recipe by 5.

×5

Water (cups)	4	20
Sugar (cups)	1	5

×5

The ratio 20 to 5 is equivalent to 4 to 1.

So, you need 5 cups of sugar.

Study Tip

In Example 2, Method 1, notice that you can eliminate a step by adding columns 2 and 3 to obtain $8 + 12 = 20$ cups of water for $2 + 3 = 5$ cups of sugar.

On Your Own

Now You're Ready
Exercises 13 and 14

3. **WHAT IF?** You use 24 cups of water. How much sugar do you need?
4. You make a sweeter mixture of sugar water for your hummingbird feeder using 3 parts of water for every 1 part of sugar. You use 9 quarts of water. How much sugar do you need?

EXAMPLE 3 Using a Ratio Table

The nutrition facts label on a box of crackers shows that there are 240 milligrams of sodium in every 36 crackers.

a. You eat 15 crackers. How much sodium do you consume?

The ratio of sodium to crackers is 240 to 36. Use a ratio table to find an equivalent ratio with 15 crackers.

$\div 2$ $\div 6$ $\times 5$

Sodium (milligrams)	240	120	20	100
Crackers	36	18	3	15

$\div 2$ $\div 6$ $\times 5$

The ratio 100 to 15 is equivalent to 240 to 36.

So, you consume 100 milligrams of sodium.

b. You eat 21 crackers. How much sodium do you consume?

Notice that you can add the two middle columns in the table above.

So, you consume $120 + 20 = 140$ milligrams of sodium in $18 + 3 = 21$ crackers.

Study Tip

In Example 3, notice that you could use one step in the ratio table: multiply by

$\frac{1}{2} \cdot \frac{1}{6} \cdot 5 = \frac{5}{12}$.

EXAMPLE 4 Using a Ratio Table

You mix 3 pints of yellow paint for every 4 pints of blue paint to make green paint. You use 10 pints of blue paint. How much green paint do you make?

The ratio of yellow paint to blue paint is 3 to 4. Use a ratio table to find an equivalent ratio with 10 parts blue paint.

$\div 2$ $\times 5$

Yellow (pints)	3	$\frac{3}{2}$	$7\frac{1}{2}$
Blue (pints)	4	2	10

$\div 2$ $\times 5$

You use $7\frac{1}{2}$ pints of yellow paint and 10 pints of blue paint.

So, you make $7\frac{1}{2} + 10 = 17\frac{1}{2}$ pints of green paint.

Study Tip

In Example 4, notice that you could use one step in the ratio table: multiply by

$\frac{1}{2} \cdot 5 = \frac{5}{2}$.

On Your Own

Now You're Ready Exercises 15 and 16

5. **WHAT IF?** In Example 3, you eat 24 crackers. How much sodium do you consume?
6. **WHAT IF?** In Example 4, you mix 2 pints of yellow paint for every 3 pints of blue paint. You use 5 pints of yellow paint. How much green paint do you make?

Laurie's Notes

Example 3

- Ask a volunteer to read the problem. The answer may not be obvious because 36 is not evenly divisible by 15.
- ? **MP2 Reason Abstractly and Quantitatively:** "Can you estimate the amount of sodium in 15 crackers? Explain." Listen for something similar to: "I know that 240 : 36 = 120 : 18, so 18 crackers have 120 milligrams of sodium. For 15 crackers, the answer must be somewhat less than 120, maybe 100 or 110 milligrams.
- Students may perform several steps before arriving at the answer, rather than seeing the one-step solution mentioned in the Study Tip. After time, they should become more efficient in their solution methods.
- Draw a ratio table with a sufficient number of columns. Label the rows and write 240, 36, and 15 in the appropriate places as shown.
- ? "What would you do to get the numbers in the second column?" *Sample answer:* Divide 240 and 36 by 2.
- ? Record the sample answer results and ask, "Can you subtract 3 from 120 and from 18? Explain." No, when you use repeated subtraction, you have to subtract a *ratio*.
- Discuss strategies to get from 18 to 15, suggesting that it may involve more than one step.
- **FYI:** Subtracting the second and third columns also gives the fourth column.
- ? "How much sodium do you consume when you eat 21 crackers?" 140 milligrams Students might notice that the sum of the middle two columns will give them the answer, though it may not be obvious.
- Mention that students can add or subtract any of the equivalent ratios in the table to obtain a new equivalent ratio.

Example 4

- Ask a volunteer to read the problem.
- Draw a ratio table with a sufficient number of columns. Label the rows and write 3, 4, and 10 in the appropriate places as shown.
- Give students time to think about strategies for finding the solution.
- Solve the problem as shown or use a strategy suggested by a student.
- The problem asks for the amount of green paint, so after you find the amount of yellow paint, add it to the given amount of blue paint.

On Your Own

- To answer Question 5, students can add the amount of sodium in 3 crackers to the amount of sodium in 21 crackers found in Example 3(b).
- **MP3:** Students can use different methods to answer these questions. Ask them to share their methods with the class. It is important for students to gain confidence in explaining their reasoning aloud to classmates.

Closure

- **Exit Ticket:** Return to the example in the Motivate section. Ask students to use annotated arrows to show how they found the missing table entries.

Extra Example 3

The nutrition facts label on a bag of pretzel rods shows that there are 450 milligrams of sodium in every 3 pretzel rods.

a. You eat 2 pretzel rods. How much sodium do you consume?
300 milligrams

b. You eat 5 pretzel rods. How much sodium do you consume?
750 milligrams

Extra Example 4

You mix 2 pints of blue paint for every 5 pints of red paint to make purple paint. You use 18 pints of red paint. How much purple paint do you make? $25\frac{1}{5}$ pints

On Your Own

5. 160 milligrams

6. $12\frac{1}{2}$ pints

Vocabulary and Concept Check

1. Two ratios are equivalent if they can be written as the same ratio.
2. no; You can only create an equivalent ratio by adding or subtracting corresponding parts, or by multiplying or dividing each part of the ratio by the same number.
3. 12 : 15; all other ratios are equivalent.

Practice and Problem Solving

4. The ratio of baseballs to gloves can be described by 8 : 4, 4 : 2, or 2 : 1.
5. The ratio of ladybugs to bees can be described by 12 : 4, 6 : 2, or 3 : 1.
6.

Boys	1	2
Girls	5	10

1 : 5 and 2 : 10

7.

Violins	8	24
Cellos	3	9

8 : 3 and 24 : 9

8.

Taxis	6	18	36
Buses	5	15	30

6 : 5, 18 : 15, and 36 : 30

9.

Burgers	3	6	9
Hot dogs	5	10	15

3 : 5, 6 : 10, and 9 : 15

10.

Towels	14	7	28
Blankets	8	4	16

14 : 8, 7 : 4, and 28 : 16

11.

Forks	16	8	48
Spoons	10	5	30

16 : 10, 8 : 5, and 48 : 30

12. $68

Assignment Guide and Homework Check

Level	Day 1 Activity Assignment	Day 2 Lesson Assignment	Homework Check
Basic	4, 5, 35–38	1–3, 7–25 odd	9, 13, 15, 19
Average	4, 5, 35–38	1–3, 8, 9, 12–20 even, 21–25	8, 12, 14, 16, 20
Advanced	4, 5, 35–38	1–3, 10–34 even	10, 14, 18, 20, 30

Common Errors

- **Exercises 4 and 5** Students may write an incorrect ratio. For example, in Exercise 4, students may write the ratio of gloves to baseballs, baseballs to baseballs and gloves, etc. Remind students to read the problem carefully to determine what is being asked.
- **Exercises 6–11** Students may add or subtract the same quantity to each number in the ratio, saying, for example, that 2 : 3 and 4 : 5 are equivalent. Make it very clear that the numbers you add or subtract must be in the same ratio. You cannot add or subtract the same number, as when multiplying or dividing.

5.2 Record and Practice Journal

Find the missing value(s) in the ratio table. Then write the equivalent ratios.

1.

Kids	3	9
Adults	1	3

3 : 1 and 9 : 3

2.

Basketballs	5	10
Footballs	10	20

5 : 10 and 10 : 20

3.

Apples	4	16
Oranges	5	20

4 : 5 and 16 : 20

4.

CDs	10	30
DVDs	9	27

10 : 9 and 30 : 27

5.

Regular	2	8	32
Decaf	3	12	48

2 : 3, 8 : 12, and 32 : 48

6.

Scooters	1	5	25
Bikes	3	15	75

1 : 3, 5 : 15, and 25 : 75

7. You read 1 chapter every hour. You read for 3 hours after school. How many chapters did you read?

3 chapters

5.2 Exercises

Check It Out
Help with Homework
BigIdeasMath.com

Vocabulary and Concept Check

1. **VOCABULARY** How can you tell whether two ratios are equivalent?

2. **NUMBER SENSE** Consider the ratio 3 : 5. Can you create an equivalent ratio by adding the same number to each quantity in the ratio? Explain.

3. **WHICH ONE DOESN'T BELONG?** Which ratio does *not* belong with the other three? Explain your reasoning.

3 : 4   9 : 12   12 : 15   12 : 16

Practice and Problem Solving

Write several ratios that describe the collection.

4. baseballs to gloves

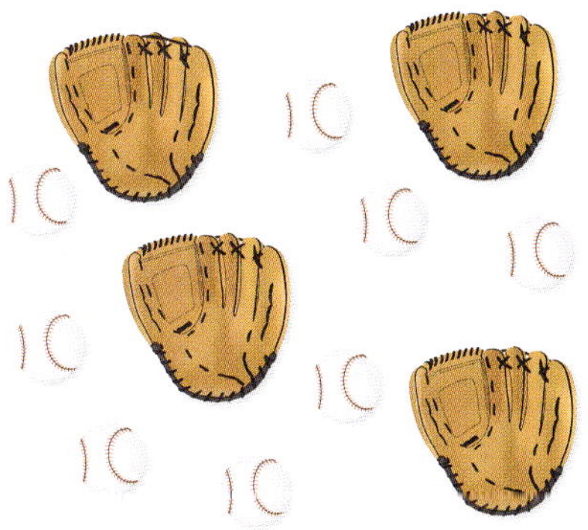

5. ladybugs to bees

Find the missing value(s) in the ratio table. Then write the equivalent ratios.

1 6.

Boys	1	
Girls	5	10

7.

Violins	8	24
Cellos	3	

8.

Taxis	6		36
Buses	5	15	

9.

Burgers	3		9
Hot Dogs	5	10	

10.

Towels	14	7	
Blankets	8		16

11.

Forks	16	8	
Spoons	10		30

12. **WORK** Your neighbor pays you \$17 for every 2 hours you work. You work for 8 hours on Saturday. How much does your neighbor owe you?

Complete the ratio table to solve the problem.

2 **13.** For every 3 tickets you sell, your friend sells 4. You sell a total of 12 tickets. How many does your friend sell?

You	3			12
Friend	4			

14. A store sells 2 printers for every 5 computers. The store sells 40 computers. How many printers does the store sell?

Printers	2		8	
Computers	5	10		40

3 **15.** First and second place in a contest use a ratio to share a cash prize. When first place pays \$100, second place pays \$60. How much does first place pay when second place pays \$36?

First	100		
Second	60		36

16. A grade has 81 girls and 72 boys. The grade is split into groups that have the same ratio of girls to boys as the whole grade. How many girls are in a group that has 16 boys?

Girls	81		
Boys	72		16

ERROR ANALYSIS **Describe and correct the error in making the ratio table.**

17. ✗

A	3	8	13
B	7	12	17

18. ✗

A	5	25	125
B	3	9	27

19. **DONATION** A sports store donates basketballs and soccer balls to the boys and girls club. The ratio of basketballs to soccer balls is 7 : 6. The store donates 24 soccer balls. How many basketballs does the store donate?

20. **DOWNLOAD** You are downloading songs to your MP3 player. The ratio of pop songs to rock songs is 5 : 4. You download 40 pop songs. How many rock songs do you download?

SCRAMBLED EGGS **In Exercises 21–25, use the ratio table showing different batches of the same recipe for scrambled eggs.**

Recipe	A	B	C	D	E	F
Servings	4	2	6	3	5	9
Eggs	8	4	12	6	10	18
Milk (cups)	$\frac{1}{2}$	$\frac{1}{4}$	$\frac{3}{4}$	$\frac{3}{8}$	$\frac{5}{8}$	$1\frac{1}{8}$

21. How can you use Recipes B and D to create Recipe E?

22. How can you use Recipes C and D to create Recipe F?

23. How can you use Recipes B and C to create Recipe A?

24. How can you use Recipes C and F to create Recipe D?

25. Describe one way to use the recipes to create a batch with 11 servings.

Common Errors

- **Exercises 13–16** There are many ways that students can use a ratio table to solve the problem. Emphasize to students that the tables shown are just templates that are similar to the examples.
- **Exercises 21–25** It is important that students realize that they can create equivalent ratios by adding or subtracting corresponding quantities in ***any*** other equivalent ratio.
- **Exercise 34** Students may have difficulty solving this problem. The key to solving this problem is recognizing that the ratio changes from 8 : 3 to 4 : 7 after giving away half of the tokens in the first quantity.

Practice and Problem Solving

13.

You	3	6	9	12
Friend	4	8	12	16

16 tickets

14.

Printers	2	4	8	16
Computers	5	10	20	40

16 printers

15.

First	100	10	60
Second	60	6	36

$60

16.

Girls	81	9	18
Boys	72	8	16

18 girls

17–25. See Additional Answers.

Differentiated Instruction

Kinesthetic

Provide students with counters and two sheets of paper. Draw a line on each paper to divide it into two parts. Each piece of paper will represent a ratio. Place counters on one piece of paper to represent a ratio. Rearrange the counters in stacks of equal number. Place counters on the second piece of paper to represent another ratio. Rearrange the counters in stacks of equal number. If the stacks in each individual ratio are the same size and there are the same number of stacks in each ratio, then the ratios are equal.

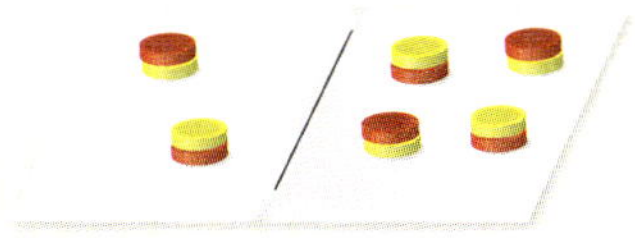

Stacks of 2, 4 counters to 8 counters
Ratio is 2 to 4.

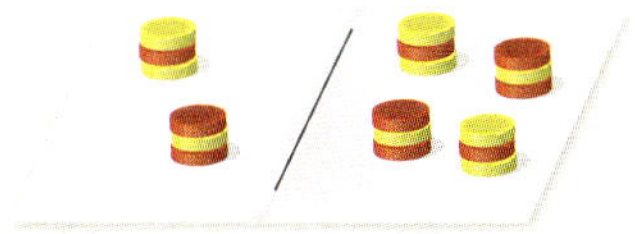

Stacks of 3, 6 counters to 12 counters
Ratio is 2 to 4.

Practice and Problem Solving

26. $A = 12, B = 18$

27. $A = 16, B = 28$

28. $A = 33, B = 15$

29. $A = 65, B = 40$

30. **a.** 69 calories

b. 21 cashews

31. 32

32. 96 students

33. 36 bugs

34. See *Taking Math Deeper*.

Fair Game Review

35. $27(2 + 1)$

36. $12(5x - 7)$

37. $14(3x + 2y)$

38. D

Mini-Assessment

Find the missing values in the ratio table. Then write the equivalent ratios.

1.

Shirts	3		12
Shorts	5	15	

9, 20; 3 : 5, 9 : 15, 12 : 20

2.

Shoes	36	9	
Socks	48		24

12, 18; 36 : 48, 9 : 12, 18 : 24

3. A soccer team has 12 girls and 18 boys. The team is split into groups that have the same ratio of girls to boys as that of the whole team. How many girls are in a group that has 9 boys? 6 girls

4. In Question 3, how many girls are in a group that has 12 boys? 8 girls

Taking Math Deeper

Exercise 34

For this exercise, you can use a tape diagram to help visualize the relationship between the numbers of tokens you and your friend have.

1. Use a tape diagram to express the ratio *for every 8 tokens you had, your friend had 3.*

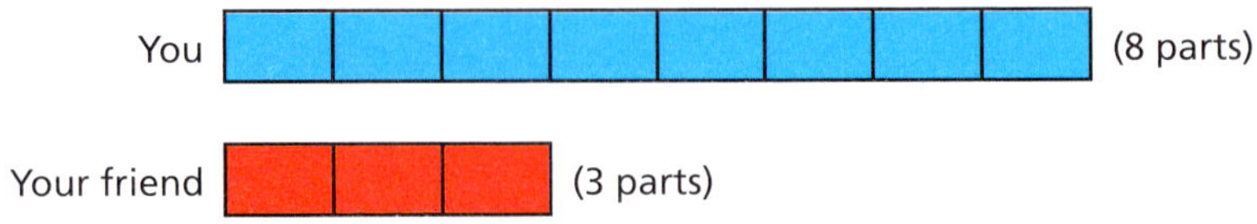

2. Draw a new tape diagram that represents this situation after you give one-half of your tokens to your friend.

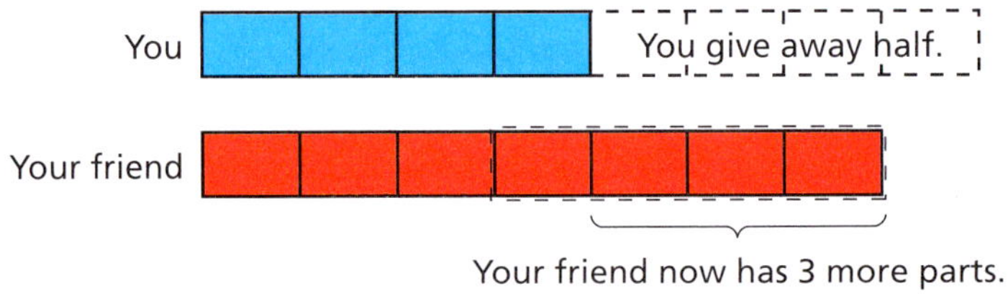

You now have 4 parts and your friend now has 7 parts.

Because your friend now has 18 more tokens than you and 3 more parts in the tape diagram, each part must represent $18 \div 3 = 6$ tokens

3. Answer the question.

Look back at the original tape diagram. You had five more parts than your friend.

So, initially, you had $5 \cdot 6 = 30$ more tokens.

Reteaching and Enrichment Strategies

If students need help. . .	If students got it. . .
Resources by Chapter • Practice A and Practice B • Puzzle Time Record and Practice Journal Practice Differentiating the Lesson Lesson Tutorials Skills Review Handbook	Resources by Chapter • Enrichment and Extension • Technology Connection Start the next section

Two whole numbers A and B satisfy the following conditions. Find A and B.

26. $A + B = 30$
$A : B$ is equivalent to $2 : 3$.

27. $A + B = 44$
$A : B$ is equivalent to $4 : 7$.

28. $A - B = 18$
$A : B$ is equivalent to $11 : 5$.

29. $A - B = 25$
$A : B$ is equivalent to $13 : 8$.

Nutrition Facts
Serving Size: 1 ounce (28g)

Amount Per Serving	
Calories 161	Calories from Fat 109
	% Daily Value*
Total Fat 13g	20%
Saturated Fat 3g	13%
Trans Fat	
Cholesterol 0mg	0%
Sodium 4mg	0%
Total Carbohydrate 9g	3%
Dietary Fiber 1g	3%
Sugars 1g	
Protein 4g	
Vitamin A 0% • Vitamin C	0%
Calcium 1% • Iron	9%

*Percent Daily Values are based on a 2,000 calorie diet. Your daily values may be higher or lower depending on your calorie needs.

30. CASHEWS The nutrition facts label on a container of dry roasted cashews indicates there are 161 calories in 28 grams. You eat 9 cashews totaling 12 grams.

a. How many calories do you consume?

b. How many cashews are in one serving?

31. REASONING The ratio of three numbers is $4 : 3 : 1$. The sum of the numbers is 64. What is the greatest number?

32. SURVEY Seven out of every 8 students surveyed owns a bike. The difference between the number of students who own a bike and those who do not is 72. How many students were surveyed?

33. BUG COLLECTION You and a classmate have a bug collection for science class. You find 5 out of every 9 bugs in the collection. You find 4 more bugs than your classmate. How many bugs are in the collection?

34. Problem Solving You and a friend each have a collection of tokens. Initially, for every 8 tokens you had, your friend had 3. After you give half of your tokens to your friend, your friend now has 18 more tokens than you. Initially, how many more tokens did you have than your friend?

Fair Game Review What you learned in previous grades & lessons

Factor the expression using the GCF. *(Section 3.4)*

35. $54 + 27$

36. $60x - 84$

37. $42x + 28y$

38. MULTIPLE CHOICE Which expression does *not* give the area of the shaded figure? *(Section 4.3)*

Ⓐ $2(6) + 2\left(\frac{1}{2}(6)(2)\right)$

Ⓑ $2\left(\frac{1}{2}(3)(2 + 6)\right)$

Ⓒ $6(6) - 4\left(\frac{1}{2}(3)(2)\right)$

Ⓓ $6(6) - \frac{1}{2}(6)(2)$

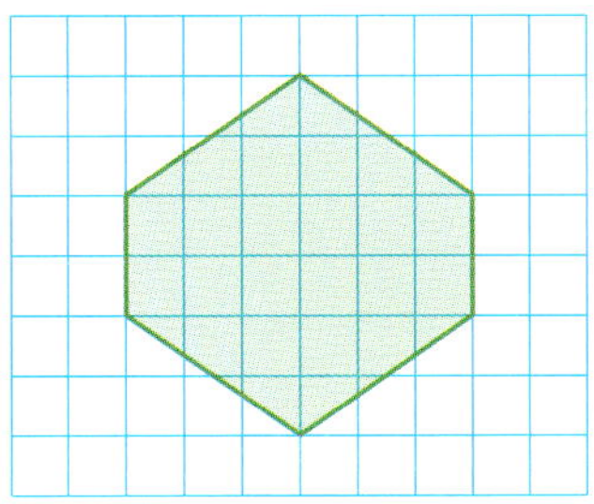

5.3 Rates

Essential Question How can you use rates to describe changes in real-life problems?

1 ACTIVITY: Stories Without Words

Work with a partner. Each diagram shows a story problem.

- **Describe the story problem in your own words.**
- **Write the rate indicated by the diagram. What are the units?**

a.

b.

c.

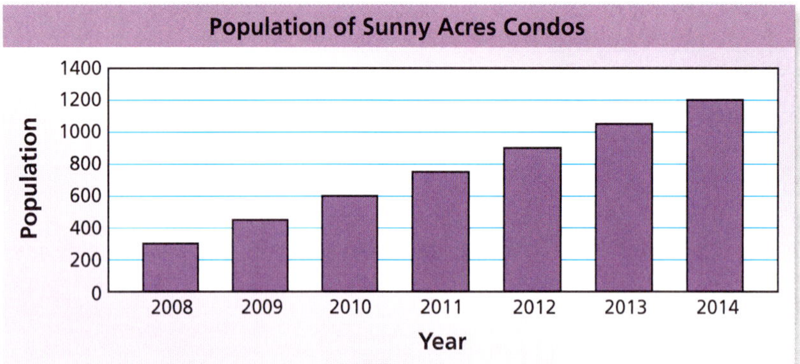

d.

Common Core

Rates

In this lesson, you will

- understand the concepts of rates and unit rates.
- write unit rates.
- solve real-life problems.

Learning Standards
6.RP.2
6.RP.3a
6.RP.3b

Laurie's Notes

Introduction

Standards for Mathematical Practice

- **MP6 Attend to Precision:** In working with rates, it is very important for students to specify the units of measure. They will have plenty of practice specifying units in the problems they work.

Motivate

- Help students calculate their heart rate, the number of beats per unit of time.
- First have students practice finding their pulse, with wrist or neck.
- Count the number of beats in a 6 second period. Record the information as a rate. Example: $\frac{7 \text{ beats}}{6 \text{ seconds}}$.

? Ask what they would predict their answer to be if they found their heart rate for one minute. They should recognize that by multiplying both the beats and seconds by 10, the rate is $\frac{70 \text{ beats}}{60 \text{ seconds}}$ which is the same as 70 beats per minute.

? "If you had found the number of beats in 10 seconds, how would you find the beats per minute?" Multiply by a factor of 6.

- Discuss the idea that they could find their heart rate in beats per hour; however that is not a common rate.

Activity Notes

Activity 1

- Students may not know a formal definition for rates, but it is a familiar idea. They just found their heart rate and have heard rates such as 65 miles per hour, 32 miles per gallon, $7.50 per hour, and $1.99 per pound.
- Explain to students that they are going to work with a partner and write a story about information given in a diagram.
- Encourage students to be creative in writing the story problem. You want students to have fun and think about each context and what the diagram is telling them.
- Give sufficient time for students to write their story problems. Encourage students to use proper grammar, spelling, and sentence structure.
- Have several groups share their stories.
- **Summarize:** All four problems involve a unit of time as the second term in the rate.

? "Do all rates involve time? Explain." No, although rates involving time are common, there are other kinds of rates, such as miles per gallon and dollars per pound.

Common Core State Standards

6.RP.2 Understand the concept of a unit rate a/b associated with a ratio $a : b$ with $b \neq 0$, and use rate language in the context of a ratio relationship.

6.RP.3a Make tables of equivalent ratios relating quantities with whole-number measurements, find missing values in the tables . . .. Use tables to compare ratios.

6.RP.3b Solve unit rate problems including those involving unit pricing and constant speed.

Previous Learning

Students know benchmark conversion facts, and how to simplify fractions.

Technology for the Teacher

Dynamic Classroom

Lesson Plans

Complete Materials List

5.3 Record and Practice Journal

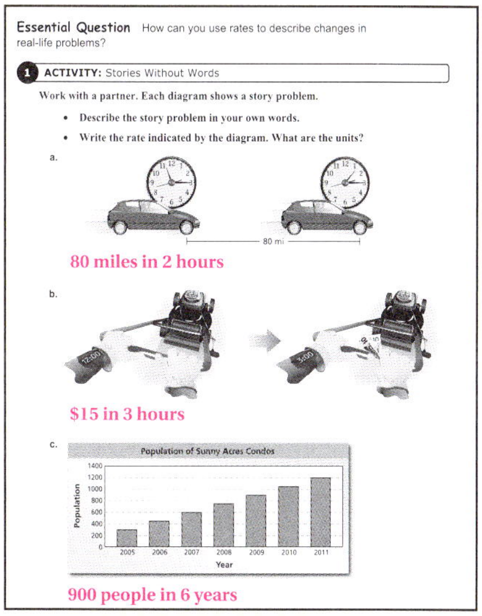

Essential Question How can you use rates to describe changes in real-life problems?

1 ACTIVITY: Stories Without Words

Work with a partner. Each diagram shows a story problem.

- Describe the story problem in your own words.
- Write the rate indicated by the diagram. What are the units?

a.

80 miles in 2 hours

b.

$15 in 3 hours

c.

900 people in 6 years

Differentiated Instruction

Auditory

Ask students to recall when they have heard or used expressions with the word *per*, such as dollars per gallon. Make a list of the expressions. Explain that these are all rates and have a denominator of 1. They are called *unit rates*. Ask students how they would find the unit rate from a rate such as $42.50 for 10 gallons of gasoline. Conversely, ask how they would use a unit rate of 25 miles per gallon to determine the number of miles traveled using 10 gallons. Motivate the students by posting the list and adding to it as you work through the chapter.

5.3 Record and Practice Journal

d.

4 feet in 4 years

2 ACTIVITY: Finding Equivalent Rates

Work with a partner. Use the diagrams in Activity 1. Explain how you found each answer.

a. How many miles does the car travel in 1 hour?

40 miles

b. How much money does the person earn every hour?

$5

c. How much does the population of Sunny Acres Condos increase each year?

150 people

d. How many feet does the alligator grow per year?

1 foot

3 ACTIVITY: Using a Double Number Line

Work with a partner. Count the number of times you can clap your hands in 12 seconds. Have your partner keep track of the time and record your results. Check students' work.

a. Use the results to complete the double number line.

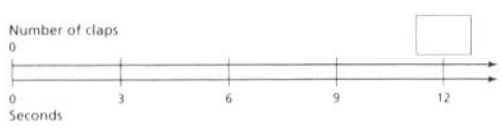

b. Explain how to use the double number line to find the number of times you clap your hands in 6 seconds and in 4 seconds.

c. Find the number of times you can clap your hands in 1 minute. Explain how you found your answer.

d. How can you find the number of times you can clap your hands in 2 minutes? 3 minutes? Explain.

What Is Your Answer?

4. **IN YOUR OWN WORDS** How can you use rates to describe changes in real-life problems? Give examples to support your explanation.

show how one quantity changes compared to another

5. **MODELING** Use a double number line to model each story in Activity 1. Show how to use the double number line to answer each question in Activity 2. Why is a double number line a good problem-solving tool for these types of problems?

See Additional Answers.

Laurie's Notes

Activity 2

- This activity is related to Activity 1. It is possible that in the stories students wrote, they may have found a **unit rate** without knowing the term.
- Give sufficient time for students to write their explanations.
- **MP3 Construct Viable Arguments and Critique the Reasoning of Others:** Ask students to share their explanations with the class. The message you are sending is that it is not only the answer you care about. You want to hear the reasoning behind their explanations.

Activity 3

- There will be some noise during this activity, but it will be good noise that engages students in the activity.
- It may be easier for students to count their number of claps using groups of 10. You may wish to be the "timer."
- Some students may want to try it a second time if their result is an "ugly" number such as 31. I always tell students that real data is not always "pretty," so they have to work with what they have.
- Depending upon their result, answering the questions may be more challenging for some groups. Remind students to explain how they are getting their answers.
- **MP3:** Students can use different methods to answer the questions. Ask them to share their methods. If their reasoning is not clear, or is incorrect, then classmates should be expected to question the explanation given.
- **Big Idea:** The double number line is another model (as is the ratio table) for finding equivalent ratios.
- **MP4 Model with Mathematics:** Mathematically proficient students will make this connection and hopefully state it for their classmates.

What Is Your Answer?

- **Think-Pair-Share:** Students should read each question independently and then work in pairs to answer the questions. When they have answered the questions, the pair should compare their answers with another group and discuss any discrepancies.

Closure

- Refer to the heart rate problem in the Motivate section. Ask students to complete the double number line for this problem.

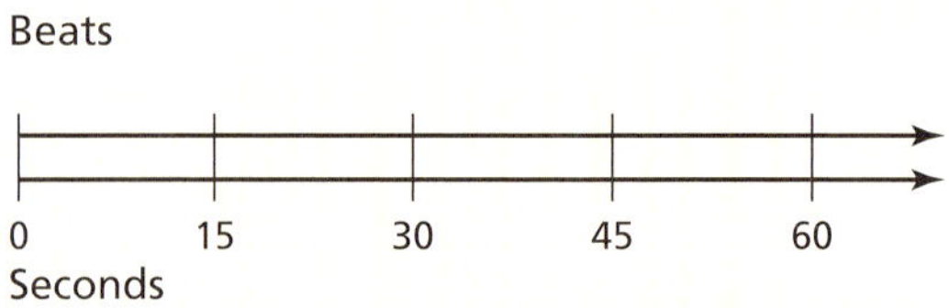

2 ACTIVITY: Finding Equivalent Rates

Math Practice 6

Specify Units

How do the given units help you find the units for your answer?

Work with a partner. Use the diagrams in Activity 1. Explain how you found each answer.

a. How many miles does the car travel in 1 hour?

b. How much money does the person earn every hour?

c. How much does the population of Sunny Acres Condos increase each year?

d. How many feet does the alligator grow per year?

3 ACTIVITY: Using a Double Number Line

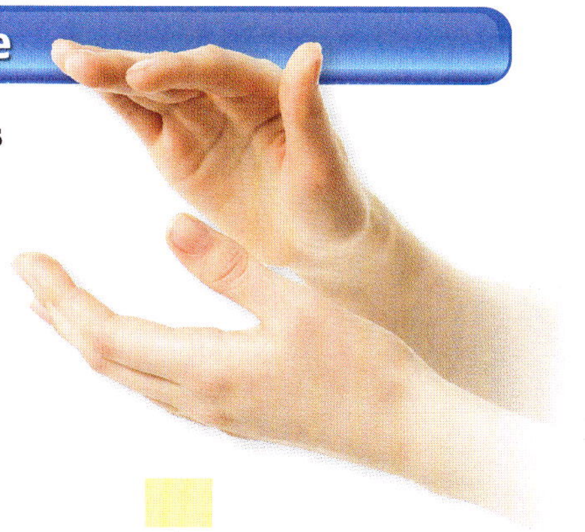

Work with a partner. Count the number of times you can clap your hands in 12 seconds. Have your partner keep track of the time and record your results.

a. Use the results to complete the double number line.

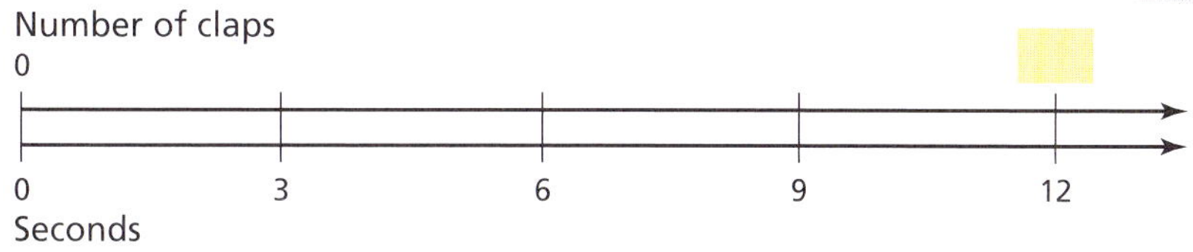

b. Explain how to use the double number line to find the number of times you clap your hands in 6 seconds and in 4 seconds.

c. Find the number of times you can clap your hands in 1 minute. Explain how you found your answer.

d. How can you find the number of times you can clap your hands in 2 minutes? 3 minutes? Explain.

What Is Your Answer?

4. **IN YOUR OWN WORDS** How can you use rates to describe changes in real-life problems? Give examples to support your explanation.

5. **MODELING** Use a double number line to model each story in Activity 1. Show how to use the double number line to answer each question in Activity 2. Why is a double number line a good problem-solving tool for these types of problems?

Practice Use what you learned about rates to complete Exercises 3 and 4 on page 208.

5.3 Lesson

Check It Out
Lesson Tutorials
BigIdeasMath.com

Key Vocabulary

rate, *p. 206*
unit rate, *p. 206*
equivalent rates, *p. 206*

Key Idea

Rate and Unit Rate

Words A **rate** is a ratio of two quantities using different units. A **unit rate** compares a quantity to one unit of another quantity. **Equivalent rates** have the same unit rate.

Numbers You pay \$27 for 3 pizzas.

Rate: \$27 : 3 pizzas

Unit rate: \$9 : 1 pizza

Algebra Rate: a units : b units Unit rate: $\frac{a}{b}$ units : 1 unit

Study Tip

In a rate $a : b$, you can divide both a and b by b to find the unit rate.

EXAMPLE 1 Writing a Rate

The double number line shows the rate at which you earn points for successfully hitting notes in a music video game. Write a rate that represents this situation.

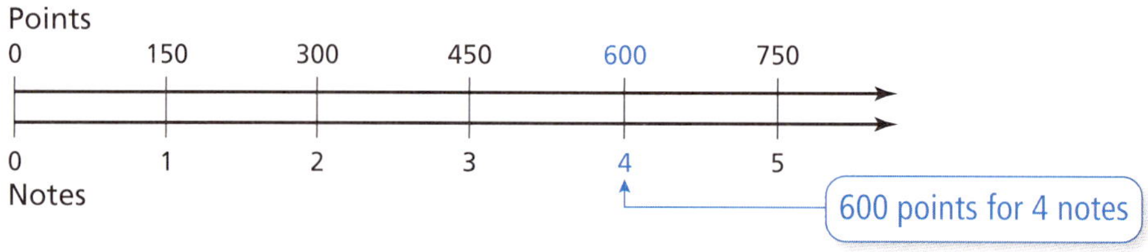

∴ One possible rate is 600 points for every 4 notes.

EXAMPLE 2 Finding a Unit Rate

A piece of space junk travels 5 miles in 8 seconds. How far does it travel per second?

Use a ratio table and divide by 8 to write an equivalent rate in which the time is 1 second.

	÷ 8 →	
Distance (miles)	5	$\frac{5}{8}$
Time (seconds)	8	1
	÷ 8 →	

The rate 5 miles : 8 seconds is equivalent to $\frac{5}{8}$ mile : 1 second.

∴ So, the space junk travels $\frac{5}{8}$ mile per second.

Multi-Language Glossary at BigIdeasMath.com

Laurie's Notes

Introduction

Connect

- **Yesterday:** Students explored how to find a rate that describes changes in real-life problems (MP3, MP4, MP6).
- **Today:** Students find rates, unit rates, and equivalent rates.

Motivate

- Pair students; give them preliminary think time and then have them write as many rates as possible in two minutes. The rates need to have an *action verb* at the beginning of the rate and must be realistic.
- Example: *Sleep* 8 hours per night; *blink* 5 times per 20 seconds; *purchase* 1 ball per 4 people.

Lesson Notes

Key Idea

- Write the Key Idea.
- When students are writing rates it is very important to write the related units. The units tell the context for the rate.
- Without the units the problem becomes $\frac{27}{3} = 9$. The 9 doesn't convey the context of $9 per pizza. Note that the word "per" is used in reading unit rates.
- Discuss the algebra connection involved in writing a rate as a unit rate. Remind students that the process they used yesterday in Activity 2 involved division.
- **Representation:** Be sure students recall that $\frac{a}{b}$ means $a \div b$.

Example 1

- The problem asks for a rate, not a unit rate.
- ? Refer to the double number line and ask, "What is one rate that represents this situation?" Listen for a rate equivalent to 150 points for every 1 note.
- **Extension:** Ask for a rate not shown on the double number line.

Example 2

- When spacecraft or parts of spacecraft no longer work, they are left to float around space indefinitely. It is usually too expensive to retrieve these objects, so they are left to orbit Earth until they fall back down or collide with other space junk.
- **MP4 Model with Mathematics:** A ratio table models the situation.
- ? "How can you find the unit rate?" Divide distance and time by 8.
- Relate the unit rate back to the algebraic representation in the Key Idea.
- Take time for students to consider how fast $\frac{5}{8}$ mile per second is. Give an example of a distance of $\frac{5}{8}$ mile that would be familiar to students.

Goal Today's lesson is finding **rates**.

Technology for the Teacher

Dynamic Classroom

Lesson Tutorials
Lesson Plans
Answer Presentation Tool

Extra Example 1

The double number line shows the rate at which your friend runs. Write a rate that represents this situation.

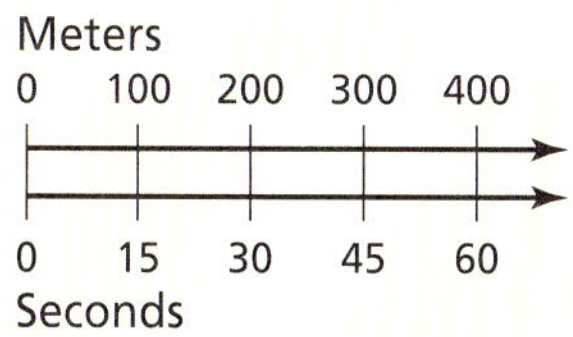

Sample answer: 200 meters in 30 seconds

Extra Example 2

At mid-latitudes, Earth spins 3 miles in 10 seconds. How far does it spin per second? $\frac{3}{10}$ mile per second

On Your Own

1. *Sample answer:* 150 points per note
2. 186 miles
3. \$0.50 : 1 ounce

Extra Example 3

a. A chef buys 6 pounds of shrimp for \$39. How much will the chef pay for 11 more pounds of shrimp? \$71.50

b. You buy 4 pounds of catfish fillets for \$22. What is the cost for 9 pounds of catfish fillets? \$49.50

On Your Own

4. **a.** 45 megabytes

   b.

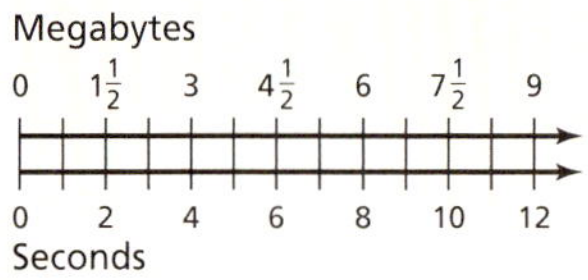

$7\frac{1}{2}$ megabytes

English Language Learners

Labels

English learners may recognize the fraction bar as division, but may not be familiar with its use in the concept of rate. The following unit rates are equivalent.

$\frac{3\text{ m}}{1\text{ h}}$, 3 m/h, 3 meters per hour

Each of these rates can be read as "three meters *for every* hour."

Laurie's Notes

On Your Own

- **Neighbor Check:** Have students work independently and then have their neighbors check their work. Have students discuss any discrepancies.

Example 3

- Ask a volunteer to read the problem.
- ? "How many of you like salmon?" Answers will vary.
- ? "What strategy would you use to answer part (a)?" Find the unit rate first, then use the unit rate to find the cost of 9 pounds.
- Draw a ratio table with a sufficient number of columns. Ask students what information is known and how many columns will be necessary to answer the question.
- ? "How do you find the unit rate?" Divide by 6. "What is the unit rate?" \$8.50 per pound.
- **MP6 Attend to Precision:** Be sure students specify the units.
- When working part (b), explain why a double number line is an appropriate model. Students should understand the reasoning behind the model selected.
- Step through the creation of the double number line. Begin incrementing by \$16 and 2 pounds. Use the strategy shown to find the solution. Ask for costs of other amounts, such as 7.5 pounds, 7.75 pounds, etc. This may help students see why using a double number line works well for the given rate. The numbers work out nicely.
- ? "Can you think of an alternate method to find the answer to part (b)?" Answers will vary.

On Your Own

- Ask a volunteer to explain his or her work.

Closure

- **Exit Ticket:** Write the following as a unit rate in miles per minute and miles per hour: You run 2 miles in 20 minutes. 0.1 mile per minute; 6 miles per hour
- ? "Which method did you use to find the answer and why?" Answers will vary.

On Your Own

Now You're Ready
Exercises 3–14

1. Write another rate that represents the situation in Example 1.
2. A Japanese bullet train travels 558 miles in 3 hours. How far does it travel every hour?
3. You pay \$8 for 16 ounces of sliced turkey. Write a rate that gives the price for each ounce of turkey.

EXAMPLE 3 Finding Equivalent Rates

a. A chef buys 6 pounds of salmon fillets for \$51. How much will the chef pay for 9 more pounds of salmon fillets?

Using a ratio table, divide to find the unit rate and then multiply to find the cost for 9 pounds of salmon fillets.

÷ 6 ×9

Cost (dollars)	51	8.5	76.5
Salmon (pounds)	6	1	9

÷ 6 ×9

So, the chef will pay \$76.50 for 9 more pounds of salmon fillets.

Study Tip

The unit rate of cost per unit is called *unit cost*. The unit cost of the salmon in Example 3(a) is \$8.50 per pound.

b. You buy 2 pounds of tilapia fillets for \$16. What is the cost for 7 pounds of tilapia fillets?

Because \$16 is easily divided into halves, fourths, and eighths, it is appropriate to model the rate using a double number line.

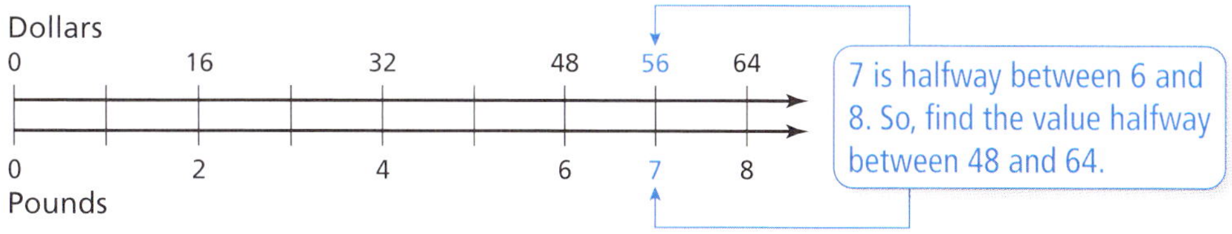

So, the cost for 7 pounds of tilapia fillets is \$56.

On Your Own

Now You're Ready
Exercises 17 and 18

4. Your download speed is 3 megabytes every 4 seconds.
   a. How many megabytes can you download in 1 minute?
   b. Construct a double number line that represents the situation. How many megabytes can you download in 10 seconds?

5.3 Exercises

Check It Out
Help with Homework
BigIdeasMath.com

Vocabulary and Concept Check

1. **WRITING** Describe a unit rate that you use in real life.
2. **DIFFERENT WORDS, SAME QUESTION** Which is different? Find "both" answers.

What is the cost per bagel?	What is the cost per dozen bagels?
What is the unit cost of a bagel?	How much does each bagel cost?

Practice and Problem Solving

Write a rate that represents the situation.

1 **3.**

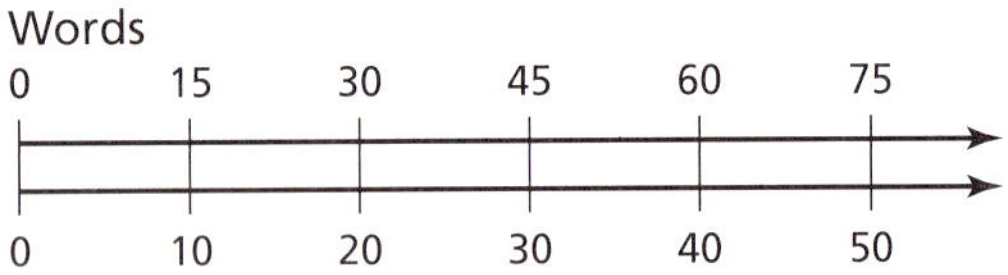

4.

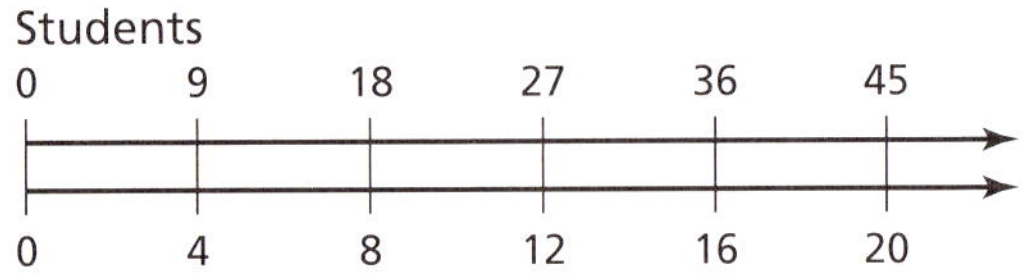

5.

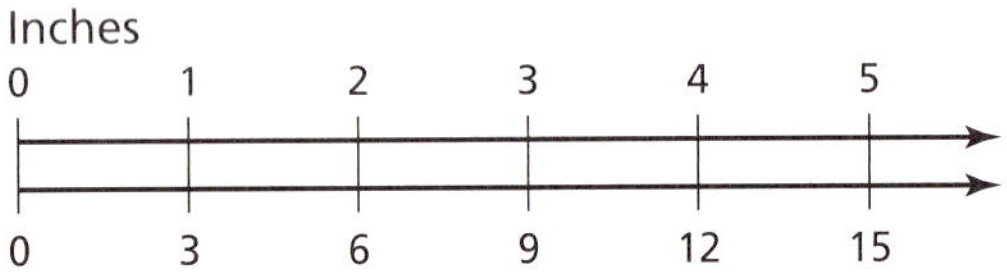

6.

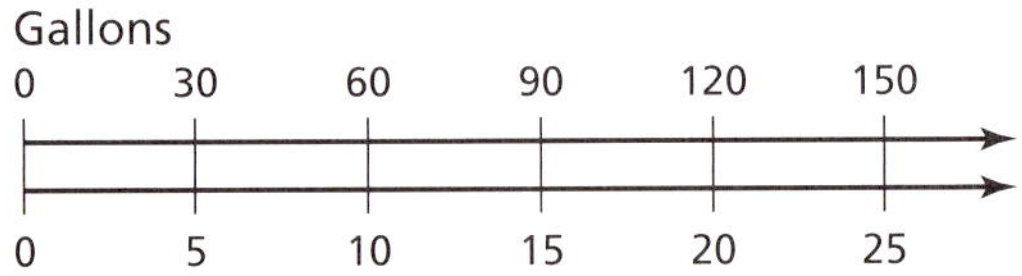

Write a unit rate for the situation.

2 **7.** \$28 saved in 4 weeks

8. 18 necklaces made in 3 hours

9. 270 miles in 6 hours

10. 228 students in 12 classes

11. 2520 kilobytes in 18 seconds

12. 880 calories in 8 servings

13. 1080 miles on 15 gallons

14. \$12.50 for 5 ounces

15. **LIGHTNING** Lightning strikes Earth 1000 times in 10 seconds. How many times does lightning strike per second?

16. **HEART RATE** Your heart beats 240 times in 4 minutes. How many times does your heart beat each minute?

3 **17.** **CAR WASH** You earn \$35 for washing 7 cars. How much do you earn for washing 4 cars?

18. **5K RACE** You jog 2 kilometers in 12 minutes. At this rate, how long will it take you to complete a 5-kilometer race?

Assignment Guide and Homework Check

Level	Day 1 Activity Assignment	Day 2 Lesson Assignment	Homework Check
Basic	3, 4, 28–32	1, 2, 5–25 odd	5, 9, 11, 17, 23
Average	3, 4, 28–32	1, 2, 5–10, 15–18, 21–27 odd	5, 10, 18, 21, 25
Advanced	3, 4, 28–32	1, 2, 6–22 even, 23–27	6, 12, 16, 22, 24

Common Errors

- **Exercise 27** Students may struggle with this exercise. Encourage them to use the first two sentences to write individual work rates. Then ask how they can find the combined work rate (by adding the individual work rates). Using the combined work rate $\frac{5}{6}$ of the job per hour, they will have to realize that they need to turn $\frac{5}{6}$ into 1 in the "portion completed" row of the ratio table, and the corresponding time will be the answer.

5.3 Record and Practice Journal

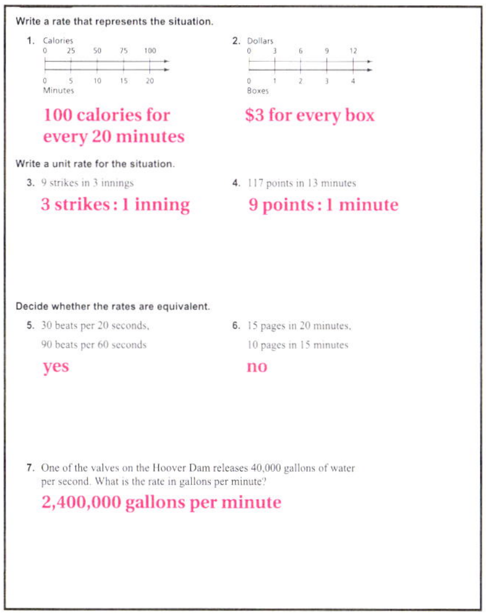

Write a rate that represents the situation.

1. Calories 0, 25, 50, 75, 100; Minutes 0, 5, 10, 15, 20
   100 calories for every 20 minutes
2. Dollars 0, 3, 6, 9, 12; Boxes 0, 1, 2, 3, 4
   $3 for every box

Write a unit rate for the situation.

3. 9 strikes in 3 innings
   3 strikes : 1 inning
4. 117 points in 13 minutes
   9 points : 1 minute

Decide whether the rates are equivalent.

5. 30 beats per 20 seconds, 90 beats per 60 seconds
   yes
6. 15 pages in 20 minutes, 10 pages in 15 minutes
   no

7. One of the valves on the Hoover Dam releases 40,000 gallons of water per second. What is the rate in gallons per minute?
   2,400,000 gallons per minute

Vocabulary and Concept Check

1. *Sample answer:* You walk at a rate of 2 blocks per minute, so you walk 12 blocks in 6 minutes.
2. What is the cost per dozen bagels?; $15; $1.25

Practice and Problem Solving

3. *Sample answer:* 45 words for every 30 minutes
4. *Sample answer:* 18 students for every 8 computers
5. *Sample answer:* 4 inches for every 12 years
6. *Sample answer:* 150 gallons for every 25 seconds
7. $7 per week
8. 6 necklaces per hour
9. 45 miles per hour
10. 19 students per class
11. 140 kilobytes per second
12. 110 calories per serving
13. 72 miles per gallon
14. $2.50 per ounce
15. 100 times per second
16. 60 beats per minute
17. $20
18. 30 min

Practice and Problem Solving

19. equivalent

20. equivalent

21. not equivalent

22. equivalent

23–24. See Additional Answers.

25. See *Taking Math Deeper.*

26. **a.** about 0.12 mile per minute

b. about 8.0 minutes per mile

c. The runner is talking about the rate in part (b) because "10-minute miles" is a way of talking about the rate in minutes per mile.

27. 1.2 h

Fair Game Review

28–31. Sample answers are given.

28. $\frac{2}{6}, \frac{3}{9}$ **29.** $\frac{10}{12}, \frac{20}{24}$

30. $\frac{4}{10}, \frac{8}{20}$ **31.** $\frac{8}{18}, \frac{12}{27}$

32. B

Mini-Assessment

Write a unit rate for the situation.

1. $350 saved in 5 months
 $70 per month
2. 605 miles in 11 hours 55 miles per hour
3. 1246 calories in 7 servings
 178 calories per serving
4. You earn $60 for mowing 5 lawns. How much do you earn for mowing 15 lawns? $180

Taking Math Deeper

Exercise 25

The problem is interesting because it could be solved more simply using algebra. However, we don't know how to use algebra at this stage. So, we need to use another problem-solving strategy, like *Guess, Check, and Revise.*

1 Make a table.

	Homeroom A	Homeroom B
Students	24	16
Canned Food	30	22
Dry Food	42	24
Additional	21	?

2 Find a rate for Homeroom A.

$$\frac{30 + 42 + 21}{24} \approx 3.88 \text{ items per student}$$

3 *Guess, Check, and Revise* a rate for Homeroom B.

$$\frac{22 + 24 + 15}{16} \approx 3.81 \text{ items per student}$$

$$\frac{22 + 24 + 16}{16} \approx 3.88 \text{ items per student}$$

$$\frac{22 + 24 + \text{(17)}}{16} \approx 3.94 \text{ items per student}$$

Project

Organize a food drive at your school. Determine the rates per student for each class. Make a poster to award the class with the highest rate.

Reteaching and Enrichment Strategies

If students need help. . .	If students got it. . .
Resources by Chapter • Practice A and Practice B • Puzzle Time Record and Practice Journal Practice Differentiating the Lesson Lesson Tutorials Skills Review Handbook	Resources by Chapter • Enrichment and Extension • Technology Connection Start the next section

Decide whether the rates are equivalent.

19. 24 laps in 6 minutes
72 laps in 18 minutes

20. 126 points every 3 games
210 points every 5 games

21. 15 breaths every 36 seconds
90 breaths every 3 minutes

22. \$16 for 4 pounds
\$1 for 4 ounces

23. **PRINTER** A printer prints 28 photos in 8 minutes.

a. How many minutes does it take to print 21 more photos?

b. Construct a double number line diagram that represents the situation. How many minutes does it take to print 35 more photos?

24. **SUN VISOR** An athletic director pays \$90 for 12 sun visors for the softball team.

a. How much will the athletic director pay to buy 15 more sun visors?

b. Construct a double number line diagram that represents the situation. What is the cost of 16 sun visors?

25. **FOOD DRIVE** The table shows the amounts of food collected by two homerooms. Homeroom A collects 21 additional items of food. How many more items does Homeroom B need to collect to have more items per student?

	Homeroom A	Homeroom B
Students	24	16
Canned Food	30	22
Dry Food	42	24

26. **MARATHON** A runner completed a 26.2-mile marathon in 210 minutes.

a. Estimate the unit rate, in miles per minute.

b. Estimate the unit rate, in minutes per mile.

c. Another runner says, "I averaged 10-minute miles in the marathon." Is this runner talking about the kind of rate described in part (a) or in part (b)? Explain your reasoning.

27. **Logic** You can do one-half of a job in an hour. Your friend can do one-third of the same job in an hour. How long will it take to do the job if you work together?

Fair Game Review *What you learned in previous grades & lessons*

Write two fractions that are equivalent to the given fraction. *(Skills Review Handbook)*

28. $\frac{1}{3}$ **29.** $\frac{5}{6}$ **30.** $\frac{2}{5}$ **31.** $\frac{4}{9}$

32. **MULTIPLE CHOICE** Which expression is equivalent to $6(x) - 6(2)$? *(Section 3.4)*

Ⓐ $2(x - 6)$ Ⓑ $6(x - 2)$ Ⓒ $12(x - 1)$ Ⓓ $36(x - 2)$

5.4 Comparing and Graphing Ratios

Essential Question How can you compare two ratios?

1 ACTIVITY: Comparing Ratio Tables

Work with a partner.

- **You make purple frosting by adding 1 drop of red food coloring for every 3 drops of blue food coloring.**
- **Your teacher makes purple frosting by adding 3 drops of red food coloring for every 5 drops of blue food coloring.**

a. Copy and complete the ratio table for each frosting mixture.

Your Frosting	
Drops of Red	Drops of Blue
1	
2	
3	
4	
5	

Your Teacher's Frosting	
Drops of Red	Drops of Blue
3	
6	
9	
12	
15	

b. Whose frosting is bluer? Whose frosting is redder? Justify your answers.

c. **STRUCTURE** Insert and complete a new column for each ratio table above that shows the total number of drops. How can you use this column to answer part (b)?

2 ACTIVITY: Graphing from a Ratio Table

Work with a partner.

a. Explain how you can use the values from the ratio table for your frosting to create a graph in the coordinate plane.

b. Use the values in the table to plot the points. Then connect the points and describe the graph. What do you notice?

c. What does the line represent?

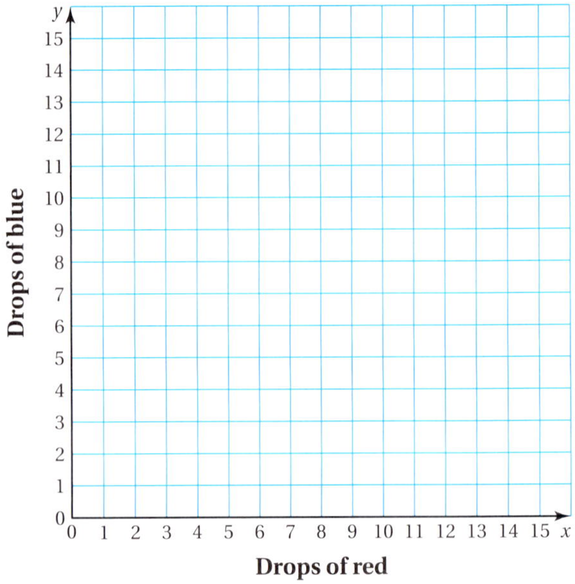

COMMON CORE

Ratios and Rates

In this lesson, you will

- compare ratios.
- compare unit rates.
- graph ordered pairs to compare ratios and rates.

Learning Standards
6.RP.2
6.RP.3a

Laurie's Notes

Introduction

Standards for Mathematical Practice

- **MP3 Construct Viable Arguments and Critique the Reasoning of Others:** This is not a hands-on activity. Instead, students work through a set of directions, and make and justify conjectures. You should develop a culture in your classroom where students are expected to reason, make conjectures, and be able to justify them in a way that is understandable to their peers. Too often, when you ask a student how they arrived at a conjecture, they shrug and say, "I don't know. I just think I'm right."

Motivate

? "What's your favorite color?" If no one says purple, pretend that it is your favorite color and you want to share the origin of purple.

- The original word was most likely used by the Semites. In Greek, the word is *porphrya*, named after a shellfish-derived dye. It was so costly to make that only the privileged could afford to wear purple-colored clothing. Purple was the color of kings through the Middle Ages. The Anglo-Saxons called purple dye *purpul*. At the beginning of the 14th century, English speaking people started using the word purple to refer to the dye and to the color.

Activity Notes

Activity 1

- Students should work with partners. Circulate to check that students are proceeding correctly.
- When students have finished part (b), stop and discuss their answers. Students should understand that because each row of a ratio table represents equivalent ratios, any entry in one table can be compared with any entry in the other table. In their answers, students may compare the third row in the student table with the first row in the teacher table.
- **MP3:** Give students the opportunity to share their answers to part (c).

Activity 2

- Before beginning, review how to plot ordered pairs in the coordinate plane.
- **MP1 Make Sense of Problems and Persevere in Solving Them:** Give sufficient time for students to complete all parts of the activity. Resist the temptation to tell them too much.

? "How can you use the information in the ratio tables in Activity 1 to make ordered pairs?" (drops of red, drops of blue)

? "What do the ordered pairs represent?" equivalent ratios

? "What does (1, 3) represent?" 1 drop of red and 3 drops of blue

? "What does (0, 0) represent?" no drops of either color

- **Extension:** You may want to discuss ordered pairs that do not have whole number coordinates. For instance, you could ask students to estimate an ordered pair such as (1.5, 4.5).

Common Core State Standards

6.RP.2 Understand the concept of a unit rate a/b associated with a ratio $a : b$ with $b \neq 0$, and use rate language in the context of a ratio relationship.

6.RP.3a Make tables of equivalent ratios relating quantities with whole-number measurements, find missing values in the tables, and plot the pairs of values on the coordinate plane. Use tables to compare ratios.

Previous Learning

Students should know how to complete ratio tables, and plot ordered pairs.

Technology for the Teacher

Dynamic Classroom

Lesson Plans
Complete Materials List

5.4 Record and Practice Journal

Essential Question How can you compare two ratios?

1 ACTIVITY: Comparing Ratio Tables

Work with a partner.

- You make purple frosting by adding 1 drop of red food coloring for every 3 drops of blue food coloring.
- Your teacher makes purple frosting by adding 3 drops of red food coloring for every 5 drops of blue food coloring.

a. Complete the ratio table for each frosting mixture.

Your Frosting	
Drops of Red	Drops of Blue
1	3
2	6
3	9
4	12
5	15

Your Teacher's Frosting	
Drops of Red	Drops of Blue
3	5
6	10
9	15
12	20
15	25

b. Whose frosting is bluer? Whose frosting is redder? Justify your answers.
your frosting; your teacher's frosting

c. **STRUCTURE** Insert and complete a new column for each ratio table above that shows the total number of drops. How can you use this column to answer part (b)?
See Additional Answers.

Differentiated Instruction

Visual

Have students create a chart to reinforce the relationship between distance, rate, and time.

distance = rate × time

$d = rt$, where d is in miles, kilometers, feet, and so on

$$\text{time} = \frac{\text{distance}}{\text{rate}}$$

$t = \frac{d}{r}$, where t is in seconds, minutes, hours, years, and so on

$$\text{rate} = \frac{\text{distance}}{\text{time}}$$

$r = \frac{d}{t}$, where r is in feet per second, kilometers per hour, miles per minute, and so on

5.4 Record and Practice Journal

2 ACTIVITY: Graphing from a Ratio Table

Work with a partner.

a. Explain how you can use the values from the ratio table for your frosting to create a graph in the coordinate plane.

ordered pairs (drops of red, drops of blue)

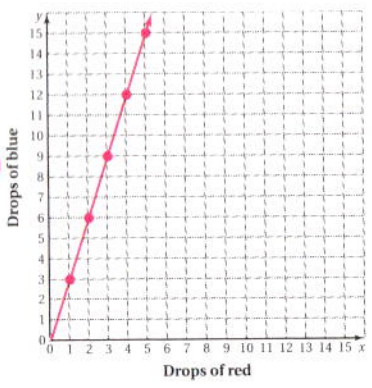

b. Use the values in the table to plot the points. Then connect the points and describe the graph. What do you notice?

points lie on a line that passes through the origin

c. What does the line represent?

all equivalent ratios

3 ACTIVITY: Comparing Graphs From Ratio Tables

Work with a partner. The graph on the next page shows the values from the ratio table for your teacher's frosting.

a. Complete the table and the graph on the next page.

b. Explain the relationship between the entries in the ratio table and the points on the graph.

increases in table are horizontal and vertical increases in graph

Your Teacher's Frosting	
Drops of Red	Drops of Blue
3	5
6	10
9	15
12	20
15	25

(+3 between successive Drops of Red entries; +5 between successive Drops of Blue entries)

c. How is this graph similar to the graph in Activity 2? How is it different?

increasing and through the origin; Activity 2 graph is steeper.

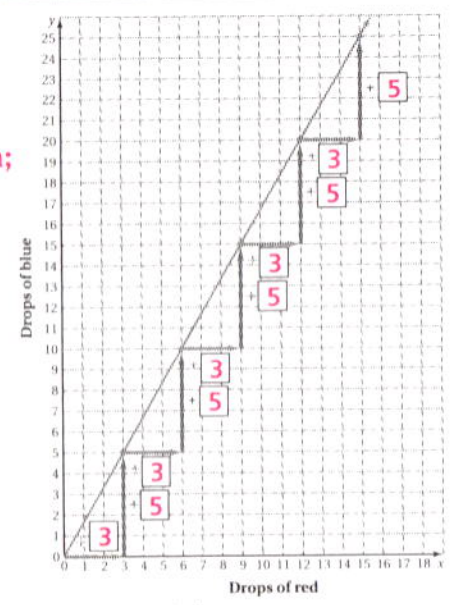

d. How can you use the graphs to determine whose frosting has more red or blue in it? Explain.

See Additional Answers.

What Is Your Answer?

4. **IN YOUR OWN WORDS** How can you compare two ratios?

ratio table or graphs

5. **PRECISION** Your teacher's frosting mixture has 7 drops of red in it. How can you use the graph to find how many drops of blue are needed to make the purple frosting? Is your answer exact? Explain.

about 12 drops of blue

Laurie's Notes

Activity 3

- This activity helps to develop the slope triangle, a key concept in algebra.
- Students should be comfortable in recognizing the additive nature of the ratio table.
- **Teaching Tip:** Ask a volunteer to show his or her work using a document camera or an overhead projector.
- Ask a volunteer to explain the answer to part (b).
- **MP3:** In listening to part (c) responses, there are many valid responses that students may offer. There are also invalid responses! Expect students to listen carefully and critique the explanations of their classmates (**MP3**). If students do not discuss the difference in steepness of the two graphs, then probe this idea.
- **Teaching Tip:** If possible, have two students plot the two graphs on transparencies and then overlay the transparencies. This allows students to visually compare the steepnesses and "read" the number of blue drops for a given number of red drops. "On which graph is there a greater vertical rise for 5 drops of red?" the graph for the teacher's frosting

What Is Your Answer?

- Question 4 is the essential question. Its answer summarizes the two techniques in this activity.

Closure

- A classmate mixes 2 drops of red food coloring for every 4 drops of blue food coloring.
  a. Create a ratio table with 5 entries to represent this situation.
  b. Write the entries of the ratio table as ordered pairs. Then plot and connect the points.
  c. Compare the graph to the graphs in Activities 2 and 3.

3 ACTIVITY: Comparing Graphs from Ratio Tables

Math Practice 7

Look for Patterns

What patterns do you notice in the graph? What does this tell you about the problem?

Work with a partner. The graph shows the values from the ratio table for your teacher's frosting.

a. Complete the table and the graph.

Your Teacher's Frosting	
Drops of Red	Drops of Blue
3	
6	
9	
12	
15	

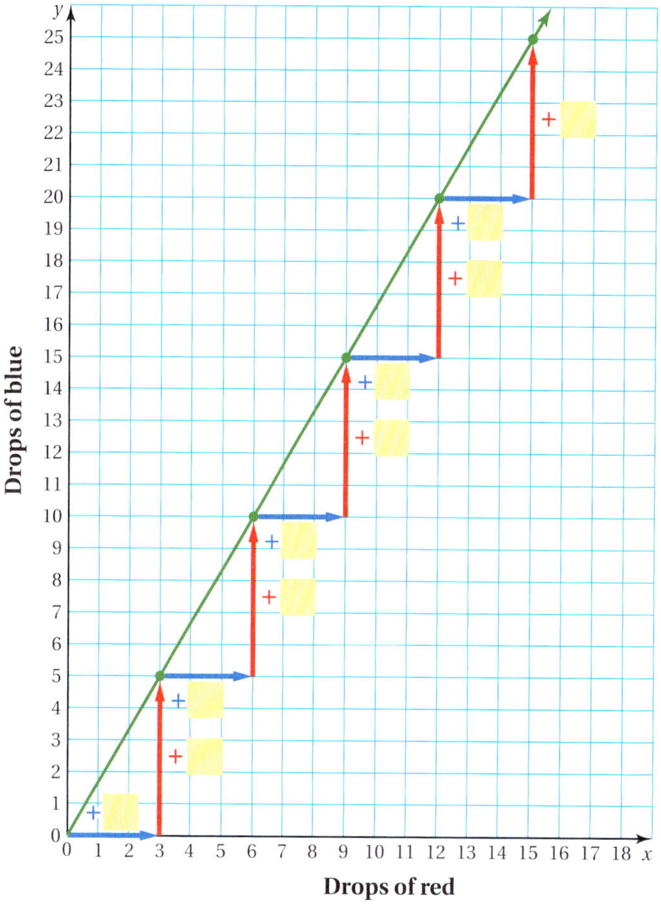

b. Explain the relationship between the entries in the ratio table and the points on the graph.

c. How is this graph similar to the graph in Activity 2? How is it different?

d. How can you use the graphs to determine whose frosting has more red or blue in it? Explain.

What Is Your Answer?

4. **IN YOUR OWN WORDS** How can you compare two ratios?

5. **PRECISION** Your teacher's frosting mixture has 7 drops of red in it. How can you use the graph to find how many drops of blue are needed to make the purple frosting? Is your answer exact? Explain.

Practice Use what you learned about comparing ratios to complete Exercises 3 and 4 on page 214.

5.4 Lesson

Check It Out
Lesson Tutorials
BigIdeasMath.com

One way to compare ratios is by using ratio tables.

EXAMPLE 1 Comparing Ratios

You mix 8 tablespoons of hot sauce and 3 cups of salsa in a green bowl. You mix 12 tablespoons of hot sauce and 4 cups of salsa in an orange bowl. Which mixture is hotter?

Use ratio tables to compare the mixtures. Find a larger batch of each mixture in which the amount of hot sauce or salsa is the same.

Green Bowl

$\times 4$

Hot Sauce (tablespoons)	8	32
Salsa (cups)	3	12

$\times 4$

Orange Bowl

$\times 3$

Hot Sauce (tablespoons)	12	36
Salsa (cups)	4	12

$\times 3$

The tables show that for a larger batch of each mixture using 12 cups of salsa, the orange bowl would have $36 - 32 = 4$ more tablespoons of hot sauce.

So, the mixture in the orange bowl is hotter.

EXAMPLE 2 Comparing Unit Rates

Which bag of dog food is the better buy?

Use ratio tables to find and compare the unit costs.

20-Pound Bag

$\div 20$

Cost (dollars)	17.20	0.86
Food (pounds)	20	1

$\div 20$

30-Pound Bag

$\div 30$

Cost (dollars)	25.20	0.84
Food (pounds)	30	1

$\div 30$

The 20-pound bag costs \$0.86 per pound, and the 30-pound bag costs \$0.84 per pound.

Because \$0.84 is less than \$0.86, the 30-pound bag is the better buy.

On Your Own

Now You're Ready
Exercises 3–10

1. In Example 1, you mix 10 tablespoons of hot sauce and 3 cups of salsa in a red bowl. Which mixture is the mildest? Explain.
2. A 30-pack of paper towels costs \$48.30. A 32-pack costs \$49.60. Which is the better buy? Explain.

Laurie's Notes

Introduction

Connect

- **Yesterday:** Students explored how to compare two ratios (MP1, MP3).
- **Today:** Students will extend their understanding of comparing ratios.

Motivate

? "What are some of the ingredients in salsa?" chili peppers for one

- In 1912, pharmacist Wilbur Scoville created a pepper hotness scale based on how much dilution is necessary for the taste to go unnoticed.
- If time allows, have students research the numbers of Scoville units for different types of peppers and use unit rates to determine how many times hotter one type of pepper is than another.

Lesson Notes

Example 1

- Make sure they understand the strategy being used in the Example, which is to find equivalent ratios for the mixtures that have the same amount of salsa, and then compare the amounts of hot sauce.

? "To make larger batches using the same number of cups of salsa in each mixture, what would you multiply each table by?" Multiply the first table by 4 and the second table by 3.

- **Big Idea:** The least common multiple of 3 and 4 is 12.
- Show other methods and/or ask students for other methods of solving the problem. They may divide to find unit rates, which would force them to compare fractions. In making the amount of hot sauce the same, they would see that the mixture in the orange bowl has less salsa for the same amount of hot sauce, interpreting it as being hotter.

Example 2

- Tell students that one of the most common applications of comparing rates is finding the better buy.

? "What is the size and price of each bag of dog food?" 20 pounds for $17.20 and 30 pounds for $25.20.

- Work through the problem as shown. Be sure to interpret the unit rate as the cost per pound. You want to find the lower cost per pound.
- **Extension:** Use an alternate method. Compare the costs of 60 pounds of dog food using (a) three 20-pound bags and (b) two 30-pound bags.

On Your Own

- In Question 2, students often ask whether they should divide units by cost (30 ÷ $48.30) or cost by units ($48.30 ÷ 30). Remind them what *unit cost* represents—the cost for 1 unit. So, you want to divide by the number of units. Dividing by the cost gives how many rolls of paper towels can be purchased for $1.

Goal Today's lesson is comparing and graphing **ratios**.

Technology for the Teacher

Dynamic Classroom

Lesson Tutorials
Lesson Plans
Answer Presentation Tool

Extra Example 1

You mix 4 drops of flavoring into 12 ounces of water. You mix 7 drops of flavoring into 18 ounces of water. Which mixture has more flavor? the 18 ounce mixture

Extra Example 2

A 6-pound bag of cat food costs $12.84. A 16-pound bag of cat food costs $33.92. Which bag of cat food is the better buy? the 16-pound bag

On Your Own

1. The one in the green bowl; For a batch of each mixture using 12 cups of salsa, the green bowl would have the least amount of hot sauce.
2. 32-pack; The unit cost of the 32-pack ($1.55) is less than the unit cost of the 30 pack ($1.61).

Extra Example 3

An eagle eats about 1 pound of food in 1 day. A crow eats about 3 pounds of food in 2 days.

a. Complete the ratio table for each bird. Which bird eats more food per day?

Eagle	
Days	Pounds Eaten
1	1
2	2
3	3
4	4

Crow	
Days	Pounds Eaten
2	3
4	6
6	9
8	12

the crow

b. Graph the ordered pairs (day, pounds eaten) from the tables in part (a). What can you conclude?

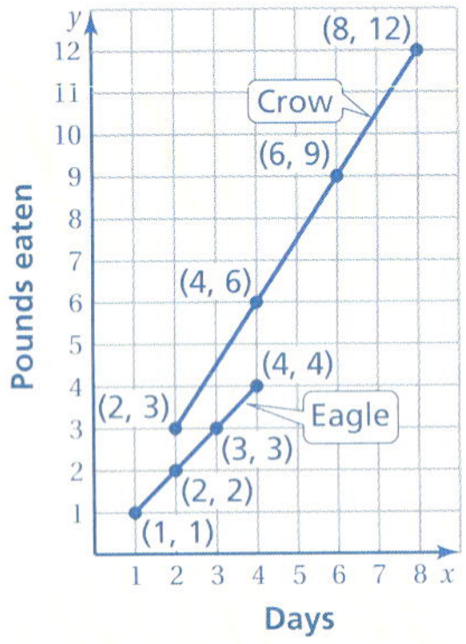

The graph for the crow is steeper, so the crow eats more food per day than the eagle does.

On Your Own

3. The two graphs now coincide because the rates are equivalent. So, both rise at the same rate.

English Language Learners

Vocabulary

Different words are used to describe rates, for example: $2 a bushel, $7.50 for every sheet, and 6 inches per year. Prepositions are often hard for English learners to master. Encourage them to learn prepositions, but remind them that in this context they are often interchangeable.

Laurie's Notes

Example 3

? "What is the rate of the hot air balloon?" 9 meters every 3 seconds

? "What is the rate of the blimp?" 7 meters every 2 seconds

? "Do you know which aircraft is rising faster?" Some students may have a sense of which is faster.

- **MP3 Construct Viable Arguments and Critique the Reasoning of Others:** Have students use the table and graph to support their answers about which aircraft is rising faster.
- Explain that they will be completing a ratio table for each aircraft and then making a graph of each table.
- Complete the ratio tables in part (a). Take time to discuss the color-coded multiplication arrows for each. The connection you want students to make is that multiplication is repeated addition. Adding four 9s is equal to 9×4. If time permits, then you could extend this idea by adding another row to the ratio tables that is not consecutive. For instance, a new row for the balloon could have 21 and 63. (Multiply each entry in the first ratio by 7.)

? "Using the ratio tables, which aircraft rises faster?" After 6 seconds, the blimp is at 21 meters and the balloon is at 18 meters. The blimp rises faster.

? "If the tables were extended, would there be other equivalent ratios that would help answer the question? Explain." Yes, after 12 seconds, the blimp would be at 42 meters and the balloon would be at 36 meters.

- **MP6: Attend to Precision** In completing part (b), it is important for students to be precise in their graphing. The scale on the y-axis needs to be something other than 1 in order to make a reasonably sized graph. Also, this is an example where the scale on each axis is different, which may not be familiar to students. Be sure they understand that the scales on the axes do not need to be the same, but each must be consistent.
- Ask students to write the ordered pairs associated with each ratio table.
- If possible, have students use colored pencils to draw the graphs.
- Have students plot and connect the ordered pairs.

? "Why does it make sense to connect the ordered pairs?" Listen for the idea that the devices keep moving as time increases. This is an informal discussion of continuity. Interpret a point such as (1, 3.5) for the blimp.

? "(0, 0) is not in the ratio table. Does it make sense to include it in the graph? Explain." Yes; (0, 0) is when both aircraft are on the ground.

? "How can you use the graphs to tell which aircraft rises faster?" Listen for mention of the steepness of the lines, or reference to two ordered pairs that are vertically aligned.

- Discuss the Study Tip. Point out that height is a distance.

On Your Own

- **Neighbor Check:** Have students work independently and then have their neighbors check their work. Have students discuss any discrepancies.

Closure

- **Writing Prompt:** To compare two ratios you can . . .

EXAMPLE 3 Graphing Values from Ratio Tables

A hot-air balloon rises 9 meters every 3 seconds. A blimp rises 7 meters every 2 seconds.

Rises 9 meters every 3 seconds.

a. Complete the ratio table for each aircraft. Which rises faster?

Balloon	
Time (seconds)	Height (meters)
3	9
6	18
9	27
12	36

(×2, ×3, ×4)

Blimp	
Time (seconds)	Height (meters)
2	7
4	14
6	21
8	28

(×2, ×3, ×4)

Every 6 seconds, the balloon rises 18 meters and the blimp rises 21 meters.

So, the blimp rises faster.

b. Graph the ordered pairs (time, height) from the tables in part (a). What can you conclude?

Write the ordered pairs.

Balloon: (3, 9), (6, 18), (9, 27), (12, 36)

Blimp: (2, 7), (4, 14), (6, 21), (8, 28)

Study Tip

When graphing speed, you often place time on the horizontal axis and distance on the vertical axis.

Plot and label each set of ordered pairs. Then draw a line through each set of points.

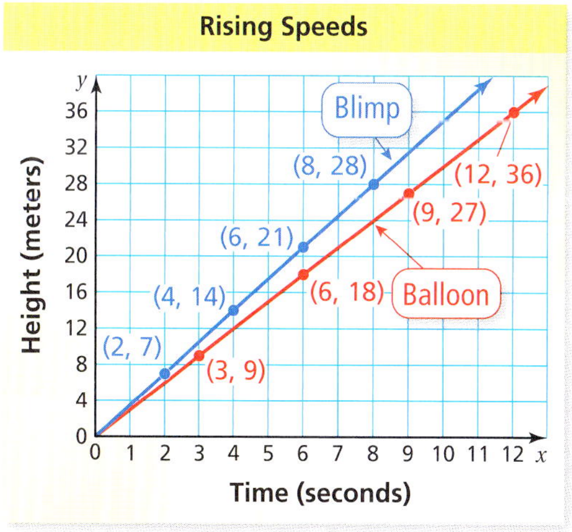

Both graphs begin at (0, 0). The graph for the blimp is steeper, so the blimp rises faster than the hot-air balloon.

On Your Own

Now You're Ready Exercises 12 and 13

3. **WHAT IF?** The blimp rises 6 meters every 2 seconds. How does this affect your conclusion?

5.4 Exercises

Check It Out
Help with Homework
BigIdeasMath.com

Vocabulary and Concept Check

1. **WRITING** Explain how to use tables to compare ratios.

2. **NUMBER SENSE** Just by looking at the graph, determine who earns a greater hourly wage. Explain.

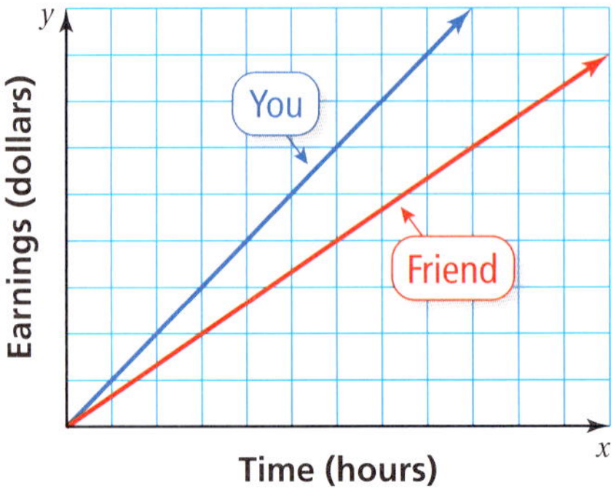

Practice and Problem Solving

Determine which car gets the better gas mileage.

1 3.

Car	A	B
Distance (miles)	125	120
Gallons Used	5	6

4.

Car	A	B
Distance (miles)	300	320
Gallons Used	8	10

5.

Car	A	B
Distance (miles)	450	405
Gallons Used	15	12

6.

Car	A	B
Distance (miles)	360	270
Gallons Used	20	18

Determine which is the better buy.

2 7.

Air Freshener	A	B
Cost (dollars)	6	12
Refills	2	3

8.

Kitten Food	A	B
Cost (dollars)	15	9
Cans	18	12

9.

Ham	A	B
Cost (dollars)	5.70	8.75
Pounds	3	5

10.

Cheese	A	B
Cost (dollars)	3.59	5.12
Slices	10	16

11. **SALT WATER GARGLE** Salt water gargle can temporarily relieve a sore throat. One recipe calls for $\frac{3}{4}$ teaspoon of salt in 1 cup of water. A second recipe calls for 1 teaspoon of salt in 2 cups of water. Which recipe will taste saltier?

Assignment Guide and Homework Check

Level	Day 1 Activity Assignment	Day 2 Lesson Assignment	Homework Check
Basic	3, 4, 19–22	1, 2, 5–15 odd, 16, 17	5, 7, 11, 13, 15
Average	3, 4, 19–22	1, 2, 5–16	5, 8, 11, 12, 14
Advanced	3, 4, 19–22	1, 2, 5–18	5, 8, 11, 13, 15

Common Errors

- **Exercises 3–10** Students may find the reciprocal of the unit rate and come up with the incorrect conclusion. Remind them that the unit rate represents the rate for one unit, so they should divide by the number of units.
- **Exercise 18** Students may not realize that they can find another point on the graph of values from a ratio table by multiplying both coordinates by the same number k. So, in general, another point on the graph is (kp, kq). Lead students to this conclusion without giving away the answer.

5.4 Record and Practice Journal

Determine which is the better buy.

1.

Iced Tea	A	B
Cost (dollars)	3	4
Refills	2	3

B

2.

Lunch Meat	A	B
Cost (dollars)	10	6
Pounds	2	1

A

3.

Movie Rental	A	B
Cost (dollars)	12	11
Rentals	4	3

A

4.

Buns	A	B
Cost (dollars)	6	5
Packages	3	2

A

5.

CDs	A	B
Cost (dollars)	60	72
CDs	6	8

B

6.

Flash Drive	A	B
Cost (dollars)	18	35
Flash Drives	3	5

A

7. You are making cookies. One recipe calls for 4 cups of chocolate morsels for 3 batches of cookies. A second recipe calls for 5 cups of chocolate morsels for 4 batches of cookies. Which cookies will contain more chocolate morsels?

first recipe

Vocabulary and Concept Check

1. Use ratio tables to write equivalent ratios where one part from each ratio has the same numerical value. Compare the other part.
2. You; The graph that represents your earnings is steeper, so you have a greater hourly wage.

Practice and Problem Solving

3. A
4. A
5. B
6. A
7. A
8. B
9. B
10. B
11. the first recipe

Practice and Problem Solving

12–13. See Additional Answers.

14. whole milk

15. See *Taking Math Deeper*.

16–17. See Additional Answers.

18. In general, all points on the graph will be of the form (kp, kq) where k is any positive number.
Sample answer: $(2p, 2q)$

Fair Game Review

19. 16 **20.** 18 R26

21. 34 R109

22. B

Mini-Assessment

1. Determine which brand of turkey is the better buy.

Brand of Turkey	A	B
Cost (dollars)	8.10	16.60
Pounds	2	4

Brand A

2. Using the table in Question 1, determine the cost for 10 pounds of each brand of turkey. Brand A: $40.50, Brand B: $41.50

3. Determine which building company has the better buy for a house.

Company	A	B
Cost of a House (thousands of dollars)	150	175
Living Space (square feet)	1250	1400

Company A

4. Using the table in Question 3, determine the cost for a 1600-square-foot house from each building company. Company A: $192,000, Company B: $200,000

Taking Math Deeper

Exercise 15

For this exercise, you can use double number lines to help visualize the relationship between the two heart rates.

1. Construct a double number line that represents the horse's heart rate.

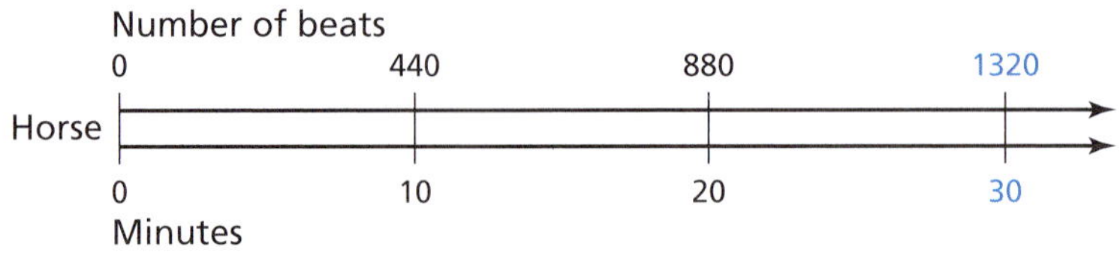

2. Construct a double number line that represents the cow's heart rate.

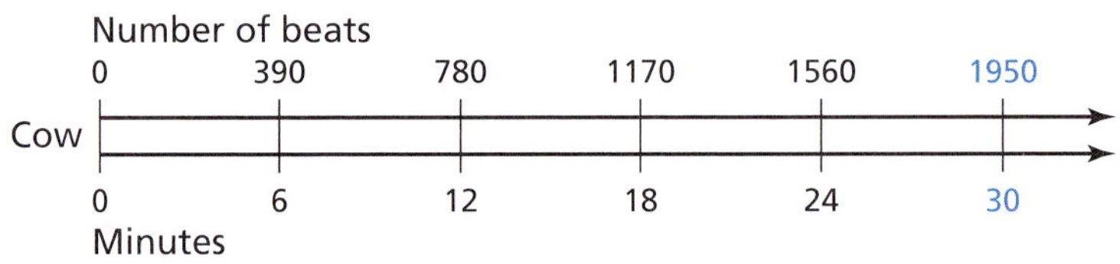

3. Answer the question.

Compare the two double number lines.

The double number lines show that the horse's heart beats 1320 times in 30 minutes and the cow's heart beats 1950 times in 30 minutes.

Because 1950 > 1320, the cow has the greater heart rate.

Project

List two other ways you can use double number lines to find the solution in Exercise 15. Compare the methods.

Reteaching and Enrichment Strategies

If students need help. . .	If students got it. . .
Resources by Chapter • Practice A and Practice B • Puzzle Time Record and Practice Journal Practice Differentiating the Lesson Lesson Tutorials Skills Review Handbook	Resources by Chapter • Enrichment and Extension • Technology Connection Start the next section

Complete the ratio tables and graph the ordered pairs from the tables. What can you conclude?

③ **12.**

Water Tank	
Time (min)	Liters Leaked
2	4
4	
6	
8	

Swimming Pool	
Time (min)	Liters Leaked
3	2
6	
9	
12	

13.

Zoo	
People	Cost (dollars)
4	60
8	
12	
16	

Museum	
People	Cost (dollars)
5	95
10	
15	
20	

14. **MILK** In whole milk, 13 parts out of 400 are milk fat. In 2% milk, 1 part out of 50 is milk fat. Which type of milk has more milk fat per cup?

15. **HEART RATE** A horse's heart beats 440 times in 10 minutes. A cow's heart beats 390 times in 6 minutes. Which animal has a greater heart rate?

16. **CHOOSE TOOLS** A chemist prepares two acid solutions.

a. Use a ratio table to determine which solution is more acidic.

b. Use a graph to determine which solution is more acidic.

c. Which method do you prefer? Explain.

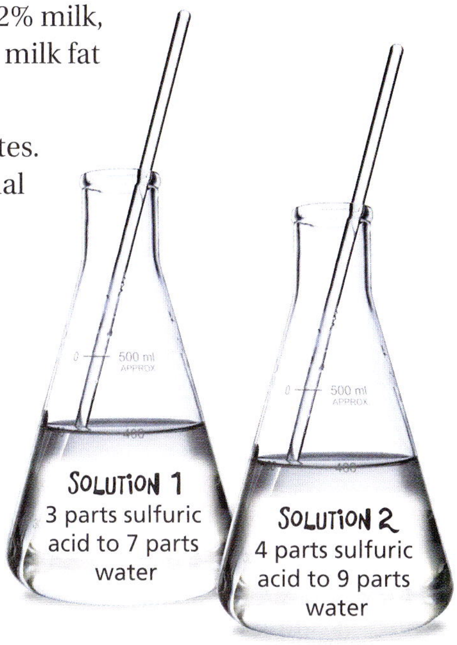

17. **NUT MIXTURE** A company offers a nut mixture with 7 peanuts for every 4 almonds. The company changes the mixture to have 8 peanuts for every 5 almonds, but the number of nuts per container does not change.

a. Create a ratio table for each mixture. How many nuts are in the smallest possible container?

b. Graph the ordered pairs from the tables. What can you conclude?

c. Almonds cost more than peanuts. Should the company charge more or less for the new mixture? Explain your reasoning.

18. **Structure** The point (p, q) is on the graph of values from a ratio table. What is another point on the graph?

Fair Game Review What you learned in previous grades & lessons

Divide. *(Section 1.1)*

19. $544 \div 34$ **20.** $1520 \div 83$ **21.** $8439 \div 245$

22. **MULTIPLE CHOICE** Which of the following numbers is equal to 9.32 when you increase it by 4.65? *(Section 2.4)*

Ⓐ 4.33 Ⓑ 4.67 Ⓒ 5.67 Ⓓ 13.97

5 Study Help

Check It Out
Graphic Organizer
BigIdeasMath.com

You can use a **definition and example chart** to organize information about a concept. Here is an example of a definition and example chart for ratio.

Ratio: a comparison of two quantities. Ratios can be part-to-part, part-to-whole, or whole-to-part comparisons.

- Example: 4 to 5
- Example: 2 : 5
- Example: teachers : students

On Your Own

Make definition and example charts to help you study these topics.

1. equivalent ratios
2. ratio table
3. rate
4. unit rate
5. equivalent rates

After you complete this chapter, make definition and example charts for the following topics.

6. percent
7. U.S. customary system
8. metric system
9. conversion factor
10. unit analysis

"My math teacher taught us how to make a **definition and example chart**."

Sample Answers

1.

Equivalent ratios: two ratios that describe the same relationship.

Example

3 : 4 and 9 : 12

Example

50 to 40 and 5 to 4

Example

1 : 12 and 5 to 60

2.

Ratio table: a table used to find and organize equivalent ratios.

Example

Girls	5	10	20
Boys	3	6	12

Equivalent ratios: 5 : 3, 10 : 6, 20 : 12

Example

Flour (cups)	3	6	9	12
Water (cups)	1	2	3	4

Equivalent ratios: 3 : 1, 6 : 2, 9 : 3, 12 : 4

Example

Red	3	18
Yellow	7	42

Equivalent ratios: 3 : 7, 18 : 42

3.

Rate: a ratio of two quantities using different units.

Example

$\frac{\$135}{\text{3 hours}}$

Example

$\frac{\text{63 feet}}{\text{12 seconds}}$

Example

$\frac{\text{8 hits}}{\text{25 at bats}}$

4–5. Available at *BigIdeasMath.com*.

List of Organizers

Available at *BigIdeasMath.com*

Comparison Chart
Concept Circle
Definition (Idea) and Example Chart
Example and Non-Example Chart
Formula Triangle
Four Square
Information Frame
Information Wheel
Notetaking Organizer
Process Diagram
Summary Triangle
Word Magnet
Y Chart

About this Organizer

A **Definition and Example Chart** can be used to organize information about a concept. Students fill in the top rectangle with a term and its definition or description. Students fill in the rectangles that follow with examples to illustrate the term. Each sample answer shows 3 examples, but students can show more or fewer examples. Definition and example charts are useful for concepts that can be illustrated with more than one type of example.

Technology for the Teacher

Editable Graphic Organizer

Answers

1. 3 to 2, or 3 : 2; For every 3 tulips, there are 2 lilies.
2. 8 to 4, or 8 : 4; For every 8 crayons, there are 4 markers.
3.

Shoes	7	28	49
Boots	2	4	14

7 : 2, 28 : 4, and 49 : 14

4.

Trains	3	12	18
Airplanes	8	32	48

3 : 8, 12 : 32, and 18 : 48

5. *Sample answer:* 20 liters for every 6 minutes
6. *Sample answer:* 15 tickets for every 60 points
7. 2 touchdowns per game
8. 3 text messages per minute
9. 20 entries per contest
10. 3 questions per minute
11. $60
12. 24-fluid-ounce shampoo; *Sample answer:* Because $0.20 is less than $0.22, the 24-fluid-ounce shampoo is the better buy.
13. **a.** 17 to 16, or 17 : 16; For every 17 Celtics championships, there are 16 Lakers championships.

    b. 3 to 4, or 3 : 4; For every 3 Pistons championships, there are 4 Spurs championships.

    c. 6 to 16, or 6 : 16; For every 6 Bulls championships, there are 16 Lakers championships.

Technology for the Teacher

Online Assessment
Assessment Book
ExamView® Assessment Suite

Alternative Quiz Ideas

100% Quiz	Math Log
Error Notebook	**Notebook Quiz**
Group Quiz	Partner Quiz
Homework Quiz	Pass the Paper

Notebook Quiz

A notebook quiz is used to check students' notebooks. Students should be told at the beginning of the course what the expectations are for their notebooks: notes, class work, homework, date, problem number, goals, definitions, or anything else that you feel is important for your class. They also need to know that it is their responsibility to obtain the notes when they miss class.

1. On a certain day, what was the answer to the warm up question?
2. On a certain day, how was this vocabulary term defined?
3. In Section 5.2, what is the answer to On Your Own Question 1?
4. In Section 5.3, what is the answer to the Essential Question?
5. On a certain day, what was the homework assignment?

Give students 5 minutes to answer these questions.

Reteaching and Enrichment Strategies

If students need help. . .	If students got it. . .
Resources by Chapter • Practice A and Practice B • Puzzle Time Lesson Tutorials *BigIdeasMath.com*	Resources by Chapter • Enrichment and Extension • Technology Connection Game Closet at *BigIdeasMath.com* Start the next section

5.1–5.4 Quiz

Check It Out
Progress Check
BigIdeasMath.com

Write the ratio. Explain what the ratio means. *(Section 5.1)*

1. tulips to lilies

2. crayons to markers

Find the missing values in the ratio table. Then write the equivalent ratios. *(Section 5.2)*

3.

Shoes	7		49
Boots	2	8	

4.

Trains	3	12	
Airplanes	8		48

Write a rate that represents the situation. *(Section 5.3)*

5.

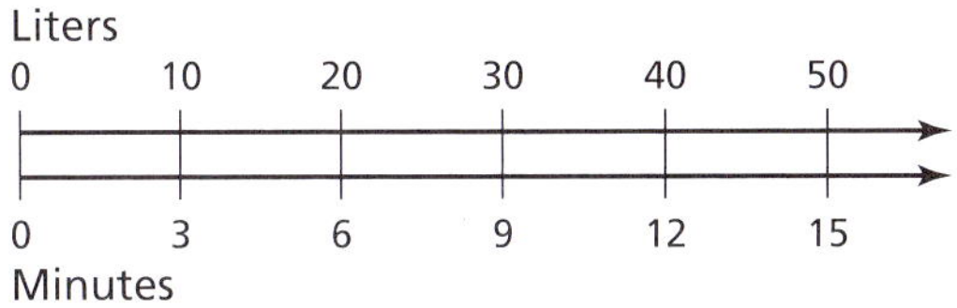

6.

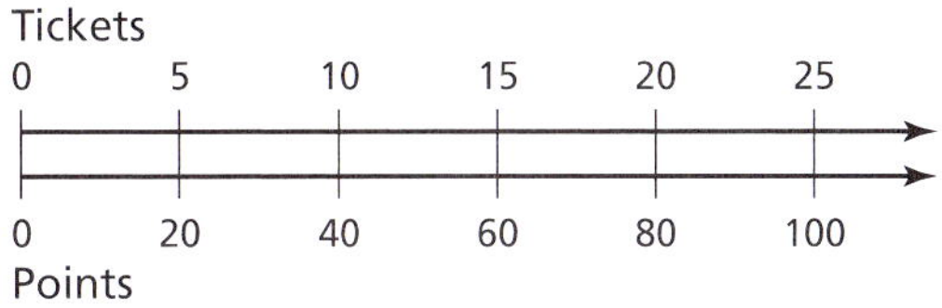

Write a unit rate for the situation. *(Section 5.3)*

7. 12 touchdowns in 6 games

8. 15 text messages in 5 minutes

9. 80 entries in 4 contests

10. 75 questions in 25 minutes

11. DOWNLOADS Three album downloads cost \$36. How much do 5 album downloads cost? *(Section 5.3)*

12. SHAMPOO You can buy 20 fluid ounces of shampoo for \$4.40 or 24 fluid ounces for \$4.80. Which is the better buy? Explain. *(Section 5.4)*

13. NBA CHAMPIONSHIPS Write each ratio. Explain what the ratio means. *(Section 5.1)*

a. Celtics championships to Lakers championships

b. Pistons championships to Spurs championships

c. Bulls championships to Lakers championships

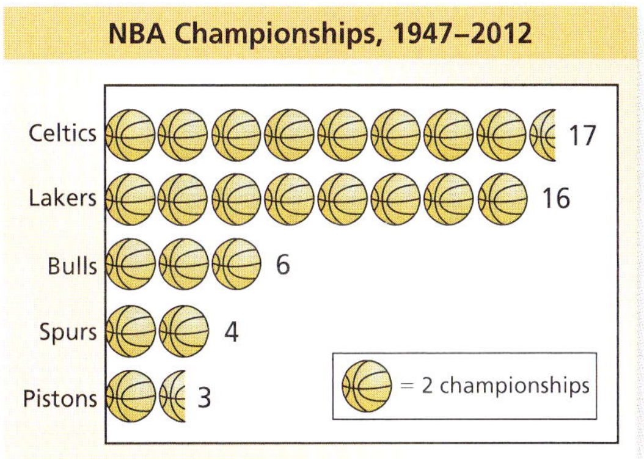

5.5 Percents

Essential Question What is the connection between ratios, fractions, and percents?

1 ACTIVITY: Writing Ratios

Work with a partner.

- **Write the fraction of the squares that are shaded.**
- **Write the ratio of the number of shaded squares to the total number of squares.**
- **How are the ratios and the fractions related?**
- **When can you write ratios as fractions?**

a. 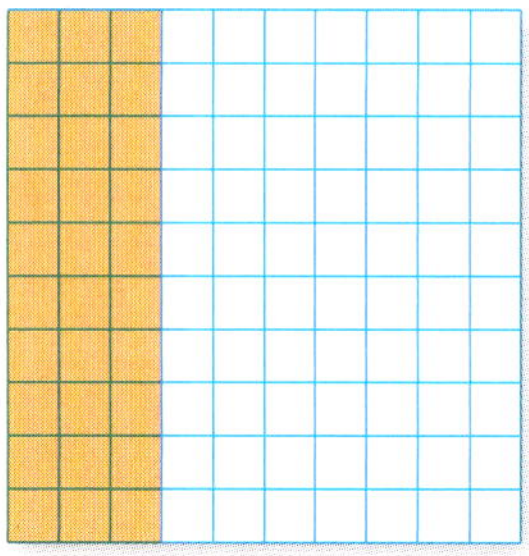

b. 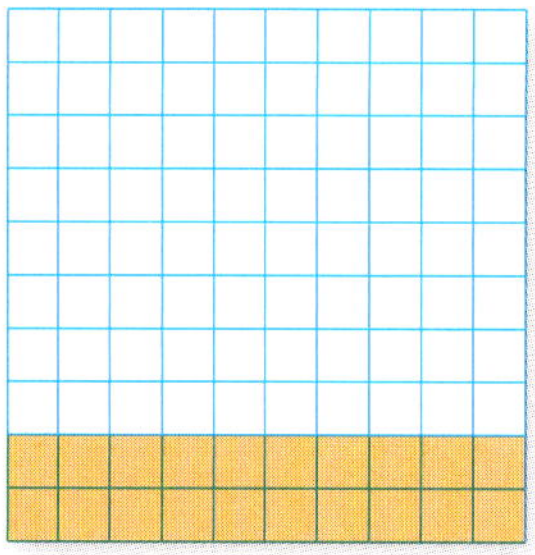

c. 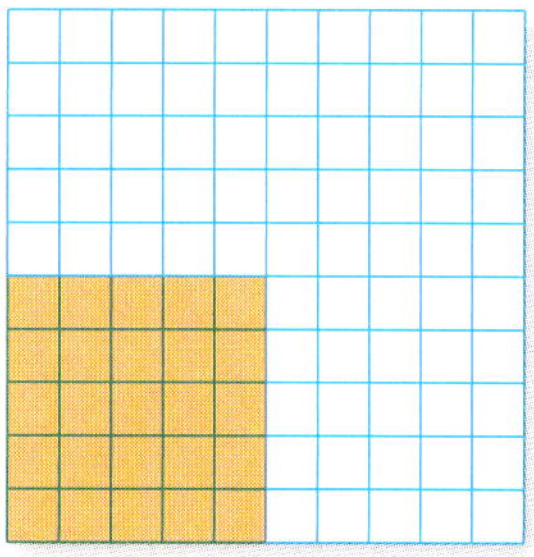

The Meaning of a Word ● Percent

A century is 100 years.

A cent is one hundredth of a dollar.

In Mexico, a centavo is one hundredth of a peso.

Cent means *one hundred*, so **percent** means *per one hundred*. The symbol for percent is %.

COMMON CORE

Percents

In this lesson, you will

- write percents as fractions with denominators of 100.
- write fractions as percents.

Learning Standard 6.RP.3c

Laurie's Notes

Introduction

Standards for Mathematical Practice

- **MP4 Model with Mathematics:** Using the 100-grid model, students make connections between ratios, fractions, and percents. Students should be able to model these relationships and read the 100-grid model.

Motivate

- Shake a few coins in your hand to draw attention to the money.
- ? Ask students to write each U.S. coin as a fractional amount of $1.00.
  penny $\frac{1}{100}$, nickel $\frac{5}{100}$, dime $\frac{10}{100}$, quarter $\frac{25}{100}$, half-dollar $\frac{50}{100}$, dollar $\frac{100}{100}$
- Be sure to include the presidential $1 coins.

Activity Notes

Activity 1

- Caution students to read the directions and questions carefully. I tell students to read directions twice, the second time aloud, before asking for help.
- It is not necessary for students to simplify the fractions.
- The ratios should be written as *a* to *b* or *a* : *b*.
- For the question in the third bullet, students may need some prompting to recall part-to-part and part-to-whole comparisons.
- **Big Idea:** When a ratio is a part-to-whole comparison, it is equivalent to the fractional representation.
- **MP4:** The 100-grid represents, or models, the whole.

Discuss

- The word *percent* is used in common language. Discuss applications and contexts.
- ? "Where do you hear the word *percent* used? Give an example of how it is used." *Sample answers:* 80% on a test; 50% chance of rain; 25% off on a sale item; 6% interest rate
- For each example given, ask students to explain.
- Discuss the examples in the book (century, cent, and centavo) and the notation.

Common Core State Standards

6.RP.3c Find a percent of a quantity as a rate per 100; solve problems involving finding the whole, given a part and the percent.

Previous Learning

Students should know how to write halves, fourths, tenths, and hundredths as percents. Students also need to be able to convert between fractions and decimals, and to simplify and write equivalent fractions.

Technology for the Teacher

Dynamic Classroom

Lesson Plans
Complete Materials List

5.5 Record and Practice Journal

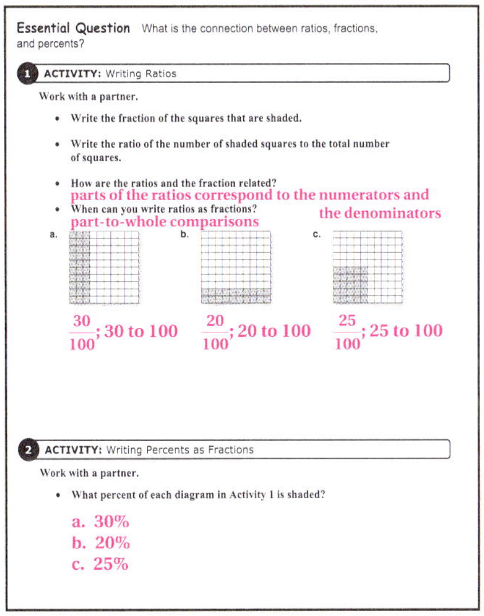

Essential Question What is the connection between ratios, fractions, and percents?

1 ACTIVITY: Writing Ratios

Work with a partner.

- Write the fraction of the squares that are shaded.
- Write the ratio of the number of shaded squares to the total number of squares.
- How are the ratios and the fraction related?
  parts of the ratios correspond to the numerators and the denominators
- When can you write ratios as fractions?
  part-to-whole comparisons

a. $\frac{30}{100}$; 30 to 100 b. $\frac{20}{100}$; 20 to 100 c. $\frac{25}{100}$; 25 to 100

2 ACTIVITY: Writing Percents as Fractions

Work with a partner.

- What percent of each diagram in Activity 1 is shaded?

a. 30%
b. 20%
c. 25%

English Language Learners

Visual

In their notebooks, have students make a table showing the fractional equivalents of common percents.

$10\% = \frac{1}{10}$	$20\% = \frac{1}{5}$	$25\% = \frac{1}{4}$
$30\% = \frac{3}{10}$	$40\% = \frac{2}{5}$	$50\% = \frac{1}{2}$
$60\% = \frac{3}{5}$	$70\% = \frac{7}{10}$	$75\% = \frac{3}{4}$
$80\% = \frac{4}{5}$	$90\% = \frac{9}{10}$	$100\% = 1$

5.5 Record and Practice Journal

- What percent of each diagram below is shaded? Write each percent as a fraction in simplest form.

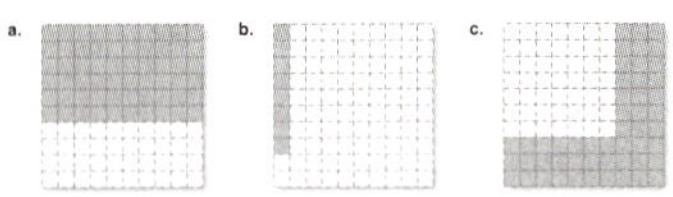

a. $60\%; \frac{3}{5}$ b. $8\%; \frac{2}{25}$ c. $51\%; \frac{51}{100}$

3 ACTIVITY: Writing Fractions as Percents

Work with a partner. Draw a model to represent the fraction. How can you write the fraction as a percent?

a. $\frac{2}{5}$

$\frac{2}{5} = \frac{40}{100} = 40\%$

b. $\frac{7}{10}$

$\frac{7}{10} = \frac{70}{100} = 70\%$

c. $\frac{3}{5}$

$\frac{3}{5} = \frac{60}{100} = 60\%$

d. $\frac{3}{4}$

$\frac{3}{4} = \frac{75}{100} = 75\%$

e. $\frac{3}{25}$

$\frac{3}{25} = \frac{12}{100} = 12\%$

What Is Your Answer?

4. IN YOUR OWN WORDS What is the connection between ratios, fractions, and percents? Give an example with your answer.

Fractions can be written as percents and vise versa, which represent part-to-whole ratios.

5. REASONING Your score on a test is 110%. What does this mean?

You earned more points on the test than it was worth.

Laurie's Notes

Activity 2

- Make sure students understand the 100-grid model. Ten rows and ten columns make 100 parts.
- ? "Where have you seen this model before?" decimals; Some students may say it looks like Base 10 pieces for whole numbers—flats, strips, units.
- Check answers to each problem.
- ? "Does it matter where the squares are shaded?" No; Could be checkerboard style or random.
- **Summarize:** Be sure to tell students that when the percent is given, for instance 40%, the percent symbol (%) means writing the number as a fraction with a denominator of 100. Write $\frac{40}{100}$, then say "to complete, be sure to simplify the fraction to $\frac{2}{5}$."

Activity 3

- This activity reverses the process of what students did in Activity 2. In Activity 2, they wrote a percent as a fraction and now they are going to write a fraction as a percent. Have students sketch a model of the fraction. Then have them write an equivalent fraction with a denominator of 100. Then write the percent.
- **FYI:** Sketch fractions on a transparency using a square model that is the same size as a square 100 grid. After shading the fraction $\left(\text{say } \frac{3}{5}\right)$, overlay the transparency that has the 100 grid. Students can see that the area shaded has not changed, only the size of the pieces.

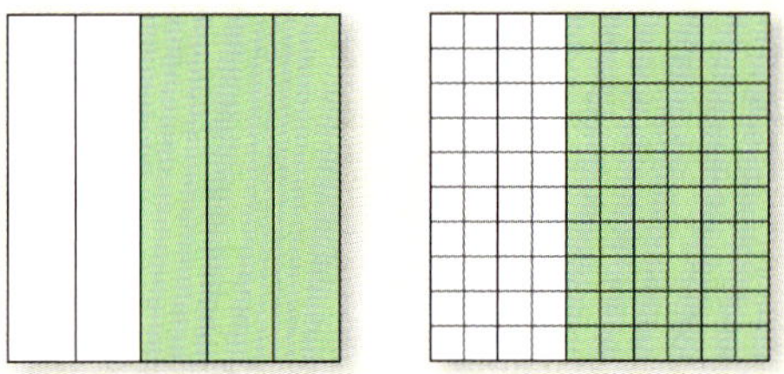

? **Extension:** "You sketched models for $\frac{3}{5}$, $\frac{3}{4}$, and $\frac{3}{25}$, and were able to write them as percents. If you wanted to write a fraction as a percent, which would be easier, $\frac{3}{10}$ or $\frac{3}{7}$ and why?" $\frac{3}{10}$; 10 is a factor of 100 and 7 is not.

What Is Your Answer?

- For Question 5, listen for an answer that includes scoring more than the whole (what the test was worth).

Closure

- Write this list on the board and ask students to put the numbers in order from least to greatest.

$\frac{1}{2}$, 60%, $\frac{2}{5}$, 75%, $\frac{3}{10}$, 90%, $\frac{3}{20}$, 25% $\frac{3}{20}$, 25%, $\frac{3}{10}$, $\frac{2}{5}$, $\frac{1}{2}$, 60%, 75%, 90%

2 ACTIVITY: Writing Percents as Fractions

Work with a partner.

- What percent of each diagram in Activity 1 is shaded?
- What percent of each diagram below is shaded? Write each percent as a fraction in simplest form.

a. 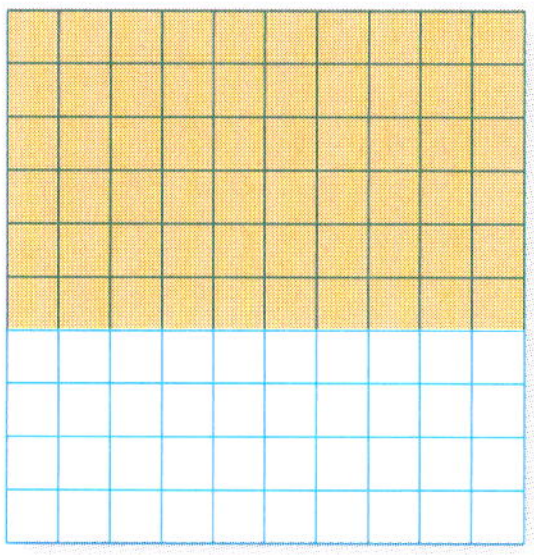

b. 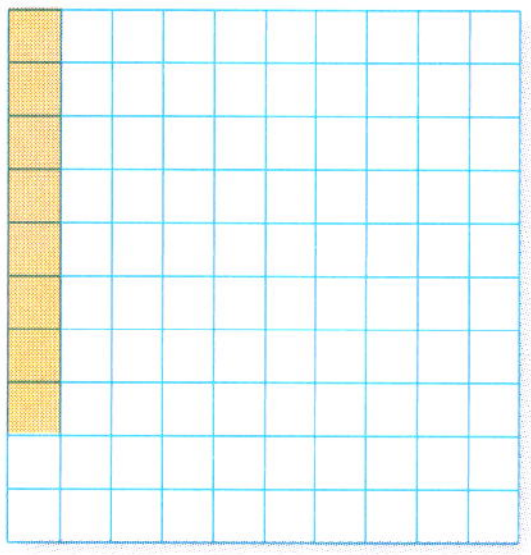

c. 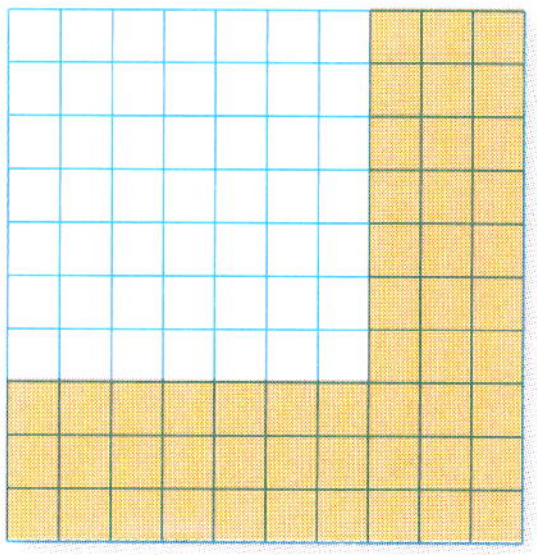

3 ACTIVITY: Writing Fractions as Percents

Work with a partner. Draw a model to represent the fraction. How can you write the fraction as a percent?

a. $\frac{2}{5} = \frac{\square}{100} = \square\%$

b. $\frac{7}{10}$

c. $\frac{3}{5}$

d. $\frac{3}{4}$

e. $\frac{3}{25}$

Math Practice 1

Consider Similar Problems

How is this problem similar to ones you have seen before? How does this help you find the solution?

What Is Your Answer?

4. **IN YOUR OWN WORDS** What is the connection between ratios, fractions, and percents? Give an example with your answer.

5. **REASONING** Your score on a test is 110%. What does this mean?

Practice Use what you learned about percents to complete Exercises 5–7 on page 222.

5.5 Lesson

Check It Out
Lesson Tutorials
BigIdeasMath.com

Key Vocabulary
percent, *p. 220*

Key Idea

Writing Percents as Fractions

Words A **percent** is a part-to-whole ratio where the whole is 100. So, you can write a percent as a fraction with a denominator of 100.

Numbers $60\% = 60 \text{ out of } 100 = \frac{60}{100}$

part (60), per (fraction bar), one hundred (whole) (100)

Algebra $n\% = \frac{n}{100}$

EXAMPLE 1 Writing Percents as Fractions

Study Tip
Equivalent fractions and percents represent the same number using different notations.

a. Write 35% as a fraction in simplest form.

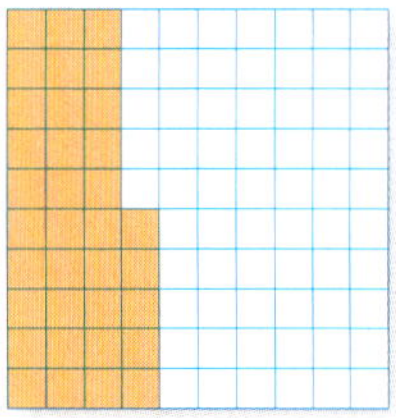

$35\% = \frac{35}{100}$ Write as a fraction with a denominator of 100.

$= \frac{7}{20}$ Simplify.

So, $35\% = \frac{7}{20}$.

b. Write 100% as a fraction in simplest form.

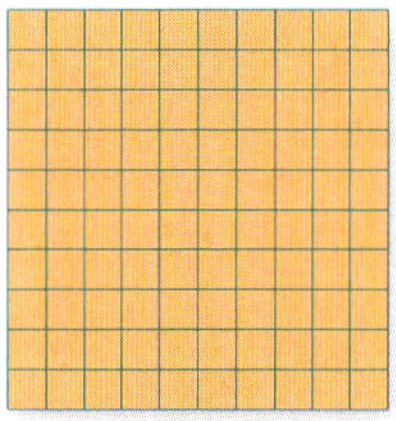

$100\% = \frac{100}{100}$ Write as a fraction with a denominator of 100.

$= 1$ Simplify.

So, $100\% = 1$.

c. Write 174% as a mixed number in simplest form.

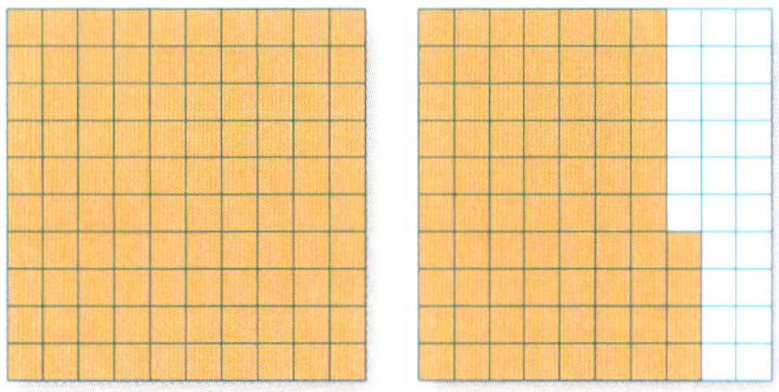

$174\% = \frac{174}{100}$ Write as a fraction with a denominator of 100.

$= \frac{87}{50}$, or $1\frac{37}{50}$ Simplify.

So, $174\% = 1\frac{37}{50}$.

On Your Own

Now You're Ready
Exercises 8–19

Write the percent as a fraction or mixed number in simplest form.

1. 5%
2. 168%
3. 36%
4. 83%

Multi-Language Glossary at BigIdeasMath.com

Laurie's Notes

Introduction

Connect

- **Yesterday:** Students used visual models to represent percents, and then wrote the percents as fractions (MP4).
- **Today:** Students will convert between percents and fractions.

Motivate

- Review *common fractions* and *percents*.
- Students should start to memorize, and have quick recall of common fractions and percents: halves, fourths, fifths, and tenths.
- Students have little difficulty writing a percent as a fraction. Writing fractions as percents is more challenging.
- As a motivator today, use any game format *Matching*, *Memory*, *I Have, Who Has?*, and so on, to review these common fractions and percents.

Lesson Notes

Key Idea

- Write the Key Idea.
- Discuss the algebraic representation.

Example 1

- Work through all three parts of the example.

? "What is 100% as a fraction?" 1

? "If the percent is greater than 100%, what do you know about the equivalent fraction?" It will be greater than 1.

? "Can you think of a context where the percent would be greater than 100%?" *Sample answer:* When you give 110%, it means that you give more than expected.

- Discuss the Study Tip. Students need to understand that fractions and percents are not different types of numbers, just different representations. For example, 30% and $\frac{3}{10}$ represent the same number. This concept is important throughout future mathematics.

On Your Own

- **Think-Pair-Share:** Students should read each question independently and then work in pairs to answer the questions. When they have answered the questions, the pair should compare their answers with another group and discuss any discrepancies.

Goal Today's lesson is converting between **percents** and fractions.

Technology for the Teacher

Dynamic Classroom

Lesson Tutorials
Lesson Plans
Answer Presentation Tool

Extra Example 1

a. Write 76% as a fraction in simplest form. $\frac{19}{25}$

b. Write 118% as a mixed number in simplest form. $1\frac{9}{50}$

On Your Own

1. $\frac{1}{20}$
2. $1\frac{17}{25}$
3. $\frac{9}{25}$
4. $\frac{83}{100}$

Laurie's Notes

Key Idea

- ? "What are equivalent fractions? For instance, $\frac{2}{5}$ and $\frac{4}{10}$ are equivalent fractions. $\frac{2}{5} \cdot \frac{20}{20} = \frac{40}{100}$ and $\frac{4}{10} \cdot \frac{10}{10} = \frac{40}{100}$." two fractions that represent the same amount
- ? "How do you write equivalent fractions with a denominator of 100?" Find a number you can multiply the denominator by so that it equals 100. Multiply the numerator by the same number.
- Students may ask about fractions that cannot be written as equivalent fractions with denominators of 100. Assure students that they will learn other techniques for writing such fractions as percents.

Example 2

- **Connection:** The visual model for $\frac{3}{50}$ and $\frac{6}{100}$ would be equivalent. Use a simpler problem to demonstrate this. $\frac{1}{4} = \frac{4}{16}$.

- ? "How many hundredths would $\frac{1}{4}$ equal?" 25

Example 3

- ? "Can you name the Midwestern states?"
- ? "Is 9 out of 12 a part-to-part comparison or part-to-whole comparison? Explain." part-to-whole; The whole is all 12 Midwestern states, and the part is the 9 states affected by the drought.
- ? "Is 12 a factor of 100?" no
- Explain that by simplifying $\frac{9}{12}$, you obtain a new denominator that is a factor of 100.
- Continue to solve the problem as shown. Point out that they are answering the question, "9 is what percent of 12?"

On Your Own

- **Neighbor Check:** Have students work independently and then have their neighbors check their work. Have students discuss any discrepancies.
- **MP3 Construct Viable Arguments and Critique the Reasoning of Others:** To get students thinking critically, after Question 9 ask them, "What if it rains in 4 Midwestern states? Can you tell what percent, if any, of the states affected by drought got rain? Explain."

Closure

- **Writing Prompts:**
  To write a percent as a fraction, . . .
  To write a fraction as a percent, . . .

Extra Example 2

Write $\frac{13}{20}$ as a percent. 65%

Extra Example 3

You delete 3 out of 16 pictures on your digital camera. What percent of the pictures do you delete? 18.75%

On Your Own

5. 62% **6.** 28%

7. 95% **8.** 150%

9. 100%

Differentiated Instruction

Auditory

Give students fractions that have denominators of 100 or any fraction with a denominator that is a factor of 100, such as $\frac{3}{50}$ or $\frac{3}{10}$. Ask students to give the percent equivalents of the fractions.

Key Idea

Writing Fractions as Percents

Words Write an equivalent fraction with a denominator of 100. Then write the numerator with the percent symbol.

Numbers $\dfrac{1}{4} = \dfrac{25}{100} = 25\%$ (× 25 numerator and denominator)

EXAMPLE 2 Writing a Fraction as a Percent

Write $\dfrac{3}{50}$ as a percent.

$\dfrac{3}{50} = \dfrac{6}{100} = 6\%$ (× 2 numerator and denominator)

Because $50 \times 2 = 100$, multiply the numerator and denominator by 2. Write the numerator with a percent symbol.

EXAMPLE 3 Real-Life Application

A drought affects 9 out of 12 midwestern states. What percent of the midwestern states are affected by the drought?

Midwestern United States

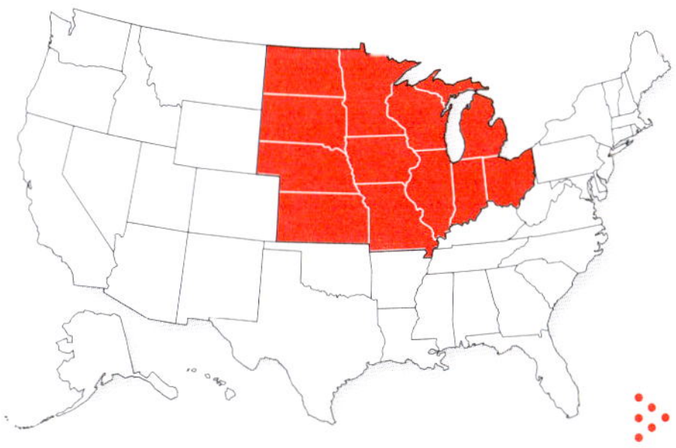

$\dfrac{9}{12} = \dfrac{3}{4}$ Simplify.

$= \dfrac{75}{100}$ ← $\dfrac{3 \times 25}{4 \times 25} = \dfrac{75}{100}$

$= 75\%$ Write the numerator with a percent symbol.

So, 75% of the midwestern states are affected by the drought.

On Your Own

Now You're Ready
Exercises 21–28

Write the fraction or mixed number as a percent.

5. $\dfrac{31}{50}$ **6.** $\dfrac{7}{25}$ **7.** $\dfrac{19}{20}$ **8.** $1\dfrac{1}{2}$

9. WHAT IF? In Example 3, it rains in all the midwestern states. In what percent of the states affected by drought does it rain?

5.5 Exercises

Check It Out
Help with Homework
BigIdeasMath.com

Vocabulary and Concept Check

1. **WRITING** Explain how you can use a 10-by-10 grid to model 42%.

2. **WHICH ONE DOESN'T BELONG?** Which one does *not* have the same value as the other three? Explain your reasoning.

   $\frac{10}{100}$ 10% $\frac{1}{10}$ 0.01

3. **OPEN-ENDED** Write three different fractions that are less than 40%.

4. **NUMBER SENSE** Can $1\frac{1}{4}$ be written as a percent? Explain.

Practice and Problem Solving

Use a 10-by-10 grid to model the percent.

5. 10%   6. 55%   7. 35%

Write the percent as a fraction or mixed number in simplest form.

1 8. 45%   9. 90%   10. 15%   11. 7%

12. 34%   13. 79%   14. 77.5%   15. 188%

16. 8%   17. 224%   18. 0.25%   19. 0.4%

20. **ERROR ANALYSIS** Describe and correct the error in writing 225% as a fraction.

   ✗ $225\% = \frac{225}{1000} = \frac{9}{40}$

Write the fraction or mixed number as a percent.

2 21. $\frac{1}{10}$ 22. $\frac{1}{5}$ 23. $\frac{11}{20}$ 24. $\frac{2}{25}$

25. $\frac{27}{50}$ 26. $\frac{18}{25}$ 27. $1\frac{17}{20}$ 28. $2\frac{41}{50}$

29. **ERROR ANALYSIS** Describe and correct the error in writing $\frac{14}{25}$ as a percent.

   ✗ $\frac{14}{25} = \frac{14 \times 4}{25 \times 4} = \frac{56}{100} = 0.56\%$

30. **LEFT-HANDED** Of the students in your class, 12% are left-handed. What *fraction* of the students are left-handed? Are there more right-handed or left-handed students? Explain.

31. **ARCADE** You have 125% of the tickets required for a souvenir. What *fraction* of the required tickets do you have? Do you need more tickets for the souvenir? Explain.

Assignment Guide and Homework Check

Level	Day 1 Activity Assignment	Day 2 Lesson Assignment	Homework Check
Basic	5–7, 41–45	1–4, 9–15 odd, 20, 21–25 odd, 29, 31	11, 13, 21, 23, 31
Average	5–7, 41–45	12–24 even, 29, 30, 34, 37	12, 14, 22, 24, 30
Advanced	5–7, 41–45	16–20 even, 26–29, 30, 34–40 even	18, 26, 28, 30, 34

Common Errors

- **Exercises 5–7** Students may not shade the correct number of squares in the 10-by-10 grid. Remind them to check their work.
- **Exercises 8–19** Students may write the percent as a fraction using the wrong denominator, or try to remove the decimal point when writing it over 100. For example, a student may write $\frac{775}{100}$ instead of $\frac{77.5}{100}$. Ask students how many percents are in one whole (100). Remind students that the denominator represents the whole and the denominator will always be 100. Also tell students that the number to the left of the percent symbol will be the numerator.
- **Exercises 21–28** Students may not know when it is appropriate to use an equivalent fraction to write a fraction as a percent. While there is no error in writing the fraction as a decimal first, sometimes it is easier to just write an equivalent fraction. Brainstorm some denominators that are easily rewritten as denominators of 100 (such as 4, 5, 10, 20, 25, etc.).

5.5 Record and Practice Journal

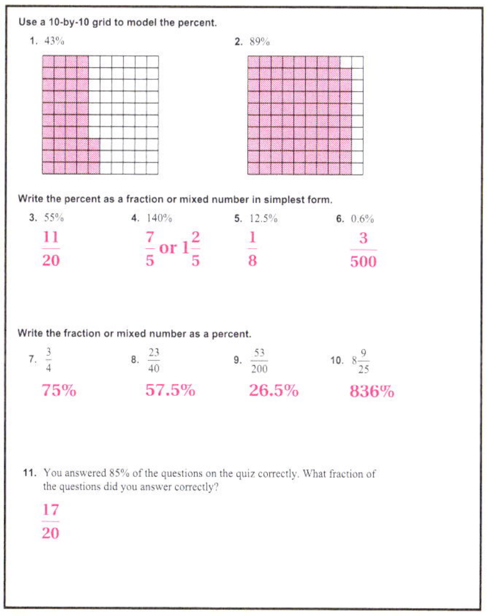

Use a 10-by-10 grid to model the percent.

1. 43% 2. 89%

Write the percent as a fraction or mixed number in simplest form.

3. 55% $\frac{11}{20}$ 4. 140% $\frac{7}{5}$ or $1\frac{2}{5}$ 5. 12.5% $\frac{1}{8}$ 6. 0.6% $\frac{3}{500}$

Write the fraction or mixed number as a percent.

7. $\frac{3}{4}$ 75% 8. $\frac{23}{40}$ 57.5% 9. $\frac{53}{200}$ 26.5% 10. $8\frac{9}{25}$ 836%

11. You answered 85% of the questions on the quiz correctly. What fraction of the questions did you answer correctly?

$\frac{17}{20}$

Vocabulary and Concept Check

1. You can shade 42 out of 100 squares to model 42%.
2. 0.01, because it is equal to 1%, and the others are equal to 10%.
3. *Sample answer:* $\frac{3}{20}, \frac{23}{100}, \frac{1}{8}$
4. yes; You can rewrite $1\frac{1}{4}$ as the improper fraction $\frac{125}{100}$, which is equal to 125%.

Practice and Problem Solving

5–7. See Additional Answers.

8. $\frac{9}{20}$ **9.** $\frac{9}{10}$

10. $\frac{3}{20}$ **11.** $\frac{7}{100}$

12. $\frac{17}{50}$ **13.** $\frac{79}{100}$

14. $\frac{31}{40}$ **15.** $1\frac{22}{25}$

16. $\frac{2}{25}$ **17.** $2\frac{6}{25}$

18. $\frac{1}{400}$ **19.** $\frac{1}{250}$

20. The 225 should be over 100.

$225\% = \frac{225}{100} = \frac{9}{4} = 2\frac{1}{4}$

21. 10% **22.** 20%

23. 55% **24.** 8%

25. 54% **26.** 72%

27. 185% **28.** 282%

29. The decimal point should not have been added to the percent expression.

$\frac{14}{25} = \frac{14 \times 4}{25 \times 4} = \frac{56}{100} = 56\%$

30. See Additional Answers.

31. $\frac{5}{4}$; No, you have more than you need.

Practice and Problem Solving

32. 37.5% **33.** 81.25%

34. 56.25% **35.** 82.5%

36. **a.** $\frac{7}{16}$

b. 27.5%

37. See *Taking Math Deeper.*

38. See Additional Answers.

39. You can shade half of one of the squares.

40. See Additional Answers.

Fair Game Review

41. $\frac{1}{2}$ **42.** 12

43. 16 **44.** $\frac{1}{12}$

45. D

Mini-Assessment

1. Use a 10-by-10 grid to model 38%.

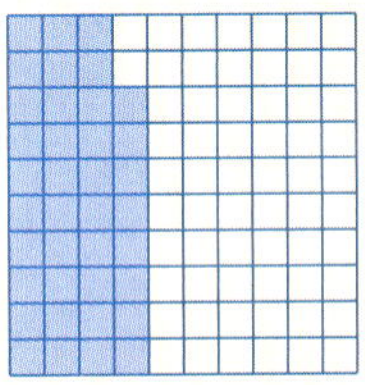

2. Identify the percent shown by the model.

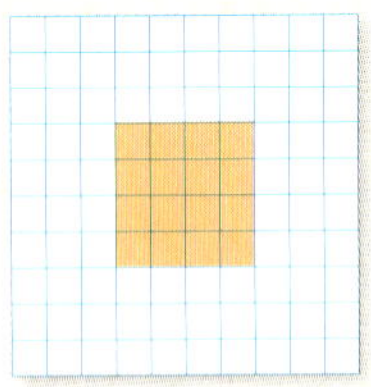

16%

3. Write 54% as a fraction in simplest form. $\frac{27}{50}$

4. Write $\frac{3}{25}$ as a percent. 12%

5. You ate $\frac{4}{5}$ of a loaf of bread. What percent of the bread did you eat? 80%

Taking Math Deeper

Exercise 37

Have you ever tried to find a U.S. map that shows Alaska in proportion to the lower 48 states? Most maps show Alaska much smaller than it is.

c.

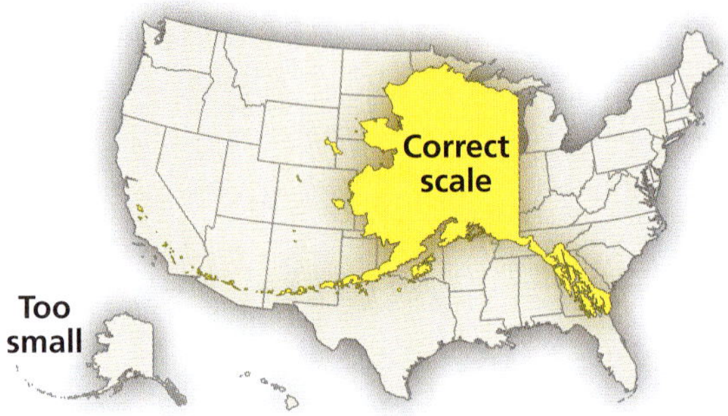

1 **a.** Make a table.

State	Area as Portion of Total U.S. Area	
	as a Percent	as a Fraction
Alaska	17.5%	$\frac{17.5}{100} = \frac{175}{1000} = \frac{7}{40}$
Florida	1.7%	$\frac{1.7}{100} = \frac{17}{1000}$
Hawaii	0.288%	$\frac{0.288}{100} = \frac{288}{100,000} = \frac{9}{3125}$
Illinois	1.5%	$\frac{1.5}{100} = \frac{15}{1000} = \frac{3}{200}$

2 Use division to compare Illinois to Hawaii.

b. $\frac{3}{200} \div \frac{9}{3125} = \frac{\overset{1}{\cancel{3}}}{\underset{8}{\cancel{200}}} \cdot \frac{\overset{125}{\cancel{3125}}}{\underset{3}{\cancel{9}}} = \frac{125}{24} \approx$ 5.2 times larger

3 **d.** 24 states are larger than Illinois. In order: AK, TX, CA, MT, NM, AZ, NV, CO, OR, WY, MI, MN, UT, ID, KS, NE, SD, WA, ND, OK, MO, FL, WI, GA

Project

How was the western border of Illinois determined? Draw a poster showing how far the borders of Illinois would need to be moved so that the area of Illinois would be approximately equal to the area of Alaska.

Reteaching and Enrichment Strategies

If students need help. . .	If students got it. . .
Resources by Chapter • Practice A and Practice B • Puzzle Time Record and Practice Journal Practice Differentiating the Lesson Lesson Tutorials Skills Review Handbook	Resources by Chapter • Enrichment and Extension • Technology Connection Start the next section

Find the percent.

32. 3 is what percent of 8?

33. 13 is what percent of 16?

34. 9 is what percent of 16?

35. 33 is what percent of 40?

36. **SOCIAL NETWORKING** A survey asked students to choose their favorite social networking website. The results are shown in the table.

a. What fraction of the students chose Website A?

b. What percent of the students chose Website C?

Social Networking Website	Number of Students
Website A	35
Website B	13
Website C	22
Website D	10

37. **GEOGRAPHY** The percent of the total area of the United States that is in each of four states is shown.

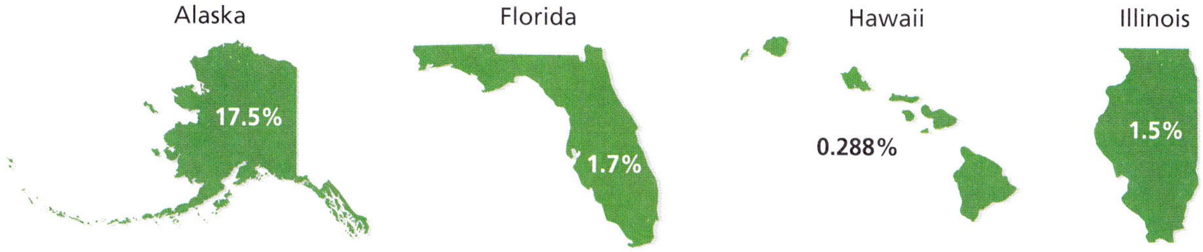

a. Write the percents as fractions in simplest form.

b. How many times larger is Illinois than Hawaii?

c. Compared to the map of Florida, is the map of Alaska the correct size? Explain your reasoning.

d. **RESEARCH** Which of the 50 states are larger than Illinois?

38. **CRITICAL THINKING** A school fundraiser raised 120% of its goal last year and 125% of its goal this year. Did the fundraiser raise more money this year? Explain your reasoning.

39. **CRITICAL THINKING** How can you use a 10-by-10 grid to model $\frac{1}{2}$%?

40. **Reasoning** Write $\frac{1}{12}$ as a percent. Explain how you found your answer.

Fair Game Review What you learned in previous grades & lessons

Divide. Write the answer in simplest form. *(Section 2.2)*

41. $\frac{1}{6} \div \frac{1}{3}$

42. $9 \div \frac{3}{4}$

43. $10 \div \frac{5}{8}$

44. $\frac{1}{6} \div 2$

45. **MULTIPLE CHOICE** Which of the following is *not* equal to 15? *(Section 2.1)*

Ⓐ $\frac{3}{4} \cdot 20$

Ⓑ $\frac{5}{9} \cdot 27$

Ⓒ $35 \cdot \frac{3}{7}$

Ⓓ $28 \cdot \frac{5}{7}$

5.6 Solving Percent Problems

Essential Question How can you use mental math to find the percent of a number?

"I have a secret way for finding 21% of 80." "10% is 8, and 1% is 0.8." "So, 21% is 8 + 8 + 0.8 = 16.8."

1 ACTIVITY: Finding 10% of a Number

Work with a partner.

a. How did Newton know that 10% of 80 is 8?

Write 10% as a fraction. $10\% = \dfrac{\square}{\square} = \dfrac{1}{\square}$ (10 per cent)

Method 1: Use a model.

0%	10%	20%	30%	40%	50%	60%	70%	80%	90%	100%
0	$\square$	$\square$	$\square$	$\square$	$\square$	$\square$	$\square$	$\square$	$\square$	80

Method 2: Use multiplication.

$$10\% \text{ of } 80 = \frac{\square}{10} \text{ of } 80 = \frac{\square}{10} \times \square = \frac{\square}{10} = \square$$

b. How do you move the decimal point to find 10% of a number?

Move the decimal point one place to the $\square$. **10% of 80. =** $\square$

COMMON CORE

Percents

In this lesson, you will

- find percents of numbers.
- find the whole given the part and the percent.

Learning Standard
6.RP.3c

2 ACTIVITY: Finding 1% of a Number

Work with a partner.

a. How did Newton know that 1% of 80 is 0.8?

b. How do you move the decimal point to find 1% of a number?

Laurie's Notes

Introduction

Standards for Mathematical Practice

- **MP1 Make Sense of Problems and Persevere in Solving Them:** The percent bar model provides a visual means to think about percents. Later, when students are solving percent problems, this will help students determine the reasonableness of their answers.

Motivate

- Draw a 24-tower showing different ways 24 can be divided into groups of equal size.

  ? Use this model to ask questions such as:

  "What is $\frac{1}{2}$ of 24? What is $\frac{3}{4}$ of 24?

  What is $\frac{5}{6}$ of 24?" 12; 18; 20

12			12		
8		8		8	
6	6		6		6
4	4	4	4	4	4

Activity Notes

Activity 1

? This example uses a percent bar model to find 10% of 80. Introduce the idea with a series of questions.

- "If 10 friends share \$80, how much will each person receive?" \$8
- "What is $\frac{1}{10}$ of \$80? What is 10% of \$80?" \$8; \$8

- In part (b), you are asking students to make the observation that they can find 10% of 80 by moving the decimal point one place to the left. This observation is sometimes an "aha" moment for students. Remind them of this observation throughout the lesson.

Activity 2

- Help students develop an understanding of the relationship between 1%, 10%, and 100%.

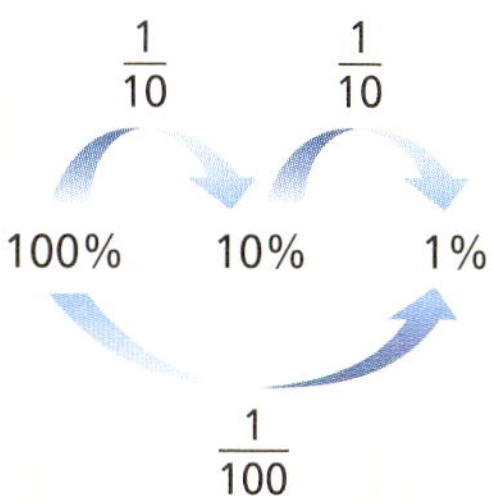

? "Finding $\frac{1}{10}$ of a number moves the decimal point one place to the left, so finding $\frac{1}{100}$ of a number should move the decimal point how many places to the left? Why?"

2; $\frac{1}{100}$ is $\frac{1}{10} \cdot \frac{1}{10}$ or $\frac{1}{10}$ times itself.

? "What is $\frac{1}{100}$ of \$80?" \$0.80 "What is 1% of \$80?" \$0.80

- **Connection:** Percents of the same number can be added or subtracted. For example, 12% of 30 could be found by finding 10% of 30 = 3 and 1% of 30 = 0.3 and adding 3 + 0.3 + 0.3 = 3.6. Discuss this concept with the students, because it is not obvious to all students.

Common Core State Standards

6.RP.3c Find a percent of a quantity as a rate per 100; solve problems involving finding the whole, given a part and the percent.

Previous Learning

Students should know how to write halves, fourths, tenths, and hundredths as percents. Students should also be able to multiply decimals and fractions, and write percents as fractions or decimals.

Technology for the Teacher

Dynamic Classroom

Lesson Plans
Complete Materials List

5.6 Record and Practice Journal

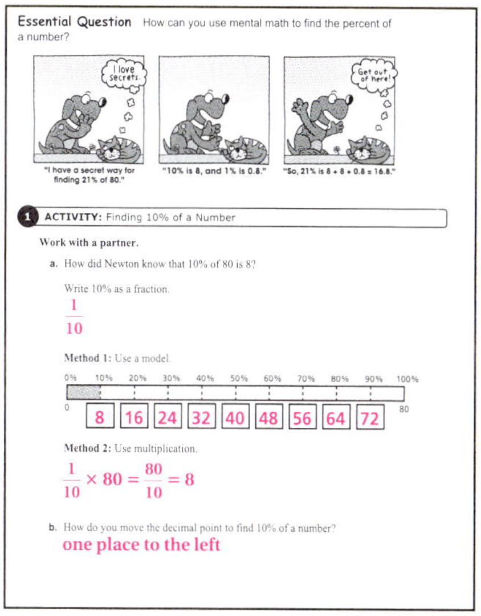

Essential Question How can you use mental math to find the percent of a number?

"I have a secret way for finding 21% of 80."

"10% is 8, and 1% is 0.8."

"So, 21% is 8 + 8 + 0.8 = 16.8."

1 ACTIVITY: Finding 10% of a Number

Work with a partner.

a. How did Newton know that 10% of 80 is 8?

Write 10% as a fraction.

$\frac{1}{10}$

Method 1: Use a model.

0% 10% 20% 30% 40% 50% 60% 70% 80% 90% 100%

0 8 16 24 32 40 48 56 64 72 80

Method 2: Use multiplication.

$\frac{1}{10} \times 80 = \frac{80}{10} = 8$

b. How do you move the decimal point to find 10% of a number?

one place to the left

English Language Learners

Vocabulary

English learners may be unfamiliar with the concepts of *tip*, *sales tax*, *service charge*, *delivery fee*, or *bonus* in Activity 4. Discuss their meanings before students begin.

5.6 Record and Practice Journal

2 ACTIVITY: Finding 1% of a Number

Work with a partner.

a. How did Newton know that 1% of 80 is 0.8?

$1\% \text{ of } 80 = \frac{1}{100} \text{ of } 80$

$= \frac{1}{100} \times 80 = \frac{80}{100} = 0.8$

b. How do you move the decimal point to find 1% of a number?

move decimal point 2 places left

3 ACTIVITY: Using Mental Math

Work with a partner. Use mental math to find each percent of a number.

a. 12% of 40 — 4.8

b. 19% of 50 — 9.5

4 ACTIVITY: Using Mental Math

Work with a partner. Use mental math to find each percent of a number.

a. 20% tip for a \$30 meal — \$6.00

b. 18% tip for a \$30 meal — \$5.40

c. 6% sales tax on a \$20 shirt — \$1.20

d. 9% sales tax on a \$20 shirt — \$1.80

e. 6% service charge for a \$200 boxing ticket — \$12

f. 2% delivery fee for a \$200 boxing ticket — \$4

g. 21% bonus on a total of 40,000 points — 8400 points

h. 38% bonus on a total of 80,000 points — 30,400 points

What Is Your Answer?

5. IN YOUR OWN WORDS How can you use mental math to find the percent of a number?

Sample answer: Think of the problem as a sum or difference of percents that are powers of 10.

6. Describe two real-life examples of finding a percent of a number.

Check students' work.

7. How can you use 10% of a number to find 20% of the number? 30%? Explain your reasoning.

double it; triple it

Laurie's Notes

Activity 3

? "What is 10% of 40? What is 1% of 40?" 4; 0.4

- Some students may need to write down each percent as they find it instead of trying to hold more than one answer in their heads.
- Work through each example.

? **Extension:** Try to help students develop number sense with percents by asking the following questions.

  - "What is the relationship between 10% of 40 and 10% of 80?" 4 is half of 8.
  - "10% of 40 is 4. What is 5% of 40?" half of 4; 2
  - "19% of 50 is 9.5. What is 19% of 100?" two times 9.5; 19
  - "19% of 50 is 9.5. What is 50% of 19? half of 19; 9.5

Activity 4

- Students will use mental math to compute the percents, but will need to record answers as they are done.
- **Example:** Find 38% of 80,000.

Find 10% of 80,000.	8,000
Quadruple 8,000, which is 40% of 80,000.	32,000
Find 1% of 80,000.	800
Double 800, which is 2% of 80,000.	1600
Find 38%, which is 40% − 2%.	32,000 − 1600

  Answer: So, 38% of 80,000 is 30,400.
- Students should work in pairs. When everyone has finished, ask for volunteers to explain their work.
- **Common Difficulty:** When numbers become fairly large, a student's ability to use mental math seems to diminish.
- **MP1:** Remind students that they can use the percent bar model as a visual aid to check the reasonableness of their answers. For example, the following can be used to check part (a):

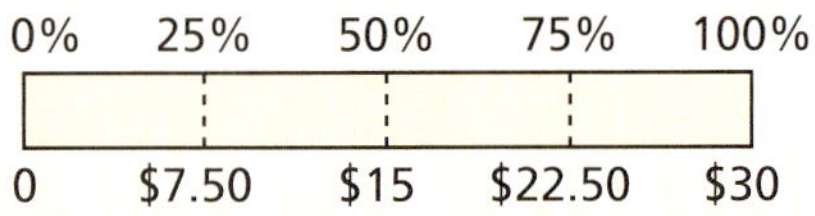

Because 25% of \$30 is \$7.50, 20% of \$30 should be less than \$7.50.

What Is Your Answer?

- Ask students to restate how to find 10% of a number.

Closure

- Use mental math to find the following percents of 120: 10%, 15%, and 20%. 12, 18, 24

3 ACTIVITY: Using Mental Math

Math Practice 8

Evaluate Results

Does your answer seem reasonable? How can you check your answer?

Work with a partner. Use mental math to find each percent of a number.

a. 12% of 40

Think: 12% = 10% + 1% + 1%

10% of 40 = ___ 1% of 40 = ___

___ + ___ + ___ = ___

b. 19% of 50

Think: 19% = 10% + 10% − 1%

10% of 50 = ___ 1% of 50 = ___

___ + ___ − ___ = ___

4 ACTIVITY: Using Mental Math

Work with a partner. Use mental math to find each percent of a number.

a. 20% tip for a $30 meal

b. 18% tip for a $30 meal

c. 6% sales tax on a $20 shirt

d. 9% sales tax on a $20 shirt

e. 6% service charge for a $200 boxing ticket

f. 2% delivery fee for a $200 boxing ticket

g. 21% bonus on a total of 40,000 points

h. 38% bonus on a total of 80,000 points

What Is Your Answer?

5. **IN YOUR OWN WORDS** How can you use mental math to find the percent of a number?
6. Describe two real-life examples of finding a percent of a number.
7. How can you use 10% of a number to find 20% of the number? 30%? Explain your reasoning.

Practice Use what you learned about finding the percent of a number to complete Exercises 3–10 on page 229.

5.6 Lesson

Check It Out
Lesson Tutorials
BigIdeasMath.com

Key Idea

Finding the Percent of a Number

Words Write the percent as a fraction. Then multiply by the whole. The percent times the whole equals the part.

Numbers 20% of 60 is 12.

$$\frac{1}{5} \times 60 = 12$$

Model

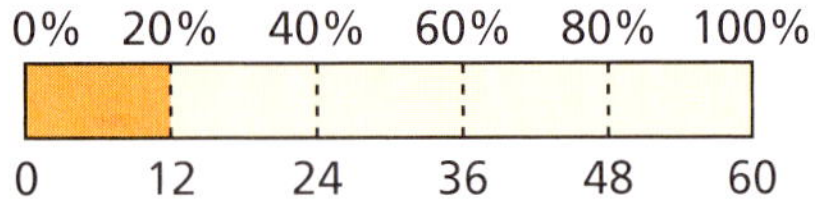

EXAMPLE 1 Finding the Percent of a Number

25% of 40 is what number?

$$25\% \text{ of } 40 = \frac{1}{4} \cdot 40$$ Write the percent as a fraction and multiply.

$$= \frac{1 \cdot \cancel{40}^{10}}{_{1}\cancel{4}}$$ Divide out the common factor.

$$= 10$$ Simplify.

So, 25% of 40 is 10.

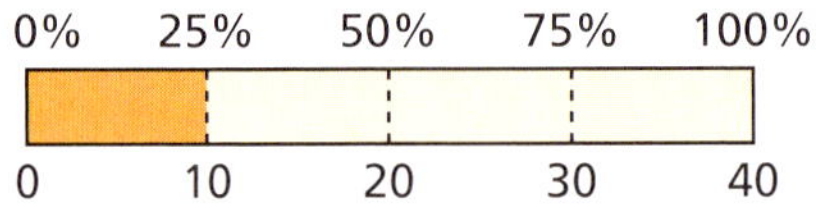

Study Tip

You can use mental math to check your answer in Example 1.

10% of 40 = 4

5% of 40 = 2

So, 25% of 40 is 4 + 4 + 2 = 10.

You can also use a ratio table to find the percent of a number.

EXAMPLE 2 Finding the Percent of a Number Using a Ratio Table

60% of 150 is what number?

Use a ratio table to find the part. Let one row be the *part*, and let the other be the *whole*. Find an equivalent ratio with 150 as the whole.

The first column represents the percent.

$$\frac{\text{part}}{\text{whole}} = \frac{60}{100} = 60\%$$

	÷2	×3	
Part	60	30	90
Whole	100	50	150

(÷2, ×3 applied to both rows)

So, 60% of 150 is 90.

On Your Own

Now You're Ready
Exercises 3–22

Find the percent of the number. Explain your method.

1. 90% of 20
2. 75% of 32
3. 10% of 110
4. 30% of 75

Laurie's Notes

Introduction

Connect

- **Yesterday:** Students used mental math to find the percent of a number (MP1).
- **Today:** Students will use multiplication to find the percent of a number and division to find the whole given the part and the percent.

Motivate

- Play a game of *Find Your Partner*. Make cards where one-half of the cards have a fraction problem and one-half of the cards have a percent problem.
- Each pair of cards has the same solution. Use mental math to solve.
- Ask each pair to explain how they found the value of their card. Connect yesterday's investigation to today's introduction.

Lesson Notes

Key Idea

- In the last lesson, students wrote percents as fractions and simplified. This section shows percents written as simplified fractions in one step. Spend time showing the simplification if necessary.
- Be sure to emphasize "the *percent* times the *whole* equals the *part*."
- **Connection:** Connect this idea to multiplying a fraction by a whole number.
- This is a long lesson. Avoid trying to rush through the content.

Example 1

? "What is 25% as a fraction? How do you find $\frac{1}{4}$ of 40?" $\frac{1}{4}$; You multiply.

- Discuss the visual model. Each part, $\frac{1}{4}$ or 25%, represents 10.
- Discuss the Study Tip.

Example 2

- This example helps students apply and connect the skill of working with ratio tables to finding the percent of a number.
- Draw the ratio table and identify the first column as 60% (60 = part, 100 = whole). The goal is to get the last ratio to have a whole of 150.
- Point out that because the columns of the tables show equivalent ratios, the top number in each column is 60% of the bottom number.
- **MP5 Use Appropriate Tools Strategically:** Have students check Example 1 using a ratio table and Example 2 using a fraction.

On Your Own

- **Think-Pair-Share:** Students should read each question independently and then work in pairs to answer the questions. When they have answered the questions, the pair should compare their answers with another group and discuss any discrepancies.

Goal Today's lesson is solving percent problems.

Technology for the Teacher

Dynamic Classroom

Lesson Tutorials
Lesson Plans
Answer Presentation Tool

Extra Example 1

Use a fraction to find 80% of 75. 60

Extra Example 2

Use a table to find 40% of 200. 80

On Your Own

1. 18
2. 24
3. 11
4. $22\frac{1}{2}$

Laurie's Notes

Key Idea

- **CCSS Connection:** The following standard justifies the reasoning used. **3.OA.7** Fluently multiply and divide within 100, using strategies such as the relationship between multiplication and division or properties of operations. . . .
- ? "What are related division equations for $3 \times 12 = 36$?" $36 \div 3 = 12$, $36 \div 12 = 3$
- Say, "In Examples 1 and 2, you found the percent of a number using *percent* × *whole* = *part*. Now, you will find the whole using the related division equation, *part* ÷ *percent* = *whole*."
- Write the Key Idea. Emphasize the original multiplication equation and the related division equation.
- Students should still recognize that the percent is represented as a fraction in each case.

Example 3

- ? "Will the answer be greater than or less than 48? Explain." greater than 48; If 75% of a number is 48, then the number must be greater than 48.
- ? "What is 75% written as a fraction?" $\frac{3}{4}$
- Write the word equation *part* ÷ *percent* = *whole*. Then work through the problem as shown. Before writing the final answer, you may want to show an extra step in which you divide out the common factor of 3.
- **MP4 Model with Mathematics:** Draw the percent bar model. Each part (25%) represents 16, so three parts (75%) represent 48. The whole (100%) is 64. So the answer is reasonable.

Example 4

- This example helps students apply and connect the skill of working with ratio tables to finding the whole given the part and the percent.
- Take a moment to estimate the answer. Because 100% of 72 is 72, 120% of a number less than 72 is 72. So, the answer will be less than 72.
- Draw the ratio table and identify the first column as 120% (120 = part, 100 = whole). The goal is to get the last ratio to have a part of 72.
- Work through the problem as shown, or have students suggest ways to create the ratio table.
- Point out that because the columns of the tables show equivalent ratios, the top number in each column is 120% of the bottom number.
- **MP5:** Have students check Example 3 using a ratio table and Example 4 using a fraction.

On Your Own

- **Neighbor Check:** Have students work independently and then have their neighbors check their work. Have students discuss any discrepancies.

Extra Example 3

60% of what number is 42? Use a fraction to find the answer. 70

Extra Example 4

140% of what number is 126? Use a table to find the answer. 90

On Your Own

5. 200

6. 50

Differentiated Instruction

Kinesthetic

Use a 100 grid to illustrate 80% of 60. Have students shade 60 of the 100 squares. Ask students for strategies for finding 80% or $\frac{4}{5}$ of the shaded squares by just counting. One strategy would be to repeatedly shade 4 squares with a second color and skip 1 square.

You can use a related division equation to find the whole given the part and the percent.

Key Idea

Finding the Whole

Write the percent as a fraction. Then divide the part by the fraction.

Words The part divided by the percent equals the whole.

Numbers 20% of 60 is 12.

$$\frac{1}{5} \times 60 = 12 \longrightarrow 12 \div \frac{1}{5} = 60$$

Multiplication equation → Related division equation

EXAMPLE 3 Finding the Whole

75% of what number is 48?

$$48 \div 75\% = 48 \div \frac{3}{4}$$ Write the percent as a fraction and divide.

$$= 48 \cdot \frac{4}{3}$$ Multiply by the reciprocal.

$$= 64$$ Simplify.

So, 75% of 64 is 48.

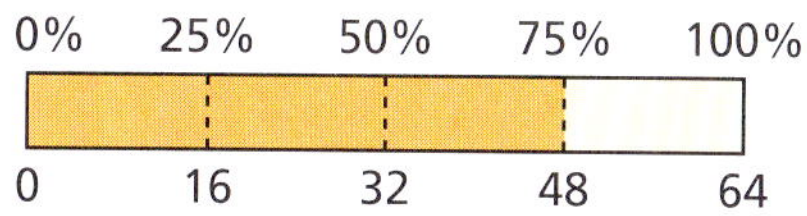

EXAMPLE 4 Finding the Whole Using a Ratio Table

120% of what number is 72?

Use a ratio table to find the whole. Find an equivalent ratio with 72 as the part.

The first column represents the percent.

$$\frac{\text{part}}{\text{whole}} = \frac{120}{100} = 120\%$$

Part	120	6	72
Whole	100	5	60

(÷ 20 from the first column to the second; × 12 from the second column to the third.)

So, 120% of 60 is 72.

On Your Own

Now You're Ready
Exercises 27–36

Find the whole. Explain your method.

5. 5% of what number is 10?

6. 62% of what number is 31?

EXAMPLE 5 Real-Life Application

The width of a rectangular room is 80% of its length. What is the area of the room?

Find 80% of 15 feet.

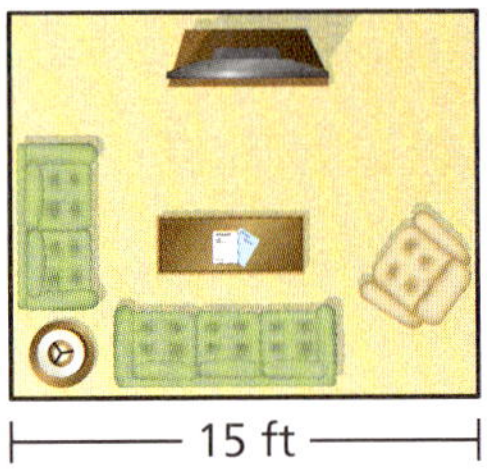

$$80\% \text{ of } 15 = \frac{4}{5} \times 15$$

$$= \frac{4 \times \cancel{15}^{3}}{\cancel{5}_{1}}$$

$$= 12$$

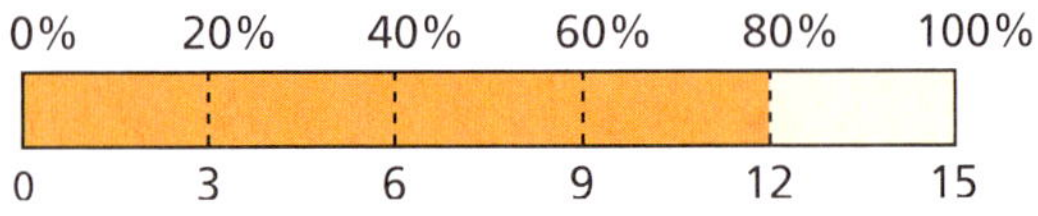

The width is 12 feet.

Use the formula for the area A of a rectangle.

$$A = 15 \times 12 = 180$$

So, the area of the room is 180 square feet.

On Your Own

7. The width of a rectangular stage is 55% of its length. The stage is 120 feet long. What is the area?

EXAMPLE 6 Real-Life Application

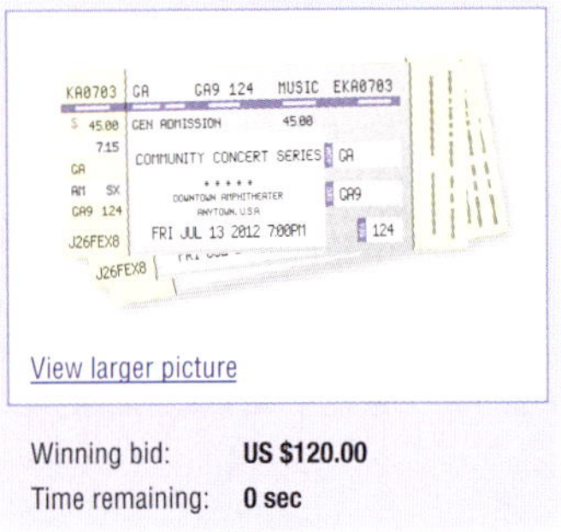

You win an online auction for concert tickets. Your winning bid is 60% of your maximum bid. How much more were you willing to pay for the tickets than you actually paid?

(A) \$72 (B) \$80 (C) \$120 (D) \$200

Your maximum bid is the *whole*, and your winning bid is the *part*. Find your maximum bid by dividing the part by the percent.

$$120 \div 60\% = 120 \div \frac{3}{5}$$ Divide the part by the percent.

$$= 120 \cdot \frac{5}{3}$$ Multiply by the reciprocal.

$$= 200$$ Simplify.

Your maximum bid is \$200, and your winning bid is \$120. So, you were willing to pay 200 − 120 = \$80 more for the tickets.

The correct answer is (B).

On Your Own

8. **WHAT IF?** Your winning bid is 96% of your maximum bid. How much more were you willing to pay for the tickets than you actually paid?

Laurie's Notes

Example 5

? "What information is given?" From the diagram, the length is given as 15 feet. From the wording of the problem, the width is 80% of the length.

- Write "80% of 15 = ?" and ask students what method(s) could be used to answer the question.
- **MP1 Make Sense of Problems and Persevere in Solving Them:** Students should know a variety of ways to find 80% of 15. It is important for students to take a lead on finding an approach to solving a problem that makes sense.
- Be sure that students have a basic understanding that if the width is 80% of the length, the width is shorter than the length. This may not be obvious to all students.
- The percent bar model visually shows the 5 equal parts of 15. Students could use this model to find 20%, 40%, 60%, or 80% of 15.
- **MP6 Attend to Precision:** Be sure that students label the answer with the correct units. Note that the question asks for the area.

Example 6

- It helps to know a little about online bidding. Bidders enter secret maximum bids. Through automated bidding, the winning bid is greater than the second greatest maximum bid, but often less than the winner's maximum bid. This may help students understand that the winning bid is the part and the maximum bid is the whole.
- Read the problem and ask what is known and what they need to find.
- **Teaching Tip:** You could draw a percent bar model to represent the information known.

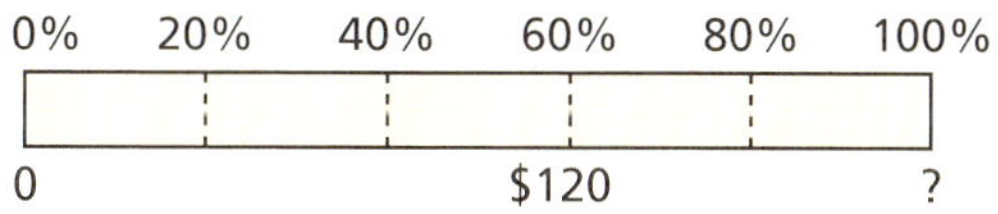

- Say, "$120 is the part and your maximum bid is the whole."
- **MP1:** Students should know a variety of ways to find a number such that 60% is 120. It is important for students to take a lead on finding an approach to solving a problem that makes sense.

? "Is $200 the answer to the question? Explain." No; The question asks how much *more* you were willing to pay for the tickets than you actually paid.

Closure

- **Writing Prompts:** Have students work in pairs to complete the following.
  To find the percent of a number, . . .
  To find the whole given the part and the percent, . . .

Extra Example 5

The length of a garden is 225% of its width. If the width of the garden is 4 meters, what is the area of the garden? 36 m^2

On Your Own

7. 7920 ft^2

Extra Example 6

Using coupons, you spend $135 grocery shopping. This is 75% of the total retail price of the groceries. How much would you have spent if you had not used the coupons? $180

On Your Own

8. $5

Vocabulary and Concept Check

1. Twenty percent of what number is 30?; 6; 150
2. less than 52; If you take more than 100% of a number, you get a greater number. So, 52 is greater than the original number.

Practice and Problem Solving

3–22. Explanations will vary.

3. 12 **4.** 4

5. 35 **6.** 9

7. 9 **8.** 3

9. 12.5 **10.** 3

11. 21 **12.** 36

13. 20.25 **14.** 17.48

15. 24 **16.** 8.36

17. 14 **18.** 47.85

19. 84 **20.** 39.6

21. 94.5 **22.** 365

23. The percent was not written as a fraction before multiplying;

$$40\% \times 75 = \frac{2}{5} \times 75 = 30$$

24. about 275.6 acres

25. 35.2 in.

26. **a.** \$3.15

b. \$48.15

Assignment Guide and Homework Check

Level	Day 1 Activity Assignment	Day 2 Lesson Assignment	Homework Check
Basic	3–10, 54–58	1, 2, 11–31 odd, 37	11, 21, 25, 27, 31
Average	3–10, 54–58	1, 2, 16–22 even, 23, 24, 29–32, 37–40	16, 20, 24, 29, 32
Advanced	3–10, 54–58	1, 2, 14, 16, 22, 23, 26, 35–37, 44–50 even, 51–53	16, 22, 36, 44, 50

Common Errors

- **Exercises 3–22** If the number ends in a zero and students multiply by a decimal, they may be off by a place value. This is because they have forgotten about the zero at the end. Encourage students to estimate their answers first. For example, 8% of 90 should be less than 10% of 90, so the answer should be less than 9.
- **Exercises 19–22** Students often struggle with percents greater than 100%. They may find a number significantly greater than or less than what they should. Remind students of how to move the decimal point. Encourage students to estimate their answers. Review that percents greater than 100% will have answers that are greater than the original number.

5.6 Record and Practice Journal

Find the percent of the number.

1. 30% of 50 **15**
2. 12% of 85 **10.2**
3. 2% of 96 **1.92**
4. 150% of 66 **99**
5. 7% of 120 **8.4**
6. 3% of 15 **0.45**

Complete the statement using <, >, or =.

7. 70% of 80 **=** 80% of 70
8. 92% of 30 **<** 48% of 75

9. You make homemade lip balm. About 11% of the lip balm is made from beeswax. You make $4\frac{2}{5}$ teaspoons of the lip balm. About how many teaspoons of beeswax do you need? Round your answer to the nearest tenth.

0.5 teaspoon

5.6 Exercises

Check It Out
Help with Homework
BigIdeasMath.com

Vocabulary and Concept Check

1. **DIFFERENT WORDS, SAME QUESTION** Which is different? Find "both" answers.

What is twenty percent of 30?	What is one-fifth of 30?
Twenty percent of what number is 30?	What is two-tenths of 30?

2. **NUMBER SENSE** If 52 is 130% of a number, is the number greater or less than 52? Explain.

Practice and Problem Solving

Find the percent of the number. Explain your method.

1 2

3. 20% of 60	**4.** 10% of 40	**5.** 50% of 70	**6.** 30% of 30
7. 10% of 90	**8.** 15% of 20	**9.** 25% of 50	**10.** 5% of 60
11. 30% of 70	**12.** 75% of 48	**13.** 45% of 45	**14.** 92% of 19
15. 40% of 60	**16.** 38% of 22	**17.** 70% of 20	**18.** 87% of 55
19. 140% of 60	**20.** 120% of 33	**21.** 175% of 54	**22.** 250% of 146

23. **ERROR ANALYSIS** Describe and correct the error in finding 40% of 75.

✗ $40\% \text{ of } 75 = 40\% \times 75 = 3000$

24. **PINE TREES** A town had about 2120 acres of pine trees 40 years ago. Only about 13% of the pine trees remain. How many acres of pine trees remain?

25. **SPIDER MONKEY** The tail of the spider monkey is 64% of the length shown. What is the length of its tail?

26. **CABLE** A family pays $45 each month for cable television. The cost increases 7%.
    a. How many dollars is the increase?
    b. What is the new monthly cost?

Find the whole. Explain your method.

3 4 **27.** 10% of what number is 14?

28. 20% of what number is 18?

29. 25% of what number is 21?

30. 75% of what number is 27?

31. 15% of what number is 12?

32. 85% of what number is 17?

33. 140% of what number is 35?

34. 160% of what number is 32?

35. 125% of what number is 25?

36. 175% of what number is 42?

37. ERROR ANALYSIS Describe and correct the error in finding the whole.

✗ *20% of what number is 5?*

$$5 \div 20\% = \frac{5}{20} = \frac{1}{4}$$

38. COUPON You have a coupon for a restaurant. You save \$3 on a meal. What was the original cost of the meal?

39. SURVEY The results of a survey are shown at the right. In the survey, 12 students said that they would like to learn French.

a. How many students were surveyed?

b. How many of the students surveyed would like to learn Spanish?

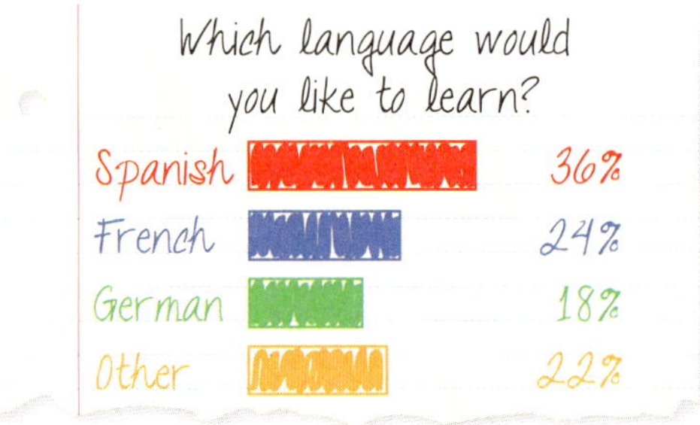

40. WEIGHT A sixth grader weighs 90 pounds, which is 120% of what he weighed in fourth grade. How much did he weigh in fourth grade?

41. PARKING LOT In a parking lot, 16% of the cars are blue. There are 4 blue cars in the parking lot. How many cars in the parking lot are *not* blue?

42. LOTION A bottle contains 20 fluid ounces of lotion and sells for \$5.80. The 20-fluid-ounce bottle contains 125% of the lotion in the next smallest size, which sells for \$5.12. Which is the better buy? Explain.

Common Errors

- **Exercises 47 and 48** Students may forget to convert the units after they have found the percent. Remind them to read the question carefully.

Practice and Problem Solving

27–36. Explanations will vary.

27. 140 **28.** 90

29. 84 **30.** 36

31. 80 **32.** 20

33. 25 **34.** 20

35. 20 **36.** 24

37. The percent should be written as a fraction before dividing;
$5 \div 20\% = 5 \div \frac{1}{5} = 25$

38. \$12

39. **a.** 50 students

b. 18 students

40. 75 pounds

41. 21 cars

42. the 20-fluid-ounce bottle; Because \$0.29 is less \$0.32, the 20-fluid-ounce bottle is the better buy.

English Language Learners

Vocabulary

English learners may have trouble identifying whether a number is the *whole* or the *part of the whole* in a percent problem. Have students identify the whole and the part of the whole in statements such as "20% of 300 is 60" and "125% of 50 is 62.5." Suggest that in most cases, the whole is the number after the word *of*, and the part of the whole is the remaining number (without the percent symbol).

Practice and Problem Solving

43. =
44. <
45. >
46. =
47. 48 min
48. 28.08 in.
49. a. 432 in.^{2}
    b. 37.5%; Because the length is doubled, the width of the rectangle is now half of 75% of its length, or 37.5%.
50. yes; To pass inspection, the ball must bounce back to between 68% and 75% of the starting height, or between 4.08 feet and 4.5 feet. It bounced back to $4.08\overline{3}$ feet, so it passes.
51. *Sample answer:* Because 30% of n is equal to 2 times 15% of n and 45% of n is equal to 3 times 15% of n, you can write 30% of $n = 2 \times 12 = 24$ and 45% of $n = 3 \times 12 = 36$.
52. See *Taking Math Deeper*.
53. 97.2%

Fair Game Review

54. 4.8
55. 16.5
56. 6.66
57. 26.28
58. D

Mini-Assessment

Find the percent of the number.

1. 15% of 70 10.5
2. 65% of 50 32.5
3. 32% of 88 28.16
4. 6% of 32 1.92
5. How many inches is 40% of 5 feet? 24 in.

Taking Math Deeper

Exercise 52

Sale items can be discounted in different ways. In this problem, students can learn that sequential discounting can produce a lesser discount than if each discount is taken off the original price.

1. Organize the given information.

Regular Price $150

Coupon 10% off Sale price

Sale: 30% off

2. Find the sale price before and after using the coupon.

   30% of \$150 = 0.3(150) = \$45

   So, the sale price is 150 − 45 = \$105.

   10% of \$105 = 0.1(105) = \$10.5

   So, after using the coupon, the sale price is 105 − 10.5 = \$94.50.

3. Compare the two discount plans.

   40% of \$150 = 0.4(150) = \$60

   Sale price = 150 − 60 = \$90

Discounts and markups will be taught in a future course. However, it is good preparation for students to think about the various ways that discounts can occur.

Project

Prepare a budget for a surfing trip or other outing. Include all expenses that might be incurred. Give some ideas as to how you might earn the money needed for the trip.

Reteaching and Enrichment Strategies

If students need help...	If students got it...
Resources by Chapter • Practice A and Practice B • Puzzle Time Record and Practice Journal Practice Differentiating the Lesson Lesson Tutorials Skills Review Handbook	Resources by Chapter • Enrichment and Extension • Technology Connection Start the next section

Copy and complete the statement using <, >, or =.

43. 80% of 60 ___ 60% of 80

44. 20% of 30 ___ 30% of 40

45. 120% of 5 ___ 0.8% of 250

46. 85% of 40 ___ 25% of 136

47. **TIME** How many minutes is 40% of 2 hours?

48. **LENGTH** How many inches is 78% of 3 feet?

49. **GEOMETRY** The width of the rectangle is 75% of its length.

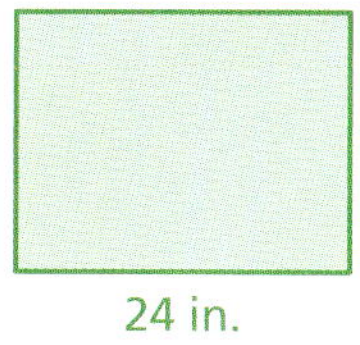

a. What is the area of the rectangle?

b. The length of the rectangle is doubled. What percent of the length is the width now? Explain your reasoning.

50. **BASKETBALL** To pass inspection, a new basketball should bounce back to between 68% and 75% of the starting height. A new ball is dropped from 6 feet and bounces back 4 feet 1 inch. Does the ball pass inspection? Explain.

51. **REASONING** You know that 15% of a number n is 12. How can you use this to find 30% of n? 45% of n? Explain.

52. **SURFBOARD** You have a coupon for 10% off the sale price of a surfboard. Which is the better buy? Explain your reasoning.

- 40% off the regular price
- 30% off the regular price and then 10% off the sale price

53. **Number Sense** On three 150-point geography tests, you earned grades of 88%, 94%, and 90%. The final test is worth 250 points. What *percent* do you need on the final to earn 93% of the total points on all tests?

Fair Game Review What you learned in previous grades & lessons

Multiply. *(Section 2.5)*

54. 0.6×8

55. 3.3×5

56. 0.74×9

57. 2.19×12

58. **MULTIPLE CHOICE** What is the quotient of 75 and 2.4? *(Section 2.6)*

Ⓐ 0.032 Ⓑ 0.3125 Ⓒ 3.2 Ⓓ 31.25

5.7 Converting Measures

Essential Question How can you compare lengths between the customary and metric systems?

1 ACTIVITY: Customary Measure History

Work with a partner.

a. Match the measure of length with its historical beginning.

Length	*Historical Beginning*
Inch	The length of a human foot
Foot	The width of a human thumb
Yard	The distance a human can walk in 1000 paces (1 pace = 2 steps)
Mile	The distance from a human nose to the end of an outstretched human arm

b. Use a ruler to measure your thumb, arm, and foot. How do your measurements compare to your answers from part (a)? Are they close to the historical measures?

You know how to convert measures within the customary and metric systems.

Equivalent Customary Lengths

1 ft = 12 in. 1 yd = 3 ft 1 mi = 5280 ft

Equivalent Metric Lengths

1 m = 1000 mm 1 m = 100 cm 1 km = 1000 m

You will learn how to convert between the two systems.

Converting Between Systems

1 in. = 2.54 cm

1 mi ≈ 1.61 km

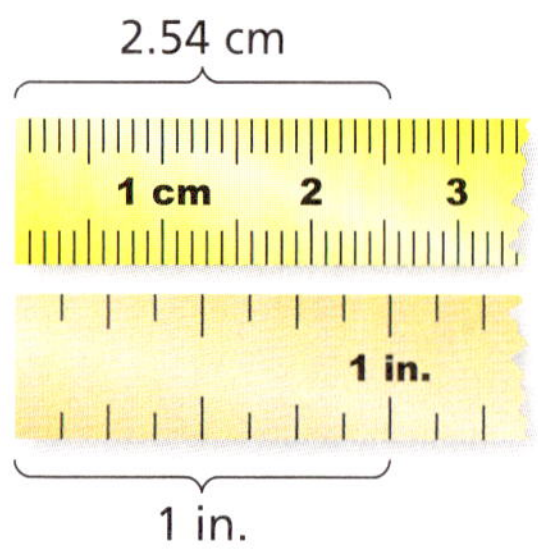

COMMON CORE

Converting Measures

In this lesson, you will
- use conversion factors (rates) to convert units of measurement.

Learning Standard 6.RP.3d

Laurie's Notes

Introduction

Standards for Mathematical Practice

- **MP6 Attend to Precision:** Mathematically proficient students are careful about specifying units of measure. In today's activities, students will need to pay close attention to units. If you sense that students are not as fluent with units as they need to be, then provide them with a review.

Discuss

- Discuss what linear measurement means—the distance between two points. Draw the following examples:

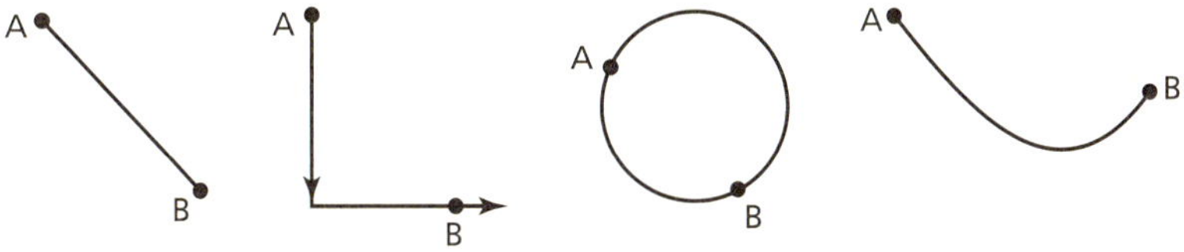

- The idea is that linear measurement does not mean it has to involve a straight line. Techniques can be used to find the lengths of curves, rivers, arteries, and so on.
- The first activity looks at the history of customary measure. Someone had to define how long an inch would be, and it wasn't a math teacher!

Activity Notes

Activity 1

- A transparent ruler for the overhead is helpful. Lay small objects on the overhead and have a student come to the overhead to measure them.
- **Common Error:** Not all rulers begin scaling with 0 at the left edge. Check rulers that will be used by students and if possible, make an overhead version for students to see the difference.
- **MP6:** Students need to examine measuring devices and be observant of their similarities and differences.
- Have students work with partners to answer part (a), and then measure and record the length in part (b).
- If comfortable, students should remove their shoes to measure the length of their feet. They do not need to step directly onto the ruler.
- **Safety:** It is a wise idea to use a disinfectant on all measuring devices on a regular basis.
- Discuss the need for *standard* units of measure that are not related to body parts or other distances that are subject to change.
- **FYI:** We generally do not convert between systems of measurement; however, there are a few conversions that are considered *common knowledge* that people encounter in daily life. These common conversions are the focus of this investigation and the lesson.
- Write the two common conversions relating inches to centimeters and miles to kilometers.

Common Core State Standards

6.RP.3d Use ratio reasoning to convert measurement units; manipulate and transform units appropriately when multiplying or dividing quantities.

Previous Learning

Students have studied measurement systems. They should be familiar with metric prefixes such as kilo, centi, and milli.

Technology for the Teacher

Dynamic Classroom

Lesson Plans
Complete Materials List

5.7 Record and Practice Journal

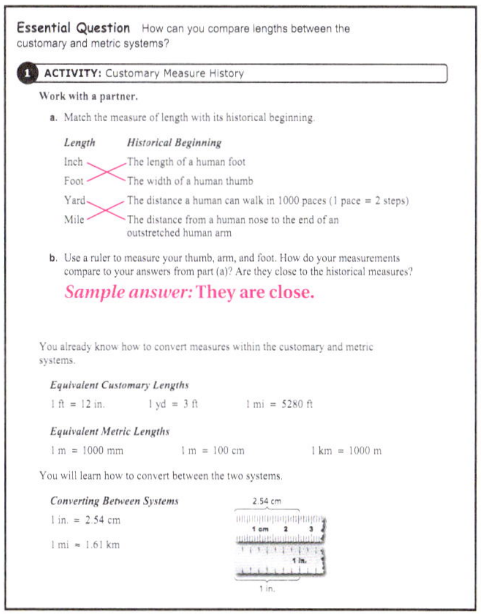

Essential Question How can you compare lengths between the customary and metric systems?

1 ACTIVITY: Customary Measure History

Work with a partner.

a. Match the measure of length with its historical beginning.

Length	*Historical Beginning*
Inch	The length of a human foot
Foot	The width of a human thumb
Yard	The distance a human can walk in 1000 paces (1 pace = 2 steps)
Mile	The distance from a human nose to the end of an outstretched human arm

b. Use a ruler to measure your thumb, arm, and foot. How do your measurements compare to your answers from part (a)? Are they close to the historical measures?

Sample answer: They are close.

You already know how to convert measures within the customary and metric systems.

Equivalent Customary Lengths

1 ft = 12 in. 1 yd = 3 ft 1 mi = 5280 ft

Equivalent Metric Lengths

1 m = 1000 mm 1 m = 100 cm 1 km = 1000 m

You will learn how to convert between the two systems.

Converting Between Systems

1 in. = 2.54 cm

1 mi ≈ 1.61 km

English Language Learners

Words and Abbreviations

Make a poster to display both customary and metric units of measure. Show the full word, both singular and plural, and the abbreviation.

Word		
Singular	**Plural**	**Abbreviation**
inch	inches	in.
foot	feet	ft
meter	meters	m

5.7 Record and Practice Journal

2 ACTIVITY: Comparing Measures

Work with a partner. Answer each question. Explain your answer. Use a diagram in your explanation.

	Metric	*Customary*
a. Car Speed: Which is faster?	80 km/h	60 mi/h
b. Trip Distance: Which is farther?	200 km	200 mi
c. Human Height: Who is taller?	180 cm	5 ft 8 in.
d. Wrench Width: Which is wider?	8 mm	5/16 in.
e. Swimming Pool Depth: Which is deeper?	1.4 m	4 ft

a. 60 mi/h

b. 200 mi

c. 180 cm

d. 8 mm

e. 1.4 m

3 ACTIVITY: Changing Units in a Rate

Work with a partner. Change the units of the rate by multiplying by a "Magic One." Write your answer as a unit rate. Show your work.

	Original Rate	*Magic One*	*New Units*	*Unit Rate*
a.	$\frac{\$120}{\text{h}}$	$\times \frac{1\text{ h}}{60\text{ min}}$	$= \frac{\$120}{60\text{ min}}$	$= \frac{\$2}{1\text{ min}}$
b.	$\frac{\$3}{\text{min}}$	$\times \frac{60\text{ min}}{1\text{ h}}$	$= \frac{\$180}{1\text{ h}}$	$= \frac{\$180}{1\text{ h}}$
c.	$\frac{12\text{ in.}}{\text{ft}}$	$\times \frac{3\text{ ft}}{1\text{ yd}}$	$= \frac{36\text{ in.}}{1\text{ yd}}$	$= \frac{36\text{ in.}}{1\text{ yd}}$
d.	$\frac{2\text{ ft}}{\text{week}}$	$\times \frac{52\text{ weeks}}{1\text{ yr}}$	$= \frac{104\text{ ft}}{1\text{ yr}}$	$= \frac{104\text{ ft}}{1\text{ yr}}$

What Is Your Answer?

4. One problem-solving strategy is called *Working Backwards*. What does this mean? How can you use this strategy to find the rates in Activity 3?
Check students' work.

5. **IN YOUR OWN WORDS** How can you compare lengths between the customary and the metric systems? Give examples with your description.
Convert one length into the units of the other length and then compare.

Laurie's Notes

Activity 2

? "Which diving board is higher, a 3-meter diving board or a 3-foot diving board?" Listen for reasoning that 1 meter is a little longer than 1 yard and because 1 yard equals 3 feet, the 3-meter diving board is about 3 times higher.

- A diagram would be helpful. It could look something like the following:

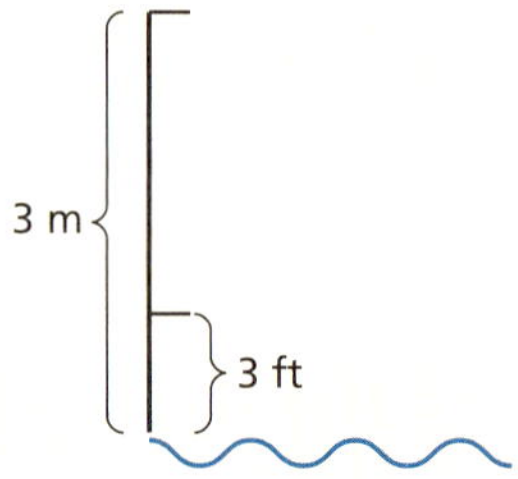

- Students could guess at each of these problems.
- The diagram provides evidence that they understand the relationship between the two units of measure.

Activity 3

- **Explain:** Students should recognize that 60 seconds = 1 minute. The fractions $\frac{60\text{ seconds}}{1\text{ minute}}$ and $\frac{1\text{ minute}}{60\text{ seconds}}$ are equivalent to 1. This is the definition of Magic One. Multiplying by 1 doesn't change the rate's value, only the units that are used to describe it.
- **Common Error:** Students multiply by the reciprocal of the "Magic One." Tell students to look at the units they have (i.e., dollars per hour) and the units they want (i.e., dollars per minute). Then ask in what order should they write the fraction so that the unwanted units are divided out and the unit they want is left. This is working backwards.
- **Connection:** When multiplying fractions, common factors are divided out. When rewriting a rate as a unit rate, common units are divided out. In some examples the numerical portion must be simplified as shown in the sample ($120 \div 60 = 2$).
- Have different pairs of students share their results at the board.

What Is Your Answer?

? "Why are there two different systems of measurement?" Because the measures in the customary system were not the same in all countries, scientists created the metric system of measures to ensure uniformity in measurements.

Closure

- By eyesight, estimate the length and width of the classroom in feet and then in meters. Answers will vary depending upon classroom size.

2 ACTIVITY: Comparing Measures

Math Practice 1

Analyze Givens

What is the relationship between the given quantities? What are you trying to find?

Work with a partner. Answer each question. Explain your answer. Use a diagram in your explanation.

		Metric	*Customary*
a.	**Car Speed:** Which is faster?	80 km/h	60 mi/h
b.	**Trip Distance:** Which is farther?	200 km	200 mi
c.	**Human Height:** Who is taller?	180 cm	5 ft 8 in.
d.	**Wrench Width:** Which is wider?	8 mm	5/16 in.
e.	**Swimming Pool Depth:** Which is deeper?	1.4 m	4 ft

3 ACTIVITY: Changing Units in a Rate

Work with a partner. Change the units of the rate by multiplying by a "Magic One." Write your answer as a unit rate. Show your work.

	Original Rate		*Magic One*		*New Units*		*Unit Rate*
a. **Sample:**	$\frac{\$120}{\cancel{h}}$	×	$\frac{1\cancel{h}}{60 \text{ min}}$	=	$\frac{\$120}{60 \text{ min}}$	=	$\frac{\$2}{1 \text{ min}}$
b.	$\frac{\$3}{\text{min}}$	×				=	$\$\frac{\square}{1 \text{ h}}$
c.	$\frac{12 \text{ in.}}{\text{ft}}$	×				=	$\frac{\square \text{ in.}}{1 \text{ yd}}$
d.	$\frac{2 \text{ ft}}{\text{week}}$	×				=	$\frac{\square \text{ ft}}{1 \text{ yr}}$

What Is Your Answer?

4. One problem-solving strategy is called *Working Backwards.* What does this mean? How can you use this strategy to find the rates in Activity 3?

5. **IN YOUR OWN WORDS** How can you compare lengths between the customary and the metric systems? Give examples with your description.

Practice Use what you learned about converting measures between systems to complete Exercises 4 and 5 on page 236.

5.7 Lesson

Check It Out
Lesson Tutorials
BigIdeasMath.com

Key Vocabulary
U.S. customary system, *p. 234*
metric system, *p. 234*
conversion factor, *p. 234*
unit analysis, *p. 234*

The **U.S. customary system** is a system of measurement that contains units for length, capacity, and weight. The **metric system** is a decimal system of measurement, based on powers of 10, that contains units for length, capacity, and mass.

To convert from one unit of measure to another, multiply by one or more *conversion factors*. A conversion factor can be written using fraction notation.

Key Idea

Conversion Factor

A **conversion factor** is a rate that equals 1.

	Relationship	*Conversion Factors*
Example	$1 \text{ m} \approx 3.28 \text{ ft}$	$\dfrac{1 \text{ m}}{3.28 \text{ ft}}$ and $\dfrac{3.28 \text{ ft}}{1 \text{ m}}$

You can use **unit analysis** to decide which conversion factor will produce the appropriate units.

EXAMPLE 1 Converting Units

a. Convert 36 quarts to gallons.

Use a conversion factor.

1 gal = 4 qt

$$36 \text{ qt} \cdot \frac{1 \text{ gal}}{4 \text{ qt}} = \frac{36 \cdot 1 \text{ gal}}{4}$$

$$= 9 \text{ gal}$$

So, 36 quarts is 9 gallons.

b. Convert 20 centimeters to inches.

Use a conversion factor.

1 in. = 2.54 cm

$$20 \text{ cm} \cdot \frac{1 \text{ in.}}{2.54 \text{ cm}} \approx 7.87 \text{ in.}$$

So, 20 centimeters is about 7.87 inches.

On Your Own

Now You're Ready
Exercises 6–17

Copy and complete the statement. Round to the nearest hundredth if necessary.

1. 48 ft = ___ yd
2. 7 lb = ___ oz
3. 5 g = ___ mg
4. 7 mi ≈ ___ km
5. 12 qt ≈ ___ L
6. 25 kg ≈ ___ lb

Multi-Language Glossary at BigIdeasMath.com

Laurie's Notes

Introduction

Connect

- **Yesterday:** Students explored linear measurement in customary and metric units, and converted between the two systems (MP6).
- **Today:** Students will convert units of length, capacity, and weight/mass between systems (customary and metric).

Motivate

- **History of Customary Length:**
  - The Egyptian cubit was developed about 3000 B.C. The unit was based on the length of the arm (from the elbow to the extended finger).
  - The Romans adopted the foot from the Greeks and divided it into 12 sections called *unicae*, which came to be known as an inch.
- **History of Metric:**
  - The word *meter* is from the Greek word *metron*, which means a *measure*. In 1793, the meter was defined to be one ten-millionth of *the length of the earth's meridian along a quadrant*. This is the same as one ten-millionth of the distance from the North Pole to the equator, along the meridian running near Dunkirk in France.
  - The United States is the only industrialized country that does not use the metric system as its predominant system of measurement.

Lesson Notes

Key Idea

- Write the Key Idea. Define *conversion factor* and give several examples. Point out that each conversion factor can be written two ways, which are reciprocals of each other.
- **Big Idea:** The technique that is used today is called **unit analysis**. You multiply the known amount by the conversion factor, written in the form that allows the known units to divide out and the desired units to remain.

Example 1

- **MP6 Attend to Precision:** For part (a), some students may ask, "Why not just divide by 4 since there are 4 quarts in a gallon?" That is indeed what happens, however paying attention to the units is a good habit to develop.
- Model part (b) with a ruler.
- When converting between systems, answers may vary, depending on which conversion factor is used. For instance, in part (b), if you use the conversion factor $\frac{0.39 \text{ in.}}{1 \text{ cm}}$, you will get an answer of 7.8 inches.

On Your Own

- **Think-Pair-Share:** Students should read each question independently and then work in pairs to answer the questions. When they have answered the questions, the pair should compare their answers with another group and discuss any discrepancies.

Goal Today's lesson is converting units of measure between systems.

Technology for the Teacher

Dynamic Classroom

Lesson Tutorials
Lesson Plans
Answer Presentation Tool

Extra Example 1

a. Convert 48 quarts to gallons. 12 gal
b. Convert 14 centimeters to inches. about 5.51 in.

On Your Own

1. 16
2. 112
3. 5000
4. 11.27
5. 11.4
6. 55

Extra Example 2

Copy and complete the statement using < or > : 2 ft _?_ 55 cm. 2 ft > 55 cm

Extra Example 3

Convert the flow rate of 3 liters per minute to quarts per minute. about 3.16 quarts per minute

Extra Example 4

Convert the speed 3 kilometers per hour to meters per minute. 50 meters per minute

On Your Own

7. <
8. <
9. >
10. 5 gallons per second
11. 176 feet per second

Differentiated Instruction

Visual

Create a poster to display the units of 1 (or ratios) that are used in converting units of measure between the customary and metric systems.

Length

$\frac{1 \text{ in.}}{2.54 \text{ cm}}, \frac{2.54 \text{ cm}}{1 \text{ in.}}, \frac{1 \text{ mi}}{1.61 \text{ km}}, \frac{1.61 \text{ km}}{1 \text{ mi}}$

Capacity

$\frac{1 \text{ qt}}{0.95 \text{ L}}, \frac{0.95 \text{ L}}{1 \text{ qt}}$

Weight and Mass

$\frac{1 \text{ lb}}{0.45 \text{ kg}}, \frac{0.45 \text{ kg}}{1 \text{ lb}}$

Laurie's Notes

Example 2

- This is a multi-step problem that requires students to possess knowledge of common equivalences, such as 1 pound equals 16 ounces. Before looking at these problems, it would be helpful to make a list of measurement facts. Ask students for examples.
- **MP1 Make Sense of Problems and Persevere in Solving Them:** A common question is how to recognize when to use $\frac{1 \text{ pound}}{16 \text{ ounces}}$ and when to use $\frac{16 \text{ ounces}}{1 \text{ pound}}$. Because you are beginning with 25 ounces, the unit of ounces must be in the denominator of the ratio in order for the units to divide out. Then, because pounds are in the numerator of that ratio, pounds must be in the denominator of the next ratio in order for the units of pounds to divide out. In other words, start with the quantity you're trying to convert, then decide on a conversion factor based on the units.

Example 3

- Write the equation as shown.
- ? "What are the units for the rate given?" quarts per minute "What units are desired for the rate?" liters per minute
- The strategy is to look at the units given (qt/min) and the units desired (L/min). Because you need to eliminate quarts and introduce liters, write the conversion factor with liters in the numerator and quarts in the denominator as shown. The common units of quarts divide out, leaving L/min.

Example 4

- ? "What is a zip line? How does it work?" A zip line is a wire, usually hung high off of the ground between trees. It is hung at a diagonal, so you use gravity to slide down. You wear a harness and climb to the top of a tower, push off and slide either to the ground or to another platform.
- ? "What are the units for the known rate and what units do you want?" known rate: mi/h; want rate: ft/sec
- In this example, the linear measure must be converted to feet and the time must be converted to seconds. So, you must multiply by two conversion factors.

On Your Own

- These questions will take some time. Students should work together to set up an appropriate expression that uses unit analysis. Use of calculators would be appropriate **(MP5 Use Appropriate Tools Strategically).**

Closure

- A major league pitcher throws a 90-mile-per-hour fastball. How many feet per second is this? 132 feet per second

EXAMPLE 2 Comparing Units

Copy and complete the statement using < or >: 25 oz ▢ 2 kg.

Convert 25 ounces to kilograms.

1 lb = 16 oz   1 lb ≈ 0.45 kg

$$25\ \text{oz} \times \frac{1\ \text{lb}}{16\ \text{oz}} \times \frac{0.45\ \text{kg}}{1\ \text{lb}} = \frac{25 \cdot 1 \cdot 0.45\ \text{kg}}{16 \cdot 1} \approx 0.70\ \text{kg}$$

Because 0.70 kilogram is less than 2 kilograms, 25 oz < 2 kg.

EXAMPLE 3 Converting a Rate: Changing One Unit

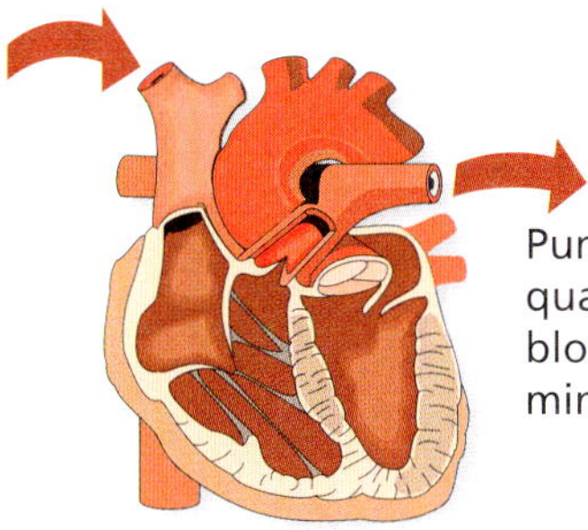

Pumps 5 quarts of blood per minute

How many liters does the human heart pump per minute?

1 qt ≈ 0.95 L

$$\frac{5\ \text{qt}}{1\ \text{min}} \cdot \frac{0.95\ \text{L}}{1\ \text{qt}} \approx \frac{4.75\ \text{L}}{1\ \text{min}}$$

The rate of 5 quarts per minute is about 4.75 liters per minute.

EXAMPLE 4 Converting a Speed: Changing Both Units

You are riding on a zip line. Your speed is 15 miles per hour. What is your speed in feet per second?

$$\frac{15\ \text{mi}}{1\ \text{h}}\left(\frac{5280\ \text{ft}}{1\ \text{mi}}\right)\left(\frac{1\ \text{h}}{3600\ \text{sec}}\right) = \frac{15 \cdot 5280\ \text{ft}}{3600\ \text{sec}}$$

1 mi = 5280 ft   1 h = 3600 sec

$$= \frac{79{,}200\ \text{ft}}{3600\ \text{sec}}$$

$$= \frac{22\ \text{ft}}{1\ \text{sec}}$$

Your speed is 22 feet per second.

On Your Own

Now You're Ready
Exercises 20–31

Copy and complete the statement using < or >.

7. 7 cm ▢ 3 in.   **8.** 8 c ▢ 2 L   **9.** 3 oz ▢ 70 g

10. An oil tanker is leaking oil at a rate of 300 gallons per minute. What is this rate in gallons per second?

11. A tennis ball travels at a speed of 120 miles per hour. What is this rate in feet per second?

5.7 Exercises

Check It Out
Help with Homework
BigIdeasMath.com

Vocabulary and Concept Check

1. **VOCABULARY** Is $\frac{10 \text{ mm}}{1 \text{ cm}}$ a conversion factor? Explain.

2. **WRITING** Describe how to convert 2 liters per hour to milliliters per second.

3. **DIFFERENT WORDS, SAME QUESTION** Which is different? Find "both" answers.

Convert 5 inches to centimeters.	Find the number of inches in 5 centimeters.
How many centimeters are in 5 inches?	Five inches equals how many centimeters?

Practice and Problem Solving

Answer the question. Explain your answer.

4. Which juice container is larger: 2 L or 1 gal?

5. Which person is heavier: 75 kg or 110 lb?

Copy and complete the statement. Round to the nearest hundredth if necessary.

1 6. 3 pt = ___ c
7. 1500 mL = ___ L
8. 40 oz = ___ lb
9. 12 L ≈ ___ qt
10. 14 m ≈ ___ ft
11. 4 ft ≈ ___ m
12. 64 lb ≈ ___ kg
13. 0.3 km ≈ ___ mi
14. 75.2 in. ≈ ___ cm
15. 17 kg ≈ ___ lb
16. 15 cm ≈ ___ in.
17. 9 mi ≈ ___ km

18. **ERROR ANALYSIS** Describe and correct the error in converting the units.

✗
$$8 \text{ L} \approx 8 \text{ L} \cdot \frac{0.95 \text{ qt}}{1 \text{ L}}$$
$$= 8 \cancel{\text{L}} \cdot \frac{0.95 \text{ qt}}{1 \cancel{\text{L}}}$$
$$= 7.6 \text{ qt}$$

19. **BRIDGE** The Mackinac Bridge in Michigan is the third-longest suspension bridge in the United States.

    a. How high above the water is the roadway in meters?

    b. The bridge has a length of 26,372 feet. What is the length in kilometers?

Assignment Guide and Homework Check

Level	Day 1 Activity Assignment	Day 2 Lesson Assignment	Homework Check
Basic	4, 5, 37–41	1–3, 7–17 odd, 18, 19, 21–25 odd	7, 11, 19, 23, 25
Average	4, 5, 37–41	1–3, 13–19 odd, 18, 20, 27–31, 33	13, 17, 19, 20, 29
Advanced	4, 5, 37–41	1–3, 10–20 even, 24–34 even, 35, 36	14, 16, 20, 24, 34

Common Errors

- **Exercises 4, 5, 20–25** Students may try to compare the measurements without converting the units. Remind them that they cannot compare quantities unless the units are the same.
- **Exercises 6–17** Students may use the wrong conversion factor or may write the units they are trying to remove in the numerator. For example, a student may write $14 \text{ m} \cdot \frac{1 \text{ m}}{3.28 \text{ ft}}$ instead of $14 \text{ m} \cdot \frac{3.28 \text{ ft}}{1 \text{ m}}$. Remind them to put the given unit in the denominator so that the units will divide out. Also caution students to make sure that they use the correct conversion factor.
- **Exercises 26–28** Students may change the wrong units. For example, a student may change from hours to minutes instead of kilometers to miles. Remind them to read the units carefully.
- **Exercises 29–31** Students may write the units they are trying to remove in the numerator. This is common because they are converting both units. Remind them to put the given unit in the denominator so that the units will divide out. Some students may need to convert one unit, and then convert the other unit as a separate problem.

5.7 Record and Practice Journal

Complete the statement.

1. $3 \text{ m} \approx$ **9.84** ft
2. $32 \text{ cm} \approx$ **12.6** in.
3. $16 \text{ qt} \approx$ **15.2** L
4. $\frac{50 \text{ mi}}{\text{h}} \approx \frac{\boxed{80} \text{ km}}{\text{h}}$
5. $\frac{25 \text{ gal}}{\text{min}} = \frac{\boxed{1.67} \text{ qt}}{\text{sec}}$
6. $\frac{1000 \text{ m}}{\text{sec}} = \frac{\boxed{60} \text{ km}}{\text{min}}$
7. Your doctor prescribes you to take 400 milligrams of medicine every 8 hours. How many ounces of medicine do you take in a day?
   about 0.043 ounce

Vocabulary and Concept Check

1. yes; Because 1 centimeter is equal to 10 millimeters, the conversion factor equals 1.
2. First convert liters to milliliters by multiplying by 1000. Then convert hours to seconds by dividing by 3600.
3. Find the number of inches in 5 cm; $5 \text{ cm} \approx 1.97$ in.; $5 \text{ in.} = 12.7 \text{ cm}$

Practice and Problem Solving

4. 1 gal container; $1 \text{ gal} = 4 \text{ qt}$, $2 \text{ L} \approx 2.11 \text{ qt}$, and $4 \text{ qt} > 2.11 \text{ qt}$
5. person weighing 75 kg; $75 \text{ kg} \approx 166.67 \text{ lb}$ and $166.67 \text{ lb} > 110 \text{ lb}$
6. 6
7. 1.5
8. 2.5
9. 12.63
10. 45.92
11. 1.22
12. 28.8
13. 0.19
14. 190.01
15. 37.78
16. 5.91
17. 14.49
18. The conversion factor is wrong.

    $$8 \text{ L} \approx 8 \cancel{\text{L}} \cdot \frac{1 \text{ qt}}{0.95 \cancel{\text{L}}} \approx 8.42 \text{ qt}$$
19. **a.** about 60.67 m

    b. about 8.04 km

Practice and Problem Solving

20. > **21.** <

22. > **23.** >

24. < **25.** >

26. 8.07 **27.** 1320

28. 0.0175 **29.** 111.8

30. 1.13 **31.** 0.001

32. no; 2 L > 2 qt

33. See *Taking Math Deeper*.

34. **a.** spine-tailed swift; mallard

b. yes, It is faster than all of the other birds in the table. Its dive speed is about 201.25 miles per hour.

35. about 669,600,000 mi/h

36. about 22.7 gallons; So you need to buy 23 gallons of paint.

Fair Game Review

37. 30 **38.** 30.55

39. 18 **40.** 2.56

41. C

Mini-Assessment

Complete the statement.

1. 5 mi ≈ ☐ km 8
2. 16 qt ≈ ☐ L 15.2
3. 3 in. ≈ ☐ cm 7.62
4. 5 lb ≈ ☐ kg 2.25
5. A new born baby weighs 8 pounds. How much does the baby weigh in kilograms? about 3.6 kg

Taking Math Deeper

Exercise 33

This is a practical problem that students need as a real-life skill—changing between miles per hour and kilometers per hour. Most countries in the world, including our neighbors Canada and Mexico, measure driving distance in kilometers.

1 Help me see the conversion factors.

0.62 mi ≈ 1 km   1 mi ≈ 1.61 km

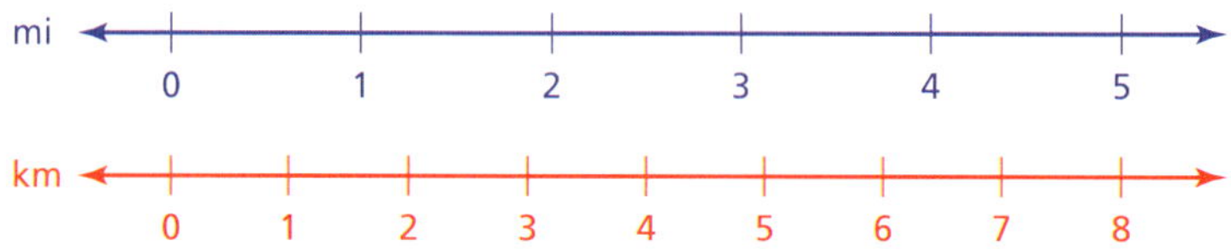

A kilometer is shorter than a mile.

2 Convert 110 miles per hour to kilometers per hour.

$$\frac{110\ \cancel{\text{mi}}}{1\ \text{h}} \cdot \frac{1.61\ \text{km}}{1\ \cancel{\text{mi}}} = \frac{177.1\ \text{km}}{1\ \text{h}}$$

a. Yes; 177.1 kilometers per hour is above the speed limit of 130 kilometers per hour.

3 Convert 80 miles per hour to kilometers per hour.

$$\frac{80\ \cancel{\text{mi}}}{1\ \text{h}} \cdot \frac{1.61\ \text{km}}{1\ \cancel{\text{mi}}} = \frac{128.8\ \text{km}}{1\ \text{h}}$$

b. Yes; 128.8 kilometers per hour is just below the speed limit of 130 kilometers per hour.

Project

Use a map to plan a family trip to a place you would like to visit outside of your state. If the car travels at an average speed of 90 kilometers per hour, then how long will your round trip take?

Reteaching and Enrichment Strategies

If students need help. . .	If students got it. . .
Resources by Chapter • Practice A and Practice B • Puzzle Time Record and Practice Journal Practice Differentiating the Lesson Lesson Tutorials Skills Review Handbook	Resources by Chapter • Enrichment and Extension • Technology Connection Start the next section

Copy and complete the statement using < or >.

2 **20.** 8 kg ☐ 30 oz **21.** 6 ft ☐ 300 cm **22.** 3 gal ☐ 6 L

23. 10 in. ☐ 200 mm **24.** 1200 g ☐ 5 lb **25.** 1500 m ☐ 3000 ft

Copy and complete the statement.

3 4 **26.** $\frac{13\text{ km}}{\text{h}} \approx \frac{\square\text{ mi}}{\text{h}}$ **27.** $\frac{22\text{ L}}{\text{min}} = \frac{\square\text{ L}}{\text{h}}$ **28.** $\frac{63\text{ mi}}{\text{h}} = \frac{\square\text{ mi}}{\text{sec}}$

29. $\frac{3\text{ km}}{\text{min}} \approx \frac{\square\text{ mi}}{\text{h}}$ **30.** $\frac{17\text{ gal}}{\text{h}} \approx \frac{\square\text{ qt}}{\text{min}}$ **31.** $\frac{6\text{ cm}}{\text{min}} = \frac{\square\text{ m}}{\text{sec}}$

32. **BOTTLE** Can you pour the water from a full 2-liter bottle into a 2-quart pitcher without spilling any? Explain.

33. **AUTOBAHN** Germany suggests a speed limit of 130 kilometers per hour on highways.

a. Is the speed shown greater than the suggested limit?

b. Suppose the speed shown drops 30 miles per hour. Is the new speed below the suggested limit?

34. **BIRDS** The table shows the flying speeds of several birds.

a. Which bird is the fastest? Which is the slowest?

b. The peregrine falcon has a dive speed of 322 kilometers per hour. Is the dive speed of the peregrine falcon faster than the flying speed of any of the birds? Explain.

Bird	Speed
Spine-tailed swift	2843.2 m/min
Spur-winged goose	129.1 ft/sec
Eider duck	31.3 m/sec
Mallard	65 mi/h

35. **SPEED OF LIGHT** The speed of light is about 300,000 kilometers per second. Convert the speed to miles per hour.

36. Critical Thinking One liter of paint covers 100 square feet. How many gallons of paint does it take to cover a room whose walls have an area of 800 square meters?

Fair Game Review What you learned in previous grades & lessons

Find the percent of the number. *(Section 5.6)*

37. 25% of 120 **38.** 65% of 47 **39.** 120% of 15 **40.** 3.2% of 80

41. **MULTIPLE CHOICE** What is the area of a parallelogram with a base of 15 centimeters and a height of 12 centimeters? *(Section 4.1)*

Ⓐ 90 cm^2 Ⓑ 175 cm^2 Ⓒ 180 cm^2 Ⓓ 205 cm^2

5.5–5.7 Quiz

Check It Out
Progress Check
BigIdeasMath.com

Write the percent as a fraction or mixed number in simplest form. *(Section 5.5)*

1. 14%

2. 124%

Write the fraction or mixed number as a percent. *(Section 5.5)*

3. $\frac{13}{20}$

4. $1\frac{1}{4}$

Find the percent of the number. Explain your method. *(Section 5.6)*

5. 25% of 64

6. 120% of 50

Find the whole. Explain your method. *(Section 5.6)*

7. 60% of what number is 24?

8. 160% of what number is 80?

Copy and complete the statement. Round to the nearest hundredth if necessary. *(Section 5.7)*

9. 6.4 in. $\approx$ ___ cm

10. 4 qt $\approx$ ___ L

11. 10 kg $\approx$ ___ lb

12. ANATOMY About 62% of the human body is composed of water. Write this percent as a fraction in simplest form. *(Section 5.5)*

13. SAVES A goalie's saves (•) and goals scored against (×) are shown. What percent of shots did the goalie save? Explain. *(Section 5.5)*

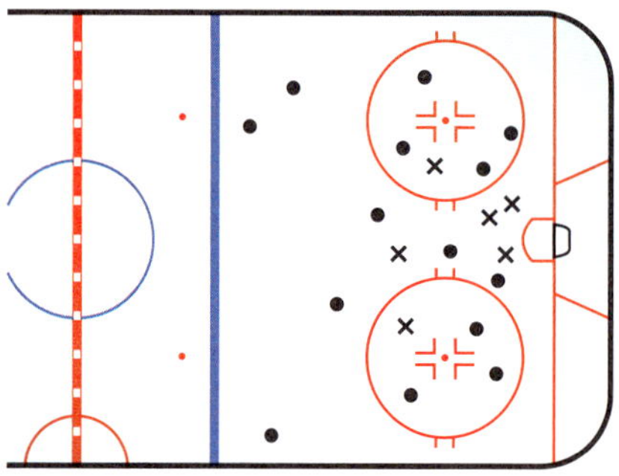

14. SHOPPING You went to the mall with $80. You spent 25% of your money on a pair of shorts and 65% of the remainder on sandals. How much did you spend on the sandals? Explain how you found your answer. *(Section 5.6)*

15. WINDSURFING Determine which windsurfer is traveling faster. Explain your reasoning. *(Section 5.7)*

Speed:
5 meters per second

Speed:
720 feet per minute

Alternative Assessment Options

Math Chat	Student Reflective Focus Question
Structured Interview	**Writing Prompt**

Writing Prompt

Ask students to draw figures that represent 100%, one less than 100%, and one greater than 100%. Remind them to label their drawings. Then have students explain their drawings in writing.

Study Help Sample Answers

Remind students to complete Graphic Organizers for the rest of the chapter.

6. Percent: a part-to-whole ratio where the whole is 100. So, you can write a percent as a fraction with a denominator of 100.

- Example: $40\% = 40 \text{ out of } 100 = \frac{40}{100} = \frac{2}{5}$
- Example: $112\% = \frac{112}{100} = 1\frac{12}{100} = 1\frac{3}{25}$
- Example: $\frac{9}{20} = \frac{9 \times 5}{20 \times 5} = \frac{45}{100} = 45\%$

7. U.S. customary system: a system of measurement that contains units for length, capacity, and weight.

- Example: Unit of length: foot
- Example: Unit of capacity: quart
- Example: Unit of weight: pound

8–10. Available at *BigIdeasMath.com.*

Reteaching and Enrichment Strategies

If students need help. . .	If students got it. . .
Resources by Chapter • Practice A and Practice B • Puzzle Time Lesson Tutorials *BigIdeasMath.com*	Resources by Chapter • Enrichment and Extension • Technology Connection Game Closet at *BigIdeasMath.com* Start the Chapter Review

Answers

1. $\frac{7}{50}$
2. $1\frac{6}{25}$
3. 65%
4. 125%
5. 16; Explanations will vary.
6. 60; Explanations will vary.
7. 40; Explanations will vary.
8. 50; Explanations will vary.
9. 16.26
10. 3.8
11. 22.22
12. $\frac{31}{50}$
13. 70%; There are 14 saves and 6 goals scored against for a total of 20 shots.
    $\frac{14}{20} = \frac{70}{100} = 70\%$
14. \$39; To find the answer, find the amount spent on shorts, subtract this from the amount you took, and multiply the result by 65%.
15. the blue windsurfer; blue speed $= 5 \text{ m/s} > 3.66 \text{ m/s}$ = red speed

Technology for the Teacher

Online Assessment
Assessment Book
ExamView® Assessment Suite

For the Teacher
Additional Review Options

- *BigIdeasMath.com*
- Online Assessment
- Game Closet at *BigIdeasMath.com*
- Vocabulary Help
- Resources by Chapter

Answers

1. 3 : 2, 3 to 2; There are 3 butterflies for every 2 caterpillars.
2. 2 : 1, 2 to 1; There are 2 saxophones for every trumpet.

Review of Common Errors

Exercises 1 and 2

- Students may write the ratios in the wrong order. Remind them that the ratio a to b is not the same as the ratio b to a.

Exercises 3 and 4

- Students may add (or subtract) the same quantity to each number in the ratio and say that 6 : 3 is equivalent to 9 : 6. Remind students that they can multiply to obtain equivalent ratios. For example, 6 : 3 is equivalent to $6 \times 2 : 3 \times 2$, or 12 : 6.

Exercises 5 and 6

- Students may not write the units on the rates. Compare rates to ratios, but with different units. Illustrate the importance of the units with an examination of the speed of a car. In the United States speed is measured in miles per hour, but in most of Europe speed is measured in kilometers per hour. If you say that the speed of a car is 35, you do not know if it is miles or kilometers per hour.

Exercise 7

- Students may first take the number of beats part of the rate instead of finding the unit rate to solve the problem. Encourage students to find the unit rate first to help them solve the problem without errors.

Exercise 8

- Students may find the reciprocal of the unit rate and come up with the incorrect conclusion. Remind them that the unit rate represents the rate for one unit, so they should divide by the number of units.

Exercises 9–14

- Students may put the percent over the wrong denominator or try to remove the decimal point when writing it over 100.
- Students may struggle with knowing when it is appropriate to use the equivalent fraction method of writing a fraction as a percent.

Exercises 15–19

- When the number ends in a zero, students may be off by a place value because they forgot about the zero at the end. Encourage students to estimate their answer first.

Exercises 20–22

- Students may put the unit that they are solving for in the denominator. Demonstrate to students visually that the original units will not cancel.

5 Chapter Review

Check It Out
Vocabulary Help
BigIdeasMath.com

Review Key Vocabulary

ratio, *p. 192*
equivalent ratios, *p. 198*
ratio table, *p. 198*
rate, *p. 206*
unit rate, *p. 206*
equivalent rates, *p. 206*
percent, *p. 220*
U.S. customary system, *p. 234*
metric system, *p. 234*
conversion factor, *p. 234*
unit analysis, *p. 234*

Review Examples and Exercises

5.1 Ratios *(pp. 190–195)*

Write the ratio of apples to oranges. Explain what the ratio means.

3 apples → 3 to 5 ← 5 oranges

So, the ratio of apples to oranges is 3 to 5, or 3 : 5. That means that for every 3 apples, there are 5 oranges.

Exercises

Write the ratio. Explain what the ratio means.

1. butterflies : caterpillars

2. saxophones : trumpets

5.2 Ratio Tables *(pp. 196–203)*

Find the missing values in the ratio table. Then write the equivalent ratios.

Trees	2	6	
Birds	5		30

You can use multiplication to find the missing values.

×3 ×2

Trees	2	6	12
Birds	5	15	30

×3 ×2

The equivalent ratios are 2 : 5, 6 : 15, and 12 : 30.

Exercises

Find the missing values in the ratio table. Then write the equivalent ratios.

3.

Levers	6		18
Pulleys	3	6	

4.

Cars	3	6	
Trucks	4		24

5.3 Rates (pp. 204–209)

A horse can run 165 feet in 3 seconds. At this rate, how far can the horse run in 5 seconds?

Using a ratio table, divide to find the unit rate. Then multiply to find the distance that the horse can run in 5 seconds.

Distance (feet)	165	55	275
Time (seconds)	3	1	5

(÷3, ×5)

So, the horse can run 275 feet in 5 seconds.

Exercises

Write a unit rate for the situation.

5. 12 stunts in 4 movies

6. 3600 stitches in 3 minutes

7. **MUSIC** A song has 28 beats in 4 seconds. At this rate, how many beats are there in 30 seconds?

5.4 Comparing and Graphing Ratios (pp. 210–215)

There are 24 grams of sugar in 6 fluid ounces of Soft Drink A, and there are 15 grams of sugar in 4 fluid ounces of Soft Drink B. Which soft drink contains more sugar in a 12-ounce can?

Use ratio tables to compare the soft drinks.

Soft Drink A (×2)

Sugar (grams)	24	48
Volume (fluid ounces)	6	12

Soft Drink B (×3)

Sugar (grams)	15	45
Volume (fluid ounces)	4	12

The tables show that a 12-ounce can of Soft Drink A has $48 - 45 = 3$ more grams of sugar than Soft Drink B.

So, a 12-ounce can of Soft Drink A has more sugar.

Exercises

8. **TUNA** A 5-ounce can of tuna costs $0.90. A 12-ounce can of tuna costs $2.40. Which is the better buy?

Review Game

Rates and Percent

Materials per Pair:

- copy of the rate and percent situations
- pencil
- paper

Directions:

Players take turns picking a rate and finding its unit rate or picking a percent situation and finding its number. Round answers to the nearest tenth if necessary. For a correct answer, the player is awarded the number of points equivalent to the answer. Play continues until all the situations are used. Each rate and percent situation may be used only once.

Who wins?

The student with the greater total wins.

Rate and Percent Situations

A. 630 miles in 5 hours

B. 780 miles in 6 hours

C. 952 miles in 8 hours

D. 149 miles in 1 hour

E. 308 miles in 1.5 hours

F. 246 miles in 2 hours

G. 416 miles in 3 hours

H. 557 miles in 5 hours

J. What is 116% of 116?

K. What is 24% of 764?

L. What is 60% of 195?

M. What is 35% of 317?

N. 92 is 45% of what number?

P. 37 is 44% of what number?

Q. 53 is 112% of what number?

R. 99 is 152% of what number?

For the Student Additional Practice

- Lesson Tutorials
- Multi-Language Glossary
- Self-Grading Progress Check
- *BigIdeasMath.com*
  Dynamic Student Edition
  Student Resources

Answers

3.

Levers	6	12	18
Pulleys	3	6	9

6 : 3, 12 : 6, 18 : 9

4.

Cars	3	6	18
Trucks	4	8	24

3 : 4, 6 : 8, 18 : 24

5. 3 stunts per movie

6. 1200 stitches per minute

7. 210 beats

8. 5-ounce can

9. $\frac{3}{25}$

10. $\frac{22}{25}$

11. $\frac{1}{125}$

12. 60%

13. 172%

14. 142%

15. 48

16. 44

17. 72

18. 50

19. 40

20. 3.16

21. 23.37

22. 6.75

My Thoughts on the Chapter

Teacher Tip

Not allowed to write in your teaching edition? Use sticky notes to record your thoughts.

What worked. . .

What did not work. . .

What I would do differently. . .

5.5 Percents (pp. 218–223)

Write $\frac{3}{20}$ as a percent.

$$\frac{3}{20} = \frac{15}{100} = 15\%$$

(numerator and denominator each $\times 5$)

Because $20 \times 5 = 100$, multiply the numerator and denominator by 5. Write the numerator with a percent symbol.

Exercises

Write the percent as a fraction or mixed number in simplest form.

9. 12%
10. 88%
11. 0.8%

Write the fraction or mixed number as a percent.

12. $\frac{3}{5}$
13. $\frac{43}{25}$
14. $1\frac{21}{50}$

5.6 Solving Percent Problems (pp. 224–231)

a. 75% of 80 is what number?

$$75\% \text{ of } 80 = \frac{3}{4} \times 80 = \frac{3 \times \cancel{80}^{20}}{\cancel{4}_1} = 60$$

So, 75% of 80 is 60.

b. 30% of what number is 27?

$$27 \div 30\% = 27 \div \frac{3}{10} = \cancel{27}^{9} \cdot \frac{10}{\cancel{3}_1} = 90$$

So, 30% of 90 is 27.

Exercises

Find the percent of the number. Explain your method.

15. 60% of 80
16. 80% of 55
17. 150% of 48

Find the whole. Explain your method.

18. 70% of what number is 35?
19. 140% of what number is 56?

5.7 Converting Measures (pp. 232–237)

Convert 8 kilometers to miles.

$$8 \text{ km} \times \frac{1 \text{ mi}}{1.6 \text{ km}} \approx 5 \text{ mi}$$

Because 1 mi ≈ 1.6 km, use the ratio $\frac{1 \text{ mi}}{1.6 \text{ km}}$.

Exercises

Copy and complete the statement. Round to the nearest hundredth if necessary.

20. 3 L ≈ ☐ qt
21. 9.2 in. ≈ ☐ cm
22. 15 lb ≈ ☐ kg

5 Chapter Test

Check It Out
Test Practice
BigIdeasMath.com

Write the ratio. Explain what the ratio means.

1. scooters : bikes

2. starfish : seashells

Find the missing values in the ratio table. Then write the equivalent ratios.

3.

Lemons	4		36
Limes	2	6	

4.

Rabbits	2	4	
Hamsters	9		54

Write a unit rate for the situation.

5. \$54.00 for 3 tickets

6. 210 miles in 3 hours

Write the fraction or mixed number as a percent.

7. $\frac{21}{25}$

8. $\frac{17}{20}$

9. $1\frac{2}{5}$

Find the percent of the number. Explain your method.

10. 80% of 90

11. 30% of 50

12. 120% of 75

Find the whole. Explain your method.

13. 34 is 40% of what number?

14. 52 is 130% of what number?

Copy and complete the statement. Round to the nearest hundredth if necessary.

15. 5 L ≈ ___ qt

16. 56 lb ≈ ___ kg

17. SOUP There are 600 milligrams of sodium in 4 ounces of Soup A, and there are 720 milligrams of sodium in 6 ounces of Soup B. You prepare an 18-ounce bowl of each soup. Which bowl of soup contains more sodium?

18. ORANGE JUICE A 48-fluid-ounce container of orange juice costs \$2.40. A 60-fluid-ounce container of orange juice costs \$3.60. Which is the better buy?

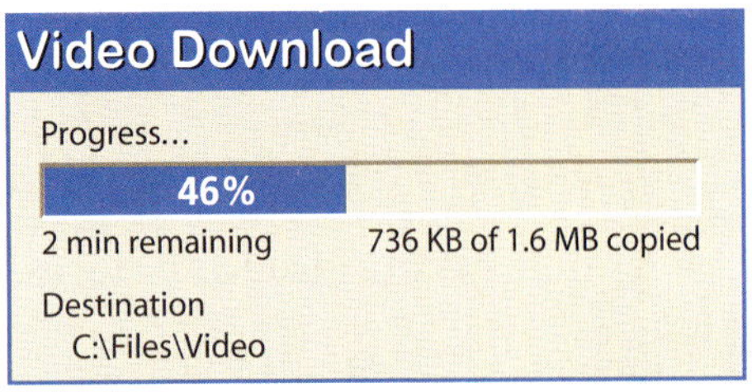

19. DOWNLOAD Your computer displays the progress of a downloading video. What fraction of the video is downloaded? Write your answer in simplest form.

20. GLASSES In a class of 20 students, 40% are boys. Twenty-five percent of the boys and 50% of the girls wear glasses. How many students wear glasses?

Test Item References

Chapter Test Questions	Section to Review	Common Core State Standards
1, 2	5.1	6.RP.1
3, 4	5.2	6.RP.1, 6.RP.3a
5, 6	5.3	6.RP.2, 6.RP.3a, 6.RP.3b
17, 18	5.4	6.RP.2, 6.RP.3a
7–9, 19	5.5	6.RP.3c
10–14, 20	5.6	6.RP.3c
15, 16	5.7	6.RP.3d

Test-Taking Strategies

Remind students to quickly look over the entire test before they start so they can budget their time. Remind them that the test is on ratios, rates, *and* percents and that they need to read the problems carefully. Students need to use the **Stop** and **Think** strategy before they answer a question. Students need to remember to think of the different representations of each number as they work through the test, such as 0.5, $\frac{1}{2}$, and 50%.

Common Errors

- **Exercises 7–9** Students may not know when it is appropriate to use an equivalent fraction to write a fraction as a percent. Point out that sometimes it is easier to write an equivalent fraction.
- **Exercises 10–12** Students might find a number that is significantly greater or less than the correct answer. Encourage students to estimate their answers. Remind them that percents greater than 100 will have answers that are greater than the original number.
- **Exercises 15 and 16** When converting units, students may use a ratio that is the reciprocal of what it should be. Remind students that the ratio should be such that the original units divide out and the new units remain.

Reteaching and Enrichment Strategies

If students need help. . .	If students got it. . .
Resources by Chapter • Practice A and Practice B • Puzzle Time Record and Practice Journal Practice Differentiating the Lesson Lesson Tutorials *BigIdeasMath.com* Skills Review Handbook	Resources by Chapter • Enrichment and Extension • Technology Connection Game Closet at *BigIdeasMath.com* Start Standards Assessment

Answers

1. 3 : 3, 3 to 3; For every 3 scooters, there are 3 bikes.
2. 2 : 5, 2 to 5; For every 2 starfish, there are 5 shells.
3.

Lemons	4	12	36
Limes	2	6	18

4 : 2, 12 : 6, 36 : 18

4.

Rabbits	2	4	12
Hamsters	9	18	54

2 : 9, 4 : 18, 12 : 54

5. \$18.00 per ticket
6. 70 miles per hour
7. 84%
8. 85%
9. 140%
10. 72; Explanations will vary.
11. 15; Explanations will vary.
12. 90; Explanations will vary.
13. 85; Explanations will vary.
14. 40; Explanations will vary.
15. 5.26
16. 25.2
17. Soup A
18. 48-fluid-ounce container
19. $\frac{23}{50}$
20. 8 students

Technology for the Teacher

Online Assessment
Assessment Book
ExamView® Assessment Suite

Test-Taking Strategies

Available at *BigIdeasMath.com*

After Answering Easy Questions, Relax
Answer Easy Questions First
Estimate the Answer
Read All Choices before Answering
Read Question before Answering
Solve Directly or Eliminate Choices
Solve Problem before Looking at Choices
Use Intelligent Guessing
Work Backwards

About this Strategy

When taking a multiple choice test, be sure to read each question carefully and thoroughly. Sometimes it is easier to solve the problem and then look for the answer among the choices.

Answers

1. A
2. G
3. B
4. G
5. C

Technology for the Teacher

Common Core State Standards Support
Performance Tasks
Online Assessment
Assessment Book
ExamView® Assessment Suite

Item Analysis

1. **A.** Correct answer

   B. The student finds the quotient of the whole numbers and the quotient of the fractions and combines these to get the answer.

   C. The student finds the product of the whole numbers and the product of the fractions and combines these to get the answer.

   D. The student rewrites the quotient as $\frac{76}{9} \div \frac{14}{3}$ but then changes this to a product without finding the reciprocal of the divisor.

2. **F.** The student does not recognize that $\frac{1}{4} = \frac{25}{100} = 25\%$.

   G. Correct answer

   H. The student does not recognize that $\frac{5}{20} = \frac{25}{100} = 25\%$.

   I. The student does not recognize that $\frac{25}{100} = 25\%$.

3. **A.** The cost for 4 pencils is \$0.10 more than \$0.10 times 4. So, the student thinks that the cost of 10 pencils is \$0.10 more than \$0.10 times 10.

   B. Correct answer

   C. The student multiplies 4 times \$0.50.

   D. The student multiplies 10 times \$0.50.

4. **F.** The student distributes the multiplication by 2 correctly, but then multiplies the resulting terms instead of adding them.

   G. Correct answer

   H. The student incorrectly distributes the multiplication by 2 by using addition and then multiplies the resulting sums.

   I. The student incorrectly distributes the multiplication by 2 by using addition.

5. **A.** The student multiplies the two ticket prices together and multiplies this product by the sum of the numbers of the two types of tickets.

   B. The student adds the two ticket prices together and multiplies this sum by the product of the numbers of the two types of tickets.

   C. Correct answer

   D. The student finds the correct expressions for the cost of each type of ticket but then multiplies these two expressions.

5 Standards Assessment

Test-Taking Strategy
Solve Problem Before Looking at Choices

"Solve the problem before looking at the choices. You know one-fourth of 20 is 5, so the answer is 5 miles."

1. What is the value of the expression below? *(6.NS.1)*

$$8\frac{4}{9} \div 4\frac{2}{3}$$

A. $1\frac{17}{21}$ **C.** $32\frac{8}{27}$

B. $2\frac{2}{3}$ **D.** $39\frac{11}{27}$

2. Which fraction is *not* equivalent to 25%? *(6.RP.3c)*

F. $\frac{1}{4}$ **H.** $\frac{5}{20}$

G. $\frac{2}{5}$ **I.** $\frac{25}{100}$

3. The school store sells 4 pencils for \$0.50. At that rate, what would be the cost of 10 pencils? *(6.RP.3b)*

A. \$1.10 **C.** \$2.00

B. \$1.25 **D.** \$5.00

4. Which expression is equivalent to the expression below? *(6.EE.4)*

$$2(m + n)$$

F. $2m \times 2n$ **H.** $(2 + m) \times (2 + n)$

G. $2m + 2n$ **I.** $(2 + m) + (2 + n)$

5. A service club wants to buy tickets to a baseball game. Tickets are available for the grandstand and for the bleachers.

Grandstand Ticket	Bleachers Ticket
\$25	\$15

Which expression represents the total cost, in dollars, for g grandstand tickets and b bleachers tickets? *(6.EE.2a)*

A. $375(g + b)$ **C.** $25g + 15b$

B. $40(g \times b)$ **D.** $25g \times 15b$

6. What property was used to simplify the expression? *(6.EE.3)*

$$\begin{aligned} 12 \times 47 &= 12 \times (40 + 7) \\ &= 12 \times 40 + 12 \times 7 \\ &= 480 + 84 \\ &= 564 \end{aligned}$$

F. Distributive Property

G. Identity Property of Addition

H. Commutative Property of Addition

I. Associative Property of Multiplication

7. What is 15% of 36? *(6.RP.3c)*

8. If 5 dogs share equally a bag of dog treats, each dog gets 24 treats. Suppose 8 dogs share equally the bag of treats. How many treats does each dog get? *(6.RP.3b)*

A. 3

B. 15

C. 21

D. 38

9. The figure below consists of a rectangle and a right triangle. *(6.G.1)*

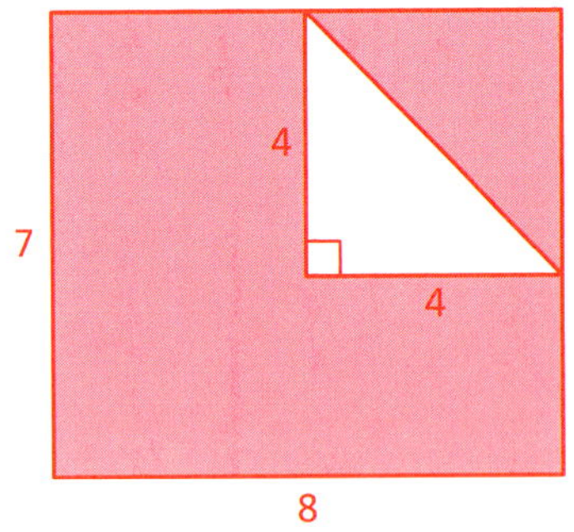

What is the area of the shaded region?

F. 23 units^2

G. 40 units^2

H. 48 units^2

I. 60 units^2

Item Analysis (continued)

6. **F.** Correct answer

   G. The student incorrectly identifies the property being used as the Identity Property of Addition.

   H. The student incorrectly identifies the property being used as the Commutative Property of Addition.

   I. The student incorrectly identifies the property being used as the Associative Property of Multiplication.

7. **Gridded Response:** Correct answer: 5.4

   Common Error: The student divides 36 by 0.15 and gets 240.

8. **A.** The student divides 24 by 8.

   B. Correct answer

   C. The student thinks that because there are 3 more dogs, each dog gets 3 fewer treats.

   D. The student uses direct variation instead of inverse variation (answer rounded to nearest whole number).

9. **F.** The student simply adds the four numbers in the figure.

   G. The student does not multiply by $\frac{1}{2}$ when finding the area of the triangle, gets 16 for the area of the triangle, and subtracts the result from 56.

   H. Correct answer

   I. The student thinks that the composite figure can be divided and recomposed as two rectangles, one 7×4 and one 8×4.

Answers

6. F
7. 5.4
8. B
9. H

Answers

10. 31.5 in.^{2}

11. A

12. *Part A:* $\frac{3}{25}$

Part B: 12%

Item Analysis (continued)

10. Gridded Response: Correct answer: Area = 31.5 in.^{2}

Common Error: The student forgets to multiply the product of the height and the sum of the bases by $\frac{1}{2}$ and finds an incorrect area of 63 in.^{2}.

11. A. Correct answer

B. The student reverses the values of k and p.

C. The student uses the values of k and p as coefficients.

D. The student incorrectly uses a decimal number division algorithm and thinks that $0.8 \times 0.5 = 4$ and that $0.5 \times 1.6 = 8$.

12. 2 points The student demonstrates a thorough understanding of multiplying fractions and writing fractions as percents. The student shows an understanding of the relationship between fractions and percents. The student correctly determines that the amount of the cherry gelatin eaten as a fraction of the total dessert is $\frac{3}{25}$ and as a percent is 12%.

1 point The student demonstrates a partial understanding of multiplying fractions, writing fractions as percents, and the relationship between fractions and percents.

0 points The student demonstrates insufficient understanding of multiplying fractions, writing fractions as percents, and the relationship between fractions and percents.

10. What is the area, in square inches, of the trapezoid-shaped award? *(6.G.1)*

11. Your friend evaluated an expression using $k = 0.5$ and $p = 1.6$ and got an answer of 12. Which expression did your friend evaluate? *(6.EE.2c)*

A. $5p + 8k$ **C.** $0.5k + 1.6p$

B. $8p + 5k$ **D.** $0.8k + 0.5p$

12. For a party, you made a gelatin dessert in a rectangular pan and cut the dessert into equal-sized pieces as shown below.

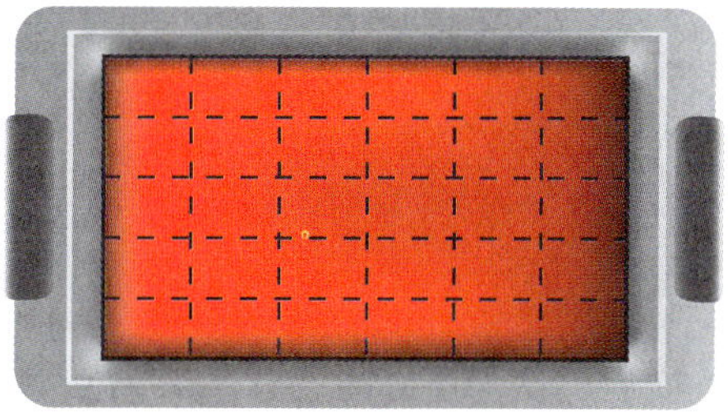

The dessert consisted of 5 layers of equal height. Each layer was a different flavor, as shown below by a side view of the pan. *(6.RP.3c)*

Your guests ate $\frac{3}{5}$ of the pieces of the dessert.

Part A Write the amount of cherry gelatin eaten by your guests as a fraction of the total dessert. Justify your answer.

Part B Write the amount of cherry gelatin eaten by your guests as a percent of the total dessert. Justify your answer.

6 Integers and the Coordinate Plane

"Don't worry. At negative 20 miles per hour, we're still under the speed limit."

"Dear Sir: You asked me to 'find' the opposite of −1."

"I didn't know it was missing."

Common Core Progression

4th Grade
• Compare and order fractions. • Compare and order decimals to hundredths place.

5th Grade
• Compare and order decimals to thousandths place. • Generate numerical patterns, identify the relationship, and form ordered pairs. • Graph ordered pairs in the first quadrant of the coordinate plane.

6th Grade
• Describe quantities with positive and negative numbers. • Compare and order integers and absolute value numbers. • Graph ordered pairs in all four quadrants of the coordinate plane.

Pacing Guide for Chapter 6

Chapter Opener	1 Day
Section 1 Activity Lesson	 1 Day 1 Day
Section 2 Activity Lesson	 1 Day 1 Day
Section 3 Activity Lesson	 1 Day 1 Day
Study Help / Quiz	1 Day
Section 4 Activity Lesson	 1 Day 1 Day
Section 5 Activity Lesson Extension	 1 Day 1 Day 1 Day
Chapter Review/ Chapter Tests	2 Days
Total Chapter 6	15 Days
Year-to-Date	97 Days

Chapter Summary

Section	Common Core State Standard	
6.1	Learning	6.NS.5, 6.NS.6a, 6.NS.6c
6.2	Learning	6.NS.6c, 6.NS.7a, 6.NS.7b
6.3	Learning	6.NS.5 ★, 6.NS.6a, 6.NS.6c, 6.NS.7a, 6.NS.7b
6.4	Learning	6.NS.7c, 6.NS.7d ★
6.5	Learning	6.NS.6b, 6.NS.6c ★, 6.NS.8 ★

Technology for the Teacher

BigIdeasMath.com
Chapter at a Glance
Complete Materials List
Parent Letters: English and Spanish

Common Core State Standards

4.NF.7 Compare two decimals to hundredths by reasoning about their size. . .. Record the results . . . and justify the conclusions . . . by using a visual model.

5.NBT.3b Compare two decimals to thousandths based on meanings of the digits in each place, using >, =, and < symbols to record the results of comparisons.

Additional Topics for Review

- Place Value
- Graphing Ordered Pairs in the First Quadrant

Try It Yourself

1. 0.01, 0.2, 0.42, 0.5
2. 0.75, 0.95, 1.01, 1.05
3. >
4. =
5. <
6. *Sample answer:* 8.5, 9, $\frac{19}{2}$
7. *Sample answer:* −1, 0, 1
8. *Sample answer:* 0.705, 0.749, 0.75

Record and Practice Journal Fair Game Review

1. 0.2, 0.4, 0.54, 0.61
2. 0.02, 0.11, 0.3, 0.45
3. 1.02, 1.24, 1.33, 1.7
4. 1.01, 1.06, 1.2, 1.42
5. 0.87, 0.9, 0.98, 1.23
6. 0.06, 0.5, 1.23, 1.4
7. 0.003, 0.03, 0.031, 0.033
8. 0.002, 0.02, 0.022, 0.2
9. group that brought fruit to lunch
10. <
11. >
12. <
13. =
14. >
15. <

16–22. See Additional Answers.

Math Background Notes

Vocabulary Review

- Common Denominator
- Mixed Number
- Improper Fraction

Ordering Decimals

- Students should know how to order decimals.

Comparing Numbers

- Students should be familiar with comparing numbers.
- Remind students that it is easier to compare numbers if they are both in the same form. In Example 3, for instance, one number is a decimal and one is a fraction. Encourage students to convert one of the numbers so that either both are decimals or both are fractions.
- **Teaching Tip:** Many students benefit from imagery when using the greater than and less than signs to compare numbers. Remind them that the greater than and less than symbols act as hungry alligators. They always want to eat the bigger number. Encourage students to be sure that the mouth of the alligator always opens towards the greater number.

Reteaching and Enrichment Strategies

If students need help. . .	If students got it. . .
Record and Practice Journal • Fair Game Review Skills Review Handbook Lesson Tutorials	Game Closet at *BigIdeasMath.com* Start the next section

What You Learned Before

"The Dog License Department wrote and said that it has run out of positive license numbers. So, they assigned me negative 3."

Ordering Decimals (4.NF.7)

Example 1 Use a number line to order 0.25, 1.15, 0.2, and 0.34 from least to greatest.

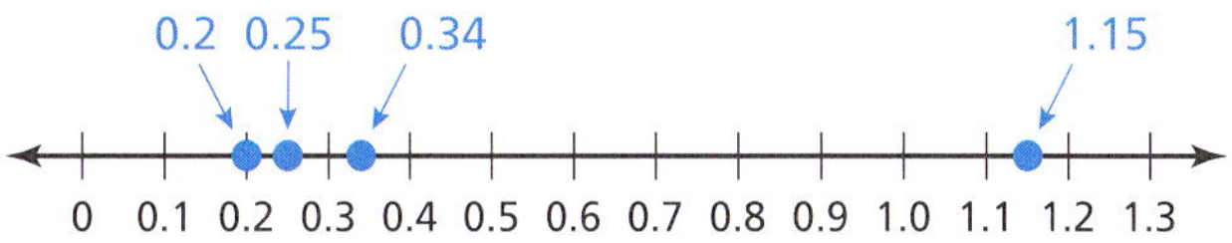

Try It Yourself

Use a number line to order the numbers from least to greatest.

1. 0.01, 0.42, 0.2, 0.5

2. 1.05, 0.95, 0.75, 1.01

Comparing Numbers (5.NBT.3b)

Complete the number sentence with <, >, or =.

Example 2 10 ▭ 15

On a number line, 10 is closer to zero than 15.

So, $10 < 15$.

Example 3 0.875 ▭ $\frac{7}{8}$

$$0.875 = \frac{875}{1000} = \frac{875 \div 125}{1000 \div 125} = \frac{7}{8}$$

So, $0.875 = \frac{7}{8}$.

Example 4 Find three numbers that make the number sentence $1\frac{2}{5} \leq$ ▭ true.

Sample answer: $1\frac{3}{5}, \frac{5}{2}, 2$

Try It Yourself

Complete the number sentence with <, >, or =.

3. 2.01 ▭ 2.001

4. 4.5 ▭ $\frac{9}{2}$

5. 3.18 ▭ 3.2

Find three numbers that make the number sentence true.

6. $\frac{17}{2} \leq$ ▭

7. $1\frac{1}{2} >$ ▭

8. $0.75 \geq$ ▭

6.1 Integers

Essential Question How can you represent numbers that are less than 0?

1 ACTIVITY: Reading Thermometers

Work with a partner. The thermometers show the temperatures in four cities.

Honolulu, Hawaii *Anchorage, Alaska*

Death Valley, California *Seattle, Washington*

Write each temperature. Then match each temperature with its most appropriate location.

a.

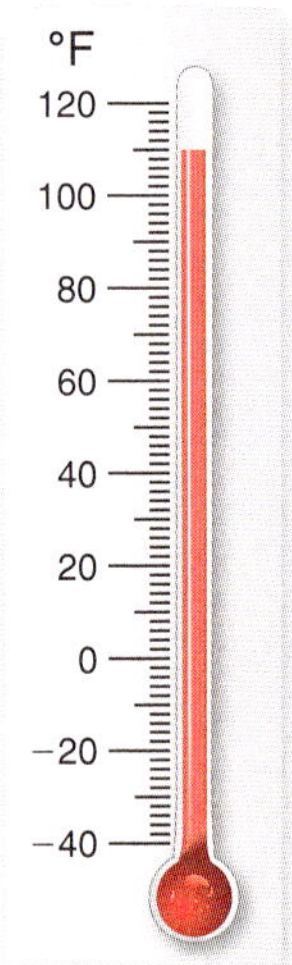

b.

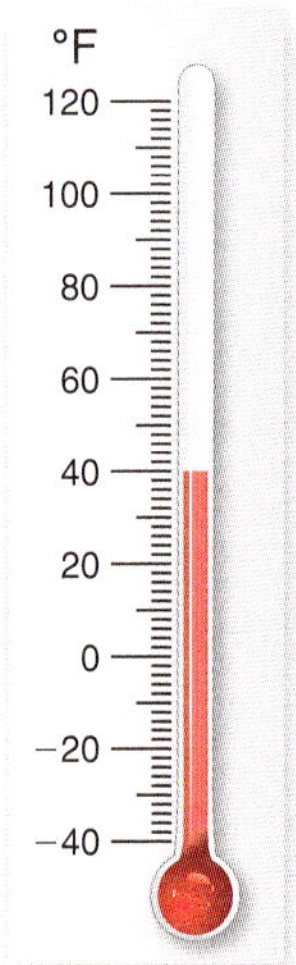

c.

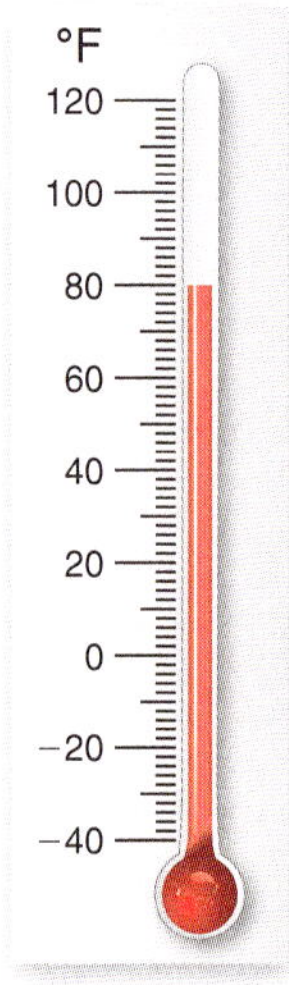

d.

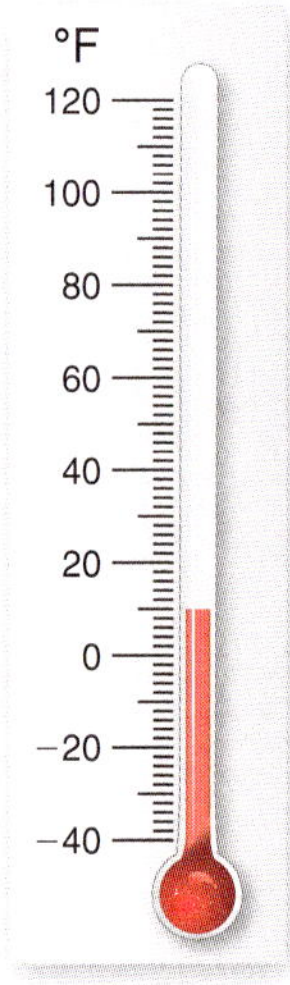

e. How would you describe all the temperatures in relation to 0°F?

2 ACTIVITY: Describing a Temperature

Work with a partner. The thermometer shows the coldest temperature ever recorded in Seattle, Washington.

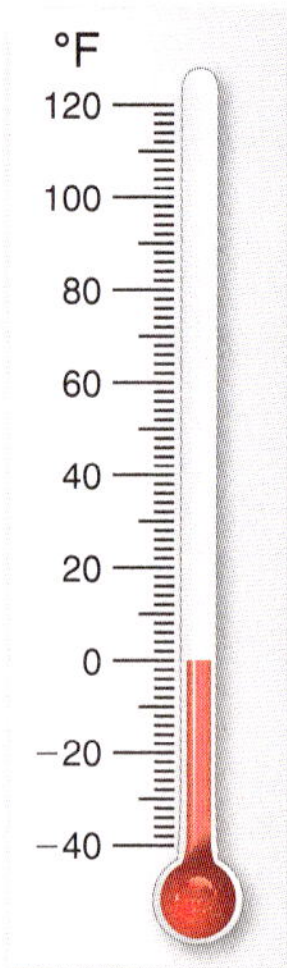

a. What is the temperature?

b. How do you write temperatures that are colder than this?

c. Suppose the record for the coldest temperature in Seattle is broken by 10 degrees. What is the new coldest temperature? Draw a thermometer that shows the new coldest temperature.

d. How is the new coldest temperature different from the temperatures in Activity 1?

COMMON CORE

Integers

In this lesson, you will
- understand positive and negative integers and use them to describe real-life situations.
- graph integers on a number line.

Learning Standards
6.NS.5
6.NS.6a
6.NS.6c

Laurie's Notes

Introduction

Standards for Mathematical Practice

- **MP5 Use Appropriate Tools Strategically:** It is important for students to create and then represent integers on a number line. A physical model that can be folded at 0 helps students to develop an understanding of the symmetric nature of integers and serves as an informal introduction to absolute value later in the chapter.

Motivate

- Have students work with partners to brainstorm types of numbers and contexts for these types of numbers. There will be three parts. Give each part a minute for brainstorming. Then, spend a minute or two for sharing.
- Part 1: Give examples of contexts where the numbers are generally greater than 100. Examples: cost of a car, population of a city
- Part 2: Give examples of contexts where the numbers are generally fractions. Examples: shoe sizes, lengths measured with a ruler
- Part 3: Give examples of contexts where the numbers are generally decimals. Examples: body temperature, track meet times

Activity Notes

Activity 1

- This activity assumes basic familiarity with different climate zones within the 50 United States.
- Working with partners, students should not have difficulty matching up the thermometers if they are familiar with the locations.
- ? "Do you think time of year has any bearing on the temperature readings? Explain." yes; Temperatures change through the seasons of the year.
- ? "What is the warmest (coldest) temperature you've experienced?" Answers will vary. This is just to start a discussion. It is likely that at least one student will mention a temperature below 0°F.
- **Discuss:** Engage students in a discussion about types of numbers. They are familiar with whole numbers, fractions, and decimals. Ask them what other types of numbers there are.

Activity 2

- This activity has students think about both sides of 0. Because the thermometer is vertical, students may start to think that there are numbers above and below zero. Number lines are generally horizontal so we also refer to left and right of zero.
- In Question 2(c), be sure that students have labeled 0 on their number line.
- Discuss Question 2(d). The new temperature, −10°F, is 10 units from 0. This idea of having two locations on the number line that are the same distance from 0 is the focus of the next activity.

Common Core State Standards

6.NS.5 Understand that positive and negative numbers are used together to describe quantities having opposite directions or values; use positive and negative numbers to represent quantities in real-world contexts, explaining the meaning of 0 in each situation.

6.NS.6a Recognize opposite signs of numbers as indicating locations on opposite sides of 0 on the number line; recognize that the opposite of the opposite of a number is the number itself, e.g., $-(-3) = 3$, and that 0 is its own opposite.

6.NS.6c Find and position integers . . . on a horizontal or vertical number line diagram; . . . .

Previous Learning

Students should know how to graph whole numbers.

Technology for the Teacher

Dynamic Classroom

Lesson Plans
Complete Materials List

6.1 Record and Practice Journal

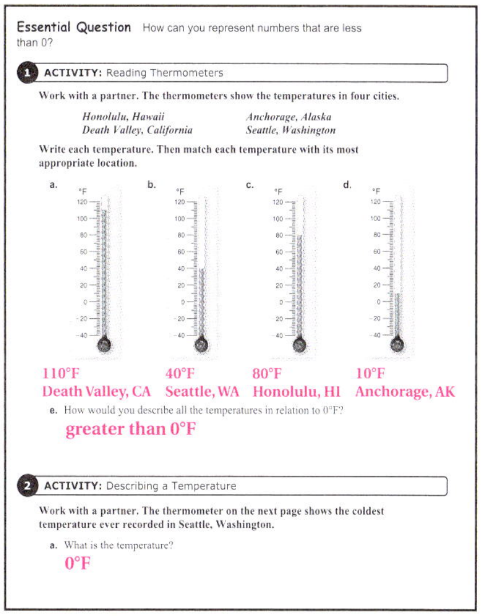

Essential Question How can you represent numbers that are less than 0?

1 ACTIVITY: Reading Thermometers

Work with a partner. The thermometers show the temperatures in four cities.

Honolulu, Hawaii *Anchorage, Alaska*
Death Valley, California *Seattle, Washington*

Write each temperature. Then match each temperature with its most appropriate location.

a. 110°F Death Valley, CA
b. 40°F Seattle, WA
c. 80°F Honolulu, HI
d. 10°F Anchorage, AK

e. How would you describe all the temperatures in relation to 0°F? greater than 0°F

2 ACTIVITY: Describing a Temperature

Work with a partner. The thermometer on the next page shows the coldest temperature ever recorded in Seattle, Washington.

a. What is the temperature? 0°F

English Language Learners

Vocabulary

Students might have difficulty understanding a negative elevation. Use a vertical scale and tell them that zero represents sea level. Locations with an elevation greater than zero are above sea level. Locations with an elevation less than zero are below sea level.

6.1 Record and Practice Journal

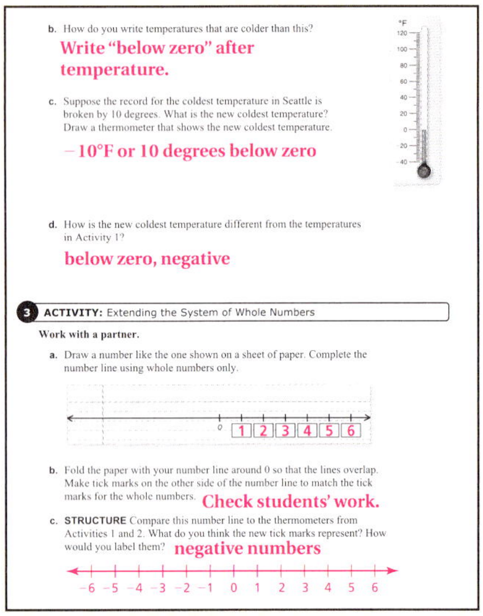

b. How do you write temperatures that are colder than this?

Write "below zero" after temperature.

c. Suppose the record for the coldest temperature in Seattle is broken by 10 degrees. What is the new coldest temperature? Draw a thermometer that shows the new coldest temperature.

−10°F or 10 degrees below zero

d. How is the new coldest temperature different from the temperatures in Activity 1?

below zero, negative

3 ACTIVITY: Extending the System of Whole Numbers

Work with a partner.

a. Draw a number like the one shown on a sheet of paper. Complete the number line using whole numbers only.

b. Fold the paper with your number line around 0 so that the lines overlap. Make tick marks on the other side of the number line to match the tick marks for the whole numbers. Check students' work.

c. **STRUCTURE** Compare this number line to the thermometers from Activities 1 and 2. What do you think the new tick marks represent? How would you label them? negative numbers

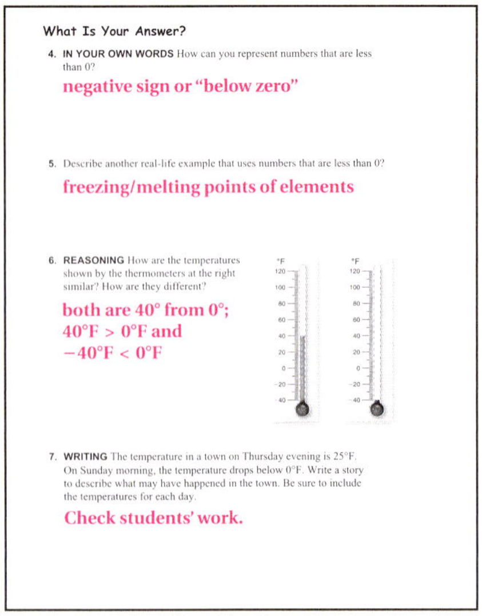

What Is Your Answer?

4. **IN YOUR OWN WORDS** How can you represent numbers that are less than 0?

negative sign or "below zero"

5. Describe another real-life example that uses numbers that are less than 0?

freezing/melting points of elements

6. **REASONING** How are the temperatures shown by the thermometers at the right similar? How are they different?

both are 40° from 0°; 40°F > 0°F and −40°F < 0°F

7. **WRITING** The temperature in a town on Thursday evening is 25°F. On Sunday morning, the temperature drops below 0°F. Write a story to describe what may have happened in the town. Be sure to include the temperatures for each day.

Check students' work.

Laurie's Notes

Activity 3

- Provide students with a ruler for this activity.
- **MP5:** As they extend their number lines, there are two key features that you want to discuss with students. The first is that the tick marks are equally spaced. The second is the symmetry about 0 that is suggested in a concrete fashion by folding the number line onto itself.
- **Common Error:** Students may equally space the tick marks and then incorrectly label the negative integers.

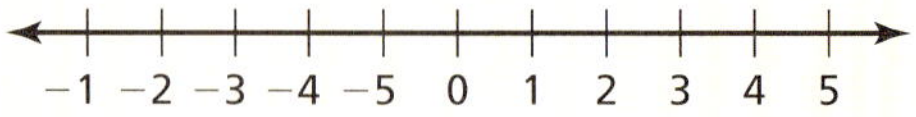

- **Extension:** Draw a number line with increments labeled as shown. Ask students what values would be used to label the remaining tick marks.

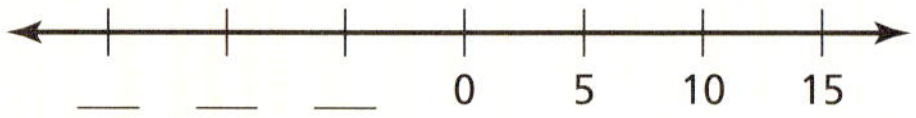

What Is Your Answer?

- For Question 7, have students share their stories. Stories should be very different. Ask some probing questions, such as whether the temperatures could have increased.

Closure

- Refer back to the opening Motivate activity and ask students to brainstorm with their partners.
- Part 4: Give examples of contexts where the numbers are generally negative numbers. Examples: winter temperatures, overdrawn checkbooks

3 ACTIVITY: Extending the System of Whole Numbers

Math Practice 8

Maintain Oversight

How does this activity help you represent numbers less than 0?

Work with a partner.

a. Copy and complete the number line using whole numbers only.

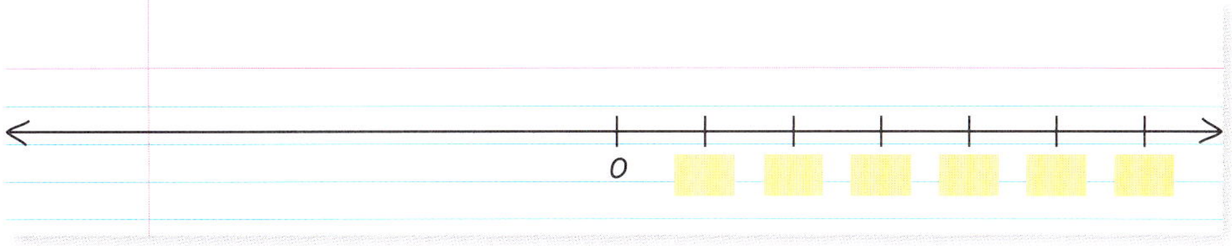

b. Fold the paper with your number line around 0 so that the lines overlap. Make tick marks on the other side of the number line to match the tick marks for the whole numbers.

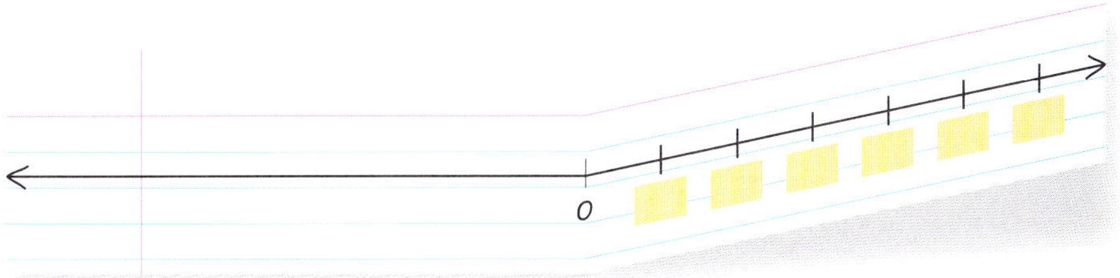

c. **STRUCTURE** Compare this number line to the thermometers from Activities 1 and 2. What do you think the new tick marks represent? How would you label them?

What Is Your Answer?

4. **IN YOUR OWN WORDS** How can you represent numbers that are less than 0?

5. Describe another real-life example that uses numbers that are less than 0.

6. **REASONING** How are the temperatures shown by the thermometers at the right similar? How are they different?

7. **WRITING** The temperature in a town on Thursday evening is 25°F. On Sunday morning, the temperature drops below 0°F. Write a story to describe what may have happened in the town. Be sure to include the temperatures for each day.

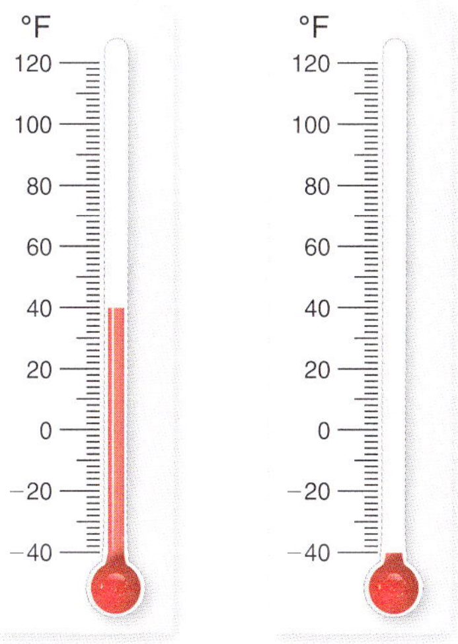

Practice Use what you learned about positive and negative numbers to complete Exercises 4–7 on page 252.

6.1 Lesson

Check It Out
Lesson Tutorials
BigIdeasMath.com

Key Vocabulary
positive numbers, *p. 250*
negative numbers, *p. 250*
opposites, *p. 250*
integers, *p. 250*

Positive numbers are greater than 0. They can be written with or without a positive sign (+).

+1   5   +20   10,000

Negative numbers are less than 0. They are written with a negative sign (−).

−1   −5   −20   −10,000

Two numbers that are the same distance from 0 on a number line, but on opposite sides of 0, are called **opposites**. The opposite of 0 is 0.

The Meaning of a Word

Opposite

When you sit across from your friend at the lunch table, you sit **opposite** your friend.

Key Idea

Integers

Words **Integers** are the set of whole numbers and their opposites.

Graph

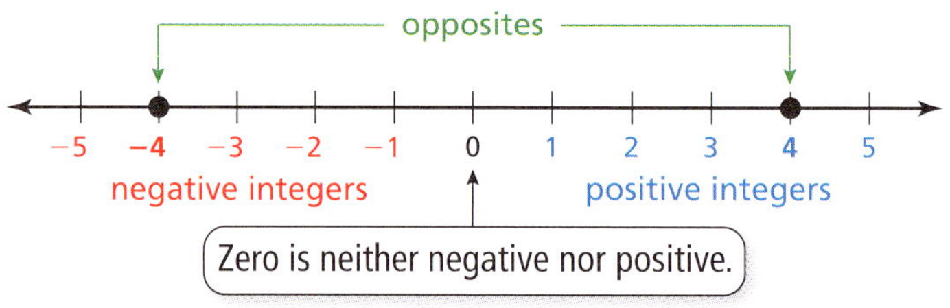

EXAMPLE 1 Writing Positive and Negative Integers

Write a positive or negative integer that represents the situation.

a. A contestant gains 250 points on a game show.

Gains indicates a number greater than 0. So, use a positive integer.

∴ +250, or 250

b. Gasoline freezes at 40 degrees below zero.

Below zero indicates a number less than 0. So, use a negative integer.

∴ −40

On Your Own

Now You're Ready
Exercises 8–13

Write a positive or negative integer that represents the situation.

1. A hiker climbs 900 feet up a mountain.
2. You have a debt of $24.
3. A student loses 5 points for being late to class.
4. A savings account earns $10.

Multi-Language Glossary at BigIdeasMath.com

Laurie's Notes

Introduction

Connect

- **Yesterday:** Students explored a real-life application of negative numbers and gained an intuitive understanding of absolute value. (MP5)
- **Today:** Students will write and graph integers.

Motivate

- Ask for 7 volunteers to hold onto a piece of rope. They should be equally spaced. Ask the person in the middle to hold a card with the number 0 written on it.
- Say to the students:
  - "If you are 2 people away from 0, raise your hand."
  - "If you are 3 people away from 0, raise your hand."
- Now hand the card with the number 1 written on it to the person on the right of the person holding the 0 card.
- Say to the students:
  - "If you are the number −1, raise your hand."
  - "If you are the number 3, raise your hand."
  - "Keep your hand in the air if you are the least of the two numbers."
  - Repeat the instructions to compare 2 and 3, and to compare −2 and −3.
- This gives students a strong visual image of how a number line looks and where positive and negative numbers are located.

Lesson Notes

Key Idea

- Define integers. Make it clear to students that there are three sets of numbers: positive integers, negative integers, and 0. Zero is neither positive nor negative.
- Remind students that whole numbers are 0, 1, 2, 3, . . .
- Draw a number line, using two colors to differentiate positive and negative integers.
- Point out the symmetry of a number line. If the number line were drawn on a piece of paper and folded at 0, opposites would fold on top of one another. Remind students of Activity 3 yesterday with the folded paper.

Example 1

- Discuss each part of the example. The context for each part should be familiar to students.
- Point out that the + sign is optional in writing a positive number. It generally is not written.

On Your Own

- **Extension:** After student work has been checked, ask a student to state the opposite of each situation. Example: A hiker descends 900 feet down a mountain.

Goal Today's lesson is reading and graphing **integers**.

Technology for the Teacher

Dynamic Classroom

Lesson Tutorials
Lesson Plans
Answer Presentation Tool

Differentiated Instruction

Kinesthetic

Have students use a compass to find opposites. Place the point of the compass at zero on a number line and adjust the compass so the pencil intersects a positive or negative integer. To find the opposite of that integer, draw an arc to the opposite side of the number line.

Extra Example 1

Write a positive or negative integer that represents the situation.

a. A balloon floats 5 feet above the ground. 5

b. An anchor is 15 feet underwater. −15

On Your Own

1. 900
2. −24
3. −5
4. 10

Laurie's Notes

Extra Example 2

Graph each integer and its opposite.

a. 4

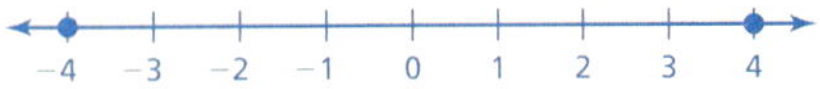

b. −10

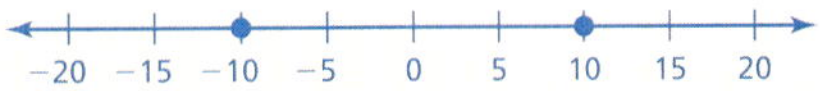

Extra Example 3

You start a game with 10 points. You gain 4 points in the 1st round. In the 2nd round, you lose 7 points.

a. Write an integer that represents each point level.

You start a game. 10
You gain 4 points. +4
You lose 7 points. −7

b. After the 3rd round, your total score is 10 points. Write an integer that represents how many points you received in the 3rd round. +3

On Your Own

5.

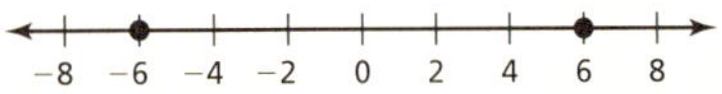

6.

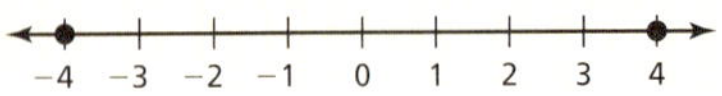

7.

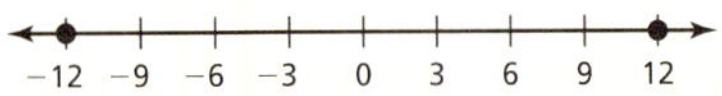

8.

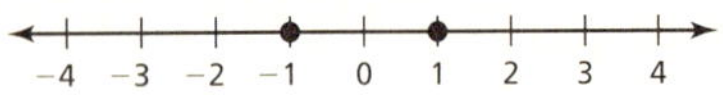

9. −7

Example 2

? "What is the opposite of 5?" −5 "What is the opposite of −13?" 13

- **Common Error:** When asked to graph the numbers 2 and −2, students may draw a number line and label −2, 0, and 2 as shown, and write nothing else.

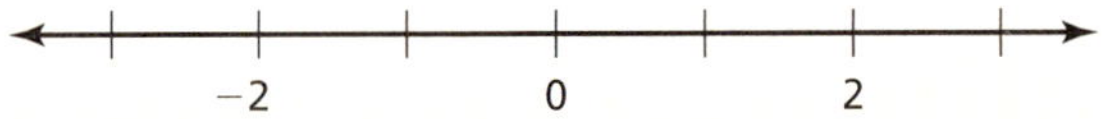

Make it clear to students that they should decide on a scale and then put a closed circle on the number line to graph a number.

? "What number is halfway between a number and its opposite?" 0

- **MP6 Attend to Precision:** Discuss the Reading Note in the margin. Tell students that −2 is read "negative 2" or "the opposite of 2." It should not be read "minus 2." Minus is an operation.

Example 3

- Draw a vertical number line on the board as a visual aid for students as you work through this problem.
- **FYI:** Students may ask if underground floors are really numbered using negative numbers. It is most common to use a letter such as L or P to denote a lower level or a parking level.

On Your Own

- **Think-Pair-Share:** Students should read each question independently and then work in pairs to answer the questions. When they have answered the questions, the pair should compare their answers with another group and discuss any discrepancies.

Closure

- **Exit Ticket:** Graph each integer on the same number line and write a context for each: −6 and 4.

−6 A student missed 6 questions on a quiz.
4 There are 4 new students in the class.

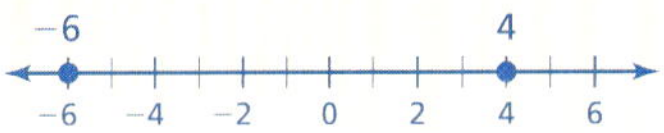

EXAMPLE 2 Graphing Integers

Graph each integer and its opposite.

a. 3

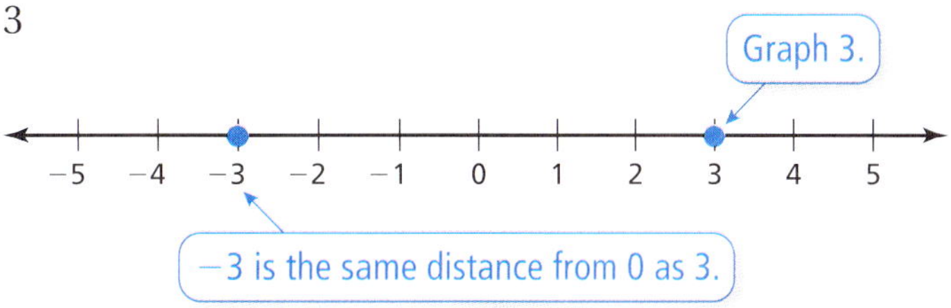

b. −2

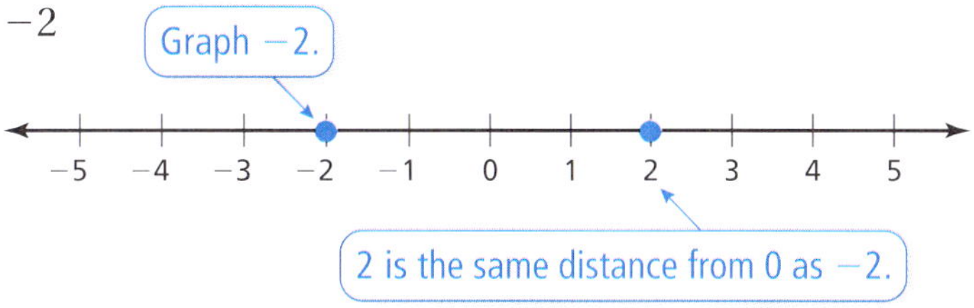

Reading

You can think of the negative sign (−) as referring to the opposite of a number. In Example 2(b), you can read −2 as "the opposite of 2."

EXAMPLE 3 Real-Life Application

You deliver flowers to an office building. You enter at ground level and go down 2 floors to make the first delivery. Then you go up 7 floors to make the second delivery.

a. Write an integer that represents each position.

Position	*Integer*
You enter at ground level.	0
You go down 2 floors.	−2
You go up 7 floors.	+7

b. Write an integer that represents how you return to ground level.

Use a number line to model your movement, as shown.

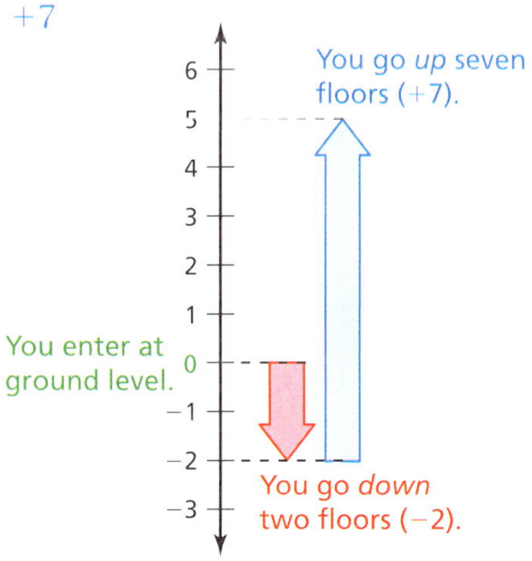

The second delivery is on the fifth floor. You must go down 5 floors to return to ground level.

∴ The integer representing "down 5 floors" is −5.

On Your Own

Now You're Ready
Exercises 16–23

Graph the integer and its opposite.

5. 6 **6.** −4 **7.** −12 **8.** 1

9. WHAT IF? In Example 3, you go up 9 floors to make the second delivery. Write an integer that represents how you return to ground level.

6.1 Exercises

Check It Out
Help with Homework
BigIdeasMath.com

Vocabulary and Concept Check

1. **VOCABULARY** Which of the following numbers are integers?

$$8, \ -4.1, \ -9, \ \frac{1}{6}, \ 1.75, \ 22$$

2. **OPEN-ENDED** Describe a real-life example that you can represent by -1200.

3. **VOCABULARY** List three words or phrases used in real life that indicate negative numbers.

Practice and Problem Solving

Graph the number that represents the situation on a number line.

4. A football team loses 3 yards.
5. The temperature is 6 degrees below zero.
6. A person climbs 600 feet up a mountain.
7. You earn $15 raking leaves.

Write a positive or negative integer that represents the situation.

(1) 8. You withdraw $42 from an account.
9. An airplane climbs to 37,500 feet.
10. The temperature rises 17 degrees.
11. You lose 56 points in a video game.
12. A ball falls 350 centimeters.
13. You receive 5 bonus points in class.

14. **STOCK MARKET** A stock market gains 83 points. The next day, the stock market loses 47 points. Write each amount as an integer.

15. **SCUBA DIVING** The world record for scuba diving is 318 meters below sea level. Write this as an integer.

Graph the integer and its opposite.

(2) 16. -5
17. -8
18. 14
19. 9
20. 30
21. -150
22. -32
23. 400

24. **ERROR ANALYSIS** Describe and correct the error in describing positive integers.

✗ The positive integers are 0, 1, 2, 3, ...

25. **TEMPERATURE** The highest temperature in February is 25°F. The lowest temperature in February is the opposite of the highest temperature. Graph both temperatures.

Assignment Guide and Homework Check

Level	Day 1 Activity Assignment	Day 2 Lesson Assignment	Homework Check
Basic	4–7, 33–35	1–3, 9–13 odd, 14, 15–23, odd, 24, 25, 27	9, 14, 21, 24, 27
Average	4–7, 33–35	1–3, 12–15, 16–24 even, 27, 29, 30, 31	14, 22, 24, 27, 30
Advanced	4–7, 33–35	1–3, 8–30 even, 31, 32	14, 22, 24, 26, 30

Common Errors

- **Exercises 8–13** Students may use the wrong sign when writing the integer. Tell them to look for key words, such as rise and fall, to help them determine the sign.
- **Exercises 16–23** Students may think that *opposites* are only negative numbers. For example, the opposite of −3 is −3 and the opposite of 3 is −3. Remind them that the opposite of a negative number is positive.

6.1 Record and Practice Journal

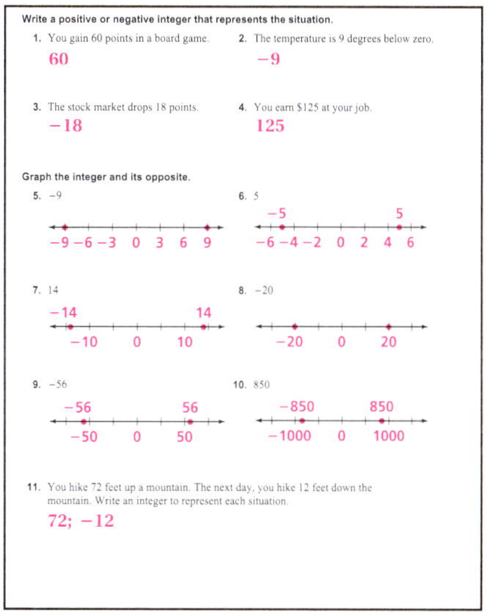
Write a positive or negative integer that represents the situation.

1. You gain 60 points in a board game. **60**
2. The temperature is 9 degrees below zero. **−9**
3. The stock market drops 18 points. **−18**
4. You earn $125 at your job. **125**

Graph the integer and its opposite.

5. −9
6. 5
7. 14
8. −20
9. −56
10. 850

11. You hike 72 feet up a mountain. The next day, you hike 12 feet down the mountain. Write an integer to represent each situation. **72; −12**

Vocabulary and Concept Check

1. 8, −9, 22
2. *Sample answer:* The status of a fundraiser's goal is −$1200.
3. *Sample answer:* below, under, lose

Practice and Problem Solving

4.
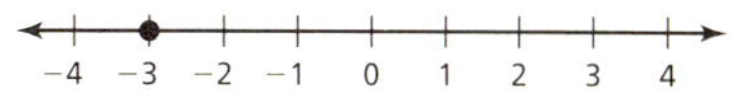

5.
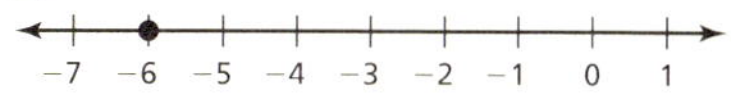

6.
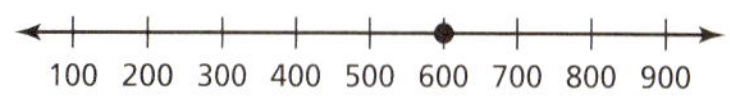

7.

8. −42
9. 37,500
10. 17
11. −56
12. −350
13. 5
14. 83; −47
15. −318

16.
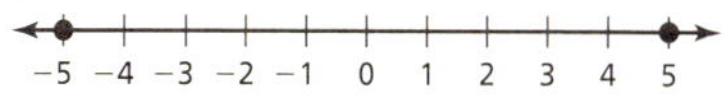

17.

18.
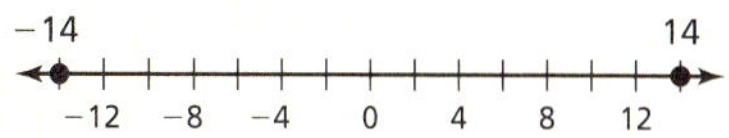

19.
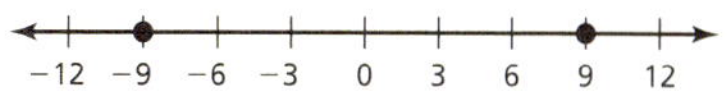

20.
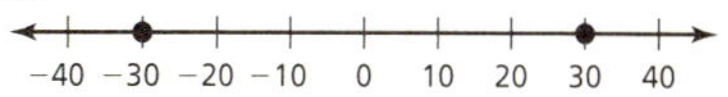

21.
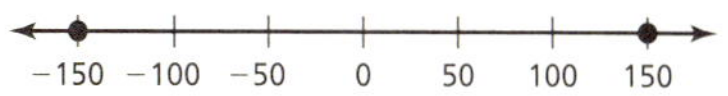

22.
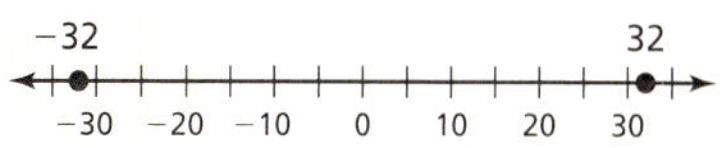

Practice and Problem Solving

23. 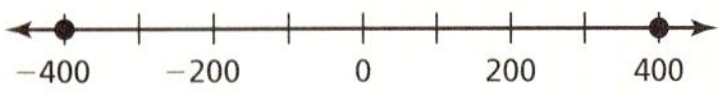

24. The numbers shown are whole numbers. Positive integers are 1, 2, 3, . . .

25. 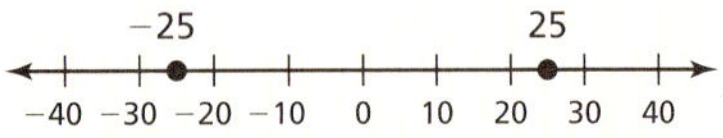

26. 5 **27.** −8

28. −15 **29.** 18

30. See *Taking Math Deeper*.

31. See Additional Answers.

32. no; The flag starts at 0 and moves 8 feet right to 8 on the number line. Then the flag moves 12 feet left to −4 on the number line. Then the flag moves 13 feet right to 9 on the number line. Nine is less than 10. So, a team has not won yet.

Fair Game Review

33. $\frac{3}{8}, \frac{1}{2}, \frac{3}{4}, \frac{7}{8}$

34. 4.312, 4.316, 4.32, 4.5

35. B

Mini-Assessment

Write a positive or negative integer that represents the situation.

1. A fish is 12 meters underwater. −12
2. You deposit $25 into your savings account. 25

Graph the integer and its opposite.

3. 7

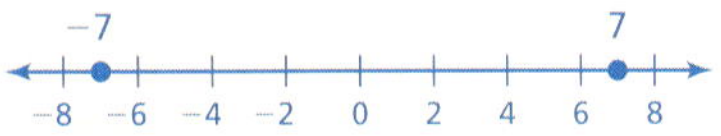

4. −15

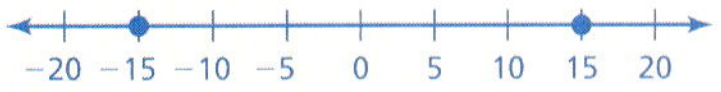

5. While playing a game, you move 6 steps forward and then 10 steps backward. Write each number as an integer. 6; −10

Taking Math Deeper

Exercise 30

This exercise involves a subtle, yet very important, use of negative numbers. The concept is that in business and science, we frequently use the number 0 as a convenient point on the number line to represent time or physical location.

① Draw a diagram.

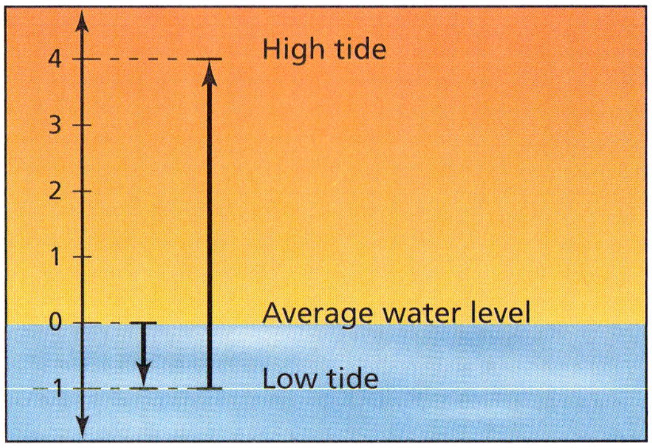

② It is not *required* that we use the number 0 to represent the average water level. However, if we do, then low tide is represented by the number −1 because it is 1 foot below the average water level. High tide is represented by the number 4 because it is 5 feet higher than low tide.

High Tide = 4
Average Water Level = 0
Low tide = −1

The average water level is 4 feet below high tide. So, the average water level relative to high tide is represented by the number −4.

③ If we choose to let 0 represent high tide, then the numbering system changes. Now, we have the following.

High Tide = 0
Average Water Level = −4
Low tide = −5

Reteaching and Enrichment Strategies

If students need help. . .	If students got it. . .
Resources by Chapter • Practice A and Practice B • Puzzle Time Record and Practice Journal Practice Differentiating the Lesson Lesson Tutorials Skills Review Handbook	Resources by Chapter • Enrichment and Extension • Technology Connection Start the next section

Identify the integer represented by the point on the number line.

26. A **27.** B **28.** C **29.** D

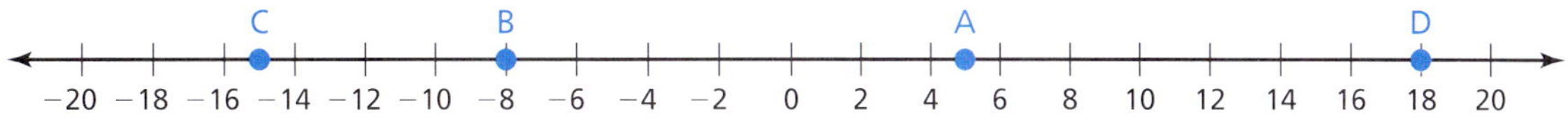

30. **TIDES** Use the information below.

- Low tide is 1 foot below the average water level.
- High tide is 5 feet higher than low tide.

Write an integer that represents the average water level relative to high tide.

31. **REPEATED REASONING** Choose any positive integer.

a. Find the opposite of the integer. **b.** Find the opposite of the integer in part (a).

c. What can you conclude about the opposite of the opposite of the integer? Is this true for all integers? Use a number line to justify your answer.

d. Describe the meaning of $-(-(-6))$. Find its value.

32. Number Sense In a game of tug-of-war, a team wins by pulling the flag over its goal line. The flag begins at 0. During a game, the flag moves 8 feet to the right, 12 feet to the left, and 13 feet back to the right. Did a team win? Explain.

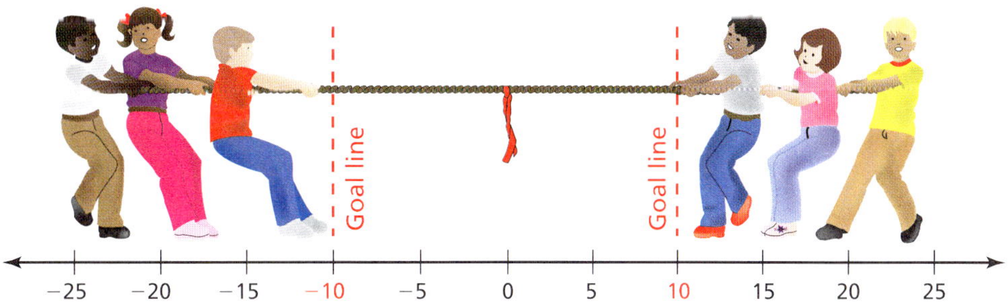

Fair Game Review What you learned in previous grades & lessons

Order the numbers from least to greatest. *(Skills Review Handbook)*

33. $\frac{7}{8}, \frac{1}{2}, \frac{3}{8}, \frac{3}{4}$ **34.** 4.5, 4.316, 4.32, 4.312

35. **MULTIPLE CHOICE** The height of a statue is 276 inches. What is the height of the statue in meters? Round your answer to the nearest hundredth. *(Section 5.7)*

(A) 1.09 m (B) 7.01 m (C) 108.66 m (D) 701.04 m

6.2 Comparing and Ordering Integers

Essential Question How can you use a number line to order real-life events?

1 ACTIVITY: Seconds to Takeoff

Work with a partner. You are listening to a command center before the liftoff of a rocket.

You hear the following:

> "T minus 10 seconds . . . go for main engine start . . . T minus 9 . . . 8 . . . 7 . . . 6 . . . 5 . . . 4 . . . 3 . . . 2 . . . 1 . . . we have liftoff."

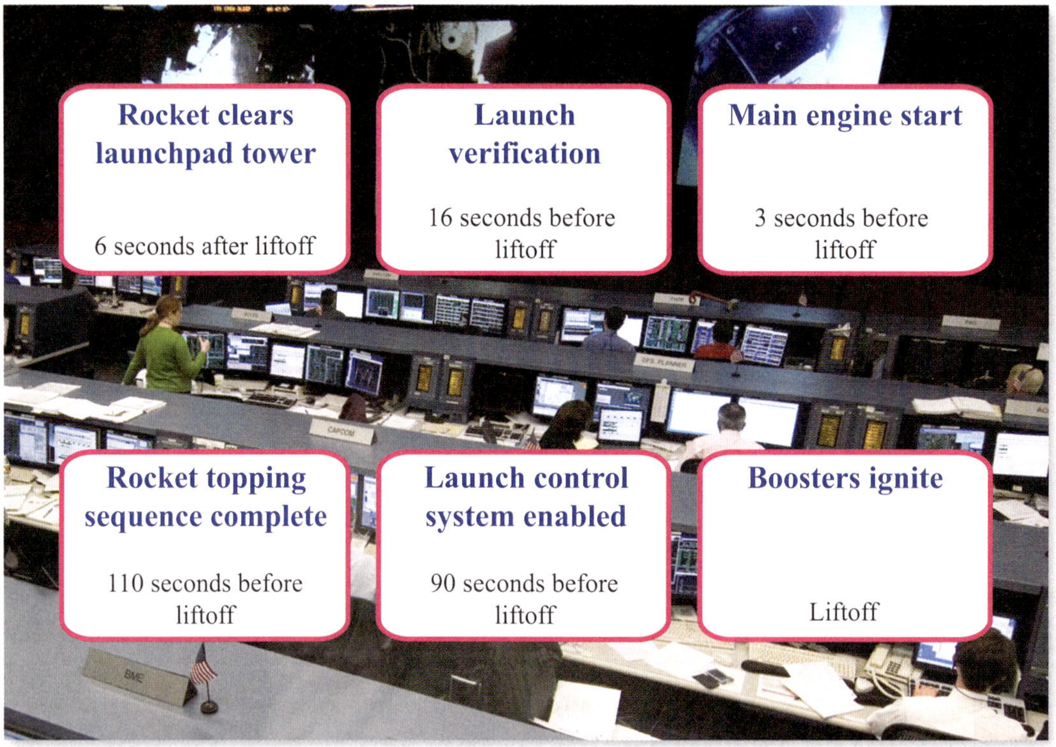

a. Draw a number line. Then locate the events shown above at appropriate points on the number line.

b. Which event occurs at zero on your number line? Explain.

c. Which of the events occurs first? Which of the events occurs last? How do you know?

d. List the events in the order they occurred.

COMMON CORE

Integers

In this lesson, you will

- use a number line to compare positive and negative integers.
- use a number line to order positive and negative integers for real-life situations.

Learning Standards
6.NS.6c
6.NS.7a
6.NS.7b

Laurie's Notes

Introduction

Standards for Mathematical Practice

- **MP5 Use Appropriate Tools Strategically:** It is important for students to create and then represent integers on a number line. A physical model that can be folded at 0 helps students to develop an understanding of the symmetric nature of integers and serves as an informal introduction to absolute value later in the chapter.

Motivate

? 3-2-1-Blast Off! Ask the NASA space trivia questions and have students write their answers on white boards or raise their hands.

- "Who was the first person in space?" Major Yuri Gagarin—Russian
- "Who was the first American woman in space?" Sally Ride; June 18, 1983
- "Who made the first U.S. spacewalk?" Ed White; June 3, 1965; Gemini 4
- "Where is the Sea of Tranquility?" moon; site of the first lunar landing
- "What is the name of the largest volcano in the solar system?" Olympus Mons; located on Mars; bigger than any mountain on Earth

Activity Notes

Activity 1

- Have students draw a number line displaying both positive and negative numbers. Suggest that students not label every number between -31 and 0. Every other number or every third number should be labeled.
- **Common Error:** It is very common for students to place the negative numbers starting with -1 to the left as shown:

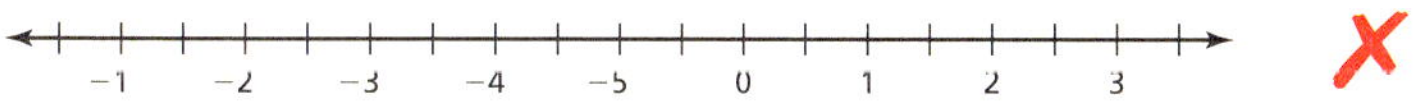

- Explain to students what it means to *locate* an event on the number line. They should put a dot on the number line and a letter above it that corresponds to one of the six events. Let rocket clears launchpad tower = A, launch verification = B, and so on.
- **MP6 Attend to Precision:** Discuss with students that even though negative numbers are *not* read as "minus 3," it is common language at a space launch to use the phrase "minus 3" instead of "negative 3."
- Discuss the order of the events. Those that are *before* launch would be to the left of 0.
- **Extension:** If you have a long piece of rope, you could have students *act out* this problem. Two students stretch the rope across the front of the classroom. Six students *locate themselves* on the rope number line. You can hang a large paper clip to identify zero. Then students place themselves in approximately the correct location.
- **Teaching Tip:** The number line doesn't have to have integers labeled. The number line could have 0 labeled and then tic marks with the events labeled.

Common Core State Standards

6.NS.6c Find and position integers and other rational numbers on a horizontal or vertical number line diagram; find and position pairs of integers . . .

6.NS.7a Interpret statements of inequality as statements about the relative position of two numbers on a number line diagram.

6.NS.7b Write, interpret, and explain statements of order for rational numbers in real-world contexts.

Previous Learning

Students should know how to compare whole numbers and graph integers.

Technology for the Teacher

Dynamic Classroom

Lesson Plans
Complete Materials List

6.2 Record and Practice Journal

Essential Question How can you use a number line to order real-life events?

1 ACTIVITY: Seconds to Takeoff

Work with a partner. You are listening to a command center before the liftoff of a rocket.

You hear the following:

"T minus 10 seconds...go for main engine start...T minus 9...8...7...6...5...4...3...2...1...we have liftoff."

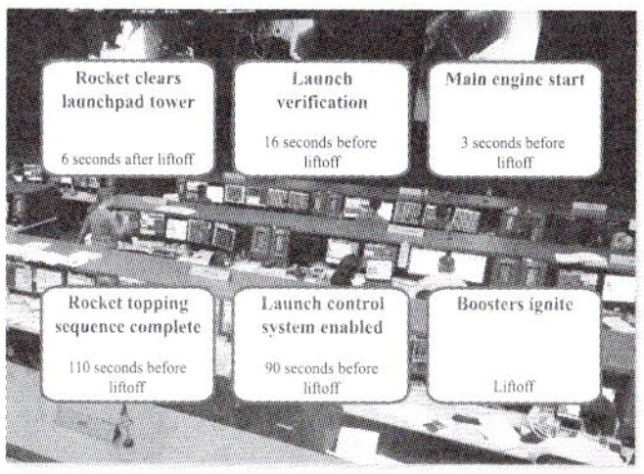

a. Draw a number line. Then locate the events shown above at appropriate points on the number line.
See Additional Answers.

b. Which event occurs at zero on your number line? Explain.
boosters ignite/liftoff

English Language Learners

Vocabulary

Integers may be a new term for English learners. Review vocabulary often with students. For example, you may say, "Name a negative integer less than negative two." Encourage your students to reply in a complete sentence, "A negative integer less than negative two is negative five."

6.2 Record and Practice Journal

c. Which of the events occurs first? Which of the events occurs last? How do you know?

rocket topping sequence complete; rocket clears launchpad tower

d. List the events in the order they occurred.

rocket topping sequence complete, launch control system enabled, launch verification, main engine start, boosters ignite, rocket clears launchpad tower

2 ACTIVITY: Being Careful with Terminology

Work with a partner.

a. Use a number line to show that the phrase "3 seconds away from liftoff" can have two meanings.

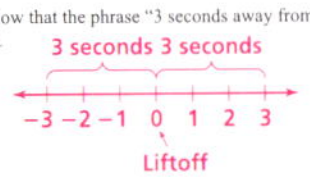

b. Reword the phrase "3 seconds away from liftoff" in two ways so that each meaning is absolutely clear.

3 seconds before liftoff; 3 seconds after liftoff

c. Explain why you must be very careful with terminology if you are working in the command center for a rocket launch.

Check students' work.

3 ACTIVITY: A Day in the Life of an Astronaut

Make a time line that shows a day in the life of an astronaut. Use the Internet or another reference source to gather information. Check students' work.

- Use a number line with units representing hours. Start at 12 hours before liftoff and end at 12 hours after liftoff. Locate the liftoff at 0. Assume liftoff occurs at noon.
- Include at least five events before liftoff, such as when the astronauts suit up.
- Include at least five events after liftoff, such as when the rocket enters Earth's orbit.
- How do you determine where each event occurs on the number line?

What Is Your Answer?

4. **IN YOUR OWN WORDS** How can you use a number line to order real-life events?

to represent the order in which real-life events occur

5. Describe how you can use a number line to create a time line.

Let units on the number line represent time.

Laurie's Notes

Activity 2

- The term *opposite* was introduced in the previous lesson and is revisited here in the context of pre- and post-launch events.
- This activity introduces the idea that every number, other than 0, has an opposite. In other words, the number line is symmetric. Starting at 0, there is a number 4 units to the left and 4 units to the right; there is a number 7 units to the left and 7 units to the right.
- Listen to students' terminology while they discuss this problem.

Activity 3

- **Research:** This is an activity that is most easily accomplished in a library, computer lab, or as a homework assignment.
- Discuss with students the need to make their units clear on the number line. Even though the liftoff may not occur at noon, for the purposes of this example, it is the centering point. So, pre-launch events happen in the morning (A.M.) and post-launch events happen in the afternoon and evening (P.M.).

What Is Your Answer?

- **Think-Pair-Share:** Students should read each question independently and then work in pairs to answer the questions. When they have answered the questions, the pair should compare their answers with another group and discuss any discrepancies.

Closure

- **Exit Ticket:** Make a number line to represent three events prior to math class and three events after math class. The events do not need to take place at school. Answers will vary.

2 ACTIVITY: Being Careful with Terminology

Work with a partner.

a. Use a number line to show that the phrase "3 seconds away from liftoff" can have two meanings.

b. Reword the phrase "3 seconds away from liftoff" in two ways so that each meaning is absolutely clear.

c. Explain why you must be very careful with terminology if you are working in the command center for a rocket launch.

3 ACTIVITY: A Day in the Life of an Astronaut

Math Practice 5

Recognize the Usefulness of Tools

Which sources would give you the most accurate information? How do you know you can trust the information you find?

Make a time line that shows a day in the life of an astronaut. Use the Internet or another reference source to gather information.

- Use a number line with units representing hours. Start at 12 hours before liftoff and end at 12 hours after liftoff. Locate the liftoff at 0. Assume liftoff occurs at noon.
- Include at least five events before liftoff, such as when the astronauts suit up.
- Include at least five events after liftoff, such as when the rocket enters Earth's orbit.
- How do you determine where each event occurs on the number line?

What Is Your Answer?

4. **IN YOUR OWN WORDS** How can you use a number line to order real-life events?

5. Describe how you can use a number line to create a time line.

Practice Use what you learned about number lines to complete Exercises 4–7 on page 258.

6.2 Lesson

Check It Out
Lesson Tutorials
BigIdeasMath.com

On a horizontal number line, numbers to the left are less than numbers to the right. Numbers to the right are greater than numbers to the left.

EXAMPLE 1 Comparing Integers on a Horizontal Number Line

Compare 2 and −6.

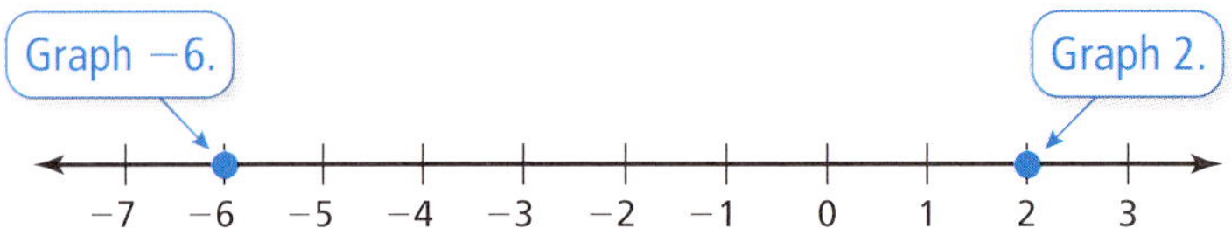

∴ 2 is to the right of −6. So, $2 > -6$.

On a vertical number line, numbers below are less than numbers above. Numbers above are greater than numbers below.

EXAMPLE 2 Comparing Integers on a Vertical Number Line

Compare −5 and −3.

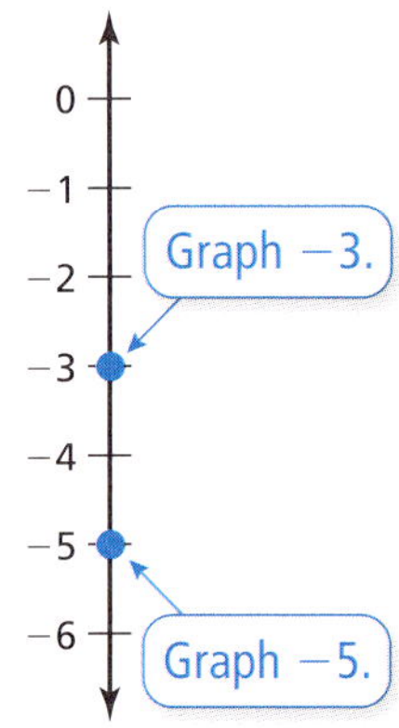

∴ −5 is below −3. So, $-5 < -3$.

On Your Own

Now You're Ready
Exercises 4–11

Copy and complete the statement using $<$ or $>$.

1. 0 ___ −4 **2.** −5 ___ 5 **3.** −8 ___ −7

EXAMPLE 3 Ordering Integers

Order −4, 3, 0, −1, −2 from least to greatest.

Graph each integer on a number line.

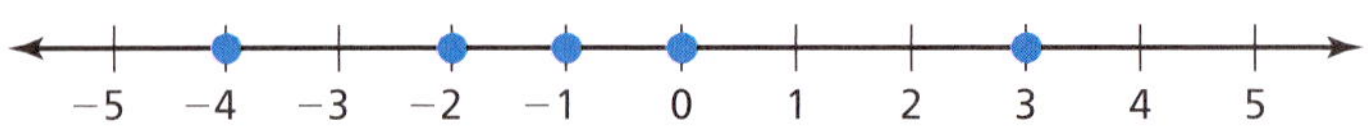

Write the integers as they appear on the number line from left to right.

∴ So, the order from least to greatest is −4, −2, −1, 0, 3.

Laurie's Notes

Introduction

Connect

- **Yesterday:** Students explored a real-life application of integers and gained an intuitive understanding of comparing integers. (MP5, MP6)
- **Today:** Students will compare and order integers.

Motivate

- Place a piece of masking tape on the floor to represent a number line. Identify the location of 0 by having a student stand at zero.
- Have two more students stand at two other integers. Have the students change their locations several times to represent the following scenarios: two positive integers, one positive integer and one negative integer, and both negative integers.

Lesson Notes

Example 1

- Draw a number line and graph −6 and 2.
- Discuss with students that 0 does not have to be in the middle of the number line. Sometimes number lines do not show 0.
- **Big Idea:** The further to the left on the number line, the less the number becomes. The further to the right on the number line, the greater the number becomes.
- **Extension:** Ask the students how far −6 and 2 are from 0. −6 is 6 units from 0, 2 is 2 units from 0.

Example 2

- **MP4 Model with Mathematics:** There is an intentional effort in this lesson to display number lines in two orientations—horizontal and vertical. These two models will help students with their graphing in the coordinate plane. They also make connections to common contexts, such as a thermometer.
- Draw a vertical number line and graph −5 and −3.
- A vertical number line should remind students of the thermometer in Activity 6.1.
- **Extension:** Ask the students how far −5 and −3 are from 0. −5 is 5 units from 0, and −3 is 3 units from 0.

On Your Own

- Check students' answers to Question 3. Comparing two negative numbers is a difficult concept for some students. Encourage them to graph the numbers on a number line.

Example 3

- Draw a number line and graph the integers. Discuss with students the difference between labeling the number line with integers and graphing integers with closed circles.

Goal Today's lesson is on comparing and ordering integers.

Technology for the Teacher

Dynamic Classroom

Lesson Tutorials
Lesson Plans
Answer Presentation Tool

Extra Example 1

Compare −2 and −10. −10 < −2

Extra Example 2

Compare −1 and −2. −1 > −2

On Your Own

1. >
2. <
3. <

Extra Example 3

Order −3, 9, 0, −7, 1 from least to greatest. −7, −3, 0, 1, 9

Differentiated Instruction

Auditory

Have a discussion about positive and negative integers. Talk about examples of when we use negative numbers in everyday life. Some examples include temperature and elevation. Have students discuss what a negative value indicates for each example.

Extra Example 4

A number is between −7 and 3. What is the least possible integer value of this number? −6

Extra Example 5

In Example 5, Wallops Island recorded a new record low last night. The new record low is greater than the record low in Roanoke. What integers can represent the new record low on Wallops Island? −10, −9, −8, −7, −6, −5, −4, −3, −2, −1

On Your Own

4. $-3, -2, -1, 1, 3$
5. $-8, -7, 1, 4, 6$
6. -7
7. $-9, -8, -7, -6, -5, -4$

Laurie's Notes

Example 4

- Drawing a number line, and plotting the possible answers, is a helpful aid for students.
- Discuss with students that the phrase "between −8 and 0" does not include the integers −8 and 0.
- **Common Error:** Students will often say −7 is the greatest integer between −8 and 0. They are still uncertain why moving to the left on the number line gives numbers that are getting smaller (less than).

Example 5

- Students should notice that all of the temperatures are negative, with the exception of Wallops Island.
- **Extension:** Have students research the coldest and warmest recorded temperatures for their town or city.

On Your Own

- Check to see that students have read Question 7 carefully.

Closure

- Explain why 2 > −42. 2 is to the right of −42 on a number line.

EXAMPLE 4 Reasoning with Integers

A number is greater than −8 and less than 0. What is the greatest possible integer value of this number?

Ⓐ −10 Ⓑ −7 Ⓒ −1 Ⓓ 2

Study Tip

In Example 4, you can eliminate Choices A and D because −10 is to the left of −8 and 2 is to the right of 0.

The number is greater than −8 and less than 0. So, the number must be to the right of −8 and to the left of 0 on a horizontal number line.

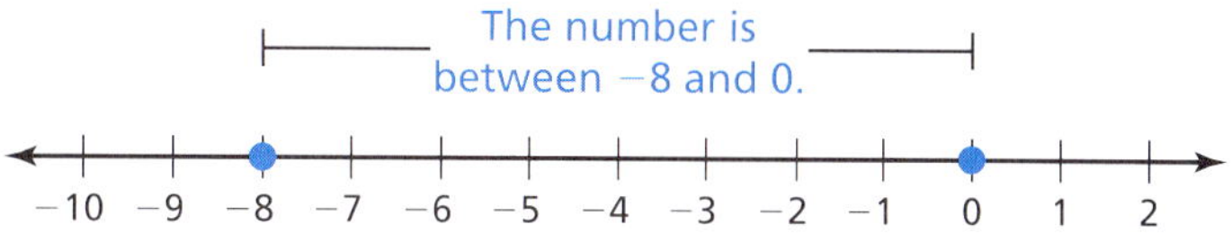

The greatest possible integer value between −8 and 0 is the integer farthest to the right on the number line between these values, which is −1.

So, the correct answer is Ⓒ.

EXAMPLE 5 Real-Life Application

The diagram shows the coldest recorded temperatures for several cities in Virginia.

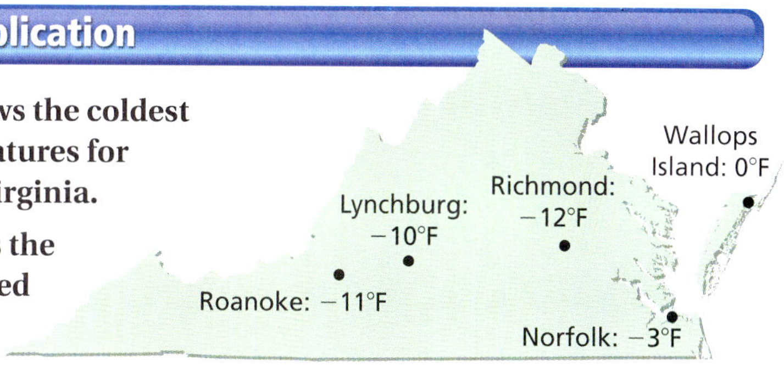

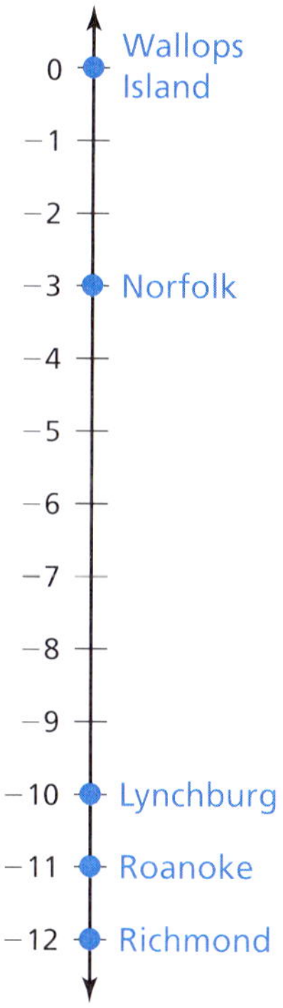

a. Which city has the coldest recorded temperature?

Graph each integer on a vertical number line.

−12 is the lowest on the number line. So, Richmond has the coldest recorded temperature.

b. Has a negative Fahrenheit temperature ever been recorded on Wallops Island? Explain.

The coldest recorded temperature on Wallops Island is 0°F, which is greater than every negative temperature. So, a negative temperature has never been recorded on Wallops Island.

On Your Own

Now You're Ready Exercises 14–19 and 22

Order the integers from least to greatest.

4. −2, −3, 3, 1, −1

5. 4, −7, −8, 6, 1

6. In Example 4, what is the least possible integer value of the number?

7. In Example 5, Norfolk recorded a new record low last night. The new record low is greater than the record low in Lynchburg. What integers can represent the new record low in Norfolk?

6.2 Exercises

Check It Out
Help with Homework
BigIdeasMath.com

Vocabulary and Concept Check

1. **WRITING** Explain how to use a number line to compare two integers.
2. **REASONING** The positions of four fish are shown.
   - **a.** Use red, blue, yellow, and green dots to graph the positions of the fish on a horizontal number line and a vertical number line.
   - **b.** Explain how to use the number lines from part (a) to order the positions from least to greatest.

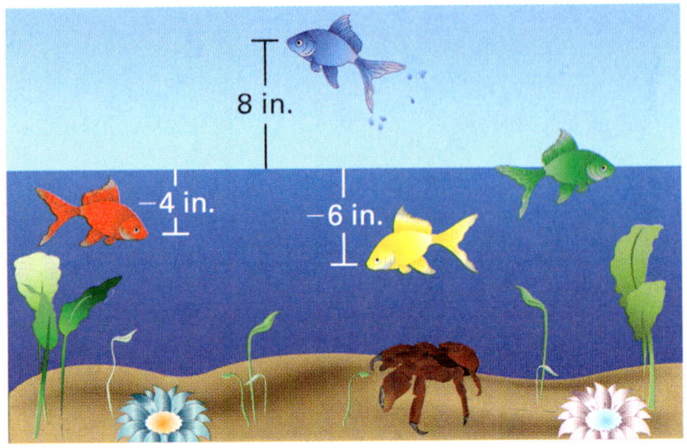

3. **NUMBER SENSE** a and b are negative integers. Compare a and b. Explain your reasoning.

Practice and Problem Solving

Copy and complete the statement using < or >.

1 2 **4.** $3 \; \square \; 0$ **5.** $-2 \; \square \; 0$ **6.** $6 \; \square \; -6$ **7.** $3 \; \square \; -4$

8. $-1 \; \square \; 4$ **9.** $-7 \; \square \; -8$ **10.** $-3 \; \square \; -2$ **11.** $-5 \; \square \; -10$

ERROR ANALYSIS **Describe and correct the error in comparing the negative numbers.**

12. ✗ Compare −3 and −1.
3 > 1. So, −3 > −1.

13. ✗ Compare −7 and −3.
Because −7 < −3, −7 is to the right of −3 on a number line.

Order the integers from least to greatest.

3 **14.** 0, −1, 2, 3, −3 **15.** −4, −2, −3, 2, 1 **16.** −2, 3, −3, −4, 4

17. −7, 2, 6, −4, 3 **18.** 10, −10, 30, −30, −50 **19.** −5, 15, −10, −20, 25

20. **ARCHAEOLOGY** An archaeologist discovers the two artifacts shown.
   - **a.** What integer represents ground level?
   - **b.** A dinosaur bone is found 42 centimeters below ground level. Is it deeper than both of the artifacts?

21. **TEMPERATURE** The freezing temperature of nitrogen is −210°C, and the freezing temperature of oxygen is −223°C. Which temperature is colder?

Assignment Guide and Homework Check

Level	Day 1 Activity Assignment	Day 2 Lesson Assignment	Homework Check
Basic	4–7, 29–33	1–3, 8, 11–23 odd	8, 13, 17, 19, 21
Average	4–7, 29–33	1–3, 8–12 even, 13–21 odd, 22–26	10, 17, 19, 21, 25
Advanced	4–7, 29–33	1–3, 10, 12, 16, 18, 20–28	10, 18, 24, 26, 27

Common Errors

- **Exercises 14–19** Students may ignore the signs on the integers and order them incorrectly. Encourage them to use a number line to help them order the integers.

6.2 Record and Practice Journal

Complete the statement using $<$ or $>$.

1. $0 \; \underline{<} \; 9$
2. $8 \; \underline{>} \; -3$
3. $-5 \; \underline{>} \; -7$
4. $-12 \; \underline{<} \; 0$
5. $-6 \; \underline{<} \; -2$
6. $-21 \; \underline{>} \; -40$

Order the integers from least to greatest.

7. $1, -3, -6, 5, 0$
   $-6, -3, 0, 1, 5$
8. $2, -4, -9, 3, -1$
   $-9, -4, -1, 2, 3$
9. $11, -11, 18, -18, -8$
   $-18, -11, -8, 11, 18$
10. $21, -14, -35, 28, -7$
   $-35, -14, -7, 21, 28$
11. Your miniature golf scores for the first half of a course are $-3, 7, -2, -5, 3, -1, 0, 4, -4$. Order the scores from least to greatest.
   $-5, -4, -3, -2, -1, 0, 3, 4, 7$

Vocabulary and Concept Check

1. On a number line, numbers to the left are less than numbers to the right. Numbers to the right are greater than numbers to the left.
2. See Additional Answers.
3. The value of a is less than the value b because a is to the left of b.

Practice and Problem Solving

4. $>$
5. $<$
6. $>$
7. $>$
8. $<$
9. $>$
10. $<$
11. $>$
12. The student compares 3 and 1, not -3 and -1. So, $-3 < -1$.
13. The explanation about where the integers are located on a number line is incorrect; $-7 < -3$; So, -7 is to the left of -3 on a number line.
14. $-3, -1, 0, 2, 3$
15. $-4, -3, -2, 1, 2$
16. $-4, -3, -2, 3, 4$
17. $-7, -4, 2, 3, 6$
18. $-50, -30, -10, 10, 30$
19. $-20, -10, -5, 15, 25$
20. **a.** 0

    b. no
21. oxygen
22. least: -9; greatest: -3
23. always; The opposite of a positive integer is a negative integer. Positive integers are greater than negative integers.

Practice and Problem Solving

24. never; If an integer is less than its opposite, it must be a negative integer, which is never greater than 0.

25. **a.** Florida, Louisiana, Arkansas, Tennessee, California

b. California, Louisiana, Florida, Arkansas, Tennessee

c. An elevation of 0 feet represents sea level.

26. See *Taking Math Deeper.*

27. no; In order for the median to be below 0°F, at least 6 of the temperatures must be below 0°F.

28. See Additional Answers.

Fair Game Review

29.

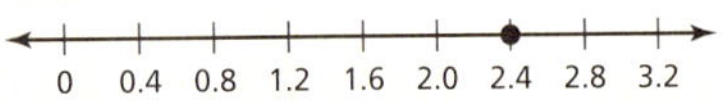

30.

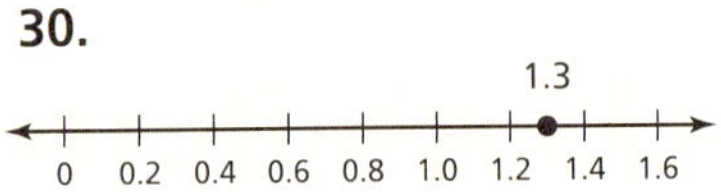

31.

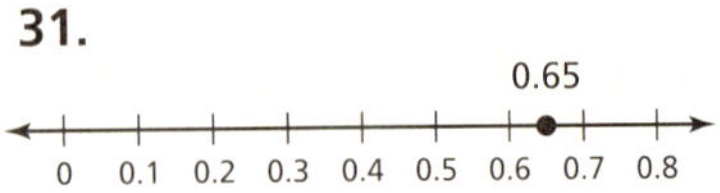

32.

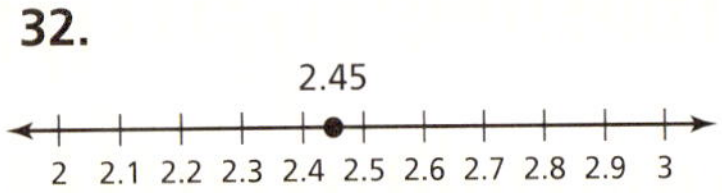

33. B

Mini-Assessment

Copy and complete the statement.

1. −5 ___ 5 $-5 < 5$
2. −4 ___ −3 $-4 < -3$

Order the integers from least to greatest.

3. −8, 4, −1, −2, 1 −8, −2, −1, 1, 4
4. 0, 5, −3, −5, −9 −9, −5, −3, 0, 5
5. In miniature golf, the person with the least score wins. Jeremy has a score of −5 and Mike has a score of −6. Who is the winner? Mike

Taking Math Deeper

Exercise 26

This exercise is a nice example of a nonstandard problem. It can be solved algebraically or graphically.

1 Draw a number line that shows −17 and 5.

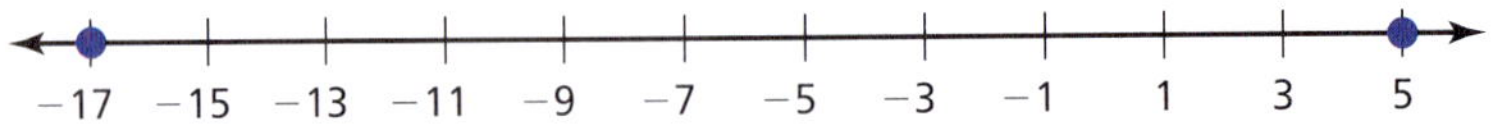

2 Find the halfway point.

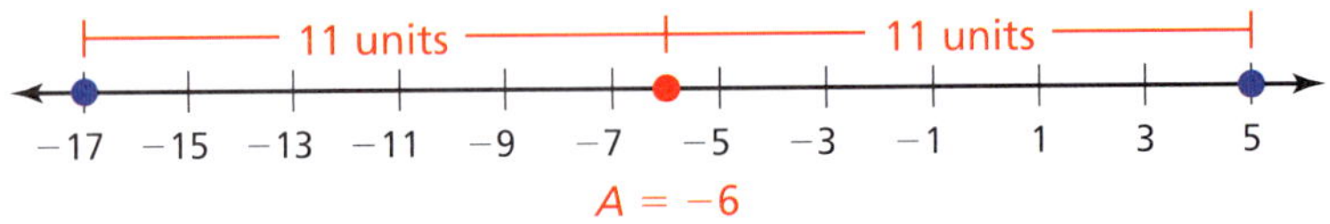

3 Find the point halfway between A and 0.

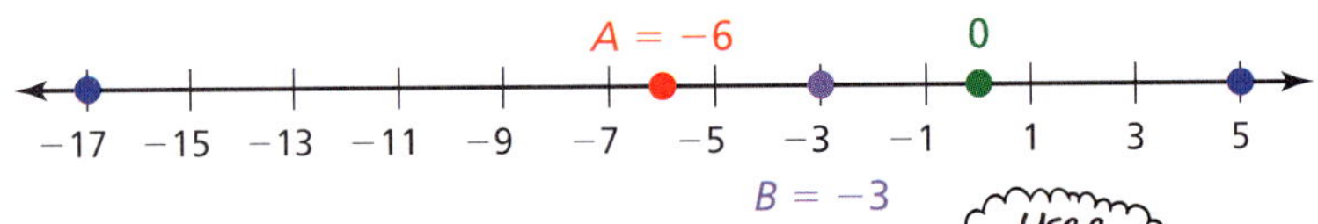

So, $A = -6$ and $B = -3$.

Project

Research deep sea diving. Find out how deep divers can go and the intervals needed for the ascent back to the surface.

Reteaching and Enrichment Strategies

If students need help. . .	If students got it. . .
Resources by Chapter • Practice A and Practice B • Puzzle Time Record and Practice Journal Practice Differentiating the Lesson Lesson Tutorials Skills Review Handbook	Resources by Chapter • Enrichment and Extension • Technology Connection Start the next section

4 **22.** **REASONING** A number is between −2 and −10. What is the least possible integer value of this number? What is the greatest possible integer value of this number?

Tell whether the statement is *always*, *sometimes*, or *never* true. Explain.

23. A positive integer is greater than its opposite.

24. An integer is less than its opposite and greater than 0.

25. **ELEVATION** The table shows the highest and lowest elevations for five states.

a. Order the states by their highest elevations, from least to greatest.

b. Order the states by their lowest elevations, from least to greatest.

c. What does the lowest elevation for Florida represent?

State	Highest Elevation (feet)	Lowest Elevation (feet)
Arkansas	2,753	55
California	14,494	−282
Florida	345	0
Louisiana	535	−8
Tennessee	6,643	178

26. **NUMBER LINE** Point A is on a number line halfway between −17 and 5. Point B is halfway between point A and 0. What integer does point B represent?

27. **TEMPERATURE** Eleven Fahrenheit temperatures are shown on a map during a weather report. When the temperatures are ordered from least to greatest, the middle temperature is below 0°F. Do you know exactly how many of the temperatures are represented by negative numbers? Explain.

28. Puzzle Nine students choose integers. Here are seven of them:

5, −8, 10, −1, −12, −20, and 1.

a. When all nine integers are ordered from least to greatest, the middle integer is 1. Describe the integers chosen by the other two students.

b. When all nine integers are ordered from least to greatest, the middle integer is −3. Describe the integers chosen by the other two students.

Fair Game Review What you learned in previous grades & lessons

Graph the decimal on a number line. *(Skills Review Handbook)*

29. 2.4 **30.** 1.3 **31.** 0.65 **32.** 2.45

33. **MULTIPLE CHOICE** What is the area of the trapezoid? *(Section 4.3)*

Ⓐ 6.3 ft^2 Ⓑ 44.1 ft^2 Ⓒ 50.4 ft^2 Ⓓ 88.2 ft^2

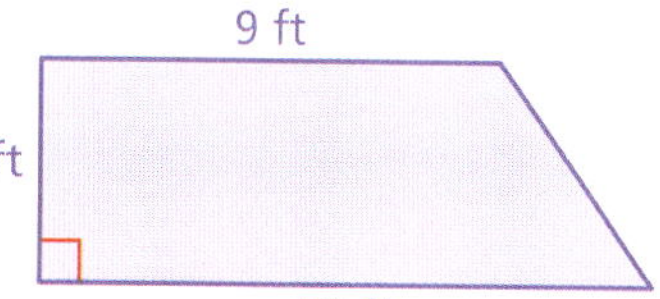

6.3 Fractions and Decimals on the Number Line

Essential Question How can you use a number line to compare positive and negative fractions and decimals?

1 ACTIVITY: Locating Fractions on a Number Line

On your time line for "A Day in the Life of an Astronaut" from Activity 3 in Section 6.2, include the following events. Represent each using a fraction or a mixed number.

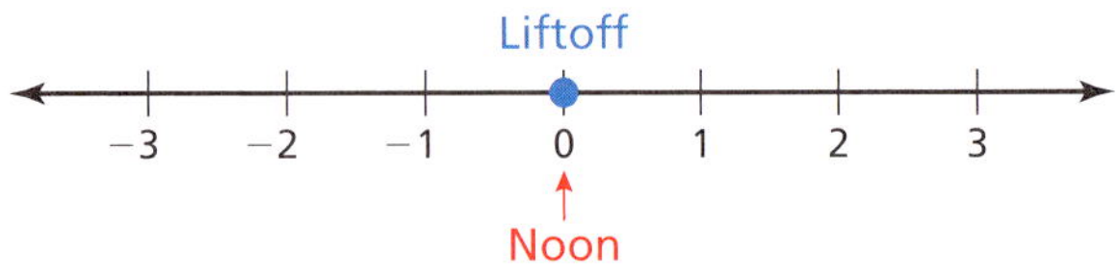

a. Radio Transmission: 10:30 A.M.

b. Space Walk: 7:30 P.M.

c. Physical Exam: 4:45 A.M.

d. Photograph Taken: 3:15 A.M.

e. Float in the Cabin: 6:20 P.M.

f. Eat Dinner: 8:40 P.M.

COMMON CORE

Fractions and Decimals

In this lesson, you will
- understand positive and negative numbers and use them to describe real-life situations.
- graph numbers on a number line.

Learning Standards
6.NS.5
6.NS.6a
6.NS.6c
6.NS.7a
6.NS.7b

Laurie's Notes

Introduction

Standards for Mathematical Practice

- **MP6 Attend to Precision:** Mathematically proficient students are able to communicate precisely to others. In open-ended questions that have more than one possible answer, expect students to be attentive to all students offering solutions. The process students use to arrive at an answer may differ, and students should listen critically to the explanations offered.

Motivate

- Draw a number line on the board from −8 to 8. There should be sufficient space to locate fractions. Under the 0 write *Noon* as is shown in the text.

? Ask students a series of questions about events in their day. Tell students that zero represents noon.
  - "Where is 9 A.M. located?" −3 "How many hours is 9 A.M. from noon?" 3 "What were you doing 3 hours ago?" Answers will vary.
  - "What time is an hour and a half before noon? 10:30 A.M. "To what fractional number does this correspond?" $-1\frac{1}{2}$
  - Continue to ask similar questions about different times.
- **Common Error:** Students may change between decimal hours and hours and minutes. Students incorrectly say 3 hours 15 minutes is 3.15 hours. Remind students that there are 60 minutes in an hour and $3\frac{15}{60}$ is equal to $3\frac{1}{4}$ or 3.25.
- **Common Error:** Students may locate negative mixed numbers incorrectly. Remind students that for negative numbers they start at 0 and move left.

Activity Notes

Activity 1

? Ask questions to check students' understanding.
  - "What type of numbers are the A.M. events?" negative numbers; "the P.M. events?" positive numbers
  - "Are there any events that are the same distance from liftoff?" yes; physical exam and space walk "Where are the events located on the number line?" $-7\frac{1}{2}$ and $7\frac{1}{2}$ "What do you know about the absolute value of these two numbers?" They are the same.
  - "Explain how you wrote 6:20 P.M. as a mixed number and how you located it on the number line." Listen for an explanation that includes $\frac{20}{60} = \frac{1}{3}$ hour and therefore, $\frac{1}{3}$ beyond 6 on the number line.

Common Core State Standards

6.NS.5 . . . use positive and negative numbers to represent quantities in real-world contexts, ...

6.NS.6a Recognize opposite signs of numbers as indicating locations on opposite sides of 0 on the number line; recognize that the opposite of the opposite of a number is the number itself, ...

6.NS.6c Find and position . . . rational numbers on a horizontal or vertical number line diagram; . . ..

6.NS.7a Interpret statements of inequality as statements about the relative position of two numbers on a number line diagram.

6.NS.7b Write, interpret, and explain statements of order for rational numbers in real-world contexts.

Previous Learning

Students should know how to convert between mixed numbers and improper fractions.

Technology for the Teacher

Dynamic Classroom

Lesson Plans
Complete Materials List

6.3 Record and Practice Journal

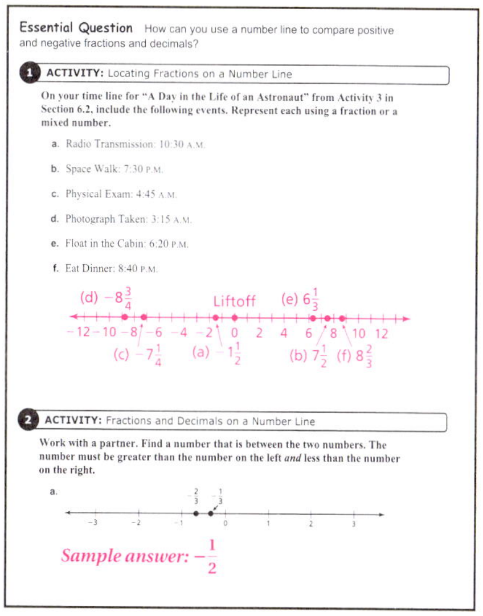

Essential Question How can you use a number line to compare positive and negative fractions and decimals?

1 ACTIVITY: Locating Fractions on a Number Line

On your time line for "A Day in the Life of an Astronaut" from Activity 3 in Section 6.2, include the following events. Represent each using a fraction or a mixed number.

a. Radio Transmission: 10:30 A.M.

b. Space Walk: 7:30 P.M.

c. Physical Exam: 4:45 A.M.

d. Photograph Taken: 3:15 A.M.

e. Float in the Cabin: 6:20 P.M.

f. Eat Dinner: 8:40 P.M.

2 ACTIVITY: Fractions and Decimals on a Number Line

Work with a partner. Find a number that is between the two numbers. The number must be greater than the number on the left *and* less than the number on the right.

a.

Sample answer: $-\frac{1}{2}$

Differentiated Instruction

Vocabulary

Write a table of opposites on the board. Encourage students to add to the list.

Word	Opposite
little	big
forward	backward

Ask the English learners to write the words in their native language in another column and to share them with the class. Explain to the class that in mathematics, every nonzero number has an opposite. Every pair of opposites consists of a positive number and a negative number. Ask students to name some pairs of opposite numbers. Write opposite numbers in the table and the words that represent them.

6.3 Record and Practice Journal

b.

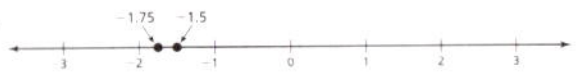

Sample answer: -1.6

c.

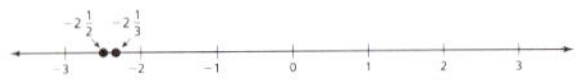

Sample answer: $-2\frac{5}{12}$

3 ACTIVITY: Decimals on a Number Line

Work with a partner.

Snorkeling: −5 meters

Scuba diving: −50 meters

Deep sea diving: −700 meters

a. Write the position of each diver in kilometers.

−0.005 km; −0.05 km; −0.7 km

b. **CHOOSE TOOLS** Would a horizontal or a vertical number line be more appropriate for representing these data? Why?

vertical; to visualize the position or depth of the diver

c. Use a number line to order the positions from deepest to shallowest.

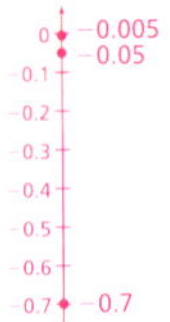

What Is Your Answer?

4. **IN YOUR OWN WORDS** How can you use a number line to compare positive and negative fractions and decimals?

Graphing fractions and decimals on a number line is the same as graphing integers. So, to order positive and negative fractions and decimals, you just read from left to right on the number line.

5. Draw a number line. Graph and label three values between −2 and −1.

Sample answer:

Laurie's Notes

Activity 2

? Prepare students for this activity by asking a few questions related to the activity.

- "Which is greater, $1\frac{1}{2}$ or $1\frac{2}{3}$? How do you know?" Students might reason that using a common denominator you have $1\frac{3}{6}$ and $1\frac{4}{6}$.
- "Can you name a number between $1\frac{3}{6}$ and $1\frac{4}{6}$? Explain." Listen for equivalent fractions, probably $1\frac{6}{12}$ and $1\frac{8}{12}$. One fraction between these is $1\frac{7}{12}$.

- Answers will vary for these problems so it is important to give several students the opportunity to share their answers and their thinking.
- **Common Error:** When locating negative fractions, students often move left to right instead of right to left. For example, students incorrectly locate $-1\frac{3}{4}$ closer to −1 than −2. Help students remember to start at −1 and then move $\frac{3}{4}$ of the way towards −2.
- **Big Idea:** If two numbers A and B are positive and $A < B$, then $-B < -A$. Discuss this idea on a number line using numbers from the three problems just completed.

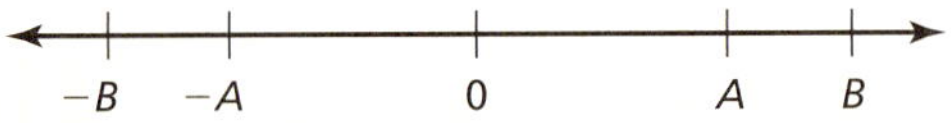

Activity 3

- **MP6:** Mathematically proficient students are careful in specifying units of measure.
- Review meters and kilometers with students. Try to relate the units to common lengths first, such as, the classroom is 6 meters wide; the distance from school to [insert a location] is 6 kilometers. Ask, "How many meters in 6 kilometers?" 6000 "How many kilometers in 6 meters?" 0.006
- Caution students to read carefully. The information is given in meters and the position is to be written in kilometers.

? "How should the number line be labeled?" −1 to 0 (in kilometers) will include all three positions.

- The vertical number line model is a good visual for the context of this problem.

What Is Your Answer?

- If time permits, have several students share their answers to Question 5.

Closure

- Draw a number line and locate the following fractions: $-2\frac{1}{4}, 1\frac{1}{3}, -1\frac{7}{8}, \frac{5}{6}$

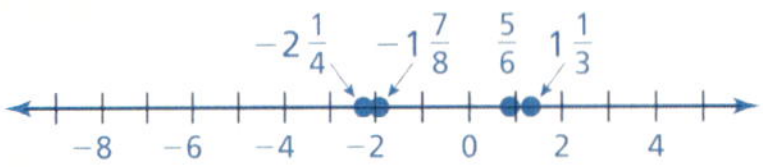

2 ACTIVITY: Fractions and Decimals on a Number Line

Math Practice 1

Make a Plan

How can you find a number between two given numbers?

Work with a partner. Find a number that is between the two numbers. The number must be greater than the green number *and* less than the blue number.

a.

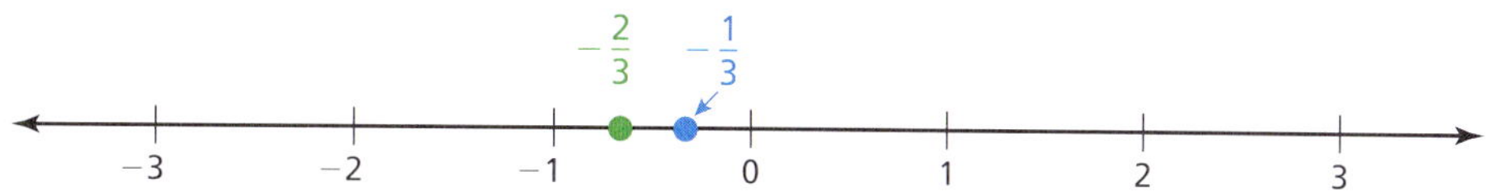

b.

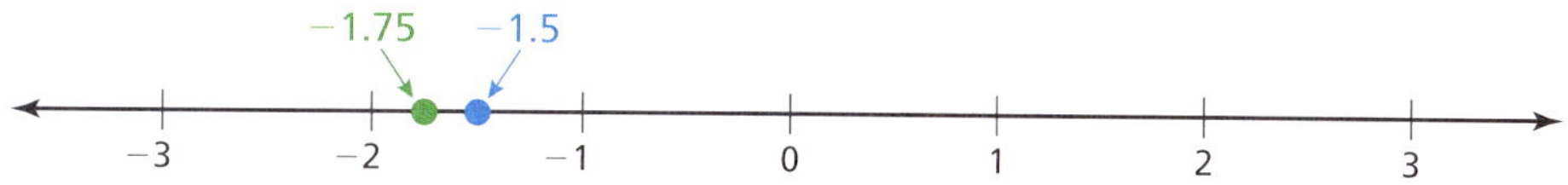

c.

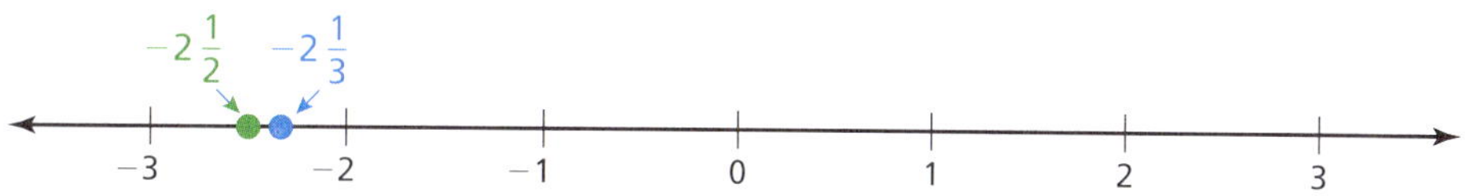

3 ACTIVITY: Decimals on a Number Line

Work with a partner.

Snorkeling: −5 meters	Scuba diving: −50 meters	Deep-sea diving: −700 meters

a. Write the position of each diver in kilometers.

b. **CHOOSE TOOLS** Would a horizontal or a vertical number line be more appropriate for representing these data? Why?

c. Use a number line to order the positions from deepest to shallowest.

What Is Your Answer?

4. **IN YOUR OWN WORDS** How can you use a number line to compare positive and negative fractions and decimals?

5. Draw a number line. Graph and label three values between −2 and −1.

Practice Use what you learned about fractions and decimals on a number line to complete Exercises 4 and 5 on page 264.

6.3 Lesson

Check It Out
Lesson Tutorials
BigIdeasMath.com

In Section 6.1, you learned that integers can be negative. Fractions and decimals can also be negative.

EXAMPLE 1 Graphing Negative Fractions and Decimals

Graph each number and its opposite.

a. $\frac{3}{4}$

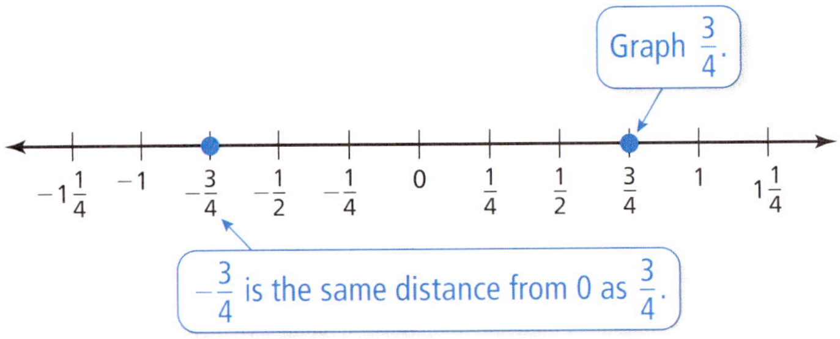

b. -1.6

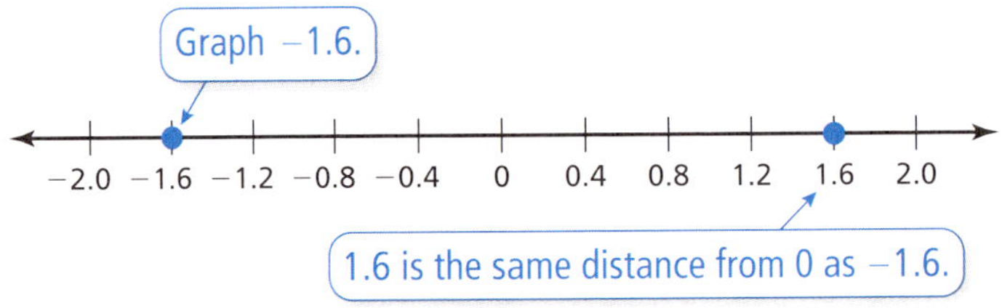

On Your Own

Now You're Ready
Exercises 6–9

Graph the number and its opposite.

1. $2\frac{1}{2}$ **2.** $-\frac{4}{5}$ **3.** -3.5 **4.** 5.25

EXAMPLE 2 Comparing Fractions and Mixed Numbers

a. Compare $-\frac{1}{2}$ and $-\frac{3}{4}$.

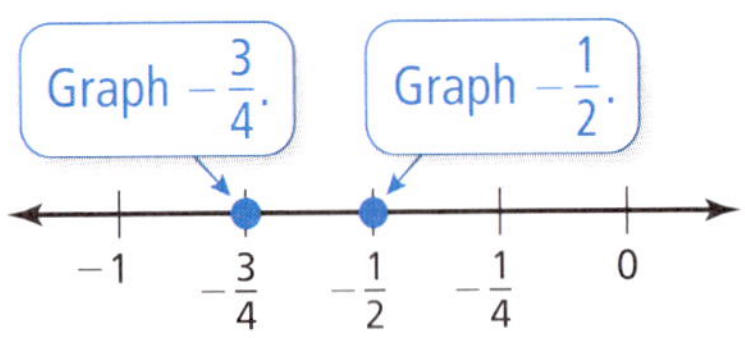

$-\frac{1}{2}$ is to the right of $-\frac{3}{4}$.

So, $-\frac{1}{2} > -\frac{3}{4}$.

b. Compare $-4\frac{5}{6}$ and $-4\frac{1}{6}$.

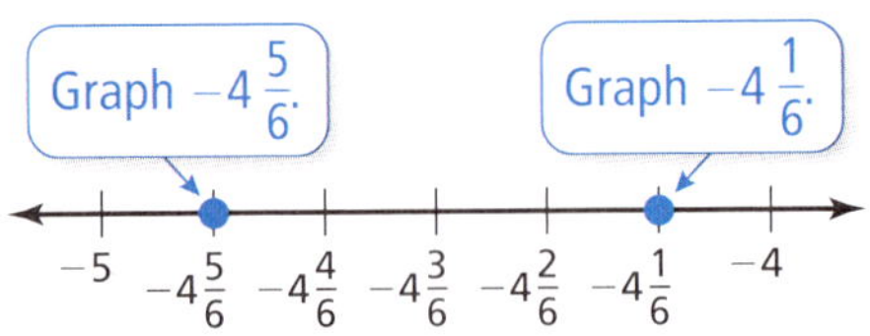

$-4\frac{5}{6}$ is to the left of $-4\frac{1}{6}$.

So, $-4\frac{5}{6} < -4\frac{1}{6}$.

Laurie's Notes

Introduction

Connect

- **Yesterday:** Students developed an understanding of how to locate and compare positive and negative fractions and decimals on a number line. (MP6)
- **Today:** Students will use a number line to compare rational numbers.

Motivate

- Place a piece of masking tape on the floor for a number line.
- Have volunteers hold index cards with $0, \frac{1}{4}, \frac{1}{2}, \frac{3}{4}$, or 1 written on them and stand at the correct places on the number line.
- Say, "Reading left to right, $0 < \frac{1}{4} < \frac{1}{2} < \frac{3}{4} < 1$."
- Use a marker to write a negative sign in front of all the numbers except 0.

? "Are the numbers still in the correct order as we look at the number line? Explain." No; 0 and −1 need to swap and $-\frac{1}{4}$ and $-\frac{3}{4}$ need to swap.

Discuss

- **Big Idea:** The big idea for students to remember is that the farther to the right a number is on the number line, the greater it is. Likewise, the farther to the left a number is on the number line, the less it is.

Lesson Notes

Example 1

? "In part (a), how far is each number from 0?" $\frac{3}{4}$ unit

? "In part (b), how far is each number from 0?" 1.6 units

- **MP6 Attend to Precision:** Students may incorrectly believe that the opposite of a number is negative. If the original number is negative, its opposite is positive. Vocabulary: opposite doesn't mean negative.

On Your Own

- **Neighbor Check:** Have students work independently and then have their neighbors check their work. Have students discuss any discrepancies.

Example 2

- In part (a), remind students that for negative numbers, you work from right to left: $0, -\frac{1}{4}, -\frac{1}{2}, -\frac{3}{4}, -1$.
- In part (b), remind students that for negative numbers, you work from right to left: $-4, -4\frac{1}{6}, -4\frac{2}{6}$, and so on.

? "Which number is farther from 0, $-4\frac{5}{6}$ or $-4\frac{1}{6}$?" $-4\frac{5}{6}$

Goal Today's lesson is comparing rational numbers on a number line.

Technology for the Teacher

Dynamic Classroom

Lesson Tutorials
Lesson Plans
Answer Presentation Tool

Extra Example 1

Graph each number and its opposite.

a. $1\frac{1}{3}$

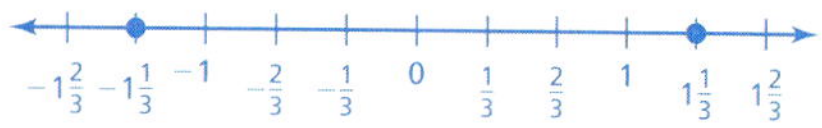

b. −0.7

On Your Own

1.

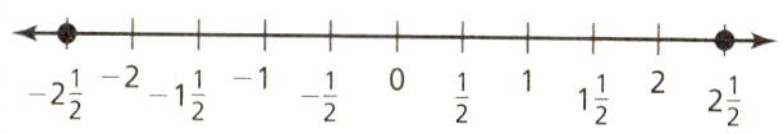

2.

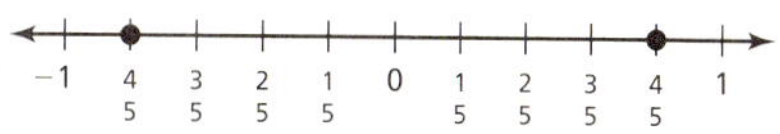

3.

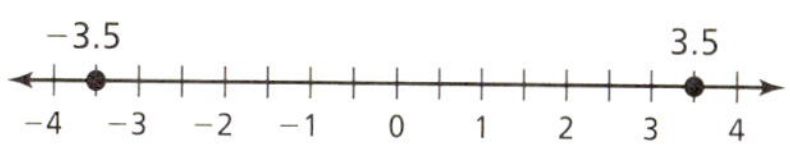

4.

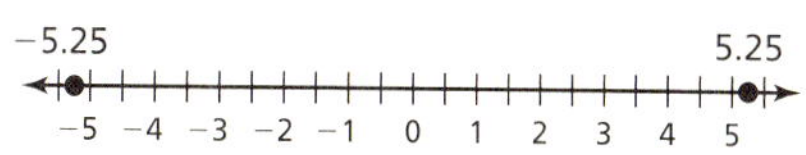

Extra Example 2

a. Compare $-\frac{5}{8}$ and $-\frac{3}{8}$.

$-\frac{5}{8} < -\frac{3}{8}$

b. Compare $-2\frac{2}{3}$ and $-3\frac{1}{3}$.

$-2\frac{2}{3} > -3\frac{1}{3}$

Extra Example 3

Compare 0.3 and −1.2. 0.3 > −1.2

Extra Example 4

Which is the least, $-1\frac{3}{4}$, $-1\frac{9}{16}$, or $-1\frac{5}{8}$?

$-1\frac{3}{4}$

On Your Own

5. $-\frac{4}{7} < -\frac{1}{7}$
6. $-1\frac{2}{3} > -1\frac{5}{6}$
7. $-0.5 < 0.3$
8. A temperature change of $-3\frac{2}{5}$°F per minute is the farthest to the left on the number line. So, the temperature dropped fastest on this day.

English Language Learners

Vocabulary

Let English language learners know that the pronunciation of rational numbers that end in –th such as fourths and fifths is sometimes difficult for native English speakers.

Laurie's Notes

Example 3

- **MP6:** Use correct language in reading the decimals: negative 3 and 8 hundredths, and negative 3 and 8 tenths.
- **Teaching Tip:** To help students compare decimals with different place values, write trailing zeros as shown:

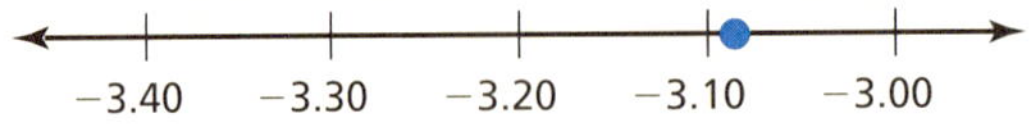

It is not obvious to all students that −3.1 is equivalent to −3.10, or in words, negative 3 and 1 tenth is equivalent to negative 3 and 10 hundredths. The trailing zeros help students recognize where −3.08 is in relation to −3.1 (or −3.10).

Example 4

- Even though the numbers can be compared without graphing on the number line, it is good reinforcement of estimating the location of the numbers.

? "Refer to the diagram. Between what two tick marks is $-\frac{5}{8}$ located? $-2\frac{1}{5}$? $-3\frac{1}{10}$?" $-\frac{3}{4}$ and $-\frac{1}{2}$; $-2\frac{1}{4}$ and −2; $-3\frac{1}{4}$ and −3

- Explain that even though $-3\frac{1}{10}$ is the least of the three numbers, it has the greatest absolute value and therefore represents the greatest drop in temperature.
- Absolute value has not been defined formally, yet students intuitively understand that because $3\frac{1}{10} > 2\frac{1}{5} > \frac{5}{8}$, when you have the opposite of each of these numbers, $-3\frac{1}{10} < -2\frac{1}{5} < -\frac{5}{8}$.

On Your Own

- **Think-Pair-Share:** Students should read each question independently and then work in pairs to answer the questions. When they have answered the questions, the pair should compare their answers with another group and discuss any discrepancies.

Closure

- **Exit Ticket:** Which is greater, $-1\frac{2}{5}$ or $-1\frac{3}{4}$? Explain. $-1\frac{2}{5}$; It is farther to the right on the number line.

EXAMPLE 3 Comparing Decimals

Compare −3.08 and −3.8.

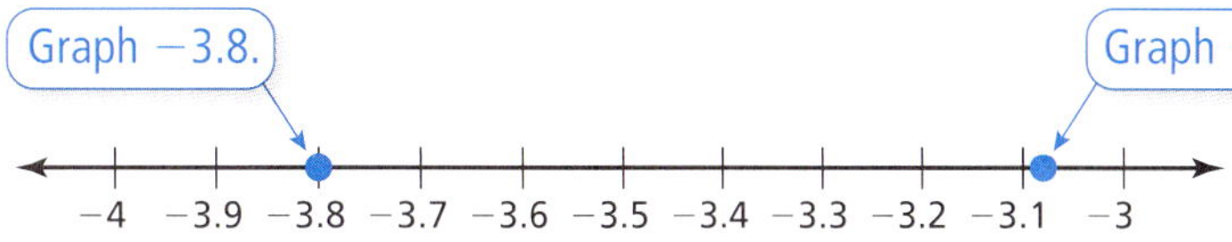

−3.08 is to the right of −3.8.

∴ So, −3.08 > −3.8.

EXAMPLE 4 Real-Life Application

A *Chinook wind* is a warm mountain wind that can cause rapid temperature changes. The table shows three of the greatest temperature drops ever recorded after a Chinook wind occurred. On which date did the temperature drop the fastest? Explain.

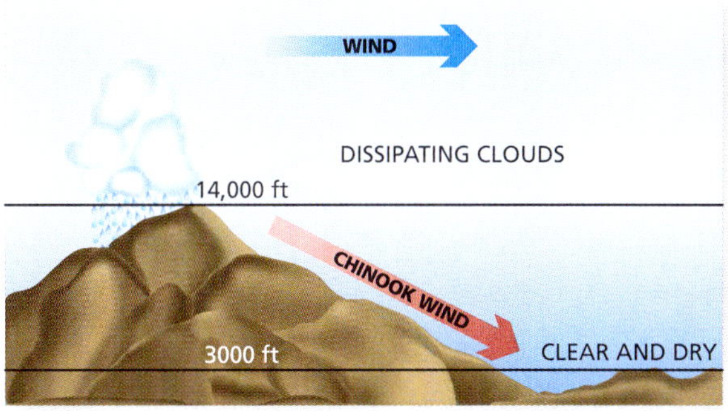

Date	Temperature Change
January 10, 1911	$-3\frac{1}{10}$°F per minute
November 10, 1911	$-\frac{5}{8}$°F per minute
January 22, 1943	$-2\frac{1}{5}$°F per minute

Graph the numbers on a number line.

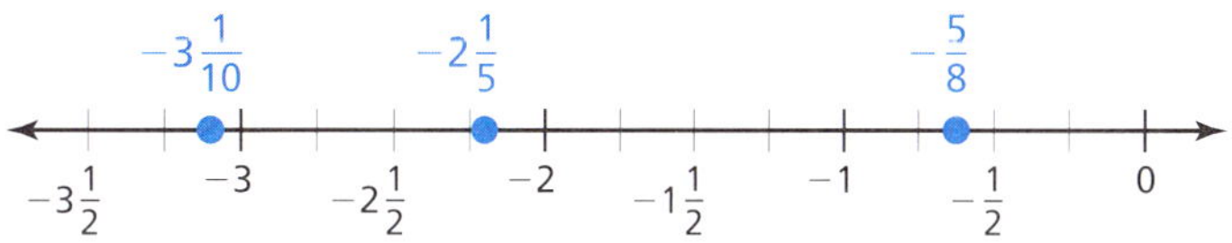

$-3\frac{1}{10}$ is farthest to the left.

∴ So, the temperature dropped the fastest on January 10, 1911.

On Your Own

Now You're Ready
Exercises 10–18 and 20–23

Copy and complete the statement using < or >.

5. $-\frac{4}{7}$ ▢ $-\frac{1}{7}$

6. $-1\frac{2}{3}$ ▢ $-1\frac{5}{6}$

7. -0.5 ▢ 0.3

8. **WHAT IF?** In Example 4, a temperature change of $-3\frac{2}{5}$°F per minute is recorded. How does this temperature change compare with the other temperature changes? Explain.

6.3 Exercises

Check It Out
Help with Homework
BigIdeasMath.com

Vocabulary and Concept Check

1. **NUMBER SENSE** Which statement is *not* true?

   a. On a number line, $-2\frac{1}{6}$ is to the left of $-2\frac{2}{3}$.

   b. $-2\frac{2}{3}$ is less than $-2\frac{1}{6}$.

   c. $-2\frac{1}{6}$ is greater than $-2\frac{2}{3}$.

   d. On a number line, $-2\frac{2}{3}$ is to the left of $-2\frac{1}{6}$.

2. **NUMBER SENSE** Is a negative decimal *always, sometimes,* or *never* equal to a positive decimal? Explain.

3. **NUMBER SENSE** On a number line, is -2.06 or -2.6 farther to the left?

Practice and Problem Solving

Find a fraction or mixed number that is between the two numbers.

4.

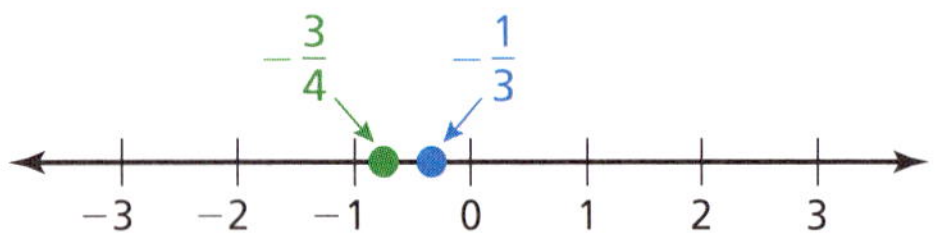

5.

Graph the number and its opposite.

1 6. $\frac{2}{3}$ 7. $-2\frac{1}{4}$ 8. -3.8 9. 2.15

Copy and complete the statement using < or >.

2 3 10. $-3\frac{1}{3}$ $\square$ $-3\frac{2}{3}$ 11. $-\frac{1}{2}$ $\square$ $-\frac{1}{6}$ 12. $-\frac{3}{4}$ $\square$ $\frac{5}{8}$

13. $-2\frac{2}{3}$ $\square$ $-2\frac{1}{2}$ 14. $-1\frac{5}{6}$ $\square$ $-1\frac{3}{4}$ 15. -4.6 $\square$ -4.8

16. -0.12 $\square$ -0.05 17. 2.41 $\square$ -3.16 18. -3.524 $\square$ -3.542

19. **SAND DOLLARS** In rough water, a small sand dollar burrows $-\frac{1}{2}$ centimeter into the sand. A larger sand dollar burrows $-1\frac{1}{4}$ centimeters into the sand. Which sand dollar burrowed farther?

Assignment Guide and Homework Check

Level	Day 1 Activity Assignment	Day 2 Lesson Assignment	Homework Check
Basic	4, 5, 28–32	1–3, 7–23 odd, 24	7, 11, 15, 19, 21
Average	4, 5, 28–32	1–3, 8–16 even, 17, 19, 21–25	8, 12, 16, 19, 22
Advanced	4, 5, 28–32	1–3, 8, 14–27	8, 14, 18, 19, 22

Common Errors

- **Exercises 10–18, 20–23** Students may ignore the sign of the number and order the numbers incorrectly. Encourage them to use a number line to determine which number is farthest to the right.
- **Exercises 11–14, 20, 21** Students may place the fractions in the wrong order. Encourage them to find a common denominator before ordering.

6.3 Record and Practice Journal

Complete the statement using < or >.

1. $3.7 \underline{\ >\ } -3.2$
2. $-10.4 \underline{\ <\ } -10.04$
3. $-\frac{2}{3} \underline{\ <\ } \frac{1}{4}$
4. $-2\frac{2}{3} \underline{\ <\ } -2\frac{1}{2}$

Order the numbers from least to greatest.

5. $-2, 0, -4, 2, 3$

   $-4, -2, 0, 2, 3$
6. $-6.3, 4.2, 7.7, -3.9, 3.4$

   $-6.3, -3.9, 3.4, 4.2, 7.7$
7. $1\frac{4}{5}, \frac{1}{2}, -3\frac{7}{8}, -\frac{7}{9}, -\frac{3}{5}$

   $-3\frac{7}{8}, -\frac{7}{9}, -\frac{3}{5}, \frac{1}{2}, 1\frac{4}{5}$
8. $-3, 0.6, \frac{1}{4}, 0, -1\frac{2}{3}$

   $-3, -1\frac{2}{3}, 0, \frac{1}{4}, 0.6$
9. An archaeologist discovers two artifacts. Compare the positions of the artifacts.

   $-44\frac{3}{5} < -38$

   The pot is deeper than the arrowhead.

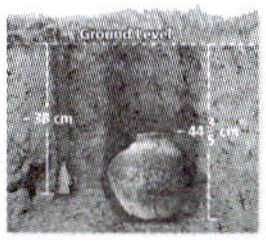

Vocabulary and Concept Check

1. a
2. never; A negative decimal is to the left of 0 on a number line and a positive decimal is to the right of 0 on a number line.
3. −2.6

Practice and Problem Solving

4. *Sample answer:* $-\frac{1}{2}$
5. *Sample answer:* $-2\frac{1}{4}$
6. 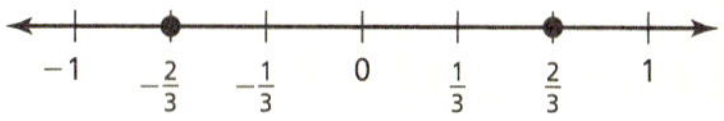

7. 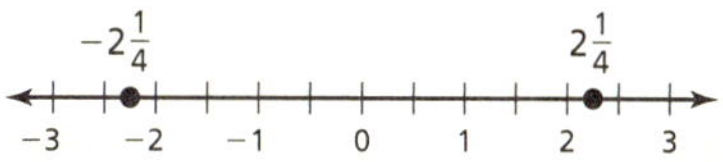

8. 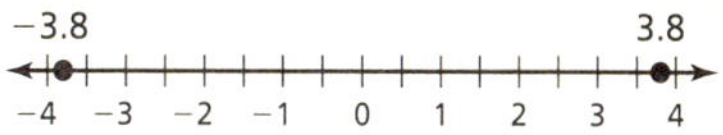

9. 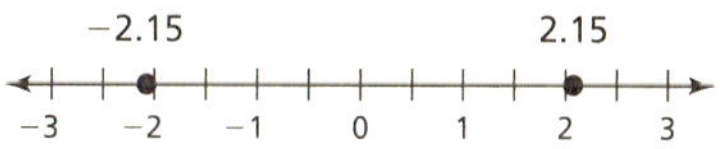

10. >
11. <
12. <
13. <
14. <
15. >
16. <
17. >
18. >
19. the larger sand dollar
20. $-3, -2\frac{1}{2}, -2\frac{2}{5}, -2\frac{3}{10}, -2$
21. $-1, -\frac{3}{4}, -\frac{5}{8}, -\frac{1}{20}, 0$
22. −2, −1.8, −1.75, 0, 1.3
23. −5, −4.9, −4.35, −4.3, −4
24. Sirius

Practice and Problem Solving

25. See *Taking Math Deeper*.

26. **a.** 0

b. Strings 1, 4, and 6

c. String 6

d. String 1

e. String 6

27. 1, 2, and any integer less than -3

Fair Game Review

28.

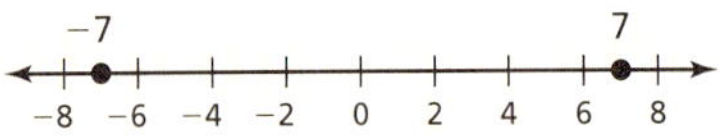

29.

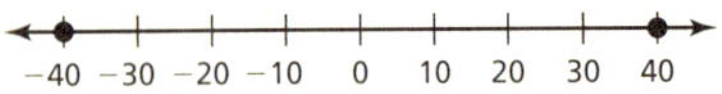

30.

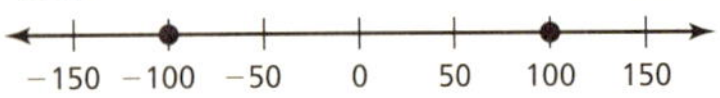

31.

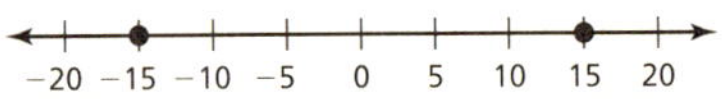

32. D

Mini-Assessment

Which number is greater?

1. 2.3, -1.4 2.3

2. $-\frac{4}{5}, \frac{6}{7}$ $\frac{6}{7}$

3. $-1\frac{1}{2}, -1\frac{1}{3}$ $-1\frac{1}{3}$

4. $-0.06, -0.10$ -0.06

5. In a pond, a school of bass is $-2\frac{1}{3}$ feet from the surface of the pond. A school of trout is $-2\frac{3}{4}$ feet from the surface of the pond. Which school of fish is farther from the surface of the pond? school of trout

Taking Math Deeper

Exercise 25

It would be difficult to compare all of the water levels in fraction form. (The least common denominator is 300.) So, for this problem it is probably better to rewrite each of the water levels in decimal form.

1 Rewrite the given facts.

Sunday	-0.12
Monday	-0.35
Tuesday	-0.54
Wednesday	-0.65
Thursday	-0.64
Friday	-0.53
Saturday	$-0.333\ldots$

2 Is the level lower or higher?

Sunday	-0.12	
Monday	-0.35	Lower
Tuesday	-0.54	Lower
Wednesday	-0.65	Lower
Thursday	-0.64	Higher
Friday	-0.53	Higher
Saturday	$-0.333\ldots$	Higher

3 On which days is the level higher than the day before? On which days is it lower?

Higher on: Thursday, Friday, Saturday
Lower on: Monday, Tuesday, Wednesday

Project

Research the changing of the tides. What makes them change? How frequently do they change? Where in the world do they have the most dramatic changes in the tides?

Reteaching and Enrichment Strategies

If students need help. . .	If students got it. . .
Resources by Chapter • Practice A and Practice B • Puzzle Time Record and Practice Journal Practice Differentiating the Lesson Lesson Tutorials Skills Review Handbook	Resources by Chapter • Enrichment and Extension • Technology Connection Start the next section

Order the numbers from least to greatest.

4 **20.** $-2\frac{3}{10}, -2\frac{2}{5}, -2, -2\frac{1}{2}, -3$

21. $-\frac{1}{20}, -\frac{5}{8}, 0, -1, -\frac{3}{4}$

22. $1.3, -2, -1.8, 0, -1.75$

23. $-4, -4.35, -4.9, -5, -4.3$

24. **STARS** The *apparent magnitude* of a star measures how bright the star appears as seen from Earth. The brighter the star, the lesser the number. Which star is the brightest?

Star	Alpha Centauri	Antares	Canopus	Deneb	Sirius
Apparent Magnitude	−0.27	0.96	−0.72	1.25	−1.46

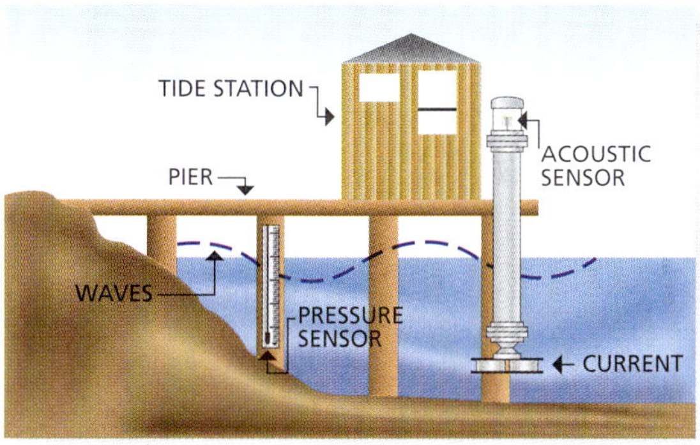

25. **LOW TIDE** The daily water level is recorded for seven straight days at a tide station on the Big Marco River in Florida. On which days is the water level higher than on the previous day? On which days is it lower?

Day	Sun.	Mon.	Tues.	Wed.	Thurs.	Fri.	Sat.
Water Level of the Day (feet)	$-\frac{3}{25}$	$-\frac{7}{20}$	$-\frac{27}{50}$	$-\frac{13}{20}$	$-\frac{16}{25}$	$-\frac{53}{100}$	$-\frac{1}{3}$

26. **PROBLEM SOLVING** A guitar tuner allows you to tune a guitar string to its correct pitch. The units on a tuner are measured in *cents*. The units tell you how far the string tone is above or below the correct pitch.

Guitar String	6	5	4	3	2	1
Number of Cents Away from the Correct Pitch	−0.3	1.6	−2.3	2.8	2.4	−3.6

a. What number on the tuner represents a correctly tuned guitar string?

b. Which strings have a pitch below the correct pitch?

c. Which string has a pitch closest to its correct pitch?

d. Which string has a pitch farthest from its correct pitch?

e. The tuner is rated to be accurate to within 0.5 cent of the true pitch. Which string could possibly be correct?

27. Number Sense What integer values of x make the statement $-\frac{3}{x} < -\frac{x}{3}$ true?

Fair Game Review What you learned in previous grades & lessons

Graph the integer and its opposite. *(Section 6.1)*

28. −7 **29.** 40 **30.** 100 **31.** −15

32. **MULTIPLE CHOICE** You pay $48 for 8 pounds of chicken. Which is an equivalent rate? *(Section 5.3)*

Ⓐ $44 for 4 pounds Ⓑ $28 for 4 pounds

Ⓒ $15 for 3 pounds Ⓓ $30 for 5 pounds

6 Study Help

Check It Out
Graphic Organizer
BigIdeasMath.com

You can use a **summary triangle** to explain a concept. Here is an example of a summary triangle for integers.

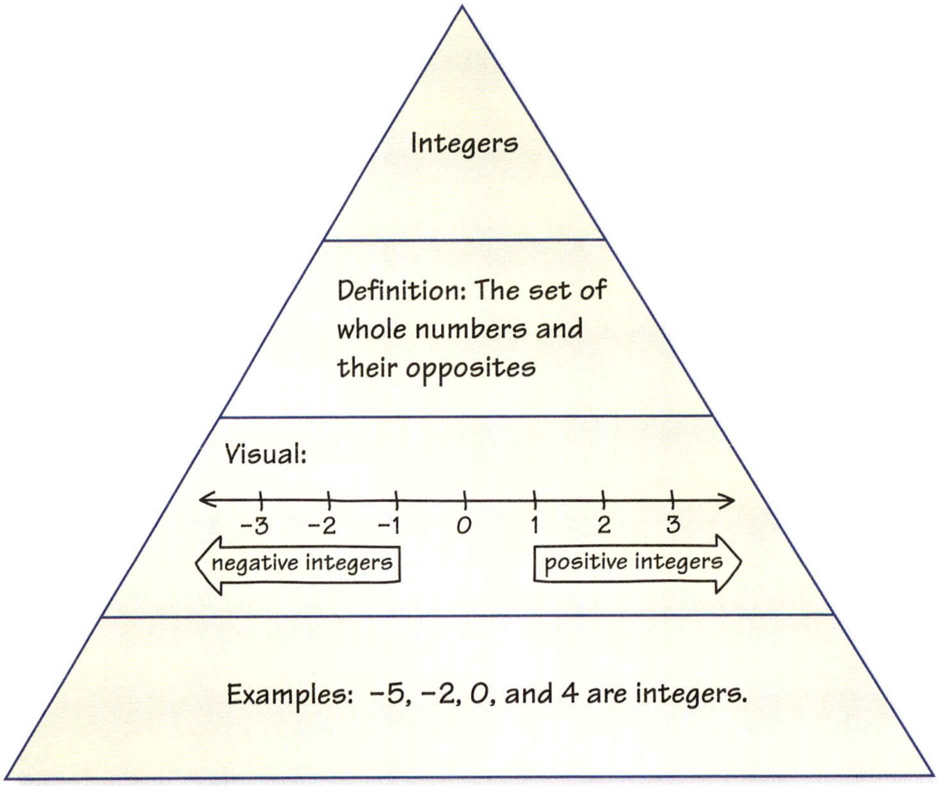

On Your Own

Make summary triangles to help you study these topics.

1. positive integers
2. negative integers
3. opposites

After you complete this chapter, make summary triangles for the following topics.

4. absolute value
5. coordinate plane
6. origin
7. quadrants

"I'm posting my new summary triangle on my daily blog. Do you think it will get me more hits?"

Sample Answers

1.

Positive integers

Definition: The whole numbers that are greater than 0; They can be written with or without a positive sign (+).

Key words: gain, up, earn, climb, rise, ascend, receive, deposit, above zero, above sea level

Examples:
+2; 9; +36; 500,000

2.

Negative integers

Definition: The opposites of the positive integers; They are written with a negative sign (−).

Key words: lose, down, fall, debt, withdraw, below zero, below sea level, descend

Examples:
−2; −9; −36; −500,000

3.

Opposites

Definition: Two numbers that are the same distance from 0 on a number line, but on opposite sides of 0

The opposite of 0 is 0.

Example:

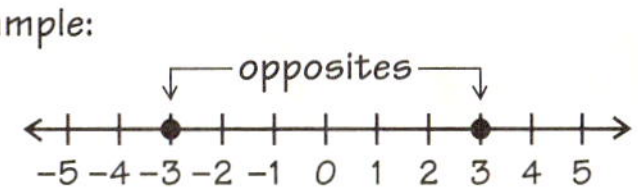

List of Organizers

Available at *BigIdeasMath.com*

Comparison Chart
Concept Circle
Definition (Idea) and Example Chart
Example and Non-Example Chart
Formula Triangle
Four Square
Information Frame
Information Wheel
Notetaking Organizer
Process Diagram
Summary Triangle
Word Magnet
Y Chart

About this Organizer

A **Summary Triangle** can be used to explain a concept. Typically, the summary triangle is divided into 3 or 4 parts. In the top part, students write the concept being explained. In the middle part(s), students write any procedure, explanation, description, definition, theorem, and/or formula(s). In the bottom part, students write an example to illustrate the concept. A summary triangle can be used as an assessment tool, in which blanks are left for students to complete. Also, students can place their summary triangles on note cards to use as a quick study reference.

Technology for the Teacher

Editable Graphic Organizer

Answers

1. 2
2. −15
3. 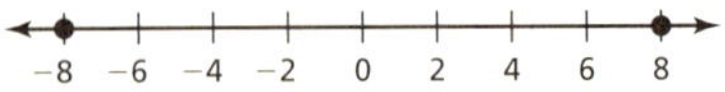
4. 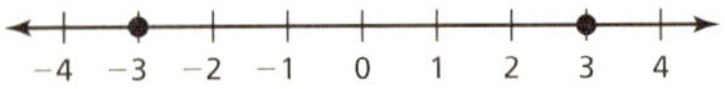
5. <
6. >
7. −5, −2, −1, 0, 3
8. −8, −6, 3, 5, 7
9. 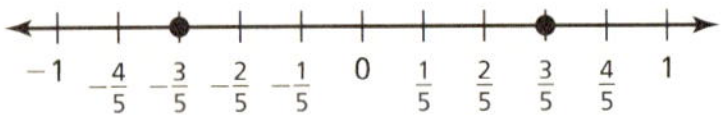
10. 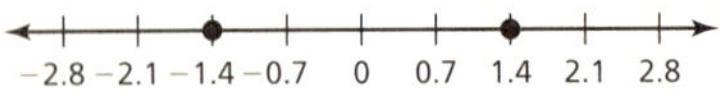
11. <
12. >
13. −195
14. Neptune; Its temperature lies farthest to the left on a number line.
15. −0.45, −0.42, 0.26, 0.37

Technology for the Teacher

Online Assessment
Assessment Book
ExamView® Assessment Suite

Alternative Quiz Ideas

100% Quiz	Math Log
Error Notebook	Notebook Quiz
Group Quiz	Partner Quiz
Homework Quiz	Pass the Paper

100% Quiz

This is a quiz where students are given the answers and then they have to explain and justify each answer.

Reteaching and Enrichment Strategies

If students need help...	If students got it...
Resources by Chapter • Practice A and Practice B • Puzzle Time Lesson Tutorials *BigIdeasMath.com*	Resources by Chapter • Enrichment and Extension • Technology Connection Game Closet at *BigIdeasMath.com* Start the next section

6.1–6.3 Quiz

Check It Out
Progress Check
BigIdeasMath.com

Write a positive or negative integer that represents the situation. *(Section 6.1)*

1. The price of a stock goes up \$2.

2. You descend 15 feet.

Graph the integer and its opposite. *(Section 6.1)*

3. 8

4. -3

Copy and complete the statement using < or >. *(Section 6.2)*

5. -5 ▭ 0

6. -7 ▭ -9

Order the integers from least to greatest. *(Section 6.2)*

7. $3, -2, 0, -1, -5$

8. $-6, 5, 3, -8, 7$

Graph the number and its opposite. *(Section 6.3)*

9. $\frac{3}{5}$

10. -1.4

Copy and complete the statement using < or >. *(Section 6.3)*

11. $-2\frac{2}{5}$ ▭ $-2\frac{1}{4}$

12. -3.28 ▭ -3.72

13. **ROLLER COASTER** At the top of a roller coaster hill, you are 210 feet above ground. At the bottom of the hill, you are 15 feet above ground. Write an integer that represents the change in height from the top to the bottom. *(Section 6.1)*

14. **PLANETS** The table shows the average surface temperatures of four planets. Which planet is the coldest? Explain. *(Section 6.2)*

Planet	Jupiter	Neptune	Saturn	Uranus
Temperature (°C)	-150	-220	-180	-214

15. **STOCK** The table shows the changes in the value of a stock over several days. Order the numbers from least to greatest. *(Section 6.3)*

Day	Change (dollars)
1	-0.42
2	0.26
3	-0.45
4	0.37

6.4 Absolute Value

Essential Question How can you describe how far an object is from sea level?

1 ACTIVITY: Sea Level

Work with a partner. Write an integer that represents the elevation of each object. How far is each object from sea level? Explain your reasoning.

a. Boeing 747

b. Seaplane

c. Bald eagle

d. Leatherback turtle

e. U.S.S. Dolphin

f. Whale

g. Jason Jr.

h. Alvin

i. Kaiko

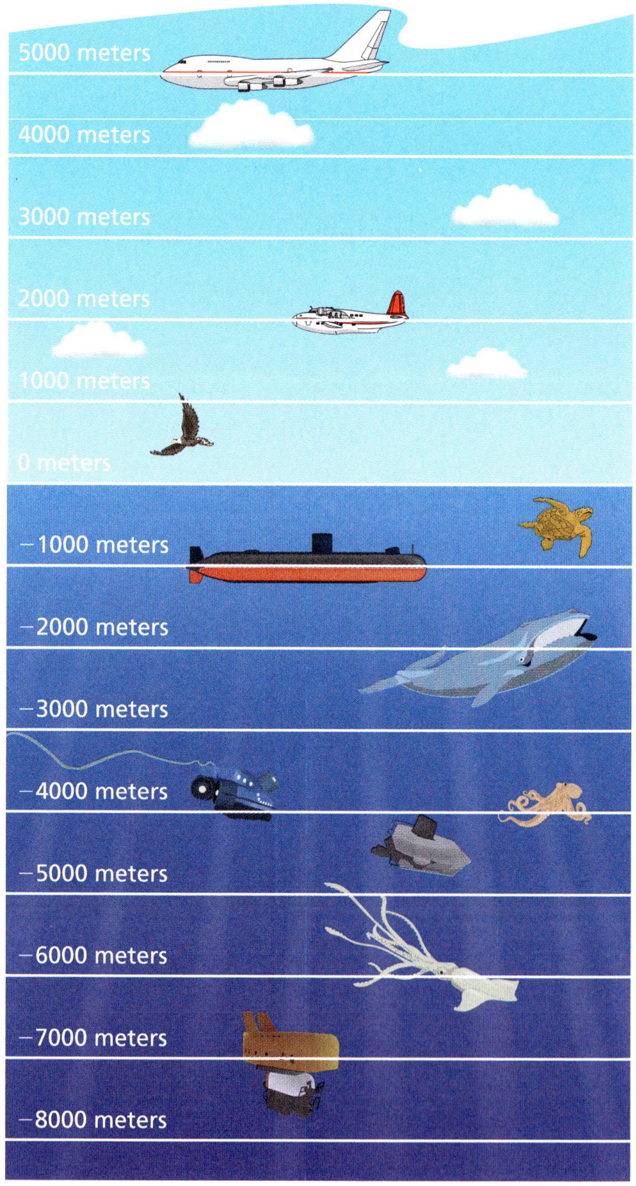

Common Core

Absolute Value

In this lesson, you will
- find the absolute value of numbers.
- use absolute value to compare numbers in real-life situations.

Learning Standards
6.NS.7c
6.NS.7d

Laurie's Notes

Introduction

Standards for Mathematical Practice

- **MP6 Attend to Precision:** Absolute value is introduced in this section. Students may incorrectly say, "Absolute value just makes the number positive." Be sure that students correctly refer to the absolute value of a number as the distance the number is from 0.

Motivate

- **FYI:** Share information with students about NASA aquanauts.
  - Aquanauts live in an undersea habitat to simulate conditions in space. The extensive training, being submerged in one of the world's most hostile environments, is designed to prepare humans and technology for the rigors of life in a vacuum. Aquanauts say living under the sea is the best way to train for space and to study the effects of isolated missions.
  - Aquanaut crews live for days or weeks at the bottom of the ocean. Aquarius, an underwater ocean laboratory, descends about 60 feet to the sea floor. The aquanauts assist scientists with research into coral health and conduct excursions that mimic spacewalks. These excursions are critical to the International Space Station.

Activity Notes

Activity 1

? "What does 500 feet below sea level mean?" The surface of the water is sea level. So, 500 feet below sea level means 500 feet below the surface of the water.

- Remind students that if distances are below sea level, the numbers will be negative.
- After students have finished, discuss their explanations.
- You could create a vertical model, using masking tape (placed on the wall) or string (attached to the ceiling), and index cards with the names of the objects written on the cards. Deciding on an appropriate scale is a good exercise for students. Have a student place the cards in approximately the correct positions and then ask general questions.

? "Which objects are closest to sea level?" bald eagle and leatherback turtle "Which objects are closest together?" leatherback turtle and U.S.S. Dolphin "As objects descend into the water, how do their elevations change?" Listen for language that suggests that the numbers are increasing, but they are negative.

Common Core State Standards

6.NS.7c Understand the absolute value of a rational number as its distance from 0 on the number line; interpret absolute value as magnitude for a positive or negative quantity in a real-world situation.

6.NS.7d Distinguish comparisons of absolute value from statements about order.

Previous Learning

Students should know how to compare and order integers.

Technology for the Teacher

Dynamic Classroom

Lesson Plans
Complete Materials List

6.4 Record and Practice Journal

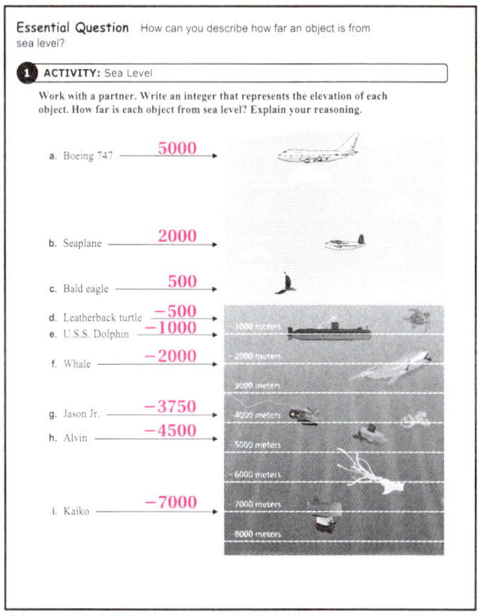

Essential Question How can you describe how far an object is from sea level?

1 ACTIVITY: Sea Level

Work with a partner. Write an integer that represents the elevation of each object. How far is each object from sea level? Explain your reasoning.

a. Boeing 747 5000
b. Seaplane 2000
c. Bald eagle 500
d. Leatherback turtle −500
e. U.S.S. Dolphin −1000
f. Whale −2000
g. Jason Jr. −3750
h. Alvin −4500
i. Kaiko −7000

English Language Learners

Vocabulary

In Activity 1, make sure that students understand that the words on the left of the figure are the names of the objects, and that they are determining the elevation of each.

6.4 Record and Practice Journal

2 ACTIVITY: Finding a Distance

Work with a partner. Use the diagram in Activity 1.

a. What integer represents sea level?

0

b. The vessel *Kaiko* ascends to the same depth as the U.S.S. *Dolphin*. About how many meters did *Kaiko* travel? Explain how you found your answer.

about 6000 meters

c. The vessel *Jason Jr.* descends to the same depth as the *Alvin*. About how many meters did *Jason Jr.* travel? Explain how you found your answer.

about 750 meters

d. **REASONING** Which pairs of objects are the same distance from sea level? How do you know?

seaplane and whale; bald eagle and leatherback turtle

e. **REASONING** An airplane is the same distance from sea level as the *Kaiko*. How far is the airplane from sea level?

7000 meters

3 ACTIVITY: Oceanography Project

Work with a partner. Use the Internet or some other resource to write a report that describes two ways in which mathematics is used in oceanography. Check students' work.

Here are two possible ideas. You can use one or both of these, or you can use other ideas.

Diving Bell

Mine Neutralization Vehicle

What Is Your Answer?

4. **IN YOUR OWN WORDS** How can you describe how far an object is from sea level?

find the distance between the number and 0 (sea level)

5. **PRECISION** In Activity 1, an object has an elevation of −7500 meters. Is −7500 greater than or less than −7000? Does this object have a depth greater than or less than 7000 meters? Explain your reasoning.

less than; greater than

Laurie's Notes

Activity 2

- Students have a vertical number line to use as a model in answering the questions about how far a vessel traveled.
- **MP3 Construct Viable Arguments and Critique the Reasoning of Others:** Students will have different ways in which they explain how far a vessel traveled. Elicit different methods from students. Knowing that their answer is correct, students are more apt to share their reasoning with classmates.
- **Extension:** You could tell students the end location and how far it moved, and then ask where the vessel started. Note, there are two answers—one above and one below the ending position.

Activity 3

? Ask questions to determine students' knowledge of oceanography. The main connection you want them to make to mathematics is the use of negative numbers to represent depths below sea level.

- Explain that oceanography is the study of Earth's oceans. Oceanographers study everything about the ocean, including marine animals, currents and waves, and the ocean floor.
- Discuss the photos shown.
- Students will need access to a library or the Internet to find information on oceanography.

What Is Your Answer?

- **Think-Pair-Share:** Students should read the question independently and then work in pairs to answer the question. When they have answered the question, the pair should compare their answer with another group and discuss any discrepancies.

Closure

- **Exit Ticket:** Fill in the blank with < or >. Explain your answer.

$$-15 \underline{\quad > \quad} -20$$

2 ACTIVITY: Finding a Distance

Work with a partner. Use the diagram in Activity 1.

a. What integer represents sea level?

b. The vessel *Kaiko* ascends to the same depth as the U.S.S. *Dolphin*. About how many meters did *Kaiko* travel? Explain how you found your answer.

c. The vessel *Jason Jr.* descends to the same depth as the *Alvin*. About how many meters did *Jason Jr.* travel? Explain how you found your answer.

d. **REASONING** Which pairs of objects are the same distance from sea level? How do you know?

e. **REASONING** An airplane is the same distance from sea level as the *Kaiko*. How far is the airplane from sea level?

3 ACTIVITY: Oceanography Project

Math Practice 5

Use Technology to Explore

How can you find more information on oceanography? What information is useful to your report?

Work with a partner. Use the Internet or some other resource to write a report that describes two ways in which mathematics is used in oceanography.

Here are two possible ideas. You can use one or both of these, or you can use other ideas.

Diving Bell

Mine Neutralization Vehicle

What Is Your Answer?

4. **IN YOUR OWN WORDS** How can you describe how far an object is from sea level?

5. **PRECISION** In Activity 1, an object has an elevation of -7500 meters. Is -7500 greater than or less than -7000? Does this object have a depth greater than or less than 7000 meters? Explain your reasoning.

Practice

Use what you learned about elevation and sea level to complete Exercises 4–6 on page 272.

6.4 Lesson

Check It Out
Lesson Tutorials
BigIdeasMath.com

Key Vocabulary
absolute value, *p. 270*

Key Idea

Absolute Value

Words The **absolute value** of a number is the distance between the number and 0 on a number line. The absolute value of a number a is written as $|a|$.

Numbers $|-2| = 2$ $|2| = 2$

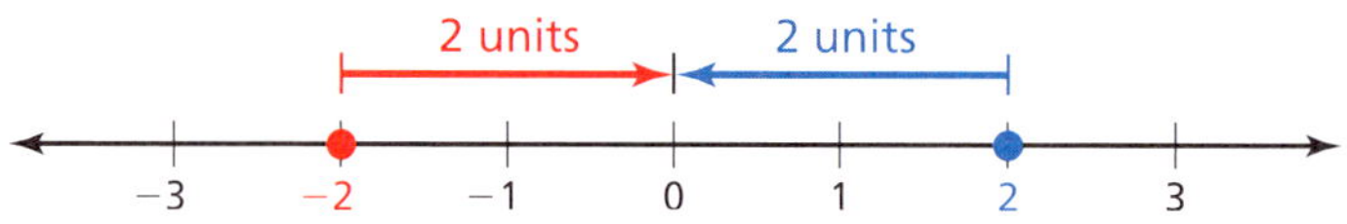

EXAMPLE 1 Finding Absolute Value

a. Find the absolute value of 3.

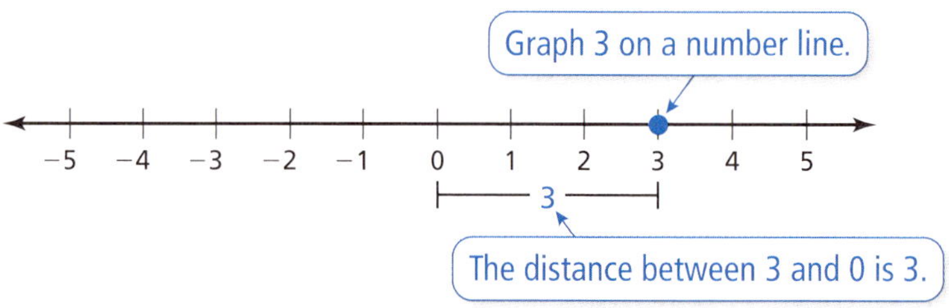

So, $|3| = 3$.

b. Find the absolute value of $-2\frac{1}{2}$.

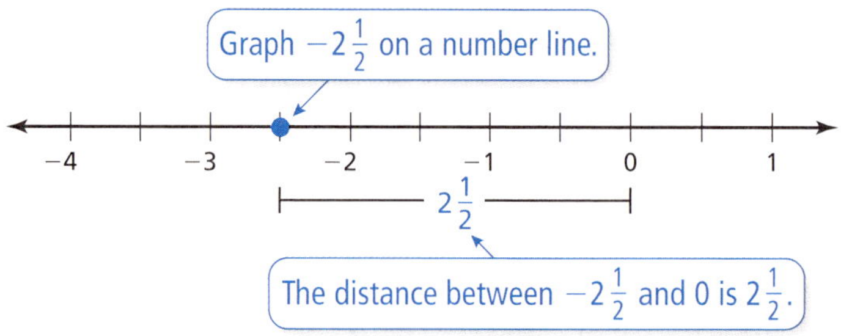

So, $\left|-2\frac{1}{2}\right| = 2\frac{1}{2}$.

On Your Own

Now You're Ready
Exercises 7–14

Find the absolute value.

1. $|8|$ **2.** $|-6|$ **3.** $|0|$

4. $\left|\frac{1}{4}\right|$ **5.** $\left|-7\frac{1}{3}\right|$ **6.** $|-12.9|$

Multi-Language Glossary at BigIdeasMath.com

Laurie's Notes

Introduction

Connect

- **Yesterday:** Students developed an understanding of absolute value. (MP3, MP6)
- **Today:** Students will find absolute values of rational numbers.

Motivate

- Have two students (A and B) stand at the front of the room holding a piece of string between them. You hold a piece of paper with the number 0 written on it. Share that the distance between the two students is 10 units.

? Position yourself at various points along the line so that A and B have various values, such as A = 3, 5, or 8 and B = −7, −5, or −2, respectively.
  - "If A is _____ [3, 5, 8], what number does B represent?" −7, −5, −2
  - "Who is closer to me (meaning 0)?" A; Neither; B
  - "How far away from me is each person?" Answers will vary depending on the numbers chosen.

? Without the string, ask students "What number or numbers are 4 units from 0?" 4 and −4

- Discuss positive and negative directions from 0 on a number line. For instance, it is possible to be 5 units from 0 in either direction.

Lesson Notes

Key Idea

- Write the Key Idea on the board.
- $|a|$ is read, "the absolute value of *a*."
- Stress that the **absolute value** of a number is the *distance* from 0. Distance is a positive number or 0. Although directed distances are addressed in science, the concept of distance here is that it is a positive number or 0.

Example 1

- Work through each example.
- **Common Misconception:** Students may say, "Absolute values are always positive." This statement can cause problems when students find $|0|$, which is 0.

? After part (b) ask, "What is the absolute value of −2.4?" 2.4

Words of Wisdom

- Point out that when you write the notation for the absolute value, it means *take the absolute value of the number inside the symbols.*

On Your Own

- **Common Error:** Students may incorrectly find that the absolute value of a negative number is negative.

Goal Today's lesson is finding the **absolute value** of a number.

Technology for the Teacher

Dynamic Classroom

Lesson Tutorials
Lesson Plans
Answer Presentation Tool

Extra Example 1

a. Find the absolute value of −10. 10

b. Find the absolute value of 8.3. 8.3

On Your Own

1. 8
2. 6
3. 0
4. $\frac{1}{4}$
5. $7\frac{1}{3}$
6. 12.9

Extra Example 2

Compare $|-10|$ and 4. $|-10| > 4$

On Your Own

7. $>$
8. $<$
9. $<$
10. $=$

Extra Example 3

The table shows the elevations of a bird and a fish. Is the bird or the fish closer to sea level? fish

Animal	Elevation (ft)
Bird	10
Fish	-9

On Your Own

11. the seagull; Because $|56| < |-65|$, the seagull is closer to 0 feet, or sea level.

Differentiated Instruction

Auditory

If students have difficulty remembering how to find the absolute value of a number, give them a mnemonic device: *I am absolutely positive that absolute value is positive.* Make sure, however, that students understand the one exception, that the absolute value of zero is zero.

Laurie's Notes

Example 2

- Explain that one way to compare numbers is to graph the numbers on a number line. The farther to the right a number is, the greater it is. The farther to the left a number is, the less it is.

? "What is $|-5|$?" 5

- **Common Error:** Students may graph $|-5|$ at -5. Explain that you find the absolute value of the number first and then graph it.
- Note that it would also be correct to conclude that $|-5| > 2$.

On Your Own

- Draw number lines on the board and ask volunteers to solve the problems by graphing.
- **MP6 Attend to Precision:** To practice re-writing inequalities, for each of the problems you could have students write two equivalent inequalities. For Problem 7, they would write $|-4| > -2$ and $-2 < |-4|$.

Example 3

- Discuss the number line. Make sure students observe that, for instance, -22 is below -20. It is not uncommon for students to graph -22 above -20 because 22 is above 20.
- Work through each part of the problem.

On Your Own

- **Think-Pair-Share:** Students should read the question independently and then work in pairs to answer the question. When they have answered the question, the pair should compare their answer with another group and discuss any discrepancies.

Closure

- **Exit Ticket:** Write these numbers in order from least to greatest.

$$|-12|,\ -8,\ -10,\ |6|,\ |-4|$$

$-10, -8, 4, 6, 12$

EXAMPLE 2 Comparing Values

Compare 2 and $|-5|$.

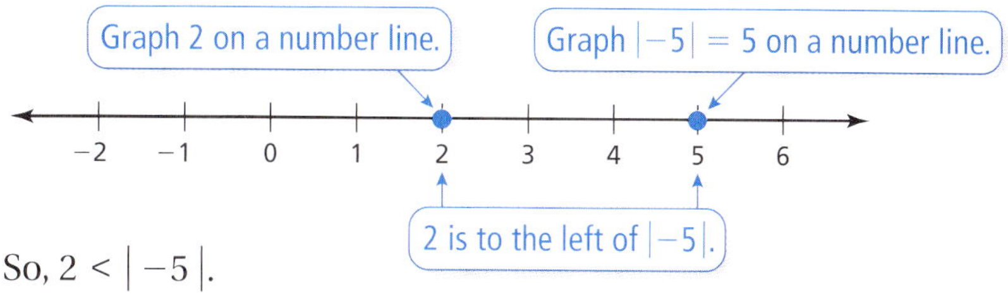

So, $2 < |-5|$.

On Your Own

Now You're Ready
Exercises 17–22

Copy and complete the statement using <, >, or =.

7. $|-4|$ ▢ -2

8. -5 ▢ $|5|$

9. $|9|$ ▢ 10

10. 3.9 ▢ $|-3.9|$

EXAMPLE 3 Real-Life Application

Animal	Elevation (ft)
Shark	−4
Sea lion	5
Seagull	56
Shrimp	−65
Turtle	22

The table shows the elevations of several animals.

a. Which animal is the deepest? Explain.

Graph each elevation.

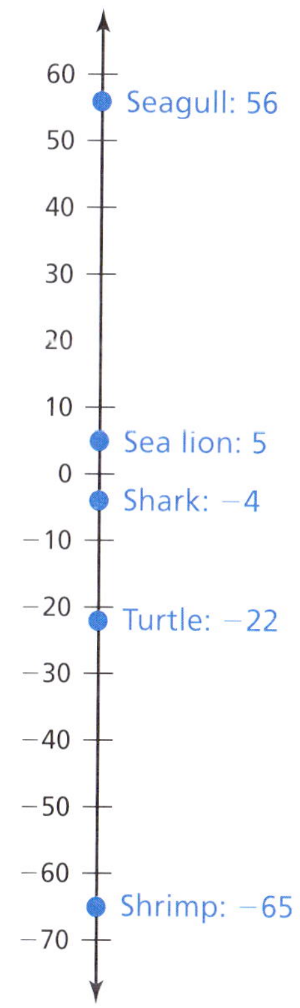

The lowest elevation represents the animal that is the deepest. The integer that is lowest on the number line is −65.

So, the shrimp is the deepest.

b. Is the shark or the sea lion closer to sea level?

Because sea level is at 0 feet, use absolute values.

Shark: $|-4| = 4$ **Sea lion:** $|5| = 5$

Because 4 is less than 5, the shark is closer to sea level than the sea lion.

On Your Own

11. Is the seagull or the shrimp closer to sea level? Explain your reasoning.

6.4 Exercises

Check It Out
Help with Homework
BigIdeasMath.com

Vocabulary and Concept Check

1. **VOCABULARY** Explain how to find the absolute value of an integer.
2. **REASONING** Which integer is greater, -50 or 25? Which has the greater absolute value? Explain.
3. **DIFFERENT WORDS, SAME QUESTION** Which is different? Find "both" answers.

How far is -3 from 0?	What integer is 3 units to the left of 0?
What is the absolute value of -3?	What is the distance between -3 and 0?

Practice and Problem Solving

Use a vertical number line to graph the location of each object. Then tell which object is farther from sea level.

4. Scuba diver: -15 m
   Dolphin: -22 m
5. Seagull: 12 m
   School of fish: -4 m
6. Shark: -40 m
   Flag on a ship: 32 m

Find the absolute value.

1 7. $|-2|$
8. $|23|$
9. $|-8.35|$
10. $\left|\frac{1}{6}\right|$
11. $\left|-3\frac{2}{5}\right|$
12. $|11|$
13. $|14.06|$
14. $|-68|$

15. **REASONING** Write two integers that have an absolute value of 10.

16. **ERROR ANALYSIS** Describe and correct the error in finding the absolute value.

✗ $|14| = -14$

Copy and complete the statement using <, >, or =.

2 17. $6 \;\square\; |-8|$
18. $|-3| \;\square\; 3$
19. $|-5.5| \;\square\; |-3.1|$
20. $\frac{3}{4} \;\square\; \left|-\frac{2}{5}\right|$
21. $|-6.8| \;\square\; |8.25|$
22. $-12 \;\square\; |12|$

23. **CAVES** Three scientists explore a cave. Which scientist is farthest underground?

Scientist A: -48 ft   Scientist B: -62 ft   Scientist C: -53 ft

MATCHING Match the account balance with the debt that it represents. Explain your reasoning.

24. account balance $= -\$25$
25. account balance $< -\$25$
26. account balance $> -\$25$

A. debt $> \$25$
B. debt $= \$25$
C. debt $< \$25$

Assignment Guide and Homework Check

Level	Day 1 Activity Assignment	Day 2 Lesson Assignment	Homework Check
Basic	4–6, 40–44	1–3, 7–15 odd, 16, 17–23 odd	11, 13, 19, 23
Average	4–6, 40–44	1–3, 8–14 even, 15, 16–22 even, 23–31 odd	14, 20, 23, 27
Advanced	4–6, 40–44	1–3, 12–22 even, 26–38 even, 39	14, 20, 28, 32, 34

For Your Information

- **Exercise 34** The Kelvin temperature scale does not use a degree symbol.
- **Exercise 38** Spaces and punctuation can be adjusted in a palindrome.

Common Errors

- **Exercises 7–14, 37** Students may think the absolute value of a number is its opposite. For example, they may think $|8| = -8$. Use a number line to remind students that absolute value is a number's distance from zero, so it is always a positive number or zero.
- **Exercises 17–22, 27–30** Students may ignore the absolute value bars when comparing or ordering values. Encourage them to graph the values on a number line, as shown in Example 3, to help them correctly compare or order values.
- **Exercise 33** Students may treat the absolute value bars as parentheses and write $-|-1| = 1$. Remind them of the definition of absolute value, and that they should find the absolute value first.

6.4 Record and Practice Journal

Find the absolute value.

1. $|-5|$
   5
2. $|7|$
   7
3. $|0|$
   0
4. $|-31|$
   31

Complete the statement using <, >, or =.

5. 7 **>** $|-4|$
6. -9 **<** $|-10|$
7. $|8|$ **=** $|-8|$

Order the values from least to greatest.

8. $0, -5, |-6|, |-2|, 4$
   $-5, 0, |-2|, 4, |-6|$
9. $|-12|, -21, |25|, |-31|, -14, 33$
   $-21, -14, |-12|, |25|, |-31|, 33$
10. You go to a store in a mall that is on the fourth floor above the main level. Your friend goes to a store that is two floors below the main level.
    a. Write an integer for the position of each person relative to the main level.
    4; −2
    b. Find the absolute value of each integer.
    4; 2
    c. Who is farther from the main level? Explain.
    you

Vocabulary and Concept Check

1. Find the distance between the number and 0 on a number line.
2. 25; −50; 25 is to the right of −50 on a number line, so $25 > -50$. $|-50|$ is farther from 0 on a number line than $|25|$, so $|-50|$ has the greater absolute value.
3. What integer is 3 units to the left of 0?; −3; 3

Practice and Problem Solving

4–6. See Additional Answers.

7. 2
8. 23
9. 8.35
10. $\frac{1}{6}$
11. $3\frac{2}{5}$
12. 11
13. 14.06
14. 68
15. −10, 10
16. The absolute value of a number cannot be negative. $|14| = 14$
17. $<$
18. $=$
19. $>$
20. $>$
21. $<$
22. $<$
23. Scientist B
24. B; You owe \$25, so debt = \$25.
25. A; You owe more than \$25, so debt > \$25.
26. C; You owe less than \$25, so debt < \$25.
27. $-2, 0, |-1|, |4|, 5$
28. $-4, -3, |-3|, |-4|, |5|$
29. $-11, 0, |3|, |-6|, 9, 10$
30. $-20, -19, -18, |-18|, |-22|, |30|$

Practice and Problem Solving

31. 0 **32.** −6

33. −1

34. See *Taking Math Deeper*.

35. sometimes; If the number is negative then its absolute value is greater, but if the number is positive or zero then it is equal to its absolute value.

36. always; The absolute value is the positive distance from zero on a number line.

37. never; The absolute value of a positive number is the number itself.

38. **a.**

b. racecar; yes

c. *Sample answer:* Madam, I'm Adam.

39. *Sample answer:* $x = -2$, $y = -3$

Fair Game Review

40.

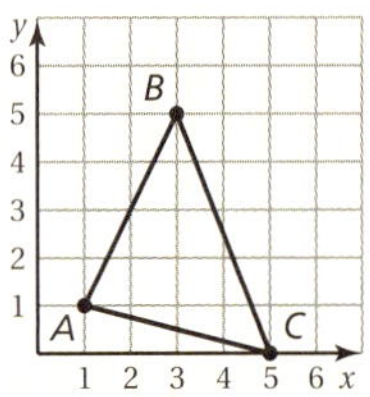

41–44. See Additional Answers.

Mini-Assessment

Find the absolute value.

1. $|-83|$ 83 **2.** $|38|$ 38

Copy and complete the statement using <, >, or =.

3. -2 ▢ $|-2|$ < **4.** 8 ▢ $|-8|$ =

5. The freezing point of airplane fuel is −53°C and the freezing point of candle wax is 55°C. Which is closer to the freezing point of water, 0°C?
the freezing point of airplane fuel

Taking Math Deeper

Exercise 34

1 The Kelvin temperature scale and the Kelvin are named after the British physicist William Thomson, Lord Kelvin (1824–1907). The Kelvin scale is a temperature scale referenced to absolute zero, the absence of all energy. Scientists believe that it is not possible to reach absolute zero. In 2003, researchers at the Massachusetts Institute of Technology produced a gas that was within one-billionth of a degree of absolute zero.

Note that the Kelvin is not referred to as a "degree," and it is not written with a degree symbol.

2 **a.** Use the graph in the exercise to determine that 32°F = 0°C and 0 K = −273°C.

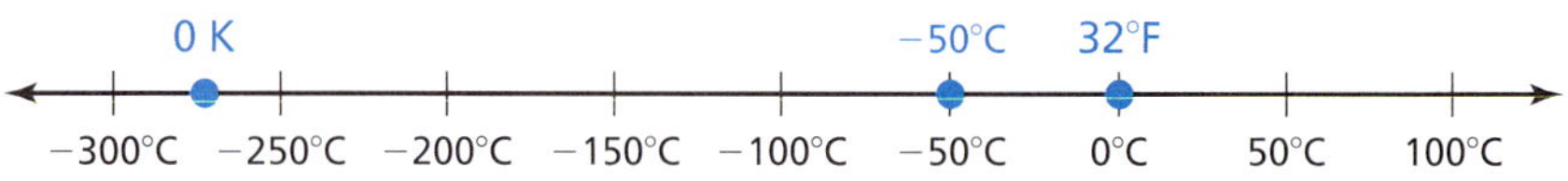

−50°C is closer to 0 K than 32°F.

3 **b.** One thing that the Kelvin scale and absolute values have in common is that neither can have *negative* numbers. For both, the least possible number is 0.

In fact, the Kelvin temperature scale is often called the *absolute temperature scale*.

Reteaching and Enrichment Strategies

If students need help. . .	If students got it. . .
Resources by Chapter • Practice A and Practice B • Puzzle Time Record and Practice Journal Practice Differentiating the Lesson Lesson Tutorials Skills Review Handbook	Resources by Chapter • Enrichment and Extension • Technology Connection Start the next section

Order the values from least to greatest.

27. $5,\ 0,\ |-1|,\ |4|,\ -2$

28. $|-3|,\ |5|,\ -3,\ -4,\ |-4|$

29. $10,\ |-6|,\ 9,\ |3|,\ -11,\ 0$

30. $-18,\ |30|,\ -19,\ |-22|,\ -20,\ |-18|$

Simplify the expression.

31. $|0|$

32. $-|6|$

33. $-|-1|$

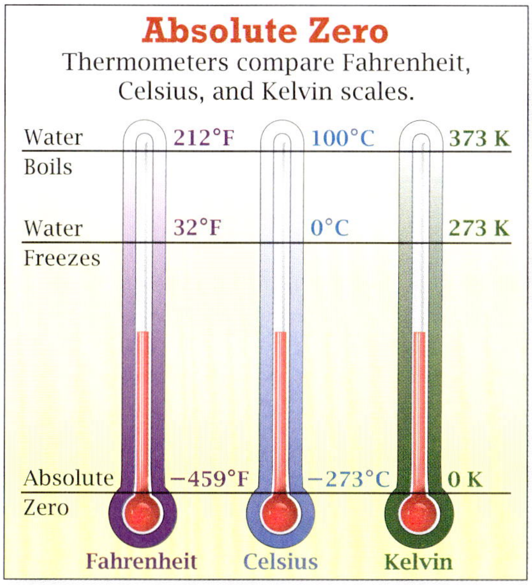

34. **ABSOLUTE ZERO** The coldest possible temperature is called *absolute zero*. It is represented by 0 K on the Kelvin temperature scale.

a. Which temperature is closer to 0 K: 32°F or −50°C?

b. What do absolute values and temperatures on the Kelvin scale have in common?

Tell whether the statement is *always, sometimes,* or *never* true. Explain.

35. The absolute value of a number is greater than the number.

36. The absolute value of a negative number is positive.

37. The absolute value of a positive number is its opposite.

38. **PALINDROME** A *palindrome* is a word or sentence that reads the same forward as it does backward.

a. Graph and label the following points on a number line: $A = -2$, $C = -1$, $E = 0$, $R = -3$. Then graph and label the absolute value of each point on the *same* number line.

b. What word do the letters spell? Is this a palindrome?

c. Make up your own palindrome.

39. Critical Thinking Find values of x and y so that $|x| < |y|$ and $x > y$.

Fair Game Review What you learned in previous grades & lessons

Draw the polygon with the given vertices in a coordinate plane. *(Section 4.4)*

40. $A(1, 1)$, $B(3, 5)$, $C(5, 0)$

41. $D(0, 6)$, $E(2, 1)$, $F(6, 3)$

42. $P(2, 1)$, $Q(4, 4)$, $R(8, 4)$, $S(6, 1)$

43. $W(1, 6)$, $X(9, 6)$, $Y(9, 1)$, $Z(4, 1)$

44. **MULTIPLE CHOICE** Which expression represents "6 less than the product of 4 and a number x"? *(Section 3.2)*

Ⓐ $(6 - 4)x$ Ⓑ $6 - 4x$ Ⓒ $\frac{6}{4x}$ Ⓓ $4x - 6$

6.5 The Coordinate Plane

Essential Question How can you graph and locate points that contain negative numbers in a coordinate plane?

You have already graphed points and polygons in one part of the coordinate plane. In Activity 1, you will form the *entire* coordinate plane.

1 ACTIVITY: Forming the Entire Coordinate Plane

Work with a partner.

a. In the middle of a sheet of grid paper, construct a horizontal number line as shown. Label the tick marks. On a different sheet of grid paper, construct and label a similar vertical number line.

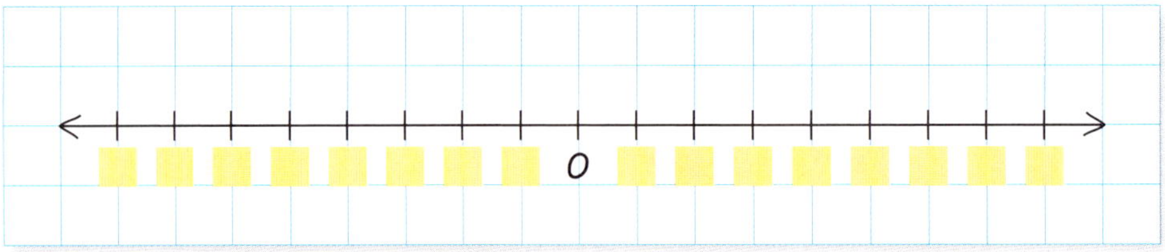

b. Cut out the vertical number line and tape it on top of the horizontal number line so that the zeros overlap. Make sure the number lines are perpendicular to one another. How many regions did you form by doing this?

c. **REASONING** What ordered pair represents the point where the number lines intersect? Why do you think this point is called the *origin*? Explain.

2 ACTIVITY: Describing Points in the Coordinate Plane

Work with a partner. Use your perpendicular number lines from Activity 1.

a. Plot and label (3, 2) on your coordinate plane. Shade this region in your coordinate plane. What do you notice about the integers along the number lines that surround (3, 2)?

b. Can you plot a point in your coordinate plane so that it is surrounded by negative numbers on the axes? If so, where is this point? Use a different color to shade this region in your coordinate plane.

c. What do you notice about the integers along the number lines for points in the regions that are not shaded?

d. **STRUCTURE** Describe how you would plot $(-3, -2)$. How is plotting this point similar to plotting (3, 2)? Plot $(-3, -2)$ in your coordinate plane.

e. **REASONING** Where in your coordinate plane do you plot $(2, -4)$? Where do you plot $(-2, 4)$? Explain your reasoning.

COMMON CORE

Coordinate Plane

In this lesson, you will

- describe the locations of points in the coordinate plane.
- plot points in the coordinate plane given ordered pairs.
- find distances between points in the coordinate plane.

Learning Standards
6.NS.6b
6.NS.6c
6.NS.8

Laurie's Notes

Introduction

Standards for Mathematical Practice

- **MP7 Look for and Make Use of Structure:** Mathematically proficient students make use of the patterns when plotting ordered pairs. Students recognize the relationship between plotting (a, b) and $(-a, -b)$.

Motivate

- **Kinesthetic Class Activity:** Create a coordinate grid on your classroom floor or school foyer. Use two strips of masking tape laid perpendicular to one another. Use a thick marker to mark the axes, leaving 1–1.5 feet between each label. Write each of the following points on a piece of paper large enough for the class to see: (3, 2), (−3, 2), (−3, −2), (3, −2)
- Ask four students to volunteer. Have each student start at the origin and face towards the positive y-axis. Give the following directions:
  - **Student 1:** Move 3 spaces to the right and forward 2 spaces.
  - **Student 2:** Move 3 spaces to the left and forward 2 spaces.
  - **Student 3:** Move 3 spaces to the left and back 2 spaces.
  - **Student 4:** Move 3 spaces to the right and back 2 spaces.

? "What figure would be formed if you connected each student in order with a piece of string?" a rectangle

Discuss

- Students should recall plotting ordered pairs in the first quadrant. This lesson expands the coordinate plane to include negative coordinates.
- Work with horizontal and vertical number lines should help students extend their understanding to plotting points in all four quadrants.

Activity Notes

Activity 1

- Circulate to be sure that the number lines are labeled correctly. In particular, check for symmetry of opposites about 0.
- Students can use a protractor or the corner of a piece of paper to check that their number lines are perpendicular.

? "Describe the coordinate grid." The number lines represent the axes of the coordinate plane: the horizontal number is the x-axis and the vertical number line is the y-axis.

Activity 2

- The goal of this activity is for students to recognize the sign of the coordinates in each quadrant, referred to as regions in this activity.

? "What do all of the coordinates of the ordered pairs have in common for the shaded region in part (a)?" all positive "in part (b)?" all negative

- **MP7:** In part (d), students should recognize that plotting (−3, −2) is similar to plotting (3, 2) except the directions are opposite. Instead of 3 units right it is 3 units left. Instead of 2 units up it is 2 units down.

Common Core State Standards

6.NS.6b Understand signs of numbers in ordered pairs as indicating locations in quadrants of the coordinate plane; recognize that when two ordered pairs differ only by signs, the locations of the points are related by reflections across one or both axes.

6.NS.6c Find and position . . . pairs of integers and other rational numbers on a coordinate plane.

6.NS.8 Solve real-world and mathematical problems by graphing points in all four quadrants of the coordinate plane. Include use of coordinates and absolute value to find distances between points with the same first coordinate or the same second coordinate.

Previous Learning

Students should be able to graph integers on a number line.

Technology for the Teacher

Dynamic Classroom

Lesson Plans

Complete Materials List

6.5 Record and Practice Journal

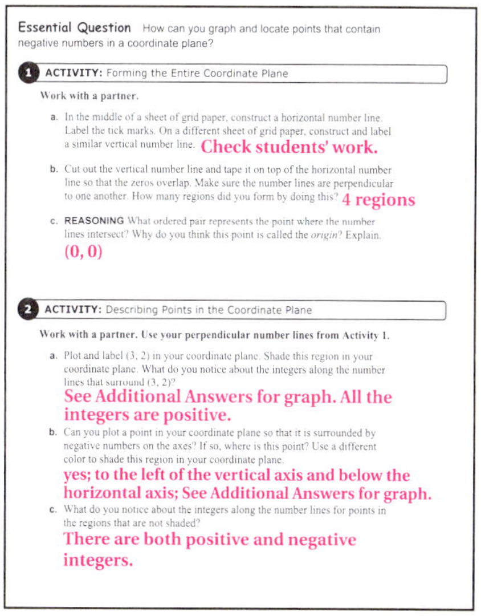

Essential Question How can you graph and locate points that contain negative numbers in a coordinate plane?

1 ACTIVITY: Forming the Entire Coordinate Plane

Work with a partner.

a. In the middle of a sheet of grid paper, construct a horizontal number line. Label the tick marks. On a different sheet of grid paper, construct and label a similar vertical number line. **Check students' work.**

b. Cut out the vertical number line and tape it on top of the horizontal number line so that the zeros overlap. Make sure the number lines are perpendicular to one another. How many regions did you form by doing this? **4 regions**

c. **REASONING** What ordered pair represents the point where the number lines intersect? Why do you think this point is called the *origin*? Explain.
(0, 0)

2 ACTIVITY: Describing Points in the Coordinate Plane

Work with a partner. Use your perpendicular number lines from Activity 1.

a. Plot and label (3, 2) in your coordinate plane. Shade this region in your coordinate plane. What do you notice about the integers along the number lines that surround (3, 2)?
See Additional Answers for graph. All the integers are positive.

b. Can you plot a point in your coordinate plane so that it is surrounded by negative numbers on the axes? If so, where is this point? Use a different color to shade this region in your coordinate plane.
yes; to the left of the vertical axis and below the horizontal axis; See Additional Answers for graph.

c. What do you notice about the integers along the number lines for points in the regions that are not shaded?
There are both positive and negative integers.

Differentiated Instruction

Kinesthetic

For kinesthetic learners who have difficulty plotting points in the coordinate plane, suggest they use a finger for tracing. Have students place a finger at the origin and trace left or right along the *x*-axis to the first coordinate, then trace up or down to the second coordinate. Students should also practice writing the coordinates of a plotted point. Guide students with questions such as, "Should you move left or right? How far? Should you move up or down? How far?"

6.5 Record and Practice Journal

d. **STRUCTURE** Describe how you would plot (−3, −2). How is plotting this point similar to plotting (3, 2)? Plot (−3, −2) in your coordinate plane.

Move 3 units to the left of the origin, and 2 units down; These movements are opposite of plotting (3, 2). See Additional Answers for graph.

e. **REASONING** Where in your coordinate plane do you plot (2, −4)? Where do you plot (−2, 4)? Explain your reasoning.

See Additional Answers.

3 **ACTIVITY:** Plotting Points in a Coordinate Plane

Work with a partner. Plot and connect the points to make a picture. Describe and color the picture when you are done.

1(6, 9) **2**(4, 11) **3**(2, 12) **4**(0, 11) **5**(−2, 9)
6(−6, 2) **7**(−9, 1) **8**(−11, −3) **9**(−7, 0) **10**(−5, −1)
11(−5, −5) **12**(−4, −8) **13**(−6, −10) **14**(−3, −9) **15**(−3, −10)
16(−4, −11) **17**(−4, −12)
18(−3, −11) **19**(−2, −12)
20(−2, −11) **21**(−1, −12)
22(−1, −11) **23**(−2, −10)
24(−2, −9) **25**(1, −9)
26(2, −8) **27**(2, −10)
28(1, −11) **29**(1, −12)
30(2, −11) **31**(3, −12)
32(3, −11) **33**(4, −12)
34(4, −11) **35**(3, −10)
36(3, −8) **37**(4, −6)
38(6, 0) **39**(9, −3)
40(9, −1) **41**(8, 1)
42(5, 3) **43**(3, 6)
44(3, 7) **45**(4, 8)

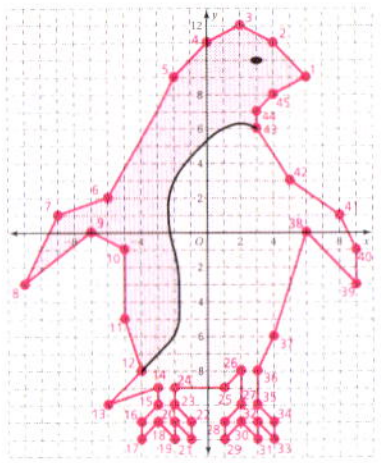

What Is Your Answer?

4. **IN YOUR OWN WORDS** How can you graph and locate points that contain negative numbers in a coordinate plane?

The first coordinate tells you how far to travel from the origin along the horizontal axis and in what direction. The second coordinate tells you how far to travel along the vertical axis and in what direction.

5. Make up your own "dot-to-dot" picture. Use at least 20 points. Your picture should have at least two points in each quadrant. **Check students' work.**

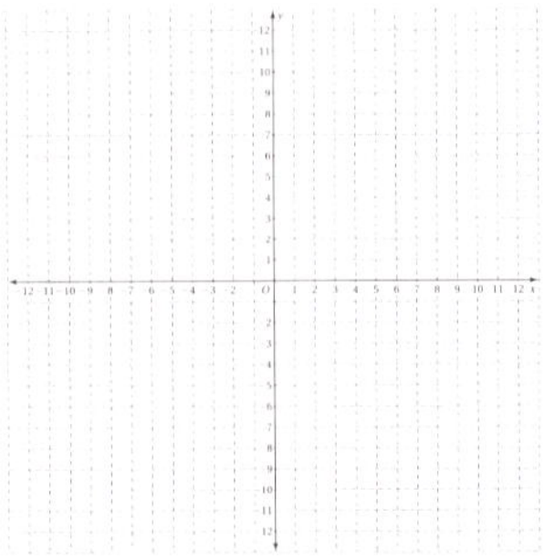

Laurie's Notes

Activity 3

- Stress that the order in which you plot the points is important.
  - The *x*-coordinate is always first. It tells us how far to go horizontally, and in which direction.
  - The *y*-coordinate is always second. It tells us how far to go vertically, and in which direction.
- Leave a summary on the board for students to reference as they work through the example.
- Students should work with partners and check their work.

Words of Wisdom

- Discuss difficulties students had in completing their pictures. Common problems/errors include carelessness, interchanging *x* and *y*, and moving in the opposite direction.
- Another error may occur when students are plotting a point. They may start at the last plotted point, rather than starting at the origin.

What Is Your Answer?

- Students should understand that each point in the coordinate plane has a unique name. The *x*- and *y*-coordinates specify an exact spot, similar to latitude and longitude on maps. Students who finish early could start a dot-to-dot design of their own.

? **Extension:** Think of some real-life situations that use coordinates to locate points. *Sample answers:* maps often use letters and numbers; stadiums give section-row-seat numbers, games (like Battleship) use numbers and letters; some maps use longitude and latitude

Closure

- "Today you learned about plotting points. You did an activity and summary that should help you remember and reinforce this concept."

3 ACTIVITY: Plotting Points in a Coordinate Plane

Math Practice 1

Check Progress

How can you check your progress to make sure you are accurately drawing the picture?

Work with a partner. Plot and connect the points to make a picture. Describe and color the picture when you are done.

1(6, 9)	2(4, 11)	3(2, 12)	4(0, 11)	5(−2, 9)
6(−6, 2)	7(−9, 1)	8(−11, −3)	9(−7, 0)	10(−5, −1)
11(−5, −5)	12(−4, −8)	13(−6, −10)	14(−3, −9)	15(−3, −10)
16(−4, −11)	17(−4, −12)	18(−3, −11)	19(−2, −12)	20(−2, −11)
21(−1, −12)	22(−1, −11)	23(−2, −10)	24(−2, −9)	25(1, −9)
26(2, −8)	27(2, −10)	28(1, −11)	29(1, −12)	30(2, −11)
31(3, −12)	32(3, −11)	33(4, −12)	34(4, −11)	35(3, −10)
36(3, −8)	37(4, −6)	38(6, 0)	39(9, −3)	40(9, −1)
41(8, 1)	42(5, 3)	43(3, 6)	44(3, 7)	45(4, 8)

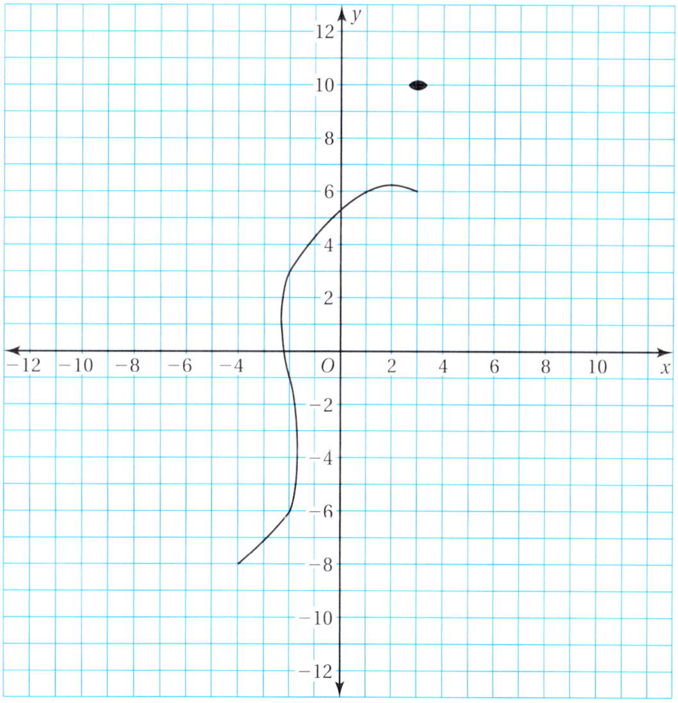

What Is Your Answer?

4. **IN YOUR OWN WORDS** How can you graph and locate points that contain negative numbers in a coordinate plane?

5. Make up your own "dot-to-dot" picture. Use at least 20 points. Your picture should have at least two points in each region of the coordinate plane.

Practice Use what you learned about the coordinate plane to complete Exercise 4 on page 279.

6.5 Lesson

Check It Out
Lesson Tutorials
BigIdeasMath.com

Key Vocabulary
coordinate plane, *p. 276*
origin, *p. 276*
quadrants, *p. 276*

Previously, you plotted points with positive coordinates. Now you will plot points with positive and negative coordinates.

Key Idea

The Coordinate Plane

A **coordinate plane** is formed by the intersection of a horizontal number line and a vertical number line. The number lines intersect at the **origin** and separate the coordinate plane into four regions called **quadrants**.

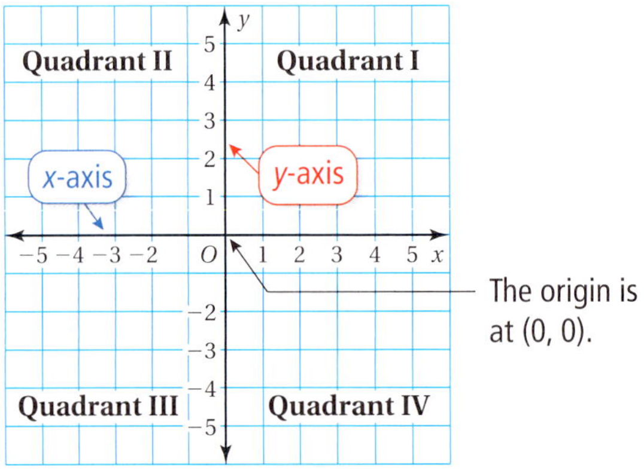

An *ordered pair* is used to locate a point in a coordinate plane.

ordered pair: $(4, -2)$

x-coordinate: 4   *y*-coordinate: -2

EXAMPLE 1 Identifying an Ordered Pair

Which ordered pair corresponds to point *T*?

Ⓐ $(-3, -3)$ Ⓑ $(-3, 3)$

Ⓒ $(3, -3)$ Ⓓ $(3, 3)$

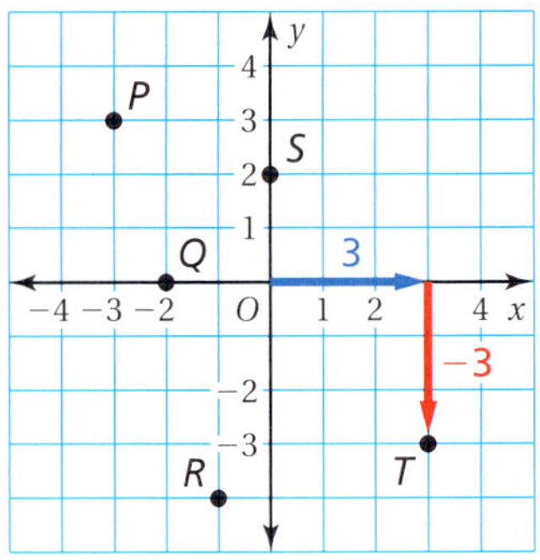

Point T is 3 units to the right of the origin and 3 units down. So, the x-coordinate is 3 and the y-coordinate is -3.

∴ The ordered pair $(3, -3)$ corresponds to point T. The correct answer is Ⓒ.

On Your Own

Now You're Ready
Exercises 5–14

Use the graph in Example 1 to write an ordered pair corresponding to the point.

1. Point P **2.** Point Q **3.** Point R **4.** Point S

Multi-Language Glossary at BigIdeasMath.com

Laurie's Notes

Introduction

Connect

- **Yesterday:** Students should be comfortable with plotting ordered pairs from the dot-to-dot activity they completed yesterday. (MP7)
- **Today:** The vocabulary and concepts are presented more formally.

Motivate

- Share a fictitious story with your students. "Last night when I was out shopping I got hungry, so I decided to get a snack from the vending machine. I decided that I was going to get crackers. To select them, I had to enter the row (a letter from A to E) and a column (a number from 1 to 6). I entered B-4 instead of C-4 and ended up with popcorn instead of crackers. But, I was hungry, so I still ate it!"
- ? "How is the vending machine like what you did yesterday?" Students should see that the crackers were in a particular spot associated with the ordered pair (C, 4). When a different ordered pair is entered, you plot a different point. (You get an unexpected item, the popcorn.)

Lesson Notes

Key Idea

- It is important to have a model of the coordinate *grid* versus only a model of scaled axes. The grid is essential in helping the students understand that a point plotted by moving in two directions (horizontal and vertical). You may want to project a coordinate grid, if possible, on the wall or board.
- Use the model of the coordinate grid to identify key vocabulary: coordinate plane, origin, quadrants, *x*-axis, *y*-axis, and ordered pair.
- **Connections:** To help students remember which way is horizontal, hold your arms out horizontally and relate them to the horizon. Origin means "where something starts." If your students have played the game of *four square*, relate the quadrants to the four square court.
- Stress that the ordered pairs (2, 4) and (4, 2) are not the same. The order matters and the ordered pairs are always (*x*, *y*). If it helps, tell students that the order of the coordinates is in alphabetical order, *x* and then *y*.

Example 1

- The colored arrows on the diagram will help students plot in the *x*-direction first, followed by the *y*-direction.

On Your Own

- Ask if there are any questions. Then have students work independently to write the ordered pairs for the remaining points, *P*, *Q*, *R*, and *S*. Ask volunteers to present their answers to the class.

Goal Today's lesson is identifying and plotting ordered pairs in a **coordinate plane**.

Technology for the Teacher

Dynamic Classroom

Lesson Tutorials
Lesson Plans
Answer Presentation Tool

Extra Example 1

Write the ordered pair represented by point *A*, point *B*, point *C*, and point *D*.

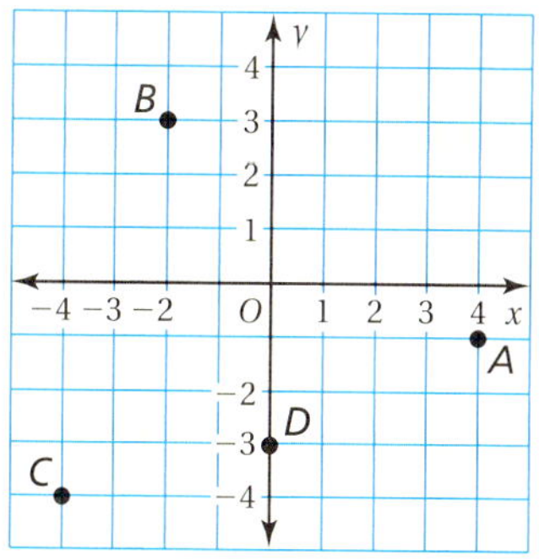

A(4, −1), *B*(−2, 3), *C*(−4, −4), *D*(0, −3)

On Your Own

1. (−3, 3) **2.** (−2, 0)

3. (−1, −4) **4.** (0, 2)

Laurie's Notes

Extra Example 2

Plot $(2, -1.5)$ in a coordinate plane. Describe the location of the point.

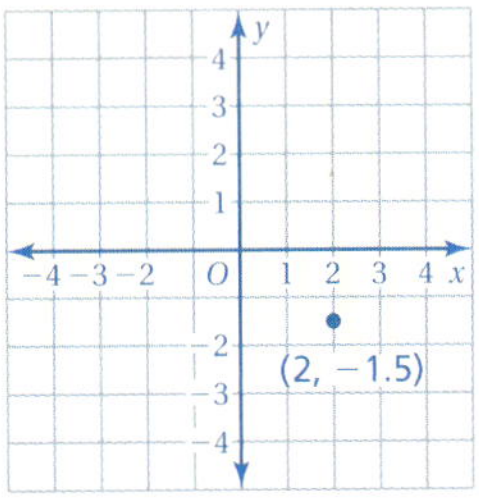

The point is in Quadrant IV.

On Your Own

5–8.

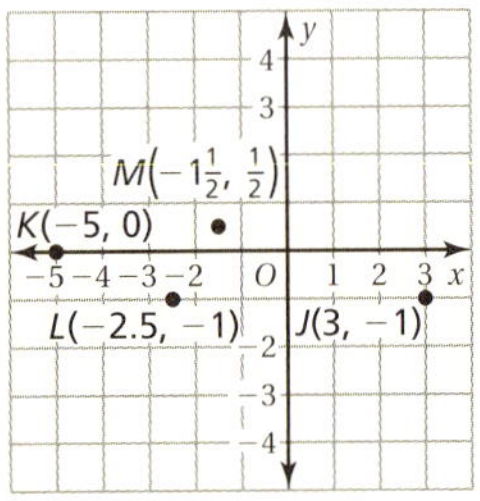

5. Quadrant IV

6. x-axis

7. Quadrant III

8. Quadrant II

Extra Example 3

An interior designer maps out a room design using a coordinate plane in which each unit represents 1 yard. The corners of the room are located at $(-3, 2)$, $(5, 2)$, $(5, -3)$, and $(-3, -3)$. What are the dimensions of the room?
8 yards long and 5 yards wide

On Your Own

9. 1 meter

English Language Learners

Visual

English learners may confuse the words *horizontal* and *vertical*. To help students distinguish between the words, draw a picture of the ocean with a ship on the horizon. Connect the word horizontal to the word horizon.

Example 2

? Ask questions about how to position the point in part (a):

- "What is the x-coordinate?" -2 "Which direction do you move from (0, 0)?" Left 2 units
- "What is the y-coordinate?" 3 "Which direction do you move from (0, 0)?" Up 3 units

- **Common Error:** Students often plot points on the axes incorrectly. Make sure the plotted point in part (b) is $(0, -3.5)$
- Explain to students that ordered pairs are in one of the four quadrants, on an axis, or at the origin. The ordered pair in part (b) is on an axis, not in a quadrant.

On Your Own

- **Common Error:** Students may move the wrong direction, or mix up the x-and y-coordinates and plot the wrong point.

Example 3

- **Connection:** Point out that the absolute values for the length come from the x-coordinates of the points. The absolute values for the width come from the y-coordinates of the points. The absolute values represent the distance from 0 for that coordinate, i.e. the distance from the axis.
- Note that you can also use the points $(-4, 1)$ and $(2, 1)$ to find the length and the points $(-4, 5)$ and $(-4, 1)$ to find the width.
- Explain why you add the absolute values to find the length (points are in different quadrants), but subtract to find the width (points are in the same quadrant).
- Explain to students that even though two of the ordered pairs are in Quadrant II where x-coordinates are negative, the dimensions of the rectangle are positive.

On Your Own

- **Neighbor Check:** Have students work independently and then have their neighbors check their work. Have students discuss any discrepancies.

EXAMPLE 2 Plotting Ordered Pairs

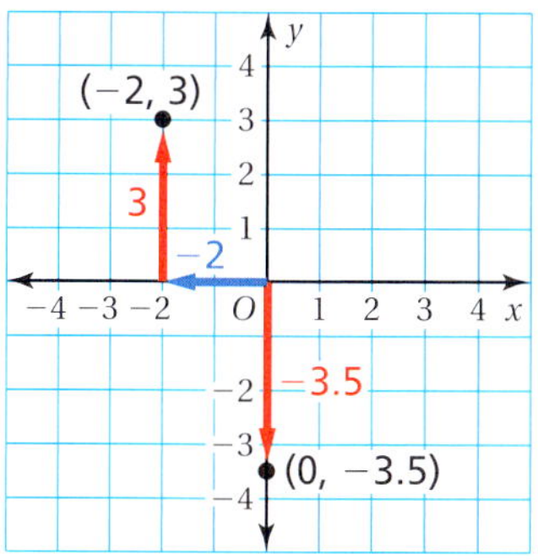

Plot (a) (−2, 3) and (b) (0, −3.5) in a coordinate plane. Describe the location of each point.

a. Start at the origin. Move 2 units left and 3 units up. Then plot the point.

∴ The point is in Quadrant II.

b. Start at the origin. Move 3.5 units down. Then plot the point.

∴ The point is on the y-axis.

On Your Own

Now You're Ready
Exercises 15–22

Plot the ordered pair in a coordinate plane. Describe the location of the point.

5. (3, −1) **6.** (−5, 0) **7.** (−2.5, −1) **8.** $\left(-1\frac{1}{2}, \frac{1}{2}\right)$

EXAMPLE 3 Finding Distances in the Coordinate Plane

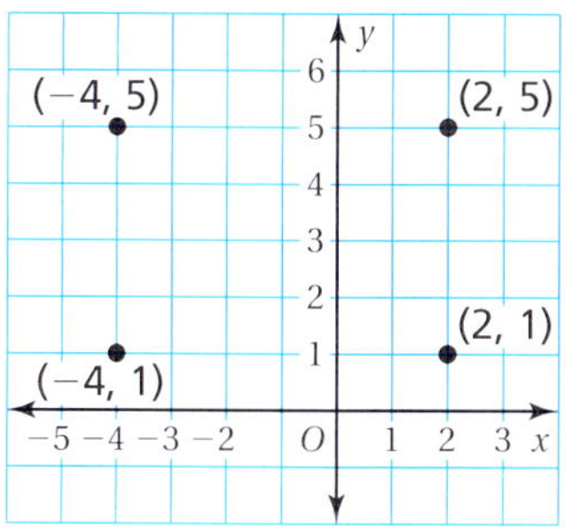

An *archaeologist* divides an area using a coordinate plane in which each unit represents 1 meter. The corners of a secret chamber are shown in the graph. What are the dimensions of the secret chamber?

The length of the chamber is the distance between (−4, 5) and (2, 5).
The width of the chamber is the distance between (2, 5) and (2, 1).

You can use absolute values to find the distances between the points.

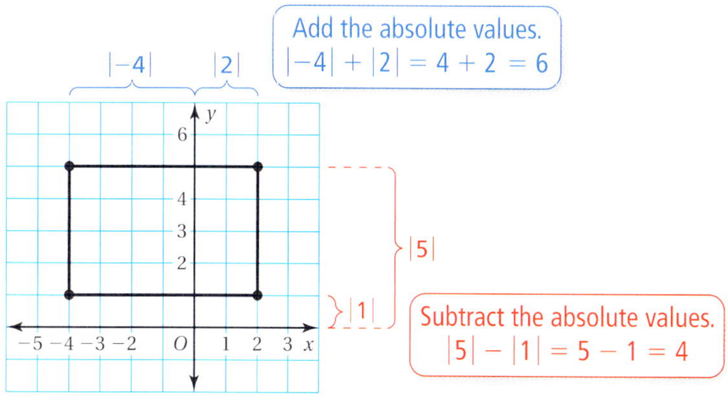

Reading

An **archaeologist** studies ancient ruins and objects to learn about people and cultures.

∴ The secret chamber is 6 meters long and 4 meters wide.

On Your Own

Now You're Ready
Exercises 25–30

9. In Example 3, the archaeologist finds a gold coin at (−1, 4), a silver coin at (−4, 2), and pottery at (−4, 4). How much closer is the pottery to the silver coin than to the gold coin?

You can use line graphs to display data that is collected over a period of time. Graphing and connecting the ordered pairs can show patterns or trends in the data. This type of line graph is also called a *time series graph*.

EXAMPLE 4 Real-Life Application

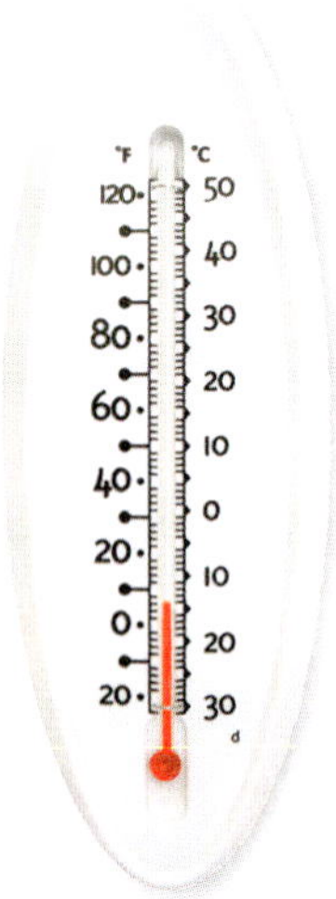

A blizzard hits a town at midnight. The table shows the hourly temperatures from midnight to 8:00 A.M.

Hours after Midnight, x	0	1	2	3	4	5	6	7	8
Temperature, y	7°F	5°F	3°F	0°F	−1°F	−4°F	−5°F	−2°F	2°F

a. Display the data in a line graph.

Write the ordered pairs.

(0, 7) (1, 5) (2, 3)

(3, 0) (4, −1) (5, −4)

(6, −5) (7, −2) (8, 2)

Plot and label the ordered pairs. Then connect the ordered pairs with line segments.

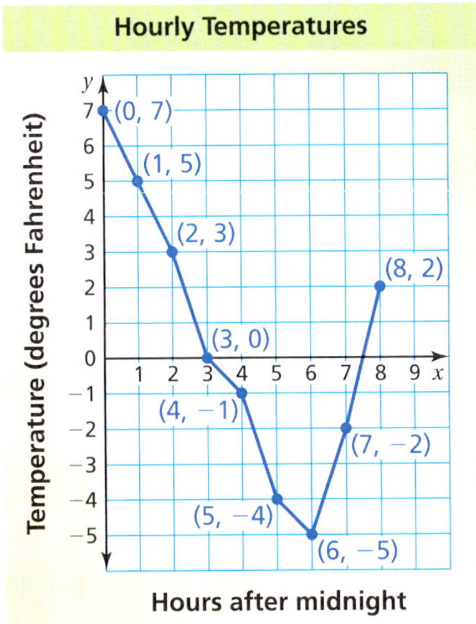

b. Make three observations from the graph.

Three possible observations follow:

- The hourly temperatures decrease from midnight to 6:00 A.M.
- The hourly temperatures increase from 6:00 A.M. to 8:00 A.M.
- The greatest decrease in hourly temperatures from one hour to the next is 3°F. This happens twice: from 2:00 A.M. to 3:00 A.M. and from 4:00 A.M. to 5:00 A.M.

Study Tip

The observations given in Example 4(b) are sample answers. You can make many other correct observations.

On Your Own

10. In Example 4, the blizzard hits another town at noon. The table shows the hourly temperatures from noon to 6:00 P.M.

Hours after Noon	0	1	2	3	4	5	6
Temperature	6°F	7°F	5°F	1°F	1°F	0°F	−3°F

a. Display the data in a line graph.

b. Make three observations from the graph.

Laurie's Notes

Example 4

- Ask students to describe the types of graphs and plots with which they are familiar.
- Define a *time series graph.* a line graph that connects ordered pairs to show patterns and trends in data over a period of time
- ? "What does the *x*-coordinate represent?" number of hours since midnight
- ? "What does the *y*-coordinate represent?" temperature
- Write the data from the table as ordered pairs.
- ? "What does (5, −4) represent?" At 5 A.M., the temperature was −4°F.
- ? "The *x*-coordinates are all positive, and the *y*-coordinates are positive and negative. What quadrants will the graph be in?" Quadrants I and IV
- Draw and label the axes so that the context is known.
- Plot the ordered pairs. Connect the ordered pairs with line segments. The segments allow the trend in the data to be more obvious.
- **MP3 Construct Viable Arguments and Critique the Reasoning of Others** and **MP6 Attend to Precision:** Ask students to make observations about the graph. Students should offer adequate detail in their observations. For instance, it would not be sufficient to simply say, "The graph goes down then up." Observations should give details and reference the context. An improved observation would be, "The temperatures decreased from midnight to 6 A.M., and then they increased for the next two hours until 8 A.M.
- After you have modeled making an observation that is detailed, give students time to brainstorm with their neighbors about other observations.
- **Extension:** Record additional data for temperatures prior to midnight. Ask students what changes are needed to the graph.

On Your Own

- **Think-Pair-Share:** Students should read the question independently and then work in pairs to answer the question. When they have answered the question, the pair should compare their answer with another group and discuss any discrepancies.

Closure

- Graph a rectangle with coordinates in Quadrants III and IV. Name the coordinates.

Extra Example 4

The table shows the hourly temperatures on a winter morning from 6 A.M. to 11 A.M.

Hours after 6 A.M., *x*	0	1	2
Temperature, *y*	−2°C	−5°C	−1°C
Hours after 6 A.M., *x*	3	4	5
Temperature, *y*	2°C	5°C	10°C

a. Display the data in a line graph.

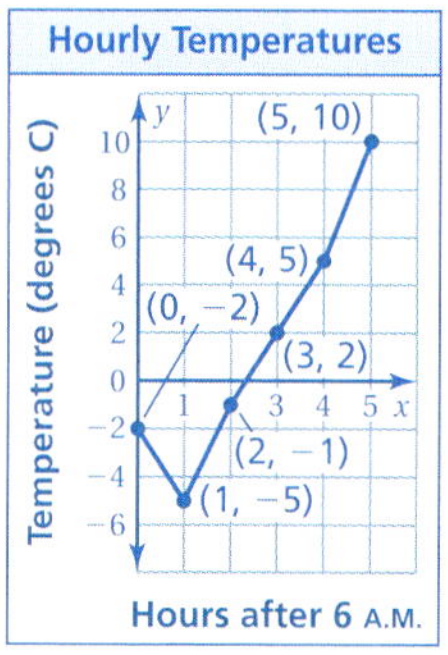

b. Make three observations from the graph. *Sample answers:* The hourly temperatures decrease from 6 A.M. to 7 A.M.; The hourly temperatures increase from 7 A.M. to 11 A.M.; The greatest increase in temperature from one hour to the next is 5°C from 10 A.M. to 11 A.M.

On Your Own

10. a.

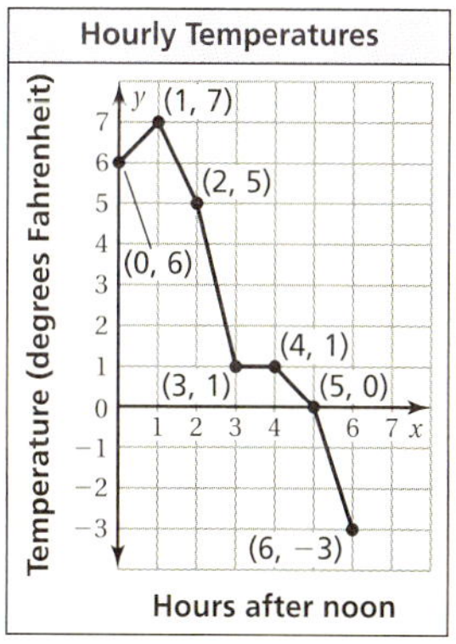

b. *Sample answers:* The greatest hourly temperature occurs at 1 P.M.; The greatest decrease in temperature from one hour to the next is 4°F; Most of the hourly temperatures are less than the previous hourly temperature.

Vocabulary and Concept Check

1. 4
2. y-axis
3. $(2, -3)$; $(2, -3)$ is in Quadrant IV. The other three points are in Quadrant II.

Practice and Problem Solving

4. 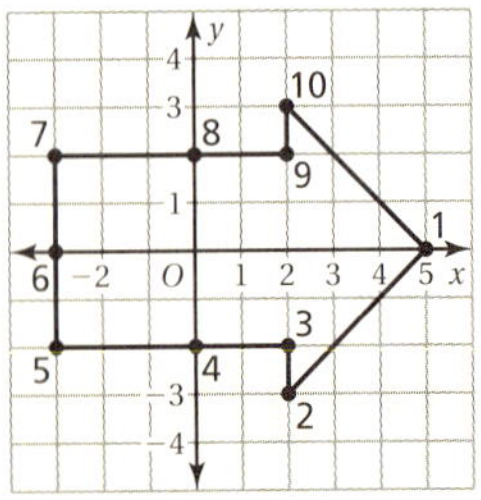

5. $(3, 1)$
6. $(-3, -2)$
7. $(-2, 4)$
8. $(1, 2)$
9. $(2, -2)$
10. $(0, -4)$
11. $(-4, 2)$
12. $(-4, -4)$
13. $(4, 0)$
14. $(4, -4)$

15–22.

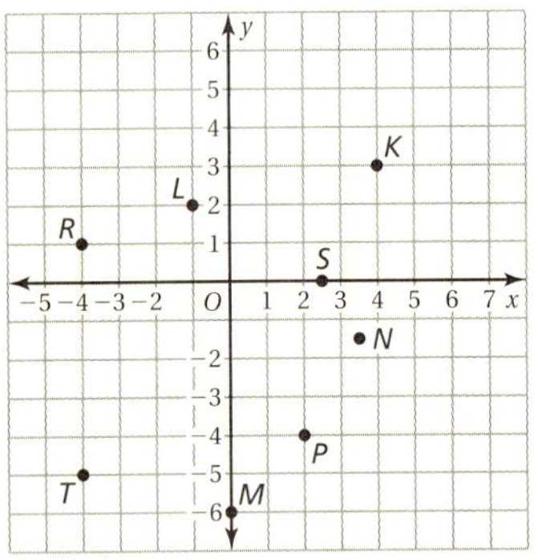

15. Quadrant I
16. Quadrant II
17. y-axis
18. Quadrant IV
19. Quadrant IV
20. Quadrant II
21. x-axis
22. Quadrant III

Assignment Guide and Homework Check

Level	Day 1 Activity Assignment	Day 2 Lesson Assignment	Homework Check
Basic	4, 55–59	1–3, 5–13 odd, 19–25 odd, 26, 31	7, 13, 21, 25
Average	4, 55–59	1–3, 11, 13, 19–23 odd, 28–36	13, 21, 28, 32, 35
Advanced	4, 55–59	1–3, 22–54 even	22, 24, 28, 34, 36

Common Errors

- **Exercises 5–14** Students may write the y-coordinate first and then the x-coordinate for the ordered pair. Tell them that x comes before y in the alphabet, so the x-coordinate must come before the y-coordinate in the ordered pair.
- **Exercises 15–22** Students may plot the x-coordinate vertically instead of horizontally and the y-coordinate horizontally instead of vertically. Remind them that x is horizontal and y is vertical.

6.5 Record and Practice Journal

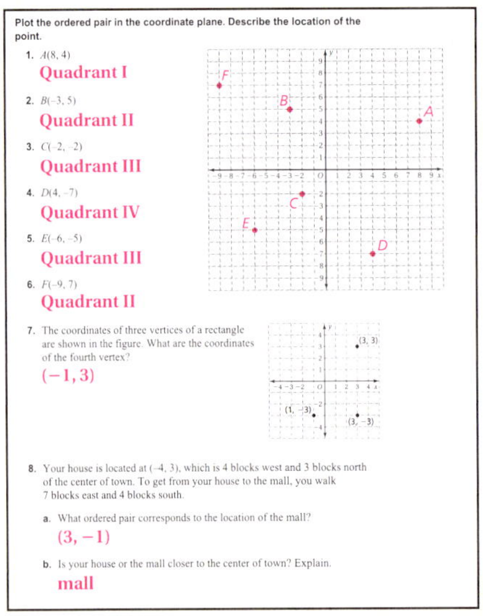

Plot the ordered pair in the coordinate plane. Describe the location of the point.

1. $A(8, 4)$ Quadrant I
2. $B(-3, 5)$ Quadrant II
3. $C(-2, -2)$ Quadrant III
4. $D(4, -7)$ Quadrant IV
5. $E(-6, -5)$ Quadrant III
6. $F(-9, 7)$ Quadrant II
7. The coordinates of three vertices of a rectangle are shown in the figure. What are the coordinates of the fourth vertex? $(-1, 3)$
8. Your house is located at $(-4, 3)$, which is 4 blocks west and 3 blocks north of the center of town. To get from your house to the mall, you walk 7 blocks east and 4 blocks south.
   a. What ordered pair corresponds to the location of the mall? $(3, -1)$
   b. Is your house or the mall closer to the center of town? Explain. mall

6.5 Exercises

Check It Out
Help with Homework
BigIdeasMath.com

Vocabulary and Concept Check

1. **VOCABULARY** How many quadrants are in a coordinate plane?

2. **VOCABULARY** Is the point $(0, -7)$ on the x-axis or the y-axis?

3. **WHICH ONE DOESN'T BELONG?** Which point does *not* belong with the other three? Explain your reasoning.

$(-2, 1)$	$(-4, 5)$	$(2, -3)$	$(-1, 3)$

Practice and Problem Solving

4. Plot and connect the points to make a picture.

1$(5, 0)$	**2**$(2, -3)$	**3**$(2, -2)$	**4**$(0, -2)$	**5**$(-3, -2)$
6$(-3, 0)$	**7**$(-3, 2)$	**8**$(0, 2)$	**9**$(2, 2)$	**10**$(2, 3)$

Write an ordered pair corresponding to the point.

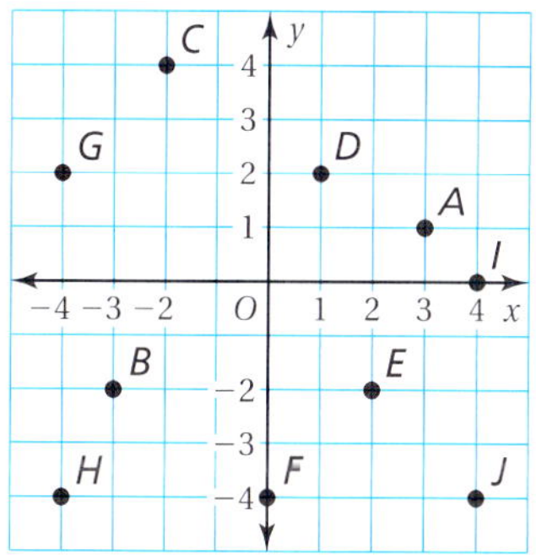

1 5. Point A 6. Point B

7. Point C 8. Point D

9. Point E 10. Point F

11. Point G 12. Point H

13. Point I 14. Point J

Plot the ordered pair in a coordinate plane. Describe the location of the point.

2 15. $K(4, 3)$ 16. $L(-1, 2)$ 17. $M(0, -6)$ 18. $N(3.5, -1.5)$

19. $P(2, -4)$ 20. $R(-4, 1)$ 21. $S\left(2\frac{1}{2}, 0\right)$ 22. $T(-4, -5)$

ERROR ANALYSIS **Describe and correct the error in the solution.**

23. ✗ To plot (4, 5), start at (0, 0) and move 5 units right and 4 units up.

24. ✗ To plot (−6, 3), start at (0, 0) and move 6 units right and 3 units down.

Plot the points and find the distance between the points.

3 25. $(2, -3), (6, -3)$ 26. $(4, 2), (4, -1)$

27. $(-1, 1), (-1, 7)$ 28. $(-5, -2), (4, -2)$

29. $(-3, 4), (5, 4)$ 30. $(-2, -4), (-2, 1)$

31. **REASONING** The coordinates of three vertices of a square are shown in the figure. What are the coordinates of the fourth vertex?

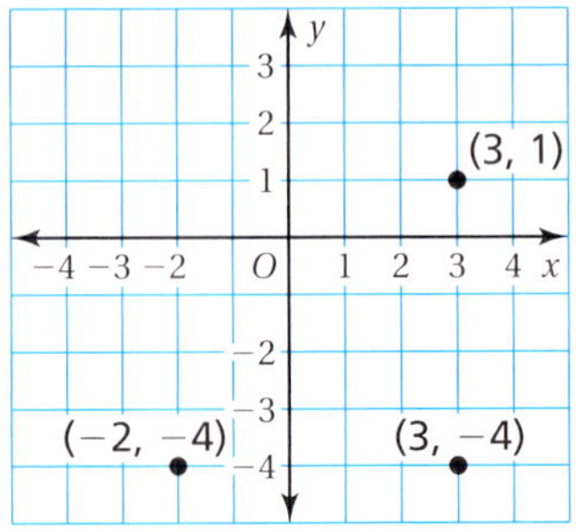

Draw the figure with the given vertices in a coordinate plane. Find the perimeter and the area of the figure.

32. $D(1, 1)$, $E(1, -2)$, $F(-2, -2)$, $G(-2, 1)$

33. $P(-2, 3)$, $Q(5, 3)$, $R(5, -1)$, $S(-2, -1)$

34. $W(-3, 2)$, $X(2, 2)$, $Y(2, -7)$, $Z(-3, -7)$

35. **POPULATION** The line graph shows the population of a city from 2005 to 2013.

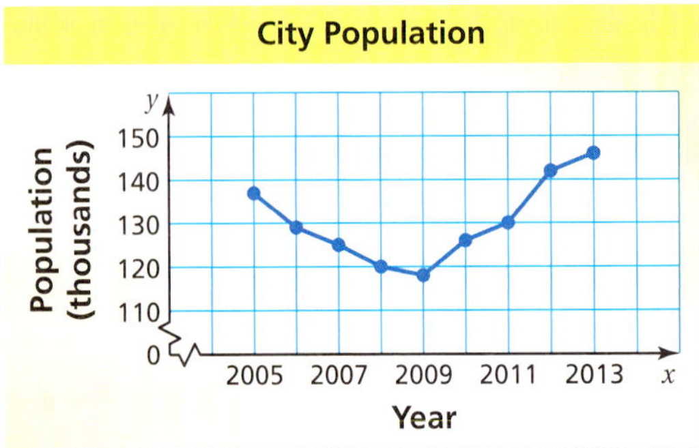

a. Estimate the population of the city in 2012.

b. Between which two years did the population increase the most?

c. Estimate the total change in population from 2005 to 2013.

36. **MODELING** The table shows the total miles run through 18 weeks for a marathon training program.

Week	1	2	3	4	5	6	7	8	9
Total Miles	22	46	72	96	124	151	181	211	244

Week	10	11	12	13	14	15	16	17	18
Total Miles	279	317	357	397	437	473	506	530	544

a. Create a table for the distance run during each week of training.

b. Display the data from part (a) in a line graph.

c. Make three observations from the graph.

d. Explain the pattern shown in the graph.

37. **PROFITS** The table shows the profits of a company from 2007 to 2013.

Years since 2000, x	7	8	9	10	11	12	13
Profit (millions of dollars), y	0.6	−0.2	−1.2	1.2	0.8	1	−0.6

a. Display the data in a line graph.

b. Make three observations from the graph.

c. What was the total profit from 2007 to 2013?

d. How could you include profits from the years 1990 to 2006 on your graph? Explain.

Describe the possible location(s) of the point (x, y).

38. $x > 0, y > 0$

39. $x < 0, y < 0$

40. $x > 0, y < 0$

41. $x > 0$

42. $y < 0$

43. $x = 0, y = 0$

Common Errors

- **Exercises 32–34** Students may plot the x-coordinate vertically instead of horizontally and the y-coordinate horizontally instead of vertically. Remind them that x is horizontal and y is vertical.
- **Exercise 44** Students may not think about the origin, which is the reason the statement is *sometimes* true. Tell students to think about the x-axis as a number line and ask if there is any place where x would be 0.

Practice and Problem Solving

23. The numbers are reversed. To plot (4, 5), start at (0, 0) and move 4 units right and 5 units up.

24. The directions are reversed. To plot (−6, 3), start at (0, 0) and move 6 units left and 3 units up.

25.

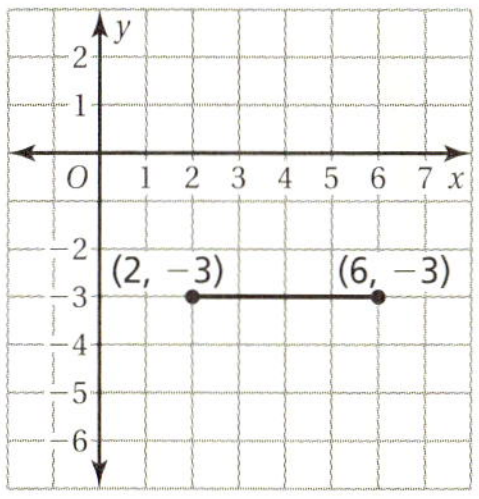

4

26.

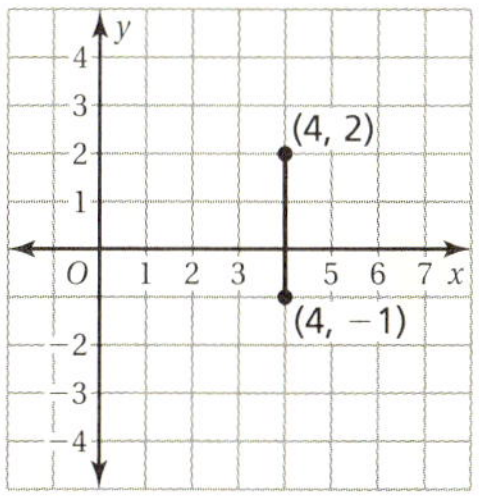

3

27.

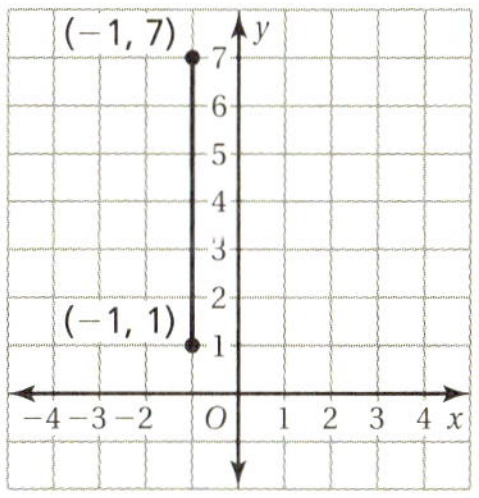

6

28–44. See Additional Answers.

English Language Learners

Vocabulary

Tell students that different words may be used to describe a reflection in a coordinate plane. For example, a figure is a reflection *in* the x-axis, *about* the x-axis, *across* the x-axis, or *over* the x-axis. The same words can be used to describe a reflection in the y-axis as well.

Practice and Problem Solving

45. never; All points in Quadrant III have negative y-coordinates.

46. always; The x-coordinate of a point in Quadrant II is negative, and so is the y-coordinate of a point in Quadrant IV.

47–51. See *Taking Math Deeper.*

52. (2, 2)

53. *Sample answer:* $(-6, 3)$, $(-2, 3)$, $(-2, -9)$, $(2, -9)$

54. **a.** $(7, -3)$

b. school; $(2, -1)$ is closer to $(0, 0)$ than $(7, -3)$.

c. 70%

Fair Game Review

55. $y - 4$ **56.** $18b$

57. $x + 9$ **58.** $\frac{w}{3}$

59. C

Mini-Assessment

The points $A(-2, 3)$, $B(4, 3)$, $C(-2, -4)$, and $D(4, -4)$ represent the vertices of a garden.

1. Plot the ordered pairs in a coordinate plane.

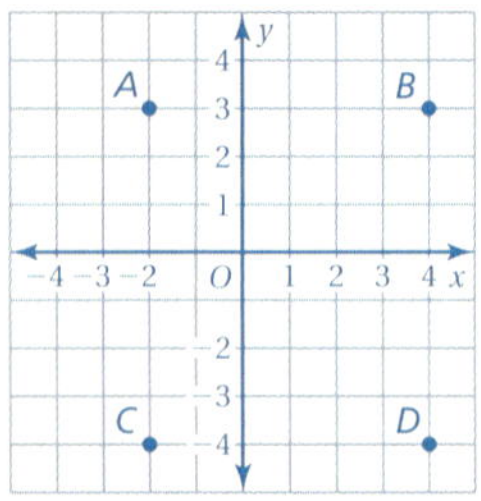

2. What shape does the garden form? a rectangle
3. Describe the location of each point. *A:* Quadrant II, *B:* Quadrant I, *C:* Quadrant III, *D:* Quadrant IV

Taking Math Deeper

Exercises 47–51

This problem shows the common practice of placing a coordinate plane over a real-life diagram or map. The decision for the location of the origin is arbitrary. The scale may also be arbitrary.

1 Use the coordinate grid to answer the questions.

(48) Flamingo Café is on positive y-axis. (47) Reptiles is at (2, 1).

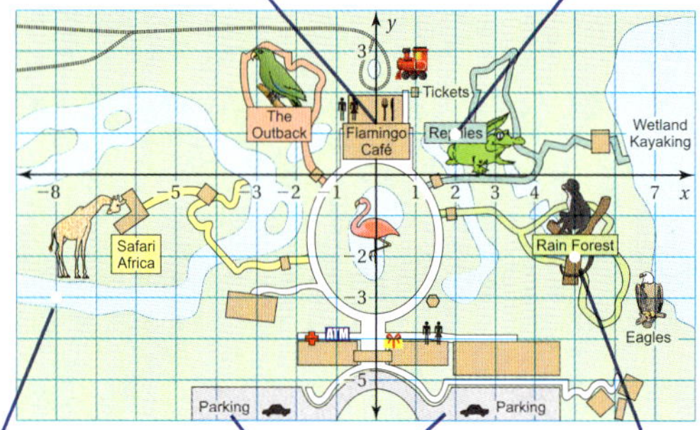

(49) Parking only in Quadrants III and IV.

(50) Many answers. Two are $(5, -1)$ and $(5, -2)$.

(51) Safari Africa is closest to $(-8, -3)$.

2 The National Zoo in Washington, D.C., is 163 acres (over 7 million square feet). It was created by Congress in 1889. The zoo includes about 400 different species, including giant pandas, which are a symbol of the zoo's conservation efforts.

Project

Use a grid to draw a map of your school or town. Include important buildings or rooms, parks, ball fields, playground equipment, and other key landmarks in your school or town.

Reteaching and Enrichment Strategies

If students need help. . .	If students got it. . .
Resources by Chapter • Practice A and Practice B • Puzzle Time Record and Practice Journal Practice Differentiating the Lesson Lesson Tutorials Skills Review Handbook	Resources by Chapter • Enrichment and Extension • Technology Connection Start the next section

Tell whether the statement is *sometimes*, *always*, or *never* true. Explain your reasoning.

44. The x-coordinate of a point on the x-axis is zero.

45. The y-coordinates of points in Quadrant III are positive.

46. The x-coordinate of a point in Quadrant II has the same sign as the y-coordinate of a point in Quadrant IV.

ZOO **In Exercises 47–51, use the map of the zoo.**

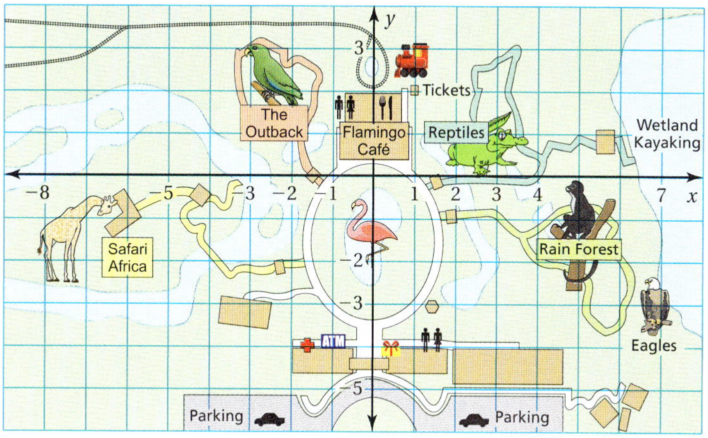

47. Which exhibit is located at (2, 1)?

48. Name an attraction on the positive y-axis.

49. Is parking available in Quadrant II? If not, name a quadrant in which you can park.

50. Write two different ordered pairs that represent the location of the Rain Forest.

51. Which exhibit is closest to $(-8, -3)$?

52. **NUMBER SENSE** Name the ordered pair that is 5 units right and 2 units down from $(-3, 4)$.

53. **OPEN-ENDED** The vertices of triangle ABC are $A(-6, -3)$ and $B(2, -3)$. List four possible coordinates of the third vertex so that the triangle has an area of 24 square units.

54. **Reasoning** Your school is located at $(2, -1)$, which is 2 blocks east and 1 block south of the center of town. To get from your house to the school, you walk 5 blocks west and 2 blocks north.

a. What ordered pair corresponds to the location of your house?

b. Is your house or your school closer to the center of town? Explain.

c. You can only walk along streets that are north and south or streets that are east and west. You are at the center of town and decide to take the shortest path home that passes by the school. When you are at the school, what percent of the walk home remains?

Fair Game Review What you learned in previous grades & lessons

Write the phrase as an expression. *(Section 3.2)*

55. 4 less than a number y

56. the product of 18 and a number b

57. a number x increased by 9

58. a number w divided by 3

59. **MULTIPLE CHOICE** What is the ratio of ducks to swans? *(Section 5.1)*

Ⓐ 4 : 9 Ⓑ 4 : 5

Ⓒ 5 : 4 Ⓓ 5 : 9

Extension 6.5 Reflecting Points in the Coordinate Plane

Check It Out
Lesson Tutorials
BigIdeasMath.com

You can *reflect* a point in the x-axis, in the y-axis, or in both axes.

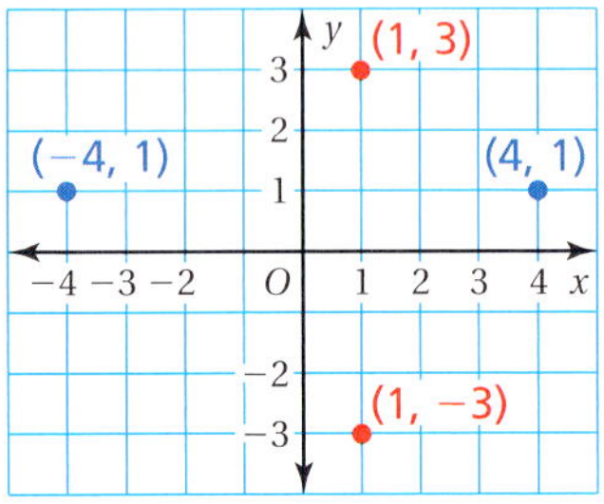

The red points are mirror images of each other in the x-axis because the x-coordinates are the same and the y-coordinates are opposites. So, the red points are 3 units from the x-axis in opposite directions. The red points represent a *reflection in the x-axis.*

The blue points are mirror images of each other in the y-axis because the y-coordinates are the same and the x-coordinates are opposites. So, the blue points are 4 units from the y-axis in opposite directions. The blue points represent a *reflection in the y-axis.*

Key Idea

Reflecting a Point in the Coordinate Plane

- To reflect a point in the x-axis, use the same x-coordinate and take the opposite of the y-coordinate.
- To reflect a point in the y-axis, use the same y-coordinate and take the opposite of the x-coordinate.

EXAMPLE 1 Reflecting Points in One Axis

a. Reflect $(-2, 4)$ in the x-axis.

Plot $(-2, 4)$.

To reflect $(-2, 4)$ in the x-axis, use the same x-coordinate, -2, and take the opposite of the y-coordinate. The opposite of 4 is -4.

So, the reflection of $(-2, 4)$ in the x-axis is $(-2, -4)$.

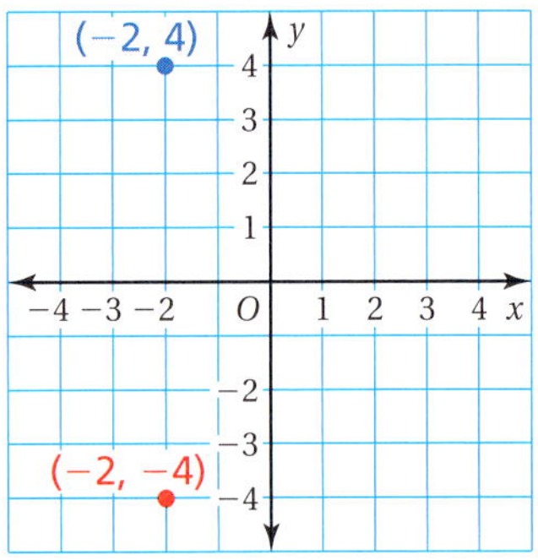

b. Reflect $(-3, -1)$ in the y-axis.

Plot $(-3, -1)$.

To reflect $(-3, -1)$ in the y-axis, use the same y-coordinate, -1, and take the opposite of the x-coordinate. The opposite of -3 is 3.

So, the reflection of $(-3, -1)$ in the y-axis is $(3, -1)$.

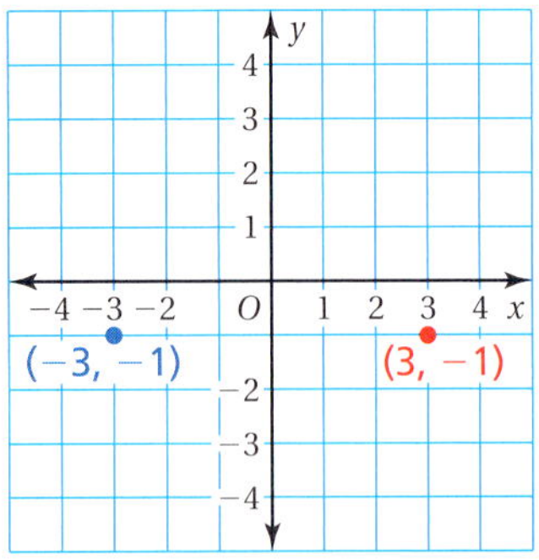

Common Core
Coordinate Plane
In this extension, you will
- understand reflections of points in the coordinate plane.

Learning Standard
6.NS.6b

Laurie's Notes

Introduction

Connect

- **Yesterday:** Students learned to plot ordered pairs in the coordinate plane in all four quadrants. (MP3, MP6)
- **Today:** Students will plot ordered pairs that are reflections of each other.

Motivate

- Take a plain piece of paper and draw two dark, perpendicular lines (*x*- and *y*-axes) that intersect near the center of the paper. Fold the paper, creasing on one of the dark lines. Fold again, creasing on the other dark line. You have now folded the paper to create the four quadrants.
- Take a paper punch and tell students that you're going to punch your paper once, going through all layers. Punch in a spot that is not on the diagonal.

? "Can you describe what the paper will look like when I unfold it?" Listen for several ideas. 4 holes, same distance from the creases; Students may use language such as "the holes are across from one another."

- **Extension:** If you can find a paper punch that is not circular, it makes it more interesting! A primary level teacher may have one.

Lesson Notes

Discuss

- Draw a coordinate grid. Plot and label ordered pairs: two points should be reflections in the *x*-axis and two should be reflections in the *y*-axis. Color code as shown.
- Ask about features of a point reflected in the *x*-axis. Students should mention that the points are the same distance from the *x*-axis, the *x*-coordinates are the same, and the *y*-coordinates are opposites.
- Repeat for the two points reflected in the *y*-axis.

Key Idea

- Refer to the coordinate grid drawn as it provides a good visual reference.
- **Common Error:** Students think that when you take the opposite of a number, it becomes negative. If the original number is negative, the opposite of it is positive.

Example 1

? "What quadrant is $(-2, 4)$ in?" Quadrant II "If you reflect $(-2, 4)$ in the *x*-axis, what quadrant will the reflection be in?" Quadrant III

- Because $(-2, 4)$ is 4 units above the *x*-axis its reflection will be 4 units below the *x*-axis. Plot $(-2, -4)$.
- Work through part (b) as shown.

Common Core State Standards

6.NS.6b Understand signs of numbers in ordered pairs as indicating locations in quadrants of the coordinate plane; recognize that when two ordered pairs differ only by signs, the locations of the points are related by reflections across one or both axes.

Goal Today's lesson is plotting ordered pairs that are reflections of each other in the coordinate plane.

Technology for the Teacher

Dynamic Classroom

Lesson Tutorials
Lesson Plans
Answer Presentation Tool

Extra Example 1

a. Reflect $(-2, -1)$ in the *x*-axis.

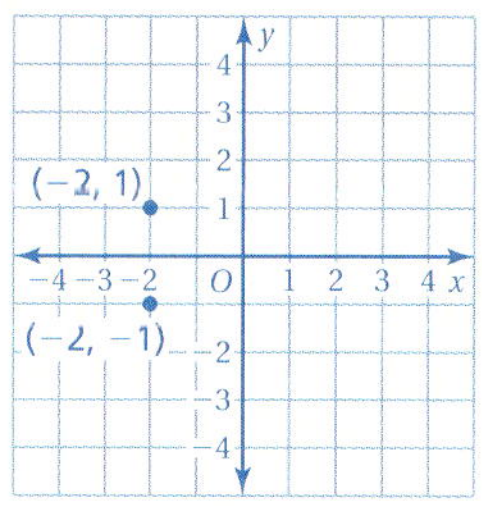

b. Reflect $(3, 2)$ in the *y*-axis.

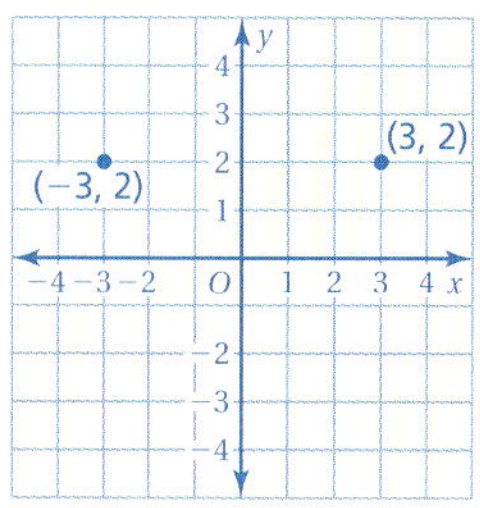

Record and Practice Journal Extension 6.5 Practice

1–11. See Additional Answers.

Extra Example 2

Reflect $(-2, -4)$ in the x-axis followed by the y-axis.

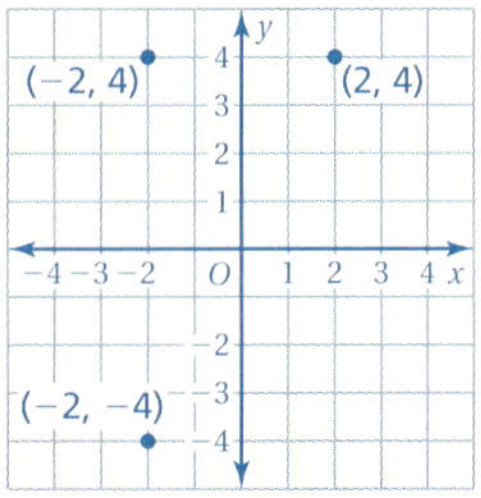

Practice

1. **a.** $(3, -2)$ **b.** $(-3, 2)$
2. **a.** $(-4, -4)$ **b.** $(4, 4)$
3. **a.** $(-5, 6)$ **b.** $(5, -6)$
4. **a.** $(4, 7)$ **b.** $(-4, -7)$
5. **a.** $(0, 1)$ **b.** $(0, -1)$

6–16. See Additional Answers.

Mini-Assessment

1. Reflect $(2, -3)$ in the x-axis.

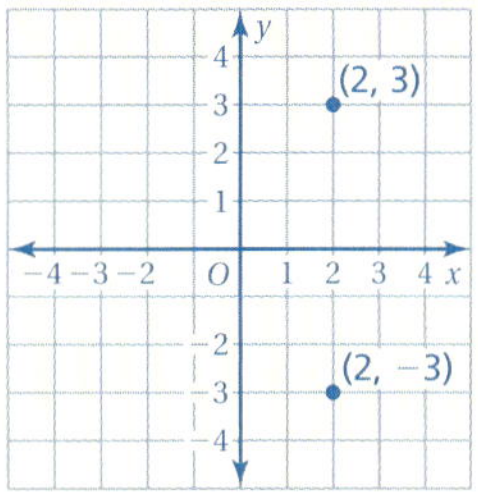

2. Reflect $(4, 3)$ in the y-axis.

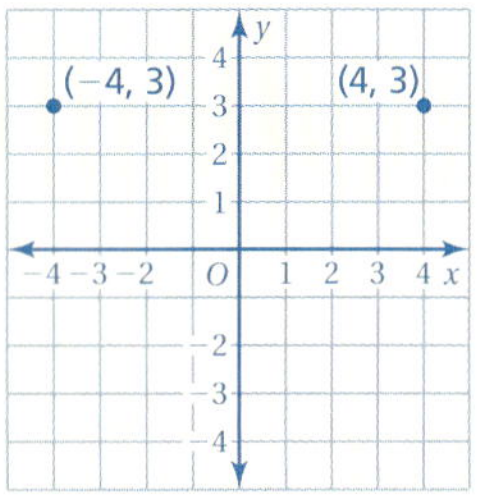

3. Reflect $(2, -2)$ in the x-axis followed by the y-axis.

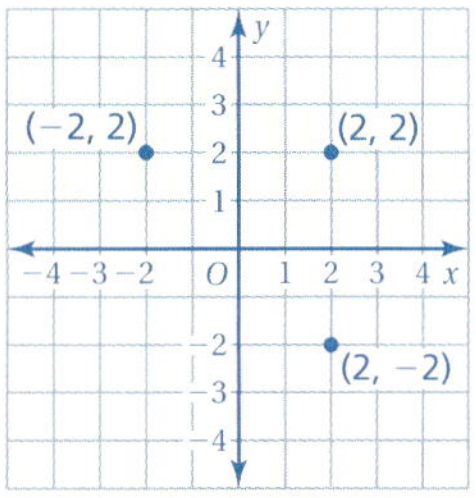

Laurie's Notes

Example 2

- **MP8 Look for and Express Regularity in Repeated Reasoning:** When students reflect a point in one axis followed by the other axis, they must repeat their thinking.
- ? "What quadrant is the point (2, 1) in?" Quadrant I
- ? "What is the reflection of (2, 1) in the x-axis?" $(2, -1)$
- At this point tell students that they are reflecting $(2, -1)$ in the y-axis and not the original point (2, 1).
- ? "What is the reflection of $(2, -1)$ in the y-axis?" $(-2, -1)$
- **Extension (MP7 Look for and Make Use of Structure):** "If you reflect (2, 1) first in the y-axis and then in the x-axis, will you end at the same location as in the example? In other words, does the order in which you do the reflections matter?" no; The order doesn't matter because you end up at the same spot.

Closure

- Plot $(-3, 4)$.
  - **a.** Reflect $(-3, 4)$ in the x-axis. $(-3, -4)$
  - **b.** Reflect $(-3, 4)$ in the y-axis. $(3, 4)$
  - **c.** If these are three of the vertices of a rectangle, what is the fourth vertex? Explain how you know. $(3, -4)$; The fourth vertex can be found by reflecting $(-3, -4)$ in the y-axis or reflecting $(3, 4)$ in the x-axis.

EXAMPLE 2 Reflecting a Point in Both Axes

Reflect (2, 1) in the x-axis followed by the y-axis.

Step1: First, plot (2, 1).

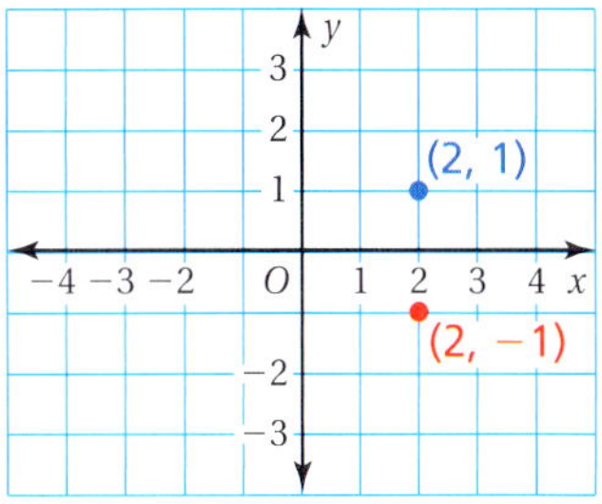

Step 2: Next, reflect (2, 1) in the x-axis. Use the same x-coordinate, 2, and take the opposite of the y-coordinate. The opposite of 1 is −1.

The point (2, 1) reflected in the x-axis is (2, −1).

Common Error

When reflecting a second time, be sure to use the reflected point and not the original point.

Step 3: Finally, reflect (2, −1) in the y-axis. Use the same y-coordinate, −1, and take the opposite of the x-coordinate. The opposite of 2 is −2.

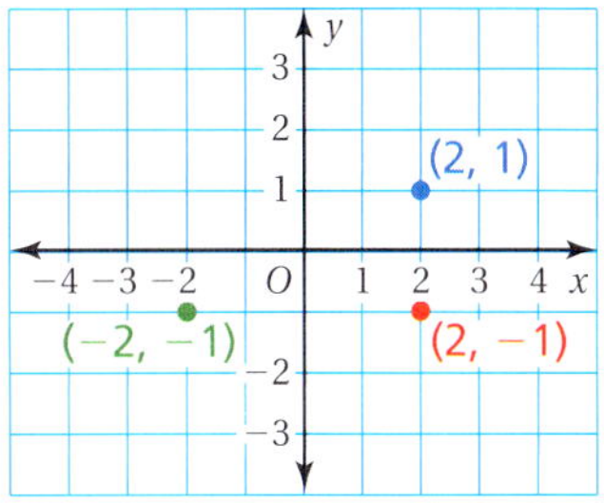

The point (2, −1) reflected in the y-axis is (−2, −1).

So, (2, 1) reflected in the x-axis followed by the y-axis is (−2, −1).

Practice

Reflect the point in (a) the x-axis and (b) the y-axis.

1. (3, 2)
2. (−4, 4)
3. (−5, −6)
4. (4, −7)
5. (0, −1)
6. (8, 0)
7. (2.5, 4.5)
8. $\left(5\frac{1}{2}, 3\right)$

Reflect the point in the x-axis followed by the y-axis.

9. (4, 5)
10. (−1, 7)
11. (−2, −2)
12. (6.5, −10.5)
13. **REASONING** A point is reflected in the x-axis. The reflected point is (3, −9). What is the original point? What is the distance between the points?
14. **REASONING** A point is reflected in the y-axis. The reflected point is (5.75, 0). What is the original point? What is the distance between the points?
15. a. **STRUCTURE** In Exercises 9–12, reflect the point in the y-axis followed by the x-axis. Do you get the same results? Explain.

    b. **LOGIC** Make a conjecture about how to use the coordinates of a point to find its reflection in both axes.
16. **GEOMETRY** The vertices of a triangle are (−1, 3), (−5, 3), and (−5, 7). How would you reflect the triangle in the x-axis? in the y-axis? Give the coordinates of the reflected triangle for each case.

6.4–6.5 Quiz

Check It Out
Progress Check
BigIdeasMath.com

Find the absolute value. *(Section 6.4)*

1. $|-12|$
2. $|4|$

Copy and complete the statement using <, >, or =. *(Section 6.4)*

3. 5 ___ $|-9|$
4. $|-11|$ ___ $|-10|$

Write an ordered pair corresponding to the point. *(Section 6.5)*

5. Point A
6. Point B
7. Point C
8. Point D

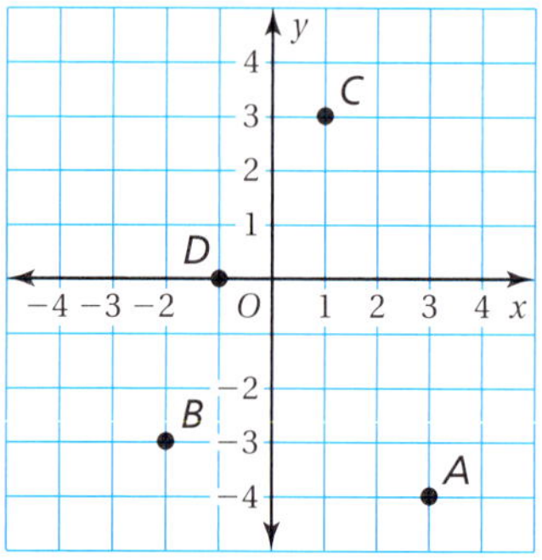

Plot the ordered pair in a coordinate plane. Describe the location of the point. *(Section 6.5)*

9. $Q(2, 5)$
10. $R(1, -4)$
11. $S(-2.5, 3.5)$
12. $T\left(0, -1\frac{1}{2}\right)$

Reflect the point in (a) the x-axis and (b) the y-axis. *(Section 6.5)*

13. $(1, 3)$
14. $(-2, 6)$

Reflect the point in the x-axis followed by the y-axis. *(Section 6.5)*

15. $(3, -2)$
16. $(-4, -5)$

17. **HIKING** The table shows the elevations of several checkpoints along a hiking trail. *(Section 6.4)*

    a. Which checkpoint is farthest from sea level?

    b. Which checkpoint is closest to sea level?

    c. Is Checkpoint 2 or Checkpoint 3 closer to sea level? Explain.

Checkpoint	Elevation (feet)
1	110
2	38
3	−24
4	12
5	−142

18. **GEOMETRY** The points $A(-4, 2)$, $B(1, 2)$, $C(1, -1)$, and $D(-4, -1)$ are the vertices of a figure. *(Section 6.5)*

    a. Draw the figure in a coordinate plane.

    b. Find the perimeter of the figure.

    c. Find the area of the figure.

Alternative Assessment Options

Math Chat	Student Reflective Focus Question
Structured Interview	Writing Prompt

Math Chat

Ask students to use their own words to summarize what they know about comparing and ordering integers, fractions, and decimals on the number line. Be sure that they include examples. Select students at random to present their summary to the class.

Study Help Sample Answers

Remind students to complete Graphic Organizers for the rest of the chapter.

4.

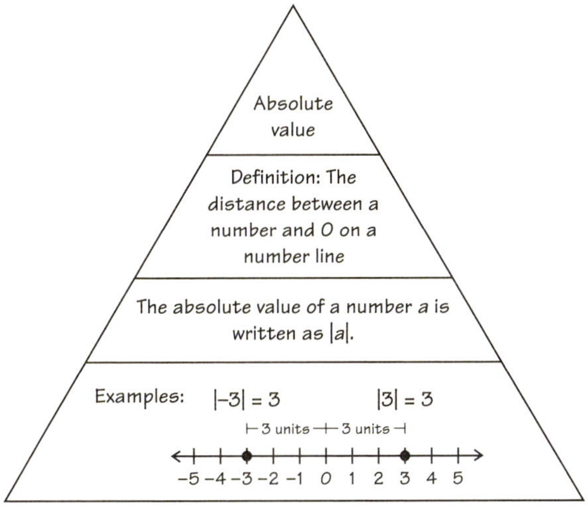

5.

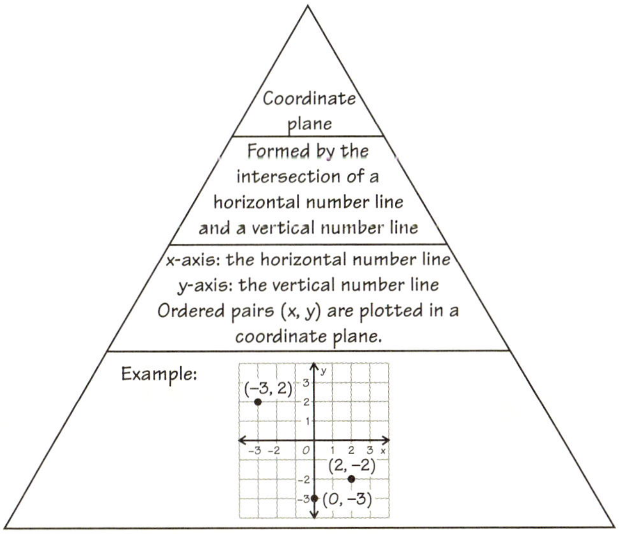

6–7. Available at *BigIdeasMath.com.*

Reteaching and Enrichment Strategies

If students need help. . .	If students got it. . .
Resources by Chapter • Practice A and Practice B • Puzzle Time Lesson Tutorials *BigIdeasMath.com*	Resources by Chapter • Enrichment and Extension • Technology Connection Game Closet at *BigIdeasMath.com* Start the Chapter Review

Answers

1. 12 **2.** 4

3. $5 < |-9|$

4. $|-11| > |-10|$

5. $(3, -4)$ **6.** $(-2, -3)$

7. $(1, 3)$ **8.** $(-1, 0)$

9–12.

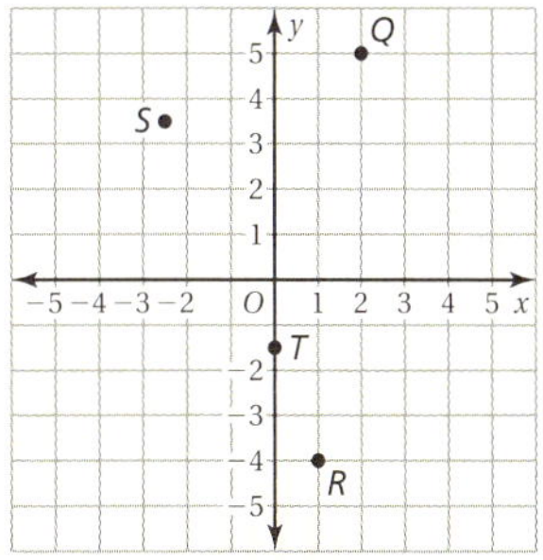

9. Quadrant I

10. Quadrant IV

11. Quadrant II

12. y-axis

13. a. $(1, -3)$ **b.** $(-1, 3)$

14. a. $(-2, -6)$ **b.** $(2, 6)$

15. $(-3, 2)$ **16.** $(4, 5)$

17. a. Checkpoint 5

b. Checkpoint 4

c. Checkpoint 3, because $|-24| < |38|$ and the absolute value of the elevation represents its distance from sea level.

18. a.

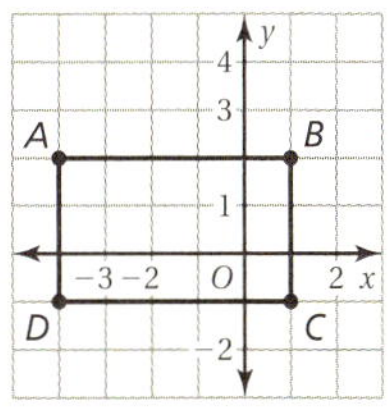

b. 16 units

c. 15 units^2

Technology for the Teacher

Online Assessment
Assessment Book
ExamView® Assessment Suite

For the Teacher
Additional Review Options

- *BigIdeasMath.com*
- Online Assessment
- Game Closet at *BigIdeasMath.com*
- Vocabulary Help
- Resources by Chapter

Answers

1. -8

2. 12

3.
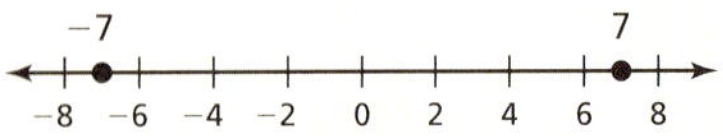

4.
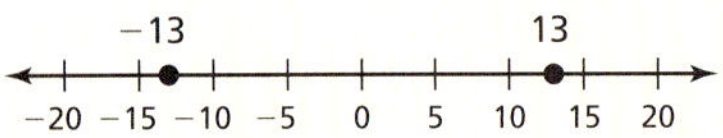

5.
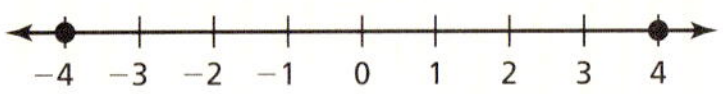

6.
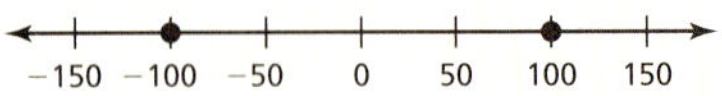

7. $-5, -3, -1, 2, 4$

8. $-20, -10, 5, 10, 15$

9. $-12°C, -7°C, -3°C, 0°C, 8°C$

Review of Common Errors

Exercises 1 and 2

- Students may use the wrong sign when writing the integer. Tell them to look for key words to help them determine the sign.

Exercises 3–6

- Students may think that *opposites* are only negative numbers. For example, the opposite of -8 is -8 and the opposite of 8 is -8. Remind them that the opposite of a negative number is positive.

Exercises 7–9

- Students may ignore the signs on the integers and order them incorrectly. Encourage students to use a number line to help them order the integers.

Exercises 14–16

- Students may ignore the sign of the number and order the numbers incorrectly. Encourage them to use a number line to determine which number is farthest to the right.

Exercise 15

- Students may place the fractions in the wrong order on the number line. Encourage them to find a common denominator before graphing on the number line.

Exercises 17–20

- Students may think the absolute value of a number is its opposite. For example, they may think $|6| = -6$. Use a number line to remind students that absolute value is a number's distance from zero, so it is always a positive number or zero.

Exercises 24–27

- Students may plot the x-coordinate vertically instead of horizontally and the y-coordinate horizontally instead of vertically. Remind students that x is horizontal and y is vertical.

6 Chapter Review

Check It Out
Vocabulary Help
BigIdeasMath.com

Review Key Vocabulary

positive numbers, *p. 250*
negative numbers, *p. 250*
opposites, *p. 250*
integers, *p. 250*
absolute value, *p. 270*
coordinate plane, *p. 276*
origin, *p. 276*
quadrants, *p. 276*

Review Examples and Exercises

6.1 Integers *(pp. 248–253)*

Write a positive or negative integer to represent losing 150 points in a pinball game.

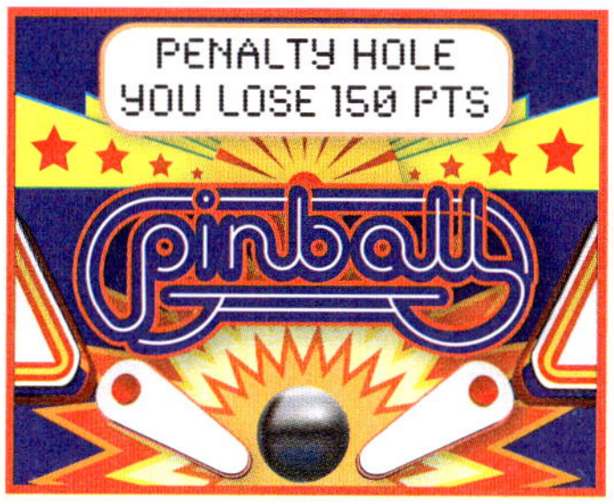

"Lose" indicates a number less than 0. So, use a negative integer.

∴ -150

Exercises

Write a positive or negative integer that represents the situation.

1. An elevator goes down 8 floors. **2.** You earn $12.

Graph the integer and its opposite.

3. -7 **4.** 13 **5.** 4 **6.** -100

6.2 Comparing and Ordering Integers *(pp. 254–259)*

Order $-3, -4, 2, 0, -1$ from least to greatest.

Graph each integer on a number line.

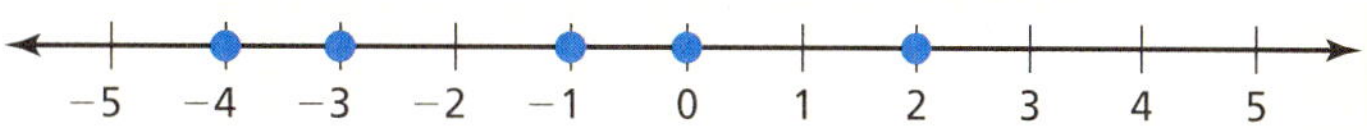

Write the integers as they appear on the number line from left to right.

∴ So, the order from least to greatest is $-4, -3, -1, 0, 2$.

Exercises

Order the integers from least to greatest.

7. $-5, 4, 2, -3, -1$ **8.** $5, -20, -10, 10, 15$

9. Order the temperatures -3°C, 8°C, -12°C, -7°C, and 0°C from coldest to warmest.

6.3 Fractions and Decimals on the Number Line (pp. 260–265)

Compare $-3\frac{7}{8}$ and $-3\frac{3}{8}$.

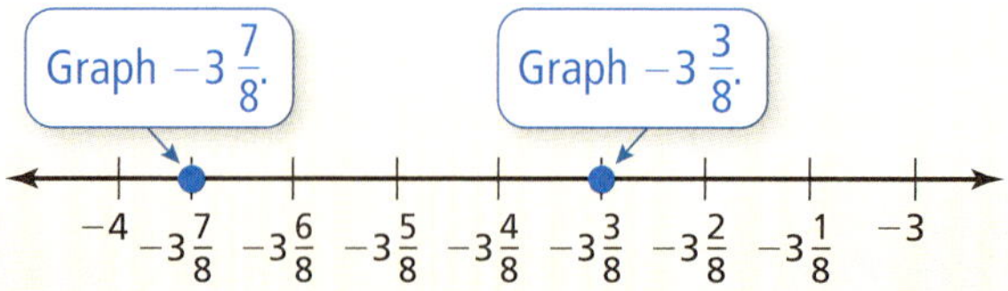

$-3\frac{7}{8}$ is to the left of $-3\frac{3}{8}$.

So, $-3\frac{7}{8} < -3\frac{3}{8}$.

Exercises

Graph the number and its opposite.

10. $-\frac{2}{5}$ **11.** $1\frac{3}{4}$ **12.** -1.6 **13.** 2.75

Copy and complete the statement using < or >.

14. $-2\frac{1}{6}$ ▭ $-2\frac{5}{6}$ **15.** $-\frac{1}{3}$ ▭ $-\frac{1}{8}$ **16.** -3.27 ▭ -2.68

6.4 Absolute Value (pp. 268–273)

Find the absolute value of -3.

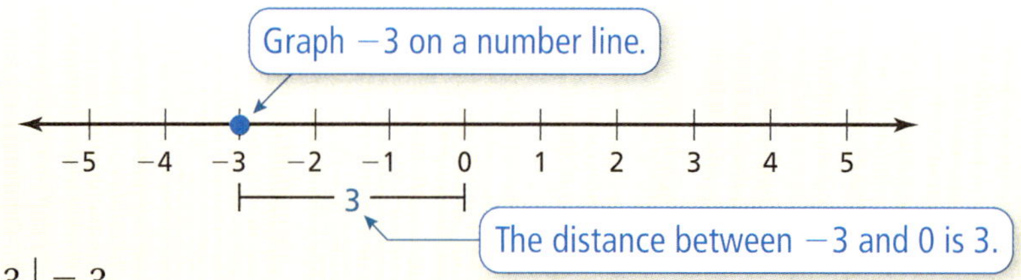

So, $|-3| = 3$.

Exercises

Find the absolute value.

17. $|-8|$ **18.** $|13|$ **19.** $\left|3\frac{6}{7}\right|$ **20.** $|-1.34|$

Copy and complete the statement using <, >, or =.

21. $|-2|$ ▭ 2 **22.** $|4.4|$ ▭ $|-2.8|$ **23.** $\left|\frac{1}{6}\right|$ ▭ $\left|-\frac{2}{9}\right|$

Review Game

Number Lines

Materials

- 1 deck of cards with the jacks, queens, kings, aces, and jokers removed
- 2 wooden craft sticks
- paper
- pencil

Players: 3

Directions:

Divide the class into groups of three. Each group draws a number line labeled −10 to 10 in increments of 1. Two students each draw a card and lay a stick on the number represented by the card. Black cards represent positive integers and red cards represent negative integers. The third student is turned so he or she cannot see the two integers being plotted. The third student says "greatest" or "least." If the third student says "greatest," then the student who plotted the greatest integer gets 1 point. If the third student says "least," then the student who plotted the least integer gets 1 point. Continue playing until one of the students reaches 10 points. That student wins. Now the three students can switch roles and play again.

Who Wins?

The student that reaches 10 points the most.

For the Student Additional Practice

- Lesson Tutorials
- Multi-Language Glossary
- Self-Grading Progress Check
- *BigIdeasMath.com*
  Dynamic Student Edition
  Student Resources

Answers

10. 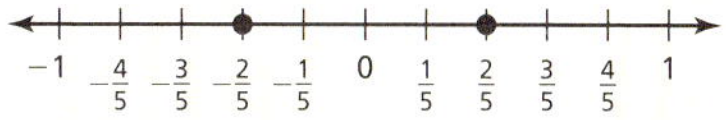

11. 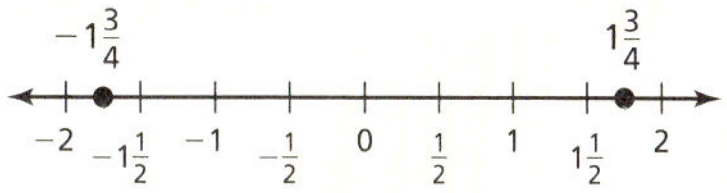

12. 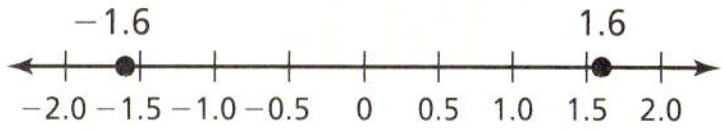

13. 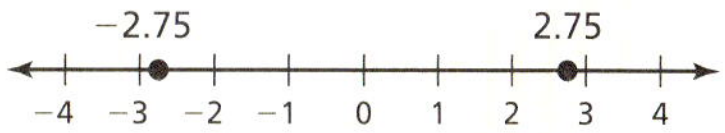

14. $>$ **15.** $<$

16. $<$ **17.** 8

18. 13 **19.** $3\frac{6}{7}$

20. 1.34 **21.** $=$

22. $>$ **23.** $<$

24–27. 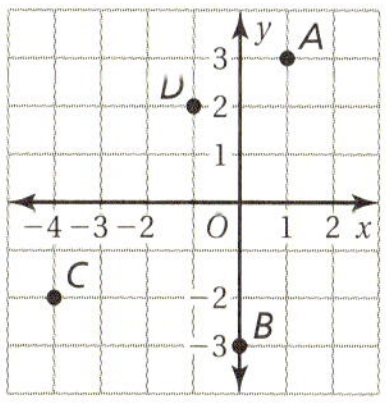

24. Quadrant I

25. y-axis

26. Quadrant III

27. Quadrant II

28. **a.** $(4, -1)$ **b.** $(-4, 1)$

29. **a.** $(-2, -3)$ **b.** $(2, 3)$

30. **a.** $(2, 5)$ **b.** $(-2, -5)$

31. **a.** $(-3.5, 2.5)$

b. $(3.5, -2.5)$

32. $(-1, -2)$ **33.** $(4, -6)$

34. $(-3, 4)$ **35.** $(3, 3)$

My Thoughts on the Chapter

Teacher Tip

Not allowed to write in your teaching edition? Use sticky notes to record your thoughts.

What worked. . .

What did not work. . .

What I would do differently. . .

6.5 The Coordinate Plane (pp. 274–283)

a. Plot (−3, 0) and (4, −4) in a coordinate plane. Describe the location of each point.

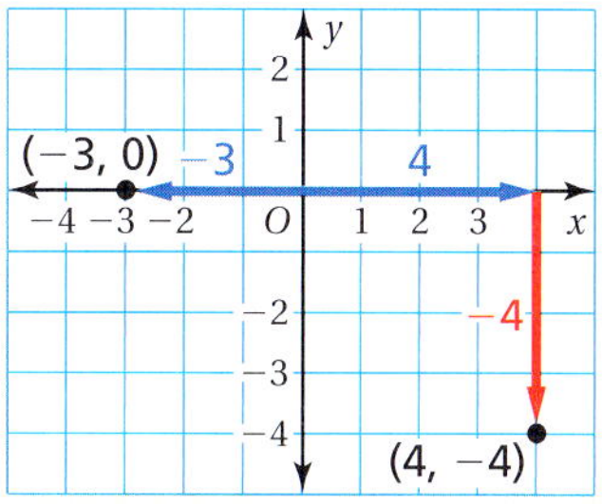

To plot (−3, 0), start at the origin. Move 3 units left. Then plot the point.

To plot (4, −4), start at the origin. Move 4 units right and 4 units down. Then plot the point.

∴ The point (−3, 0) is on the x-axis.
The point (4, −4) is in Quadrant IV.

b. Reflect (2, −3) in the x-axis.

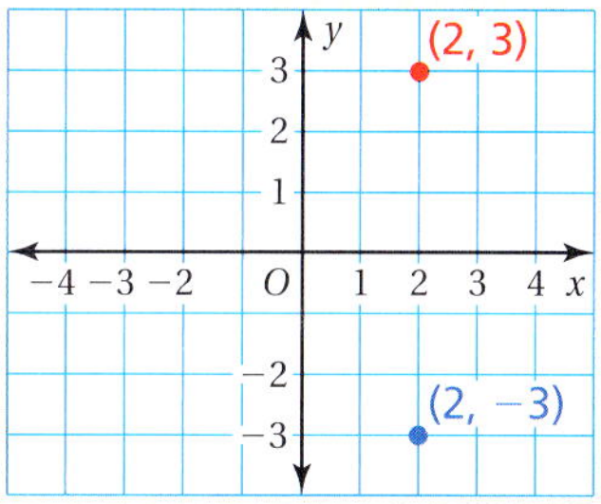

Plot (2, −3).

To reflect (2, −3) in the x-axis, use the same x-coordinate, 2, and take the opposite of the y-coordinate. The opposite of −3 is 3.

∴ So, the reflection of (2, −3) in the x-axis is (2, 3).

c. Reflect (2, −3) in the y-axis.

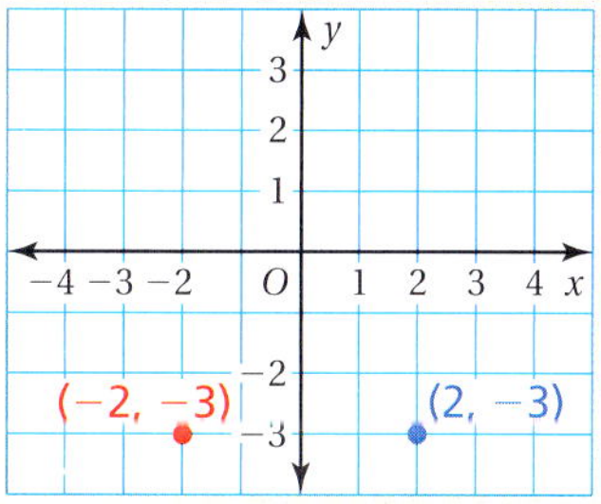

Plot (2, −3).

To reflect (2, −3) in the y-axis, use the same y-coordinate, −3, and take the opposite of the x-coordinate. The opposite of 2 is −2.

∴ So, the reflection of (2, −3) in the y-axis is (–2, –3).

Exercises

Plot the ordered pair in a coordinate plane. Describe the location of the point.

24. $A(1, 3)$

25. $B(0, -3)$

26. $C(-4, -2)$

27. $D(-1, 2)$

Reflect the point in (a) the x-axis and (b) the y-axis.

28. (4, 1)

29. (−2, 3)

30. (2, −5)

31. (−3.5, −2.5)

Reflect the point in the x-axis followed by the y-axis.

32. (1, 2)

33. (−4, 6)

34. (3, −4)

35. (−3, −3)

6 Chapter Test

Check It Out
Test Practice
BigIdeasMath.com

Order the integers from least to greatest.

1. $0, -2, 3, 1, -4$

2. $-8, -3, 5, 4, -5$

Graph the number and its opposite.

3. 14

4. -40

5. $-1\frac{1}{3}$

6. 1.75

Find the absolute value.

7. $|-7|$

8. $|-11|$

Copy and complete the statement using <, >, or =.

9. $-\frac{2}{3}$ ▭ $-\frac{3}{5}$

10. 1.55 ▭ -2.46

11. $|-6|$ ▭ -3

12. -2.5 ▭ $|2.5|$

Plot the ordered pair in a coordinate plane. Describe the location of the point.

13. $J(4, 0)$

14. $K(-3, 5)$

15. $L(1.5, -3.5)$

16. $M(-2, -3)$

Reflect the point in the x-axis followed by the y-axis.

17. $(2, 4)$

18. $(-5, 1)$

19. POOL A diver is on a springboard that is 3 meters above the surface of a pool. Another diver is 2 meters below the surface of the pool.

a. Write an integer for the position of each diver relative to the surface of the pool.

b. Find the absolute value of each integer.

c. Who is farther from the surface of the pool?

20. OPEN-ENDED Two vertices of a triangle are $F(1, -4)$ and $G(6, -4)$. List two possible coordinates of the third vertex so that the triangle has an area of 20 square units.

21. MELTING POINT The table shows the melting points (in degrees Celsius) of several elements. Compare the melting point of mercury to the melting point of each of the other elements.

Element	Mercury	Radon	Bromine	Cesium	Francium
Melting Point (°C)	−38.83	−71	−7.2	28.5	27

Test Item References

Chapter Test Questions	Section to Review	Common Core State Standards
3, 4	6.1	6.NS.5, 6.NS.6a, 6.NS.6c
1, 2	6.2	6.NS.6c, 6.NS.7a, 6.NS.7b
5, 6, 9, 10, 21	6.3	6.NS.5, 6.NS.6a, 6.NS.6c, 6.NS.7a, 6.NS.7b
7, 8, 11, 12, 19	6.4	6.NS.7c, 6.NS.7d
13–18, 20	6.5	6.NS.6b, 6.NS.6c, 6.NS.8

Test-Taking Strategies

Remind students to quickly look over the entire test before they start so that they can budget their time. On this test, it is very important for them to **Stop** and **Think**. When students hurry on a test dealing with signed numbers, they often make 'sign errors.' Encourage them to represent problems with a number line, if appropriate, to ensure that they think through the process.

Common Errors

- **Exercises 1, 2, 9–12** Students may ignore the signs on the integers and order them incorrectly. Encourage students to use a number line to help them order the integers.
- **Exercises 3–6** Students may think that *opposites* are only negative numbers. For example, the opposite of −3 is −3 and the opposite of 3 is −3. Remind them that the opposite of a negative number is positive.
- **Exercises 13–16** Students may plot the *x*-coordinate vertically instead of horizontally and the *y*-coordinate horizontally instead of vertically. Remind them that *x* is horizontal and *y* is vertical.

Reteaching and Enrichment Strategies

If students need help. . .	If students got it. . .
Resources by Chapter • Practice A and Practice B • Puzzle Time Record and Practice Journal Practice Differentiating the Lesson Lesson Tutorials *BigIdeasMath.com* Skills Review Handbook	Resources by Chapter • Enrichment and Extension • Technology Connection Game Closet at *BigIdeasMath.com* Start Standards Assessment

Answers

1. −4, −2, 0, 1, 3
2. −8, −5, −3, 4, 5
3. 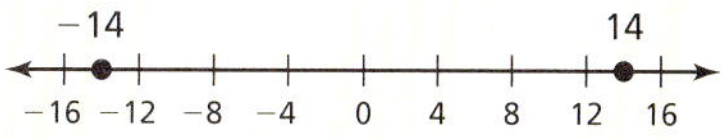

4. 

5. 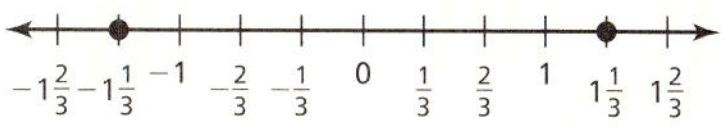

6. 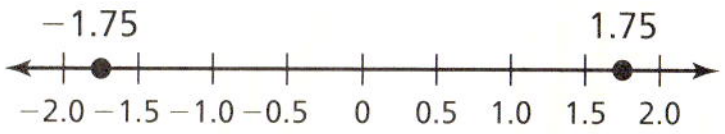

7. 7
8. 11
9. <
10. >
11. >
12. <

13–16. 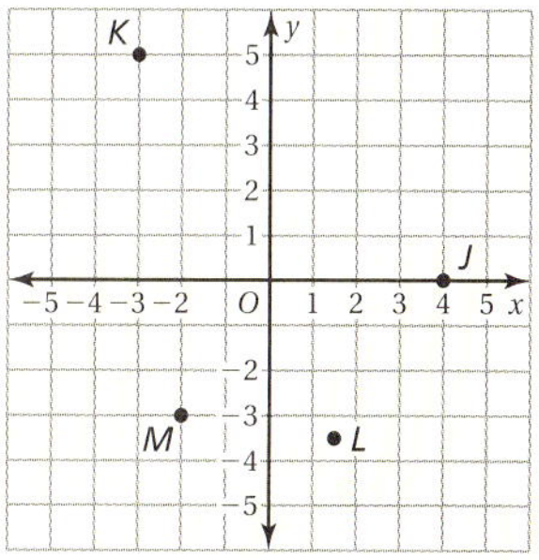

13. *x*-axis
14. Quadrant II
15. Quadrant IV
16. Quadrant III
17. (−2, −4)
18. (5, −1)
19. **a.** 3; −2
    b. 3; 2
    c. the diver on the springboard
20. *Sample answer:* (1, 4), (6, 4)
21. Mercury has a higher melting point than Radon and a lower melting point than Bromine, Cesium, and Francium.

Technology for the Teacher

Online Assessment
Assessment Book
ExamView® Assessment Suite

Test-Taking Strategies

Available at *BigIdeasMath.com*

After Answering Easy Questions, Relax
Answer Easy Questions First
Estimate the Answer
Read All Choices before Answering
Read Question before Answering
Solve Directly or Eliminate Choices
Solve Problem before Looking at Choices
Use Intelligent Guessing
Work Backwards

About this Strategy

When taking a multiple choice test, be sure to read each question carefully and thoroughly. It is also very important to read each answer choice carefully. Do not pick the first answer that you think is correct! If two answer choices are the same, eliminate them both. There can only be one correct answer.

Answers

1. C
2. G
3. $\frac{3}{4}$
4. D
5. H

Technology for the Teacher

Common Core State Standards Support
Performance Tasks
Online Assessment
Assessment Book
ExamView® Assessment Suite

Item Analysis

1. **A.** The student makes an order of operations error, adding $3c$ to $5b$ and then subtracting this sum from $8a$.

   B. The student substitutes 5 for c and 4 for b.

   C. Correct answer

   D. The student makes two-digit numbers out of the coefficients and the variables, such as 43 instead of 4×3.

2. **F.** The student incorrectly thinks that points in Quadrant II have negative y-coordinates.

   G. Correct answer

   H. The student incorrectly thinks that points in Quadrant II have negative y-coordinates. The student also reverses the x- and y-coordinates.

   I. The student reverses the x- and y-coordinates.

3. **Gridded Response:** Correct answer: $\frac{3}{4}$

   Common Error: The student divides the greater number by the lesser number, getting an answer of $\frac{4}{3}$ or equivalent.

4. **A.** The student orders the numbers by magnitude.

   B. The student incorrectly thinks $|4| = -4$

   C. The student ignores the absolute value bars.

   D. Correct answer

5. **F.** The student multiplies the numerator and denominator.

   G. The student uses the digits from the numerator and denominator.

   H. Correct answer

   I. The student divides 5 by 4 and converts the quotient into a percent.

6 Standards Assessment

Test-Taking Strategy

Read All Choices Before Answering

"Reading all choices before answering can get you a lot more yummy treats!"

1. What is the value of the expression below when $a = 6$, $b = 5$, and $c = 4$? *(6.EE.2c)*

$$8a - 3c + 5b$$

A. 11 **C.** 61

B. 53 **D.** 107

2. Point P is plotted in the coordinate plane below.

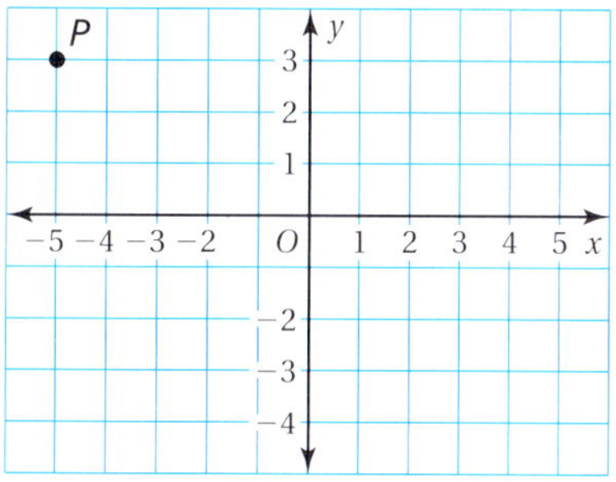

What are the coordinates of point P? *(6.NS.8)*

F. $(-5, -3)$ **H.** $(-3, -5)$

G. $(-5, 3)$ **I.** $(3, -5)$

3. What is the value of the expression below? *(6.NS.1)*

$$4\frac{1}{8} \div 5\frac{1}{2}$$

4. Which list of numbers is in order from least to greatest? *(6.NS.7)*

A. $2, |-3|, |4|, -6$ **C.** $-6, |-3|, 2, |4|$

B. $-6, |4|, 2, |-3|$ **D.** $-6, 2, |-3|, |4|$

5. Which percent is equivalent to $\frac{4}{5}$? *(6.RP.3c)*

F. 20% **H.** 80%

G. 45% **I.** 125%

6. Which property is illustrated by the statement below? *(6.EE.3)*

$$4 + (6 + n) = (4 + 6) + n$$

A. Associative Property of Addition

B. Commutative Property of Addition

C. Associative Property of Multiplication

D. Distributive Property

7. You bought 0.875 kilogram of mixed nuts. What was the total cost, in dollars, of the mixed nuts that you bought? *(6.NS.3)*

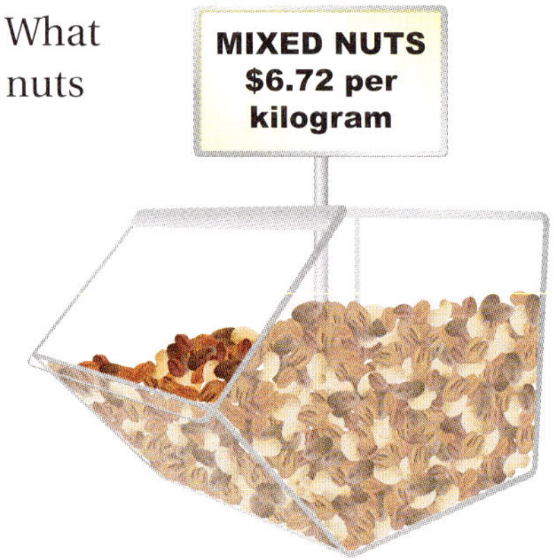

8. On Saturday, you earned $35 mowing lawns. This was x dollars more than you earned on Thursday. Which expression represents the amount, in dollars, you earned mowing lawns on Thursday? *(6.EE.6)*

F. $35x$

G. $x + 35$

H. $x - 35$

I. $35 - x$

9. Helene was finding the percent of a number in the box below.

75% of 24 is what number?

$$75\% \text{ of } 24 = 24 \div \frac{3}{4}$$

$$= 32$$

What should Helene do to correct the error that she made? *(6.RP.3c)*

A. Divide 24 by 75.

B. Divide $\frac{3}{4}$ by 24.

C. Multiply 24 by 75.

D. Multiply 24 by $\frac{3}{4}$.

Item Analysis (continued)

6. **A.** Correct answer

   B. The student misidentifies the property being used as the Commutative Property of Addition.

   C. The student misidentifies the property being used as the Associative Property of Multiplication.

   D. The student misidentifies the property being used as the Distributive Property.

7. **Gridded Response:** Correct answer: 5.88

   Common Error: The student incorrectly places the decimal point in the product.

8. **F.** The student misinterprets "more than" as meaning multiplication.

   G. The student misinterprets the problem, thinking that you received more on Thursday than Saturday.

   H. The student subtracts in the wrong order and writes an expression that represents $35 less than the difference of the amounts earned on Thursday and Saturday.

   I. Correct answer

9. **A.** The student divides by 100 times the percent.

   B. The student switches the order of division instead of changing to multiplication.

   C. The student multiplies by 100 times the percent.

   D. Correct answer

Answers

6. A
7. 5.88
8. I
9. D

Answers

10. F

11. *Part A* and *Part B*

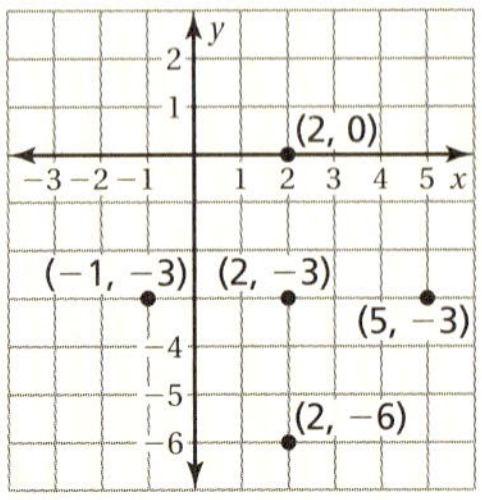

12. A

Item Analysis (continued)

10. **F.** Correct answer

G. The student chooses a benchmark percent that visually approximates the number of squares shaded red.

H. The student thinks that the quantity of 48 squares is the same as 48%.

I. The student chooses a benchmark percent that approximates the number of squares shaded red.

11. **2 points** The student demonstrates a thorough understanding of plotting points in all four quadrants of the coordinate plane. The student correctly plots and labels points at $(2, -3)$, $(2, -6)$, $(-1, -3)$, $(2, 0)$, and $(5, -3)$.

1 point The student demonstrates a partial understanding of plotting points in all four quadrants of the coordinate plane. The student shows some knowledge of how to plot points, but does not successfully plot all points.

0 points The student demonstrates insufficient understanding of plotting points in all four quadrants of the coordinate plane. The student makes many errors in plotting points and/or is unable to correctly draw an x-axis and y-axis.

12. **A.** Correct answer

B. The student does not find the reciprocal of $\frac{10}{3}$.

C. The student does not change division to multiplication.

D. The students incorrectly rewrites $3\frac{1}{3}$ as $\frac{7}{3}$ by adding 3, 3, and 1 to get 7.

10. In the mural below, the squares that are painted red are marked with the letter R.

R	R	R															R	R	R
R	R	R															R	R	R
R	R	R															R	R	R
							R	R	R	R	R	R							
							R	R	R	R	R	R							
R	R	R															R	R	R
R	R	R															R	R	R
R	R	R															R	R	R

What percent of the mural is painted red? *(6.RP.3c)*

F. 24%

G. 25%

H. 48%

I. 50%

11. Use grid paper to complete the following. *(6.NS.8)*

Think
Solve
Explain

Part A Draw an x-axis and a y-axis in the coordinate plane. Then plot and label the point $(2, -3)$.

Part B Plot and label *four* points that are 3 units away from $(2, -3)$.

12. Which expression is equivalent to the expression below? *(6.NS.1)*

$$k \div 3\frac{1}{3}$$

A. $k \cdot \frac{3}{10}$

B. $k \cdot \frac{10}{3}$

C. $k \div \frac{3}{10}$

D. $k \div \frac{7}{3}$

7 Equations and Inequalities

Common Core Progression

4th Grade
• Use and interpret simple equations.

5th Grade
• Generate numerical patterns, identify the relationship, and form ordered pairs.

6th Grade
• Determine if a value is a solution. • Solve one-step equations. • Represent constraints with inequalities and recognize they can have infinitely many solutions. • Solve one-step inequalities.

Pacing Guide for Chapter 7

Chapter Opener	1 Day
Section 1 Activity Lesson	 1 Day 1 Day
Section 2 Activity Lesson	 1 Day 1 Day
Section 3 Activity Lesson	 1 Day 1 Day
Section 4 Activity Lesson	 1 Day 1 Day
Study Help / Quiz	1 Day
Section 5 Activity Lesson	 1 Day 1 Day
Section 6 Activity Lesson	 1 Day 1 Day
Section 7 Activity Lesson	 1 Day 1 Day
Chapter Review/ Chapter Tests	2 Days
Total Chapter 7	18 Days
Year-to-Date	115 Days

Chapter Summary

Section	Common Core State Standard	
7.1	Learning	6.EE.6 ★
7.2	Learning	6.EE.5, 6.EE.7
7.3	Learning	6.EE.5, 6.EE.7 ★
7.4	Learning	6.EE.9 ★
7.5	Learning	6.EE.5, 6.EE.8
7.6	Applying	6.EE.5, 6.EE.8
7.7	Applying	6.EE.5 ★, 6.EE.8 ★

★ Teaching is complete. Standard can be assessed.

Technology for the Teacher

BigIdeasMath.com
Chapter at a Glance
Complete Materials List
Parent Letters: English and Spanish

Common Core State Standards

6.EE.2c Evaluate expressions at specific values of their variables. Include expressions that arise from formulas used in real-world problems. Perform arithmetic operations, including those involving whole-number exponents, in the conventional order when there are no parentheses to specify a particular order (Order of Operations).

6.EE.2a Write expressions that record operations with numbers and with letters standing for numbers.

Additional Topics for Review

- Order of Operations
- Ordering Fractions
- Comparing Numbers

Try It Yourself

1. 21 **2.** 1

3. 11 **4.** 46

5. $3w + 6$ **6.** $\frac{7}{p}$

7. $t - 2$ **8.** $5x$

9. $\frac{6}{r} + 5$ **10.** $3b - 4$

Record and Practice Journal Fair Game Review

1. 30 **2.** 10

3. 17 **4.** 6

5. 6 **6.** 44

7. 16 **8.** 72

9. 85 **10.** $2k + 3$

11. $\frac{q}{2} + 8$ **12.** $p - 6$

13. $9x$ **14.** $\frac{5}{n}$

15. $1 + 3y$ **16.** 2c

17. $\frac{p}{2} + 6$

Math Background Notes

Vocabulary Review

- Evaluate
- Substitute
- Expression
- Order of Operations

Evaluating Expressions

- Students should know how to evaluate expressions. Because this skill is fairly new to them, you may want to allot slightly more time for practice.
- Remind students that *evaluate* means to find a value.
- To evaluate an expression, students will need to substitute values into the expression.
- **Teaching Tip:** English Language Learners may find the vocabulary and writing involved in these opening exercises challenging. To ease the transition, help them make analogies. In the same way a sugar substitute replaces sugar, substituting numbers into an expression will replace the variables until you are able to evaluate.
- After they substitute, remind students that they must follow the correct order of operations as they simplify. You may wish to review order of operations prior to completing the examples.

Writing Expressions

- Students should know how to write expressions.
- Review key words with students. For example, words such as sum and product alert students to use addition and multiplication.
- **Common Error:** Many students will read Example 3, part (b) and incorrectly write the expression $12 - 4y$. Remind students that they cannot subtract 12 from a number unless they have a number to start with. Always write the number first and then add or subtract as indicated.

Reteaching and Enrichment Strategies

If students need help. . .	If students got it. . .
Record and Practice Journal • Fair Game Review Skills Review Handbook Lesson Tutorials	Game Closet at *BigIdeasMath.com* Start the next section

What You Learned Before

Evaluating Expressions (6.EE.2c)

"Dear Sir: I have the answer to the contest question, 'How many seconds are in a year?' There are 12: January 2nd, February 2nd, ..."

Example 1 Evaluate $7x + 3y$ when $x = 2$ and $y = 4$.

$7x + 3y = 7 \cdot 2 + 3 \cdot 4$	Substitute 2 for x and 4 for y.
$= 14 + 12$	Using order of operations, multiply from left to right.
$= 26$	Add 14 and 12.

Example 2 Evaluate $5x^2 - 2(y + 1) + 9$ when $x = 2$ and $y = 1$.

$5x^2 - 2(y + 1) + 9 = 5(2)^2 - 2(1 + 1) + 9$	Substitute 2 for x and 1 for y.
$= 5(2)^2 - 2 \cdot 2 + 9$	Using order of operations, evaluate within the parentheses.
$= 5 \cdot 4 - 2 \cdot 2 + 9$	Using order of operations, evaluate the exponent.
$= 20 - 4 + 9$	Using order of operations, multiply from left to right.
$= 25$	Subtract 4 from 20. Add the result to 9.

Try It Yourself

Evaluate the expression when $a = \frac{1}{2}$ and $b = 7$.

1. $6ab$ **2.** $16a - b$ **3.** $3b - 2a - 9$ **4.** $b^2 - 16a + 5$

Writing Expressions (6.EE.2a)

Example 3 Write the phrase as an expression.

a. the sum of twice a number n and five

$2n + 5$

b. twelve less than four times a number y

$4y - 12$

Try It Yourself

Write the phrase as an expression.

5. six more than three times a number w

6. the quotient of seven and a number p

7. two less than a number t

8. the product of a number x and five

9. five more than six divided by a number r

10. four less than three times a number b

7.1 Writing Equations in One Variable

Essential Question How does rewriting a word problem help you solve the word problem?

1 ACTIVITY: Rewriting a Word Problem

Work with a partner. Read the problem several times. Think about how you could rewrite the problem. Leave out information that you do not need to solve the problem.

Given Problem (63 words)

Your minivan has a flat, rectangular area in the back. When you fold down the rear seats of the van and move them forward, the width of the rectangular area in the van is increased by 2 feet, as shown in the diagram.

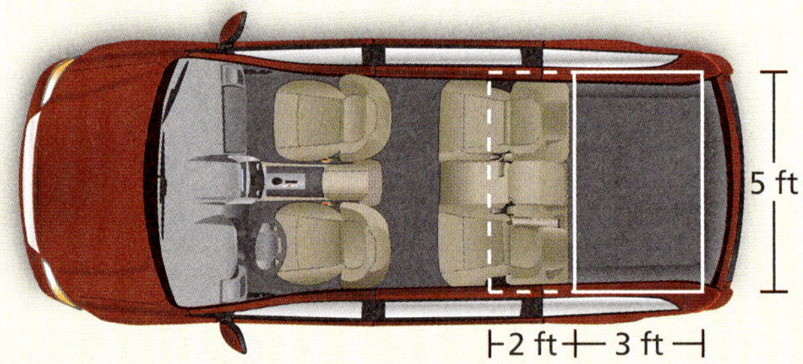

By how many square feet does the rectangular area increase when the rear seats are folded down and moved forward?

Rewritten Problem (28 words)

When you fold down the back seats of a minivan, the added area is a 5-foot by 2-foot rectangle. What is the area of this rectangle?

Can you make the problem even simpler?

Rewritten Problem (words)

Explain why your rewritten problem is easier to read.

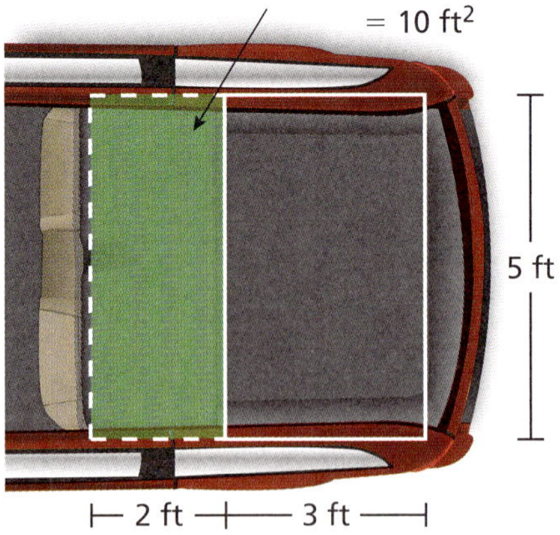

Common Core
Writing Equations
In this lesson, you will
- write word sentences as equations.

Learning Standard
6.EE.6

Laurie's Notes

Introduction

Standards for Mathematical Practice

- **MP1 Make Sense of Problems and Persevere in Solving Them:** The first step in solving a problem is to understand the language and what the problem is about. A student has to be able to explain the meaning of a problem in order to find an entry point to its solution. In the activities, students will read through problems multiple times and try to eliminate extraneous words. Stating the problem more simply aids in attempting to solve the problem.

Motivate

- Because this investigation is about words in math problems, try a version of Word Scrabble as a warm-up. Start with a 4-letter word. Each player in succession changes just one letter to make a new word. The player may also rearrange the letters. Sample:
  MATH → MOTH → BOTH → BOAT → COAT → SCAT → MATS → MATH
- This sample comes back to the word it began with, but that is not required.
- Try any of these words for starters: FOUR, MADE, BOOK, MALL, TALK.

Activity Notes

Activity 1

- Students may become intimidated by problems that have a lot of words. In this example, they have the opportunity to see the stages needed to reduce the number of words to get at the real question.
- Explain that additional words are often added to a problem for interest, and/or to help clarify the problem. Once the problem is clear, it is rewritten using fewer words.

? "What words and information are not necessary in the original form of the problem?" Listen to a series of responses.

- Pause at the end of the second section to discuss what the words tell you and what is being asked. Discuss how the diagram aids understanding of the problem.
- Ask different volunteers to read their final questions. If possible, write the questions on individual white boards and have students display their questions to the class. Discuss each question to be sure the essence of the question is not lost.

? "What is the answer to the question?" 10 ft^2

Common Core State Standards

6.EE.6 Use variables to represent numbers and write expressions when solving a real-world or mathematical problem; understand that a variable can represent an unknown number, or, depending on the purpose at hand, any number in a specified set.

Previous Learning

Students should be familiar with the vocabulary associated with the four operations.

Technology for the Teacher

Dynamic Classroom

Lesson Plans
Complete Materials List

7.1 Record and Practice Journal

Essential Question How does rewriting a word problem help you solve the word problem?

1 ACTIVITY: Rewriting a Word Problem

Work with a partner. Read the problem several times. Think about how you could rewrite the problem, leaving out information that you do not need to solve the problem.

Given Problem (63 words)

Your minivan has a flat, rectangular area in the back. When you fold down the rear seats of the van and move them forward, the width of the rectangular area in the van is increased by 2 feet, as shown in the diagram.

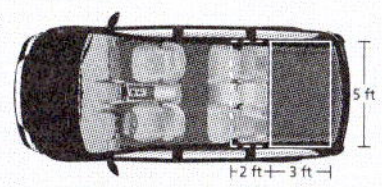

By how many square feet does the rectangular area increase when the rear seats are folded down and moved forward?

Rewritten Problem (28 words)

When you fold down the back seats of a minivan, the added area is a 5-foot by 2-foot rectangle. What is the area of this rectangle?

Can you make the problem even simpler?

What is the area of a rectangle that is 5 feet long and 2 feet wide?

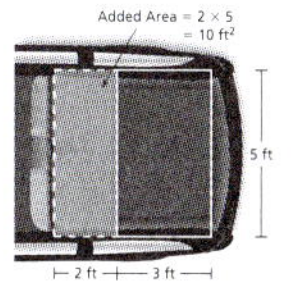

Explain why your rewritten problem is easier to read.

Check students' work.

English Language Learners

Vocabulary

Have students create a table in their notebooks of common words used to indicate addition, subtraction, multiplication, and division. *Sample:*

Addition	Subtraction
added to	subtracted from
sum of	difference of
more than	less than
increased by	decreased by
total of	fewer than
and	take away
Multiplication	**Division**
multiplied by	divided by
times	quotient of
product of	
twice	
of	

7.1 Record and Practice Journal

2 ACTIVITY: Rewriting a Word Problem

Work with a partner. Rewrite each problem using fewer words. Leave out information that you do not need to solve the problem. Then solve the problem.

a. **(63 words)**

A supermarket is having its grand opening on Saturday morning. Every fifth customer will receive a $10 coupon for a free turkey. Every seventh customer will receive a $3 coupon for 2 gallons of ice cream. You are the manager of the store and you expect 400 customers. How many of each type of coupon should you plan to give away?

80 $10 coupons; 57 $3 coupons

b. **(71 words)**

You and your friend are at a football game. The stadium is 4 miles from your home. You each brought $5 to spend on refreshments. During the third quarter of the game, you say, "I read that the greatest distance that a baseball has been thrown is 445 feet 10 inches." Your friend says, "That's about one and a half times the length of the football field." Is your friend correct?

yes

c. **(90 words)**

You are visiting your cousin who lives in the city. To get back home, you take a taxi. The taxi charges $2.10 for the first mile and $0.90 for each additional mile. After riding 13 miles, you decide that the fare is going to be more than the $20 you have with you. So, you tell the driver to stop and let you out. Then you call a friend and ask your friend to come pick you up. After paying the driver, how much of your $20 is left?

$7.10

What Is Your Answer?

3. **IN YOUR OWN WORDS** How does rewriting a word problem help you solve the word problem? Make up a word problem that has more than 50 words. Then show how you can rewrite the problem using at most 25 words.

Leaves out information you do not need and is easier to read.

Laurie's Notes

Activity 2

- **MP1:** Understanding the problem certainly is an important part of problem solving. Reading through the problem and eliminating extraneous words help students find entry points into the solution.
- **Teaching Tip:** It would be helpful to provide a transparency and pen, or individual white board, for each pair so that they can share their sample rewrites easily.
- Remind students to keep track of the number of words used in each successive rewrite. Encourage them to use a diagram.
- **FYI:** There is additional (extraneous) information in each problem. Some students may find it helpful to make a list of extraneous information.
- **Part (a):** The question pares down to multiples of 5 and 7 and how many times each divides into 400. The question could be extended to ask how many people will receive both coupons. multiples of 35
- **Part (b):** Students need to know that a football field is 100 yards or 300 feet.
- **Part (c):** Students need to be familiar with how taxis charge their fares—a base fare plus so much for each mile (or portion of a mile).
- Have students share their rewritten problems and their answers. Using transparencies or white boards makes it possible to share more than one version of each problem.

What Is Your Answer?

- **FYI:** Read the cartoon. Students may not be aware of how much sap it takes to make a gallon of maple syrup!

Closure

- How do adverbs and adjectives in an English sentence relate to today's activities? Sometimes you can remove adverbs and adjectives from a sentence without changing its meaning. For example, the two sentences below have the same meaning.

  I rode my green bike to school today.

  I rode my bike to school today.

Math Practice 1

Analyze Givens

What information do you need to solve the problem?

2 ACTIVITY: Rewriting a Word Problem

Work with a partner. Rewrite each problem using fewer words. Leave out information that you do not need to solve the problem. Then solve the problem.

a. (63 words)

A supermarket is having its grand opening on Saturday morning. Every fifth customer will receive a $10 coupon for a free turkey. Every seventh customer will receive a $3 coupon for 2 gallons of ice cream. You are the manager of the store and you expect to have 400 customers. How many of each type of coupon should you plan to give away?

b. (71 words)

You and your friend are at a football game. The stadium is 4 miles from your home. You each brought $5 to spend on refreshments. During the third quarter of the game, you say, "I read that the greatest distance that a baseball has been thrown is 445 feet 10 inches." Your friend says, "That's about one and a half times the length of the football field." Is your friend correct?

c. (90 words)

You are visiting your cousin who lives in the city. To get back home, you take a taxi. The taxi charges $2.10 for the first mile and $0.90 for each additional mile. After riding 13 miles, you decide that the fare is going to be more than the $20 you have with you. So, you tell the driver to stop and let you out. Then you call a friend and ask your friend to come and pick you up. After paying the driver, how much of your $20 is left?

What Is Your Answer?

3. **IN YOUR OWN WORDS** How does rewriting a word problem help you solve the word problem? Make up a word problem that has more than 50 words. Then show how you can rewrite the problem using at most 25 words.

"Solving a math word problem is like making maple syrup." "You need to boil down 40 gallons of sap from a sugar maple tree to get 1 gallon of syrup."

Practice Use what you learned about writing equations to complete Exercises 4 and 5 on page 298.

7.1 Lesson

Check It Out
Lesson Tutorials
BigIdeasMath.com

Key Vocabulary
equation, *p. 296*

An **equation** is a mathematical sentence that uses an equal sign, $=$, to show that two expressions are equal.

Expressions	*Equations*
$4 + 8$	$4 + 8 = 12$
$x + 8$	$x + 8 = 12$

To write a word sentence as an equation, look for key words or phrases such as *is*, *the same as*, or *equals* to determine where to place the equal sign.

EXAMPLE 1 Writing Equations

Write the word sentence as an equation.

a. The sum of a number n and 7 is 15.

The sum of a number n and 7 is 15.

$n + 7 \quad = 15$ — *Sum of* means *addition*.

An equation is $n + 7 = 15$.

b. A number y decreased by 4 is 3.

A number y decreased by 4 is 3.

$y - 4 \quad = 3$ — *Decreased by* means *subtraction*.

An equation is $y - 4 = 3$.

c. 12 times a number p equals 48.

12 times a number p equals 48.

$12p \quad = \quad 48$ — *Times* means *multiplication*.

An equation is $12p = 48$.

On Your Own

Now You're Ready
Exercises 6–13

Write the word sentence as an equation.

1. 9 less than a number b equals 2.
2. The product of a number g and 5 is 30.
3. A number k increased by 10 is the same as 24.
4. The quotient of a number q and 4 is 12.

Multi-Language Glossary at BigIdeasMath.com

Laurie's Notes

Introduction

Connect

- **Yesterday:** Students practiced rewriting word problems using fewer words. (MP1)
- **Today:** Students will translate words into mathematical equations.
- Previously, students learned to translate words into algebraic expressions. In this section, students take the process one step further and write an equation.

Motivate

- Ask four students to volunteer. Two of them will hold sentence strip cards with English phrases and two will hold sentence strip cards with English sentences. Don't tell them the difference. Simply say that two of them are similar in some way and the other two are similar in some way. They should sort themselves and explain their grouping.

  Examples: WINTER IS HERE.
  THE LONG COLD WINTER (PHRASE)
  ORANGE JUICE FOR HEALTHY LIVING (PHRASE)
  I LIKE ORANGE JUICE.
- **Discuss:** Discuss the difference between expressions (English phrases) and equations (English sentences). Note that variables are used in each, but equations and expressions can be written without variables. A phrase is a group of words that has meaning but is not a full sentence. A sentence is a group of words that must include a subject and a verb.

Lesson Notes

Example 1

- Note the use of color to connect the words and the operation. If possible, change chalk or marker color when writing the examples on the board.
- **Common Error:** Subtraction is not commutative so the order in which the terms are written does matter. *A number decreased by 4* is different from *4 decreased by a number*.
- **Representation:** The notation $12p$ is still somewhat new to students so be sure to ask, "What does $12p$ mean?" Multiply 12 by the value of p.

? Ask a few questions about part (a) and part (c).
  - "Would $7 + n = 15$ be correct for part (a)? Explain." yes; Addition is commutative.
  - "Would $p12 = 48$ be correct for part (c)? Explain." yes; Multiplication is commutative. It is standard notation, however, to write the number (coefficient) before the variable.

On Your Own

- **Think-Pair-Share:** Students should read each question independently and then work in pairs to answer the questions. When they have answered the questions, the pair should compare their answers with another group and discuss any discrepancies.
- **Common Error:** In Question 4, division is not commutative. The quotient of a number q and 4, $q \div 4$, is not the same as the quotient of 4 and a number q, $4 \div q$.

Goal Today's lesson is reading and translating a word problem into a mathematical **equation**.

Technology for the Teacher

Dynamic Classroom

Lesson Tutorials
Lesson Plans
Answer Presentation Tool

Extra Example 1

Write the word sentence as an equation.

a. A number m added to 3 equals 15. $3 + m = 15$

b. The difference of a number w and 6 is the same as 12. $w - 6 = 12$

c. 22 is twice a number h. $22 = 2h$

On Your Own

1. $b - 9 = 2$
2. $5g = 30$
3. $k + 10 = 24$
4. $q \div 4 = 12$

Laurie's Notes

Extra Example 2

Ten servers decorate 20 tables for a party. Each table has 4 red candles and 3 white candles. Let c be the total number of red and white candles. Write an equation that can be used to find c. $c = 20 \times (4 + 3)$

Extra Example 3

After three rounds, 15 students are eliminated from a geography bee. There are 55 students remaining. Write an equation you can use to find the number of students that started the geography bee. $s - 15 = 55$

On Your Own

5. $x - 7 = 10$
6. $28 + x = 52$
7. $24{,}000 = (24x)(60)$

Differentiated Instruction

Auditory

On the board, write a sentence that translates to an algebraic expression. Discuss with students what each word in the sentence means and how it relates to an algebraic equation. Sentences similar to Example 1:

The quotient of 8 and a number p is 2.
The difference of a number s and 6 is 8.
Twice a number z is 12.
The total of a number y and 4 is 11.

Example 2

- Students may need to read the problem more than once. Just as with the activity yesterday, there is additional information that is not important in solving the problem.
- Model rewriting the problem using fewer words.
- ? When the problem is finished ask, "Would $25(4 + 6) = c$ also be correct?" yes
- **MP6 Attend to Precision:** Be mindful of the examples you do in class. You do not want to mislead students into thinking that all of the arithmetic occurs on the left side of the equation, and the answer is on the right side of the equation. This is often referred to as the "rightward arrow." You do not want students to view the equation sign as an arrow. ($3 + 4 \rightarrow 7$ or $3 + x \rightarrow 7$): Mathematically proficient students understand the symbols they use, including the equal sign!
- ? **Extension:** Ask how many of each color of candle is needed. 100 white and 150 purple

Example 3

- Ask a volunteer to read through the problem. Note the *Reading* tip at the left.
- **Common Error:** Students may subtract 24 from 96.
- **MP4 Model with Mathematics:** The words written are an example of a *verbal model*. The verbal model says in words how the problem can be solved. As I say to my students, "If you can't say in words how to solve the problem, don't try to write an equation. You need to understand the problem first." Once the problem is understood, the equation is one way to model the situation.
- **MP6:** The variable is defined to be *the number of students that started*, not simply *students*. Often students are too brief in describing what the variable represents. You want students to develop the habit of being precise with their variable definitions.

On Your Own

- For each question the students should write the question in words (verbal model), define the variable, and then translate the words into an equation.
- For Question 7, students will need to determine the number of minutes in one day before writing the equation.
- Students could work in pairs to complete the questions.
- Share solutions at the board.

Closure

- **Exit Ticket:** Translate each sentence.
  The quotient of a number and 8 is 40. $n \div 8 = 40$
  The quotient of 40 and a number is 8. $40 \div n = 8$
  40 is the quotient of 8 and a number. $40 = 8 \div n$

EXAMPLE 2 Writing an Equation

Ten servers decorate 25 tables for a wedding. Each table is decorated as shown. Let c be the total number of white and purple candles. Which equation can you use to find c?

Ⓐ $c = 25 + (4 \times 6)$ Ⓑ $c = 25(4 + 6)$

Ⓒ $c = 10(25 + 4 + 6)$ Ⓓ $c = 10(4 + 6)$

Words The total number of candles is the number of tables times the number of candles on each table.

Variable Let c be the total number of candles.

Equation $c = 25 \times (4 + 6)$

∴ The correct answer is Ⓑ.

EXAMPLE 3 Real-Life Application

After two rounds, 24 students are eliminated from a spelling bee. There are 96 students remaining. Write an equation you can use to find the number of students that started the spelling bee.

Reading

The word *eliminated* means *subtraction*.

Words The number of students that started minus the number of students eliminated is the number of students remaining.

Variable Let s be the number of students that started.

Equation $s - 24 = 96$

∴ An equation is $s - 24 = 96$.

On Your Own

5. You enter an elevator and go down 7 floors. You exit on the 10th floor. Write an equation you can use to find the floor where you entered the elevator.
6. Together you and a friend have $52. Your friend has $28. Write an equation you can use to find how much money you have.
7. A typical person takes about 24,000 breaths each day. Write an equation you can use to find the number of breaths a typical person takes each minute.

7.1 Exercises

Check It Out
Help with Homework
BigIdeasMath.com

Vocabulary and Concept Check

1. **VOCABULARY** How are expressions and equations different?
2. **DIFFERENT WORDS, SAME QUESTION** Which is different? Write "both" equations.

4 less than a number n is 8.	A number n is 4 less than 8.
A number n minus 4 equals 8.	4 subtracted from a number n is 8.

3. **OPEN-ENDED** Write a word sentence for the equation $28 - n = 5$.

Practice and Problem Solving

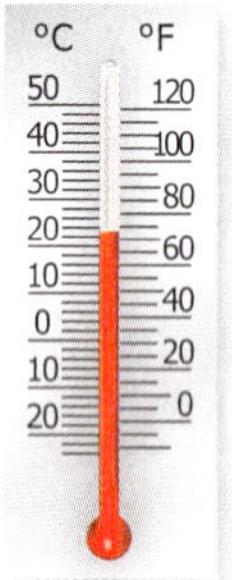

Rewrite the problem using fewer words. Leave out information that you do not need to solve the problem. Then solve the problem.

4. In a cross-country race you run at a steady rate of 7 minutes per mile. After 21 minutes, you finish in fourth place. How long is the race?
5. For a science project, you record the high temperature each day. The high temperature on Day 1 was 6° less than on Day 4 and 4° less than on Day 10. The high temperature on Day 10 was 62°F. What was the high temperature on Day 1?

Write the word sentence as an equation.

1 6. The sum of a number x and 4 equals 12.
7. A number y decreased by 9 is 8.
8. 9 times a number b is 36.
9. A number w divided by 5 equals 6.
10. 54 equals 9 more than a number t.
11. 5 is one-fourth of a number c.
12. 11 is the quotient of a number y and 6.
13. 9 less than a number n equals 27.

14. **ERROR ANALYSIS** Describe the error in writing the sentence as an equation.

✗ A number n is 5 more than 12.
$n + 5 = 12$

15. **FUNDRAISING** Students and faculty raised \$6042 for band uniforms. The faculty raised \$1780. Write an equation you can use to find the amount a raised by the students.

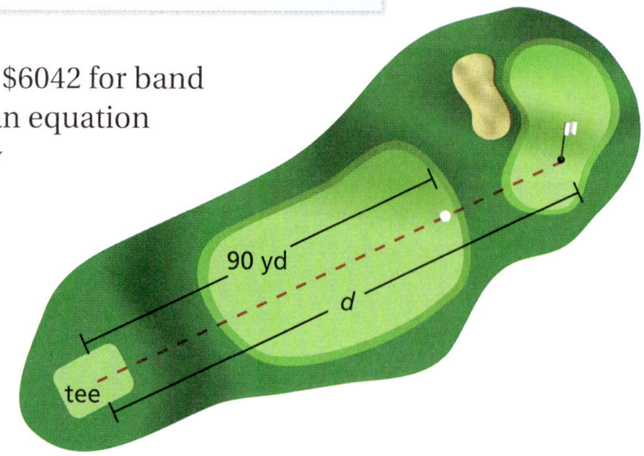

16. **GOLF** You hit a golf ball 90 yards. It travels three-fourths of the distance to the hole. Write an equation you can use to find the distance d from the tee to the hole.

Assignment Guide and Homework Check

Level	Day 1 Activity Assignment	Day 2 Lesson Assignment	Homework Check
Basic	4, 5, 23–27	1–3, 7–17 odd, 14, 16	7, 11, 16, 17
Average	4, 5, 23–27	1–3, 7–13 odd, 14–20	7, 11, 15, 17, 20
Advanced	4, 5, 23–27	1–3, 10–16 even, 17–22	12, 16, 18, 20, 21

Common Errors

- **Exercises 6–13** Students may write the equations in the wrong order or forget parts of the equation. Remind them of the guidelines for writing expressions. Remind students that these are equations, so there will be equal signs in the answer and numbers and/or variables on each side.
- **Exercise 20** Students may forget that the sales tax rate *r* is the decimal form of the percent of sales tax. Remind them that they are using the decimal form of the percent in part (a).

7.1 Record and Practice Journal

Write the word sentence as an equation.

1. 27 is 3 times a number y.
   $27 = 3y$
2. The difference of a number x and 4 is 3.
   $x - 4 = 3$
3. 8 more than a number p is 17.
   $p + 8 = 17$
4. Half of a number q is 14.
   $\frac{q}{2} = 14$

Write an equation that can be used to find the value of *x*.

5. Perimeter of rectangle: 32 cm

   $2x + 8 = 32$

6. Perimeter of triangle: 20 in.

   $3x = 20$

7. You spend $16 on 3 notebooks and x binders. Notebooks cost $2 each and binders cost $5 each. Write an equation you can use to find the number of binders you bought.
   $5x + 6 = 16$

Vocabulary and Concept Check

1. An equation has an equal sign and an expression does not.
2. A number n is 4 less than 8; $n = 8 - 4$; $n - 4 = 8$
3. *Sample answer:* A number n subtracted from 28 is 5.

Practice and Problem Solving

4. You run for 21 minutes at a rate of 7 minutes per mile. How far do you run? 3 mi
5. What was the high temperature if it was 4° less than 62°F? 58°F
6. $x + 4 = 12$
7. $y - 9 = 8$
8. $9b = 36$
9. $w \div 5 = 6$
10. $54 = t + 9$
11. $5 = \frac{1}{4}c$
12. $11 = y \div 6$
13. $n - 9 = 27$
14. The word *is* indicates an equal sign, so the equal sign is in the wrong position. $n = 12 + 5$
15. $6042 = 1780 + a$
16. $90 = \frac{3}{4}d$

Practice and Problem Solving

17. $16 = 3x$

18. $30 = 4x$

19. $326 = 12(14) + 6(5) + 16x$

20. **a.** *Answer should include, but is not limited to:* The equation should have the form total amount paid = total price + (total price × rate).

b. yes; Multiplying r by 100 gives the percent for the sales tax.

21. See *Taking Math Deeper.*

22. $84 = 14s - \frac{1}{2}\left(14 \cdot \frac{1}{2}s\right)$

Fair Game Review

23. 13 **24.** 3

25. 28 **26.** 5

27. B

Mini-Assessment

Write the word sentence as an equation.

1. The sum of a number x and 5 equals 20. $x + 5 = 20$
2. 64 equals 33 more than a number y. $64 = 33 + y$
3. A number c decreased by 12 is 4. $c - 12 = 4$
4. 10 is the quotient of a number s and 8. $\frac{s}{8} = 10$
5. You went grocery shopping with \$40.65. You left the store with \$5.10. Write an equation you can use to find the cost c of the groceries. *Sample answer:* $5.10 + c = 40.65$

Taking Math Deeper

Exercise 21

This problem is similar to real life because the given information is not presented in nice little bundles. Start by organizing the given information.

1 Organize the given information.

Information about blocks

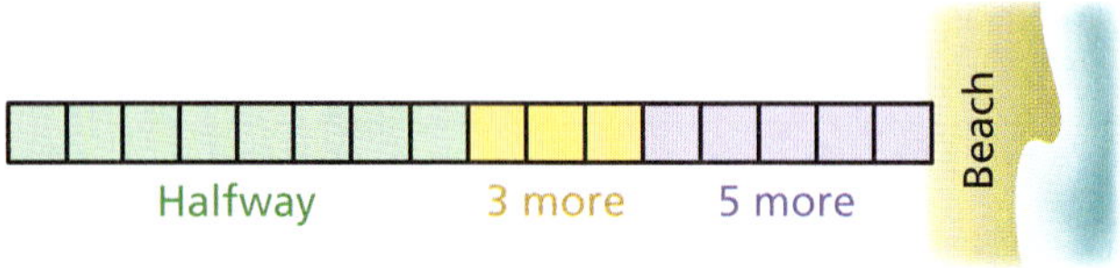

There are 16 blocks.
The time is 28 minutes.

In U.S. cities there are 12–16 blocks per mile.

Information about strawberries

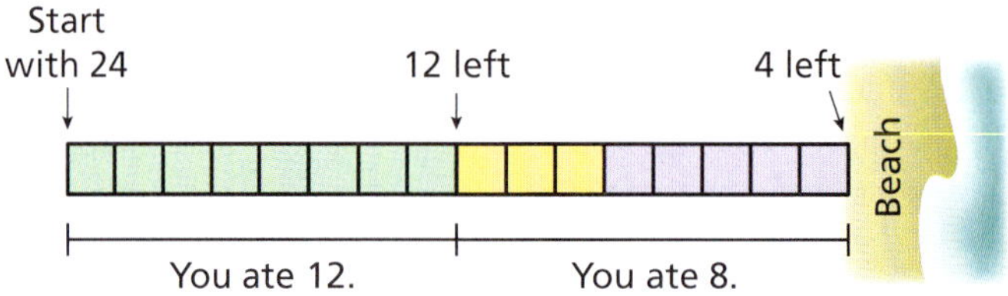

2 **a.** Yes, there is enough information to find the time it takes to walk each block. You walked 16 blocks in 28 minutes.

Time: $\frac{28}{16} = 1.75$ minutes, or 1 minute 45 seconds per block

3 **b.** No, there is not enough information to find how many strawberries you ate while walking the last block. The rate at which you are eating the strawberries is not constant. You ate 12 in the first 8 blocks, but only 8 in the last 8 blocks. On the other hand, if a student estimates that eating 8 strawberries in 8 blocks means that the person probably ate 1 strawberry while walking the last block, that is still good thinking.

Reteaching and Enrichment Strategies

If students need help. . .	If students got it. . .
Resources by Chapter • Practice A and Practice B • Puzzle Time Record and Practice Journal Practice Differentiating the Lesson Lesson Tutorials Skills Review Handbook	Resources by Chapter • Enrichment and Extension • Technology Connection Start the next section

GEOMETRY Write an equation that you can use to find the value of x.

17. Perimeter of triangle: 16 in.

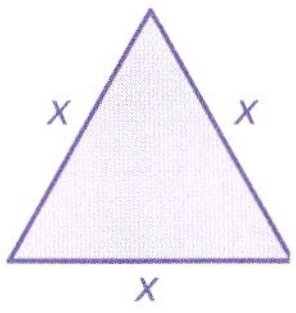

18. Perimeter of square: 30 mm

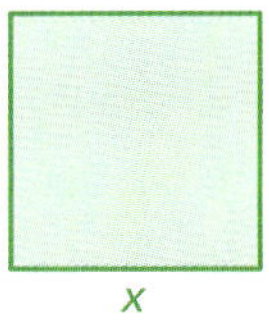

19. **MUSIC** You sell instruments at a Caribbean music festival. You earn \$326 by selling 12 sets of maracas, 6 sets of claves, and x *djembe* drums. Write an equation you can use to find the number of *djembe* drums you sold.

20. **SALES TAX** Find a sales receipt from a store that shows the total price and the total amount paid including sales tax.

a. Write an equation you can use to find the sales tax rate r.

b. Can you use r to find the *percent* for the sales tax? Explain.

21. **STRAWBERRIES** You buy a basket of 24 strawberries. You eat them as you walk to the beach. It takes the same amount of time to walk each block. When you are halfway there, half of the berries are gone. After walking 3 more blocks, you still have 5 blocks to go. You reach the beach 28 minutes after you began. One-sixth of your strawberries are left.

a. Is there enough information to find the time it takes to walk each block? Explain.

b. Is there enough information to find how many strawberries you ate while walking the last block? Explain.

22. Geometry A triangle is cut from a rectangle. The height of the triangle is half of the unknown side length s. The area of the shaded region is 84 square inches. Write an equation you can use to find the side length s.

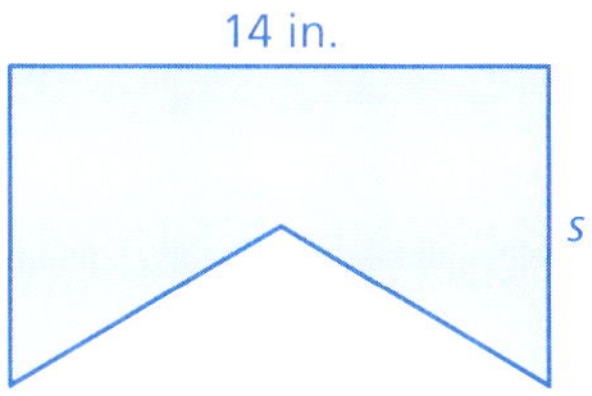

Fair Game Review What you learned in previous grades & lessons

Evaluate the expression when $a = 7$. *(Section 3.1)*

23. $6 + a$ **24.** $a - 4$ **25.** $4a$ **26.** $\frac{35}{a}$

27. **MULTIPLE CHOICE** Which expression is equivalent to $8(x + 3)$? *(Section 3.4)*

Ⓐ $8x + 3$ Ⓑ $8x + 24$ Ⓒ $8x + 11$ Ⓓ $x + 24$

7.2 Solving Equations Using Addition or Subtraction

Essential Question How can you use addition or subtraction to solve an equation?

When two sides of a scale weigh the same, the scale will balance.

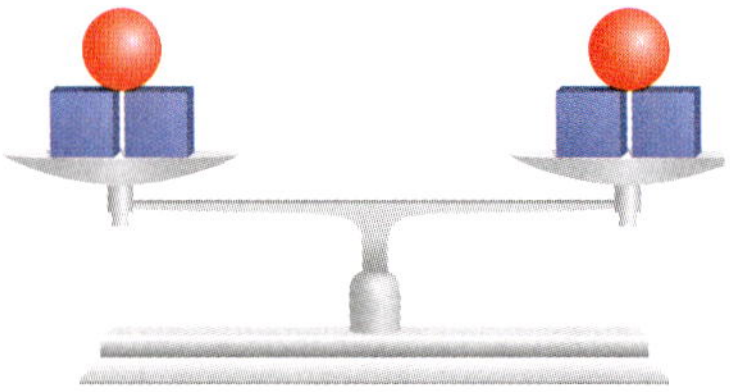

When you add or subtract the same amount on each side of the scale, it will still balance.

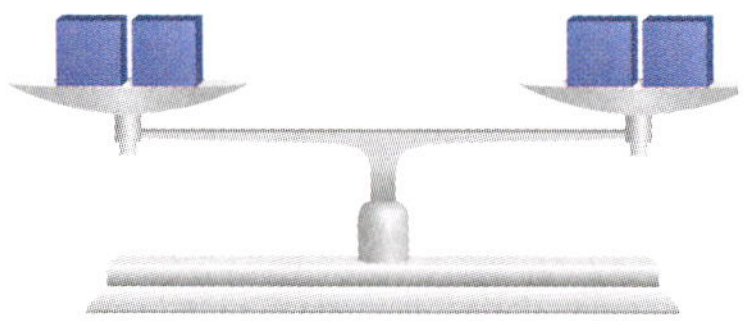

1 ACTIVITY: Solving an Equation

Work with a partner.

a. Use a model to solve $n + 3 = 7$.

- Explain how the model represents the equation $n + 3 = 7$.

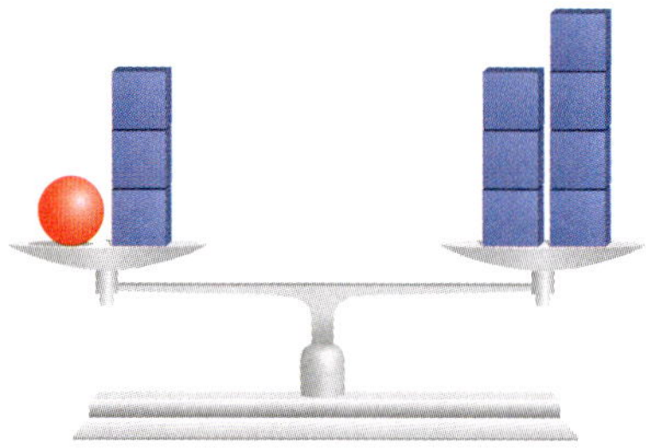

- How much does one  weigh? How do you know?

The solution is $n =$ ____.

b. Describe how you could check your answer in part (a).

c. Which model below represents the solution of $n + 1 = 9$? How do you know?

Common Core

Solving Equations

In this lesson, you will
- use addition or subtraction to solve equations.
- use substitution to check answers.
- solve real-life problems.

Learning Standards
6.EE.5
6.EE.7

Laurie's Notes

Introduction

Standards for Mathematical Practice

- **MP5 Use Appropriate Tools Strategically:** Students begin solving equations by using mental math to estimate and then check their answers to see if they make sense. Using mental math to estimate an answer is a tool that students need to utilize.

Motivate

- Ask for 6 volunteers to come to the front of the room. Three students are Team A and the other 3 are Team B.
- After they are positioned say, "I changed my mind. I want [name a 7th student] to join Team A." Team B will immediately want someone else to join their team so it will be fair (balanced). A = 4 people, B = 3 people
- Change your mind again and have two students from Team A sit down. A = 2 people, B = 3 people Now Team A will say it is unfair (not balanced).
- Students have a sense about fairness, and they want everyone to be treated the same. This idea can be used with equations. You want to treat both sides of the equation the same way.

Discuss

- Discuss the two scales in the context of being balanced. If any quantity is added to one side, the same amount must be added to the other. If any quantity is removed (subtracted) from one side, the same amount must be removed (subtracted) from the other side.

Activity Notes

Activity 1

- **Representation:** If you are familiar with algebra tiles, you could use the tiles to model these problems. You could also draw the sketch shown or model with actual materials on a pan balance.
- **Note:** Physical objects on a pan balance make more sense to students than simply seeing a variable and numbers. The equation written above the balance scale (as in Activity 2) helps students think about two sides of an equation being balanced. The physical objects help students think what must be done mathematically (operations) to solve the equation.
- The concept of what it means to *solve an equation* may be very unfamiliar to students. They have solved many problems in math, and now they are being asked to solve an equation. Problems and equations are not the same, especially if students think of the equal sign as a rightward pointing arrow. The *problem* $4 + 3 \rightarrow ?$ and the *equation* $? + 3 = 7$ are different to students.
- Discuss the example shown and model it with physical objects. Note that the goal is stated: "Get n by itself on one side." You want the variable by itself. With the physical models you want the object representing the variable by itself on one side.

Common Core State Standards

6.EE.5 Understand solving an equation . . . as a process of answering a question: which values from a specified set, if any, make the equation . . . true? Use substitution to determine whether a given number in a specified set makes an equation . . . true.

6.EE.7 Solve real-world and mathematical problems by writing and solving equations of the form $x + p = q$ and $px = q$ for cases in which p, q and x are all nonnegative rational numbers.

Previous Learning

Students should know how to evaluate expressions.

Technology for the Teacher

Dynamic Classroom

Lesson Plans

Complete Materials List

7.2 Record and Practice Journal

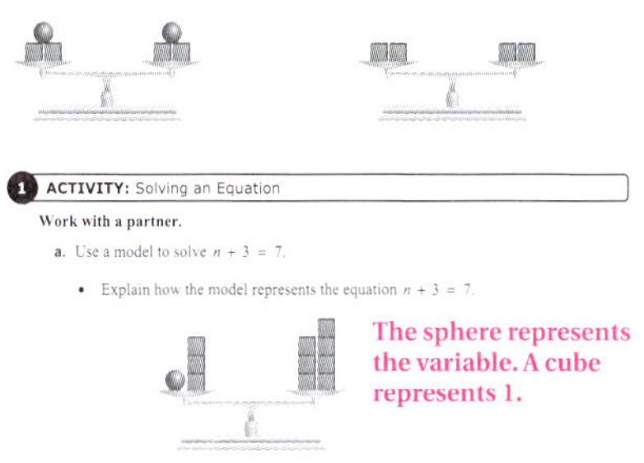

Essential Question How can you use addition or subtraction to solve an equation?

When two sides of a scale weigh the same, the scale will balance.

When you add or subtract the same amount on each side of the scale, it will still balance.

1 ACTIVITY: Solving an Equation

Work with a partner.

a. Use a model to solve $n + 3 = 7$.

- Explain how the model represents the equation $n + 3 = 7$.

The sphere represents the variable. A cube represents 1.

- How much does one sphere weigh? How do you know? 4 cubes

The solution is $n =$ 4.

b. Describe how you could check your answer in part (a).

Substitute 4 for n and evaluate. If true, then answer is correct.

Differentiated Instruction

Kinesthetic

Have students play *The Solution Game*. On index cards, write equations that can be solved using mental math. Working in small groups, one student turns over a card. The other students suggest solutions. The solution should be checked by substitution into the equation.

7.2 Record and Practice Journal

c. Which model below represents the solution of $n + 1 = 9$? How do you know? **left**

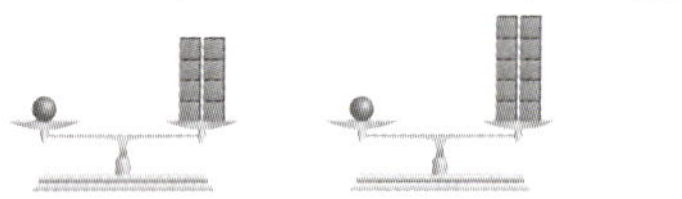

2 ACTIVITY: Solving Equations

Work with a partner. Solve the equation using the method in Activity 1.

a. $n + 5 = 10$
$n = 5$

b. $x + 2 = 11$
$x = 9$

c. $6 = y + 3$
$y = 3$

d. $8 = m + 8$
$m = 0$

3 ACTIVITY: Solving Equations Using Mental Math

Work with a partner. Write a question that represents the equation. Use mental math to answer the question. Then check your solution.

Equation	Question	Solution	Check
a. $x + 1 = 5$	a–e.	$x = 4$	$4 + 1 = 5$ ✓
b. $4 + m = 11$	See Additional	$m = 7$	$4 + 7 = 11$ ✓
c. $8 = a + 3$	Answers.	$a = 5$	$8 = 5 + 3$
d. $x - 9 = 21$		$x = 30$	$30 - 9 = 21$
e. $13 = p - 4$		$p = 17$	$13 = 17 - 4$

What Is Your Answer?

4. **REPEATED REASONING** In Activity 3, how are parts (d) and (e) different from parts (a)–(c)? Did your process to find the solution change? Explain.
(d) and (e) are subtraction equations; use addition to get the variable by itself

5. Decide whether the statement is *true* or *false*. If false, explain your reasoning.
   a. In an equation, you can use any letter as a variable.
   true
   b. The goal in solving an equation is to get the variable by itself.
   true
   c. In the solution, the variable must be on the left side of the equal sign.
   false; It can be on either side.
   d. If you add a number to one side, you should subtract it from the other side.
   false; If you add to one side, you should add to the other side.

6. **IN YOUR OWN WORDS** How can you use addition or subtraction to solve an equation? Give two examples to show how your procedure works.
Use the inverse operation on both sides to undo the operation.

7. Are the following equations equivalent? Explain your reasoning.
$x - 5 = 12$ and $12 = x - 5$
yes

Laurie's Notes

Activity 2

- In the formal lesson, students will be expected to show the mathematical process. In other words, identify which number has been added to or subtracted from each side of the equation. At this stage, you might want students to simply note what is being done to each side of the equation, and then write the solution. For example, part (c) would be "subtract 3 from each side; $y = 3$."
- **MP2 Reason Abstractly and Quantitatively:** Now students are transitioning from a physical object (sphere) to a variable and from counters (cubes) to a number. The pan scale remains the same and the concept of wanting to keep the scale balanced remains the same. Students should be thinking, "What does the variable represent in the equation?" The variable must represent a quantity to keep the scale balanced.

Activity 3

- Read the sample problem.

? "Can you ask a similar question for the equation $x + 1 = 5$?" Yes, you want students to understand that there is more than one way to phrase the question. What do you add to 1 to get 5?

- This is just mental math at this stage. Do not expect students to state inverse operations. Let students rely on their mental math.
- **Note:** Two of the problems have the variables on the right side of the equation. The question formed is still an equivalence: "8 is the sum of 3 and what number?"

? When students finish ask, "Could part (d) be phrased, what number do you subtract from 9 to get 21?" No, *subtracting a number from 9* and *subtracting 9 from a number* are not the same.

What Is Your Answer?

- Have students work with partners. To do a quick (but private) check of student understanding, have students give a thumbs up for *true* and a thumbs down for *false*. Discuss each question as needed.

Closure

- Select 3 people for Team A and 3 people for Team B and repeat the questions from the motivate activity.
  - "If I have 3 people join Team B, then …?" Students should say, "Have 3 people join Team A." 12 students are now standing.
  - "If I have 5 people from Team B sit down, then…?" Students should say, "Have 5 people from Team A sit down."

2 ACTIVITY: Solving Equations

Work with a partner. Solve the equation using the method in Activity 1.

Math Practice 2

Understand Quantities

What does the variable represent in the equation?

a. 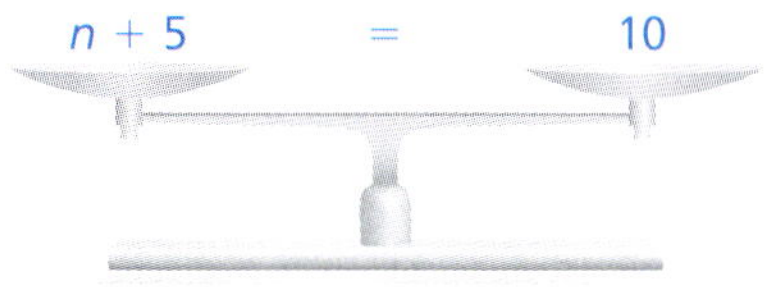

b. 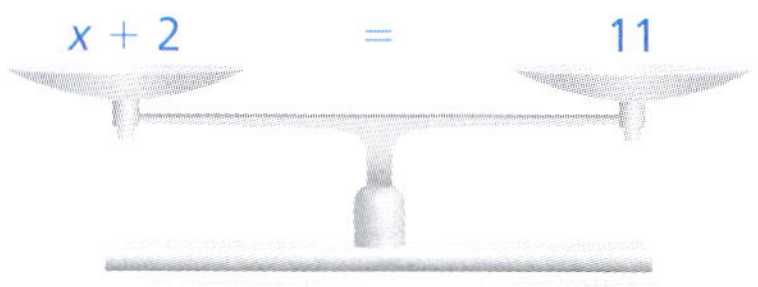

c. 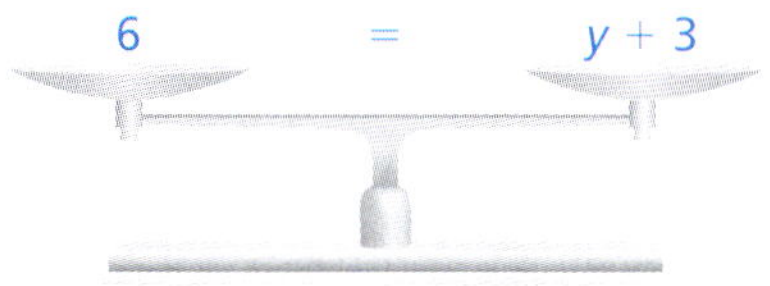

d. 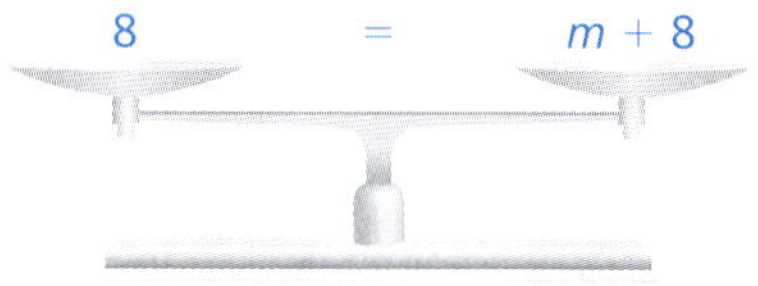

3 ACTIVITY: Solving Equations Using Mental Math

Work with a partner. Write a question that represents the equation. Use mental math to answer the question. Then check your solution.

Equation	Question	Solution	Check
a. $x + 1 = 5$			
b. $4 + m = 11$			
c. $8 = a + 3$			
d. $x - 9 = 21$			
e. $13 = p - 4$			

What Is Your Answer?

4. **REPEATED REASONING** In Activity 3, how are parts (d) and (e) different from parts (a)–(c)? Did your process to find the solution change? Explain.
5. Decide whether the statement is *true* or *false*. If false, explain your reasoning.
   - **a.** In an equation, you can use any letter as a variable.
   - **b.** The goal in solving an equation is to get the variable by itself.
   - **c.** In the solution, the variable must always be on the left side of the equal sign.
   - **d.** If you add a number to one side, you should subtract it from the other side.
6. **IN YOUR OWN WORDS** How can you use addition or subtraction to solve an equation? Give two examples to show how your procedure works.
7. Are the following equations equivalent? Explain your reasoning.

$$x - 5 = 12 \qquad \text{and} \qquad 12 = x - 5$$

Practice Use what you learned about solving equations to complete Exercises 12–17 on page 305.

7.2 Lesson

Check It Out
Lesson Tutorials
BigIdeasMath.com

Key Vocabulary
solution, *p. 302*
inverse operations, *p. 303*

Equations may be true for some values and false for others. A **solution** of an equation is a value that makes the equation true.

Value of x	$x + 3 = 7$	Are both sides equal?
3	$3 + 3 \stackrel{?}{=} 7$ $6 \neq 7$ ✗	no
4	$4 + 3 \stackrel{?}{=} 7$ $7 = 7$ ✓	yes
5	$5 + 3 \stackrel{?}{=} 7$ $8 \neq 7$ ✗	no

Reading
The symbol $\neq$ means *is not equal to*.

So, the value $x = 4$ is a solution of the equation $x + 3 = 7$.

EXAMPLE 1 Checking Solutions

Tell whether the given value is a solution of the equation.

a. $p + 10 = 38;\ p = 18$

$18 + 10 \stackrel{?}{=} 38$ — Substitute 18 for p.

$28 \neq 38$ ✗ — Sides are *not* equal.

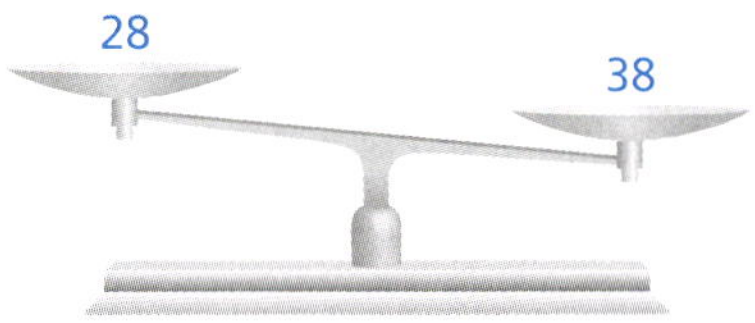

So, $p = 18$ is *not* a solution.

b. $4y = 56;\ y = 14$

$4(14) \stackrel{?}{=} 56$ — Substitute 14 for y.

$56 = 56$ ✓ — Sides are equal.

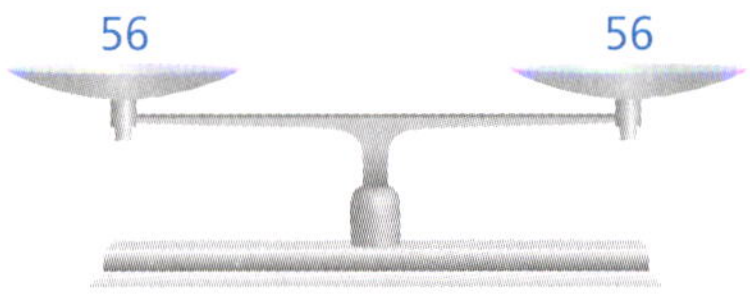

So, $y = 14$ is a solution.

On Your Own

Now You're Ready
Exercises 6–11

Tell whether the given value is a solution of the equation.

1. $a + 6 = 17;\ a = 9$

2. $9 - g = 5;\ g = 3$

3. $35 = 7n;\ n = 5$

4. $\frac{q}{2} = 28;\ q = 14$

Multi-Language Glossary at BigIdeasMath.com

Laurie's Notes

Introduction

Connect

- **Yesterday:** Students gained an intuitive understanding of how equations are solved using addition and subtraction. (MP2, MP5)
- **Today:** Students will use Properties of Equality to solve equations.

For the Teacher

- Use file folder paper to make a quick model to help students think about what it means to solve an equation.
- Make a number strip with the numbers 3, 4, and 5. Make three equation strips with equations $x + 4 = 9$, $1 = x - 3$, and $5 + x = 8$. Cut a slit in the paper on either side of the variable so that the number strip can slide through, replacing the variable with a number.

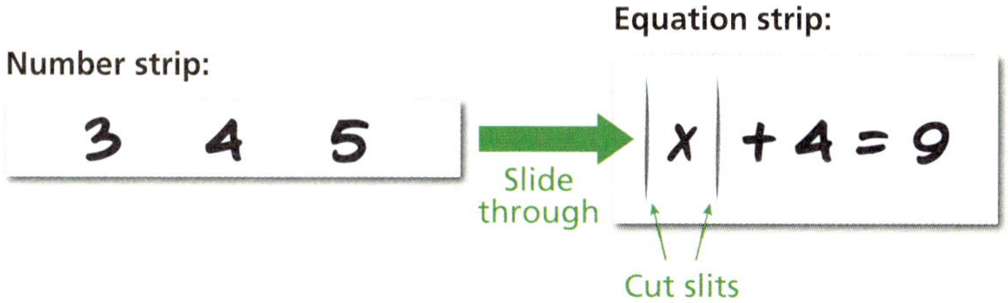

Motivate

- Tell the students you have several equations. They are to decide if the equation is *true* or *false*. They can vote with thumbs up/down or on personal white boards.
- When the number slip is woven through the opening, it will read: $5 + 4 = 9$ (true), $4 + 4 = 9$ (false), and $3 + 4 = 9$ (false).
- Point out to students that equations can have only numbers, $5 + 4 = 9$, or numbers and variables, $x + 4 = 9$.
- Use the same number strip with the other equation strips.
- The goal is for students to recognize that one of the numbers makes a true statement and the other numbers make false statements.

? "If you use the equation $x - 3 = 4$, do you think any of the numbers (3, 4, or 5) would have made a true statement?" no

- Define a solution of an equation.

Lesson Notes

Example 1

- Work through each part.
- **MP6 Attend to Precision:** Be careful to use the correct language. You *substitute* for the value of the variable. You do not *plug it in*!

? "What does $4y$ mean, forty-something?" No, it means 4 times the value of y.

On Your Own

- **Neighbor Check:** Have students work independently and then have their neighbors check their work. Have students discuss any discrepancies.
- **Common Error:** For Question 4, students will quickly look at the numbers 2, 14, and 28 and say 14 is a solution without thinking about the operation. Caution students to work slowly and actually substitute the value of the variable.

Goal Today's lesson is using the **inverse operations** of addition and subtraction to find the **solution** of an equation.

Technology for the Teacher

Dynamic Classroom

Lesson Tutorials
Lesson Plans
Answer Presentation Tool

Extra Example 1

Tell whether the given value is a solution of the equation.

a. $t - 5 = 17$; $t = 12$ no

b. $b \div 6 = 8$; $b = 48$ yes

On Your Own

1. no	**2.** no
3. yes	**4.** no

English Language Learners

Vocabulary

English learners may have difficulty understanding that addition *undoes* subtraction and that subtraction *undoes* addition. Emphasize that *undo* means to do the opposite of what was done before.

You get $x - 2$ by subtracting 2 from x.
You get $x - 2$ back to x by adding 2 to $x - 2$.
$x - 2 + 2 = x$

You get $x + 11$ by adding 11 to x.
You get $x + 11$ back to x by subtracting 11 from $x + 11$.
$x + 11 - 11 = x$

This will help students to better understand *inverse operations.*

Extra Example 2

a. Solve $p - 8 = 14$. $p = 22$

b. Solve $11 = y - 4$. $y = 15$

On Your Own

5. $k = 4$

6. $n = 14$

7. $r = 21$

Laurie's Notes

Key Ideas

- Discuss other inverse operations such as: putting your shoes on and taking them off; going to school and coming home from school; an airplane taking off and landing; turning a light switch on and off.
- Write the Key Ideas.
- Discuss the Words, Numbers, and Algebra.
- How you record what is being done mathematically to each side of the equation is very important. You should not be sloppy. Encourage students to develop good habits from the start.

Example 2

- Explain to students that although they can probably solve these problems using mental math, another equally important part of equation solving is the *process*. How can the problem be solved when the answer cannot be found using mental math?
- Write a few examples.

  (a) $x - 5 = 13$ **(b)** $x - 8.7 = 19.49$ **(c)** $x - 4\frac{3}{5} = 6\frac{2}{3}$
- You can use mental math for (a). For (b) and (c), you probably want to use paper and pencil. What is the process now?
- Again note that the variable is not always on the left side of the equation. You want students to be as comfortable solving $x - 7 = 18$ as $18 = x - 7$.
- **Note:** Some textbooks use the horizontal format as well, but using just one format this year is fine.
- Discuss with students how to determine if their answer is correct by substituting the solution for the variable in the original equation.
- **Big Idea:** Whatever number is added to one side of the equation is also added to the other. Using an alternate color (chalk or overhead pen) to show this will help the many students who are visual learners. Notice in the examples that the *process* is annotated along the right side of the equation.

On Your Own

- Ask volunteers to share their work at the board.

You can use *inverse operations* to solve equations. **Inverse operations** "undo" each other. Addition and subtraction are inverse operations.

Key Ideas

Addition Property of Equality

Words When you add the same number to each side of an equation, the two sides remain equal.

Numbers

$$\begin{aligned} 8 &= 8 \\ +5 &\quad +5 \\ \hline 13 &= 13 \end{aligned}$$

Algebra

$$\begin{aligned} x - 4 &= 5 \\ +4 &\quad +4 \\ \hline x &= 9 \end{aligned}$$

Subtraction Property of Equality

Words When you subtract the same number from each side of an equation, the two sides remain equal.

Numbers

$$\begin{aligned} 8 &= 8 \\ -5 &\quad -5 \\ \hline 3 &= 3 \end{aligned}$$

Algebra

$$\begin{aligned} x + 4 &= 5 \\ -4 &\quad -4 \\ \hline x &= 1 \end{aligned}$$

EXAMPLE 2 Solving Equations Using Addition

a. Solve $x - 2 = 6$.

$x - 2 = 6$ Write the equation.

Undo the subtraction. → $+2 \quad +2$ Addition Property of Equality

$x = 8$ Simplify.

∴ The solution is $x = 8$.

Check

$x - 2 = 6$

$8 - 2 \stackrel{?}{=} 6$

$6 = 6$ ✓

Study Tip

You can check your solution by substituting it for the variable in the original equation.

b. Solve $18 = x - 7$.

$18 = x - 7$ Write the equation.

$+7 \quad +7$ Addition Property of Equality

$25 = x$ Simplify.

∴ The solution is $x = 25$.

Check

$18 = x - 7$

$18 \stackrel{?}{=} 25 - 7$

$18 = 18$ ✓

On Your Own

Now You're Ready
Exercises 18–20

Solve the equation. Check your solution.

5. $k - 3 = 1$
6. $n - 10 = 4$
7. $15 = r - 6$

EXAMPLE 3 Solving Equations Using Subtraction

a. Solve $x + 2 = 9$.

	$x + 2 = 9$	Write the equation.
Undo the addition. →	$\underline{-2} \quad \underline{-2}$	Subtraction Property of Equality
	$x = 7$	Simplify.

The solution is $x = 7$.

Check

$x + 2 = 9$

$7 + 2 \stackrel{?}{=} 9$

$9 = 9$ ✓

b. Solve $26 = 11 + x$.

$26 = 11 + x$	Write the equation.
$\underline{-11} \quad \underline{-11}$	Subtraction Property of Equality
$15 = x$	Simplify.

The solution is $x = 15$.

Check

$26 = 11 + x$

$26 \stackrel{?}{=} 11 + 15$

$26 = 26$ ✓

EXAMPLE 4 Real-Life Application

Your parents give you \$20 to help buy the new pair of shoes shown. After you buy the shoes, you have \$5.50 left. Write and solve an equation to find how much money you had before your parents gave you \$20.

Words	The starting amount	plus	the amount your parents gave you	minus	the cost of the shoes	is	the amount left.
Variable	Let s be the starting amount.						
Equation	s	$+$	20	$-$	59.95	$=$	5.50

$s + 20 - 59.95 = 5.50$	Write the equation.
$s + 20 - 59.95 + 59.95 = 5.50 + 59.95$	Addition Property of Equality
$s + 20 = 65.45$	Simplify.
$s + 20 - 20 = 65.45 - 20$	Subtraction Property of Equality
$s = 45.45$	Simplify.

You had \$45.45 before your parents gave you money.

Study Tip

In Example 4, you can solve the problem arithmetically by working backwards from \$5.50.

$5.50 + 59.95 - 20 = 45.45$

So, your answer is reasonable.

On Your Own

Now You're Ready Exercises 21–23

Solve the equation. Check your solution.

8. $s + 8 = 17$ **9.** $9 = y + 6$ **10.** $13 + m = 20$

11. You eat 8 blueberries and your friend eats 11 blueberries from a package. There are 23 blueberries left. Write and solve an equation to find the number of blueberries in a full package.

Laurie's Notes

Example 3

? "How do you undo the operation of addition?" subtraction

- Work through the problems.

? "Why isn't 9 subtracted from each side of the equation in part (a)?" You could; however, it wouldn't achieve the goal of getting the variable by itself on one side of the equation. You need to subtract 2 from each side of the equation.

- **Big Idea:** Whatever is subtracted from one side of the equation is also subtracted from the other. Using an alternate color (chalk or overhead pen) to show this will help visual learners. Notice that the *process* is annotated along the right side of the equation.

Example 4

- This example uses skills developed in the last section. Note how the words in the verbal model come directly from the written problem.
- Information to write the verbal model must be read from the photograph.
- Students will want to skip writing the equation and simply jump to $5.50 + 59.95 - 20$ to arrive at the answer. Remind them that the question asks for an equation to be written. The equation helps to explain why subtraction is the correct operation to use in solving.
- Students should check their answers in the original equation and within the context of the problem. Does the answer make sense?

On Your Own

- Ask volunteers to share their work at the board.

Closure

- **Exit Ticket:** Solve the equations.

  $17 = 6 + t$ $t = 11$

  $m + 6 = 17$ $m = 11$

? "Are these two problems the same? Explain." Yes; Both equations could be stated as *6 added to a number is 17.*

Extra Example 3

a. Solve $z + 5 = 14$. $z = 9$

b. Solve $18 = r + 6$. $r = 12$

Extra Example 4

You climb 12 feet on a rock climbing wall that is 25 feet high. Write and solve an equation to find the distance to the top. $12 + d = 25$; $d = 13$ ft

On Your Own

8. $s = 9$

9. $y = 3$

10. $m = 7$

11. $x - 8 - 11 = 23$; $x = 42$ blueberries

Vocabulary and Concept Check

1. Substitute your solution back into the original equation and see if you obtain a true statement.
2. addition
3. subtraction
4. subtraction
5. so that x is by itself; so that the two sides remain equal

Practice and Problem Solving

6. yes
7. yes
8. no
9. no
10. no
11. yes
12. $n = 2$
13. $t = 1$
14. $c = 6$
15. What number plus 5 equals 12?; $a = 7$
16. What number plus 9 equals 18?; $v = 9$
17. 20 is what number minus 6?; $d = 26$
18. $y = 10$
19. $z = 16$
20. $r = 22$
21. $p = 3$
22. $k = 12$
23. $h = 34$
24. $f = 46$
25. $q = 11$
26. $j = 1\frac{1}{4}$
27. $x = \frac{7}{30}$
28. $m = 3.7$
29. $a = 11.8$
30. They should subtract 7 from both sides.

$$\begin{aligned} x + 7 &= 13 \\ -7 \quad & \;\; -7 \\ x &= 6 \end{aligned}$$

31. They must apply the same operation to both sides.

$$\begin{aligned} 34 &= y - 12 \\ +12 \quad & \quad\; +12 \\ 46 &= y \end{aligned}$$

Assignment Guide and Homework Check

Level	Day 1 Activity Assignment	Day 2 Lesson Assignment	Homework Check
Basic	12–17, 57–61	1–5, 7, 9, 11, 19–25 odd, 30, 31, 33	7, 19, 21, 33
Average	12–17, 57–61	1–5, 19–31 odd, 32–34, 44–46	19, 27, 32, 44
Advanced	12–17, 57–61	1–5, 30, 34, 35, 40–50 even, 49, 52–54	34, 40, 50, 52, 53

Common Errors

- **Exercises 6–11** Students may try to perform operations on both sides of the equation in order to get the same answer on both sides. Remind them that they are checking to see if the value of the variable makes the equation true, so they should focus on simplifying one side of the equation and then compare the answer to the other side.
- **Exercises 18–29** Instead of using the inverse operation, students may use the same operation in solving for x. Simplify the equation on the board or overhead to demonstrate that this will not work. Students may have ignored the side with the variable when they made this mistake. Remind them to check their answers in the original equation.
- **Exercises 18–29** Students may add or subtract the number on the side without the variable. For example, $-14 + 14 = k + 6 + 14$ instead of $-14 - 6 = k + 6 - 6$. Remind them that they are trying to get the variable by itself. They should start on the side with the variable and use the inverse of that operation.

7.2 Record and Practice Journal

Tell whether the given value is a solution of the equation.

1. $34 + x = 46; x = 12$
   yes
2. $y - 9 = 14; y = 22$
   no
3. $6d = 54; d = 9$
   yes
4. $\frac{n}{3} = 13; n = 39$
   yes

Solve the equation. Check your solution.

5. $7 + k = 11$
   $k = 4$
6. $p - 24 = 13$
   $p = 37$
7. $b - 16 = 7$
   $b = 23$
8. $\frac{2}{5} + m = \frac{5}{6}$
   $m = \frac{13}{30}$
9. In the heavyweight class of professional wrestling, the junior weight limit is 190 pounds. This is 15 pounds heavier than the light heavyweight limit. Write and solve an equation to find the weight limit of the light heavyweight class.

   $w + 15 = 190$
   $w = 175$

7.2 Exercises

Check It Out
Help with Homework
BigIdeasMath.com

Vocabulary and Concept Check

1. **WRITING** How can you check the solution of an equation?

Name the inverse operation you can use to solve the equation.

2. $x - 8 = 12$
3. $n + 3 = 13$
4. $b + 14 = 33$

5. **WRITING** When solving $x + 5 = 16$, why do you subtract 5 from the left side of the equation? Why do you subtract 5 from the right side of the equation?

Practice and Problem Solving

Tell whether the given value is a solution of the equation.

1 6. $x + 42 = 85;\ x = 43$

7. $8b = 48;\ b = 6$

8. $19 - g = 7;\ g = 15$

9. $\dfrac{m}{4} = 16;\ m = 4$

10. $w + 23 = 41;\ w = 28$

11. $s - 68 = 11;\ s = 79$

Use a scale to model and solve the equation.

12. $n + 7 = 9$
13. $t + 4 = 5$
14. $c + 2 = 8$

Write a question that represents the equation. Use mental math to answer the question. Then check your solution.

15. $a + 5 = 12$
16. $v + 9 = 18$
17. $20 = d - 6$

Solve the equation. Check your solution.

2 18. $y - 7 = 3$

19. $z - 3 = 13$

20. $8 = r - 14$

3 21. $p + 5 = 8$

22. $k + 6 = 18$

23. $64 = h + 30$

24. $f - 27 = 19$

25. $25 = q + 14$

26. $\dfrac{3}{4} = j - \dfrac{1}{2}$

27. $x + \dfrac{2}{3} = \dfrac{9}{10}$

28. $1.2 = m - 2.5$

29. $a + 5.5 = 17.3$

ERROR ANALYSIS **Describe and correct the error in solving the equation.**

30. ✗

$$\begin{array}{rcl} x + 7 & = & 13 \\ +7 & & +7 \\ \hline x & = & 20 \end{array}$$

31. ✗

$$\begin{array}{rcl} 34 & = & y - 12 \\ -12 & & +12 \\ \hline 22 & = & y \end{array}$$

32. **PENGUINS** An emperor penguin is 45 inches tall. It is 24 inches taller than a rockhopper penguin. Write and solve an equation to find the height of a rockhopper penguin. Is your answer reasonable? Explain.

33. **ELEVATOR** You get in an elevator and go down 8 floors. You exit on the 16th floor. Write and solve an equation to find what floor you got on the elevator.

34. **AREA** The area of Jamaica is 6460 square miles less than the area of Haiti. Write and solve an equation to find the area of Haiti.

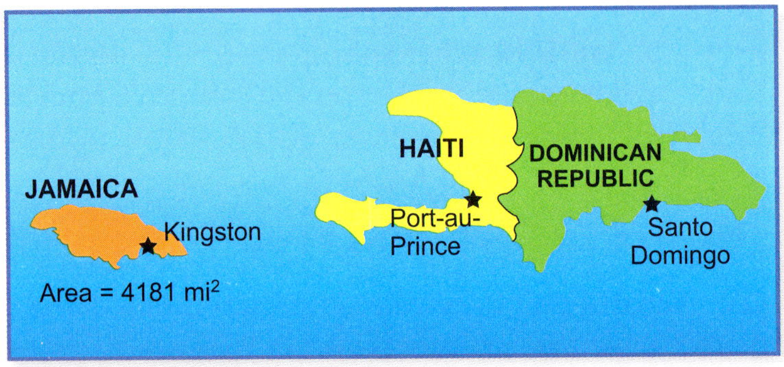

35. **REASONING** The solution of the equation $x + 3 = 12$ is shown. Explain each step. Use a property, if possible.

$$x + 3 = 12$$
$$x + 3 - 3 = 12 - 3$$
$$x + 0 = 9$$
$$x = 9$$

Write the equation.

Write the word sentence as an equation. Then solve the equation.

36. 13 subtracted from a number w is 15.

37. A number k increased by 7 is 34.

38. 9 is the difference of a number n and 7.

39. 93 is the sum of a number g and 58.

Solve the equation. Check your solution.

40. $b + 7 + 12 = 30$

41. $y + 4 - 1 = 18$

42. $m + 18 + 23 = 71$

43. $v - 7 = 9 + 12$

44. $5 + 44 = 2 + r$

45. $22 + 15 = d - 17$

GEOMETRY Write and solve an addition equation to find x.

46. Perimeter = 48 ft

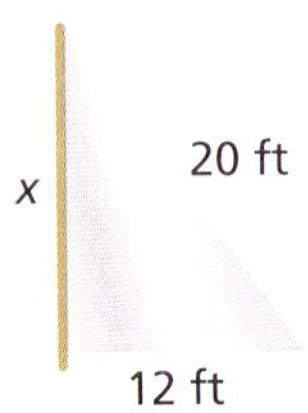

47. Perimeter = 132 in.

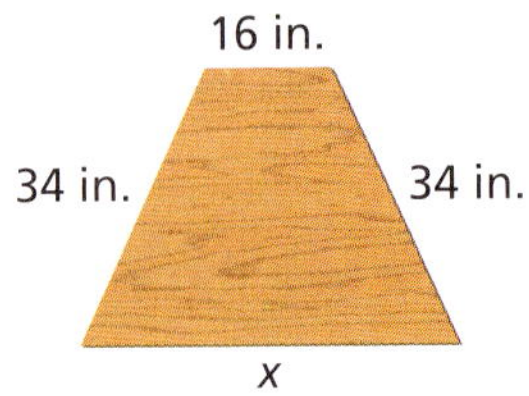

48. Perimeter = 93 ft

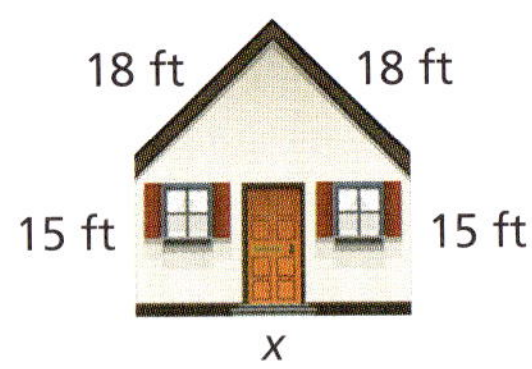

49. **REASONING** Explain why the equations $x + 4 = 13$ and $4 + x = 13$ have the same solution.

50. **REASONING** Explain why the equations $x - 13 = 4$ and $13 - x = 4$ do *not* have the same solution.

Common Errors

- **Exercises 32–34** Students may be struggling with determining what information is important in a word problem. Encourage them to simplify the word problem as they did in the previous section.
- **Exercises 36–39** Students may write the wrong equation and get an answer that is negative or have a variable that is negative. Remind them to read the word sentence carefully. Refer them to the previous section for help writing the equation.
- **Exercises 40–45** Students may lose one of the numbers when solving for the variable. For example,

$$\begin{aligned} b + 7 + 12 &= \quad 30 \\ \underline{-7} \quad\quad &\quad \underline{-7} \\ b &= \quad 23 \end{aligned}$$

Remind them to be very careful when solving for the variable. Encourage them to simplify and combine like terms before solving. Developing this skill will be valuable in future courses. Students should check their answers by substituting the value for the variable back into the original equation. This will help them to catch their mistakes.

Practice and Problem Solving

32. $45 = h + 24$; 21 in.; The answer is reasonable because the rockhopper penguin looks like it is about half the height of the emperor penguin in the picture.

33. $x - 8 = 16$; 24th floor

34. $h - 6460 = 4181$; 10,641 mi^2

35. Subtract 3 from each side. (Subtraction Property of Equality); Subtract 3 from 3. (Subtract.); Add x and 0. (Addition Property of Zero)

36. $w - 13 = 15$; $w = 28$

37. $k + 7 = 34$; $k = 27$

38. $9 = n - 7$; $n = 16$

39. $93 = g + 58$; $g = 35$

40. $b = 11$ **41.** $y = 15$

42. $m = 30$ **43.** $v = 28$

44. $r = 47$ **45.** $d = 54$

46. $x + 12 + 20 = 48$; 16 ft

47. $x + 34 + 34 + 16 = 132$; 48 in.

48. $x + 15 + 15 + 18 + 18 = 93$; 27 ft

49. Addition is commutative.

50. Subtraction is not commutative.

English Language Learners

Vocabulary

Make sure that students understand the definition of *solution*. They should understand that a solution needs to be checked in the original equation. Students should also understand the difference between solving an equation and simplifying an expression.

Practice and Problem Solving

51. See *Taking Math Deeper.*

52. $a + 53 = 75$; $b + 44 = 75$; $c + 62 = 75$
$a = 22$; $b = 31$; $c = 13$

53. $x + 15.50 + 8.75 = 66.55$; \$42.30

54. \$49

55. **a.** \$5.25

b. no; You have \$5.25 left and it costs \$9.75 to ride each ride once.

56. decreases by 3

Fair Game Review

57. 96 **58.** 208

59. 5 **60.** 24

61. C

Mini-Assessment

Solve the equation.

1. $y - 10 = 4$ $y = 14$
2. $25 = f + 13$ $f = 12$
3. $q - 67 = 3$ $q = 70$
4. $s + 24 = 35$ $s = 11$
5. You get in an elevator and go up 4 floors. You exit on the 12th floor. Write and solve an equation to find what floor you got on the elevator. $x + 4 = 12$; You got on the elevator on the 8th floor.

Taking Math Deeper

Exercise 51

It would be great if you could spend some extra time on this problem. Even in high school, many students are confused about the difference between simplifying an expression and solving an equation.

1 Simplifying an expression

Simplify the expression $2(x + 3) - 4$.

$$2(x + 3) - 4 = 2x + 6 - 4$$
$$= 2x + 2$$

Characteristics:
- Solution steps are "L"-shaped.
- Write a sequence of equivalent expressions.
- Not solving for a value of x.
- There are *not two sides* of the equation.

2 Solving an equation

Solve the equation $x + 3 = 4$.

$$\begin{aligned} x + 3 &= 4 \\ -3 &\quad -3 \\ \hline x &= 1 \end{aligned}$$

Characteristics:
- Equation has two sides.
- Want two sides to be equivalent expressions.
- Trying to find values of x that make the equation true.
- Whatever you do to one side, you need to do the same thing to the other side.

3 If I try to "solve" the equation,
$2(x + 3) - 4 = 2x + 2$
I will discover that the equation is true for *all* values of x. This happens because $2(x + 3) - 4$ and $2x + 2$ are equivalent expressions.

Reteaching and Enrichment Strategies

If students need help. . .	If students got it. . .
Resources by Chapter • Practice A and Practice B • Puzzle Time Record and Practice Journal Practice Differentiating the Lesson Lesson Tutorials Skills Review Handbook	Resources by Chapter • Enrichment and Extension • Technology Connection Start the next section

51. **SIMPLIFYING AND SOLVING** Compare and contrast the two problems.

Simplify the expression $2(x + 3) - 4$.

$$2(x + 3) - 4 = 2x + 6 - 4$$
$$= 2x + 2$$

Solve the equation $x + 3 = 4$.

$$x + 3 = 4$$
$$-3 \quad -3$$
$$x = 1$$

52. **PUZZLE** In a *magic square*, the sum of the numbers in each row, column, and diagonal is the same. Write and solve equations to find the values of a, b, and c.

a	37	16
19	25	b
34	c	28

53. **FUNDRAISER** You participate in a dance-a-thon fundraiser. After your parents pledge \$15.50 and your neighbor pledges \$8.75, you have \$66.55. Write and solve an equation to find how much money you had before your parents and neighbor pledged.

54. **MONEY** On Saturday, you spend \$33, give \$15 to a friend, and receive \$20 for mowing your neighbor's lawn. You have \$21 left. Use two methods to find how much money you started with that day.

55. **AMUSEMENT PARK** You have \$15.

a. How much money do you have left if you ride each ride once?

b. Do you have enough money to ride each ride twice? Explain.

56. **Critical Thinking** Consider the equation $x + y = 15$. The value of x increases by 3. What has to happen to the value of y so that $x + y = 15$ remains true?

Fair Game Review What you learned in previous grades & lessons

Find the value of the expression. Use estimation to check your answer. *(Section 1.1)*

57. 12×8 58. 13×16 59. $75 \div 15$ 60. $72 \div 3$

61. **MULTIPLE CHOICE** What is the area of the parallelogram? *(Section 4.1)*

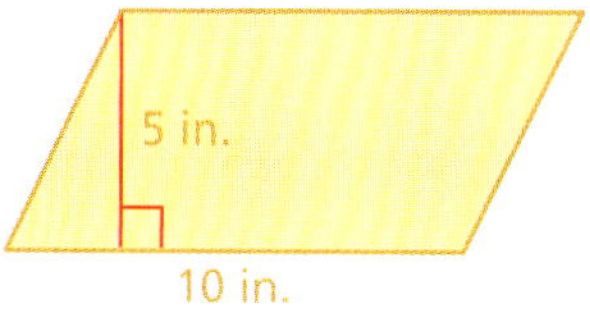

Ⓐ 25 in.^2 Ⓑ 30 in.^2

Ⓒ 50 in.^2 Ⓓ 100 in.^2

7.3 Solving Equations Using Multiplication or Division

Essential Question How can you use multiplication or division to solve an equation?

1 ACTIVITY: Finding Missing Dimensions

Work with a partner. Describe how you would find the value of *x*. Then find the value and check your result.

a. rectangle

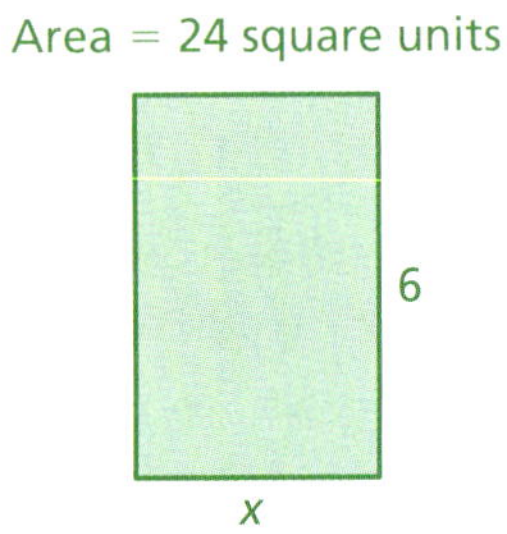

b. parallelogram

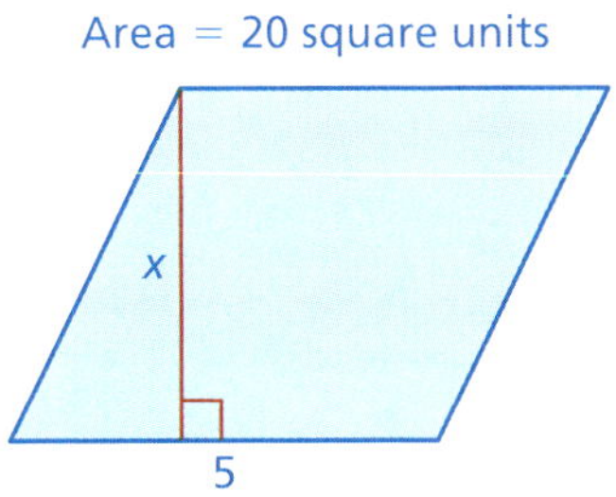

c. triangle

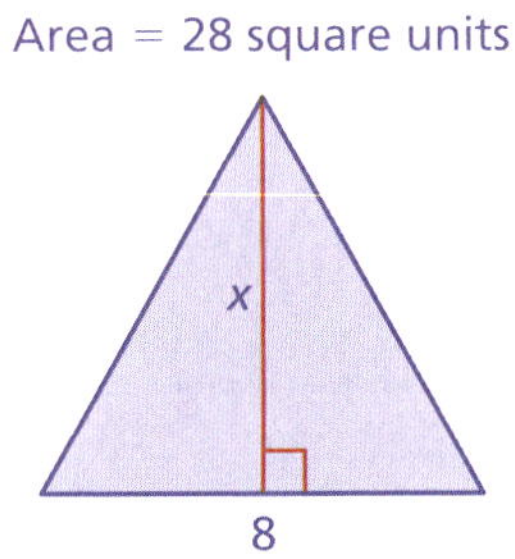

2 ACTIVITY: Using an Equation to Model a Story

Work with a partner.

a. Use a model to solve the problem.

> Three people go out to lunch. They decide to share the $12 bill evenly. How much does each person pay?

- What equation does the model represent? Explain how this represents the problem.

- How much does one  weigh? How do you know?

∴ Each person pays ____.

b. Describe how you can check your answer in part (a).

COMMON CORE

Solving Equations

In this lesson, you will

- use multiplication or division to solve equations.
- use substitution to check answers.
- solve real-life problems.

Learning Standards
6.EE.5
6.EE.7

Laurie's Notes

Introduction

Standards for Mathematical Practice

- **MP1 Make Sense of Problems and Persevere in Solving Them:** Beginning equation solving can sometimes be difficult because of the notation, representations, and symbols used. Equation solving is a simple skill for teachers, and in time it will be for students as well. It is important for us to be clear in our language and notation so that students will make sense of the problems. If we are sloppy with either, it can cause students to develop misconceptions about how equations are solved.

Motivate

- ? "Have you ever thought about building a robot to perform a simple task? What skills do you think are necessary to build a robot?"
- Share information about a middle school competition called *FIRST* LEGO League (FLL), a robotics program for students ages 9 to 14.
- **Check This Out:** To find out more about this international program visit: *www.usfirst.org*.
- Explain to students that several of today's word problems are about robots.

Activity Notes

Activity 1

- Introduce the activity. Explain that they are to solve for the missing dimension in each figure.
- ? "What is the name of each polygon?" rectangle, parallelogram, triangle
- ? Draw the parallelogram including the red line and right angle mark. You may want to make the red line dashed on the board. Then point to the right angle mark and ask, "What does this mean?" indicates a 90° angle
- The first two parts should be easy. For part (c), remind students that it is okay to use the problem solving strategy of *Guess, Check, and Revise.* In doing so, students are actually checking to see if a value is a solution to an equation. For the triangle, students are checking for a solution to $28 = \frac{1}{2} \cdot x \cdot 8$.

Activity 2

- **MP3 Construct Viable Arguments and Critique the Reasoning of Others:** This example demonstrates how an equation can be used to model a real-life problem. There are two bulleted steps, and in each, students are asked to explain their thinking. Students should be very comfortable with sharing their reasoning with classmates at this point in the year.
- Students may quickly say that the answer is \$4 without bothering to write an equation. Remind them that by showing the process they can check their thinking.

Common Core State Standards

6.EE.5 Understand solving an equation . . . as a process of answering a question: which values from a specified set, if any, make the equation . . . true? Use substitution to determine whether a given number in a specified set makes an equation . . . true.

6.EE.7 Solve real-world and mathematical problems by writing and solving equations of the form $x + p = q$ and $px = q$ for cases in which p, q and x are all nonnegative rational numbers.

Previous Learning

Students should know how to evaluate expressions and use area formulas.

Technology for the Teacher

Dynamic Classroom

Lesson Plans
Complete Materials List

7.3 Record and Practice Journal

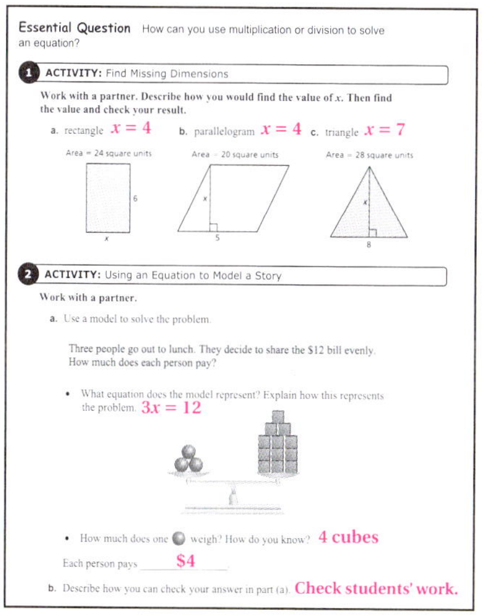

English Language Learners

Graphic Organizer

When solving an equation, students must understand that they need to isolate the variable in the equation. To do this, they use inverse operations. Have students create a reference table with examples that show the operation needed to move the numbers to the other side of the equal sign.

Equation Type	Operation to Solve	Example
Subtraction	Addition	$b - 4 = 2$
Addition	Subtraction	$h + 3 = 7$
Division	Multiplication	$w \div 5 = 5$
Multiplication	Division	$m \cdot 3 = 6$

7.3 Record and Practice Journal

3 ACTIVITY: Using Equations to Model a Story

Work with a partner.

- **What is the unknown?**
- **Write an equation that represents each problem.**
- **What does the variable in your equation represent?**
- **Explain how you can solve the equation.**
- **Answer the question.**

Problem	Equation
a. Three robots go out to lunch. They decide to share the $11.91 bill evenly. How much does each robot pay? **$3.97**	$3x = 11.91$
b. On Earth, objects weigh 6 times what they weigh on the Moon. A robot weighs 96 pounds on Earth. What does it weigh on the Moon? **16 lb**	$6x = 96$
c. At maximum speed, a robot runs 6 feet in 1 second. How many feet does the robot run in 1 minute? **360 ft**	$\frac{x}{60} = 6$
d. Four identical robots lie on the ground head-to-toe and measure 14 feet. How tall is each robot? **3.5 ft**	$4x = 14$

What Is Your Answer?

4. Complete each sentence by matching.

- The inverse operation of addition — is subtraction.
- The inverse operation of subtraction — is addition.
- The inverse operation of multiplication — is division.
- The inverse operation of division — is multiplication.

/ is division.)

5. **IN YOUR OWN WORDS** How can you use multiplication or division to solve an equation? Give two examples to show how your procedure works.

Use the inverse operation on both sides to undo the operation.

Laurie's Notes

Activity 3

- Explain to students that they are to use Activity 2 as a model for completing this activity.
- Students need practice in identifying what the variable represents. Caution them to be complete in their description of the variable. For instance, in Activity 2, the variable x was equal to the amount each person paid. It would not be sufficient to say "x equals money" or "x equals pay."
- When students finish, ask for volunteers to share their solutions.

? "Is $3x = 11.91$ the same as $11.91 = 3x$?" yes

What Is Your Answer?

- Students should know that multiplication and division are inverse operations.
- The inverse operation must be done on both sides of the equation so that the two sides remain equal.

Closure

- **Writing Prompt:** To solve $28 = 7x$, you . . . Divide both sides by 7.

  To solve $7 = \frac{x}{28}$, you . . . Multiply both sides by 28.

3 ACTIVITY: Using Equations to Model a Story

Work with a partner.

- **What is the unknown?**
- **Write an equation that represents each problem.**
- **What does the variable in your equation represent?**
- **Explain how you can solve the equation.**
- **Answer the question.**

	Problem	Equation
a.	Three robots go out to lunch. They decide to share the $11.91 bill evenly. How much does each robot pay?	
b.	On Earth, objects weigh 6 times what they weigh on the Moon. A robot weighs 96 pounds on Earth. What does it weigh on the Moon?	
c.	At maximum speed, a robot runs 6 feet in 1 second. How many feet does the robot run in 1 minute?	
d.	Four identical robots lie on the ground head-to-toe and measure 14 feet. How tall is each robot?	

Math Practice 4

Interpret Results

What does the solution represent? Does the answer make sense?

What Is Your Answer?

4. Complete each sentence by matching.

• The inverse operation of addition	• is multiplication.
• The inverse operation of subtraction	• is subtraction.
• The inverse operation of multiplication	• is addition.
• The inverse operation of division	• is division.

5. **IN YOUR OWN WORDS** How can you use multiplication or division to solve an equation? Give two examples to show how your procedure works.

Practice Use what you learned about solving equations to complete Exercises 15–18 on page 312.

7.3 Lesson

Check It Out
Lesson Tutorials
BigIdeasMath.com

Key Ideas

Remember

Inverse operations "undo" each other. Multiplication and division are inverse operations.

Multiplication Property of Equality

Words When you multiply each side of an equation by the same nonzero number, the two sides remain equal.

Numbers

$$\frac{8}{4} = 2$$

$$\frac{8}{4} \cdot 4 = 2 \cdot 4$$

$$8 = 8$$

Algebra

$$\frac{x}{4} = 2$$

$$\frac{x}{4} \cdot 4 = 2 \cdot 4$$

$$x = 8$$

Multiplicative Inverse Property

Words The product of a nonzero number n and its reciprocal, $\frac{1}{n}$, is 1.

Numbers $5 \cdot \frac{1}{5} = 1$

Algebra $n \cdot \frac{1}{n} = \frac{1}{n} \cdot n = 1, n \neq 0$

EXAMPLE 1 Solving Equations Using Multiplication

a. Solve $\frac{w}{4} = 12$.

$\frac{w}{4} = 12$	Write the equation.
$\frac{w}{4} \cdot 4 = 12 \cdot 4$	Multiplication Property of Equality
$w = 48$	Simplify.

Undo the division. →

The solution is $w = 48$.

Check

$$\frac{w}{4} = 12$$

$$\frac{48}{4} \stackrel{?}{=} 12$$

$$12 = 12 \checkmark$$

b. Solve $\frac{2}{7}x = 6$.

$\frac{2}{7}x = 6$	Write the equation.
$\frac{7}{2} \cdot \left(\frac{2}{7}x\right) = \frac{7}{2} \cdot 6$	Multiplication Property of Equality
$x = 21$	Simplify.

Use the Multiplicative Inverse Property. →

The solution is $x = 21$.

On Your Own

Now You're Ready
Exercises 7–10

Solve the equation. Check your solution.

1. $\frac{a}{8} = 6$
2. $14 = \frac{2y}{5}$
3. $3z \div 2 = 9$

Laurie's Notes

Introduction

- **Yesterday:** Students gained an intuitive understanding about how equations are solved using multiplication and division. (MP1, MP3)
- **Today:** Students will use Properties of Equality to solve equations.

Motivate

- Write the following on the board. Students fill in all ten squares using each of the digits 0 to 9. Read the first equation as "12 times a single-digit number equals a 2-digit number that ends in 0."

  $12 \times \boxed{5} = \boxed{6}0$ $\boxed{4}\boxed{8} \div 4 = 12$

  $\boxed{7} = \boxed{3}5 \div 5$ $\boxed{1} \times 0 = \boxed{0}$

  $7\boxed{2} = \boxed{9} \times 8$
- This helps thinking of multiplication and division as inverse operations.

Lesson Notes

Key Idea

- Discuss the *Words*, *Numbers*, and *Algebra*.
- **Representation:** Point out that division is often represented in the fraction format. To write and solve equations, do not write $x \div 4 = 2$.
- **Connection:** When multiplying a fraction and a whole number, write the whole number over a denominator of 1. Divide out common factors. Then multiply.

  Numbers: $\frac{8}{4} \times 4 = \frac{8}{\cancel{4}} \times \frac{\cancel{4}}{1} = 8$ *Algebra:* $\frac{x}{4} \cdot 4 = \frac{x}{\cancel{4}} \cdot \frac{\cancel{4}}{1} = x$
- Discuss the representation of multiplication. The dot is used versus $\times$.
- The Multiplicative Inverse Property mentioned in Section 2.2 is defined now. Say that $\frac{1}{5}$ and 5 are multiplicative inverses (reciprocals).

Example 1

? "How else can you write a division problem?" Rewrite it as a fraction.

- Writing division as a fraction is not obvious to students. Use a simple problem like $12 \div 4 = \frac{12}{4} = 3$, then point out that in part (b), the numerator is $2x$ and the denominator is 7.
- **MP1 Make Sense of Problems and Persevere in Solving Them:** Make the connection to multiplying a fraction times a whole number to rewrite $\frac{2x}{7}$ as $\frac{2}{7}x$, by thinking of $\frac{2x}{7}$ as the product of the factors $\frac{2}{7}$ and x.

? "What can you multiply each side of the equation by to have $1x$?" $\frac{7}{2}$

- Another way to solve this equation is to multiply by 7, then divide by 2. Look for the efficient way to "multiply by the reciprocal of the coefficient."
- Discuss how to check that their answers are correct.
- **Big Idea:** Multiply each side of the equation by the same factor. Using an alternate color to show this helps visual learners.

On Your Own

- Ask volunteers to share their work at the board.

Goal Today's lesson is using the inverse operations of multiplication and division to find the solution of an equation.

Technology for the Teacher

Dynamic Classroom

Lesson Tutorials
Lesson Plans
Answer Presentation Tool

Extra Example 1

a. Solve $\frac{t}{6} = 9$. $t = 54$

b. Solve $2y \div 8 = 7$. $y = 28$

On Your Own

1. $a = 48$
2. $y = 35$
3. $z = 6$

Differentiated Instruction

Visual

Show that the inverse operations of multiplication and division produce the multiplicative identity, 1.

$$5b = 65$$

$$1\left\{\frac{\overset{1}{\cancel{5}}b}{\underset{1}{\cancel{5}}} = \frac{65}{5}\right.$$

$$b = 13$$

Extra Example 2

Solve $7c = 42$. $c = 6$

Extra Example 3

A driver travels 2500 miles at a rate of 500 miles per day. Write and solve an equation to find the number of days it takes to complete the trip.
$500d = 2500$; 5 days

On Your Own

4. $p = 6$
5. $q = 5$
6. $r = 9$
7. $5x = 70$; \$14

Laurie's Notes

Key Idea

- Write the Key Idea.
- Discuss the *Words*, *Numbers*, and *Algebra*.
- **Representation:** Remind students that $4x$ is how multiplication is represented.
- **Common Misconception:** Students sometimes think that when $\frac{4x}{4}$ is simplified to x, the 4s have just been "crossed out" because they are the same. Explain that the 4s represent common factors in the numerator and denominator and have been *divided out*, in the same way a common factor of 4 would be divided out when simplifying the fraction $\frac{12}{16}$.

Example 2

? "How do you undo the operation of multiplication?" division

- **MP1:** Work through the problem. Being clear about notation and symbols will help students make sense of the mathematics and how to work towards the solution.
- Discuss the last step in the solution. This can be confusing for students because of the 5 on each side of the equation. On the left, $5b$ means 5 times the value of b, and on the right, 5 is the ones digit in a two-digit number. If the equation had been written $b5 = 65$, it would have been even more confusing. Remember that our comfort level with representing multiplication as $5b$ is not the same as students'. It's not uncommon to see a wrong answer here of $b = 6$. The 5s were "crossed out" on both sides! It might be helpful to write $\frac{5 \cdot b}{5} = \frac{5 \cdot 13}{5}$ as an intermediate step.

Example 3

- Ask a volunteer to read the problem.

? "What additional information do you need in order solve this problem?" You need to know how to find the area of a parallelogram.

? "How do you find the area of a parallelogram?" area = base × height

- Substitute the known information and continue to solve as shown.
- The answer is 42 feet, not just 42.

On Your Own

- Note that variables can appear on either side of the equal sign.
- Ask volunteers to share their work at the board.

Closure

- Write an equation that could be solved by multiplying both sides of the equation by $\frac{2}{3}$.
- Write an equation that could be solved by dividing both sides of the equation by 8.

Key Idea

Division Property of Equality

Words When you divide each side of an equation by the same nonzero number, the two sides remain equal.

Numbers

$8 \cdot 4 = 32$

$8 \cdot 4 \div 4 = 32 \div 4$

$8 = 8$

Algebra

$4x = 32$

$\frac{4x}{4} = \frac{32}{4}$

$x = 8$

EXAMPLE 2 Solving an Equation Using Division

Solve $5b = 65$.

	$5b = 65$	Write the equation.
Undo the multiplication. →	$\frac{5b}{5} = \frac{65}{5}$	Division Property of Equality
	$b = 13$	Simplify.

Check

$5b = 65$

$5(13) \stackrel{?}{=} 65$

$65 = 65$ ✓

The solution is $b = 13$.

EXAMPLE 3 Real-Life Application

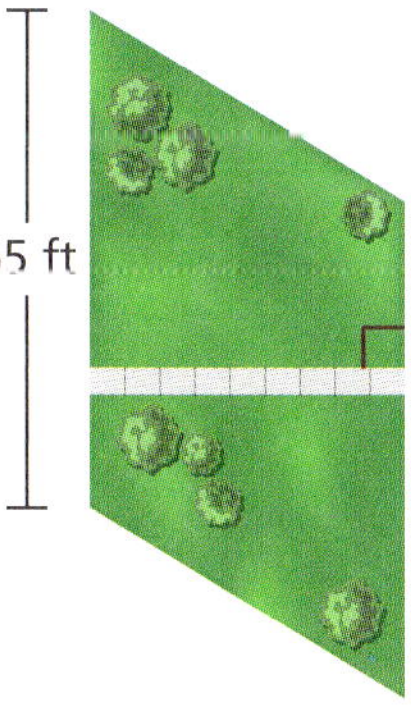

The area of the parallelogram-shaped courtyard is 2730 square feet. What is the length of the sidewalk?

The height of the parallelogram represents the length of the sidewalk.

$A = bh$	Use the formula for area of a parallelogram.
$2730 = 65h$	Substitute 2730 for A and 65 for b.
$\frac{2730}{65} = \frac{65h}{65}$	Division Property of Equality
$42 = h$	Simplify.

So, the sidewalk is 42 feet long.

On Your Own

Now You're Ready
Exercises 11–14

Solve the equation. Check your solution.

4. $p \cdot 3 = 18$ **5.** $12q = 60$ **6.** $81 = 9r$

7. You and four friends buy tickets to a baseball game. The total cost is $70. Write and solve an equation to find the cost of each ticket.

7.3 Exercises

Check It Out
Help with Homework
BigIdeasMath.com

Vocabulary and Concept Check

1. **NUMBER SENSE** What number divided by 12 equals 1?

2. **WRITING** What property of equality would you use to solve $\frac{x}{6} = 7$? Explain how you would use the property.

Copy and complete the first step in the solution.

3. $4x = 24$

   $\frac{4x}{\square} = \frac{24}{\square}$

4. $\frac{x}{3} = 11$

   $\frac{x}{3} \cdot \square = 11 \cdot \square$

5. $8 = n \div 3$

   $8 \cdot \square = (n \div 3) \cdot \square$

6. **OPEN-ENDED** Write an equation that can be solved using the Division Property of Equality.

Practice and Problem Solving

Solve the equation. Check your solution.

1 7. $\frac{s}{10} = 7$ 8. $6 = \frac{t}{5}$ 9. $5x \div 6 = 20$ 10. $24 = \frac{3r}{4}$

2 11. $3a = 12$ 12. $5 \cdot z = 35$ 13. $40 = 4y$ 14. $42 = 7k$

15. $7x = 105$ 16. $75 = 6 \cdot w$ 17. $13 = d \div 6$ 18. $9 = v \div 5$

19. $\frac{2c}{15} = 8.8$ 20. $7b \div 12 = 4.2$ 21. $12.5 \cdot n = 32$ 22. $3.4m = 20.4$

23. **ERROR ANALYSIS** Describe and correct the error in solving the equation.

✗

$$x \div 4 = 28$$

$$\frac{x \div 4}{4} = \frac{28}{4}$$

$$x = 7$$

24. **ANOTHER WAY** Show how you can solve the equation $3x = 9$ by multiplying each side by the reciprocal of 3.

25. **BASKETBALL** Forty-five basketball players participate in a tournament. Write and solve an equation to find the number of 3-person teams that they can form.

26. **THEATER** A theater has 1200 seats. Each row has 20 seats. Write and solve an equation to find the number of rows in the theater.

Assignment Guide and Homework Check

Level	Day 1 Activity Assignment	Day 2 Lesson Assignment	Homework Check
Basic	15–18, 36–38	1–6, 7–13 odd, 23–27 odd, 26	7, 11, 26, 27
Average	15–18, 36–38	1–6, 7–13 odd, 23, 27, 30, 31	7, 11, 27, 30
Advanced	15–18, 36–38	1–6, 20–26 even, 23, 27–34	20, 22, 30, 32

Common Errors

- **Exercises 7–22** Students may do the same operation on both sides instead of the opposite operation. This is most common with problems like Exercise 9, where the equation is written in division format instead of fraction format. Remind them that to get the variable alone, they need to use the inverse operation.
- **Exercises 27–29** Students may use the formula for perimeter instead of area. Remind them of the two formulas and how to find both.

7.3 Record and Practice Journal

Solve the equation. Check your solution.

1. $7k = 77$
   $k = 11$
2. $\frac{p}{5} = 10$
   $p = 50$
3. $3 = \frac{m}{12}$
   $m = 36$
4. $4a = 36$
   $a = 9$
5. $5 \bullet x = 12$
   $x = 2.4$
6. $4.2 = \frac{c}{8}$
   $c = 33.6$
7. You earn \$5 for every friendship bracelet you sell. Write and solve an equation to find the number of bracelets you have to sell to earn \$85.
   $5a = 85$
   $a = 17$
8. You practice the piano for 30 minutes each day. Write and solve an equation to find the total time t you spend practicing the piano in a week.
   $\frac{t}{7} = 30$
   $t = 210$

Vocabulary and Concept Check

1. 12
2. Multiplication Property of Equality; Multiply each side of the equation by 6.
3. $\frac{4x}{4} = \frac{24}{4}$
4. $\frac{x}{3} \bullet 3 = 11 \bullet 3$
5. $8 \bullet 3 = (n \div 3) \bullet 3$
6. *Sample answer:* $3x = 15$

Practice and Problem Solving

7. $s = 70$
8. $t = 30$
9. $x = 24$
10. $r = 32$
11. $a = 4$
12. $z = 7$
13. $y = 10$
14. $k = 6$
15. $x = 15$
16. $w = 12.5$
17. $d = 78$
18. $v = 45$
19. $c = 66$
20. $b = 7.2$
21. $n = 2.56$
22. $m = 6$
23. They should have multiplied by 4.
    $$x \div 4 = 28$$
    $$(x \div 4) \bullet 4 = 28 \bullet 4$$
    $$x = 112$$
24. $$3x = 9$$
    $$3x \bullet \frac{1}{3} = 9 \bullet \frac{1}{3}$$
    $$x = 3$$
25. $3x = 45$; 15 teams
26. $20x = 1200$; 60 rows

Practice and Problem Solving

27. 9 units

28. 11 units

29. 8 units

30. 50 points

31. 20 cards

32. See *Taking Math Deeper.*

33. $x = 6$; Because $5x$ is on both sides of the equation, $3x$ must be equal to 18 so that the equation is true.

34. $y = 5$; Because $2y$ is on both sides of the equation, $8y$ must be equal to 40 so that the equation is true.

35. length: 20 in.; width: 5 in.

Fair Game Review

36. $b + 8 = 17$

37. $\frac{t}{3} = 7$

38. C

Mini-Assessment

Solve the equation.

1. $\frac{t}{5} = 7$ $t = 35$
2. $12d = 60$ $d = 5$
3. $16 = \frac{s}{2}$ $s = 32$
4. $35 = 5g + 5$ $g = 6$
5. You pay $1.56 for 4 pounds of bananas. Write and solve an equation to find how much it costs for one pound of bananas. $4x = 1.56$; It costs $0.39 for one pound of bananas.

Taking Math Deeper

Exercise 32

The second part of this problem tends to be difficult for students. Before trying to answer part (b), ask students to decide the following.

- *The cones are smaller. So, can the machine fill more cones than cups? Or fewer cones than cups?*

1 Organize the given information.

Rate: 1440 cups in 24 hours

30 symbols represents 1440 cups

2 **a.** Write and solve an equation for c.

$30c = 1440$

$c = 48$

Each symbol represents 48 cups.

3 How many cones can be filled in 24 hours?

1440 cups　　n cones

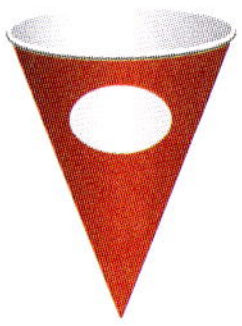

Because a cone holds 20% less than a cup, a cone holds 80% of what a cup holds. So, the number of cups must be 80% of the number n of cones. Solve a percent problem to find n.

b. 80% of what number is 1440?

$0.8 \cdot n = 1440$

$n = 1800$

The machine can fill 1800 cones in 24 hours.

Reteaching and Enrichment Strategies

If students need help. . .	If students got it. . .
Resources by Chapter • Practice A and Practice B • Puzzle Time Record and Practice Journal Practice Differentiating the Lesson Lesson Tutorials Skills Review Handbook	Resources by Chapter • Enrichment and Extension • Technology Connection Start the next section

Solve for x. Check your answer.

27. rectangle

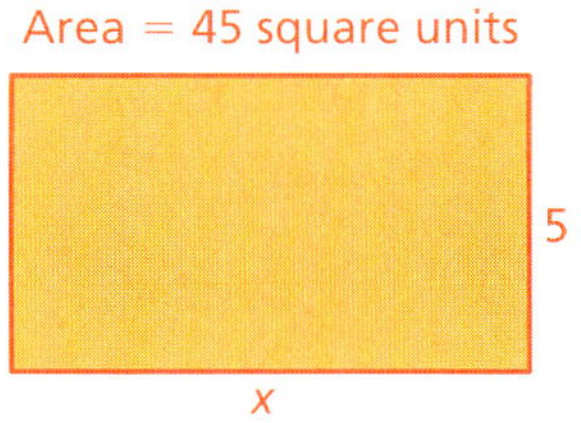

28. rectangle

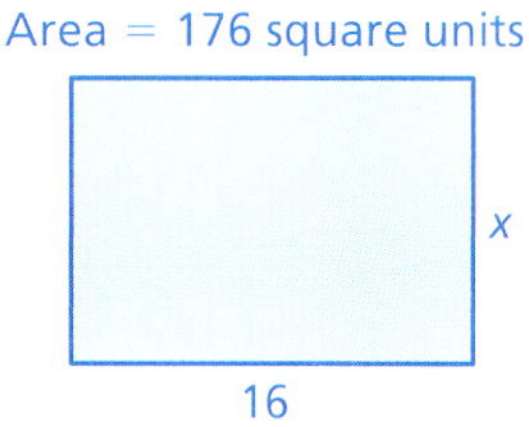

29. parallelogram

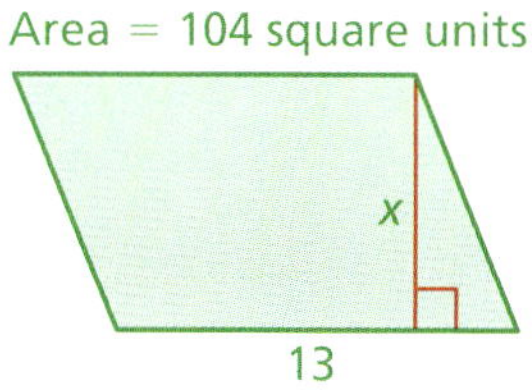

30. **TEST SCORE** On a test, you correctly answer six 5-point questions and eight 2-point questions. You earn 92% of the possible points on the test. How many points p is the test worth?

31. **CARD GAME** You use index cards to play a homemade game. The object is to be the first to get rid of all your cards. How many cards are in your friend's stack?

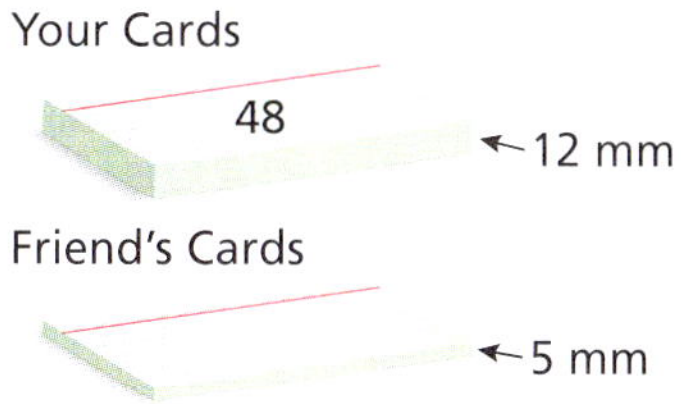

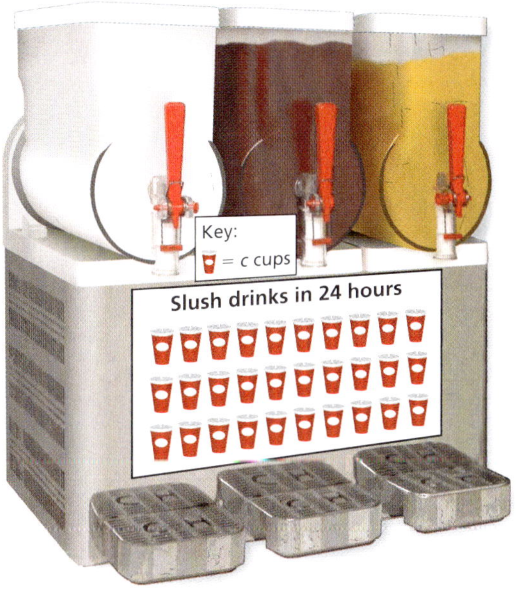

32. **SLUSH DRINKS** A slush drink machine fills 1440 cups in 24 hours.

a. Write and solve an equation to find the number c of cups each symbol represents.

b. To lower costs, you replace the cups with paper cones that hold 20% less. Write and solve an equation to find the number n of paper cones that the machine can fill in 24 hours.

STRUCTURE **Solve the equation. Explain how you found your answer.**

33. $5x + 3x = 5x + 18$

34. $8y + 2y - 2y + 40$

35. Number Sense The area of the picture is 100 square inches. The length is 4 times the width. Find the length and width of the picture.

Fair Game Review What you learned in previous grades & lessons

Write the word sentence as an equation. *(Section 7.1)*

36. The sum of a number b and 8 is 17.

37. A number t divided by 3 is 7.

38. **MULTIPLE CHOICE** What is the value of a^3 when $a = 4$? *(Section 3.1)*

Ⓐ 12 Ⓑ 43 Ⓒ 64 Ⓓ 81

7.4 Writing Equations in Two Variables

Essential Question How can you write an equation in two variables?

1 ACTIVITY: Writing an Equation in Two Variables

Work with a partner. You earn $8 per hour working part-time at a store.

a. Complete the table.

Hours Worked	Money Earned (dollars)
1	
2	
3	
4	
5	

b. Use the values from the table to complete the graph. Then answer each question below.

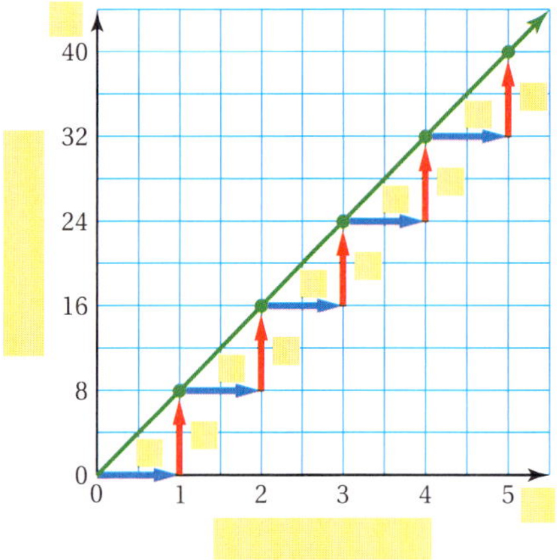

- What does the horizontal axis represent? What variable did you use to identify it?
- What does the vertical axis represent? What variable did you use to identify it?
- How are the ordered pairs in the graph related to the values in the table?
- How are the horizontal and vertical distances shown on the graph related to the values in the table?

c. How can you write an equation that shows how the two variables are related?

d. What does the green line in the graph represent?

COMMON CORE

Writing Equations

In this lesson, you will

- identify independent and dependent variables.
- write equations in two variables.
- use tables and graphs to analyze the relationship between two variables.

Learning Standard
6.EE.9

Laurie's Notes

Introduction

Standards for Mathematical Practice

- **MP8 Look for and Express Regularity in Repeated Reasoning:** Students will be creating tables and graphs in this activity. They will recognize patterns in the table of values (earnings are increasing by \$8) and the same pattern should be apparent in the graph as they are plotting points (increase hours worked by 1, increase earnings by \$8).

Motivate

? "How many of you have had a job earning money outside of your home?"

- Without asking how much money they earned, ask how they were paid, meaning by the hour, for the job, or some other manner.

Activity Notes

Activity 1

- **FYI:** The work in this section serves as the foundation for function work in later grades.
- **Financial Literacy:** You might spend some time discussing minimum wage and what that means in the context of working for an employer.
- This activity connects to two prior skills—making a table of values and graphing ordered pairs. Although the data in the table are not written as ordered pairs, you can refer to them as such.

? When students have finished creating the table of data ask, "What does the ordered pair (4, 32) mean?" For working 4 hours, you earn \$32.

- Students may label the axes with words but forget to label them with variables as well.
- **Note:** Most students will use x and y as their variables, yet it is equally correct to write $E = 8h$ where h is the number of hours worked and E is the earnings. Explain that you often use variables other than x and y in real-life problems so the variables relate to the context. My students know that I often avoid the variables "s" and "l" because they look too much like 5 and 1 when handwritten.
- In part (c), compare the equations written. Students may write $y = 8x$, $8x = y$, $8h = M$, and so on. Be sure that students understand that the choice of variables is theirs. Logical choices can aid in communication and understanding.
- **Extension:** What other types of ordered pairs can be written that are different from those in the table in part (a), and what do they represent? You want students to think about working for a portion of an hour, not just whole hours.

Common Core State Standards

6.EE.9 Use variables to represent two quantities in a real-world problem that change in relationship to one another; write an equation to express one quantity, thought of as the dependent variable, in terms of the other quantity, thought of as the independent variable. Analyze the relationship between the dependent and independent variables using graphs and tables, and relate these to the equation.

Previous Learning

Students should know how to solve equations in one variable.

Technology for the Teacher

Dynamic Classroom

Lesson Plans
Complete Materials List

7.4 Record and Practice Journal

Essential Question How can you write an equation in two variables?

1 ACTIVITY: Writing an Equation in Two Variables

Work with a partner. You earn \$8 per hour working part-time at a store.

a. Complete the table.

Hours Worked	Money Earned (dollars)
1	8
2	16
3	24
4	32
5	40

b. Use the values from the table to complete the graph. Then answer each question below and on the next page.

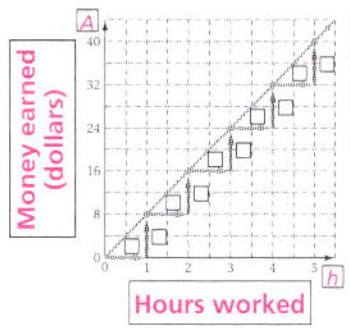

All horizontal distances are +1.
All vertical distances are +8.

- What does the horizontal axis represent? What variable did you use to identify it?

hours worked; h

Differentiated Instruction

Kinesthetic

Allow students to construct a square using one of the side lengths from their tables. Based on how they constructed the square, discuss which variable depends on the other.

7.4 Record and Practice Journal

- What does the vertical axis represent? What variable did you use to identify it?
  money earned; A
- How are the ordered pairs in the graph related to the values in the table?
  The ordered pairs (h, A) represent the values in the table.
- How are the horizontal and vertical distances shown on the graph related to the values in the table?
  See Additional Answers.

c. How can you write an equation that shows how the two variables are related?
   $A = 8h$

d. What does the line in the graph represent?
   all of the solutions of the equation

2 ACTIVITY: Describing Variables

Work with a partner. Use the equation you wrote in Activity 1.

a. How is this equation different from the equations earlier in this chapter?
   It has two variables.

b. One of the variables in this equation *depends* on the other variable. Determine which variable is which by answering the following questions:

   - Does the amount of money you earn *depend* on the number of hours you work?
     yes
   - Does the number of hours you work *depend* on the amount of money you earn?
     no

   What do you think is the significance of having two types of variables? How do you think you can use these types of variables in real life?
   An equation can now be graphed in the coordinate plane. Helps to visualize how two variables vary together.

3 ACTIVITY: Describing a Formula in Two Variables

Work with a partner. Recall that the perimeter of a square is 4 times its side length.

a. Write the formula for the perimeter of a square. Tell what each variable represents.
   $P = 4s$

b. Describe how the perimeter of a square changes as its side length increases by 1 unit. Use a table and a graph to support your answer.
   See Additional Answers.

c. In your formula, which variable depends on which?
   perimeter P depends on side length s

What Is Your Answer?

4. **IN YOUR OWN WORDS** How can you write an equation in two variables?
   See Additional Answers.

5. The equation $y = 7.75x$ shows how the number of movie tickets is related to the total amount of money spent. Describe what each part of the equation represents.
   7.75 = ticket price; x = number of tickets; y = total spent

6. **CHOOSE TOOLS** In Activity 1, you want to know the amount of money you earn after working 30.5 hours during a week. Would you use the table, the graph, or the equation to find your earnings? What are your earnings? Explain your reasoning.
   Check students' work.

7. Give an example of another real-life situation that you can model by an equation in two variables.
   Check students' work.

Laurie's Notes

Activity 2

- **Big Idea:** All of the previous equations in the chapter involved one variable. In this activity there are two variables. Say "an equation *in* two variables" versus "an equation *with* two variables."
- The language being developed in this lesson will help students understand functions and function language in later courses.

? "Does the amount of money you earn depend on the number of hours you work?" yes

? "Does the number of hours worked depend on the amount of money you earn?" no

- In discussing the last question, have students think of other common equations in two variables and how one variable depends upon the other variable. Example: The cost of the salmon depends upon how many pounds you are buying.

Activity 3

? "What is the formula for the perimeter of a square?" $P = 4s$

- **MP7 Look for and Make Use of Structure:** Students should work with their partners to repeat the same steps as in Activity 1: Make a table, graph the ordered pairs, and connect the points. The goal is to write an equation in two variables ($P = 4s$) and think about which variable depends upon the other.
- **Misconception:** Just because an equation has two variables, it is not necessarily true that one variable must depend upon the other. If there are 20 students in a class and the number of girls is known, then the number of boys can be found ($B = 20 - G$). One does not depend upon the other.

What Is Your Answer?

- When students finish, take time to review their answers, particularly Questions 6 and 7.

Closure

- Revisit Activity 1. Make a table and graph for a job paying \$9.50. How does this graph compare to the graph in Activity 1?

2 ACTIVITY: Describing Variables

Work with a partner. Use the equation you wrote in Activity 1.

a. How is this equation different from the equations earlier in this chapter?

b. One of the variables in this equation *depends* on the other variable. Determine which variable is which by answering the following questions:

- Does the amount of money you earn *depend* on the number of hours you work?
- Does the number of hours you work *depend* on the amount of money you earn?

What do you think is the significance of having two types of variables? How do you think you can use these types of variables in real life?

3 ACTIVITY: Describing a Formula in Two Variables

Math Practice 7

Look for Patterns

What pattern do you notice in the table for the perimeter of the square?

Work with a partner. Recall that the perimeter of a square is 4 times its side length.

a. Write the formula for the perimeter of a square. Tell what each variable represents.

b. Describe how the perimeter of a square changes as its side length increases by 1 unit. Use a table and a graph to support your answer.

c. In your formula, which variable depends on which?

What Is Your Answer?

4. **IN YOUR OWN WORDS** How can you write an equation in two variables?
5. The equation $y = 7.75x$ shows how the number of movie tickets is related to the total amount of money spent. Describe what each part of the equation represents.
6. **CHOOSE TOOLS** In Activity 1, you want to know the amount of money you earn after working 30.5 hours during a week. Would you use the table, the graph, or the equation to find your earnings? What are your earnings? Explain your reasoning.
7. Give an example of another real-life situation that you can model by an equation in two variables.

Practice Use what you learned about equations in two variables to complete Exercises 4 and 5 on page 319.

7.4 Lesson

Check It Out
Lesson Tutorials
BigIdeasMath.com

An **equation in two variables** represents two quantities that change in relationship to one another. A **solution of an equation in two variables** is an ordered pair that makes the equation true.

Key Vocabulary
equation in two variables, *p. 316*
solution of an equation in two variables, *p. 316*
independent variable, *p. 316*
dependent variable, *p. 316*

EXAMPLE 1 Identifying Solutions of Equations in Two Variables

Tell whether the ordered pair is a solution of the equation.

a. $y = 2x$; (3, 6)

$6 \stackrel{?}{=} 2(3)$ Substitute.

$6 = 6$ ✓ Compare.

So, (3, 6) is a solution.

b. $y = 4x - 3$; (4, 12)

$12 \stackrel{?}{=} 4(4) - 3$

$12 \neq 13$ ✗

So, (4, 12) is *not* a solution.

You can use equations in two variables to represent situations involving two related quantities. The variable representing the quantity that can change freely is the **independent variable**. The other variable is called the **dependent variable** because its value *depends* on the independent variable.

EXAMPLE 2 Using an Equation in Two Variables

The equation $y = 128 - 8x$ gives the amount y (in fluid ounces) of milk remaining in a gallon jug after you pour x cups.

a. Identify the independent and dependent variables.

Because the amount y remaining depends on the number x of cups you pour, y is the dependent variable and x is the independent variable.

b. How much milk remains in the jug after you pour 10 cups?

Use the equation to find the value of y when $x = 10$.

$y = 128 - 8x$ Write the equation.

$= 128 - 8(10)$ Substitute 10 for x.

$= 48$ Simplify.

There are 48 fluid ounces remaining.

On Your Own

Now You're Ready
Exercises 6–11 and 13–17

Tell whether the ordered pair is a solution of the equation.

1. $y = 7x$; (2, 21)
2. $y = 5x + 1$; (3, 16)
3. The equation $y = 10x + 25$ gives the amount y (in dollars) in your savings account after x weeks.
   a. Identify the independent and dependent variables.
   b. How much is in your savings account after 8 weeks?

Multi-Language Glossary at BigIdeasMath.com

Laurie's Notes

Introduction

Connect

- **Yesterday:** Students explored writing equations in two variables. (MP7, MP8)
- **Today:** Students will write equations in two variables and determine whether an ordered pair is a solution of an equation.

Motivate

- Hold up a gallon jug partially filled with water. Don't let students see the amount. Say you have been drinking water out of the jug for a few days.
- ? "How much water is left in the jug?" Listen for *"it depends on how much you already drank."* Restate and stress this statement.
- ? "What are the two quantities that are related in this problem?" amount of water you drank, amount of water remaining

Discuss

- **Define:** *Equation in two variables, solution of an equation in two variables.* Relate the definitions to the work done in yesterday's investigation.

Lesson Notes

Example 1

- ? "How can you tell whether (3, 6) is a solution of the equation in part (a)?" Substitute 3 for x and 6 for y and see if that makes the equation true.
- Write the equation in part (b). Students should notice two operations on the right side of the equation. Perform multiplication before subtraction.
- Mention using parentheses to represent multiplication when you substitute 4 for x. The dot and $\times$ symbol are not commonly used in this context.
- **Discuss:** In contextual problems, it is often understood which variable "can change freely." This is typically called x. Describe independent and dependent variables. This will become common language for the students.

Example 2

- **MP1 Make Sense of Problems and Persevere in Solving Them:** The similarity to the motivator should help students understand the context and make sense of the variables used in the equation.
- **MP6 Attend to Precision:** Students need to pay attention to the units in this problem: x is measured in cups and y is measured in fluid ounces. Recognizing the difference in units explains the "8" in the equation.
- ? "Which is the independent variable? Explain why this makes sense." The amount poured is independent because it can change. The amount of milk remaining depends upon how many cups have been poured.
- ? "How many fluid ounces are in a gallon? a cup?" 128, 8
- Discuss the equation and why it makes sense. I find it helpful to make a quick table of values off to the side beginning with $x = 0$ through $x = 3$.
- ? "How can you find the amount of milk left after pouring 10 cups?" Substitute 10 for x and solve.

On Your Own

- If time permits discuss Question 3. What do 10 and 25 represent?

Goal Today's lesson is writing and representing **equations in two variables**.

Technology for the Teacher

Dynamic Classroom

Lesson Tutorials
Lesson Plans
Answer Presentation Tool

Extra Example 1

Tell whether the ordered pair is a solution of the equation.

a. $y = 3x$; (3, 6) no

b. $y = 2x + 1$; (2, 5) yes

Extra Example 2

The equation $y = 16 - 2x$ gives the amount y (in ounces) of trail mix remaining after you eat x servings.

a. Identify the independent and dependent variables.
independent variable: x
dependent variable: y

b. How much trail mix remains after you eat 6 servings? 4 ounces

On Your Own

1. not a solution
2. solution
3. a. independent variable: x
   dependent variable: y
   b. $105

English Language Learners

Comprehension

Use the Key Idea to discuss the solutions of the given equation in two variables. *One* solution of the equation is represented by an ordered pair or a point. *All* of the solutions of the equation are represented by its graph. Can all of the solutions be represented in a table?

Extra Example 3

A teenager burns 100 calories in a Hatha Yoga class. The teen then walks on an inclined treadmill and burns 5 calories for every minute. Write and graph an equation in two variables that represents the total number of calories burned.

Let c = calories burned and m = number of minutes on the inclined treadmill; $c = 100 + 5m$

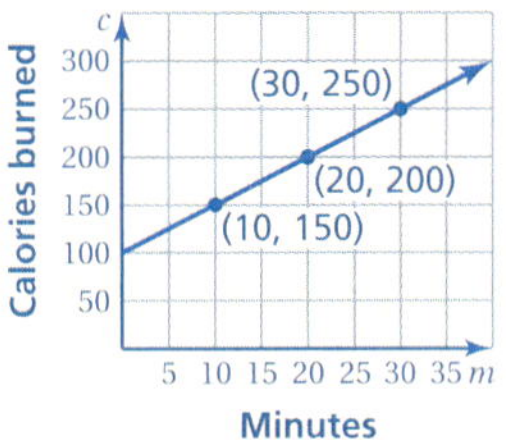

On Your Own

4. $c = 8h + 25$

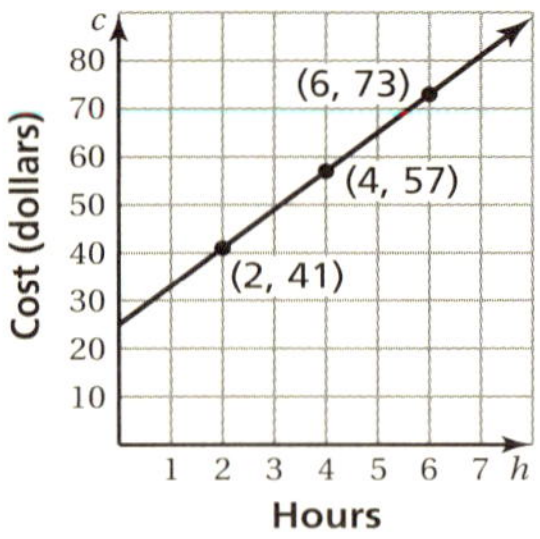

Laurie's Notes

Key Idea

- Write the Key Idea and example as shown. The goal is for students to see the connection between the different representations: equations (symbols), tables, and graphs. As students become more proficient and comfortable with these representations, they will recognize patterns, such as the constant rate of change.
- Discuss the Study Tip.

? **Extension:** "Do you think the graph of every equation will be a line?" Answers will vary. You don't need to tell the whole story of different types of functions, but make sure students realize that there are equations whose graphs are not lines.

Example 3

- Ask a volunteer to read the problem. Explain any vocabulary that is unfamiliar.
- **MP1:** Write the verbal model. The words act as a road map for how to solve the problem. You write the equation by translating the words into symbols. This is a practice that you want to develop in your students—say in words how you are going to solve the problem before you try to write an equation.

? "What are the variables in this problem?" c represents the total calories burned, m represents the minutes spent on the elliptical trainer.

- Write the equation. Students may ask whether $200 + 10m = c$ would also be correct. Explain that while it is correct, the convention is to write the form dependent variable = expression.

? "What do you know about the workout when $m = 0$?" The athlete has not spent any time on the elliptical trainer.

- Ask volunteers to help fill in the table.
- Discuss the scale on the graph. Because there are 10 calories for every minute, it makes sense to scale the vertical axis differently than the horizontal axis.

On Your Own

- Students should write a verbal model for the problem in words, define the variables, and then write the equation by translating the words into symbols.

Key Idea

Study Tip

When you draw a line through the points, you graph *all* the solutions of the equation.

Tables, Graphs, and Equations

You can use tables and graphs to represent equations in two variables. The table and graph below represent the equation $y = x + 2$.

Independent Variable, x	Dependent Variable, y	Ordered Pair, (x, y)
1	3	(1, 3)
2	4	(2, 4)
3	5	(3, 5)

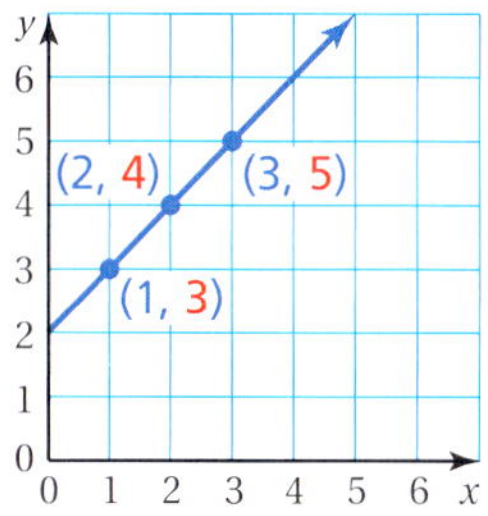

EXAMPLE 3 Writing and Graphing an Equation in Two Variables

An athlete burns 200 calories weight lifting. The athlete then works out on an elliptical trainer and burns 10 calories for every minute. Write and graph an equation in two variables that represents the total number of calories burned during the workout.

Reading

Make sure you read and understand the context of the problem. Because you cannot have a negative number of minutes, use only whole number values of *m*.

Words	The total number of calories burned	equals	calories burned weight lifting	plus	calories burned per minute	times	the number of minutes.

Variables Let c be the total number of calories burned, and let m be the number of minutes on the elliptical trainer.

Equation $c = 200 + 10 \cdot m$

To graph the equation, first make a table. Then plot the ordered pairs and draw a line through the points.

Minutes, m	$c = 200 + 10m$	Calories, c	Ordered Pair, (m, c)
10	$c = 200 + 10(10)$	300	(10, 300)
20	$c = 200 + 10(20)$	400	(20, 400)
30	$c = 200 + 10(30)$	500	(30, 500)

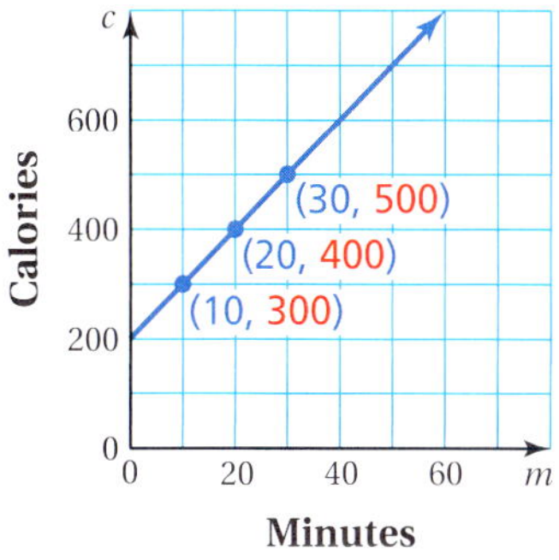

On Your Own

Now You're Ready
Exercises 22 and 23

4. It costs $25 to rent a kayak plus $8 for each hour. Write and graph an equation in two variables that represents the total cost of renting the kayak.

You can model many rate problems by using the *distance formula* $d = rt$, where d is the distance traveled, r is the speed, and t is the time. When you are given a speed, you can use the formula to write an equation in two variables that represents the situation.

Key Idea

Remember

Speed is an example of a rate.

Distance Formula

Words To find the distance traveled d, multiply the speed r by the time t.

Algebra $d = rt$

EXAMPLE 4 Real-Life Application

A train averages 40 miles per hour between two cities. Use a graph to show the relationship between the time and the distance traveled.

Method 1: Use a ratio table.

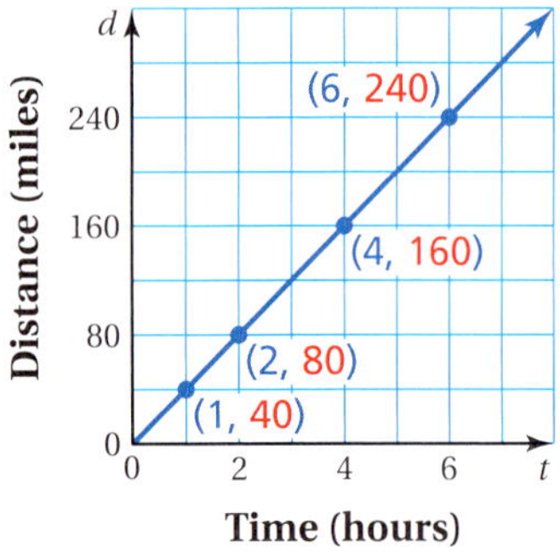

You can use a ratio table and multiplication to find equivalent rates. Then plot the ordered pairs (time, distance) from the table and draw a line through the points.

Time (hours)	1	2	4	6
Distance (miles)	40	80	160	240

×2 ×4 ×6

Method 2: Use an equation in two variables.

Use the distance formula to write the equation $d = 40t$. Use the equation to make a table. Then plot the ordered pairs and draw a line through the points, as shown in the graph above.

Time (hours), t	$d = 40t$	Distance (miles), d	Ordered Pair, (t, d)
1	$d = 40(1)$	40	(1, 40)
2	$d = 40(2)$	80	(2, 80)
4	$d = 40(4)$	160	(4, 160)
6	$d = 40(6)$	240	(6, 240)

On Your Own

Now You're Ready
Exercise 25

5. **WHAT IF?** The train averages 50 miles per hour. Use a graph to show the relationship between the time and the distance traveled.

Laurie's Notes

Key Idea

- Write the Key Idea.
- Relate this formula to work done with rates in Lesson 5.3. Speed is an example of a rate.
- It is important for students to remember that the unit of time specified in the rate needs to be the same as the unit of time in the formula. For instance, if a car travels 60 mph for 10 minutes, then you can't multiply 60×10 to find the distance.

Example 4

- Ask a volunteer to read the problem.
- The first method reviews ratio tables from Lesson 5.2.
- Complete the ratio table and discuss the ordered pairs formed, namely (hours, miles). Plot the points on a graph.
- Be sure that students label the axes with both words and variables.
- Practice interpreting the graph, left to right. As the time increases, the distance traveled increases.
- If time permits, make a ratio table using addition for the same context.
- Method 2 asks students to write an equation in two variables.

? "What equation can you write to describe the relationship between the two variables?" $d = 40t$

- Continue to work through the problem as shown.
- **MP1:** You have demonstrated at least two ways to make sense of this problem. Different methods will enable more students to make sense of the problem. Different methods provide different entry points into the problem for students.

On Your Own

- Discuss the changes in the results for each method in Example 4 when the average rate of speed is 50 miles per hour.

Closure

- A scoop of ice cream costs $1.20. Use a table and a graph to show the relationship between the number of scoops purchased and the money spent.

Extra Example 4

A train averages 60 miles per hour between two cities. Use a graph to show the relationships between the time and the distance traveled.

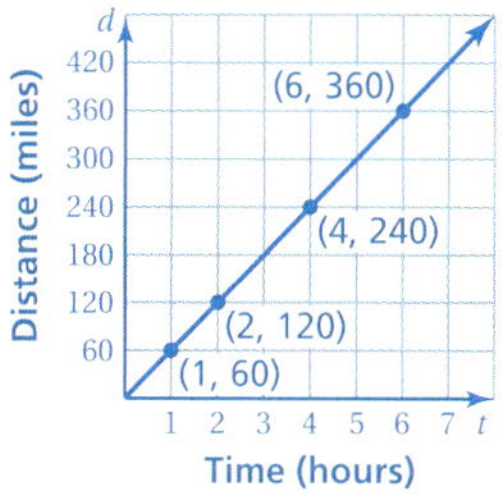

On Your Own

5.

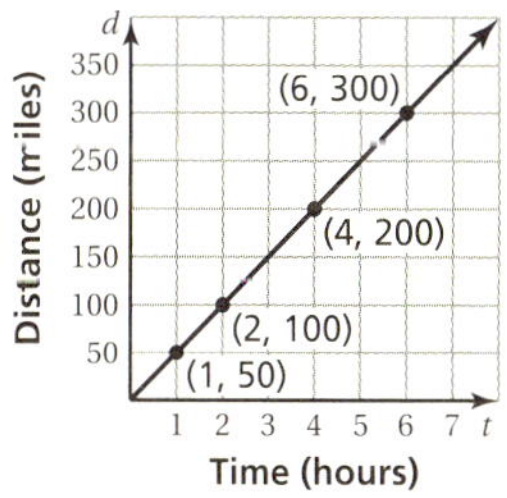

Vocabulary and Concept Check

1. *Sample answer:* An independent variable can change freely. A dependent variable depends on the independent variable.
2. Create a table. Plot the ordered pairs from the table. Draw a line through the points.
3. $n = 4n - 6$; This one is not an equation in two variables.

Practice and Problem Solving

4. $P = 2w + 10$ where P is the perimeter in inches and w is the width in inches; P depends on w.
5. $A = 9h$ where A is the area in square feet and h is the height in feet; A depends on h.
6. no
7. yes
8. yes
9. no
10. no
11. yes
12. The values for x and y were substituted in reverse order.

    $y = 3x + 2$; (5, 1)

    $1 \stackrel{?}{=} 3(5) + 2$

    $1 \neq 17$

    So, (5, 1) is not a solution.
13. w is independent and A is dependent.
14. s is independent and c is dependent.
15. p is independent and t is dependent.
16. m is independent and h is dependent.
17. $270

Assignment Guide and Homework Check

Level	Day 1 Activity Assignment	Day 2 Lesson Assignment	Homework Check
Basic	4, 5, 40–44	1–3, 7–23 odd, 12	9, 15, 17, 23
Average	4, 5, 40–44	1–3, 10–16 even, 20–29	10, 16, 22, 26
Advanced	4, 5, 40–44	1–3, 10–38 even	10, 16, 22, 28, 32

Common Errors

- **Exercises 6–11** Students may substitute incorrectly for x and y. Remind them that the first coordinate of the ordered pair is always the x-value and the second coordinate is always the y-value.
- **Exercises 13–16** Students may have trouble identifying the independent variable. Remind them that the independent variable represents the quantity that can change freely.

7.4 Record and Practice Journal

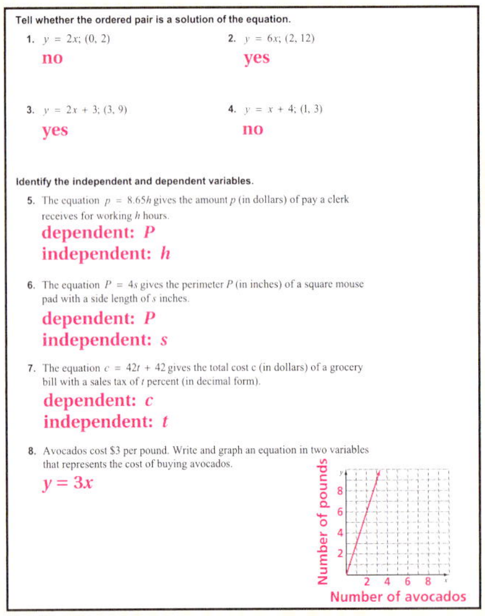

Tell whether the ordered pair is a solution of the equation.

1. $y = 2x$; (0, 2)
   no
2. $y = 6x$; (2, 12)
   yes
3. $y = 2x + 3$; (3, 9)
   yes
4. $y = x + 4$; (1, 3)
   no

Identify the independent and dependent variables.

5. The equation $p = 8.65h$ gives the amount p (in dollars) of pay a clerk receives for working h hours.
   dependent: P
   independent: h
6. The equation $P = 4s$ gives the perimeter P (in inches) of a square mouse pad with a side length of s inches.
   dependent: P
   independent: s
7. The equation $c = 42t + 42$ gives the total cost c (in dollars) of a grocery bill with a sales tax of t percent (in decimal form).
   dependent: c
   independent: t
8. Avocados cost $3 per pound. Write and graph an equation in two variables that represents the cost of buying avocados.
   $y = 3x$

7.4 Exercises

Check It Out
Help with Homework
BigIdeasMath.com

Vocabulary and Concept Check

1. **VOCABULARY** How are independent variables and dependent variables different?
2. **PRECISION** Explain how to graph an equation in two variables.
3. **WHICH ONE DOESN'T BELONG?** Which one does *not* belong with the other three? Explain your reasoning.

$y = 12x + 25$	$c = 10t - 5$	$a = 7b + 11$	$n = 4n - 6$

Practice and Problem Solving

Write a formula for the given measure. Tell what each variable represents. Identify which variable depends on which in the formula.

4. the perimeter of a rectangle with a length of 5 inches
5. the area of a trapezoid with base lengths of 7 feet and 11 feet

Tell whether the ordered pair is a solution of the equation.

1 6. $y = 4x$; (0, 4)
7. $y = 3x$; (2, 6)
8. $y = 5x - 10$; (3, 5)
9. $y = x + 7$; (1, 6)
10. $y = 7x + 2$; (2, 0)
11. $y = 2x - 3$; (4, 5)

12. **ERROR ANALYSIS** Describe and correct the error in finding a solution of the equation in two variables.

✗
$y = 3x + 2$; (5, 1)
$5 \stackrel{?}{=} 3(1) + 2$
$5 = 5$
So, (5, 1) is a solution.

Identify the independent and dependent variables.

2 13. The equation $A = 25w$ gives the area A (in square feet) of a rectangular dance floor with a width of w feet.

14. The equation $c = 0.09s$ gives the amount c (in dollars) of commission a salesperson receives for making a sale of s dollars.

15. The equation $t = 12p + 12$ gives the total cost t (in dollars) of a meal with a tip of p percent (in decimal form).

16. The equation $h = 60 - 4m$ gives the height h (in inches) of the water in a tank m minutes after it starts to drain.

17. **DRUM SET** The equation $b = 540 - 30m$ gives the balance b (in dollars) that you owe on a drum set after m monthly payments. What is the balance after 9 monthly payments?

OPEN-ENDED **Complete the table by describing possible independent or dependent variables.**

	Independent Variable	Dependent Variable
18.	The number of hours you study for a test	
19.	The speed you are pedaling a bike	
20.		Your monthly cell phone bill
21.		The amount of money you earn

3 **22.** **PIZZA** A cheese pizza costs \$5. Additional toppings cost \$1.50 each. Write and graph an equation in two variables that represents the total cost of a pizza.

23. **GYM MEMBERSHIP** It costs \$35 to join a gym. The monthly fee is \$25. Write and graph an equation in two variables that represents the total cost of a gym membership.

24. **TEXTING** The maximum size of a text message is 160 characters. A space counts as one character.

a. Write an equation in two variables that represents the remaining (unused) characters in a text message as you type.

b. Identify the independent and dependent variables.

c. How many characters remain in the message shown?

4 **25.** **CHOOSE TOOLS** A car averages 60 miles per hour on a road trip. Use a graph to show the relationship between the time and the distance traveled. What method did you use to create your graph?

Write and graph an equation in two variables that shows the relationship between the time and the distance traveled.

26.

Moves 2 meters every 3 hours.

27.

Rises 5 stories every 6 seconds.

28.

Moves 660 feet every 10 seconds.

29.

Moves 960 kilometers every 4 minutes.

Common Errors

- **Exercises 26–29** Students may not realize that the rate is given. Remind them that they need to convert this rate into a unit rate before using the distance formula $d = rt$.
- **Exercise 39** When discussing part (a), tell students that it does not make sense to draw a line between the points because you can't sell a part of a ticket. This exercise illustrates the difference between discrete and continuous domains of functions, which they will study in future lessons.

Practice and Problem Solving

18–21. See Additional Answers.

22. *Sample answer:* $c = 1.5t + 5$ where t is the number of toppings and c is the total cost of the pizza.

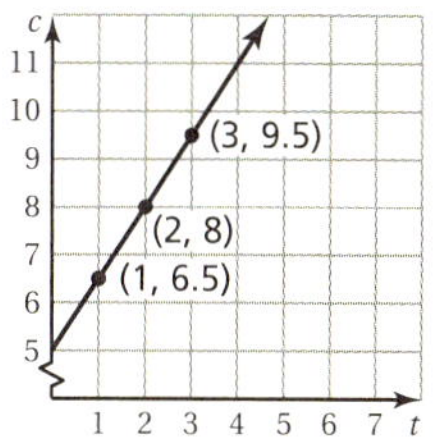

23. *Sample answer:* $c = 25m + 35$ where m is the number of months and c is the total cost of the gym membership.

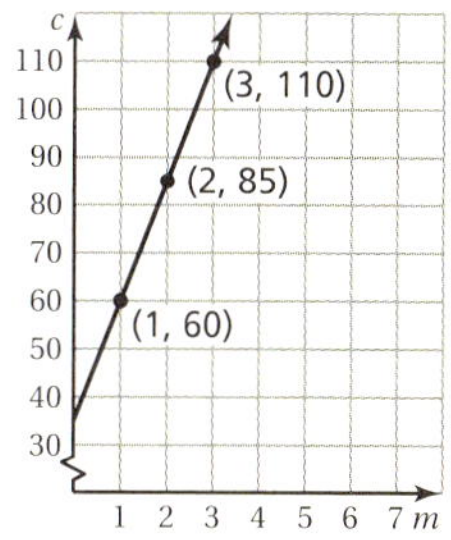

24. a. *Sample answer:* $r = 160 - x$ where x is the number of typed characters and r is the number of remaining characters.

b. x is independent and r is dependent.

c. 145 characters

25–29. See Additional Answers.

30. 11 **31.** 1

32. 2

Differentiated Instruction

Auditory

Have someone read aloud the problem statement and part (a) for Exercise 38. Discuss the concept that the question is getting at. Relative to this concept, ask how this situation is different from the situation in Exercise 28.

Practice and Problem Solving

33. no; By definition, the independent variable can change freely.

34. *Sample answer:* $y = 2x - 2$

35. 50 city blocks

36. 14 in./min

37. Methods to create the graph will vary.

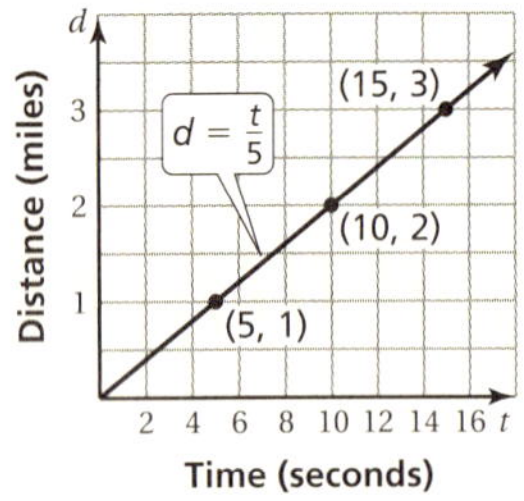

38. See *Taking Math Deeper.*

39. **a.** no; It does not make sense to draw a line between the points to show the solutions because you cannot sell part of a ticket.

b. $c = 10n$

Fair Game Review

40. 30% **41.** 80%

42. 45% **43.** 68%

44. A

Mini-Assessment

Find the distance.

1. $d = \square$, $r = 55$ mi/h, $t = 2$ h
   110 miles

2. $d = \square$, $r = 40$ mi/h, $t = 3$ h
   120 miles

Find the speed.

3. 800 feet in 20 minutes
   40 feet per minute

4. 1300 miles in 25 hours
   52 miles per hour

5. You walk 2 city blocks in 10 minutes. How many city blocks can you walk in 1 hour? 12 city blocks

Taking Math Deeper

Exercise 38

This problem keeps students on their toes regarding units. Each rate is given in terms of seconds, but each time is given as minutes. Moreover, with the distances ending up as thousands of feet, it makes sense to list the final distances in miles.

1 Make a table.

	You	Friend
Rate	$\frac{108}{8} = 13.5$ ft per sec	$\frac{63}{6} = 10.5$ ft per sec
15 min = 900 sec	$d = (13.5)(900)$ $= 12{,}150$ ft ≈ 2.30 mi	$d = (10.5)(900)$ $= 9450$ ft ≈ 1.79 mi
35 min = 2100 sec	$d = (13.5)(2100)$ $= 28{,}350$ ft ≈ 5.37 mi	
40 min = 2400 sec		$d = (10.5)(2400)$ $= 25{,}200$ ft ≈ 4.77 mi

2 After 15 minutes, your distance apart is about:
a. $2.30 + 1.79 = 4.09$ mi

3 After 40 minutes, your distance apart is about:
b. $5.37 + 4.77 = 10.14$ mi

Reteaching and Enrichment Strategies

If students need help. . .	If students got it. . .
Resources by Chapter • Practice A and Practice B • Puzzle Time Record and Practice Journal Practice Differentiating the Lesson Lesson Tutorials Skills Review Handbook	Resources by Chapter • Enrichment and Extension • Technology Connection Start the next section

Fill in the blank so that the ordered pair is a solution of the equation.

30. $y = 8x + 3$; (1, ___)

31. $y = 12x + 2$; (___ , 14)

32. $y = 22 - 9x$; (___ , 4)

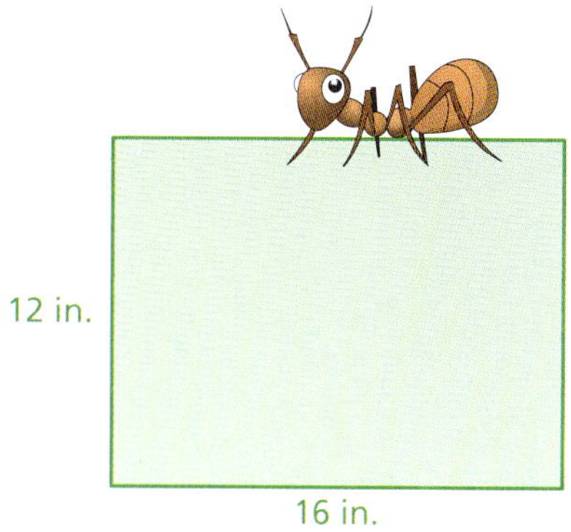

33. CRITICAL THINKING Can the dependent variable cause a change in the independent variable? Explain.

34. OPEN-ENDED Write an equation in two variables that has (3, 4) as a solution.

35. WALKING You walk 5 city blocks in 12 minutes. How many city blocks can you walk in 2 hours?

36. ANT How fast should the ant walk to go around the rectangle in 4 minutes?

37. LIGHTNING To estimate how far you are from lightning (in miles), count the number of seconds between a lightning flash and the thunder that follows. Then divide the number of seconds by 5. Use a graph to show the relationship between the time and the distance. Describe the method you used to create your graph.

38. PROBLEM SOLVING You and a friend start biking in opposite directions from the same point. You travel 108 feet every 8 seconds. Your friend travels 63 feet every 6 seconds.

a. How far apart are you and your friend after 15 minutes?

b. After 20 minutes, you take a 5-minute rest, but your friend does not. How far apart are you and your friend after 40 minutes? Explain your reasoning.

39. Reasoning The graph represents the cost c (in dollars) of buying n tickets to a baseball game.

a. Should the points be connected with a line to show all the solutions? Explain your reasoning.

b. Write an equation in two variables that represents the graph.

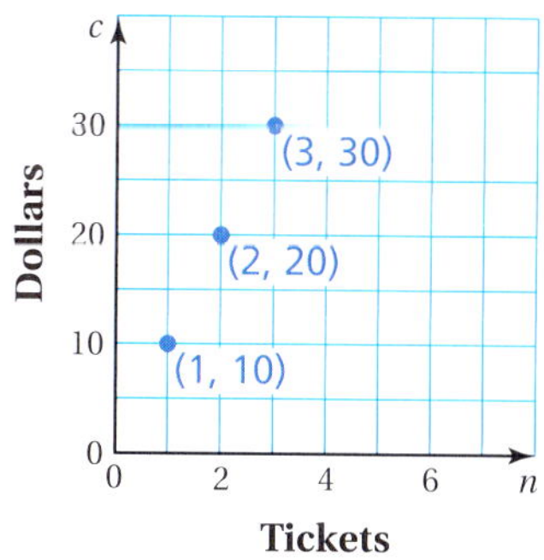

Fair Game Review What you learned in previous grades & lessons

Write the fraction as a percent. *(Section 5.5)*

40. $\frac{3}{10}$

41. $\frac{4}{5}$

42. $\frac{9}{20}$

43. $\frac{17}{25}$

44. MULTIPLE CHOICE What is the area of the triangle? *(Section 4.2)*

Ⓐ 36 cm^2 Ⓑ 68 cm^2

Ⓒ 72 cm^2 Ⓓ 76.5 cm^2

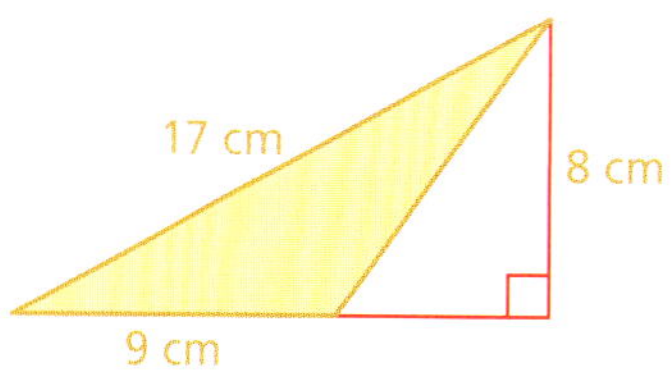

7 Study Help

Check It Out
Graphic Organizer
BigIdeasMath.com

You can use an **example and non-example chart** to list examples and non-examples of a vocabulary word or term. Here is an example and non-example chart for equations.

Equations

Examples	Non-Examples
$x = 5$	5
$2a = 16$	$2a$
$x + 4 = 19$	$x + 4$
$5 = x + 3$	$x + 3$
$12 - 7 = 5$	$12 - 7$
$\frac{3}{4}y = 6$	$\frac{3}{4}$

On Your Own

Make example and non-example charts to help you study these topics.

1. inverse operations
2. equations solved using addition or subtraction
3. equations solved using multiplication or division
4. equations in two variables

After you complete this chapter, make example and non-example charts for the following topics.

5. inequalities
6. graphs of inequalities
7. inequalities solved using addition or subtraction
8. inequalities solved using multiplication or division

"I need a good non-example of a cool animal for my **example and non-example chart**."

Sample Answers

1. Inverse Operations

Examples	Non-Examples
addition and subtraction	addition and multiplication
multiplication and division	addition and division
	subtraction and multiplication
	subtraction and division

2. Equations Solved Using Addition or Subtraction

Examples	Non-Examples
$y + 9 = 12$	$2a = 16$
$a - 11 = 3 - 2$	$0.4b = 4$
$10 = x - 19$	$\frac{3}{4}y = 6$
$25 = r + 14$	$\frac{y}{6} = 0$

3. Equations Solved Using Multiplication or Division

Examples	Non-Examples
$2a = 16$	$x - 1 = 5$
$\frac{y}{4} = 6$	$y + 9 = 12$
$0.4b = 4$	$a - 11 = 3 - 2$
$\frac{6}{7} = \frac{x}{14}$	$10 = x - 19$

4. Equations in Two Variables

Examples	Non-Examples
$y = 5x$	$12x = 3$
$y = 7 - x$	$\frac{y}{4} = 6$
$C = 50 + 3w$	$x - 1 = 5$
$d = 25t$	$a + 11 = 30$
$y = 4x - 3$	$0.4b = 4$

List of Organizers

Available at *BigIdeasMath.com*

Comparison Chart
Concept Circle
Definition (Idea) and Example Chart
Example and Non-Example Chart
Formula Triangle
Four Square
Information Frame
Information Wheel
Notetaking Organizer
Process Diagram
Summary Triangle
Word Magnet
Y Chart

About this Organizer

An **Example and Non-Example Chart** can be used to list examples and non-examples of a vocabulary word or term. Students write examples of the word or term in the left column and non-examples in the right column. This type of organizer serves as a good tool for assessing students' knowledge of pairs of topics that have subtle but important differences, such as equations and inequalities. Blank example and non-example charts can be included on tests or quizzes for this purpose.

Technology for the Teacher

Editable Graphic Organizer

Answers

1. $x - 3 = 5$
2. $a \div 7 = 14$
3. $k = 10$
4. $m = 5.7$
5. $w = 6$
6. $y = 5$
7. no
8. yes
9. $d = \frac{4}{9}t$;

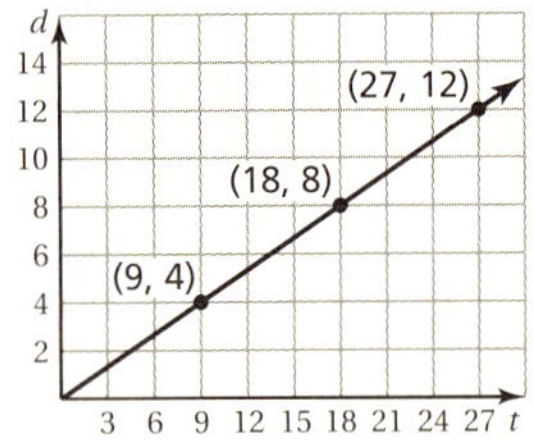

10. $d = 90t$;

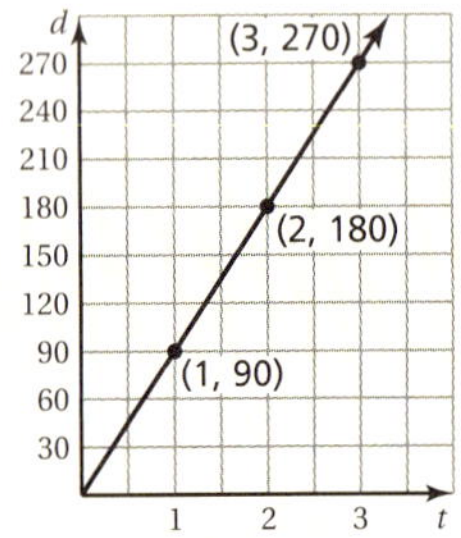

11. $\frac{2}{3}r = 30$
12. $360 = \ell - 30$; 390 m
13. $0.8m = 4(3) + 2(2)$; \$20
14. *Sample answer:* $c = 0.5t + 2$ where t is the number of toppings and c is the total cost of the sundae.

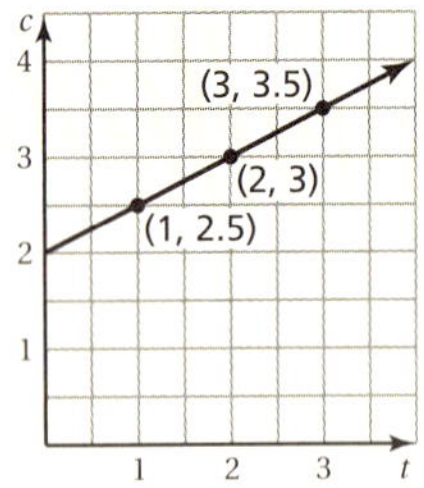

Technology for the Teacher

Online Assessment
Assessment Book
ExamView® Assessment Suite

Alternative Quiz Ideas

100% Quiz
Error Notebook
Group Quiz
Homework Quiz
Math Log
Notebook Quiz
Partner Quiz
Pass the Paper

Homework Quiz

A homework notebook provides an opportunity for teachers to check that students are doing their homework regularly. Students keep their homework in notebooks. They should be told to record the page number, problem number, and copy the problem exactly in their homework notebooks. Each day the teacher walks around and visually checks that homework is completed. Periodically, without advance notice, the teacher tells the students to put everything away except their homework notebooks.

Questions are from students' homework.

1. What is the answer to Exercise 14 on page 298?
2. What are the answers to Exercises 18–20 on page 305?
3. What are the answers to Exercises 15–18 on page 312?
4. What is the answer to Exercise 23 on page 320?

Reteaching and Enrichment Strategies

If students need help. . .	If students got it. . .
Resources by Chapter • Practice A and Practice B • Puzzle Time Lesson Tutorials *BigIdeasMath.com*	Resources by Chapter • Enrichment and Extension • Technology Connection Game Closet at *BigIdeasMath.com* Start the next section

7.1–7.4 Quiz

Check It Out
Progress Check
BigIdeasMath.com

Write the word sentence as an equation. *(Section 7.1)*

1. A number x decreased by 3 is 5.

2. A number a divided by 7 equals 14.

Solve the equation. Check your solution. *(Section 7.2 and Section 7.3)*

3. $4 + k = 14$

4. $3.5 = m - 2.2$

5. $8 = \frac{4w}{3}$

6. $31 = 6.2 \cdot y$

Tell whether the ordered pair is a solution of the equation. *(Section 7.4)*

7. $y = 6x$; $(3, 24)$

8. $y = 3x + 4$; $(4, 16)$

Write and graph an equation in two variables that shows the relationship between the time and the distance traveled. *(Section 7.4)*

9.

10.

11. RIBBON The length of the blue ribbon is two-thirds the length of the red ribbon. Write an equation you can use to find the length r of the red ribbon. *(Section 7.1)*

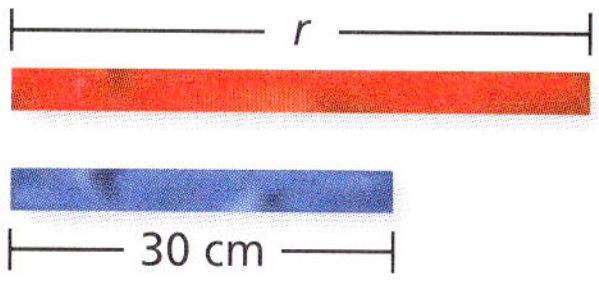

12. BRIDGES The main span of the Sunshine Skyway Bridge is 360 meters long. The Skyway's main span is 30 meters shorter than the main span of the Dames Point Bridge. Write and solve an equation to find the length ℓ of the main span of the Dames Point Bridge. *(Section 7.2)*

13. SHOPPING At a farmer's market, you buy 4 pounds of tomatoes and 2 pounds of sweet potatoes. You spend 80% of the money in your wallet. Write and solve an equation to find how much money is in your wallet before you pay. *(Section 7.3)*

14. SUNDAE A sundae costs $2. Additional toppings cost $0.50 each. Write and graph an equation in two variables that represents the total cost of a sundae. *(Section 7.4)*

7.5 Writing and Graphing Inequalities

Essential Question How can you use a number line to represent solutions of an inequality?

1 ACTIVITY: Understanding Inequality Statements

Work with a partner. Read the statement. Circle each number that makes the statement true, and then answer the questions.

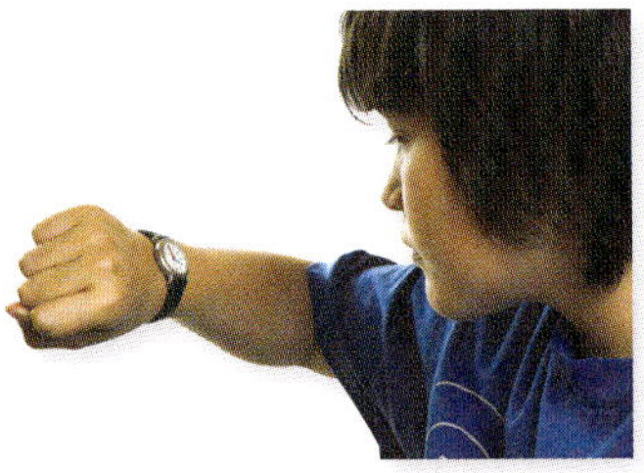

a. **"Your friend is more than 3 minutes late."**

−3 −2 −1 0 1 2 3 4 5 6

- What do you notice about the numbers that you circled?
- Is the number 3 included? Why or why not?
- Write four other numbers that make the statement true.

b. **"The temperature is at most 2 degrees."**

−5 −4 −3 −2 −1 0 1 2 3 4

- What do you notice about the numbers that you circled?
- Can the temperature be exactly 2 degrees? Explain.
- Write four other numbers that make the statement true.

c. **"You need at least 4 pieces of paper for your math homework."**

−3 −2 −1 0 1 2 3 4 5 6

- What do you notice about the numbers that you circled?
- Can you have exactly 4 pieces of paper? Explain.
- Write four other numbers that make the statement true.

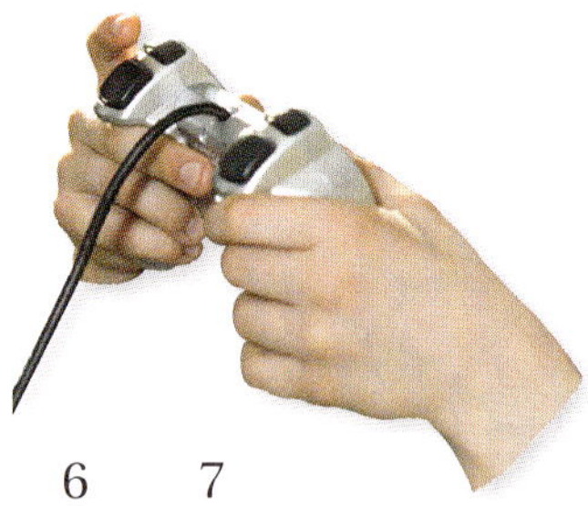

d. **"After playing a video game for 20 minutes, you have fewer than 6 points."**

−2 −1 0 1 2 3 4 5 6 7

- What do you notice about the numbers that you circled?
- Is the number 6 included? Why or why not?
- Write four other numbers that make the statement true.

Common Core

Writing Inequalities

In this lesson, you will

- write word sentences as inequalities.
- use a number line to graph the solution set of inequalities.
- use inequalities to represent real-life situations.

Learning Standards
6.EE.5
6.EE.8

Laurie's Notes

Introduction

Standards for Mathematical Practice

- **MP6 Attend to Precision:** Mathematically proficient students can state clearly the meanings of the symbols they use. Symbols used in this lesson include $<$, $\leq$, $>$, and $\geq$. Proficient students understand the difference between $x > 4$ and $x \geq 4$.

Motivate

- As a quick introduction to the vocabulary in this section, ask a variety of questions. Each student will stand up if the answer is yes, otherwise remain seated.
- ? Sample questions to ask with appropriate inequalities provided for discussion later:
  - Is your birth month before May? $m < 5$ or $m \leq 4$
  - Is your height more than 5 feet? $h > 5$
  - Do you have at most 1 sibling? $x \leq 1$
  - Do you have at least 1 pet? $p \geq 1$
- In discussing each problem, students should observe that not all students who stood up had the same "answer," meaning that for the first example birth months of 1, 2, 3, and 4 all satisfied the statement. Students should also make note of the numbers which do not work. For instance, students with a birth month of 9 (September) did not stand.

Activity Notes

Activity 1

- In this activity, students must read a statement and decide what numbers make sense in the problem.
- Students should pay particular attention to the key words in red.
- Students may need additional examples of contexts where *at most* and *at least* are used. For instance, you may have *at most* 2 cookies. You must read *at least* 20 pages tonight. Ask students for more examples.
- ? "For part (a), could your friend be 30 minutes late?" yes; "3 minutes late?" no; "$5\frac{1}{2}$ minutes late?" yes
- ? "How many solutions are there for part (a)?" Students may or may not know the word infinite but should be able to describe the concept.
- **Note:** Students may only consider whole number solutions, but you should bring up rational numbers. Also, consider the boundary point such as $x = 3$ in part (a).
- **Big Idea:** When giving additional numbers that make the statement true, students must consider the context of the problem. For instance, in part (d), it may not make sense to have a negative number of points, and for part (c) you would use whole sheets of paper.

Common Core State Standards

6.EE.5 Understand solving an . . . inequality as a process of answering a question: which values from a specified set, if any, make the . . . inequality true? Use substitution to determine whether a given number in a specified set makes an . . . inequality true.

6.EE.8 Write an inequality of the form $x > c$ or $x < c$ to represent a constraint or condition in a real-world or mathematical problem. Recognize that inequalities of the form $x > c$ or $x < c$ have infinitely many solutions; represent solutions of such inequalities on number line diagrams.

Previous Learning

Students should know the symbols and vocabulary associated with inequalities.

Technology for the Teacher

Dynamic Classroom

Lesson Plans
Complete Materials List

7.5 Record and Practice Journal

Essential Question How can you use a number line to represent solutions of an inequality?

1 ACTIVITY: Understanding Inequality Statements

Work with a partner. Read the statement. Circle each number that makes the statement true, and then answer the questions.

a. **"Your friend is *more than* 3 minutes late."**

−3 −2 −1 0 1 2 3 (4) (5) (6)

- What do you notice about the numbers that you circled?
  greater than 3
- Is the number 3 included? Why or why not?
  no
- Write four other numbers that make the statement true.
  7, 8, 9, 10

b. **"The temperature is *at most* 2 degrees."**

(−5) (−4) (−3) (−2) (−1) (0) (1) (2) 3 4

- What do you notice about the numbers that you circled?
  2 or less
- Can the temperature be exactly 2 degrees? Explain.
  yes
- Write four other numbers that make the statement true.
  −9, −8, −7, −6

c. **"You need *at least* 4 pieces of paper for your math homework."**

−3 −2 −1 0 1 2 3 (4) (5) (6)

- What do you notice about the numbers that you circled?
  greater than or equal to 4
- Can you have exactly 4 pieces of paper? Explain.
  yes
- Write four other numbers that make the statement true.
  7, 8, 9, 10

English Language Learners

Vocabulary and Symbols

Students should review the vocabulary and symbols for inequalities. Have students add a table to their notebooks of symbols and what the symbols mean. Students should add to the table as new phrases are used in the chapter.

Symbol	Phrase
=	is equal to
≠	is not equal to
<	is less than
≤	is less than or equal to
>	is greater than
≥	is greater than or equal to

7.5 Record and Practice Journal

d. "After playing a video game for 20 minutes, you have *fewer than* 6 points."

(−2) (−1) (0) (1) (2) (3) (4) (5) 6 7

- What do you notice about the numbers that you circled?
  less than 6
- Is the number 6 included? Why or why not?
  no
- Write four other numbers that make the statement true.
  −6, −5, −4, −3

2 ACTIVITY: Understanding Inequality Symbols

Work with a partner.

a. Consider the statement "x is a number such that $x < 2$."

- Can the number be exactly 2? Explain.
  no
- Circle each number that makes the statement true.
  (−5) (−4) (−1) (−2) (−1) (0) (1) 2 3 4
- Write four other numbers that make the statement true.
  −9, −8, −7, −6

b. Consider the statement "x is a number such that $x \geq 1$."

- Can the number be exactly 1? Explain.
  yes
- Circle each number that makes the statement true.
  −5 −4 −3 −2 −1 0 (1) (2) (3) (4)
- Write four other numbers that make the statement true.
  5, 6, 7, 8

3 ACTIVITY: How Close Can You Come to 0?

Work with a partner.

a. Which number line shows $x > 0$? Which number line shows $x \geq 0$? Explain your reasoning.

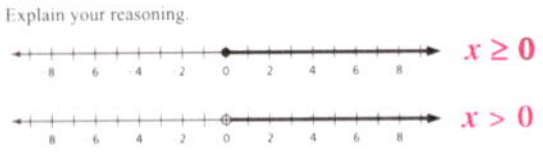

b. Write the least positive number you can think of that is still a solution of the inequality $x > 0$. Explain your reasoning.

Sample answer: 0.00001

What Is Your Answer?

4. **IN YOUR OWN WORDS** How can you use a number line to represent solutions of an inequality?
   Shade the numbers that are solutions of the inequality.

5. Write an inequality. Graph all solutions of your inequality on a number line.
   Check students' work.

6. Graph the inequalities $x > 9$ and $9 < x$ on different number lines. What do you notice?

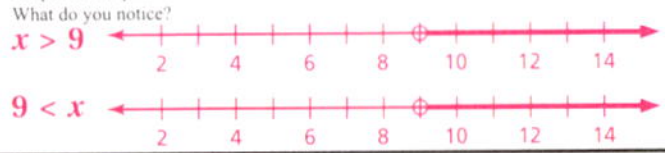

Laurie's Notes

Activity 2

- **MP6:** It is important for students to be familiar with inequality symbols. Students easily confuse them. I have a poster in my classroom with the symbols and their translations for students to refer to.
- ? Write the less than symbol < on the board and ask the following questions.
  - "What does this symbol mean? How is the symbol read?" less than; is less than
  - "What inequality symbols are you familiar with?" They may respond with >, <, ≤, ≥; They may also know ≠.
- Write a few inequality statements on the board as a visual reference for students. Use simple statements such as $4 < 5$ and $8 > 6$.
- **Note:** This activity and Activity 1 try to distinguish between $x < 2$ and $x \leq 2$.
- ? "In part (a), would 1.5 make $x < 2$ true?" yes; "would $1\frac{3}{4}$?" yes
- ? "How many numbers are there that are less than 2?" Infinite if you are not restricted to only whole numbers.

Activity 3

- Discuss the difference in the symbols $x > 0$ and $x \geq 0$. Students should be able to describe the difference in words.
- Graphing the solution of an inequality is new. If decimals and fractions have been suggested throughout, the graph should not be a big stretch for them. The open and closed circle notation will be new, but generally the notation will make sense to students.
- **Big Idea:** The graph shows that the set of solutions is infinite. It is not possible to list all of the solutions, although the . . . notation suggests that the list continues in the same fashion.
- ? Related to part (b), ask "What is the greatest number that is a solution of $x > 0$?" Compare the numbers suggested by students. Students may have heard of a googol (a 1 followed by 100 zeros). There is no greatest number.

What Is Your Answer?

- **Think-Pair-Share:** Students should read each question independently and then work in pairs to answer the questions. When they have answered the questions, the pair should compare their answers with another group and discuss any discrepancies.

Closure

- Use questions similar to those used to motivate the lesson. Ask, "How would you write (and graph) the solutions of these inequalities?"
  - You need at least 8 hours of sleep. $h \geq 8$
  - You will have at most 2 hours of homework. $h \leq 2$

2 ACTIVITY: Understanding Inequality Symbols

Work with a partner.

a. Consider the statement "x is a number such that $x < 2$."

- Can the number be exactly 2? Explain.
- Circle each number that makes the statement true.

  −5 −4 −3 −2 −1 0 1 2 3 4

- Write four other numbers that make the statement true.

Math Practice 6

State the Meaning of Symbols

What do the symbols < and ≥ mean?

b. Consider the statement "x is a number such that $x \geq 1$."

- Can the number be exactly 1? Explain.
- Circle each number that makes the statement true.

  −5 −4 −3 −2 −1 0 1 2 3 4

- Write four other numbers that make the statement true.

3 ACTIVITY: How Close Can You Come to 0?

Work with a partner.

a. Which number line shows $x > 0$? Which number line shows $x \geq 0$? Explain your reasoning.

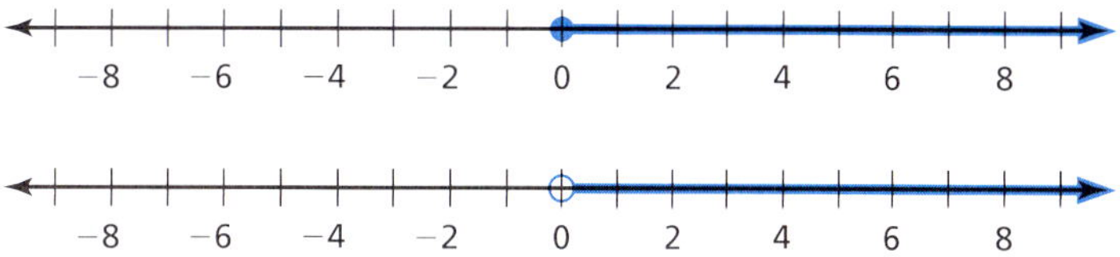

b. Write the least positive number you can think of that is still a solution of the inequality $x > 0$. Explain your reasoning.

What Is Your Answer?

4. **IN YOUR OWN WORDS** How can you use a number line to represent solutions of an inequality?
5. Write an inequality. Graph all solutions of your inequality on a number line.
6. Graph the inequalities $x > 9$ and $9 < x$ on different number lines. What do you notice?

Practice Use what you learned about graphing inequalities to complete Exercises 17–20 on page 329.

7.5 Lesson

Check It Out
Lesson Tutorials
BigIdeasMath.com

Key Vocabulary
inequality, *p. 326*
solution of an inequality, *p. 327*
solution set, *p. 327*
graph of an inequality, *p. 328*

An **inequality** is a mathematical sentence that compares expressions. It contains the symbols <, >, ≤, or ≥. To write an inequality, look for the following phrases to determine where to place the inequality symbol.

Inequality Symbols				
Symbol	<	>	≤	≥
Key Phrases	• is less than • is fewer than	• is greater than • is more than	• is less than or equal to • is at most • is no more than	• is greater than or equal to • is at least • is no less than

EXAMPLE 1 Writing Inequalities

Write the word sentence as an inequality.

a. A number c is less than -4.

A number c | is less than | -4.

c $<$ -4

An inequality is $c < -4$.

b. A number k plus 5 is greater than or equal to 8.

A number k plus 5 | is greater than or equal to | 8.

$k + 5$ $\geq$ 8

An inequality is $k + 5 \geq 8$.

c. Four times a number q is at most 16.

Four times a number q | is at most | 16.

$4q$ $\leq$ 16

An inequality is $4q \leq 16$.

On Your Own

Now You're Ready
Exercises 5–10

Write the word sentence as an inequality.

1. A number n is greater than 1.
2. Twice a number p is fewer than 7.
3. A number w minus 3 is less than or equal to 10.
4. A number z divided by 2 is at least -6.

Multi-Language Glossary at BigIdeasMath.com

Laurie's Notes

Introduction

Connect

- **Yesterday:** Students used a variety of contexts to explore inequalities. (MP6)
- **Today:** Students will translate words into mathematical inequalities and determine what numbers are solutions of an inequality.

Motivate

- Make eleven number cards using 3×5 index cards. Write the integers from -5 to 5 on the cards.
- Ask 11 students to stand at the front of the room, each holding one card and facing their classmates who are seated. Tell them you will describe a set of numbers. If the number the student is holding is in the set described, he or she should take one step forward.
- "If your number is [sample] take one step forward on the word go."
- Samples:
  - greater than 2
  - less than or equal to 1
  - at least -1
  - at most 3
- This introduction helps to connect the vocabulary to the graphs that students will be drawing today.

Lesson Notes

Example 1

- Discuss the definition of inequality and the vocabulary associated with each symbol.
- If possible, change chalk or marker color as you do the problems. This will help students view the different components of writing an inequality.
- Work through each example. Note that the inequalities are *not* being solved.
- Part (c) will be the most challenging. Students may struggle with determining the correct inequality symbol for *at most*. Give them a context to relate to such as, "The wait time to enter is at most 15 minutes." This will help students translate the words using the correct inequality symbol, ≤.

On Your Own

- **Think-Pair-Share:** Students should read each question independently and then work in pairs to answer the questions. When they have answered the questions, the pair should compare their answers with another group and discuss any discrepancies.
- Ask volunteers to write their inequalities at the board.

? "In Question 4, the inequality is $\frac{z}{2} \geq -6$. Could $z = 4$? How do you know?" yes; $4 \div 2$ is greater than -6. "Could $z = 0$? How do you know?" yes; $0 \div 2$ is greater than -6.

Goal Today's lesson is graphing and determining **solutions of an inequality**.

Technology for the Teacher

Dynamic Classroom

Lesson Tutorials
Lesson Plans
Answer Presentation Tool

Extra Example 1

Write the word sentence as an inequality.

a. A number t is no more than 7. $t \leq 7$

b. The difference of a number p and 3 is at least 10. $p - 3 \geq 10$

c. One-fourth of a number x is greater than 2. $\frac{1}{4}x > 2$

On Your Own

1. $n > 1$
2. $2p < 7$
3. $w - 3 \leq 10$
4. $\frac{z}{2} \geq -6$

Extra Example 2

Tell whether the given value is a solution of the inequality.

a. $d - 5 \leq 7$; $d = 9$ yes

b. $4y \geq 24$; $y = 6$ yes

c. $\frac{m}{5} < 6$; $m = 30$ no

On Your Own

5. no **6.** yes

7. yes

Differentiated Instruction

Auditory

Stress to students the importance of reading a statement and translating it into an expression, equation, or inequality. The word "is" plays an important role in the meaning of the statement. For instance, *six less than a number* translates to $x - 6$. While *six **is** less than a number* translates to $6 < x$.

Laurie's Notes

Example 2

- **Vocabulary:** Explain the vocabulary words *solution of an inequality* and *solution set.*
- **Representation:** Discuss the notation used to show that 8 is not less than or equal to 7 ($8 \not\leq 7$).
- Use the samples in the textbook or Example 1 to show the substitution for the value of the variable. If the inequality statement is true, the value of the variable is a solution, otherwise it is not a solution.
- Work through each part.
- **MP6 Attend to Precision:** In each part, the *value of the variable is substituted* for the variable. Avoid using language such as, "Plug in the number for *x*."
- Note that in part (c), 5 is greater than or equal to 5.

? "Would 15 be a solution of the inequality $\frac{z}{3} > 5$? Explain."
no; 5 is not greater than 5.

On Your Own

- **Neighbor Check:** Have students work independently and then have their neighbors check their work. Have students discuss any discrepancies.

A **solution of an inequality** is a value that makes the inequality true. An inequality can have more than one solution. The set of all solutions of an inequality is called the **solution set**.

Value of x	$x + 3 \le 7$	Is the inequality true?
3	$3 + 3 \overset{?}{\le} 7$ $6 \le 7$ ✓	yes
4	$4 + 3 \overset{?}{\le} 7$ $7 \le 7$ ✓	yes
5	$5 + 3 \overset{?}{\le} 7$ $8 \not\le 7$ ✗	no

Reading

The symbol $\not\le$ means *is not less than or equal to.*

EXAMPLE 2 Checking Solutions

Tell whether the given value is a solution of the inequality.

a. $x + 1 > 7;\ x = 8$

$x + 1 > 7$ Write the inequality.

$8 + 1 \overset{?}{>} 7$ Substitute 8 for x.

$9 > 7$ ✓ Add. 9 is greater than 7.

So, 8 is a solution of the inequality.

b. $7y < 27;\ y = 4$

$7y < 27$ Write the inequality.

$7(4) \overset{?}{<} 27$ Substitute 4 for y.

$28 \not< 27$ ✗ Multiply. 28 is *not* less than 27.

So, 4 is *not* a solution of the inequality.

c. $\frac{z}{3} \ge 5;\ z = 15$

$\frac{z}{3} \ge 5$ Write the inequality.

$\frac{15}{3} \overset{?}{\ge} 5$ Substitute 15 for z.

$5 \ge 5$ ✓ Divide. 5 is greater than or equal to 5.

So, 15 is a solution of the inequality.

On Your Own

Now You're Ready
Exercises 11–16

Tell whether 3 is a solution of the inequality.

5. $b + 4 < 6$ **6.** $9 - n \ge 6$ **7.** $18 \div x \le 10$

The **graph of an inequality** shows all the solutions of the inequality on a number line. An open circle ○ is used when a number is *not* a solution. A closed circle ● is used when a number is a solution. An arrow to the left or right shows that the graph continues in that direction.

EXAMPLE 3 Graphing an Inequality

Reading

The inequality $g > 2$ is the same as $2 < g$.

Graph $g > 2$.

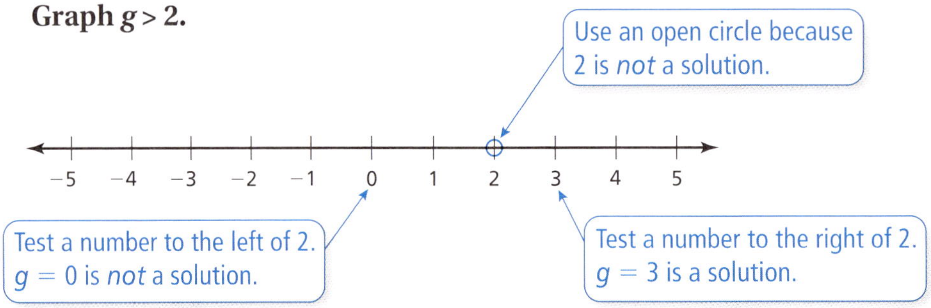

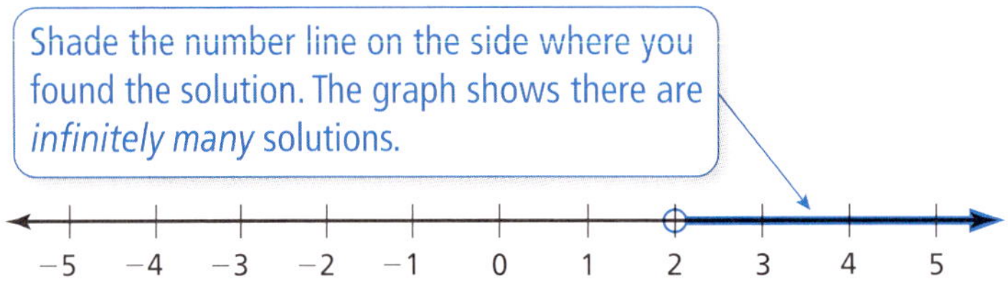

EXAMPLE 4 Real-Life Application

The NASA Solar Probe Plus can withstand temperatures up to and including 2600°F. Write and graph an inequality that represents the temperatures the probe can withstand.

Words	temperatures	up to and including	2600°F
Variable	Let t be the temperatures the probe can withstand.		
Inequality	t	$\leq$	2600

∴ An inequality is $t \leq 2600$.

On Your Own

Now You're Ready
Exercises 25–36

Graph the inequality on a number line.

8. $a < 4$
9. $f \leq 7$
10. $n > 0$
11. $p \geq -3$

Write and graph an inequality for the situation.

12. A cruise ship can carry at most 3500 passengers.
13. A board game is designed for ages 12 and up.

Laurie's Notes

Example 3

- **Vocabulary:** Explain the vocabulary word *graph of an inequality*.
- Discuss the difference between $x > 5$ and $x \geq 5$. Explain to students that they know how to describe the difference in words. Now you want to show them the differences when you graph the solution of each.
- Describe the difference between an open circle and a closed circle.
- Graph $x > 5$ and $x \geq 5$.
- Work through the example. Use the language of *test point* to describe how to determine which side of the boundary point (2) to shade.
- **Summarize:** Shading on the number line indicates a solution of the inequality, such as $g = 3$. No shading on the number line indicates that the number is not a solution, such as $g = 0$.
- Be sure to discuss the reading note with students as they will encounter inequalities with the variable on the right side of the inequality symbol in the exercise set. It is very common for students to graph this type of inequality incorrectly. They will shade in the wrong direction.

Example 4

- You may need to explain that *up to and including* means the same as *less than or equal to*.
- **Common Misconception:** Students may think that 0 must be included on the number line. In this example, the students use a scale on the number line that is not in increments of 1 and is not centered at 0.

On Your Own

- Students should work with partners to graph the inequalities.
- **Teaching Tip:** If you have small individual white boards, do these questions together and have students display their graphs simultaneously.

Closure

- Repeat the inequalities used to motivate at the beginning of the lesson. Ask students to write the symbols and then graph the solution.
  - Greater than 2 $x > 2$

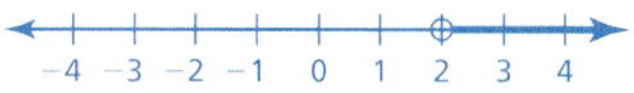

  - Less than or equal to 1 $x \leq 1$

  - At least −1 $x \geq -1$

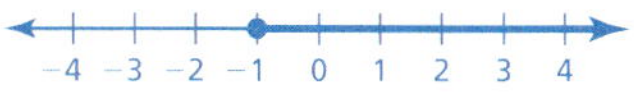

  - At most 3 $x \leq 3$

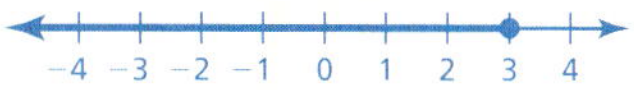

Extra Example 3

Graph $h \leq 5$.

Extra Example 4

A classroom has enough room for up to 25 students. Write and graph an inequality that represents the number s of students in the classroom. $s \leq 25$

On Your Own

8.

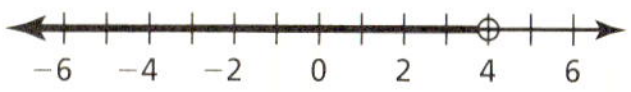

9.

10.

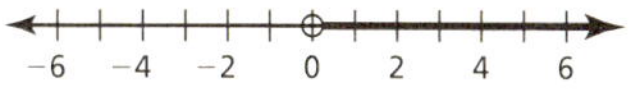

11.

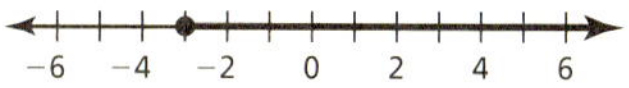

12. $p \leq 3500$;

13. $a \geq 12$;

Vocabulary and Concept Check

1. Both phrases refer to numbers that are greater than a given number. The difference is that "greater than or equal to" includes the number itself, whereas "greater than" does not.
2. A number n is at least 3; $n \geq 3$; $n \leq 3$
3. The graph of $x \leq 6$ has a closed circle at 6. The graph of $x < 6$ has an open circle at 6.
4. Both graphs are the same, because both indicate that x is less than or equal to 5.

Practice and Problem Solving

5. $k < 10$
6. $a > 6$
7. $z < \frac{3}{4}$
8. $b \geq -3$
9. $1 + y \leq -13$
10. $\frac{x}{3} \leq 5$
11. yes
12. no
13. yes
14. no
15. no
16. yes
17. B
18. A
19. D
20. C
21. $x < 1$; A number x is less than 1.
22. $x \leq 1$; A number x is at most 1.
23. $x \geq -4$; A number x is at least -4.
24. $x > 0$; A number x is more than 0.
25. 

26. 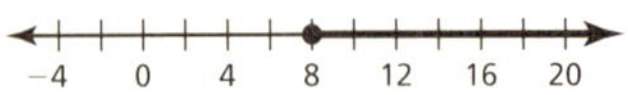

Assignment Guide and Homework Check

Level	Day 1 Activity Assignment	Day 2 Lesson Assignment	Homework Check
Basic	17–20, 47–51	1–4, 5–13 odd, 25–31 odd, 37, 40	4, 5, 11, 25, 40
Average	17–20, 47–51	1–4, 5–15 odd, 25–39 odd, 40, 41	4, 5, 11, 29, 40
Advanced	17–20, 47–51	1–4, 22–38 even, 41–46	22, 26, 42, 44

For Your Information

- **Exercise 46** Part (b) introduces the concept of compound inequalities.

Common Errors

- **Exercises 5–10** Students may struggle with knowing which symbol to use in the inequality. Encourage them to put the word sentence into a real-life context and to use the table in the lesson that explains what symbol matches each phrase.
- **Exercises 5–10** Students may write the subtraction inequalities in the wrong order. For example, a student will write $7 - x < 4$ instead of $x - 7 < 4$. Remind them that writing inequalities is similar to writing equations and expressions.
- **Exercises 11–16** Students may try to perform operations on both sides of the inequality (often not the same operation on both sides either) to get both sides to be the same answer. Remind them that they are checking to see if the value of the variable is true, so they are focusing on simplifying one side of the equation, then comparing the answer to the other side.
- **Exercises 17–20** Students may mix up the open and closed circles when matching. Remind them that the open circle means the number is not included, and a closed circle means that the number is included.

7.5 Record and Practice Journal

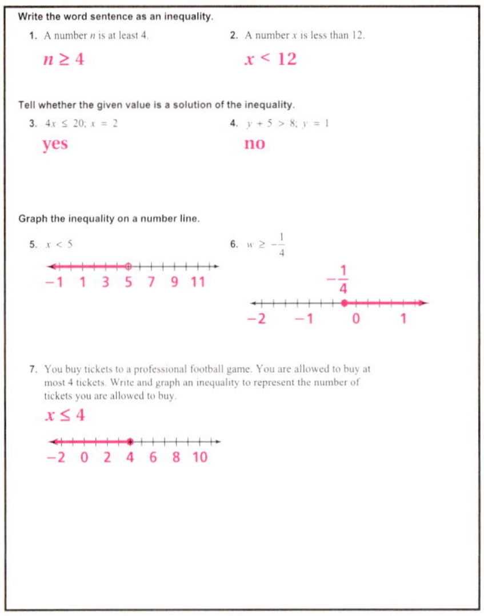

Write the word sentence as an inequality.

1. A number n is at least 4.

$n \geq 4$

2. A number x is less than 12.

$x < 12$

Tell whether the given value is a solution of the inequality.

3. $4x \leq 20$; $x = 2$

yes

4. $y + 5 > 8$; $y = 1$

no

Graph the inequality on a number line.

5. $x < 5$

6. $w \geq -\frac{1}{4}$

7. You buy tickets to a professional football game. You are allowed to buy at most 4 tickets. Write and graph an inequality to represent the number of tickets you are allowed to buy.

$x \leq 4$

7.5 Exercises

Check It Out
Help with Homework
BigIdeasMath.com

Vocabulary and Concept Check

1. **VOCABULARY** How are *greater than* and *greater than or equal to* similar? How are they different?

2. **DIFFERENT WORDS, SAME QUESTION** Which is different? Write "both" inequalities.

A number n is at most 3.	A number n is at least 3.
A number n is less than or equal to 3.	A number n is no more than 3.

3. **WRITING** Explain how the graph of $x \leq 6$ is different from the graph of $x < 6$.

4. **WRITING** Are the graphs of $x \leq 5$ and $5 \geq x$ the same or different? Explain.

Practice and Problem Solving

Write the word sentence as an inequality.

1 5. A number k is less than 10.

6. A number a is more than 6.

7. A number z is fewer than $\frac{3}{4}$.

8. A number b is at least -3.

9. One plus a number y is no more than -13.

10. A number x divided by 3 is at most 5.

Tell whether the given value is a solution of the inequality.

2 11. $x - 1 \leq 7;\ x = 6$

12. $y + 5 < 13;\ y = 17$

13. $3z > 6;\ z = 3$

14. $\frac{b}{2} \geq 6;\ b = 10$

15. $c + 2.5 < 4.3;\ c = 1.8$

16. $a \leq 0;\ a = -5$

Match the inequality with its graph.

17. $x \geq 2$

18. $x < 2$

19. $x > -2$

20. $x \leq -2$

A. 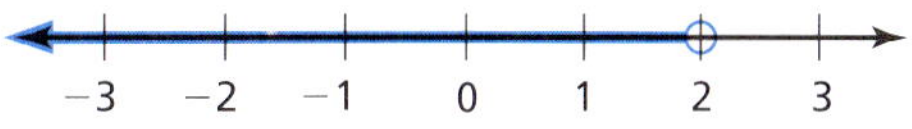

B. 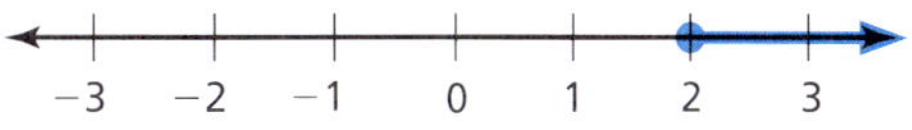

C. 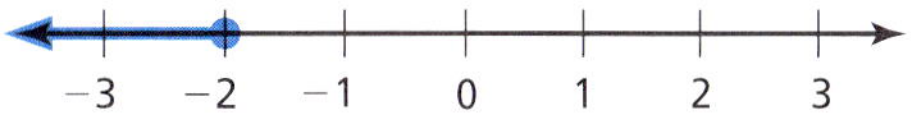

D. 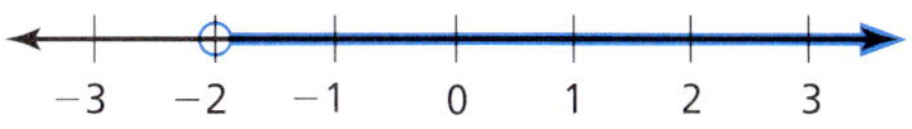

Write an inequality and a word sentence that represent the graph.

21. 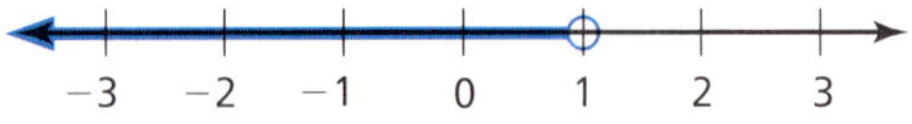

22. 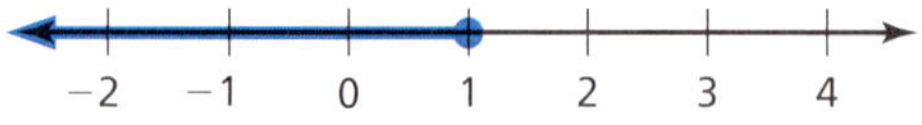

23. 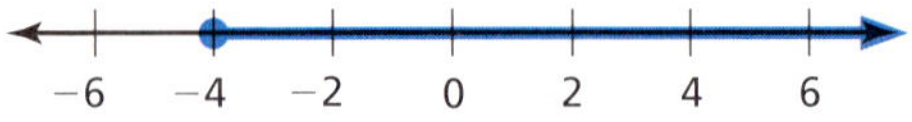

24. 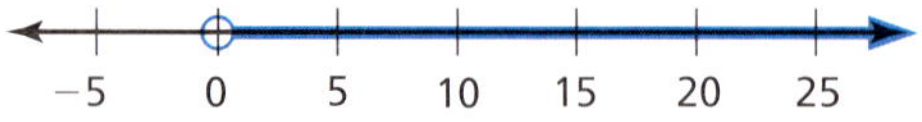

Graph the inequality on a number line.

③ **25.** $a > 4$ **26.** $n \geq 8$ **27.** $3 \geq x$ **28.** $y < \frac{1}{2}$

29. $x < \frac{2}{9}$ **30.** $-3 \geq c$ **31.** $m > -5$ **32.** $0 \leq b$

33. $1.5 > f$ **34.** $t \geq -\frac{1}{2}$ **35.** $p > -1.6$ **36.** $\frac{7}{3} \geq z$

ERROR ANALYSIS **Describe and correct the error in graphing the inequality.**

37. 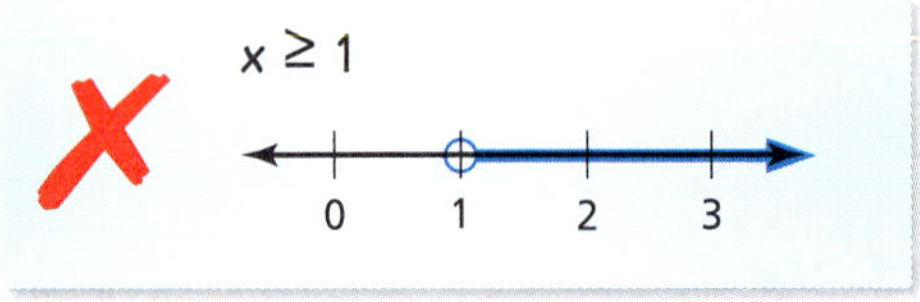

38. 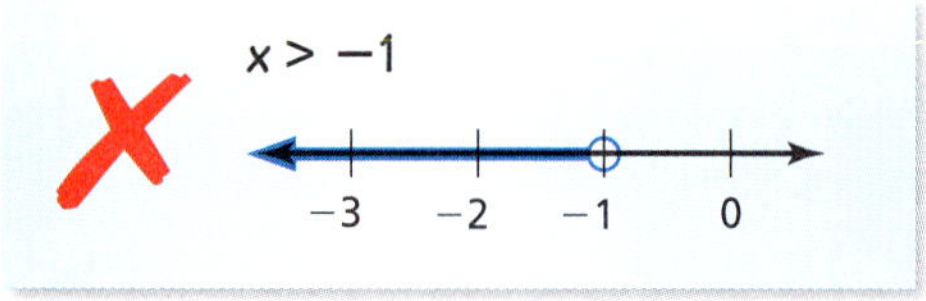

39. FISHING You are fishing and are allowed to catch at most 3 striped bass. Each striped bass must be no less than 18 inches long.

a. Write and graph an inequality to represent the number of striped bass you are allowed to catch.

b. Write and graph an inequality to represent the length of each striped bass you are allowed to catch.

40. MODELING For a food to be labeled *low sodium,* there must be no more than 140 milligrams of sodium per serving.

a. Write and graph an inequality to represent the amount of sodium in a low-sodium serving.

b. Write and graph an inequality to represent the amount of sodium in a serving that does *not* qualify as low sodium.

c. Does the food represented by the nutrition facts label qualify as a low-sodium food? Explain.

Nutrition Facts	
Serving Size ½ cup (114g)	
Servings Per Container 4	
Amount Per Serving	
Calories 90	Calories from fat 30
	% Daily Value*
Total Fat 3g	**5%**
Saturated Fat 0g	**0%**
Cholesterol 0mg	**0%**
Sodium 300mg	**13%**
Total Carbohydrate 13g	**4%**
Dietary Fiber 3g	**12%**
Sugars 3g	
Protein 3g	
Vitamin A 80% •	Vitamin C 60%
Calcium 4% •	Iron 4%

Common Errors

- **Exercises 25–36** Students may shade the number line in the wrong direction. This often happens when the variable is on the right of the inequality symbol instead of the left. Remind them to check numbers to the left and right of the given number to determine which direction to shade the number line.
- **Exercises 25–36** Students may draw a closed circle instead of an open circle and vice versa. Remind them of the definitions of open and closed circles.
- **Exercise 39** Students may mix up the inequalities given in the word problem. Remind them to read carefully and encourage them to write word sentences for the information given.

Practice and Problem Solving

27. 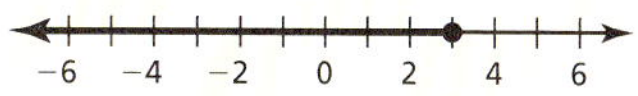

28. 

29. 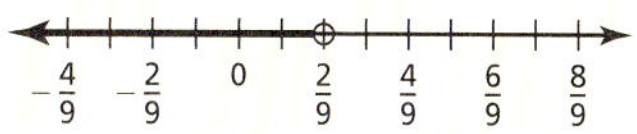

30. 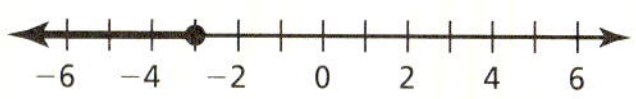

31. 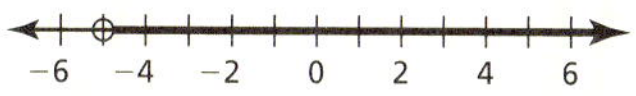

32. 

33. 

34. 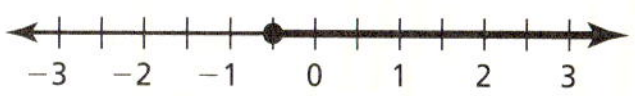

35. 

36. 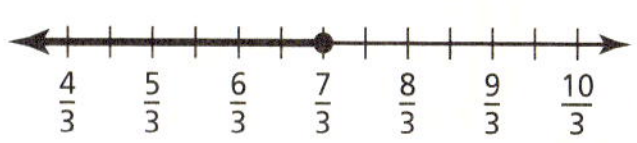

37. $x \geq 1$ means that 1 is also a solution, so a closed circle should be used.

38. The graph shows $x < -1$, not $x > -1$.

39. **a.** $b \leq 3$;

b. $\ell \geq 18$;

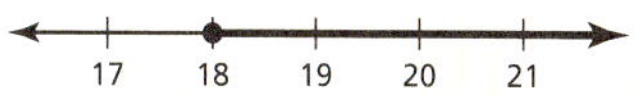

Differentiated Instruction

Visual

Discuss with students the graph of the inequality for Exercise 40(a). Have them describe numbers that are solutions, like 139.99, and numbers that are not solutions, like 140.01. Ask students if the number −1 makes sense as a solution. Point out that in application problems, students need to determine whether the solution of an equation or inequality makes sense as an answer.

Practice and Problem Solving

40. **a.** $s \leq 140$;

b. $s > 140$;

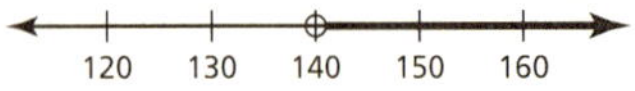

c. no; There are 300 milligrams of sodium in one serving, which is greater than 140.

41. See *Taking Math Deeper*.

42. always; A number that is greater than 5 is always greater than or equal to 5.

43. sometimes; The only time this is not true is if $x = 5$.

44. $1.50x > 36$

45. $p \leq 375$

46. See Additional Answers.

Fair Game Review

47. $x = 9$ **48.** $x = 14$

49. $x = 28$ **50.** $x = 14.1$

51. D

Mini-Assessment

Write the word sentence as an inequality.

1. A number b is at most 5. $b \leq 5$
2. Twice a number s is less than 15. $2s < 15$

Tell whether the given value is a solution of the inequality.

3. $3x > 16$; $x = 6$ yes
4. $12 - n \geq 5$; $n = 8$ no
5. No more than 10 people will go to the movies. Write and graph an inequality to represent the number of movie goers. $x \leq 10$

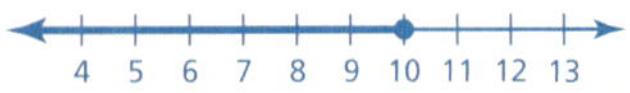

Taking Math Deeper

Exercise 41

This problem gives students an opportunity to use an inequality to describe a real-life situation.

Item	Price (with tax)
T-shirt	\$15
Book	\$20
DVD	\$13
Necklace	\$16

① You have \$33 and plan to buy a necklace for \$16. One way to approach the problem is to let x represent the cost of another item.

a. $16 + x \leq 33$

Necklace — Another item — Starting Amount

Students might also subtract the \$16 from \$33 and write the following inequality.

$x \leq 17$

② **b.** A T-shirt costs \$15, so there *is* enough money left to buy a T-shirt.

③ **c.** A book costs \$20, so there *is not* enough money left to buy a book.

Reteaching and Enrichment Strategies

If students need help. . .	If students got it. . .
Resources by Chapter • Practice A and Practice B • Puzzle Time Record and Practice Journal Practice Differentiating the Lesson Lesson Tutorials Skills Review Handbook	Resources by Chapter • Enrichment and Extension • Technology Connection Start the next section

41. **SHOPPING** You have \$33. You want to buy a necklace and one other item from the list.

Item	Price (with tax)
T-shirt	\$ 15
Book	\$ 20
DVD	\$ 13
Necklace	\$ 16

a. Write an inequality to represent the situation.

b. Can the other item be a T-shirt? Explain.

c. Can the other item be a book? Explain.

Determine whether the statement is *sometimes, always,* or *never* true. Explain your reasoning.

42. A number that is a solution of the inequality $x > 5$ is also a solution of the inequality $x \geq 5$.

43. A number that is a solution of the inequality $5 \leq x$ is also a solution of the inequality $x > 5$.

44. **BUS RIDE** A bus ride costs \$1.50. A 30-day bus pass costs \$36. Write an inequality to represent the number of bus rides you would need to take for the bus pass to be a better deal.

45. **MOVIE THEATER** Fifty people are seated in a movie theater. The maximum capacity of the theater is 425 people. Write an inequality to represent the number of additional people who can still be seated.

46. **Critical Thinking** The map shows the elevations above sea level for an area of land.

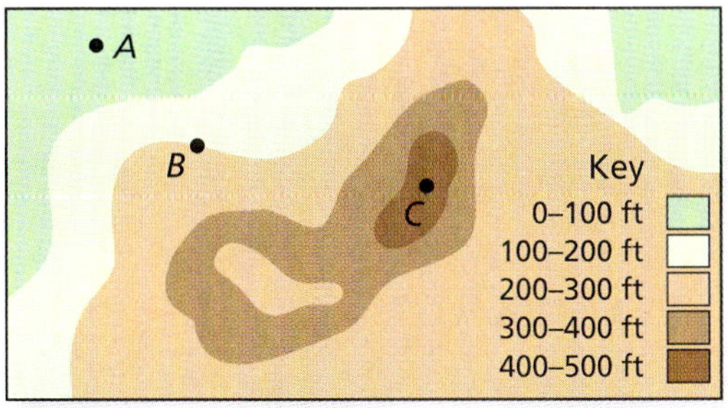

a. Graph the possible elevations of A. Write the set of elevations as two inequalities.

b. Graph the possible elevations of C. How can you write this set of elevations as a single inequality? Explain.

c. What is the elevation of B? Explain.

Fair Game Review What you learned in previous grades & lessons

Solve the equation. Check your solution. *(Section 7.2)*

47. $x + 3 = 12$
48. $x - 6 = 8$
49. $16 + x = 44$
50. $7.6 = x - 6.5$

51. **MULTIPLE CHOICE** A stack of boards is 24 inches high. Each board is $\frac{3}{8}$ of an inch thick. How many boards are in the stack? *(Section 2.2)*

Ⓐ $\frac{1}{9}$ Ⓑ $\frac{1}{6}$ Ⓒ 9 Ⓓ 64

7.6 Solving Inequalities Using Addition or Subtraction

Essential Question How can you use addition or subtraction to solve an inequality?

1 ACTIVITY: Writing an Inequality

Work with a partner. In 3 years, your friend will still not be old enough to vote.

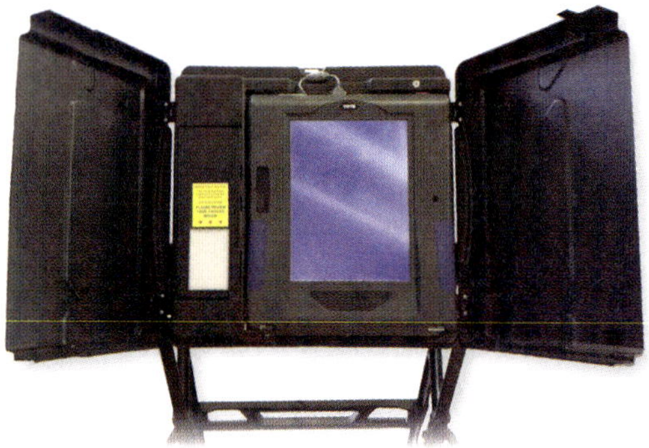

a. Which of the following represents your friend's situation? What does x represent? Explain your reasoning.

$x + 3 < 18$ $\quad$ $x + 3 \le 18$

$x + 3 > 18$ $\quad$ $x + 3 \ge 18$

b. Graph the possible ages of your friend on a number line. Explain how you decided what to graph.

2 ACTIVITY: Writing an Inequality

Work with a partner. Baby manatees are about 4 feet long at birth. They grow to a maximum length of 13 feet.

a. Which of the following can represent a baby manatee's growth? What does x represent? Explain your reasoning.

$x + 4 < 13$ $\quad$ $x + 4 \le 13$

$x - 4 > 13$ $\quad$ $x - 4 \ge 13$

b. Graph the solution on a number line. Explain how you decided what to graph.

COMMON CORE

Solving Inequalities

In this lesson, you will

- use addition or subtraction to solve inequalities.
- use a number line to graph the solution set of inequalities.
- solve real-life problems.

Applying Standards
6.EE.5
6.EE.8

Laurie's Notes

Introduction

Standards for Mathematical Practice

- **MP5 Use Appropriate Tools Strategically:** A number line can help students visualize the results of solving an inequality. Students know what the graph of $x \geq 6$ looks like. Seeing the graph of the solution to the inequality $x + 2 \geq 6$ helps students verify that $x = 4$ is a solution, but all values less than 4 are not solutions.

Motivate

- **FYI:** Share some historical notes about inequalities with students:
  - One of the earliest publications using the current day symbols < and > was in a posthumous publication by Thomas Harriot in 1631. Some claim Harriot invented these symbols while others suggest it was Harriot's editor who invented the symbols.
  - Forty years later, in 1670, John Wallis used ⩾ ("equal or greater") and ⩽ ("equal or less") in his publication.

Activity Notes

Activity 1

? "What is the legal voting age in the United States?" 18

- **FYI:** The Twenty-sixth Amendment to the U.S. Constitution was passed in 1971 to guarantee the right to vote to all 18-year-olds. World War II passions that "any American old enough to fight is old enough to vote" had resurfaced because of the Vietnam War.
- Students should be able to use their own age to think through the correct inequality for this problem.
- **MP5:** Graphing various ages on the number line is a helpful tool in writing the inequality for this problem.

Activity 2

- **FYI:** Adult manatees have been known to weigh over 3500 pounds.
- Students may use Guess and Check to begin thinking about which inequality represents the manatee's growth.
- **MP3 Construct Viable Arguments and Critique the Reasoning of Others:** Listen for clear reasoning as students describe how they decided what to graph. With an inequality involving 4 and 13, they might simply guess, placing a closed circle at 9 (13 − 4) or 17 (13 + 4). Students need to give an explanation, not a guess.

Common Core State Standards

6.EE.5 Understand solving an . . . inequality as a process of answering a question: which values from a specified set, if any, make the . . . inequality true? Use substitution to determine whether a given number in a specified set makes an . . . inequality true.

6.EE.8 Write an inequality of the form $x > c$ or $x < c$ to represent a constraint or condition in a real-world or mathematical problem. Recognize that inequalities of the form $x > c$ or $x < c$ have infinitely many solutions; represent solutions of such inequalities on number line diagrams.

Extending the Standard: Although Standard 6.EE.5 does not explicitly require solving inequalities, it follows solving equations well. Introducing this topic with calculations limited to positive number operations now will lessen the burden when the topic is covered in Grade 7 with negative numbers.

Previous Learning

Students should know how to solve equations using addition or subtraction.

Technology for the Teacher

Dynamic Classroom

Lesson Plans

Complete Materials List

7.6 Record and Practice Journal

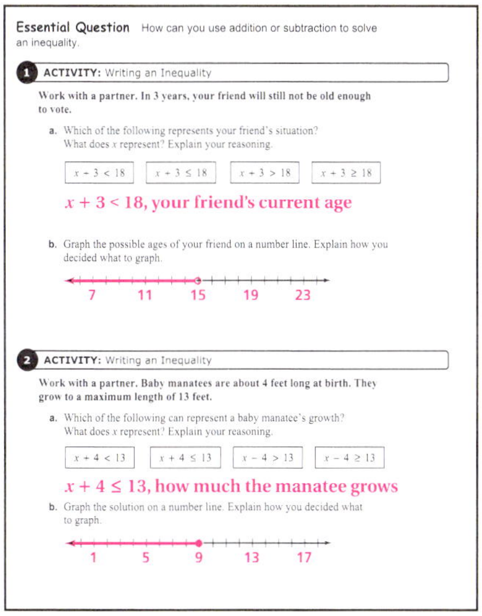

Essential Question How can you use addition or subtraction to solve an inequality.

1 ACTIVITY: Writing an Inequality

Work with a partner. In 3 years, your friend will still not be old enough to vote.

a. Which of the following represents your friend's situation? What does x represent? Explain your reasoning.

$x + 3 < 18$ | $x + 3 \leq 18$ | $x + 3 > 18$ | $x + 3 \geq 18$

$x + 3 < 18$, your friend's current age

b. Graph the possible ages of your friend on a number line. Explain how you decided what to graph.

2 ACTIVITY: Writing an Inequality

Work with a partner. Baby manatees are about 4 feet long at birth. They grow to a maximum length of 13 feet.

a. Which of the following can represent a baby manatee's growth? What does x represent? Explain your reasoning.

$x + 4 < 13$ | $x + 4 \leq 13$ | $x - 4 > 13$ | $x - 4 \geq 13$

$x + 4 \leq 13$, how much the manatee grows

b. Graph the solution on a number line. Explain how you decided what to graph.

English Language Learners

Vocabulary

English learners may benefit by restating word problems so that the wording corresponds to the wording of inequality symbols. In Activity 2, the problem could be rewritten as:

> *Baby manatees are about 4 feet long at birth. They grow a certain number of feet to a length less than or equal to 13 feet.*

This leads to writing the inequality $4 + x \leq 13$.

7.6 Record and Practice Journal

3 ACTIVITY: Solving Inequalities

Work with a partner. Complete the following steps for Activity 1. Then repeat the steps for Activity 2.

- Use your inequality from part (a). Replace the inequality symbol with an equal sign.

  Activity 1: $x + 3 = 18$
  Activity 2: $x + 4 = 13$

- Solve the equation.

  Activity 1: $x = 15$
  Activity 2: $x = 9$

- Replace the equal sign with the original inequality symbol.

  Activity 1: $x < 15$
  Activity 2: $x \leq 9$

- Graph this new inequality.

Activity 1:

Activity 2:

- Compare the graph with your graph in part (b). What do you notice?

  Graphs are the same.

4 ACTIVITY: The Triangle Inequality

Work with a partner. Draw different triangles whose sides have lengths 10 cm, 6 cm, and x cm.

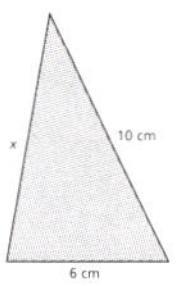

a. Which of the following describes how *small* x can be? Explain your reasoning.

$6 + x < 10$	$6 + x \leq 10$
$6 + x > 10$	$6 + x \geq 10$

$6 + x > 10$

b. Which of the following describes how *large* x can be?

$x - 6 < 10$	$x - 6 \leq 10$	$x - 6 > 10$	$x - 6 \geq 10$

$x - 6 < 10$

c. Graph the possible values of x on a number line.

What Is Your Answer?

5. **IN YOUR OWN WORDS** How can you use addition and subtraction to solve an inequality? Replace the inequality symbol with an equal sign, solve the equation, and then replace the equal sign with the original inequality symbol.
6. Describe a real-life situation that you can represent with an inequality. Write the inequality. Graph the solution on a number line.

Check students' work.

Laurie's Notes

Activity 3

- This activity steps students through the process of solving an inequality. They may indeed have an intuitive sense that solving an inequality is very much like solving an equation.
- Be sure that students work through all the solution steps for Activity 1 and Activity 2.

Activity 4

- **FYI:** The Triangle Inequality in a geometry course would state that "the sum of any two legs of a triangle must exceed the length of the last side." You can model this by using 3 pencils of lengths (approximately) 5", 6", and 8". Selecting any two pencils, the sum of the lengths is longer than the third pencil. This would not be the case if the initial pencils were 3", 4", and 8". The sum of the two shortest pencils is shorter than the longest pencil.

? If students need a hint on this activity, ask the following questions to help them think about the range of numbers for x.
  - "Can x be 2 centimeters?" no
  - "Can x be 20 centimeters?" no
- **Note:** x must be longer than 4 cm and shorter than 16 centimeters.

What Is Your Answer?

- **Neighbor Check:** For Question 6, have students work independently and then have their neighbors check their work. Have students discuss any discrepancies.

Closure

- "How would you write and graph the solution of $x + 5 \leq 13$?" $x \leq 8$

3 ACTIVITY: Solving Inequalities

Work with a partner. Complete the following steps for Activity 1. Then repeat the steps for Activity 2.

Math Practice 4

Interpret Results

What does the solution of the inequality represent?

- Use your inequality from part (a). Replace the inequality symbol with an equal sign.
- Solve the equation.
- Replace the equal sign with the original inequality symbol.
- Graph this new inequality.
- Compare the graph with your graph in part (b). What do you notice?

4 ACTIVITY: The Triangle Inequality

Work with a partner. Draw different triangles whose sides have lengths 10 cm, 6 cm, and *x* cm.

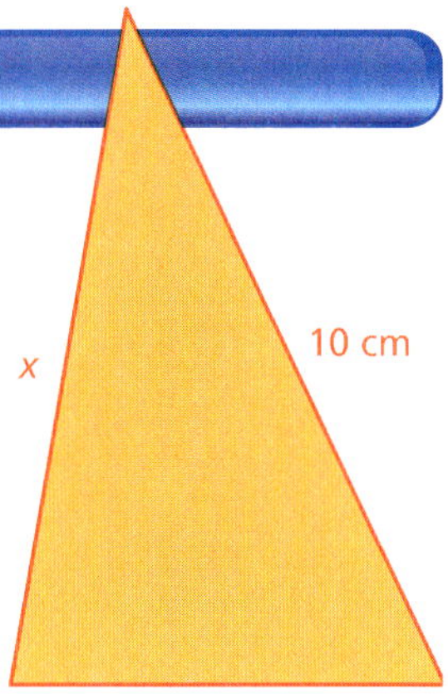

a. Which of the following describes how *small* x can be? Explain your reasoning.

$6 + x < 10$	$6 + x \leq 10$
$6 + x > 10$	$6 + x \geq 10$

b. Which of the following describes how *large* x can be? Explain your reasoning.

$x - 6 < 10$	$x - 6 \leq 10$	$x - 6 > 10$	$x - 6 \geq 10$

c. Graph the possible values of x on a number line.

What Is Your Answer?

5. **IN YOUR OWN WORDS** How can you use addition or subtraction to solve an inequality?
6. Describe a real-life situation that you can represent with an inequality. Write the inequality. Graph the solution on a number line.

Practice Use what you learned about solving inequalities to complete Exercises 5–7 on page 336.

7.6 Lesson

Check It Out
Lesson Tutorials
BigIdeasMath.com

Study Tip

You can solve inequalities the same way you solve equations. Use inverse operations to get the variable by itself.

Key Ideas

Addition Property of Inequality

Words When you add the same number to each side of an inequality, the inequality remains true.

Numbers

$$\begin{aligned} 3 &< 5 \\ +2 & \quad +2 \\ \hline 5 &< 7 \end{aligned}$$

Algebra

$$\begin{aligned} x - 4 &> 5 \\ +4 & \quad +4 \\ \hline x &> 9 \end{aligned}$$

Graph

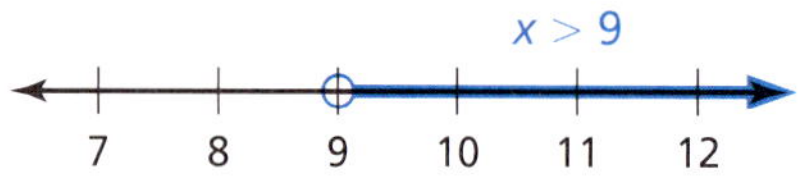

Subtraction Property of Inequality

Words When you subtract the same number from each side of an inequality, the inequality remains true.

Numbers

$$\begin{aligned} 3 &< 5 \\ -2 & \quad -2 \\ \hline 1 &< 3 \end{aligned}$$

Algebra

$$\begin{aligned} x + 4 &> 5 \\ -4 & \quad -4 \\ \hline x &> 1 \end{aligned}$$

Graph

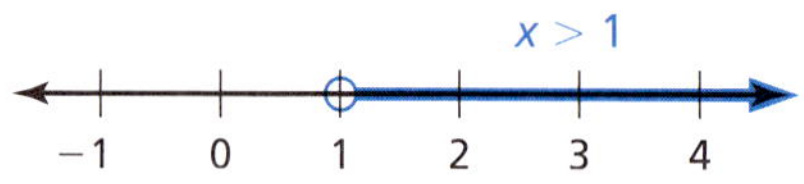

These properties are also true for ≤ and ≥.

EXAMPLE 1 Solving an Inequality Using Addition

Solve $x - 3 > 1$. Graph the solution.

$x - 3 > 1$	Write the inequality.
$+3 \quad +3$	Addition Property of Inequality
$x > 4$	Simplify.

Undo the subtraction. (→ $+3$)

∴ The solution is $x > 4$.

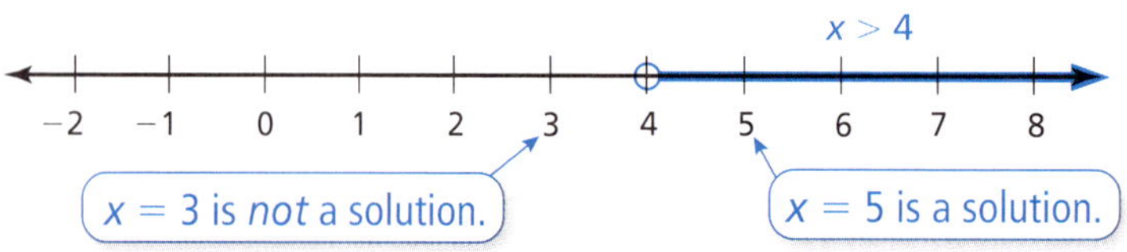

Check:

$x = 3$: $3 - 3 \overset{?}{>} 1$

$0 \not> 1$ ✗

$x = 5$: $5 - 3 \overset{?}{>} 1$

$2 > 1$ ✓

On Your Own

Solve the inequality. Graph the solution.

1. $x - 2 < 3$
2. $x - 6 \geq 4$
3. $10 \geq x - 1$

Laurie's Notes

Introduction

Connect

- **Yesterday:** Students used a variety of contexts to explore inequalities that could be solved using addition or subtraction. (MP3, MP5)
- **Today:** Students will use the Properties of Inequality to solve inequalities.

Motivate

? Ask a series of questions that suggest the Properties of Inequality that will be used today. Samples:
  - "I am older than each of you. In 3 years, who will be older?" adding 3 years to each age did not change the fact that you are older than your students.
  - "Sarah is shorter than Matt. If they both grow 2 inches, who is shorter?" adding 2 inches to each height did not change the fact that Sarah is shorter than Matt.
- This introduction helps to create a context for the Properties of Inequality.
- Point out how the Key Idea streamlines the process for solving inequalities. The properties correspond to Properties of Equality used to solve equations. When you use these properties, there is no need to temporarily replace the inequality symbol with an equal sign as in yesterday's activity.

Lesson Notes

Key Ideas

- Explain that solving inequalities is very similar to solving equations (refer to the Study Tip).
- Remind students to line up their inequality symbols as they solve vertically.
- **Teaching Tip:** Students will want to solve the inequalities in their heads and resist writing their work down. Remind students that you are trying to help them develop good problem solving techniques. The problems will become more involved later and students need to know how to record the process.

Example 1

? "How do you graph $x > 4$?" open circle at 4, shade to the right of 4

? "How do you know which side of the boundary point ($x = 4$) to shade?" test values on each side of 4 "Why is there an open circle?" 4 is not a solution.

- **MP5 Use Appropriate Tools Strategically:** In checking their solutions, have students check numbers on either side of the open (or closed) circle.

On Your Own

- **Think-Pair-Share:** Students should read each question independently and then work in pairs to answer the questions. When they have answered the questions, the pair should compare their answers with another group and discuss any discrepancies.
- **Common Error:** In Question 3, because the variable is on the right, it is common for students to shade the wrong side of the inequality when graphing. Encourage students to rewrite the solution $11 \geq x$ as $x \leq 11$.

Goal Today's lesson is solving inequalities using addition or subtraction.

Technology for the Teacher

Dynamic Classroom

Lesson Tutorials
Lesson Plans
Answer Presentation Tool

Extra Example 1

Solve $5 \leq x - 4$. Graph the solution.
$9 \leq x$ or $x \geq 9$

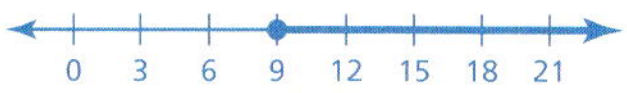

On Your Own

1. $x < 5$;

2. $x \geq 10$;

3. $11 \geq x$;

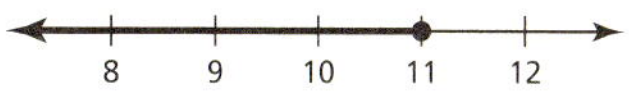

Extra Example 2

Solve $n + 7 < 8$. Graph the solution.
$n < 1$

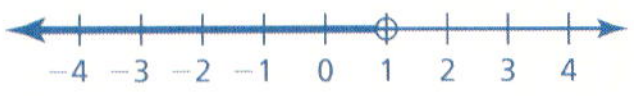

On Your Own

4. $x > 4$;

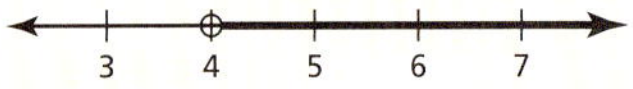

5. $y < 15$;

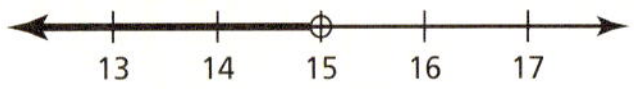

6. $7 \leq m$;

Extra Example 3

Your book report must be at least 500 words. So far, you have written 334 words. Write and solve an inequality to represent the number w of words you still need to write. $w + 334 \geq 500$; $w \geq 166$

On Your Own

7. $24 + x \geq 75$; $x \geq 51$; You must spend at least \$51 more to get free shipping.

Differentiated Instruction

Visual

To show students that the Addition and Subtraction Properties of Inequalities are true, graph two numbers, 2 and 5, on the number line. Write the ordered relationship, $2 < 5$, on the board. Now add 4 to each number. Both points move the same distance to the right, so their ordered positions remain the same, $6 < 9$. Now subtract 2 from each of the original numbers. Both points move 2 units to the left, so their ordered positions remain the same, $0 < 3$.

Laurie's Notes

Example 2

? Discuss the initial inequality with the variable written on the right side.

- "Is $15 \geq 6 + x$ equivalent to $15 \geq x + 6$? Explain."
  yes; Commutative Property
- "Is $15 \geq 6 + x$ equivalent to $6 + x \leq 15$? Explain."
  yes; If $a > b$, then $b < a$.

- Students could solve an equivalent inequality or the inequality as it is stated in the textbook. If students solve the inequality as it is stated, note that the last step involves rewriting the solution so the variable is on the left side of the inequality symbol.
- **Big Idea:** The solution $x \leq 9$ means that all numbers less than or equal to 9 are solutions. The graph conveys this idea with the left-pointing arrow and the use of negative numbers.
- Discuss with students the scale used for the graph. Remind them that it is not necessary to use an increment of 1 for the scaling on the number line.

On Your Own

- **Neighbor Check:** Have students work independently and then have their neighbors check their work. Have students discuss any discrepancies.

Example 3

- **Note:** The last sentence in the problem identifies what you need to find. While reading the problems, students should be thinking about what the question is asking them.
- **Extension:** Ask students to graph the solution.

On Your Own

- **Think-Pair-Share:** Students should read the question independently and then work in pairs to answer the question. When they have answered the question, the pair should compare their answer with another group and discuss any discrepancies.
- Ask students to think of other examples where inequalities are used. An example is unlimited text messaging up to 500 messages.

Closure

- Explain how solving $x - 8 > 14$ and $x - 8 = 14$ are the same and how they are different. Solve each. $x > 22$ and $x = 22$; *Sample answer:* For both, you add 8 to each side (Addition Property). The inequality has infinitely many solutions, but the equation has only one solution.

EXAMPLE 2 Solving an Inequality Using Subtraction

Solve $15 \geq 6 + x$. Graph the solution.

	$15 \geq 6 + x$	Write the inequality.
Undo the addition. →	$\underline{-6} \quad \underline{-6}$	Subtraction Property of Inequality
	$9 \geq x$	Simplify.

The solution is $x \leq 9$.

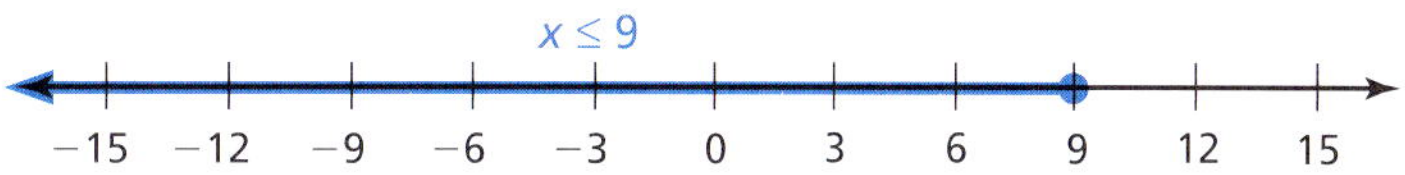

Reading

The inequality $x \leq 9$ is the same as $9 \geq x$.

On Your Own

Now You're Ready
Exercises 5–16

Solve the inequality. Graph the solution.

4. $x + 3 > 7$ **5.** $y + 2 < 17$ **6.** $16 \leq m + 9$

EXAMPLE 3 Real-Life Application

A flea market advertises that it has more than 250 vending booths. Of these, 184 are currently filled. Write and solve an inequality to represent the number of vending booths still available.

Words	The number of booths filled	plus	the number of remaining booths	is greater than	the total number of booths.
Variable	Let b be the number of remaining booths.				
Inequality	184	$+$	b	$>$	250

$184 + b > 250$	Write the inequality.
$\underline{-184} \quad \underline{-184}$	Subtraction Property of Inequality
$b > 66$	Simplify.

More than 66 vending booths are still available.

On Your Own

7. You have already spent \$24 shopping online for clothes. Write and solve an inequality to represent the additional amount you must spend to get free shipping.

7.6 Exercises

Check It Out
Help with Homework
BigIdeasMath.com

Vocabulary and Concept Check

1. **OPEN-ENDED** Write an inequality that can be solved by subtracting 7 from each side.
2. **WRITING** Explain how to solve the inequality $x - 6 > 3$.
3. **WRITING** Describe the graph of the solution of $x + 3 \leq 4$.
4. **OPEN-ENDED** Write an inequality that the graph represents. Then use the Subtraction Property of Inequality to write another inequality that the graph represents.

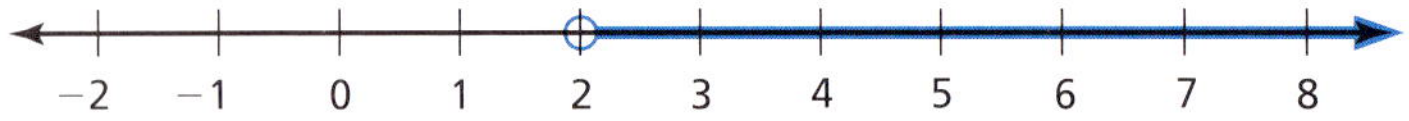

Practice and Problem Solving

Solve the inequality. Graph the solution.

1 2

5. $x - 4 < 5$
6. $5 + h > 7$
7. $3 \geq y - 2$
8. $9 \leq c + 1$
9. $18 > 12 + x$
10. $37 + z \leq 54$
11. $y - 21 < 85$
12. $g - 17 \geq 17$
13. $7.2 < x + 4.2$
14. $12.7 \geq s - 5.3$
15. $\frac{3}{4} \leq \frac{1}{2} + n$
16. $\frac{1}{3} + b > \frac{3}{4}$

17. **ERROR ANALYSIS** Describe and correct the error in solving the inequality.

$$\begin{aligned} 28 &\geq t - 9 \\ -9 &\quad\;\; -9 \\ \hline 19 &\geq t \end{aligned}$$

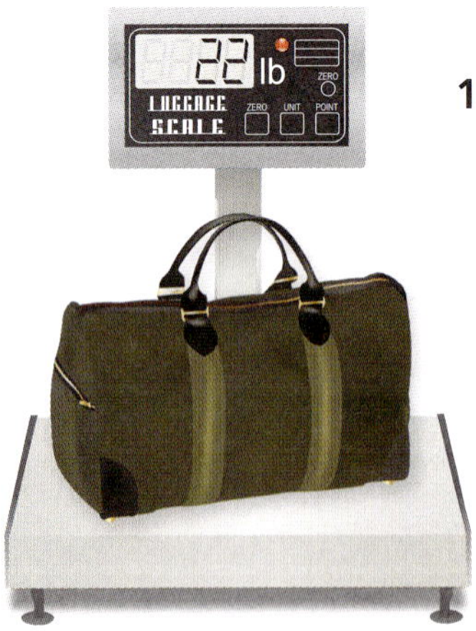

18. **AIR TRAVEL** Your carry-on bag can weigh at most 40 pounds. Write and solve an inequality to represent how much more weight you can add to the bag and still meet the requirement.

19. **SHOPPING** It costs $\$x$ for a round-trip bus ticket to the mall. You have \$24. Write and solve an inequality to represent the greatest amount of money you can spend for the bus fare and still have enough to buy the baseball cap.

Write the word sentence as an inequality. Then solve the inequality.

20. Five more than a number is less than 17.
21. Three less than a number is more than 15.

Assignment Guide and Homework Check

Level	Day 1 Activity Assignment	Day 2 Lesson Assignment	Homework Check
Basic	5–7, 29–33	1–4, 9–21 odd, 18, 20	9, 11, 18, 20
Average	5–7, 29–33	1–4, 9–17 odd, 18, 20, 21, 23, 25	9, 11, 20, 23
Advanced	5–7, 29–33	1–4, 8–28 even, 17, 27	14, 16, 22, 26

For Your Information

- **Exercise 26** In some cases, the weights of drivers and gasoline are included in a vehicle's weight.

Common Errors

- **Exercises 5–16** Students may use the same operation in solving for the variable instead of the inverse operation. Demonstrate to them that this will not work by simplifying the inequality on the board. Students most likely ignored the side with the variable when they made this mistake. Remind them to check their answers in the original inequality.
- **Exercises 5–16** Students may also switch the direction of the inequality when adding or subtracting. Remind them that the inequality does not change. Work through an example without variables to demonstrate that the inequality is still true when numbers are added or subtracted.

7.6 Record and Practice Journal

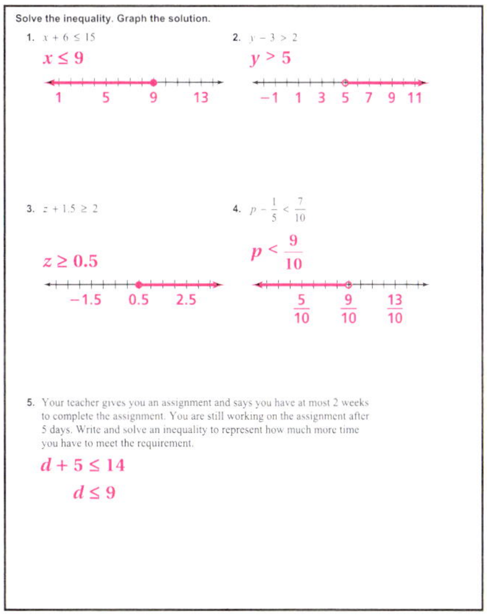
Solve the inequality. Graph the solution.

1. $x + 6 \leq 15$
   $x \leq 9$
2. $y - 3 > 2$
   $y > 5$
3. $z + 1.5 \geq 2$
   $z \geq 0.5$
4. $p - \frac{1}{5} < \frac{7}{10}$
   $p < \frac{9}{10}$
5. Your teacher gives you an assignment and says you have at most 2 weeks to complete the assignment. You are still working on the assignment after 5 days. Write and solve an inequality to represent how much more time you have to meet the requirement.
   $d + 5 \leq 14$
   $d \leq 9$

Vocabulary and Concept Check

1. *Sample answer:* $x + 7 \geq 143$
2. Add 6 to both sides of the inequality.
3. By solving the inequality to obtain $x \leq 1$, the graph has a closed circle at 1 and an arrow pointing in the negative direction.
4. *Sample answer:* $x > 2$; $x - 1 > 1$

Practice and Problem Solving

5. $x < 9$;

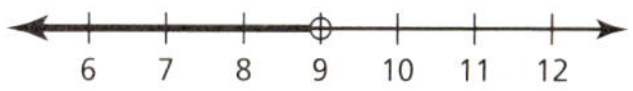

6. $h > 2$;

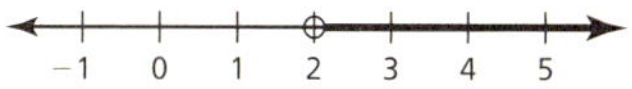

7. $5 \geq y$;

8. $8 \leq c$;

9. $6 > x$;

10. $z \leq 17$;

11. $y < 106$;

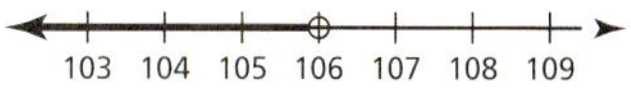

12. $g \geq 34$;

13. $3 < x$;

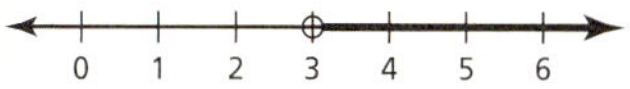

14. $18 \geq s$;

15–19. See Additional Answers.

Practice and Problem Solving

20. $x + 5 < 17; x < 12$

21. $x - 3 > 15; x > 18$

22. $x \le 8$;

23. $11 > s$;

24. $4.8 \ge c$;

25. $34{,}280 + d + 1000 > 36{,}480$; $d > 1200$ dragonflies

26. See Additional Answers.

27. See *Taking Math Deeper*.

28. 25

Fair Game Review

29. $t = 48$ **30.** $27 = s$

31. $x = 9$ **32.** $6 = z$

33. C

Mini-Assessment

Solve the inequality. Graph the solution.

1. $x + 9 \le 15$ $x \le 6$

2. $5 > 2 + y$ $y < 3$

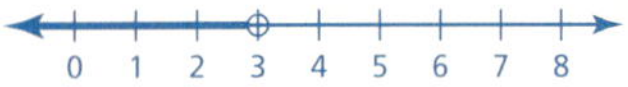

3. $z - 2.3 < 4.7$ $z < 7$

4. $6 \le t + 4$ $t \ge 2$

5. You have 70 DVDs. The DVD cabinet you own holds 150 DVDs. Write and solve an inequality to represent how many DVDs you can buy and fit into the cabinet. $70 + x \le 150$; $x \le 80$

Taking Math Deeper

Exercise 27

This is an open-ended problem that allows students to use estimation, common sense, and real-life experiences to come up with a reasonable solution.

Triathlon	
Event	**Your Time (minutes)**
Swimming	18.2
Biking	45.4
Running	?

① Write an inequality.

a. $18.2 + 45.4 + x < 100$

Swimming Biking Running Total

② $63.6 + x < 100$ Combine like terms.

$x < 36.4$ Subtract 63.6 from each side.

③ **b.** There are many ways students can estimate how long it takes to run a mile. However they do this, their estimates should be greater than 4 minutes, which is approximately the world record for running a mile.

Suppose a student estimates that it would take 8 minutes to run a mile. With this estimate, there is plenty of time to run 3.1 miles and still meet the goal of less than 100 minutes in the race.

Project

Research triathlons that are run in the United States. Report on the training, cost, and time commitments of the participants.

Reteaching and Enrichment Strategies

If students need help. . .	If students got it. . .
Resources by Chapter • Practice A and Practice B • Puzzle Time Record and Practice Journal Practice Differentiating the Lesson Lesson Tutorials Skills Review Handbook	Resources by Chapter • Enrichment and Extension • Technology Connection Start the next section

Solve the inequality. Graph the solution.

22. $x + 9 - 3 \leq 14$

23. $44 > 7 + s + 26$

24. $6.1 - 0.3 \geq c + 1$

25. **VIDEO GAME** The high score for a video game is 36,480. Your current score is 34,280. Each dragonfly you catch is worth 1 point. You also get a 1000-point bonus for reaching 35,000 points. Write and solve an inequality to represent the number of dragonflies you must catch to earn a new high score.

26. **PICKUP TRUCKS** You can register a pickup truck as a passenger vehicle if the truck is not used for commercial purposes and the weight of the truck with its contents does not exceed 8500 pounds.

a. Your pickup truck weighs 4200 pounds. Write an inequality to represent the number of pounds your truck can carry and still qualify as a passenger vehicle. Then solve the inequality.

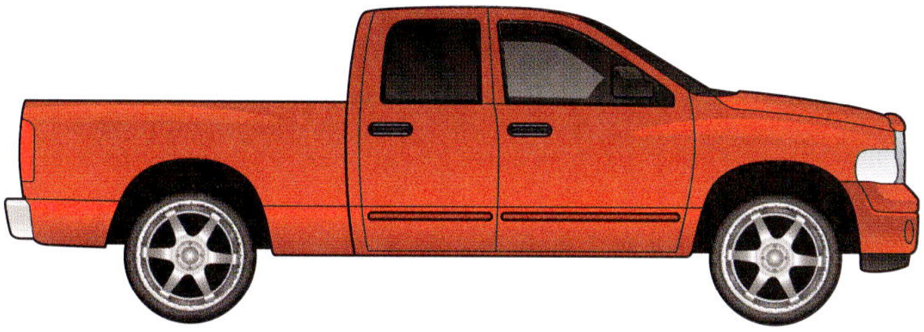

b. A cubic yard of sand weighs about 1600 pounds. How many cubic yards of sand can you haul in your truck and still qualify as a passenger vehicle? Explain your reasoning.

27. **TRIATHLON** You complete two events of a triathlon. Your goal is to finish with an overall time of less than 100 minutes.

a. Write and solve an inequality to represent how many minutes you can take to finish the running event and still meet your goal.

b. The running event is 3.1 miles long. Estimate how many minutes it would take you to run 3.1 miles. Would this time allow you to reach your goal? Explain your reasoning.

Triathlon	
Event	**Your Time (minutes)**
Swimming	18.2
Biking	45.4
Running	?

28. Number Sense The possible values of x are given by $x - 3 \geq 2$. What is the least possible value of $5x$?

Fair Game Review What you learned in previous grades & lessons

Solve the equation. Check your solution. *(Section 7.3)*

29. $\frac{t}{12} = 4$

30. $6 = \frac{2s}{9}$

31. $8x = 72$

32. $9 = 1.5z$

33. **MULTIPLE CHOICE** Which brand of turkey is the best buy? *(Section 5.4)*

Ⓐ Brand A　Ⓑ Brand B

Ⓒ Brand C　Ⓓ Brand D

Brand	A	B	C	D
Cost (dollars)	10.38	13.47	21.45	34.93
Pounds	2	3	5	7

7.7 Solving Inequalities Using Multiplication or Division

Essential Question How can you use multiplication or division to solve an inequality?

1 ACTIVITY: Writing an Inequality

Work with a partner. A store has a clearance rack of shirts that each cost the same amount. You buy 2 shirts and have money left after paying with a $20 bill.

a. Which of the following represents your purchase? What does x represent? Explain your reasoning.

$2x < 20$	$2x \le 20$
$2x > 20$	$2x \ge 20$

b. Graph the possible values of x on a number line. Explain how you decided what to graph.

c. Can you buy a third shirt? Explain your reasoning.

2 ACTIVITY: Writing an Inequality

Work with a partner. One of your favorite stores is having a 75% off sale. You have $20. You want to buy a pair of jeans.

a. Which of the following represents your ability to buy the jeans with $20? What does x represent? Explain your reasoning.

$\frac{1}{4}x < 20$	$\frac{1}{4}x \le 20$
$\frac{1}{4}x > 20$	$\frac{1}{4}x \ge 20$

b. Graph the possible values of x on a number line. Explain how you decided what to graph.

c. Can you afford a pair of jeans that originally costs $100? Explain your reasoning.

COMMON CORE

Solving Inequalities

In this lesson, you will

- use multiplication or division to solve inequalities.
- use a number line to graph the solution set of inequalities.
- solve real-life problems.

Applying Standards
6.EE.5
6.EE.8

Laurie's Notes

Introduction

Standards for Mathematical Practice

- **MP5 Use Appropriate Tools Strategically:** A number line can help students to visualize the solution of an inequality. Test values on each side of the boundary point to determine which side of the number line to shade.

Motivate

- **Story Time:** Tell students about a sale (could be Internet-based) where all of the items (name something you would buy) were 50% off. Students know I love to geocache, so I would use geocoins for my example.
- ? "What did I pay for geocoins that were normally \$12? \$8? \$6? \$4?" \$6, \$4, \$3, \$2
- ? "Suppose the geocoins were 75% off. What would I pay for coins that were normally \$12? \$8? \$3? \$2? \$3, \$2, \$0.75, \$0.50
- Discuss *the amount saved* in each instance and *the percent paid* of the original cost. For instance, for a \$12 item marked down 75%, you save \$9 and pay 25% of the original cost.

Activity Notes

Activity 1

- ? Read the problem and ask, "What is a clearance rack?" a rack of clothes priced very low so they all sell and are cleared from the store
- **MP1 Make Sense of Problems and Persevere in Solving Them:** I tell my students to read the problem, and then read the problem again. Too often students ask for assistance before making an honest effort to understand the problem. Make students try before you help them.
- ? "Can the price of each of the two shirts be \$12.50? Explain." No; The two shirts would cost \$25 and you only used \$20.
- ? "Can the price of each shirt be \$10? Explain." No; The two shirts would cost \$20, and you are supposed to have money left over from the \$20 bill.
- ? "Which inequality did you select and what does x represent?" $2x < 20$; x is the cost of each shirt.
- ? "Could you purchase a third shirt at the same price as the first two? Explain." Listen for an explanation that shows an understanding of the unit cost being less than one-third of \$20.

Activity 2

- Make sure students understand that x equals the original purchase price of the jeans.
- ? "Why is 0.25 used in this problem when the jeans are 75% off?" 25% is the amount you must pay when there is a 75% off sale; 25% = 0.25.
- **MP3 Construct Viable Arguments and Critique the Reasoning of Others:** Listen for clear reasoning as students describe how they decided what to graph.
- **Extension:** Discuss the sales tax that would be added in your state.

Common Core State Standards

6.EE.5 Understand solving an . . . inequality as a process of answering a question: which values from a specified set, if any, make the . . . inequality true? Use substitution to determine whether a given number in a specified set makes an . . . inequality true.

6.EE.8 Write an inequality of the form $x > c$ or $x < c$ to represent a constraint or condition in a real-world or mathematical problem. Recognize that inequalities of the form $x > c$ or $x < c$ have infinitely many solutions; represent solutions of such inequalities on number line diagrams.

Previous Learning

Students should know how to solve equations using multiplication or division.

Technology for the Teacher

Dynamic Classroom

Lesson Plans

Complete Materials List

7.7 Record and Practice Journal

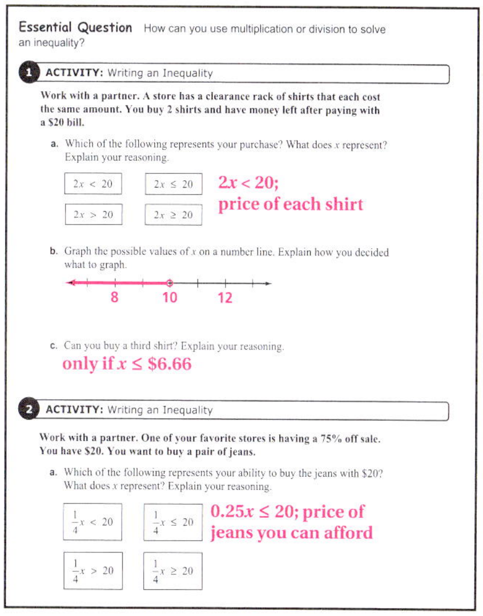

Essential Question How can you use multiplication or division to solve an inequality?

1 ACTIVITY: Writing an Inequality

Work with a partner. A store has a clearance rack of shirts that each cost the same amount. You buy 2 shirts and have money left after paying with a \$20 bill.

a. Which of the following represents your purchase? What does x represent? Explain your reasoning.

$2x < 20$ $\quad$ $2x \le 20$ $\quad$ $2x > 20$ $\quad$ $2x \ge 20$

$2x < 20$; price of each shirt

b. Graph the possible values of x on a number line. Explain how you decided what to graph.

8 $\quad$ 10 $\quad$ 12

c. Can you buy a third shirt? Explain your reasoning.

only if $x \le \$6.66$

2 ACTIVITY: Writing an Inequality

Work with a partner. One of your favorite stores is having a 75% off sale. You have \$20. You want to buy a pair of jeans.

a. Which of the following represents your ability to buy the jeans with \$20? What does x represent? Explain your reasoning.

$\frac{1}{4}x < 20$ $\quad$ $\frac{1}{4}x \le 20$ $\quad$ $\frac{1}{4}x > 20$ $\quad$ $\frac{1}{4}x \ge 20$

$0.25x \le 20$; price of jeans you can afford

English Language Learners

Vocabulary

It is important that English learners understand the difference between *less than* and *less than or equal to,* as well as *greater than* and *greater than or equal to.* Starting with 1, give each student a unique number card. Tell students to stand up if their number is *less than* (say a number) and then *less than or equal to* that same number. Students should discuss the difference between the two. This can be repeated with *greater than* and *greater than or equal to.*

7.7 Record and Practice Journal

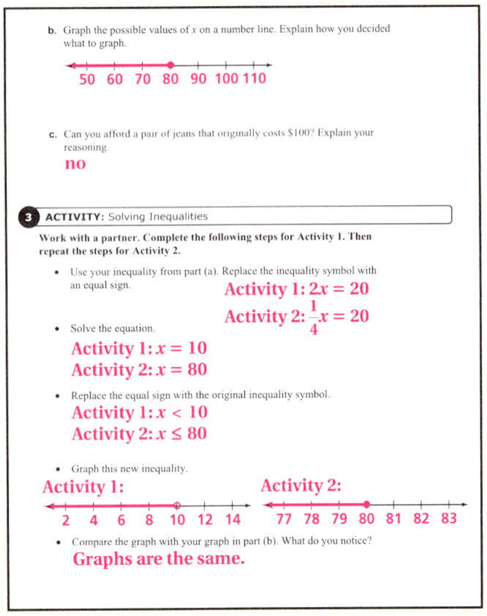

b. Graph the possible values of x on a number line. Explain how you decided what to graph.

50 60 70 80 90 100 110

c. Can you afford a pair of jeans that originally costs $100? Explain your reasoning.

no

3 ACTIVITY: Solving Inequalities

Work with a partner. Complete the following steps for Activity 1. Then repeat the steps for Activity 2.

- Use your inequality from part (a). Replace the inequality symbol with an equal sign.
  Activity 1: $2x = 20$
  Activity 2: $\frac{1}{4}x = 20$
- Solve the equation.
  Activity 1: $x = 10$
  Activity 2: $x = 80$
- Replace the equal sign with the original inequality symbol.
  Activity 1: $x < 10$
  Activity 2: $x \le 80$
- Graph this new inequality.
  Activity 1: 2 4 6 8 10 12 14
  Activity 2: 77 78 79 80 81 82 83
- Compare the graph with your graph in part (b). What do you notice?
  Graphs are the same.

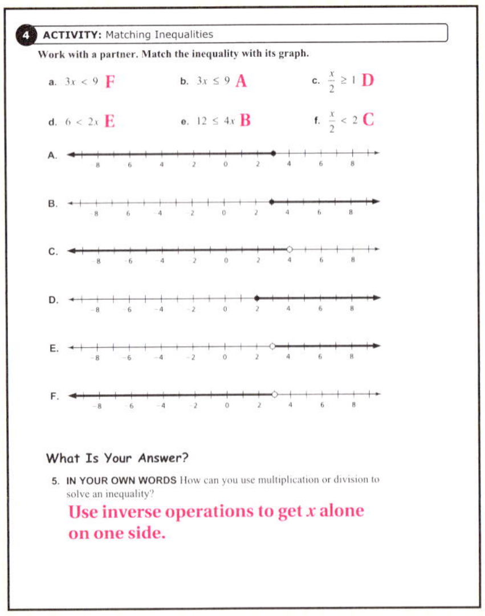

4 ACTIVITY: Matching Inequalities

Work with a partner. Match the inequality with its graph.

a. $3x < 9$ F b. $3x \le 9$ A c. $\frac{x}{2} \ge 1$ D

d. $6 < 2x$ E e. $12 \le 4x$ B f. $\frac{x}{2} < 2$ C

A. B. C. D. E. F.

What Is Your Answer?

5. IN YOUR OWN WORDS How can you use multiplication or division to solve an inequality?
Use inverse operations to get x alone on one side.

Laurie's Notes

Activity 3

- Students already have a sense of how inequalities are solved. This activity leads students through the symbolic process of solving inequalities using multiplication or division.
- Check that students use the correct circle (open or closed) to graph the inequality in each activity.

Activity 4

- Students should be able to use Guess, Check, and Revise to complete this matching activity. In the process, students should be able to confirm that you solve a multiplication inequality such as $3x < 9$ the same way you solve the equation $3x = 9$.
- Caution students to be careful with the inequalities that have the variable on the right side. Also remind students how division is represented $\left(\frac{x}{2} \text{ versus } x \div 2\right)$.

? "What are the inequalities graphed in (A) and (F)? (A) is $x \le 3$ and (F) is $x < 3$.

? "What are the inequalities graphed in (B) and (E)?" (B) is $x \ge 3$ and (E) is $x > 3$.

- Given the similarity of all 6 solutions, it is important to have students discuss how they completed the matching activity.
- **Extension:** Ask students to name a value of x that would be a solution of (C) but would not be a solution of (F). *Sample answer:* $x = 3.5$ Name a value of x that would be a solution of (D) but would not be a solution of (E). *Sample answer:* $x = 2.5$ Name a value of x that is a solution of (A) and (B). $x = 3$

What Is Your Answer?

- **Neighbor Check:** Have students work independently and then have their neighbors check their work. Have students discuss any discrepancies.
- Discuss whether it is necessary to temporarily substitute an equal sign for the inequality symbol when you solve an inequality using multiplication or division.

Closure

- **Exit Ticket:** Solve each inequality.

$4x \ge 16$ $x \ge 4$ and $\frac{x}{4} \ge 16$ $x \ge 64$

3 ACTIVITY: Solving Inequalities

Work with a partner. Complete the following steps for Activity 1. Then repeat the steps for Activity 2.

- Use your inequality from part (a). Replace the inequality symbol with an equal sign.
- Solve the equation.
- Replace the equal sign with the original inequality symbol.
- Graph this new inequality.
- Compare the graph with your graph in part (b). What do you notice?

4 ACTIVITY: Matching Inequalities

Work with a partner. Match the inequality with its graph. Explain your method.

a. $3x < 9$ **b.** $3x \le 9$ **c.** $\frac{x}{2} \ge 1$

d. $6 < 2x$ **e.** $12 \le 4x$ **f.** $\frac{x}{2} < 2$

Math Practice 1

Make a Plan

What strategy will you use to choose the correct graph?

A.

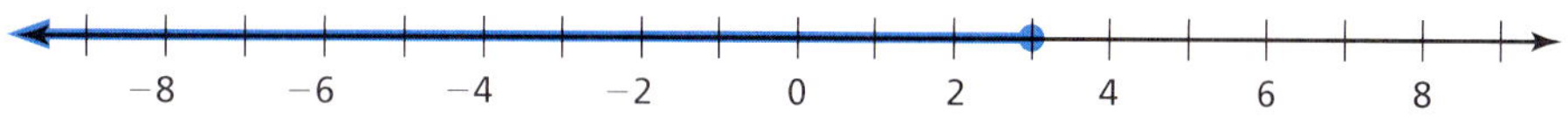

B.

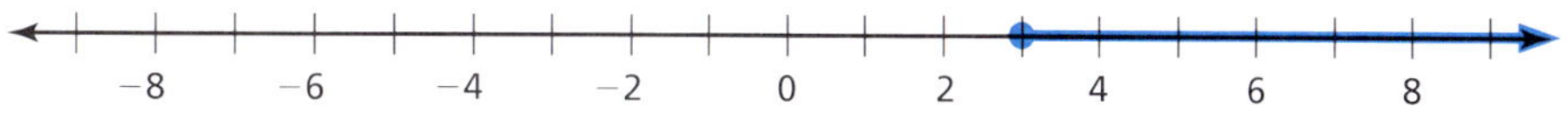

C.

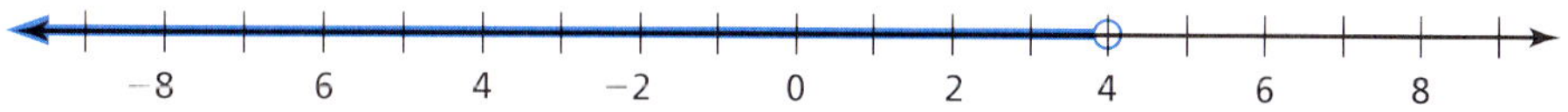

D.

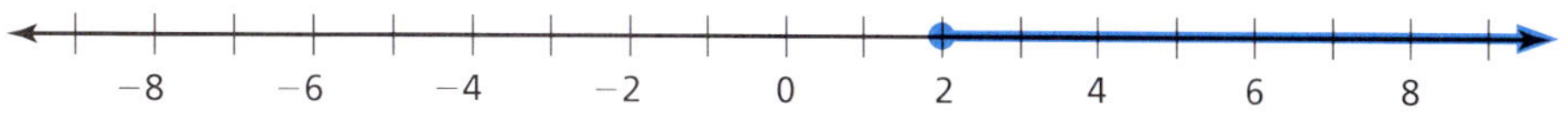

E.

F.

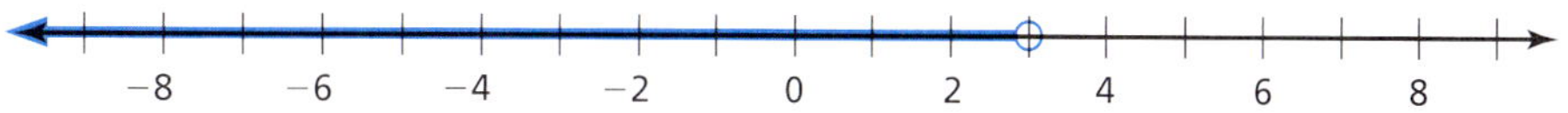

What Is Your Answer?

5. **IN YOUR OWN WORDS** How can you use multiplication or division to solve an inequality?

Practice Use what you learned about solving inequalities to complete Exercises 8–11 on page 342.

7.7 Lesson

Check It Out
Lesson Tutorials
BigIdeasMath.com

Key Ideas

Remember

Multiplication and division are inverse operations.

Multiplication Property of Inequality

Words When you multiply each side of an inequality by the same *positive* number, the inequality remains true.

Numbers

$8 > 6$

$8 \times 2 > 6 \times 2$

$16 > 12$

Algebra

$\frac{x}{4} < 2$

$\frac{x}{4} \cdot 4 < 2 \cdot 4$

$x < 8$

Division Property of Inequality

Words When you divide each side of an inequality by the same *positive* number, the inequality remains true.

Numbers

$8 > 6$

$8 \div 2 > 6 \div 2$

$4 > 3$

Algebra

$4x < 8$

$\frac{4x}{4} < \frac{8}{4}$

$x < 2$

These properties are also true for $\le$ and $\ge$.

EXAMPLE 1 Solving an Inequality Using Multiplication

Solve $\frac{x}{5} \le 2$. Graph the solution.

$\frac{x}{5} \le 2$	Write the inequality.
$\frac{x}{5} \cdot 5 \le 2 \cdot 5$	Multiplication Property of Inequality
$x \le 10$	Simplify.

Undo the division. → $\frac{x}{5} \cdot 5 \le 2 \cdot 5$

The solution is $x \le 10$.

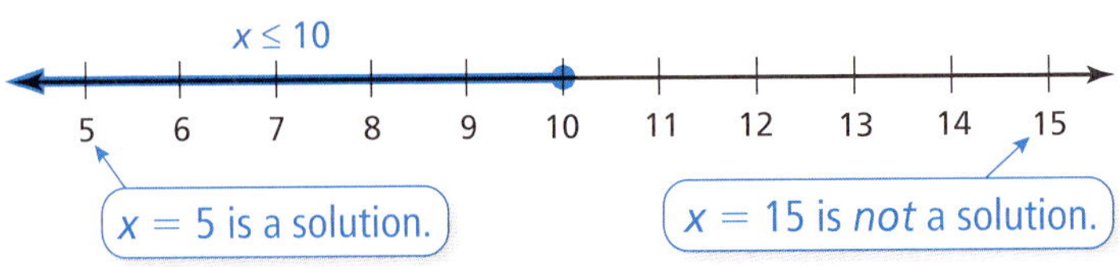

On Your Own

Now You're Ready
Exercises 6–9

Solve the inequality. Graph the solution.

1. $p \div 3 > 2$
2. $\frac{3}{5}q \le 6$
3. $1 < \frac{s}{7}$

Laurie's Notes

Introduction

Connect

- **Yesterday:** Students developed an intuitive understanding about how multiplication and division inequalities are solved. (MP1, MP3, MP5)
- **Today:** Students will use the Properties of Inequality to solve inequalities.

Motivate

- Present the following problem: If you were the owner of a theme park that included many different rides, would you sell individual tickets for each ride or would you sell a pass that allowed an unlimited number of rides?
- Allow students to discuss this problem. It is related to Example 3. The hope is that students will recognize that there is a point at which the individual tickets start to cost more for the consumer than the pass. The other factor to consider is the consumer who only wishes to take a few rides. That consumer may not want to purchase a pass.

Lesson Notes

Key Ideas

- Work through the Words, Numbers, and Algebra.
- Explain that solving inequalities is similar to solving equations.
- Today's inequalities involve multiplication and division which are inverse operations.
- **Teaching Tip:** Students will want to solve the inequalities in their heads and resist writing their work down. Remind students that you are trying to help them develop good problem solving techniques. The problems will become more involved later and students need to know how to record the process.
- The inequalities solved in this lesson do *not* involve multiplying or dividing by a negative quantity. This type of inequality is taught in a future course.

Example 1

- Work through the example.
- ? "How do you graph $x \leq 10$?" closed circle at 10, shade to the left of 10
- ? "How do you know which side of $x = 10$ to shade?" test values "Why is there a closed circle?" 10 is a solution.
- **MP5 Use Appropriate Tools Strategically:** In checking their solutions, have students check numbers on each side of the solid circle.

On Your Own

- **Think-Pair-Share:** Students should read each question independently and then work in pairs to answer the questions. When they have answered the questions, the pair should compare their answers with another group and discuss any discrepancies.
- Note that division in Question 1 is represented differently than in the next two questions.
- **Common Error:** In Question 3, because the variable is on the right, it is common for students to shade the wrong side of the inequality when graphing.

Goal Today's lesson is solving inequalities using multiplication or division.

Technology for the Teacher

Dynamic Classroom

Lesson Tutorials
Lesson Plans
Answer Presentation Tool

Extra Example 1

Solve $4 \geq r \div 2$. Graph the solution.

$8 \geq r$ or $r \leq 8$

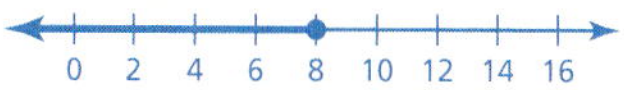

On Your Own

1. $p > 6$;

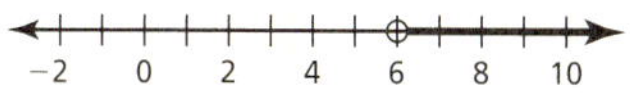

2. $q \leq 10$;

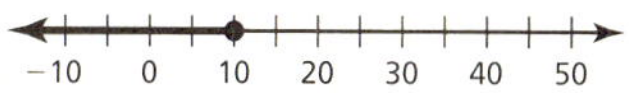

3. $7 < s$;

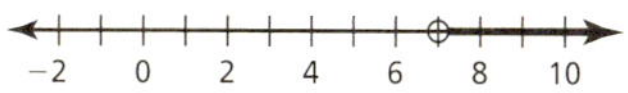

Laurie's Notes

Extra Example 2

Solve $8h < 56$. Graph the solution. $h < 7$

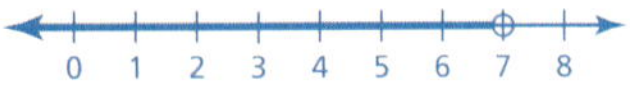

Extra Example 3

A one-way bus ride costs \$1.50. A 30-day bus pass costs \$30.

a. Write and solve an inequality to find the least number of one-way rides you must take for the 30-day pass to be a better deal. $1.5r > 30$; $r > 20$

b. You ride the bus an average of 24 times each month. Is the pass a better deal? Explain. Yes. The cost of 24 one-way rides is more than \$30. So, the pass is a better deal.

On Your Own

4. $k \leq 3$;

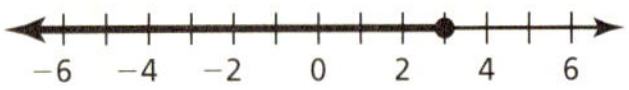

5. $j > 4$;

6. $25 \leq m$;

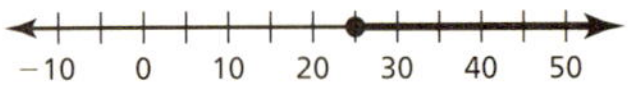

7. $2.5x \leq 15$; $x \leq 6$

Differentiated Instruction

Visual

Students may question why they are asked to graph the solution of an inequality, but not asked to graph the solution of an equation. The graph of an equation is one point on the number line. The graph of an inequality provides more information, because of the infinite number of possible solutions.

Example 2

? "How do you undo a multiplication inequality?" divide

? "Why is the circle open?" The solution does not include 8.

Example 3

? Discuss how multiple-day bus passes and single ride tickets work.

- "How many rides can you take with a 30-day bus pass?" unlimited
- "If you took 42 rides using a 30-day pass, what would the cost be per ride? Explain your thinking." \$1.00; Divide the cost of \$42 by the number of rides.
- "If you took 42 rides and paid each time, what would the total cost be? Explain your thinking." \$73.50; Multiply 42 rides by \$1.75, the cost of a single ride.

? Discuss how to find the least number of rides you would need to take in order to make the cost of the 30-day pass a better deal.

- "How many rides do you need to take before the cost of the individual tickets is at least as expensive as the 30-day pass?" Set up the inequality and solve.

- **Estimate:** Encourage students to estimate. $42 \div 1.75$ is approximately $42 \div 2 = 21$.
- **MP3 Construct Viable Arguments and Critique the Reasoning of Others:** Mathematically proficient students take into account the context of the problem. Students should consider the number of times a person needs to ride a bus monthly in deciding whether a monthly pass would save money.

On Your Own

- **Think-Pair-Share:** Students should read each question independently and then work in pairs to answer the questions. When they have answered the questions, the pair should compare their answers with another group and discuss any discrepancies.
- Note that Question 5 is written as $5 \cdot j$ instead of $5j$. The inequality is still solved by dividing both sides by 5.
- Question 6 is written with the variable on the right side. Students should rewrite the solution as $m \geq 25$.

Closure

- **Writing Prompt:** To solve inequalities, ... Make sure that students include all 4 operations in their description.

EXAMPLE 2 Solving an Inequality Using Division

Solve $4n > 32$. Graph the solution.

	$4n > 32$	Write the inequality.
Undo the multiplication. →	$\frac{4n}{4} > \frac{32}{4}$	Division Property of Inequality
	$n > 8$	Simplify.

∴ The solution is $n > 8$.

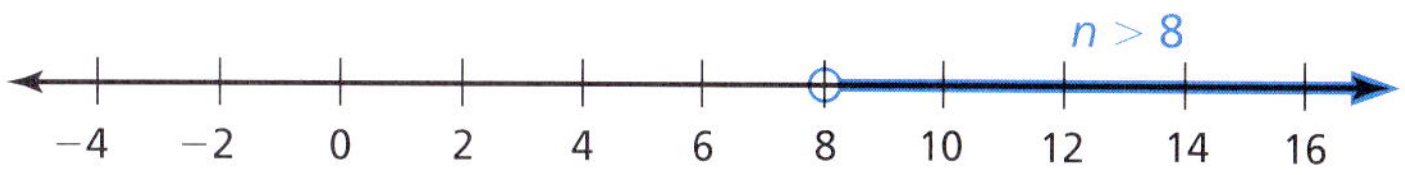

EXAMPLE 3 Real-Life Application

A one-way bus ride costs \$1.75. A 30-day bus pass costs \$42.

a. Write and solve an inequality to find the least number of one-way rides you must take for the 30-day pass to be a better deal.

b. You ride the bus an average of 20 times each month. Is the pass a better deal? Explain.

a. **Words** The price of a one-way ride times the number of one-way rides is more than \$42.

Variable Let r be the number of one-way rides.

Inequality $1.75 \quad \cdot \quad r \quad > \quad 42$

$1.75r > 42$	Write the inequality.
$\frac{1.75r}{1.75} > \frac{42}{1.75}$	Division Property of Inequality
$r > 24$	Simplify.

∴ So, you need to take more than 24 one-way rides for the pass to be a better deal.

b. No. The cost of 20 one-way rides is less than \$42. So, the pass is not a better deal.

On Your Own

Now You're Ready
Exercises 10–13

Solve the inequality. Graph the solution.

4. $11k \leq 33$
5. $5 \cdot j > 20$
6. $50 \leq 2m$
7. The sign shows the toll for driving on Alligator Alley. Write and solve an inequality to represent the number of times someone can drive on Alligator Alley with \$15.

7.7 Exercises

Check It Out
Help with Homework
BigIdeasMath.com

Vocabulary and Concept Check

1. **REASONING** How is the graph of the solution of $2x \geq 10$ different from the graph of the solution of $2x = 10$?

Name the property you should use to solve the inequality.

2. $3x \leq 27$
3. $7x > 49$
4. $\frac{x}{2} < 36$
5. **OPEN-ENDED** Write two inequalities that have the same solution set: one that you can solve using division and one that you can solve using multiplication.

Practice and Problem Solving

Solve the inequality. Graph the solution.

1
6. $\frac{m}{8} < 4$
7. $n \div 6 > 2$
8. $\frac{t}{3} \geq 15$
9. $\frac{1}{11}c \geq 9$

2
10. $12x < 96$
11. $5x \geq 25$
12. $8 \cdot w \leq 72$
13. $7p \leq 42$
14. $\frac{3}{4}b > 15$
15. $6x < 90$
16. $3s \geq 36$
17. $\frac{5}{9}v \leq 45$
18. $4t > 72$
19. $\frac{3}{4}w \leq 24$
20. $12m < 132$
21. $\frac{5x}{8} \geq 30$

22. **ERROR ANALYSIS** Describe and correct the error in solving the inequality.

✗
$$\frac{x}{6} \leq 30$$
$$\frac{x}{6} \cdot 6 \leq \frac{30}{6}$$
$$x \leq 5$$

23. **GEOMETRY** The length of a rectangle is 8 feet, and its area is less than 168 square feet. Write and solve an inequality to represent the width of the rectangle.

24. **PLAYGROUND** Students at a playground are divided into 5 equal groups with at least 6 students in each group. Write and solve an inequality to represent the number of students at the playground.

Write the word sentence as an inequality. Then solve the inequality.

25. Eight times a number n is less than 72.
26. A number t divided by 32 is at most 4.25.
27. 225 is no less than 12 times a number w.

Assignment Guide and Homework Check

Level	Day 1 Activity Assignment	Day 2 Lesson Assignment	Homework Check
Basic	8–11, 39–42	1–5, 7, 13–29 odd, 22	7, 13, 23, 25
Average	8–11, 39–42	1–5, 13–21 odd, 22, 26, 27, 29–31	7, 11, 26, 29, 31
Advanced	8–11, 39–42	1–5, 12–38 even	8, 20, 28, 32, 34

For Your Information

- **Exercises 35–38** Encourage students to use numerical values for the variables to test if the statement is true.

Common Errors

- **Exercises 6–21** Students may do the same operation on both sides instead of the opposite operation. Remind them that to get the variable alone, they need to use the opposite operation.
- **Exercises 6–21** Students may also switch the direction of the inequality when multiplying or dividing.
- **Exercises 28 and 29** Students may not understand how to graph these two inequalities together. Encourage them to draw the graph for each inequality separately, one above the other, and then lay them on top of each other and remove the parts that do not overlap.

7.7 Record and Practice Journal

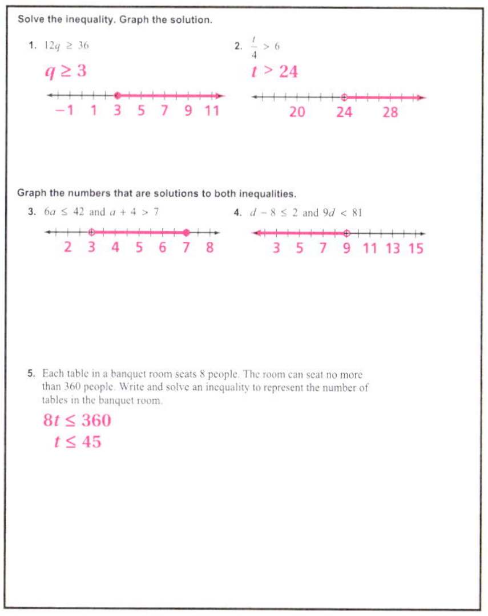
Solve the inequality. Graph the solution.

1. $12q \geq 36$
$q \geq 3$

2. $\frac{t}{4} > 6$
$t > 24$

Graph the numbers that are solutions to both inequalities.

3. $6a \leq 42$ and $a + 4 > 7$

4. $d - 8 \leq 2$ and $9d < 81$

5. Each table in a banquet room seats 8 people. The room can seat no more than 360 people. Write and solve an inequality to represent the number of tables in the banquet room.
$8t \leq 360$
$t \leq 45$

Vocabulary and Concept Check

1. The solution of $2x \geq 10$ includes the solution of $2x = 10$, $x = 5$, and all other x values that are greater than 5.
2. Div. Prop. of Ineq.
3. Div. Prop. of Ineq.
4. Mult. Prop. of Ineq.
5. *Sample answer:* $\frac{x}{2} \geq 4$, $2x \geq 16$

Practice and Problem Solving

6. $m < 32$;

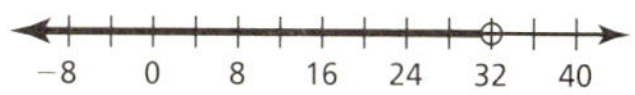

7. $n > 12$;

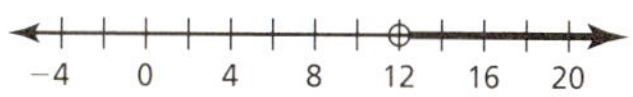

8. $t \geq 45$;

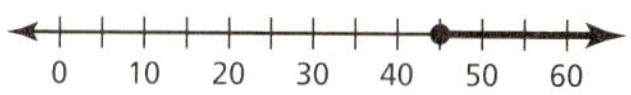

9. $c \geq 99$;

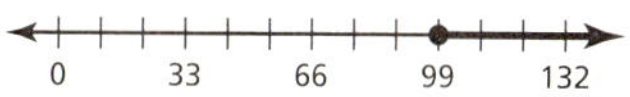

10. $x < 8$;

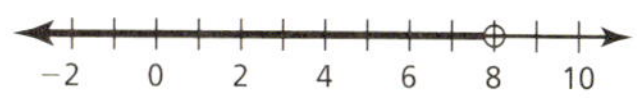

11. $x \geq 5$;

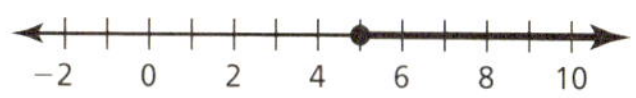

12. $w \leq 9$;

13. $p \leq 6$;

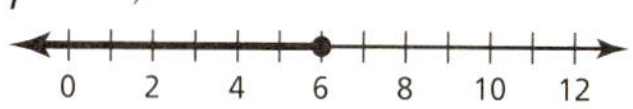

14. $b > 20$;

15. $x < 15$;

16. $s \geq 12$;

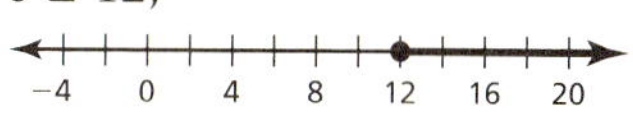

17. See Additional Answers.

Practice and Problem Solving

18. $t > 18$;

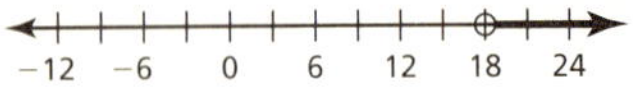

19. $w \leq 32$;

20. $m < 11$;

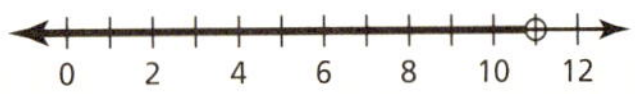

21. $x \geq 48$;

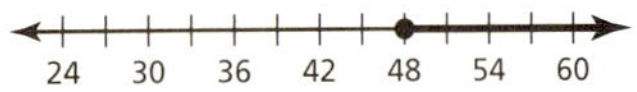

22–29. See Additional Answers.

30. See *Taking Math Deeper*.

31–38. See Additional Answers.

Fair Game Review

39. rectangle

40. trapezoid

41. parallelogram

42. C

Mini-Assessment

Solve the inequality. Graph the solution.

1. $\frac{r}{3} \geq 2$ $r \geq 6$

2. $1 > \frac{w}{7}$ $w < 7$

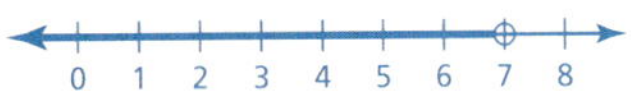

3. $5g > 25$ $g > 5$

4. $60 \geq 10s$ $s \leq 6$

5. You use a square section of your yard for a sandbox. You have at most 24 feet of wood to build the sides of the sandbox. Write and solve an inequality to represent the possible lengths of each side of the sandbox. $4x \leq 24$; $x \leq 6$

Taking Math Deeper

Exercise 30

This problem gives students an opportunity to estimate time. A good way to begin is to organize the given information.

1 Organize the given information.

12 people per ride

15,000 people per day

2 How many rides are needed for 15,000 people?

a. Let x be the number of rides per day.

$$12 \cdot x \geq 15{,}000$$
$$x \geq 1250$$

At least 1250 rides are needed.

3 The park is open for 12 hours per day. To have 1250 rides in a day, the thrill ride would have to operate

$$\frac{1250}{12} \approx 104.2 \text{ times per hour.}$$

b. This means that the thrill ride would have to operate once every 35 seconds. Even without being at the ride and seeing it, it seems unreasonable to unload 12 people, load another 12 people, and conduct the ride in only 35 seconds.

Project

There are several drop towers in amusement parks around the country. Research the heights of the different towers. Draw a bar graph that compares the heights.

Reteaching and Enrichment Strategies

If students need help. . .	If students got it. . .
Resources by Chapter • Practice A and Practice B • Puzzle Time Record and Practice Journal Practice Differentiating the Lesson Lesson Tutorials Skills Review Handbook	Resources by Chapter • Enrichment and Extension • Technology Connection Start the next section

Graph the numbers that are solutions to both inequalities.

28. $x + 7 > 9$ and $8x \leq 64$

29. $x - 3 \leq 8$ and $6x < 72$

30. THRILL RIDE A thrill ride at an amusement park holds a maximum of 12 people per ride.

a. Write and solve an inequality to find the least number of rides needed for 15,000 people.

b. Do you think it is possible for 15,000 people to ride the thrill ride in 1 day? Explain.

Park Hours
10:00 A.M.–10:00 P.M.

31. FOOTBALL A winning football team more than doubled the offensive yards gained by its opponent. The opponent gained 272 offensive yards. The winning team had 80 offensive plays. Write and solve an inequality to find the possible number of yards per play for the winning team.

32. LOGIC Explain how you know that $7x < 7x$ has no solution.

33. OPEN-ENDED Give an example of a real-life situation in which you can list all the solutions of an inequality. Give an example of a real-life situation in which you cannot list all the solutions of an inequality.

34. FUNDRAISER You are selling items from a catalog for a school fundraiser. Write and solve two inequalities to find the range of sales that will earn you between $40 and $50.

Critical Thinking **Let $a > b$ and $x > y$. Tell whether the statement is *always* true. Explain your reasoning.**

35. $a + x > b + y$

36. $a - x > b - y$

37. $ax > by$

38. $\frac{a}{x} > \frac{y}{b}$

Fair Game Review What you learned in previous grades & lessons

Classify the quadrilateral. *(Skills Review Handbook)*

39.

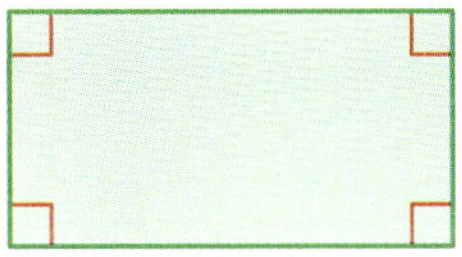

40.

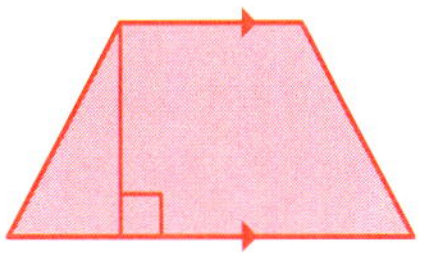

41.

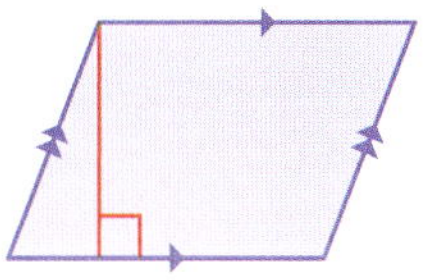

42. MULTIPLE CHOICE On a normal day, 12 airplanes arrive at an airport every 15 minutes. Which rate does not represent this situation? *(Section 5.3)*

Ⓐ 24 airplanes every 30 minutes

Ⓑ 4 airplanes every 5 minutes

Ⓒ 6 airplanes every 5 minutes

Ⓓ 48 airplanes each hour

7.5–7.7 Quiz

Check It Out
Progress Check
BigIdeasMath.com

Write the word sentence as an inequality. *(Section 7.5)*

1. A number x is greater than 0.

2. Twice a number c is at least -8.

Tell whether the given value is a solution of the inequality. *(Section 7.5)*

3. $2n > 16;\ n = 9$

4. $x - 1 \leq 9;\ x = 10$

Graph the inequality on a number line. *(Section 7.5)*

5. $y > -4$

6. $m \leq \frac{3}{5}$

Solve the inequality. Graph the solution. *(Section 7.6)*

7. $x + 4 \leq 8$

8. $18 > 16 + g$

Write the word sentence as an inequality. Then solve the inequality. *(Section 7.6)*

9. Two less than a number is more than 15.

10. Seven more than a number is less than or equal to 27.

Solve the inequality. Graph the solution. *(Section 7.7)*

11. $\frac{3a}{2} < 24$

12. $121 \geq 11s$

Write the word sentence as an inequality. Then solve the inequality. *(Section 7.7)*

13. Three times a number x is more than 18.

14. 84 is no less than 7 times a number k.

15. **WATER PARK** Each visit to a water park costs \$19.95. An annual pass to the park costs \$89.95. Write an inequality to represent the number of times you would need to visit the park for the pass to be a better deal. *(Section 7.5)*

16. **GARDEN** You want to use a square section of your yard for a garden. You have at most 52 feet of fencing to surround the garden. Write and solve an inequality to represent the possible lengths of each side of the garden. *(Section 7.7)*

17. **DELIVERY** You were planning to spend \$12 on a pizza. Write and solve an inequality to represent the additional amount you must spend to get free delivery. *(Section 7.6)*

Alternative Assessment Options

Math Chat	**Student Reflective Focus Question**
Structured Interview	Writing Prompt

Student Reflective Focus Question

Ask students to use their own words to summarize the similarities and differences between solving equations and solving inequalities. Be sure that they include examples. Select students at random to present their summary to the class.

Study Help Sample Answers

Remind students to complete Graphic Organizers for the rest of the chapter.

5. Inequalities

Examples	Non-Examples
$x - 5 \geq 8$	$6y = 12$
$4 + 13 < n + 2$	$x + 9 = 17$
$r > -\frac{3}{4}$	$a = 5$
$0 \leq k$	$10 = c - 1$

6. Graphs of Inequalities

Examples	Non-Examples
$n < 3$ 	$x = 2$ 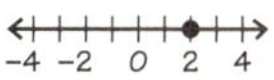
$-\frac{2}{3} \leq x$ 	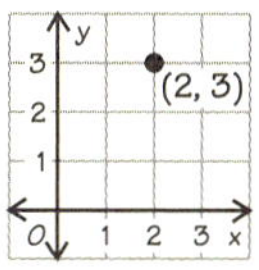
$C \leq 500$ 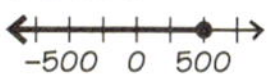	

7–8. Available at *BigIdeasMath.com*.

Reteaching and Enrichment Strategies

If students need help. . .	If students got it. . .
Resources by Chapter • Practice A and Practice B • Puzzle Time Lesson Tutorials *BigIdeasMath.com*	Resources by Chapter • Enrichment and Extension • Technology Connection Game Closet at *BigIdeasMath.com* Start the Chapter Review

Answers

1. $x > 0$
2. $2c \geq -8$
3. yes
4. yes
5. 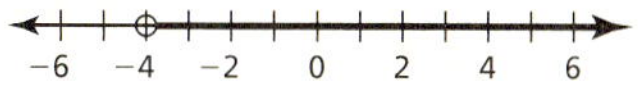

6. 

7. $x \leq 4$;
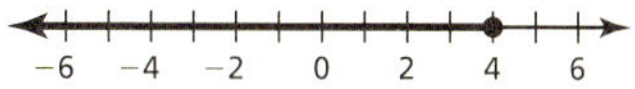

8. $g < 2$;
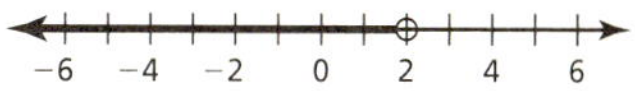

9. $x - 2 > 15$; $x > 17$
10. $7 + x \leq 27$; $x \leq 20$
11. $a < 16$;

12. $s \leq 11$;

13. $3x > 18$; $x > 6$
14. $84 \geq 7k$; $12 \geq k$
15. $19.95x > 89.95$
16. $4s \leq 52$; $s \leq 13$ ft
17. $x + 12 \geq 30$; $x \geq 18$

Technology for the Teacher

Online Assessment
Assessment Book
ExamView® Assessment Suite

For the Teacher
Additional Review Options

- *BigIdeasMath.com*
- Online Assessment
- Game Closet at *BigIdeasMath.com*
- Vocabulary Help
- Resources by Chapter

Answers

1. $m \cdot 2 = 8$
2. $t - 6 = 7$
3. $m + 5 = 7$
4. $8 = g \div 3$
5. $x = 9$
6. $m = 4$
7. $p = 33$
8. $q = 6$
9. $k = 9$
10. $a = 35$
11. yes
12. no
13. $c = 3 + 2.50m$;

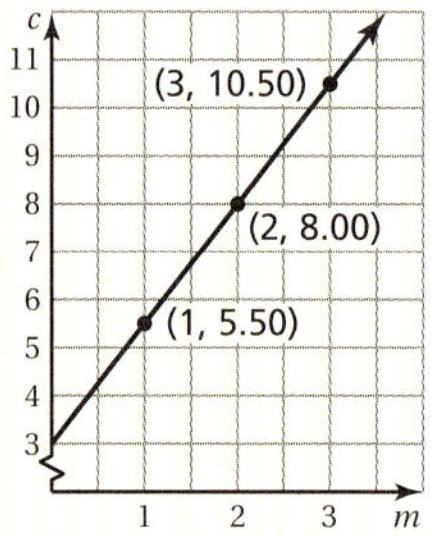

14. $m < 5$
15. $h \geq -12$
16. 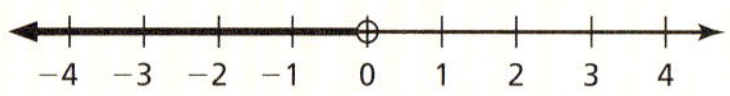

17. 

18. 

19. $x > 2$;

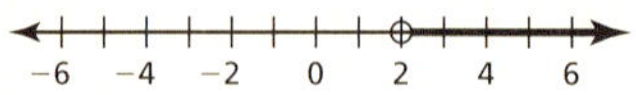

20. $k \leq 7$;

21. $y \geq 1$;

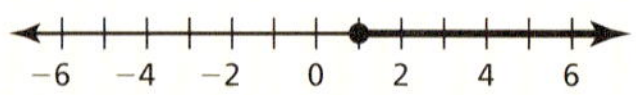

Review of Common Errors

Exercises 1–4

- Students may write the equations in the wrong order or forget parts of the equation. Remind them of the guidelines for writing expressions. Remind students that these are equations, so there will be equal signs in the answers and numbers and/or variables on each side.

Exercises 5–7

- Students may add or subtract the number on the side without the variable. For example, $-14 + 14 = k + 6 + 14$ instead of $-14 - 6 = k + 6 - 6$. Remind them that they are trying to get the variable by itself. They should start on the side with the variable and use the inverse of that operation.

Exercises 8–10

- Students may do the same operation on both sides instead of the opposite operation. This is most common with problems like Exercise 9, where the equation is written in division format instead of fraction format. Remind them that to get the variable alone, they need to use the inverse operation.

Exercises 11 and 12

- Students may substitute incorrectly for x and y. Remind them that the first coordinate of the ordered pair is always the x-value and the second coordinate is always the y-value.

Exercises 14 and 15

- Students may struggle with knowing which symbol to use in the inequality. Encourage them to put the word sentence into a real-life context and to use the table in the lesson that explains what symbol matches each phrase.

Exercises 16–18

- Students may shade the number line in the wrong direction. Remind them to check numbers to the left and right of the given number to determine which direction to shade the number line.

Exercises 16–18

- Students may draw a closed circle instead of an open circle and vice versa. Remind them of the definitions of open and closed circles.

Exercises 19–27

- Students may use the same operation in solving for the variable instead of the inverse operation. Demonstrate to them that this will not work by simplifying the inequality on the board. Students most likely ignored the side with the variable when they made this mistake. Remind them to check their answers in the original inequality.

Exercises 28–33

- Students may do the same operation on both sides instead of the opposite operation. Remind them that to get the variable alone, they need to use the opposite operation.

7 Chapter Review

Check It Out
Vocabulary Help
BigIdeasMath.com

Review Key Vocabulary

equation, *p. 296*
solution, *p. 302*
inverse operations, *p. 303*
equation in two variables, *p. 316*
solution of an equation in two variables, *p. 316*
independent variable, *p. 316*
dependent variable, *p. 316*
inequality, *p. 326*
solution of an inequality, *p. 327*
solution set, *p. 327*
graph of an inequality, *p. 328*

Review Examples and Exercises

7.1 Writing Equations in One Variable *(pp. 294–299)*

Write the word sentence "The quotient of a number b and 6 is 9" as an equation.

The quotient of a number b and 6 is 9.

$b \div 6 = 9$ — *Quotient of* means *division.*

∴ An equation is $b \div 6 = 9$.

Exercises

Write the word sentence as an equation.

1. The product of a number m and 2 is 8.
2. 6 less than a number t is 7.
3. A number m increased by 5 is 7.
4. 8 is the quotient of a number g and 3.

7.2 Solving Equations Using Addition or Subtraction *(pp. 300–307)*

Solve $z + 5 = 13$.

$$
\begin{aligned}
z + 5 &= 13 && \text{Write the equation.}\\
-5 &\quad -5 && \text{Subtraction Property of Equality}\\
z &= 8 && \text{Simplify.}
\end{aligned}
$$

Undo the addition.

∴ The solution is $z = 8$.

Check

$z + 5 = 13$

$8 + 5 \stackrel{?}{=} 13$

$13 = 13$ ✓

Exercises

Solve the equation. Check your solution.

5. $x - 1 = 8$
6. $m + 7 = 11$
7. $21 = p - 12$

7.3 Solving Equations Using Multiplication or Division (pp. 308–313)

Solve $4c = 32$.

$4c = 32$	Write the equation.
$\frac{4c}{4} = \frac{32}{4}$	Division Property of Equality
$c = 8$	Simplify.

Undo the multiplication.

Check

$4c = 32$

$4(8) \stackrel{?}{=} 32$

$32 = 32$ ✓

Exercises

Solve the equation. Check your solution.

8. $7 \cdot q = 42$ **9.** $7k \div 3 = 21$ **10.** $\frac{5a}{7} = 25$

7.4 Writing Equations in Two Variables (pp. 314–321)

Tell whether (6, 16) is a solution of the equation $y = 3x - 4$.

$16 \stackrel{?}{=} 3(6) - 4$	Substitute.
$16 \neq 14$ ✗	Compare.

So, (6, 16) is *not* a solution.

Exercises

Tell whether the ordered pair is a solution of the equation.

11. $y = 3x + 1$; (2, 7) **12.** $y = 7x - 4$; (4, 22)

13. **TAXI** A taxi ride costs \$3 plus \$2.50 per mile. Write and graph an equation in two variables that represents the total cost of a taxi ride.

7.5 Writing and Graphing Inequalities (pp. 324–331)

Write the word sentence as an inequality.

a. A number x is more than -9.

A number x | is more than | -9.

x | $>$ | -9

An inequality is $x > -9$.

b. A number r divided by 2 is at most 4.

A number r divided by 2 | is at most | 4.

$\frac{r}{2}$ | $\leq$ | 4

An inequality is $\frac{r}{2} \leq 4$.

Exercises

Write the word sentence as an inequality.

14. A number m is less than 5. **15.** A number h is at least -12.

Graph the inequality on a number line.

16. $x < 0$ **17.** $a \geq 3$ **18.** $n \leq -1$

Review Game

Equations

Materials per Group:

- a copy of the chapter test from the pupil's edition
- chalk or dry erase markers

Directions:

Divide the class into four teams. Team members gather in groups at the board. One member of each team works the first problem from the chapter test. Coaching is allowed from other team members. Check each student's work and award 2 points to the team if the work is correct. If the work needs to be corrected, award 1 point to the team after the student corrects it. Team members take turns working the problems and the process is repeated until the test is finished.

Who Wins?

The team with the most points wins.

For the Student Additional Practice

- Lesson Tutorials
- Multi-Language Glossary
- Self-Grading Progress Check
- *BigIdeasMath.com*
  Dynamic Student Edition
  Student Resources

Answers

22. $x > 13$;

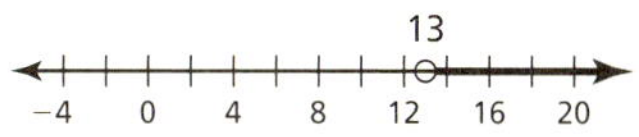

23. $n \geq 8$;

24. $x > 44$;

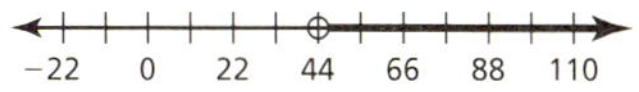

25. $b \leq 14$;

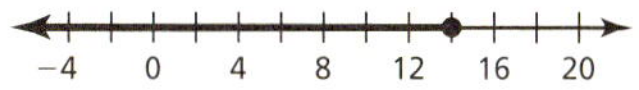

26. $s < 4$;

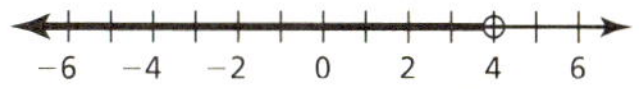

27. $m \leq \frac{1}{4}$;

28. $x < 8$;

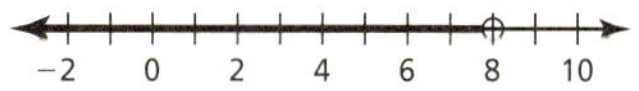

29. $n \geq 7$;

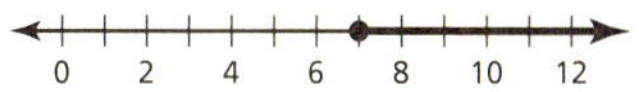

30. $x \leq 6$;

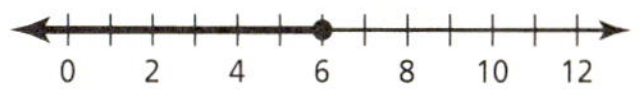

31. $b \leq 3$;

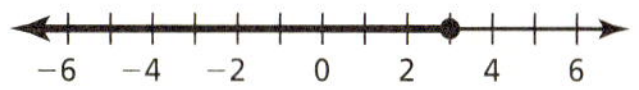

32. $p > 4$;

33. $k < 55$;

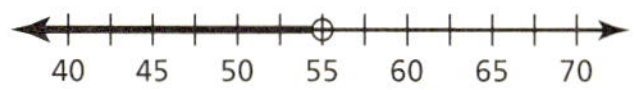

34. $3x \geq 20$; $x \geq \$6.67$

My Thoughts on the Chapter

What worked. . .

What did not work. . .

What I would do differently. . .

Teacher Tip

Not allowed to write in your teaching edition? Use sticky notes to record your thoughts.

7.6 Solving Inequalities Using Addition or Subtraction (pp. 332–337)

Solve $1 \le x - 4$. Graph the solution.

	$1 \le x - 4$	Write the inequality.
Undo the subtraction. →	$\underline{+4} \quad \underline{+4}$	Addition Property of Inequality
	$5 \le x$	Simplify.

The inequality $5 \le x$ is the same as $x \ge 5$.

∴ The solution is $x \ge 5$.

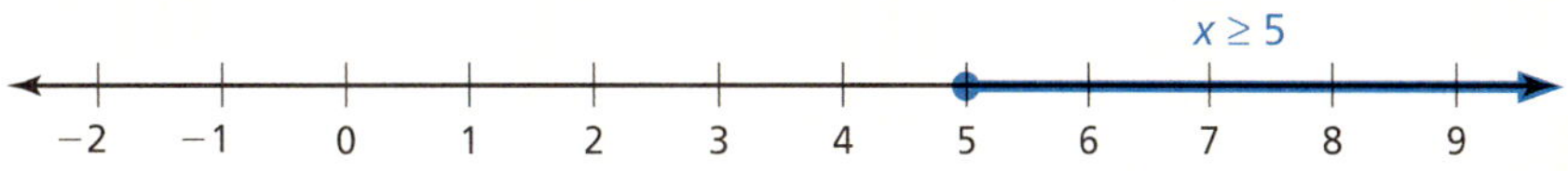

Exercises

Solve the inequality. Graph the solution.

19. $x + 1 > 3$ **20.** $k - 7 \le 0$ **21.** $y + 8 \ge 9$

22. $24 < 11 + x$ **23.** $4 \le n - 4$ **24.** $x - 20 > 24$

25. $b + 12 \le 26$ **26.** $s - 1.5 < 2.5$ **27.** $\frac{1}{4} + m \le \frac{1}{2}$

7.7 Solving Inequalities Using Multiplication or Division (pp. 338–343)

Solve $7n < 42$. Graph the solution.

	$7n < 42$	Write the inequality.
Undo the multiplication. →	$\frac{7n}{7} < \frac{42}{7}$	Division Property of Inequality
	$n < 6$	Simplify.

∴ The solution is $n < 6$.

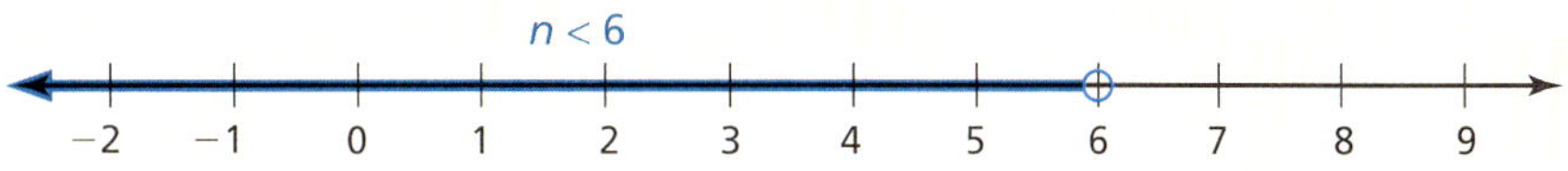

Exercises

Solve the inequality. Graph the solution.

28. $x \div 2 < 4$ **29.** $9n \ge 63$ **30.** $\frac{5}{3}x \le 10$

31. $9 \ge 3b$ **32.** $10p > 40$ **33.** $\frac{3}{11}k < 15$

34. **TICKETS** The cost of three tickets to a movie is at least \$20. Write and solve an inequality that represents the situation.

7 Chapter Test

Check It Out
Test Practice
BigIdeasMath.com

Write the word sentence as an equation.

1. 7 times a number s is 84.
2. 13 is one-third of a number m.

Solve the equation. Check your solution.

3. $15 = 7 + b$
4. $v - 6 = 16$
5. $5x = 70$
6. $3b = 45$
7. $\frac{6m}{7} = 30$
8. $\frac{8k}{3} = 32$

Tell whether the ordered pair is a solution of the equation.

9. $y = 9x$; $(3, 27)$
10. $y = 4x + 2$; $(8, 36)$

Write an inequality for the situation.

11. An MP3 player holds up to 300 songs.
12. Riders must be at least 48 inches tall.

Graph the inequality on a number line.

13. $x \geq 5$
14. $m \leq -2$

Solve the inequality. Graph the solution.

15. $x - 3 < 7$
16. $12 \geq n + 6$
17. $\frac{4}{3}b \leq 12$
18. $72 > 12p$

19. **SCHOOL DANCE** Each ticket to a school dance is \$4. The total amount collected in ticket sales is \$332. Write and solve an equation to find the number of students attending the dance.

20. **T-SHIRTS** A soccer team will sell T-shirts for a fundraiser. The company that makes the T-shirts charges \$10 per shirt plus a \$20 shipping fee per order.
    a. Write and graph an equation in two variables that represents the total cost of ordering the shirts.
    b. Choose an ordered pair that lies on your graph in part (a). Interpret it in the context of the problem.

21. **HURRICANE** A hurricane has wind speeds that are greater than or equal to 74 miles per hour. Write an inequality to represent the possible wind speeds during a hurricane.

Test Item References

Chapter Test Questions	Section to Review	Common Core State Standards
1, 2	7.1	6.EE.6
3, 4	7.2	6.EE.5, 6.EE.7
5–8, 19	7.3	6.EE.5, 6.EE.7
9, 10, 20	7.4	6.EE.9
11–14, 21	7.5	6.EE.5, 6.EE.8
15, 16	7.6	6.EE.5, 6.EE.8
17, 18	7.7	6.EE.5, 6.EE.8

Test-Taking Strategies

Remind students to quickly look over the entire test before they start so that they can budget their time. When working with equations and inequalities, they need to write all numbers and variables clearly. Some numbers and variables are easy to confuse. Have them pay attention to 6, *s*, 4, 9, 7, and *x*. Remind students to line up terms in each step and to not crowd their work. Have them use the **Stop** and **Think** strategy before they answer each question.

Common Errors

- **Exercises 1 and 2** Students may write the equations in the wrong order or forget parts of the equation. Remind them of the guidelines for writing expressions. Remind students that these are equations, so there will be equal signs in the answers and numbers and/or variables on each side.
- **Exercises 3 and 4** Students may add or subtract the number on the side without the variable. For example, $-14 + 14 = k + 6 + 14$ instead of $-14 - 6 = k + 6 - 6$. Remind them that they are trying to get the variable by itself. They should start on the side with the variable and use the inverse of that operation.
- **Exercises 15–18** Students may shade the number line in the wrong direction. This often happens when the variable is on the right of the inequality symbol instead of the left. Remind them to check numbers to the left and right of the given number to determine which direction to shade the number line.

Reteaching and Enrichment Strategies

If students need help. . .	If students got it. . .
Resources by Chapter • Practice A and Practice B • Puzzle Time Record and Practice Journal Practice Differentiating the Lesson Lesson Tutorials *BigIdeasMath.com* Skills Review Handbook	Resources by Chapter • Enrichment and Extension • Technology Connection Game Closet at *BigIdeasMath.com* Start Standards Assessment

Answers

1. $7s = 84$
2. $13 = \frac{1}{3} \cdot m$
3. $b = 8$
4. $v = 22$
5. $x = 14$
6. $b = 15$
7. $m = 35$
8. $k = 12$
9. yes
10. no
11. $m \le 300$
12. $h \ge 48$
13. 

14. 

15. $x < 10$;
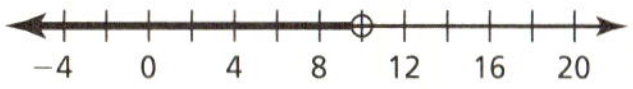

16. $n \le 6$;
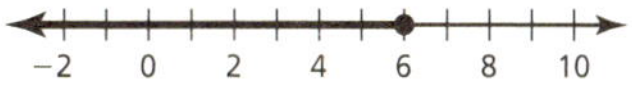

17. $b \le 9$;

18. $p < 6$;
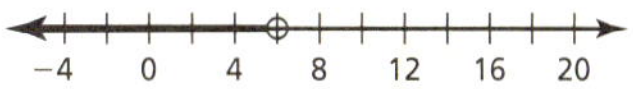

19. $4x = 332$; 83 students
20. a. $c = 20 + 10t$;
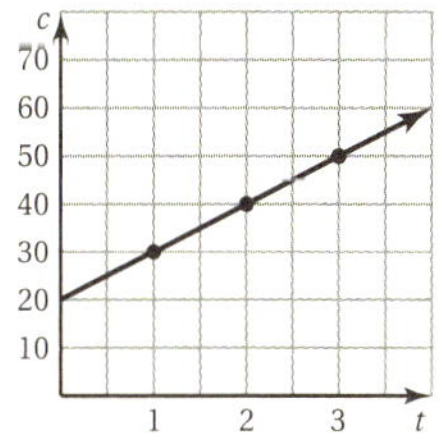

b. *Sample answer:* The ordered pair (3, 50) represents 3 shirts being ordered for \$50.
21. $w \ge 74$ miles per hour

Technology for the Teacher

Online Assessment
Assessment Book
ExamView® Assessment Suite

Test-Taking Strategies

Available at *BigIdeasMath.com*

After Answering Easy Questions, Relax
Answer Easy Questions First
Estimate the Answer
Read All Choices before Answering
Read Question before Answering
Solve Directly or Eliminate Choices
Solve Problem before Looking at Choices
Use Intelligent Guessing
Work Backwards

About this Strategy

One way to answer the question is to work backwards. Try putting the responses into the question, one at a time, and see if you get a correct solution.

Answers

1. C
2. H
3. C
4. F

Technology for the Teacher

Common Core State Standards Support Performance Tasks
Online Assessment
Assessment Book
ExamView® Assessment Suite

Item Analysis

1. **A.** The student subtracts bases instead of adding them.

   B. The student subtracts bases and fails to multiply by $\frac{1}{2}$.

   C. Correct answer

   D. The student fails to multiply by $\frac{1}{2}$.

2. **F.** The student finds a common factor of the three numbers but not the greatest common factor.

   G. The student finds a common factor of the three numbers but not the greatest common factor.

   H. Correct answer

   I. The student selects the smallest of the three numbers.

3. **A.** The student writes an expression that represents the difference of 18 and 5 instead of the sum. The student misinterprets "less than" and writes the expression on the right side of the equation in the wrong order.

   B. The student misinterprets "less than" and writes the expression on the right side of the equation in the wrong order.

   C. Correct answer

   D. The student writes an expression that represents the difference of 18 and 5 instead of the sum.

4. **F.** Correct answer

   G. The student uses the incorrect type of circle on the number line.

   H. The student shades in the wrong direction on the number line.

   I. The student uses the incorrect type of circle and shades in the wrong direction on the number line.

7 Standards Assessment

Test-Taking Strategy
Work Backwards

"**Work backwards** by trying 6, 7, 8, and 9. You will see that 3(8)=24. So, C is correct."

1. What is the area of the balcony shown below? *(6.G.1)*

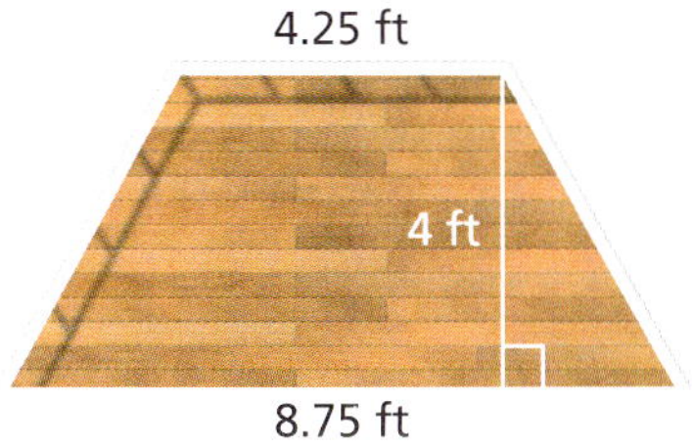

A. 9 ft^2 **C.** 26 ft^2

B. 18 ft^2 **D.** 52 ft^2

2. You are making identical fruit baskets using 16 apples, 24 pears, and 32 bananas. What is the greatest number of baskets you can make using all the fruit? *(6.NS.4)*

F. 2 **H.** 8

G. 4 **I.** 16

3. Which equation represents the word sentence below? *(6.EE.7)*

The sum of 18 and 5 is equal to 9 less than a number y.

A. $18 - 5 = 9 - y$ **C.** $18 + 5 = y - 9$

B. $18 + 5 = 9 - y$ **D.** $18 - 5 = y - 9$

4. Which number line is a graph of the solution of the inequality below? *(6.EE.8)*

$$x \geq 5$$

F. 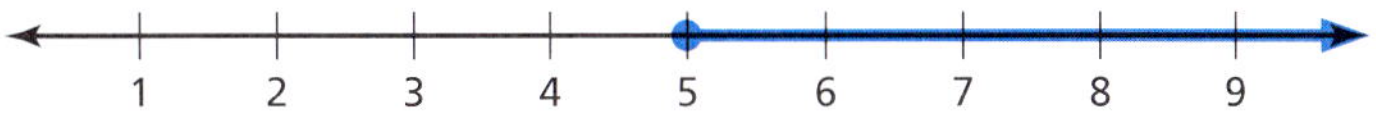

G. 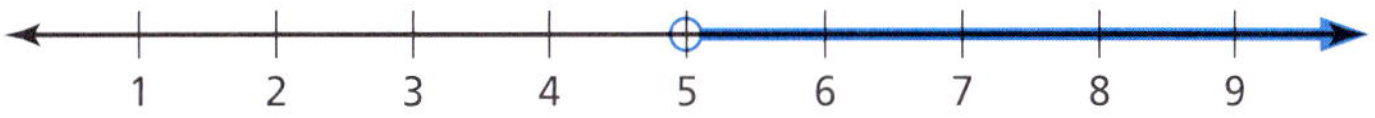

H. 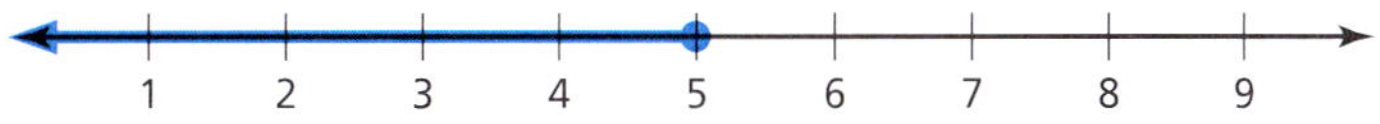

I. 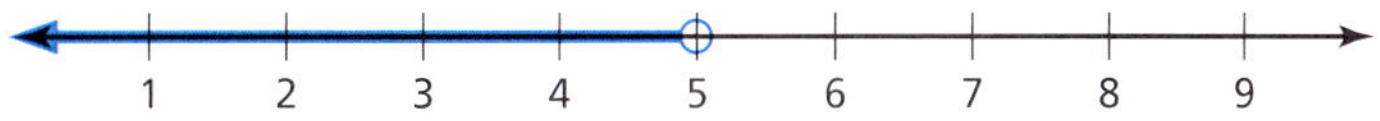

5. The steps your friend took to divide two mixed numbers are shown below.

$$3\frac{3}{5} \div 1\frac{1}{2} = \frac{18}{5} \times \frac{3}{2}$$
$$= \frac{27}{5}$$
$$= 5\frac{2}{5}$$

What should your friend change in order to divide the two mixed numbers correctly? *(6.NS.1)*

A. Find a common denominator of 5 and 2.

B. Multiply by the reciprocal of $\frac{18}{5}$.

C. Multiply by the reciprocal of $\frac{3}{2}$.

D. Rename $3\frac{3}{5}$ as $2\frac{8}{5}$.

6. An inequality is graphed on the number line below.

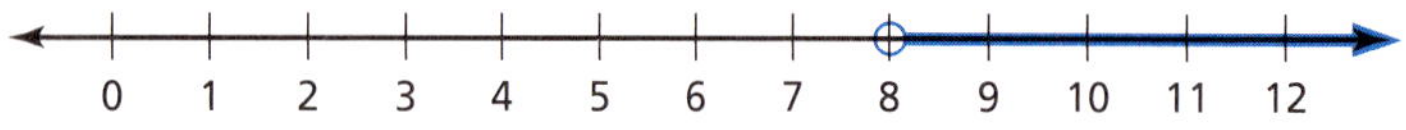

What is the least whole number value that is a solution of the inequality? *(6.EE.8)*

7. A company ordering parts receives a charge of \$25 for shipping and handling plus \$20 per part. Which equation represents the cost c of ordering p parts? *(6.EE.9)*

F. $c = 25 + 20p$

G. $c = 20 + 25p$

H. $p = 25 + 20c$

I. $p = 20 + 25c$

8. Which property is illustrated by the statement below? *(6.EE.3)*

$$5(3 + 6) = 5(3) + 5(6)$$

A. Associative Property of Multiplication

B. Commutative Property of Multiplication

C. Commutative Property of Addition

D. Distributive Property

Item Analysis (continued)

5. **A.** The student confuses the intended division algorithm with an addition/subtraction algorithm.

   B. The student finds the reciprocal of the dividend rather than the reciprocal of the divisor.

   C. Correct answer

   D. The student renames the first number of the expression as if the student were going to follow a subtraction algorithm.

6. **Gridded Response:** Correct answer: 9

   Common Error: The student misinterprets the meaning of an open circle and gets an answer of 8.

7. **F.** Correct answer

   G. The student assigns the incorrect coefficient to p.

   H. The student switches p and c.

   I. The student switches p and c and assigns the incorrect coefficient to p.

8. **A.** The student confuses the Associative Property of Multiplication with the Distributive Property.

   B. The student confuses the Commutative Property of Multiplication with the Distributive Property.

   C. The student confuses the Commutative Property of Addition with the Distributive Property.

   D. Correct answer

Answers

5. C

6. 9

7. F

8. D

Answers

9. 300
10. H
11. B
12. G
13. *Part A* $850 + 150w = 4000$; 21 weeks

    Part B $\frac{3}{4}(20a) = 150$

    Part C \$10 per hour

Item Analysis (continued)

9. Gridded Response: Correct answer: 300

Common Error: The student incorrectly places the decimal point in the product.

10. **F.** The student multiplies 24 by 0.25.

G. The student multiplies 24 by 0.75.

H. Correct answer

I. The student divides 24 by 0.25.

11. **A.** The student incorrectly thinks that points in Quadrant IV have negative x-coordinates.

B. Correct answer

C. The student incorrectly thinks that points in Quadrant IV have negative x-coordinates. The student also reverses the x- and y-coordinates.

D. The student reverses the x- and y-coordinates.

12. **F.** The student incorrectly multiplies from left to right before evaluating inside parentheses.

G. Correct answer

H. The student applies the exponent incorrectly.

I. The student applies the exponent incorrectly.

13. **4 points** The student demonstrates a thorough understanding of writing and solving equations. In Part A, the student writes and solves $850 + 150w = 4000$ or an equivalent equation to find that Jeff can afford the car in 21 weeks. In Part B, the student writes $\frac{3}{4}(20a) = 150$ or an equivalent equation to find the amount a per hour needed. In Part C, the student solves the equation correctly to find that Jeff needs to earn \$10 per hour. The student provides accurate work with clear and complete explanations.

3 points The student demonstrates an understanding of writing and solving equations, but the student's work and explanations demonstrate an essential but less than thorough understanding.

2 points The student demonstrates a partial understanding of writing and solving equations. The student's work and explanations demonstrate a lack of essential understanding.

1 point The student demonstrates a limited understanding of writing and solving equations. The student's response is incomplete and exhibits many flaws.

0 points The student provided no response, a completely incorrect or incomprehensible response, or a response that demonstrates insufficient understanding of writing and solving equations.

9. What is the value of the expression below? *(6.NS.3)*

$$46.8 \div 0.156$$

10. In a fish tank, 75% of the fish are goldfish. How many fish are in the tank if there are 24 goldfish? *(6.RP.3c)*

F. 6 **H.** 32

G. 18 **I.** 96

11. What are the coordinates of point P in the coordinate plane below? *(6.NS.6c)*

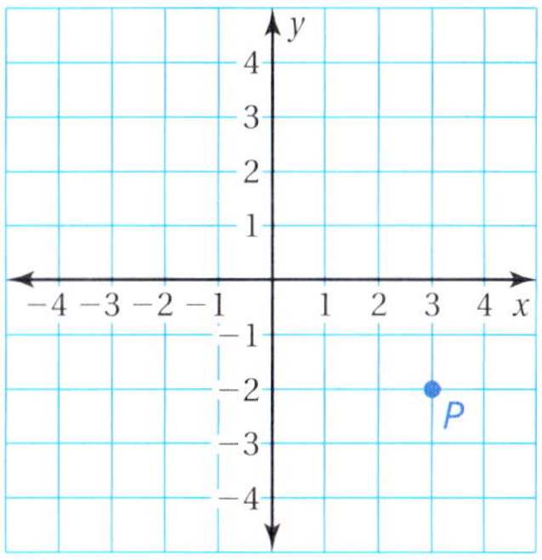

A. $(-3, -2)$ **C.** $(-2, -3)$

B. $(3, -2)$ **D.** $(-2, 3)$

12. What is the first step in evaluating the expression below? *(6.EE.1)*

$$3 \cdot (5 + 2)^2 \div 7$$

F. Multiply 3 and 5. **H.** Evaluate 5^2.

G. Add 5 and 2. **I.** Evaluate 2^2.

13. Jeff wants to save \$4000 to buy a used car. He has already saved \$850. He plans to save an additional \$150 each week. *(6.EE.7)*

Think Solve Explain

Part A Write and solve an equation to represent the number of weeks remaining until he can afford the car.

Jeff saves \$150 per week by saving $\frac{3}{4}$ of what he earns at his job each week. He works 20 hours per week.

Part B Write an equation to represent the amount per hour that Jeff must earn to save \$150 per week. Explain your reasoning.

Part C What is the amount per hour that Jeff must earn? Show your work and explain your reasoning.

8 Surface Area and Volume

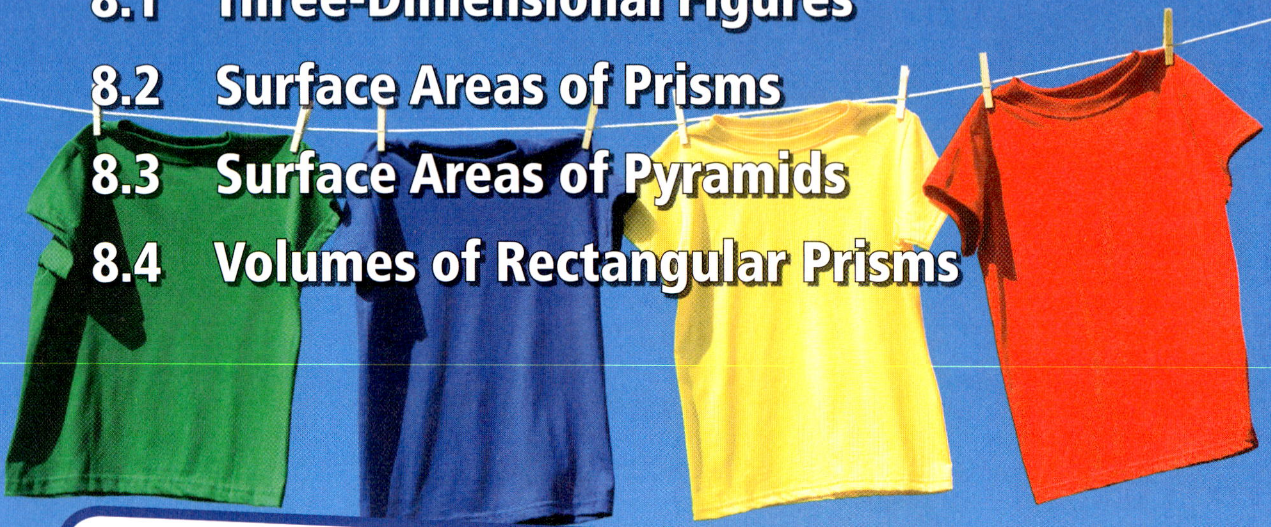

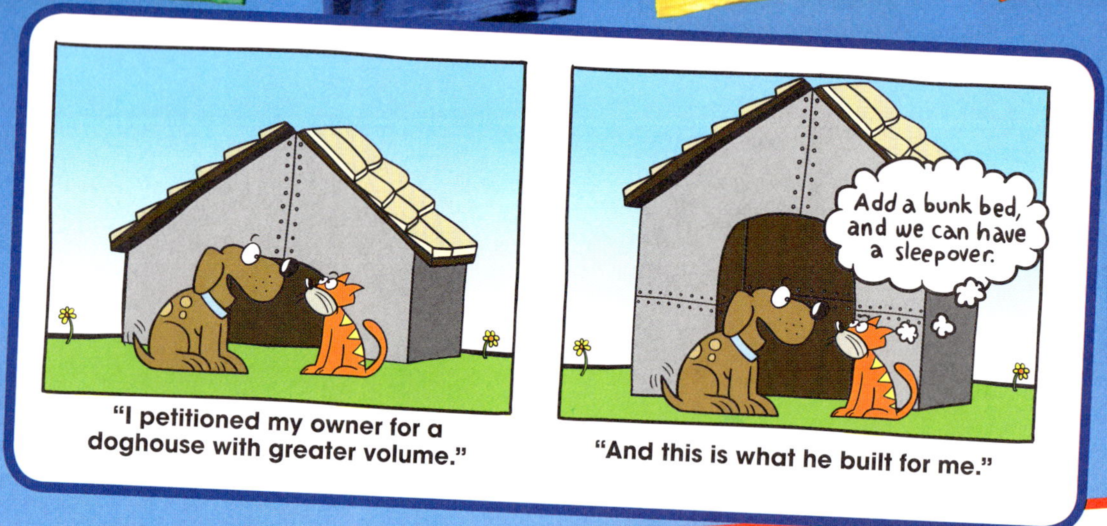

Common Core Progression

4th Grade
• Find the areas of rectangles with whole number side lengths. • Classify two-dimensional figures using angles and parallel and perpendicular lines. • Draw line segments, angles, parallel lines, and perpendicular lines.
5th Grade
• Find the areas of rectangles with fractional side lengths. • Classify two-dimensional figures into categories based on properties. • Understand volume, and measure it by counting unit cubes. • Find the volumes of rectangular prisms using the formula.
6th Grade
• Find areas of triangles, special quadrilaterals, and polygons. • Use nets made up of rectangles and triangles to find surface areas. • Find the volumes of prisms with fractional edge lengths.

Chapter Summary

Section	Common Core State Standard	
8.1	Learning	6.G.4
8.2	Learning	6.G.4
8.3	Learning	6.G.4 ★
8.4	Learning	6.G.2 ★
★ Teaching is complete. Standard can be assessed.		

Pacing Guide for Chapter 8

Chapter Opener	1 Day
Section 1 Activity Lesson	 1 Day 1 Day
Section 2 Activity Lesson	 1 Day 1 Day
Study Help / Quiz	1 Day
Section 3 Activity Lesson	 1 Day 1 Day
Section 4 Activity Lesson	 1 Day 1 Day
Chapter Review/ Chapter Tests	2 Days
Total Chapter 8	12 Days
Year-to-Date	127 Days

Technology for the Teacher

BigIdeasMath.com
Chapter at a Glance
Complete Materials List
Parent Letters: English and Spanish

Common Core State Standards

5.G.4 Classify two-dimensional figures in a hierarchy based on properties.
5.MD.5a Find the volume of a right rectangular prism with whole-number side lengths by packing it with unit cubes, and show that the volume is the same as would be found by multiplying the edge lengths, equivalently by multiplying the height by the area of the base. Represent threefold whole-number products as volumes, e.g., to represent the associative property of multiplication.

Additional Topics for Review

- Areas of Rectangles and Triangles
- Operations with Fractions
- Operations with Decimals

Try It Yourself

1. square
2. rectangle
3. trapezoid
4. 90 cubic units
5. 120 cubic units
6. 96 cubic units

Record and Practice Journal Fair Game Review

1. trapezoid
2. rectangle
3. square
4. equilateral triangle
5. square
6. 504 cubic units
7. 160 cubic units
8. 224 cubic units
9. 60 cubic units
10. 66 cubic units
11. 126 cubic units

Math Background Notes

Vocabulary Review

- Rectangle
- Square
- Parallelogram
- Right Angle
- Isosceles Triangle
- Equilateral Triangle
- Cubic Units

Classifying Figures

- Students have been defining and identifying plane figures since Grade 4.
- Review with students the marks indicating congruent segment lengths, parallel segments, and right angles.
- You may want to review the different types of triangles: right, acute, obtuse, equilateral, isosceles, and scalene.
- A flow chart is a great way to review classifying quadrilaterals.
- Students should be familiar with area formulas for squares, rectangles, and triangles.

Finding Volumes of Rectangular Prisms

- Students should have a sense that volume is a *filling process* and that it can be found by stacking equal layers on top of one another.
- In Example 3, the area of the base is 28 square units, and the height of the prism is 5 units.
- **Connection:** Students can memorize formulas and have little understanding of why the formula makes sense. It is important that students see that all volume formulas are similar. The volume is found by finding the area of the base and then multiplying by the number of layers.
- You may want to review units with students. A complete answer will include a numeric solution and the correct units. Remind students that area is always measured in square units, and volume is always measured in cubic units.

Reteaching and Enrichment Strategies

If students need help. . .	If students got it. . .
Record and Practice Journal • Fair Game Review Skills Review Handbook Lesson Tutorials	Game Closet at *BigIdeasMath.com* Start the next section

What You Learned Before

"Name these shapes."

Classifying Figures (5.G.4)

Example 1

Identify the figure.

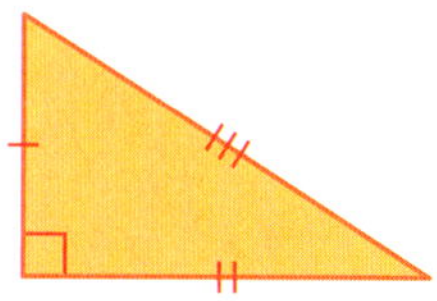

Because the figure has a right angle and three sides of different lengths, it is a right scalene triangle.

Example 2

Identify the figure.

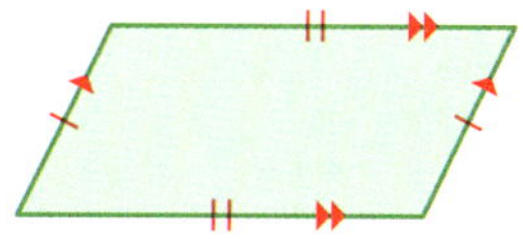

Because the figure is a quadrilateral with opposite sides that are parallel, it is a parallelogram.

Try It Yourself

Identify the figure.

1. 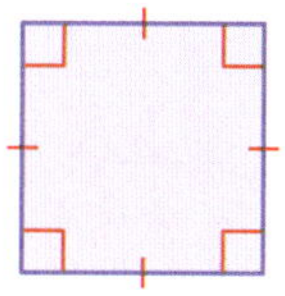

2. 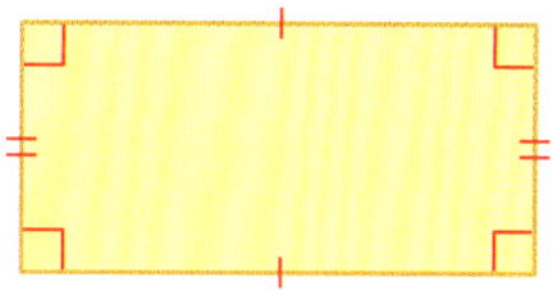

3. 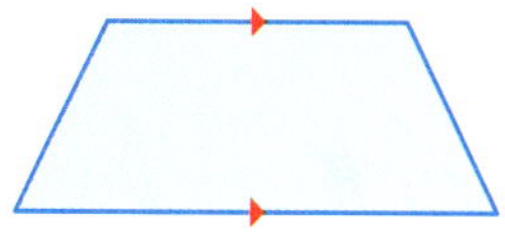

Finding Volumes of Rectangular Prisms (5.MD.5a)

Example 3

Find the volume of the rectangular prism.

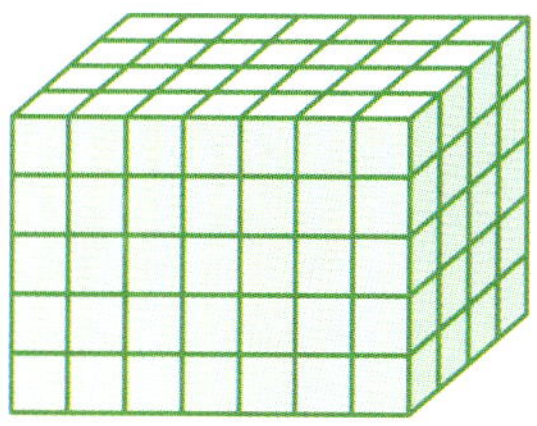

There are $4 \times 7 = 28$ unit cubes in each layer.

Because there are 5 layers, there are $5 \times 28 = 140$ unit cubes in the prism.

So, the volume is 140 cubic units.

Try It Yourself

Find the volume of the rectangular prism.

4. 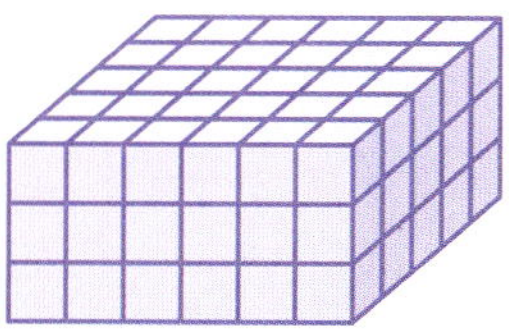

5. 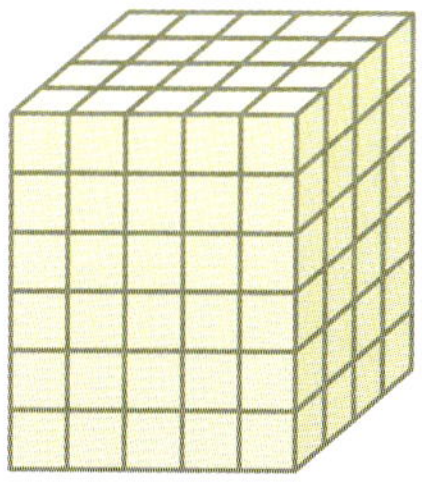

6. 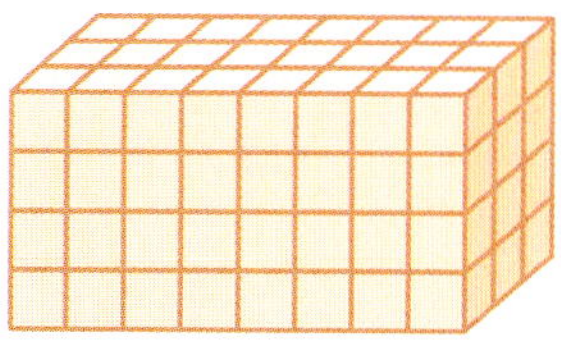

8.1 Three-Dimensional Figures

Essential Question How can you draw three-dimensional figures?

Dot paper can help you draw three-dimensional figures, or *solids.*

Square Dot Paper

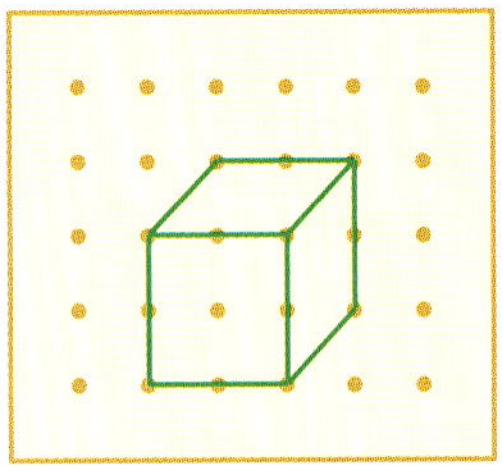

Face-On view

Isometric Dot Paper

Corner view

1 ACTIVITY: Drawing Views of a Solid

Work with a partner. Draw the front, side, and top views of each stack of cubes. Then find the number of cubes in the stack.

a. Sample:

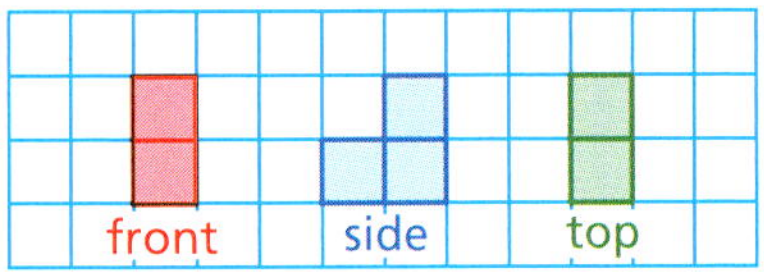

Number of cubes: 3

b.

c.

d.

e.

f.

g.

Common Core

Geometry

In this lesson, you will

- draw three-dimensional figures.
- find the number of faces, edges, and vertices of solids.

Preparing for Standard 6.G.4

Laurie's Notes

Introduction

Standards for Mathematical Practice

- **MP5 Use Appropriate Tools Strategically:** As students are developing their spatial and visual skills, having cubes to hold and view is very helpful. The samples in the book are two-dimensional representations of three-dimensional figures. Manipulating and viewing actual cubes is a helpful tool for many students.

Motivate

- Place a cube-shaped tissue box on a desk. Ask students to describe what they see when standing directly in front of the cube (front view), to the side of the cube (side view), and looking down on the cube (top view).
- Now place a rectangular prism (a shoe box) on the same surface. Describe all three views.
- Place a cube on top of a rectangular prism creating a solid that might be similar to the one shown in part (a) of Activity 1.
- Describe all three views. Students are challenged to ignore the difference in depth when describing a solid from one of the view points. They should focus on the surface that they see, not the depth of the solid. This can often be confusing to students.

Activity Notes

Activity 1

- **MP5:** If enough cubes are available, give each pair of students 10 cubes to create the models on their desks or tables. To see each view, the students need to be at "eye level" with the solid.
- **Teaching Tip:** Another way to help students think about the top view is to ask them, "If you want to paint the top, what shapes would you paint?" Two squares would be painted. Repeat this strategy for each view.
- Discuss volume. How many cubes are necessary to build the figure? Students are often confused by the cubes that are not visible in the picture. Explain that parts (b), (d), (f), and (g) would fit into the corner of a room. There are cubes behind cubes that are supporting the top cubes.
- **Scaffolding:** You may want half the class to work on parts (b), (d), and (f), while the other half works on parts (c), (e), and (g).
- Generally, it is the side view that is often challenging to draw.
- Check to see if students understand that the volume is the same number of cubes as the number of cubes used to build the figure.

Common Core State Standards

6.G.4 Represent three-dimensional figures . . ..

Previous Learning

Students should know the vocabulary of prisms.

Technology for the Teacher

Dynamic Classroom

Lesson Plans
Complete Materials List

8.1 Record and Practice Journal

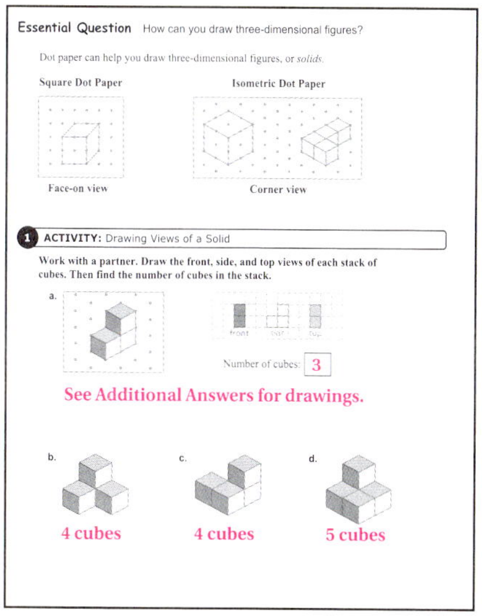

Essential Question How can you draw three-dimensional figures?

Dot paper can help you draw three-dimensional figures, or *solids*.

1 ACTIVITY: Drawing Views of a Solid

Work with a partner. Draw the front, side, and top views of each stack of cubes. Then find the number of cubes in the stack.

a. Number of cubes: 3

See Additional Answers for drawings.

b. 4 cubes

c. 4 cubes

d. 5 cubes

Differentiated Instruction

Kinesthetic

Provide building blocks or cubes for students to use. Students may work in pairs. Each student builds a solid out of sight of his/her partner and draws front, top, and side views. Students meet at a neutral site and trade drawings. At the building site, the student builds the solid shown in the drawing. Students get together to compare their drawings and models.

8.1 Record and Practice Journal

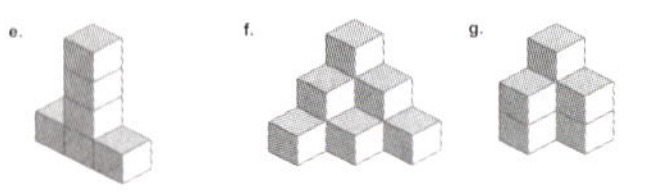

e. 6 cubes f. 10 cubes g. 7 cubes

2 ACTIVITY: Drawing Solids

Work with a partner.

a. Use isometric dot paper to draw three different solids that use the same number of cubes as the solid at the right.

b. Use square dot paper to draw a different solid that uses the same number of *prisms* as the solid at the right.

Sample answer:

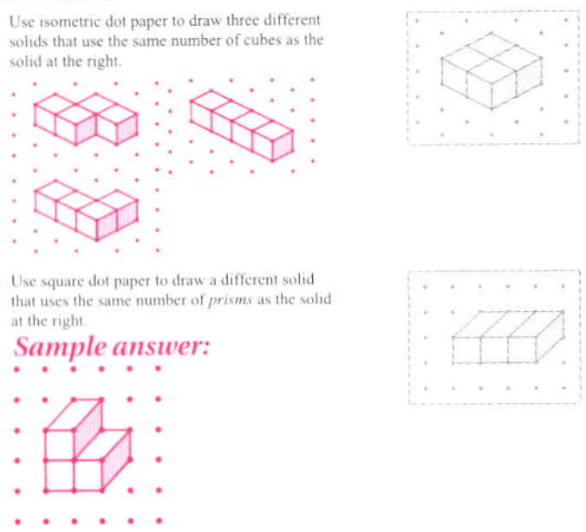

3 ACTIVITY: Exploring Faces, Edges, and Vertices

Work with a partner. Use the solid shown.

a. Match each word to the figure. Then write a definition for each word.

face *edge* *vertex*

b. Identify the number of faces, edges, and vertices in a rectangular prism.

6 faces, 12 edges, 8 vertices

c. When using dot paper to draw a solid, what represents the vertices? How do you draw edges? How do you draw faces?

the dots; connect two dots; connect at least 3 dots forming a closed figure

d. What do you think it means for lines or planes to be parallel or perpendicular in three dimensions? Use drawings to identify one pair of each of the following:

- parallel faces
- perpendicular faces
- parallel edges
- perpendicular edges
- edge parallel to a face
- edges perpendicular to a face

See Additional Answers.

What Is Your Answer?

4. IN YOUR OWN WORDS How can you draw three-dimensional figures?

Check students' work.

Laurie's Notes

Activity 2

- If time is a concern, you might elect to skip Activity 2.
- Students will use 4 cubes and arrange them to form three different solids.
- Help students understand that the orientation of the solid does not make it different, meaning 4 horizontal cubes would be the same solid as 4 vertical cubes.
- Students may need help getting started with their sketches. Suggest that they refer back to the sample. If the front cube were moved to make a number 7, what would still be the same? What would be different?

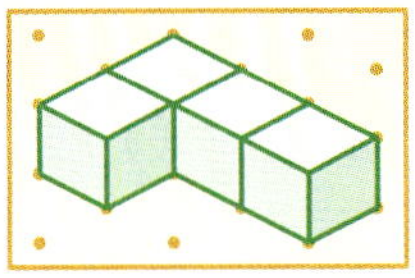

- Students should be familiar with the word *prism* from prior grades.
- In part (b), students will use 3 rectangular prisms. Primary teachers may have colored rods or similar prisms for your students to use in this activity.
- Ask volunteers to share their solutions.

Activity 3

- If students are stuck on a vocabulary word, they should ask their partners or look the word up in a glossary or a math dictionary.
- Before students go too far, ensure that they have correctly identified the vertex, edge, and face of the prism shown.
- In part (d), make sure they answer the first question for all 6 combinations: (line and line, plane and plane, plane and line) and (parallel, perpendicular)
- It is essential for students to think of an edge as a part of a line and a face as part of a plane.
- Students will likely say that lines are parallel in three dimensions if they do not intersect. However, they must not intersect *and* lie in the same plane. Allow them to struggle with this and be incorrect at first. After making this distinction, tell them that lines that do not intersect in three dimensions and are in different planes are called *skew lines*. You could now ask them to identify a pair of skew edges if desired.
- To get students away from only considering planes in which the faces of the prism lie, ask if the "top" edge of one of the "sides" is parallel to the "bottom" edge of the opposite side. Because a plane is not drawn through those lines (edges), they may think they are skew. Students should be able to visualize a plane crossing through the prism diagonally containing both lines. This is an important preview of CCSS 7.G.3, slicing three-dimensional figures with planes.

What Is Your Answer?

- For Question 4, students should refer back to their sketches from the first activity.

Closure

- How did the three views of a solid help you determine the number of cubes in the stack? Answers will vary.

2 ACTIVITY: Drawing Solids

Work with a partner.

a. Use isometric dot paper to draw three different solids that use the same number of cubes as the solid at the right.

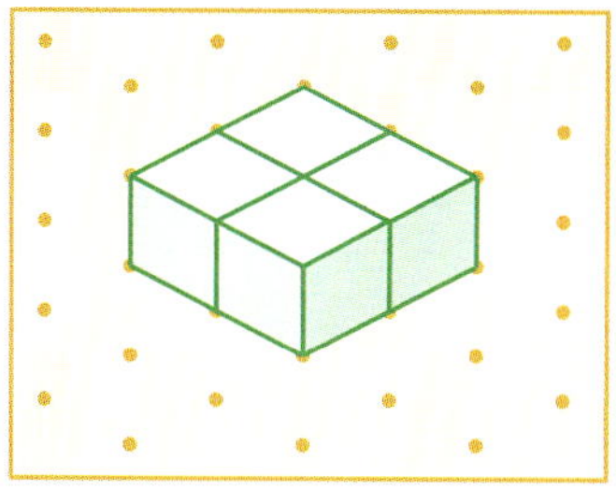

b. Use square dot paper to draw a different solid that uses the same number of *prisms* as the solid at the right.

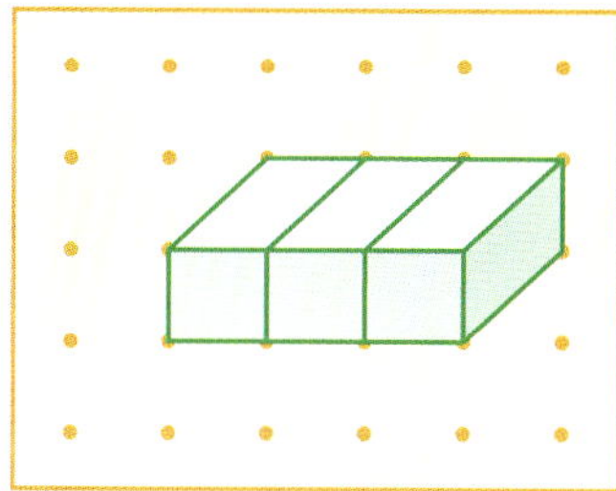

3 ACTIVITY: Exploring Faces, Edges, and Vertices

Work with a partner. Use the solid shown.

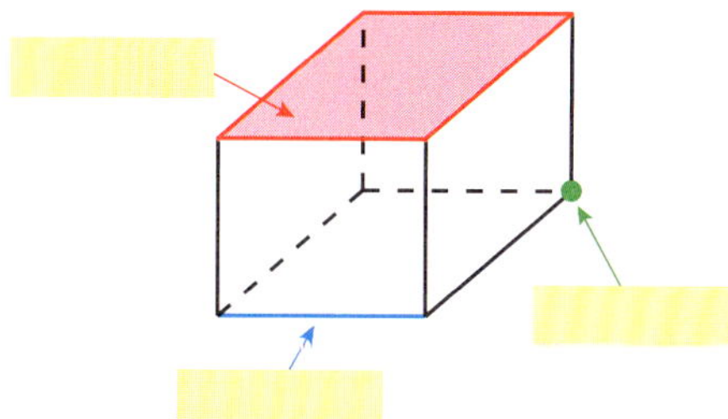

a. Match each word to the figure. Then write a definition for each word.

face *edge* *vertex*

b. Identify the number of faces, edges, and vertices in a rectangular prism.

c. When using dot paper to draw a solid, what represents the vertices? How do you draw edges? How do you draw faces?

d. What do you think it means for lines or planes to be parallel or perpendicular in three dimensions? Use drawings to identify one pair of each of the following:

- parallel faces
- parallel edges
- edge parallel to a face
- perpendicular faces
- perpendicular edges
- edge perpendicular to a face

Math Practice 7

View as Components

What are the different parts of a three-dimensional object? How can dot paper help you draw the parts of the object?

What Is Your Answer?

4. **IN YOUR OWN WORDS** How can you draw three-dimensional figures?

Practice Use what you learned about three-dimensional figures to complete Exercises 7–9 on page 358.

8.1 Lesson

Check It Out
Lesson Tutorials
BigIdeasMath.com

Key Vocabulary
solid, *p. 356*
polyhedron, *p. 356*
face, *p. 356*
edge, *p. 356*
vertex, *p. 356*
prism, *p. 356*
pyramid, *p. 356*

A **solid** is a three-dimensional figure that encloses a space. A **polyhedron** is a solid whose *faces* are all polygons.

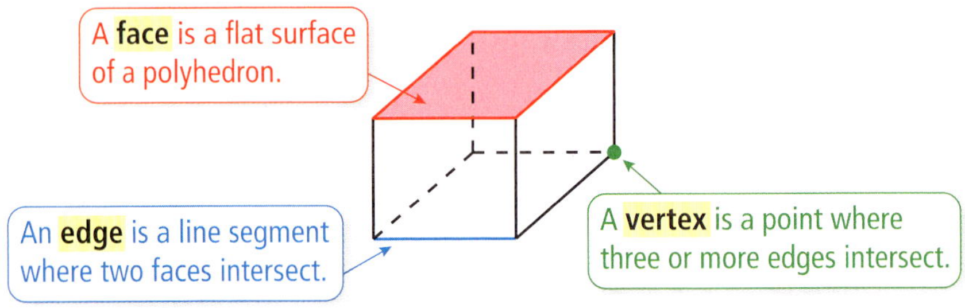

EXAMPLE 1 Finding the Number of Faces, Edges, and Vertices

Find the number of faces, edges, and vertices of the solid.

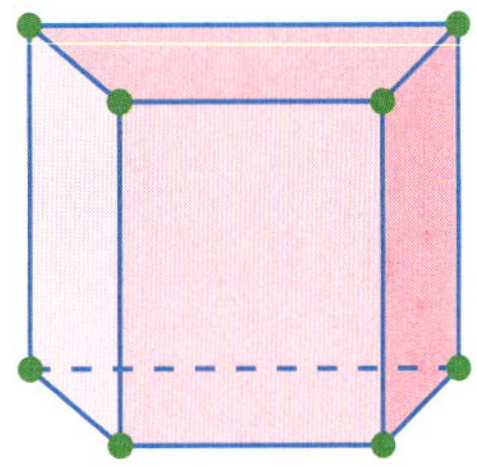

The solid has 1 face on the bottom, 1 face on the top, and 4 faces on the sides.

The faces intersect at 12 different line segments.

The edges intersect at 8 different points.

So, the solid has 6 faces, 12 edges, and 8 vertices.

On Your Own

Now You're Ready
Exercises 10–12

1. Find the number of faces, edges, and vertices of the solid.

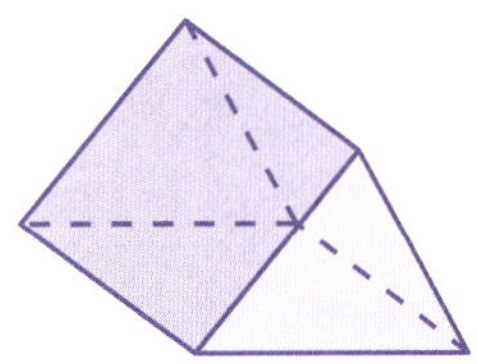

Key Ideas

Prisms

A **prism** is a polyhedron that has two parallel, identical *bases*. The *lateral faces* are parallelograms.

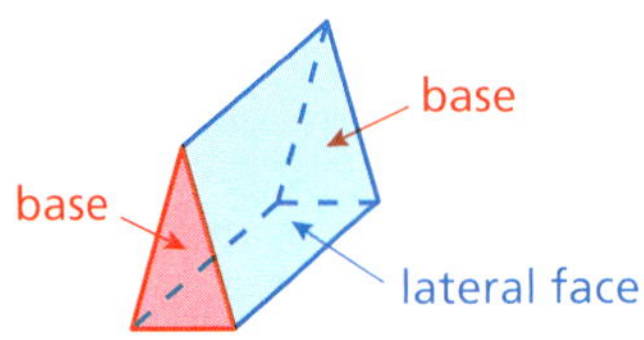

Triangular Prism

Pyramids

A **pyramid** is a polyhedron that has one base. The lateral faces are triangles.

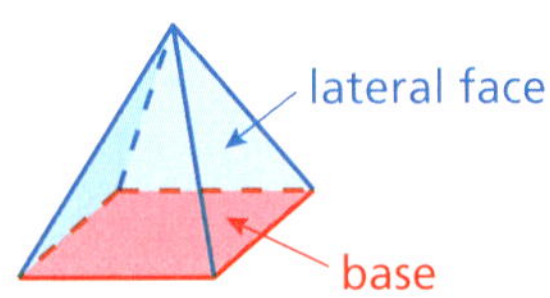

Rectangular Pyramid

The shape of the base tells the name of the prism or the pyramid.

Multi-Language Glossary at BigIdeasMath.com

Laurie's Notes

Introduction

Connect

- **Yesterday:** Students explored sketching solids on isometric dot paper. (MP5)
- **Today:** Students will be introduced to the vocabulary of prisms and pyramids and asked to sketch several solids.

Motivate

- Collect and display prisms and pyramids from school or home.
- Ask a volunteer to select one solid and describe it using mathematical vocabulary. You want students to say "prism" instead of "box of cereal." Besides naming the object, they should identify its features or attributes. They may say that the prism has 6 rectangular sides.

? "Are there any other solids in the collection that share the same features or attributes?" Listen for what attribute(s) the second solid shares with the first.

- Repeat this process for several solids.

Lesson Notes

Example 1

? "How many faces are there, and how would you describe their locations?" 6; top, bottom, 4 around the sides

? "How many edges are there, and how would you describe their locations? 12; 4 around the top, 4 around the bottom, and 4 vertical on the sides

? "How many vertices are there, and how would you describe their locations? 8; 4 on the top and 4 on the bottom

- **MP7 Look for and Make Use of Structure:** Repeat the example using prisms with different bases. Students should recognize that the number of faces will always be 2 more than the number of edges on the base. The number of vertices will be twice the number of edges on the base. The number of edges will always be 3 times the number of edges on the base.

On Your Own

- **Neighbor Check:** Have students work independently and then have their neighbors check their work. Have students discuss any discrepancies.

Key Ideas

- The point of the vocabulary is not to memorize definitions, but to have a sense as to the attributes of the solid. This will help in generalizing surface area and volume formulas later.
- Mention to students that prisms and pyramids have a qualifying name, given the type of base. A triangular prism has two bases that are triangles.
- **Common Error:** Students often think that the face that is "on the bottom" is the base. The solid does not need to be oriented so that it is resting on a base. Demonstrate this with several solids.
- It may be helpful to use vocabulary, such as vertex (vertices) and edge.

Goal Today's lesson is sketching and learning the vocabulary of **prisms** and **pyramids**.

Technology for the Teacher

Dynamic Classroom

Lesson Tutorials
Lesson Plans
Answer Presentation Tool

English Language Learners

Vocabulary

By this time, students should be able to tell the difference between a prism and a pyramid. Discuss the terms *base* and *face*. Explain that all the surfaces of prisms and pyramids are *faces*. Prisms have two bases that are congruent polygons and that are parallel to each other. Pyramids have only one face that is called the base.

Extra Example 1

Find the number of faces, edges, and vertices of the solid.

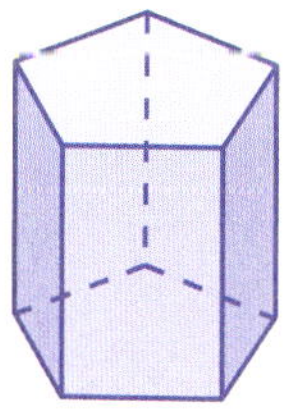

7 faces, 15 edges, 10 vertices

On Your Own

1. 5 faces, 9 edges, 6 vertices

Extra Example 2

a. Draw a hexagonal prism.

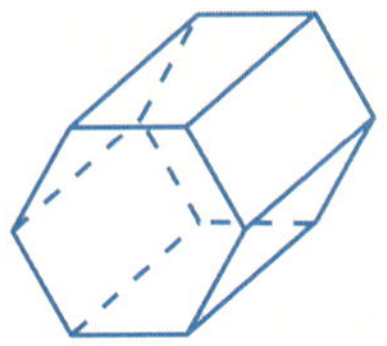

b. Draw a rectangular pyramid.

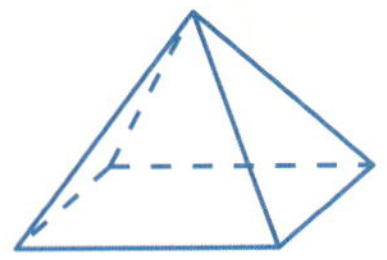

Extra Example 3

Draw the front, side, and top views of the solid.

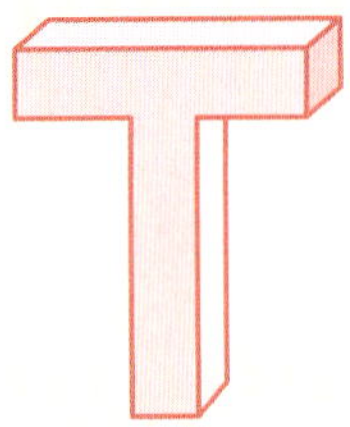

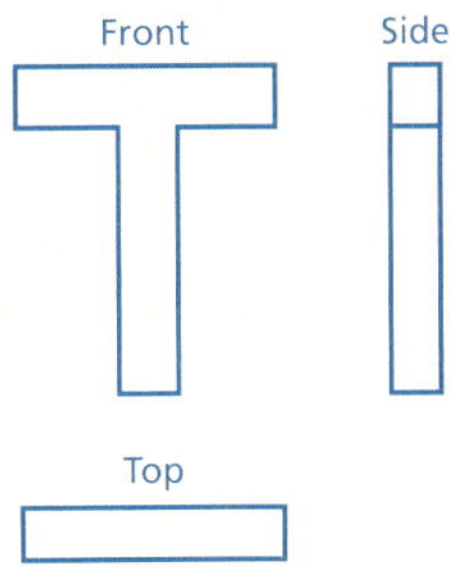

On Your Own

2.

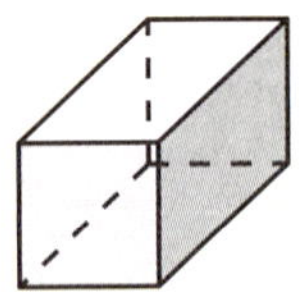

3.

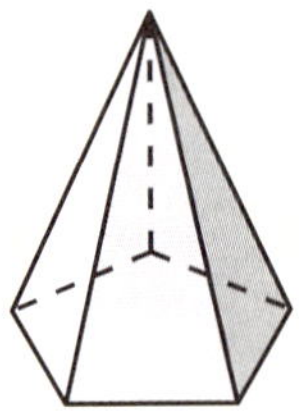

4–5. See Additional Answers.

Laurie's Notes

Example 2

- Demonstrate how to sketch a triangular pyramid.
- Drawings of objects may differ based on perspective. For instance, the point in step 1 could be placed below the triangle, or different lines in step 3 could be dashed. Each would offer a different perspective of the same object.

Example 3

- This example connects to yesterday's activity.
- Remind students that they need to take a *bird's eye view* in each of the three directions.

On Your Own

- Ask volunteers to draw their sketches at the board.

Closure

- Sketch a rectangular prism. Draw the front, side, and top views.

EXAMPLE 2 Drawing Solids

a. Draw a rectangular prism.

Step 1:

Draw identical rectangular bases.

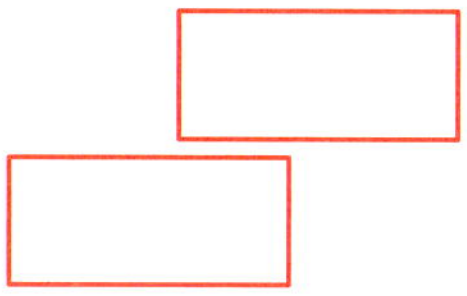

Step 2:

Connect corresponding vertices.

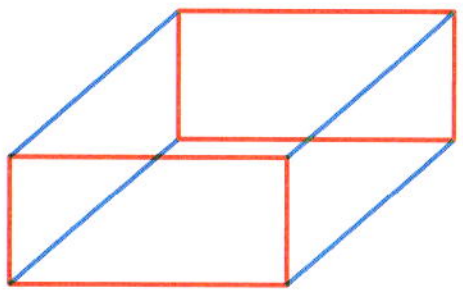

Step 3:

Change any *hidden* lines to dashed lines.

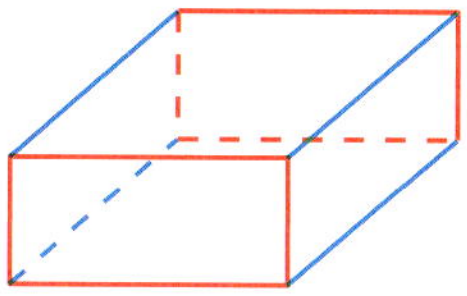

b. Draw a triangular pyramid.

Step 1:

Draw a triangular base and a point.

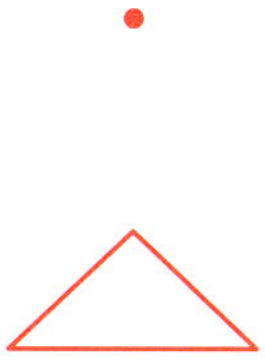

Step 2:

Connect the vertices of the triangle to the point.

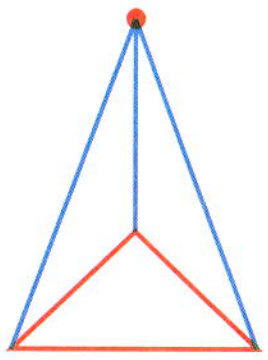

Step 3:

Change any *hidden* lines to dashed lines.

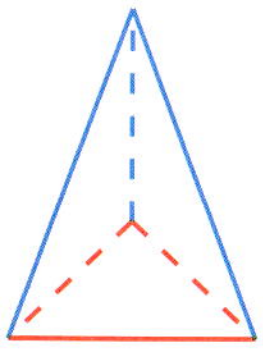

EXAMPLE 3 Drawing Views of a Solid

Draw the front, side, and top views of the eraser.

The front view is a parallelogram.

The side view is a rectangle.

The top view is a rectangle.

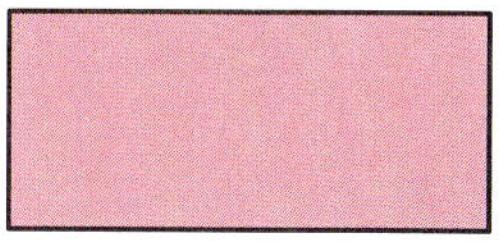

On Your Own

Now You're Ready
Exercises 13–22

Draw the solid.

2. square prism

3. pentagonal pyramid

Draw the front, side, and top views of the solid.

4. 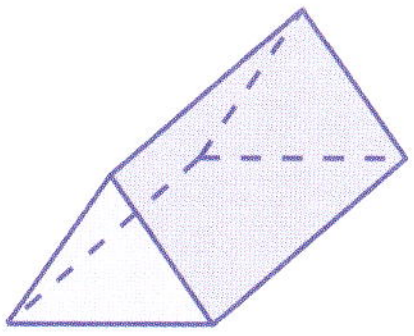

5. 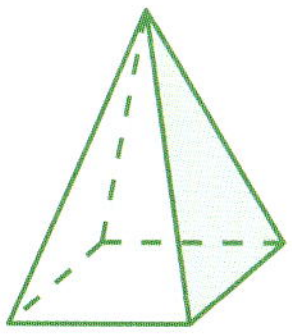

8.1 Exercises

Check It Out
Help with Homework
BigIdeasMath.com

Vocabulary and Concept Check

LOGIC **Decide whether the statement is *true* or *false*. If false, explain your reasoning.**

1. A triangular prism has three triangular faces.
2. A triangular prism has three rectangular faces.
3. A rectangular pyramid has one rectangular face.
4. A rectangular pyramid has three triangular faces.
5. All of the edges of a rectangular prism are parallel.
6. None of the edges of a rectangular pyramid are parallel.

Practice and Problem Solving

Draw the front, side, and top views of the stack of cubes. Then find the number of cubes in the stack.

7. 

8. 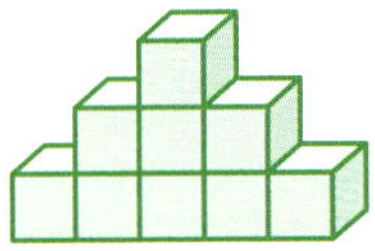

9. 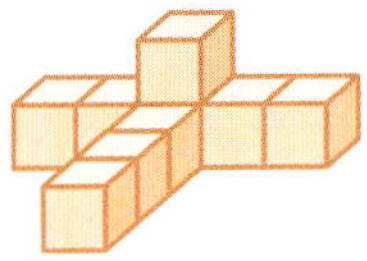

Find the number of faces, edges, and vertices of the solid.

10. 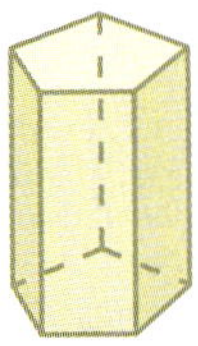

11. 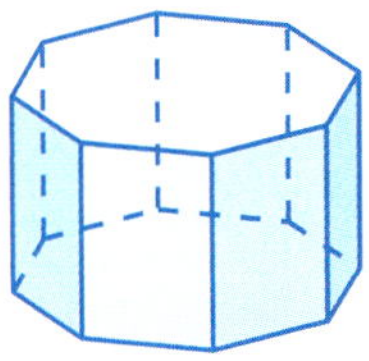

12. 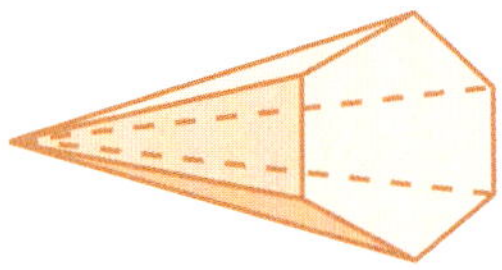

Draw the solid.

13. triangular prism
14. pentagonal prism
15. rectangular pyramid
16. hexagonal pyramid

Draw the front, side, and top views of the solid.

17. 

18. 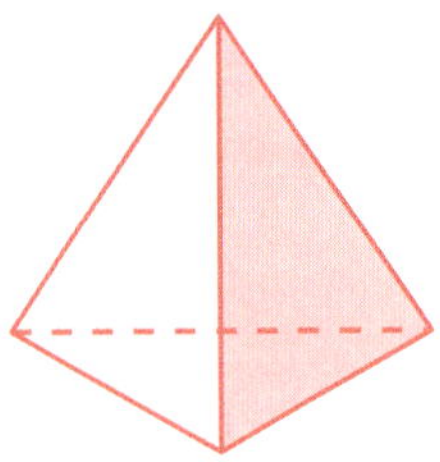

19. 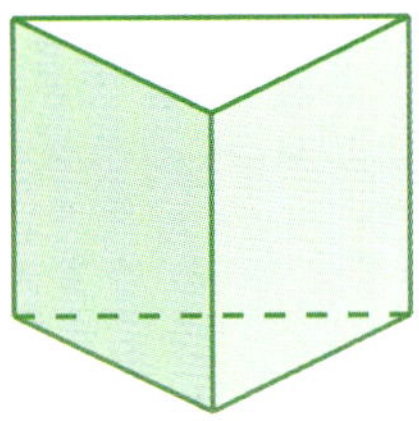

20. 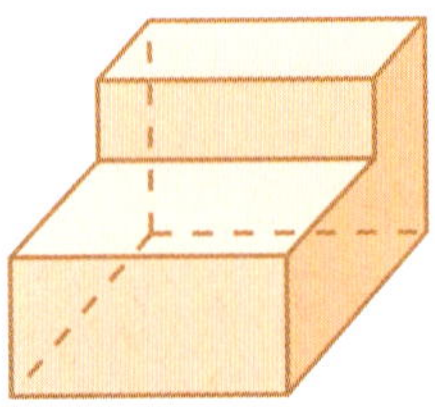

21. 

22. 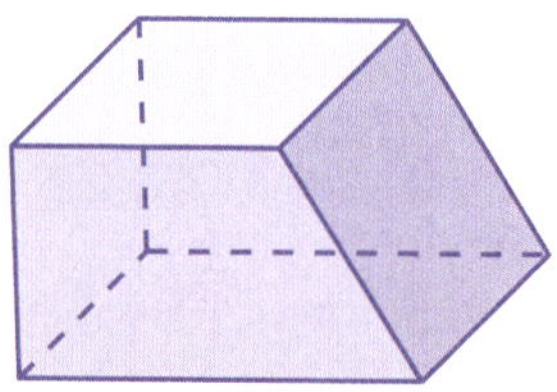

Assignment Guide and Homework Check

Level	Day 1 Activity Assignment	Day 2 Lesson Assignment	Homework Check
Basic	7–9, 30–33	1–6, 11–25 odd	11, 13, 21, 23
Average	7–9, 30–33	1–6, 11–23 odd, 24–26	11, 13, 21, 23, 26
Advanced	7–9, 30–33	1–6, 10–28 even	12, 14, 18, 26, 28

For Your Information

- **Exercise 23** The pyramid was built 18 B.C.–12 B.C. as a tomb for a Roman magistrate. It measures 100 Roman feet (22 meters) square at the base and is 125 Roman feet (27 meters) high.

Common Errors

- **Exercises 13–16** Students may mix up the different types of solids. Remind them of the definition of each solid and give a few real-life examples of each solid.
- **Exercises 17–22** Students may have difficulty visualizing the front, side, and top views of the solid. Create paper objects for those who are struggling to draw the different sides of the solid.
- **Exercises 25 and 26** Students may not be able to see how the shapes go together. Have them cut out pieces of paper or use blocks to model the solid.

8.1 Record and Practice Journal

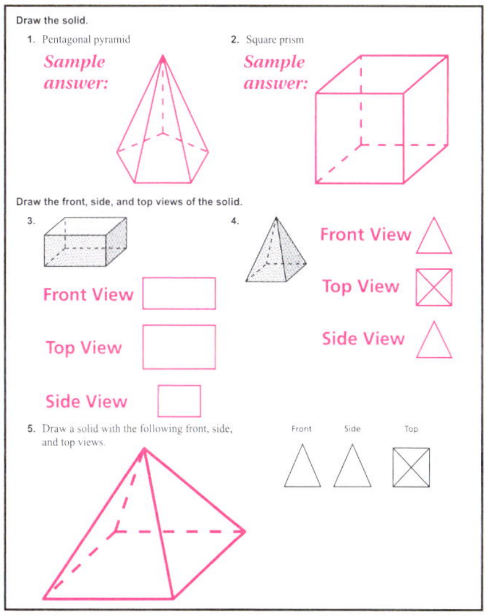

Vocabulary and Concept Check

1. false; It has two triangular faces.
2. true
3. true
4. false; It has four triangular faces.
5. false; Some are perpendicular and some are neither (skew).
6. false; Opposite edges of the base are parallel.

Practice and Problem Solving

7. front: 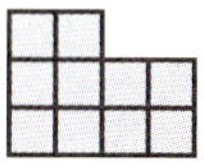

   side:  top: 

   10 cubes
8. front: 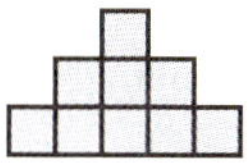

   side:  top: 

   9 cubes
9. front: 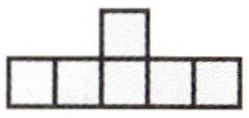

   side: 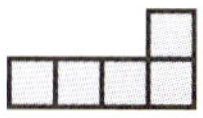

   top: 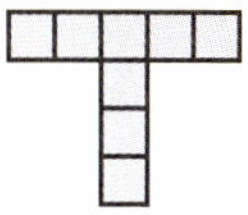

   9 cubes
10. 7 faces, 15 edges, and 10 vertices
11. 10 faces, 24 edges, and 16 vertices
12. 7 faces, 12 edges, and 7 vertices

13–26. See Additional Answers.

Practice and Problem Solving

27. *Answer should include, but is not limited to*: an original drawing of a house; a description of any solids that make up any part of the house
28. See *Taking Math Deeper*.
29. See Additional Answers.

Fair Game Review

30. 28 m^2 31. 12 cm^2
32. 15 ft^2 33. D

Mini-Assessment

1. Find the number of faces, edges, and vertices of the solid.

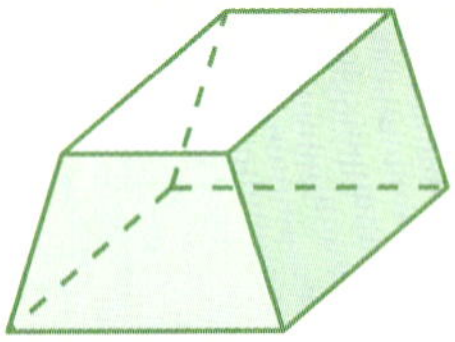

6 faces, 12 edges, 8 vertices

2. Draw a hexagonal prism.

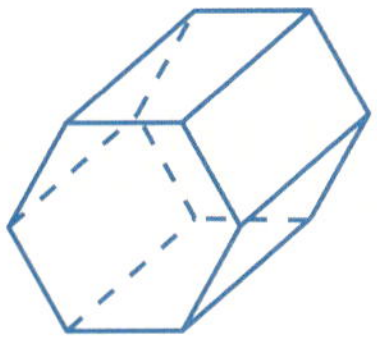

3. Draw a triangular pyramid.

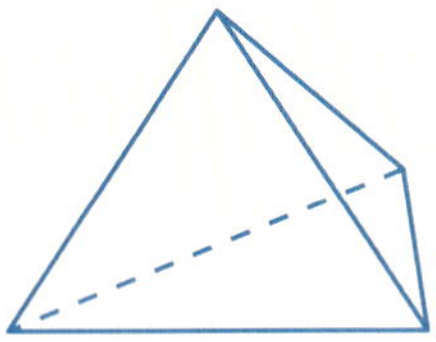

4. You and a friend attend a birthday party. Draw the front, side, and top views of the birthday cake.

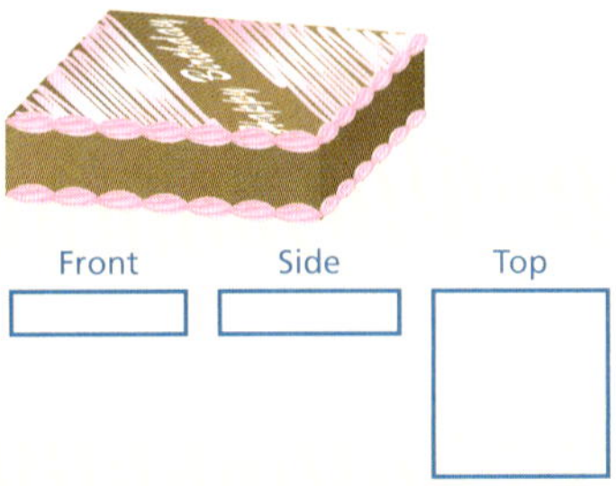

Taking Math Deeper

Exercise 28

This type of problem begs to be touched, felt, and seen. Give students 9 cubes and ask them to construct different solids that have the given top and side views.

1 Help me see it.
Build different solids out of 9 cubes.

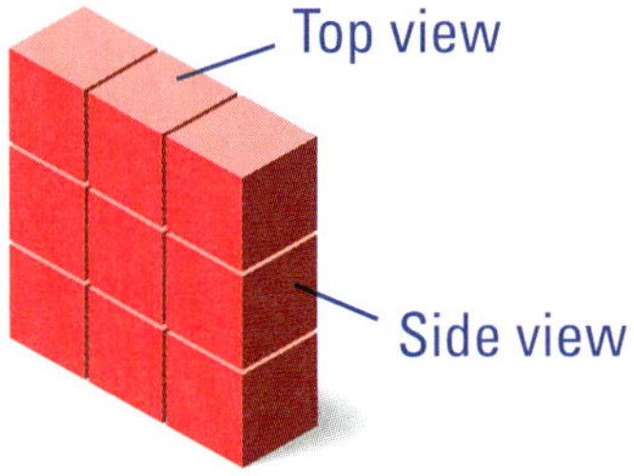

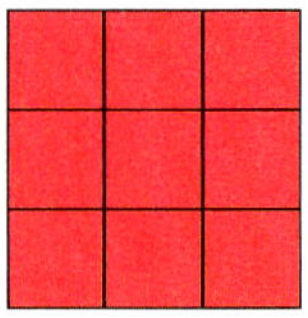

a. The greatest number of cubes is 9.

2 Here is a different one.

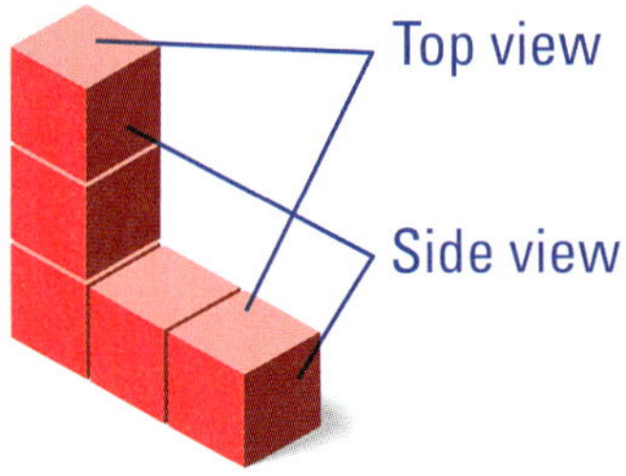

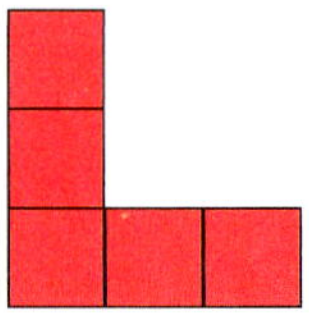

b. The least number of cubes is 5.

3 Here are the other possibilities (does not include reflections).

Reteaching and Enrichment Strategies

If students need help...	If students got it...
Resources by Chapter • Practice A and Practice B • Puzzle Time Record and Practice Journal Practice Differentiating the Lesson Lesson Tutorials Skills Review Handbook	Resources by Chapter • Enrichment and Extension • Technology Connection Start the next section

23. **PYRAMID ARENA** The Pyramid of Caius Cestius in Rome, Italy, is in the shape of a square pyramid. Draw a sketch of the pyramid.

24. **RESEARCH** Use the Internet to find a picture of the Washington Monument. Describe its shape.

Draw a solid with the following front, side, and top views.

25. 

26. 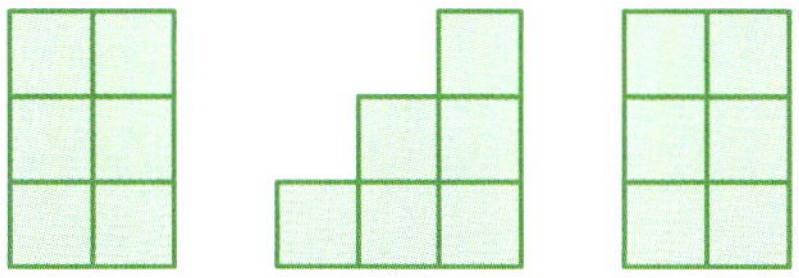

27. **PROJECT** Design and draw a house. Name the different solids that you can use to make a model of the house.

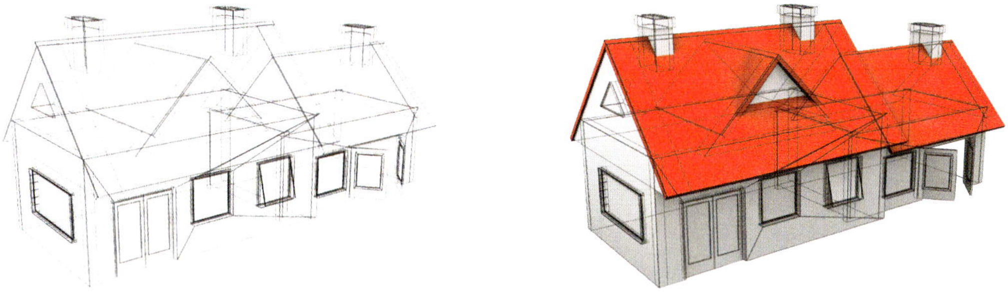

28. **REASONING** Two of the three views of a solid are shown.

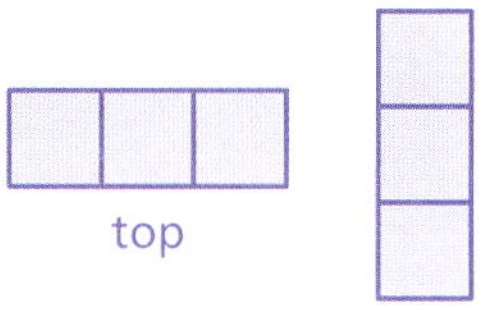

a. What is the greatest number of unit cubes in the solid?

b. What is the least number of unit cubes in the solid?

c. Draw the front views of both solids in parts (a) and (b).

29. **Reasoning** Draw two different solids with five faces.

a. Write the number of vertices and edges for each solid.

b. Explain how knowing the numbers of edges and vertices helps you draw a three-dimensional figure.

Fair Game Review What you learned in previous grades & lessons

Find the area of the figure. *(Section 4.1, Section 4.2, and Section 4.3)*

30. 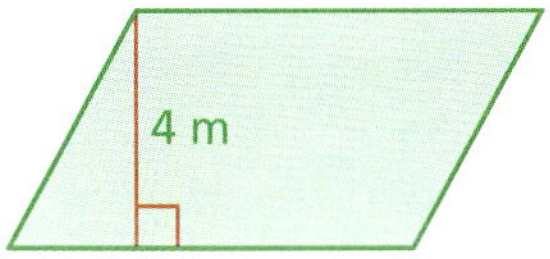

31. 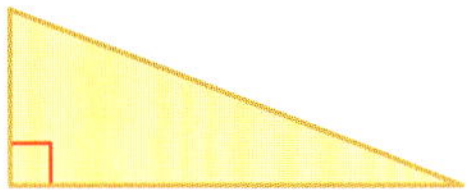

32. 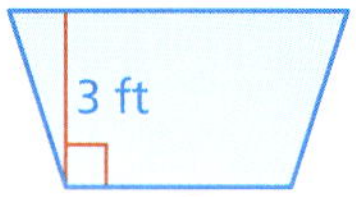

33. **MULTIPLE CHOICE** Which statement is true when $x = -2$ and $y = |-2|$? *(Section 6.4)*

Ⓐ $x = y$ Ⓑ $y < 0$ Ⓒ $x > y$ Ⓓ $y > x$

8.2 Surface Areas of Prisms

Essential Question How can you find the area of the entire surface of a prism?

1 ACTIVITY: Identifying Prisms

Work with a partner. Label one of the faces as a "base" and the other as a "lateral face." Use the shape of the base to identify the prism.

a.

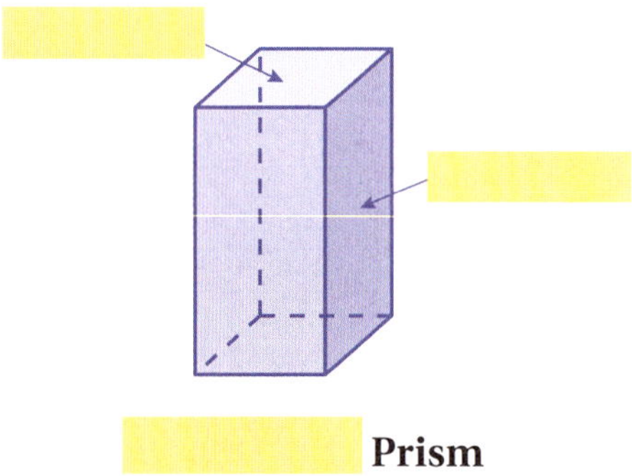

b.

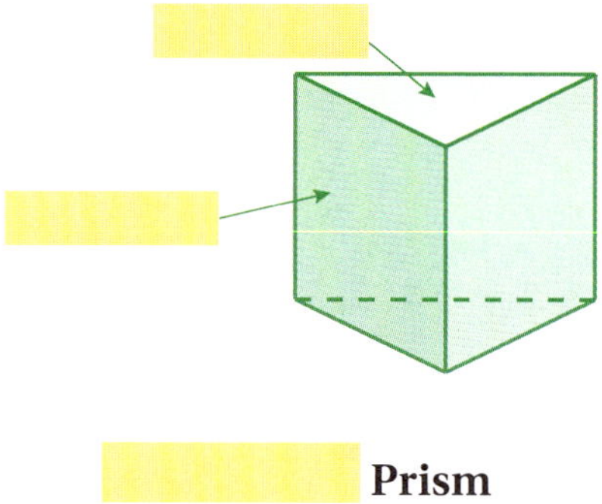

2 ACTIVITY: Using Grid Paper to Construct a Prism

Work with a partner.

a. Copy the figure shown below onto grid paper.

b. Cut out the figure and fold it to form a prism. What type of prism does it form?

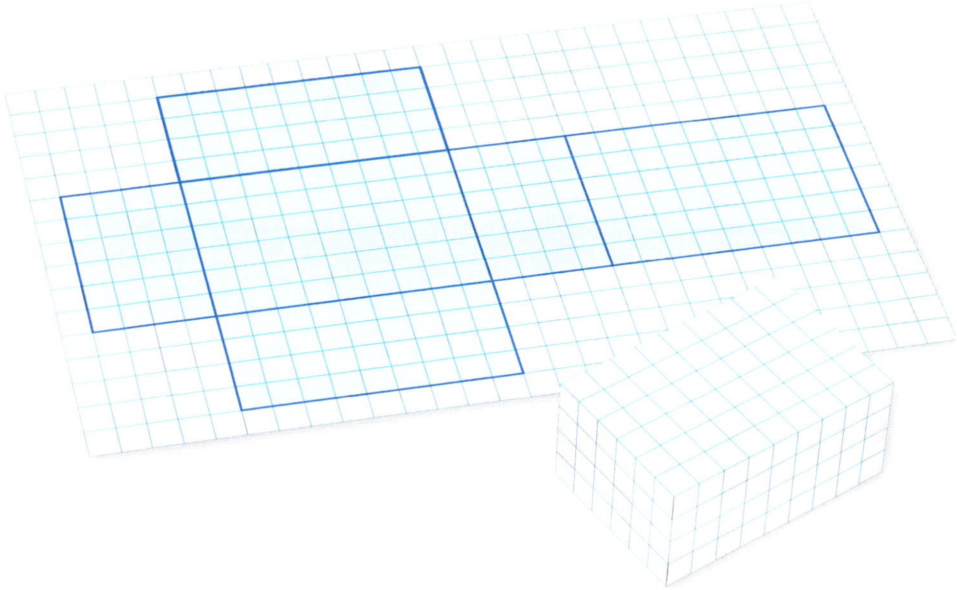

COMMON CORE

Geometry

In this lesson, you will

- use nets to represent prisms.
- find the surface area of prisms.
- solve real-life problems.

Learning Standard
6.G.4

Laurie's Notes

Introduction

Standards for Mathematical Practice

- **MP8 Look for and Express Regularity in Repeated Reasoning:** Students will be finding the area of the entire surface of several prisms. In doing so they may discover an efficient method. For instance, a rectangular prism has 3 pairs of congruent faces. Finding the area of one from each pair and then doubling is more efficient than doing six computations.

Motivate

- Hold up two different rectangular prisms made of cardboard, one clearly larger than the other.

? "Which box used more cardboard when it was made? Explain how you know." Listen for an explanation that refers to finding the area of each face and adding up all of the areas and then repeating that process for the second prism.

Discuss

- Review the vocabulary of prisms (base, lateral face).
- Point out that prisms are named by the shapes of their bases. For instance, a triangular prism has triangular bases.
- Note: The term *lateral face* is used in the activity. *Lateral surface area* is taught in Grade 7.

Activity Notes

Activity 1

- Students should work with their partners to identify the bases and lateral faces of each prism.
- In this activity, students can label the faces however they want in part (a). Any pair of opposite sides can be considered the bases. Explore this with students. Note that it is traditional to consider the *top and bottom* faces of a rectangular prism as the bases.
- Check to see that the prisms are named correctly.

Activity 2

- **Teaching Tip:** To avoid having scraps of paper on the floor, recycle plastic bags and tape them to desks around the room. There should be a bag close enough to each group so that there is no excuse for leaving a mess.
- In order to copy the figure correctly, students will need to determine the dimensions of each face, or at least do some counting. This process will be helpful in developing the concept of surface area of a solid.
- If time permits, ask students questions about the prism they folded. How many vertices, faces, and edges does the prism have? Which faces are parallel? Which faces are perpendicular?

Common Core State Standards

6.G.4 Represent three-dimensional figures using nets made up of rectangles and triangles, and use the nets to find the surface area of these figures. Apply these techniques in the context of solving real-world and mathematical problems.

Previous Learning

Students need to be familiar with prisms.

Technology for the Teacher

Dynamic Classroom

Lesson Plans
Complete Materials List

8.2 Record and Practice Journal

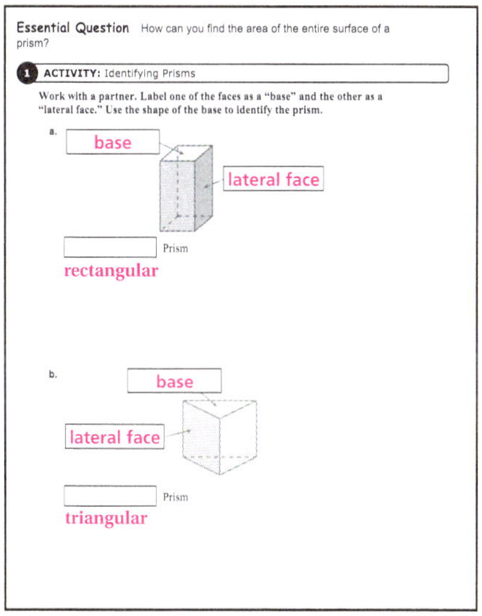
Essential Question How can you find the area of the entire surface of a prism?

1 ACTIVITY: Identifying Prisms

Work with a partner. Label one of the faces as a "base" and the other as a "lateral face." Use the shape of the base to identify the prism.

a. base; lateral face; rectangular Prism

b. base; lateral face; triangular Prism

English Language Learners

Vocabulary

English learners may be familiar with the word *net* in everyday context. In finance, the net profit describes the bottom line of a financial transaction. In fishing, a net is a collection of knotted strings used to catch fish. In mathematical context, *net* or *geometric net* is used to mean the two-dimensional representation of a solid object.

8.2 Record and Practice Journal

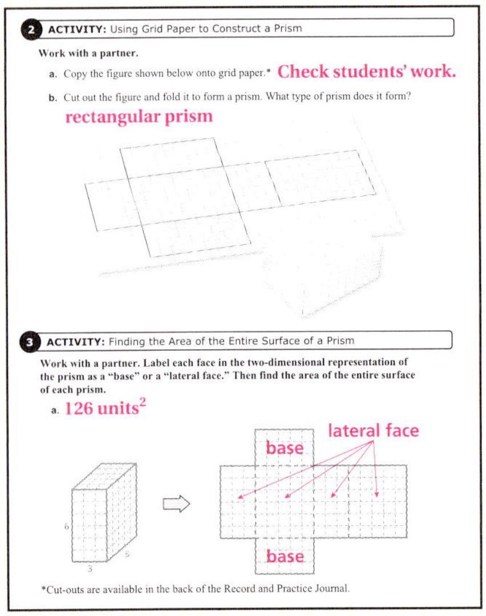

2 ACTIVITY: Using Grid Paper to Construct a Prism

Work with a partner.

a. Copy the figure shown below onto grid paper.* Check students' work.

b. Cut out the figure and fold it to form a prism. What type of prism does it form?

rectangular prism

3 ACTIVITY: Finding the Area of the Entire Surface of a Prism

Work with a partner. Label each face in the two-dimensional representation of the prism as a "base" or a "lateral face." Then find the area of the entire surface of each prism.

a. 126 units^2

*Cut-outs are available in the back of the Record and Practice Journal.

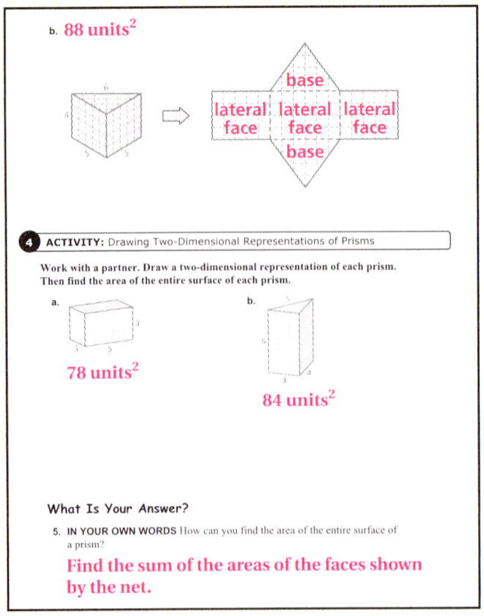

b. 88 units^2

4 ACTIVITY: Drawing Two-Dimensional Representations of Prisms

Work with a partner. Draw a two-dimensional representation of each prism. Then find the area of the entire surface of each prism.

a. 78 units^2

b. 84 units^2

What Is Your Answer?

5. **IN YOUR OWN WORDS** How can you find the area of the entire surface of a prism?

Find the sum of the areas of the faces shown by the net.

Laurie's Notes

Activity 3

- In this activity, students can label the faces however they want in part (a). Any pair of opposite sides can be considered the bases. Explore this with students. Note that it is traditional to consider the *top and bottom* faces of a rectangular prism as the bases.
- Instead of cutting and folding up a prism as they did in Activity 2, this activity reverses that process. Students have to think about how the prism is unfolded to make the template shown.
- Students may recognize that part (a) is the prism (net) from Activity 2. They have a model of the prism in front of them that they can relate to for this activity.
- **Teaching Tip:** When working through the activity, refer to the faces as the *surface*, so they are finding the area of the surface. This leads into the definition of *surface area* in the lesson.
- **MP3 Construct Viable Arguments and Critique the Reasoning of Others** and **MP8:** When students have finished, ask volunteers to share how they found the area of the entire surface of the prism. There will be different strategies so expect all students to listen carefully to the process. Does the process described for part (a) work for part (b)?
- **Big Idea:** The grid shown on the surface of the prisms helps students focus on the idea of area. The square units are visible.

Activity 4

- **Scaffolding:** If time is a concern, have half of the class work on part (a) and the other half work on part (b).
- **Teaching Tip:** When working through the activity, refer to the faces as the *surface*, so they are finding the area of the surface. This leads into the definition of *surface area* in the lesson.
- These drawings can be challenging for some students. Going from the solid to the two-dimensional representation is different than the reverse. If students are having trouble, suggest that they begin with one face and sketch it. Then move to a second face that shares an edge with the first face. The two faces are either both lateral or one is a lateral face and the second is a base. Refer back to Activity 1 or 3 for help in thinking about where to draw the second face. Repeat this process for the remaining faces.
- There are different two-dimensional representations (nets) that can be drawn for each prism. Take time for students to investigate this.

What Is Your Answer?

- Listen for a process that describes finding the sum of the areas of all of the faces.

Closure

- **Writing Prompt:** To find the total area of all faces of a tissue box I would . . . Sum the areas of the six faces.

3 ACTIVITY: Finding the Area of the Entire Surface of a Prism

Work with a partner. Label each face in the two-dimensional representation of the prism as a "base" or a "lateral face." Then find the area of the entire surface of each prism.

a.

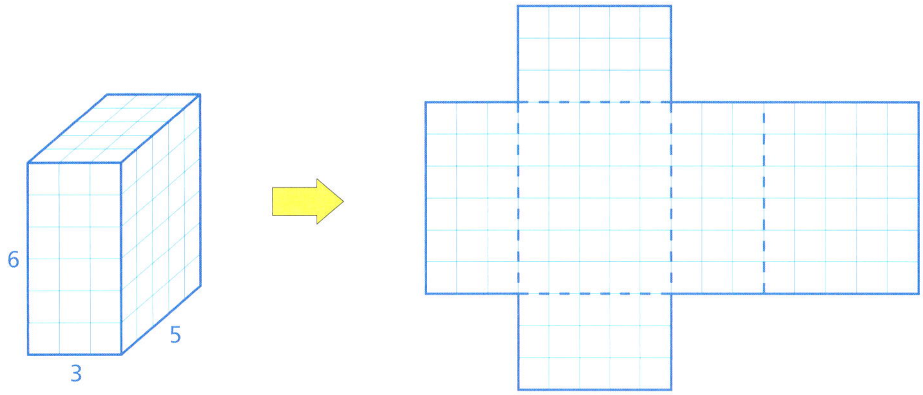

Math Practice 8

Repeat Calculations

When finding the areas of the faces, what calculations do you repeat?

b.

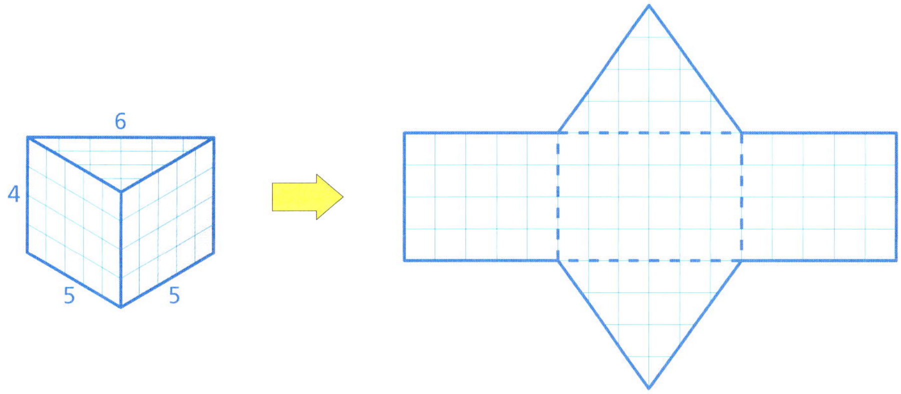

4 ACTIVITY: Drawing Two-Dimensional Representations of Prisms

Work with a partner. Draw a two-dimensional representation of each prism. Then find the area of the entire surface of each prism.

a.

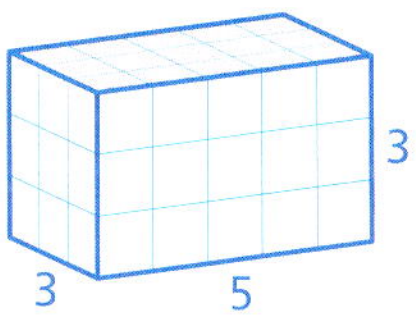

b.

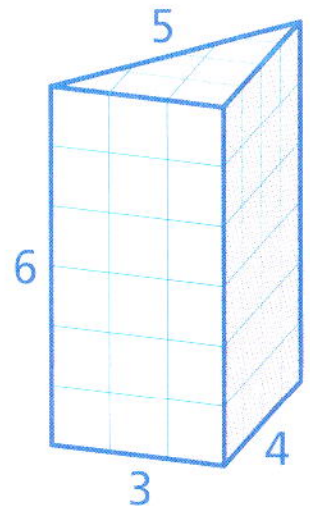

What Is Your Answer?

5. **IN YOUR OWN WORDS** How can you find the area of the entire surface of a prism?

Practice Use what you learned about the area of the entire surface of a prism to complete Exercises 3–5 on page 364.

8.2 Lesson

Check It Out
Lesson Tutorials
BigIdeasMath.com

Key Vocabulary
surface area, *p. 362*
net, *p. 362*

The **surface area** of a solid is the sum of the areas of all of its faces. You can use a two-dimensional representation of a solid, called a **net**, to find the surface area of the solid. Surface area is measured in *square units.*

Key Idea

Net of a Rectangular Prism

A *rectangular prism* is a prism with rectangular bases.

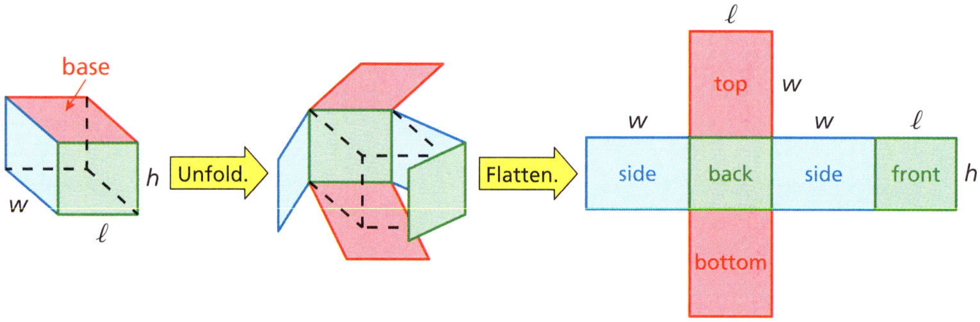

EXAMPLE 1 Finding the Surface Area of a Rectangular Prism

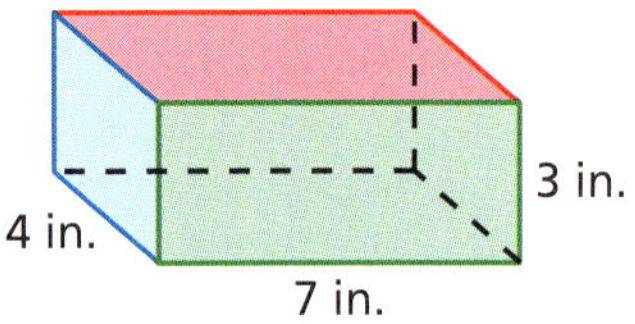

Find the surface area of the rectangular prism.

Use a net to find the area of each face.

Top: $7 \cdot 4 = 28$

Bottom: $7 \cdot 4 = 28$

Front: $7 \cdot 3 = 21$

Back: $7 \cdot 3 = 21$

Side: $4 \cdot 3 = 12$

Side: $4 \cdot 3 = 12$

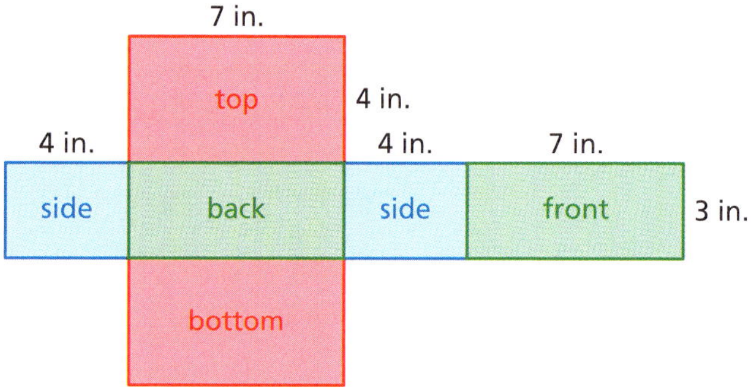

Find the sum of the areas of the faces.

$$\text{Surface Area} = \text{Area of top} + \text{Area of bottom} + \text{Area of front} + \text{Area of back} + \text{Area of a side} + \text{Area of a side}$$

$$S = 28 + 28 + 21 + 21 + 12 + 12$$

$$= 122$$

So, the surface area is 122 square inches.

Multi-Language Glossary at BigIdeasMath.com

Laurie's Notes

Introduction

Connect

- **Yesterday:** Students worked with two-dimensional representations (nets) of prisms and found the area of each face. (MP3, MP8)
- **Today:** Students will use a net to find the surface area of a prism.

Motivate

- Have available cardboard prisms that have been folded.
- Holding the assembled solids, ask students to visualize and then describe the cardboard nets that result when the solids are "unfolded."
- Unfold one or more of the solids so that students can see the net of the prism.
- Explain the connection between the cardboard net and the surface area of the prism.

Lesson Notes

Discuss

- Define surface area and net.
- Remind students that square units are used to label surface area.

Key Idea

- Write the Key Idea.
- Refer to the physical models in the Motivate activity to describe a net for a rectangular prism.

Example 1

? Hold a cardboard box and ask, "How can you find the total amount of cardboard used to make this box?" Find the sum of the areas of the six faces.

- **MP8 Look for and Express Regularity in Repeated Reasoning:** Work through the example as shown. Students may find it easier to consider the three pairs of opposite faces instead of finding the area of each of the six faces separately.

? "What units are used to label surface area?" square units

Goal Today's lesson is finding **surface areas** of prisms using **nets**.

Technology for the Teacher

Dynamic Classroom

Lesson Tutorials
Lesson Plans
Answer Presentation Tool

Extra Example 1

Find the surface area of the rectangular prism.

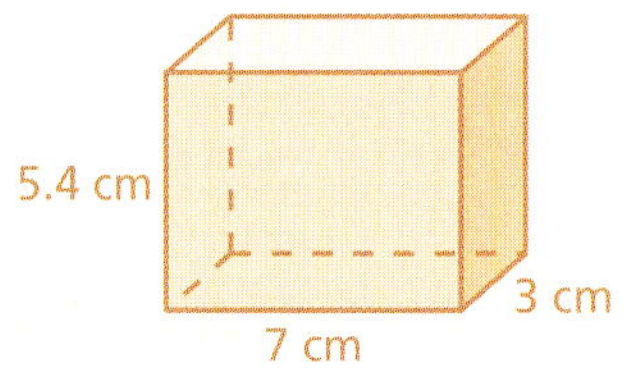

150 cm^2

Differentiated Instruction

Visual

Use a rectangular box to demonstrate two ways of finding surface area. The first method is to find the area of each face and then add areas. The second method is to open the box into a net and find the area of the net. Students should see that the two methods have the same result.

Extra Example 2

Find the surface area of the triangular prism.

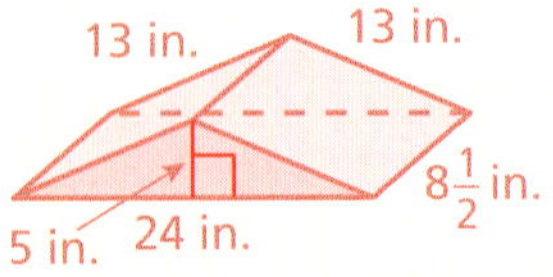

545 in.^{2}

On Your Own

1. 258 m^2
2. 180 in.^{2}
3. 286 ft^2
4. 60 yd^2
5. 420 m^2
6. 299.32 ft^2

Laurie's Notes

Key Idea

- Have a net available as a visual, whether it is a cardboard box or a triangular prism made from polygon frames.
- Discuss with students that a triangular prism has two opposite faces that are triangles, and the rest of the faces are rectangles.

Example 2

- Note that the net is a visual reminder to find the area of each face. Color-coding the faces helps students keep track of their work.

? "How many faces does a triangular prism have?" 5

? "How do you find the area of a triangle?" $A = \frac{1}{2}bh$

- Encourage students to write a verbal model for the surface area of the prism and then substitute the areas of the faces as they are computed.

On Your Own

- **MP6 Attend to Precision:** Students may forget to include the area of one or more of the faces when finding the surface area.
- Verify that fraction (Exercise 2) and decimal (Exercises 3 and 6) computations are done correctly.
- Exercise 4 involves a right triangle. Some students may need to be reminded that one side is the base and one side is the height.
- Exercise 6 includes decimal dimensions. Students may find it helpful to think of the area formula for triangles as $A = (0.5)bh$.

Closure

- Hold one of the items from the Motivate activity of today's lesson and ask students to find the surface area of the prism.

Key Idea

Net of a Triangular Prism

A *triangular prism* is a prism with triangular bases.

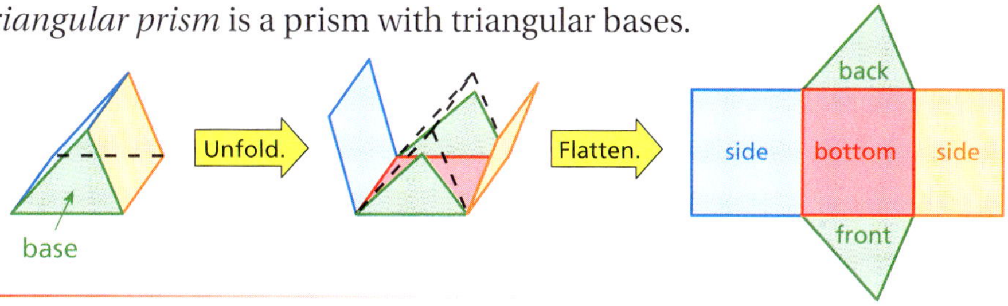

> **Remember**
>
> The area A of a triangle with base b and height h is $A = \frac{1}{2}bh$.

EXAMPLE 2 Finding the Surface Area of a Triangular Prism

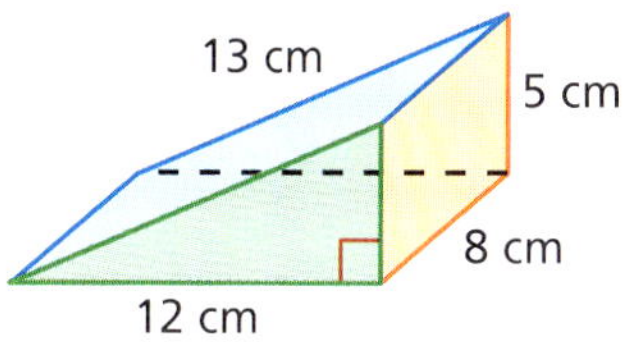

Find the surface area of the triangular prism.

Use a net to find the area of each face.

Bottom: $12 \cdot 8 = 96$

Front: $\frac{1}{2} \cdot 12 \cdot 5 = 30$

Back: $\frac{1}{2} \cdot 12 \cdot 5 = 30$

Side: $13 \cdot 8 = 104$

Side: $8 \cdot 5 = 40$

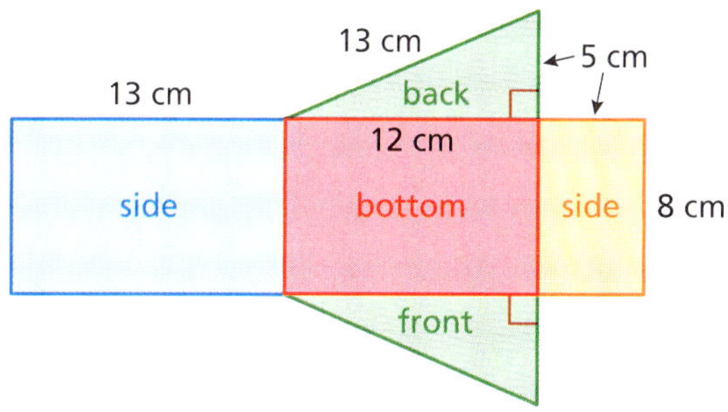

Find the sum of the areas of the faces.

$$\text{Surface Area} = \text{Area of bottom} + \text{Area of front} + \text{Area of back} + \text{Area of a side} + \text{Area of a side}$$

$$S = 96 + 30 + 30 + 104 + 40 = 300$$

So, the surface area is 300 square centimeters.

On Your Own

Now You're Ready Exercises 6–11

Find the surface area of the rectangular prism.

1. 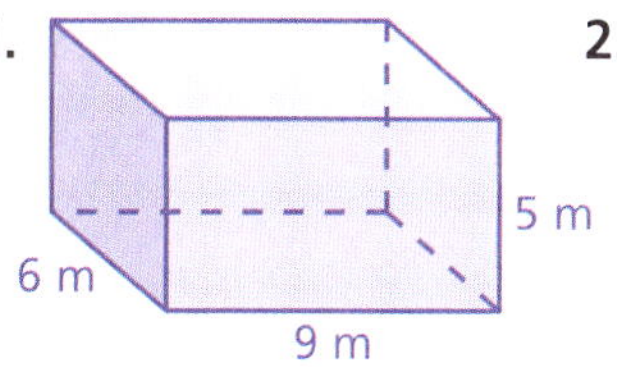

2. 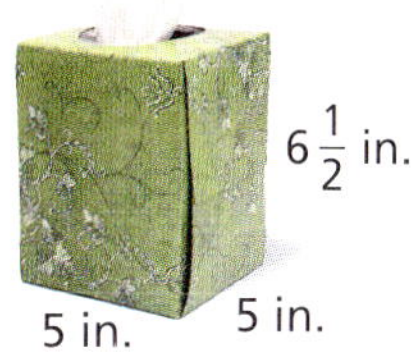

3. 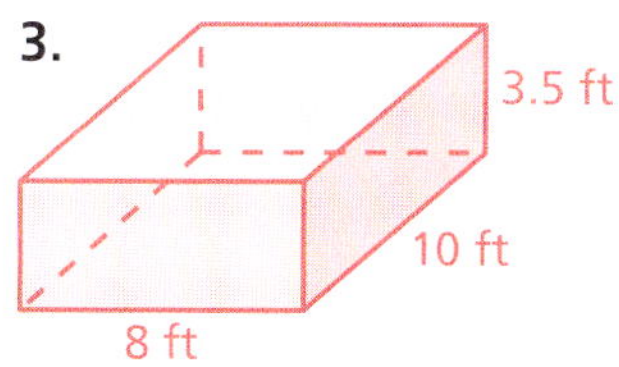

Find the surface area of the triangular prism.

4. 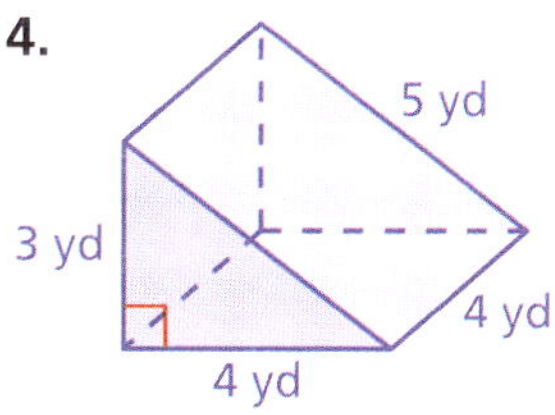

5. 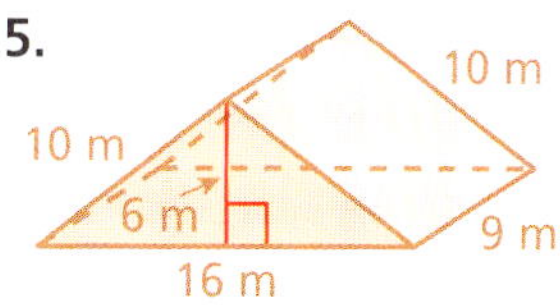

6. 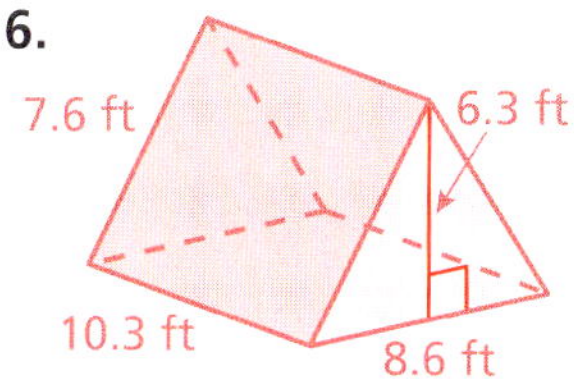

8.2 Exercises

Check It Out
Help with Homework
BigIdeasMath.com

Vocabulary and Concept Check

1. **VOCABULARY** Explain how to find the surface area of a prism.
2. **DIFFERENT WORDS, SAME QUESTION** Which is different? Find "both" answers.

What is the sum of the areas of the faces of the prism?

What is the area of the entire surface of the prism?

What is the area of the triangular faces of the prism?

What is the surface area of the prism?

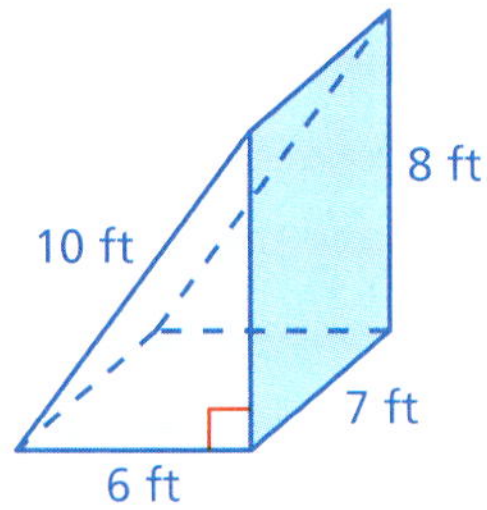

Practice and Problem Solving

Draw a two-dimensional representation of the prism. Then find the area of the entire surface of the prism.

3. 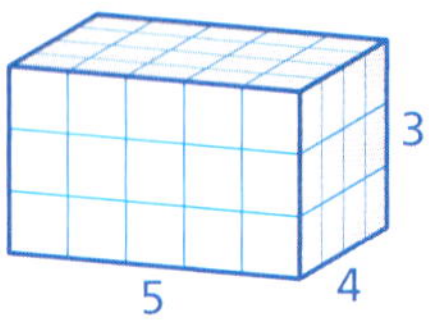

4. 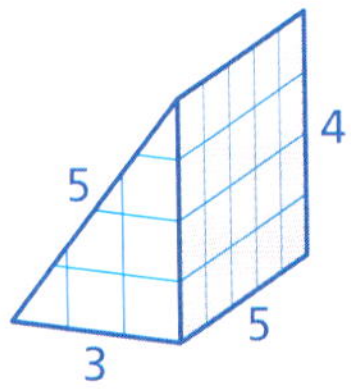

5. 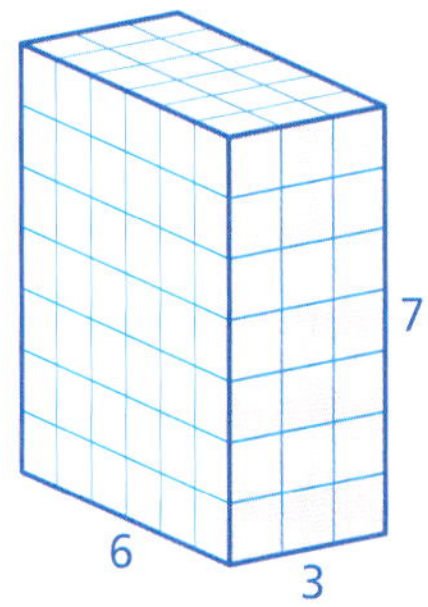

Find the surface area of the prism.

1 6. 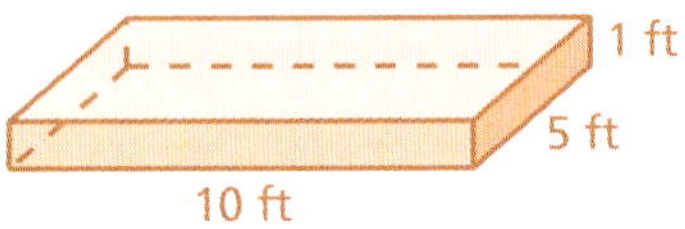

7. 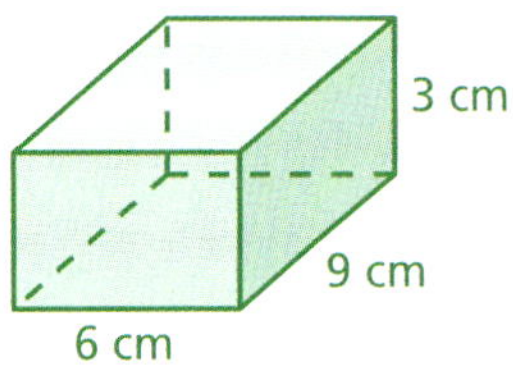

8. 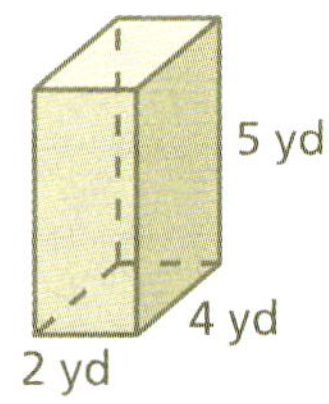

2 9. 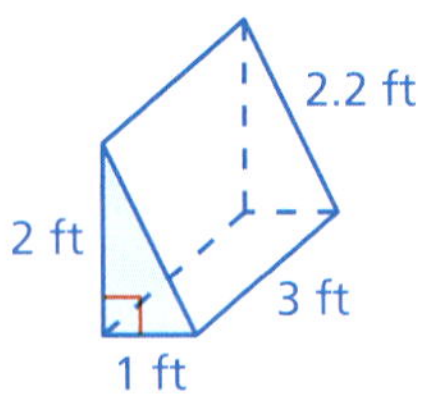

10. 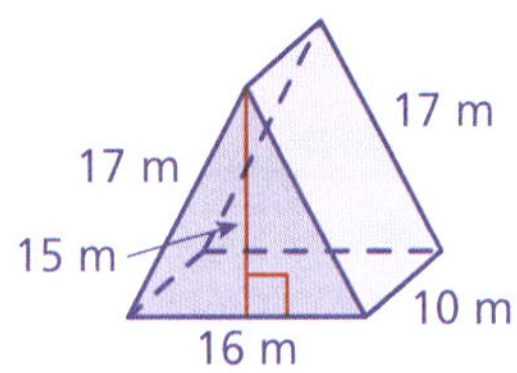

11. 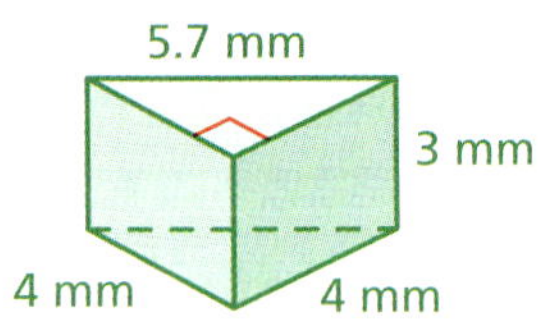

12. **GIFT BOX** A gift box in the shape of a rectangular prism measures 8 inches by 8 inches by 10 inches. What is the least amount of wrapping paper needed to wrap the gift box? Explain.

13. **TENT** What is the least amount of fabric needed to make the tent?

Assignment Guide and Homework Check

Level	Day 1 Activity Assignment	Day 2 Lesson Assignment	Homework Check
Basic	3–5, 18–21	1, 2, 7–15 odd	7, 11, 13, 15
Average	3–5, 18–21	1, 2, 6–16	7, 9, 12, 13 ,15
Advanced	3–5, 18–21	1, 2, 6–17	7, 9, 13, 15, 16

Common Errors

- **Exercises 3–11** Students may forget to count the surface area of the top and bottom faces. Remind them that they must count all of the faces, which includes the top and bottom as well as the sides.

8.2 Record and Practice Journal

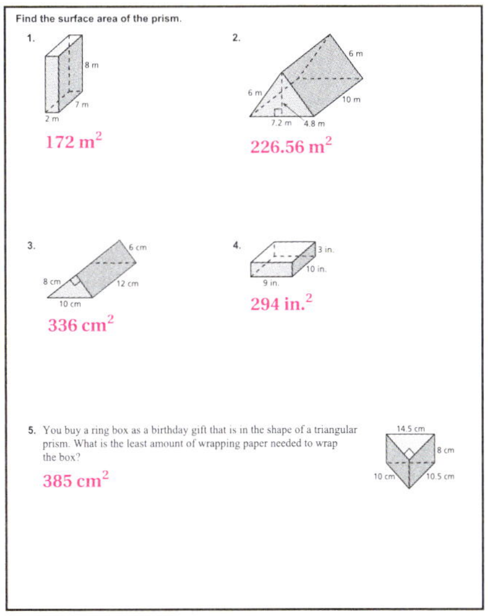
Find the surface area of the prism.

1. 8 m, 7 m, 2 m
172 m^2

2. 6 m, 10 m, 6 m, 7.2 m, 4.8 m
226.56 m^2

3. 6 cm, 8 cm, 12 cm, 10 cm
336 cm^2

4. 3 in., 10 in., 9 in.
294 $in.^2$

5. You buy a ring box as a birthday gift that is in the shape of a triangular prism. What is the least amount of wrapping paper needed to wrap the box?
14.5 cm, 8 cm, 10 cm, 10.5 cm
385 cm^2

Vocabulary and Concept Check

1. Find the sum of the areas of the faces.
2. What is the area of the triangular faces of the prism?; 48 ft^2; 216 ft^2

Practice and Problem Solving

3–5. See Additional Answers.

6. 130 ft^2
7. 198 cm^2
8. 76 yd^2
9. 17.6 ft^2
10. 740 m^2
11. 57.1 mm^2
12. 448 $in.^2$; The surface area of the box is 448 square inches, so that is the least amount of paper needed to cover the box.
13. 136 ft^2

Practice and Problem Solving

14. 83 ft^2

15. See *Taking Math Deeper.*

16. 2 qt

17. 364 ft^2

Fair Game Review

18. 48 m^2 **19.** 165 ft^2

20. 15 $in.^2$ **21.** C

Mini-Assessment

Find the surface area of the rectangular prism.

1.

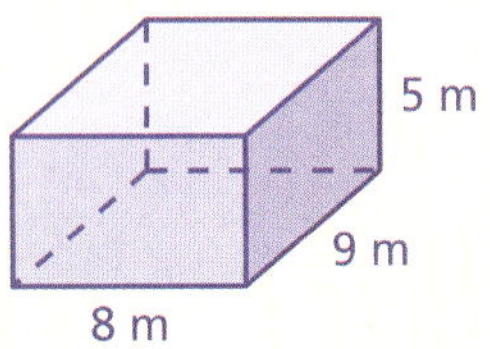

314 m^2

2.

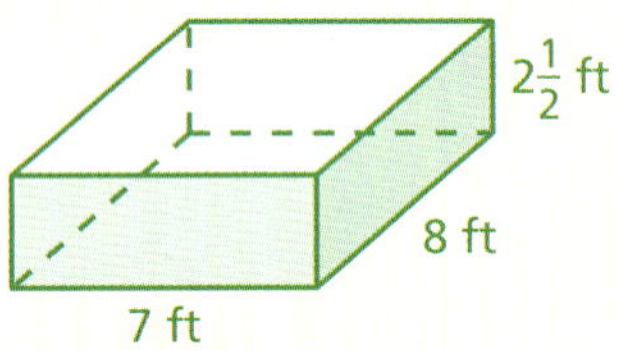

187 ft^2

Find the surface area of the triangular prism.

3.

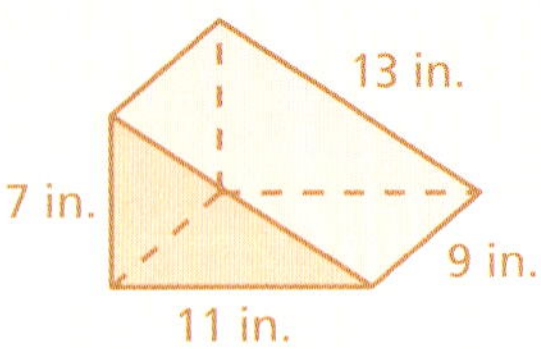

356 $in.^2$

4.

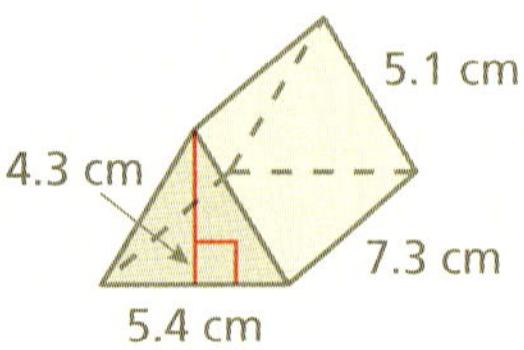

137.1 cm^2

Taking Math Deeper

Exercise 15

This is a classic type of problem in manufacturing. For a given volume, what is the least amount of material I can use? The general answer is that the more cube-like, the more efficient the use of material. For instance, a cube-like tissue box is much more cost effective than a cereal box.

1 Help me see it.
Each storage box has a volume of 480 cubic inches. However, the shapes are quite different.

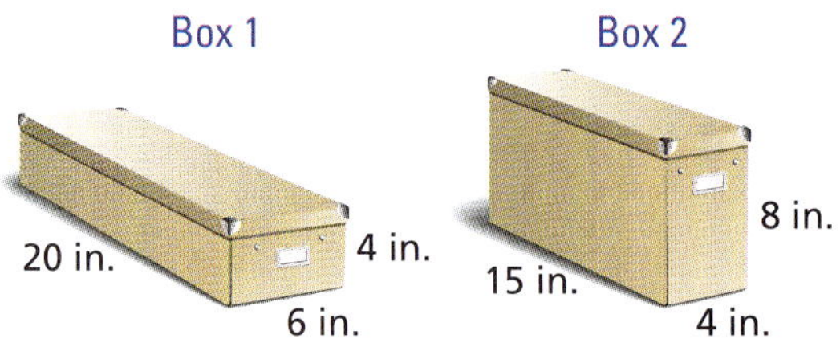

2 Find the surface area of each—in square feet.

Box 1

$2(20 \cdot 4) + 2(20 \cdot 6) + 2(4 \cdot 6) = 448 \text{ in.}^2 \approx 3.11 \text{ ft}^2$

Box 2

$2(15 \cdot 8) + 2(15 \cdot 4) + 2(8 \cdot 4) = 424 \text{ in.}^2 \approx 2.94 \text{ ft}^2$

Divide by 144 to get square feet.

3 Find the cost of each type and answer the question.

Box 1 Cost: $50(3.11 \text{ ft}^2)\left(1.25 \frac{\$}{\text{ft}^2}\right) \approx \$194.38$

Box 2 Cost: $50(2.94 \text{ ft}^2)\left(1.25 \frac{\$}{\text{ft}^2}\right) = \$183.75$

A company saves \$10.63 by using Box 2.

A more exact answer of \$10.42 can be found when the surface areas of the boxes are left in fraction form.

Project

Design a box that would have a volume of 480 cubic inches using the least possible amount of cardboard.

Reteaching and Enrichment Strategies

If students need help. . .	If students got it. . .
Resources by Chapter • Practice A and Practice B • Puzzle Time Record and Practice Journal Practice Differentiating the Lesson Lesson Tutorials Skills Review Handbook	Resources by Chapter • Enrichment and Extension • Technology Connection Start the next section

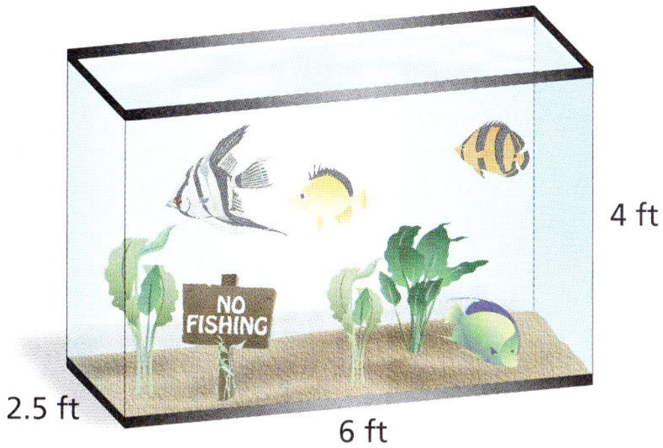

14. AQUARIUM A public library has an aquarium in the shape of a rectangular prism. The base is 6 feet by 2.5 feet. The height is 4 feet. How many square feet of glass were used to build the aquarium? (The top of the aquarium is open.)

15. STORAGE BOX The material used to make a storage box costs \$1.25 per square foot. The boxes have the same volume. How much does a company save by choosing to make 50 of Box 2 instead of 50 of Box 1?

	Length	Width	Height
Box 1	20 in.	6 in.	4 in.
Box 2	15 in.	4 in.	8 in.

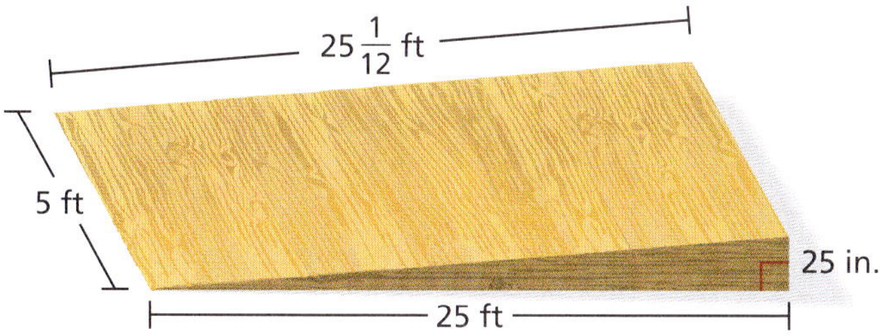

16. RAMP A quart of stain covers 100 square feet. How many quarts should you buy to stain the wheelchair ramp? (Assume you do not have to stain the bottom of the ramp.)

17. Critical Thinking A cube is removed from a rectangular prism. Find the surface area of the figure after removing the cube.

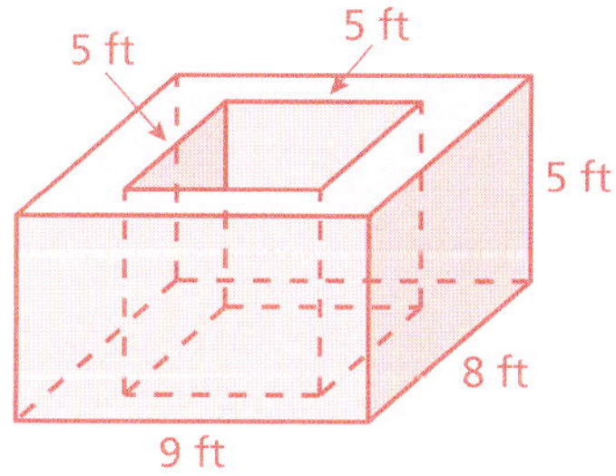

Fair Game Review What you learned in previous grades & lessons

Find the area of the triangle. *(Section 4.2)*

18.

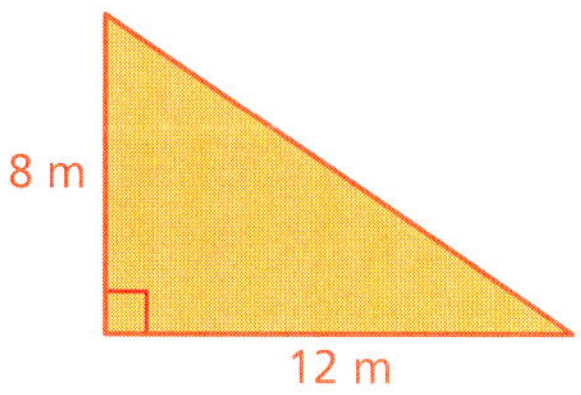

19.

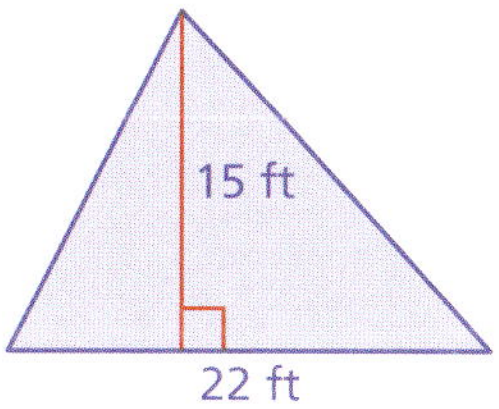

20.

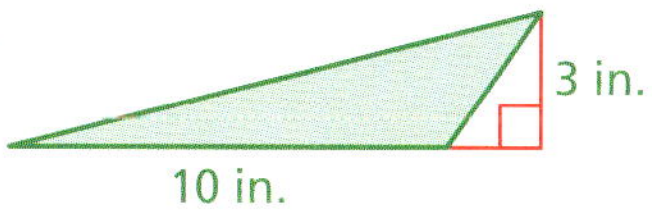

21. MULTIPLE CHOICE Which value is *not* a solution of the inequality $x - 4 \geq 2$? *(Section 7.5)*

Ⓐ $x = 10$ Ⓑ $x = 6$ Ⓒ $x = 4$ Ⓓ $x = 14$

8 Study Help

Check It Out
Graphic Organizer
BigIdeasMath.com

You can use a **process diagram** to show the steps involved in a procedure. Here is an example of a process diagram for drawing a prism.

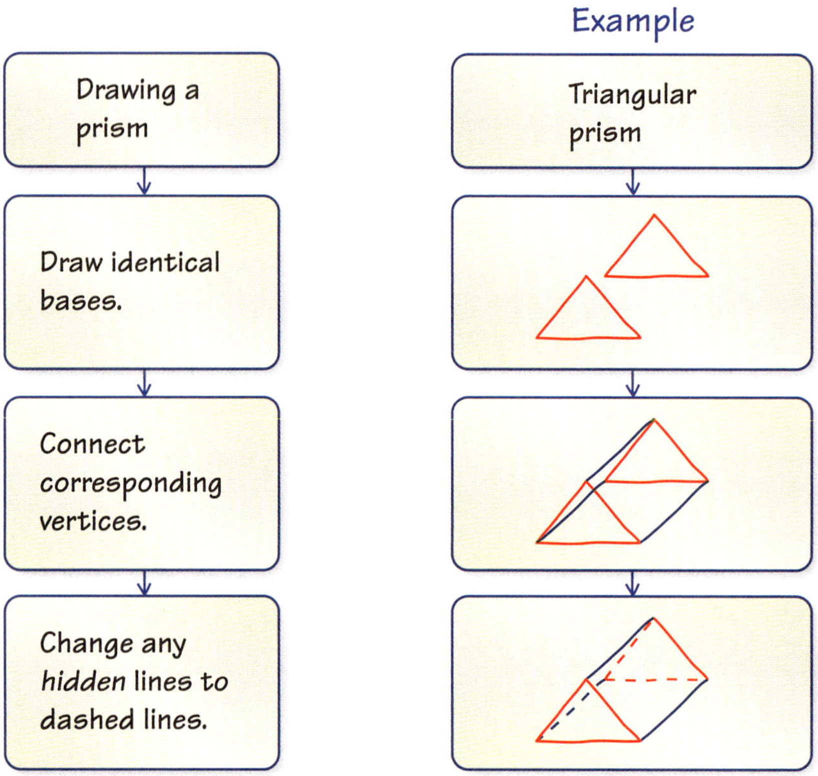

On Your Own

Make process diagrams with examples to help you study these topics.

1. drawing a pyramid
2. finding the surface area of a prism

After you complete this chapter, make process diagrams with examples for the following topics.

3. finding the surface area of a pyramid
4. finding the volume of a rectangular prism

"Descartes, you should use my process diagram when you eat your treats."

Sample Answers

1.

	Example
Drawing a pyramid	Hexagonal pyramid
Draw a base and a point.	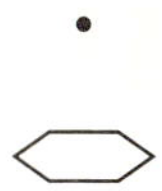
Connect the vertices of the base to the point.	
Change any hidden lines to dashed lines.	

2.

	Example
Finding the surface area of a prism	Rectangular prism 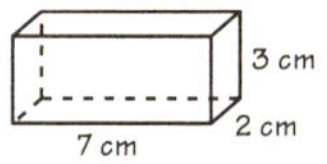
Draw a net.	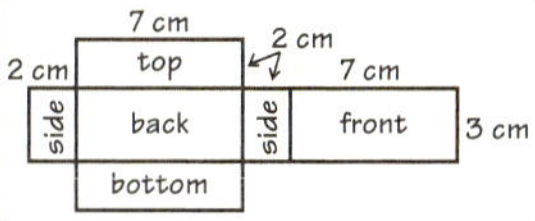
Find the areas of the faces.	Top: $7 \cdot 2 = 14$ Bottom: $7 \cdot 2 = 14$ Front: $7 \cdot 3 = 21$ Back: $7 \cdot 3 = 21$ Side: $2 \cdot 3 = 6$ Side: $2 \cdot 3 = 6$
Find the sum of the areas of the faces.	$S = 14 + 14 + 21 + 21 + 6 + 6$ $= 82 \text{ cm}^2$

List of Organizers

Available at *BigIdeasMath.com*

Comparison Chart
Concept Circle
Definition (Idea) and Example Chart
Example and Non-Example Chart
Formula Triangle
Four Square
Information Frame
Information Wheel
Notetaking Organizer
Process Diagram
Summary Triangle
Word Magnet
Y Chart

About this Organizer

A **Process Diagram** can be used to show the steps involved in a procedure. Process diagrams are particularly useful for illustrating procedures with two or more steps, and they can have one or more branches. As shown, students' process diagrams can have two parallel parts, in which the procedure is stepped out in one part and an example illustrating each step is shown in the other part. Or, the diagram can be made up of just one part, with example(s) included in the last "bubble" to illustrate the steps that precede it.

Technology for the Teacher

Editable Graphic Organizer

Answers

1. 6 faces, 12 edges, 8 vertices
2. 4 faces, 6 edges, 4 vertices
3. 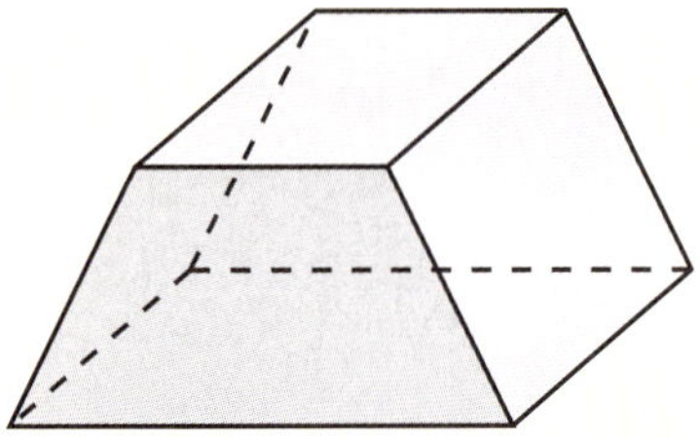
4. 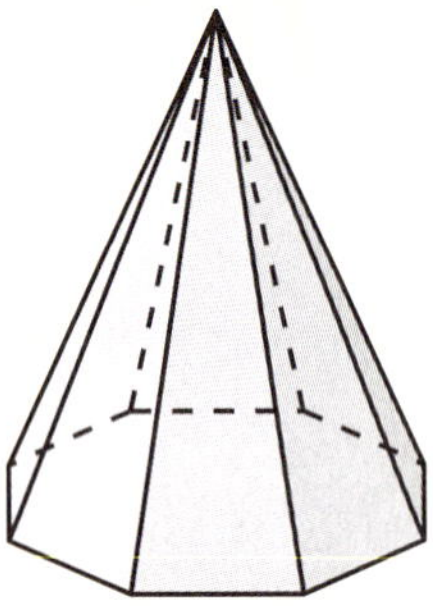
5. Front: 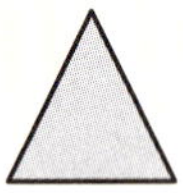 Side: 

   Top: 
6. Front:  Side: 

   Top: 
7. 132 cm^2
8. 52 $in.^2$
9. **a.** 384 $in.^2$

   b. 94 $in.^2$
10. 390 cm^2

Technology for the Teacher

Online Assessment
Assessment Book
ExamView® Assessment Suite

Alternative Quiz Ideas

100% Quiz	**Math Log**
Error Notebook	Notebook Quiz
Group Quiz	Partner Quiz
Homework Quiz	Pass the Paper

Math Log

Ask students to keep a math log for the chapter. Have them include diagrams, definitions, and examples. Everything should be clearly labeled. It might be helpful if they put the information in a chart. Students can add to the log as new topics are introduced.

Reteaching and Enrichment Strategies

If students need help. . .	If students got it. . .
Resources by Chapter • Practice A and Practice B • Puzzle Time Lesson Tutorials *BigIdeasMath.com*	Resources by Chapter • Enrichment and Extension • Technology Connection Game Closet at *BigIdeasMath.com* Start the next section

8.1–8.2 Quiz

Check It Out
Progress Check
BigIdeasMath.com

Find the number of faces, edges, and vertices of the solid. *(Section 8.1)*

1. 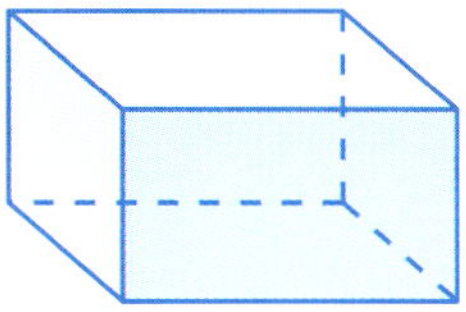

2. 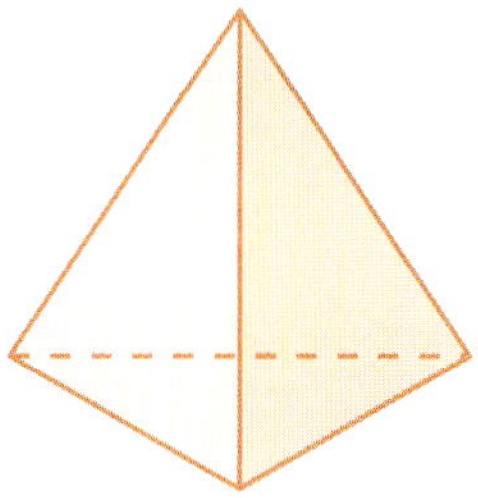

Draw the solid. *(Section 8.1)*

3. trapezoidal prism

4. octagonal pyramid

Draw the front, side, and top views of the solid. *(Section 8.1)*

5. 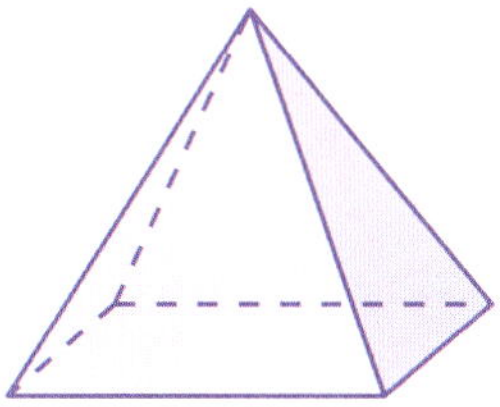

6. 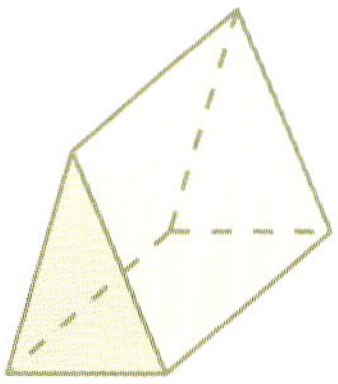

Find the surface area of the prism. *(Section 8.2)*

7. 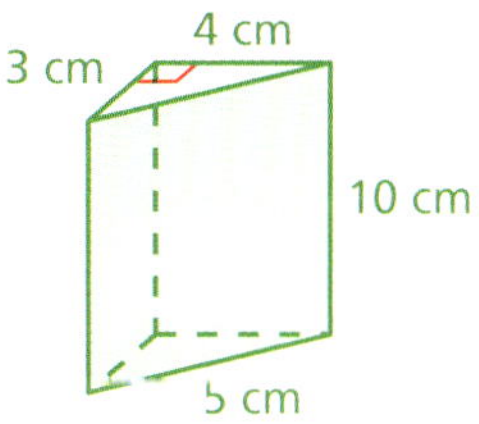

8. 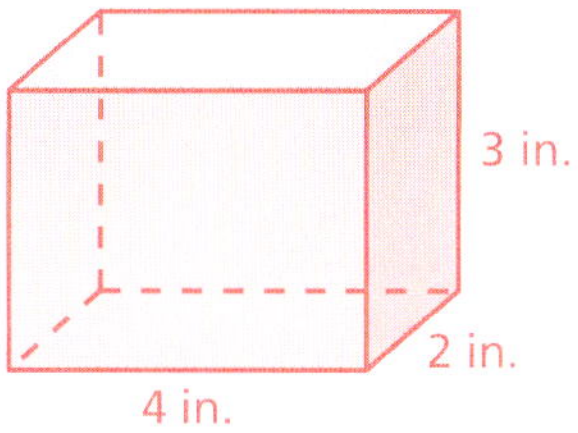

9. **CEREAL** A cereal box has the dimensions shown. *(Section 8.2)*

   a. Find the surface area of the cereal box.

   b. The manufacturer decides to decrease the size of the box by reducing each of the dimensions by 1 inch. Find the decrease in surface area.

10. **GIFT BOX** Find the surface area of the gift box. *(Section 8.2)*

8.3 Surface Areas of Pyramids

Essential Question How can you use a net to find the surface area of a pyramid?

1 ACTIVITY: Identifying Pyramids

Work with a partner. Label one of the faces as a "base" and the other as a "lateral face." Use the shape of the base to identify the pyramid.

a.

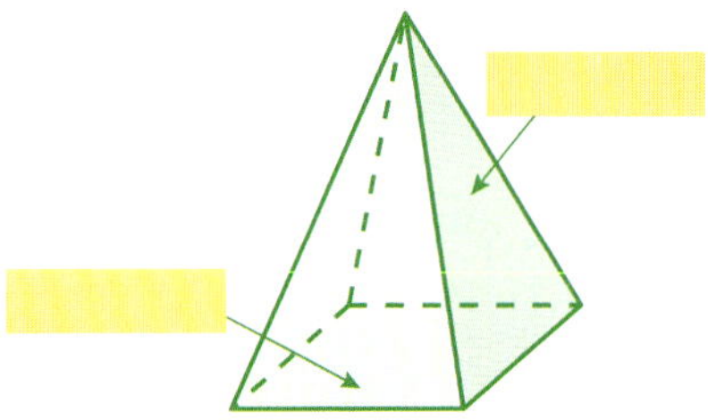

_______ **Pyramid**

b.

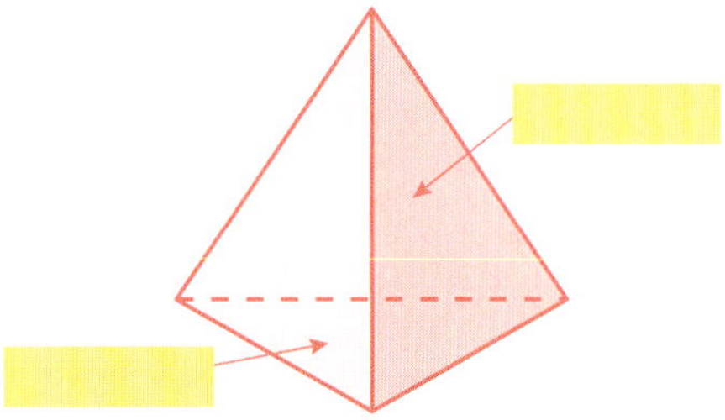

_______ **Pyramid**

2 ACTIVITY: Using a Net

Work with a partner.

a. Copy the net shown below onto grid paper.

b. Cut out the net and fold it to form a pyramid. What type of rectangle is the base? Use this shape to name the pyramid.

c. Find the surface area of the pyramid.

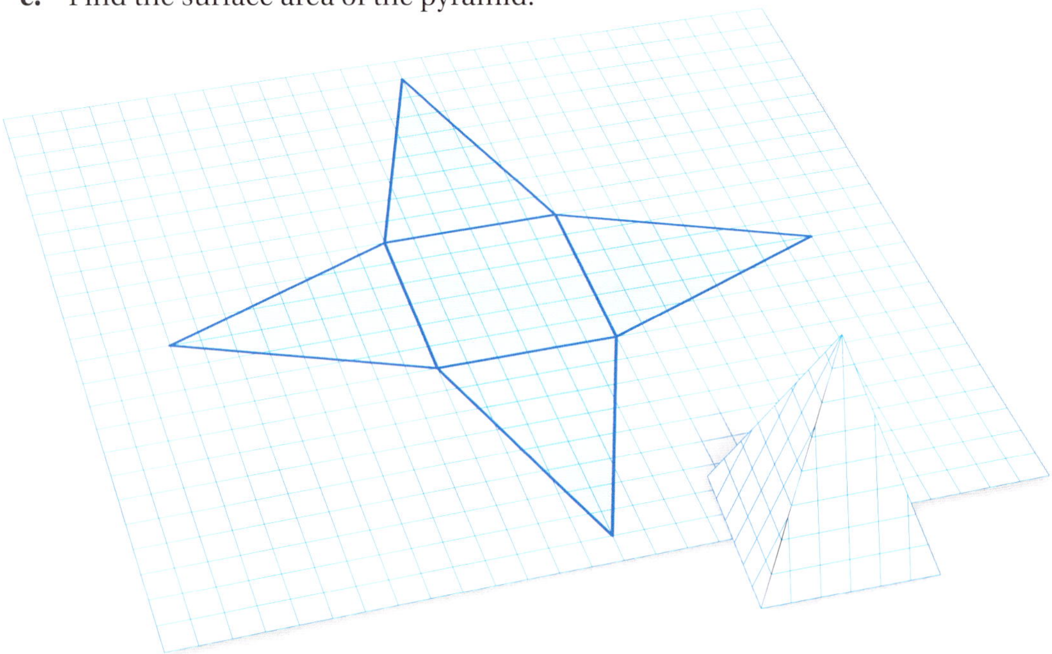

COMMON CORE

Geometry

In this lesson, you will

- use nets to represent pyramids.
- find the surface area of pyramids.
- solve real-life problems.

Learning Standard
6.G.4

Laurie's Notes

Introduction

Standards for Mathematical Practice

- **MP8 Look for and Express Regularity in Repeated Reasoning:** Students will be finding the surface areas of several pyramids. In doing so they may discover an efficient method. For instance, a pyramid with a regular base has congruent lateral faces. Finding the area of one lateral face and multiplying by the number of lateral faces is more efficient than doing multiple computations.

For Your Information

- The term *lateral face* is used in the activity. *Lateral surface area* is taught in Grade 7.
- To introduce surface area of pyramids, all pyramids in this section are right pyramids and have regular bases (squares and equilateral triangles).

Motivate

- Show a picture of the Louvre pyramid in Paris. It was designed by I.M. Pei. as the Louvre's main entrance, which is used to handle the enormous number of visitors on an everyday basis.
- I.M. Pei also designed the East Building of the National Gallery of Art in Washington, D.C., and the Rock and Roll Hall of Fame in Cleveland, Ohio.

Activity Notes

Activity 1

- Students should work with their partners to identify the base and lateral face of each pyramid.
- Like a prism, a pyramid is further identified by the type of base it has.
- Check to see that the pyramids are named correctly.

Activity 2

- **Teaching Tip:** To avoid having scraps of paper on the floor, recycle plastic bags and tape them to desks around the room. There should be a bag close enough to each group so that there is no excuse for leaving a mess.
- In order to correctly copy the figure, students will need to determine the base and height of each triangular face. In this activity, the base of the pyramid is a square meaning that all of the triangles are the same.
- **MP3 Construct Viable Arguments and Critique the Reasoning of Others:** If time permits, ask students questions about the pyramid they folded.

? "If you know the shape of the base, what do you know about the total number of faces?" total number of faces = number of sides of base + 1

? "If you know the shape of the base, what do you know about the total number of edges?" total number of edges = number of sides of base × 2

? "If you know the shape of the base, what do you know about the total number of vertices?" Total number of vertices = number of sides of base + 1

Common Core State Standards

6.G.4 Represent three-dimensional figures using nets made up of rectangles and triangles, and use the nets to find the surface area of these figures. Apply these techniques in the context of solving real-world and mathematical problems.

Previous Learning

Students should know how to find the area of a triangle and should know the general properties of squares and isosceles triangles.

Technology for the Teacher

Dynamic Classroom

Lesson Plans
Complete Materials List

8.3 Record and Practice Journal

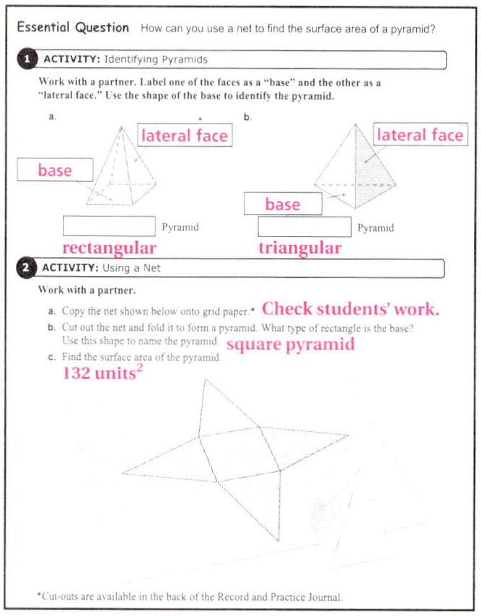

Essential Question How can you use a net to find the surface area of a pyramid?

1 ACTIVITY: Identifying Pyramids

Work with a partner. Label one of the faces as a "base" and the other as a "lateral face." Use the shape of the base to identify the pyramid.

a. lateral face; base; rectangular Pyramid

b. lateral face; base; triangular Pyramid

2 ACTIVITY: Using a Net

Work with a partner.

a. Copy the net shown below onto grid paper.* Check students' work.

b. Cut out the net and fold it to form a pyramid. What type of rectangle is the base? Use this shape to name the pyramid. square pyramid

c. Find the surface area of the pyramid. 132 units^2

*Cut-outs are available in the back of the Record and Practice Journal.

English Language Learners

Vocabulary

Have students work in pairs, one English learner and one English speaker. Have each pair write a problem involving the surface area of a pyramid. On a separate piece of paper, students should solve their own problem. Then have students exchange their problem with another pair of students. Students solve the new problem. After solving the problems, the four students discuss the problems and solutions.

8.3 Record and Practice Journal

3 ACTIVITY: Estimating the Surface Area of a Triangular Pyramid

Work with a partner. Label each face in the net of the triangular pyramid as a "base" or a "lateral face." Then estimate the surface area of the pyramid.

69 units^2

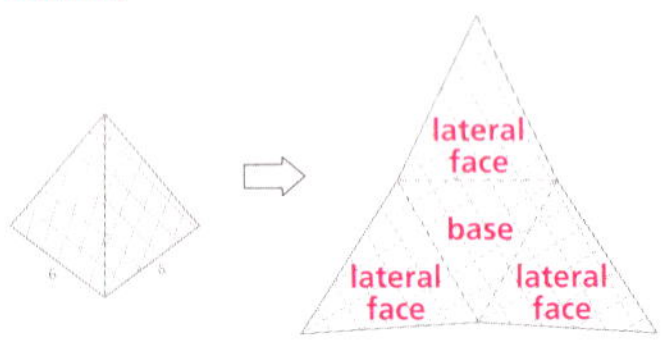

4 ACTIVITY: Finding the Surface Area of a Square Pyramid

Work with a partner. Draw a net for each square pyramid. Use the net to find the surface area of the pyramid.

a. 33 units^2

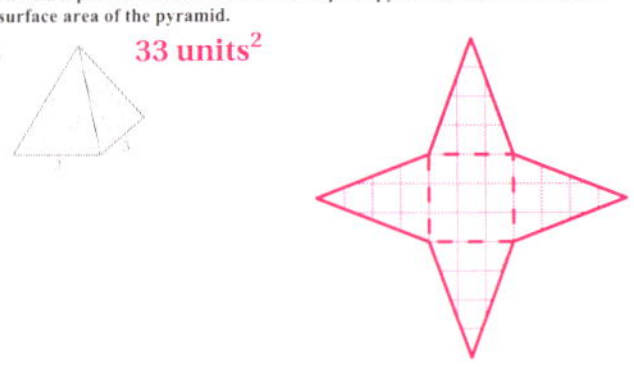

b. 75 units^2

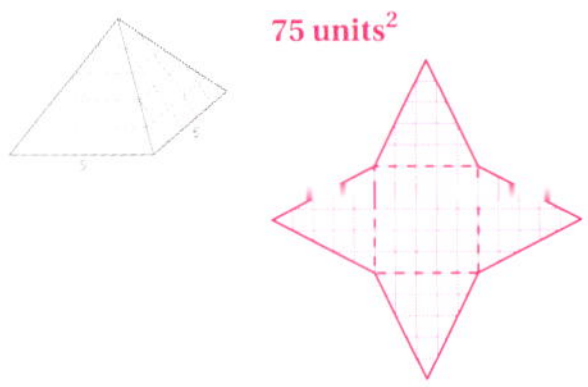

What Is Your Answer?

5. **IN YOUR OWN WORDS** How can you use a net to find the surface area of a pyramid?

Find the sum of the areas of the faces shown by the net.

6. **CONJECTURE** Make a conjecture about the lateral faces of a pyramid when the side lengths of the base have the same measure. Explain.

The faces are identical so the areas are the same.

Laurie's Notes

Activity 3

- Instead of cutting and folding up a prism as they did in Activity 2, this activity reverses that process. Students have to think about how the pyramid is unfolded to make the template shown.

? "How many faces does the pyramid have?" four "Are they all the same?" no "Describe the faces." There are 3 isosceles triangles and 1 equilateral triangle.

- The top isosceles triangle is positioned such that the height (6 units) and base (6 units) can be found easily. The side lengths of the equilateral triangle are 6 and the height is $\sqrt{27} \approx 5.2$. Because students are asked to estimate the surface area, it is okay to estimate the height as 5 units.
- **MP3** and **MP8:** When students have finished, ask volunteers to share how they found the surface area of the pyramid. There will be different strategies so expect all students to listen carefully to the process.

Activity 4

- **Scaffolding:** If time is a concern, have half of the class work on part (a) and the other half work on part (b).
- These drawings can be challenging for some students. Going from the solid to the two-dimensional representation is different than the reverse. If students are having trouble, suggest that they look back at Activity 2 for a hint as to how to sketch it.
- Although there are different two-dimensional representations (nets) that can be drawn for each pyramid, it is likely that students will draw a template similar to Activity 2. If time permits, investigate other nets. If you have snap together pieces, this is an easy way to investigate nets.

What Is Your Answer?

- Listen for a process that describes finding the sum of the areas of all of the lateral faces and the base.

Closure

- **Writing Prompt:** How is finding the surface area of a pyramid different from finding the surface area of a prism? To find the area of a pyramid you sum the area of the base and the triangular faces. To find the area of a prism you sum the area of the two bases and the rectangular lateral faces.

3 ACTIVITY: Estimating the Surface Area of a Triangular Pyramid

Work with a partner. Label each face in the net of the triangular pyramid as a "base" or a "lateral face." Then estimate the surface area of the pyramid.

Math Practice 1

Analyze Givens

What information is given in the diagram? How does this help you estimate the surface area of the pyramid?

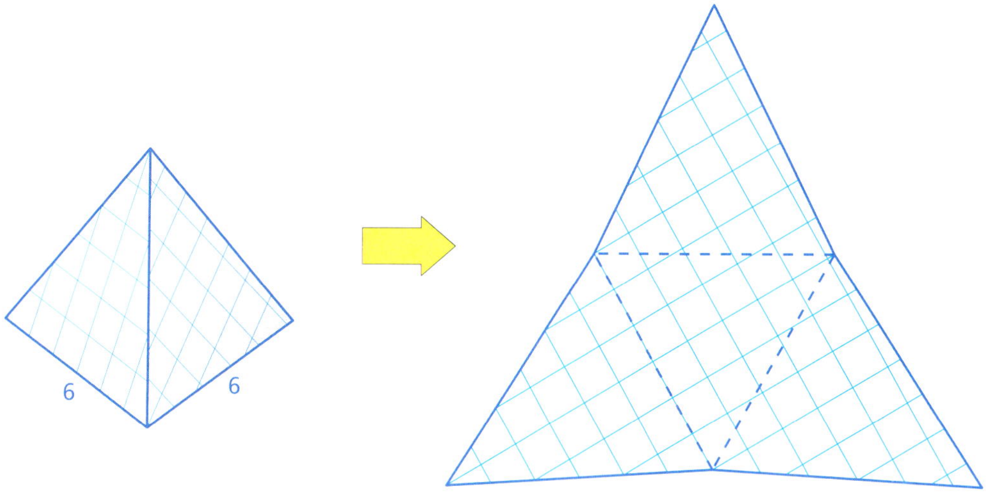

4 ACTIVITY: Finding the Surface Area of a Square Pyramid

Work with a partner. Draw a net for each square pyramid. Use the net to find the surface area of the pyramid.

a.

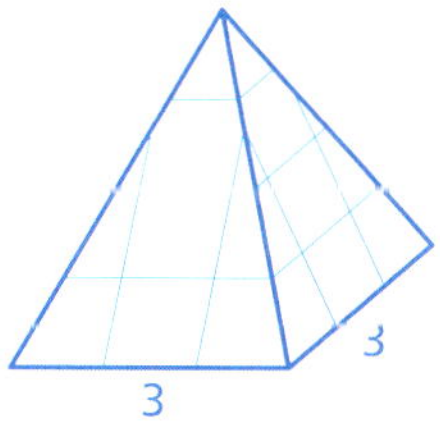

b.

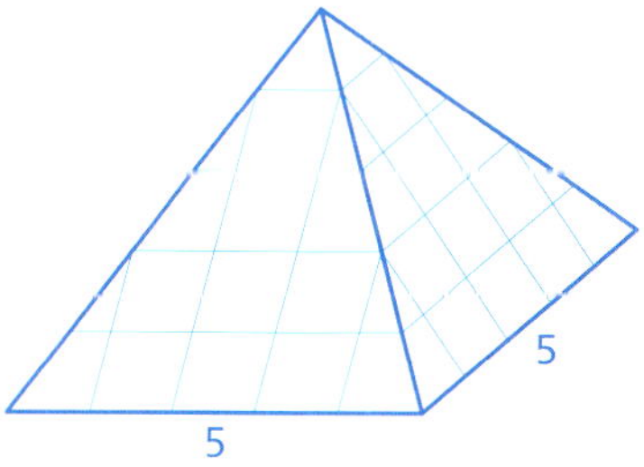

What Is Your Answer?

5. **IN YOUR OWN WORDS** How can you use a net to find the surface area of a pyramid?

6. **CONJECTURE** Make a conjecture about the lateral faces of a pyramid when the side lengths of the base have the same measure. Explain.

Practice Use what you learned about the surface area of a pyramid to complete Exercises 3–5 on page 372.

Check It Out
Lesson Tutorials
BigIdeasMath.com

Key Idea

Net of a Square Pyramid

A *square pyramid* is a pyramid with a square base.

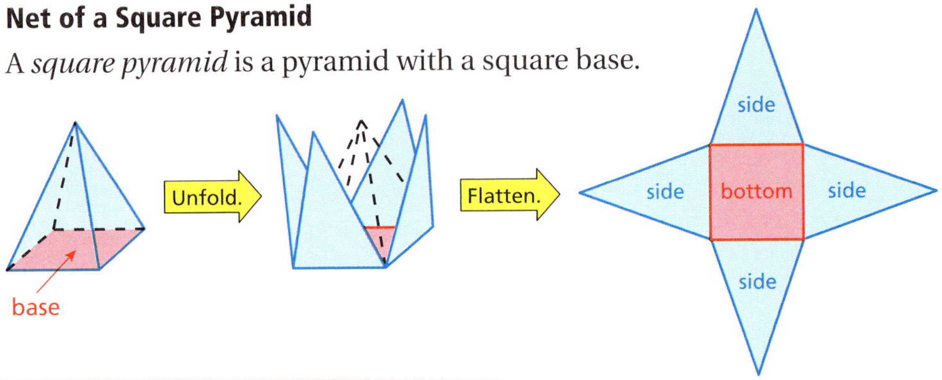

EXAMPLE 1 Finding the Surface Area of a Square Pyramid

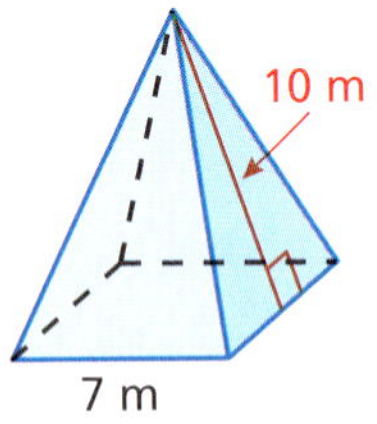

Find the surface area of the square pyramid.

Use a net to find the area of each face.

Bottom: $7 \cdot 7 = 49$

Side: $\frac{1}{2} \cdot 7 \cdot 10 = 35$

Side: $\frac{1}{2} \cdot 7 \cdot 10 = 35$

Side: $\frac{1}{2} \cdot 7 \cdot 10 = 35$

Side: $\frac{1}{2} \cdot 7 \cdot 10 = 35$

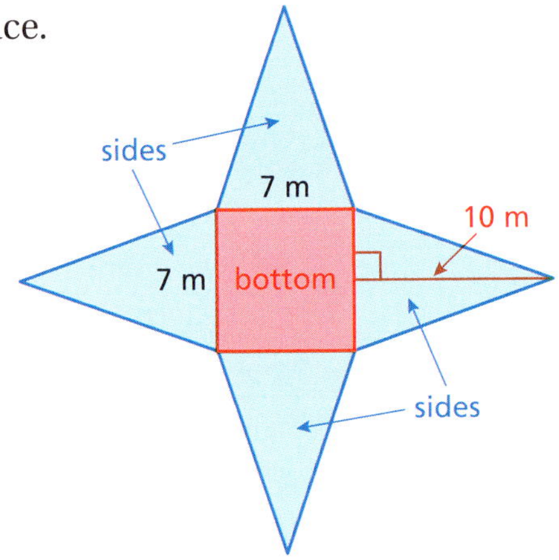

Find the sum of the areas of the faces.

Surface Area	=	Area of bottom	+	Area of a side	+	Area of a side	+	Area of a side	+	Area of a side	
S	=	49	+	35	+	35	+	35	+	35	= 189

So, the surface area is 189 square meters.

On Your Own

Now You're Ready
Exercises 6–8

Find the surface area of the square pyramid.

1. 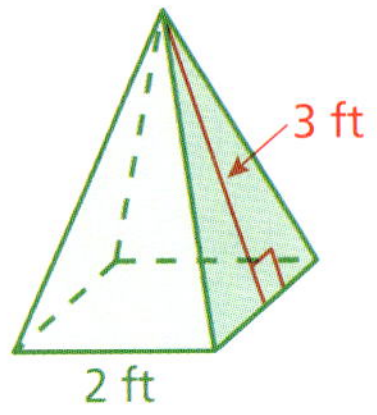

2. 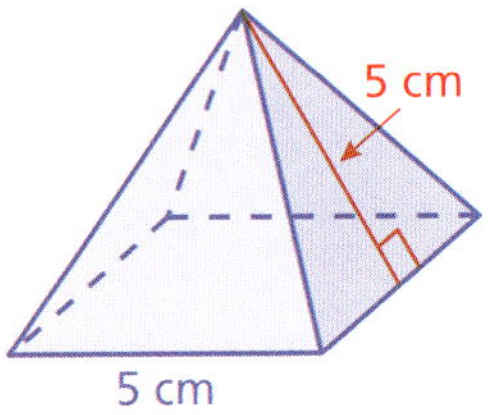

3. 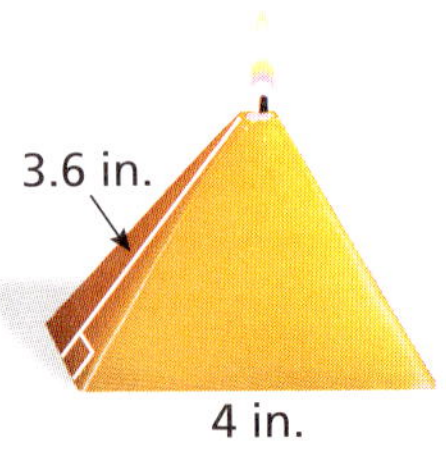

Laurie's Notes

Introduction

Connect

- **Yesterday:** Students worked with two-dimensional representations (nets) of pyramids and found the area of each face. (MP3, MP8)
- **Today:** Students will use a net to find the surface area of a pyramid.

Motivate

- Ask students where they have heard about pyramids or have seen them before. Give groups of students 3–4 minutes to brainstorm a list. They may mention the pyramid on the back of U.S. dollar bills, camping tents, roof designs, tetrahedral dice, and of course, Egyptian pyramids.

Lesson Notes

Key Idea

- Having a physical model of a square pyramid to refer to will be helpful. Use a paper model or commercially made model if you have one.
- Note that the base and lateral faces are labeled as *bottom* and *side* when the pyramid is unfolded as a net.

Example 1

- Draw the net and label the known information.
- ? "What do we need to do in order to find the surface area of this square pyramid?" Find the area of the base and each of the lateral faces.
- Write the *formula* for surface area in words as shown.
- ? "What is the area of the bottom (base)?" 49 square meters
- ? "What is the area of each side (lateral face)?" 35 square meters
- ? **Common Error:** In using the area formula for a triangle, the $\frac{1}{2}$ often produces a computation mistake. In this instance, students must multiply $\frac{1}{2} \times 7 \times 10$. Remind students that it's okay to change the order of the factors (Commutative Property). Rewriting the problem as $\frac{1}{2} \times 10 \times 7$ means that you can work with whole numbers: $\frac{1}{2} \times 10 \times 7 = 5 \times 7 = 35$.
- **MP8 Look for and Express Regularity in Repeated Reasoning:** Students may ask about the repeated steps of finding the area of each lateral face. Ask them to conjecture about an alternate way to find the surface area. Listen for (surface area) = (area of bottom) + 4 × (area of a side).

On Your Own

- Each of the problems is a square pyramid. Remind students to find the area of the square base, followed by the areas of the four triangles.
- Check to see that students label their answers the with correct units.

Goal Today's lesson is finding the surface areas of pyramids using nets.

Technology for the Teacher

Dynamic Classroom

Lesson Tutorials
Lesson Plans
Answer Presentation Tool

Extra Example 1

Find the surface area of the square pyramid.

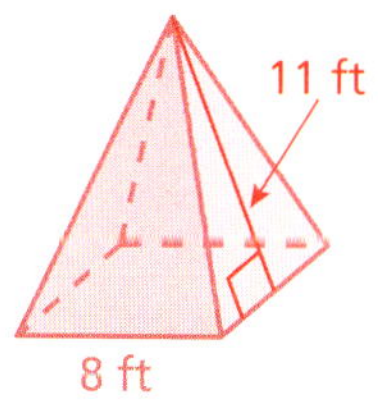

240 ft^2

On Your Own

1. 16 ft^2
2. 75 cm^2
3. 44.8 $in.^2$

Differentiated Instruction

Kinesthetic

Photocopy nets of solids for students to cut out and assemble. Then have students draw their own nets to cut out and assemble.

Extra Example 2

Find the surface area of the triangular pyramid.

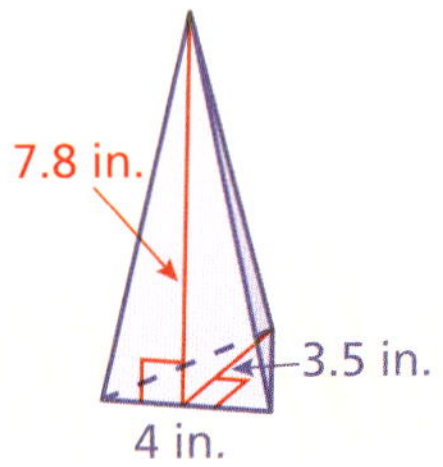

53.8 in.^2

On Your Own

4. 10.7 cm^2
5. 129.6 in.^2
6. 533 yd^2

Laurie's Notes

Key Ideas

- Write the Key Idea.
- Having a physical model of a triangular pyramid to refer to will be helpful. Use a paper model or commercially made model if you have one.
- As stated in the Study Tip, all of the triangular pyramids in this section will have an equilateral triangle for a base. The base and height of the triangular base and the lateral faces will be given.

Example 2

- Drawing the net is an important step. It allows the key dimensions to be labeled in a way that can be seen.
- Ask students to estimate the area of the base before computing. $\left(\frac{1}{2} \times 6 \times 5 = 15\right)$.
- **MP8:** Students may ask about the repeated steps of finding the area of each lateral face. Ask if their conjecture from Example 1 still holds. Listen for reasoning about the number of lateral faces being different so (surface area) = (area of bottom) + 3 × (area of a side).

On Your Own

- Each of the problems is a triangular pyramid. Remind students to find the area of the triangular base, followed by the areas of the three side triangles.
- Ask volunteers to share their work at the board.
- **Common Error:** Working with decimals may cause errors. Encourage students to estimate their answer before multiplying.
- Check to see that students label their answers with correct units.

Closure

- **Exit Ticket:** Sketch a square pyramid with a slant height of 4 centimeters and a base side length of 3 centimeters. Sketch the net and find the surface area. 33 cm^2

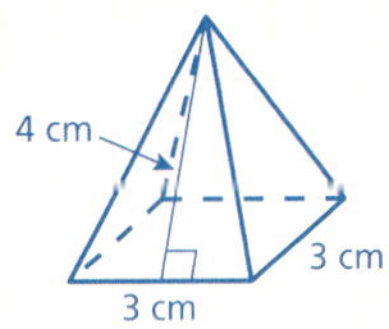

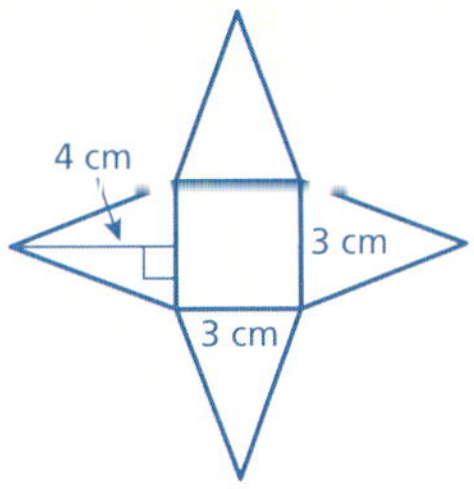

Key Idea

Net of a Triangular Pyramid

A *triangular pyramid* is a pyramid with a triangular base.

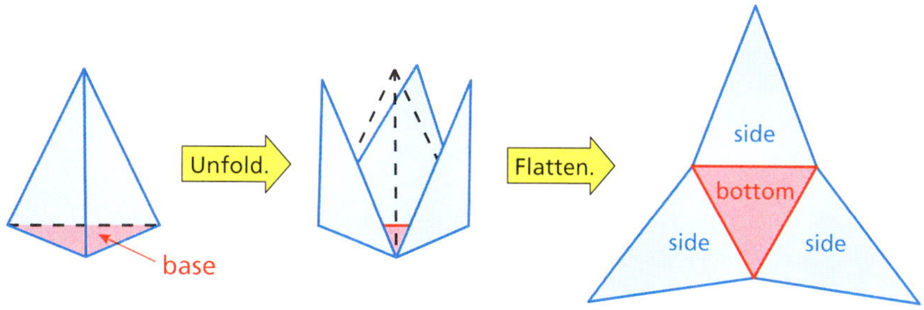

Study Tip

The base of each triangular pyramid in this section is an equilateral triangle.

EXAMPLE 2 Finding the Surface Area of a Triangular Pyramid

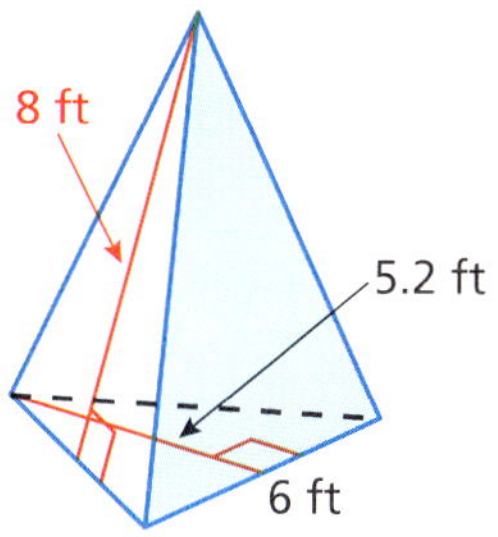

Find the surface area of the triangular pyramid.

Use a net to find the area of each face.

Bottom: $\frac{1}{2} \cdot 6 \cdot 5.2 = 15.6$

Side: $\frac{1}{2} \cdot 6 \cdot 8 = 24$

Side: $\frac{1}{2} \cdot 6 \cdot 8 = 24$

Side: $\frac{1}{2} \cdot 6 \cdot 8 = 24$

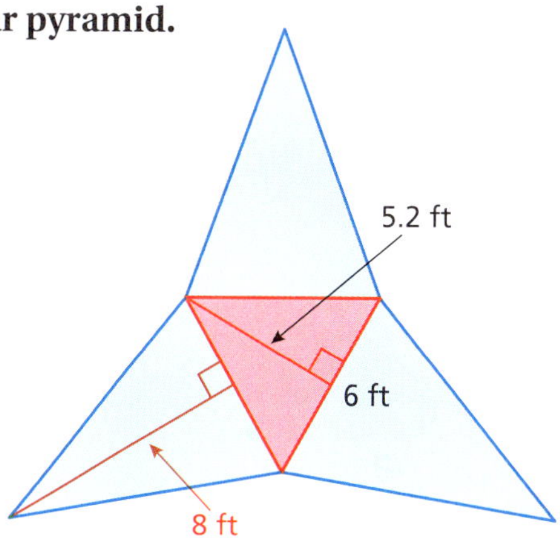

Find the sum of the areas of the faces.

Surface Area	=	Area of bottom	+	Area of a side	+	Area of a side	+	Area of a side
S	=	15.6	+	24	+	24	+	24
	= 87.6							

So, the surface area is 87.6 square feet.

On Your Own

Now You're Ready
Exercises 9–11

Find the surface area of the triangular pyramid.

4. 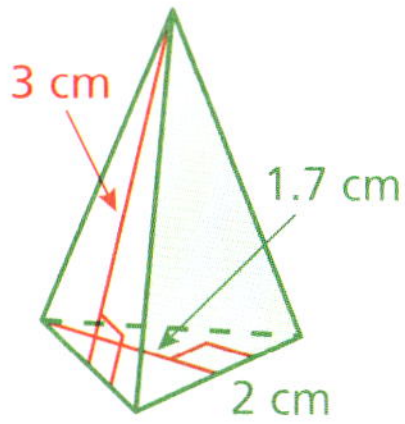

5. 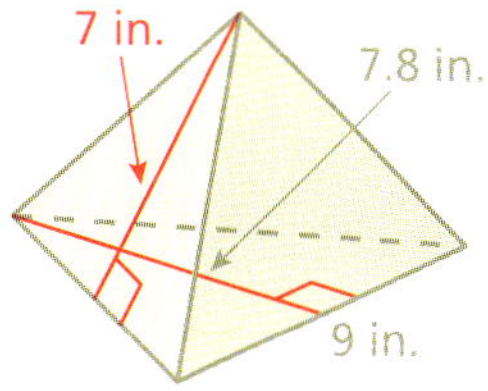

6. 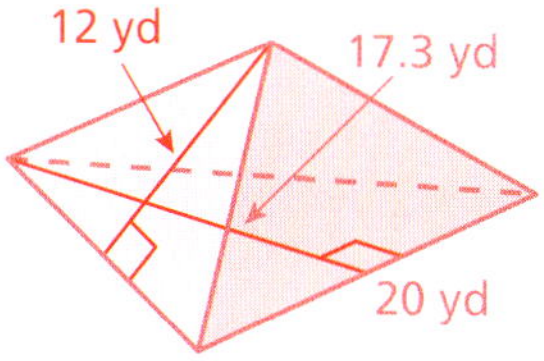

8.3 Exercises

Check It Out
Help with Homework
BigIdeasMath.com

Vocabulary and Concept Check

1. **PRECISION** Explain how to find the surface area of a pyramid.

2. **WHICH ONE DOESN'T BELONG?** Which figure does *not* belong with the other three? Explain your reasoning.

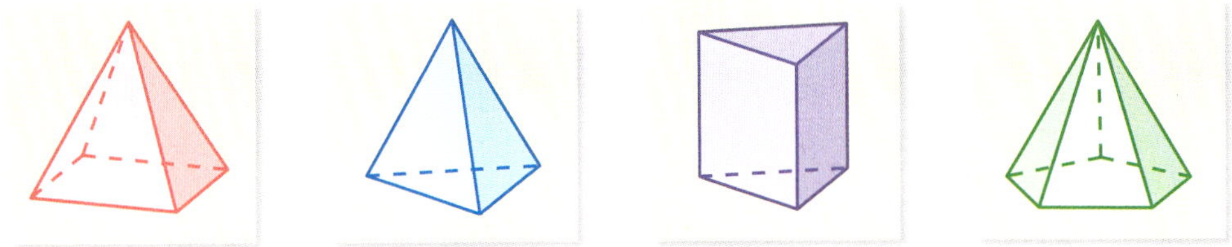

Practice and Problem Solving

Draw a net of the square pyramid. Then find the surface area of the pyramid.

3. 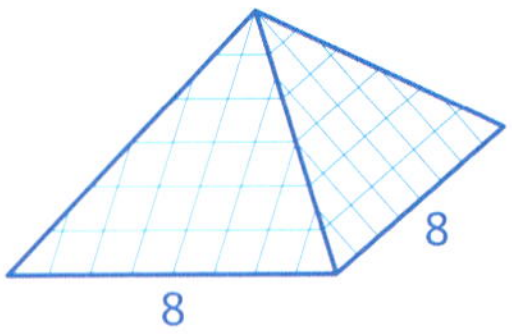

4. 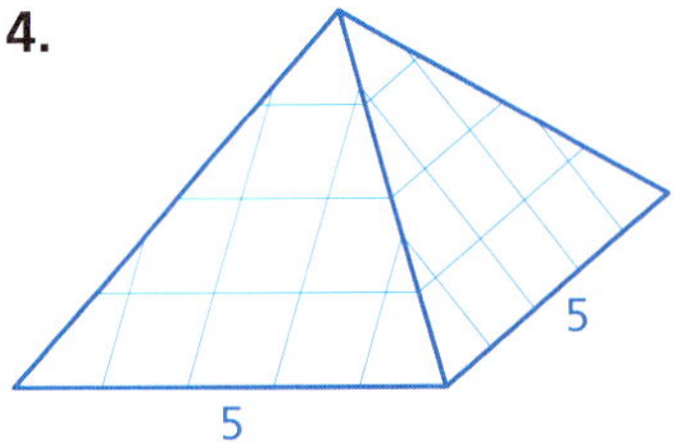

5. 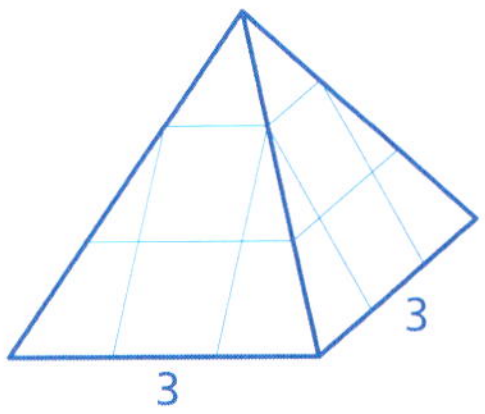

Find the surface area of the pyramid. The side lengths of the base are equal.

1 6. 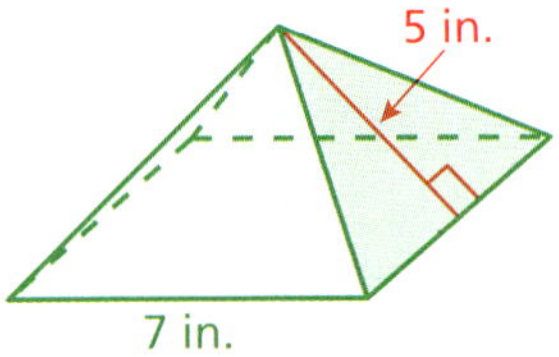

7. 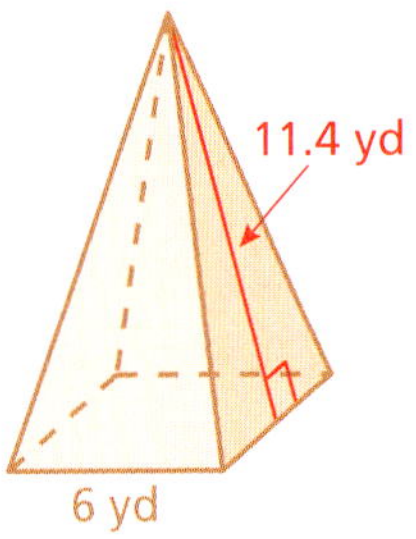

8. 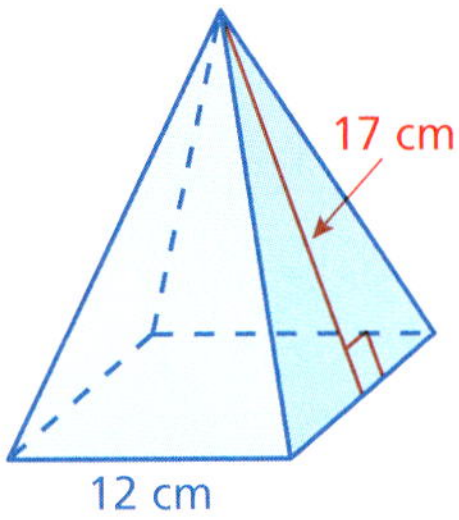

2 9. 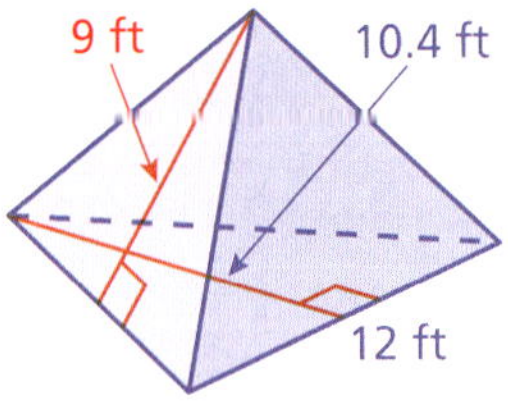

10. 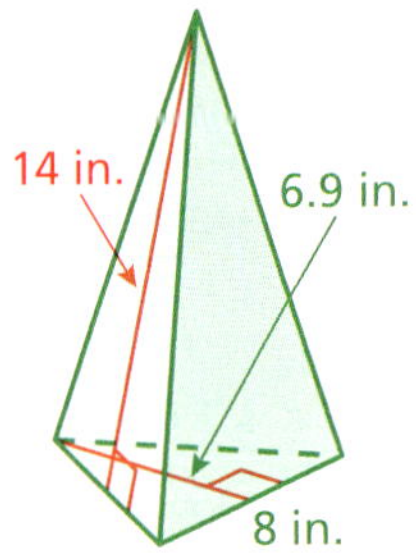

11. 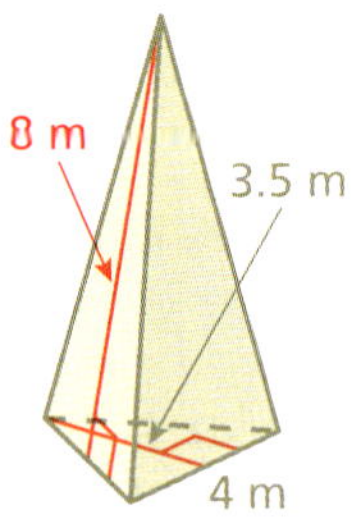

12. **PAPERWEIGHT** A paperweight is shaped like a triangular pyramid. The base is an equilateral triangle. Find the surface area of the paperweight.

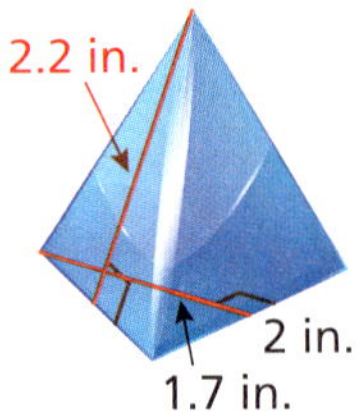

Assignment Guide and Homework Check

Level	Day 1 Activity Assignment	Day 2 Lesson Assignment	Homework Check
Basic	3–5, 18–20	1, 2, 7–15 odd	7, 11, 13, 15
Average	3–5, 18–20	1, 2, 6–12 even, 13–16	8, 10, 12, 14
Advanced	3–5, 18–20	1, 2, 6–17	8, 10, 12, 15, 16

Common Errors

- **Exercises 6–11** Students may forget to add on the area of the base when finding the surface area. Remind them that when asked to find the surface area, the base is included.
- **Exercises 6–11** Students may add the wrong number of lateral face areas to the area of the base. Examine several different pyramids with different bases and ask if they can find a relationship between the number of sides of the base and the number of lateral faces. (They are the same.) Remind students that the number of sides on the base determines how many triangles make up the lateral surface area.
- **Exercises 9–11** Students may forget to multiply by $\frac{1}{2}$ when finding the area of the base triangle. Remind them that the formula for the area of a triangle is $\frac{1}{2}bh$.

8.3 Record and Practice Journal

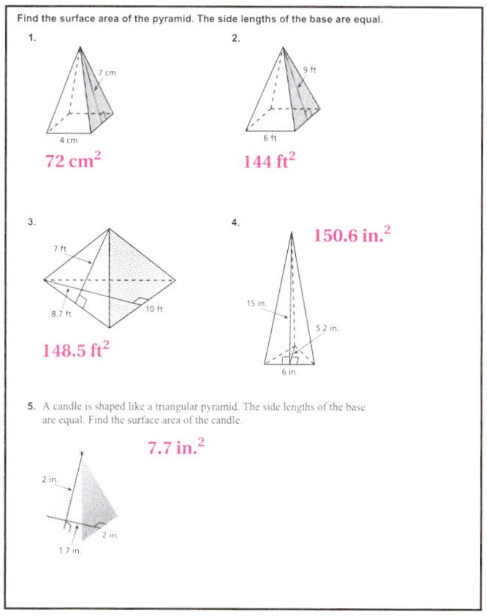
Find the surface area of the pyramid. The side lengths of the base are equal.

1. 72 cm²
2. 144 ft²
3. 148.5 ft²
4. 150.6 in.²
5. A candle is shaped like a triangular pyramid. The side lengths of the base are equal. Find the surface area of the candle.
7.7 in.²

Vocabulary and Concept Check

1. Find the sum of the areas of the faces.
2. 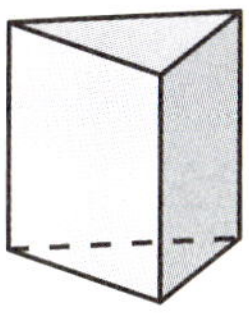
This figure is a prism and the other three figures are pyramids.

Practice and Problem Solving

3–5. See Additional Answers.

6. 119 in.^{2}
7. 172.8 yd^2
8. 552 cm^2
9. 224.4 ft^2
10. 195.6 in.^{2}
11. 55 m^2
12. 8.3 in.^{2}
13. 21,274.4 ft^2
14. yes; The weight of the glass is 29.4 pounds, which is less than the limit of 35 pounds for the chain.
15. 4
16. See *Taking Math Deeper*.
17. no; You can place the four triangles on top of the square and it covers the entire square. But when you lift up the triangles, they do not touch. So, they do not form a pyramid.

Fair Game Review

18.

Frogs	7	14	28
Turtles	3	6	12

7 : 3, 14 : 6, 28 : 12

19.

Apples	10	5	30
Oranges	4	2	12

10 : 4, 5 : 2, 30 : 12

20. B

Mini-Assessment

Find the surface area of the square pyramid.

1.

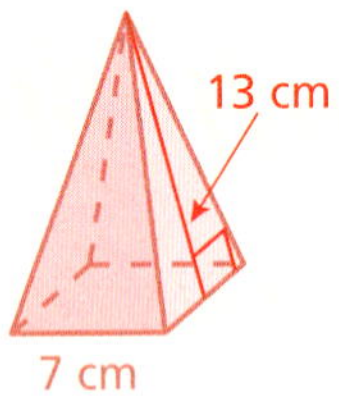

231 cm^2

2.

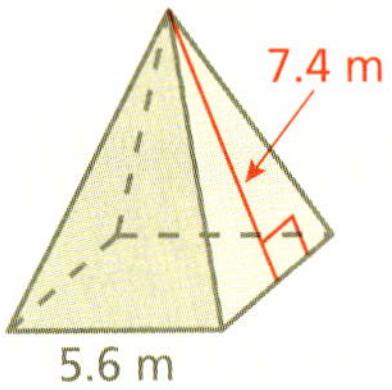

114.24 m^2

Find the surface area of the triangular pyramid.

3.

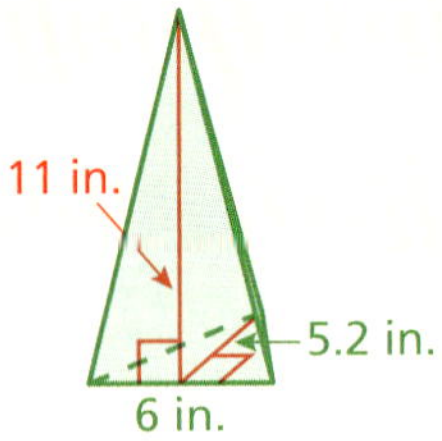

114.6 in.^{2}

4.

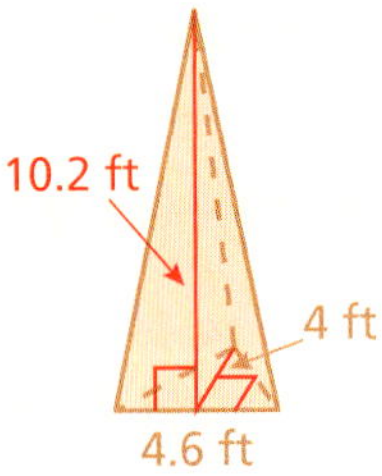

79.58 ft^2

Taking Math Deeper

Exercise 16

One way to solve this problem is to rearrange the base and the lateral faces into different shapes.

1. Find the area of the base by rearranging the triangles to form a parallelogram.

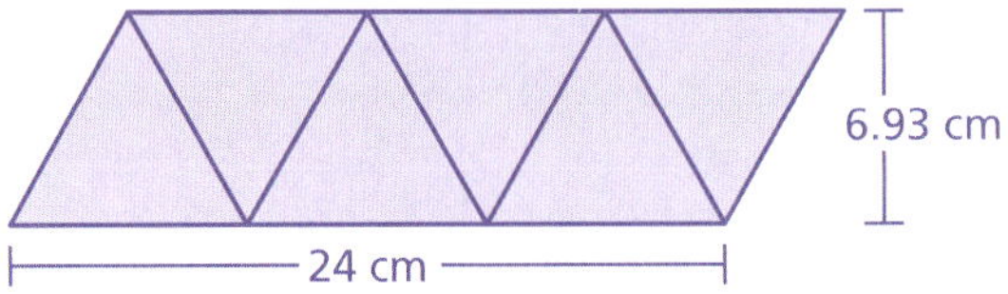

$$A = bh = 24(6.93) = 166.32 \text{ cm}^2$$

2. Find the area of the lateral faces by rearranging them to form a parallelogram.

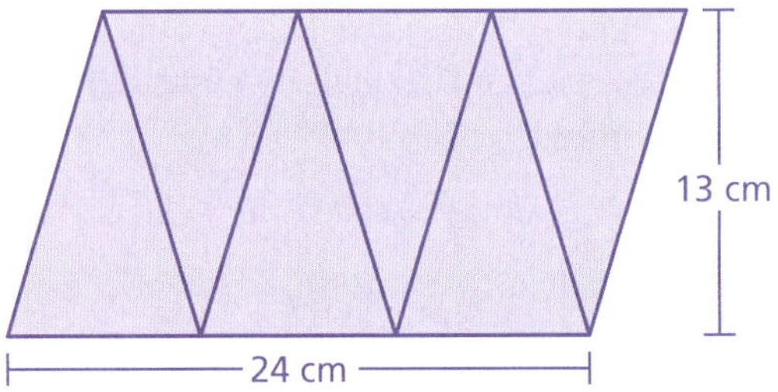

$$A = bh = 24(13) = 312 \text{ cm}^2$$

3. So, the surface area of the hexagonal pyramid is $166.32 + 312 = 478.32$ square centimeters.

Project

Use grid paper to draw a net of the pyramid in Exercise 16. Cut out the net and fold it to form the pyramid.

Reteaching and Enrichment Strategies

If students need help. . .	If students got it. . .
Resources by Chapter • Practice A and Practice B • Puzzle Time Record and Practice Journal Practice Differentiating the Lesson Lesson Tutorials Skills Review Handbook	Resources by Chapter • Enrichment and Extension • Technology Connection Start the next section

13. **LOUVRE** The entrance to the Louvre Museum in Paris, France, is a square pyramid. The side length of the base is 116 feet, and the height of one of the triangular faces is 91.7 feet. Find the surface area of the four triangular faces of the entrance to the Louvre Museum.

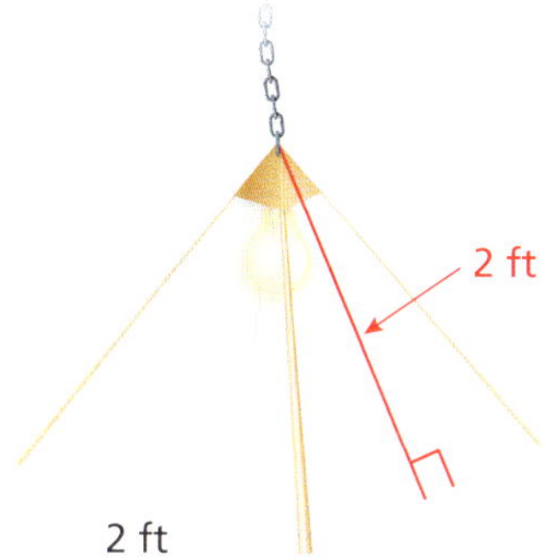

14. **LIGHT COVER** A hanging light cover made of glass is shaped like a square pyramid. The cover does not have a bottom. One square foot of the glass weighs 2.45 pounds. The chain can support 35 pounds. Will the chain support the light cover? Explain.

15. **GEOMETRY** The surface area of a square pyramid is 84 square inches. The side length of the base is 6 inches. What is the value of x?

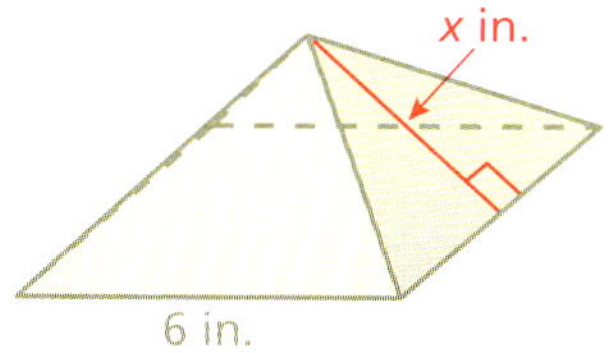

16. **STRUCTURE** In the diagram of the base of the hexagonal pyramid, all the triangles are the same. Find the surface area of the hexagonal pyramid.

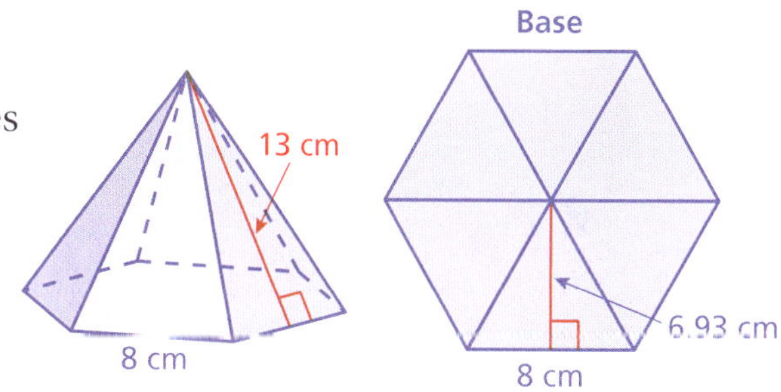

17. **Critical Thinking** Can you form a square pyramid using four of the triangles shown? Explain your reasoning.

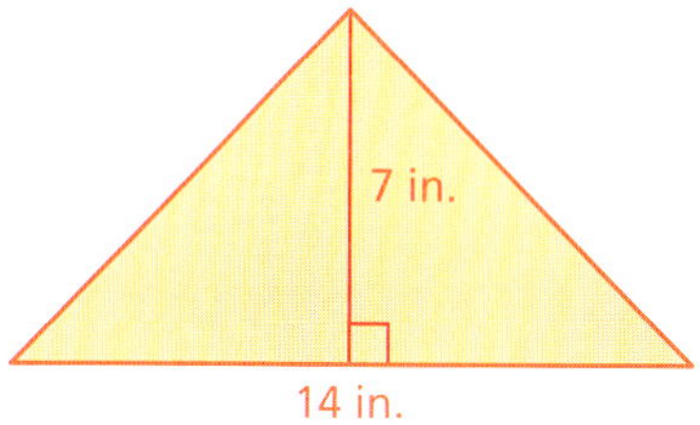

Fair Game Review What you learned in previous grades & lessons

Find the missing values in the ratio table. Then write the equivalent ratios. *(Section 5.2)*

18.

Frogs	7		28
Turtles	3	6	

19.

Apples	10	5	
Oranges	4		12

20. **MULTIPLE CHOICE** Which ordered pair is in Quadrant III? *(Section 6.5)*

Ⓐ (5, −1) Ⓑ (−2, −3) Ⓒ (2, 4) Ⓓ (−7, 1)

8.4 Volumes of Rectangular Prisms

Essential Question How can you find the volume of a rectangular prism with fractional edge lengths?

Recall that the **volume** of a three-dimensional figure is a measure of the amount of space that it occupies. Volume is measured in *cubic units*.

A *unit cube* is a cube with an edge length of 1 unit.

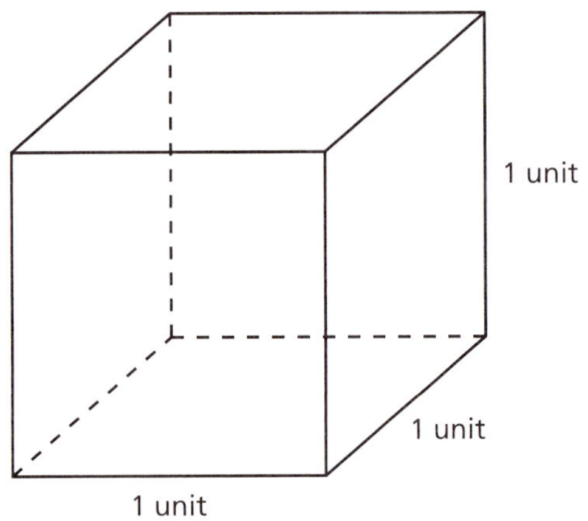

1 ACTIVITY: Using a Unit Cube

Work with a partner. The parallel edges of the unit cube have been divided into 2, 3, and 4 equal parts to create smaller rectangular prisms that are identical.

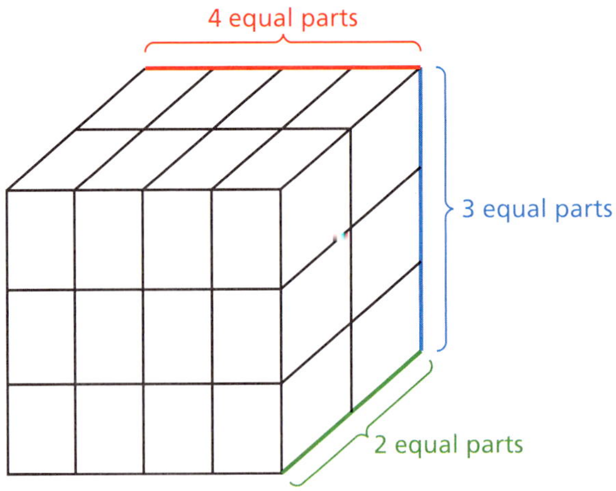

a. Draw one of these identical prisms and label its dimensions.

b. What fraction of the volume of the unit cube does one of these identical prisms represent? Use this value to find the volume of one of the identical prisms. Explain your reasoning.

COMMON CORE

Geometry

In this lesson, you will

- find the volume of prisms with fractional edge lengths by using models.
- find the volume of prisms by using formulas.

Learning Standard 6.G.2

Laurie's Notes

Introduction

Standards for Mathematical Practice

- **MP3 Construct Viable Arguments and Critique the Reasoning of Others:** The essential question for this lesson is how to find the volume of a rectangular prism with fractional dimensions. Students will work through activities that will help them to make a connection to the volume formula they studied in Grade 5. As they make conjectures, students should be expected to listen carefully and critique the reasoning of their classmates.

Motivate

- Hold up a variety of common containers and ask what is commonly found inside. Examples: egg carton (12 eggs); playing cards box (52 cards); crayon box (8 crayons)
- Discuss with students these examples of volume. Each container is filled with objects of the same size. How many eggs fit in the egg carton, or how many crayons fit in the crayon box? Because the units are different (eggs, cards, crayons), you can't compare the volumes.

Activity Notes

Activity 1

- **FYI:** In Grade, 5 students developed the volume formula $V = \ell wh$ for rectangular prisms with whole number dimensions. This activity focuses on finding the volume of a rectangular prism with fractional edge lengths.
- **Teaching Tip:** Make a unit cube out of poster board or construction paper. It's important that it not look like a cubic foot or cubic inch so that students are simply thinking of a unit cube as having all edges equal in length.
- ? Hold the unit cube and ask, "What name would you give this solid?" expect answers such as prism, cube, box
- ? "What is special about this prism?" All faces are the same size and shape. All edges have the same length.
- Say, "This is a unit cube. It is a prism with six square faces. In this first activity, you are going to divide up this cube. If helpful, think of it as a large chunk of cheese, and you're able to slice down through it!"
- When students have a sketch of part (a), ask a volunteer to share it with the class. You want to be sure that all students have the dimensions labeled correctly. In completing part (b), students will need to know how many identical prisms there are. You might suggest that they think about the layers, and how many prisms are in each layer.
- **MP3:** Ask volunteers to share their reasoning in part (b). Most students will think about the 24 identical prisms, so each must have a volume of $\frac{1}{24}$ cubic unit. Some students may think about the dimensions and use the formula they developed in Grade 5. They will also have a volume of $\frac{1}{24}$ cubic unit.
- Discuss the meaning of cubic units with students.

Common Core State Standards

6.G.2 Find the volume of a right rectangular prism with fractional edge lengths by packing it with unit cubes of the appropriate unit fraction edge lengths, and show that the volume is the same as would be found by multiplying the edge lengths of the prism. Apply the formulas $V = \ell wh$ and $V = bh$ to find volumes of right rectangular prisms with fractional edge lengths in the context of solving real-world and mathematical problems.

Previous Learning

Students should know how to distinguish between perimeter, area, and volume, and how to determine the volume of a rectangular solid using cubes or pictorial representations.

Technology for the Teacher

Dynamic Classroom

Lesson Plans
Complete Materials List

8.4 Record and Practice Journal

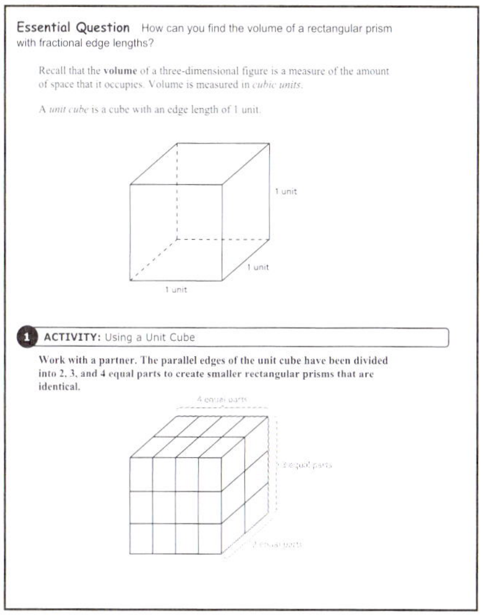

Essential Question How can you find the volume of a rectangular prism with fractional edge lengths?

Recall that the **volume** of a three-dimensional figure is a measure of the amount of space that it occupies. Volume is measured in *cubic units*.

A *unit cube* is a cube with an edge length of 1 unit.

1 unit
1 unit
1 unit

1 ACTIVITY: Using a Unit Cube

Work with a partner. The parallel edges of the unit cube have been divided into 2, 3, and 4 equal parts to create smaller rectangular prisms that are identical.

English Language Learners

Forming Answers

Encourage English learners to form complete sentences in their responses. Students can use the question to help them form the answer.

Question: If you know the height and the width of a prism, what else do you need to know to find the volume?

Response: If you know the height and width of a prism, you need to know the length of the prism to find the volume.

8.4 Record and Practice Journal

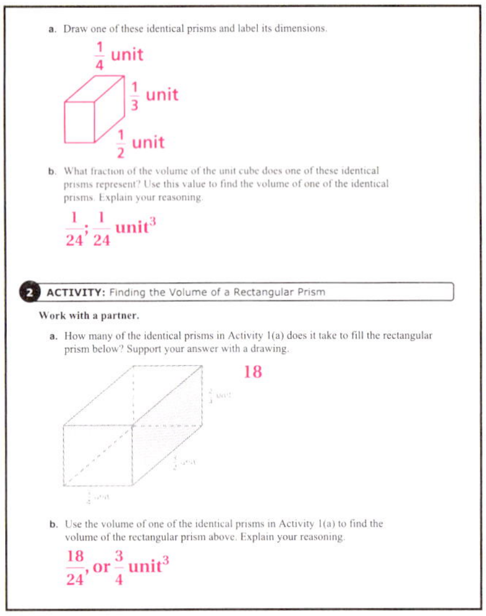

a. Draw one of these identical prisms and label its dimensions.

$\frac{1}{4}$ unit, $\frac{1}{3}$ unit, $\frac{1}{2}$ unit

b. What fraction of the volume of the unit cube does one of these identical prisms represent? Use this value to find the volume of one of the identical prisms. Explain your reasoning.

$\frac{1}{24}$; $\frac{1}{24}$ unit^3

2 ACTIVITY: Finding the Volume of a Rectangular Prism

Work with a partner.

a. How many of the identical prisms in Activity 1(a) does it take to fill the rectangular prism below? Support your answer with a drawing.

18

b. Use the volume of one of the identical prisms in Activity 1(a) to find the volume of the rectangular prism above. Explain your reasoning.

$\frac{18}{24}$, or $\frac{3}{4}$ unit^3

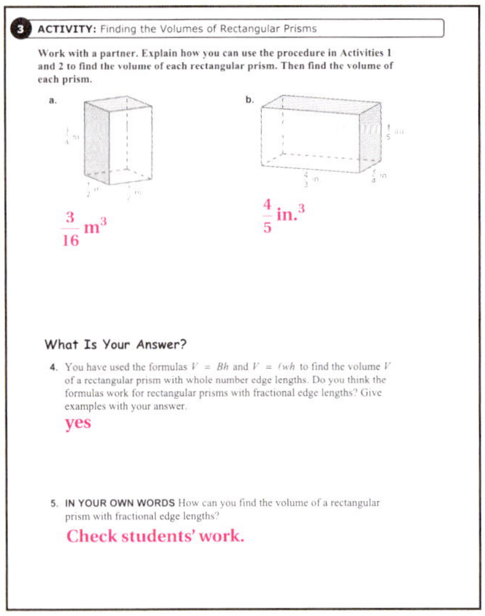

3 ACTIVITY: Finding the Volumes of Rectangular Prisms

Work with a partner. Explain how you can use the procedure in Activities 1 and 2 to find the volume of each rectangular prism. Then find the volume of each prism.

a. $\frac{3}{16}$ m^3

b. $\frac{4}{5}$ in.^{3}

What Is Your Answer?

4. You have used the formulas $V = Bh$ and $V = \ell wh$ to find the volume V of a rectangular prism with whole number edge lengths. Do you think the formulas work for rectangular prisms with fractional edge lengths? Give examples with your answer.

yes

5. **IN YOUR OWN WORDS** How can you find the volume of a rectangular prism with fractional edge lengths?

Check students' work.

Laurie's Notes

Activity 2

- Help students see the connection between the denominators of the dimensions of the prism in Activity 2 (which are 2, 3, and 4) and the number of equal parts the unit cube was divided up into in Activity 1 (2, 3, and 4).
- Say, "If the width of the prism is $\frac{3}{4}$ you can think of it as three of four equal parts. Each part would be $\frac{1}{4}$-unit long." Use the same strategy for the thirds and halves.
- Students can sketch lines parallel to the edges to reveal the smaller prisms, all of which will have the same dimensions as Activity 1: $\frac{1}{2}$ by $\frac{1}{4}$ by $\frac{1}{3}$. From Activity 1 they know the volume of the small identical prisms is $\frac{1}{24}$ cubic unit. There are 18 of these prisms, therefore the volume of the prism is $\frac{18}{24}$ cubic units.
- **MP3:** Ask volunteers to share their reasoning in part (b). Again, students may recognize that an alternate way to find the volume is to multiply the dimensions of the original prism: Volume $= \frac{3}{4} \times \frac{3}{2} \times \frac{2}{3} = \frac{18}{24}$ cubic units.

Activity 3

- In this activity, students must think about how to divide up the prism.
- **MP1 Make Sense of Problems and Persevere in Solving Them:** Do not jump in too quickly to rescue students. Tell them that these two problems are similar to Activities 1 and 2. They need to think through what information is given and how the first two activities can help them.
- **MP3:** When students have finished, ask volunteers to share how they found the volume of each prism. There may be different strategies so expect all students to listen carefully to the process.

What Is Your Answer?

- In Question 4, the volume formula may have already come up during the activities. If not, students should now be able to make the connection by looking back at their results in each activity.

Closure

- Find the volume of each prism. How are the results the same?

24 cubic inches; $\frac{24}{125}$ cubic units; The volume of each prism is 24 of its cubic units of measurement, 1 cubic inch and $\frac{1}{125}$ cubic unit.

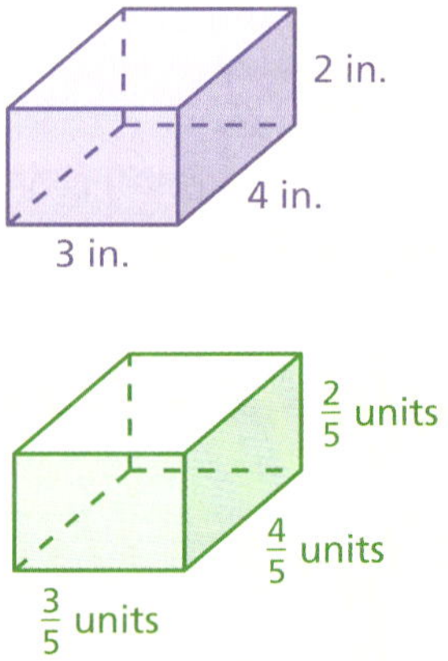

2 ACTIVITY: Finding the Volume of a Rectangular Prism

Work with a partner.

a. How many of the identical prisms in Activity 1(a) does it take to fill the rectangular prism below? Support your answer with a drawing.

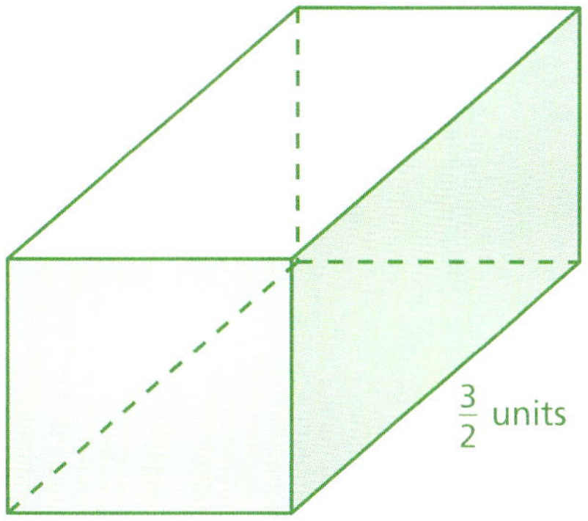

Math Practice 4

Analyze Relationships

What is the relationship between the solid shown here and the solid in the previous activity?

b. Use the volume of one of the identical prisms in Activity 1(a) to find the volume of the rectangular prism above. Explain your reasoning.

3 ACTIVITY: Finding the Volumes of Rectangular Prisms

Work with a partner. Explain how you can use the procedure in Activities 1 and 2 to find the volume of each rectangular prism. Then find the volume of each prism.

a.

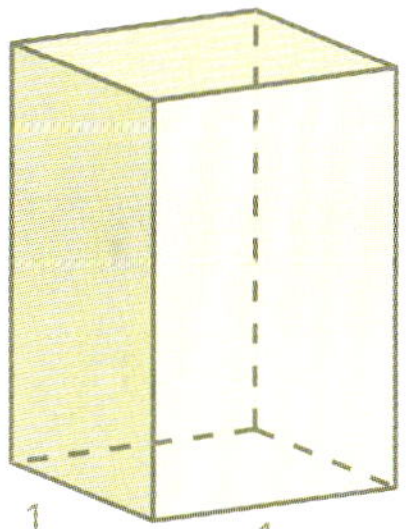

b.

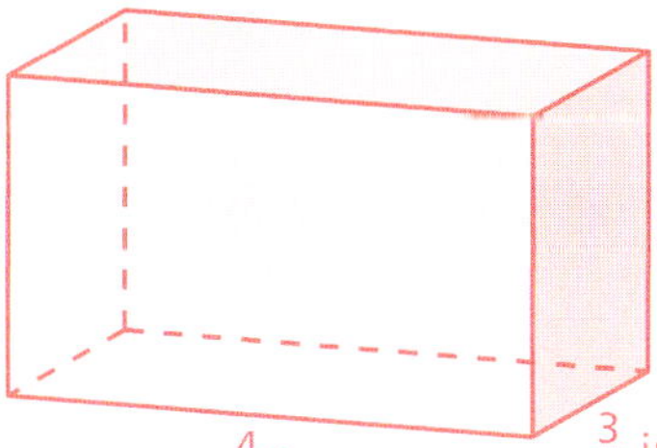

What Is Your Answer?

4. You have used the formulas $V = Bh$ and $V = \ell wh$ to find the volume V of a rectangular prism with whole number edge lengths. Do you think the formulas work for rectangular prisms with fractional edge lengths? Give examples with your answer.

5. **IN YOUR OWN WORDS** How can you find the volume of a rectangular prism with fractional edge lengths?

Practice Use what you learned about the volume of a rectangular prism to complete Exercises 4–6 on page 378.

Check It Out
Lesson Tutorials
BigIdeasMath.com

Key Vocabulary
volume, *p. 374*

Key Idea

Volume of a Rectangular Prism

Words The volume V of a rectangular prism is the product of the area of the base and the height of the prism.

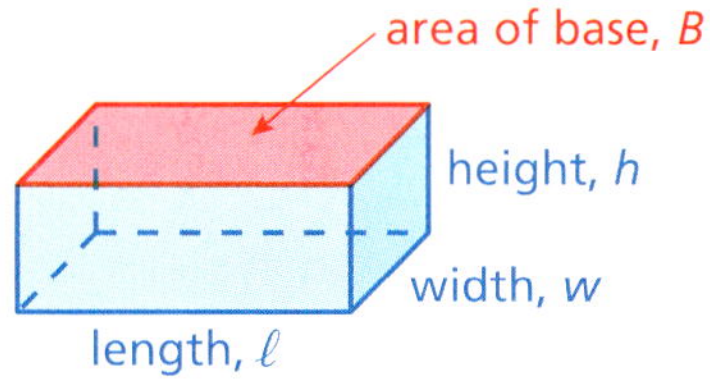

Algebra $V = Bh$ or $V = \ell wh$

EXAMPLE 1 Finding Volumes of Rectangular Prisms

Find the volume of each prism.

a.

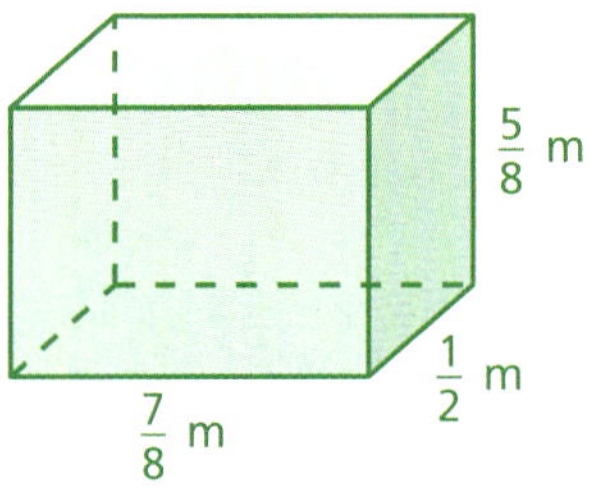

$V = \ell wh$ Write formula.

$= \frac{7}{8}\left(\frac{1}{2}\right)\left(\frac{5}{8}\right)$ Substitute values.

$= \frac{35}{128}$ Multiply.

So, the volume is $\frac{35}{128}$ cubic meter.

b.

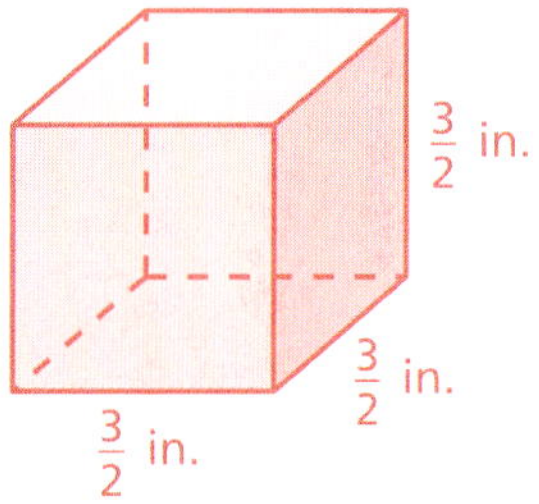

$V = \ell wh$

$= \frac{3}{2}\left(\frac{3}{2}\right)\left(\frac{3}{2}\right)$

$= \frac{27}{8}$

$= 3\frac{3}{8}$

So, the volume is $3\frac{3}{8}$ cubic inches.

Study Tip

In Example 1(b), the rectangular prism is a cube. You can use the formula $V = s^3$ to find the volume V of a cube with an edge length of s.

On Your Own

Now You're Ready
Exercises 4–9

Find the volume of the prism.

1.

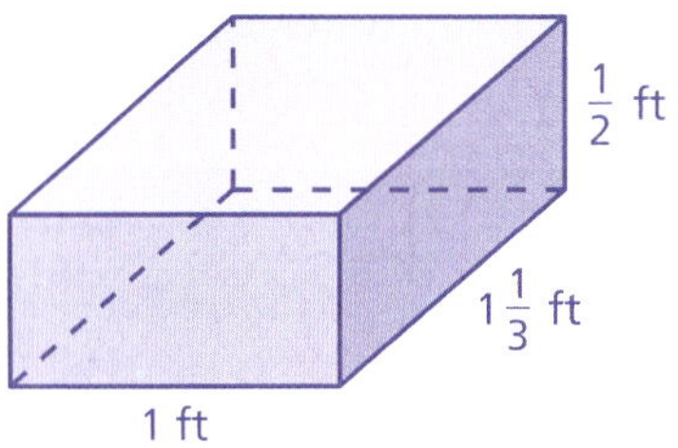

2.

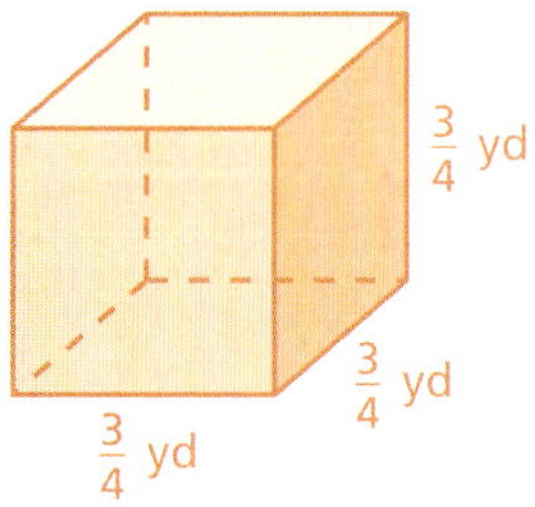

Multi-Language Glossary at BigIdeasMath.com

Laurie's Notes

Introduction

Connect

- **Yesterday:** Students explored how to find the volume of a rectangular prism with fractional dimensions. (MP1, MP3)
- **Today:** Students will use the formula for the volume of a rectangular prism to solve problems.

Motivate

- ? Hold up a tissue box and ask, "What does this prism hold?" tissues "How many?" Answers will vary.
- ? Hold up a ream of paper and ask, "What does this prism hold?" paper "How many sheets?" likely 500
- Ask students to think of other rectangular prisms whose volume is referred to by the contents, like the tissue box and ream of paper.
- Have students generate a list with their partners and then share.
- **Connection:** Similar to yesterday's Motivate, the layers help us to think about how to measure the space inside the prism.

Lesson Notes

Key Idea

- ? "What is a prism?" three-dimensional solid with two identical, parallel bases
- Point out to students that the dashed lines in the figure represent edges that cannot be seen from the perspective given.
- ? "Why is it called a *rectangular* prism?" The bases are rectangles.
- Write the formula for the volume of a rectangular prism in words and in symbols.
- **Review Vocabulary:** *Product* is the answer to a multiplication problem.

Example 1

- Write the formula for volume and substitute the value of each dimension.
- Note that parentheses are used to denote multiplication.
- After students have worked through each problem, ask them to use their hands to demonstrate the size of a cubic meter and a cubic inch.
- ? "How different are the two units?" open-ended

On Your Own

- ? "In Question 1, can the 1 ft and $\frac{1}{2}$ ft dimensions be the length and width, and the $1\frac{1}{3}$ ft dimension be the height? Explain." Yes, the bases of the prism would be the 1 ft by $\frac{1}{2}$ ft faces.
- **Extension:** Point out to students that in Question 2, all of the dimensions are in yards. "What would the dimensions be if they were labeled in feet?" $\frac{9}{4}$ feet "How would you find the volume?" $\left(\frac{9}{4}\right)^3$ "What units would be used in the answer?" cubic feet

Goal Today's lesson is finding the **volume** of a prism using a formula.

Technology for the Teacher

Dynamic Classroom

Lesson Tutorials
Lesson Plans
Answer Presentation Tool

Extra Example 1

Find the volume of each prism.

a.

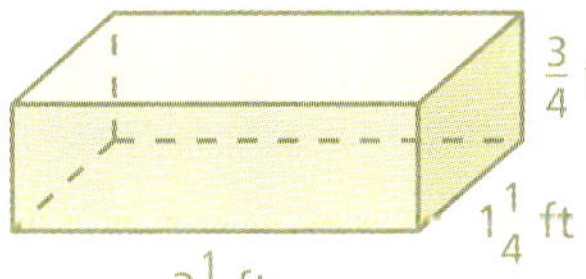

$2\frac{11}{32}$ ft^3

b.

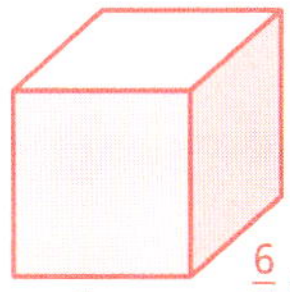

$\frac{216}{343}$ cm^3

On Your Own

1. $\frac{2}{3}$ ft^3
2. $\frac{27}{64}$ yd^3

Extra Example 2

In Example 2, the length of the dump truck is 22 feet. How many pounds of dirt can the dump truck haul when it is full?
58,520 lb

Extra Example 3

An MP3 player has a depth (length) of 0.6 centimeter, a width of 4 centimeters, and a volume of 21.6 cubic centimeters. Write and solve an equation to find the height of the MP3 player.
$21.6 = 0.6(4)h$, 9 cm

On Your Own

3. 53,200 lb

4. $72 = \ell(6)(2)$; 6 in.

5. $1375 = 20(w)\left(5\frac{1}{2}\right)$; $12\frac{1}{2}$ cm

Differentiated Instruction

Kinesthetic

Divide the class into two groups. Give each group a piece of yarn or string 36 inches long. Have the first group use all of the yarn to form a rectangle with the largest possible area. Have the second group use all of the yarn to form a rectangle with the smallest possible area, with one side at least 1 inch long. Ask students:

What are the dimensions of your rectangle? 9 in. by 9 in.; 1 in. by 17 in.

If the rectangle was the base of a prism having a height of 5 inches, what would the volume of the prism be? 405 in.^{3}; 85 in.^{3}

What is the difference between the two volumes? 320 in.^{3}

Laurie's Notes

Example 2

- It may be helpful to sketch and label the dump truck in the horizontal position.

? "Would a cubic foot of dirt fit in the wastebasket?" Answers will vary depending upon size of the wastebasket in the classroom.

- If the wastebasket is close to a cubic foot, relate this to the statement that a cubic foot of dirt weighs about 70 pounds.
- **MP6 Attend to Precision:** Note the use of dimensional analysis to show that the answer has units of pounds.

? **Extension:** "A ton is 2000 pounds. How many tons of dirt can the dump truck haul when it is full?" 22.61 tons

Example 3

- Students worked with equation solving in the last chapter, so they should have little trouble with this problem.

? "What is known in this problem?" volume, length, and width "What are you trying to find out?" height

? "Is the answer reasonable? Explain." yes; Answers will vary.

On Your Own

- **Neighbor Check:** Have students work independently and then have their neighbors check their work. Have students discuss any discrepancies.

Closure

- **Exit Ticket:** Sketch a rectangular prism. Label the dimensions $\frac{3}{4}$ foot, 2 feet, and $\frac{3}{2}$ feet. Find the volume of the prism.

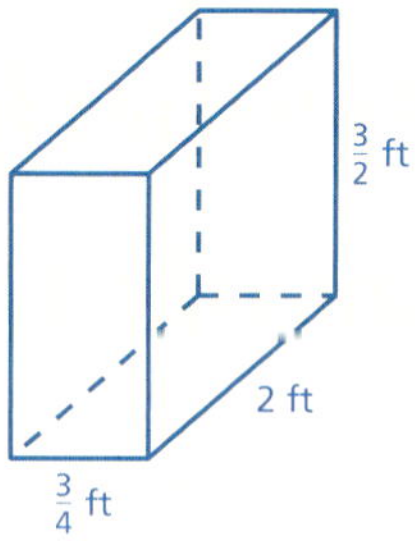

$V = \frac{3}{4} \cdot 2 \cdot \frac{3}{2} = 2\frac{1}{4}$ ft^3

EXAMPLE 2 Using the Volume of a Rectangular Prism

One cubic foot of dirt weighs about 70 pounds. How many pounds of dirt can the dump truck haul when it is full?

Find the volume of dirt that the dump truck can haul when it is full.

$V = \ell wh$	Write formula for volume.
$= 17(8)\left(4\frac{3}{4}\right)$	Substitute values.
$= 646$	Multiply.

So, the dump truck can haul 646 cubic feet of dirt when it is full. To find the weight of the dirt, multiply by $\frac{70 \text{ lb}}{1 \text{ ft}^3}$.

$$646 \cancel{\text{ft}^3} \times \frac{70 \text{ lb}}{1 \cancel{\text{ft}^3}} = 45{,}220 \text{ lb}$$

The dump truck can haul about 45,220 pounds of dirt when it is full.

EXAMPLE 3 Finding a Missing Dimension of a Rectangular Prism

Write and solve an equation to find the height of the computer tower.

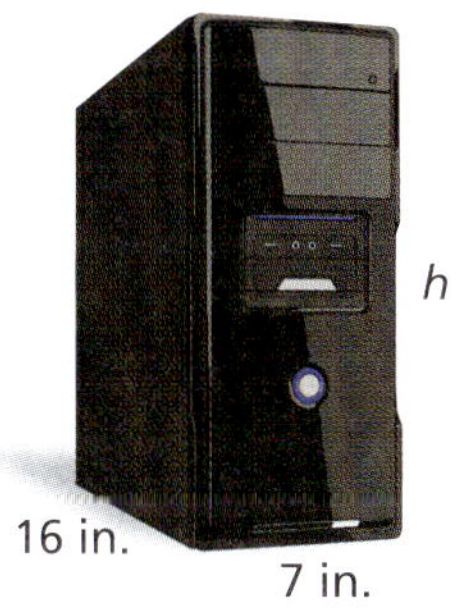

Volume = 1792 in.^{3}

$V = \ell wh$	Write formula for volume.
$1792 = 16(7)h$	Substitute values.
$1792 = 112h$	Simplify.
$\frac{1792}{112} = \frac{112h}{112}$	Division Property of Equality
$16 = h$	Simplify.

So, the height of the computer tower is 16 inches.

On Your Own

3. **WHAT IF?** In Example 2, the length of the dump truck is 20 feet. How many pounds of dirt can the dump truck haul when it is full?

Now You're Ready
Exercises 10–12

Write and solve an equation to find the missing dimension of the prism.

4. Volume = 72 in.^{3}

5. Volume = 1375 cm^3

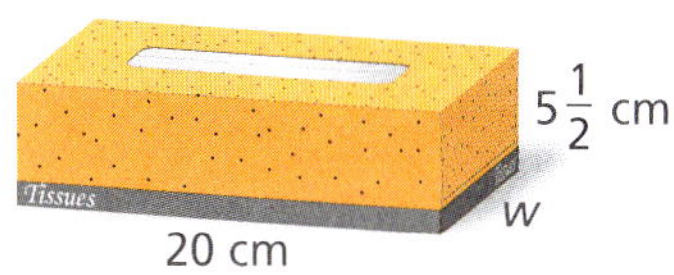

8.4 Exercises

Check It Out
Help with Homework
BigIdeasMath.com

Vocabulary and Concept Check

1. **CRITICAL THINKING** Explain how volume and surface area are different.

2. **REASONING** Will the formulas for volume work for rectangular prisms with decimal edge lengths? Explain.

3. **DIFFERENT WORDS, SAME QUESTION** Which is different? Find "both" answers.

How much does it take to fill the rectangular prism?

What is the capacity of the rectangular prism?

How much does it take to cover the rectangular prism?

How much does the rectangular prism contain?

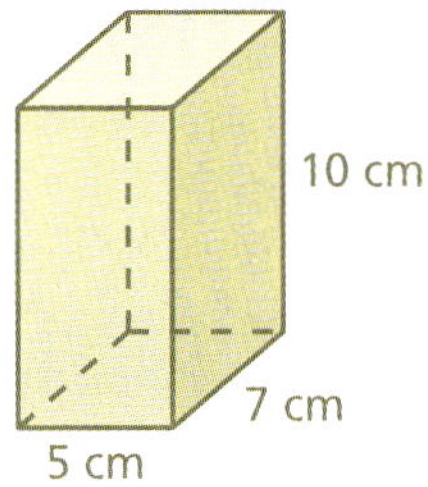

Practice and Problem Solving

Find the volume of the prism.

1 4.

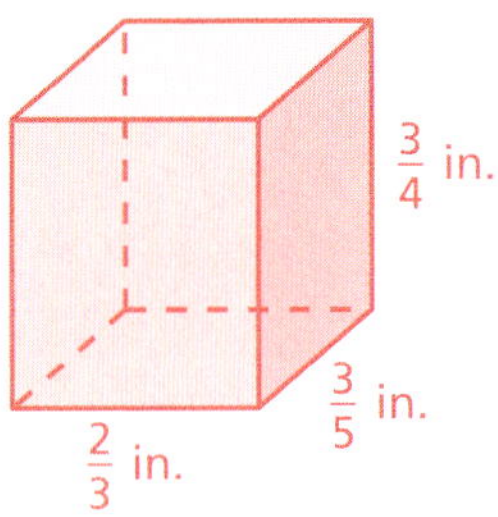

5.

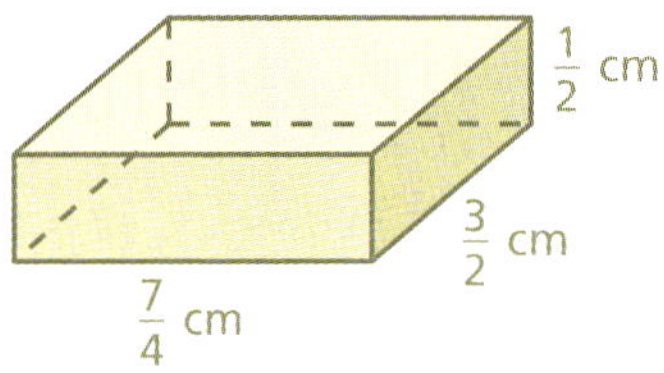

6.

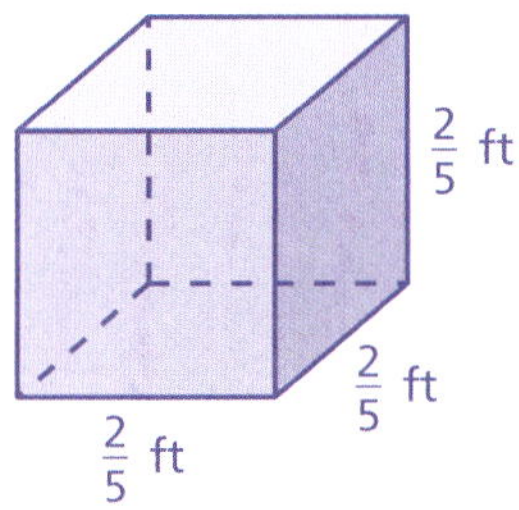

7.

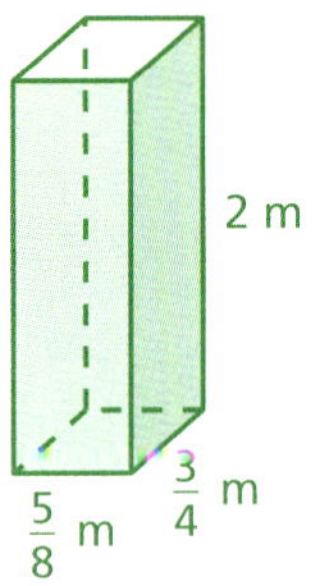

8.

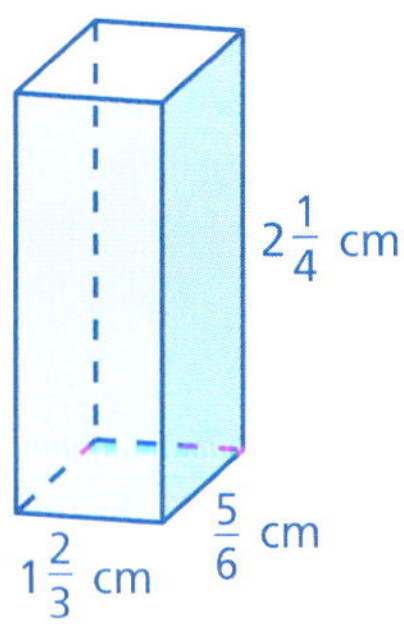

9.

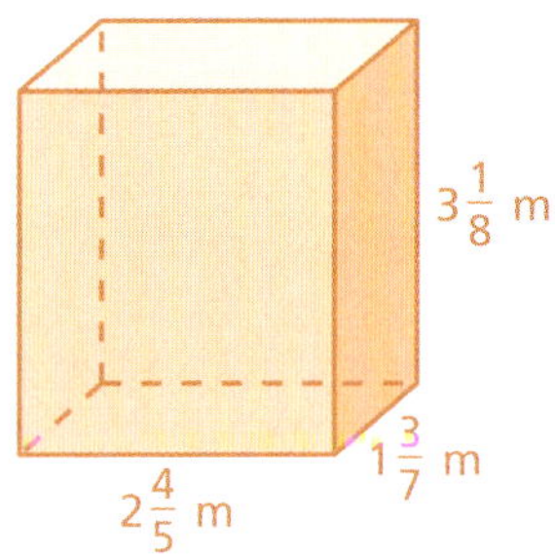

Write and solve an equation to find the missing dimension of the prism.

3 10. Volume = 1620 cm^3

11. Volume = 220.5 cm^3

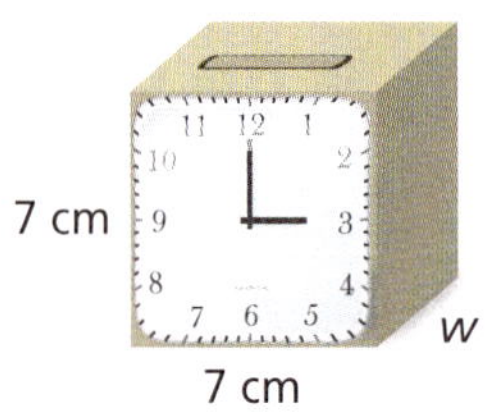

12. Volume = 532 in.^{3}

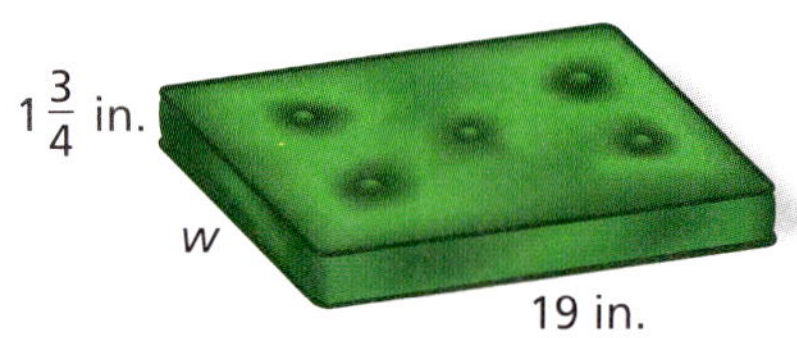

Assignment Guide and Homework Check

Level	Day 1 Activity Assignment	Day 2 Lesson Assignment	Homework Check
Basic	4–6, 19–22	1–3, 7–15 odd	7, 9, 11, 13
Average	4–6, 19–22	1–3, 7–14, 16	7, 9, 11, 13, 14
Advanced	4–6, 19–22	1–3, 7–18	8, 10, 12, 14, 16

Common Errors

- **Exercises 4–9** Students may write the incorrect units. Remind them that volume has units cubed because there are three dimensions.
- **Exercises 10–13** Students may forget to simplify before solving for the variable and may lose or forget about one of the dimensions. For example:

$$\frac{1620}{9} = \frac{(9)(9)h}{9}$$
$$180 = h \ \text{✗}$$

$$1620 = (9)(9)h$$
$$\frac{1620}{81} = \frac{81h}{81}$$
$$20 = h$$

Remind them to simplify their equations before solving for a variable.

8.4 Record and Practice Journal

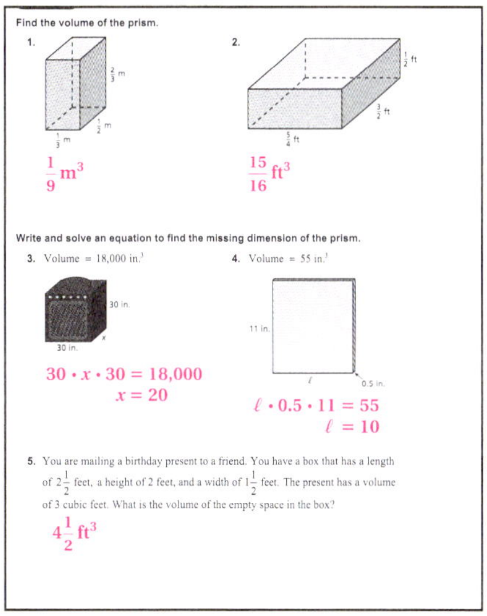

Find the volume of the prism.

1. $\frac{1}{9}$ m^3

2. $\frac{15}{16}$ ft^3

Write and solve an equation to find the missing dimension of the prism.

3. Volume = 18,000 in.^{3}

$30 \cdot x \cdot 30 = 18{,}000$
$x = 20$

4. Volume = 55 in.^{3}

$\ell \cdot 0.5 \cdot 11 = 55$
$\ell = 10$

5. You are mailing a birthday present to a friend. You have a box that has a length of $2\frac{1}{2}$ feet, a height of 2 feet, and a width of $1\frac{1}{2}$ feet. The present has a volume of 3 cubic feet. What is the volume of the empty space in the box?

$4\frac{1}{2}$ ft^3

Vocabulary and Concept Check

1. The volume of an object is the amount of space it occupies. The surface area of an object is the sum of the areas of all its faces.
2. yes; You just substitute the decimal edge lengths into the formula and multiply.
3. How much does it take to cover the rectangular prism?; 310 cm^2; 350 cm^3

Practice and Problem Solving

4. $\frac{3}{10}$ in.^{3}
5. $1\frac{5}{16}$ cm^3
6. $\frac{8}{125}$ ft^3
7. $\frac{15}{16}$ m^3
8. $3\frac{1}{8}$ cm^3
9. $12\frac{1}{2}$ m^3
10. $1620 = 9 \cdot 9 \cdot h$; 20 cm
11. $220.5 = 7 \cdot w \cdot 7$; 4.5 cm
12. $532 = 19 \cdot w \cdot 1\frac{3}{4}$; 16 in.

Practice and Problem Solving

13. See *Taking Math Deeper*.

14. $4096\frac{3}{4}$-centimeter cubes

15. See Additional Answers.

16. **a.** *Sample answer:* 297 in.^{3}

b. no; The container only holds 196 cubic inches.

17. 1152 cm^3

18. See Additional Answers.

Fair Game Review

19. yes **20.** no

21. no **22.** C

Mini-Assessment

Find the volume of the rectangular prism.

1.

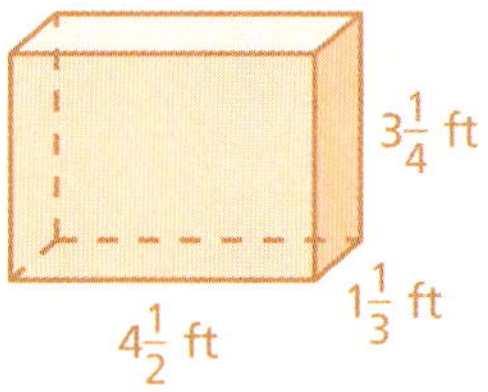

$19\frac{1}{2}$ ft^3

2.

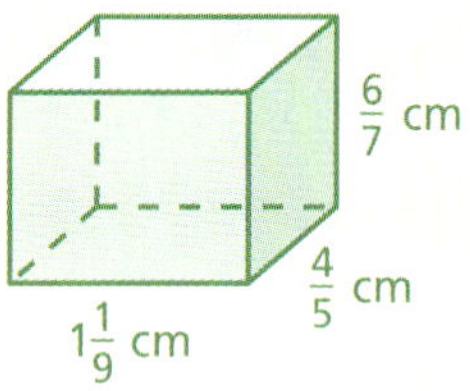

$\frac{16}{21}$ cm^3

3. The volume of the cell phone is 75 cubic centimeters. Write and solve an equation to find the height of the cell phone. Round your answer to the nearest tenth. $75 = 10(4.7)h$; 1.6 cm

Taking Math Deeper

Exercise 13

You can solve this problem visually by reasoning about cubes.

1 One cubic foot of water weighs about 62.4 pounds. You can visualize that a 1-foot cube fits in the fish tank, but the tank is 0.5 feet taller. So, imagine slicing a 1-foot cube of water in half horizontally. Each half weighs $62.4 \div 2 = 31.2$ pounds.

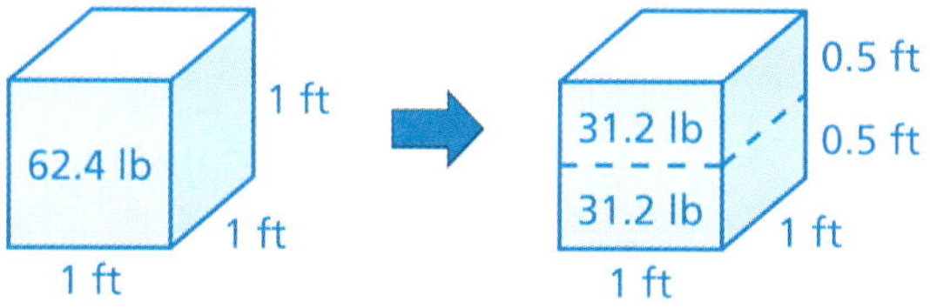

2 When you put one of these halves on top of a 1-foot cube, the new prism weighs $31.2 + 62.4 = 93.6$ pounds. It fits inside the tank without extra space along the width or height.

You can visualize that two of the 93.6-pound prisms fit inside the tank, but the tank is 0.5 feet *longer*. So, imagine slicing the prism in half vertically. Each half weighs $93.6 \div 2 = 46.8$ pounds.

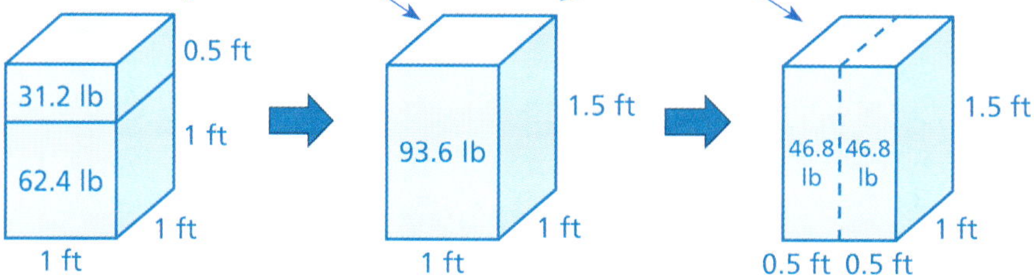

3 The tank will hold two of the 93.6-pound prisms of water and one of the 46.8-pound prisms of water.

So, the fish tank can hold $46.8 + 93.6 + 93.6 = 234$ pounds of water when full.

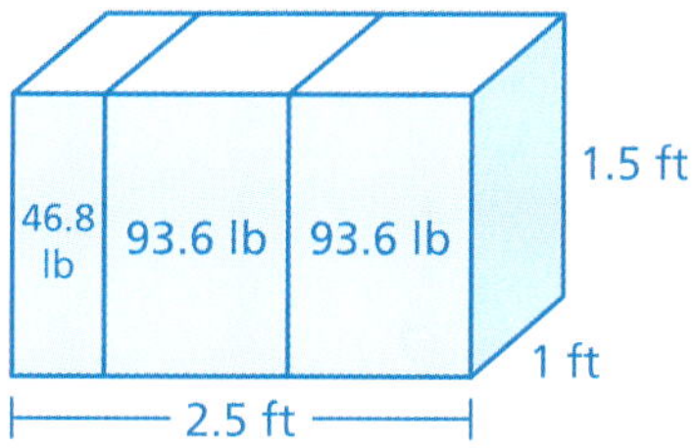

Reteaching and Enrichment Strategies

If students need help. . .	If students got it. . .
Resources by Chapter • Practice A and Practice B • Puzzle Time Record and Practice Journal Practice Differentiating the Lesson Lesson Tutorials Skills Review Handbook	Resources by Chapter • Enrichment and Extension • Technology Connection Start the next section

13. **FISH TANK** One cubic foot of water weighs about 62.4 pounds. How many pounds of water can the fish tank hold when it is full?

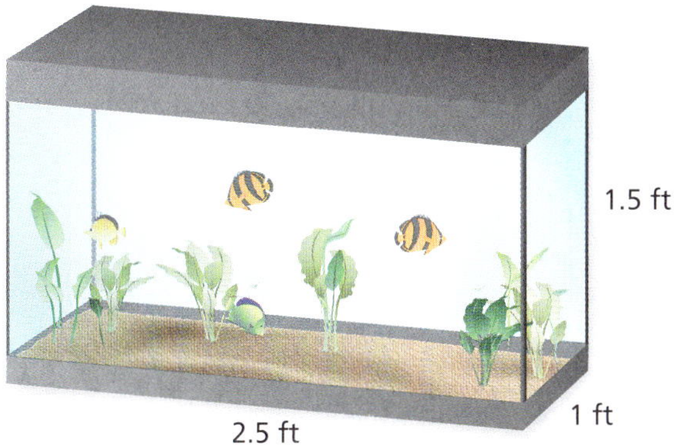

14. **CUBE** How many $\frac{3}{4}$-centimeter cubes do you need to create a cube with an edge length of 12 centimeters?

15. **REASONING** How many 1-inch cubes do you need to fill a cube that has an edge length of 1 foot? How can this result help you convert a volume from cubic inches to cubic feet? from cubic feet to cubic inches?

16. **FOOD STORAGE**
   a. Estimate the amount of casserole left in the dish.
   b. Will the casserole fit in the storage container? Explain your reasoning.

17. **PROBLEM SOLVING** The area of the shaded face is 96 square centimeters. What is the volume of the rectangular prism?

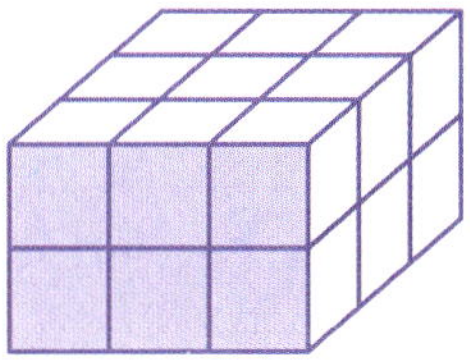

18. **Project** You have 1400 square feet of boards to use for a new tree house.
   a. Design a tree house that has a volume of at least 250 cubic feet. Include sketches of your tree house.
   b. Are your dimensions reasonable? Explain your reasoning.

Fair Game Review What you learned in previous grades & lessons

Tell whether the given value is a solution of the equation. *(Section 7.2)*

19. $x + 17 = 24;\ x = 7$

20. $\frac{x}{5} = 6;\ x = 35$

21. $x - 19 = 42;\ x = 21$

22. **MULTIPLE CHOICE** Which set of integers is ordered from least to greatest? *(Section 6.2)*

   (A) $-1, 3, -5, -8, 12$ (B) $-1, 3, -5, -8, 12$

   (C) $-4, -2, 1, 7, 10$ (D) $-14, -9, 6, -4, 2$

8.3–8.4 Quiz

Check It Out
Progress Check
BigIdeasMath.com

Find the surface area of the pyramid. The side lengths of the base are equal. *(Section 8.3)*

1.

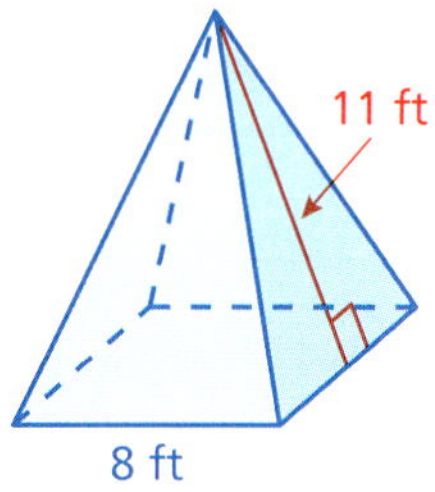

2.

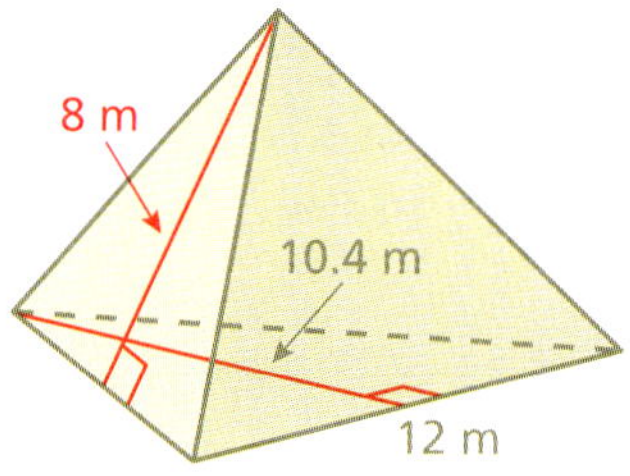

Find the volume of the prism. *(Section 8.4)*

3.

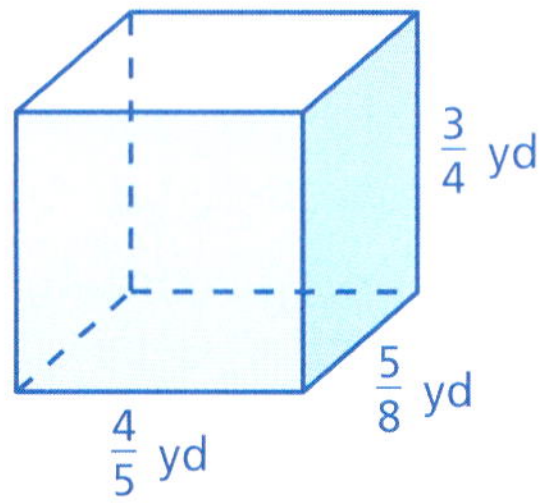

4.

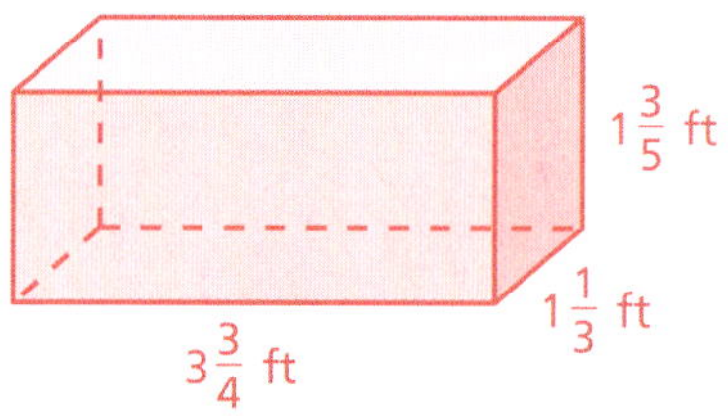

Write and solve an equation to find the missing dimension of the prism. *(Section 8.4)*

5. Volume = 1620 in.^{3}

6. Volume = 154 in.^{3}

7. Volume = 4250 in.^{3}

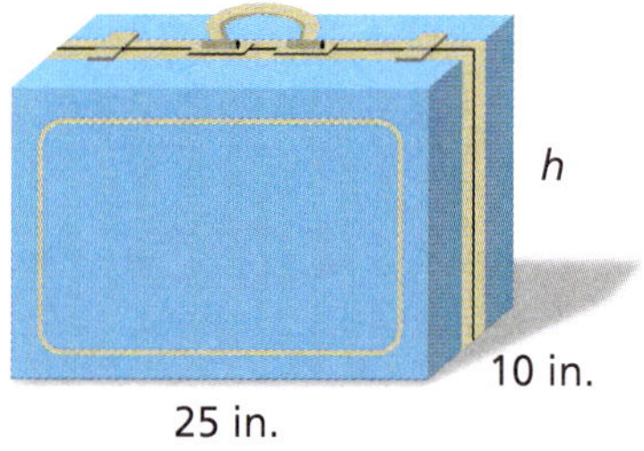

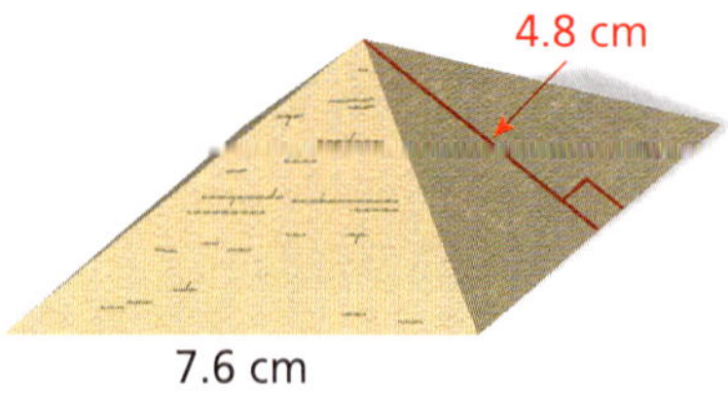

8. GREAT PYRAMID The Great Pyramid of Giza is a square pyramid. A gift shop sells miniature models of this pyramid. Find the surface area of the model shown at the left. *(Section 8.3)*

9. CUBE How many 1-inch cubes do you need to create a cube with an edge length of 7 inches? *(Section 8.4)*

10. TOY CHEST A toy company sells two different toy chests. The toy chests have different dimensions, but the same volume. What is the width w of Toy Chest 2? *(Section 8.4)*

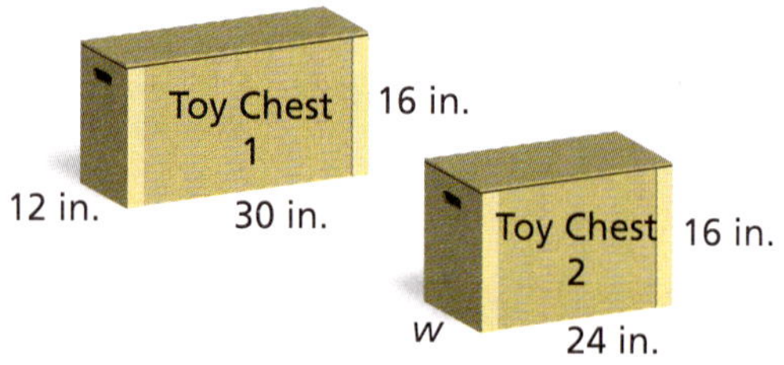

Alternative Assessment Options

Math Chat — Student Reflective Focus Question
Structured Interview — Writing Prompt

Math Chat

Ask students to use their own words to summarize what they know about surface areas of rectangular prisms and pyramids, and volumes of rectangular prisms. Be sure that they include examples. Select students at random to present their summaries to the class.

Study Help Sample Answers

Remind students to complete Graphic Organizers for the rest of the chapter.

3.

	Example
Finding the surface area of a pyramid	Triangular pyramid 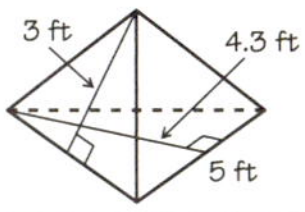
Draw a net.	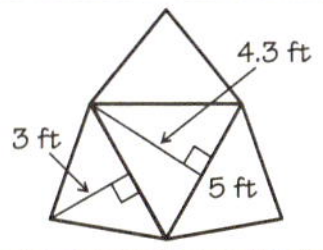
Find the areas of the faces.	Bottom: $\frac{1}{2} \cdot 5 \cdot 4.3 = 10.75$ Side: $\frac{1}{2} \cdot 5 \cdot 3 = 7.5$ Side: $\frac{1}{2} \cdot 5 \cdot 3 = 7.5$ Side: $\frac{1}{2} \cdot 5 \cdot 3 = 7.5$
Find the sum of the areas of the faces.	$S = 10.75 + 7.5 + 7.5 + 7.5$ $= 33.25 \text{ ft}^2$

4. Available at *BigIdeasMath.com*.

Reteaching and Enrichment Strategies

If students need help. . .	If students got it. . .
Resources by Chapter • Practice A and Practice B • Puzzle Time Lesson Tutorials *BigIdeasMath.com*	Resources by Chapter • Enrichment and Extension • Technology Connection Game Closet at *BigIdeasMath.com* Start the Chapter Review

Answers

1. 240 ft^2
2. 206.4 m^2
3. $\frac{3}{8} \text{ yd}^3$
4. 8 ft^3
5. $1620 = 15 \cdot w \cdot 12$; 9 in.
6. $154 = \ell \cdot 2 \cdot 11$; 7 in.
7. $4250 = 25 \cdot 10 \cdot h$; $h = 17$ in.
8. 130.72 cm^2
9. 343 cubes
10. 15 in.

Technology for the Teacher

Online Assessment
Assessment Book
ExamView® Assessment Suite

For the Teacher
Additional Review Options

- *BigIdeasMath.com*
- Online Assessment
- Game Closet at *BigIdeasMath.com*
- Vocabulary Help
- Resources by Chapter

Answers

1. 6 faces, 12 edges, 8 vertices
2. 6 faces, 10 edges, 6 vertices
3. 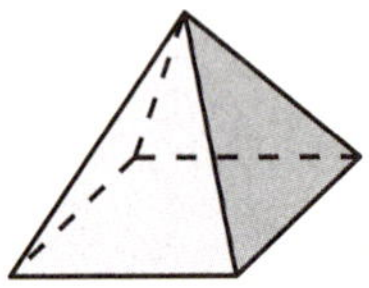
4. 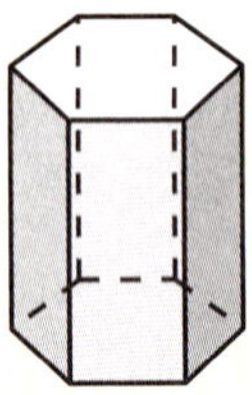

Review of Common Errors

Exercises 3 and 4

- Students may mix up the different types of solids. Remind them of the definition of each solid and give them a few real-life examples of each.

Exercises 5–7

- Students may sum the areas of only 3 of the faces of the rectangular prism instead of all 6. Remind them that a rectangular prism has 6 faces.

Exercises 11–13

- Students may forget to include the area of the base when finding the surface area of a pyramid. Remind them that when asked to find the surface area, the base is included.
- Students may add the wrong number of lateral face areas to the area of the base. Remind them that they must add the area of a lateral face as many times as there are sides on the base of the pyramid.

Exercises 14 and 15

- Students may write the units incorrectly, often writing square units instead of cubic units. Remind them that they are working in three dimensions, so the units are cubed. Give an example showing the formula as three units multiplied together. For example, write the volume of Exercise 14 as $V = \left(\frac{5}{2}\text{ ft}\right)\left(\frac{3}{2}\text{ ft}\right)\left(\frac{4}{3}\text{ ft}\right)$.

8 Chapter Review

Check It Out
Vocabulary Help
BigIdeasMath.com

Review Key Vocabulary

solid, *p. 356*	vertex, *p. 356*	net, *p. 362*
polyhedron, *p. 356*	prism, *p. 356*	volume, *p. 374*
face, *p. 356*	pyramid, *p. 356*	
edge, *p. 356*	surface area, *p. 362*	

Review Examples and Exercises

8.1 Three-Dimensional Figures *(pp. 354–359)*

a. Find the number of faces, edges, and vertices of the solid.

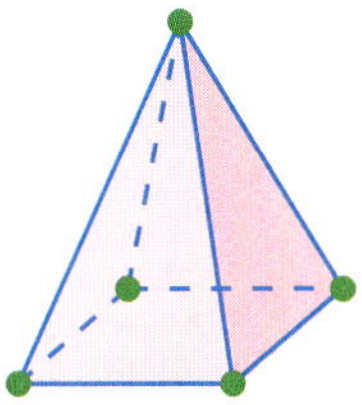

The solid has 1 face on the bottom and 4 faces on the sides.

The faces intersect at 8 different line segments.

The edges intersect at 5 different points.

So, the solid has 5 faces, 8 edges, and 5 vertices.

b. Draw a triangular prism.

Draw identical triangular bases.

Connect corresponding vertices.

Change any *hidden* lines to dashed lines.

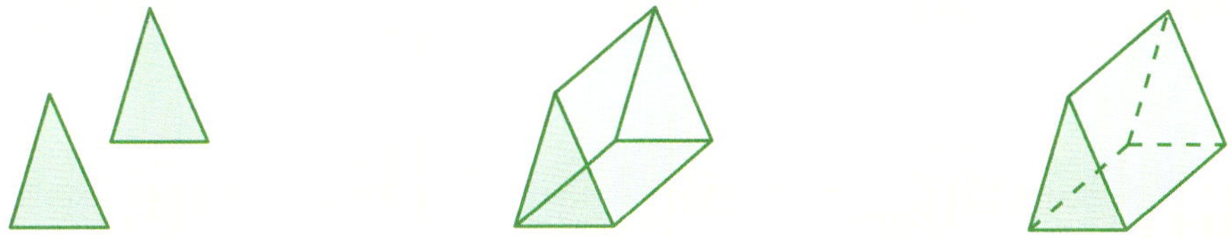

Exercises

Find the number of faces, edges, and vertices of the solid.

1. 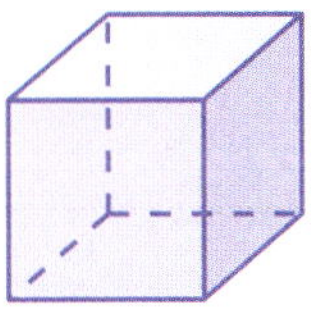

2. 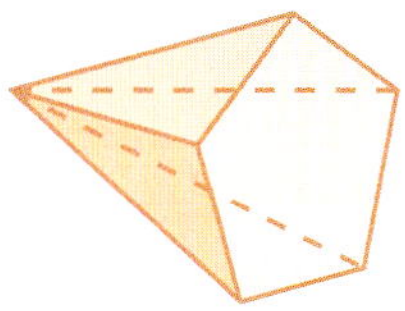

Draw the solid.

3. square pyramid

4. hexagonal prism

8.2 Surface Areas of Prisms (pp. 360–365)

Find the surface area of the rectangular prism.

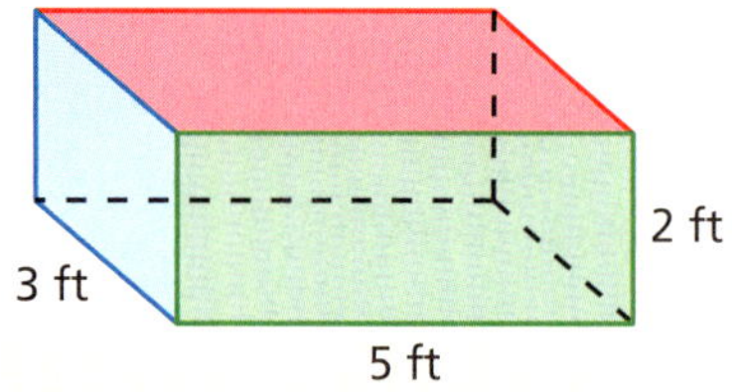

Use a net to find the area of each face.

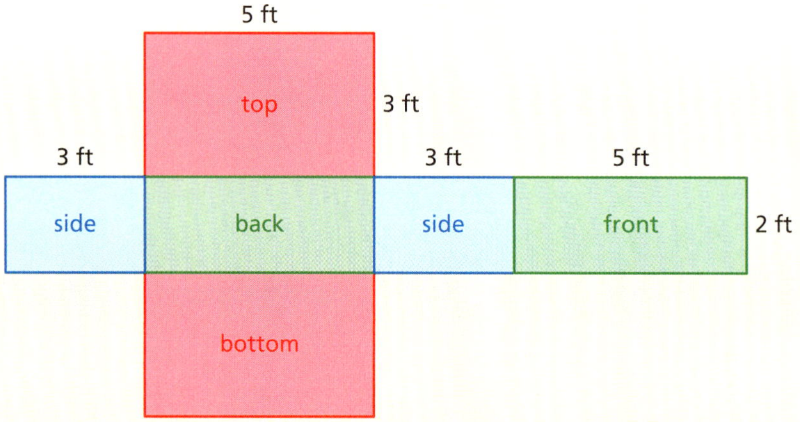

Top: $5 \cdot 3 = 15$	Front: $5 \cdot 2 = 10$	Side: $3 \cdot 2 = 6$
Bottom: $5 \cdot 3 = 15$	Back: $5 \cdot 2 = 10$	Side: $3 \cdot 2 = 6$

Find the sum of the areas of the faces.

$$S = 15 + 15 + 10 + 10 + 6 + 6$$

$$= 62$$

The surface area is 62 square feet.

Exercises

Find the surface area of the prism.

5. 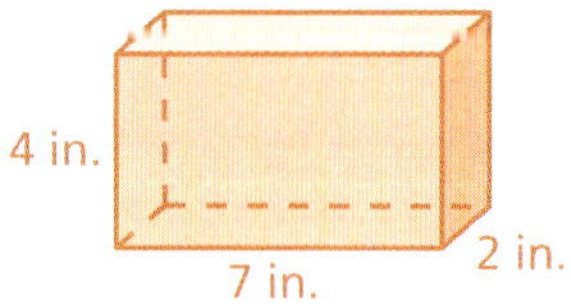

6. 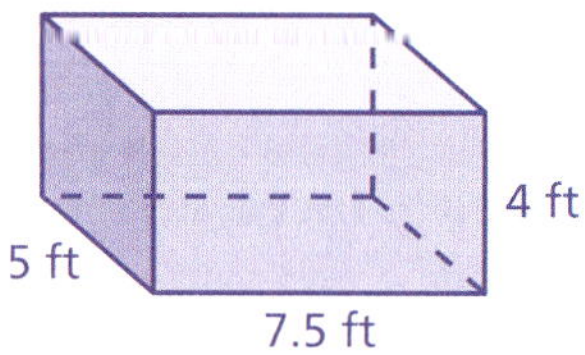

7. 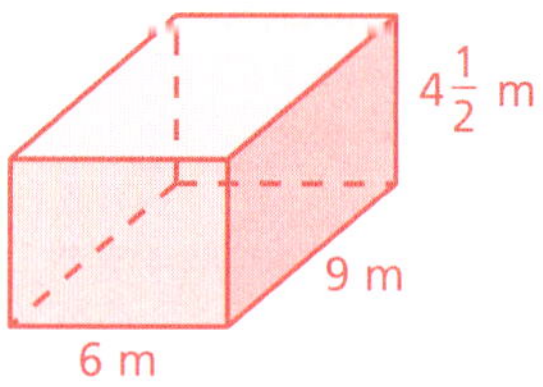

8. 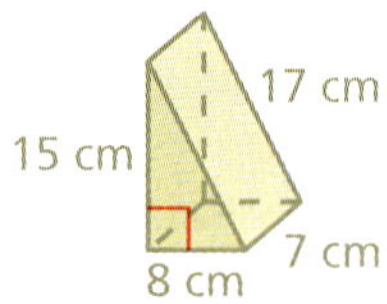

9. 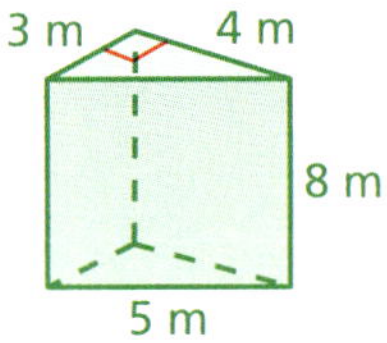

10. 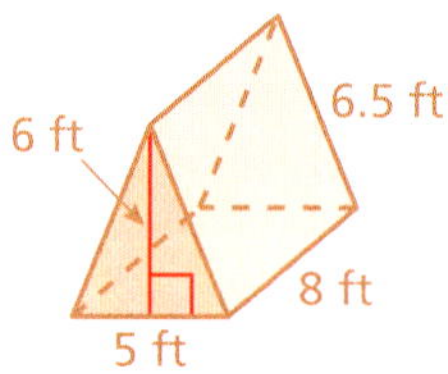

Review Game

Surface Areas of Rectangular Prisms

Materials per Pair:

- paper
- pencils (colored pencils optional)

Directions:

Each group of four divides into teams of two. Each team competes with the other team. The class is directed to draw and label a rectangular prism with a predetermined surface area. For example, the teacher could say, "Draw and label a rectangular prism with a surface area of 50 square units." Each team then draws their prism, and calculates the surface area. Papers are exchanged within the group and teams check each other's work. The team whose correctly-drawn figure has an area that is closest to the area specified receives 1 point. If both teams are equally close, both teams receive 1 point. Play for a set amount of time.

Who Wins?

When time is up, the team with the most points wins.

For the Student Additional Practice

- Lesson Tutorials
- Multi-Language Glossary
- Self-Grading Progress Check
- *BigIdeasMath.com*
  Dynamic Student Edition
  Student Resources

Answers

5. 100 in.^2
6. 175 ft^2
7. 243 m^2
8. 400 cm^2
9. 108 m^2
10. 174 ft^2
11. 16 in.^2
12. 147.6 m^2
13. 180.6 cm^2
14. 5 ft^3
15. $\frac{11}{18} \text{ cm}^3$

My Thoughts on the Chapter

Teacher Tip

Not allowed to write in your teaching edition? Use sticky notes to record your thoughts.

What worked. . .

What did not work. . .

What I would do differently. . .

8.3 Surface Areas of Pyramids (pp. 368–373)

Find the surface area of the triangular pyramid.

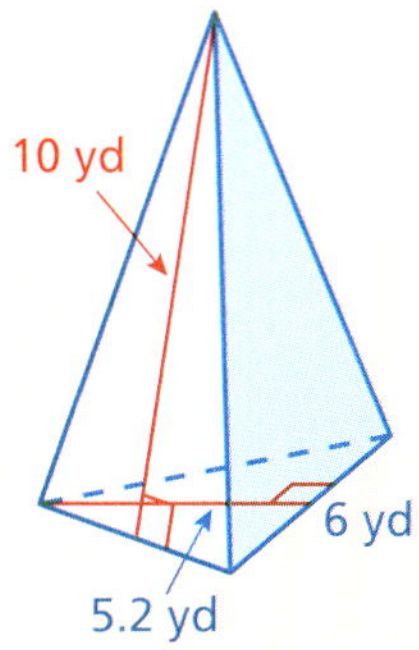

Use a net to find the area of each face.

Bottom: $\frac{1}{2} \cdot 6 \cdot 5.2 = 15.6$ Side: $\frac{1}{2} \cdot 6 \cdot 10 = 30$

Side: $\frac{1}{2} \cdot 6 \cdot 10 = 30$ Side: $\frac{1}{2} \cdot 6 \cdot 10 = 30$

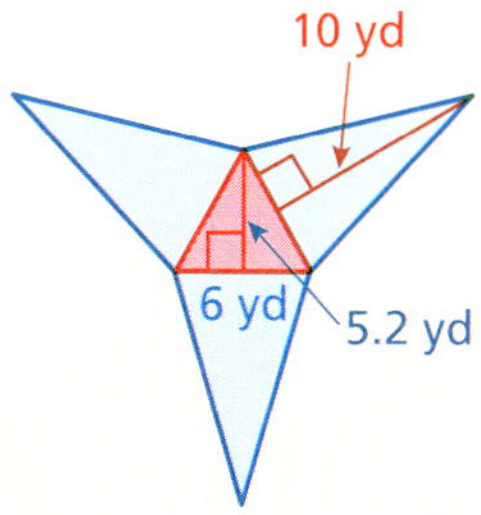

Find the sum of the areas of the faces.

$S = 15.6 + 30 + 30 + 30 = 105.6$

The surface area is 105.6 square yards.

Exercises

Find the surface area of the pyramid. The side lengths of the base are equal.

11.

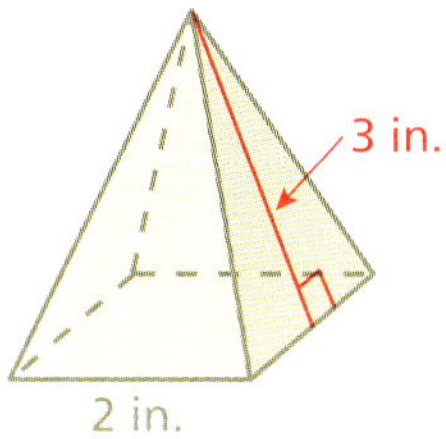

12.

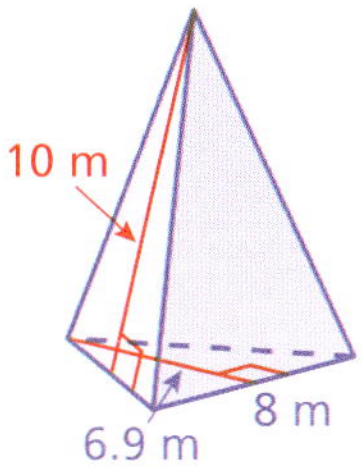

13.

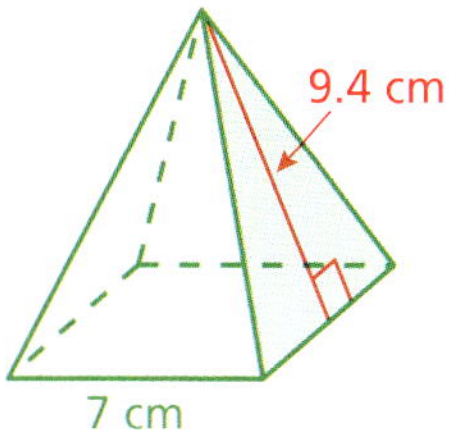

8.4 Volumes of Rectangular Prisms (pp. 374–379)

Find the volume of the prism.

$V = \ell wh$ Write formula for volume.

$= \frac{5}{6}\left(\frac{3}{4}\right)\left(\frac{4}{5}\right)$ Substitute values.

$= \frac{1}{2}$ Multiply.

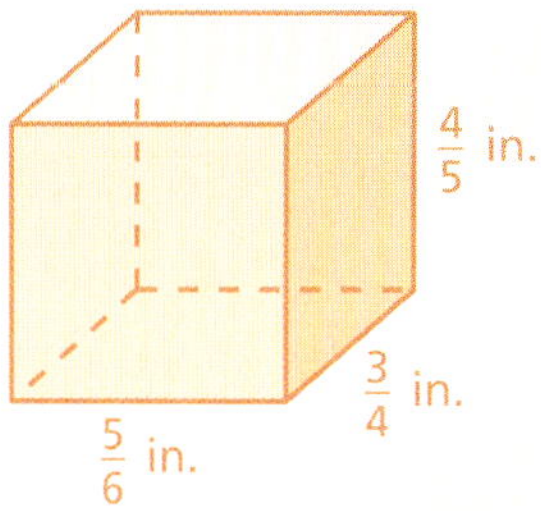

The volume is $\frac{1}{2}$ cubic inch.

Exercises

Find the volume of the prism.

14.

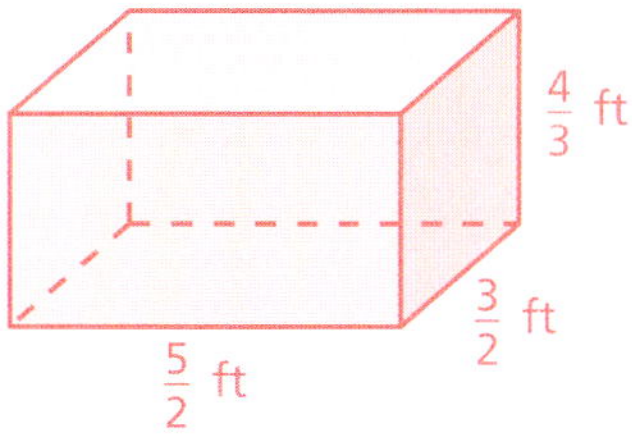

15.

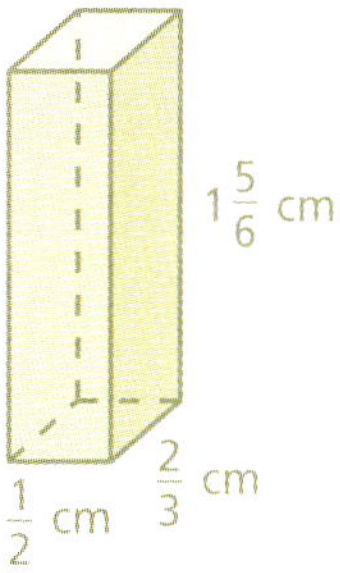

8 Chapter Test

Check It Out
Test Practice
BigIdeasMath.com

Find the number of faces, edges, and vertices of the solid.

1.

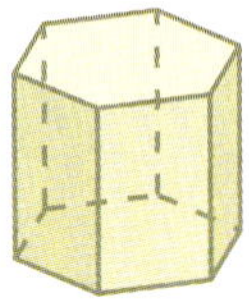

2.

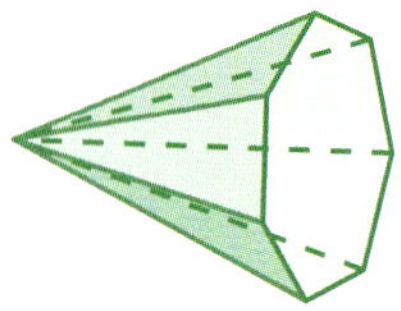

Find the surface area of the prism.

3.

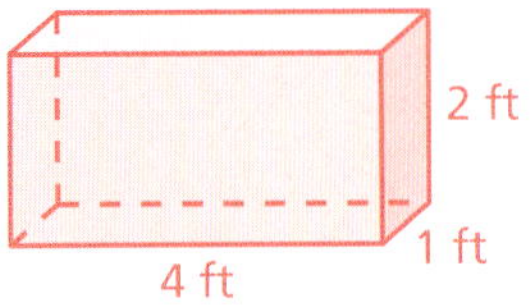

4.

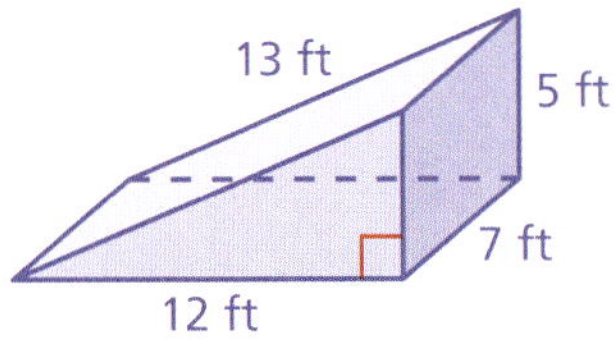

Find the surface area of the pyramid. The side lengths of the base are equal.

5.

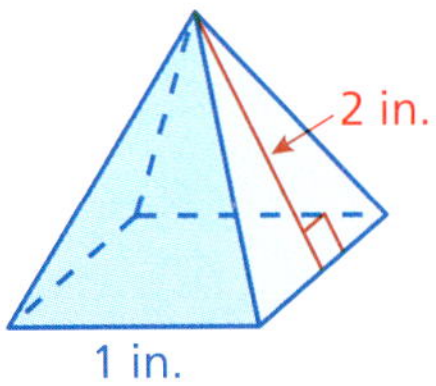

6.

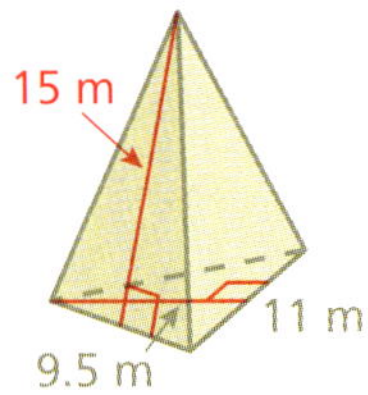

Find the volume of the prism.

7.

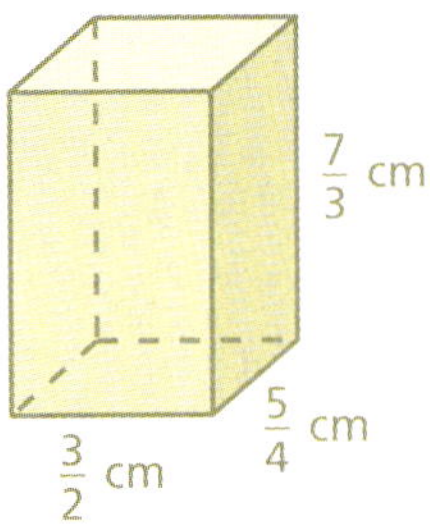

8.

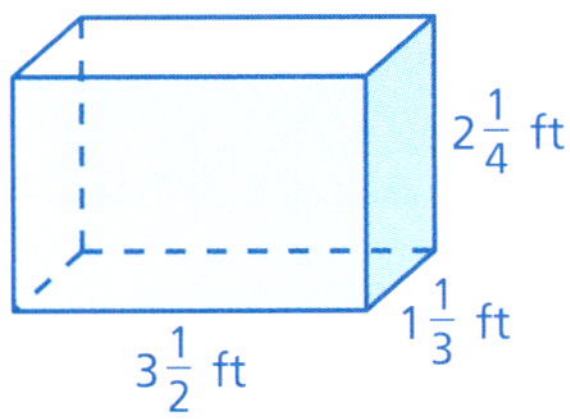

9. DRAWING A SOLID Draw an octagonal prism.

10. DVD COLLECTION You are wrapping the boxed DVD collection as a present. What is the least amount of wrapping paper needed to wrap the box?

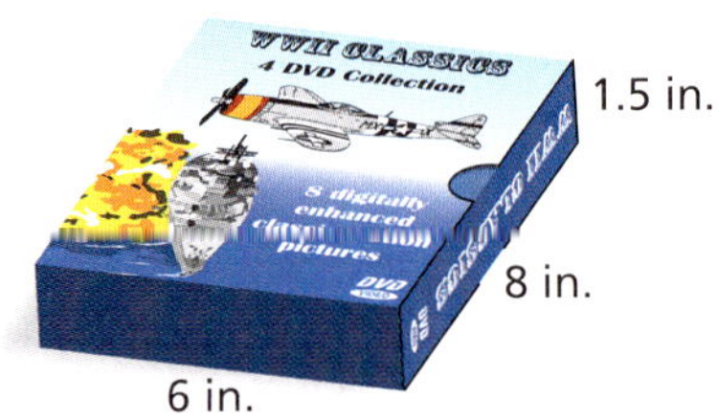

11. SKATEBOARD RAMP A quart of paint covers 80 square feet. How many quarts should you buy to paint the ramp with two coats? (Assume you will not paint the bottom of the ramp.)

12. CUBE A cube has an edge length of 4 inches. You double the side lengths. How many times greater is the volume of the new cube?

Test Item References

Chapter Test Questions	Section to Review	Common Core State Standards
1, 2, 9	8.1	6.G.4
3, 4, 10, 11	8.2	6.G.4
5, 6	8.3	6.G.4
7, 8, 12	8.4	6.G.2

Test-Taking Strategies

Remind students to quickly look over the entire test before they start so that they can budget their time. This test is very visual and requires that students remember many terms. It might be helpful for them to jot down some of the terms on the back of the test before they start.

Common Errors

- **Exercise 3** Students may sum the areas of only 3 of the faces of the rectangular prism instead of all 6. Remind them that a rectangular prism has 6 faces.
- **Exercises 5 and 6** Students may forget to include the area of the base or add the wrong number of lateral face areas when finding the surface area of a pyramid. Remind them that the base is included as part of the surface area, and that they must add the area of each lateral face.
- **Exercises 7 and 8** Students may write the units incorrectly, often writing square units instead of cubic units. Remind them that they are working in three dimensions, so the units are cubed.

Reteaching and Enrichment Strategies

If students need help. . .	If students got it. . .
Resources by Chapter • Practice A and Practice B • Puzzle Time Record and Practice Journal Practice Differentiating the Lesson Lesson Tutorials *BigIdeasMath.com* Skills Review Handbook	Resources by Chapter • Enrichment and Extension • Technology Connection Game Closet at *BigIdeasMath.com* Start Standards Assessment

Answers

1. 8 faces, 18 edges, 12 vertices
2. 8 faces, 14 edges, 8 vertices
3. 28 ft^2
4. 270 ft^2
5. 5 $in.^2$
6. 299.75 m^2
7. $\frac{35}{8}$ cm^3, or $4\frac{3}{8}$ cm^3
8. $\frac{21}{2}$ ft^3, or $10\frac{1}{2}$ ft^3
9. 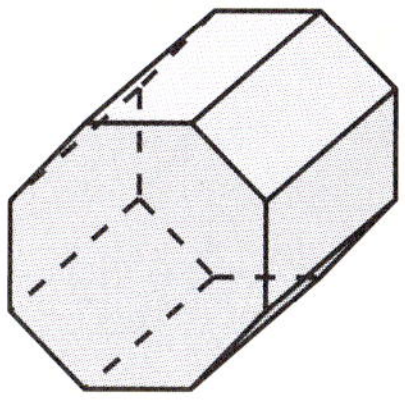
10. 138 $in.^2$
11. 13 quarts of paint
12. 8 times greater

Technology for the Teacher

Online Assessment
Assessment Book
ExamView® Assessment Suite

Test-Taking Strategies

Available at *BigIdeasMath.com*

After Answering Easy Questions, Relax
Answer Easy Questions First
Estimate the Answer
Read All Choices before Answering
Read Question before Answering
Solve Directly or Eliminate Choices
Solve Problem before Looking at Choices
Use Intelligent Guessing
Work Backwards

About this Strategy

When taking a multiple choice test, be sure to read each question carefully and thoroughly. When taking a timed test, it is often best to skim the test and answer the easy questions first. Be careful that you record your answer in the correct position on the answer sheet.

Answers

1. B
2. H
3. C
4. F

Technology for the Teacher

Common Core State Standards Support Performance Tasks
Online Assessment
Assessment Book
ExamView® Assessment Suite

Item Analysis

1. **A.** The student misinterprets "never been above" to mean "always been below."

   B. Correct answer

   C. The student confuses less than and greater than symbols, and uses strict inequality inappropriately.

   D. The student confuses less than and greater than symbols.

2. **F.** The student does not follow the order of operations and adds 6 to $3 \cdot 4^2$ instead of dividing 6 by 2.

   G. The student does not follow the order of operations and adds before multiplying and dividing.

   H. Correct answer

   I. The student does not follow the order of operations and evaluates the expression as $[(3 \cdot 4)^2 + 6] \div 2$

3. **A.** The student multiplies the volume by $\frac{1}{2}$.

   B. The student calculates the surface area instead of the volume.

   C. Correct answer

   D. The student multiplies the volume by 2.

4. **F.** Correct answer

   G. The student adds 60 and 8 together.

   H. The student interchanges the roles of 60 and 8 in the equation.

   I. The student adds 8 and y together instead of multiplying them.

8 Standards Assessment

1. The temperature in a town has never been above 38 degrees Fahrenheit. Let t represent the temperature, in degrees Fahrenheit. Which inequality represents the temperature in the town? *(6.EE.8)*

   A. $t < 38$ **C.** $t > 38$

   B. $t \leq 38$ **D.** $t \geq 38$

2. Which number is equivalent to the expression below? *(6.EE.1)*

$$3 \cdot 4^2 + 6 \div 2$$

   F. 27 **H.** 51

   G. 33 **I.** 75

3. What is the volume of the package shown below? *(6.G.2)*

   A. 240 in.^{3} **C.** 480 in.^{3}

   B. 376 in.^{3} **D.** 960 in.^{3}

4. A housing community started with 60 homes. In each of the following years, 8 more homes were built. Let y represent the number of years that have passed since the first year, and let n represent the number of homes. Which equation describes the relationship between n and y? *(6.EE.9)*

   F. $n = 8y + 60$ **H.** $n = 60y + 8$

   G. $n = 68y$ **I.** $n = 60 + 8 + y$

5. What is the value of m that makes the equation below true? *(6.EE.7)*

$$4m = 6$$

6. A square pyramid is shown below.

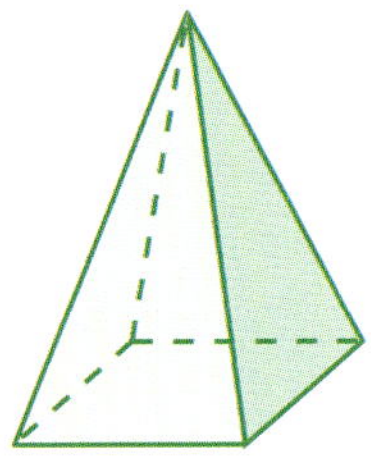

The square base and one of the triangular faces of the square pyramid are shown below with their dimensions.

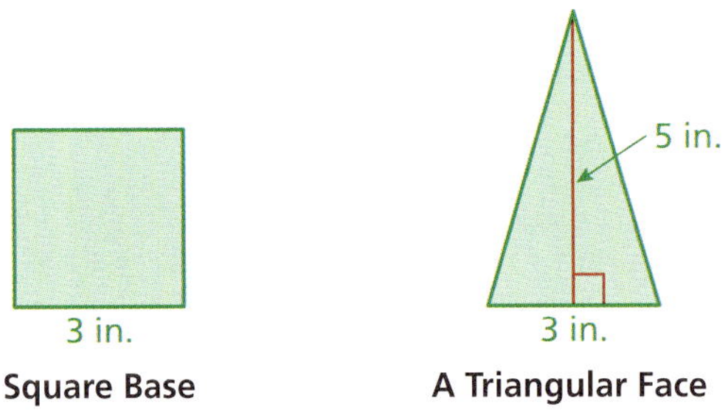

What is the total surface area of the square pyramid? *(6.G.4)*

A. 16.5 in.^{2}

B. 31.5 in.^{2}

C. 39 in.^{2}

D. 69 in.^{2}

7. A wooden box has a length of 12 inches, a width of 6 inches, and a height of 8 inches. *(6.G.2, 6.G.4)*

Think Solve Explain

Part A Draw and label a rectangular prism with the dimensions of the wooden box.

Part B What is the surface area, in square inches, of the wooden box? Show your work.

Part C You have a 2-ounce sample of wood stain that covers 900 square inches. Is this enough to give the entire box two coats of stain? Show your work and explain your reasoning.

Item Analysis (continued)

5. **Gridded Response:** Correct answer: 1.5 or $\frac{3}{2}$

   Common Error: The student subtracts 4 from 6 and gets an answer of 2.

6. **A.** The student includes the area of only one triangular face.

   B. The student includes the area of only three triangular faces.

   C. Correct answer

   D. The student does not multiply by $\frac{1}{2}$ when determining the area of a triangular face.

7. **4 points** The student demonstrates a thorough understanding of the shape and attributes of a rectangular prism. The student draws and correctly labels the dimensions of the box and calculates the surface area as 432 square inches. The student is able to explain why the 2-ounce sample of stain is enough for two coats. The student shows accurate, complete work for all parts and provides clear and complete explanations.

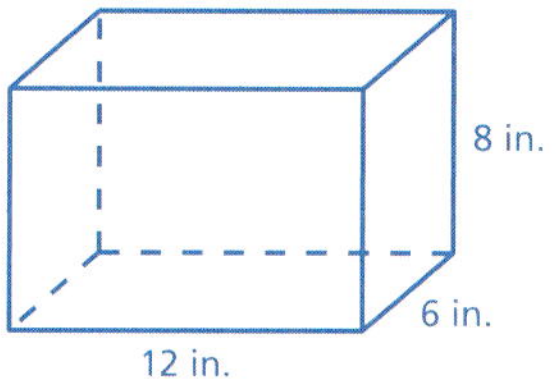

3 points The student demonstrates an understanding of the shape and attributes of a rectangular prism, but the student's work and explanations demonstrate an essential but less than thorough understanding.

2 points The student demonstrates a partial understanding of the shape and attributes of a rectangular prism. The student's work and explanations demonstrate a lack of essential understanding.

1 point The student demonstrates very limited understanding of the shape and attributes of a rectangular prism. The student's response is incomplete and exhibits many flaws.

0 points The student provided no response, a completely incorrect or incomprehensible response, or a response that demonstrates insufficient understanding of the shape and attributes of a rectangular prism.

Answers

5. 1.5, or $\frac{3}{2}$

6. C

7. *Part A*

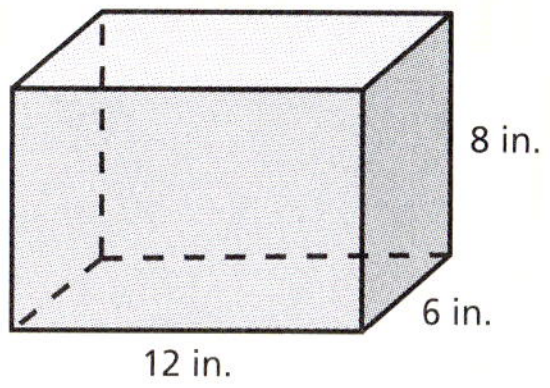

Part B

Surface Area = Top + Bottom + Front + Back + Side + Side

= 12 • 6 + 12 • 6 + 12 • 8 + 12 • 8 + 6 • 8 + 6 • 8

= 432 square inches

Part C yes; 900 ÷ 432 ≈ 2.1, so you should have just enough to cover the box twice.

Answers

8. I
9. 6
10. A

Item Analysis (continued)

8. **F.** The student multiplies the two lengths instead of dividing.

G. The student subtracts the two lengths instead of dividing.

H. The student divides 6 by 3, and then divides 32 by the result.

I. Correct answer

9. **Gridded Response:** Correct answer: 6

Common error: The student adds 1 to 2 in the top row to get 3, or the student multiplies 2 by 2 to get 4 because the second entry is double the first entry.

10. **A.** Correct answer

B. The student counts all squares with any shading.

C. The student uses $A = bh$ when finding the area of the triangle.

D. The student multiplies the height of the figure by the base of the triangle.

8. A biologist measures the lengths of a crazy ant and a green anole that he has in his laboratory. His measurements are shown below.

Not drawn to scale

The length of the green anole is how many times greater than the length of the crazy ant? *(6.NS.1)*

F. $\frac{9}{16}$

G. $5\frac{29}{32}$

H. 16

I. 64

9. What is the missing value in the ratio table? *(6.RP.3a)*

Castles	1	2		12
Towers	4	8	24	48

10. What is the area of the shaded figure shown below? *(6.G.1)*

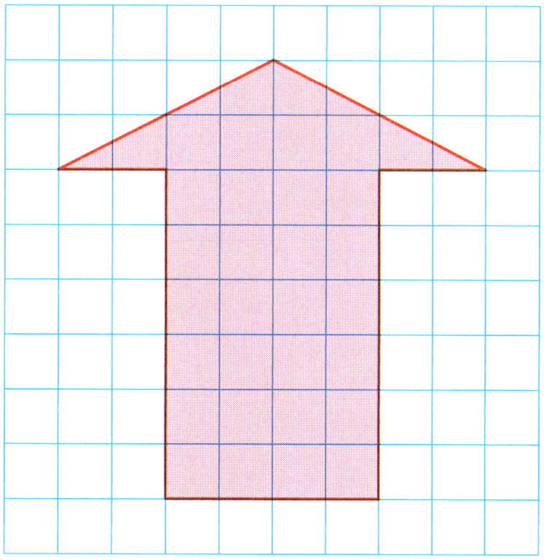

A. 32 units^2

B. 36 units^2

C. 40 units^2

D. 64 units^2

9 Statistical Measures

"Mom, my owner, and Fluffy have agreed to participate in my survey. Will you be my fourth participant?"

"Please hold still. I am trying to find the mean of 6, 8, and 10 by dividing their sum into three equal piles."

Common Core Progression

4th Grade

- Make a line plot with data in fractions of a unit.
- Use line plots to solve problems involving addition and subtraction of fractions.

5th Grade

- Use line plots to solve problems involving operations on fractions.

6th Grade

- Recognize statistical questions as ones anticipating variability.
- Understand that data used to answer statistical questions has a distribution that can be described by center and spread.
- Use measures of center to summarize all of the values in a data set with a single number.
- Use measures of variation to summarize how all of the values in a data set vary with a single number.

Chapter Summary

Section	Common Core State Standard	
9.1	Learning	6.SP.1 ★, 6.SP.4
9.2	Learning	6.SP.2, 6.SP.3, 6.SP.5a, 6.SP.5c
9.3	Learning	6.SP.2, 6.SP.3, 6.SP.5c
9.4	Learning	6.SP.2, 6.SP.3, 6.SP.5c
9.5	Learning	6.SP.2, 6.SP.3 ★, 6.SP.5c

★ Teaching is complete. Standard can be assessed.

Pacing Guide for Chapter 9

Chapter Opener	1 Day
Section 1 Activity Lesson	 1 Day 1 Day
Section 2 Activity Lesson	 1 Day 1 Day
Section 3 Activity Lesson	 1 Day 1 Day
Study Help / Quiz	1 Day
Section 4 Activity Lesson	 1 Day 1 Day
Section 5 Activity Lesson	 1 Day 1 Day
Chapter Review/ Chapter Tests	2 Days
Total Chapter 9	14 Days
Year-to-Date	141 Days

Technology for the Teacher

BigIdeasMath.com
Chapter at a Glance
Complete Materials List
Parent Letters: English and Spanish

Common Core State Standards

4.NF.7 Compare two decimals to hundredths by reasoning about their size. . . . justify the conclusions, e.g., by using a visual model.
3.MD.3 Draw a . . . scaled bar graph to represent a data set with several categories. Solve one- and two-step "how many more" and "how many less" problems using information presented in scaled bar graphs.
4.NBT.4 Fluently add and subtract multi-digit whole numbers using the standard algorithm.

Additional Topics for Review

- Making Line Plots
- Operations with Decimals

Try It Yourself

1.
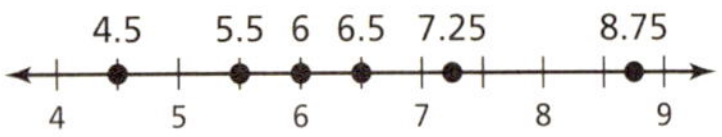

2.
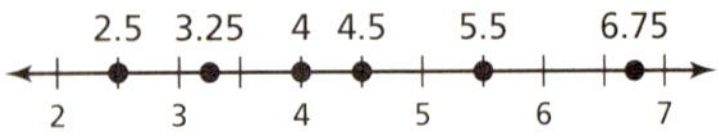

3.
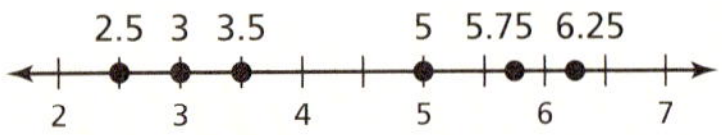

4.
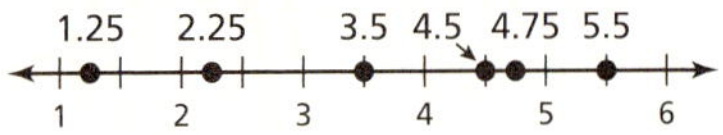

5. 2131 more female athletes

6. *Sample answer:* There are more females and fewer males in 2004 than in 2000.

Record and Practice Journal Fair Game Review

1.
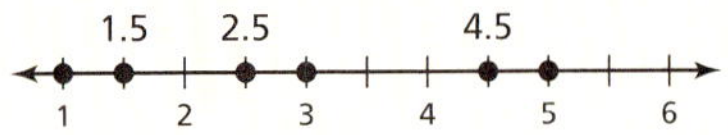

2.
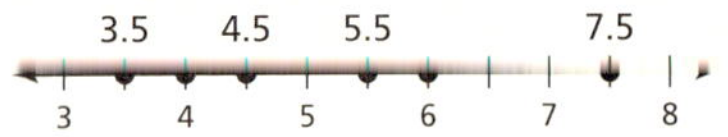

3.
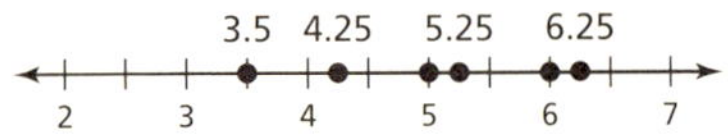

4.
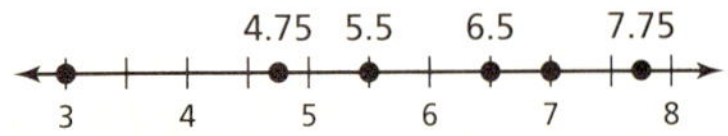

5.
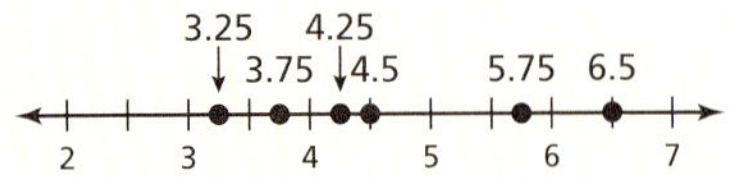

6. See Additional Answers.

7. \$122

8. pants **9.** pants

Math Background Notes

Vocabulary Review

- Decimal
- Number Line
- Double Bar Graph

Ordering Decimals

- Students should know how to use a number line to order decimals.

Analyzing Double Bar Graphs

- Students should know how to read and draw double bar graphs. Because this is a fairly recent skill, students may still need practice to achieve mastery.
- In single bar graphs, each category contains one data set (one bar). Remind students that in a double bar graph each category contains two data sets (two bars).
- **Teaching Tip:** This is an excellent opportunity to review the parts of a graph with students. Sample questions could include: What is the title of the graph? Where is the legend? What does it tell you? What are the categories? What data sets are being compared? Or, what scale is being used to measure the axes?

Reteaching and Enrichment Strategies

If students need help. . .	If students got it. . .
Record and Practice Journal • Fair Game Review Skills Review Handbook Lesson Tutorials	Game Closet at *BigIdeasMath.com* Start the next section

What You Learned Before

"Our mean weight is 18 pounds."

Ordering Decimals (4.NF.7)

Example 1 Use a number line to order 8, 6.5, 7.25, 5.5, 4.25, and 7 from least to greatest.

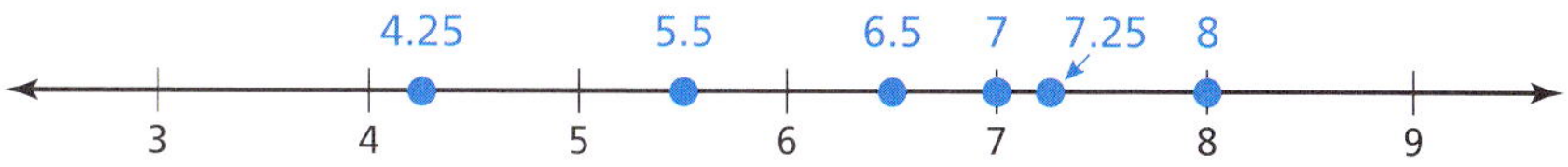

Try It Yourself

Use a number line to order the numbers from least to greatest.

1. 7.25, 4.5, 6.5, 6, 5.5, 8.75

2. 4, 2.5, 3.25, 5.5, 4.5, 6.75

3. 6.25, 3, 2.5, 3.5, 5.75, 5

4. 1.25, 5.5, 4.75, 4.5, 3.5, 2.25

Analyzing Double Bar Graphs (3.MD.3, 4.NBT.4)

Example 2 How many more male athletes than female athletes participated in the 1992 Summer Olympics?

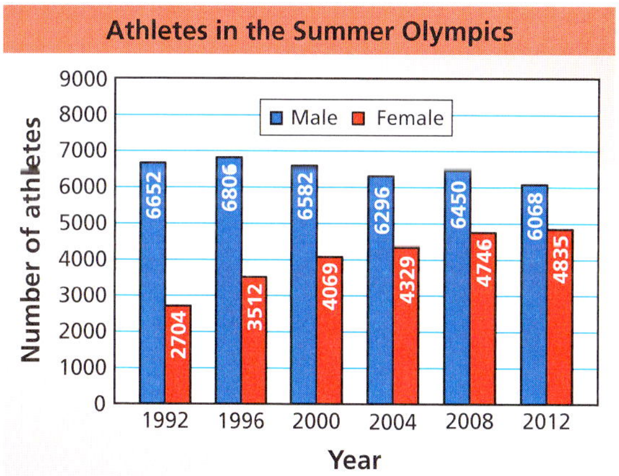

$6652 - 2704 = 3948$

6652: Number of male athletes in 1992

2704: Number of female athletes in 1992

∴ 3948 more male athletes participated.

Example 3 How many athletes participated in the 2000 Summer Olympics?

∴ $6582 + 4069 = 10{,}651$ athletes participated in the 2000 Summer Olympics.

Try It Yourself

5. How many more female athletes participated in the 2012 Summer Olympics than in the 1992 Summer Olympics?

6. Describe the relationship between the number of athletes in the 2000 Summer Olympics and the number of athletes in the 2004 Summer Olympics.

9.1 Introduction to Statistics

Essential Question How can you tell whether a question is a statistical question?

Your heart rate is the number of times your heart beats in a certain time period, such as 1 minute. To measure your heart rate, you can check your pulse. The illustration shows how to check your pulse by pressing lightly on your wrist.

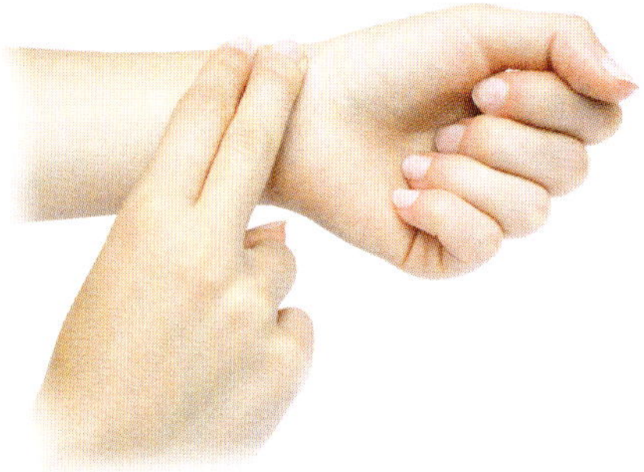

Here are other places to check your pulse:

- inside your elbow
- side of your neck
- top of your foot

1 ACTIVITY: Using Data to Answer Questions

Work with a partner.

a. Find your pulse by counting the number of beats in 10 seconds. Have your partner keep track of the time. Write a rate to describe your result.

b. Complete the ratio table. What is your heart rate in beats per minute?

Time (seconds)	10	30	60
Number of Beats			

c. Collect the recorded heart rates (in beats per minute) of the students in your class, including yourself. Compare the heart rates.

d. **MODELING** Make a *line plot* of your data. Then answer the following questions:

- How many values are in your data set?
- Do the heart rates *cluster* around a particular value or values?
- Are there any *peaks* or *gaps* in the data?
- Are there any unusual heart rates that are far removed from the other values?

e. **REASONING** How would you answer the following question by using only one value? Explain your reasoning.

"What is the heart rate of sixth grade students?"

f. **REASONING** Read and compare the following questions. How did you answer each question? Could the answer be the same for both questions? Explain.

- *What is your heart rate?*
- *What is the heart rate of sixth grade students?*

COMMON CORE
Statistics
In this lesson, you will
- recognize statistical questions.
- use dot plots to display numerical data.

Learning Standards
6.SP.1
6.SP.4

Laurie's Notes

Introduction

Standards for Mathematical Practice

- **MP3 Construct Viable Arguments and Critique the Reasoning of Others:** Mathematically proficient students make conjectures and justify their explanations or conclusions to others. At this point, students should understand the need to give logical support for their statements.

Motivate

- **MP5 Use Appropriate Tools Strategically:** You may want students to use a calculator to quickly explore your question.

? When students settle down ask, "Have you had your millionth heartbeat?" You may need to explain the question.

- Assuming a rate of 60 beats per minute: 60 bpm × (60 min/1 hour) × (24 hours/1 day) × (365 days/1 year) × (11 years old) = 346,896,000; A 12-year-old will have had about 378,432,000 beats.
- Students who are older than 12 may wish to figure out a closer approximation using the months and days since their last birthday.

Activity Notes

Activity 1

- This activity provides some groundwork for statistics: students are collecting, analyzing, and interpreting data to answer a question.
- Provide directions for finding your pulse.
  - Face the palm of one of your hands upward.
  - Place the tips of your fingers on your wrist below the base of your thumb as shown.
  - Press lightly until you feel the pulsing beneath your fingers.
- Students may need to feel around on their wrists to find a pulse. If unsuccessful, they can try at the elbow, neck, or foot, as shown in the chart.
- Have students count the number of beats for 10 seconds, then add or multiply to find the number of beats per minute. Use a ratio table.
- In part (c), students will find that heart rates vary from person to person.
- In part (d), students are familiar with line plots from previous grades. Data are shown as marks above a number line. A line plot should have at least 15 data values in order to show trends. For a class of fewer than 15 students, gather additional data from another class.
- **MP3:** In part (e), students understand that they can describe a set of values using only one number. Students should explain why they chose the value they did.

Common Core State Standards

6.SP.1 Recognize a statistical question as one that anticipates variability in the data related to the question and accounts for it in the answers.
6.SP.4 Display numerical data in plots on a number line, including dot plots, . . .

Previous Learning

Students should know how to make line plots and ratio tables.

Technology for the Teacher

Dynamic Classroom

Lesson Plans
Complete Materials List

9.1 Record and Practice Journal

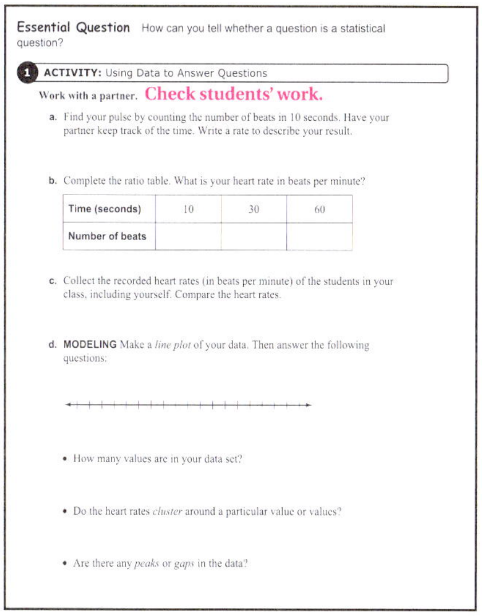

Essential Question How can you tell whether a question is a statistical question?

1 ACTIVITY: Using Data to Answer Questions

Work with a partner. Check students' work.

a. Find your pulse by counting the number of beats in 10 seconds. Have your partner keep track of the time. Write a rate to describe your result.

b. Complete the ratio table. What is your heart rate in beats per minute?

Time (seconds)	10	30	60
Number of beats			

c. Collect the recorded heart rates (in beats per minute) of the students in your class, including yourself. Compare the heart rates.

d. MODELING Make a *line plot* of your data. Then answer the following questions.

- How many values are in your data set?
- Do the heart rates *cluster* around a particular value or values?
- Are there any *peaks* or *gaps* in the data?

Differentiated Instruction

Auditory

Some students may benefit by hearing how to construct a line plot. Ask students questions and have them explain how they got their answers. Should you draw a horizontal or vertical line plot? What increments should you use? What scale should you use and where should it start and end? How can you identify clusters, peaks, and gaps in the data?

9.1 Record and Practice Journal

Check students' work.

- Are there any unusual heart rates that are far removed from the other values?

e. **REASONING** How would you answer the following question by using only one value? Explain your reasoning.

"What is the heart rate of sixth grade students?"

f. **REASONING** Read and compare the following questions. How did you answer each question? Could the answer be the same for both questions? Explain.

- *What is your heart rate?*
- *What is the heart rate of sixth grade students?*

2 ACTIVITY: Identifying Types of Questions

Work with a partner. Check students' work.

a. Answer each question below on your own. Then compare your answers with your partner's answers. For which questions should your answers be the same? For which questions might your answers be different?

Same: 2, 4
Different: 1, 3, 5, 6

1. What is your shoe size?
2. How many states are in the United States?
3. How many brothers and sisters do you have?
4. How many U.S. presidents have been in office?
5. What is your favorite type of movie?
6. How tall are you?

b. **CONJECTURE** Some of the questions on the previous page are considered *statistical* questions. Which ones do you think they are? Why?
1, 3, 5, 6; Answers will most likely vary.

3 ACTIVITY: Analyzing a Question in a Survey

Work with a partner. A student asks the following question in a survey.

"Do you prefer salty potato chips or healthy granola bars to be sold in the school's vending machines?"

a. Do you think this is a fair question to ask in a survey? Explain.
no

b. **LOGIC** Identify the words in the question that may influence someone's response. Then explain how you can reword the question.
salty, healthy

c. How might the results of the survey differ when the student asks the original question and your reworded question in part (b)?
Check students' work.

What Is Your Answer?

4. **REASONING** What do you think "statistics" means?
Science of collecting, organizing, analyzing, and interpreting data
5. **IN YOUR OWN WORDS** How can you tell whether a question is a statistical question? Give examples to support your explanation.
when it anticipates variability
6. Find the least and the greatest heart rates in your class. How can you use these two values to answer the question in Activity 1(e)?
Check students' work.
7. Create a one-question survey. Explain why your question is a statistical question. Then conduct your survey and organize your results in a line plot. Make three observations about your data set.
Check students' work.

Laurie's Notes

Activity 2

- "How many of you have ever answered a survey?" Most say yes.
- "Does everyone answer the same on a survey?" no
- "Why?" Surveys are designed to gather information that varies from person to person.
- In this activity, students should develop an understanding of what *statistical questions* are.
- **MP3:** Ask volunteers to explain what they think statistical questions are. Listen for the idea that statistical questions have answers that vary.
- For a *factual* question such as "How many states are there in the U.S.?" there is one correct answer, and it does not vary across respondents.
- For a *statistical* question such as "What is your shoe size?" there is a correct answer for each respondent, but the answer will vary across respondents.

Activity 3

- In this activity, students must think about how survey questions are worded. Could the language used influence the respondent?
- **MP3:** When students have finished, ask volunteers to share how they reworded the question in part (b). Ask students to explain how the wording of a question can influence responses in a survey.
- **Teaching Tip:** Return to Activity 1(f) and ask why the first question is not a statistical question. Then ask how you can change it to make it a statistical question. For example, "What is your heart rate?" has only one answer or data value. On the other hand, "What is your heart rate over the course of a month?" is a statistical question because it leads to a variety of heart rate data.

What Is Your Answer?

- In Question 4, you could share that the word *statistics* is most likely derived from the Latin word *status*, meaning "state."
- For Question 7, encourage students to identify the number of people in their survey, the units of measurement (if any), and so on. The work can be done in class, if time permits, or it could be homework.

Closure

- Is the question, "Have you had your millionth heartbeat?" a statistical question? Explain. yes; There is a correct answer for each respondent, but the answer will vary across respondents.

2 ACTIVITY: Identifying Types of Questions

Work with a partner.

a. Answer each question below on your own. Then compare your answers with your partner's answers. For which questions should your answers be the same? For which questions might your answers be different?

1. What is your shoe size?

2. How many states are in the United States?

3. How many brothers and sisters do you have?

4. How many U.S. presidents have been in office?

5. What is your favorite type of movie?

6. How tall are you?

Math Practice 3

Build Arguments

How can comparing your answers help you support your conjecture?

b. **CONJECTURE** Some of the questions above are considered *statistical* questions. Which ones do you think they are? Why?

3 ACTIVITY: Analyzing a Question in a Survey

Work with a partner. A student asks the following question in a survey:

"Do you prefer salty potato chips or healthy granola bars to be sold in the school's vending machines?"

a. Do you think this is a fair question to ask in a survey? Explain.

b. **LOGIC** Identify the words in the question that may influence someone's response. Then explain how you can reword the question.

c. How might the results of the survey differ when the student asks the original question and your reworded question in part (b)?

What Is Your Answer?

4. **REASONING** What do you think "statistics" means?

5. **IN YOUR OWN WORDS** How can you tell whether a question is a statistical question? Give examples to support your explanation.

6. Find the least and the greatest heart rates in your class. How can you use these two values to answer the question in Activity 1(e)?

7. Create a one-question survey. Explain why your question is a statistical question. Then conduct your survey and organize your results in a line plot. Make three observations about your data set.

Practice Use what you learned about different types of questions to complete Exercises 4–7 on page 394.

9.1 Lesson

Check It Out
Lesson Tutorials
BigIdeasMath.com

Key Vocabulary
statistics, *p. 392*
statistical question, *p. 392*

Statistics is the science of collecting, organizing, analyzing, and interpreting data. A **statistical question** is one for which you do not expect to get a single answer. Instead, you expect a variety of answers, and you are interested in the distribution and tendency of those answers.

Recall that a dot plot uses a number line to show the number of times each value in a data set occurs. Dot plots show the *spread* and the *distribution* of a data set.

EXAMPLE 1 Answering a Statistical Question

You conduct a science experiment on house mice. Your teacher asks you, "What is the weight of a mouse?"

a. Is this a statistical question? Explain.

Because you can anticipate that the weights of mice will vary, it is a statistical question.

b. You weigh some mice and record the weights (in grams) in the table. Display the data in a dot plot. Identify any clusters, peaks, or gaps in the data.

Weights (grams)			
20	19	21	20
18	20	27	21
28	23	20	19
20	21	18	27
19	22	21	20

Draw a number line that includes the least value, 18, and greatest value, 28. Then place a dot above the number line for each data value.

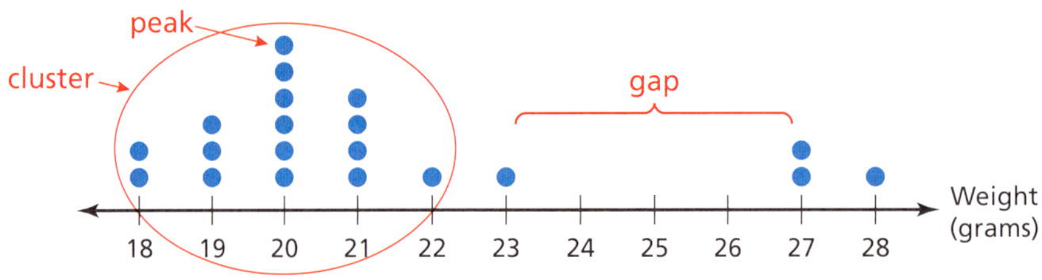

Most of the data are clustered around 20. There is a peak at 20 and a gap between 23 and 27.

c. Use the distribution of the data to answer the question.

Most mice weigh about 20 grams.

Study Tip
Dot plots are sometimes called *line plots*. It is easy to see clusters, peaks, and gaps in a dot plot.

On Your Own

Now You're Ready
Exercises 8–16

1. The table shows the ages of some people who retired early. You are asked, "How old are people who retire early?"
   a. Is this a statistical question? Explain.
   b. Display the data in a dot plot. Identify any clusters, peaks, or gaps in the data.
   c. Use the distribution of the data to answer the question.

Ages			
60	61	59	60
62	56	64	59
58	60	61	60
59	60	58	61

Multi-Language Glossary at BigIdeasMath.com

Laurie's Notes

Introduction

Connect

- **Yesterday:** Students explored the idea of a statistical question. (MP3, MP5)
- **Today:** Students will use the formal definition of a statistical question and plot the responses to a statistical question.

Motivate

- Give each student a sticky note when class begins.
- ? "How many times have you eaten pizza (or another popular food) in the last month? If you can't remember, an estimate is fine. Write your answer on the sticky note."
- Use student responses to make a line plot on the board. Draw a horizontal number line with a scale that fits the range of the data.
- Have students go to the board to place their sticky notes above the appropriate values.

Discuss

- ? "What do you know about the word *statistics*?" Answers will vary.
- Define *statistics* and relate it to the answers given by students.
- Define *statistical question* and relate it to yesterday's activities. You are interested in the distribution and tendency of the answers.
- Ask whether the "Motivate" question about pizza is a statistical question.
- Review *dot plot*. The plot made by students can be called a dot plot.

Lesson Notes

Example 1

- Help students understand that a survey question is not itself a statistical question, but is used to help answer one. The survey question, "How old are you?" is used to gather data to answer the statistical question, "How old are the students in your school?"
- The terms *cluster*, *peak*, and *gap* describe the distribution of data. A cluster is when several data points lie in a small interval. A peak is the data value that occurs most often in an interval. A gap is an interval with no data. A data set can have *zero*, *one*, or *more* clusters, peaks, or gaps.
- ? "What is the least weight of the mice?" 18 grams "What is the greatest weight of the mice?" 28 grams
- Draw a number line with whole numbers from 18 to 28. Do not skip 24, 25, and 26. These numbers must be included to see the gap in the data.
- Have students describe the dot plot in terms of clusters, peaks, and gaps.
- Ask the original (statistical) question, "What is the weight of a mouse?"

On Your Own

- **Neighbor Check:** Have students work independently and then have their neighbors check their work. Have students discuss any discrepancies.

Goal Today's lesson is using dot plots to answer **statistical questions**.

Technology for the Teacher

Dynamic Classroom

Lesson Tutorials
Lesson Plans
Answer Presentation Tool

Extra Example 1

The table shows the weights of the students in a class. You ask, "What is the weight of a student in the class?"

Weights (lbs)			
83	101	94	92
85	93	101	84
95	92	94	90
95	91	95	92
91	92	100	92

a. Is this a statistical question? Explain. Because you can anticipate that the weights of students will vary, it is a statistical question.

b. Display the data in a dot plot. Identify any clusters, peaks, or gaps in the data.

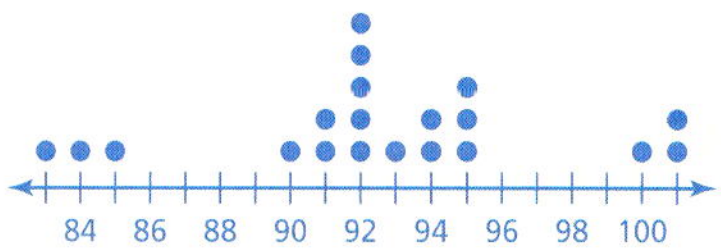

Many of the data are clustered around 92. There is a peak at 92. There are gaps between 85 and 90 and between 95 and 100.

c. Use the distribution of the data to answer the question. Most students in the class weigh about 92 pounds.

On Your Own

1. See Additional Answers.

English Language Learners

Vocabulary

Emphasize the relationship between a survey question and a statistical question:

A survey question is asked to one person in a group. The answers from each person are used to answer a related **statistical question** about the whole group.

Extra Example 2

The dot plot shows the heights of sixth grade students in a gym class.

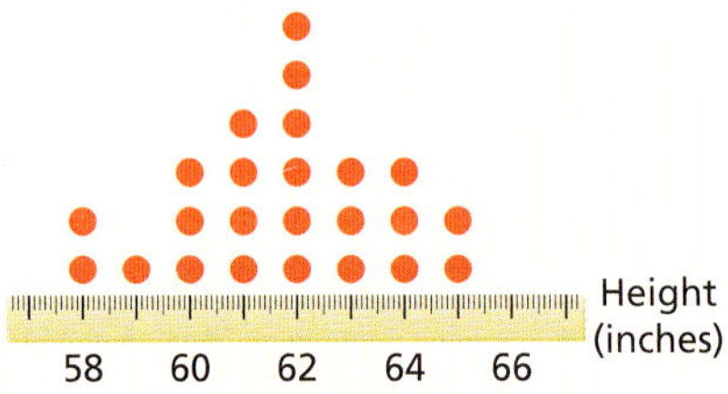

a. How many students are in the class? 24

b. How can you collect these data? What are the units? Use a measuring tape; inches

c. Write a statistical question that you can answer using the dot plot. Then answer the question. *Sample answer:* How tall are the students in the gym class? Most of the students are about 62 inches tall.

On Your Own

2. a. 20

b. Time the students using a stopwatch; seconds (in tenths)

c. *Sample answer:* How long does it take a sixth grade student to run 100 meters? It takes most sixth graders about 13.8 seconds to run 100 meters.

Laurie's Notes

Example 2

- This is a nice example of a vertical dot plot.
- Thermometers are not scaled as shown. This is done for illustrative purposes.
- Ask students whether they think they could use negative values in a dot plot (negative temperatures, golf scores).
- In part (a), point out how a conversion factor is used to convert days to weeks.
- **MP5 Use Appropriate Tools Strategically:** Students should be familiar with Fahrenheit and Celsius thermometers. Ask students to estimate the range of numbers on a Celsius thermometer that would correspond to those shown.
- "What is the daily high temperature in August?" is an example of a statistical question that does not have a single answer. The temperatures are spread out from 80° to 87° with two peaks occurring at 81° and 86°.
- **Extension:** Ask another statistical question: "How often does the daily high temperature exceed 85°F in August?" It makes sense to answer this question for a specific location as in Example 2. You would expect different answers for Seattle and Miami.

On Your Own

- The units of time are not specified, but students should know that reasonable units for this question would be seconds.

Closure

- Write a statistical question that you can answer using the dot plot made with the sticky notes at the beginning of class. Then answer the question.

EXAMPLE 2 Using a Dot Plot

You record the high temperature every day while at summer camp in August. Then you create the vertical dot plot.

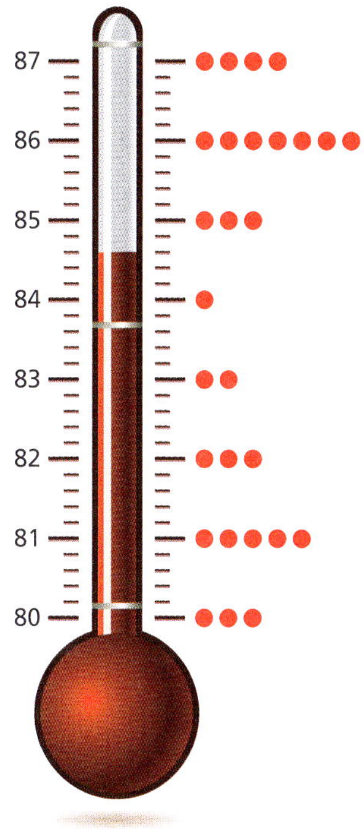

a. How many weeks were you at summer camp?

Because there are 28 data values on the dot plot, you were at camp 28 days.

$$28 \text{ days} \cdot \frac{1 \text{ week}}{7 \text{ days}} = 4 \text{ weeks}$$

So, you were at summer camp for 4 weeks.

b. How can you collect these data? What are the units?

You can collect these data with a thermometer. The units are degrees Fahrenheit (°F).

c. Write a statistical question that you can answer using the dot plot. Then answer the question.

One possible statistical question is:

What is the daily high temperature in August?

The high temperatures are spread out with about half of the temperatures around 81°F and half of the temperatures around 86°F.

On Your Own

Now You're Ready
Exercises 17 and 18

2. The dot plot shows the times of sixth grade students in a 100-meter race.

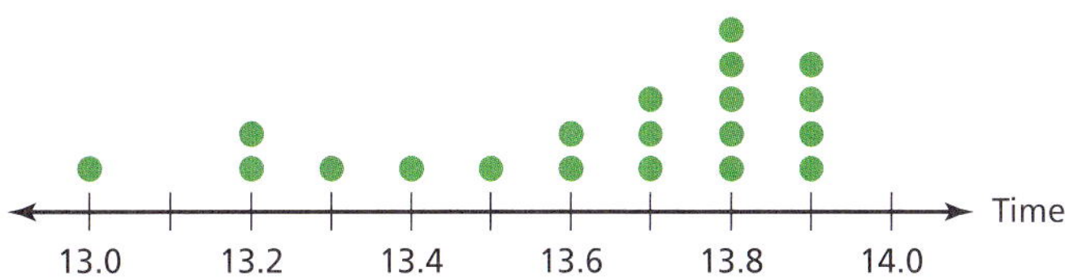

a. How many students ran in the race?

b. How can you collect these data? What are the units?

c. Write a statistical question that you can answer using the dot plot. Then answer the question.

9.1 Exercises

Check It Out
Help with Homework
BigIdeasMath.com

Vocabulary and Concept Check

1. **VOCABULARY** What is a statistical question? Give an example.
2. **CRITICAL THINKING** What process can you use to answer a statistical question?
3. **NUMBER SENSE** The results of a survey are shown in the table. Did the survey ask a statistical question? Explain.

Miles			
6	1	9	2
2	5	4	9
8	10	6	6
5	1	8	1

Practice and Problem Solving

Answer the question. Tell whether your answer would be the same as your classmates'.

4. How many inches are in 1 foot?
5. How many pets do you have?
6. On what day of the month were you born?
7. How many senators are in Congress?

Determine whether the question is a statistical question. Explain.

1 8. What is the eye color of sixth grade students?
9. At what temperature (in degrees Fahrenheit) does water freeze?
10. How many pages are in the favorite books of students your age?
11. How many hours do sixth grade students use the Internet each week?

Display the data in a dot plot. Identify any clusters, peaks, or gaps in the data.

12.

Number of Fouls					
2	1	2	0	0	2
2	1	6	1	1	0

13.

Camper Registrations				
21	25	25	22	21
23	24	26	25	16
24	26	22	25	22

14.

Years			
2011	2008	2013	2009
2009	2010	2010	2009
2010	2012	2009	2010

15.

Test Scores				
85	80	83	90	88
82	83	81	80	89
89	84	86	87	83

16. **SURVEY** You conduct a survey to answer: "How many hours does a sixth grade student spend on homework during a school night?" The table shows the results.

Hours of Homework			
2	4	3	2
1	2	2	1
2	3	5	2

a. Is this a statistical question? Explain.
b. Display the data in a dot plot. Identify any clusters, peaks, or gaps in the data.
c. Use the distribution of the data to answer the question.

Assignment Guide and Homework Check

Level	Day 1 Activity Assignment	Day 2 Lesson Assignment	Homework Check
Basic	4–7, 26–29	1–3, 9–23 odd	9, 13, 15, 17
Average	4–7, 26–29	1–3, 8–16 even, 17–23 odd	10, 14, 16, 17, 23
Advanced	4–7, 26–29	1–3, 8–16 even, 17–25	10, 14, 16, 18, 23

Common Errors

- **Exercises 12–15** Students may miss gaps in the data because they include only data values along the number line for the dot plot. Remind students to include all the integers from the least data value to the greatest data value along the number line.
- **Exercises 17 and 18** Students will have to use reasonableness to determine a method for collecting the data and the units of measurement used. For instance, it is not reasonable for the units to be feet in Exercises 17 and 18.

9.1 Record and Practice Journal

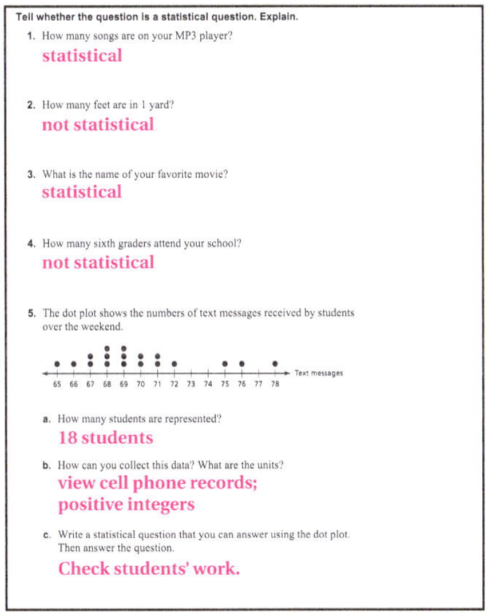
Tell whether the question is a statistical question. Explain.

1. How many songs are on your MP3 player?
   statistical
2. How many feet are in 1 yard?
   not statistical
3. What is the name of your favorite movie?
   statistical
4. How many sixth graders attend your school?
   not statistical
5. The dot plot shows the numbers of text messages received by students over the weekend.

   a. How many students are represented?
   18 students

   b. How can you collect this data? What are the units?
   view cell phone records; positive integers

   c. Write a statistical question that you can answer using the dot plot. Then answer the question.
   Check students' work.

Vocabulary and Concept Check

1. A statistical question is one for which you don't expect to get a single answer. Instead, you expect a variety of answers, and you are interested in the distribution and tendency of those answers. *Sample question:* How old are the teachers in middle school?
2. You need to collect, organize, analyze, and interpret data related to the question.
3. yes; There are many different answers.

Practice and Problem Solving

4. 12 in.; yes
5. *Sample answer:* 2 pets; no
6. *Sample answer:* 9th; no
7. 100 senators; yes
8. statistical; There are many different answers.
9. not statistical; There is only one answer.
10. statistical; There are many different answers.
11. statistical; There are many different answers.
12. 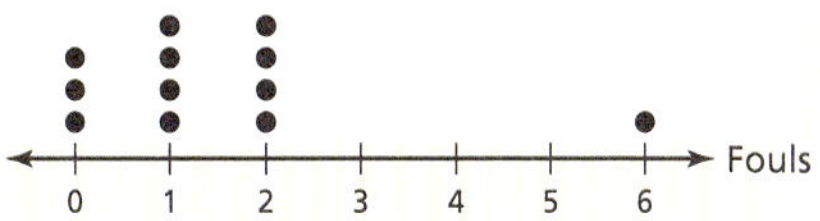

    Most of the data are clustered around 1. There are peaks at 1 and 2. There is a gap between 2 and 6.

13–16. See Additional Answers.

Practice and Problem Solving

17–24. See Additional Answers.

25. See *Taking Math Deeper.*

Fair Game Review

26. yes **27.** no

28. yes **29.** D

Mini-Assessment

1. You conduct a survey in your third class of the day. You ask, "How many hours of television does a student in this class watch in a week?" Is this a statistical question? Explain.
   Because you can anticipate that the number of hours will vary, it is a statistical question.
2. The table shows the results of the survey. Display the data in a dot plot. Identify any clusters, peaks, or gaps in the data.

Time Spent Watching TV (hours)					
5	7	10	10	13	14
14	6	17	18	5	11
12	10	11	12	12	13
15	17	18	11	18	11
13	12	13	12	12	

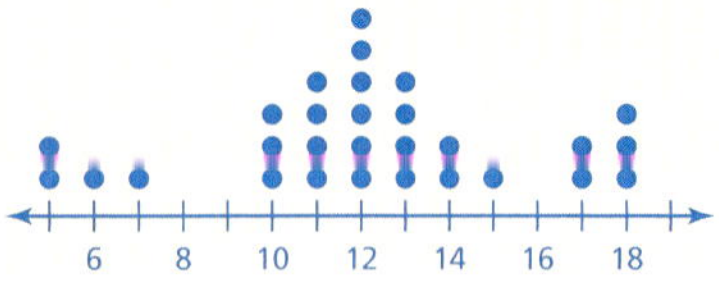

   Most of the data are clustered around 12. There is a peak at 12. There are gaps between 7 and 10 and between 15 and 17.
3. Use the distribution to answer the question. Most students spend 12 hours each week watching television.
4. How many students were in the third class of the day with you? 29

Taking Math Deeper

Exercise 25

Understanding the difference between qualitative and quantitative data will help students solve this problem.

1 One way to begin to solve the problem is to make up data and create a bar graph.

- Choose a few colors: red (R), blue (B), orange (O), yellow (Y), purple (P), green (G).
- Create a bar graph representing 30 people.
- Include a "gap" in the data.

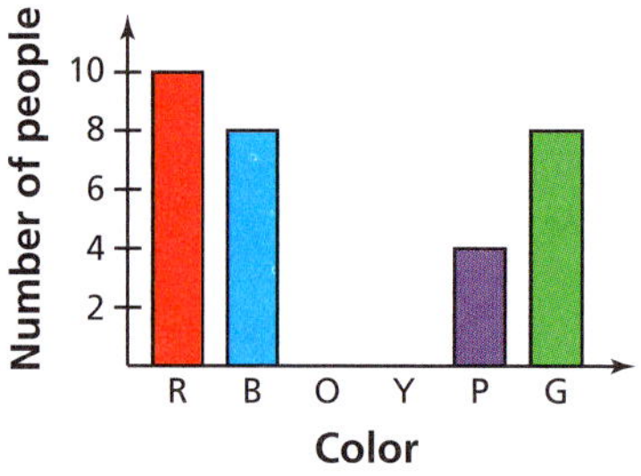

2 Does it make sense to say that there is a gap in the data between blue and purple? Consider changing the order of the colors.

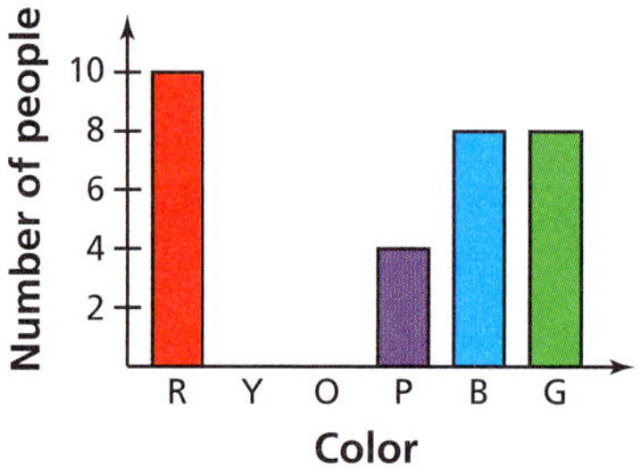

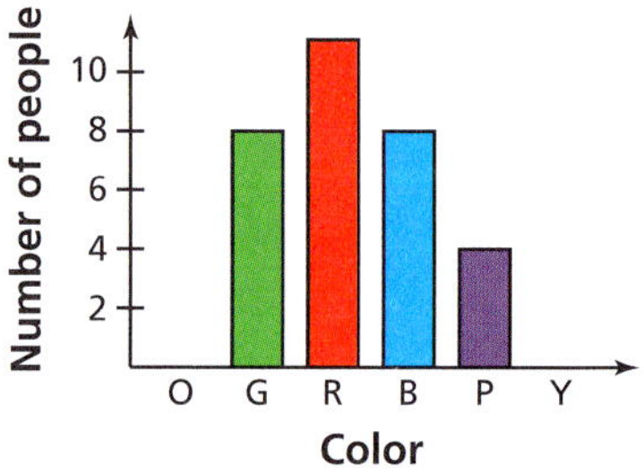

The "distribution" changes when you change the order of the colors.

3 The order of the colors is arbitrary because the data is qualitative. So, it does not make sense to describe the distribution of these data.

Reteaching and Enrichment Strategies

If students need help. . .	If students got it. . .
Resources by Chapter • Practice A and Practice B • Puzzle Time Record and Practice Journal Practice Differentiating the Lesson Lesson Tutorials Skills Review Handbook	Resources by Chapter • Enrichment and Extension • Technology Connection Start the next section

2 **17. EARTHWORMS** The dot plot shows the lengths of earthworms.

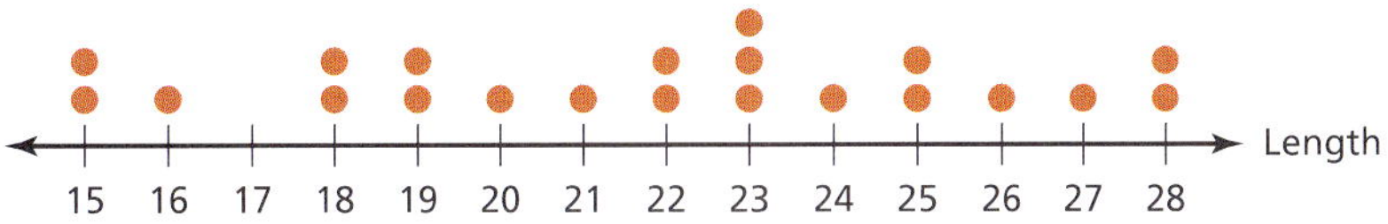

a. How many earthworms does it represent?

b. How can you collect these data? What are the units?

c. Write a statistical question that you can answer using the dot plot. Then answer the question.

18. BASKETBALL The vertical dot plot shows the heights of the players on a recent NBA championship team.

a. How many players were on the team?

b. How can you collect these data? What are the units?

c. Write a statistical question that you can answer using the dot plot. Then answer the question.

Use the Internet to research and identify the method of measurement and the units used when collecting data about the topic.

19. wind speed **20.** amount of rainfall **21.** earthquake intensity

The dot plot shows the speeds of cars in a traffic study. Estimate the speed limit. Explain your reasoning.

22.

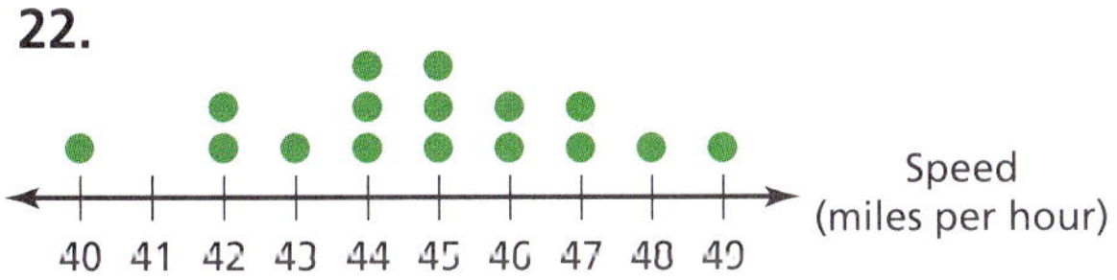

23.

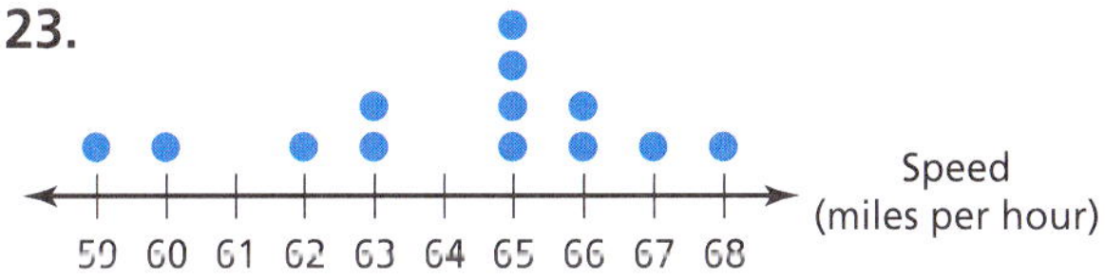

24. REASONING "How many letters are in the English alphabet?" is *not* a statistical question. Write a question about letters that is a statistical question. Explain your reasoning.

25. Reasoning A bar graph shows the favorite colors of 30 people. Does it make sense to describe the distribution of these data? Explain.

Fair Game Review What you learned in previous grades & lessons

Tell whether the ordered pair is a solution of the equation. *(Section 7.4)*

26. $y = 4x$; (2, 8) **27.** $y = 3x + 5$; (3, 15) **28.** $y = 6x - 15$; (4, 9)

29. MULTIPLE CHOICE A point is reflected in the x-axis. The reflected point is $(4, -3)$. What is the original point? *(Section 6.5)*

Ⓐ $(-3, 4)$ Ⓑ $(-4, 3)$ Ⓒ $(-4, -3)$ Ⓓ $(4, 3)$

9.2 Mean

Essential Question How can you find an average value of a data set?

1 ACTIVITY: Finding a Balance Point

Work with a partner. Discuss the distribution of the data. Where on the number line do you think the data set is *balanced*? Is this a good representation of the average? Explain.

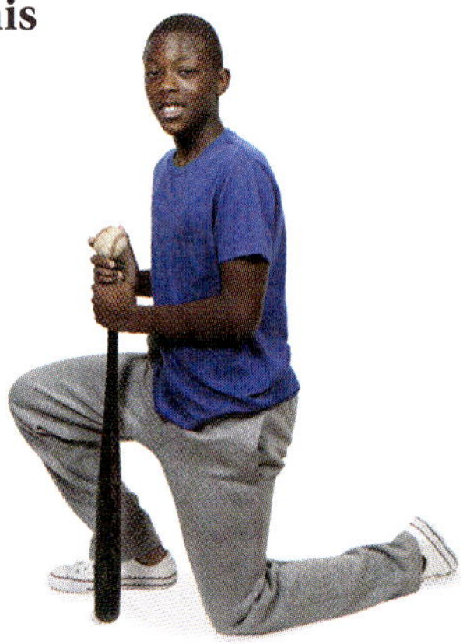

a. number of quarters brought to a batting cage

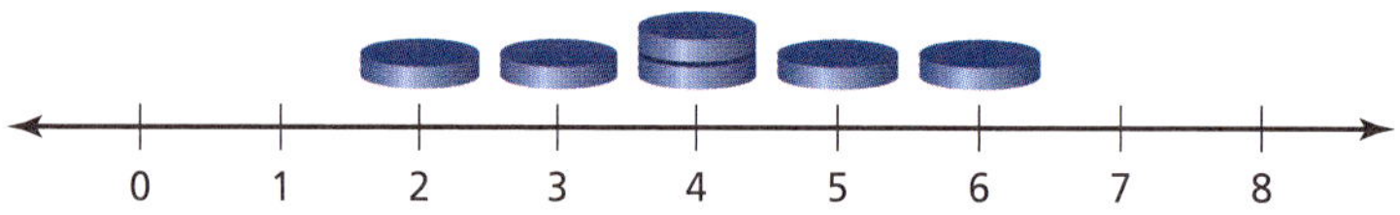

b. annual income of recent graduates (in thousands of dollars)

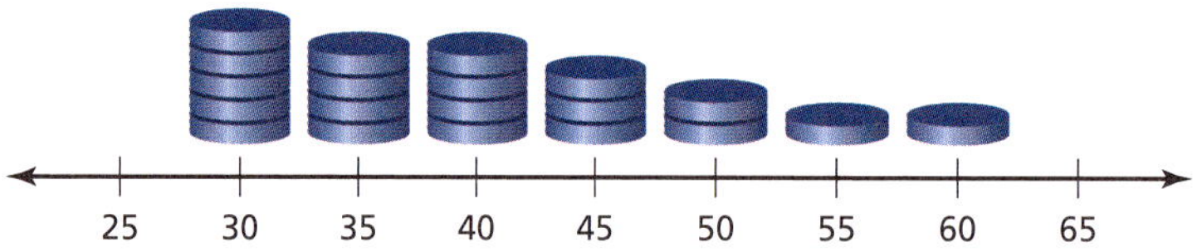

c. hybrid fuel economy (miles per gallon)

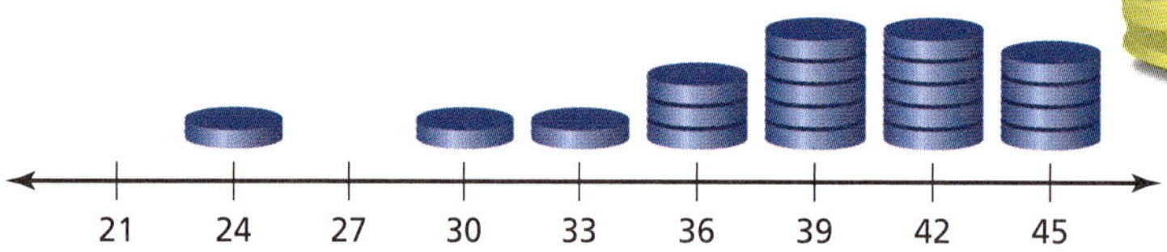

2 ACTIVITY: Finding a Fair Share

Work with a partner. It costs $0.25 to hit 12 baseballs in a batting cage. The table shows the numbers of quarters six friends bring to the batting cage. They want to group the quarters so that everyone has the same amount.

Quarters					
John	Lisa	Miguel	Matt	Cheryl	Jean
6	3	4	5	2	4

Use counters to represent each number in the table. How can you use the counters to determine how many times each friend can use the batting cage? Explain how this procedure results in a "fair share."

Common Core

Statistics

In this lesson, you will

- understand the concept of the mean of data sets.
- find the mean of data sets.
- compare and interpret the means of data sets.

Learning Standards
6.SP.2
6.SP.3
6.SP.5a
6.SP.5c

Laurie's Notes

Introduction

Standards for Mathematical Practice

- **MP2 Reason Abstractly and Quantitatively:** Students use counters to develop the concept of a mean abstractly. Then they use quantitative reasoning to determine the mathematical processes for finding a mean.

Motivate

- Give each student a sticky note, with different colors for girls and boys. Ask students to write their shoe sizes on the note. Make a quick bar graph on the board representing shoe size for each gender.
- Ask students to make observations about each bar graph. Some vocabulary to listen for: *least*, *greatest*, *typical*, *most common*, and so on.
- Ask, "What is the typical, or average, shoe size for girls? For boys?"

Activity Notes

Activity 1

- To introduce this activity, balance a 12-inch ruler at the tip of your index finger. Place the same number of pennies at each end of the ruler.

? "Why is the ruler balanced?" The finger is in the center, and the weight is the same at each end.

? "What will happen if more pennies are added to one end but not the other end?" The ruler will become unbalanced, and the pennies will fall off.

- Probe at the idea of a balancing point. Ask, "How far from the center of the ruler are the piles of pennies?" A likely answer is 5 inches. "Will the ruler be balanced if I place the pennies at the 5-inch and 7-inch marks?" yes "at the 4-inch and 9-inch marks?" no
- Have students work through each part of the activity with partners.
- **MP3 Construct Viable Arguments and Critique the Reasoning of Others:** Have a class discussion about how students decided where on the number line the data set is balanced. Students may start to talk about a *fair share* process, that is, an "equal dividing" of the data.
- Each part of the activity can be modeled using your ruler and some pennies. In part (a), you can place 2 pennies at the 6-inch mark and 1 penny each at the 4-inch, 5-inch, 7-inch, and 8-inch marks.

Activity 2

- **MP5 Use Appropriate Tools Strategically:** The quarters (or counters) are used as a physical model to show the "leveling out" process for finding the "fair share" portion (the mean).
- **MP3:** Six volunteers could model this activity. Have them talk aloud as they share the quarters (or counters) evenly. Relate this to the pennies balancing on the ruler at the beginning of class.
- **Extension:** Ask your students how they would share three more quarters.

Common Core State Standards

6.SP.2 Understand that a set of data collected to answer a statistical question has a distribution which can be described by its center, ...

6.SP.3 Recognize that a measure of center for a numerical data set summarizes all of its values with a single number, ...

6.SP.5a Summarize numerical data sets in relation to their context, such as by reporting the number of observations.

6.SP.5c Summarize numerical data sets in relation to their context, such as by giving quantitative measures of center (. . . mean) . . ., as well as describing any overall pattern and any striking deviations from the overall pattern with reference to the context in which the data were gathered.

Previous Learning

Students should know how to make dot plots to answer statistical questions.

Technology for the Teacher

Dynamic Classroom

Lesson Plans

Complete Materials List

9.2 Record and Practice Journal

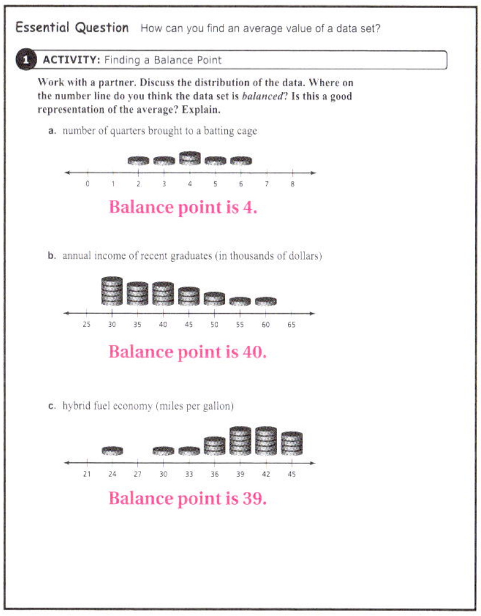

Essential Question How can you find an average value of a data set?

1 ACTIVITY: Finding a Balance Point

Work with a partner. Discuss the distribution of the data. Where on the number line do you think the data set is *balanced*? Is this a good representation of the average? Explain.

a. number of quarters brought to a batting cage

Balance point is 4.

b. annual income of recent graduates (in thousands of dollars)

Balance point is 40.

c. hybrid fuel economy (miles per gallon)

Balance point is 39.

English Language Learners

Vocabulary

English learners may be confused by the word *average*. In math they may read phrases such as "the average rate" or "the average number of participants." Explain to students that the word *average* can be used to describe a common outcome or result. Looking up *average* in the dictionary and writing synonyms may help students to better understand the concept.

9.2 Record and Practice Journal

2 ACTIVITY: Finding a Fair Share

Work with a partner. It costs $0.25 to hit 12 baseballs in a batting cage. The table shows the numbers of quarters six friends bring to the batting cage. They want to group the quarters so that everyone has the same amount.

Quarters					
John	Lisa	Miguel	Matt	Cheryl	Jean
6	3	4	5	2	4

Use counters to represent each number in the table. How can you use the counters to determine how many times each friend can use the batting cage? Explain how this procedure results in a fair share.

4 times

3 ACTIVITY: Finding an Average

Work with a partner. Use the information in Activity 2.

a. What is the total number of quarters the group of friends brought to the batting cage?

24 quarters

b. **REASONING** How can you use math to find the average number of quarters that each friend brought to the batting cage? Find the average number of quarters. Why do you think this average represents a "fair share?"

Divide the total number of quarters by the number of friends.

4 ACTIVITY: Answering a Statistical Question

Work with a partner. The table shows the numbers of quarters several people bring to a batting cage. You want to answer the question:

"How many quarters do people bring to the batting cage?"

Quarters			
6	8	8	12
8	12	8	4
8	6	6	10
7	10	7	8

a. Explain why this question is a statistical question.

Number of quarters will vary from person to person.

b. **MODELING** Make a dot plot of the data. Use the distribution of the data to answer the question. Explain your reasoning.

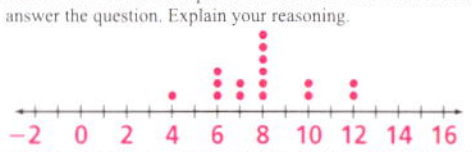

Most people brought between 6 and 8 quarters.

c. **REASONING** Use an average to answer the question. Explain your reasoning.

8 quarters

What Is Your Answer?

5. **IN YOUR OWN WORDS** How can you find an average value of a data set?
Sample answer: You can show the distribution of a data set and find the balance point.

6. Give two real-life examples of averages.
Check students' work.

7. Explain what it means to say the average of a data set is the point on a number line where the data set is balanced.
Sample answer: The sum of the distances from the average of all points above the average is equal to the sum of the distances of all points below the average.

8. There are 5 students in the cartoon. Four of the students are 66 inches tall. One is 96 inches tall.

"Yup, the average height in our class is 6 feet."

a. How do you think the students decided their average height is 6 feet?
Sample answer: Added all heights and divided by 5.

b. Does a height of 6 feet seem like a good representation of the average height of the 5 students? Explain why or why not.
no

Laurie's Notes

Activity 3

- Using counters will help students to see the total number of quarters the six friends bring to the batting cage. The friends then divide the total "fairly" so that each person has the same number of quarters.
- **MP3:** After students finish, ask a volunteer to share how to find the average mathematically and why it represents a "fair share."
- Be careful not to imply that the sharing process is the only way to find an average. There are different types of averages, and the mean is only one.

Activity 4

? "Why is this question a statistical question?" You anticipate that there will be different answers; The answer will vary among different people.

- The dot plot in part (b) will show a cluster and some gaps. Ask a student to describe the plot. Most of the data are clustered around 8 quarters.
- Finding the average is another way to answer the question without using a dot plot. Using the average is a numerical approach and using a dot plot is a graphical approach.

What Is Your Answer?

- In Question 5, students have the opportunity to describe several ways to find an average value of a numeric data set.
- For Question 8, chances are someone in the class knows how to find a mean—add all numbers and divide by the number of values.
- The bigger idea of this question is to help students develop a sense about outliers and how they affect the mean of a data set.

Closure

- **Exit Ticket:** How would you find the *average* time it takes students in your school to commute from home to school? *Sample answer:* You could conduct a survey to see how long it takes each student in the class to commute from home to school. Add all the commute times together and divide by the total number of students who responded to the survey.

3 ACTIVITY: Finding an Average

Work with a partner. Use the information in Activity 2.

a. What is the total number of quarters the group of friends brought to the batting cage?

b. **REASONING** How can you use math to find the average number of quarters that each friend brought to the batting cage? Find the average number of quarters. Why do you think this average represents a fair share?

4 ACTIVITY: Answering a Statistical Question

Work with a partner. The table shows the numbers of quarters several people bring to a batting cage. You want to answer the question:

> ***"How many quarters do people bring to a batting cage?"***

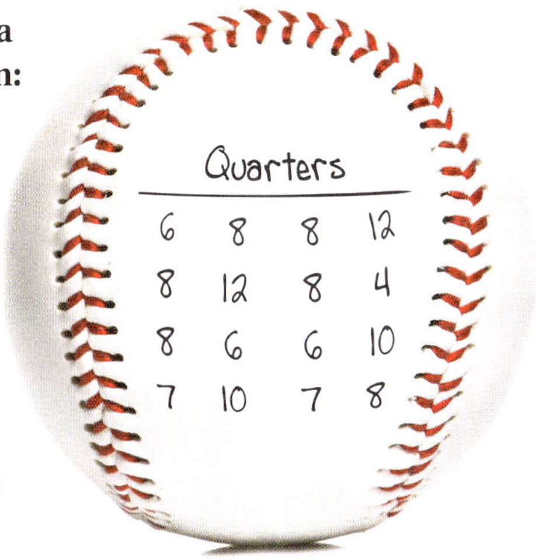

Quarters			
6	8	8	12
8	12	8	4
8	6	6	10
7	10	7	8

a. Explain why this question is a statistical question.

b. **MODELING** Make a dot plot of the data. Use the distribution of the data to answer the question. Explain your reasoning.

c. **REASONING** Use an average to answer the question. Explain your reasoning.

Math Practice 6

Use Clear Definitions

What does it mean for data to have an average? How does this help you answer the question?

What Is Your Answer?

5. **IN YOUR OWN WORDS** How can you find an average value of a data set?

6. Give two real-life examples of averages.

7. Explain what it means to say the average of a data set is the point on a number line where the data set is balanced.

8. There are 5 students in the cartoon. Four of the students are 66 inches tall. One is 96 inches tall.

a. How do you think the students decided that their average height is 6 feet?

b. Does a height of 6 feet seem like a good representation of the average height of the 5 students? Explain why or why not.

"Yup, the average height in our class is 6 feet."

Practice Use what you learned about averages to complete Exercises 4 and 5 on page 400.

Check It Out
Lesson Tutorials
BigIdeasMath.com

Key Vocabulary
mean, *p. 398*
outlier, *p. 399*

A mean is a type of average.

Key Idea

Mean

Words The **mean** of a data set is the sum of the data divided by the number of data values.

Numbers **Data:** 8, 5, 6, 9 **Mean:** $\frac{8+5+6+9}{4} = \frac{28}{4} = 7$

4 data values

EXAMPLE 1 Finding the Mean

Text Messages Sent
Mark: 120
Laura: 95
Stacy: 101
Josh: 125
Kevin: 82
Maria: 108
Manny: 90

The table shows the number of text messages sent by a group of friends over 1 week. What is the mean number of messages sent?

Ⓐ 100 Ⓑ 102 Ⓒ 103 Ⓓ 104

$$\text{mean} = \frac{120 + 95 + 101 + 125 + 82 + 108 + 90}{7}$$

Sum of the data

Number of values

$$= \frac{721}{7}, \text{ or } 103$$

Simplify.

∴ The mean number of text messages sent is 103. The correct answer is Ⓒ.

EXAMPLE 2 Comparing Means

The double bar graph shows the monthly rainfall amounts for two cities over a six-month period. Compare the mean monthly rainfalls.

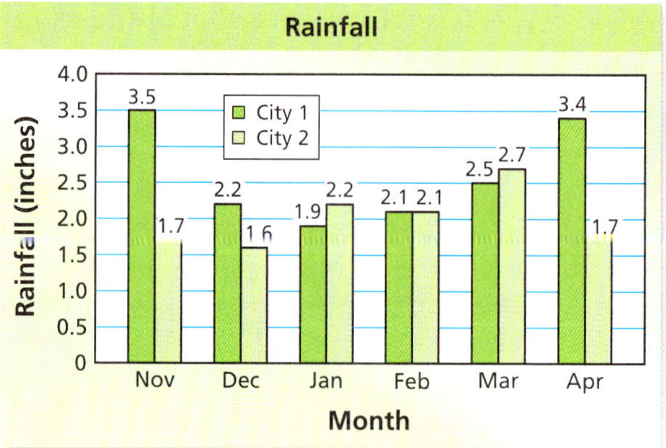

City 1 mean: $\frac{3.5 + 2.2 + 1.9 + 2.1 + 2.5 + 3.4}{6} = \frac{15.6}{6}$, or 2.6

City 2 mean: $\frac{1.7 + 1.6 + 2.2 + 2.1 + 2.7 + 1.7}{6} = \frac{12}{6}$, or 2

∴ Because 2.6 is greater than 2, City 1 averaged more rainfall.

Laurie's Notes

Introduction

Connect

- **Yesterday:** Students developed an intuitive understanding about averages. (MP2, MP3, MP5)
- **Today:** Students will use the formal definition of a mean to answer real-life problems.

Motivate

- Use 24 pennies or counters and sort them into 3 stacks of height 3, 7, and 14. Ask how to find the *average* number of pennies in a stack.
- Ask a volunteer to show how it could be done. Perhaps the student will "share" some pennies among the stacks. The stacks will even out to 8 per stack, which is the mean.
- **Big Idea:** Finding the mean of a data set is a sharing process. By summing the data set and dividing by the number of data values you even out the data, so they are all the same.
- Dividing the pennies into 3 unequal stacks model the process: the sum (24) divided by the number of data values (3) gives you even stacks (8).

Lesson Notes

Key Idea

- There are three *measures of central tendency* or *averages*. Today's lesson is about the mean.
- Write the Key Idea.
- Discuss the example. For the data values of 5, 6, 8, and 9, the mean is somewhere between the low (5) and high (9) data values. After sharing some from the 9 to the 5 and the 8 to the 6, the mean is 7.

Example 1

? "What is the greatest value?" 125 "the least value?" 82

? "What do you know about the mean?" It is a number between 82 and 125.

- Work through the problem.

? "Is 103 a reasonable answer? How can you tell?" Yes, because 103 falls between 82 (the low) and 125 (the high).

- **Caution:** If students use a calculator and enter all the operations in one step, they must remember to place the numerator in parentheses and then divide by the denominator: $\frac{(120 + 95 + 101 + 125 + 82 + 108 + 90)}{7} = 103.$

Example 2

? "Which city is represented by the dark green bar?" City 1

? "In what month was the rainfall the same in the two cities?" February

- Work through the computation for each city.

Goal Today's lesson is finding the **mean** of data sets and discovering the relationship between the mean and **outliers**.

Technology for the Teacher

Dynamic Classroom

Lesson Tutorials
Lesson Plans
Answer Presentation Tool

Extra Example 1

Find the mean of the data.

30, 81, 50, 24, 15, 64 44

Extra Example 2

Compare the mean monthly rainfall (in inches) for the two cities.

City A: 2.5, 4.3, 4.8, 2.7, 1.2

City B: 1.7, 4.1, 5.5, 3.2, 0.5

Because 3.1 is greater than 3, City A averaged more rainfall.

On Your Own

1. 57 2. 7.5

Extra Example 3

The table shows the weights of several kittens.

Kitten Weights (pounds)				
4.5	5.7	4.4	4.45	5.5
5.6	4.7	4.9	7.25	5

a. Identify the outlier. 7.25 pounds

b. Find the mean with and without the outlier. mean with outlier: 5.2 pounds, mean without outlier: about 4.97 pounds

c. Describe how the outlier affects the mean. With the outlier, the mean is greater than 6 of the values (that's over half!). Without the outlier, the mean better represents the weights.

On Your Own

3. The weight of 8 pounds is the outlier, and it makes the mean less than all but two of the weights.
4. The price of $900 is the outlier, and it makes the mean greater than all of the other prices.

Differentiated Instruction

Vocabulary

Have students add a glossary to their math notebooks. Key vocabulary words should be added as they are introduced. Mean, median, and mode are often confusing to students. Illustrations next to the vocabulary words will help in understanding and reinforcing their meanings.

Laurie's Notes

On Your Own

- **Neighbor Check:** Have students work independently and then have their neighbors check their work. Have students discuss any discrepancies.

Example 3

- Ask students to examine the height of each pony.
- ? "Is there one pony that is noticeably different in size from the others? Describe how you can tell." Yes. There is one pony that is 28 inches in height. This is much smaller compared to the other ponies.
- Define outlier. In the pony example, 28 inches is the outlier.
- **Multiple Representations:** Some students will be able to find the outlier simply by reading through the chart. Others may struggle to find the outlier. Have these students graph the heights on a dot plot as shown in part (a). This will visually reinforce why 28 inches is the outlier.
- Calculate the mean of the data.
- ? "Are any of the ponies exactly 37.9 inches tall? How many ponies are taller than the mean?" No. There are 7 ponies that are more than 37.9 inches tall.
- ? "Does the mean do a good job of describing the average pony height?" No.
- Ask students to calculate the mean *without* the outlier.
- **Common Error:** Some students will forget to subtract 1 from the total number of data values. Remind students that they are using one less value and so the denominator of their calculation should decrease by 1.
- ? "Are any ponies exactly 39 inches tall? How many ponies are taller than the mean? Shorter than the mean?" Yes. There are 4 ponies taller than 39 inches, and 4 ponies shorter than 39 inches (ignoring the outlier).
- ? "Which mean better describes the average pony height?" 39 inches. Listen for: the heights are more balanced (4 greater than the mean, 4 less than the mean).
- **Big Idea:** An **outlier** that is very low compared to the rest of the data values will cause the mean to be too low. By eliminating the outlier, you can calculate a mean that better demonstrates the average of the data.
- **MP2 Reason Abstractly and Quantitatively:** Discuss the question, "What effect would an extremely tall pony have on the mean? What could you do to correct this?" It would cause the mean to be too high. You could eliminate the outlier and calculate the mean without it.

On Your Own

- **MP7 Look for and Make Use of Structure:** You could ask students to estimate the mean with and without the outlier. Ask how they found their estimates.

Closure

- **Exit Ticket:** Compute the mean of points scored by a football team in their first 5 games one season: 16, 21, 6, 30, 27 20 points

On Your Own

Now You're Ready
Exercises 6–9

Find the mean of the data.

1. 49, 62, 52, 54, 61, 70, 55, 53
2. 7.2, 8.5, 7.0, 8.1, 6.7

An **outlier** is a data value that is much greater or much less than the other values. When included in a data set, it can affect the mean.

EXAMPLE 3 Finding the Mean With and Without an Outlier

Shetland Pony Heights (inches)				
40	37	39	40	42
38	38	37	28	40

The table shows the heights of several Shetland ponies.

a. Identify the outlier.

b. Find the mean with and without the outlier.

c. Describe how the outlier affects the mean.

a. Display the data in a dot plot.

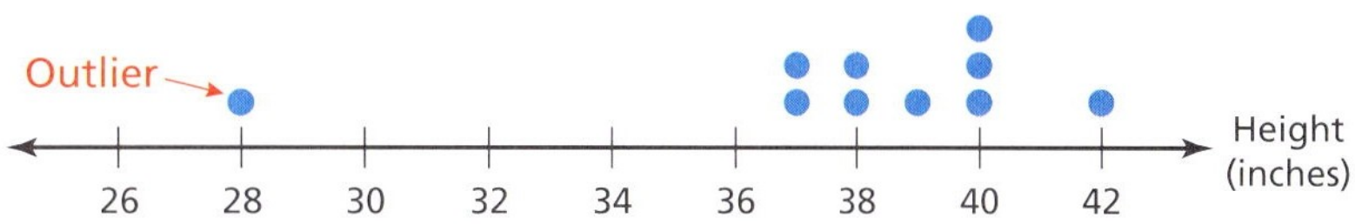

The height of 28 inches is much less than the other heights. So, it is an outlier.

b. Mean with outlier:

$$\frac{40 + 37 + 39 + 40 + 42 + 38 + 38 + 37 + 28 + 40}{10} = \frac{379}{10}, \text{ or } 37.9$$

Mean without outlier:

$$\frac{40 + 37 + 39 + 40 + 42 + 38 + 38 + 37 + 40}{9} = \frac{351}{9}, \text{ or } 39$$

c. With the outlier, the mean is less than all but three of the heights. Without the outlier, the mean better represents the heights.

On Your Own

Now You're Ready
Exercise 14

For each data set, identify the outlier. Then describe how it affects the mean.

3. Weights (in pounds) of dogs at a kennel:

   48, 50, 55, 60, 8, 37, 50

4. Prices for flights from Miami, Florida, to San Juan, Puerto Rico:

   $456, $512, $516, $900, $436, $516

9.2 Exercises

Check It Out
Help with Homework
BigIdeasMath.com

Vocabulary and Concept Check

1. **VOCABULARY** Arrange the words to explain how to find a mean.

   the data values   divide by   the number of data values   add   then

2. **NUMBER SENSE** Is the mean always equal to a value in the data set? Explain.
3. **REASONING** Can you use the mean to answer a statistical question? Explain.

Practice and Problem Solving

Describe an average value of the data.

4. Ages in a class: 11, 12, 12, 12, 12, 12, 13
5. Movies seen this week: 0, 0, 0, 1, 1, 2, 3

Find the mean of the data.

(1) 6.

Pets Owned	
Brandon	I
Jill	III
Mark	II
Nicole	IIII
Steve	0

7.

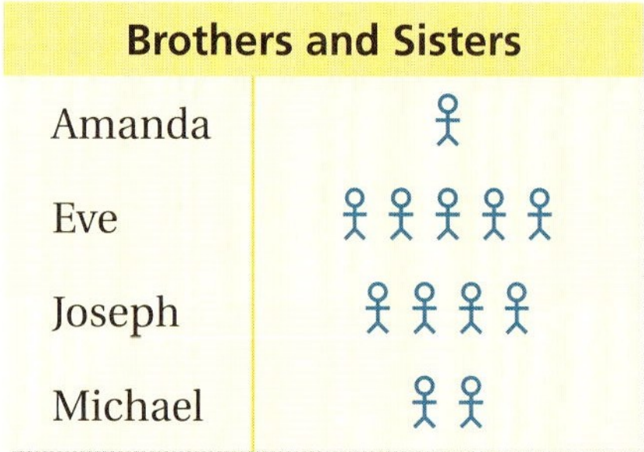

8.

Sit-ups		
108	85	94
103	112	115
98	119	126
105	82	89

9.

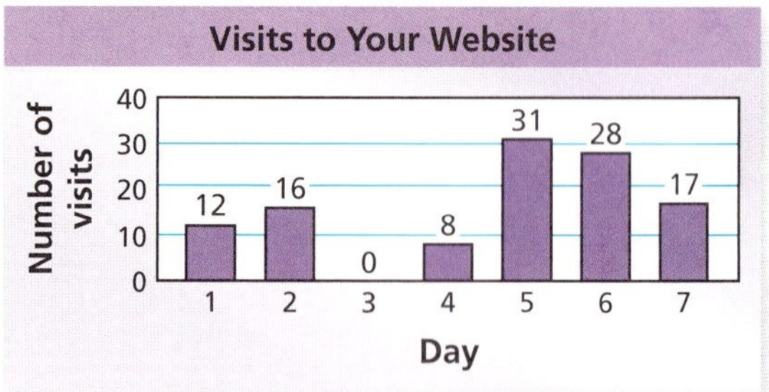

10. **GOLF** The table shows tournament finishes for a golfer.

    a. What was the golfer's mean finish?

    b. Identify two outliers for the data.

Tournament Finishes							
1	1	2	1	1	12	6	2
15	37	1	2	1	26	9	1

11. **COMMERCIALS** You and your friends are watching a television show. One of your friends asks, "How long are the commercial breaks during this show?"

    a. Is this a statistical question? Explain.

    b. Use the mean of the values in the table to answer the question.

Time (minutes)				
4.2	3.5	4.55	2.75	2.25

Assignment Guide and Homework Check

Level	Day 1 Activity Assignment	Day 2 Lesson Assignment	Homework Check
Basic	4, 5, 19–23	1–3, 7, 9, 10, 11, 12	2, 7, 10, 12
Average	4, 5, 19–23	1–3, 7, 9, 11–14	2, 7, 12, 13
Advanced	4, 5, 19–23	1–3, 6, 8, 13, 15–18	2, 6, 13, 17

Common Errors

- **Exercises 6–9** Students may forget to count the total to divide by, and instead use the maximum number. Tell students it is as though they are dividing the total evenly among the number of groups that there are. For example, the mean number of pets owned describes how many pets each person would have if all the pets were divided evenly among each person.
- **Exercise 10** Students may have difficulty finding the outliers just by looking at the table. Ask them to place all the values on a number line. This will help make the two outliers more obvious.

9.2 Record and Practice Journal

Find the mean of the data.

1. Emails sent in the last 4 hours:

   2, 5, 4, 5

   4 emails

2. Magazine subscriptions sold this week:

   3, 6, 7, 6, 7, 9, 11

   7 subscriptions

3.

Books Brought Home	
Monday	|
Tuesday	|||
Wednesday	||||
Thursday	||||
Friday	|

   2.6 books

4. The table shows the number of points scored by your team in each quarter of a football game. What is the mean number of points scored in a quarter?

Quarter	1	2	3	4
Points	3	14	10	0

   6.75 points

Vocabulary and Concept Check

1. Add the data values then divide by the number of data values.
2. No; Dividing the sum of the data by the number of data values to find the mean does not necessarily result in one of the data values.
3. yes; Because of the variability of the answers to a statistical question, the mean gives an average of the answers. That way, you can use only one value, the mean, to answer the statistical question.

Practice and Problem Solving

4. 12 years old; Find the sum of the ages and divide by the number of students in the class.
5. 1 movie seen this week; Find the total number of movies and divide by the number of people.
6. 2 pets
7. 3 brothers and sisters
8. 103 sit-ups
9. 16 visits
10. **a.** about 7th

    b. 26 and 37 are outliers because they are much greater than the other values.
11. **a.** yes; There will be variability in the lengths of the commercial breaks.

    b. 3.45 minutes

Practice and Problem Solving

12. 3.245 inches

13. *Sample answer:*
20, 21, 21, 21, 21, 22
20, 20.5, 20.5, 21.5, 21.5, 22

14. **a.** The 288 minutes used in September is much less than the other values, so it is an outlier.

b. With outlier: 488.4
Without outlier: 538.5
The outlier caused the mean to be about 50 minutes less.

c. *Sample answer:* School could have caused you to spend less time talking on your cell phone.

15. 3.9 inches; No, neither team has a height that is much shorter or taller than the other heights. So, you can say that the Tigers are taller than the Dolphins on average.

16. See Additional Answers.

17. See *Taking Math Deeper.*

18. See Additional Answers.

Fair Game Review

19. 9 **20.** 30

21. 18.5 **22.** 15.5

23. B

Mini-Assessment

Find the mean of the data set.

1. 2, 5, 7, 2, 4 4
2. 22, 29, 18, 31 25
3. 11, 15, 8, 14, 18, 10, 9, 19 13
4. 35, 56, 41, 59, 37, 48, 53 47
5. In your class, 7 students do not receive a weekly allowance, 8 students receive $8, 6 students receive $5, and 10 students receive $3. What is the mean weekly allowance? $4

Taking Math Deeper

Exercise 17

This is an important type of problem in statistics. The key to understanding this problem is to realize that each of 5 different allowance amounts are represented more than once.

1 Make a table.

Students	Allowances	Total
7	$0	7 • 0 = $0
5	$3	5 • 3 = $15
7	$5	7 • 5 = $35
3	$6	3 • 6 = $18
2	$8	2 • 8 = $16

24 Students $84

2 Summarize the table.
So, 24 students receive a weekly total of $84.

Mean: $\frac{84}{24} = \$3.50$

3 According to National Longitudinal Survey of Youth, allowance for 12- to 17-year-olds varies. 50% reported receiving no regular allowance. However, the remaining 50% reported a median allowance of $50 a week. In spite of this, we kept the amounts low to avoid making those with no or low allowances feel bad.

Project

Assume you receive an allowance of $7.00 per week. Keep a budget for one week. In one week, how much money did you spend on food? Calculate the average amount of money you spent on food per day.

Reteaching and Enrichment Strategies

If students need help. . .	If students got it. . .
Resources by Chapter • Practice A and Practice B • Puzzle Time Record and Practice Journal Practice Differentiating the Lesson Lesson Tutorials Skills Review Handbook	Resources by Chapter • Enrichment and Extension • Technology Connection Start the next section

Month	Rainfall (inches)	Month	Rainfall (inches)
Jan	2.22	Jul	3.27
Feb	1.51	Aug	5.40
Mar	1.86	Sep	5.45
Apr	2.06	Oct	4.34
May	3.48	Nov	2.64
Jun	4.57	Dec	2.14

12. **RAINFALL** The table shows the monthly rainfall at a measuring station. What is the mean monthly rainfall?

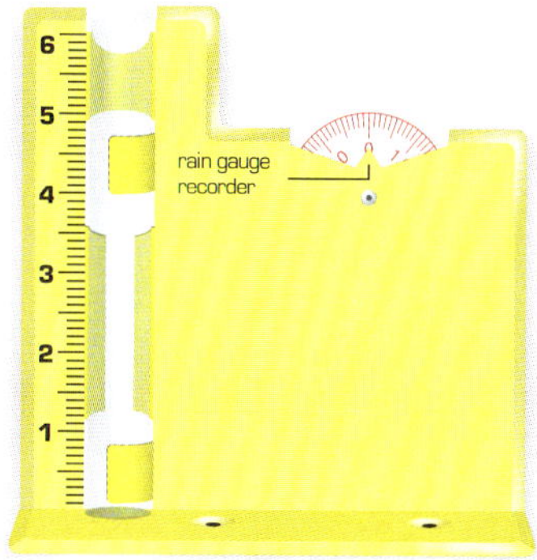

13. **OPEN-ENDED** Create two different sets of data that have six values and a mean of 21.

③ **14.** **CELL PHONE** The bar graph shows your cell phone usage for five months.

a. Which data value is an outlier? Explain.

b. Find the mean with and without the outlier. Then describe how the outlier affects the mean.

c. Describe a situation that could have caused the outlier in this problem.

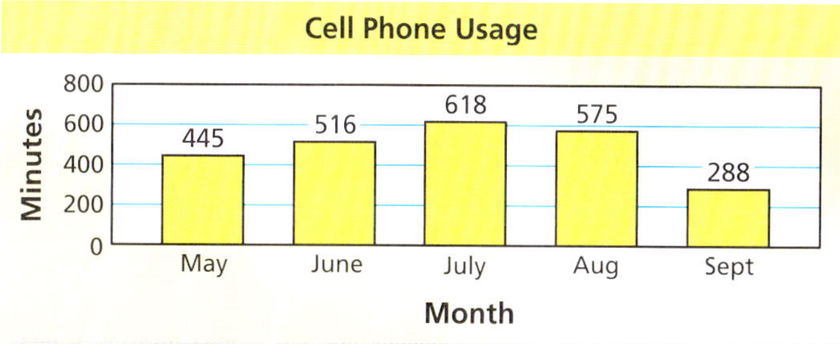

15. **HEIGHT** The table shows the heights of the volleyball players from two schools. What is the difference between the mean heights of the two teams? Do outliers affect either mean? Explain.

	Player Height (inches)														
Dolphins	59	65	53	56	58	61	64	68	51	56	54	57			
Tigers	63	68	66	58	54	55	61	62	53	70	64	64	62	67	69

16. **REASONING** Make a dot plot of the data set 11, 13, 17, 15, 12, 18, and 12. Use the dot plot to explain how the mean is the point where the data set is balanced.

17. **ALLOWANCE** In your class, 7 students do not receive a weekly allowance, 5 students receive \$3, 7 students receive \$5, 3 students receive \$6, and 2 students receive \$8. What is the mean weekly allowance? Explain how you found your answer.

18. **Precision** A collection of 8 backpacks has a mean weight of 14 pounds. A different collection of 12 backpacks has a mean weight of 9 pounds. What is the mean weight of the 20 backpacks? Explain how you found your answer.

Fair Game Review What you learned in previous grades & lessons

Evaluate the expression. *(Section 1.3)*

19. $\dfrac{8 + 10}{2}$ **20.** $\dfrac{26 + 34}{2}$ **21.** $\dfrac{18 + 19}{2}$ **22.** $\dfrac{14 + 17}{2}$

23. **MULTIPLE CHOICE** 60% of what number is 105? *(Section 5.6)*

Ⓐ 63 Ⓑ 175 Ⓒ 630 Ⓓ 1750

9.3 Measures of Center

Essential Question In what other ways can you describe an average of a data set?

1 ACTIVITY: Finding a Median

Work with a partner.

a. Write the total number of letters in the first and last names of 19 celebrities, historical figures, or people you know. Organize your data in a table. One person is already listed for you.

Person	Number of letters in first and last name
Abraham Lincoln	14

b. Order the values in your data set from least to greatest. Then write the data on a strip of grid paper with 19 boxes.

c. Place a finger on the square at each end of the strip. Move your fingers toward the center of the ordered data set until your fingers touch. On what value do your fingers touch?

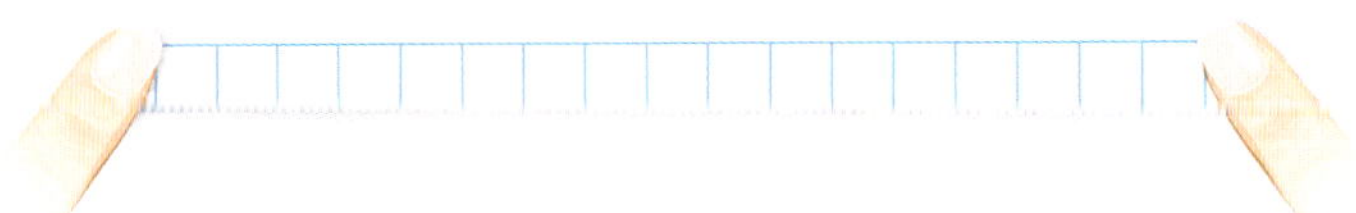

d. Now take your strip of grid paper and fold it in half. On what number is the crease? What do you notice? This value is called the *median*. How would you describe to another student what the median of a data set represents?

e. How many values are greater than the median? How many are less than the median?

f. Why do you think the median is considered an average of a data set?

COMMON CORE

Statistics

In this lesson, you will

- understand the concept of measures of center.
- find the median and mode of data sets.

Learning Standards
6.SP.2
6.SP.3
6.SP.5c

Laurie's Notes

Introduction

Standards for Mathematical Practice

- **MP5 Use Appropriate Tools Strategically:** Sorting data from least to greatest or organizing data in a dot plot can help to determine the median and mode of a data set efficiently.

Motivate

- Play Quiz Show with the students. You could pair up students or have them play individually. Tell them you will ask 5 questions, and they will have 15 seconds to answer each question.
  1. What are the two continents with the longest names? North America, South America (12 letters)
  2. What are the three U.S. states with the longest names? North Carolina, South Carolina, Massachusetts (13 letters)
  3. What are the two U.S. presidents who had the longest last names? Washington and Eisenhower (10 letters)
  4. What are the two Great Lakes with the longest names? Lake Superior and Lake Michigan (12 letters)
  5. What is the state capital with the longest name? Jefferson City, MO (13 letters)
- Award something silly to the winner(s) such as a pencil eraser.

Activity Notes

Activity 1

- **Discuss:** The mean is not the only way to describe a data set. The mean is one of many types of averages used in statistics. Two more averages are the median and mode. You will work with these today.
- Let students freely choose the names of 19 people, as long as they can spell the names correctly. People from the past are fair game.
- In part (c), students should move both fingers at the same time towards the center of the strip.
- After students finish the activity, ask a volunteer to describe what the median of a data set is. The student may fail to mention that the data needs to be ordered from least to greatest. Discuss whether the middle value in their *original* list of 19 values is the median. (Most likely it is not.)
- Students may ask what to do if there is an even number of values. This occurs in Activity 2.

Common Core State Standards

6.SP.2 Understand that a set of data collected to answer a statistical question has a distribution which can be described by its center, ...

6.SP.3 Recognize that a measure of center for a numerical data set summarizes all of its values with a single number, ...

6.SP.5c Summarize numerical data sets in relation to their context, such as by giving quantitative measures of center (median and/or mean) . . ., as well as describing any overall pattern and any striking deviations from the overall pattern with reference to the context in which the data were gathered.

Previous Learning

Students should know how to find a mean.

Technology for the Teacher

Dynamic Classroom

Lesson Plans

Complete Materials List

9.3 Record and Practice Journal

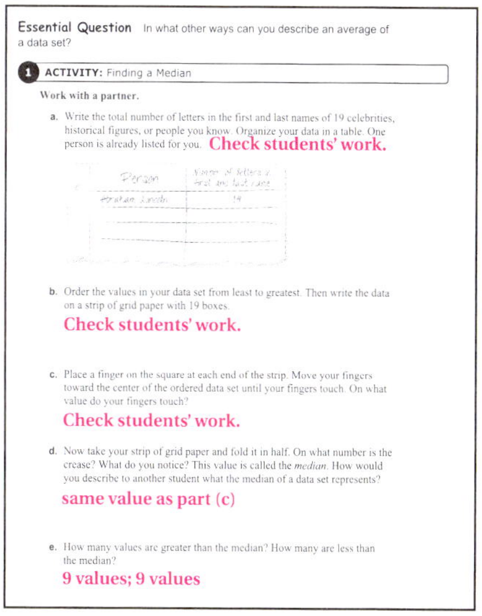

Essential Question In what other ways can you describe an average of a data set?

1 ACTIVITY: Finding a Median

Work with a partner.

a. Write the total number of letters in the first and last names of 19 celebrities, historical figures, or people you know. Organize your data in a table. One person is already listed for you. **Check students' work.**

Person	Number of letters in first and last name
Abraham Lincoln	14

b. Order the values in your data set from least to greatest. Then write the data on a strip of grid paper with 19 boxes.
Check students' work.

c. Place a finger on the square at each end of the strip. Move your fingers toward the center of the ordered data set until your fingers touch. On what value do your fingers touch?
Check students' work.

d. Now take your strip of grid paper and fold it in half. On what number is the crease? What do you notice? This value is called the *median*. How would you describe to another student what the median of a data set represents?
same value as part (c)

e. How many values are greater than the median? How many are less than the median?
9 values; 9 values

English Language Learners

Graphic Organizer

Provide small groups with different data sets and a copy of the organizer. Have each group find the mean, median, and mode of their data set and present their work to the class.

Data set:	
Mean	
Median	
Mode	

9.3 Record and Practice Journal

f. Why do you think the median is considered an average of a data set?

It measures the midpoint of a data set.

2 ACTIVITY: Adding a Value to a Data Set

Work with a partner.

a. How many total letters are in your first name and last name? Add this value to the ordered data set in Activity 1. How many values are now in your data set?

20

b. Write the ordered data, including your new value from part (a), on a strip of grid paper. Check students' work.

c. Repeat parts (c) and (d) from Activity 1. Explain your findings. How do you think you can find the median of this data set?

Find the mean of the middle two values.

d. Compare the medians in Activities 1 and 2. Then answer the following questions. Explain your reasoning.

- Do you think the median always has to be a value in the data set?

  no

- Do you think the median always has to be a whole number?

  no

3 ACTIVITY: Finding a Mode

Work with a partner.

a. Make a dot plot for the data set in Activity 2. Describe the distribution of the data. Check students' work.

b. Which value occurs most often in the data set? This value is called the *mode*.

Check students' work.

c. Do you think a data set can have no mode or more than one mode? Explain.

yes

d. Do you think the mode always has to be a value in the data set? Explain.

yes

e. Why do you think the mode is considered an average of a data set?

Sample answer: In some cases, the data value that occurs most frequently may also be the most typical number in a data set.

What Is Your Answer?

4. **IN YOUR OWN WORDS** In what other ways can you describe an average of a data set?

use median or mode

5. Find the mean of your data set in Activity 2. Then compare the mean, median, and mode. Is there one measure that you think best represents your data set? Explain your reasoning.

Check students' work.

Laurie's Notes

Activity 2

- The number of letters in the student's first and last names is now added to the list of 19 values.
- Give time for the students to try to find the median of the new data set. Eventually you should explain that they need to find the mean of the middle two numbers.
- **MP2 Reason Abstractly and Quantitatively:** A variety of outcomes are possible in this activity. Discuss whether the median changed and how. Is it a whole number? Is it the same as one of the data? Is it greater or less than the median in Activity1? Why?

Activity 3

- Students should be comfortable making a dot plot. Check that their number lines are scaled correctly and no values are skipped.
- In part (b), students may have more than one value that occurs most often. Discuss the possible numbers of modes: there may be no mode, one mode, or more than one mode.
- After students finish, have them compare their dot plots with others. Have them describe the distributions of their data to each other. The distributions and dot plots will vary. This will allow them to see different types of distributions.
- Ask volunteers to share their mode(s). Because there are 20 values in the data set, it is unlikely that a student will have a data set with no mode.
- **MP2:** A variety of outcomes are possible. Discuss whether any of the following occurred or could occur. In each case, ask students to explain their reasoning.
  - There was more than one mode.
  - There was no mode.
  - The median was the same as the mode.
  - The median was different from the mode.

What Is Your Answer?

- In Question 4 you are listening for students to name all three averages they have investigated: mean, median, and mode.
- After students finish Question 5, have them discuss and explain different circumstances when each average (mean, median, and mode) will represent a set of data better. For instance, the mode is best for a set of non-numerical data.

Closure

- Ask students to find the median and mode of the data set that contains the numbers of letters in the names of the Great Lakes.
  - Lake Superior (12 letters)
  - Lake Michigan (12 letters)
  - Lake Ontario (11 letters)
  - Lake Huron (9 letters)
  - Lake Erie (8 letters)

  median: 11; mode: 12

2 ACTIVITY: Adding a Value to a Data Set

Work with a partner.

a. How many total letters are in your first name and last name? Add this value to the ordered data set in Activity 1. How many values are now in your data set?

b. Write the ordered data, including your new value from part (a), on a strip of grid paper.

c. Repeat parts (c) and (d) from Activity 1. Explain your findings. How do you think you can find the median of this data set?

d. Compare the medians in Activities 1 and 2. Then answer the following questions. Explain your reasoning.

- Do you think the median always has to be a value in the data set?
- Do you think the median always has to be a whole number?

3 ACTIVITY: Finding a Mode

Work with a partner.

Math Practice 4

Use a Graph

How can you use the dot plot to find the mode?

a. Make a dot plot for the data set in Activity 2. Describe the distribution of the data.

b. Which value occurs most often in the data set? This value is called the *mode*.

c. Do you think a data set can have no mode or more than one mode? Explain.

d. Do you think the mode always has to be a value in the data set? Explain.

e. Why do you think the mode is considered an average of a data set?

What Is Your Answer?

4. **IN YOUR OWN WORDS** In what other ways can you describe an average of a data set?

5. Find the mean of your data set in Activity 2. Then compare the mean, median, and mode. Is there one measure that you think best represents your data set? Explain your reasoning.

Practice Use what you learned about the median of a data set to complete Exercises 5 and 6 on page 407.

9.3 Lesson

Check It Out
Lesson Tutorials
BigIdeasMath.com

Key Vocabulary
measure of center, *p. 404*
median, *p. 404*
mode, *p. 404*

A **measure of center** is a measure that describes the typical value of a data set. The mean is one type of measure of center. Here are two others.

Key Ideas

Median

Words Order the data. For a set with an odd number of values, the **median** is the middle value. For a set with an even number of values, the **median** is the mean of the two middle values.

Numbers **Data:** 5, 8, 9, 12, 14 — The median is 9.

Data: 2, 3, 5, 7, 10, 11 — The median is $\frac{5+7}{2}$, or 6.

Mode

Words The **mode** of a data set is the value or values that occur most often. Data can have one mode, more than one mode, or no mode. When all values occur only once, there is no mode.

Numbers **Data:** 11, 13, 15, 15, 18, 21, 24, 24

The modes are 15 and 24.

Study Tip

The mode is the only measure of center that you can use to describe a set of data that is *not* made up of numbers.

EXAMPLE 1 Finding the Median and Mode

Bowling Scores				
120	135	160	125	90
205	160	175	105	145

Find the median and mode of the bowling scores.

90, 105, 120, 125, 135, 145, 160, 160, 175, 205 — Order the data.

Median: $\frac{135+145}{2} = \frac{280}{2}$, or 140 — Add the two middle values and divide by 2.

Mode: 90, 105, 120, 125, 135, 145, 160, 160, 175, 205

The value 160 occurs most often.

∴ The median is 140. The mode is 160.

On Your Own

Now You're Ready
Exercises 7–12

Find the median and mode of the data.

1. 20, 4, 17, 8, 12, 9, 5, 20, 13
2. 100, 75, 90, 80, 110, 102

Laurie's Notes

Introduction

Connect

- **Yesterday:** Students developed a conceptual understanding of the median and mode for a data set. (MP2, MP5)
- **Today:** Students will find the median and mode for a data set.

Motivate

? Ask a few questions that don't necessarily need to be answered. The questions are intended to get students thinking about another type of average (mode).

- "What is the most common (average) color of car in the parking lot?"
- "What is the typical (average) eye color of students in the class?"
- "What is the typical (average) style of shoe worn by students in the class?"

Lesson Notes

Key Ideas

- Define **measure of center**. The three types (mean, median, and mode) describe how the data *tends*, its center.
- Write the Key Ideas.
- Discuss the median and mode and how each is determined.
- **FYI:** The median household income in the U.S. is $50,500 (U.S. Census Bureau 2011). Discuss with students that this means half of all households have an income greater than the median and half have an income less than the median.
- To help students remember the difference between mean and median, think about the median on a highway.
- Note the Study Tip: The mode is the only measure of center that can be used to describe non-numeric data sets such as the questions in the beginning of the lesson.

Example 1

? "What is the first step in finding the median?" order the data

- **Common Error:** Students forget to arrange the data in order before finding the median.
- **Note:** Sorting the data also makes it easier to determine the mode(s).
- **Connection:** Adding two middle values and dividing by 2 is the same as finding the mean of the middle two data values.
- If time permits, find the mean. Discuss the three measures of center found and if all are representative of the data.
- **Extension:** "Is it possible for the median and the mode to be the same?" yes

On Your Own

- Ask students to explain how to find the median for a data set.

Goal Today's lesson is finding the **median** and **mode** and identifying which **measure of center** best describes a data set.

Technology for the Teacher

Dynamic Classroom

Lesson Tutorials
Lesson Plans
Answer Presentation Tool

Extra Example 1

Find the median and mode of the data.
95, 80, 85, 86, 89, 100, 95
median: 89, mode: 95

On Your Own

1. median: 12; mode: 20
2. median: 95; no mode

Extra Example 2

The list shows the favorite colors for students in a class. Organize the data in a frequency table. Then find the mode.

blue, green, purple, blue, green, pink, purple, blue, green, purple, blue, blue, green, pink, blue, blue, green, blue, green

Color	Tally	Frequency
Blue	卌 III	8
Green	卌 I	6
Pink	II	2
Purple	III	3

mode: blue

On Your Own

3. yes; Comedy and horror are both modes now that horror has the same number of votes as comedy.

Extra Example 3

Find the mean, median, and mode of the data. Choose the measure that best represents the data. Explain your reasoning.

85, 70, 60, 10, 89, 85, 63

mean: 66, median: 70, mode: 85

Sample: The mean or the median could be used to represent the data. The mode is greater than most of the data.

On Your Own

4. mean: 67.25; median: 51.5; no mode The median best represents the data. It is closer to more of the data because the mean is influenced by outliers.
5. mean: 110; median: 99; mode: 150 The median best represents the data because the mode is the greatest data value and the mean is also greater than most of the data.

Laurie's Notes

Example 2

- **Common Error:** When making the frequency table students can forget to make a tally for the data on the first entry; for instance, in this example they would write comedy as the first type of movie and then forget to make a tally mark for that entry.
- **MP5 Use Appropriate Tools Strategically:** The frequency table is an effective way to determine the mode for this non-numeric data set.

Example 3

- **Discuss:** When you shop for sneakers, there is a range of prices in the store. How do you decide what the *typical* price is of sneakers?

? Work through the example, asking the following questions.
  - "How do you find the mean for a set of data?" Add the data values and divide by the total number of data values.
  - "How do you find the median for a set of data?" Order the data and find the mean of the middle values.
  - "How do you find the mode? Find the most frequent data valve.
- Note that by sorting the data first, the mode is easier to identify.
- In analyzing the data, ask why the mean is so high.

? "What effect did the outlier have on the data?" pulls the mean up
- **MP2 Reason Abstractly and Quantitatively:** The units (dollars) are not written in computing the mean, median, and mode. A mathematically proficient student will consider the units, however, when they talk about the values of these measures of center.

On Your Own

- **Neighbor Check:** Have students work independently and then have their neighbors check their work. Have students discuss any discrepancies.
- Have students share their reasoning about their choices.

EXAMPLE 2 Finding the Mode

Favorite Types of Movies

Comedy	Drama	Horror
Horror	Drama	Horror
Comedy	Comedy	Action
Action	Comedy	Action
Horror	Drama	Comedy
Comedy	Comedy	Horror
Horror	Comedy	Action
Horror	Action	Drama

The list shows the favorite types of movies for students in a class. Organize the data in a frequency table. Then find the mode.

Type	Tally	Frequency
Action	卌	5
Comedy	卌 III	8
Drama	IIII	4
Horror	卌 II	7

The number of tally marks is the frequency.

Make a tally for each vote.

Comedy received the most votes.

So, the mode is comedy.

On Your Own

Now You're Ready
Exercises 14–15

3. One member of the class was absent and ends up voting for horror. Does this change the mode? Explain.

EXAMPLE 3 Choosing the Best Measure of Center

Find the mean, median, and mode of the sneaker prices. Which measure best represents the data?

Mean: $\dfrac{20 + 31 + 122 + 48 + 37 + 20 + 45 + 65}{8} = \dfrac{388}{8}$, or 48.5

Median: 20, 20, 31, 37, 45, 48, 65, 122 — Order from least to greatest.

$\dfrac{37 + 45}{2} = \dfrac{82}{2}$, or 41

Mode: 20, 20, 31, 37, 45, 48, 65, 122 — The value 20 occurs most often.

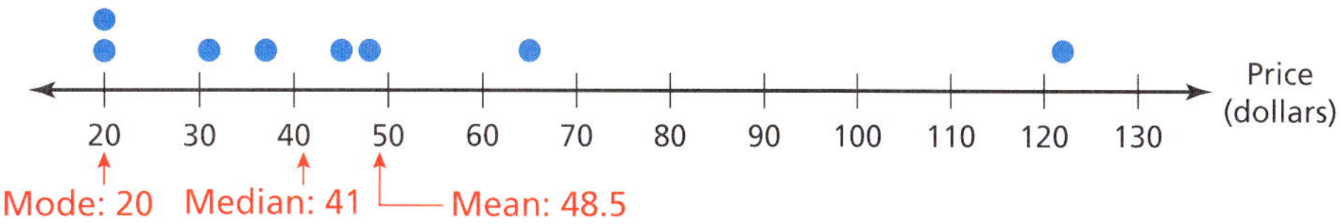

The median best represents the data. The mode is less than most of the data, and the mean is greater than most of the data.

On Your Own

Now You're Ready
Exercises 17–20

Find the mean, median, and mode of the data. Choose the measure that best represents the data. Explain your reasoning.

4. 1, 93, 46, 48, 34, 194, 67, 55

5. 96, 150, 102, 87, 150, 75

EXAMPLE 4 Removing an Outlier

Identify the outlier in Example 3. Find the mean, median, and mode without the outlier. Which measure does the outlier affect the most?

The price of $122 is much greater than any other price. So, it is the outlier.

	Mean	Median	Mode
With Outlier (Example 3)	48.5	41	20
Without Outlier	38	37	20

∴ The mean is affected the most by the outlier.

On Your Own

Now You're Ready
Exercises 21–22

6. The times (in minutes) it takes six students to travel to school are 8, 10, 10, 15, 20, and 45. Identify the outlier. Find the mean, median, and mode with and without the outlier. Which measure does the outlier affect the most?

EXAMPLE 5 Changing the Values of a Data Set

The prices of six video games at an online store are shown in the table. The price of each game increases by $4.98 when a shipping charge is included. How does this increase affect the mean, median, and mode?

Video Game Prices	
$53.42	$35.69
$18.99	$25.13
$27.97	$53.42

Make a new table by adding $4.98 to each price. Then find the mean, median, and mode of both data sets.

Video Game Prices with Shipping Charge	
$58.40	$40.67
$23.97	$30.11
$32.95	$58.40

	Mean	Median	Mode
Original Price	35.77	31.83	53.42
Price with Shipping Charge	40.75	36.81	58.4

Compare:

Mean: $40.75 - 35.77 = 4.98$

Median: $36.81 - 31.83 = 4.98$

Mode: $58.4 - 53.42 = 4.98$

∴ By increasing each video game price by $4.98 for shipping, the mean, median, and mode all increase by $4.98.

On Your Own

7. **WHAT IF?** The store decreases the price of each video game by $3. How does this decrease affect the mean, median, and mode?

Laurie's Notes

Example 4

- ? "Which of the three measures does the outlier impact the most?" mean
- ? "If $122 is not the *typical* price for a pair of sneakers, would it be okay to eliminate it from the data set?" Yes. If you're trying to describe the typical price, eliminating the outlier is fine.
- Computing the mean, median, and mode should be efficient due to the work previously done.
- ? "Why does the median change?" There are an odd number of data values now, and the middle two were not the same in the original set of data.
- **Big Idea:** The mean of a data set is affected the most by an outlier.
- ? "If the outlier of a data set is a least value, what effect does it have on the mean?" pulls the mean down

On Your Own

- Question 6 will take time for students to complete. They could work with partners.

Example 5

- The big idea in this example is how the mean, median, and mode change when every data value increases by the same amount. You could pose the idea by asking the following question.
- ? "When I graded your tests for Chapter 8, I computed the mean, median, and mode of the scores for the class. How would those three averages change if I give everyone 3 bonus points?" Student responses will vary.
- Like the 3 bonus points, in Example 5 a shipping charge of $4.98 is added to each price. Ask a student to read the problem.
- Give students time to compute the mean, median, and mode of the original set of data. Remind them to sort the data before finding the median.
- ? "How does the total cost of each video game change when you add for shipping and handling?" The total cost of each game increases by $4.98.
- Give students time to compute the mean, median, and mode of the new set of data.
- ? "How do the mean, median, and mode change?" each increases by $4.98
- **Big Idea:** When all values of a data set increase by an amount n, the mean, median, and mode will each increase by n.
- **Extension:** If time permits, you could explore the question of what happens to the mean, median, and mode when all of the prices increase by 10%.

Closure

- **Writing Prompt:** To determine the median of one dozen data values, … *Sample answer:* Start by sorting the data values. Then find the mean of the two middle values by adding them and dividing by two.

Extra Example 4

Identify the outlier in Extra Example 3. Find the mean, median, and mode without the outlier. Which measure does the outlier affect the most? outlier: 10; mean: $75.\overline{3}$, median: 77.5, mode: 85; The mean is affected most by the outlier.

On Your Own

6. 45 is the outlier.
With outlier: mean: 18; median: 12.5; mode: 10
Without outlier: mean: 12.6; median: 10; mode: 10
The mean is affected most by the outlier.

Extra Example 5

The prices of six digital e-readers at an online store are shown in the table. The price of each e-reader increases by $6.27 when a shipping charge is included. How does this increase affect the mean, median, and mode?

e-Reader Prices	
$59.99	$128.99
$69.99	$99.99
$159.99	$128.99

The mean, median, and mode all increase by $6.27.

On Your Own

7. The mean, median, and mode all decrease by $3.

Differentiated Instruction

Visual

To give students a visual sense of an outlier, have them construct a line plot of the data in Example 3 on page 405. Then have them eliminate the outlier, redo the line plot, and recalculate the mean, median, and mode. They should notice that the new mean is more central to the data. You can mention that sometimes outliers are eliminated to get a truer picture of the data.

Vocabulary and Concept Check

1. *Sample answer:* 1, 2, 3, 4, 5, 6
2. It depends on the data. The mean is often affected more by an outlier when the middle values of the ordered data are relatively close, but if they are not close, the median may be affected more.
3. outlier; The other three are measures of center.
4. The number 8 must be in the data set at least twice, because the mode is the data value that occurs most often.

Practice and Problem Solving

5. 5.5
6. 24
7. median: 7; mode: 3
8. median: 15; modes: 14, 16
9. median: 92.5; mode: 94
10. median: 33; no mode
11. median: 17; mode: 12
12. median: 51.5; modes: 44, 55
13. The data were not ordered from least to greatest; The median is 55. 49, 50, 51, 55, 58, 59, 63
14. black, blue
15. singing
16. No, only the mode can describe a set of data that is not made up of numbers.

Assignment Guide and Homework Check

Level	Day 1 Activity Assignment	Day 2 Lesson Assignment	Homework Check
Basic	5, 6, 33–37	1–4, 7–27 odd	2, 7, 15, 25
Average	5, 6, 33–37	1–4, 7–15 odd, 18–24 even, 25–29	2, 7, 15, 22, 29
Advanced	5, 6, 33–37	1–4, 12–14, 16–32 even	2, 14, 16, 22, 30

Common Errors

- **Exercises 7–12** Students may try to identify the median without ordering the data first. Remind them that it is essential to order the data first and then find the median. This makes finding the mode easier as well.
- **Exercises 14 and 15** Students may make an educated guess of which color or act appears the most, instead of organizing the data. Remind them of the importance of organizing data, that it not only helps them find the mode, but also helps answer questions such as "how many students wore black?"

9.3 Record and Practice Journal

Find the median and mode(s) of the data.

1. 3, 2, 3, 6, 7, 5, 9
   median: 5
   mode: 3

2. 17, 21, 30, 17, 28, 21
   median: 21
   modes: 17 and 21

Find the mean, median, and mode(s) of the data with and without the outlier. Describe the effect of the outlier on the measures of center.

3. 4, 15, 6, 12, 68, 12
   See Additional Answers.
   The outlier increases the mean.

4. 0, 54, 62, 64, 55, 55, 54, 62
   See Additional Answers.
   The outlier decreases the mean.

5. The data show your strokes for 18 holes of miniature golf.

   4, 5, 3, 3, 1, 2, 3, 2, 4, 8, 2, 4, 4, 5, 2, 3, 6, 2

   Find the mean, median, and mode(s) of the data. Which measure best represents the data? Explain your reasoning.

   mean: 3.5
   median: 3
   mode: 2
   The median best represents the data because 8 affects the mean.

9.3 Exercises

Check It Out
Help with Homework
BigIdeasMath.com

Vocabulary and Concept Check

1. **NUMBER SENSE** Give an example of a data set that has no mode.
2. **WRITING** Which is affected most by an outlier: the mean, median, or mode? Explain.
3. **WHICH ONE DOESN'T BELONG** Which word does *not* belong with the other three? Explain.

median	outlier	mode	mean

4. **NUMBER SENSE** A data set has a mean of 7, a median of 5, and a mode of 8. Which of the numbers 7, 5, and 8 *must* be in the data set? Explain.

Practice and Problem Solving

Use grid paper to find the median of the data.

5. 9, 7, 2, 4, 3, 5, 9, 6, 8, 0, 3, 8
6. 16, 24, 13, 36, 22, 26, 22, 28, 25

Find the median and mode(s) of the data.

1 7. 3, 5, 7, 9, 11, 3, 8
8. 14, 19, 16, 13, 16, 14
9. 93, 81, 94, 71, 89, 92, 94, 99
10. 44, 13, 36, 52, 19, 27, 33
11. 12, 33, 18, 28, 29, 12, 17, 4, 2
12. 55, 44, 40, 55, 48, 44, 58, 67
13. **ERROR ANALYSIS** Describe and correct the error in finding the median of the data.

✗ The median is 58.

63, 55, 49, 58, 50, 59, 51

Find the mode(s) of the data.

2 14.

Shirt Color		
Black	Blue	Red
Pink	Black	Black
Gray	Green	Blue
Blue	Blue	Red
Yellow	Blue	Blue
Black	Orange	Black
Black		

15.

Talent Show Acts		
Singing	Dancing	Comedy
Singing	Singing	Dancing
Juggling	Dancing	Singing
Singing	Poetry	Dancing
Comedy	Magic	Dancing
Poetry	Singing	Singing

16. **REASONING** In Exercises 14 and 15, can you find the mean and median of the data? Explain.

Find the mean, median, and mode(s) of the data. Choose the measure that best represents the data. Explain your reasoning.

3 **17.** 48, 12, 11, 45, 48, 48, 43, 32

18. 12, 13, 40, 95, 88, 7, 95

19. 2, 8, 10, 12, 56, 9, 5, 2, 4

20. 126, 62, 144, 81, 144, 103

Find the mean, median, and mode(s) of the data with and without the outlier. Describe the effect of the outlier on the measures of center.

4 **21.** 45, 52, 17, 63, 57, 42, 54, 58

22. 85, 77, 211, 88, 91, 84, 85

Find the mean, median, and mode(s) of the data.

23. 4.7, 8.51, 6.5, 7.42, 9.64, 7.2, 9.3

24. $8\frac{1}{2}, 6\frac{5}{8}, 3\frac{1}{8}, 5\frac{3}{4}, 6\frac{5}{8}, 5\frac{1}{4}, 10\frac{5}{8}, 4\frac{1}{2}$

25. WEATHER The weather forecast for a week is shown.

	Sun	Mon	Tue	Wed	Thu	Fri	Sat

High	90° F	91° F	89° F	97° F	101° F	99° F	91° F
Low	74° F	78° F	77° F	77° F	83° F	78° F	72° F

a. Find the mean, median, and mode(s) of the high temperatures. Which measure best represents the data? Explain your reasoning.

b. Repeat part (a) for the low temperatures.

26. RESEARCH Find the unit costs of 10 different kinds of cereal. Choose one cereal whose unit cost will be an outlier.

a. Find the mean, median, and mode(s) of the data. Which measure best represents the data? Explain your reasoning.

b. Identify the outlier in the data set. Find the mean, median, and mode(s) of the data set without the outlier. Which measure does the outlier affect the most?

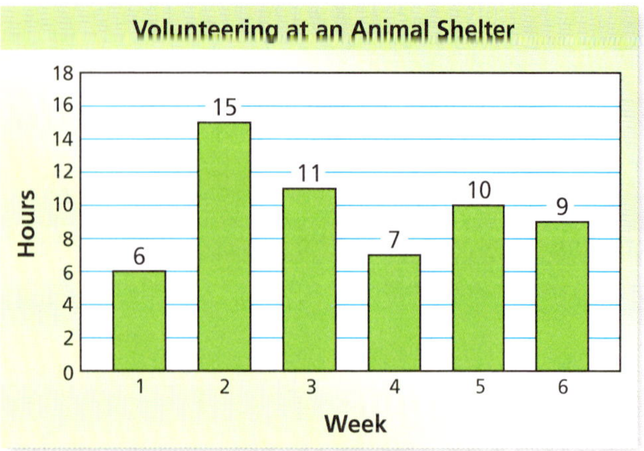

27. PROBLEM SOLVING The bar graph shows the numbers of hours you volunteered at an animal shelter. What is the minimum number of hours you need to work in the seventh week to justify that you worked an average of 10 hours for the 7 weeks? Explain your answer using measures of center.

28. REASONING Why do you think the mode is the least frequently used measure to describe a data set? Explain.

Common Errors

- **Exercises 17–20** Students may only find the mean, median, and mode instead of putting all the numbers on a number line before deciding which measure of center is the best representative. Even if students are correct, encourage them to draw number lines to check their answer. Creating a visual of the data helps students to understand, and it is also a way for them to explain why that measure of center is the best representative.
- **Exercise 21 and 22** When finding the mean for the data after eliminating the outlier, students may forget to subtract one from the denominator. Remind them that when the outlier is removed, there is one less data value, so divide by a lesser number.
- **Exercises 23 and 24** Students may have difficulties ordering the data before finding the median because the numbers are harder to work with. Remind them to take one step at a time. Tell students to write the numbers as they are and then cross out the numbers as they re-write them in order from least to greatest.

Practice and Problem Solving

17. mean: 35.875; median: 44
mode: 48
Sample answer: The median is probably best, because it is close to most of the data. The mean is less than most of the data and the mode is the greatest value.

18. mean: 50; median: 40
mode: 95
Sample answer: The mean is probably best, because the mode is the greatest value and the median is too far from the greater values.

19. mean: 12; median: 8
mode: 2
Sample answer: The median is the best measure, because the mean is greater than most of the data and the mode is the least value.

20. mean: 110; median: 114.5
mode: 144
Sample answer: Either the mean or median is best, because they are both at the middle of the data, while the mode is the greatest value.

21–28. See Additional Answers.

Differentiated Instruction

Visual

Some students may incorrectly find the median by counting the number of values and dividing by 2. Have students write the numbers in increasing order. Then cross the numbers out in pairs, one on the left end and one on the right end, until 1 or 2 numbers remain. If there is one number remaining, the median is that number. If two numbers remain, then the median is the mean of those two numbers.

Practice and Problem Solving

29. See *Taking Math Deeper.*

30. **a.** no; The price is the mode, but it is the lowest price. Most cameras cost more.

b. By advertising the lowest price, they are likely to draw more customers to the store.

c. *Sample answer:* Knowing all the measures can help you know whether the store has many models in your price range.

31–32. See Additional Answers.

Fair Game Review

33. 13 **34.** 65

35. 119 **36.** 2875

37. D

Mini-Assessment

Find the median and mode(s) of the data.

1. 56, 27, 35, 27, 44, 25 31, 27
2. 65, 2, 23, 41, 11, 7, 55 23, no mode
3. The data are the number of times you win playing a video game.

Mon	Tues	Wed	Thu	Fri	Sat
2	10	12	14	2	11

a. Find the mean, median, and mode(s) of the data. Choose the measure that best represents the data. Explain your reasoning. 8.5, 10.5, 2; The median best represents the data. The mean and mode are less than most of the data.

b. Suppose you win 33 times on Sunday. Find the mean, median, and mode(s) of the data with this outlier. Which measure does the outlier affect the most? 12, 11, 2; The mean is affected the most by the outlier.

Taking Math Deeper

Exercise 29

This problem shows how the change of a single number in a data set can affect the mean, median, and mode. Of these, the mode is the least likely to be affected.

1. Measures with the 30-year-old:

14, 14, 17, 20, 21, 22, 25, 25, 26, 27, 29, 30

Mean: $\frac{270}{12} = 22.5$ years

Median: $\frac{22 + 25}{2} = 23.5$ years

Modes: 14 and 25

2. Measures with the 15-year-old:

14, 14, 15, 17, 20, 21, 22, 25, 25, 26, 27, 29

Mean: $\frac{255}{12} = 21.25$ years

Median: $\frac{21 + 22}{2} = 21.5$ years

Modes: 14 and 25

3. Compare the measures.

The mean and median were lowered.
The modes are unchanged.

Project

Research motocross racing in the United States. Write a report that shows when it started, costs involved, number of participants, and other interesting facts.

Reteaching and Enrichment Strategies

If students need help. . .	If students got it. . .
Resources by Chapter • Practice A and Practice B • Puzzle Time Record and Practice Journal Practice Differentiating the Lesson Lesson Tutorials Skills Review Handbook	Resources by Chapter • Enrichment and Extension • Technology Connection Start the next section

29. **MOTOCROSS** The ages of the racers in a bicycle motocross race are 14, 22, 20, 25, 26, 17, 21, 30, 27, 25, 14, and 29. The 30-year-old drops out of the race and is replaced with a 15-year-old. How are the mean, median, and mode of the ages affected?

30. **CAMERAS** The data are the prices of several digital cameras at a store.

$130 $170 $230 $130
$250 $275 $130 $185

a. Does the price shown in the advertisement represent the prices well? Explain.

b. Why might the store use this advertisement?

c. In this situation, why might a person want to know the mean? the median? the mode? Explain.

31. **SALARIES** The table shows the monthly salaries for employees at a company.

Monthly Salaries (dollars)				
1940	1660	1860	2100	1720
1540	1760	1940	1820	1600

a. Find the mean, median, and mode of the data.

b. Each employee receives a 5% raise. Find the mean, median, and mode of the data with the raise. How does this increase affect the mean, median, and mode of the data?

c. Use the original monthly salaries to calculate the annual salaries. Find the mean, median, and mode of the annual salaries. How are these values related to the mean, median, and mode of the monthly salaries?

32. Critical Thinking Consider the algebraic expressions $3x$, $9x$, $4x$, $23x$, $6x$, and $3x$. Assume $x > 0$.

a. Find the mean, median, and mode.

b. Is there an outlier? If so, what is it?

Fair Game Review What you learned in previous grades & lessons

Find the value of the expression. *(Section 1.1)*

33. 48 − 35 **34.** 188 − 123 **35.** 416 − 297 **36.** 6249 − 3374

37. **MULTIPLE CHOICE** A shelf in your room can hold at most 30 pounds. There are 12 pounds of books already on it. Which inequality represents the number of pounds you can add to the shelf? *(Section 7.6)*

Ⓐ $x < 18$ Ⓑ $x \geq 18$ Ⓒ $x \leq 42$ Ⓓ $x \leq 18$

9 Study Help

Check It Out
Graphic Organizer
BigIdeasMath.com

You can use a **concept circle** to organize information about a concept. Here is an example of a concept circle for a statistical question.

Statistical Question

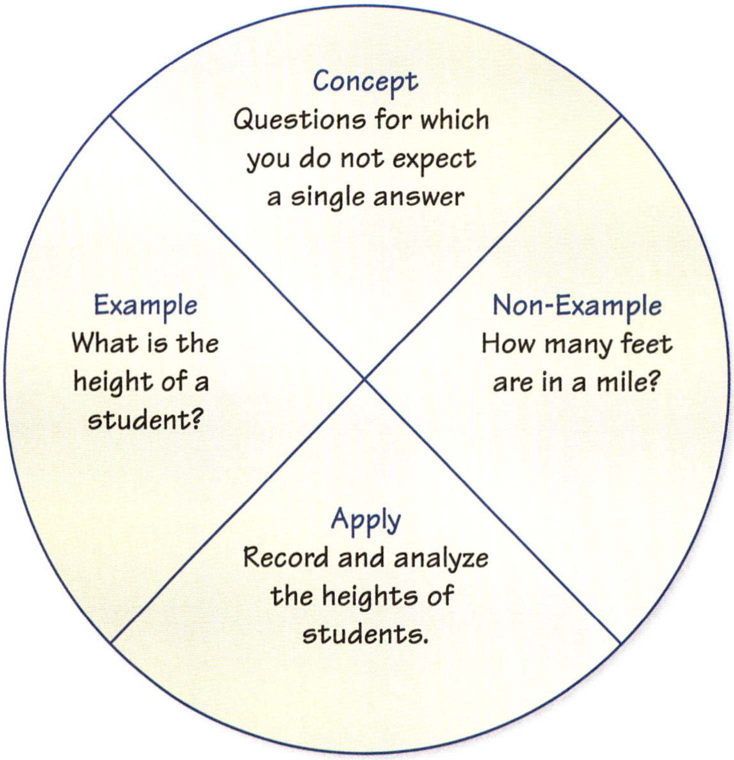

On Your Own

Make concept circles to help you study these topics.

1. mean
2. outlier
3. measures of center
4. median
5. mode

After you complete this chapter, make concept circles for the following topics.

6. measures of variation
7. range
8. quartiles
9. interquartile range
10. mean absolute deviation

"Do you think this concept circle will help my owner understand that 'Speak' and 'Sit' need motivation?"

Sample Answers

1. Mean

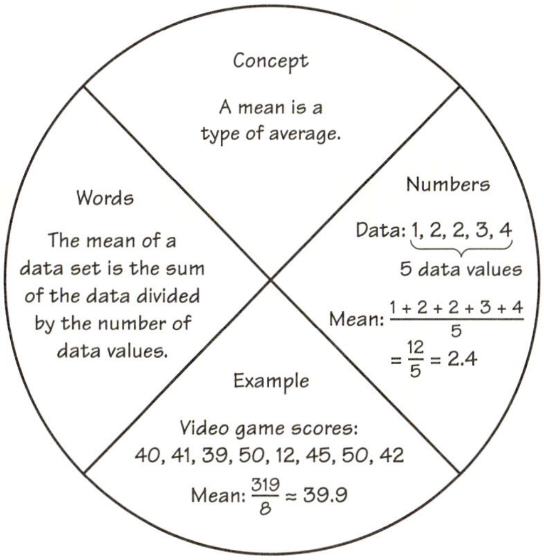

2. Outlier

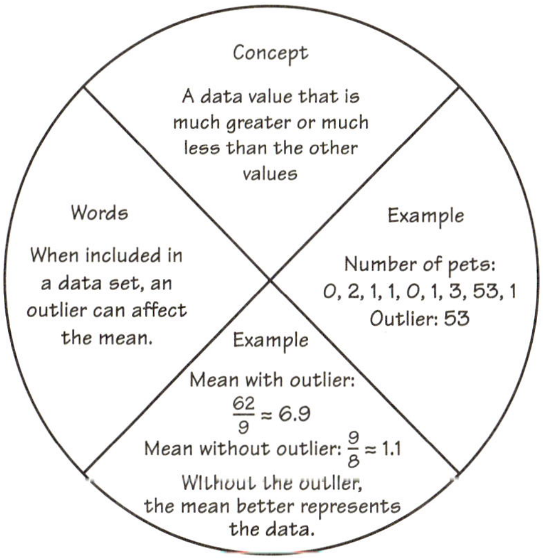

3. Measures of Center

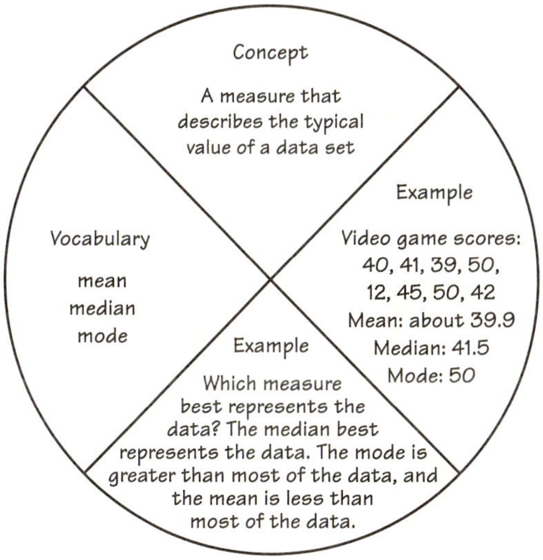

4–5. Available at *BigIdeasMath.com.*

List of Organizers

Available at *BigIdeasMath.com*

Comparison Chart
Concept Circle
Example and Non-Example Chart
Definition (Idea) and Example Chart
Formula Triangle
Four Square
Information Frame
Information Wheel
Notetaking Organizer
Process Diagram
Summary Triangle
Word Magnet
Y Chart

About this Organizer

A **Concept Circle** can be used to organize information about a concept. Students write the concept above the circle. Then students write associated information in the sectors of the circle. Associated information can include (an explanation of the) *Concept, Apply, Solve, Check, Example, Non-Example, Vocabulary, Property, Visual, Words, Algebra,* and *Justify.* Concept circles can have more or fewer than four sectors. Students can place their concept circles on note cards to use as a quick study reference.

Technology for the Teacher

Editable Graphic Organizer

Answers

1–2. See Additional Answers.

3. 10 **4.** 9

5. median: 5; mode: 3

6. median: 37; mode: 56

7. mean: 57.5; median: 44.5; mode: 47; *Sample answer:* The median best represents the data because the mean and mode are greater than most of the data.

8. mean: 45.2; median: 56; mode: none; *Sample answer:* The median best represents the data because there is no mode, the mean is less than most of the data, and the median is very close to most of the data.

9. See Additional Answers.

10. 8 emails

11. median: 18.5; mode: 20

12. *Sample answer:* With the outlier, the median (2.8) best represents the data because the mean (4) is greater than most of the data and the mode (2.2) is less than the rest of the data. Without the outlier, the mean (2.725) or the median (2.7) best represent the data because the mode (2.2) is less than the rest of the data.

Technology for the Teacher

Online Assessment
Assessment Book
ExamView® Assessment Suite

Alternative Quiz Ideas

100% Quiz	Math Log
Error Notebook	Notebook Quiz
Group Quiz	Partner Quiz
Homework Quiz	**Pass the Paper**

Pass the Paper

- Work in groups of four. The first student copies the problem and completes the first step, explaining his or her work.
- The paper is passed, and the second student works through the next step, also explaining his or her work.
- This process continues until the problem is completed.
- The second member of the group starts the next problem. Students should be allowed to question and debate as they are working through the quiz.
- Student groups can be selected by the teacher, by students, through a random process, or any way that works for your class.
- The teacher walks around the classroom listening to the groups and asks questions to ensure understanding.

Reteaching and Enrichment Strategies

If students need help. . .	If students got it. . .
Resources by Chapter • Practice A and Practice B • Puzzle Time Lesson Tutorials *BigIdeasMath.com*	Resources by Chapter • Enrichment and Extension • Technology Connection Game Closet at *BigIdeasMath.com* Start the next section

9.1–9.3 Quiz

Check It Out
Progress Check
BigIdeasMath.com

Display the data in a dot plot. Identify any clusters, peaks, or gaps in the data. *(Section 9.1)*

1.

Weight (grams)			
42	40	37	42
43	41	42	43
37	41	41	42

2.

Time (seconds)				
63	66	65	60	59
59	64	58	65	58
64	60	59	64	63

Find the mean of the data. *(Section 9.2)*

3.

Tour Dates	
May	III
June	卌 卌
July	卌 卌 IIII
August	卌 卌 卌 III
September	卌

4.

Scores	
Judge 1	8.9
Judge 2	9.4
Judge 3	8.6
Judge 4	9.1

Find the median and the mode(s) of the data. *(Section 9.3)*

5. 3, 5, 9, 11, 3

6. 24, 4, 37, 56, 6, 56, 45

Find the mean, median, and mode(s) of the data. Choose the measure that best represents the data. Explain your reasoning. *(Section 9.3)*

7. 47, 147, 24, 47, 38, 42

8. 34, 57, 58, 56, 21

Hours of Exercise				
5	1	5	3	5
4	5	2	5	4
3	4	6	5	6

9. EXERCISE You conduct a survey to answer: "How many hours does a sixth-grade student spend exercising during a week?" The table shows the results. *(Section 9.1)*

a. Is this a statistical question? Explain.

b. Display the data in a dot plot. Identify any clusters, peaks, or gaps in the data.

c. Use the distribution of the data to answer the question.

10. EMAILS The number of emails you received in 5 days is shown. What is the mean number of emails you received per day? *(Section 9.2)*

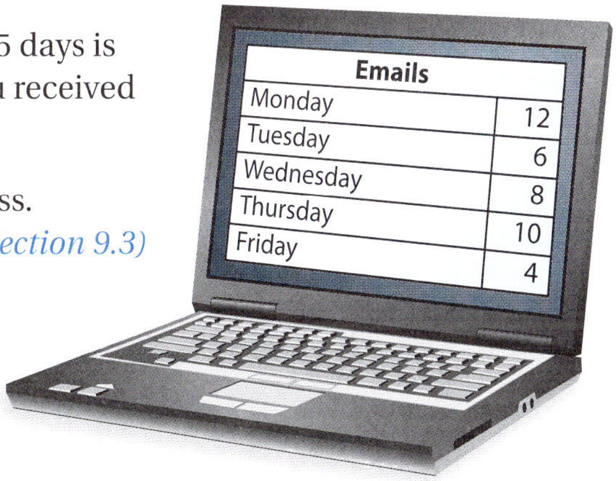

Emails	
Monday	12
Tuesday	6
Wednesday	8
Thursday	10
Friday	4

11. QUIZZES The data are your quiz scores for a class. Find the median and the mode of the data. *(Section 9.3)*

18, 19, 17, 14, 20, 20, 15, 21

12. MUSIC The data are the lengths of the songs (in minutes) on your new CD. Which measure of center best represents the data with and without the outlier? Explain. *(Section 9.3)*

2.2, 2.2, 2.4, 2.6, 2.8, 3.0, 3.2, 3.4, 14.2

9.4 Measures of Variation

Essential Question How can you describe the spread of a data set?

1 ACTIVITY: Interpreting Statements

Work with a partner. There are 24 students in your class. Your teacher makes the following statements:

- ***"The exam scores range from 75% to 96%."***
- ***"Most of the students received high scores."***

a. What do you think the first statement means? Explain.

b. In the first statement, is your teacher describing the center of the data set? If not, what do you think your teacher is describing?

c. What do you think the scores are for most of the students in the class? Explain your reasoning.

d. Use your teacher's statements to make a dot plot that can represent the distribution of the exam scores of the class.

2 ACTIVITY: Grouping Data

Work with a partner. The numbers of U.S. states visited by each student in a sixth grade class are shown.

Number of States Visited

1	7	5	2
11	6	3	20
4	18	1	6
2	7	1	8
10	2	12	5
	3	21	

a. Between what values do the data range?

b. Write the ordered data values on a strip of grid paper and fold it to find the median. How many values are greater than the median? How many are less than the median?

c. **REPEATED REASONING** Fold the strip in half again. On what values are the two new creases? What do you think these values represent?

d. Into how many parts did you divide the data set? How many data values are in each part?

e. Graph the median and the values you found in parts (a) and (c) on a number line. Are the distances the same between these points?

f. How can you use these values to describe the spread of the data?

COMMON CORE

Statistics

In this lesson, you will

- find the range of data sets.
- find the interquartile range of data sets.
- check for outliers in data sets.

Learning Standards
6.SP.2
6.SP.3
6.SP.5c

Laurie's Notes

Introduction

Standards for Mathematical Practice

- **MP8 Look for and Express Regularity in Repeated Reasoning:** Before students found the median for a data set by folding strips of grid paper. Now they will make a second fold to divide the data into 4 parts.

Motivate

- Write the word RANGE on the board and ask what it means.
- Allow discussion of the many meanings. This lesson concerns the difference between the greatest and least values in a data set—the range is a single number that describes the spread of the data.
- Give an example: "The range of ages at the dance is 5 years."
- You have been studying measures of center. Now you will explore how spread out data are. There are several ways to measure variation.

Activity Notes

Activity 1

- Give students time to work. Encourage reading carefully and critically.
- ? "What does the first statement mean?" The lowest score was 75%, and the highest score was 96%.
- ? "Do you know how many students scored either of these two grades?" no
- **MP3 Construct Viable Arguments and Critique the Reasoning of Others:** Ask students to share and compare dot plots, and explain why they drew their dot plots as they did. Each plot should have 24 dots.

Activity 2

- Begin with a discussion about trips the students have made.
- **MP8:** Students fold once as before to divide the 22 data values into two halves of 11 values. They fold again to find the median of the lower half (6th data value) and the median of the upper half (17th data value).
- ? "In part (d), what fraction of the data values are in each part?" about $\frac{1}{4}$
- In part (e), students find that the differences of the least and greatest data values in each of the four groups of data are not equal.
- Point out that the three medians formed may or may not all be data values. In this case, one median is not a data value and the others are.

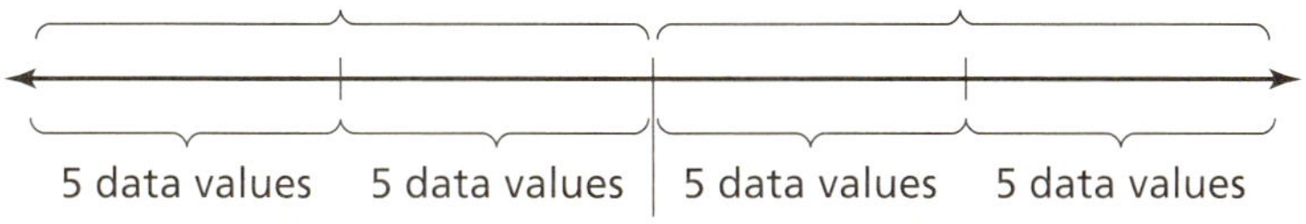

- **FYI:** In large data sets it is okay for the number of values to not be the same in each part. The goal is to get a sense of how the data vary.

Common Core State Standards

6.SP.2 Understand that a set of data collected to answer a statistical question has a distribution which can be described by its center, spread . . ..

6.SP.3 Recognize that . . . a measure of variation describes how its values vary with a single number.

6.SP.5c Summarize numerical data sets in relation to their context, such as by giving quantitative measures of center and variability (interquartile range . . .), as well as describing any overall pattern and any striking deviations from the overall pattern with reference to the context in which the data were gathered.

Previous Learning

Students should know how to find a median.

Technology for the Teacher

Dynamic Classroom

Lesson Plans

Complete Materials List

9.4 Record and Practice Journal

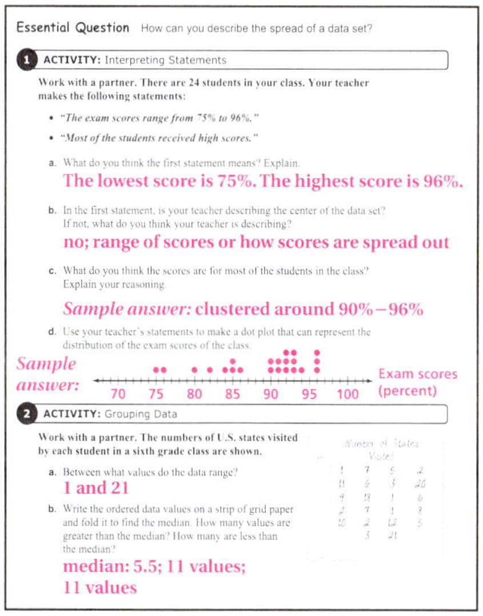

Essential Question How can you describe the spread of a data set?

1 ACTIVITY: Interpreting Statements

Work with a partner. There are 24 students in your class. Your teacher makes the following statements:

- *"The exam scores range from 75% to 96%."*
- *"Most of the students received high scores."*

a. What do you think the first statement means? Explain.
The lowest score is 75%. The highest score is 96%.

b. In the first statement, is your teacher describing the center of the data set? If not, what do you think your teacher is describing?
no; range of scores or how scores are spread out

c. What do you think the scores are for most of the students in the class? Explain your reasoning.
Sample answer: clustered around 90%–96%

d. Use your teacher's statements to make a dot plot that can represent the distribution of the exam scores of the class.

2 ACTIVITY: Grouping Data

Work with a partner. The numbers of U.S. states visited by each student in a sixth grade class are shown.

Number of States Visited

1	7	5	2
11	6	3	20
4	18	1	6
2	7	1	8
10	2	12	5
	3	21	

a. Between what values do the data range?
1 and 21

b. Write the ordered data values on a strip of grid paper and fold it to find the median. How many values are greater than the median? How many are less than the median?
median: 5.5; 11 values; 11 values

English Language Learners

Visual

Make a dot plot of the data in Activity 2, and shade the interval from the first dot to the last dot to help students visualize what the range of a set of data represents. Then plot a dot to represent the outlier described in Activity 3. Extend the shading to that dot using a different color to show the effect of the outlier on the range.

9.4 Record and Practice Journal

c. **REPEATED REASONING** Fold the strip in half again. On what values are the two new creases? What do you think these values represent?

2 and 10; median of lower half and median of upper half

d. Into how many parts did you divide the data set? How many data values are in each part?

4; 5

e. Graph the median and the values you found in parts (a) and (c) on a number line. Are the distances the same between these points? no

f. How can you use these values to describe the spread of the data?

subtract least value from the greatest value

3 ACTIVITY: Adding a Value to a Data Set

Work with a partner. A new student joins the class in Activity 2. The new student has visited 41 states.

a. Add this value to the ordered data set in Activity 2. Does your answer to part (a) change? Explain.

yes; data set range from 1 to 41

b. How does the distribution of the data change when this value is added? Explain your reasoning.

The distribution is more spread out.

c. How does adding this value affect the values on your number line in part (e) of Activity 2? Greatest value is now 41.

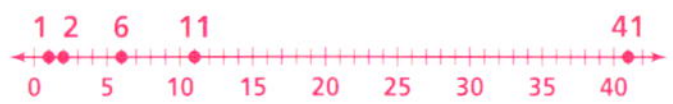

4 ACTIVITY: Analyzing Data Sets

Work with a partner. Identify the data set that is the least spread out and the data set that is the most spread out. Explain your reasoning.

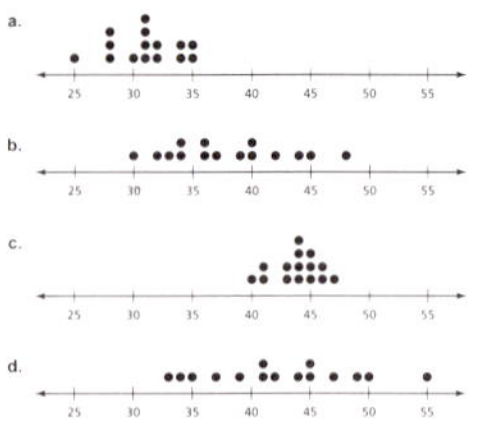

least: c
most: d

What Is Your Answer?

5. **IN YOUR OWN WORDS** How can you describe the spread of a data set?

You can find the difference between the greatest and least values in a data set.

6. Make a dot plot of the data set in Activity 2. Describe any similarities between the dot plot and the number line in part (e).

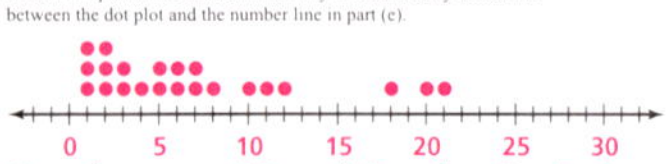

Sample answer: Most of the values in the data set are between 0 and 10.

Laurie's Notes

Activity 3

- This activity adds an outlier to the data set. In Section 9.3, students investigated the effect of an outlier on the measures of center. Now they investigate the effect of an outlier on the spread of the data.
- Students should quickly notice that the range has increased.
- ? "How many data values are there?" 23
- ? "Is the median of this data set one of the values in the set? Explain." yes; Because the number (23) of data values is odd, the median is the middle number when the data is ordered.
- The goal is for students to see that the range increases quite a bit, but the three median values change very little.
- **Extension:** Although the students do not find the mode or mean, ask how they think these two averages change when 41 is added to the data set.

Activity 4

- Mention to students that the terms *spread*, *spread out*, *variation*, and *variability* all have similar meanings when used to describe data sets.
- **MP3:** Give students time to develop their reasoning. To determine the spread, students are actually finding the range of the data set.
- ? "To compare data sets, is it important that the data sets have the exact same number of data values?" Answers may vary; You can compare two data sets even if they do not have the same number of data values; teachers often compare the variability of two classes where one has 20 students and the other has 23.

What Is Your Answer?

- In Question 5, listen for students to describe the range even if they do not use this word.

Closure

- Give an example of a data set with 6 data values that has a median of 20 and varies by 20 from the least to the greatest value. *Sample answer:* 4, 5, 18, 22, 23, 24

3 ACTIVITY: Adding a Value to a Data Set

Work with a partner. A new student joins the class in Activity 2. The new student has visited 41 states.

a. Add this value to the ordered data set in Activity 2. Does your answer to part (a) change? Explain.

b. How does the distribution of the data change when this value is added? Explain your reasoning.

c. How does adding this value affect the values on your number line in part (e) of Activity 2?

4 ACTIVITY: Analyzing Data Sets

Work with a partner. Identify the data set that is the least spread out and the data set that is the most spread out. Explain your reasoning.

a.

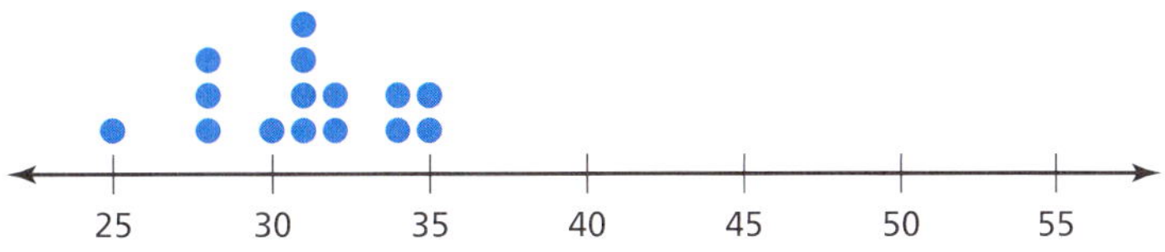

b.

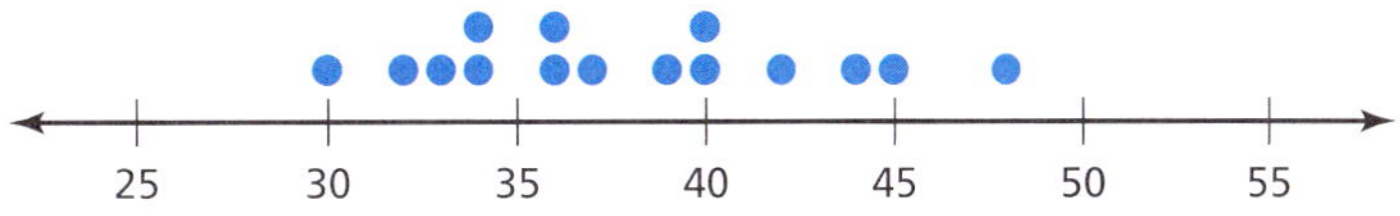

c.

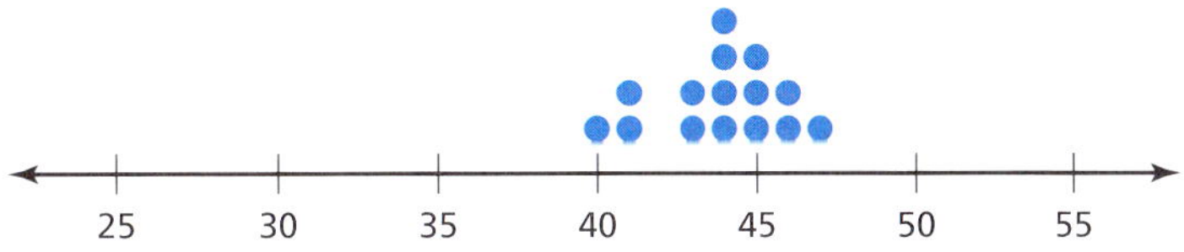

d.

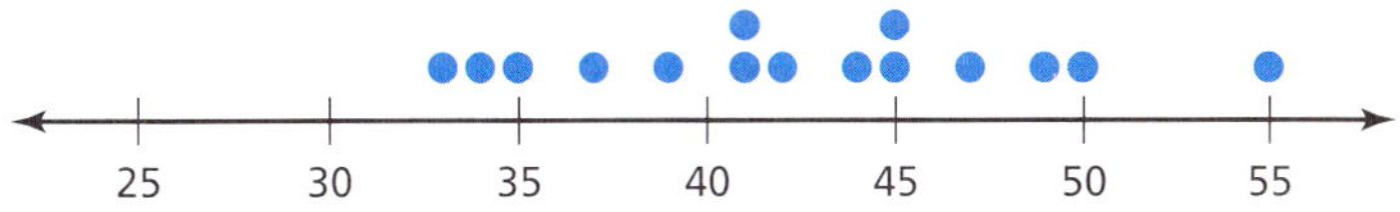

Math Practice 1

Analyze Givens

How can you use the given information to determine how spread out the data are?

What Is Your Answer?

5. **IN YOUR OWN WORDS** How can you describe the spread of a data set?

6. Make a dot plot of the data set in Activity 2. Describe any similarities between the dot plot and the number line in part (e).

Practice Use what you learned about variation to complete Exercises 4 and 5 on page 416.

9.4 Lesson

Check It Out
Lesson Tutorials
BigIdeasMath.com

A **measure of variation** is a measure that describes the distribution of a data set. A simple measure of variation to find is the *range*. The **range** of a data set is the difference between the greatest value and the least value.

Key Vocabulary
measure of variation, *p. 414*
range, *p. 414*
quartiles, *p. 414*
first quartile, *p. 414*
third quartile, *p. 414*
interquartile range, *p. 414*

EXAMPLE 1 Finding the Range

The table shows the lengths of several Burmese pythons captured for a study. Find and interpret the range of their lengths.

Lengths (feet)	
18.5	8
11	10
14	15.5
12.5	6.25
16.25	5

To find the least and the greatest values, order the lengths from least to greatest.

5, 6.25, 8, 10, 11, 12.5, 14, 15.5, 16.25, 18.5

The least value is 5. The greatest value is 18.5.

So, the range of the lengths is 18.5 − 5, or 13.5 feet. This means that the lengths vary by no more than 13.5 feet.

On Your Own

Now You're Ready
Exercises 6–9

1. The ages of people in line for a roller coaster are 15, 17, 21, 32, 41, 30, 25, 52, 16, 39, 11, and 24. Find and interpret the range of their ages.

Key Ideas

Quartiles

The **quartiles** of a data set divide the data into four equal parts. Recall that the median (second quartile) divides the data set into two halves.

lower half Median = 29 upper half

18 21 **22** 24 28 30 31 **32** 36 37

The median of the lower half is the **first quartile**, Q_1.

The median of the upper half is the **third quartile**, Q_3.

Reading
The first quartile can also be called the *lower quartile*. The third quartile can also be called the *upper quartile*.

Interquartile Range (IQR)

The difference between the third quartile and the first quartile is called the **interquartile range**. The IQR represents the range of the middle half of the data and is another measure of variation.

18 21 **22** 24 28 30 31 **32** 36 37

$$\text{IQR} = Q_3 - Q_1$$
$$= 32 - 22$$
$$= 10$$

Multi-Language Glossary at BigIdeasMath.com

Laurie's Notes

Introduction

Connect

- **Yesterday:** Students developed a conceptual understanding of how to describe the spread of a set of data. (MP3, MP8)
- **Today:** Students will use the range and the quartiles of a data set to describe the spread of the data.

Motivate

- **Discussion:** In August 2012, a Burmese python found in Everglades National Park had a length of 17 feet, 7 inches, weighed 164 pounds, and carried 87 eggs. Because of their invasive impact in Florida, scientists are conducting studies to gather data and statistics on these snakes.

Discuss

- Define and discuss *measure of variation*. Refer to the Activity 2 data set to help the students understand.
- Define *range*.
- **Common Error:** Students may write the range as an interval such as 5–22 rather than the difference between the greatest and least value, 17.

Lesson Notes

Example 1

- In finding the range, ordering the data helps to ensure that you find the least and greatest data values correctly.
- **MP2 Reason Abstractly and Quantitatively:** To interpret the range of 13.5 feet, state what it means in the context of the problem: "The lengths vary by no more than 13.5 feet."

On Your Own

- Check that students give both the range and the interpretation of what the range means in the context of the problem.

Key Ideas

- Say, "Recall folding a strip of grid paper twice yesterday to divide a data set into four parts. The folds corresponded to the *quartiles* of the data."
- Write the Key Ideas.
- Explain that the median of the lower half is called the *first quartile* (Q_1) and the median of the upper half is the *third quartile* (Q_3). The overall median is also known as the *second quartile* (Q_2).
- Define and discuss the *interquartile range* (IQR), which measures the spread of the middle 50% of the data. Because the IQR tends to ignore outliers, it may give a clearer picture of a data set than the range.
- **MP2:** Ask what fraction or percent of the data set is within the IQR and what fraction or percent is greater than the third quartile.

Goal Today's lesson is finding the **range**, **quartiles**, and **interquartile range** of a data set.

Technology for the Teacher

Dynamic Classroom

Lesson Tutorials
Lesson Plans
Answer Presentation Tool

Extra Example 1

The table shows the lengths of great white sharks at a local aquarium. Find and interpret the range of their lengths.

Lengths (feet)	
15	16.5
14	18
19	15
17.5	16
14	15.5

The range of the lengths is 5 feet, which means that the lengths vary by no more than 5 feet.

On Your Own

1. 41; The ages of people in line vary by no more than 41 years.

Differentiated Instruction

Auditory

Remind students of other words that have the same root as *quartile:* quarter and quart, for example. Define the words as *four* or *fourths*. Mention that one-fourth is the same as 25%.

Extra Example 2

The numbers of pages in eight different sixth grade math textbooks are shown below. Find and interpret the interquartile range of the data.

560, 532, 466, 728, 636, 512, 698, 444

178; The middle half of the numbers of pages vary by no more than 178.

Extra Example 3

Check for outliers in the data set in Extra Example 2. There are no books with fewer than 222 pages or more than 934 pages, so the data set has no outliers.

On Your Own

2. a. 38; The middle half of the numbers of pages vary by no more than 38.

   b. yes; The boundaries for outliers are 303 and 455. None of the numbers of pages are less than 303. One value, 468, is greater than 455. So, 468 is an outlier.

Laurie's Notes

Example 2

- Ask a volunteer to read the problem.
- ? "What measure of center can you find quickly from the dot plot?" the mode
- Some students will use the dot plot to find the quartiles without rewriting the data set in an ordered list. Suggest that they will make fewer errors when they rewrite the data.
- Write the list and show the location of the median (Q_2).
- ? "How do you find the first quartile, (Q_1)?" Find the median of the lower half.
- ? "How do you find the third quartile (Q_3)?" Find the median of the upper half.
- Remind students to never include the median in the lower half or the upper half of data. This way they will know how to handle a data set with an odd number of data.
- Find the IQR. Interpret what it means in the context of the problem.

Example 3

- You can use quartiles and the IQR to check for outliers. An outlier is any data value that is more than 1.5(IQR) below Q_1 or above Q_3. The boundaries are $Q_1 - 1.5(\text{IQR})$ and $Q_3 + 1.5(\text{IQR})$.
- **Discuss:** Compare checking for outliers by using the IQR formulas with using the definition of an outlier as a number that *is much greater or much less than the other data values.*
- Continue to work through the problem as shown.

On Your Own

- Remind students not to include the median in the lower half or the upper half of the data.
- **Neighbor Check:** Have students work independently and then have their neighbors check their work. Have students discuss any discrepancies.

Closure

- **Writing Prompt:** To describe the variability of a data set . . .
  Sample answer: Use a measure of variation, such as the range or interquartile range (IQR). The range is the difference between the greatest value and the least value. The IQR is the difference between the third quartile and the first quartile.

EXAMPLE 2 Finding the Interquartile Range

The dot plot shows the top speeds of 12 sports cars. Find and interpret the interquartile range of the data.

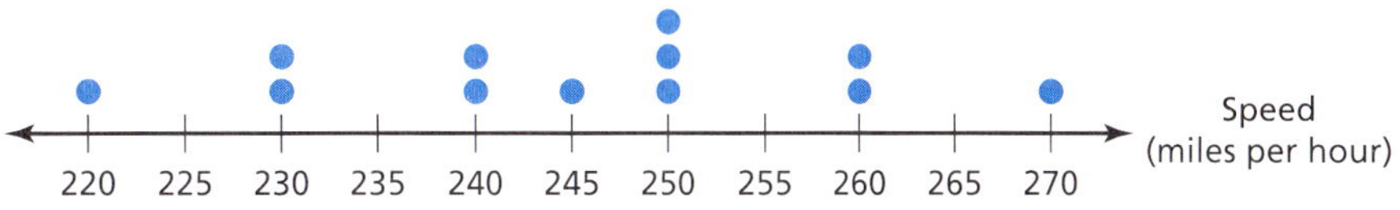

Order the speeds from slowest to fastest. Find the quartiles.

Median: $\frac{245 + 250}{2} = 247.5$

lower half: 220 230 230 240 240 245 | upper half: 250 250 250 260 260 270

Q_1: $\frac{230 + 240}{2} = 235$ Q_3: $\frac{250 + 260}{2} = 255$

∴ So, the interquartile range is $255 - 235 = 20$. This means that the middle half of the speeds vary by no more than 20 miles per hour.

You can use the quartiles and the interquartile range to check for outliers.

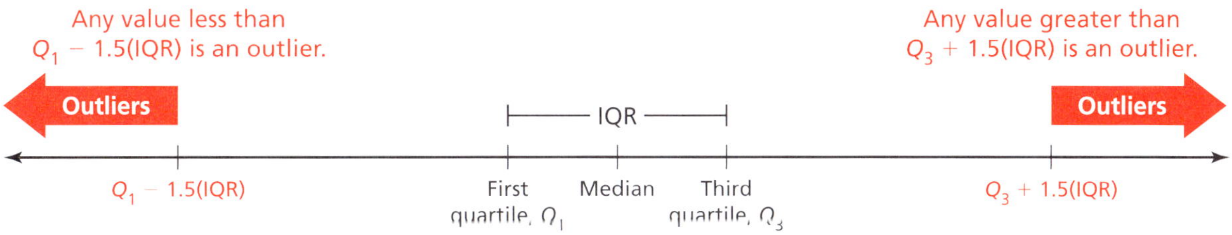

EXAMPLE 3 Checking for Outliers

Check for outliers in the data set in Example 2.

$Q_1 - 1.5(\text{IQR})$	Outlier boundaries	$Q_3 + 1.5(\text{IQR})$
$235 - 1.5(20)$	Substitute values.	$255 + 1.5(20)$
205	Simplify.	285

∴ There are no speeds less than 205 miles per hour or greater than 285 miles per hour. So, the data set has no outliers.

On Your Own

Now You're Ready
Exercises 11–14 and 17

2. The number of pages in each of an author's novels is shown.

356, 364, 390, 468, 400, 382, 376, 396, 350

a. Find and interpret the interquartile range of the data.

b. Does this data set contain any outliers? Justify your answer.

9.4 Exercises

Check It Out
Help with Homework
BigIdeasMath.com

Vocabulary and Concept Check

1. **VOCABULARY** How are measures of center different from measures of variation?

2. **VOCABULARY** How many quartiles does a data set have?

3. **DIFFERENT WORDS, SAME QUESTION** Which is different? Find "both" answers.

53, 47, 60, 45, 62, 59, 65, 50, 56, 48

What is the interquartile range of the data?	What is the range of the data?
What is the range of the middle half of the data?	What is the difference between the third quartile and the first quartile?

Practice and Problem Solving

Use grid paper to find the median of the data. Then find the median of the lower half and the median of the upper half of the data. Describe the spread of the data.

4. 5, 8, 10, 1, 7, 6, 15, 8, 6

5. 82, 62, 95, 81, 89, 51, 72, 56, 97, 98, 79, 85

Find the range of the data.

1 6. 26, 21, 27, 33, 24, 29

7. 52, 40, 49, 48, 62, 54, 44, 58, 39

8. 133, 117, 152, 127, 168, 146, 174

9. 4.8, 5.5, 4.2, 8.9, 3.4, 7.5, 1.6, 3.8

10. **ERROR ANALYSIS** Describe and correct the error in finding the range of the data.

✗ 49, 48, 51, 41, 35, 44, 38

The range is 49 − 38, or 11.

Find the median, first quartile, third quartile, and interquartile range of the data.

2 11. 40, 33, 37, 54, 41, 34, 27, 39, 35

12. 84, 75, 90, 87, 99, 91, 85, 88, 76, 92, 94

13. 132, 127, 106, 140, 158, 135, 129, 138

14. 38, 55, 61, 56, 46, 67, 59, 75, 65, 58

15. **PAPER AIRPLANE** The table shows the distances traveled by a paper airplane. Find and interpret the range and the interquartile range of the distances.

Distances (feet)			
$13\frac{1}{2}$	$21\frac{1}{2}$	21	$16\frac{3}{4}$
$10\frac{1}{4}$	19	32	$26\frac{1}{2}$
29	$16\frac{1}{4}$	$28\frac{1}{2}$	$18\frac{1}{2}$

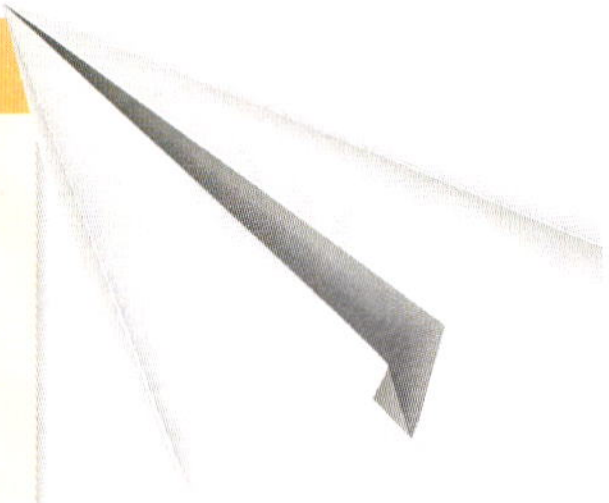

Assignment Guide and Homework Check

Level	Day 1 Activity Assignment	Day 2 Lesson Assignment	Homework Check
Basic	4, 5, 23–25	1–3, 7–9 odd, 10, 11–21 odd	7, 13, 15, 19
Average	4, 5, 23–25	1–3, 6–18 even, 19–21	8, 14, 19, 21
Advanced	4, 5, 23–25	1–3, 6–18 even, 19–22	8, 14, 19, 20, 21

Common Errors

- **Exercises 4 and 5** Students may fail to order the data before writing them on grid paper. Tell them it is important to order the data to get the correct values for the medians.
- **Exercises 6–9** Students may write the range as an interval such as 5–22 rather than as the distance between the greatest and least values, 17. Emphasize that the range is a value, not an interval.
- **Exercises 11–14** Students may include the median in the two halves of data when finding the quartiles. Remind them never to include the median in the upper or lower half of the data.
- **Exercise 16** Students may fail to realize that the range must be greater than the IQR. Point out that for the range and IQR to be equal, either each data value must be the same, or all of the data values consist of only two different numbers. In both cases, the data set will have a mode. Otherwise, the quartiles must be between the least and greatest data values.

9.4 Record and Practice Journal

Find the range of the data.

1. 3, 1, 6, 10, 12, 2
   11
2. 9, 13, 8, 7, 14, 16
   9
3. 11, 17, 21, 23, 19, 16
   12
4. 29, 37, 27, 28, 31, 33
   10

Find the median, first quartile, third quartile, and interquartile range of the data.

5. 20, 12, 14, 22, 18, 21, 24, 15
   median: 19
   Q_1: 14.5
   Q_3: 21.5
   IRQ: 7
6. 37, 28, 30, 40, 31, 27, 33, 42, 43
   median: 33
   Q_1: 29
   Q_3: 41
   IRQ: 12
7. The table shows the number of song downloads a group of friends made on Saturday.

Numbers of Songs			
12	6	9	14
15	16	8	10

a. Find and interpret the range of the numbers of song downloads.
10; The number of song downloads vary by no more than 10 downloads.

b. Find and interpret the interquartile range of the numbers of song downloads.
6; The middle half of the song downloads vary by no more than 6 downloads.

Vocabulary and Concept Check

1. A measure of center represents the center of a data set, but a measure of variation describes the distribution of a data set.
2. 3 quartiles
3. What is the range of the data?; 20; 12

Practice and Problem Solving

4. median = 7, median of lower half = 5.5, median of upper half = 9; The data are close together.
5. median = 81.5; median of lower half = 67; median of upper half = 92; The data are spread out.
6. 12
7. 23
8. 57
9. 7.3
10. The data were not ordered from least to greatest; 35, 38, 41, 44, 48, 49, 51; The range is 16.
11. median = 37; $Q_1 = 33.5$; $Q_3 = 40.5$; IQR = 7
12. median = 88; $Q_1 = 84$; $Q_3 = 92$; IQR = 8
13. median = 133.5; $Q_1 = 128$; $Q_3 = 139$; IQR = 11
14. median = 58.5; $Q_1 = 55$; $Q_3 = 65$; IQR = 10
15. range = $21\frac{3}{4}$ ft; The distances traveled by the paper airplane vary by no more than $21\frac{3}{4}$ feet; IQR = 11 ft; The middle half of the distances traveled by the paper airplane vary by no more than 11 feet.

Practice and Problem Solving

16. range; The range is the difference between the greatest value and the least value. The interquartile range is the range of the middle half of the data. So, the range is greater than the interquartile range.

17. Exercise 11: 54
Exercise 12: none
Exercise 13: 106 and 158
Exercise 14: 38

18. *Sample answer:* An outlier increases the range of a data set because there is a wider spread between the greatest and least values.

19. **a.** range = 172 points; IQR = 42 points

b. The outlier is 193 points; range = 101; IQR = 34; range

20–21. See Additional Answers.

22. See *Taking Math Deeper.*

Fair Game Review

23. 11 **24.** 56

25. D

Mini-Assessment

Find the range and the interquartile range of the data. Determine whether there are any outliers.

1. 8, 9, 1, 5, 5, 7, 9, 12, 15
range: 14, interquartile range: 5.5, no outliers

2. 43, 25, 39, 45, 57, 44, 48
range: 32, interquartile range: 9, outlier: 25

3. 110, 115, 116, 118, 121
range: 11, interquartile range: 7, no outliers

4. 359, 496, 482, 456, 498
range: 139, interquartile range: 89.5, no outliers

Taking Math Deeper

Exercise 22

It is important for students to make a plan for solving this exercise rather than simply jumping in and trying numbers.

1. Summarize what you know about the structure of the data set.
   - The median must be one of the 7 values.
   - The IQR is determined by 2 other values.
   - The range is determined by 2 other values.

2. Draw a diagram that represents the data set, marking the value that represents the median and the values that determine the IQR and the range.

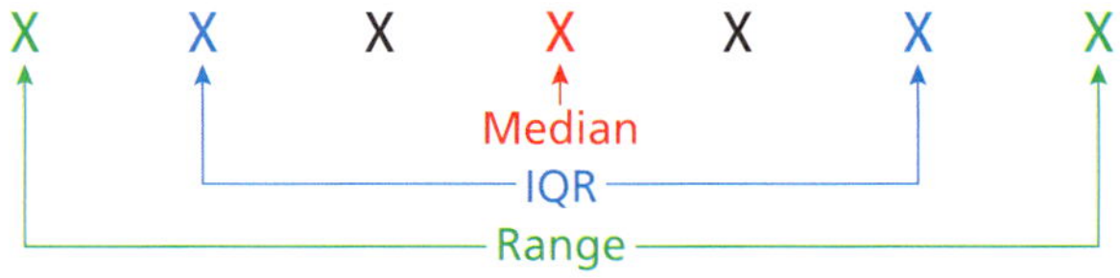

3. Look for an entry pint. The only measure affected by the third and fifth values is the mean. So, save these for last when you can choose them in a way that gives the desired mean of 30.

Start from the inside and work your way out.

The median is 26.

Let the first quartile be 14. So, the third quartile must be 14 + 36 = 50.

Let the least value be 10. So, the greatest value must be 10 + 50 = 60.

10 14 X 26 X 50 60

The mean is 30, so the sum of the 7 values must be 7 • 30 = 210. The sum of the values you have found is 10 + 14 + 26 + 50 + 60 = 160. Choose the last two values so that one is between 14 and 26, the other is between 26 and 50, and their sum is 210 − 160 = 50. Two possibilities are 20 and 30.

So, one possible data set is 10, 14, 20, 26, 30, 50, 60.

Project

Tell students to write a problem with different criteria for a data set that ends up being impossible to create. Have students exchange problems and explain why it is not possible to create the data set.

Reteaching and Enrichment Strategies

If students need help...	If students got it...
Resources by Chapter • Practice A and Practice B • Puzzle Time Record and Practice Journal Practice Differentiating the Lesson Lesson Tutorials Skills Review Handbook	Resources by Chapter • Enrichment and Extension • Technology Connection Start the next section

16. **WRITING** Consider a data set that has no mode. Which measure of variation is greater, the range or the interquartile range? Explain your reasoning.

3 17. **OUTLIERS** Use the interquartile range to identify any outliers in Exercises 11–14.

18. **REASONING** How does an outlier affect the range of a data set? Explain.

19. **BASKETBALL** The table shows the numbers of points scored by players on a basketball team.

Points Scored					
21	53	74	82	84	93
103	108	116	122	193	

a. Find the range and the interquartile range of the data.

b. Use the interquartile range to identify the outlier(s) in the data set. Find the range and the interquartile range of the data set without the outlier(s). Which measure did the outlier(s) affect more?

20. **STRUCTURE** Two data sets have the same range. Can you assume that the interquartile ranges of the two data sets are about the same? Give an example to justify your answer.

21. **SINGING** The tables show the ages of the finalists for two reality singing competitions.

a. Find the mean, median, range, and interquartile range of the ages for each show. Compare the results.

b. A 21-year-old is voted off Show A, and the 36-year-old is voted off Show B. How do these changes affect the measures in part (a)? Explain.

Ages for Show A	
18	17
15	21
22	16
18	28
24	21

Ages for Show B	
21	20
23	13
15	18
17	22
36	25

22. **Open-Ended** Create a set of data with 7 values that has a mean of 30, a median of 26, a range of 50, and an interquartile range of 36.

Fair Game Review What you learned in previous grades & lessons

Find the mean of the data. *(Section 9.2)*

23. 8, 14, 22, 7, 2, 11, 25, 7, 5, 9

24. 55, 64, 58, 43, 49, 67

25. **MULTIPLE CHOICE** What is the surface area of the rectangular prism? *(Section 8.2)*

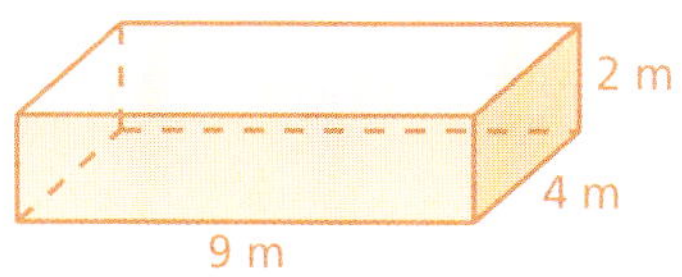

(A) $62\ m^2$ (B) $72\ m^2$

(C) $88\ m^2$ (D) $124\ m^2$

9.5 Mean Absolute Deviation

Essential Question How can you use the distances between each data value and the mean of a data set to measure the spread of a data set?

The Meaning of a Word ● Deviate

When you **deviate** from something, you stray or depart from the normal course of action.

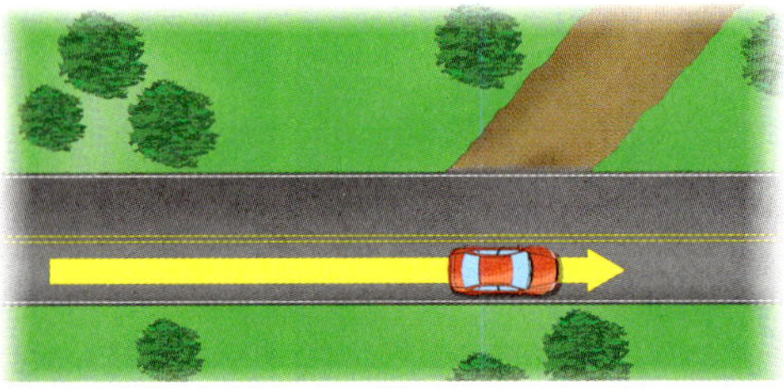

1 ACTIVITY: Finding Distances From the Mean

Work with a partner. The table shows the exam scores of 14 students in your class.

Exam Scores			
Ben	89	Mike	95
Emma	86	Hong	96
Jeremy	80	Rob	92
Pete	80	Amy	90
Ryan	96	Sue	76
Dan	94	Kim	84
Lucy	89	Heather	85

a. What is the mean exam score?

b. Make a dot plot of the data. Place an "X" on the number line to represent the mean.

c. Is the number of exam scores that are greater than the mean equal to the number of exam scores that are less than the mean? Explain.

d. Which exam score *deviates* the most from the mean? Which exam score *deviates* the least from the mean? Explain how you found your answers.

e. Overall, do you think the exam scores are *close* to the mean or *far away* from the mean? Explain your reasoning.

COMMON CORE

Statistics

In this lesson, you will

- understand the meaning of *mean absolute deviation.*
- find the mean absolute deviation of data sets.

Learning Standards
6.SP.2
6.SP.3
6.SP.5c

Laurie's Notes

Introduction

Standards for Mathematical Practice

- **MP2 Reason Abstractly and Quantitatively:** For a set of test scores, students compute how much each score deviates from the mean, then they sum all the deviations and divide by the number of scores. They interpret the meaning of the value they found.

Motivate

- Write the number 20 on a piece of paper. Fold it and place it on a front desk before students enter the classroom.
- Give each student 2 number cubes. Tell them that you wrote a special number on the piece of paper. Whoever rolls a total closest to the special number in three rolls of the number cubes is the Big Kahuna of the day! (My students love the title "Big Kahuna." You may want to use something different.)
- Each student quickly records three rolls of the number cubes and finds the sum.
- Reveal your special number. One or more students may hit 20.
- For several of the totals that are not 20, say, "You were off, or deviated from 20 by *x*." Do this for numbers both less than and greater than 20.

Discuss

- In this chapter, students have looked at measures of central tendency and measures of variability. In this lesson, students will investigate mean absolute deviation, which represents how much the data values differ from the mean. A formal definition follows in the lesson.
- Discuss the word "deviate" and give examples of its use, as in the motivator.

Activity Notes

Activity 1

- Give partners time to work through parts (a) through (c).
- Students may ask what part (d) means and how to find the amount of deviation. Tell them to recall how much they deviated from the special number in the Motivator, and they should be able to answer part (d).
- Ask students to share their answers to each part, including their explanations. In part (e), some students may think the scores are close to the mean while others may not. Some students may observe that no student failed, so everyone did okay!

? "What is the median?" 89 "What is the mode?" there are three modes: 80, 89, and 96

- Unlike the median, the mean does not necessarily divide the data set into two sets with equal numbers of values.

Common Core State Standards

6.SP.2 Understand that a set of data collected to answer a statistical question has a distribution which can be described by its center, spread . . ..

6.SP.3 Recognize that . . . a measure of variation describes how its values vary with a single number.

6.SP.5c Summarize numerical data sets in relation to their context, such as by giving quantitative measures of center and variability (. . . mean absolute deviation), as well as describing any overall pattern and any striking deviations from the overall pattern with reference to the context in which the data were gathered.

Previous Learning

Students should be able to find a mean, range, and interquartile range.

Technology for the Teacher

Dynamic Classroom

Lesson Plans
Complete Materials List

9.5 Record and Practice Journal

Essential Question How can you use the distance between each data value and the mean of a data set to measure the spread of a data set?

1 ACTIVITY: Finding Distances from the Mean

Work with a partner. The table shows the exam scores of 14 students in your class.

Exam Scores			
Ben	89	Mike	95
Emma	86	Hong	96
Jeremy	80	Rob	92
Pete	80	Amy	90
Ryan	96	Sue	76
Dan	94	Kim	84
Lucy	89	Heather	85

a. What is the mean exam score?

88

b. Make a dot plot of the data. Place an "X" on the number line to represent the mean.

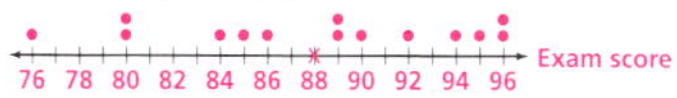

c. Is the number of exam scores that are greater than the mean equal to the number of exam scores that are less than the mean? Explain.

no

d. Which exam score *deviates* the most from the mean? Which exam score *deviates* the least from the mean? Explain how you found your answers.

76 (Sue's); 89 (Ben's and Lucy's)

e. Overall, do you think the exam scores are *close to* the mean or *far away* from the mean? Explain your reasoning.

far away

English Language Learners

Vocabulary

Help students connect the word *deviate* with the *distance from the mean*. Read the definition, "When you deviate from something, you stray or depart from the normal course of action." Then explain that each "distance from the mean" found in Activity 2 is the *amount* that a score *deviates* from the mean.

9.5 Record and Practice Journal

2 ACTIVITY: Using Distances From the Mean

Work with a partner. Use the information in Activity 1.

a. Complete the table below. Add rows if needed. Be sure to find the sum of the values in the last column of the table.

Student with Score *Less Than* the Mean	Exam Score	Distance from the Mean
Emma	86	2
Jeremy	80	8
Pete	80	8
Sue	76	12
Kim	84	4
Heather	85	3
	Sum:	37

b. Complete the table for students with scores greater than the mean.

Student with Score *Greater Than* the Mean	Exam Score	Distance from the Mean
Ben	89	1
Ryan	96	8
Dan	94	6
Lucy	89	1
Mike	95	7
Hong	96	8
Rob	92	4
Amy	90	2
	Sum:	37

c. **LOGIC** What do you notice about the sums you found in your tables? Why do you think this happens?

The sums are both equal to 37. This shows why the mean is the balance point.

3 ACTIVITY: Interpreting Distances From the Mean

Work with a partner.

a. **LOGIC** Add the sums you found in your tables in Activity 2. Divide that amount by the total number of students. Round your result to the nearest tenth.

$37 + 37 = 74; \frac{74}{14} \approx 5.3$

In your own words, what do you think this value represents?

average distance from the mean

b. **REASONING** In a data set, what do you think it means when the value you found in part (a) is close to 0? Explain.

The values in the data set are more consistent.

What Is Your Answer?

4. **IN YOUR OWN WORDS** How can you use the distances between each data value and the mean of a data set to measure the spread of a data set?

Find the sum of the distances and divide by the number of values in the data set.

5. **REASONING** Find the range and the interquartile range of the data set in Activity 1. What do you think it means when these values are close to 0? Explain.

range: 12

IQR: 10

Laurie's Notes

Activity 2

- Have students read through the directions for part (a) and record the names and exam scores of students who scored less than the mean.
- **MP2:** The data represent exam scores. Students need to find the distance from the mean of each score.

? "How can you find the distance from the mean?" Subtract the exam score from the mean.

- Have students make a second table for exam scores greater than the mean.

? "How can you find the distance from the mean?" Subtract the mean from the exam score.

- Students may be confused about the order of subtraction. Tell them to always subtract the smaller number from the greater number so that the difference will be positive.
- Students should find the sums for each table.
- **MP2:** Students may not understand why the sums are the same. Explain that for only 2 exam scores (say 80 and 90), the mean (85) is halfway between the scores. It is the balancing point. As additional scores are added, the mean is still that balancing point between all of the values.

Activity 3

- In part (a), students add the sums from each of the tables: $37 + 37 = 74$.
- Students may struggle to describe in the context of the problem what they have computed. They may say something such as, "It is the average amount of distance that the data values are from the mean." Remind students that they have studied three different averages.
- In discussing part (b), it might be useful to review the different measures of variation they have studied.

What Is Your Answer?

- **MP3 Construct Viable Arguments and Critique the Reasoning of Others:** In discussing Question 5, listen for the idea that a small amount of variation means that the data is not spread out much.

Closure

- Write the steps performed in finding the distance of each data value from the mean. *Sample answer:* (1) Find the mean of the data set. (2) For data values less than or equal to the mean, find the difference between the mean and the data value. (3) For data values greater than the mean, find the difference between the data value and the mean.

2 ACTIVITY: Using Distances from the Mean

Work with a partner. Use the information in Activity 1.

a. Complete the table below. Add rows if needed. Be sure to find the sum of the values in the last column of the table.

Math Practice 2

Use Operations

What operation can you use to find the distance from the mean? Explain.

Student with Score *Less Than* the Mean	Exam Score	Distance from the Mean
	Sum:	

b. Create a table similar to the one above for students with scores *greater than* the mean.

c. **LOGIC** What do you notice about the sums you found in your tables? Why do you think this happens?

3 ACTIVITY: Interpreting Distances from the Mean

Work with a partner.

a. **LOGIC** Add the sums you found in your tables in Activity 2. Divide that amount by the total number of students. Round your result to the nearest tenth.

In your own words, what do you think this value represents?

b. **REASONING** In a data set, what do you think it means when the value you found in part (a) is close to 0? Explain.

What Is Your Answer?

4. **IN YOUR OWN WORDS** How can you use the distances between each data value and the mean of a data set to measure the spread of a data set?

5. **REASONING** Find the range and the interquartile range of the data set in Activity 1. What do you think it means when these values are close to 0? Explain.

Practice Use what you learned about distances from the mean to complete Exercises 3 and 4 on page 422.

9.5 Lesson

Check It Out
Lesson Tutorials
BigIdeasMath.com

Key Vocabulary
mean absolute deviation, *p. 420*

Another measure of variation is the *mean absolute deviation*. The **mean absolute deviation** is an average of how much data values differ from the mean.

Key Idea

Finding the Mean Absolute Deviation (MAD)

Step 1: Find the mean of the data.

Step 2: Find the distance between each data value and the mean.

Step 3: Find the sum of the distances in Step 2.

Step 4: Divide the sum in Step 3 by the total number of data values.

EXAMPLE 1 Finding the Mean Absolute Deviation

You record the numbers of raisins in 8 scoops of cereal. Find and interpret the mean absolute deviation of the data.

1, 2, 2, 2, 4, 4, 4, 5

Step 1: Mean $= \frac{1+2+2+2+4+4+4+5}{8} = \frac{24}{8} = 3$

Step 2: You can use a dot plot to organize the data. Replace each dot with its distance from the mean.

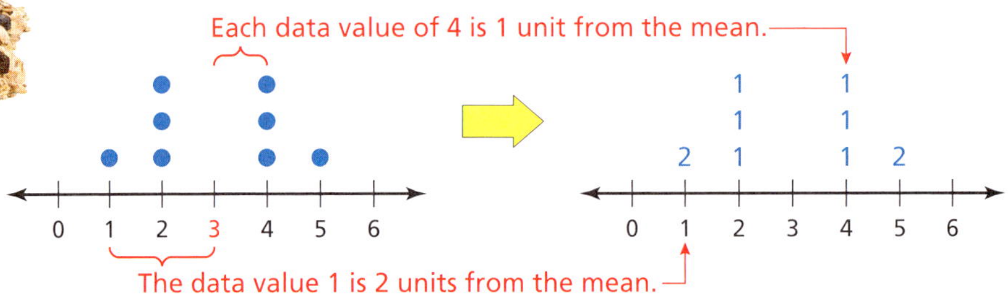

Step 3: The sum of the distances is $2 + 1 + 1 + 1 + 1 + 1 + 1 + 2 = 10$.

Step 4: The mean absolute deviation is $\frac{10}{8} = 1.25$.

So, the data values differ from the mean by an average of 1.25 raisins.

On Your Own

Now You're Ready
Exercises 5–8

1. Find and interpret the mean absolute deviation of the data.

   5, 8, 8, 10, 13, 14, 16, 22

Multi-Language Glossary at BigIdeasMath.com

Laurie's Notes

Introduction

Connect

- **Yesterday:** Students developed a conceptual understanding of describing how far data values are from the mean. (MP2, MP3)
- **Today:** Students will use the mean absolute deviation of a data set to describe the spread of the data.

Motivate

? "Does anyone know what the Cy Young Award is?" Someone may know that it is a pitching award for baseball.

- The **Cy Young Award** is given each year to the best pitcher in each league of Major League Baseball, both the National League and the American League. The award is voted on by the Baseball Writers Association of America.
- Cy Young was a pitcher in Major League Baseball from 1890 to1911. He held numerous pitching records, and some still stand. He died in 1955.
- Explain that in one of the examples in today's lesson, they will work with a data set involving baseball pitchers.

Key Idea

- Write the steps for finding the Mean Absolute Deviation (MAD).
- **Note:** The steps talk about "distance" rather than "deviation." See the Teacher Note in Example 1.

Example 1

- **MP1 Make Sense of Problems and Persevere in Solving Them:** Read the problem. Make sure students understand the context and what the data represent. You examine 8 different scoops of cereal to see how many raisins are in each scoop.

? "What is the first step?" Find the mean of the 8 data values.

? "What is the second step?" Find the distance between each data value and the mean.

- **Teacher Note:** In Step 2, notice that the *deviations* on the left side of the mean are negative: $1 - 3 = -2$, $2 - 3 = -1$, and so on. Students have not performed operations with negative numbers yet. This is why *distance* is used instead of *absolute deviation*. Students can simply count how far away each value is from the mean without performing integer operations.
- **MP5 Use Appropriate Tools Strategically:** A dot plot is a helpful tool for organizing the data. It can then be used to record the distance each data value is from the mean.
- Continue to work through the problem as shown.
- The mean absolute deviation is 1.25 raisins, which is how much the number of raisins in the average scoop differs from the mean of 3 raisins.

On Your Own

- **Neighbor Check:** Have students work independently and then have their neighbors check their work. Have students discuss any discrepancies.

Goal Today's lesson is finding the **mean absolute deviation** of a set of data.

Technology for the Teacher

Dynamic Classroom

Lesson Tutorials
Lesson Plans
Answer Presentation Tool

Extra Example 1

You record the numbers of pieces of orange colored candies in ten bags of candy of assorted colors. Find and interpret the mean absolute deviation of the data.

5, 5, 5, 6, 7, 7, 7, 9, 9, 10

1.4; The data values differ from the mean by an average of 1.4 pieces of candy.

On Your Own

1. 4.25; The data values differ from the mean by an average of 4.25.

Extra Example 2

The numbers of interceptions thrown by two quarterbacks in six seasons are shown below.

Quarterback 1: 11, 9, 19, 14, 15, 13

Quarterback 2: 12, 19, 7, 15, 8, 20

a. Find the mean, median and mean absolute deviation of the numbers of interceptions thrown.
Quarterback 1: mean: 13.5, median: 13.5, MAD: 2.5
Quarterback 2: mean: 13.5, median: 13.5, MAD: 4.5

b. Which measure can you use to distinguish the data? What can you conclude about the quarterbacks from this measure?
the mean absolute deviation; The numbers of interceptions deviate less for Quarterback 1.

On Your Own

2. The mean absolute deviation will decrease. The distance between the new value and the mean is less than the mean absolute deviation of 2.4.

Differentiated Instruction

Visual

Use a vibrating string (such as a guitar string or rubber band) to demonstrate different amounts of deviation about the same mean. When plucked, the string always vibrates in waves about the middle point (the mean), but how far it deviates from the middle point depends on how hard the string is plucked.

Laurie's Notes

Example 2

- Ask a volunteer to read the problem. Be sure that students understand the context. Pitchers want a low number of runs scored against them, and coaches want consistent performances from pitchers.
- You could scaffold this problem to have half of the class work with one data set and the other half of the class work with the other data set.
- ? "What is the mean for each pitcher?" each has a mean of 3.5 runs
- ? "What is the median for each pitcher?" each has a median of 4 runs
- ? "What can you conclude about the two pitchers from their means and medians?" They have the same measures of center, but you do not know how spread out the data are.
- **MP5:** A dot plot is again used to organize the data. It is then modified to record the distance of each data value from the mean.
- ? "After Step 2, what else must you do to find the mean absolute deviation for each pitcher?" Find the sum of the differences and divide by 10, the number of games.
- ? "What is different about the pitchers based on the statistical measures found? Explain." *Sample Answer:* You do not learn anything different about the pitchers from the mean and median because each are the same. Because Rodriguez has a lower mean absolute deviation, however, his performances are more consistent than Mendoza's.
- **MP3 Construct Viable Arguments and Critique the Reasoning of Others:** Listen for an explanation that states that the mean absolute deviation is less for Rodriquez, therefore there is less variation for him as a pitcher in the last 10 starts.

On Your Own

- **Neighbor Check:** Have students work independently and then have their neighbors check their work. Have students discuss any discrepancies.

Closure

- Explain how the range and the mean absolute deviation are similar. Explain how they are different. *Sample answer:* The range and the mean absolute deviation measure the variability of the data set. The range is determined using only two values of the data set. The mean standard deviation is determined using all of the values of the data set.

EXAMPLE 2 Real-Life Application

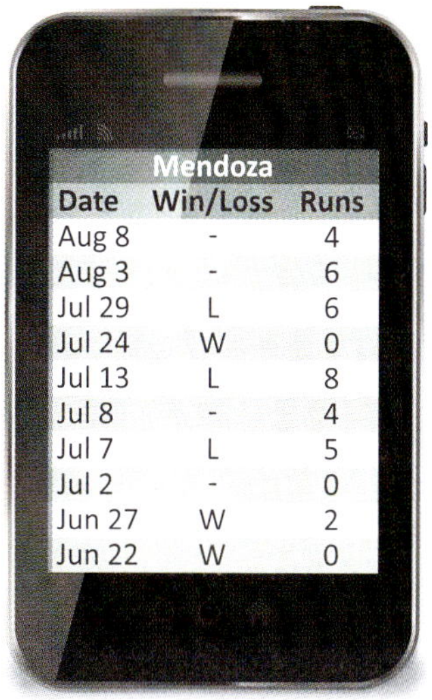

Mendoza		
Date	Win/Loss	Runs
Aug 8	-	4
Aug 3	-	6
Jul 29	L	6
Jul 24	W	0
Jul 13	L	8
Jul 8	-	4
Jul 7	L	5
Jul 2	-	0
Jun 27	W	2
Jun 22	W	0

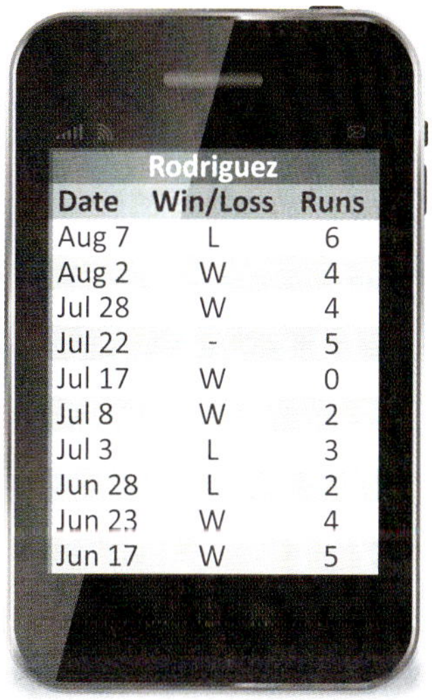

Rodriguez		
Date	Win/Loss	Runs
Aug 7	L	6
Aug 2	W	4
Jul 28	W	4
Jul 22	-	5
Jul 17	W	0
Jul 8	W	2
Jul 3	L	3
Jun 28	L	2
Jun 23	W	4
Jun 17	W	5

The smartphones show the numbers of runs allowed by two pitchers in their last 10 starts.

a. Find the mean, median, and mean absolute deviation of the numbers of runs allowed for each pitcher.

Order the runs allowed for Mendoza:
0, 0, 0, 2, 4, 4, 5, 6, 6, 8.

Mean $= \frac{35}{10} = 3.5$ Median $= \frac{4+4}{2} = 4$

Mean absolute deviation:

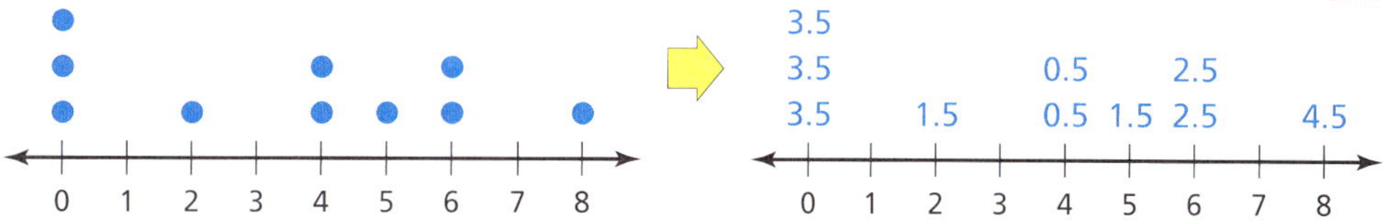

The mean absolute deviation is $\frac{24}{10} = 2.4$

Order the runs allowed for Rodriguez: 0, 2, 2, 3, 4, 4, 4, 5, 5, 6.

Mean $= \frac{35}{10} = 3.5$ Median $= \frac{4+4}{2} = 4$

Mean absolute deviation:

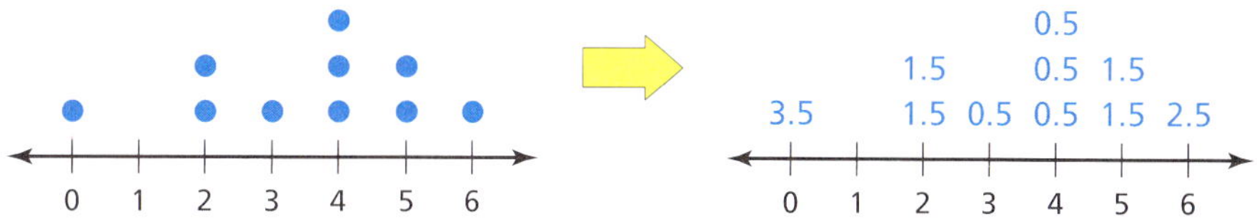

The mean absolute deviation is $\frac{14}{10} = 1.4$.

b. Which measure can you use to distinguish the data? What can you conclude about the pitchers from this measure?

You cannot use the measures of center to distinguish the data because they are the same for each data set. The measure of variation, MAD, is 2.4 for Mendoza and 1.4 for Rodriguez. This indicates that the data for Rodriguez has less variation.

∴ Using the MAD to distinguish the data, you can conclude that Rodriguez is more consistent than Mendoza.

Study Tip

The greater the mean absolute deviation, the greater the variation of the data.

On Your Own

2. **WHAT IF?** Mendoza allows 4 runs in the next game. How would you expect the mean absolute deviation to change? Explain.

9.5 Exercises

Check It Out
Help with Homework
BigIdeasMath.com

Vocabulary and Concept Check

1. **REASONING** Describe a data set that has a mean absolute deviation of 0.

2. **WHICH ONE DOESN'T BELONG?** Which one does *not* belong with the other three? Explain your reasoning.

range	interquartile range	mean	mean absolute deviation

Practice and Problem Solving

Find the average distance each data value in the set is from the mean. Round your answer to the nearest tenth, if necessary.

3. Model years of used cars on a lot: 2010, 2002, 2005, 2007, 2001

4. Prices of kites at a shop: \$7, \$20, \$9, \$35, \$12, \$15, \$7, \$10, \$20, \$25

Find and interpret the mean absolute deviation of the data. Round your answer to the nearest tenth, if necessary.

1 5.

Prices of Microphones (dollars)				
25	28	20	22	32
28	35	34	30	36

6.

Heights of 10-Year-Old Octuplets (inches)			
61	61	61	61
61	61	61	61

7.

Capacities of Stadiums (thousands of people)		
101.5	95.4	109.8
98.7	92.3	104.7

8.

Numbers of Visitors to a Website During a Week			
103	115	124	125
171	165	170	

9. **ERROR ANALYSIS** Describe and correct the error in finding the mean absolute deviation of the data set 35, 40, 38, 32, 42, and 41.

✗

$$\text{mean} = \frac{35 + 40 + 38 + 32 + 42 + 41}{6} = 38$$

$$\text{MAD} = \frac{3 + 2 + 0 + 4 + 3}{5} = 3.6$$

So, the values differ from the mean by an average of 3.6.

10. **MUSEUMS** The data set shows the admission prices at several museums.

\$20, \$20, \$16, \$12, \$15, \$25, \$11

Find and interpret the range, interquartile range, and mean absolute deviation of the data.

Assignment Guide and Homework Check

Level	Day 1 Activity Assignment	Day 2 Lesson Assignment	Homework Check
Basic	3, 4, 18–20	1, 2, 5–9 odd, 10, 11–15 odd	5, 7, 11, 13
Average	3, 4, 18–20	1, 2, 7–11, 12–16 even	7, 8, 10, 11, 12
Advanced	3, 4, 18–20	1, 2, 7–17	7, 11, 13, 14, 16

Common Errors

- **Exercises 3, 4** Students may find the distance from the median instead of the mean. Remind students to read the problem carefully.
- **Exercises 5–8** Students may find the mean absolute deviation using the median instead of the mean. Remind students that the MAD measures the variation from the mean.
- **Exercise 15** Students may be confused about units. Discuss using the correct units for this exercise. Rainfall is most likely measured in inches. Water usage in a home is most likely measured in gallons.

9.5 Record and Practice Journal

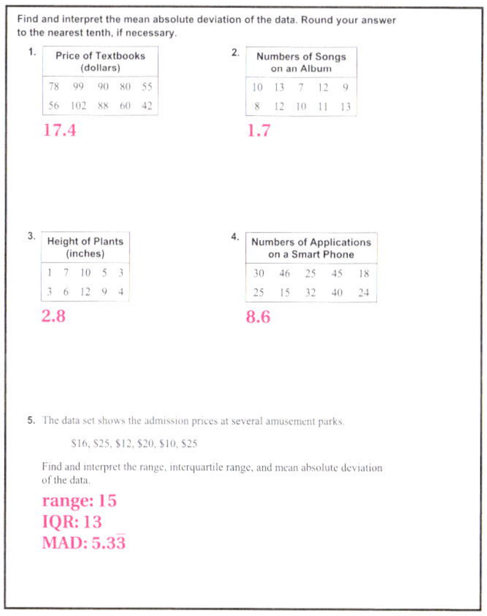

Find and interpret the mean absolute deviation of the data. Round your answer to the nearest tenth, if necessary.

1.

Price of Textbooks (dollars)				
78	99	90	80	55
56	102	88	60	42

17.4

2.

Numbers of Songs on an Album				
10	13	7	12	9
8	12	10	11	13

1.7

3.

Height of Plants (inches)				
1	7	10	5	3
3	6	12	9	4

2.8

4.

Numbers of Applications on a Smart Phone				
30	46	25	45	18
25	15	32	40	24

8.6

5. The data set shows the admission prices at several amusement parks.

$16, $25, $12, $20, $10, $25

Find and interpret the range, interquartile range, and mean absolute deviation of the data.

range: 15
IQR: 13
MAD: $5.3\overline{3}$

Vocabulary and Concept Check

1. All the values in the data set are the same.
2. mean; It is a measure of center. All the other measures are measures of variation.

Practice and Problem Solving

3. 2.8 years
4. $7.20
5. 4.4; The prices differ from the mean price by an average of $4.40.
6. 0; The heights are the same, so the mean absolute deviation is 0.
7. 4.9; The capacities differ from the mean capacity by an average of 4.9 thousand, or 4900 people.
8. 25.4; The numbers of visitors differ from the mean by an average of 25.4, or about 25 visitors.
9. When calculating the mean absolute deviation, you need to divide by 6, not 5. Even though the distance from the mean of one of the values (38) is 0, it is still included in the calculation.

   mean absolute deviation $= \dfrac{3 + 2 + 0 + 6 + 4 + 3}{6} = 3$

   So, the values differ from the mean by an average 3.0.
10. range: 14; The prices vary by no more than $14.

    IQR: 8; The middle half of the prices vary by no more than $8.

    mean absolute deviation: 4; The admission prices differ from the mean price by an average of $4.

Practice and Problem Solving

11. The MAD of the five most-expensive dishes is 3.6. The MAD of the five least-expensive dishes is 1.76. The MAD of the five least-expensive dishes is much less than the MAD of the five most-expensive dishes. So, the data for the five least-expensive dishes is closer together compared to the five most-expensive dishes.

12–13. See Additional Answers.

14. See *Taking Math Deeper.*

15–16. See Additional Answers.

17. **a.** 50%; 87.5%; 2 and 15

b. *Sample answer:* A good portion of a data set is within one MAD of the mean and most of the data set is within 2 MADs of the mean. As you get more and more MADs away from the mean, the percent increases because more and more data are included in the interval.

Fair Game Review

18. mean: 6.5; median: 6; mode: 6

19. mean: 1.6; median: 1.7; modes: 1.2, 1.7

20. C

Mini-Assessment

Find the mean, median, and mean absolute deviation of the data.

1. 2, 2, 3, 3, 4, 5, 7, 7, 8, 9
Mean: 5, median: 4.5, MAD: 2.2

2. 7, 9, 10, 12, 12, 13, 14
Mean: 11, median: 12, MAD: 2

3. 40, 8, 29, 9, 41, 30, 45, 42
Mean: 30.5, median: 35, MAD: 11.5

4. 189, 161, 283, 499, 348
Mean: 296, median: 283, MAD: 102

Taking Math Deeper

Exercise 14

The key assumption in this problem is that the jars are reasonable sizes. Neither is much larger or smaller than the other jar. Students can solve the problem by visualizing the jars and analyzing their own guesses.

1. Visualize the situation using two jars that are about the same size.

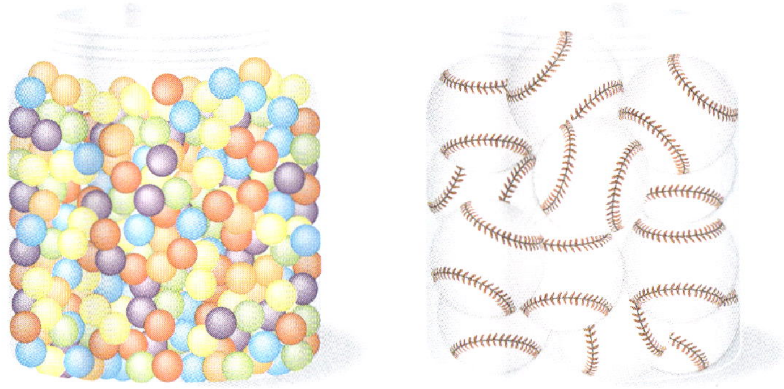

2. Make 10 guesses for the number of objects in each jar. To increase your chances of guessing correctly, you guess a few numbers that are greater than and less than what you think the actual number is.

   Gumball guesses: 130, 140, 145, 150, 155, 170, 180, 195, 210, 225

   Baseball guesses: 7, 8, 10, 11, 13, 14, 15, 16, 18, 20

3. The guesses for gumballs are greater and have a much larger range than the guesses for baseballs. So, you can reason that the MAD for gumballs is greater than the MAD for baseballs. In general, guesses for gumballs will *deviate* more because there are many more in the jar, making it harder to guess and producing a larger range of guesses.

Project

Design an experiment using two different-sized spherical objects. Fill two identical jars with the objects. Collect guesses from at least 20 people. Compare the MADs of the guesses for the jars.

Reteaching and Enrichment Strategies

If students need help. . .	If students got it. . .
Resources by Chapter • Practice A and Practice B • Puzzle Time Record and Practice Journal Practice Differentiating the Lesson Lesson Tutorials Skills Review Handbook	Resources by Chapter • Enrichment and Extension • Technology Connection Start the next section

11. **MENU** The table shows the prices of the five most-expensive and least-expensive dishes on a menu. Find the MAD of each data set. Then compare their variations.

Five Most-Expensive Dishes					Five Least-Expensive Dishes				
$28	$30	$28	$39	$25	$7	$7	$10	$8	$12

12. **COINS** The data sets show the years of the coins in two collections.

Derek's collection: 1950, 1952, 1908, 1902, 1955, 1954, 1901, 1910

Paul's collection: 1929, 1935, 1928, 1930, 1925, 1932, 1933, 1920

Find the measures of center and the measures of variation for each data set. Compare the measures. What can you conclude?

13. **PROBLEM SOLVING** You survey students in your class about the number of movies they watched last month. The results are shown in the table.

Movies Watched			
7	5	14	5
6	9	10	12
15	4	5	8
11	10	9	2

a. Find the measures of center and the measures of variation for the data.

b. A new student joins the class who watched 21 movies last month. Is 21 an outlier? How does including this value affect the measures of center and the measures of variation? Explain.

REASONING Which data set do you think would have the greater mean absolute deviation? Explain your reasoning.

14. guesses for number of gumballs in a jar
guesses for number of baseballs in a jar

15. monthly rainfall amounts in a city
monthly amounts of water used in a home

16. **REASONING** The MAD of a data set is considered a more reliable measure of variation than the range or the interquartile range. Why do you think this is true?

17. **Critical Thinking** Add and subtract the MAD from the mean in the original data set in Exercise 13.

a. What percent of the values are within one MAD of the mean? two MADs of the mean? Which values are more than twice the MAD from the mean?

b. What do you notice as you get more and more MADs away from the mean? Explain.

Fair Game Review What you learned in previous grades & lessons

Find the mean, median, and mode(s) of the data. *(Section 9.2 and Section 9.3)*

18. 4, 6, 7, 9, 6, 4, 5, 6, 8, 10

19. 1.2, 1.7, 1.7, 2.1, 1.4, 1.2, 1.9

20. **MULTIPLE CHOICE** What is the surface area of the square pyramid? *(Section 8.3)*

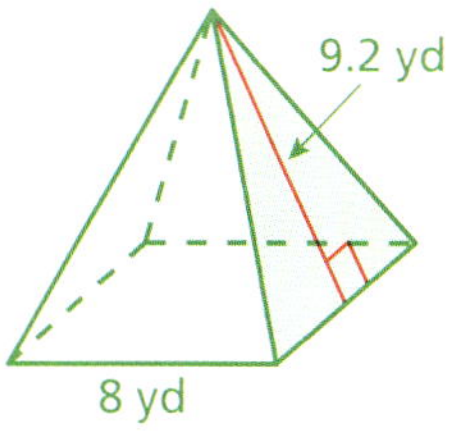

(A) 100.8 yd^2 (B) 147.2 yd^2

(C) 211.2 yd^2 (D) 368 yd^2

9.4–9.5 Quiz

Check It Out
Progress Check
BigIdeasMath.com

Find the range of the data. *(Section 9.4)*

1. 35, 76, 43, 58, 34, 67

2. 19, 21, 22, 22, 19, 25, 24, 23, 24

Find the median, first quartile, third quartile, and interquartile range of the data. *(Section 9.4)*

3. 56, 48, 72, 37, 35, 42, 48, 33, 28

4. 95, 14, 86, 62, 55, 46, 28, 37, 33, 70, 31

Find and interpret the mean absolute deviation of the data. Round your answer to the nearest tenth if necessary. *(Section 9.5)*

5.

Ages of Television Show Viewers (years)			
29	18	26	33
33	22	34	26

6.

Prices of Houses (thousands of dollars)				
80	120	95	240	140
75	135	110	90	125

7. AMUSEMENT PARKS The data set shows the admission prices at several amusement parks.

$65, $70, $40, $55, $35, $40, $60

Find and interpret the range, interquartile range, and mean absolute deviation of the data. *(Section 9.4 and Section 9.5)*

8. TEACHING EXPERIENCE The tables show the years of teaching experience of faculty members at two schools. *(Section 9.4)*

a. Find the mean, median, range, and interquartile range of the years of experience for each school. Compare the results.

b. The teacher with 11 years of experience leaves School A, and the teacher with 33 years of experience retires from School B. How does this affect the measures in part (a)? Explain.

School A: Teaching Experience (years)	
5	11
10	22
7	8
8	6
10	35

School B: Teaching Experience (years)	
4	15
6	12
10	33
12	20
8	7

9. BOOK CLUB You survey the students in your book club about the number of books they read last summer. The results are shown in the table. *(Section 9.4 and Section 9.5)*

Books Read			
8	14	15	9
6	12	9	13
11	11	7	5
12	6	10	8

a. Find the measures of center and the measures of variation for the data.

b. A new student who read 18 books last summer joins the club. Is 18 an outlier? How does adding this value to the data set affect the measures of center and variation? Explain.

Alternative Assessment Options

Math Chat	Student Reflective Focus Question
Structured Interview	Writing Prompt

Structured Interview

Interviews can occur formally or informally. Ask a student to perform a task. While working, have the student explain the task and his or her thought process. Probe the student for more information. Do not ask leading questions. Keep a rubric or notes.

Teacher Prompts	Student Answers	Teacher Notes
Describe and give possible test scores for Amy, given that she has the same mean but a lower mean absolute deviation than Don, who has scores of 80, 85, 90, 95, and 100.	Because his scores are evenly spread around 90, Don's mean score is 90. So Amy's scores are evenly spread around 90, but are closer to 90. Possible scores: 88, 89, 90, 91, 92	Student understands that data are evenly spread about the mean and that data with a lower mean absolute deviation are closer to the mean. Constructs a feasible set of data.
Add a statement about how the ranges and interquartile ranges of Don and Amy's test scores compare.	Don's range (20) and interquartile range (15) are greater than Amy's range (4) and interquartile range (3).	Student understands and correctly calculates the range and interquartile range.

Study Help Sample Answers

Remind students to complete Graphic Organizers for the rest of the chapter.

6–10. Available at *BigIdeasMath.com.*

Reteaching and Enrichment Strategies

If students need help...	If students got it...
Resources by Chapter • Practice A and Practice B • Puzzle Time Lesson Tutorials *BigIdeasMath.com*	Resources by Chapter • Enrichment and Extension • Technology Connection Game Closet at *BigIdeasMath.com* Start the Chapter Review

Answers

1. 42 **2.** 6

3. median = 42, $Q_1 = 34$, $Q_3 = 52$, IQR = 18

4. median = 46, $Q_1 = 31$, $Q_3 = 70$, IQR = 39

5. about 4.6; The ages differ from the mean by an average of about 4.6 years.

6. 31.2; The prices differ from the mean by an average of 31.2 thousand dollars, or \$31,200.

7. range: 35; The prices vary by no more than \$35.

IQR: 25; The middle half of the prices vary by no more than \$25.

MAD: about 11.8; The admission prices differ from the mean price by an average of about \$11.80.

8. See Additional Answers.

9. a. mean: 9.75; median: 9.5; modes: 8, 6, 11, 12, 9 range: 10; IQR: 4.5; MAD: about 2.5

b. no; Using the interquartile range, 18 is inside the outlier boundaries.
mean: about 10.2;
median: 10;
modes: 8, 6, 11, 12, 9
range: 13;
interquartile range: 5;
mean absolute deviation: about 2.8

The range increased by 3. The modes stayed the same. The mean, median, IQR, and MAD all increased slightly.

Technology for the Teacher

Online Assessment
Assessment Book
ExamView® Assessment Suite

For the Teacher
Additional Review Options

- *BigIdeasMath.com*
- Online Assessment
- Game Closet at *BigIdeasMath.com*
- Vocabulary Help
- Resources by Chapter

Answers

1–2. See Additional Answers.

3. 11

4. 8.1

Review of Common Errors

Exercises 1 and 2

- Students may miss gaps in the data because they include only data values along the number line for their dot plots. Remind students to include all the integers from the least data value to the greatest data value along the number line.

Exercises 3 and 4

- Students may add to find the total for the data and then incorrectly divide by the maximum data value. Remind students to divide by the total number of data values to find the mean.

Exercises 5 and 6

- Students may try to identify the median without ordering the data first. Remind them that it is essential to order the data first and then find the median. This makes finding the mode easier as well.

Exercises 7 and 8

- Students may write the range as an interval such as 5–22 rather than as the distance between the greatest and least values, 17. Emphasize that the range is a value, not an interval.

Exercises 9 and 10

- Students may include the median in the two halves of data when finding the quartiles. Remind them never to include the median in the upper or lower half of the data.

Exercises 11 and 12

- Students may find the mean absolute deviation using the median instead of the mean. Remind students that the MAD measures the variation from the mean.

9 Chapter Review

Check It Out
Vocabulary Help
BigIdeasMath.com

Review Key Vocabulary

statistics, *p. 392*
statistical question, *p. 392*
mean, *p. 398*
outlier, *p. 399*
measure of center, *p. 404*
median, *p. 404*
mode, *p. 404*
measure of variation, *p. 414*
range, *p. 414*
quartiles, *p. 414*
first quartile, *p. 414*
third quartile, *p. 414*
interquartile range, *p.414*
mean absolute deviation, *p. 420*

Review Examples and Exercises

9.1 Introduction to Statistics *(pp. 390–395)*

Display the data in a dot plot. Identify any clusters, peaks, or gaps in the data.

Heights (inches)				
70	71	70	72	69
68	69	71	64	70
64	63	72	70	67

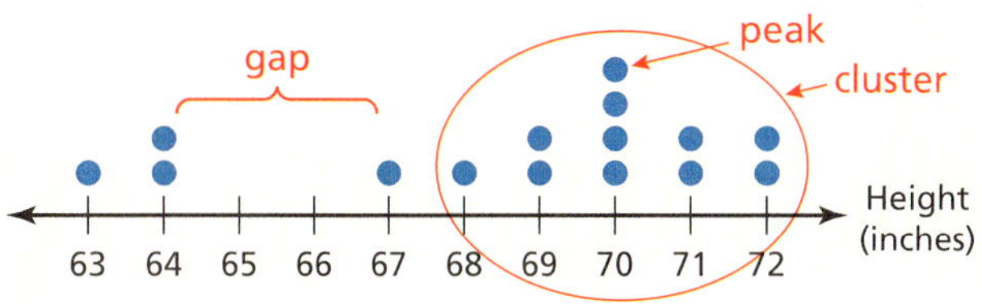

Exercises

Display the data in a dot plot. Identify any clusters, peaks, or gaps in the data.

1.

Distance (feet)			
56	55	56	57
58	54	51	55
51	56	49	56

2.

Weight (pounds)				
83	88	89	90	89
91	89	84	90	92
90	88	89	83	88

9.2 Mean *(pp. 396–401)*

Find the mean of 5, 9, 10, 6, 6, and 12.

$$\text{mean} = \frac{5 + 9 + 10 + 6 + 6 + 12}{6}$$

sum of the data (numerator); number of values (denominator)

$$= \frac{48}{6}, \text{ or } 8$$ Simplify.

Exercises

Find the mean of the data.

3. 4, 5, 7, 14, 17, 12, 18

4. 15, 5, 8, 12, 5, 9, 4, 10, 2, 11

9.3 Measures of Center (pp. 402–409)

Find the median and the mode of the movie lengths in the table.

Movie Lengths (minutes)		
91	112	126
142	122	112
92	144	

Order the data from least to greatest.

Median:

91, 92, 112, 112, 122, 126, 142, 144

$$\frac{112 + 122}{2} = \frac{234}{2}, \text{ or } 117$$

Mode:

91, 92, 112, 112, 122, 126, 142, 144

The value 112 occurs most often.

The median is 117 minutes, and the mode is 112 minutes.

Exercises

Find the median and the mode(s) of the data.

5. 8, 8, 6, 8, 4, 5, 6

6. 24, 74, 61, 29, 38, 27, 68, 54

9.4 Measures of Variation (pp. 412–417)

The table shows the weights of several adult emperor penguins. (a) Find and interpret the range. (b) Find and interpret the interquartile range. (c) Check for outliers.

Weights (kilograms)	
25	27
36	23.5
33.5	31.25
30.75	32
24	29.25

a. Ordered from least to greatest, the weights are 23.5, 24, 25, 27, 29.25, 30.75, 31.25, 32, 33.5, and 36.

So, the range of the weights is 36 − 23.5, or 12.5 kilograms. The weights vary by no more than 12.5 kilograms.

b. Find the quartiles.

Median: $\frac{29.25 + 30.75}{2} = 30$

lower half: 23.5, 24, 25, 27, 29.25,

upper half: 30.75, 31.25, 32, 33.5, 36

Q_1: 25

Q_3: 32

So, the interquartile range is 32 − 25 = 7. This means that the middle half of the weights vary by no more than 7 kilograms.

c. Calculate the outlier boundaries.

$$Q_1 - 1.5(\text{IQR}) = 25 - 1.5(7) = 14.5$$

$$Q_3 + 1.5(\text{IQR}) = 32 + 1.5(7) = 42.5$$

There are no weights less than 14.5 kilograms or greater than 42.5 kilograms. So, the data set has no outliers.

Review Game

Calculating Statistical Measures

Materials

- 2 spinners labeled with the integers 0 through 12 (as shown below)
- 1 spinner labeled with *Mean, Median, Mode, Range, Interquartile Range,* and *Mean Absolute Deviation* (as shown below)
- 2 pencils
- paper

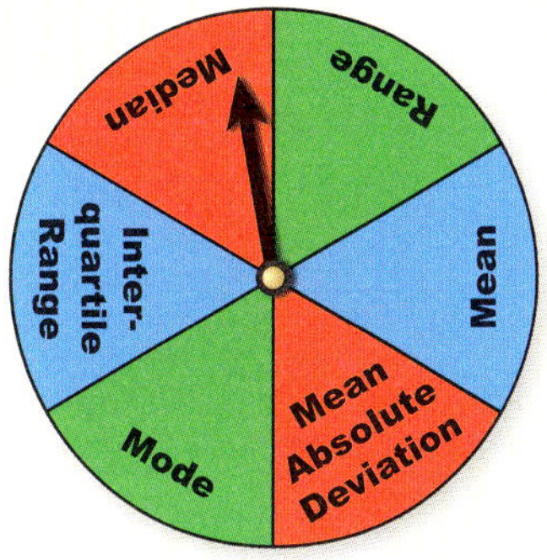

Players: 2

Directions

1. One student spins to determine whether to calculate mean, median, mode, range, interquartile range, or mean absolute deviation.
2. Each player spins his or her integer spinner 10 times, recording the results.
3. Each player calculates the mean, median, mode, range, interquartile range, or mean absolute deviation, as determined by the spinner, of his or her numbers.

Who Wins?

The player with the greater mean, median, mode, range, interquartile range, or mean absolute deviation wins a point for the round. Students can play several rounds or for a predetermined amount of time.

For the Student
Additional Practice

- Lesson Tutorials
- Multi-Language Glossary
- Self-Grading Progress Check
- *BigIdeasMath.com*
  Dynamic Student Edition
  Student Resources

Answers

5. median: 6; mode: 8
6. median: 46; mode: none
7. 77
8. 87
9. median = 44; $Q_1 = 25$; $Q_3 = 61.5$; IQR = 36.5
10. median = 55; $Q_1 = 26.5$; $Q_3 = 93$; IQR = 66.5
11. about 1.3; The shoe sizes differ from the mean size by an average of about 1.3.
12. 26; The prices differ from the mean price by an average of $26.

My Thoughts on the Chapter

Teacher Tip

Not allowed to write in your teaching edition? Use sticky notes to record your thoughts.

What worked. . .

What did not work. . .

What I would do differently. . .

Exercises

Find the range of the data.

7. 45, 76, 98, 21, 52, 39

8. 95, 63, 52, 8, 93, 16, 42, 37, 62

Find the median, first quartile, third quartile, and interquartile range of the data.

9. 28, 46, 25, 76, 18, 25, 47, 83, 44

10. 14, 25, 97, 55, 66, 28, 92, 38, 94

9.5 Mean Absolute Deviation *(pp. 418–423)*

You record the prices of 8 printers. Find and interpret the mean absolute deviation of the data.

\$120, \$150, \$90, \$110,

\$140, \$120, \$140, \$90

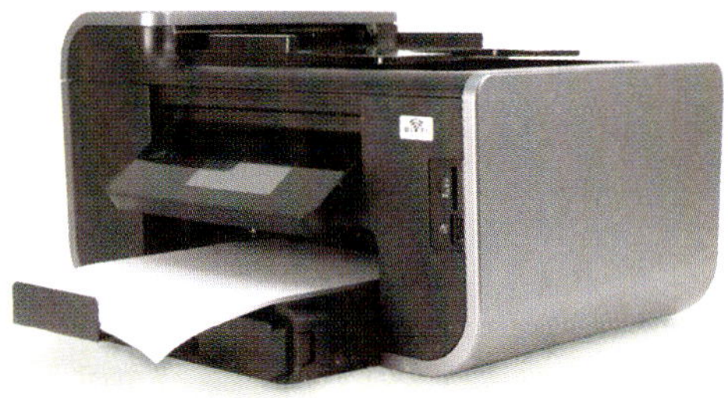

Step 1: Mean $= \dfrac{120 + 150 + 90 + 110 + 140 + 120 + 140 + 90}{8} = \dfrac{960}{8} = 120$

Step 2: Use a dot plot to organize the data. Replace each dot with its distance from the mean.

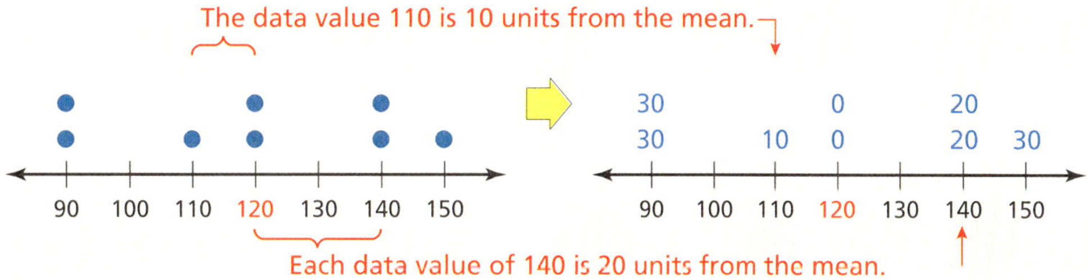

Step 3: The sum of the distances is $30 + 30 + 10 + 0 + 0 + 20 + 20 + 30 = 140$.

Step 4: The mean absolute deviation is $\dfrac{140}{8} = 17.50$.

∴ The data values differ from the mean by an average of \$17.50.

Exercises

Find and interpret the mean absolute deviation of the data. Round your answer to the nearest tenth if necessary.

11.

Shoe Sizes			
6	8.5	6	9
10	7	8	9.5

12.

Prices of Monitors (dollars)				
130	150	190	100	175
120	165	140	180	190

9 Chapter Test

Check It Out
Test Practice
BigIdeasMath.com

Display the data in a dot plot. Identify any clusters, peaks, or gaps in the data.

1.

Time (minutes)			
33	40	32	40
39	38	40	39
38	39	39	33

2.

Temperature (°F)				
81	81	80	82	81
83	76	83	76	80
75	83	82	82	81

Find the mean, median, and mode(s) of the data.

3. 2, 7, 7, 12, 4

4. 4, 5, 7, 5, 9, 9, 10, 6

Find the mean, median, and mode(s) of the data. Choose the measure that best represents the data. Explain your reasoning.

5. 5, 6, 4, 24, 18

6. 46, 27, 94, 56, 53, 65, 43

Find the range of the data.

7. 24, 56, 9, 83, 77, 14

8. 43, 12, 55, 91, 25, 86, 84, 23, 1

Find the median, first quartile, third quartile, and interquartile range of the data.

9. 32, 58, 19, 36, 44, 57, 11, 26, 74

10. 36, 24, 49, 32, 37, 28, 38, 40, 39

Find and interpret the mean absolute deviation of the data. Round your answer to the nearest tenth if necessary.

11.

Distances Driven (miles)			
312	286	196	201
158	225	206	192

12.

Prices of Sunglasses (dollars)				
15	8	19	20	18
20	22	14	10	15

13. **HOTEL** The table shows the numbers of guests at a hotel on different days.

Numbers of Guests					
66	58	90	57	63	55
60	62	56	54	72	

a. Find the range and the interquartile range of the data.

b. Use the interquartile range to identify the outlier(s) in the data set. Find the range and the interquartile range of the data set without the outlier(s). Which measure did the outlier(s) affect more?

14. **JOBS** The data sets show the numbers of hours worked each week by two friends for several weeks.

Greg's hours: 9, 18, 12, 6, 9, 21, 3, 12
Tom's hours: 12, 18, 15, 16, 14, 12, 15, 18

Find the measures of center and the measures of variation for each data set. Compare the measures. What can you conclude?

Test Item References

Chapter Test Questions	Section to Review	Common Core State Standards
1, 2	9.1	6.SP.1, 6.SP.4
3, 4	9.2	6.SP.2, 6.SP.3, 6.SP.5a, 6.SP.5c
3–6	9.3	6.SP.2, 6.SP.3, 6.SP.5c
7–10, 13	9.4	6.SP.2, 6.SP.3, 6.SP.5c
11, 12, 14	9.5	6.SP.2, 6.SP.3, 6.SP.5c

Test-Taking Strategies

Remind students to quickly look over the entire test before they start so that they can budget their time. Students should jot down the definitions of mean, median, mode, range, interquartile range, and mean absolute deviation on the back of the test before they start. By doing this, they will not become confused when they are under pressure. Students need to use the **Stop** and **Think** strategy before answering questions.

Common Errors

- **Exercises 1 and 2** Students may miss gaps in the data because they include only data values along the number line for their dot plots. Remind students to include all the integers from the least data value to the greatest data value along the number line.
- **Exercises 3–6** Students may try to identify the median without ordering the data first. Remind them that it is essential to order the data first and then find the median. This makes finding the mode easier as well.
- **Exercises 7 and 8** Students may write the range as an interval such as 5–22 rather than as the distance between the greatest and least values, 17. Emphasize that the range is a value, not an interval.
- **Exercises 9 and 10** Students may include the median in the two halves of data when finding the quartiles. Remind them never to include the median in the upper or lower half of the data.
- **Exercises 11 and 12** Students may find the mean absolute deviation using the median instead of the mean. Remind students that the MAD measures the variation from the mean.

Reteaching and Enrichment Strategies

If students need help. . .	If students got it. . .
Resources by Chapter • Practice A and Practice B • Puzzle Time Record and Practice Journal Practice Differentiating the Lesson Lesson Tutorials *BigIdeasMath.com* Skills Review Handbook	Resources by Chapter • Enrichment and Extension • Technology Connection Game Closet at *BigIdeasMath.com* Start Standards Assessment

Answers

1–3. See Additional Answers.

4. mean: 6.875; median: 6.5; modes: 5 and 9; range: 6

5. mean: 11.4; median: 6; mode: none; *Sample answer:* The median is best because there is no mode, the mean is greater than most of the data, and the median is close to 3 of 5 data values.

6. mean: about 54.86; median: 53; mode: none; The mean or median is best because there is no mode and the mean and median are close to each other.

7. 74 **8.** 90

9. median = 36; $Q_1 = 22.5$; $Q_3 = 57.5$; IQR = 35

10. median = 37; $Q_1 = 30$; $Q_3 = 39.5$; IQR = 9.5

11. about 39.3; The distances driven differ from the mean distance by an average of about 39.3 miles.

12. 3.7; The prices of sunglasses differ from the mean price by an average of $3.70.

13. **a.** range = 36 guests; IQR = 10 guests

b. The outlier is 90 guests; range = 18; IQR = 7; range

14. See Additional Answers.

Technology for the Teacher

Online Assessment
Assessment Book
ExamView® Assessment Suite

Test-Taking Strategies
Available at *BigIdeasMath.com*

After Answering Easy Questions, Relax
Answer Easy Questions First
Estimate the Answer
Read All Choices before Answering
Read Question before Answering
Solve Directly or Eliminate Choices
Solve Problem before Looking at Choices
Use Intelligent Guessing
Work Backwards

About this Strategy
When taking a multiple choice test, be sure to read each question carefully and thoroughly. Sometimes you do not know the answer. So . . . guess intelligently! Look at the choices and choose the ones that are possible answers.

Answers
1. A
2. I
3. 24
4. A

Technology for the Teacher
Common Core State Standards Support Performance Tasks
Online Assessment
Assessment Book
ExamView® Assessment Suite

Item Analysis

1. **A.** Correct answer
   B. The student finds the quotient of the whole numbers and the quotient of the fractions and combines these to get the answer.
   C. The student finds the product of the whole numbers and the product of the fractions and combines these to get the answer.
   D. The student rewrites the quotient as $\frac{76}{9} \div \frac{14}{3}$ but then changes this to a product without finding the reciprocal of the divisor.

2. **F.** The student does not carry any 1s.
   G. The student forgets to carry a 1.
   H. The student forgets to carry a 1.
   I. Correct answer

3. **Gridded Response:** Correct answer: 24
   Common Error: The student may forget that there can be a mode and writes 23.

4. **A.** Correct answer
   B. The student does not list the data in order and finds the median using 12 and 3, getting 7.5, which is less than the mean of 8.
   C. The student does not correctly determine that the mode is 10 and instead incorrectly determines a value that is less than the other measures.
   D. The student does not list the data in order and finds the range using 7 and 6, getting 1, which is less than the mean of 8.

9 Standards Assessment

1. What is the value of the expression below? *(6.NS.1)*

$$8\frac{4}{9} \div 4\frac{2}{3}$$

A. $1\frac{17}{21}$

B. $2\frac{2}{3}$

C. $32\frac{8}{27}$

D. $39\frac{11}{27}$

2. What is the value of the expression below? *(6.NS.3)*

$$4.18 + 6.225 + 5.7$$

F. 15.005

G. 15.105

H. 16.005

I. 16.105

3. One number is missing from the data set in the box below.

18, 24, 22, 30, 26, ____, 25

The median of the data set is 24. What is the greatest possible value of the missing number? *(6.SP.5c)*

4. The number of hours that each of 6 students spent reading last week is shown in the bar graph below.

For the data in the bar graph, which measure is the *least*? *(6.SP.5c)*

A. mean

B. median

C. mode

D. range

5. You go to a beach and collect buckets of shells. Of the many shells you have collected, you notice the following.

- 9% of the seashells are auger shells.
- $\frac{1}{8}$ of the seashells are coquina shells.
- 11% of the seashells are rough scallop shells.
- 0.1 of the seashells are fighting conch shells.

Which list correctly shows the types of shells in order from least to greatest? *(6.NS.6c)*

F. auger, coquina, rough scallop, fighting conch

G. fighting conch, coquina, auger, rough scallop

H. fighting conch, auger, rough scallop, coquina

I. auger, fighting conch, rough scallop, coquina

6. What is the mean absolute deviation of the data shown in the line plot, rounded to the nearest tenth? *(6.SP.5c)*

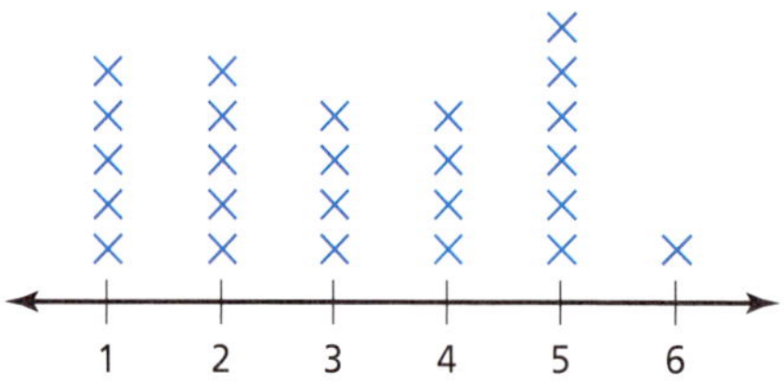

A. 1.4

B. 3

C. 3.2

D. 5

7. A family wants to buy tickets to a theme park. There are separate ticket prices for adults and children.

Which expression represents the total cost, in dollars, for a adult tickets and c child tickets? *(6.EE.2a)*

F. $600(a + c)$

G. $50(a \times c)$

H. $30a + 20c$

I. $30a \times 20c$

Item Analysis (continued)

5. **F.** The student lists the types of seashells in the order given.
   G. The student uses only the numbers given and ignores the symbols, ordering the numbers 1, 8, 9, and 11.
   H. The student thinks that 0.1 is less than 9% and reverses the order of the fighting conch and auger.
   I. Correct answer

6. **A.** Correct answer
   B. The student confuses the median for the mean absolute deviation.
   C. The student confuses the mean for the mean absolute deviation.
   D. The student confuses the mode for the mean absolute deviation.

7. **F.** The student multiplies the product of the prices by the sum of the variables.
   G. The student multiplies the sum of the prices by the product of the variables.
   H. Correct answer
   I. The student finds the product instead of the sum.

Answers

5. I
6. A
7. H

Answers

8. 390

9. C

10. G

11. *Sample answer:* 2, 3, 9, 10,11; numbers sum to 35.

12. B

Item Analysis (continued)

8. Gridded Response: Correct answer: 330

Common Error: The student does not add a zero as a place holder, rewriting the quotient as $528 \div 16$ instead of $5280 \div 16$, getting an answer of 33.

9. A. The student uses 0.6 instead of 6 for a and 0.14 instead of 14 for b and finds $0.8 \times 0.6 + 0.02 \times 0.14$.

B. The student uses 6 for the hundredths digit in $0.8a$ and 14 for the thousandths and ten-thousandths digits in $0.02b$ and finds $0.86 + 0.0214$.

C. Correct answer

D. The student finds the sum of the coefficients and the sum of the values of the variables and then finds the product of these two sums.

10. F. The student does not recognize that the Distributive Property is used to write the expression in the second line.

G. Correct answer

H. The student does not recognize that the Identity Property of Multiplication is used to find the final answer.

I. The student does not recognize that the Commutative Property of Multiplication is used in the first line.

11. 2 points The student demonstrates a thorough understanding of analyzing sets of data with respect to the mean and median. The student correctly determines a data set that has a mean of 7 and a median of 9. One possible answer is the data set of 2, 3, 9, 10, and 11. The student clearly explains the steps used, and the method used is unambiguous. The student correctly shows that the data set has a mean of 7 and a median of 9.

1 point The student demonstrates a partial understanding of analyzing sets of data with respect to the mean and median. The student shows some knowledge of how to determine the mean and/or median but does not successfully answer the question.

0 points The student demonstrates insufficient understanding of analyzing sets of data with respect to the mean and median. The student makes many errors and shows little or no ability to work with the mean or the median.

12. A. The student divides 400 by 25.

B. Correct answer

C. The student expresses 25% as 2.5.

D. The student expresses 25% as 25.

8. What is the value of the expression below? *(6.NS.3)*

$$52.8 \div 0.16$$

9. What is the value of the expression below when $a = 6$ and $b = 14$? *(6.EE.2c)*

$$0.8a + 0.02b$$

A. 0.4828

B. 0.8814

C. 5.08

D. 16.4

10. Which property was *not* used in the box below to simplify the expression? *(6.EE.3)*

$$\begin{aligned} 0.3 \times 53 + 53 \times 0.7 &= 53 \times 0.3 + 53 \times 0.7 \\ &= 53 \times (0.3 + 0.7) \\ &= 53 \times 1 \\ &= 53 \end{aligned}$$

F. Distributive Property

G. Associative Property of Addition

H. Identity Property of Multiplication

I. Commutative Property of Multiplication

11. Determine a data set of 5 numbers that has the following measures:

Think Solve Explain

- a mean of 7 and
- a median of 9.

Explain how you determined your data set. Then demonstrate that the mean of your data set is 7 and the median is 9. *(6.SP.5c)*

12. What is the value of the expression below? *(6.RP.3c)*

25% of 400

A. 16

B. 100

C. 1000

D. 10,000

10 Data Displays

"I took a survey of pet owners on how many times per day you should treat your dog to a biscuit."

"What do you think?"

"I've completed a circle graph analyzing what you do each day."

Common Core Progression

4th Grade

- Make a line plot with data in fractions of a unit.
- Use line plots to solve problems involving addition and subtraction of fractions.
- Identify line-symmetric figures and draw lines of symmetry.

5th Grade

- Use line plots to solve problems involving operations on fractions.
- Graph ordered pairs in the first quadrant of the coordinate plane.

6th Grade

- Display data on a number line, including line (dot) plots, stem-and-leaf plots, histograms, and box-and-whisker plots.
- Recognize that a measure of center for a numerical data set summarizes all of its values with a single number, and a measure of variation describes how its values vary with a single number.
- Understand that data used to answer statistical questions has a distribution that can be described by center, spread, and shape.
- Choose appropriate measures of center and variation based on shape.

Pacing Guide for Chapter 10

Chapter Opener	1 Day
Section 1 Activity Lesson	 1 Day 1 Day
Section 2 Activity Lesson	 1 Day 1 Day
Study Help / Quiz	1 Day
Section 3 Activity Lesson Extension	 1 Day 1 Day 1 Day
Section 4 Activity Lesson	 1 Day 1 Day
Chapter Review/ Chapter Tests	2 Days
Total Chapter 10	13 Days
Year-to-Date	154 Days

Chapter Summary

Section	Common Core State Standard	
10.1	Applying	6.SP.4
10.2	Learning	6.SP.2, 6.SP.4
10.3	Learning	6.SP.2, 6.SP.4, 6.SP.5d
10.4	Learning	6.SP.2 ★, 6.SP.4 ★, 6.SP.5c ★

★ Teaching is complete. Standard can be assessed.

Technology for the Teacher

BigIdeasMath.com
Chapter at a Glance
Complete Materials List
Parent Letters: English and Spanish

Common Core State Standards

3.MD.3 . . . Solve one- and two-step "how many more" and "how many less" problems using information presented in scaled bar graphs.

6.RP.3c Find a percent of a quantity as a rate per 100 . . ..

Additional Topics for Review

- Multiplying Fractions and Whole Numbers
- Line (Dot) Plots
- Ordering Whole Numbers and Integers
- Outliers
- Measures of Center
- Measures of Variation
- Quartiles
- Interquartile Range

Try It Yourself

1. orange
2. 11 students
3. 21 students
4. 25 students
5. 8 students
6. 7 students

Record and Practice Journal Fair Game Review

1. Ranch
2. 8 students
3. 18 students
4. 24 students
5. 9 students
6. 12 students
7. 6 students

Math Background Notes

Vocabulary Review

- Bar Graph
- Circle Graph
- Percent

Analyzing Bar Graphs

- Students should know how to read and draw bar graphs. Even though this is a skill from a previous grade, students may still need practice to achieve mastery.
- **Teaching Tip:** This is an excellent opportunity to review the parts of a bar graph with students. Sample questions could include: What is the title of the graph? What are the categories? What data are being compared? What scale is being used to measure the axes?

Finding Percents

- Students should be able to read and obtain information from a circle graph. Students should also know how to find the percent of a number. They learned this skill in Chapter 5.
- Take time to review the information that is given in a circle graph:
  - The circle represents all of the data.
  - The percents total 100%.
  - Each section represents part of the data.
- If necessary, review the information in Section 5.5 on how to write a percent as a fraction and how to find the percent of a number.
- **Teaching Tip:** When going over the solution of Example 2, in the simplification before the final answer, avoid using terms such as "cancel out." The factors *divide* out. Instilling good habits, such as using correct terminology, will help students later in their mathematical careers. For instance, in algebra, when students are faced with simplifying rational expressions, they are more likely to get correct answers when they look for common factors to divide out instead of common terms to "cancel" out.

Reteaching and Enrichment Strategies

If students need help. . .	If students got it. . .
Record and Practice Journal • Fair Game Review Skills Review Handbook Lesson Tutorials	Game Closet at *BigIdeasMath.com* Start the next section

What You Learned Before

"Okay, I have the box. But I need your help to complete my box-and-whisker plot."

Analyzing Bar Graphs (3.MD.3)

Example 1 **The bar graph shows the favorite colors of the students in a class. How many students said their favorite color is blue?**

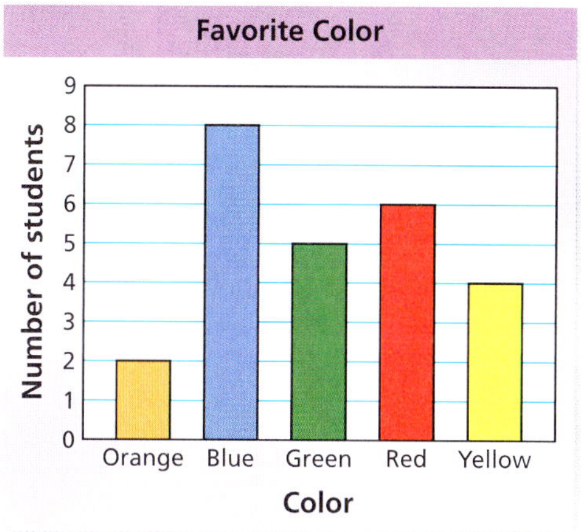

The height of the bar labeled "Blue" is 8.

∴ So, 8 students said their favorite color is blue.

Try It Yourself

1. What color was chosen the least?
2. How many students said green or red is their favorite color?
3. How many students did *not* choose yellow as their favorite color?
4. How many students are in the class?

Finding Percents (6.RP.3c)

Example 2 **The circle graph shows the favorite fruits of the students in a class. There are 20 students in the class. How many students said their favorite fruit is an orange?**

Find 25% of 20.

$$25\% \text{ of } 20 = \frac{1}{4} \cdot 20 = \frac{1 \cdot \overset{5}{\cancel{20}}}{\underset{1}{\cancel{4}}} = 5$$

∴ So, 5 students said their favorite fruit is an orange.

Try It Yourself

5. How many students said their favorite fruit is an apple?
6. How many students said their favorite fruit is a banana?

10.1 Stem-and-Leaf Plots

Essential Question How can you use place values to represent data graphically?

1 ACTIVITY: Making a Data Display

Work with a partner. The list below gives the ages of these women when they became first ladies of the United States.

THE WHITE HOUSE

WASHINGTON

Frances Cleveland – 21	Mamie Eisenhower – 56
Caroline Harrison – 56	Jacqueline Kennedy – 31
Ida McKinley – 49	Claudia Johnson – 50
Edith Roosevelt – 40	Patricia Nixon – 56
Helen Taft – 48	Elizabeth Ford – 56
Ellen Wilson – 52	Rosalynn Carter – 49
Florence Harding – 60	Nancy Reagan – 59
Grace Coolidge – 44	Barbara Bush – 63
Lou Hoover – 54	Hillary Clinton – 45
Eleanor Roosevelt – 48	Laura Bush – 54
Elizabeth Truman – 60	Michelle Obama – 45

a. The incomplete data display shows the ages of the first ladies in the left column of the list above.

What do the numbers to the left of the line represent? What do the numbers to the right of the line represent?

Ages of First Ladies

2	1
3	
4	0 4 8 8 9
5	2 4 6
6	0 0

b. This data display is called a *stem-and-leaf plot.* What numbers do you think represent the *stems? leaves?* Explain your reasoning.

c. Complete the stem-and-leaf plot using the remaining ages in the right column. Order the numbers to the right of the line in numerical order.

d. **REASONING** Write a question about the ages of first ladies that would be easier to answer using a stem-and-leaf plot than a dot plot.

COMMON CORE

Data Displays

In this lesson, you will

- make and interpret stem-and-leaf plots.

Applying Standard 6.SP.4

Laurie's Notes

Introduction

Standards for Mathematical Practice

- **MP4 Model with Mathematics:** Students will construct stem-and-leaf plots to represent and describe features of a data set, including measures of center and measures of variation.

Motivate

- Make a stem-and-leaf plot using sticky notes. As students enter, hand each one a sticky note. On the board draw a vertical line and use stems of 0, 1, 2, and 3. Ask students to write down the day of the month they were born, making their numbers large enough and dark enough to be seen from a distance.
- Ask those students born before the 10th of the month to come forward and put their sticky notes in a row adjacent to the 0 stem.
- Repeat for each of the stems; 10th–19th, 20th–29th, and 30th–31st.
- ? "What observations can you make about the data?" most likely there are fewer in the 30s; would expect randomness and balanced distribution over other 3 stems, but it might not be!
- Do not sort—this will be done at the end of the class. Explain that today they will explore a new type of plot similar to what they have made with their sticky notes.

Activity Notes

Activity 1

- The stem-and-leaf plot is fairly simple for students to understand. The motivator should give them a good understanding of how to construct one.
- **FYI:** The ages given are how old these women were when they *first* became first ladies, which was not necessarily on the day of inauguration. For instance, Frances and Grover Cleveland were married on June 2, 1886, during his first term. Also, Woodrow Wilson's second wife, Edith, is not listed. His first wife, Ellen, died while he was in office.
- Students find this data set interesting. Most students know that you must be at least 35 years old to run for president, so they realize that some first ladies were quite young. As an extension to this activity, in Activity 2 students will make a back-to-back stem-and-leaf plot with presidents' and first ladies' ages. The distributions are different!
- ? What do the stems represent?" the tens digit of the first ladies' ages
- Note that the leaves are in numerical order. Students may not notice this.
- When students have finished, display a completed plot. Have a few students share their answers to part (d).
- **Big Idea:** A stem-and-leaf plot is very similar to a dot plot, but the stem-and-leaf plot gives additional information. The stem-and-leaf plot uses the ones digit of the data values instead of dots, so you can see the distribution within each group of ten.

Common Core State Standards

6.SP.4 Display numerical data in plots on a number line, including dot plots, histograms, and box plots.

Previous Learning

Students should know how to use line plots (or dot plots) to display and analyze data.

Technology for the Teacher

Dynamic Classroom

Lesson Plans
Complete Materials List

10.1 Record and Practice Journal

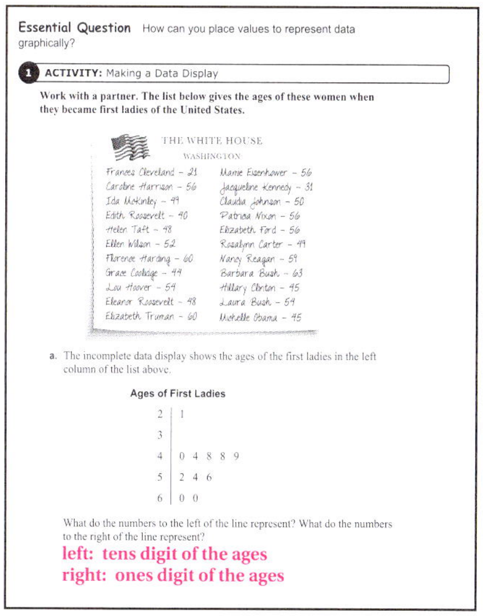

Essential Question How can you place values to represent data graphically?

1 ACTIVITY: Making a Data Display

Work with a partner. The list below gives the ages of these women when they became first ladies of the United States.

THE WHITE HOUSE
WASHINGTON

Frances Cleveland – 21	Mamie Eisenhower – 56
Caroline Harrison – 56	Jacqueline Kennedy – 31
Ida McKinley – 49	Claudia Johnson – 50
Edith Roosevelt – 40	Patricia Nixon – 56
Helen Taft – 48	Elizabeth Ford – 56
Ellen Wilson – 52	Rosalynn Carter – 49
Florence Harding – 60	Nancy Reagan – 59
Grace Coolidge – 44	Barbara Bush – 63
Lou Hoover – 54	Hillary Clinton – 45
Eleanor Roosevelt – 48	Laura Bush – 54
Elizabeth Truman – 60	Michelle Obama – 45

a. The incomplete data display shows the ages of the first ladies in the left column of the list above.

Ages of First Ladies

Stem	Leaf
2	1
3	
4	0 4 8 8 9
5	2 4 6
6	0 0

What do the numbers to the left of the line represent? What do the numbers to the right of the line represent?

left: tens digit of the ages
right: ones digit of the ages

Differentiated Instruction

Kinesthetic

To make the display in Activity 2, copy the data onto individual pieces of paper. Then sort the data into piles based on the stem value and whether the number is a first lady's age or a president's age. For each stem value, cut one of the pieces to separate the stem and leaf. Tape the stems to the board in increasing order from top to bottom. Cut the remaining leaves with the same stem and tape them to the board in ascending order. Place the leaves for the first ladies' ages to the right of the stem and order them from left to right. Place the leaves for the presidents' ages to the left of the stem and order them from right to left.

10.1 Record and Practice Journal

b. This data display is called a *stem-and-leaf plot*. What numbers do you think represent the *stems*? *leaves*? Explain your reasoning.

stems: numbers on left
leaves: numbers on right

c. Complete the stem-and-leaf plot on the previous page using the remaining ages in the right column. Order the numbers to the right of the line in numerical order. **See Additional Answers.**

d. Write a question about the ages of first ladies that would be easier to answer using a stem-and-leaf plot than a dot plot.

Sample answer: **In which age range were there more first ladies,40s or 50s?**

2 ACTIVITY: Making a Back-to-Back Stem-and-Leaf Plot

Work with a partner. The table below shows the ages of presidents of the United States from 1885 to 2009 on their first inauguration day.

Ages of Presidents										
47	55	54	42	51	56	55	51	54	51	60
62	43	55	56	61	52	69	64	46	54	47

a. On your stem-and-leaf plot from Activity 1(c), draw a vertical line to the left of the display. Represent the ages of the presidents by including numbers to the left of the line. **See Additional Answers.**

b. Find the median ages of both the first ladies and presidents of the United States.

first ladies: 51
presidents: 54

c. Compare the distribution of each data set.

Sample answer: **Distributions are similar, with most ages in the 40s, 50s, and 60s, but presidents appear to be older than first ladies, in general.**

3 ACTIVITY: Conducting an Experiment

Work with a partner. Use two number cubes to conduct the following experiment.

- Toss the cubes and find the product of the resulting numbers.
- Repeat this process 30 times. Record your results.

Toss	1	2	3	4	5	6	7	8	9	10
Product										
Toss	11	12	13	14	15	16	17	18	19	20
Product										
Toss	21	22	23	24	25	26	27	28	29	30
Product										

a. Use a stem-and-leaf plot to organize your results.

Check students' work.

b. Describe the distribution of the data.

Distributions should have more values as the product decreases.

What Is Your Answer?

4. IN YOUR OWN WORDS How can you use place values to represent data graphically?

Sample answer: **Choose digit(s) according to place values and display the data in a stem-and-leaf plot.**

5. How can you display data in a stem-and-leaf plot whose values range from 82 through 129?

Use 8, 9, 10, 11, and 12 as the stems.

Laurie's Notes

Activity 2

- Have students read through the directions to make sure they understand how to make the back-to-back stem-and-leaf plot.
- In part (a), have students decide how to list the leaves (ascending or descending order).
- In addition to finding the median, you could also ask students to find the mean to help compare the data.
- **MP3 Construct Viable Arguments and Critique the Reasoning of Others:** Ask students to compare the two distributions. Listen for evidence that supports their description.

Activity 3

- The first two activities are time consuming, so there may not be time to complete Activity 3. You could assign Activity 3 as homework.

? "What is the range of this data set?" $36 - 1 = 35$

What Is Your Answer?

- Have students work in pairs.

Closure

- Return to the sticky note plot made when motivating today's lesson. Ask a volunteer to sort the data. Now ask students to describe the plot, along with finding the median, mode, and range.

2 ACTIVITY: Making a Back-to-Back Stem-and-Leaf Plot

Work with a partner. The table below shows the ages of presidents of the United States from 1885 to 2009 on their first inauguration day.

Ages of Presidents										
47	55	54	42	51	56	55	51	54	51	60
62	43	55	56	61	52	69	64	46	54	47

a. On your stem-and-leaf plot from Activity 1(c), draw a vertical line to the left of the display. Represent the ages of the presidents by including numbers to the left of the line.

b. Find the median ages of both the first ladies and the presidents of the United States.

c. Compare the distribution of each data set.

3 ACTIVITY: Conducting an Experiment

Math Practice 4

Interpret Results

How can you use the stem-and-leaf plot to interpret your results? Explain.

Work with a partner. Use two number cubes to conduct the following experiment.

- Toss the cubes and find the product of the resulting numbers.
- Repeat this process 30 times. Record your results.

a. Use a stem-and-leaf plot to organize your results.

b. Describe the distribution of the data.

What Is Your Answer?

4. **IN YOUR OWN WORDS** How can you use place values to represent data graphically?

5. How can you display data in a stem-and-leaf plot whose values range from 82 through 129?

Practice Use what you learned about stem-and-leaf plots to complete Exercises 4 and 5 on page 438.

10.1 Lesson

Check It Out
Lesson Tutorials
BigIdeasMath.com

Key Vocabulary

stem-and-leaf plot, *p. 436*
stem, *p. 436*
leaf, *p. 436*

Key Idea

Stem-and-Leaf Plots

A **stem-and-leaf plot** uses the digits of data values to organize a data set. Each data value is broken into a **stem** (digit or digits on the left) and a **leaf** (digit or digits on the right).

A stem-and-leaf plot shows how data are distributed.

Stem	Leaf
2	0 0 1 2 5 7
3	1 4 8
4	2
5	8 9

Key: 2 | 0 = 20

The *key* explains what the stems and leaves represent.

EXAMPLE 1 Making a Stem-and-Leaf Plot

	A	B
1	DATE	MINUTES
2	JULY 9	55
3	JULY 9	3
4	JULY 9	6
5	JULY 10	14
6	JULY 10	18
7	JULY 10	5
8	JULY 10	23
9	JULY 11	30
10	JULY 11	23
11	JULY 11	10
12	JULY 11	2
13	JULY 11	36

Make a stem-and-leaf plot of the length of the 12 cell phone calls.

Step 1: Order the data.

2, 3, 5, 6, 10, 14, 18, 23, 23, 30, 36, 55

Step 2: Choose the stems and the leaves. Because the data values range from 2 to 55, use the *tens* digits for the stems and the *ones* digits for the leaves. Be sure to include the key.

Step 3: Write the stems to the *left* of the vertical line.

Step 4: Write the leaves for each stem to the *right* of the vertical line.

Cell Phone Call Lengths

Stem	Leaf
0	2 3 5 6
1	0 4 8
2	3 3
3	0 6
4	
5	5

Key: 1 | 4 = 14 minutes

Order the stems vertically. The stem for data values less than 10 is 0.

Include stems without leaves.

Write the leaves horizontally.

On Your Own

Now You're Ready
Exercises 4–9

1. Make a stem-and-leaf plot of the hair lengths.

Hair Length (centimeters)									
5	1	20	12	27	2	30	5	7	38
40	47	1	2	1	32	4	44	33	23

Multi-Language Glossary at BigIdeasMath.com

Laurie's Notes

Introduction

Connect

- **Yesterday:** Students explored the features of a stem-and-leaf plot. (MP3, MP4)
- **Today:** Students will construct and interpret stem-and-leaf plots.

Motivate

- Share a little trivia about cell phones, the context for Example 1.
  - 1973—Dr. Martin Cooper is considered the inventor of the first portable handset. Dr. Cooper, former general manager for the systems division at Motorola, was the first person to make a call on a cell phone.
  - 1977—Cell phones go public. The first trials of cell phone testing began in the city of Chicago with 2000 customers. Japan began testing cellular phone service in 1979.

Lesson Notes

Key Idea

- A common question that students ask is "what if the data are not two-digit numbers?"
  - If there are 3-digit numbers within a small range (i.e., 431–476), the stem can be two digits (43, 44, 45, 46, 47).
  - If there are decimals (i.e., 3.4), the stem can be the whole number portion and the leaf is the decimal portion.

Example 1

- By sorting the data first, the leaves are arranged in order when written to the right of the stem.
- **MP6 Attend to Precision:** Discuss with students the need to have a key that describes how to read the data in the plot.
- **Common Error:** When a data value repeats, remind students that the leaf must be listed again. The number of leaves must equal the number of data in the set.
- ? "Describe the stem-and-leaf plot. What does the plot tell you about the data?" Data are skewed towards the lower data. It tapers at the upper end. There is a gap from 36 to 55.
- **Big Idea**: Because the stem-and-leaf plot shows how data are distributed, a stem must be included in the plot even if there are no data values in that interval. In this example the stem of "4" is still included to show the gap in the data.

On Your Own

- **Think-Pair-Share:** Students should read the question independently and then work in pairs to answer the question. When they have answered the question, the pair should compare their answers with another group and discuss any discrepancies.
- ? "Describe what the plot tells you about the data." There are some smaller data values and larger data values with few in the middle.
- ? "Explain why this might be the case." perhaps girls versus boys

Goal Today's lesson is constructing and interpreting **stem-and-leaf plots**.

Technology for the Teacher

Dynamic Classroom

Lesson Tutorials
Lesson Plans
Answer Presentation Tool

Extra Example 1

Make a stem-and-leaf plot of the length of the eleven fish, in inches.
7, 12, 20, 14, 20, 25, 8, 18, 16, 20, 14

Fish Length

Stem	Leaf
0	7 8
1	2 4 4 6 8
2	0 0 0 5

Key: 1 | 8 = 18 inches

On Your Own

1. **Hair Length**

Stem	Leaf
0	1 1 1 2 2 4 5 5 7
1	2
2	0 3 7
3	0 2 3 8
4	0 4 7

Key: 1 | 2 = 12 cm

Extra Example 2

Use the stem-and-leaf plot of student test scores in Example 2.

a. How many students scored more than 85 points? 8

b. How many students scored at most 75 points? 3

c. What is the median test score? 84

On Your Own

2. a. 9 students

   b. 4 students

Extra Example 3

Use the stem-and-leaf plot in Example 2. Which statement is *not* true? B

A. The range of test scores is 34.

B. The mean test score is 85.

C. The test scores that occur most often are 81, 84, and 88.

D. Most of the test scores are over 80.

On Your Own

3. yes; From the stem-and-leaf plot, there are plants that are 22 inches, 25 inches, and 36 inches tall.

English Language Learners

Pair Activity

Form pairs of English learners and English speakers. Have them consider different types of data such as golf scores, the number of text messages each person in class sends in a week, the number of sixth, seventh, and eighth graders that ride the school bus, the numbers rolled on a number cube, and the number of hours that students in a class slept last night. Instruct students to discuss whether a stem-and-leaf plot would be an appropriate way to display the data.

Laurie's Notes

Example 2

- **Big Idea**: The leaves in a stem-and-leaf are sorted from least to greatest. The range can be found quickly, along with the median. To find the median, alternate counting in from each end of the data (the least and the greatest).

? "How many students are represented in the plot?" 18

? "What is the median? Explain how you decided." 84; There are 18 numbers so the middle will be the average of the 9th and 10th numbers which are both 84.

On Your Own

- **Extension:** Ask students to describe the plot.

Example 3

- This example reviews data analysis concepts from Chapter 9. You may need to remind students how to find the mean, median, mode, and range.

On Your Own

- **Neighbor Check:** Have students work independently and then have their neighbors check their work. Have students discuss any discrepancies.

Closure

- Explain how a stem-and-leaf plot is similar to a bar graph and how it differs from a bar graph. *Sample answer:* A stem-and-leaf plot shows the frequency of different intervals for the data similar to a bar graph as well as displaying the distribution of the data. However, all of the data are displayed in a stem-and-leaf plot, but not in a bar graph.

EXAMPLE 2 Interpreting a Stem-and-Leaf Plot

Test Scores

Stem	Leaf
6	6
7	0 5 7 8
8	1 1 3 4 4 6 8 8 9
9	0 2 9
10	0

Key: 9 | 2 = 92 points

The stem-and-leaf plot shows student test scores. (a) How many students scored less than 80 points? (b) How many students scored at least 90 points? (c) How are the data distributed?

a. There are five scores less than 80 points: 66, 70, 75, 77, and 78.

∴ Five students scored less than 80 points.

b. There are four scores of at least 90 points: 90, 92, 99, and 100.

∴ Four students scored at least 90 points.

c. There are few low test scores and few high test scores. So, most of the scores are in the middle.

On Your Own

Now You're Ready
Exercises 12–15

2. Use the grading scale at the right.

a. How many students received a B on the test?

b. How many students received a C on the test?

A: 90–100
B: 80–89
C: 70–79
D: 60–69
F: 59 and below

EXAMPLE 3 Making Conclusions from a Stem-and-Leaf Plot

Which statement is *not* true?

Ⓐ Most of the plants are less than 20 inches tall.

Ⓑ The median plant height is 11 inches.

Ⓒ The range of the plant heights is 35 inches.

Ⓓ The plant height that occurs most often is 11 inches.

Plant Heights

Stem	Leaf
0	1 2 4 5 6 8 9
1	0 1 1 5 7
2	2 5
3	6

Key: 1 | 5 = 15 inches

There are 15 plant heights. So, the median is the eighth data value, 10 inches.

∴ The correct answer is Ⓑ.

On Your Own

3. You are told that three plants are taller than 20 inches. Is the statement true? Explain.

10.1 Exercises

Check It Out
Help with Homework
BigIdeasMath.com

Vocabulary and Concept Check

1. **VOCABULARY** The key for a stem-and-leaf plot is $3|4 = 34$. Which number is the stem? Which number is the leaf?
2. **WRITING** Describe how to make a stem-and-leaf plot of the data values 14, 22, 9, 13, 30, 8, 25, and 29.
3. **WRITING** How does a stem-and-leaf plot show the distribution of data?

Practice and Problem Solving

Make a stem-and-leaf plot of the data.

1 4.

Books Read			
26	15	20	9
31	25	29	32
17	26	19	40

5.

Hours Online			
8	12	21	14
18	6	15	24
12	17	2	0

6.

Test Scores (%)				
87	82	95	91	69
88	68	87	65	81
97	85	80	90	62

7.

Points Scored				
58	50	42	71	75
45	51	43	38	71
42	70	56	58	43

8.

Bikes Sold			
78	112	105	99
86	96	115	100
79	81	99	108

9.

Minutes in Line			
4.0	2.6	1.9	3.1
3.6	2.2	2.7	3.8
1.6	2.0	3.1	2.9

10. **ERROR ANALYSIS** Describe and correct the error in making a stem-and-leaf plot of the data.

    51, 25, 47, 42, 55, 26, 50, 44, 55

Stem	Leaf
2	5 6
4	2 4 7
5	0 1 5 5

Key: $4|2 = 42$

11. **PUPPIES** The weights (in pounds) of eight puppies at a pet store are 12, 24, 17, 8, 18, 31, 24, and 15. Make a stem-and-leaf plot of the data. Describe the distribution of the data.

Assignment Guide and Homework Check

Level	Day 1 Activity Assignment	Day 2 Lesson Assignment	Homework Check
Basic	4, 5, 19–21	1–3, 8–13, 15	8, 9, 10, 11, 13
Average	4, 5, 19–21	1–3, 9–12, 13, 15, 16–18	9, 12, 16, 17
Advanced	4, 5, 19–21	1–3, 8–18	9, 12, 14, 16, 17

For Your Information

- **Exercises 12–15** A dig is a forearm pass of a hard-driven ball, such as a spike.

Common Errors

- **Exercises 4–9** Students may forget to include the numbers that have zeros in the ones place in the leaf part of the plot. Remind them that they should be able to read the numbers in the data set by reading stem *and* leaf.
- **Exercises 4–9** Students may not include repeats of numbers. Remind them that the plot represents all of the data values, so they should be able to count the values in the leaf part and have all of the data accounted for.
- **Exercises 4–9** Students may forget to include stems that have no data values. Remind them of Example 1. It is necessary to include the stems with no data to help answer questions about the data set.
- **Exercises 12–15** Students may need to be reminded of the definitions of some of the terms so that they can answer the questions. Give an example of each term and how to find it using a stem-and-leaf plot.

10.1 Record and Practice Journal

Make a stem-and-leaf plot of the data.

1. Class Sizes

12	10	21	28
9	16	19	16
25	32	14	21

0	9
1	0 2 4 6 6 9
2	1 1 5 8
3	2

2. Minutes Spent on Homework

75	82	91	68
92	86	79	76
75	81	88	60

6	0 8
7	5 5 6 9
8	1 2 6 8
9	1 2

3. The number of text messages from eight phones are 8, 11, 14, 22, 5, 15, 7, and 20. Make a stem-and-leaf plot of the data. Describe the distribution of the data.

0	5 7 8
1	1 4 5
2	0 2

The data are evenly distributed.

4. The number of minutes seven members spent at band practice are 57, 49, 55, 62, 78, 72, and 75. Make a stem-and-leaf plot of the data. Describe the distribution of the data.

4	9
5	5 7
6	2
7	2 5 8

The data shows that few students do less than 50 minutes of practice.

5. The stem-and-leaf plot shows the numbers of miles students travel to get to school.

Stem	Leaf
0	5 7
1	2 4 8
2	0 1 5 7
3	3

Key: 1 | 4 = 14 miles

a. How many students travel more than 15 miles?

6 students

b. Find the mean, median, mode, range, and interquartile range of the data.

mean: 18.2 range: 28
median: 19 IQR: 13
mode: none

Vocabulary and Concept Check

1. 3 is the stem.; 4 is the leaf.
2. Write the data set in order from least to greatest.
8, 9, 13, 14, 22, 25, 29, 30
Use the tens digits for the stems and the ones digits for the leaves. Write the stems to the left of the vertical line. Then write the leaves to the right of the vertical line.

0	8 9
1	3 4
2	2 5 9
3	0

Key: 1 | 3 = 13

3. From the leaves, you can see where most of the data lies and whether there are many values that are low or high.

Practice and Problem Solving

4. **Books Read**

Stem	Leaf
0	9
1	5 7 9
2	0 5 6 6 9
3	1 2
4	0

Key: 1 | 5 = 15 books

5. **Hours Online**

Stem	Leaf
0	0 2 6 8
1	2 2 4 5 7 8
2	1 4

Key: 2 | 1 = 21 hours

6–11. See Additional Answers.

Practice and Problem Solving

12. 6 players

13. mean: 56.6; median: 53; modes: 41, 43, 63; range: 56; IQR = 20

14. Most of the data are in the 40s, 50s, and 60s.

15. 97; It increases the mean.

16–17. See Additional Answers.

18. See *Taking Math Deeper.*

Fair Game Review

19.

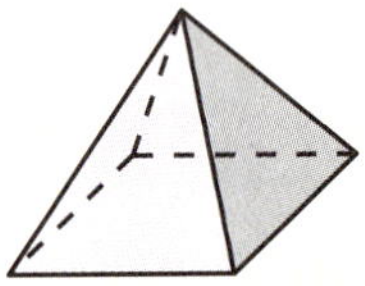

20.

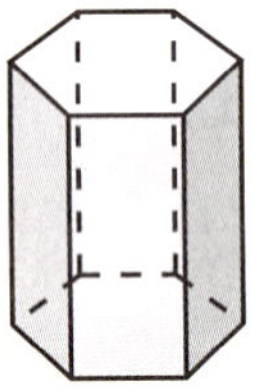

21. B

Mini-Assessment

The table shows the number of hours 15 students were online this week.

Hours Online				
21	14	8	13	17
18	9	12	7	21
15	12	21	15	7

1. Make a stem-and-leaf plot of the data.

Stem	Leaf
0	7 7 8 9
1	2 2 3 4 5 5 7 8
2	1 1 1

Key: 1 | 8 = 18 hours

2. Find the mean, median, mode, and range of the data. mean = 14, median = 14, mode = 21, range = 14

Taking Math Deeper

Exercise 18

With many problems involving data analysis, it helps to start by listing the data in order.

1. Order the scores for each golfer in the back-to-back stem-and-leaf plot.

   Rich: 35, 37, 41, 42, 43, 44, 45, 48, 50, 55
   Will: 42, 43, 44, 44, 46, 47, 47, 48, 49, 50

2. Use a table to organize the measures of center and the measures of variation.

	Mean	Median	Mode	Range	IQR	MAD
Rich	44	43.5	no mode	20	7	4.4
Will	46	46.5	44 and 47	8	4	2.2

3. You can cite reasons for choosing either golfer.

   You could choose Rich because he has the three lowest scores, a lower mean, and a lower median.

   You could choose Will because he is more consistent. He has a lower range, a lower interquartile range, and a lower mean absolute deviation.

Project

Research the meaning of a golfer's *handicap*. Which golfer would you choose if the tournament was a 4-day, 18-hole tournament based on handicap? Explain.

Reteaching and Enrichment Strategies

If students need help. . .	If students got it. . .
Resources by Chapter • Practice A and Practice B • Puzzle Time Record and Practice Journal Practice Differentiating the Lesson Lesson Tutorials Skills Review Handbook	Resources by Chapter • Enrichment and Extension • Technology Connection Start the next section

VOLLEYBALL The stem-and-leaf plot shows the number of *digs* for the top 15 players at a volleyball tournament.

Stem	Leaf
4	1 1 3 3 5
5	0 2 3 4
6	2 3 3 7
7	5
8	
9	7

Key: 5 | 0 = 50 digs

2 **12.** How many players had more than 60 digs?

13. Find the mean, median, mode, range, and interquartile range of the data.

14. Describe the distribution of the data.

15. Which data value is the outlier? Describe how the outlier affects the mean.

16. **REASONING** Each stem-and-leaf plot below has a mean of 39. Without calculating, determine which stem-and-leaf plot has the lesser mean absolute deviation. Explain your reasoning.

Stem	Leaf
2	3 7
3	0 2 6 9
4	1 2 5 8
5	1 4

Key: 4 | 1 = 41

Stem	Leaf
2	2 4 5 8 9
3	3 8
4	5
5	3 6 7 8

Key: 5 | 3 = 53

17. **TEMPERATURE** The stem-and-leaf plot shows the daily high temperatures (in degrees Fahrenheit) for the first 15 days of a month.

Stem	Leaf
6	7 8
7	0 0 3 4 6 8 9
8	2 3 6 7 8 9

Key: 6 | 7 = 67°F

a. Find and interpret the mean absolute deviation of the data.

b. After you include the daily high temperatures for the rest of the month in the stem-and-leaf plot, the mean absolute deviation increases. Where do you think most of the data values for the rest of the month are located in the stem-and-leaf plot? Explain.

18. Critical Thinking The back-to-back stem-and-leaf plot shows the 9-hole golf scores for two golfers. Only one of the golfers can compete in a tournament. Use measures of center and measures of variation to give reasons why you would choose each golfer.

Rich		Will
7 5	3	
8 5 4 3 2 1	4	2 3 4 4 6 7 7 8 9
5 0	5	0

Key: 1 | 4 | 2 = 41 and 42 strokes

Fair Game Review What you learned in previous grades & lessons

Draw the solid. *(Section 8.1)*

19. square pyramid

20. hexagonal prism

21. **MULTIPLE CHOICE** In a bar graph, what determines the length of each bar? *(Skills Review Handbook)*

Ⓐ frequency Ⓑ data value Ⓒ leaf Ⓓ change in data

10.2 Histograms

Essential Question How can you use intervals, tables, and graphs to organize data?

1 ACTIVITY: Conducting an Experiment

Work with a partner.

a. Roll a number cube 20 times. Record your results in a tally chart.

b. Make a bar graph of the totals.

c. Go to the board and enter your totals in the class tally chart.

d. Make a second bar graph showing the class totals. Compare and contrast the two bar graphs.

	Tally Chart
1	
2	
3	
4	
5	
6	

Key: | = 1   ~~||||~~ = 5

2 ACTIVITY: Using Intervals to Organize Data

Work with a partner. You are judging a paper airplane contest. A contestant flies a paper airplane 20 times. You record the following distances:

20.5 ft, 24.5 ft, 18.5 ft, 19.5 ft, 21.0 ft, 14.0 ft, 12.5 ft, 20.5 ft, 17.5 ft, 24.5 ft, 19.5 ft, 17.0 ft, 18.5 ft, 12.0 ft, 21.5 ft, 23.0 ft, 13.5 ft, 19.0 ft, 22.5 ft, 19.0 ft

a. Complete the tally chart and the bar graph of the distances.

Tally Chart		
Interval	Tally	Total
10.0–12.9		
13.0–15.9		
16.0–18.9		
19.0–21.9		
22.0–24.9		

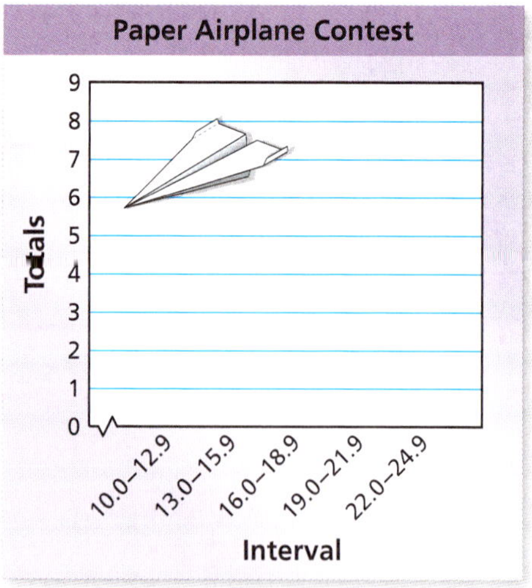

b. Make a different tally chart and bar graph of the distances. Use the following intervals:

10.0–11.9, 12.0–13.9, 14.0–15.9, 16.0–17.9, 18.0–19.9, 20.0–21.9, 22.0–23.9, 24.0–25.9

c. Which graph do you think represents the distances better? Explain.

COMMON CORE

Data Displays

In this lesson, you will

- make histograms.
- use histograms to analyze data.

Learning Standards
6.SP.2
6.SP.4

Laurie's Notes

Introduction

Standards for Mathematical Practice

- **MP4 Model with Mathematics:** Students will construct histograms to represent data sets. Such graphical representation enables students to describe features of the data set.

Motivate

- Ask if students have ever made paper airplanes. If so, how long did their planes usually stay in flight before crashing?

Activity Notes

Activity 1

? "When you roll a standard number cube, what are the possible outcomes?" 1, 2, 3, 4, 5, 6

- Discuss part (d). Depending upon the number of students, the class results should be close to the theoretical probability for this event, whereas each student's results may not be.

? **MP4:** "Was it helpful to use a chart and tally marks to keep track of the results from your experiment? Explain." Answers will vary.

Activity 2

- Discuss the context of this problem. Ask for the range of the data. Compare the range to the distances in the classroom to help students learn to judge distances.
- After students finish part (a), ask them to compare the column of tally marks in the chart with the graph at the right known as a histogram. Some might observe that if the tally marks are rotated 90° counterclockwise, the shape resembles the histogram.

? "How are the intervals in part (b) different from those in part (a)?" Smaller; these intervals have a width of 1.9 versus 2.9.

? "How will the chart and graph be different?" It's ok for students not to have an answer – it's a big idea!

- **Big Idea:** When the interval is smaller, you have *more* intervals and the tallies become more *spread out*, meaning the number of tallies in each interval will be smaller.
- **MP1 Make Sense of Problems and Persevere in Solving Them:** The size of the interval influences the shape of the histogram. Too many or too few intervals *may* mean that some features of the data go unnoticed.

? "Looking only at the tally chart or graph, did the person fly the airplane 20.0 feet?" cannot determine

- **Big Idea:** Once the tallies or graph have been made, the actual identity of the data is lost. Only in looking at the raw data can the question above be answered. The interval 19.0–21.9 contains 20.0 but you do not know if it is actually one of the data values.

Common Core State Standards

6.SP.2 Understand that a set of data collected to answer a statistical question has a distribution which can be described by its center, spread, and overall shape.

6.SP.4 Display numerical data in plots on a number line, including . . . histograms . . . .

Previous Learning

Students know how to find the mean, median, and mode and describe the range of a data set.

Technology for the Teacher

Dynamic Classroom

Lesson Plans

Complete Materials List

10.2 Record and Practice Journal

Essential Question How can you use intervals, tables, and graphs to help organize data?

1 ACTIVITY: Conducting an Experiment

Work with a partner. Check students' work.

a. Roll a number cube 20 times. Record your results in a tally chart.

	1	2	3	4	5	6
Tally						

Key: | = 1 𝍸 = 5

b. Make a bar graph of the totals.

c. Go to the board and enter your totals in the class tally chart.

d. Make a second bar graph showing the class totals. Compare and contrast the two bar graphs.

Differentiated Instruction

Auditory

Some students may benefit by hearing how to construct a bar graph. Ask students questions and have them explain how they got their answers. Should you draw a horizontal or vertical bar graph? What increments should you use? What scale should you use and where should it start and end? What is the title of the bar graph?

10.2 Record and Practice Journal

2 ACTIVITY: Using Intervals to Organize Data

Work with a partner. You are judging a paper airplane contest. A contestant flies a paper airplane 20 times. You record the following distances:

20.5 ft, 24.5 ft, 18.5 ft, 19.5 ft, 21.0 ft, 14.0 ft, 12.5 ft, 20.5 ft, 17.5 ft, 24.5 ft, 19.5 ft, 17.0 ft, 18.5 ft, 12.0 ft, 21.5 ft, 23.0 ft, 13.5 ft, 19.0 ft, 22.5 ft, 19.0 ft

a. Complete the tally chart and the bar graph of the distances.

Interval	Tally	Total
10.0–12.9	II	2
13.0–15.9	II	2
16.0–18.9	IIII	4
19.0–21.9	~~IIII~~ III	8
22.0–24.9	IIII	4

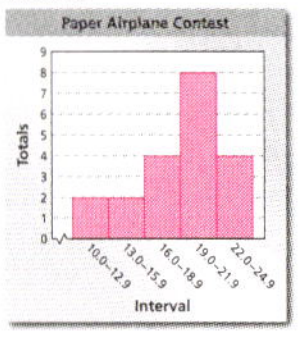

b. Make a different tally chart and bar graph of the distances. Use the following intervals:

10.0–11.9, 12.0–13.9, 14.0–15.9, 16.0–17.9, 18.0–19.9, 20.0–21.9, 22.0–23.9, 24.0–25.9

See Additional Answers.

c. Which graph do you think represents the distances better? Explain.

the new graph

3 ACTIVITY: Developing an Experiment

Work with a partner. **Check students' work.**

a. Make the airplane shown in your textbook from a single sheet of $8\frac{1}{2}$-by-11 inch paper. Then design and make your own paper airplane.

b. PRECISION Fly each airplane 20 times. Keep track of the distance flown each time.

Flight	1	2	3	4	5	6	7	8	9	10
Plane A										
Plane B										
Flight	11	12	13	14	15	16	17	18	19	20
Plane A										
Plane B										

c. **MODELING** Organize the results of the flights using frequency tables and graphs. Which airplane flies farther? Explain your reasoning.

What Is Your Answer?

4. **IN YOUR OWN WORDS** How can you use intervals, tables, and graphs to organize data? ***Sample answer:*** **Choose intervals and find the frequency for each, then graph the data using the intervals and frequencies.**

5. What intervals could you use in a graph that displays data whose values range from 40 through 59?

Sample answer: **Intervals of 40–44, 45–49, 50–54, 55–59**

Laurie's Notes

Activity 3

- This activity allows students to collect their own data. It is always more engaging and interesting for students to work with data they generated.
- Use scrap paper and allow only one sheet per student. They need to think before they fold!
- **Management Tip:** Lay tape measures out in a parallel fashion about 6 feet apart on the soccer field or other school grounds.
- Two students at a time fly their planes. They must make a quick read of distance to the nearest foot and call out their distances to their partners who records them.
- Review how to round to the nearest foot before going outside, if necessary.
- Students continue in pairs until all have completed 20 trials. The recorder and pilot switch roles and repeat the process.
- To save class time, you could decrease the number of trials from 20 to, say, 10, but keep in mind that a larger data set is often better for drawing conclusions.
- If the planes hook left or right too far and make a reading difficult, use your judgment to call out a quick measurement.
- **Safety:** Students should fly their planes in the same direction so that no one is accidentally hit in the face.
- **Management Tip:** You may choose to complete this activity inside. If so, it is helpful to mark the floor prior to the start of the activity. For instance, have students measure and mark distances 2 feet apart. When students test their planes, they only have to measure between the marks. This will allow students to review their measuring skills as well as make the activity more time efficient.
- Once the data is collected, complete part (c) inside. To save time, do not have students do any conversions to total inches. The intervals are generally a width of about 2 feet.
- **MP1** and **MP6 Attend to Precision:** Students have fun folding and flying the airplane, but don't forget the mathematics. Raw data does not tell much of a story until it is organized in some fashion.

? "Where are the mean, median, and mode located in the chart and graph?"

- **Extension:** You are the executive of a paper airplane business. Use the charts and graphs to decide which airplane you would put into production and explain the reasoning of your choice. Answers will vary.

What Is Your Answer?

- Have students share their answers.

Closure

- Explain how the intervals in a tally chart help you organize data. The interval with the most tallies helps you determine the most common answer (the mode).

The tally chart in Activity 2 is also called a *frequency table*. A **frequency table** groups data values into intervals. The **frequency** is the number of values in an interval.

3 ACTIVITY: Developing an Experiment

Math Practice 6

Specify Units

What units will you use to measure the distance flown each time? Will the units you use affect the results in your frequency table? Explain.

Work with a partner.

a. Make the airplane shown from a single sheet of $8\frac{1}{2}$-by-11-inch paper. Then design and make your own paper airplane.

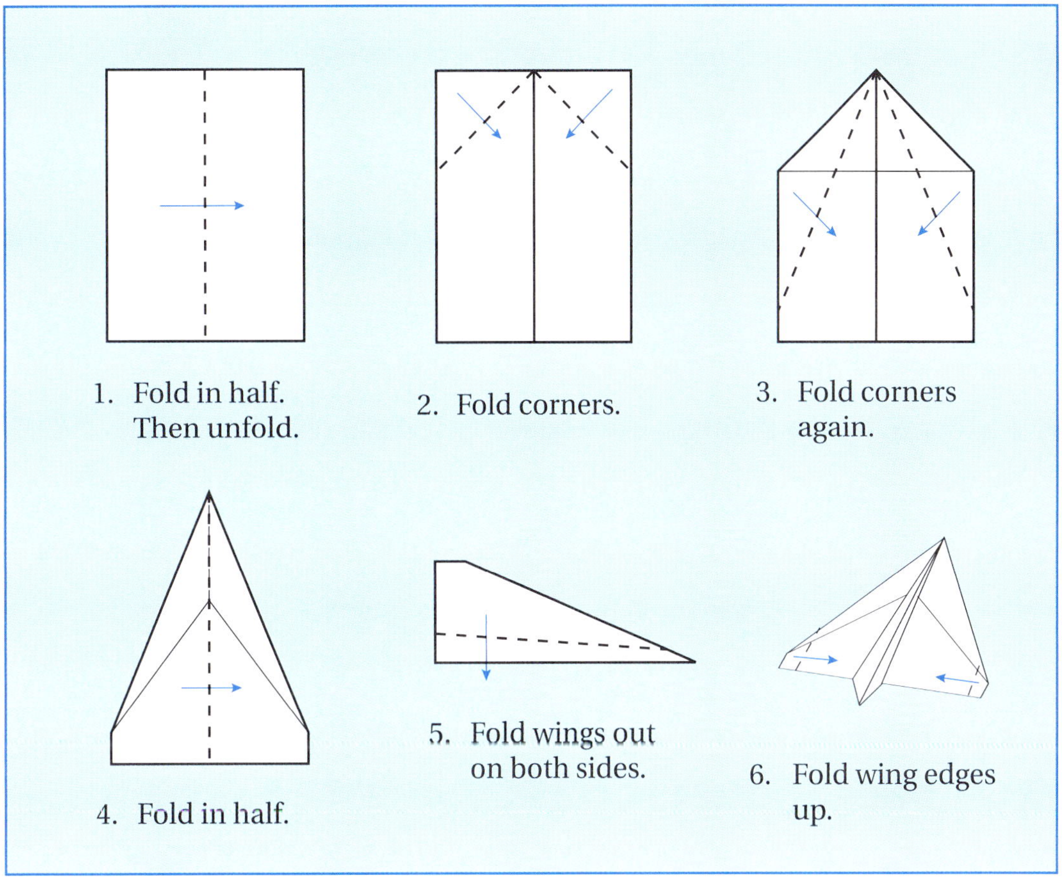

b. **PRECISION** Fly each airplane 20 times. Keep track of the distance flown each time.

c. **MODELING** Organize the results of the flights using frequency tables and graphs. Which airplane flies farther? Explain your reasoning.

What Is Your Answer?

4. **IN YOUR OWN WORDS** How can you use intervals, tables, and graphs to organize data?

5. What intervals could you use in a graph that displays data whose values range from 40 through 59?

Practice Use what you learned about organizing data into intervals to complete Exercises 4 and 5 on page 445.

10.2 Lesson

Check It Out
Lesson Tutorials
BigIdeasMath.com

Key Vocabulary
frequency table, *p. 441*
frequency, *p. 441*
histogram, *p. 442*

Key Idea

Histograms

A **histogram** is a bar graph that shows the frequency of data values in intervals of the same size.

The height of a bar represents the frequency of the values in the interval.

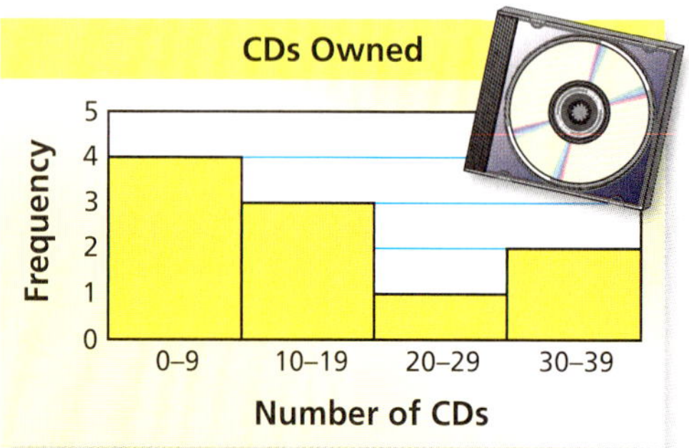

EXAMPLE 1 Making a Histogram

The frequency table shows the numbers of laps that people in a swimming class completed today. Display the data in a histogram.

Number of Laps	Frequency
1–3	11
4–6	4
7–9	0
10–12	3
13–15	6

Step 1: Draw and label the axes.

Step 2: Draw a bar to represent the frequency of each interval.

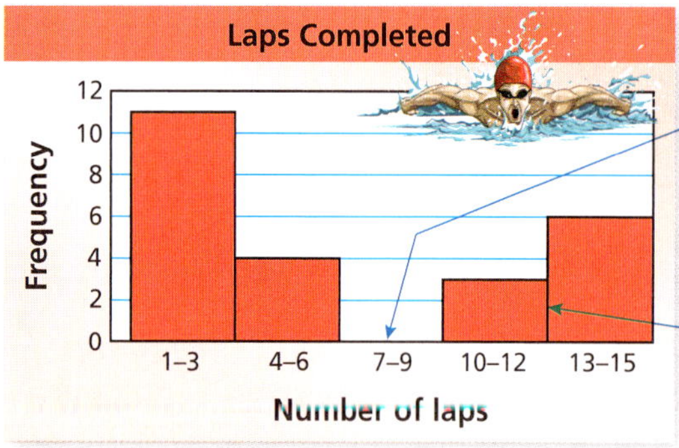

Include any interval with a frequency of 0. The bar height is 0.

There is no space between the bars of a histogram.

On Your Own

Now You're Ready
Exercises 6–8

1. The frequency table shows the ages of people riding a roller coaster. Display the data in a histogram.

Age	10–19	20–29	30–39	40–49	50–59
Frequency	16	11	5	2	4

Multi-Language Glossary at BigIdeasMath.com

Laurie's Notes

Introduction

Connect

- **Yesterday:** Students explored how to use tables and graphs to organize data. (MP1, MP4, MP6)
- **Today:** Students will construct and interpret a histogram.

Motivate

? "How many of you wear swimming caps in a pool?" Answers will vary.
- Competitive swimmers wear swimming caps.
- The largest gathering of people wearing swimming caps was 2,049 at Tybee Island, Georgia's annual Polar Bear Plunge on January 1, 2012.

Lesson Notes

Key Idea

- Students have constructed bar graphs. A histogram is a particular type of bar graph where the data is numeric, and the data is grouped into intervals of equal size. A bar graph includes categorical data (i.e., favorite vegetable) and numeric data.

? Use the sample shown to ask questions, checking students' understanding of how to read a histogram.
  - "How many people were polled?" 10
  - "How many people polled had 9 or fewer CDs?" 4
  - "How many people polled had 15 CDs?" cannot determine

Example 1

- The intervals for the histogram have been pre-determined. The size of each interval is 3. Students should notice the pattern in the right-end value of each interval: 3, 6, 9, 12, 15.
- Labeling the axes can often present a challenge for students. Students may write a number below the hash mark instead of writing an interval between the hash marks.
- Explain that no space is left between the bars because the intervals are continuous.
- Remind students that the axes are labeled and the histogram is given a title explaining what the data is about.
- Ask questions about the completed histogram.

On Your Own

- Students should recognize that the data is about people ages 10 to 59. It would not be correct to leave a gap for people ages 0 to 9.
- Ask a volunteer to share his or her histogram at the board.

Goal Today's lesson is constructing and interpreting a **histogram**.

Technology for the Teacher

Dynamic Classroom

Lesson Tutorials
Lesson Plans
Answer Presentation Tool

Extra Example 1

The frequency table shows the numbers of T-shirts each person in a class owns. Display the data in a histogram.

T-shirts	Frequency
1–4	3
5–8	6
9–12	13
13–16	0
17–20	4

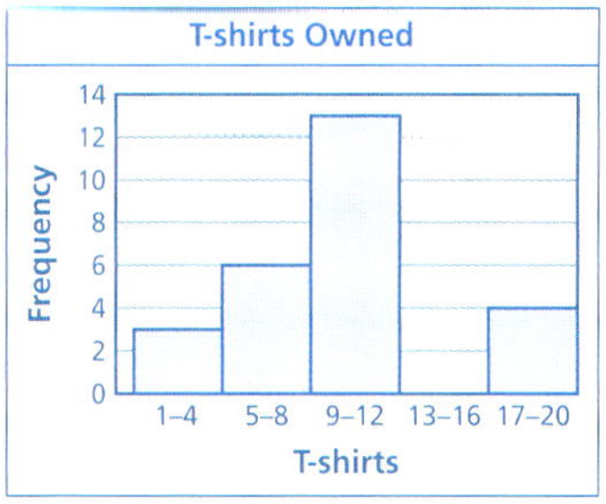

On Your Own

1.

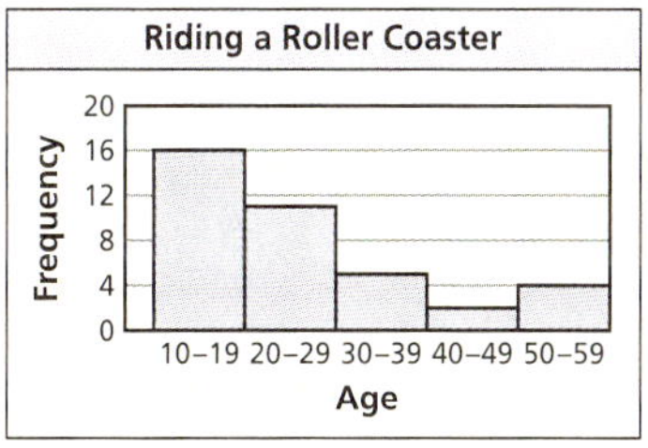

Extra Example 2

The histogram shows the numbers of different instruments each member of a jazz band can play.

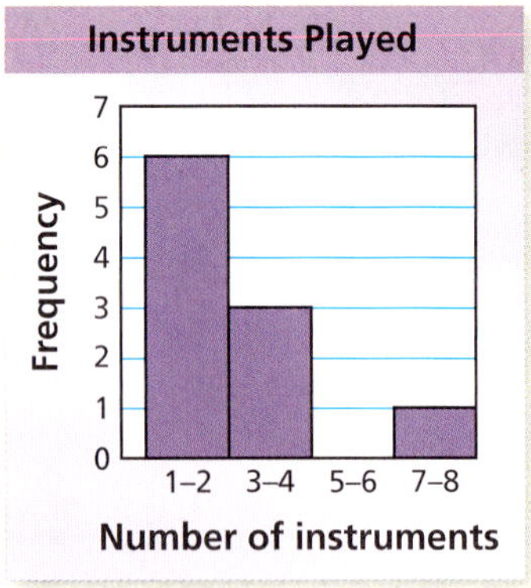

a. Which interval contains the most data values? 1–2

b. How many band members can play more than 2 instruments? 4

c. How many band members can play at most 2 instruments? 6

On Your Own

2. a. 11 students

   b. 18 students

Differentiated Instruction

Visual

Give students the following set of data.

1, 2, 2, 4, 5, 7, 7, 8, 9, 11, 13, 14, 14, 16, 17, 19, 20, 20, 21, 22, 24, 24, 25, 28, 28, 29

Instruct half of the class to create a histogram using intervals of length 5 and the other half of the class to create a histogram using intervals of length 10. Compare the shapes of the two graphs and discuss how the choice of intervals affects the interpretation of the data.

Laurie's Notes

Example 2

- The focus of this example is to interpret information given in a histogram. Review the vocabulary *less than*, *at least*, *at most*, and *more than*.
- In addition to the questions posed, ask students to describe the distribution of the data. Also ask about the number of races shown in the graph.
- **MP3 Construct Viable Arguments and Critique the Reasoning of Others:** Listen for student reasoning as they interpret the histogram.

On Your Own

- This graph is almost symmetric. Ask students what it means for a graph to be symmetric.

EXAMPLE 2 Using a Histogram

The histogram shows the winning speeds at the Daytona 500. (a) Which interval contains the most data values? (b) How many of the winning speeds are less than 140 miles per hour? (c) How many of the winning speeds are at least 160 miles per hour?

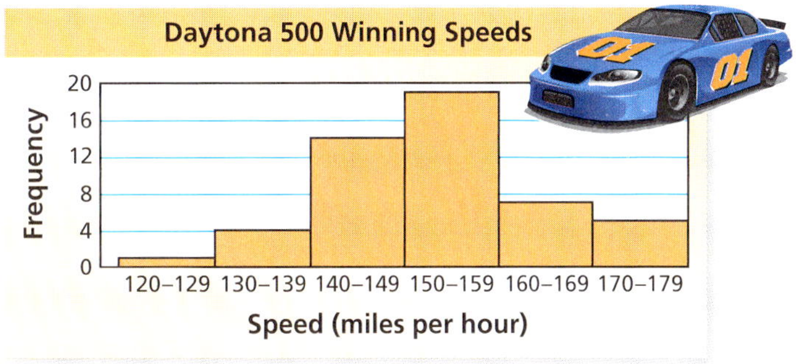

a. The interval with the tallest bar contains the most data values.

So, the 150–159 miles per hour interval contains the most data values.

b. One winning speed is in the 120–129 miles per hour interval, and four winning speeds are in the 130–139 miles per hour interval.

So, $1 + 4 = 5$ winning speeds are less than 140 miles per hour.

c. Seven winning speeds are in the 160–169 miles per hour interval, and five winning speeds are in the 170–179 miles per hour interval.

So, $7 + 5 = 12$ winning speeds are at least 160 miles per hour.

On Your Own

Now You're Ready
Exercises 10–13

2. The histogram shows the numbers of hours that students in a class slept last night.

   a. How many students slept at least 8 hours?

   b. How many students slept less than 12 hours?

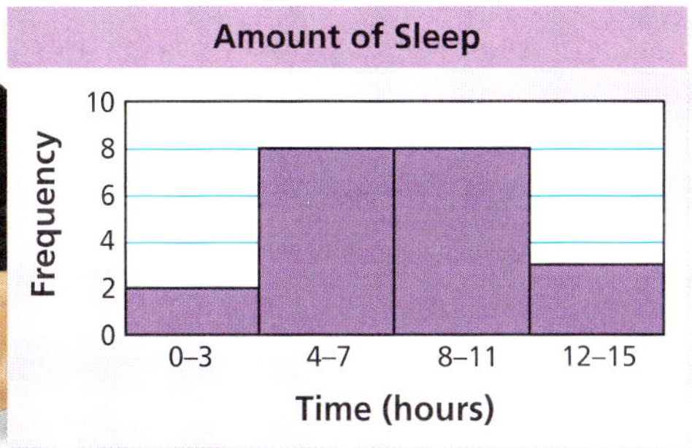

EXAMPLE 3 Comparing Data Displays

The data displays show how many push-ups students in a class completed for a physical fitness test. Which data display can you use to find how many students are in the class? Explain.

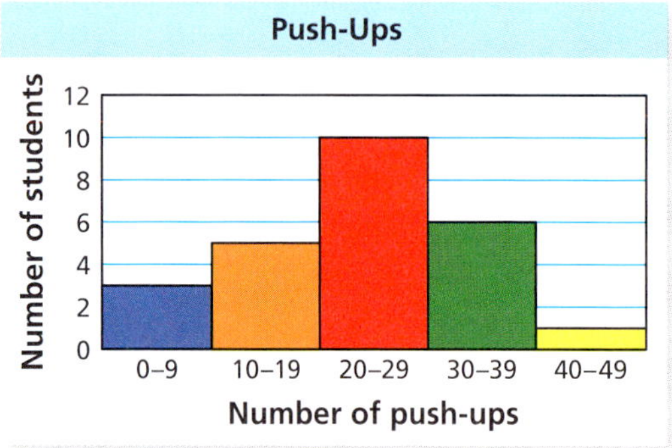

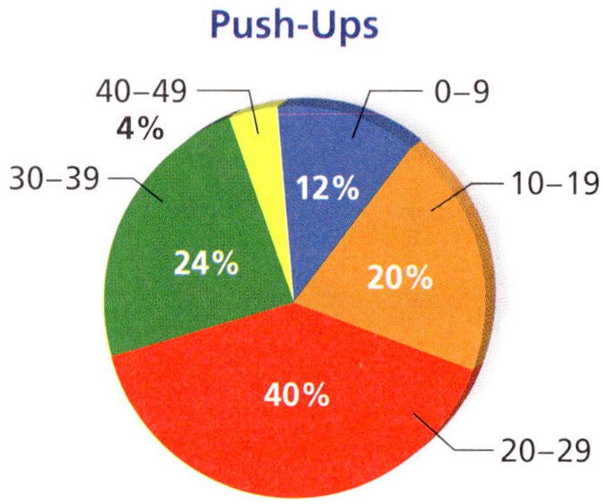

∴ You can use the histogram because it shows the number of students in each interval. The sum of these values represents the number of students in the class. You cannot use the circle graph because it does not show the number of students in each interval.

EXAMPLE 4 Making Conclusions from Data Displays

Which statement *cannot* be made using the data displays in Example 3?

Ⓐ Twelve percent of the class completed less than 10 push-ups.

Ⓑ Five students completed at least 10 and at most 19 push-ups.

Ⓒ At least one student completed more than 39 push-ups.

Ⓓ Twenty-nine percent of the class completed 30 or more push-ups.

The circle graph shows that 12% completed 0–9 push-ups. So, Statement A can be made.

In the histogram, the bar height for the 10–19 interval is 5, and the bar height for the 40–49 interval is 1. So, Statements B and C can be made.

The circle graph shows that 24% completed 30–39 push-ups, and 4% completed 40–49 push-ups. So, 24% + 4% = 28% completed 30 or more push-ups. Statement D cannot be made.

∴ The correct answer is Ⓓ.

On Your Own

Now You're Ready
Exercises 14 and 15

3. In Example 3, which data display should you use to describe the portion of the entire class that completed 30–39 push-ups?

4. Make two more conclusions from the data displays in Example 3.

Laurie's Notes

Example 3

? "When the sections of a circle graph are labeled with percents, what do you know about the sum of these percents?" They add up to the whole amount, 100% of the data.

- **Connection:** Students should recognize that the colors in each data display represent the same interval of push-ups. Doing 20–29 push-ups was the most frequent amount done by students. The histogram shows that there were 10 students who did 20–29 push-ups, which represents 40% of the class.

? "Could the data set have a mode? Explain." It is possible but you cannot tell for sure. Neither display gives individual values.

Example 4

- **MP3:** Ask different students to read each statement and explain whether or not it can be made using one of the data displays.

? "Can you determine how many students did 15 push-ups from either data display? Explain." no; The histogram only tells how many students did 10–19 push-ups and the circle graph tells the percent of students who did 10–19 push-ups.

- Justify to students that you can add percents using fractions: $\frac{24}{100} + \frac{4}{100} = \frac{28}{100} = 28\%$.

On Your Own

- Have students share the conclusions they wrote for Question 4.

Closure

- Explain how a histogram is similar to a bar graph and how it differs from a bar graph. *Sample answer:* Each bar in a histogram represents the frequency of data values in intervals of the same size. Each bar in a bar graph represents the frequency of a single data value.

Extra Example 3

In Example 3, which data display can you use to compare the number of students that completed 10–19 push-ups with the number of students that completed 20–29 push-ups? Explain. both; The 20–29 interval bar height is two times taller than the 10–19 interval bar height in the histogram, and the percent of the 20–29 section is twice the percent of the 10–19 section in the circle graph. So, twice as many students completed 20–29 push-ups as 10–19 push-ups.

Extra Example 4

Using the data displays in Example 3, can you conclude that everyone in the class did at least 1 push-up? Explain. no; Three students completed 0–9 push-ups. It is possible that at least one of these students completed 0 push-ups.

On Your Own

3. circle graph
4. *Sample answer:* 16 students completed 20–39 push-ups; 88% of the class completed 10 or more push-ups.

Vocabulary and Concept Check

1. The *Test Scores* graph is a histogram because the number of students (frequency) achieving the test scores are shown in intervals of the same size (20).
2. The scores falling into the interval 1–20 are outliers because most of the scores are between 41 and 100.
3. No bar is shown on that interval.

Practice and Problem Solving

4–6. See Additional Answers.

7.

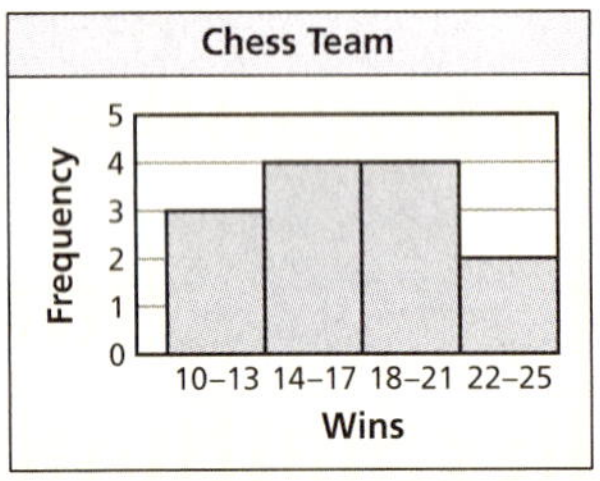

8.

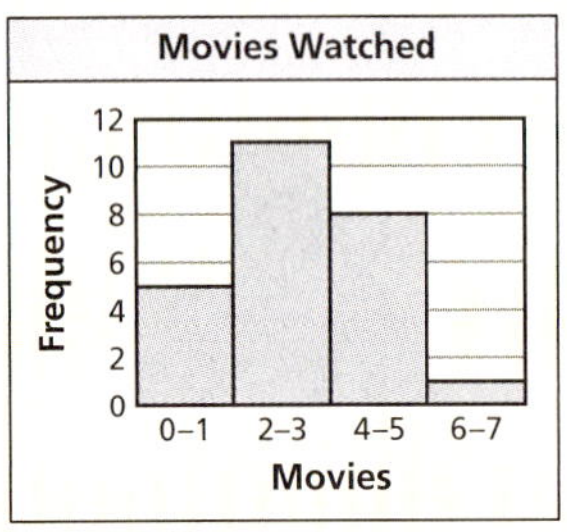

9. There should not be space between the bars of the histogram.

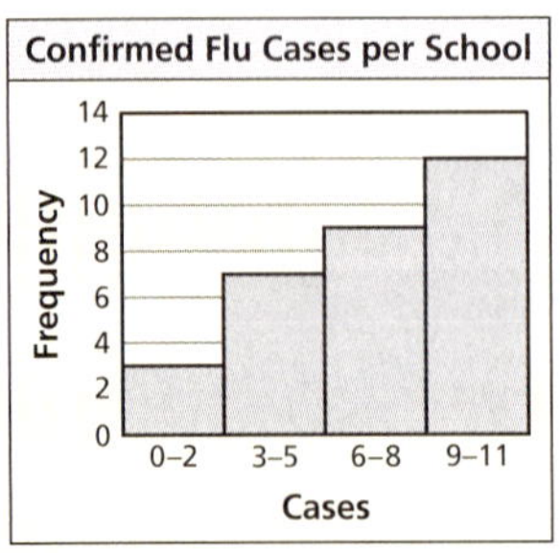

Assignment Guide and Homework Check

Level	Day 1 Activity Assignment	Day 2 Lesson Assignment	Homework Check
Basic	4, 5, 19–23	1–3, 6–11, 13	2, 8–11
Average	4, 5, 19–23	1–3, 7, 9–15	2, 7, 10, 11, 14
Advanced	4, 5, 19–23	1–3, 8, 10–18	2, 11–13, 15

Common Errors

- **Exercises 6–8** Students may struggle with determining how to scale the vertical axis of the histogram. Remind them to use consistent intervals and to base their decision on the frequency of the data. For example, if there is a high frequency, they should count by 5s or 10s, but if there is a low frequency, then they should count by 1s or 2s.

10.2 Record and Practice Journal

Display the data in a histogram.

1.

Birthdays	
Months	Frequency
Jan–Mar	15
Apr–June	9
Jul–Sept	6
[illegible]	[illegible]

2.

Goals Scored	
Goals	Frequency
0–2	6
3–5	8
6–8	2
[illegible]	[illegible]

1–2. See Additional Answers.

3.

Height Jumped	
Inches	Frequency
0–11	7
12–23	10
24–35	5
36–47	2

4.

Money Spent	
Dollars	Frequency
0–19	3
20–39	8
40–59	8
60–79	15

3–4. See Additional Answers.

5. The histogram shows the times students ran the mile in gym class.

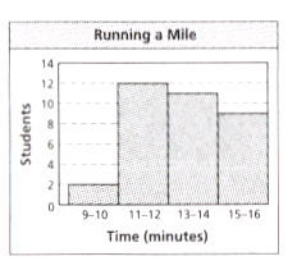

a. Which interval contains the fewest data values?
9–10 minutes

b. How many students are in the class?
34 students

c. What percent of students ran the mile in 12 minutes or less?
about 41%

10.2 Exercises

Check It Out
Help with Homework
BigIdeasMath.com

Vocabulary and Concept Check

1. **VOCABULARY** Which graph is a histogram? Explain your reasoning.

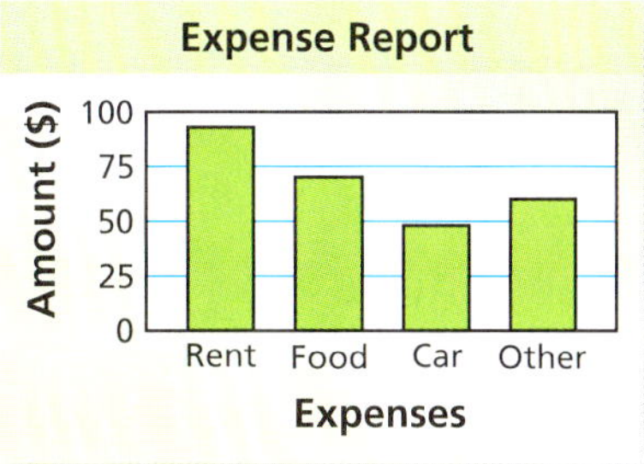

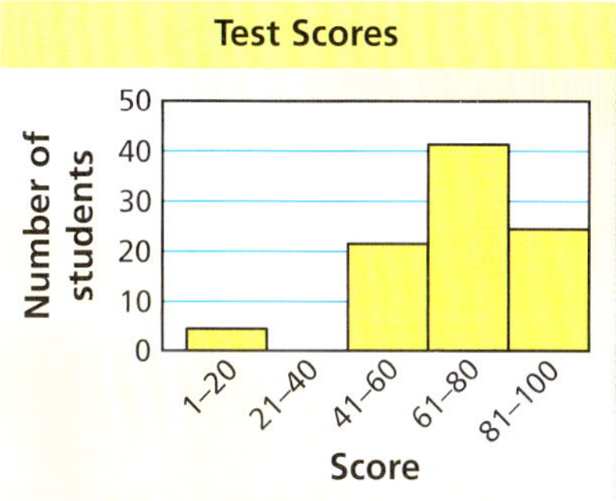

2. **REASONING** Describe the outliers in the histogram.

3. **REASONING** How can you tell when an interval of a histogram has a frequency of zero?

Practice and Problem Solving

Make a tally chart and a bar graph of the data.

4.

Members of Book Clubs			
6	17	13	19
13	9	18	24
11	15	21	14

5.

Points Scored				
42	45	57	39	55
38	48	36	48	46
51	29	45	54	42

Display the data in a histogram.

1 6.

States Visited	
States	Frequency
1–5	12
6–10	14
11–15	6
16–20	3

7.

Chess Team	
Wins	Frequency
10–13	3
14–17	4
18–21	4
22–25	2

8.

Movies Watched	
Movies	Frequency
0–1	5
2–3	11
4–5	8
6–7	1

9. **ERROR ANALYSIS** Describe and correct the error made in displaying the data in a histogram.

Confirmed Flu Cases per School	
Cases	Frequency
0–2	3
3–5	7
6–8	9
9–11	12

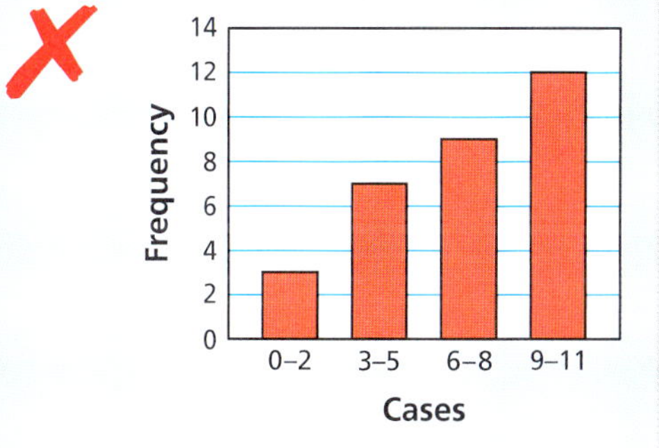

2 10. **MAGAZINES** The histogram shows the number of magazines read last month by the students in a class.

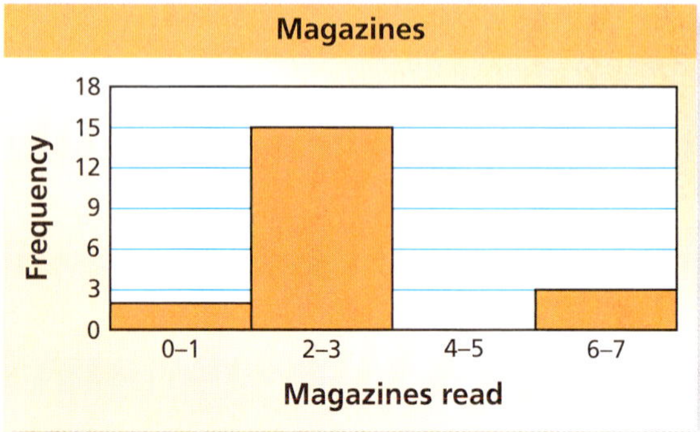

a. Which interval contains the fewest data values?

b. How many students are in the class?

c. What percent of the students read less than six magazines?

d. Can you find the mean or the median of the data? Explain.

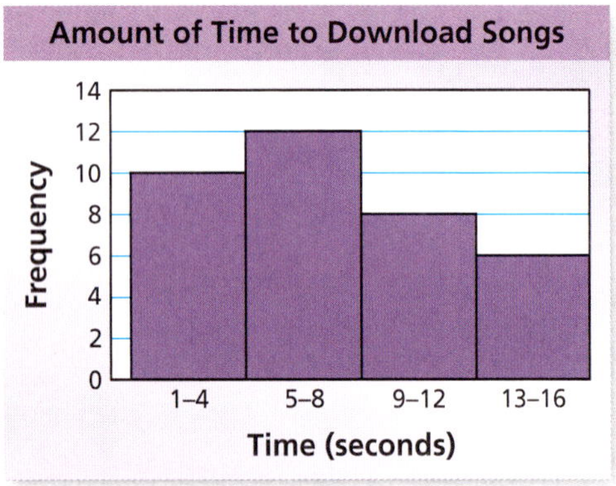

11. **ERROR ANALYSIS** Describe and correct the error made in reading the histogram.

✗ 12% of the songs took 5–8 seconds to download.

12. **VOTING** The histogram shows the percent of the voting-age population that voted in a recent presidential election. Explain whether the graph supports each statement.

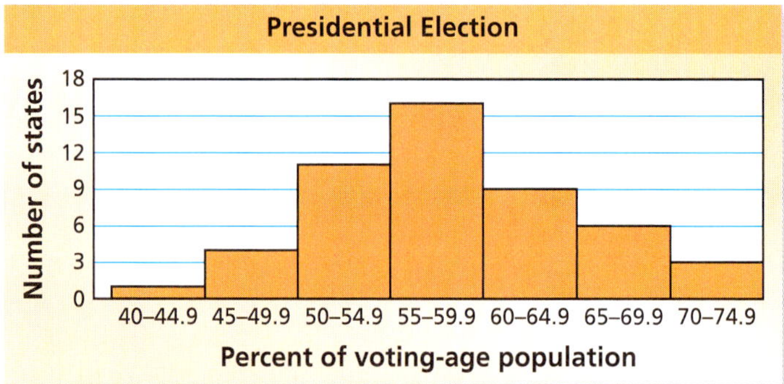

a. Only 40% of one state voted.

b. In most states, between 50% and 64.9% voted.

c. The mode of the data is between 55 and 59.9.

13. **PROBLEM SOLVING** The histograms show the areas of counties in Pennsylvania and Indiana. Which state do you think has the greater area? Explain.

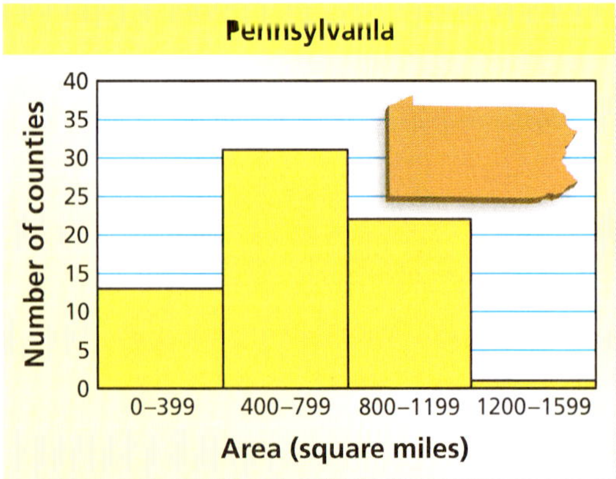

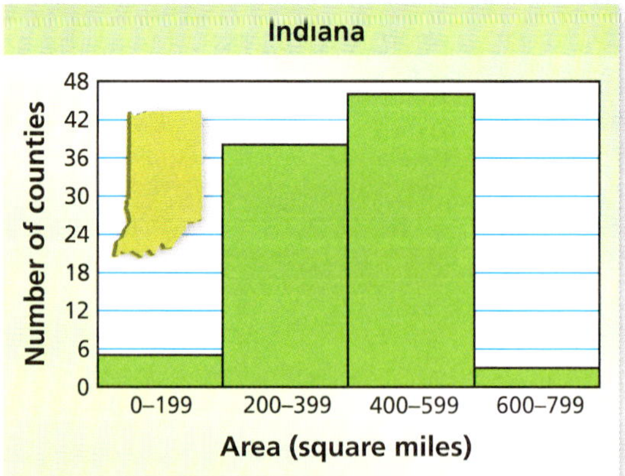

For Your Information

- **Exercise 13** Pennsylvania has a total area of 46,055 square miles. Indiana has a total area of 36,418 square miles.

Common Errors

- **Exercise 10** Students may say that the interval 0–1 has the fewest data values, but it is really 4–5. To help them answer this question, ask them to label each interval with the frequency of that interval. Writing the frequency helps students to read the histogram.
- **Exercise 13** Students may only look at the frequency for each graph and say that Indiana has the greater area. Encourage them to look at the intervals as well. The two graphs are drawn using different intervals and cannot be compared strictly by the heights of the bars.

Practice and Problem Solving

10. **a.** 4–5

b. 20 students

c. 85%

d. no; You only know what interval each of the data values falls into, not the specific data values.

11. The frequency is the number of songs not the percent of songs. The statement should be "12 of the songs took 5–8 seconds to download."

12. **a.** no; The histogram shows that only one state fell in the interval of 40–44.9%. This state did not necessarily have 40% of possible voters vote.

b. yes; 36 states are between 50% and 64.9%.

c. no; The 55–59.9 interval has the highest frequency, but does not necessarily contain the mode of the data.

13. Pennsylvania; You can see from the intervals and frequencies that Pennsylvania counties are greater in area, which makes up for it having fewer counties.

English Language Learners

Visual

Histograms and bar graphs have a similar appearance. Students may confuse the two. Compare and contrast the two types of data displays. Both use the lengths of bars to represent data. The bars of a histogram represent numerical intervals and are touching. The bars of a bar graph represent categories and are separated.

Practice and Problem Solving

14. stem-and-leaf plot; You need to know the specific data values, the intervals in the histogram do not give enough information.

15. See Additional Answers.

16. no; You only know what interval each of the data values falls into, not the specific data values.

17. See *Taking Math Deeper*.

18. 2, 2.5, 3

Fair Game Review

19. 45 **20.** 27

21. 22.4 **22.** 51.2

23. D

Mini-Assessment

The table shows the number of songs downloaded last month by your friends.

Songs Downloaded	
Songs	**Frequency**
0–1	4
2–3	12
4–5	8
6–7	5

1. Display the data in a histogram.

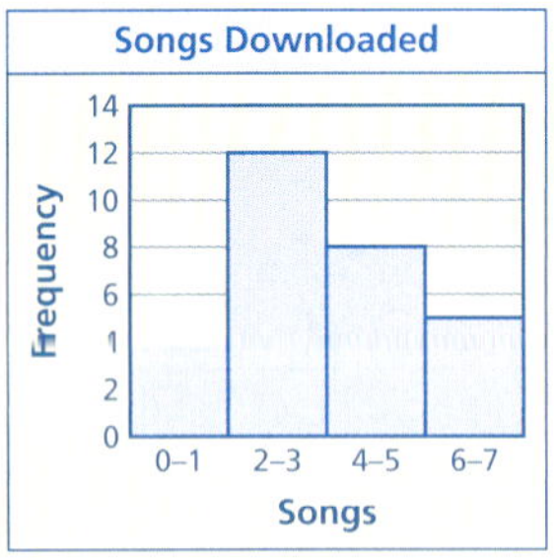

2. Which interval contains the most data values? The 2–3 songs interval contains the most data values.
3. Which interval contains the fewest data values? The 0–1 songs interval contains the fewest data values.
4. What is the difference between the 2–3 and 4–5 songs intervals? 4

Taking Math Deeper

Exercise 17

This exercise points out to students that making a histogram is as much an *art* as a *science*. The big question lies in "how many intervals should I use?"

With too many intervals, most distributions will appear flat. With too few intervals, most distributions will have only a few tall spikes.

1 Order the data.

51, 54, 55, 57, 59, 67, 68, 70, 71, 73, 75, 76, 77, 77, 78, 78, 79, 79, 80, 80, 81, 81, 82, 82, 83, 83, 84, 85, 85, 88

2 **a.** Make a frequency distribution and histogram.

Interval	Tally
51–55	III
56–60	II
61–65	
66–70	III
71–75	III
76–80	~~IIII~~ IIII
81–85	~~IIII~~ IIII
86–90	I

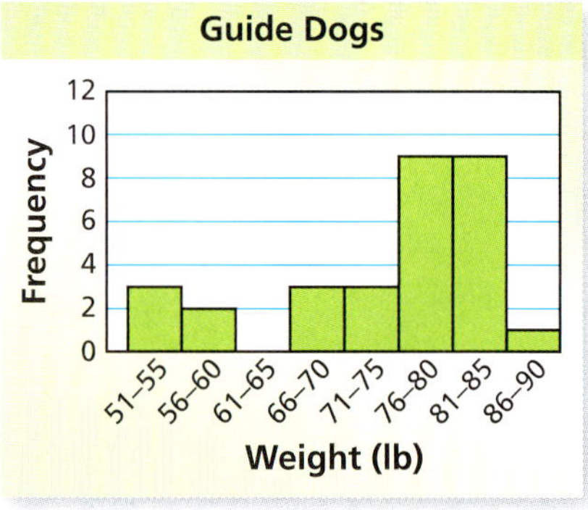

3 **b.** Make a frequency distribution and histogram using different-sized intervals.

Interval	Tally
51–60	~~IIII~~
61–70	III
71–80	~~IIII~~ ~~IIII~~ II
81–90	~~IIII~~ ~~IIII~~

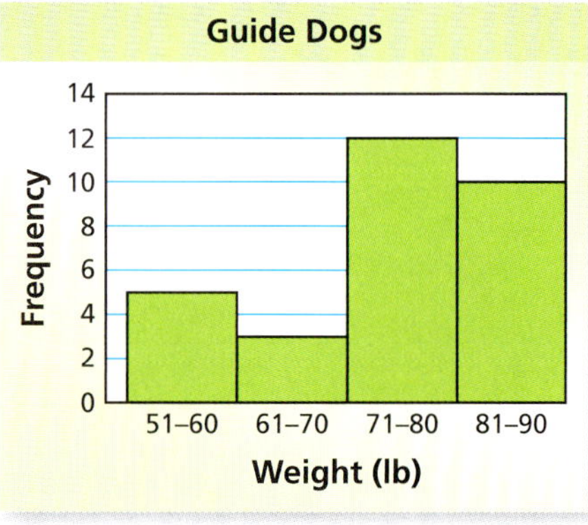

c. The second histogram has four intervals, and it does not have a gap as in the first histogram.

Reteaching and Enrichment Strategies

If students need help. . .	If students got it. . .
Resources by Chapter • Practice A and Practice B • Puzzle Time Record and Practice Journal Practice Differentiating the Lesson Lesson Tutorials Skills Review Handbook	Resources by Chapter • Enrichment and Extension • Technology Connection Start the next section

3 **14. GARBAGE** The data displays show how many pounds of garbage apartment residents produced in 1 week. Which data display can you use to find how many residents produced more than 25 pounds of garbage? Explain.

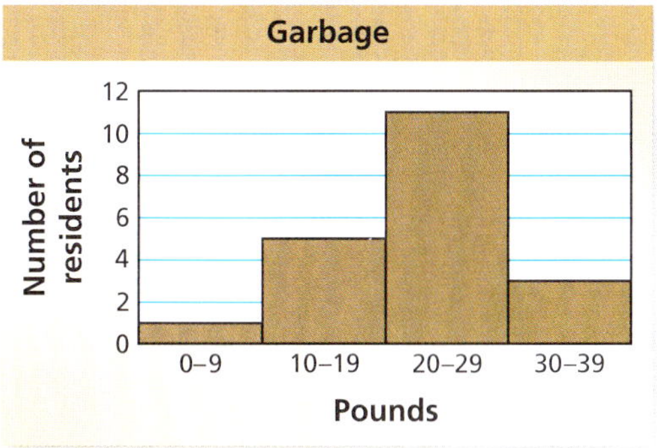

Garbage

Stem	Leaf
0	9
1	0 5 8 8 9
2	1 2 5 5 6 7 7 7 9 9 9
3	2 3 3

Key: 1 | 5 = 15 pounds

15. REASONING Determine whether you can make each statement by using the data displays in Exercise 14. Explain your reasoning.

a. One resident produced 10 pounds of garbage.

b. Twelve residents produced between 20 and 29 pounds of garbage.

16. NUMBER SENSE Can you find the range and the interquartile range of the data in Exercise 7? If so, find them. If you cannot find them, explain why not.

17. CRITICAL THINKING The table shows the weights of guide dogs enrolled in a training program.

a. Make a histogram of the data starting with the interval 51–55.

b. Make another histogram of the data using different-sized intervals.

c. Compare and contrast the two histograms.

Weights (pounds)					
81	88	57	82	70	85
71	51	82	77	79	77
83	80	54	80	81	73
59	84	75	76	68	78
83	78	55	67	85	79

18. **Logic** What are the possible values for the median in Exercise 10?

Fair Game Review What you learned in previous grades & lessons

Find the percent of the number. *(Section 5.6)*

19. 25% of 180 **20.** 30% of 90 **21.** 16% of 140 **22.** 64% of 80

23. MULTIPLE CHOICE Which is the solution of the inequality represented by "Four times a number n is at least 28"? *(Section 7.7)*

Ⓐ $n < 7$ Ⓑ $n > 7$ Ⓒ $n \leq 7$ Ⓓ $n \geq 7$

10 Study Help

Check It Out
Graphic Organizer
BigIdeasMath.com

You can use a **word magnet** to organize information associated with a vocabulary word. Here is an example of a word magnet for histogram.

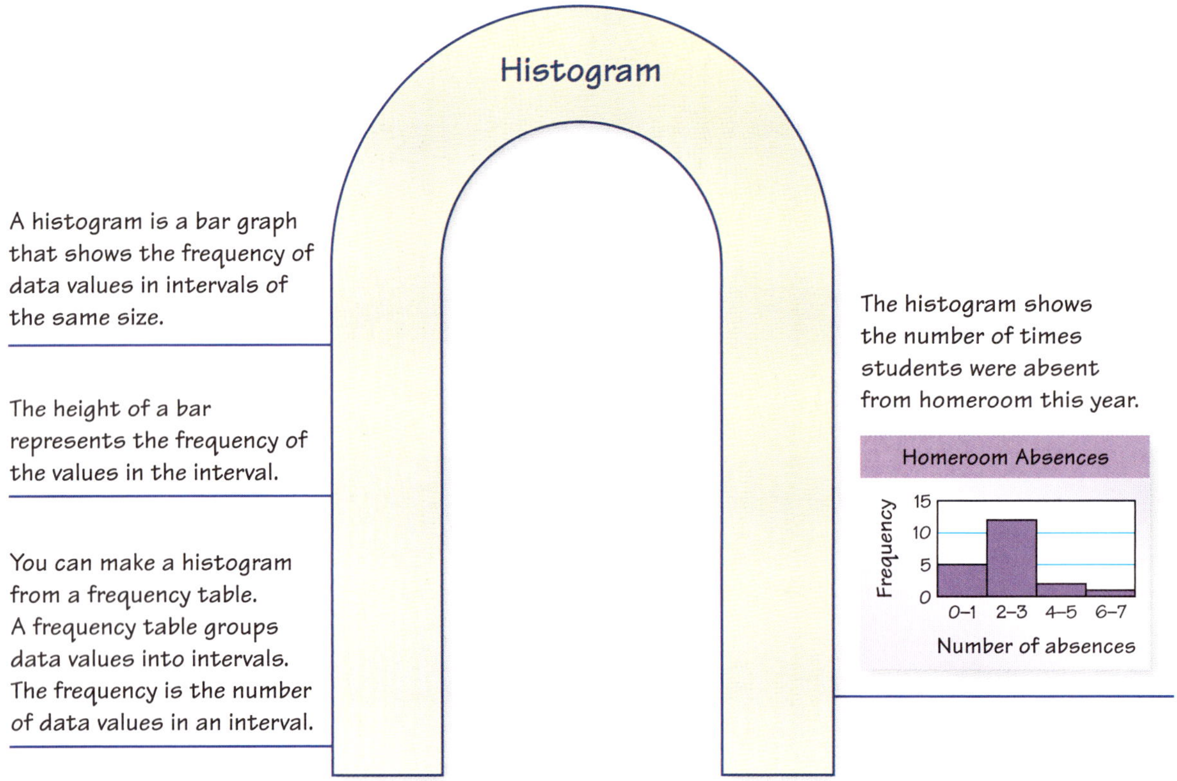

On Your Own

Make a word magnet to help you study this topic.

1. stem-and-leaf plot

After you complete this chapter, make word magnets for the following topics.

2. shapes of distributions
3. box-and-whisker plot
4. Choose three other topics that you studied earlier in this course. Make a word magnet for each topic.

"How do you like the word magnet I made for 'Beagle'?"

Sample Answer

1.

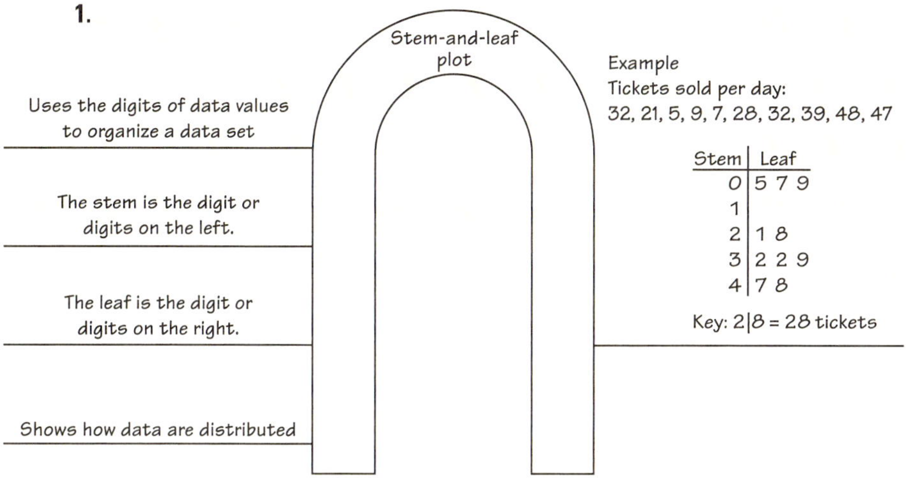

List of Organizers

Available at *BigIdeasMath.com*

Comparison Chart
Concept Circle
Example and Non-Example Chart
Formula Triangle
Four Square
Idea (Definition) and Examples Chart
Information Frame
Information Wheel
Notetaking Organizer
Process Diagram
Summary Triangle
Word Magnet
Y Chart

About this Organizer

A **Word Magnet** can be used to organize information associated with a vocabulary word or term. As shown, students write the word or term inside the magnet. Students write associated information on the blank lines that "radiate" from the magnet. Associated information can include, but is not limited to: other vocabulary words or terms, definitions, formulas, procedures, examples, and visuals. This type of organizer serves as a good summary tool because any information related to a topic can be included.

Technology for the Teacher

Editable Graphic Organizer

Answers

1. **Cans Collected Each Month**

Stem	Leaf
5	5 8 9
6	1 8
7	6 9
8	0 3 4
9	0 2

Key: 6 | 1 = 61 cans

2. **Miles Driven Each Day**

Stem	Leaf
0	9
1	0 1 2 5 6 6 8
2	0 0 1 8
3	5 7
4	
5	0

Key: 3 | 5 = 35 miles

3. **Age of Tortoises**

Stem	Leaf
8	3 5 6
9	2 9
10	0 4
11	0 5
12	4 9
13	0

Key: 9 | 2 = 92 years

4. **Kilometers Run Each Day**

Stem	Leaf
2	0 5
3	0 1 1 5 8 9
4	5
5	0 6
6	0 1 2 2

Key: 2 | 0 = 2.0 kilometers

5–10. See Additional Answers.

Technology for the Teacher

Online Assessment
Assessment Book
ExamView® Assessment Suite

Alternative Quiz Ideas

100% Quiz
Error Notebook
Group Quiz
Homework Quiz
Math Log
Notebook Quiz
Partner Quiz
Pass the Paper

Group Quiz

Students work in groups. Give each group a large index card. Each group writes five questions that they feel evaluate the material they have been studying. On a separate piece of paper, students solve the problems. When they are finished, they exchange cards with another group. The new groups work through the questions on the card.

Reteaching and Enrichment Strategies

If students need help. . .	If students got it. . .
Resources by Chapter • Practice A and Practice B • Puzzle Time Lesson Tutorials *BigIdeasMath.com*	Resources by Chapter • Enrichment and Extension • Technology Connection Game Closet at *BigIdeasMath.com* Start the next section

10.1–10.2 Quiz

Check It Out
Progress Check
BigIdeasMath.com

Make a stem-and-leaf plot of the data. *(Section 10.1)*

1.

Cans Collected Each Month			
80	90	84	92
76	83	79	59
68	55	58	61

2.

Miles Driven Each Day				
21	18	12	16	10
16	9	15	20	28
35	50	37	20	11

3.

Ages of Tortoises			
86	99	100	124
92	85	110	130
115	129	83	104

4.

Kilometers Run Each Day				
6.0	5.6	6.2	3.0	2.5
3.5	2.0	5.0	3.9	3.1
6.2	3.1	4.5	3.8	6.1

Display the data in a histogram. *(Section 10.2)*

5. Soccer Team Goals

Goals per Game	Frequency
0–1	5
2–3	4
4–5	0
6–7	1

6. Minutes Practiced

Minutes	Frequency
0–19	8
20–39	10
40–59	11
60–79	2

7. Poems Written for Class

Poems	Frequency
0–4	6
5–9	16
10–14	4
15–19	2
20–24	2

8. WEIGHTS The weights (in ounces) of nine packages are 7, 22, 16, 12, 6, 18, 15, 13, and 25. Make a stem-and-leaf plot of the data. Describe the distribution of the data. *(Section 10.1)*

9. REBOUNDS The histogram shows the number of rebounds per game for a middle school basketball player this season. *(Section 10.2)*

a. Which interval contains the most data values?

b. How many games did the player play this season?

c. What percent of the games did the player have 4 or more rebounds?

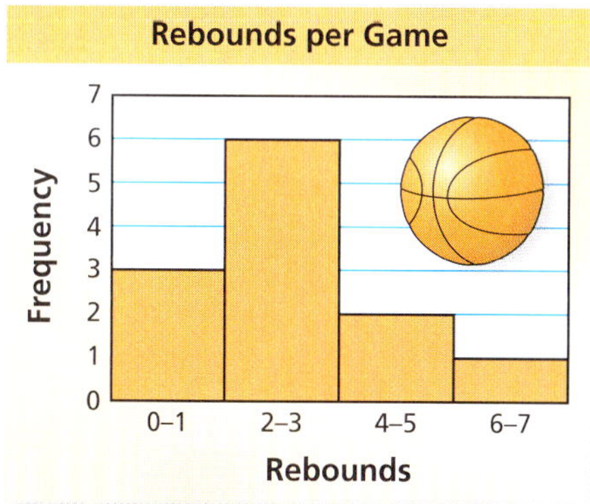

Stem	Leaf
0	6 8 8 9
1	0 1 2 3 7 8
2	0

Key: 0 | 9 = 9 hours

10. STAGE CREW The stem-and-leaf plot shows the number of hours 11 stage crew members spent building sets. Find the mean, median, mode, range, and interquartile range of the data. *(Section 10.1)*

10.3 Shapes of Distributions

Essential Question How can you describe the shape of the distribution of a data set?

1 ACTIVITY: Describing the Shape of a Distribution

Work with a partner. The lists at the left show the last four digits of a set of phone numbers in a phone book.

-7253
-7290
-7200
-1192
-1142

-3500
-2531
-2079
-5897
-5341
-1392
-5406
-7875
-7335
-0494
-9018
-2184
-2367

-8678
-2063
-2911
-2103
-4328
-7826
-7957
-7246
-2119
-7845
-1109
-9154

a. Create a list that represents the last digit of each phone number shown. Make a dot plot of the data.

b. In your own words, how would you describe the shape of the distribution? What single word do you think you can use to identify this type of distribution? Explain your reasoning.

2 ACTIVITY: Describing the Shape of a Distribution

Work with a partner. The lists at the right show the first three digits of a set of phone numbers in a phone book.

a. Create a list that represents the first digit of each phone number shown. Make a dot plot of the data.

b. In your own words, how would you describe the shape of the distribution? What single word do you think you can use to identify this type of distribution? Explain your reasoning.

c. In your dot plot, draw a vertical line through the middle of the data set. What do you notice?

d. Repeat part (c) for the dot plot you constructed in Activity 1. What do you notice? Compare the distributions from Activities 1 and 2.

538-
438-
664-
761-
868-

735-
694-
599-
725-
556-
555-
456-
736-
664-
576-
664-
664-
725-

664-
664-
538-
855-
664-
538-
654-
654-
725-
538-
799-
764-

COMMON CORE

Data Displays

In this lesson, you will

- describe shapes of distributions.

Learning Standards
6.SP.2
6.SP.4

Laurie's Notes

Introduction

Standards for Mathematical Practice

- **MP3 Construct Viable Arguments and Critique the Reasoning of Others:** Students will describe the shape of a distribution and make a connection to the measures of center and variation studied in Chapter 9.

Motivate

- ? "Who said, 'Mr. Watson, come here; I want to see you'?" Alexander Graham Bell
- ? When did Alexander Graham Bell utter those words?" March 10, 1876
- Have a general discussion about the impact of telephones, and now, cellular phones, on society.

Activity Notes

Activity 1

- Ask a volunteer to read the problem.
- **Management Tip:** Have one student read the numbers from the lists and one student fill in the dot plot. At this point, students should not be overwhelmed by the amount of data. Each value is a single digit and students have been constructing line/dot plots now for several grades.
- Discuss telephone numbers. The area code and exchanges for a particular area are not randomly distributed. The last digits of the given telephone numbers, which can be considered random, are evenly distributed. So, the dot plot appears flat or rectangular.

Activity 2

- This activity is related to Activity 1 and involves exchanges. A telephone book will have a limited number of exchanges so the dot plot will look different than that of Activity 1. Some numbers (0, 1, 2, 3, and 9 in this activity) do not appear at all.
- **MP3:** Listen to student reasoning as they compare the distributions from Activities 1 and 2.
- Review lines of symmetry to extend students' knowledge of symmetric shapes. Also, review the discussion in Section 9.1 regarding the basic characteristics of a distribution such as peaks, gaps, and clusters.
- For homework, you could have students repeat Activities 1 and 2 using a sample of 100 phone numbers.

Common Core State Standards

6.SP.2 Understand that a set of data collected to answer a statistical question has a distribution which can be described by its center, spread, and overall shape.

6.SP.4 Display numerical data in plots on a number line, including dot plots, histograms . . ..

Previous Learning

Students should know how to use line plots (or dot plots), stem-and-leaf plots, and histograms to display and analyze data.

Technology for the Teacher

Dynamic Classroom

Lesson Plans

Complete Materials List

10.3 Record and Practice Journal

Essential Question How can you describe the shape of the distribution of a data set?

1 ACTIVITY: Describing the Shape of a Distribution

Work with a partner. The lists at the right show the last four digits of a set of phone numbers in a phone book.

-7253, -7290, -7200, -1192, -1142, -3500, -2531, -2079, -5897, -5341, -1392, -5406, -7875, -7335, -0494, -9018, -2184, -2367

-8678, -2063, -2911, -2103, -4328, -7826, -7957, -7246, -2119, -7845, -1109, -9154

a. Create a list that represents the last digit of each phone number shown. Make a dot plot of the data.

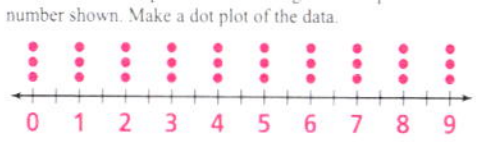

b. In your own words, how would you describe the shape of the distribution? What single word do you think you can use to identify this type of distribution? Explain your reasoning.

flat or rectangular

2 ACTIVITY: Describing the Shape of a Distribution

Work with a partner. The lists at the right show the first three digits of a set of phone numbers in a phone book.

538-, 438-, 664-, 761-, 868-, 735-, 694-, 599-, 725-, 556-, 555-, 456-, 736-, 664-, 576-, 664-, 664-, 725-

664-, 664-, 538-, 855-, 664-, 538-, 654-, 654-, 725-, 538-, 799-, 764-

a. Create a list that shows the first digit of each phone number shown. Make a dot plot of the data.

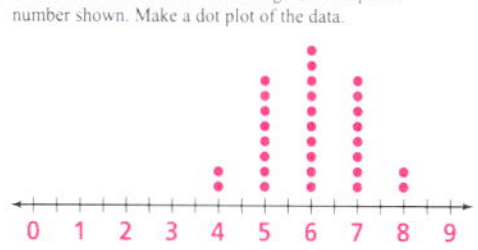

Differentiated Instruction

Inclusion

In this section, students are expected to make histograms. Provide a resource page of blank axes with horizontal lines and vertical lines. Students add the vertical and horizontal scales and labels and color the bars to represent the data.

10.3 Record and Practice Journal

b. In your own words, how would you describe the shape of the distribution? What single word do you think you can use to identify this type of distribution? Explain your reasoning.

symmetric or reflective

c. In your dot plot, draw a vertical line through the middle of the data set. What do you notice?

The line divides the data set into two halves that are mirror images of each other.

d. Repeat part (c) for the dot plot you constructed in Activity 1. What do you notice? Compare the distributions from Activities 1 and 2.

You can draw a vertical line through the middle of the data to get two halves that are mirror images of each other.

3 ACTIVITY: Describing the Shape of a Distribution

Work with a partner. The table shows the ages of cellular phones owned by a group of students.

Ages of Cellular Phones (years)				
0	1	0	6	4
2	3	5	1	1
0	1	2	3	1
0	0	1	1	1
7	1	4	2	2
0	2	0	1	2

a. Make a dot plot of the data.

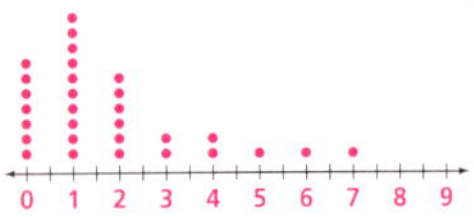

b. In your own words, how would you describe the shape of the distribution? Compare it to the distributions in Activities 1 and 2.

Most of the data are on the left.

c. Why do you think this type of distribution is called a *skewed distribution*?

The distribution is "slanted" in a certain direction.

4 ACTIVITY: Finding Measures of Center

Work with a partner.

a. Find the mean and median of the data sets in Activities 1–3.

Activity 1	Activity 2	Activity 3
mean: 4.5	mean: 6	mean: 1.8
median: 4.5	median: 6	median: 1

b. What do you notice about the means and medians of the data sets and the shapes of the distributions? Explain.

symmetric - Mean and median are the same.
skewed - Mean and median will vary.

c. Which measure of center do you think best describes the data set in Activity 2? in Activity 3? Explain your reasoning.

Activity 2: mean
Activity 3: median

d. Using your answers to part (c), decide which measure of variation you think best describes the data set in Activity 2. Which measure of variation do you think best describes the data set in Activity 3? Explain your reasoning.

Activity 2: MAD
Activity 3: IQR

What Is Your Answer?

5. **IN YOUR OWN WORDS** How can you describe the shape of the distribution of a data set?

identify peaks, gaps, and outliers; look at the distribution

6. Name two other ways you can describe the distribution of a data set.

You can describe the center and the variation of a data distribution.

Laurie's Notes

Activity 3

- Discuss the meaning of *skewed*. Look around the classroom for an example of something that is skewed.

? "What does an age of 0 mean?" a fairly new cell phone

- After students finish the activity, ask a volunteer to share his or her dot plot with the class.
- **MP3:** Listen to student reasoning regarding why this type of distribution is called a *skewed distribution*.

? "Can you think of other data sets that have skewed distributions?" Answers will vary.

- Mention that distributions can have many different shapes. Point out that the first three activities identify some of the more common shapes.

Activity 4

- This activity could be optional if you are not teaching Extension 10.3.
- Review mean and median from Chapter 9.
- Encourage students to recognize and use shortcuts to find the means. For instance, in Activity 1, because there are 3 of each number, students can multiply each value by 3 and then add each product to find the sum of the data. For the data in Activity 2, ask whether anyone can tell "by inspection" that the mean is 6.

? "How can you find the median in a dot plot?" Because the data values in a dot plot are in order, students should describe moving from the least and greatest numbers toward the center of the distribution.

- In answering part (c), students may notice that the mean and median are equal for the data sets in Activities 1 and 2. If students are stuck trying to come up with an answer, then suggest that the mean uses all of the data values in its calculation, so it could be considered more reliable. In Activity 3, as well as the median being equal to 1, the mode is 1 and most of the data are clustered around 1. The mean is more affected by the outliers than the median, so the median should be used.
- Student answers to part (c) will influence their answers to part (d).
- To help students with part (d), review interquartile range and mean absolute deviation from Chapter 9.

What Is Your Answer?

- Question 6 helps students understand that a distribution has a visual component in addition to numeric values.

Closure

- Sketch a dot plot that has a symmetric distribution and a mean of 4.

Sample answer:

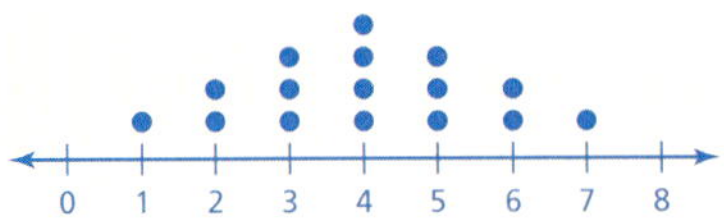

The Meaning of a Word • Skewed

When something is **skewed**, it has a slanted direction or position.

3 ACTIVITY: Describing the Shape of a Distribution

Work with a partner. The table shows the ages of cellular phones owned by a group of students.

Ages of Cellular Phones (years)				
0	1	0	6	4
2	3	5	1	1
0	1	2	3	1
0	0	1	1	1
7	1	4	2	2
0	2	0	1	2

a. Make a dot plot of the data.

b. In your own words, how would you describe the shape of the distribution? Compare it to the distributions in Activities 1 and 2.

c. Why do you think this type of distribution is called a *skewed distribution*?

4 ACTIVITY: Finding Measures of Center

Math Practice 3

Use Prior Results

How is the distribution of the data related to the mean and the median?

Work with a partner.

a. Find the means and the medians of the data sets in Activities 1–3.

b. What do you notice about the means and the medians of the data sets and the shapes of the distributions? Explain.

c. Which measure of center do you think best describes the data set in Activity 2? in Activity 3? Explain your reasoning.

d. Using your answers to part (c), decide which measure of variation you think best describes the data set in Activity 2. Which measure of variation do you think best describes the data set in Activity 3? Explain your reasoning.

What Is Your Answer?

5. **IN YOUR OWN WORDS** How can you describe the shape of the distribution of a data set?

6. Name two other ways you can describe the distribution of a data set.

Practice Use what you learned about shapes of distributions to complete Exercises 3 and 4 on page 454.

10.3 Lesson

Check It Out
Lesson Tutorials
BigIdeasMath.com

You can use dot plots and histograms to identify shapes of distributions.

Key Ideas

Symmetric and Skewed Distributions

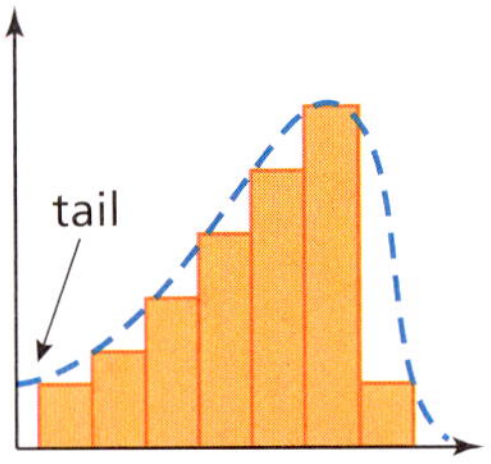

Skewed left

- The "tail" of the graph extends to the left.
- Most data are on the right.

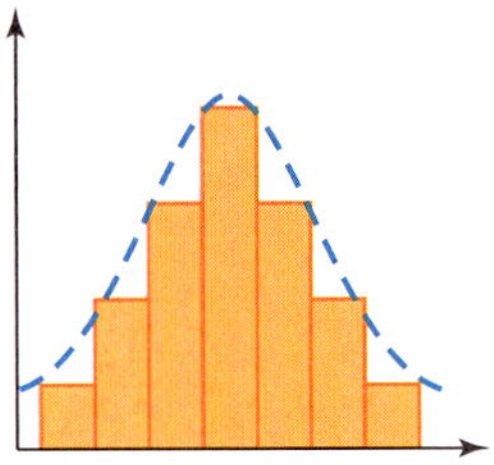

Symmetric

- The left side of the graph is a mirror image of the right side of the graph.

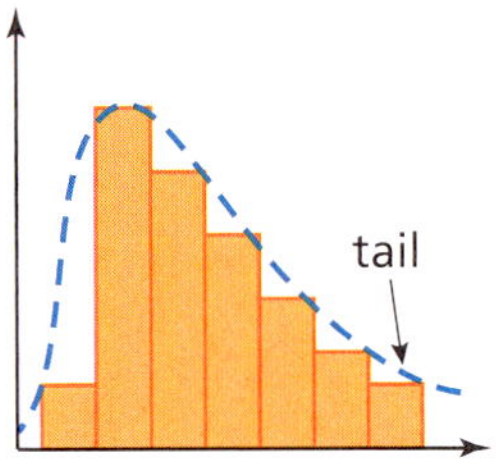

Skewed right

- The "tail" of the graph extends to the right.
- Most data are on the left.

Study Tip

If all the dots of a dot plot or bars of a histogram are about the same height, then the distribution is a *flat*, or *uniform*, distribution. A uniform distribution is also symmetric.

EXAMPLE 1 Describing the Shapes of Distributions

Describe the shape of each distribution.

a.

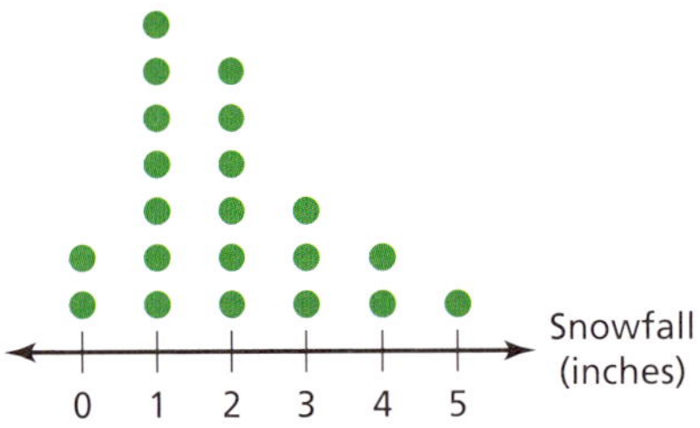

Most of the data are on the left, and the tail extends to the right.

So, the distribution is skewed right.

b.

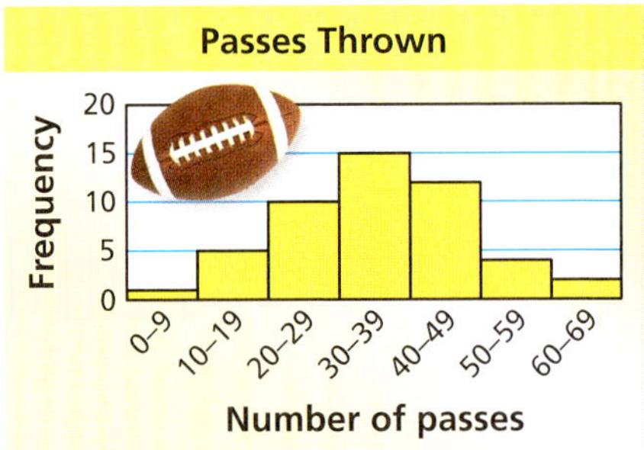

The left side of the graph is approximately a mirror image of the right side of the graph.

So, the distribution is symmetric.

On Your Own

Now You're Ready
Exercises 5–8

1. Describe the shape of the distribution.

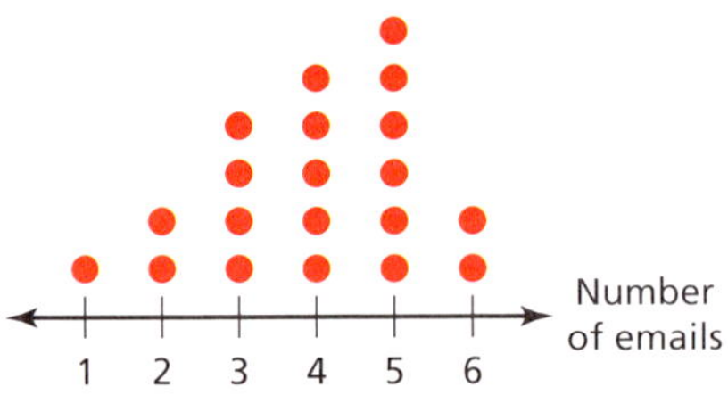

Laurie's Notes

Introduction

Connect

- **Yesterday:** Students made dot plots and explored the shapes of distributions. (MP3)
- **Today:** Students will describe and compare shapes of distributions.

Motivate

- **Story time:** Tell the students that yesterday in the teacher's lounge, three teachers were describing the results of a recent test they gave.
  - **Teacher A:** Lots of low scores; Ds and Fs, some Cs, a few Bs, and one A
  - **Teacher B:** Lots of high scores; As and Bs, some Cs, a few Ds, and one F
  - **Teacher C:** Lots of average scores; Cs, some Bs and Ds, and a few Fs and As
- Ask students to quickly sketch a histogram for each of the three sets of test scores. They should sketch something similar to the three distributions in the Key Ideas.
- Discuss the attributes of a histogram: similar to a bar graph, intervals are all the same size, and frequencies of data values are displayed in the vertical direction by the bar heights.

Lesson Notes

Key Ideas

- Draw a sketch of each type of distribution and label it: skewed left, symmetric, and skewed right.
- Connect the distributions to the test score descriptions above.
- Explain that a dashed line can be drawn to help identify the greatest frequency and to help highlight the basic shape of a distribution.

? "Do all distributions fall into one of these three categories?" no; There are many different types of distributions.

- Skewed left is also called *negatively* skewed and skewed right is also called *positively* skewed. This may help students remember by relating skewness to the number line: the negative direction is to the left and the positive direction is to the right.

Example 1

? "How many days are represented by the distribution in part (a)?" 21

- **MP4 Model with Mathematics:** Ask students to make a statement about the data represented by the histogram in part (b). Answers will vary.

On Your Own

- **Neighbor Check:** Have students work independently and then have their neighbors check their work. Have students discuss any discrepancies.
- **Extension:** If time allows, ask questions about the mean, median, mode, and number of pieces of data.

Goal Today's lesson is describing and comparing shapes of distributions.

Extra Example 1

Describe the shape of each distribution.

a. Monthly Rainfall Amounts

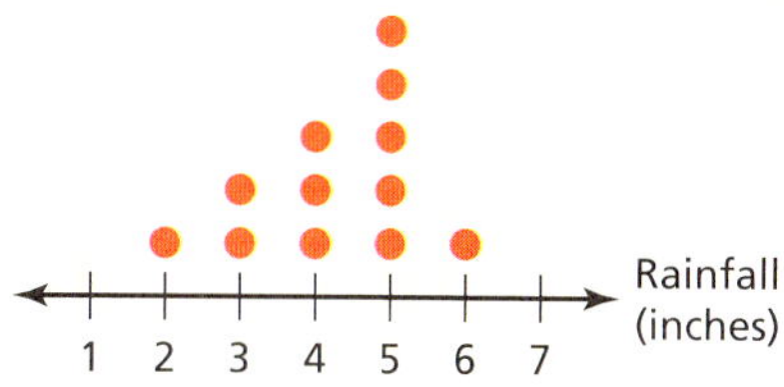

skewed left

b.

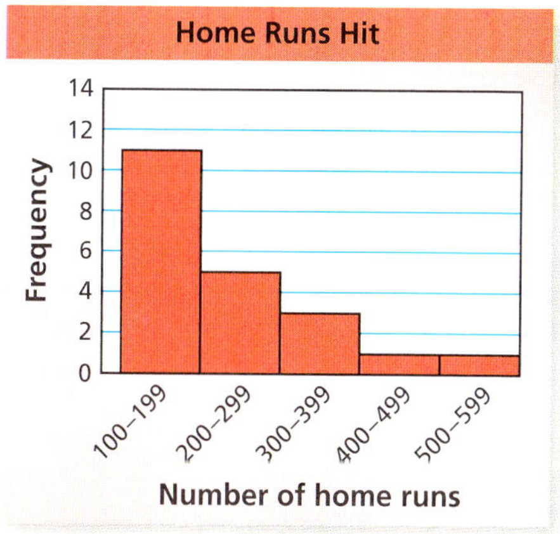

skewed right

On Your Own

1. skewed left

English Language Learners

Notebook Development

Provide English language learners with a handout of a simplified version of the notes from this lesson. This will allow the student to focus on the concepts being taught. Students should highlight information and add additional notes to the handout during the classroom discussion. They can refer to the page as they work through the exercises.

Extra Example 2

The frequency table shows the quiz scores (in percent) in a science class. Display the data in a histogram. Describe the shape of the distribution.

Scores (percent)	Frequency
75–79	4
80–84	6
85–89	11
90–94	7
95–100	4

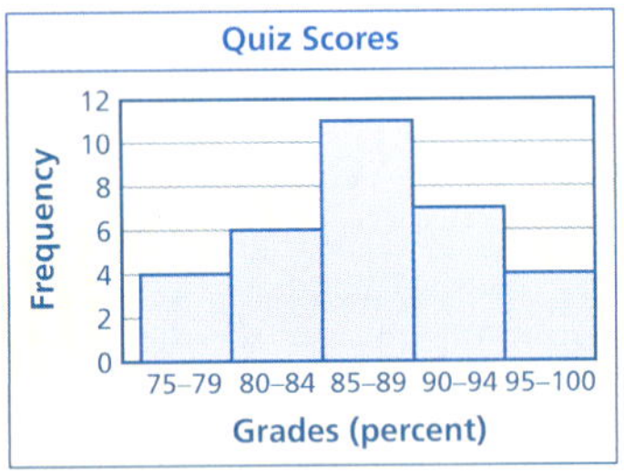

symmetric distribution

Extra Example 3

The histogram shows the quiz scores (in percent) in a social studies class in the same school as in Extra Example 2.

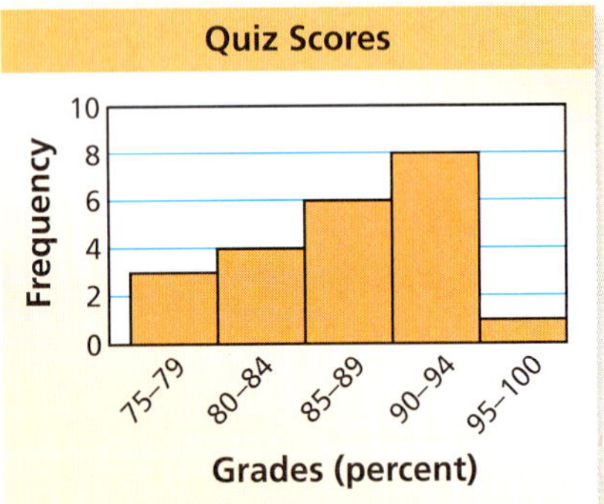

a. Describe the shape of the distribution. skewed left

b. Which class has the better grade? the social studies class

On Your Own

2. See Additional Answers.

Laurie's Notes

Example 2

- This is a good review of drawing a histogram. The intervals are given, which are often difficult for students to determine.
- Ask for a volunteer to read the values in the frequency table as you write them. Have students describe the process for creating the histogram.
- Ask a few students to do their work on transparencies if a document camera is not available.

Example 3

- Discuss the data represented in the histogram.
- **MP3 Construct Viable Arguments and Critique the Reasoning of Others:** The explanations that students offer provide additional practice with constructing viable arguments. Students should mention that the intervals in both histograms are the same.

On Your Own

- After students have finished, ask a volunteer to share his or her histogram at the board or under a document camera.
- Students should observe that the intervals are different in Question 2 than in Examples 2 and 3.

Closure

- **Exit Ticket:** Return to the example in the Motivate section. A fourth teacher describes the results of a test: lots of As and Fs, some Bs and Ds, and a few Cs. Sketch a histogram for the test scores. How would you describe the distribution? *Sample answer:* The distribution is symmetric, but most of the data does not lie near the mean.

EXAMPLE 2 Describing the Shape of a Distribution

Ages	Frequency
10–13	1
14–17	3
18–21	7
22–25	12
26–29	20
30–33	18
34–37	3

The frequency table shows the ages of people watching a comedy in a theater. Display the data in a histogram. Describe the shape of the distribution.

Draw and label the axes. Then draw a bar to represent the frequency of each interval.

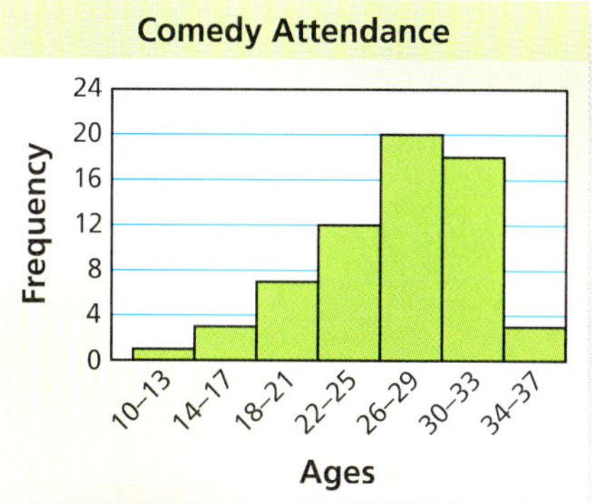

Most of the data are on the right, and the tail extends to the left.

So, the distribution is skewed left.

EXAMPLE 3 Comparing Shapes of Distributions

The histogram shows the ages of people watching an animated movie in the same theater as in Example 2.

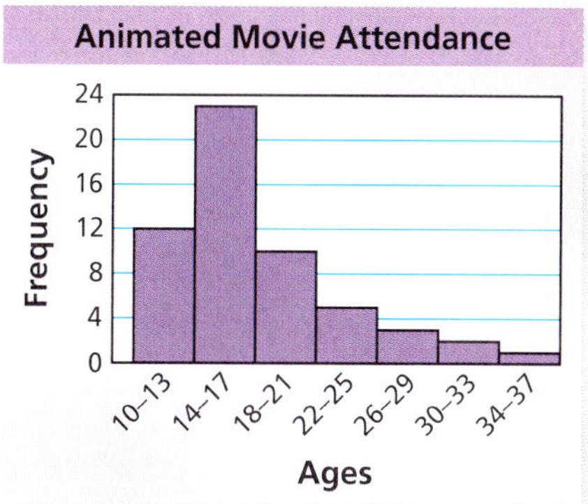

a. Describe the shape of the distribution.

Most of the data are on the left, and the tail extends to the right.

So, the distribution is skewed right.

b. Which movie has an older audience?

The intervals in the histograms are the same. Most of the data for the animated movie are on the left, while most of the data for the comedy are on the right. This means that the people watching the comedy are generally older than the people watching the animated movie.

So, the comedy has an older audience.

On Your Own

Now You're Ready
Exercise 9

2. The frequency table shows the ages of people watching a historical movie in a theater.

Ages	10–19	20–29	30–39	40–49	50–59	60–69
Frequency	3	18	36	40	14	5

a. Display the data in a histogram. Describe the shape of the distribution.

b. Compare the distribution of the data to the distributions in Examples 2 and 3. What can you conclude?

10.3 Exercises

Check It Out
Help with Homework
BigIdeasMath.com

Vocabulary and Concept Check

1. **VOCABULARY** How does the shape of a symmetric distribution differ from the shape of a skewed distribution?

2. **VOCABULARY** For a distribution that is skewed right, which direction does the tail extend? Where do most of the data lie?

Practice and Problem Solving

Make a dot plot of the data. In your own words, how would you describe the shape of the distribution?

3.

Miles Run per Day										
1	4	2	0	3	2	1	2	4	2	3
2	1	6	3	2	4	0	5	3	1	5

4.

Raffle Tickets Sold							
15	12	16	15	13	14	16	13
13	16	14	12	15	12	14	

Describe the shape of each distribution.

1 5.

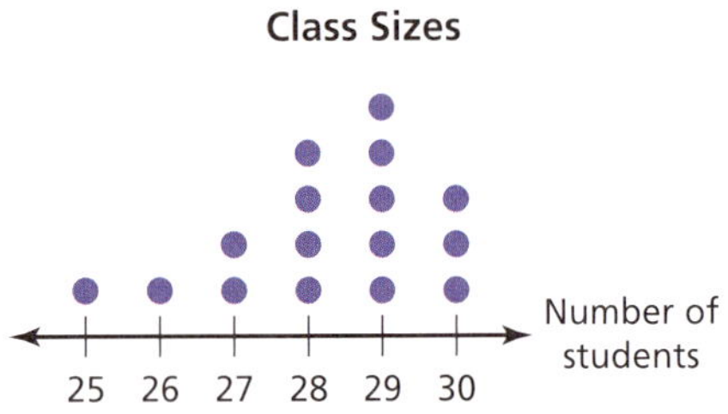

6.

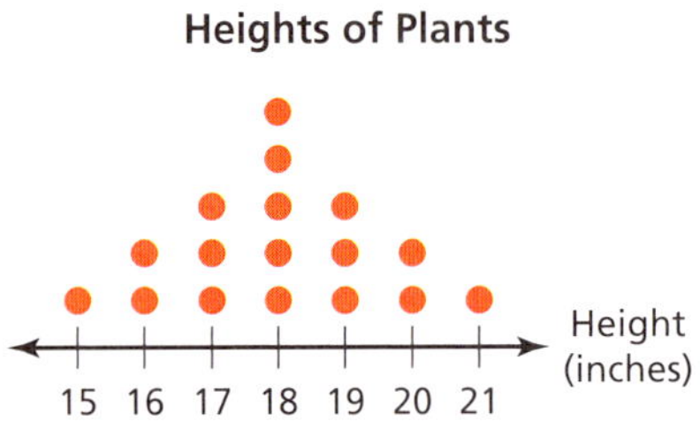

7.

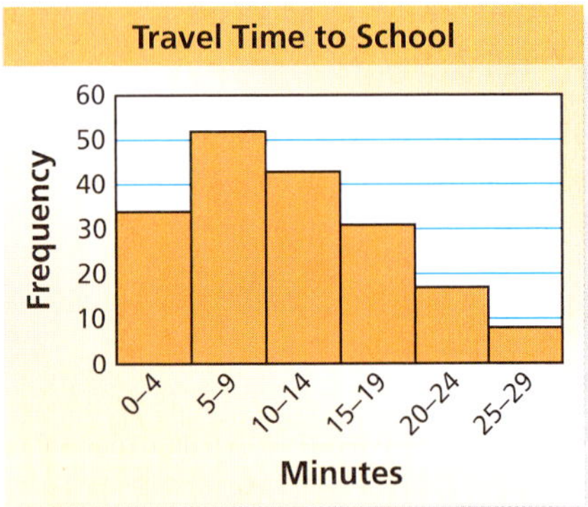

8.

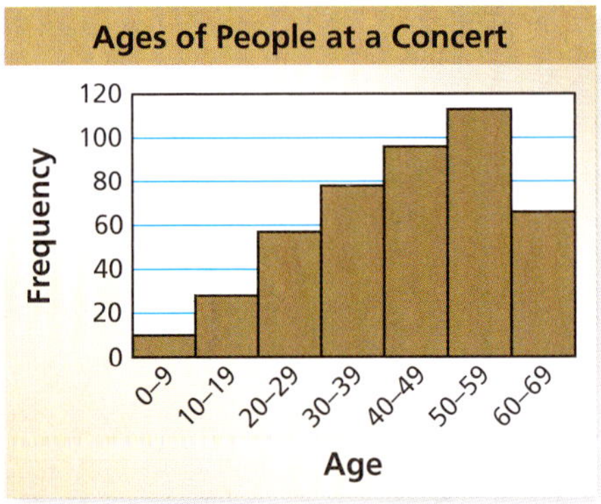

2 3 9. **POLICE** The frequency table shows the years of service for the police officers of Jones County and Pine County. Display the data for each county in a histogram. Describe the shape of each distribution. Which county's police force has less experience? Explain.

Years of Service	0–3	4–7	8–11	12–15	16–19	20–23	24–27
Frequency for Jones County	7	15	17	12	8	5	3
Frequency for Pine County	3	5	9	14	10	6	2

Assignment Guide and Homework Check

Level	Day 1 Activity Assignment	Day 2 Lesson Assignment	Homework Check
Basic	3, 4, 15–17	1, 2, 5–13 odd	5, 7, 9, 11
Average	3, 4, 15–17	1, 2, 5–9 odd, 10–14	5, 7, 12, 13
Advanced	3, 4, 15–17	1, 2, 5–14	6, 7, 8, 11, 13

Common Errors

- **Exercises 3–10** Students may confuse "skewed left" and "skewed right." Point out that the "tail" of the graph of a distribution determines the direction of the skew. That is, in a skewed left distribution, the tail extends to the left, and in a skewed right distribution, the tail extends to the right.

10.3 Record and Practice Journal

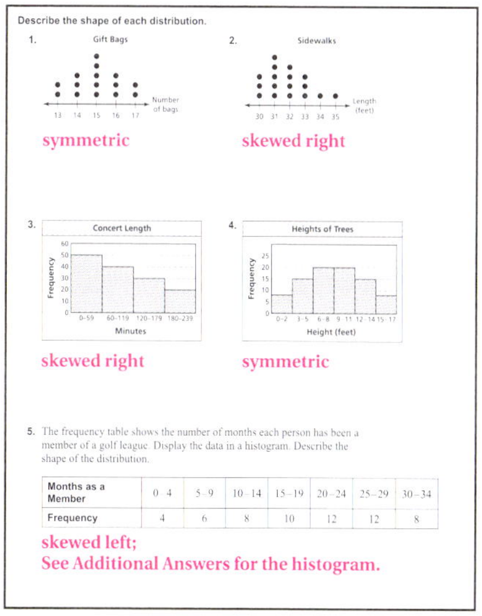
Describe the shape of each distribution.

1. Gift Bags — symmetric

2. Sidewalks — skewed right

3. Concert Length — skewed right

4. Heights of Trees — symmetric

5. The frequency table shows the number of months each person has been a member of a golf league. Display the data in a histogram. Describe the shape of the distribution.

Months as a Member	0–4	5–9	10–14	15–19	20–24	25–29	30–34
Frequency	4	6	8	10	12	12	8

skewed left;
See Additional Answers for the histogram.

Vocabulary and Concept Check

1. The shape of a skewed distribution will have a tail on one side. The shape of a symmetric distribution is when the data on the left are a mirror image of the data on the right.
2. to the right; on the left

Practice and Problem Solving

3–4. See Additional Answers.

5. skewed left
6. symmetric
7. skewed right
8. skewed left
9.

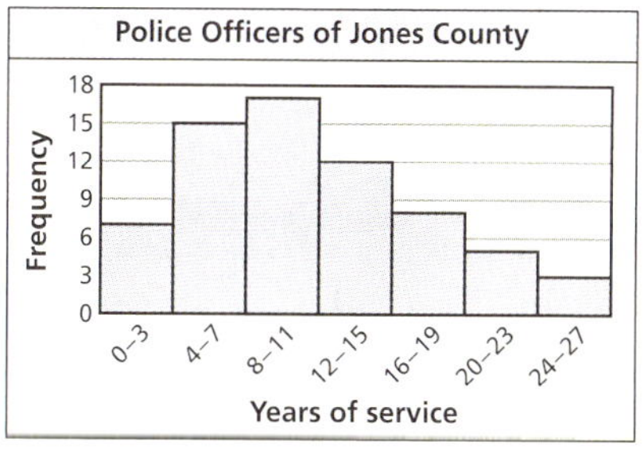

skewed right

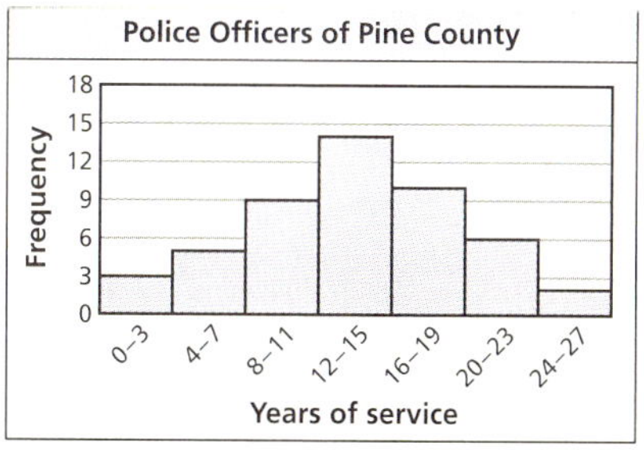

symmetric

Jones County; The distribution of Jones County is skewed right, so most of the data values are on the left.

Practice and Problem Solving

10. symmetric; The data on the left is a mirror image of the data on the right.

11–13. See Additional Answers.

14. See *Taking Math Deeper.*

Fair Game Review

15. median = 70; $Q_1 = 65.5$; $Q_3 = 75$; IQR = 9.5

16. median = 37.5; $Q_1 = 30$; $Q_3 = 44$; IQR = 14

17. A

Mini-Assessment

The frequency table shows the ages of people in attendance at a baseball game.

Ages	Frequency
0–9	25
10–19	30
20–29	58
30–39	60
40–49	32
50–59	21
60–69	16
70–79	8
80–89	4

1. Display the data in a histogram.

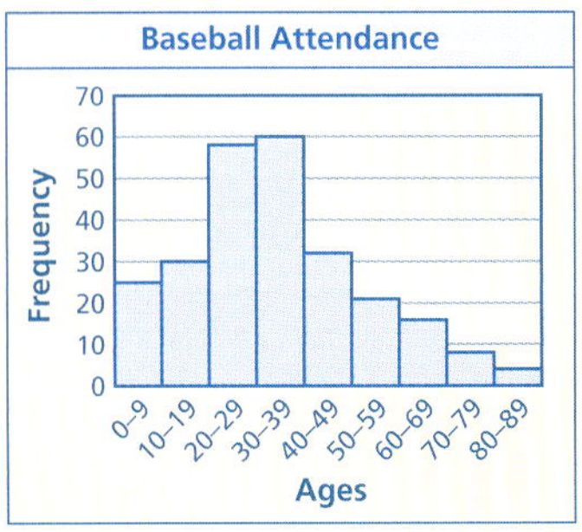

2. Describe the shape of the distribution. skewed right

Taking Math Deeper

Exercise 14

The key to this problem is distinguishing between the mean and the median in the two skewed distributions.

1 Identify the mode of each data set.

The mode is the value that occurs most often. So, the tallest bar in a bar graph represents the mode. The tallest bar is represented by the letter A in part (a) and by the letter C in part (b).

2 Identify the mean and the median of each data set.

In part (a), there are 71 values in the data set. The median is the 36th value, which occurs in the third bar and is represented by the letter B. The letter C represents the mean because the greater values on the tail extending to the right lead to a mean that is greater than the median.

In part (b), there are 107 values in the data set. The median is the 54th value, which occurs in the seventh bar and is represented by the letter B. The letter A represents the mean because the lesser values on the tail extending to the left lead to a mean that is less than the median.

3 Answer the question.

a. Most of the data are on the left, and the tail extends to the right. So, the distribution is skewed right.

A: Mode B: Median C: Mean

b. Most of the data are on the right, and the tail extends to the left. So, the distribution is skewed left.

A: Mean B: Median C: Mode

Project

Does the tail of the data in a skewed distribution "pull" the median and mean towards the tail or away from it? Which measure does it pull more? Provide data to justify your reasoning.

Reteaching and Enrichment Strategies

If students need help. . .	If students got it. . .
Resources by Chapter • Practice A and Practice B • Puzzle Time Record and Practice Journal Practice Differentiating the Lesson Lesson Tutorials Skills Review Handbook	Resources by Chapter • Enrichment and Extension • Technology Connection Start the next section

10. **REASONING** What is the shape of the distribution of the restaurant waiting times? Explain your reasoning.

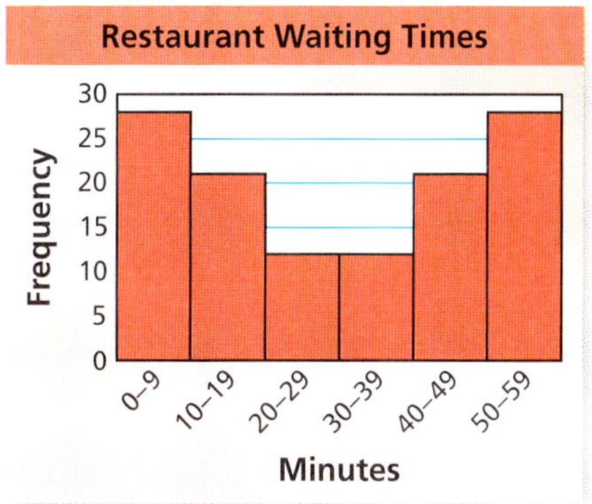

11. **LOGIC** Are all distributions either approximately symmetric or skewed? Explain. If not, give an example.

12. **REASONING** Can you use a stem-and-leaf plot to describe the shape of a distribution? Explain your reasoning.

13. **CHARITY** The table shows the donation amounts received by a charity in one day.

Donations (dollars)												
20	15	40	70	20	5	25	50	47	20	62	55	40
10	50	18	20	100	40	80	60	20	80	3	30	50
25	30	10	33	20	50	7	35	40	25	70		

a. Make a histogram of the data starting with the interval 0–14. Describe the shape of the distribution.

b. A company adds $5 to each donation. Make another histogram starting with the same first interval as in part (a). Compare the shape of this distribution with the distribution in part (a). Explain any differences in the distributions.

14. Critical Thinking Describe the shape of the distribution of each bar graph. Match the letters A, B, and C with the mean, the median, and the mode of the data set. Explain your reasoning.

a.

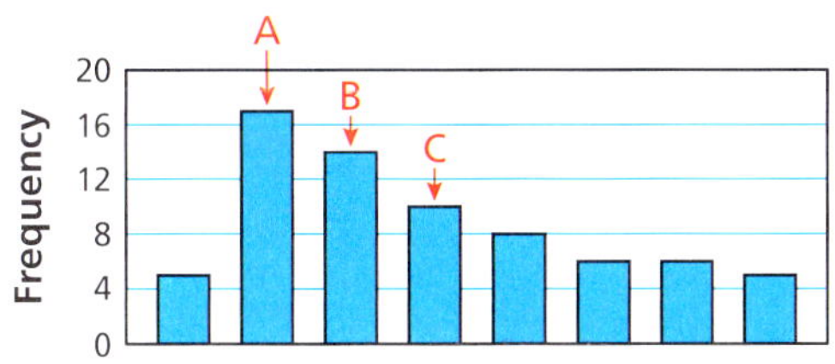

b.

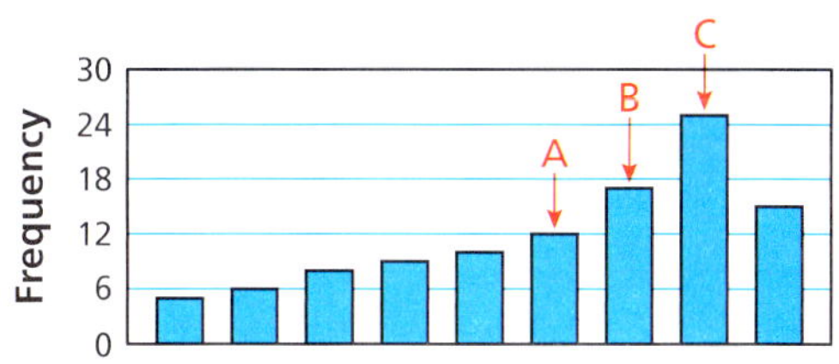

Fair Game Review What you learned in previous grades & lessons

Find the median, first quartile, third quartile, and interquartile range of the data. *(Section 9.4)*

15. 68, 74, 67, 72, 63, 70, 78, 64, 76

16. 39, 48, 33, 24, 30, 44, 36, 41, 28, 53

17. **MULTIPLE CHOICE** Sixty people participate in a trivia contest. How many four-person teams can be formed? *(Section 7.3)*

Ⓐ 15 Ⓑ 56 Ⓒ 64 Ⓓ 240

Extension 10.3 Choosing Appropriate Measures

Check It Out
Lesson Tutorials
BigIdeasMath.com

You can use a measure of center and a measure of variation to describe the distribution of a data set. The shape of the distribution can help you choose which measures are the most appropriate to use.

Key Idea

Choosing Appropriate Measures

The mean absolute deviation (MAD) uses the mean in its calculation. So, when a data distribution is *symmetric*,

- use the mean to describe the center and
- use the MAD to describe the variation.

The interquartile range (IQR) uses quartiles in its calculation. So, when a data distribution is *skewed*,

- use the median to describe the center and
- use the IQR to describe the variation.

Math Practice 2

Understand Quantities

What effect can outliers have on the mean? on the median? Explain.

EXAMPLE 1 Choosing Appropriate Measures

The dot plot shows the average number of hours students in a class sleep each night.

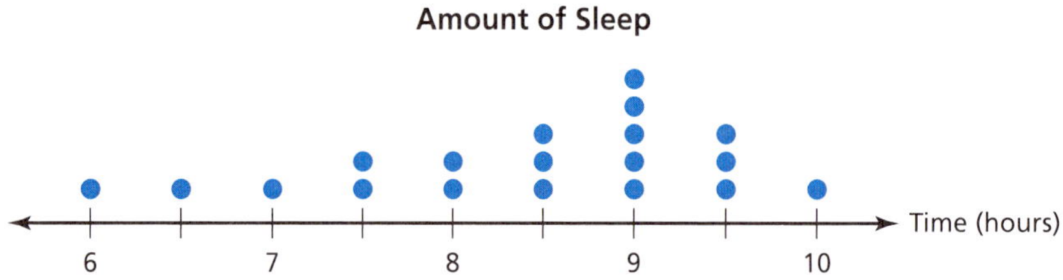

a. What are the most appropriate measures to describe the center and the variation?

Most of the data values are on the right clustered around 9, and the tail extends to the left. The distribution is skewed left.

So, the median and the interquartile range are the most appropriate measures to describe the center and the variation.

b. Describe the center and the variation of the data set.

The median is 8.5 hours. The first quartile is 7.5, and the third quartile is 9. So, the interquartile range is $9 - 7.5 = 1.5$ hours.

The data are centered around 8.5 hours. The middle half of the data varies by no more than 1.5 hours.

Common Core

Data Displays

In this extension, you will

- choose appropriate measures of center and variation to represent data sets.

Learning Standard
6.SP.5d

Laurie's Notes

Introduction

Connect

- **Yesterday:** Students described and compared shapes of distributions. (MP3, MP4)
- **Today:** Students will use the shape of a distribution to choose the most appropriate measures to describe the center and the variation.

Motivate

? "How many states does [insert your state] border?" Answers will vary.

? "How many states border 0 or 1 other states? Explain." 3; Hawaii and Alaska border 0 other states and Maine borders 1 other state.

- Explain that one of the examples in today's lesson will explore a data set involving state borders.

Discuss

- Introduce this lesson by asking how students can describe the distribution of a data set. Listen for an understanding of the measure of center (mean, median, mode), the measure of variation (mean absolute deviation, interquartile range), and the shape of the distribution (symmetric, skewed).
- Acknowledge the need to keep track of lots of vocabulary and definitions.
- In this lesson, the goal is to use the shape of a distribution to choose the most appropriate measures to describe the center and the variation.

Lesson Notes

Key Idea

- Write the Key Idea. In choosing the appropriate measures, first determine the shape of the distribution, symmetric or skewed. Students should know that the shape does not need to be either of these, but only these types of shapes are considered here.
- **MP2 Reason Abstractly and Quantitatively:** Discuss how outliers can affect mean and median. Both symmetric and skewed distributions can have outliers.

Example 1

? "What is the average number of hours of sleep you get each night?" Answers will vary.

- Sketch the dot plot shown.

? "How many students does this plot represent?" 19 "What is the range?" 4 hours "What is the mode?" 9 hours

? "What are the most appropriate measures to describe the center and variation? Explain." median and interquartile range; The distribution is skewed (left).

- Work through the steps in finding the median and the interquartile range. Summarize the results as shown.

Common Core State Standards

6.SP.5d Summarize numerical data sets in relation to their context, such as by relating the choice of measures of center and variability to the shape of the data distribution and the context in which the data were gathered.

Goal Today's lesson is choosing the most appropriate measures to describe the center and the variation.

Technology for the Teacher

Dynamic Classroom

Lesson Tutorials
Lesson Plans
Answer Presentation Tool

Extra Example 1

The dot plot shows the average number of hours students in a class exercise each day.

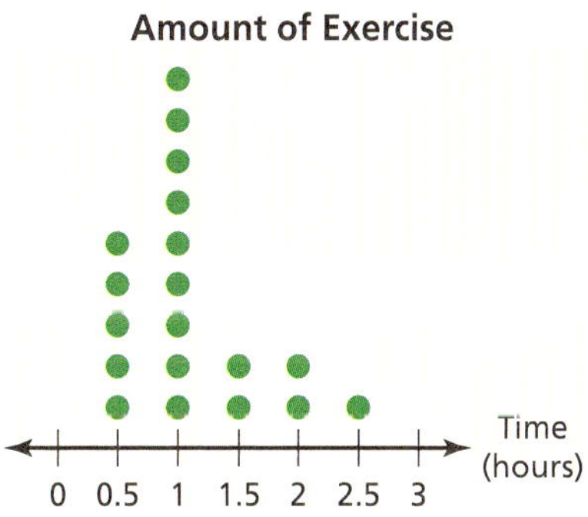

a. What are the most appropriate measures to describe the center and the variation? median and interquartile range

b. Describe the center and the variation of the data set. The data are centered around 1 hour. The middle half of the data vary by no more than 1 hour.

Record and Practice Journal Extension 10.3 Practice

1. The distribution is skewed left.; median and interquartile range; median: $19.5; interquartile range: $2

2–5. See Additional Answers.

Extra Example 2

The frequency table shows the gas mileages of several vehicles made by a company.

Mileage (miles per gallon)	Frequency
15–17	2
18–20	4
21–23	3
24–26	2
27–29	1

a. Display the data in a histogram.

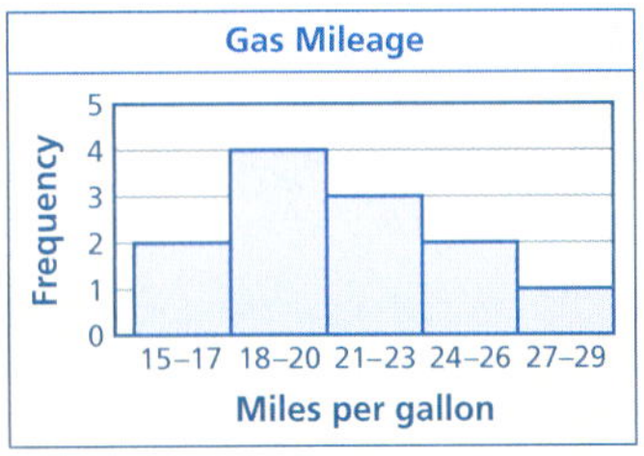

b. What are the most appropriate measures to describe the center and the variation? median and interquartile range

Practice

1–5. See Additional Answers.

Mini-Assessment

The dot plot shows the average number of minutes students in a class exercise every day.

Amount of Exercise

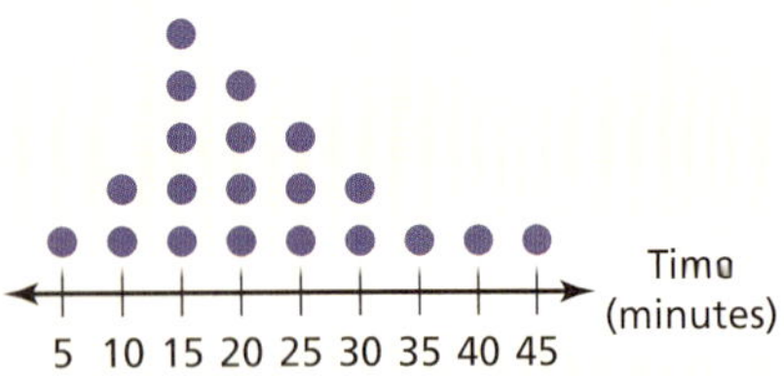

1. What are the most appropriate measures to describe the center and the variation? median and interquartile range

2. Describe the center and the variation of the data set. The data are centered around 20 minutes. The middle half of the data vary by no more than 12.5 minutes.

Laurie's Notes

Example 2

- Draw the frequency table.
- ? "Where does [insert your state] fall in the frequency table?" Answers will vary.
- ? "How many values does the data set have?" 50
- **FYI:** Tennessee and Missouri both border 8 other states.
- Ask students to construct the corresponding histogram.
- ? "What are the most appropriate measures to describe the center and variation? Explain." mean and mean absolute deviation; The distribution is symmetric.
- **MP2:** Students should be able to explain why it is not possible to find exact values of the mean and the mean absolute deviation for the data. Question 3 in the Practice Exercises asks students to explain whether they can find these values.

Practice

- **Think-Pair-Share:** Students should read each question independently and then work in pairs to answer the questions. When they have answered the questions, the pair should compare their answers with another group and discuss any discrepancies.

Closure

- Summarize how to use the shape of a distribution to choose the most appropriate measures to describe the center and the variation. When the data is symmetric, use the mean and the mean absolute deviation. When the data is skewed, use the median and interquartile range.

EXAMPLE 2 Choosing Appropriate Measures

Bordering States	Frequency
0–1	3
2–3	13
4–5	22
6–7	10
8–9	2

The frequency table shows the number of states that border each state in the United States.

a. Display the data in a histogram.

Draw and label the axes. Then draw a bar to represent the frequency of each interval.

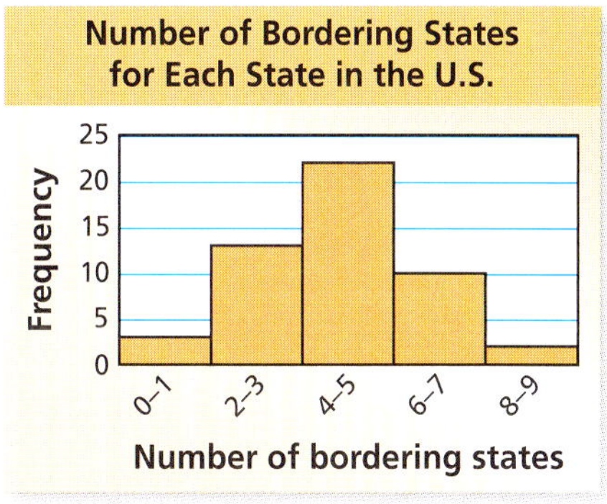

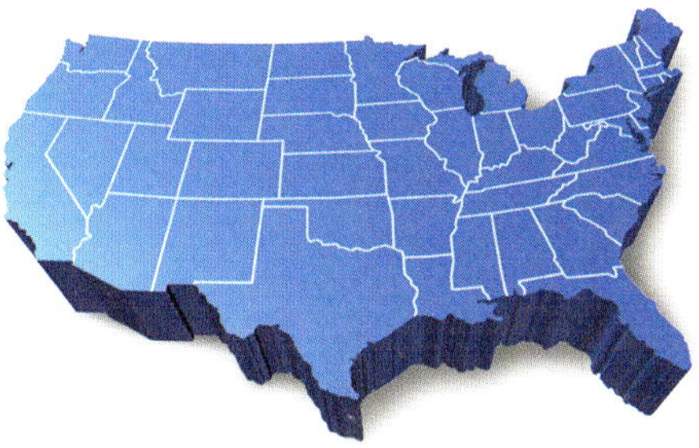

b. What are the most appropriate measures to describe the center and the variation?

The left side of the graph is approximately a mirror image of the right side of the graph. The distribution is symmetric.

So, the mean and the mean absolute deviation are the most appropriate measures to describe the center and the variation.

Practice

Choose the most appropriate measures to describe the center and the variation. Find the measures you chose.

1.

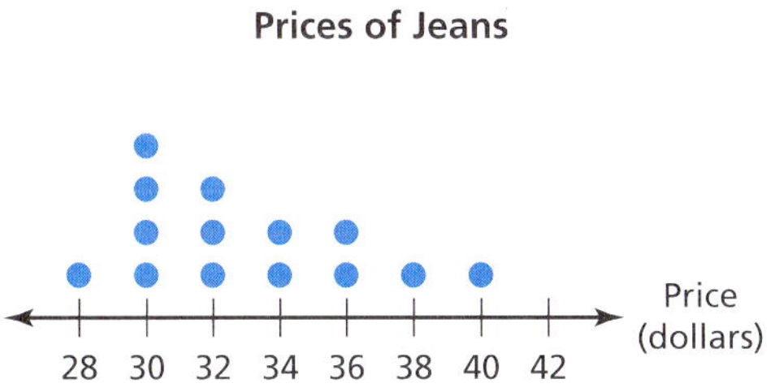

2.

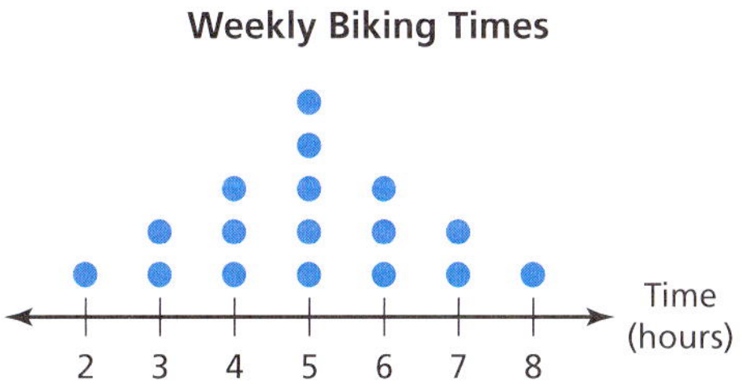

3. REASONING Can you find the exact values of the mean and the mean absolute deviation for the data in Example 2? Explain.

4. GAS MILEAGE The frequency table shows the gas mileages of several vehicles made by a company.

a. What are the most appropriate measures to describe the center and the variation?

b. What conclusions can you make?

Mileage (miles per gallon)	Frequency
10–14	2
15–19	1
20–24	6
25–29	8
30–34	10
35–39	3

5. OPEN-ENDED Construct a dot plot for which the mean is the most appropriate measure to describe the center of the distribution.

10.4 Box-and-Whisker Plots

Essential Question How can you use quartiles to represent data graphically?

1 ACTIVITY: Drawing a Box-and-Whisker Plot

Work with a partner.

The numbers of pairs of footwear owned by each student in a sixth grade class are shown.

Numbers of Pairs of Footwear			
2	5	12	3
7	2	4	6
14	10	6	28
5	3	2	4
9	25	4	10
8	15	5	8

a. Order the data set from least to greatest. Then write the data on a strip of grid paper with 24 boxes.

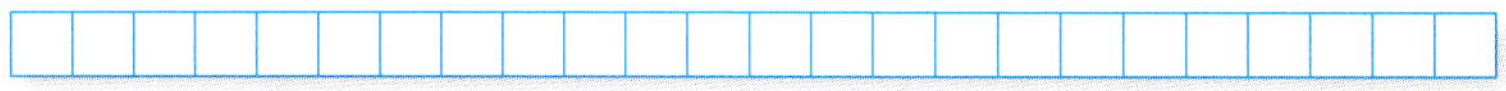

b. Use the strip of grid paper to find the median, the first quartile, and the third quartile. Identify the least value and the greatest value in the data set.

c. Graph the five numbers that you found in part (b) on the number line below.

d. The data display shown below is called a *box-and-whisker plot*. Fill in the missing labels and numbers. Explain how a box-and-whisker plot uses quartiles to represent the data.

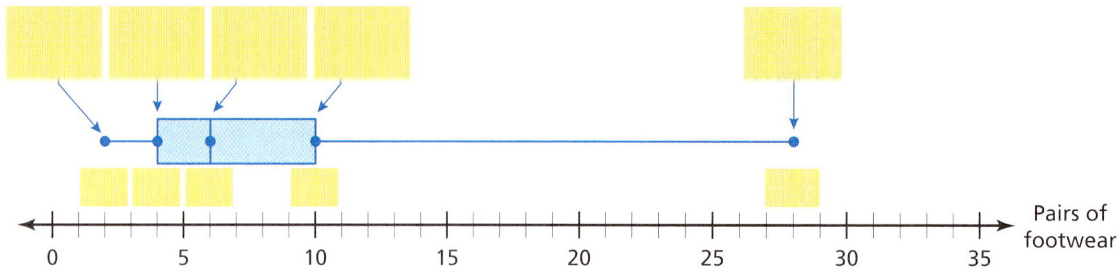

e. Using only the box-and-whisker plot, which measure(s) of center can you find for the data set? Which measure(s) of variation can you find for the data set? Explain your reasoning.

f. Why do you think this type of data display is called a box-and-whisker plot? Explain.

COMMON CORE

Data Displays

In this lesson, you will

- make and interpret box-and-whisker plots.
- compare box-and-whisker plots.

Learning Standards
6.SP.2
6.SP.4
6.SP.5c

Laurie's Notes

Introduction

Standards for Mathematical Practice

- **MP3a Construct Viable Arguments:** Students create box-and-whisker plots to display data sets. Multiple box-and-whisker plots can be displayed on the same number line, therefore students are asked to construct viable arguments in comparing data sets.

Motivate

- Collect class data about the numbers of minutes it took your students to commute to school this morning, from the time they left their front doors until they walked into the school. If this is awkward data to collect, change the question. Data can be collected on slips of paper.
- Record the data on the board and leave it for later. You may want to take time to have students make comments about the data set.

Activity Notes

Discuss

- The box-and-whisker plot is used to display information about very large data sets. However, in order to make learning the technique manageable for students, the actual data sets used today are small.
- Students quickly recognize that box-and-whisker plots involve finding five key values. Constructing the plot is not difficult. Analyzing the plot is the challenging part for students.
- **Connection:** When discussing the meaning of percentiles and quartiles with my students, I had one boy who said, "It's like at the doctor's office when they tell me what percentile I'm in for height and weight." The connection was immediate for the class and *they got it*!

Activity 1

- Explain that today they are going to construct a box-and-whisker plot, a data display that is generally used for very large data sets. Data are *not* graphed, but characteristics of the data set are still conveyed. For instance, the results of a state test for all 6th graders could be displayed using a box-and-whisker plot.
- The box-and-whisker plot uses a number line to visually represent the data.
- **Big Idea:** The 5 numbers graphed summarize the entire data set. The least and greatest values are the boundaries. The median separates the data into two parts. The first (or lower) quartile is the median of the lower half. The third (or upper) quartile is the median of the upper half. The box encloses the middle 50% of the data.

? "What percent of the data is represented by each whisker?" 25%
"How many data values are in each whisker?" 6

- **MP2 Reason Abstractly and Quantitatively:** The above question provides students with the opportunity to reason quantitatively.

Common Core State Standards

6.SP.2 Understand that a set of data collected to answer a statistical question has a distribution which can be described by its center, spread, and overall shape.
6.SP.4 Display numerical data in plots on a number line, including . . . box plots.
6.SP.5c Summarize numerical data sets in relation to their context, such as by giving quantitative measures of center (median and/or mean) and variability (interquartile range and/or mean absolute deviation), as well as describing any overall pattern and any striking deviations from the overall pattern with reference to the context in which the data were gathered.

Previous Learning

Students should know how to find the median, quartiles, and interquartile range of a data set.

Technology for the Teacher

Dynamic Classroom

Lesson Plans
Complete Materials List

10.4 Record and Practice Journal

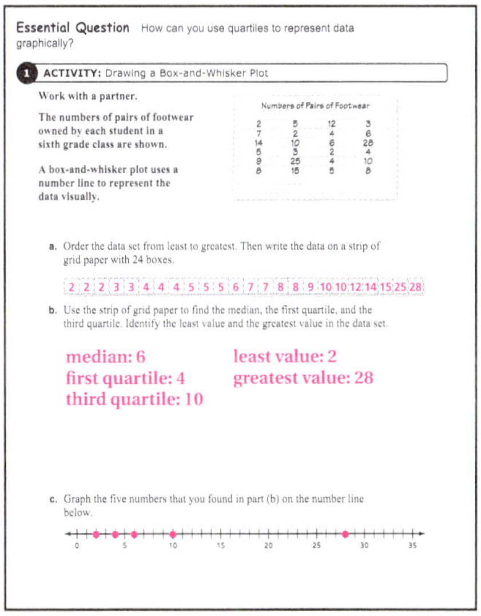

Essential Question How can you use quartiles to represent data graphically?

1 ACTIVITY: Drawing a Box-and-Whisker Plot

Work with a partner.

The numbers of pairs of footwear owned by each student in a sixth grade class are shown.

A box-and-whisker plot uses a number line to represent the data visually.

Numbers of Pairs of Footwear

2	5	12	3
7	2	4	6
14	10	6	28
5	3	2	4
9	25	4	10
8	15	5	8

a. Order the data set from least to greatest. Then write the data on a strip of grid paper with 24 boxes.

2 2 2 3 3 4 4 4 5 5 5 6 7 7 8 8 9 10 10 12 14 15 25 28

b. Use the strip of grid paper to find the median, the first quartile, and the third quartile. Identify the least value and the greatest value in the data set.

median: 6
first quartile: 4
third quartile: 10

least value: 2
greatest value: 28

c. Graph the five numbers that you found in part (b) on the number line below.

English Language Learners

Visual

Use a diagram of a generic box-and-whisker plot on an overhead as a visual aid for English learners. Have students identify the parts of the box-and-whisker plot: *median, first quartile, third quartile, least value,* and *greatest value.* Make sure students understand that they can interpret a box-and-whisker plot that does not have a scale.

10.4 Record and Practice Journal

d. The data display shown below is called a *box-and-whisker plot*. Fill in the missing labels and numbers. Explain how the box-and-whisker plot uses quartiles to represent the data.

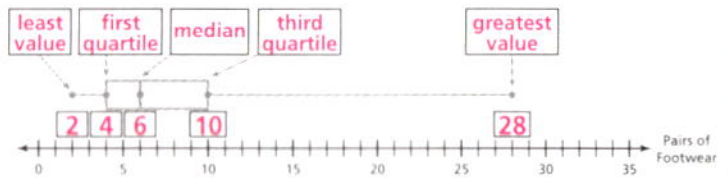

e. Using only the box-and-whisker plot, which measure(s) of center can you find for the data set? Which measure(s) of variation can you find for the data set? Explain your reasoning.

median; range and interquartile range

f. Why do you think this type of data display is called a box-and-whisker plot? Explain.

Rectangle represents the "box."
Line segments represent the "whiskers."

2 ACTIVITY: Conducting a Survey

Have your class conduct a survey. Each student will write on the chalkboard the number of pairs of footwear that he or she owns.

Now, work with a partner to draw a box-and-whisker plot of the data.

Check students' work.

3 ACTIVITY: Reading a Box-and-Whisker Plot

Work with a partner. The box-and-whisker plots show the test score distributions of two sixth grade achievement tests. The same group of students took both tests. The students took one test in the fall and the other in the spring.

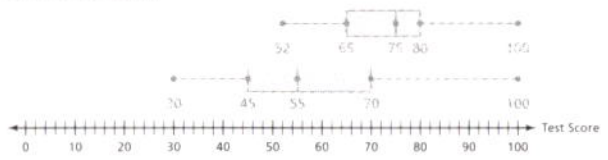

a. Compare and contrast the test results.

Upper plot shows higher scores.

b. Decide which box-and-whisker plot represents the results of which test. How did you make your decision?

spring; fall

What Is Your Answer?

4. IN YOUR OWN WORDS How can you use quartiles to represent data graphically?

You can display how the data are distributed, using the quartiles, the median, and the range.

5. Describe who might be interested in test score distributions like those shown in Activity 3. Explain why it is important for such people to know test score distributions.

Sample answer: **teachers; if they know when scores are lower, they can determine the reason and address it.**

Laurie's Notes

Activity 2

- Explain that you want to practice making a box-and-whisker plot with data collected from students. You will gather and record information about the number of pairs of footwear each of your students owns.
- Students should follow the steps from Activity 1 to construct this plot. If there is an odd number of students in the class, the median is the middle value of the sorted data. To find the first quartile, exclude the median and find the median of the lower half of the data. To find the third quartile, exclude the median and find the median of the upper half of the data.
- As students are making the plot, make the same plot at the board for discussion purposes.
- Ask questions about the plot: median, range, number of data values considered versus number of data values graphed.

Activity 3

- One advantage of box-and-whisker plots is that multiple plots can be displayed and analyzed using the same number line. For instance, state test scores for 5 different schools could be displayed on the same number line.
- Read the information given and analyze the two plots.
- **MP3 Construct Viable Arguments and Critique the Reasoning of Others:** Students should be listening to the analysis offered by their partners or other classmates and critiquing their reasoning.

? "Which test is represented by which plot? Explain." Listen for students discussing the location of the median and the third quartile for each plot. The spring test is the top plot.

? "True or false: 50% of the top plot is greater than 75% of the bottom plot." true

What Is Your Answer?

- **Think-Pair-Share:** Students should read each question independently and then work in pairs to answer the questions. When they have answered the questions, the pair should compare their answers with another group and discuss any discrepancies.

Closure

- Make a box-and-whisker plot of the data collected at the beginning of class. Write one or two observations about the plot.

2 ACTIVITY: Conducting a Survey

Have your class conduct a survey. Each student will write on the chalkboard the number of pairs of footwear that he or she owns.

Now, work with a partner to draw a box-and-whisker plot of the data.

3 ACTIVITY: Reading a Box-and-Whisker Plot

Math Practice 7

View as Components

What do the different components of a box-and-whisker plot represent?

Work with a partner. The box-and-whisker plots show the test score distributions of two sixth grade achievement tests. The same group of students took both tests. The students took one test in the fall and the other in the spring.

a. Compare and contrast the test results.

b. Decide which box-and-whisker plot represents the results of which test. How did you make your decision?

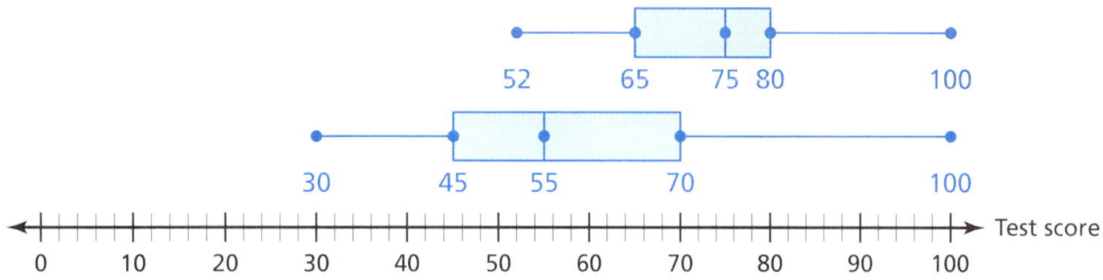

What Is Your Answer?

4. **IN YOUR OWN WORDS** How can you use quartiles to represent data graphically?

5. Describe who might be interested in test score distributions like those shown in Activity 3. Explain why it is important for such people to know test score distributions.

Practice Use what you learned about box-and-whisker plots to complete Exercise 4 on page 463.

10.4 Lesson

Check It Out
Lesson Tutorials
BigIdeasMath.com

Key Vocabulary
box-and-whisker plot, *p. 460*
five-number summary, *p. 460*

Key Idea

Box-and-Whisker Plot

A **box-and-whisker plot** represents a data set along a number line by using the least value, the greatest value, and the quartiles of the data. A box-and-whisker plot shows the *variability* of a data set.

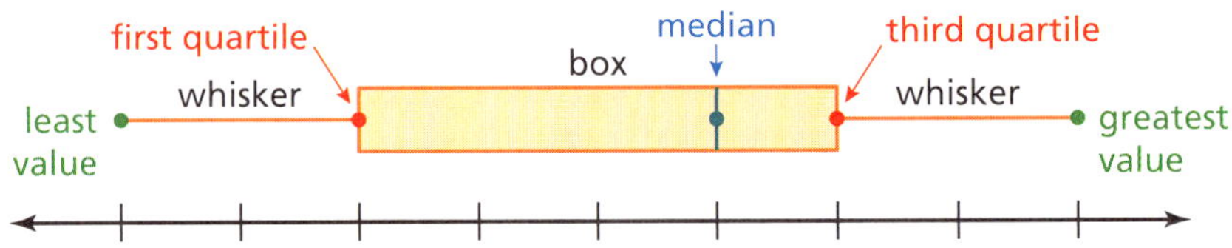

The five numbers that make up the box-and-whisker plot are called the **five-number summary** of the data set.

EXAMPLE 1 Making a Box-and-Whisker Plot

Make a box-and-whisker plot for the ages (in years) of the spider monkeys at a zoo:

15, 20, 14, 38, 30, 36, 30, 30, 27, 26, 33, 35

Step 1: Order the data. Find the median and the quartiles.

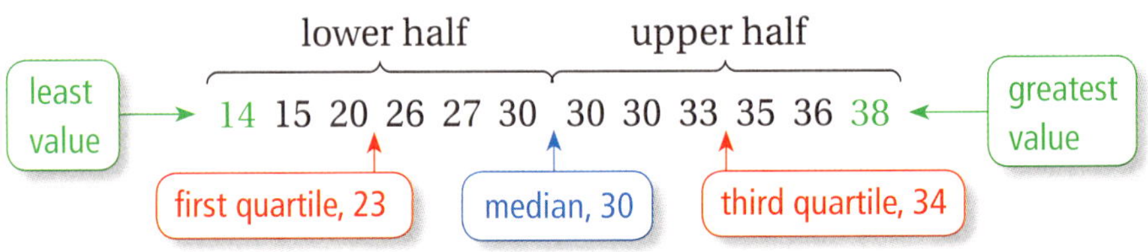

Step 2: Draw a number line that includes the least and greatest values. Graph points above the number line that represent the five-number summary.

Step 3: Draw a box using the quartiles. Draw a line through the median. Draw whiskers from the box to the least and the greatest values.

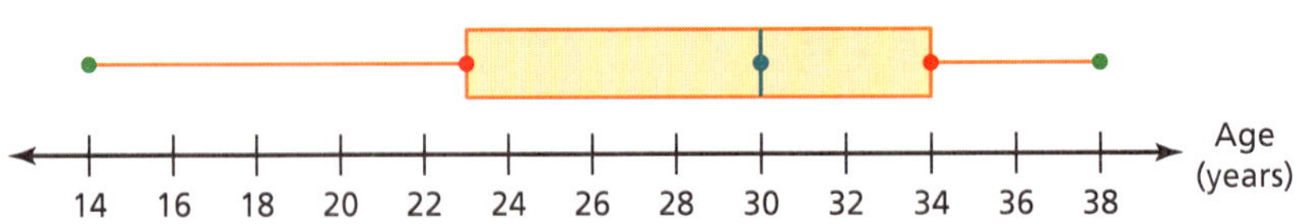

On Your Own

Now You're Ready
Exercises 5–8

1. A group of friends spent 1, 0, 2, 3, 4, 3, 6, 1, 0, 1, 2, and 2 hours online last night. Make a box-and-whisker plot for the data.

Multi-Language Glossary at BigIdeasMath.com

Laurie's Notes

Introduction

Connect

- **Yesterday:** Students gained a general understanding of how a box-and-whisker plot is constructed. (MP2, MP3)
- **Today:** Students will construct and analyze box-and-whisker plots.

Motivate

- The physical involvement of making a human box-and-whisker plot makes a lasting impression on students.
- **Preparation:** Give each student an index card with a number written on it. Include an outlier or two on one end of the data.

? "What is the first step in making a box-and-whisker plot?" sort the data

- Students should stand up and sort themselves. Have the median, the first and third quartiles, and the least and greatest data values take one step forward. If there is an even number of data values, the middle two students must figure out how to represent the mean and take a step forward.
- Make a number line on the floor or on the board. Position the 5 key values. If the plot is done on the floor, use string to form the whiskers. Students will have to visualize the box.
- **MP2 Reason Abstractly and Quantitatively:** Discuss features of the plot. If the plot includes an outlier, the length of the string becomes an instant topic of conversation. Students recognize that the same number of data values (about 25% of the class) is being represented by each whisker, yet the lengths of string are very different.

Lesson Notes

Key Idea

- Define a box-and-whisker plot. Draw the sample plot and discuss the process and vocabulary.
- **Discuss:** The box-and-whisker plot shows the *variability* of the data. Refer to this idea in each example done today.

Example 1

- Point out that ordering the data from least to greatest makes it possible to find the least value, greatest value, and quartiles.

? "How many data values are there?" 12 "When there are 12 data values, how do you find the median?" Find the mean of the 6th and 7th data values.

- There are six data values in the lower half and six data values in the upper half. The first quartile and third quartile are the mean of the middle two data values in each half.

On Your Own

- **Common Error:** Students may forget to order the data before creating the box-and-whisker plot.

Goal Today's lesson is constructing and analyzing a **box-and-whisker plot**.

Technology for the Teacher

Dynamic Classroom

Lesson Tutorials
Lesson Plans
Answer Presentation Tool

Extra Example 1

Make a box-and-whisker plot for the numbers of text messages received over several days.

32, 18, 11, 28, 42, 33, 40, 21, 37, 24

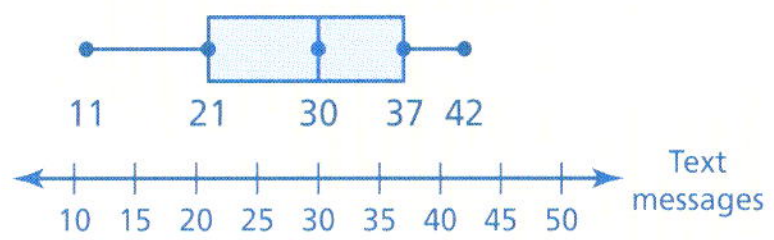

On Your Own

1.

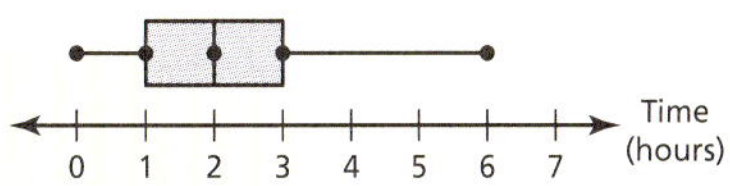

Extra Example 2

The box-and-whisker plot shows the times that the students in a gym class spent on an obstacle course.

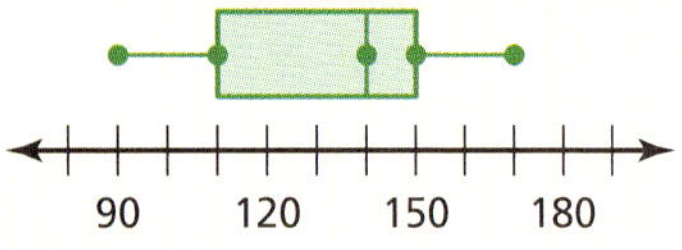

a. What fraction of the class spent 110 seconds or less on the obstacle course? about $\frac{1}{4}$

b. Are the data more spread out below the first quartile or above the third quartile? Explain. neither; Both whiskers are the same length.

c. Find and interpret the interquartile range of the data. 40; The middle half of the times varies by no more than 40 seconds.

On Your Own

2. a. about $\frac{3}{4}$

b. neither; The data are evenly spread out below and above the median.

c. 80; The middle half of the heights varies by no more than 80 feet.

Differentiated Instruction

Auditory

Remind students of other words that have the same root as *quartile:* quarter and quart, for example. Define the words as *four* or *fourths*. Mention that one-fourth is the same as 25%.

Laurie's Notes

Discuss

- The five key values plotted in a box-and-whisker plot divide the data into four parts.
- ? "What fraction of the data in the box are to the left of the median? to the right?" about $\frac{1}{4}$; about $\frac{1}{4}$
- Relate the definition of interquartile range to the box-and-whisker plot, indicating that it represents the "length" of the box.

Example 2

- **FYI:** Body Mass Index (BMI) is a numerical value determined using a person's weight and height. For most people, it provides a reliable indication of body "fatness" and can be used to screen for health problems.
- ? "Do you know from the box-and-whisker plot how many students are in the class?" no
- It is important for students to remember that you cannot determine the number of data points from a given box-and-whisker plot. You can only determine the five-number summary points and of those, only two *must* be values in the data set (the least value and the greatest value).
- ? "What is the range of the data and what does it mean in the context of the problem?" $28 - 17 = 11$; This means that the difference between the highest and lowest BMIs is 11.
- ? "What does the length of each part of the box-and-whisker plot tell you about the distribution of the data?" *Sample answers:* About 50% of the BMIs are between 17 and 21, and about 50% are between 21 and 28.
- In part (c), the students should be focusing on the middle half of the data. The IQR is 3.

On Your Own

- **MP3a Construct Viable Arguments:** Ask different students to explain their answers. Check for correct language and valid arguments.
- **Common Error:** Students may read the box-and-whisker plot incorrectly and find the range instead of the interquartile range.

The figure shows how data are distributed in a box-and-whisker plot.

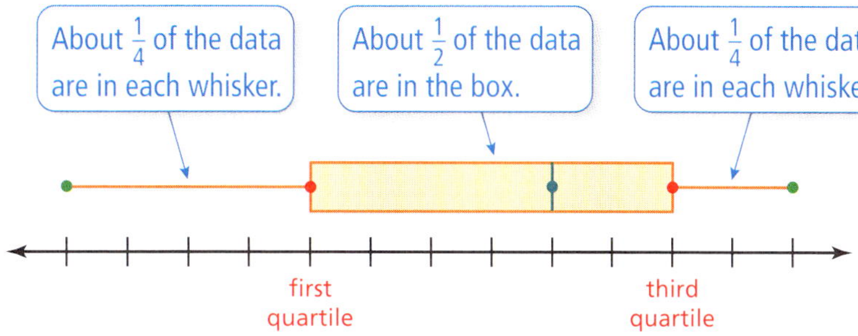

Study Tip

A long whisker or box indicates that the data are more spread out.

EXAMPLE 2 Analyzing a Box-and-Whisker Plot

The box-and-whisker plot shows the body mass index (BMI) of a sixth grade class.

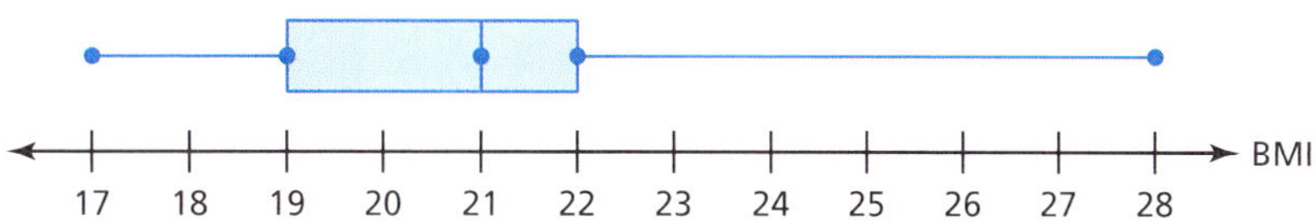

a. What fraction of the students have a BMI of at least 22?

The right whisker represents students who have a BMI of at least 22.

So, about $\frac{1}{4}$ of the students have a BMI of at least 22.

b. Are the data more spread out below the first quartile or above the third quartile? Explain.

The right whisker is longer than the left whisker.

So, the data are more spread out above the third quartile than below the first quartile.

c. Find and interpret the interquartile range of the data.

$$\text{interquartile range} = \text{third quartile} - \text{first quartile}$$
$$= 22 - 19 = 3$$

So, the middle half of the students' BMIs varies by no more than 3.

On Your Own

Now You're Ready Exercises 11 and 12

2. The box-and whisker plot shows the heights of the roller coasters at an amusement park. (a) What fraction of the roller coasters are between 120 feet tall and 220 feet tall? (b) Are the data more spread out below or above the median? Explain. (c) Find and interpret the interquartile range of the data.

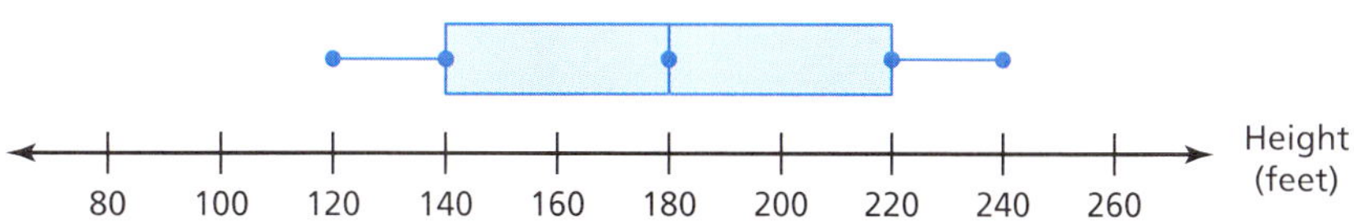

A box-and-whisker plot also shows the shape of a distribution.

Key Ideas

> **Study Tip**
> If you can draw a line through the median of a box-and-whisker plot, and each side is a mirror image of the other, then the distribution is symmetric.

Shapes of Box-and-Whisker Plots

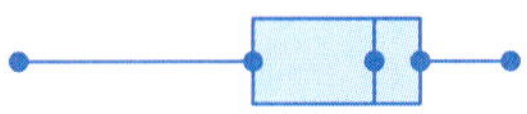

Skewed left

- Left whisker longer than right whisker
- Most data on the right

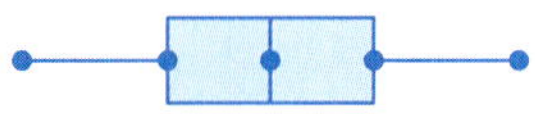

Symmetric

- Whiskers about same length
- Median in the middle of the box

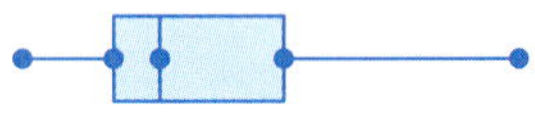

Skewed right

- Right whisker longer than left whisker
- Most data on the left

EXAMPLE 3 Comparing Box-and-Whisker Plots

The double box-and-whisker plot represents the prices of snowboards at two stores.

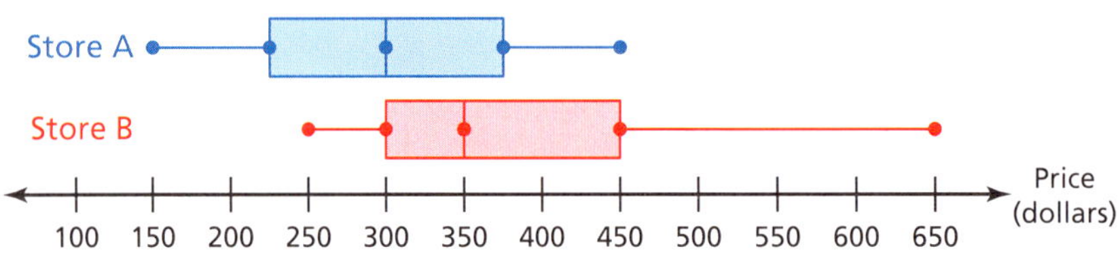

a. Identify the shape of each distribution.

For Store A, the whisker lengths are equal. The median is in the middle of the box. The data on the left are the mirror image of the data on the right. So, the distribution is symmetric.

For Store B, the right whisker is longer than the left whisker, and most of the data are on the left side of the display. So, the distribution is skewed right.

b. Which store's prices are more spread out? Explain.

Both boxes appear to be the same length. So, the interquartile range of each data set is equal. However, the range of the prices in Store B is greater than the range of the prices in Store A. So, the prices in Store B are more spread out.

On Your Own

Now You're Ready Exercises 13–17

3. The double box-and-whisker plot represents the life spans of crocodiles and alligators at a zoo. Identify the shape of each distribution. Which reptile's life spans are more spread out? Explain.

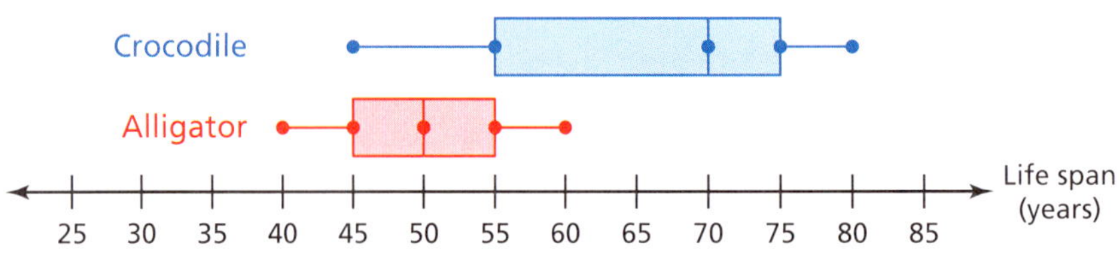

Laurie's Notes

Key Ideas

- Sketch each of the box-and-whisker plots. Discuss what the shape of the plot implies about the distribution of the data.

? "If the box-and-whisker plots represent the prices of jeans at three different stores and the range is the same for all three plots, what does the shape tell you about the distribution of the prices at all three stores?" Listen for student understanding of the spread of the data, referencing percentages and relative price.

Example 3

- One advantage of a box-and-whisker plot is that multiple data sets can be shown in the same display, as with double bar graphs and double line graphs. The data sets can also have different numbers of data values.
- Part (a) asks students to identify the shape. A student might comment that a high price in Store B could be the reason for it being skewed. Without the outlier it could be a symmetric graph.
- **MP2:** Supplement the discussion in part (b) with quantitative comparisons, such as 50% of the prices of snowboards in Store A are as low as, or lower than, the lowest 25% of the prices of snowboards in Store B.

On Your Own

- **Neighbor Check:** Have students work independently and then have their neighbors check their work. Have students discuss any discrepancies.
- Ask others quantitative questions about the two box-and-whisker plots.

Closure

- **Exit Ticket:**
  - What are the 5 key values that are graphed in a box-and-whisker plot? least value, first quartile, median, third quartile, greatest value
  - How does an outlier affect a box-and-whisker plot? *Sample answer:* increases the length of one of the whiskers
  - Explain why two data sets of different sizes can be graphed on the same number line. Box-and-whisker plots show the distribution of the data, not individual data points.

Extra Example 3

The double box-and-whisker plot represents the length of time (in minutes) it takes students in your gym class and your friend's gym class to run 1 mile.

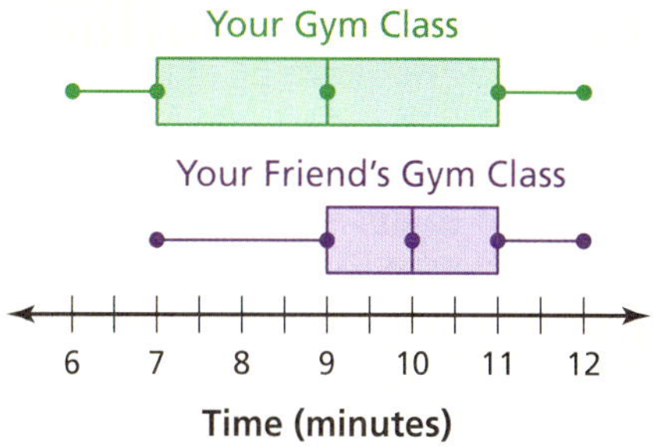

a. Identify the shape of each distribution. For your class, the whisker lengths are equal. The median is in the middle of the box. The data appear to be evenly distributed on both sides of the median. So, the distribution is symmetric. For your friend's class, the left whisker is longer than the right whisker, and most of the data are on the right side of the display. So, the distribution is skewed left.

b. Which lengths of time are more spread out? Explain. The range of the times in your class is greater than the range in your friend's class. Also, because the box for your class is longer than the box for your friend's class, the interquartile range is also greater. So, the times in your class are more spread out.

On Your Own

3. Crocodiles: skewed left, alligators: symmetric; crocodile; The box for crocodiles is longer than the box for alligators. So, the IQR for crocodiles is greater than the IQR for alligators. Also, the range of life spans for crocodiles is greater than the range of life spans for alligators.

Vocabulary and Concept Check

1. Order the data. The first number is the *least value* and the last number is the *greatest value*. The middle value is the *median*. The middle value of the lower half of the data is the *first quartile*. The middle value of the upper half of the data is the *third quartile*.
2. $\frac{3}{4}$
3. Is the distribution skewed right?; yes; no

Practice and Problem Solving

4. *Sample answer:* The attendance for Beach 2 is more spread out than the attendance for Beach 1. Beach 1 seems to be more popular because its daily attendance is higher, on average, than Beach 2.

5–10. See Additional Answers.

11. **a.** about $\frac{1}{2}$

b. The right whisker is longer than the left whisker. So the data are more spread out above the third quartile than below the first quartile.

c. 150; The middle half of the data varies by no more than 150 gallons.

Assignment Guide and Homework Check

Level	Day 1 Activity Assignment	Day 2 Lesson Assignment	Homework Check
Basic	4, 27–30	1–3, 5–17 odd	5, 7, 9, 11
Average	4, 27–30	1–3, 7–14, 17–23 odd	7, 8, 10, 12
Advanced	4, 27–30	1–3, 6, 8, 9, 10–26 even	6, 8, 10, 12, 14

Common Errors

- **Exercise 2** Students may have trouble corresponding fractions of the data with quartiles. Remind them that they can obtain this information by studying the Key Idea on page 460 as well as the figure at the top of page 461. Also, a quick review of the definition of *quartiles* can be helpful.
- **Exercises 5–8** Students may have difficulty creating the box-and-whisker plots. Remind them of the five-number summary of a data set.

10.4 Record and Practice Journal

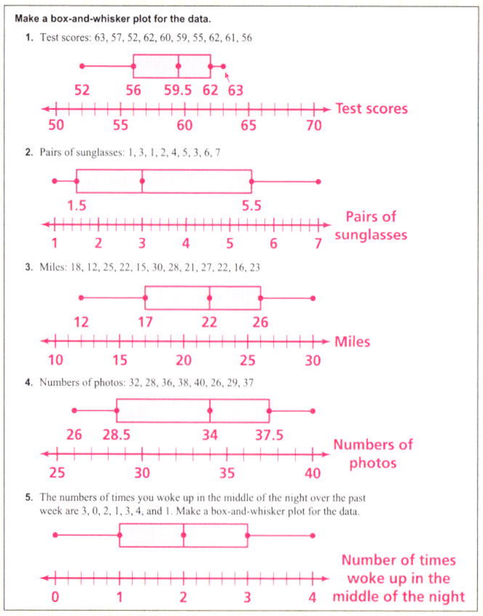
Make a box-and-whisker plot for the data.

1. Test scores: 63, 57, 52, 62, 60, 59, 55, 62, 61, 56

2. Pairs of sunglasses: 1, 3, 1, 2, 4, 5, 3, 6, 7

3. Miles: 18, 12, 25, 22, 15, 30, 28, 21, 27, 22, 16, 23

4. Numbers of photos: 32, 28, 36, 38, 40, 26, 29, 37

5. The numbers of times you woke up in the middle of the night over the past week are 3, 0, 2, 1, 3, 4, and 1. Make a box-and-whisker plot for the data.

10.4 Exercises

Check It Out
Help with Homework
BigIdeasMath.com

Vocabulary and Concept Check

1. **VOCABULARY** Explain how to find the five-number summary of a data set.

2. **NUMBER SENSE** In a box-and-whisker plot, what fraction of the data is greater than the first quartile?

3. **DIFFERENT WORDS, SAME QUESTION** Which is different? Find "both" answers.

Is the distribution skewed right?

Is the left whisker longer than the right whisker?

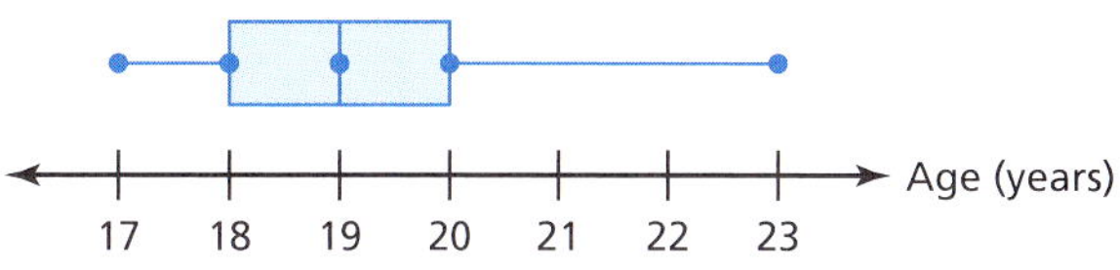

Are the data more spread out below the first quartile than above the third quartile?

Does the lower fourth of the data vary more than the upper fourth of the data?

Practice and Problem Solving

4. The box-and-whisker plots represent the daily attendance at two beaches during July. Compare and contrast the attendances for the two beaches.

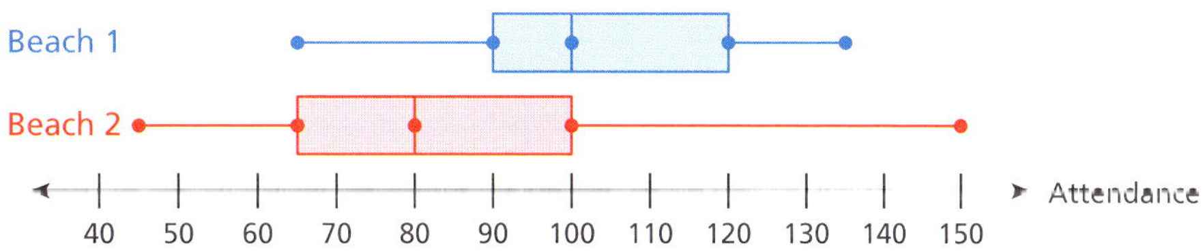

Make a box-and-whisker plot for the data.

1 5. Ages of teachers (in years): 30, 62, 26, 35, 45, 22, 49, 32, 28, 50, 42, 35

6. Quiz scores: 8, 12, 9, 10, 12, 8, 5, 9, 7, 10, 8, 9, 11

7. Donations (in dollars): 10, 30, 5, 15, 50, 25, 5, 20, 15, 35, 10, 30, 20

8. Ski lengths (in centimeters): 180, 175, 205, 160, 210, 175, 190, 205, 190, 160, 165, 195

9. **ERROR ANALYSIS** Describe and correct the error in making a box-and-whisker plot for the data.

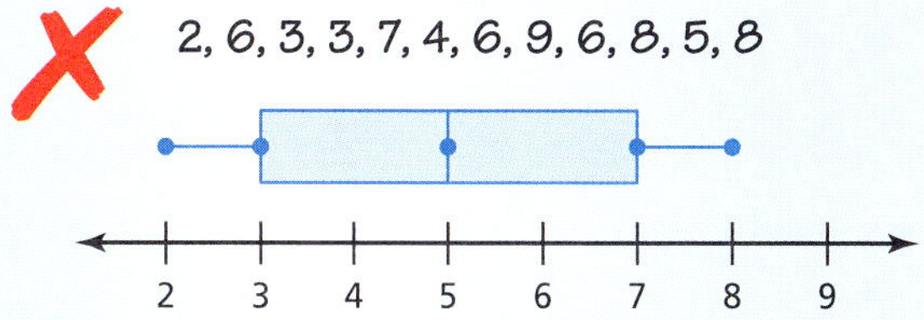

10. **CAMPING** The numbers of days 12 friends went camping during the summer are 6, 2, 0, 10, 3, 6, 6, 4, 12, 0, 6, and 2. Make a box-and-whisker plot for the data. What is the range of the data?

2 **11. DUNK TANK** The box-and-whisker plot represents the numbers of gallons of water needed to fill different types of dunk tanks offered by a company.

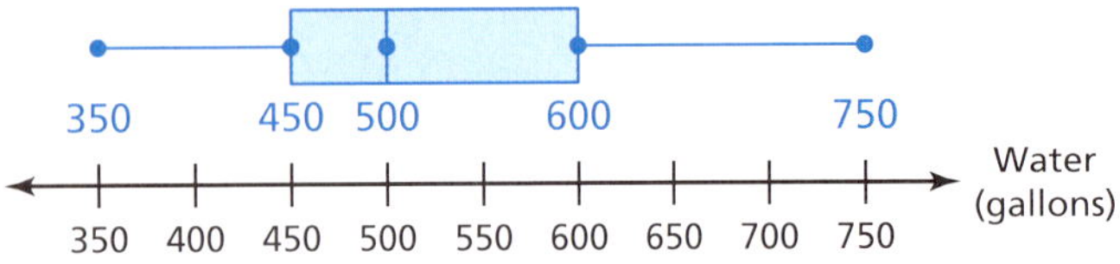

a. What fraction of the dunk tanks require at least 500 gallons of water?

b. Are the data more spread out below the first quartile or above the third quartile? Explain.

c. Find and interpret the interquartile range of the data.

12. BUILDINGS The box-and-whisker plot represents the heights (in meters) of the tallest buildings in Chicago.

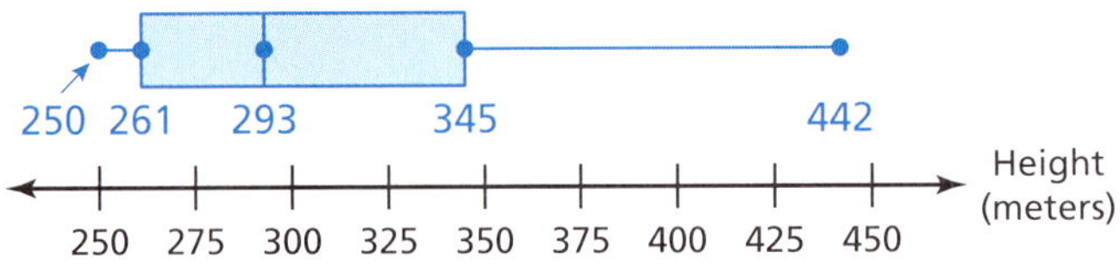

a. What percent of the buildings are no taller than 345 meters?

b. Is there more variability in the heights above 345 meters or below 261 meters? Explain.

c. Find and interpret the interquartile range of the data.

Identify the shape of the distribution. Explain.

3 **13.** 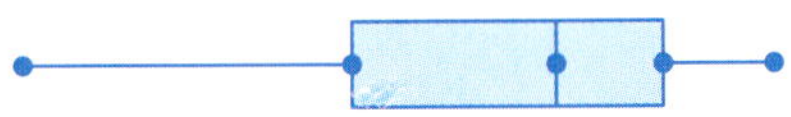

14. 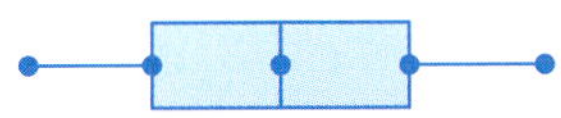

15. 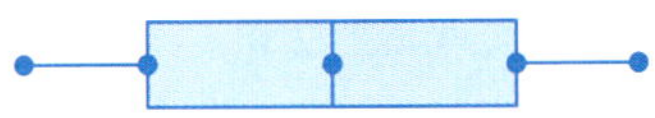

16. 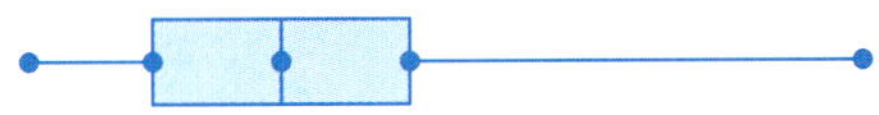

17. RECESS The double box-and-whisker plot represents the start times of recess for two schools.

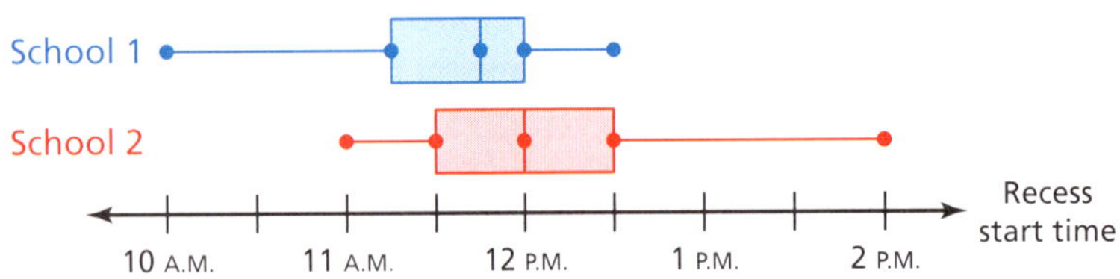

a. Identify the shape of each distribution.

b. Which school's start times for recess are more spread out? Explain.

c. Which school is more likely to have recess before lunch? Explain.

Common Errors

- **Exercises 10–12** Students may confuse the range and interquartile range of a data set. Remind them of these definitions.
- **Exercises 13–16** When identifying the shape of the distribution, students may confuse the meanings of skewed left and skewed right. Tell them that the shape is determined by the longer whisker.

Practice and Problem Solving

12. **a.** about 75%

b. The right whisker is longer than the left whisker. So there is more variability in the heights above 345 meters than the heights below 261 meters.

c. 84; The middle half of the data varies by no more than 84 meters.

13. skewed left; The left whisker is longer than the right whisker, and most of the data are on the right.

14. symmetric; The whiskers are about the same length, and the median is in the middle of the box.

15. symmetric; The whiskers are about the same length, and the median is in the middle of the box.

16. skewed right; The right whisker is longer than the left whisker, and most of the data are on the left.

17. **a.** School 1 is skewed left and School 2 is skewed right.

b. School 2; The range for School 2 is a half hour greater than the range for School 1. Also, the IQR of School 2 is greater than the IQR of School 1.

c. School 1; School 1 has more data on the left than School 2. So, School 1 is more likely to have recess before lunch.

Differentiated Instruction

Kinesthetic

To make ordering data easier, have students write each data value on a sticky note. Using their desks or notebook paper, students can move the notes to order the data and divide them into quartiles.

Practice and Problem Solving

18–20. See Additional Answers.

21. Use the median to describe the center and the interquartile range to describe the variation.

22. See *Taking Math Deeper*.

23. When the least value and the first quartile are equal, there is no whisker on the left. When the greatest value and the third quartile are equal, there is no whisker on the right.

24–26. See Additional Answers.

Fair Game Review

27. >	**28.** <
29. >	**30.** B

Mini-Assessment

The list represents the numbers of DVDs owned by the students in your class.

25, 31, 27, 36, 19, 22, 20, 24, 30, 32, 29, 27

1. Make a box-and-whisker plot for the data.

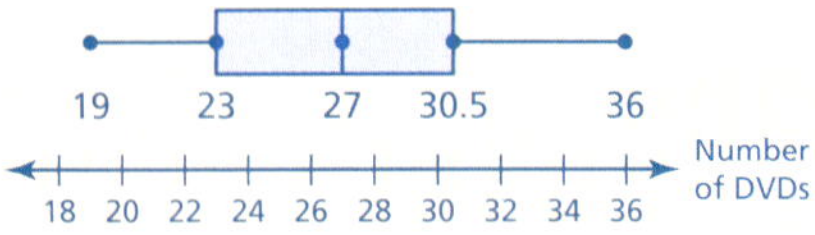

2. What is the range of the data? 17
3. What fraction of students owns at least 23 DVDs? about $\frac{3}{4}$
4. Find and interpret the interquartile range of the data. 7.5; The middle half of the numbers of DVDs varies by no more than 7.5.

Taking Math Deeper

Exercise 22

It is important for students to make a plan for solving this exercise rather than simply jumping in and trying numbers.

1 Summarize what you know about the structure of the data set.

- The data set has 12 values.
- The quartiles break the data set up into 4 parts. Each part has 3 values.

2 Draw a symmetric box-and-whisker plot using a simple scale on a number line. Indicate where the 12 values in the data set are represented in the box-and-whisker plot.

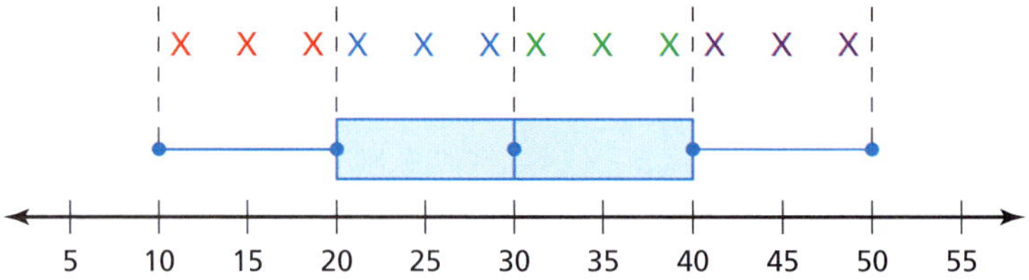

3 Begin finding values for the Xs.

- The least value is 10 and the greatest value is 50.
- Be careful choosing adjacent Xs that have different colors. The mean of each of these three pairs of values must equal the associated quartile. One way to do this is to make each value in the pair equal to the quartile. For example, the first quartile is 20. So, you can let the third and fourth values be 20.

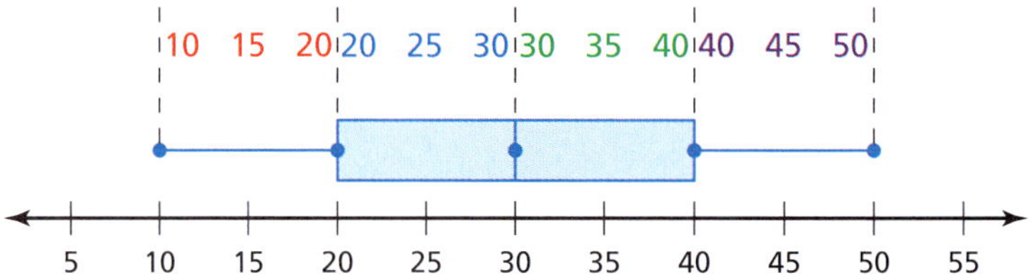

So, one possible data set is 10, 15, 20, 20, 25, 30, 30, 35, 40, 40, 45, 50.

Reteaching and Enrichment Strategies

If students need help...	If students got it...
Resources by Chapter • Practice A and Practice B • Puzzle Time Record and Practice Journal Practice Differentiating the Lesson Lesson Tutorials Skills Review Handbook	Resources by Chapter • Enrichment and Extension • Technology Connection Start the next section

Make a box-and-whisker plot for the data.

18. Temperatures (in °C): 5, 1, 4, 0, 9, 0, −8, 5, 2, 4, −1, 10, 7, −5

19. Checking account balances (in dollars): 30, 0, −10, 50, 20, 90, −15, 40, 100, 45, −20, 70

20. **REASONING** The data set in Exercise 18 has an outlier. Describe how removing the outlier affects the box-and-whisker plot.

21. **CHOOSE TOOLS** What are the most appropriate measures to describe the center and the variation of the distribution in Exercise 12?

22. **OPEN-ENDED** Write a data set with 12 values that has a symmetric box-and-whisker plot.

23. **CRITICAL THINKING** When would a box-and-whisker plot *not* have one or both whiskers?

24. **STRUCTURE** Draw a histogram that could represent the distribution shown in Exercise 15.

25. **REASONING** The double box-and-whisker plot represents the runs scored per game by two softball teams during a 32-game season.

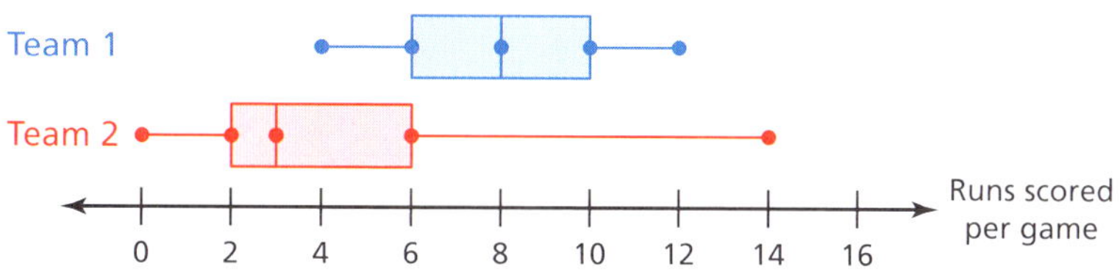

a. Which team is more consistent at scoring runs? Explain.

b. In how many games did Team 2 score 6 runs or less?

c. Team 1 played Team 2 once during the season. Which team do you think won? Explain.

d. Which team do you think has the greater mean? Explain.

26. Choose Tools A market research company wants to summarize the variability of the SAT scores of graduating seniors in the United States. Do you think the company should use a stem-and-leaf plot, a histogram, or a box-and-whisker plot? Explain.

Fair Game Review What you learned in previous grades & lessons

Copy and complete the statement using < or >. *(Section 6.3)*

27. $-\frac{2}{3}$ ▢ $-\frac{3}{4}$

28. $-2\frac{1}{5}$ ▢ $-2\frac{1}{6}$

29. -5.3 ▢ -5.5

30. **MULTIPLE CHOICE** Which of the following items is most likely represented by a rectangular prism with a volume of 1785 cubic inches? *(Section 8.4)*

Ⓐ closet

Ⓑ computer tower

Ⓒ filing cabinet

Ⓓ your math book

10.3–10.4 Quiz

Check It Out
Progress Check
BigIdeasMath.com

Describe the shape of each distribution. *(Section 10.3)*

1.

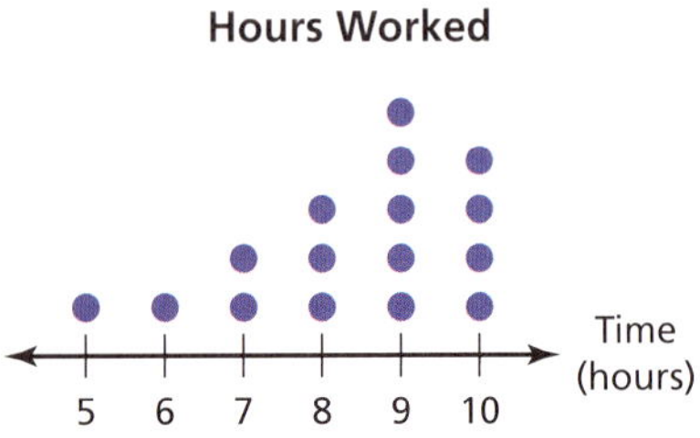

2.

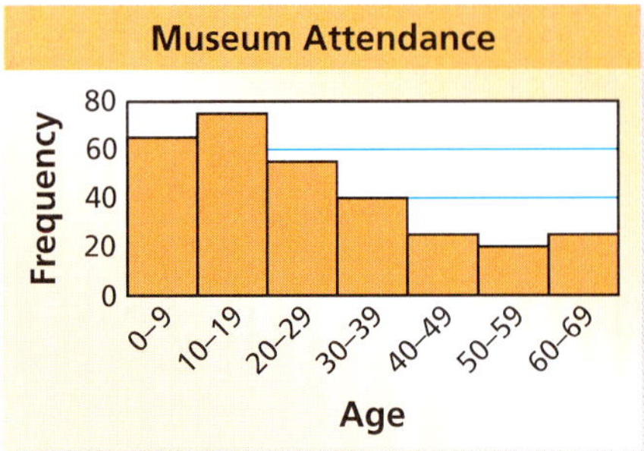

Choose the most appropriate measures to describe the center and the variation. Find the measures you chose. *(Section 10.3)*

3.

4.

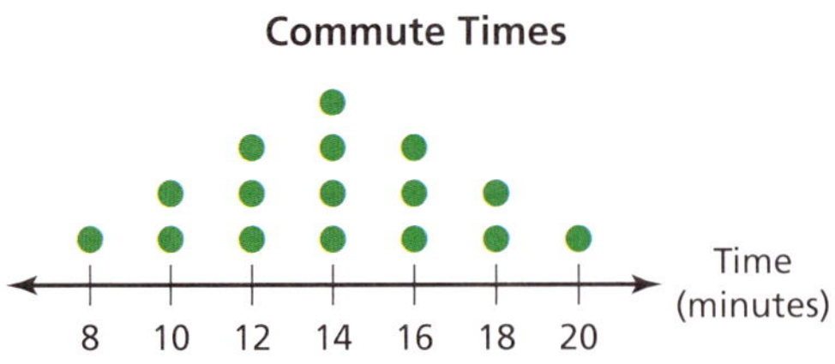

Make a box-and-whisker plot for the data. *(Section 10.4)*

5. Science test scores: 85, 76, 99, 84, 92, 95, 68, 100, 93, 88, 87, 85

6. Shoe sizes: 12, 8.5, 9, 10, 9, 11, 11.5, 9, 9, 10, 10, 10.5, 8

7. **MOVIES** The box-and-whisker plot represents the lengths (in minutes) of movies being shown at a theater. *(Section 10.4)*

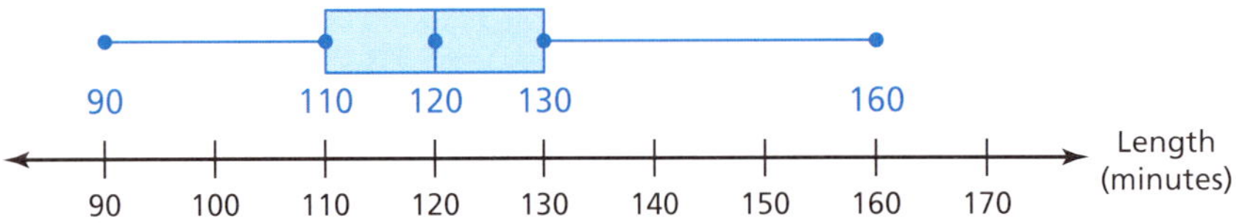

a. What percent of the movies are no longer than 120 minutes?

b. Is there more variability in the movie lengths longer than 130 minutes or shorter than 110 minutes? Explain.

c. Find and interpret the interquartile range of the data.

8. **EXPERIENCE** The frequency table shows the years of experience of employees at two branches of a company. Display the data for each branch in a histogram. Describe the shape of each distribution. Which branch has less experience? Explain. *(Section 10.3)*

Years of Experience	0–2	3–6	7–10	11–14	15–18	19–22	23–26
Frequency at Branch A	10	25	14	20	8	5	2
Frequency at Branch B	3	6	8	10	15	25	8

Alternative Assessment Options

Math Chat
Structured Interview
Student Reflective Focus Question
Writing Prompt

Student Reflective Focus Question

Ask students to use their own words to summarize the similarities and differences among line plots, stem-and-leaf plots, histograms, and box-and-whisker plots. Be sure that they include examples. Select students at random to present their summaries to the class.

Study Help Sample Answers

Remind students to complete Graphic Organizers for the rest of the chapter.

2.

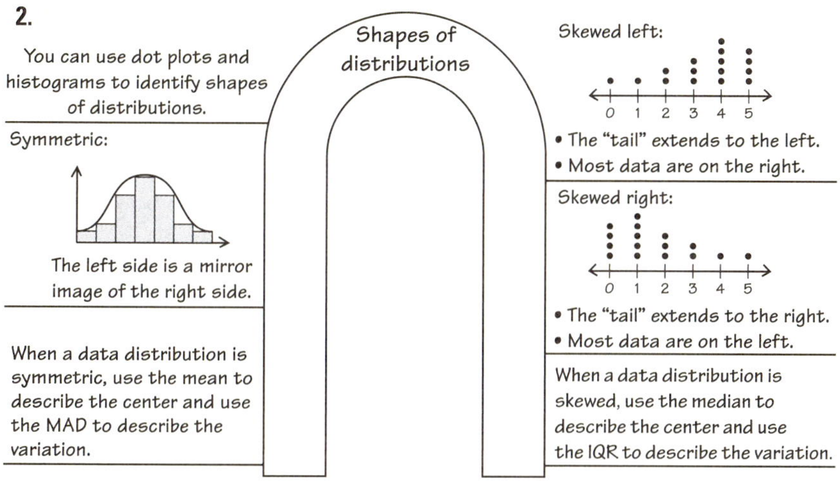

3–4. Available at *BigIdeasMath.com*.

Reteaching and Enrichment Strategies

If students need help. . .	If students got it. . .
Resources by Chapter • Practice A and Practice B • Puzzle Time Lesson Tutorials *BigIdeasMath.com*	Resources by Chapter • Enrichment and Extension • Technology Connection Game Closet at *BigIdeasMath.com* Start the Chapter Review

Answers

1. skewed left **2.** skewed right

3. median and IQR; median = 40, IQR = 12

4. mean and MAD; mean = 14, MAD = 2.5

5.

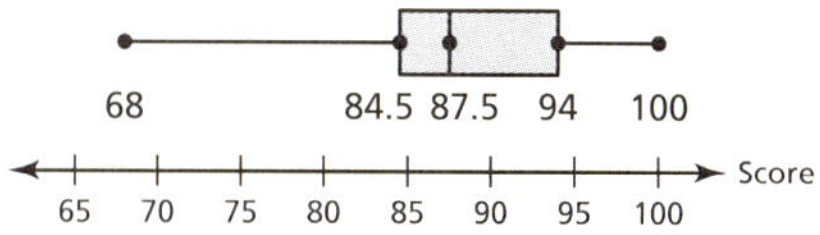

6.

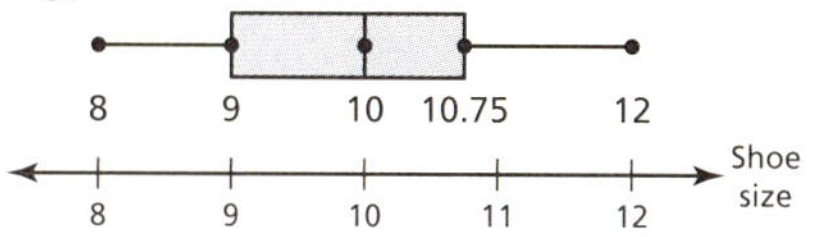

7. a. about 50%

b. longer than 130 minutes, because the right whisker is longer

c. 20 minutes; The middle half of the data vary by no more than 20 minutes.

8.

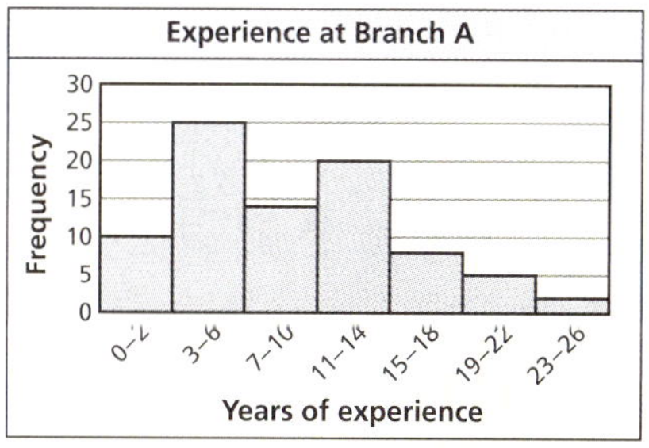

skewed right

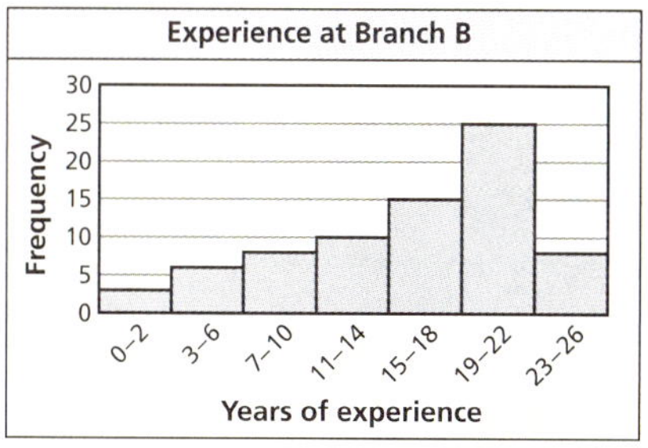

skewed left

Branch A; The distribution of Branch A is skewed right, so most of the data values are on the left.

For the Teacher
Additional Review Options

- *BigIdeasMath.com*
- Online Assessment
- Game Closet at *BigIdeasMath.com*
- Vocabulary Help
- Resources by Chapter

Answers

1. **Hats Sold Each Day**

Stem	Leaf
0	5 8 9
1	2 2 3 4 5 8
2	1 5
3	0

Key: 2 | 1 = 21 hats

2. **Ages of Park Volunteers**

Stem	Leaf
1	3 3 5 7 9
2	0 1
3	
4	0 8
5	2 5
6	0

Key: 1 | 3 = 13 years old

3. 6 tuna
4. 87 pounds

Review of Common Errors

Exercises 1 and 2

- Students may forget to include the zeros in the ones place as leaves or forget to include repeats of numbers. Remind them that *all* of the numbers in a data set should be represented in a stem-and-leaf plot.
- Students may forget to include stems that do not have leaves. Remind them that a stem-and-leaf plot shows how data are distributed, so it is important to include all of the stems in a range of data.

Exercise 3

- Students may misinterpret the question and count the data values represented with a stem of 9. Point out that neither of the values represented with this stem is less than 90.

Exercise 4

- Students may forget how to find a median, or confuse the median with the mean or mode. Remind them how to find the median of a data set.

Exercises 5 and 6

- Students may have difficulty determining how to scale the vertical axis of the histogram. Suggest that they find the interval with the greatest frequency and scale the axis so that this interval will fit.

Exercise 7

- Students may confuse "skewed left" and "skewed right." Point out that the "tail" of the graph of a distribution determines the direction of the skew. That is, in a skewed left distribution, the tail extends to the left, and in a skewed right distribution, the tail extends to the right.

Exercise 8

- Students may not recognize the distribution as symmetric and subsequently use the wrong measures to describe the center and the variation. Review with students the different types of distributions.

Exercises 9 and 10

- Students may have difficulty creating the box-and-whisker plots. Remind them of the five-number summary of a data set.

10 Chapter Review

Check It Out
Vocabulary Help
BigIdeasMath.com

Review Key Vocabulary

stem-and-leaf plot, *p. 436*
stem, *p. 436*
leaf, *p. 436*
frequency table, *p. 441*
frequency, *p. 441*
histogram, *p. 442*
box-and-whisker plot, *p. 460*
five-number summary, *p. 460*

Review Examples and Exercises

10.1 Stem-and-Leaf Plots *(pp. 434–439)*

Day	DVDs Rented
Sun	50
Mon	19
Tue	25
Wed	28
Thu	39
Fri	53
Sat	50

Make a stem-and-leaf plot of the number of DVDs rented each day at a store.

Step 1: Order the data. 19, 25, 28, 39, 50, 50, 53

Step 2: Choose the stems and the leaves. Because the data range from 19 to 53, use the *tens* digits for the stems and the *ones* digits for the leaves. Be sure to include the key.

Step 3: Write the stems to the *left* of the vertical line.

Step 4: Write the leaves for each stem to the *right* of the vertical line.

DVDs Rented

Stem	Leaf
1	9
2	5 8
3	9
4	
5	0 0 3

Key: 2 | 5 = 25 DVDs

Order the stems vertically. The stem for data values less than 10 is 0.

Include stems without leaves.

Write the leaves horizontally.

Exercises

Make a stem-and-leaf plot of the data.

1.

Hats Sold Each Day			
5	18	12	15
21	30	8	12
13	9	14	25

2.

Ages of Park Volunteers			
13	17	40	15
48	21	19	52
13	55	60	20

The stem-and-leaf plot shows the weights (in pounds) of yellowfin tuna caught during a fishing contest.

Stem	Leaf
7	6
8	0 2 5 7 9
9	5 6
10	2

Key: 8 | 5 = 85 pounds

3. How many tuna weigh less than 90 pounds?

4. What is the median weight of the tuna?

10.2 Histograms (pp. 440–447)

The frequency table shows the number of crafts each member of a craft club made for a fundraiser. Display the data in a histogram.

Crafts	Frequency
0–2	10
3–5	8
6–8	5
9–11	0
12–14	2

Step 1: Draw and label the axes.

Step 2: Draw a bar to represent the frequency of each interval.

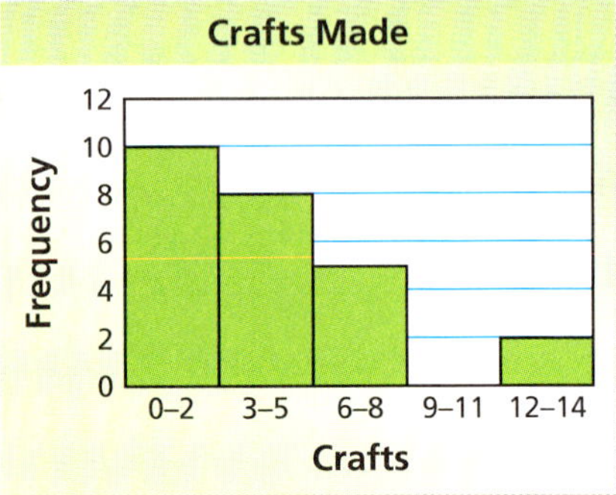

Exercises

Display the data in a histogram.

5.

Heights of Gymnasts	
Heights (in.)	Frequency
50–54	1
55–59	8
60–64	5
65–69	2

6.

Minutes Studied	
Minutes	Frequency
0–19	5
20–39	9
40–59	12
60–79	3

10.3 Shapes of Distributions (pp. 450–457)

Describe the shape of each distribution.

a.

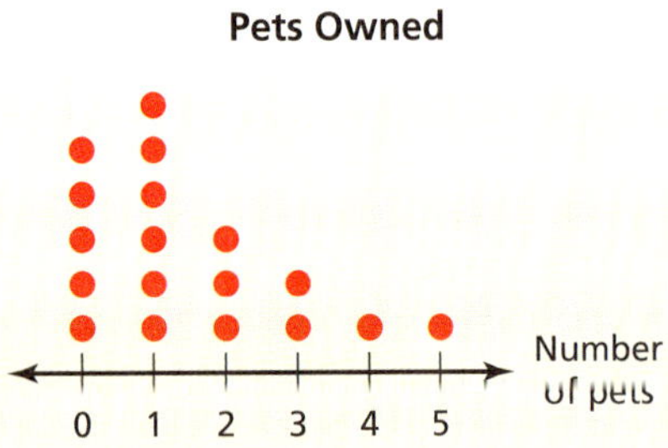

Most of the data are on the left, and the tail extends to the right.

So, the distribution is skewed right.

b.

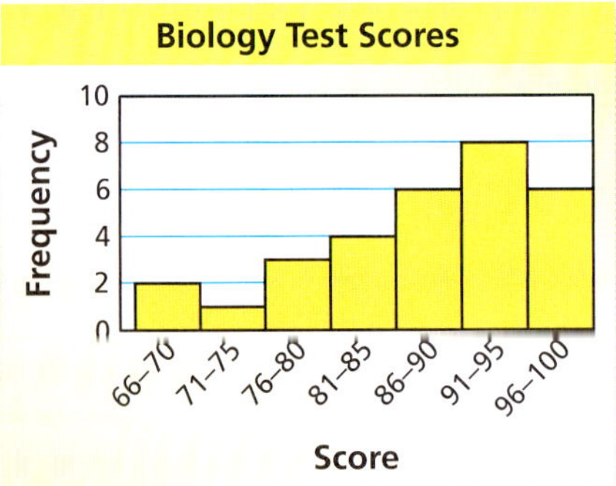

Most of the data are on the right, and the tail extends to the left.

So, the distribution is skewed left.

Review Game

Sampling and Predictions

Materials per group:

- pencils
- paper

Directions:

Divide the class into groups of 2 to 4 students. Each group comes up with a survey question. For instance, a group's survey question could be "how many pets does a student have?" Make sure each group has a different question.

Each group surveys 25% (or some other predetermined percentage) of the class and makes a display of the results, choosing the most appropriate data display. Groups then use their displays to predict the results for the entire class.

Each group then surveys the entire class and makes a display of these results, using the same type of data display chosen to analyze the sample. Groups then write a short paragraph in which they compare their actual results for their entire class to the predicted results for their entire class.

The larger the sample size, the more accurate the predictions may be. If possible, arrange to involve more students, in addition to the students in your class, in the surveys.

Who Wins?

The group whose actual results are closest to the predicted results wins.

For the Student Additional Practice

- Lesson Tutorials
- Multi-Language Glossary
- Self-Grading Progress Check
- *BigIdeasMath.com*
  Dynamic Student Edition
  Student Resources

Answers

5.

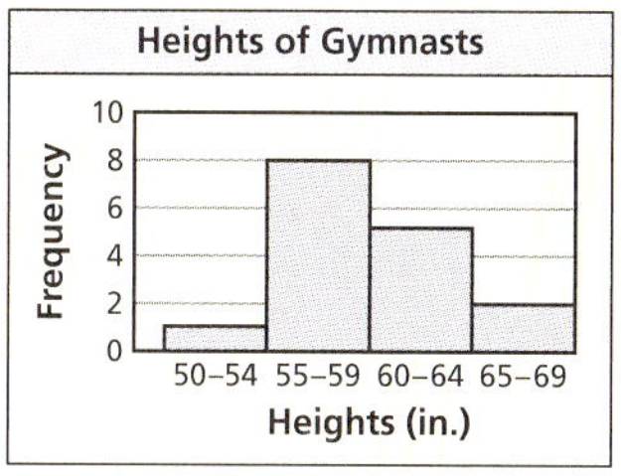

6.

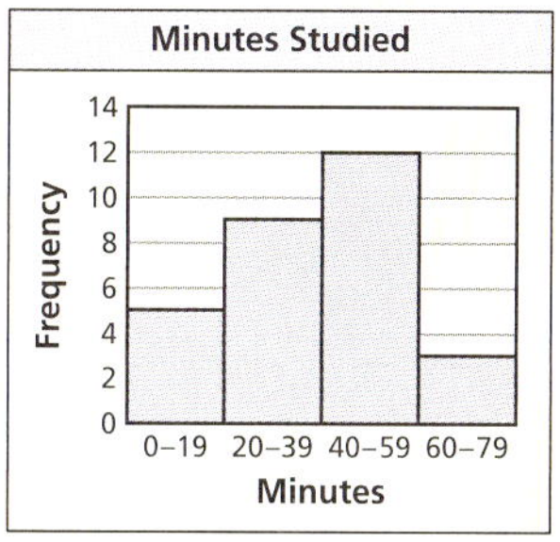

7. skewed left

8. mean, MAD; mean = 60, MAD ≈ 0.9

9.

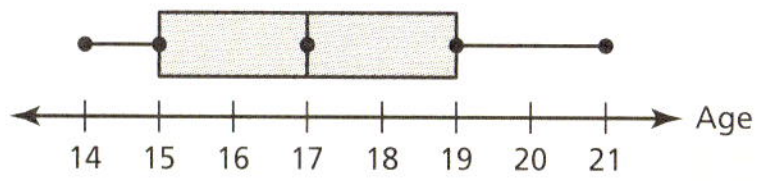

10.

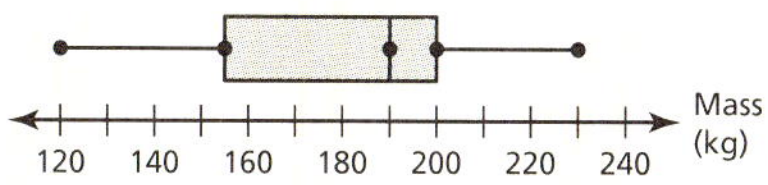

My Thoughts on the Chapter

What worked. . .

Teacher Tip

Not allowed to write in your teaching edition? Use sticky notes to record your thoughts.

What did not work. . .

What I would do differently. . .

Exercises

7. Describe the shape of the distribution.

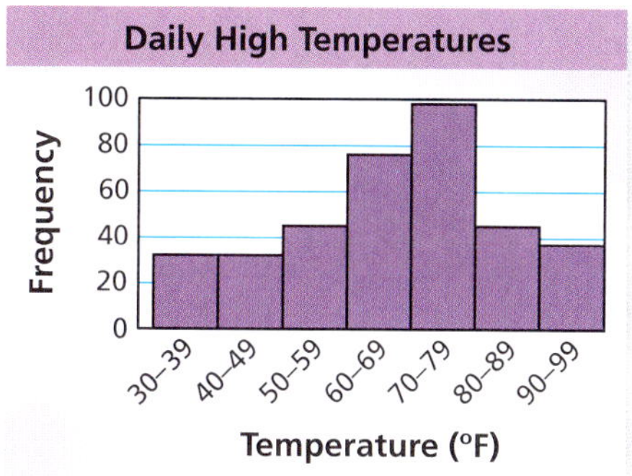

8. Choose the most appropriate measures to describe the center and the variation. Find the measures you chose.

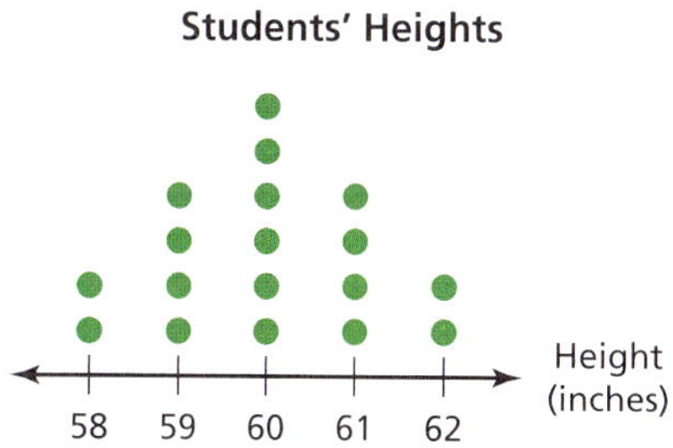

10.4 Box-and-Whisker Plots *(pp. 458–465)*

Make a box-and-whisker plot for the weights (in pounds) of pumpkins sold at a market.

16, 20, 14, 15, 12, 8, 8, 19, 14, 10, 8, 16

Step 1: Order the data. Find the median and the quartiles.

lower half: 8 8 8 10 12 14 — upper half: 14 15 16 16 19 20

least value → 8 8 8 10 12 14 14 15 16 16 19 20 ← greatest value

first quartile, 9 — median, 14 — third quartile, 16

Step 2: Draw a number line that includes the least and the greatest values. Graph points above the number line that represent the five-number summary.

Step 3: Draw a box using the quartiles. Draw a line through the median. Draw whiskers from the box to the least and the greatest values.

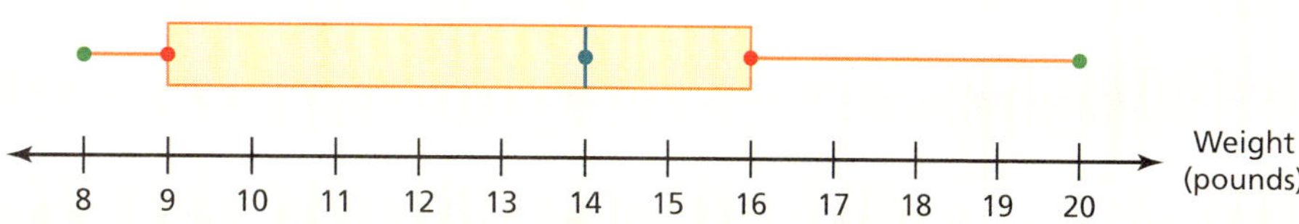

Exercises

Make a box-and-whisker plot for the data.

9. Ages of volunteers at a hospital: 14, 17, 20, 16, 17, 14, 21, 18

10. Masses (in kilograms) of lions: 120, 200, 180, 150, 200, 200, 230, 160

10 Chapter Test

Check It Out
Test Practice
BigIdeasMath.com

Make a stem-and-leaf plot of the data.

1.

Quiz Scores (%)			
96	88	80	72
80	94	92	100
76	80	68	90

2.

CDs Sold Each Day				
45	31	29	38	38
67	40	62	45	60
40	39	60	43	48

3. Find the mean, median, mode, range, and interquartile range of the data.

Cooking Time (minutes)

Stem	Leaf
3	5 8
4	0 1 8
5	0 4 4 4 5 9
6	0

Key: 4 | 1 = 41 minutes

4. Display the data in a histogram.

Television Watched Per Week	
Hours	**Frequency**
0–9	14
10–19	16
20–29	10
30–39	8

5. **WATER** The dot plot shows the number of glasses of water that the students in a class drink in one day.

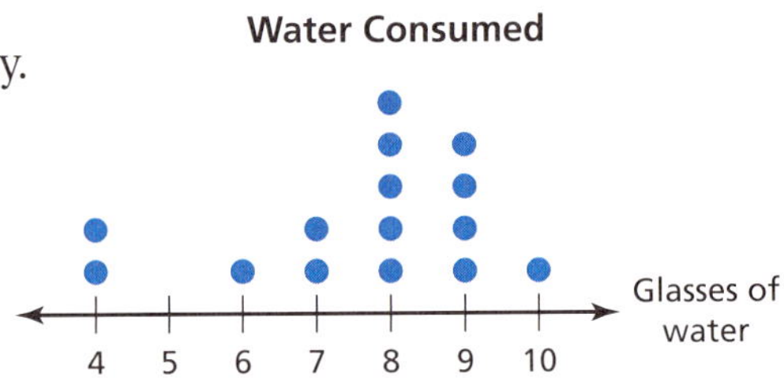

a. Describe the shape of the distribution.

b. Choose the most appropriate measures to describe the center and the variation. Find the measures you chose.

Make a box-and-whisker plot for the data.

6. Ages (in years) of dogs at a vet's office: 1, 3, 5, 11, 5, 7, 5, 9

7. Lengths (in inches) of fish in a pond: 12, 13, 7, 8, 14, 6, 13, 10

8. Hours practiced each week: 7, 6, 5, 4.5, 3.5, 7, 7.5, 2, 8, 7, 7.5, 6.5

9. **CELL PHONES** The double box-and-whisker plot compares the battery life (in hours) of two brands of cell phones.

a. What is the range of the upper 75% of each brand?

b. Which battery has a longer battery life? Explain.

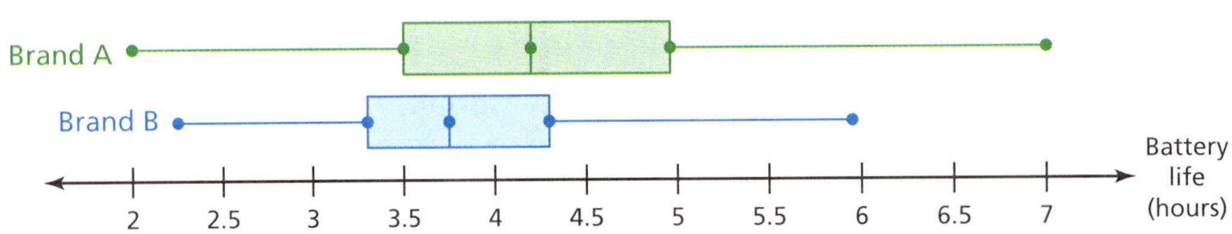

Test Item References

Chapter Test Questions	Section to Review	Common Core State Standards
1–3	10.1	6.SP.4
4	10.2	6.SP.2, 6.SP.4
5	10.3	6.SP.2, 6.SP.4, 6.SP.5d
6–9	10.4	6.SP.2, 6.SP.4, 6.SP.5c

Test-Taking Strategies

Remind students to quickly look over the entire test before they start so that they can budget their time. On this test, students are asked to display as well as analyze data. It can be difficult for students to determine how to begin. So, remind students to use the **Stop** and **Think** strategy before they answer a question.

Common Errors

- **Exercises 1 and 2** Students may forget to include the zeros in the ones place as leaves, or forget to include repeats of numbers. Remind them that *all* of the numbers in a data set should be represented in a stem-and-leaf plot.
- **Exercise 2** Students may forget to include 5 as a stem because it does not have leaves. Remind them that a stem-and-leaf plot shows how data are distributed, so it is important to include all of the stems in a range of data.
- **Exercise 3** Students may need to be reminded of the definitions of some of the terms so that they can answer the questions. Give an example of each term and how to find it using a stem-and-leaf plot.
- **Exercise 4** Students may have difficulty determining how to scale the vertical axis of the histogram. Suggest that they find the interval with the greatest frequency and scale the axis so that this interval will fit.
- **Exercise 5** Students may not recognize the distribution as skewed left and subsequently use the wrong measures to describe the center and the variation. Review with students the different types of distributions.
- **Exercises 6–8** Students may have difficulty creating the box-and-whisker plots. Remind them of the five-number summary of a data set.

Reteaching and Enrichment Strategies

If students need help. . .	If students got it. . .
Resources by Chapter • Practice A and Practice B • Puzzle Time Record and Practice Journal Practice Differentiating the Lesson Lesson Tutorials *BigIdeasMath.com* Skills Review Handbook	Resources by Chapter • Enrichment and Extension • Technology Connection Game Closet at *BigIdeasMath.com* Start Standards Assessment

Answers

1. **Quiz Scores**

Stem	Leaf
6	8
7	2 6
8	0 0 0 8
9	0 2 4 6
10	0

Key: 7 | 2 = 72%

2. **CDs Sold Each Day**

Stem	Leaf
2	9
3	1 8 8 9
4	0 0 3 5 5 8
5	
6	0 0 2 7

Key: 4 | 0 = 40 CDs

3. mean: 49; median: 52; mode: 54; range: 25; IQR: 14

4. 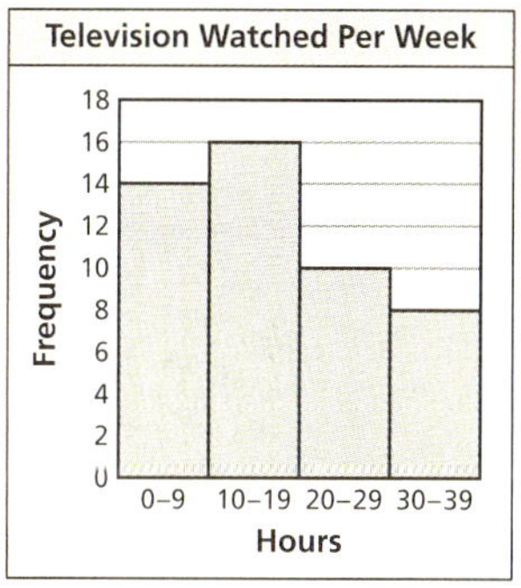

5. **a.** skewed left
   b. median, IQR; median = 8, IQR = 2

6. 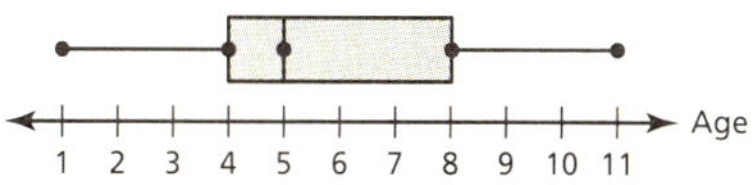

7–9. See Additional Answers.

Technology for the *Teacher*

Online Assessment
Assessment Book
ExamView® Assessment Suite

Test-Taking Strategies

Available at *BigIdeasMath.com*

After Answering Easy Questions, Relax
Answer Easy Questions First
Estimate the Answer
Read All Choices before Answering
Read Question before Answering
Solve Directly or Eliminate Choices
Solve Problem before Looking at Choices
Use Intelligent Guessing
Work Backwards

About this Strategy

When taking a multiple choice test, be sure to read each question carefully and thoroughly. Look closely for words that change the meaning of the question, such as *not, never, all, every,* and *always.*

Answers

1. D
2. F
3. C

Technology for the Teacher

Common Core State Standards Support
Performance Tasks
Online Assessment
Assessment Book
ExamView® Assessment Suite

Item Analysis

1. **A.** The student misreads the plot, thinking that the median is the number at its right end.

   B. The student misreads the plot, thinking that 8 is the median.

   C. The student misreads the plot, thinking that the lower whisker represents 50% of the data.

   D. Correct answer

2. **F.** Correct answer

   G. The student does not correctly compare the unit cost in answer choice G, \$0.325, with the unit cost in answer choice F, \$0.30.

   H. The student does not correctly compare the unit cost in answer choice H, \$0.375, with the unit cost in answer choice F, \$0.30.

   I. The student does not correctly compare the unit cost in answer choice I, \$0.33, with the unit cost in answer choice F, \$0.30.

3. **A.** The student rewrites the divisor as $\frac{17}{3}$ but does not find the reciprocal.

   B. The student only takes the reciprocal of the fraction part of the divisor and rewrites it as $5\frac{3}{2}$, which the student then rewrites as $\frac{13}{2}$.

   C. Correct answer

   D. The student only takes the reciprocal of the fraction part of the divisor and ignores the whole number part.

10 Standards Assessment

1. Research scientists are measuring the number of days lettuce seeds take to germinate. In a study, 500 seeds were planted. Of these, 473 seeds germinated. The box-and-whisker plot summarizes the number of days it took the seeds to germinate. What can you conclude from the box-and-whisker plot? *(6.SP.4, 6.SP.5c)*

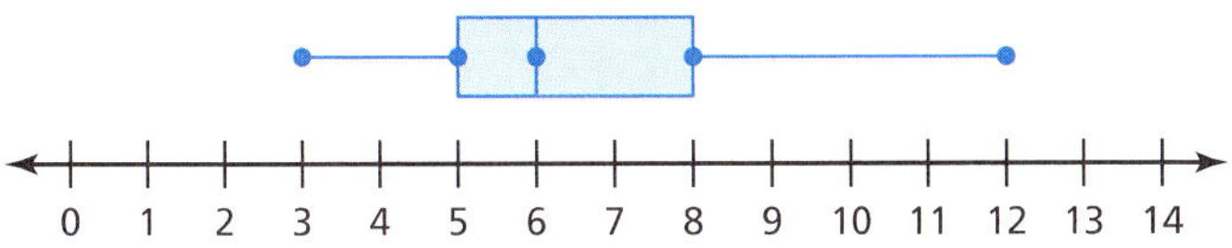

   A. The median number of days for the seeds to germinate is 12.

   B. 50% of the seeds took more than 8 days to germinate.

   C. 50% of the seeds took less than 5 days to germinate.

   D. The median number of days for the seeds to germinate was 6.

Test-Taking Strategy
Read Question Before Answering

"Be sure to read the question before choosing your answer. You may find a word that changes the meaning."

2. You are comparing the costs of buying bottles of water at the supermarket. Which of the following has the least cost per liter? *(6.RP.3b)*

   F. six 1-liter bottles for \$1.80

   G. one 2-liter bottle for \$0.65

   H. eight $\frac{1}{2}$-liter bottles for \$1.50

   I. twelve $\frac{1}{2}$-liter bottles for \$1.98

3. What number belongs in the box to make the equation true? *(6.NS.1)*

$$3\frac{1}{2} \div 5\frac{2}{3} = \frac{7}{2} \times \square$$

   A. $\frac{17}{3}$

   B. $\frac{13}{2}$

   C. $\frac{3}{17}$

   D. $\frac{3}{2}$

4. What is the mean number of seats? *(6.SP.5c)*

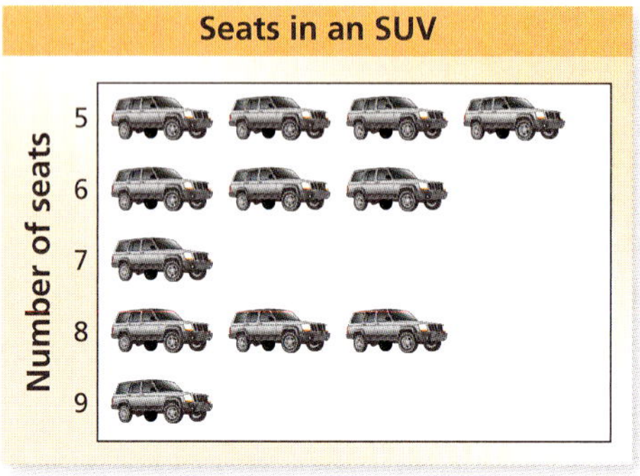

F. 2.4 seats

G. 5 seats

H. 6.5 seats

I. 7 seats

5. On Wednesday, the town of Mims received 17 millimeters of rain. This was x millimeters more rain than the town received on Tuesday. Which expression represents the amount of rain, in millimeters, the town received on Tuesday? *(6.EE.2a, 6.EE.6)*

A. $17x$

B. $17 - x$

C. $x + 17$

D. $x - 17$

6. One of the leaves is missing in the stem-and-leaf plot.

The median of the data set represented by the stem-and-leaf plot is 38. What is the value of the missing leaf? *(6.SP.4, 6.SP.5c)*

Stem	Leaf
1	3 4
2	
3	4 5 7 7 7 ? 9
4	0 1 1 4
5	0 2 3

Key: 1 | 4 = 14

7. Which property is demonstrated by the equation below? *(6.EE.3)*

$$723 + (884 + 277) = 723 + (277 + 884)$$

F. Associative Property of Addition

G. Commutative Property of Addition

H. Distributive Property

I. Identity Property of Addition

Item Analysis (continued)

4. **F.** The student adds the frequencies and divides by 5 to find the mean.

   G. The student thinks mean is mode.

   H. Correct answer.

   I. The student chooses the middle row.

5. **A.** The student misinterprets "more than" as meaning multiplication.

   B. Correct answer

   C. The student misinterprets the inverse relationship in the problem, thinking that the town received more rain on Tuesday than on Wednesday.

   D. The student subtracts in the wrong order and writes an expression that represents 17 millimeters less rain than the difference of the rain amounts on Tuesday and Wednesday.

6. **Gridded Response**: Correct answer: 7

   Common Error: The student thinks that the missing leaf represents the ones digit of the actual median, getting an answer of 8.

7. **F.** The student does not recognize the equation as an example of the Commutative Property of Addition.

   G. Correct answer

   H. The student does not recognize the equation as an example of the Commutative Property of Addition.

   I. The student does not recognize the equation as an example of the Commutative Property of Addition.

Answers

4. H

5. B

6. 7

7. G

Answers

8. C

9. 14 students

10. *Part A* 22, 24, 25, 28, 28, 30, 31, 37, 37, 39, 40, 40, 44, 51, 58, 62

Part B 37

Part C

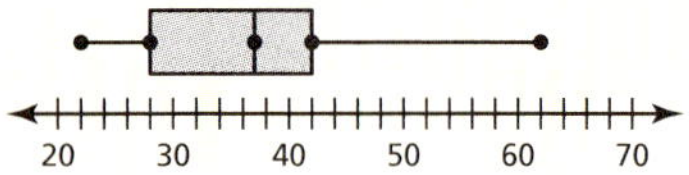

Item Analysis (continued)

8. **A.** The student thinks mean is mode.

B. The student finds the mean of the four scores (91) and adds 1 to 92 because 91 is 1 less than 92.

C. Correct answer

D. The student makes an arithmetic error finding the sum.

9. **Gridded Response:** Correct answer: 14 students

Common Error: The student only considers the bar immediately to the left of 10, getting an answer of 6.

10. **4 points** The student demonstrates a thorough understanding of how to make a box-and-whisker plot. The data is ordered correctly. The median of 37, first quartile of 28, third quartile of 42, least value of 22, and greatest value of 62 are identified correctly and graphed accurately.

3 points The student demonstrates an essential but less than thorough understanding of how to make a box-and-whisker plot. The data is ordered correctly and the median is found correctly. But, there may be a small mistake made subsequently, e.g., when finding the third quartile or when trying to accurately match the plot to the number line given.

2 points The student demonstrates a partial understanding of how to make a box-and-whisker plot. The student's work and explanations demonstrate a lack of essential understanding. For example, the data is ordered correctly, but the quartiles are misidentified.

1 point The student demonstrates very limited understanding of making a box-and-whisker plot. E.g., the only key value that the student can accurately find is the median, and the data may not be correctly ordered.

0 points The student provides no response, a completely incorrect or incomprehensible response, or a response that demonstrates insufficient understanding of making a box-and-whisker plot.

8. A student took 5 tests this marking period and had a mean score of 92. Her scores on the first 4 tests were 90, 96, 86, and 92. What was her score on the fifth test? *(6.SP.5c)*

A. 92 **C.** 96

B. 93 **D.** 98

9. At the end of the school year, your teacher counted up the number of absences for each student. The results are shown in the histogram below.

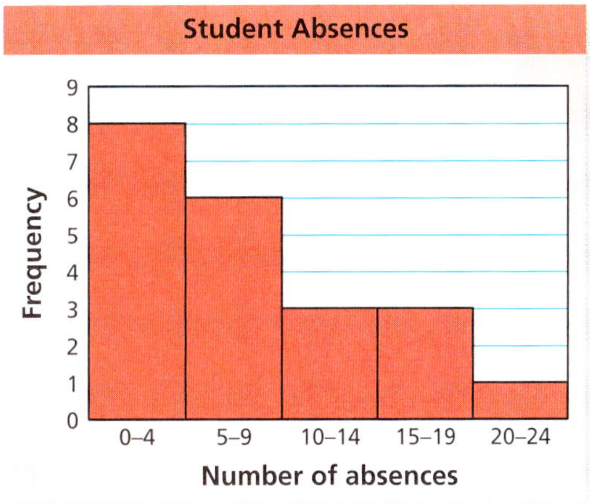

Based on the histogram, how many students had fewer than 10 absences? *(6.SP.4)*

10. The 16 members of a camera club have the ages listed below. *(6.SP.4, 6.SP.5c)*

40, 22, 24, 58, 30, 31, 37, 25, 62, 40, 39, 37, 28, 28, 51, 44

Part A Order the ages from least to greatest.

Part B Find the median of the ages.

Part C Make a box-and-whisker plot for the ages of the camera club members.

Appendix A
My Big Ideas Projects

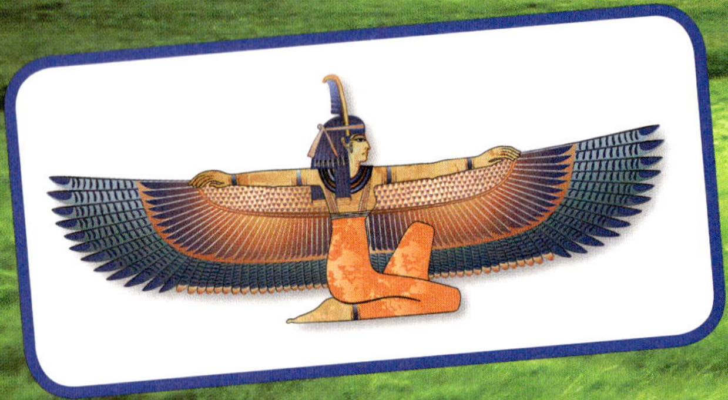

About the Appendix

- The interdisciplinary projects can be used anytime throughout the year.
- The projects offer students an opportunity to build on prior knowledge, to take mathematics to a deeper level, and to develop organizational skills.
- Students will use the Essential Questions to help them form "need to knows" to focus their research.

Essential Question

- **Literature Project**
  How does mathematical problem solving influence a story plot?
- **History Project**
  How is mathematical knowledge that was originally discovered by the ancient Egyptians used today?

Additional Resources

BigIdeasMath.com

Additional Resources

BigIdeasMath.com

Essential Question

- **Art Project**
  Do polyhedra influence the design of games and architecture?
- **Science Project**
  How does floating ice affect you and the world in which you live?

My Big Ideas Projects

A.1 Literature Project

Alice's Adventures in Wonderland

1 Getting Started

Lewis Carroll was a mathematician who also wrote literature books for children.

Essential Question How does mathematical problem solving influence a story plot?

Read *Alice's Adventures in Wonderland* by Lewis Carroll. In each chapter, rewrite one of the episodes so that it contains some of the math that you have studied this year.

Sample: You could rewrite Alice's shrinking adventure as follows.

Alice's normal height was 4′7″ or 55 inches. After drinking the bottle marked "Drink Me," Alice could feel herself shrinking more and more. After several minutes, Alice was only two-elevenths of her original height. Would Alice have to bend over to walk through a door that is only 1 foot high?

Project Notes

Introduction

For the Teacher

- **Goal:** Students will rewrite episodes from *Alice's Adventures in Wonderland* using some of the mathematics they studied this year. Samples from each chapter are included below.
- **Management Tip:** You may want to have students work together in groups. Depending on the number of groups you have, you can assign one or more chapters to each group.

Essential Question

- How does mathematical problem solving influence a story plot?

Things to Think About

- **Chapter 2: The Pool of Tears** When Alice was caught in the hallway, she had grown to be over 9 feet tall. She began "shedding gallons of tears, until there was a large pool all round her, about four inches deep and reaching half down the hall." She held onto a fan and began shrinking. How tall was Alice when she fell into the pool if she was "up to her chin in salt water?" (Her head is about 12.5% of her height.) approximately 4.57 inches
- **Chapter 3: A Caucus-Race and a Long Tale** The race course for the Caucus-Race was circular. Alice and her friends ran for about 30 minutes. If they ran approximately 1000 meters per hour and made 50 laps around the course, how long would it take them to run once across the center of the circle? approximately 11.5 seconds
- **Chapter 4: The Rabbit Sends in a Little Bill** Alice was stuck in the Rabbit's house when the Rabbit and his friends decided to send Bill down the chimney. They got two ladders, tied them together, and put them at the corner of the house. Even so, they did not reach half high enough. They had enough rope to enable Bill to get up to the roof and down the chimney. The ladders were each 2.4 feet tall. What other information would you need to find the length of the rope? Answers will vary.
- **Chapter 5: Advice from a Caterpillar** Alice began her day about 4 feet 7 inches tall. She began growing and shrinking and growing and shrinking, so much so that she was quite confused. As she happened upon the Caterpillar, she was approximately 3 inches tall. After eating from one side of the mushroom, she was as short as she could be, with her chin upon her feet. Eating from the other side of the mushroom resulted in her being taller than the tallest tree in the woods. She nibbled her way back to her original height and was quite happy until she happened upon a house that was four feet tall. She then began nibbling the mushroom until she was 9 inches tall. Overall, how much did Alice's height change after she began nibbling the mushroom and why did she decide that 9 inches was a good height? 6 inches; She did not want to frighten the owner of the house.

References

Go to *BigIdeasMath.com* to access links related to this project.

Cross-Curricular Instruction

Meet with a reading or language arts teacher and review curriculum maps to identify whether students have read *Alice's Adventures in Wonderland.* If the book has been read, you may want to discuss the work students have completed and then review the book with them. If the book has not been read, perhaps you can both work simultaneously and share notes. Or, you may want to explore activities that the reading or language arts teacher has done in the past to support student learning in this particular area.

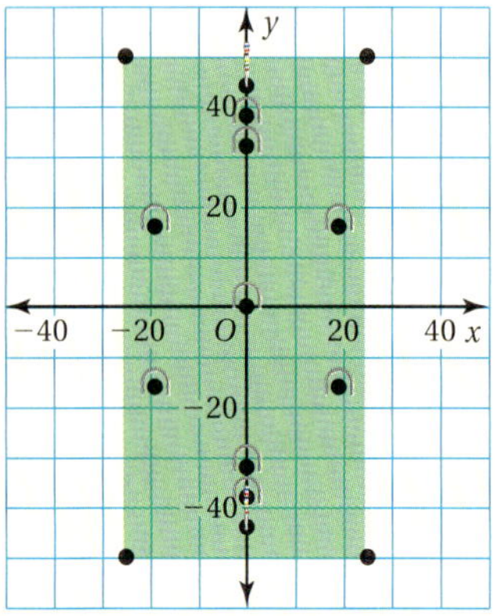

Project Notes

- **Chapter 6: Pig and Pepper** The Cheshire Cat used reasoning to prove to Alice that he was mad. Alice used reasoning to decide that visiting the March Hare was the better choice. She also used reasoning to determine when she had arrived at the March Hare's house. What was Alice's reasoning? Do you agree with her? She thought that visiting the March Hare would be more interesting than visiting the Hatter because she had seen hatters before; She thought it must be the right house, because the chimneys were shaped like ears and the roof was thatched with fur; *Answers will vary.*
- **Chapter 7: A Mad Tea-Party** During the tea-party, all the attendees moved one place on. The result of the translation was that the Hatter was the only one who had a clean cup. What translation would have been necessary for all of the guests to have a clean cup? four places on
- **Chapter 8: The Queen's Croquet Ground** The Queen's croquet court was quite different as, among other things, the soldiers were used as the arches and they kept moving about. Use the Internet or some other reference to find what a croquet court looks like. Using a coordinate plane with the center of the court as the origin, draw and label the location of each arch (wicket) and stake as well as the ordered pairs that would mark the corners of the court. See the diagram at the left.
- **Chapter 9: The Mock Turtle's Story** When the Mock Turtle went to school, his lessons were quite unusual, as was the length of the school day. The first day was 10 hours long, the next was 9 hours long, and so on to the end. How many hours were the lessons on the 5th day? How many hours were the lessons all together? What do you think happened on the 12th day? 6 hours; 55 hours; *Answers will vary.*
- **Chapter 10: The Lobster Quadrille** The Lobster Quadrille was a very curious dance. Find the definition of the word *quadrille.* Name and describe as many quadrilaterals as possible. Explain how a quadrille is related to a quadrilateral. *Sample answers:* a square dance for 4 couples (see illustration); *parallelogram:* a quadrilateral with 2 pairs of parallel sides; *rectangle:* a parallelogram with 4 right angles; *rhombus:* a parallelogram with 4 sides of equal length; *square:* a parallelogram with 4 right angles and 4 sides of equal length; *trapezoid:* a quadrilateral with one pair of parallel sides; A quadrille is a square dance and a square is a type of quadrilateral. Also, a quadrille requires 4 couples and a quadrilateral has 4 sides.
- **Chapter 11: Who Stole the Tarts?** During court, Alice began to grow. If she was a foot tall as court began and she grew at a rate of 2 inches per minute, how long did it take for her to be back to her original height of 4 feet 7 inches? 21.5 minutes

Closure

- **Rubric** An editable rubric for this project is available at *BigIdeasMath.com.*
- Students may present their reports to the class or school as a television report or public information broadcast.

The illustrations on these two pages are from the 1907 edition of *Alice's Adventures in Wonderland*. They were drawn by Arthur Rackham (1867–1939), a book illustrator from England.

2 Things to Remember

- You can download each chapter of the book at *BigIdeasMath.com*.
- Add your own illustrations to your project.
- Try to include as many different math concepts as possible. Your goal is to include at least one concept from each of the chapters you studied this year.
- Organize your math stories in a folder, and think of a title for your report.

A.2 History Project

Mathematics in Ancient Egypt

1 Getting Started

The ancient Egyptian civilization in North Africa began around 3150 B.C. with the union of Upper and Lower Egypt under the first pharaoh. The rule of the pharaohs ended in 31 B.C. when the Romans conquered Egypt and made it a province.

The ancient Egyptians' achievements included a system of mathematics, glass technology, medicine, literature, irrigation and agricultural techniques, and art. Their mastery of surveying, quarrying, and construction techniques allowed them to build pyramids, temples, and obelisks.

Essential Question How is mathematical knowledge that was originally discovered by the ancient Egyptians used today?

Sample: To multiply two whole numbers, the ancient Egyptians repeatedly multiplied by 2. The "doubling lists" used to multiply 53 by 85 are shown. The lists stop when the last number in the first list is greater than the first number in the second list.

•	1	53
	2	106
•	4	212
	8	424
•	16	848
	32	1696
•	64	3392

Because $85 = 1 + 4 + 16 + 64$,
$85 \cdot 53 = 53 + 212 + 848 + 3392 = 4505$.

Ancient Papyrus

The Blue Sphinx

Ancient Egyptian Numbers

Project Notes

Introduction

For the Teacher

- **Goal:** Students will discover how ancient Egyptians wrote and used mathematics.
- **Management Tip:** Students can work in groups to create reports about the mathematics used in ancient Egypt.

Essential Question

- How is mathematical knowledge that was originally discovered by the ancient Egyptians used today?

Things to Think About

? How did the ancient Egyptians multiply two whole numbers?

- Egyptians only knew how to multiply and divide by 2. For example, in order to find 27 • 5, they did the following:

$$(27 \cdot 2) + (27 \cdot 2) + (27 \cdot 1) = 54 + 54 + 27 = 135.$$

- For more complicated numbers, they used powers of 2 and doubling. For example, the doubling pattern below shows how they found 45 • 67.

1	2	4	8	16	32	64	128
67	134	268	536	1072	2144	4288	8576

Because $45 = 1 + 4 + 8 + 32$, $45 \cdot 67 = 67 + 268 + 536 + 2144 = 3015$.

? How did the ancient Egyptians divide two whole numbers?

- The process the Egyptians used to divide was similar to the process they used to multiply. Suppose they wanted to divide 308 by 28. They thought about this problem as "28 times some number equals 308."

1	2	4	8	16	32
28	56	112	224	448	896

Because $308 = 28 + 56 + 224$ and $1 + 2 + 8 = 11$, $308 \div 28 = 11$.

? How did the ancient Egyptians write whole numbers?

- The numbers we use today are far different from the ones used by the Egyptians. The Egyptians used hieroglyphics to represent numbers. They used a vertical line (a staff) to represent 1, an upside-down U (a whale bone) to represent 10, a coil (a coil of rope) to represent 100, and a flower (a lotus flower) to represent 1000.
- To write the number 234, draw, from right to left, two coils of rope, three whale bones, and four staffs.

234 = 

References

Go to *BigIdeasMath.com* to access links related to this project.

Cross-Curricular Instruction

Meet with a history teacher and review curriculum maps to identify whether students have covered ancient Egypt. If the topic has been covered, you may want to discuss the work students have completed and review prior knowledge with them. If the history teacher has not discussed these concepts, perhaps you can both work simultaneously on these concepts and share notes. Or, you may want to explore activities that the history teacher has done in the past to support student learning in this particular area.

Project Notes

- To write the number 1302, draw, from right to left, one lotus flower, three coils of rope, and two staffs.

1302 = 

? How did the ancient Egyptians write fractions?

- Egyptians understood the concept of fractions but did not write them as we do today. They wrote most fractions with a numerator of 1. To do this, they wrote the number the same as the whole number, but placed an oval symbol above the number to show that it was a fraction.

$\frac{1}{10}$ = 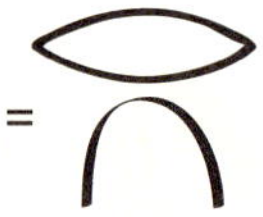

- There were three exceptions to this rule:

$\frac{1}{2}$ = , $\frac{2}{3}$ = , and $\frac{3}{4}$ = .

? How did the ancient Egyptians use ratios to measure the height of a pyramid?

- Thales, an ancient philosopher, discovered a way to measure the heights of very tall objects, such as the pyramids, using shadows. He noticed that the ratio of his height to the length of his shadow was equal to the ratio of a pyramid's height to the length of its shadow at the same time of day. So, to find the height of a pyramid, he used the proportion

$$\frac{\text{Height of Thales}}{\text{Length of Thales' Shadow}} = \frac{\text{Height of Pyramid}}{\text{Length of Pyramid's Shadow}}.$$

? How did the ancient Egyptians use mathematics?

- The ancient Egyptians did not have electric clocks or wrist watches, so they developed a way to measure time based on shadow clocks. They built obelisks and used the moving shadows created by the obelisks to help tell time.
- Instead of using yardsticks and measuring tape to measure lengths, the ancient Egyptians used parts of their body. The length of a person's forearm was called a royal cubit. A handbreadth (the length of a palm) was considered to be one seventh of a royal cubit. A fingerbreadth (the length of a finger) was even smaller; it was one fourth of a handbreadth. They even had a measurement for the volume of grains called a hekat.
- Because rulers did not exist, ancient Egyptians used rods with markings on them to measure the depth of the Nile River. They also invented a calendar to help them predict when the Nile would flood.

Closure

- **Rubric** An editable rubric for this project is available at *BigIdeasMath.com.*
- You may hold a class debate where students compare, defend, and discuss their findings with other students.

2 Things to Include

- Explain how the ancient Egyptians multiplied two whole numbers. Give an example.
- Explain how the ancient Egyptians divided two whole numbers. Give an example.
- How did the ancient Egyptians write whole numbers? Give some examples.
- How did the ancient Egyptians write fractions? Give some examples.
- How did the ancient Egyptians use ratios to measure the height of a pyramid? Give an example.
- Write your name using ancient Egyptian hieroglyphics.
- Describe how ancient Egyptians used mathematics. How does this compare with the ways in which mathematics is used today?

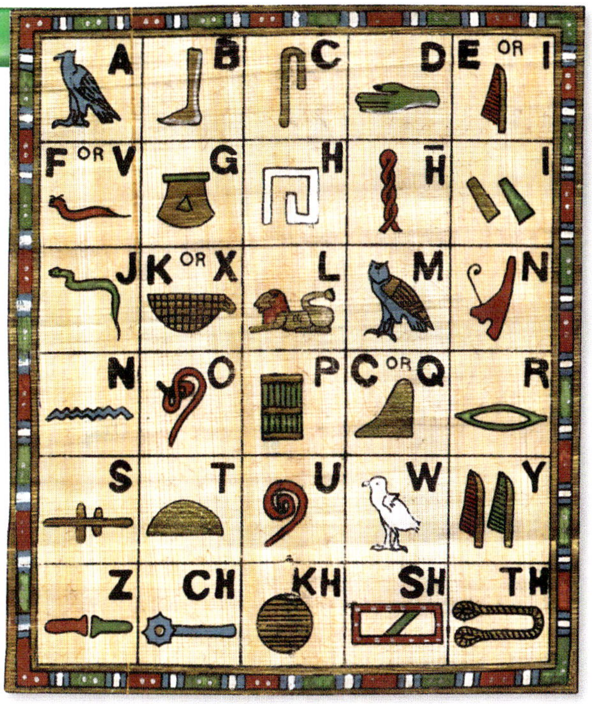

Ancient Egyptian Alphabet

3 Things to Remember

- Add your own illustrations to your project.
- Try to include as many different math concepts as possible. Your goal is to include at least one concept from each of the chapters you studied this year.
- Organize your report in a folder, and think of a title for your report.

What does this spell?

A.3 Art Project

Polyhedra in Art

1 Getting Started

Polyhedra is the plural of *polyhedron*. Polyhedra have been used in art for many centuries, in cultures all over the world.

Essential Question Do polyhedra influence the design of games and architecture?

Some of the most famous polyhedra are the five Platonic solids shown at the right.

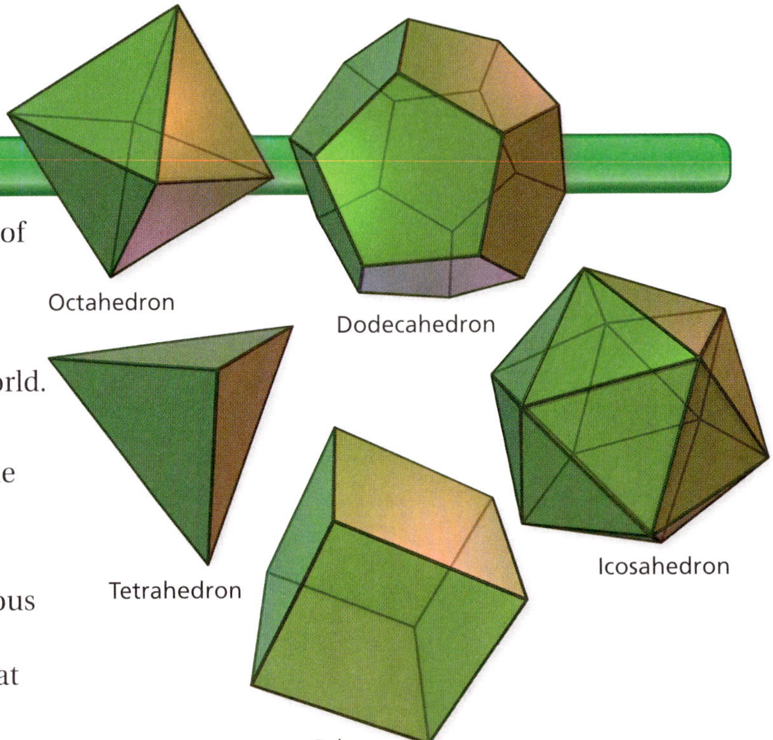

Sample: Three of the Platonic solids have faces that are equilateral triangles. One of these is the octahedron. Here is a net for an octahedron.

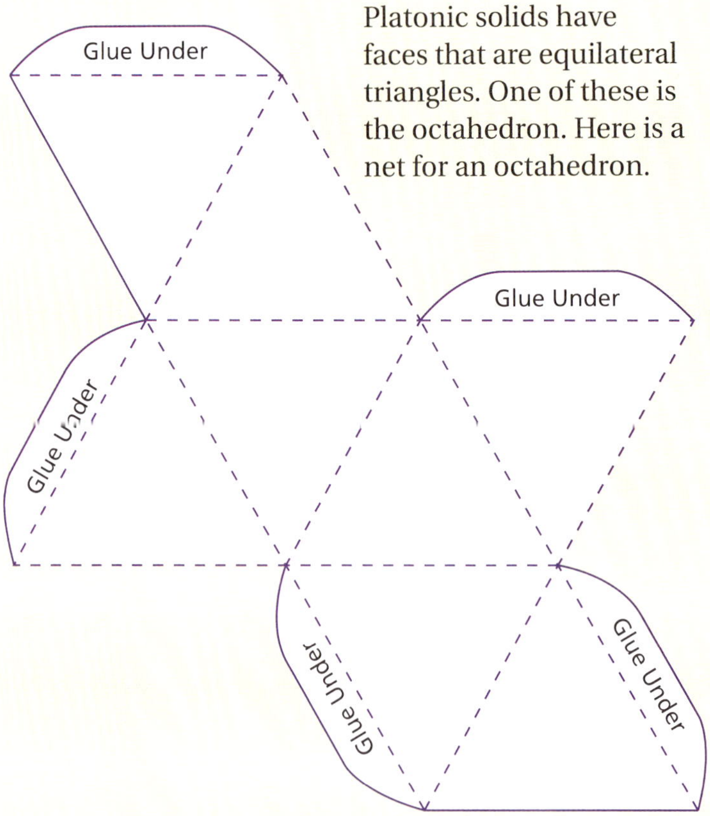

Mosaic by Paolo Uccello, 1430 A.D.

Project Notes

Introduction

For the Teacher

- **Goal:** Students will research and discover how polyhedra are used in games and architecture. They will also create their own polyhedra.
- **Management Tip:** Students may wish to search online or visit *BigIdeasMath.com* for links to websites containing instructions on how to make their own icosahedrons or dodecahedrons.

Essential Question

- Do polyhedra influence the design of games and architecture?

Things to Think About

? What are polyhedra?

- Polyhedra are three-dimensional shapes that have faces (sides) that are polygons. The polygons meet along straight line segments (edges) and the edges meet at vertices.

? What did Plato associate with the five Platonic solids?

- The Platonic solids, named for Plato, are a prominent feature of his philosophy. In his dialogue *Timaeus c. 360 B.C.*, Plato associated each of the four classical elements with a regular solid.

  "There was intuitive justification for these associations: the heat of fire feels sharp and stabbing (like little tetrahedra). Air is made of the octahedron; its minuscule components are so smooth that one can barely feel it. Water the icosahedron, flows out of one's hand when picked up, as if it is made of tiny little balls. By contrast, a highly un-spherical solid, the hexahedron (cube) represents earth. These clumsy little solids cause dirt to crumble and break when picked up, in stark difference to the smooth flow of water."

- Because there were only the four elements, there was need for only the four polyhedra. However, a fifth polyhedra, the dodecahedron, was used for the universe. Plato remarked, "*...the god used for arranging the constellations on the whole heaven.*"

? Why are polyhedra used in architecture?

- From the Renaissance to present time, many artists saw the use of polyhedra as a challenging way to demonstrate the mastery of perspective. Others, such as Plato, thought that polyhedra symbolized philosophical truths or religious beliefs. Other artists used polyhedra simply because of their symmetrical beauty. From the Great Pyramids in Egypt to the Spaceship Earth geosphere at Epcot, polyhedra have been a part of architecture since prehistoric times.

References

Go to *BigIdeasMath.com* to access links related to this project.

Cross-Curricular Instruction

Meet with an art teacher and review curriculum maps to identify whether students have covered or have yet to discuss polyhedra. If the topic has been covered, you may want to discuss the work students have completed and review prior knowledge with them. If the art teacher has not discussed these concepts, perhaps you can both work simultaneously on these concepts and share notes. Or, you may want to explore activities that the art teacher has done in the past to support student learning in this particular area.

Project Notes

? What is one of the most popular uses of polyhedra in games?

- Polyhedra dice games are a popular way to explore mathematics. The following are examples of polyhedra dice.

? What are the Archimedean Solids?

- The Archimedean Solids are symmetric, semi-regular convex polyhedra. They are composed of two or more types of regular polygons meeting at identical vertices. Three are listed below. Visit *BigIdeasMath.com* to view all 13 Archimedean Solids.

Name	rhombicubocta-hedron	truncated cuboctahedron	rhombicosidodeca-hedron
Solid		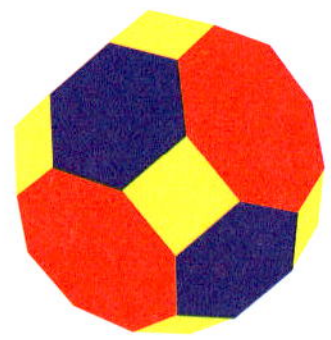	
Net			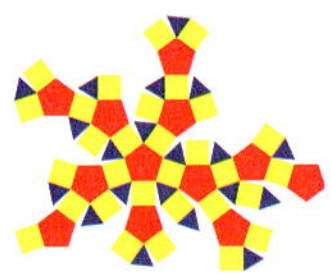
Faces	26	26	62
Faces (by type)	8 triangles; 18 squares	12 squares; 8 hexagons; 6 octagons	20 triangles; 30 squares; 12 pentagons
Edges	48	72	120

Closure

- **Rubric** An editable rubric for this project is available at *BigIdeasMath.com*.
- Students may present their reports to a parent panel or community members.

2 Things to Include

- Explain why the platonic solids are sometimes referred to as the cosmic figures.
- Draw a net for an icosahedron or a dodecahedron. Cut out the net and fold it to form the polyhedron.
- Describe the 13 polyhedra that are called Archimedean solids. What is the definition of this category of polyhedra? Draw a net for one of them. Then cut out the net and fold it to form the polyhedron.
- Find examples of polyhedra in games and architecture.

Faceted Cut Gem

Origami Polyhedron

3 Things to Remember

- Add your own illustrations or paper creations to your project.
- Organize your report in a folder, and think of a title for your report.

Concrete Tetrahedrons by Ocean

Bulatov Sculpture

A.4 Science Project

Why Does Ice Float?

1 Getting Started

Among the planets in our solar system, Earth is clearly the "water planet." Water occurs on Earth's surface as a liquid, solid, and gas. Ocean waters cover about 70% of Earth's surface. Fresh water in lakes and rivers covers less than 1%. Thick sheets of ice permanently cover Earth's polar regions, and glaciers occur in its higher mountains. Water in the form of clouds covers about half of Earth's surface at any time.

Essential Question How does floating ice affect you and the world in which you live?

There are three parts to this question. First, you have to understand why things float in water. Make lists of things that float in water and things that do not float in water. How can you describe, scientifically, the difference between these two types of objects?

Second, you have to discover what is special about water. Almost all other elements and compounds have the property that their solid forms sink in their liquid forms.

Third, you have to research how water in solid form influences the environment and living organisms. You may want to explore ocean currents, weather, and various cycles including the water cycle.

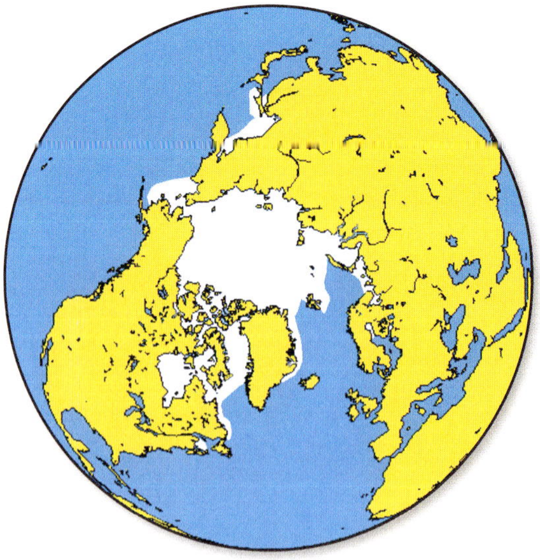

Earth's North Polar Ice Cap

Project Notes

Introduction

For the Teacher

- **Goal:** Students will discover why objects sink or float in water.
- **Management Tip:** You can have students discover other items that float in water. Allow students to bring items to school for this experiment. You can use a sink or bucket to collect the water.

Essential Question

- How does floating ice affect you and the world in which you live?

Things to Think About

? Why do some objects float in water and others do not?

- Objects float in water because they are less dense than water. Some objects may float in salt water and not in tap water, because salt makes water more dense.

Do these things float in water?	
Yes	**No**
bread	egg
apple	chocolate chip
vegetable oil	rock
pepper	penny
aluminum can	spaghetti
flour	key

? Do all compounds in a solid state float in their liquid state?

- With gold, nitrogen, solid ethanol, and most other compounds, the solid form is smaller than the liquid form. So the solid form is more dense and will sink when placed in the liquid form.
- Water is the opposite. It expands when it goes from liquid to solid, so ice is less dense and floats in water.

? Why does ice float in water?

- Even though ice is a solid state, ice (density $= 0.92$ g/mL) is less dense than water (density $= 1.0$ g/mL). Objects less dense than 1.0 g/mL will float. Objects more dense than 1.0 g/mL sink. Ice is less dense than water because water expands as it freezes.

? Does ice float in other liquids?

- Ice will float in vegetable oil because ice is less dense than vegetable oil.

References

Go to *BigIdeasMath.com* to access links related to this project.

Cross-Curricular Instruction

Meet with a science teacher and review curriculum maps to identify whether students have covered density. If the topic has already been covered, you may want to discuss the work students have completed and review prior knowledge with them. If the science teacher has not discussed these concepts, perhaps you can both work simultaneously on these concepts and share notes. Or, you may want to explore activities that the science teacher has done in the past to support student learning in this particular area.

Project Notes

? Are the majority of icebergs below the surface?

- The phrase "just the tip of the iceberg" came from the fact that most of an iceberg is below water. Research indicates that 80%-90% of an iceberg is below water.
- Very dense icebergs are deep blue. Less dense icebergs are white because they contain many trapped air bubbles. Icebergs that are greenish-black have calved off the bottom of a glacier. Dark brown stripes in an iceberg are an indication that the iceberg came from two glaciers that joined together.

? What would happen if ice did not float?

- If icebergs did not float, the oceans and waterways would freeze from the bottom up. The result would be a loss of biological life throughout the world.
- If icebergs were completely under water the sea level would rise and there would be great flooding, particularly in the low coastal regions. This would be devastating to the plants and animals in those areas.

? How does floating ice keep the oceans warm?

- Floating ice acts as a blanket in the winter, keeping the warm ocean from releasing all of its heat into the colder atmosphere.

Closure

- **Rubric** An editable rubric for this project is available at *BigIdeasMath.com.*
- Students may present their reports to the class or compare their reports with other students' reports.

2 Things to Include

- Does vegetable cooking oil float in water? Does ice float in vegetable oil?
- Compare the density of water and the density of ice. How do these densities relate to the fact that ice floats in water?
- What percent of an iceberg is above the water? What percent is below?
- What environmental and biological systems would not work if ice sunk in the oceans instead of floating?
- Explain how the floating of ice helps keep the oceans warm and therefore helps aquatic life exist in Earth's polar oceans.

3 Things to Remember

- Add your own illustrations to your project.
- Try to include as many different math concepts as possible. Your goal is to include at least one concept from each of the chapters you studied this year.
- Organize your report in a folder, and think of a title for your report.

Floating Iceberg

Key Vocabulary Index

Mathematical terms are best understood when you see them used and defined *in context*. This index lists where you will find key vocabulary. A full glossary is available in your Record and Practice Journal and at *BigIdeasMath.com*.

Student Index

This student-friendly index will help you find vocabulary, key ideas, and concepts. It is easily accessible and designed to be a reference for you whether you are looking for a definition, real-life application, or help with avoiding common errors.

A

B

C

D

Student Index

E

N

O

Student Index

Student Index

S

T

U

V

W

Additional Answers

Chapter 1

Section 1.2

Practice and Problem Solving

37.

Power	4^6	4^5	4^4	4^3	4^2	4^1
Value	4096	1024	256	64	16	4

As the exponent decreases, the value of the power is divided by 4. $4^0 = 1$

Session 1.4

Practice and Problem Solving

25.

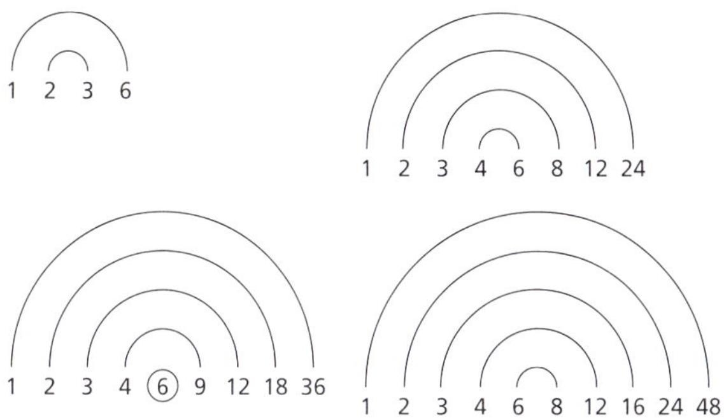

39. 6 prisms; There are 6 unique arrangements of length, width, and height using the factors of 40. (Note that $1 \times 1 \times 40$ names the same prism as $40 \times 1 \times 1$.); $1 \times 1 \times 40$, $1 \times 2 \times 20$, $1 \times 4 \times 10$, $1 \times 5 \times 8$, $2 \times 2 \times 10$, $2 \times 4 \times 5$

Section 1.5

Record and Practice Journal

4. *Sample answer:* There are two methods you can use: (1) organize the factors of the two numbers using a Venn diagram and identify the greatest common factor, or (2) write the prime factorizations of the two numbers and find the product of their common prime factors.

Practice and Problem Solving

32. a.

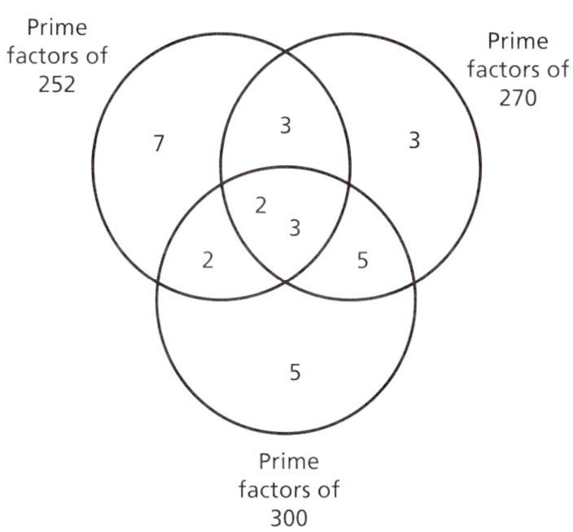

b. 6

c. 18; 12; The GCF of two numbers is the product of the prime factors in the overlap of the circles representing the numbers in the Venn diagram.

Section 1.6

Record and Practice Journal

3. *Sample answer:* There are two methods you can use: (1) organize the multiples of the two numbers using a Venn diagram and identify the least common multiple, or (2) write the prime factorizations of the two numbers and find the product of all their prime factors.

Practice and Problem Solving

35. a.

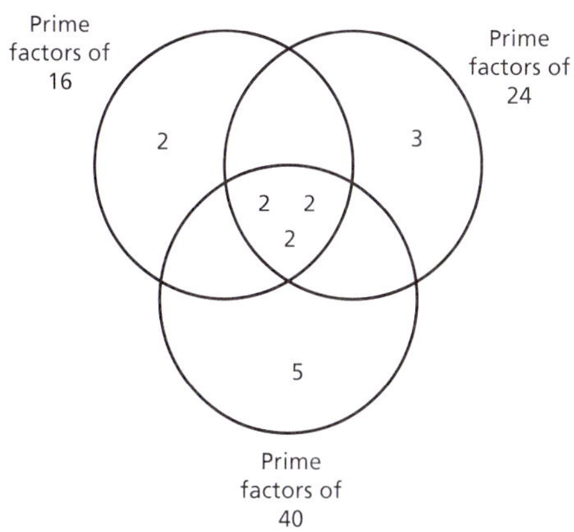

Chapter 2

Section 2.1

Practice and Problem Solving

25. $\frac{5}{6} = \left(\frac{5}{6} \times \frac{7}{7}\right)$; Because $\frac{7}{7} = 1$, by the Multiplication Property of One, the two expressions are equal.

43. You must first rewrite the mixed numbers as improper fractions and then multiply.

$$2\frac{1}{2} \times 7\frac{4}{5} = \frac{5}{2} \times \frac{39}{5}$$
$$= \frac{{}^{1}\cancel{5} \times 39}{2 \times \cancel{5}_{1}}$$
$$= \frac{39}{2}, \text{ or } 19\frac{1}{2}$$

44. $\frac{3}{80}$ g

45. a. 7 ft^2 **b.** $10\frac{1}{3}$ ft^2

46. less than; greater than; Because $\frac{4}{5} < 1$, the product will be less than the other factor, $1\frac{1}{6}$. Because $1\frac{1}{6} > 1$, the product will be greater than the other factor, $\frac{4}{5}$.

Section 2.5

Practice and Problem Solving

68. *Answers should include, but are not limited to:*

- **a.** menu with main items, desserts, beverages, and prices
- **b.** guest check for 5 people showing items, prices, and subtotal
- **c.** tax and total with tax are shown
- **d.** amount rounded to nearest dollar, 20% tip, and total amount paid are shown

Chapter 3

Record and Practice Journal Fair Game Review

1. $2 \times (126 + 2566)$ is 2 times as large as $126 + 2566$.

2. $4 \times (6425 + 25)$ is 4 times as large as $6425 + 25$.

3. $(65 - 23) + 3$ is 3 more than $65 - 23$.

4. $(65{,}000 - 5169) + 58$ is 58 more than $65{,}000 - 5169$.

5. $(890 \div 2) \div 2$ is half as large as $890 \div 2$.

6. $(65 \times 6) \div 3$ is one-third as large as 65×6.

7. *Sample answer:* You and two friends each buy an amusement park ticket for \$20 and game tickets for \$6. What is the total amount spent?

8. 0 **9.** 9

10. 12 **11.** 21

12. 20 **13.** 20

14. 38 **15.** 8

16. \$105

Section 3.1

Practice and Problem Solving

14. There is only one term, not three.
Term: $2x^2y$
Coefficient: 2
Constant: none

15. a. Terms: 2ℓ, $2w$; Coefficients: 2, 2; Constant: none

b. The coefficient 2 of ℓ represents that there are 2 lengths on the rectangle.
The coefficient 2 of w represents that there are 2 widths on the rectangle.

54. a. \$20 for an adult; \$13 for a student; The coefficient of a represents the cost per adult. The coefficient of c represents the cost per student.

b. \$392

c. yes; *Sample answer:* For 8 adults and 48 students, the cost is $20(8) + 13(48) = \$784$. Because $2(392) = 784$, the cost doubles when the number of adults and students in the group both double.

d. no; *Sample answer:* For 2 adults and 48 students, the cost is $20(2) + 13(48) = \$664$. The cost is not the same.

Section 3.2

Practice and Problem Solving

31. a.

Game	1	2	3	4	5
Cost	\$5	\$8	\$11	\$14	\$17

b. $2 + 3g$ **c.** \$26

Section 3.3

Record and Practice Journal

5. *Sample answer:* You can use the Commutative Property of Addition to reorder the numbers: $11 + 9 + 7 + 13 + 12 + 8$. Then you can use the Associative Property of Addition to group the numbers and add:
$(11 + 9) + (7 + 13) + (12 + 8)$
$= 20 + 20 + 20 = 60$. Because each group of numbers adds to 20, it is easier to add the sum in your head.

Practice and Problem Solving

17. $(0 + a) + 8 = a + 8$ Add. Prop. of Zero

18. $9 \cdot c \cdot 4 = 9 \cdot 4 \cdot c$ Comm. Prop. of Mult.
$= (9 \cdot 4) \cdot c$ Assoc. Prop. of Mult.
$= 36c$ Multiply 9 and 4.

19. $(18.6 \cdot d) \cdot 1 = 18.6 \cdot (d \cdot 1)$ Assoc. Prop. of Mult.
$= 18.6\,d$ Mult. Prop. of One

20. $\left(3k + 4\frac{1}{5}\right) + 8\frac{3}{5} = 3k + \left(4\frac{1}{5} + 8\frac{3}{5}\right)$ Assoc. Prop. of Add.
$= 3k + 12\frac{4}{5}$ Add $4\frac{1}{5}$ and $8\frac{3}{5}$.

21. $(2.4 + 4n) + 9 = (4n + 2.4) + 9$ Comm. Prop. of Add.
$= 4n + (2.4 + 9)$ Assoc. Prop. of Add.
$= 4n + 11.4$ Add 2.4 and 9.

22. $(3s) \cdot 8 = (s \cdot 3) \cdot 8$ Comm. Prop. of Mult.
$= s \cdot (3 \cdot 8)$ Assoc. Prop. of Mult.
$= s \cdot 24$ Multiply 3 and 8.
$= 24s$ Comm. Prop. of Mult.

23. $z \cdot 0 \cdot 12 = (z \cdot 0) \cdot 12$ Assoc. Prop. of Mult.
$= 0 \cdot 12$ Mult. Prop. of Zero
$= 0$ Mult. Prop. of Zero

24. $x + 16$

25. a. x represents the cost of a box of cookies.

b. $120x$

27. $7 + (x + 5) = x + 12$

28. $8 \cdot (y \cdot 9) = 72y$

29. $(7 \cdot 2) \cdot y$

30. $13.2 \cdot (1 \cdot x)$

31. $(17 + 6) + 2x$

32. $2 + c$

33. $w \cdot 16$

34. a. $x + 37$

b. In the expression $37(14) + 10x$, 37 represents the total number of hats you sold, 14 represents the price per hat you charged, 10 represents the price per hat your friend charged, and x represents the number of hats sold by your friend. Because \$14 > \$10, your friend was selling the hats at a discounted price.

c. $x \leq 51$

Section 3.4

Practice and Problem Solving

63. a. 6.2

b. 14

Sample answer: The preferred method is not the same for both expressions. For part (a), evaluating inside the parentheses first makes for easier and fewer calculations. For part (b), using the Distributive Property will eliminate the fractions.

64. 0; For every positive value of x, the perimeters are equal.

65. $7(x + 3) + 8 \cdot x + 3 \cdot x + 8 - 9 = 2(9x + 10)$

Extension 3.4

Practice

20. The first solution calculates the total earned from selling the tickets and the total paid for the tickets, and then subtracts to find the profit. The second method applies the Distributive Property first to generate an equivalent expression which represents the profit per ticket times the number of tickets; *Sample answer:* The second method because it is easier to calculate.

Chapter 4

Record and Practice Journal Fair Game Review

7. 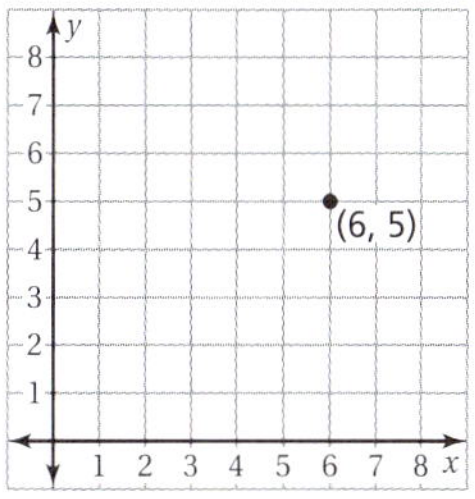

8. 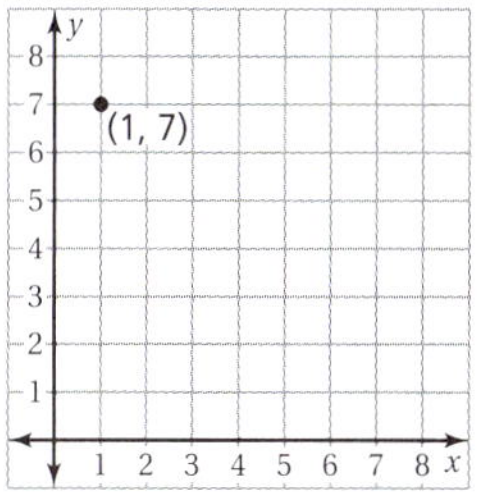

9. 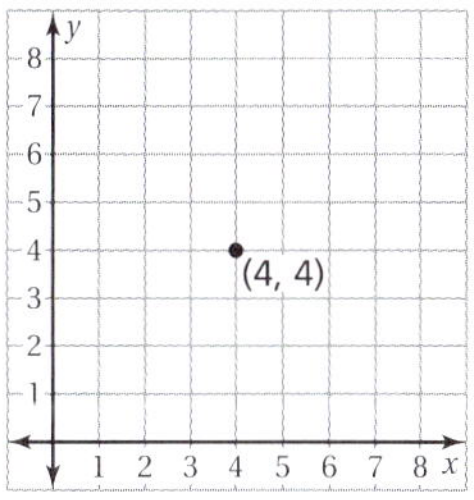

10. 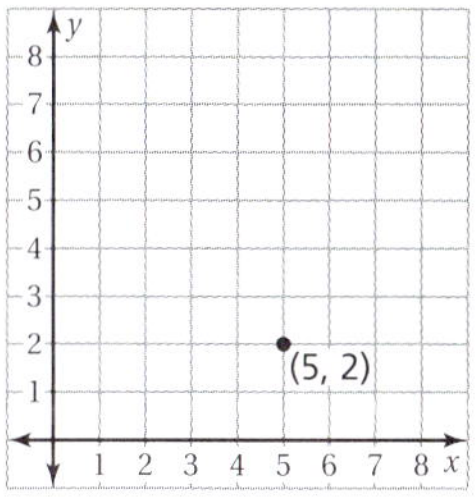

11. 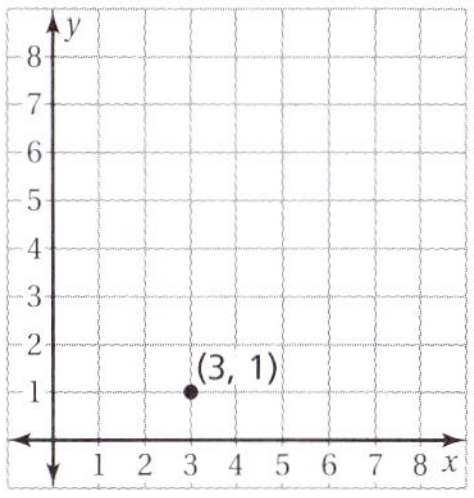

Section 4.1

Practice and Problem Solving

20.

Parallelogram	Base	Height	Area
A	$x + 4$	5	$5x + 20$
B	$x - 3$	8	$8x - 24$
C	6	$2x + y$	$12x + 6y$

21. n^2bh where b represents the base and h represents the height of the original parallelogram, or n^2A where A represents the area of the original parallelogram.

Section 4.3

Fair Game Review

23–26.

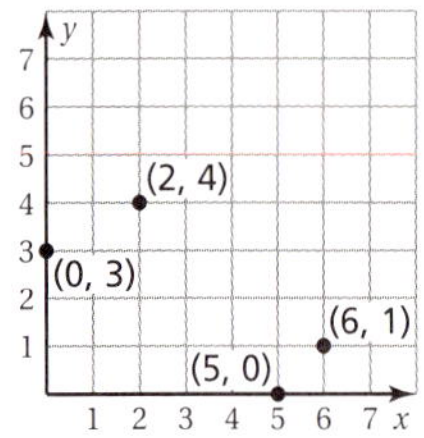

Extension 4.3

Practice

7. 3400 yd^2; *Sample answer:* Separate the figure into a trapezoid and a square.

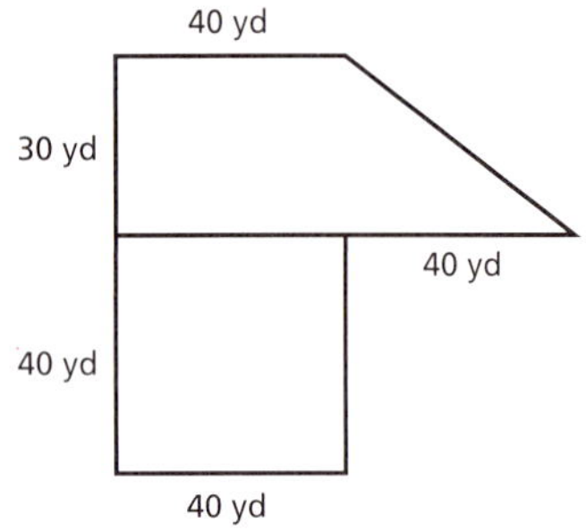

Section 4.4

Practice and Problem Solving

7.

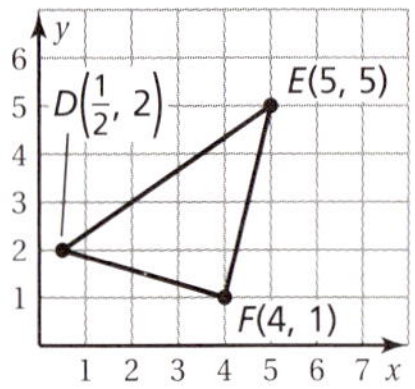

8.

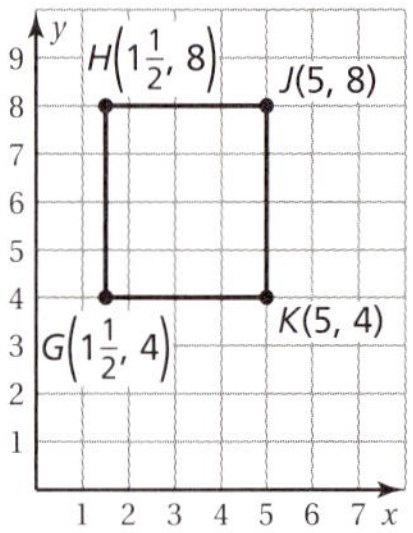

9.

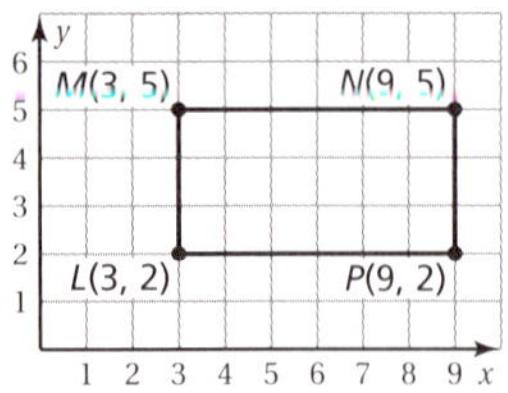

10.

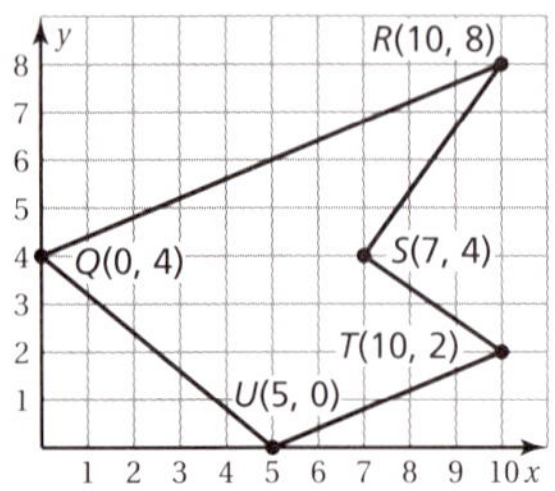

11.

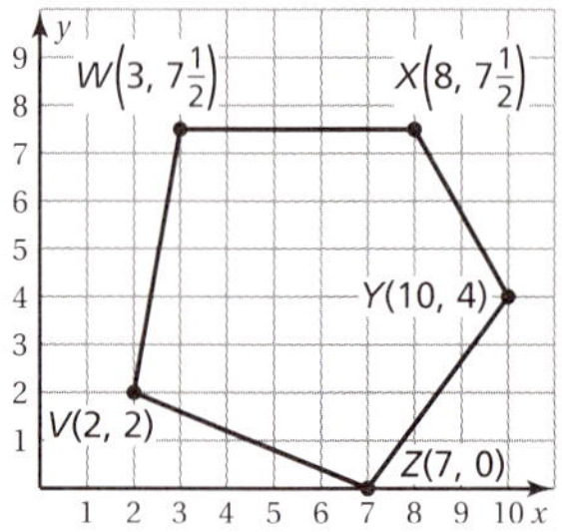

12. 12 units; 9 units^2

13. 24 units; 36 units^2

14. 16 units; 15 units^2

15. 28 units; 45 units^2

16. The x- and y-coordinates are reversed.

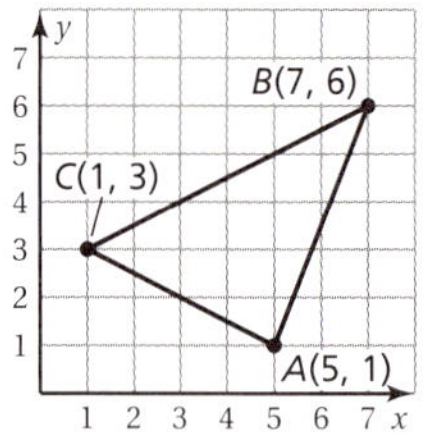

17. **a.** square **b.** 28 ft; 49 ft^2

18. *Sample answer:*

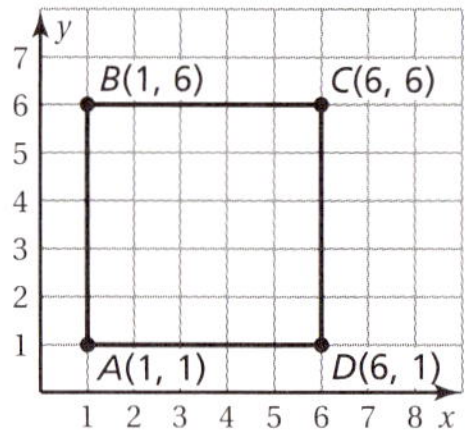

19. *Sample answer:*

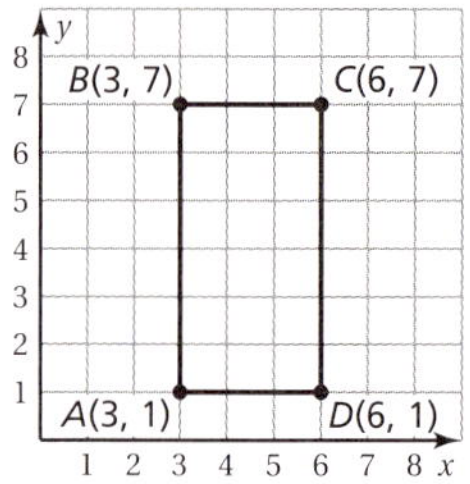

20. *Sample answer:*

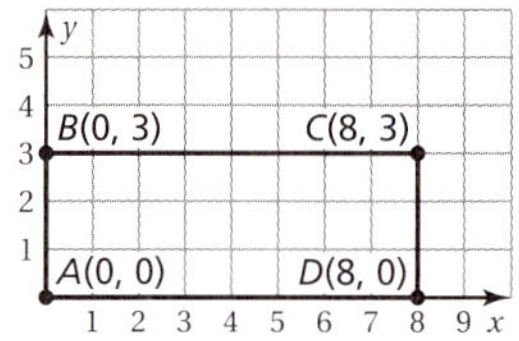

21. *Sample answer:*

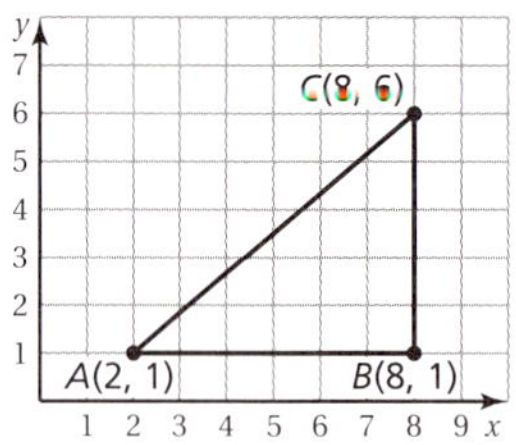

22. *Sample answer:* (6, 1) and (8, 9)

23. 27 miles; There are only two ways to go from station P to station L. Traveling from station P to N to M to L is 27 miles. Traveling from station P to J to K to L is 33 miles.

24. 41 mi^2

25. 2.5 times larger

Chapter 5

Record and Practice Journal Fair Game Review

6. $\frac{5}{27}$

7. $\frac{2}{5}$

8. $\frac{1}{72}$

9. $\frac{20}{27}$

10. $\frac{14}{11}$

11. $\frac{3}{4}$

12. $7\frac{1}{12}$ c

Section 5.1

Practice and Problem Solving

4.

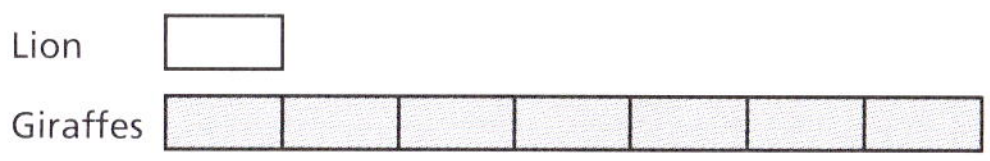

5.

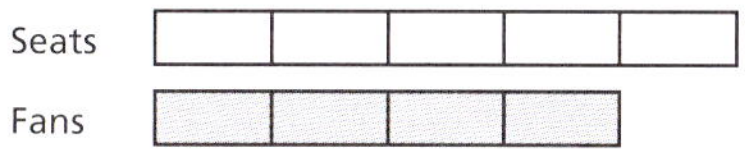

23. 4 pints of soda water, 8 pints of fruit punch concentrate, 20 pints of ginger ale; Yes; *Sample answer:* There is twice as much fruit punch as soda water (as in the original ratio). There is 5 times as much ginger ale as soda water (as in the original ratio).

24. 5 girls; *Sample answer:* Using a tape diagram with 6 parts (boys) and 5 parts (girls), each part represents 2 people. By adding 6 boys, each part now represents 3 people (one extra person than before), so there would then need to be 15 girls total (5 more) for the ratio to stay the same.

Section 5.2

Practice and Problem Solving

17. Adding the same number, 5 in this case, to each part of the ratio does not create equivalent ratios. You can add corresponding parts of equivalent ratios to create new equivalent ratios.

Sample answer:

A	3	6	9
B	7	14	21

18. Each part of the original ratio was not multiplied by the same number.

Sample answers:

A	5	25	125
B	3	15	75

or

A	5	15	45
B	3	9	27

19. 28 basketballs

20. 32 rock songs

21. Add the corresponding quantities of Recipes B and D to create Recipe E.

22. Add the corresponding quantities of Recipes C and D to create Recipe F.

23. Subtract the corresponding quantities of Recipe B from Recipe C to create Recipe A.

24. Subtract the corresponding quantities of Recipe C from Recipe F to create Recipe D.

25. *Sample answer:* Add the corresponding quantities of Recipes B and F to create a batch with 11 servings.

Section 5.3

Record and Practice Journal

5. a. Miles driven

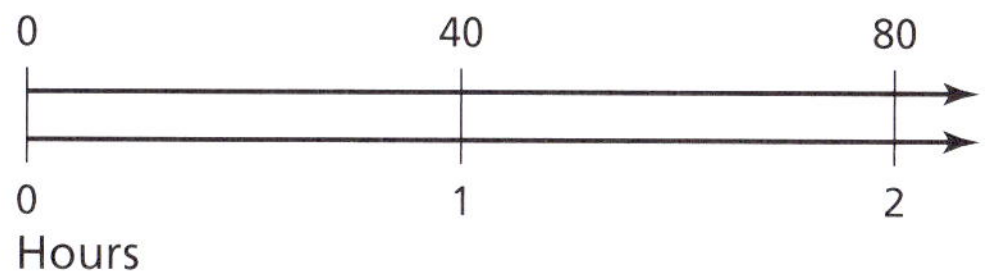

b. Money earned

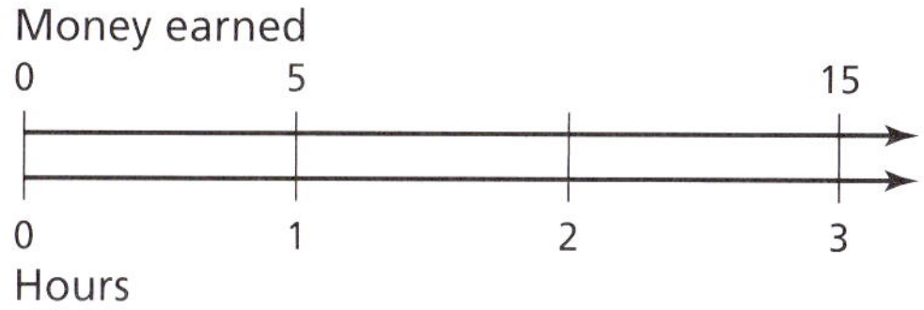

c. People

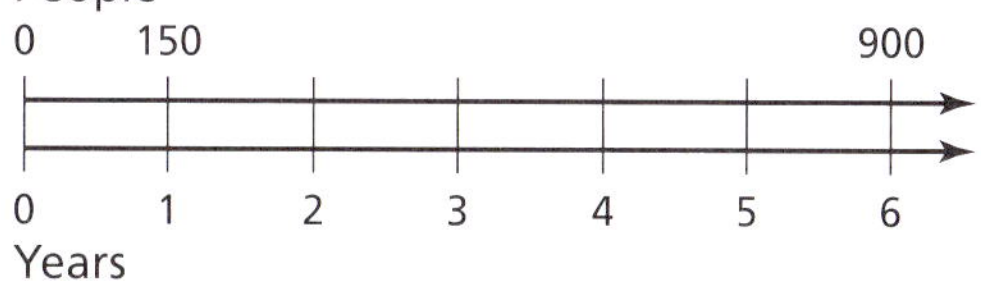

d. Feet

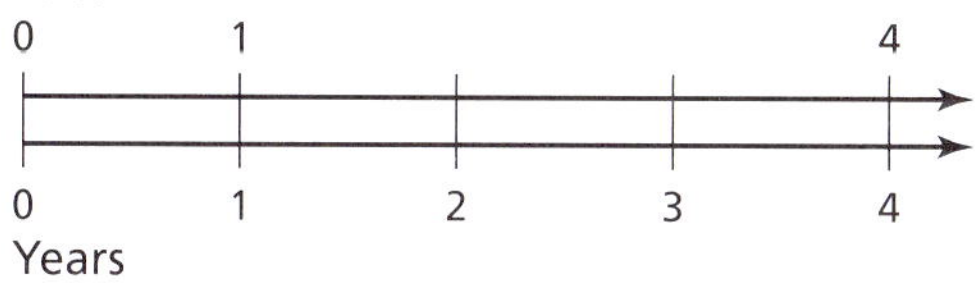

Sample answer: It is a good problem-solving strategy because you are comparing two quantities with different units.

Practice and Problem Solving

23. a. 6 min

b. Photos

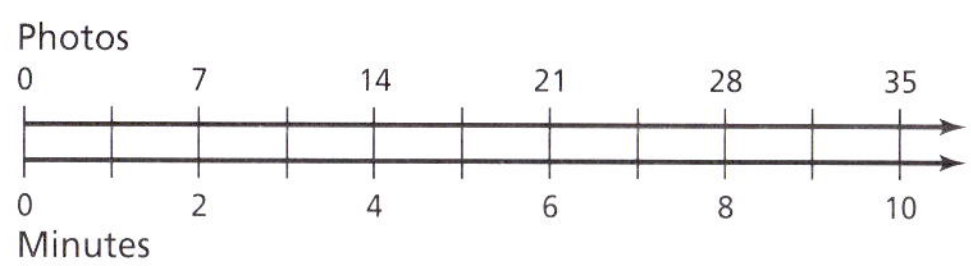

10 min

Additional Answers

24. **a.** $112.50

b.

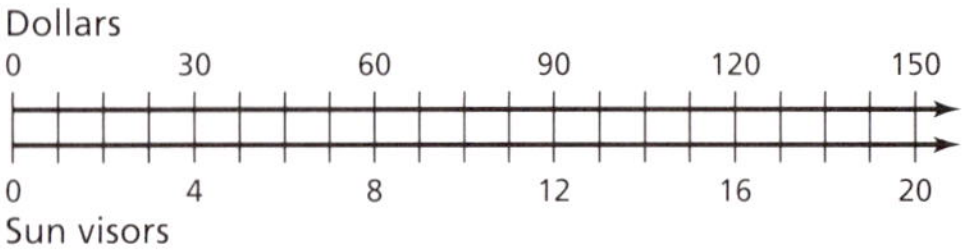

$120

Section 5.4

Record and Practice Journal

1. **c.**

Your Frosting		
Drops of Red	Drops of Blue	Total Drops
1	3	4
2	6	8
3	9	12
4	12	16
5	15	20

Your Teacher's Frosting		
Drops of Red	Drops of Blue	Total Drops
3	5	8
6	10	16
9	15	24
12	20	32
15	25	40

3. **d.** *Sample answer:* Moving vertically from the horizontal axis, you intersect your teacher's graph first, so you can conclude that your frosting is bluer, and your teacher's frosting is redder. The same conclusions can be drawn when moving horizontally from the vertical axis.

Practice and Problem Solving

12.

Water Tank	
Time (min)	Liters Leaked
2	4
4	8
6	12
8	16

Swimming Pool	
Time (min)	Liters Leaked
3	2
6	4
9	6
12	8

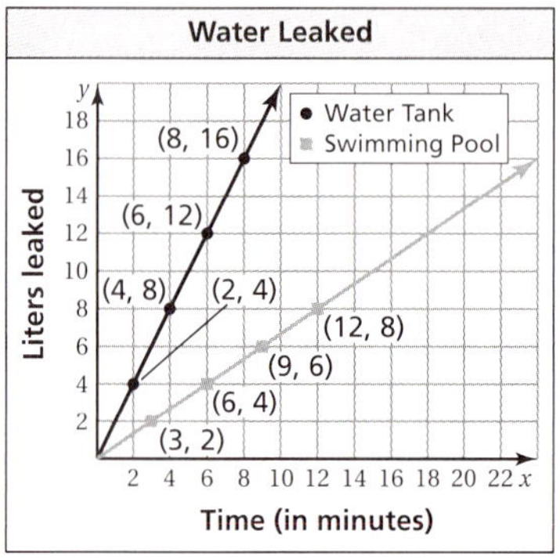

Both graphs begin at (0, 0). The graph for the water tank is steeper, so the water tank leaks faster than the swimming pool.

13.

Zoo	
People	Cost (dollars)
4	60
8	120
12	180
16	240

Museum	
People	Cost (dollars)
5	95
10	190
15	285
20	380

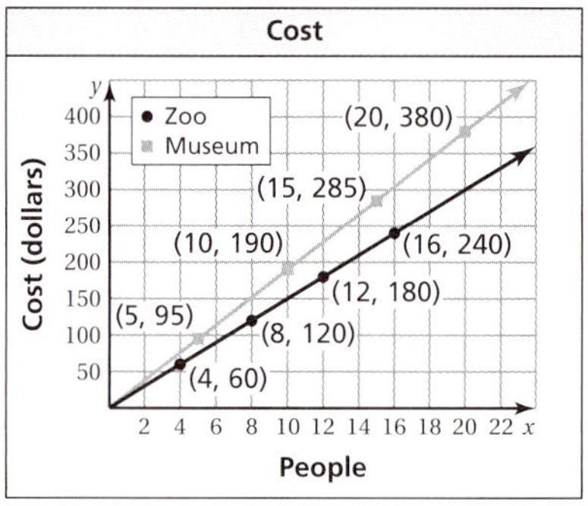

Both graphs begin at (0, 0). The graph for the museum is steeper, so the cost to attend the museum is greater than the cost to attend the zoo.

16. **a.** *Sample answer:*

Solution 1	
Water	Sulfuric Acid
7	3
21	9
42	18
63	27

Solution 2	
Water	Sulfuric Acid
9	4
18	8
45	20
63	28

Solution 2 has the greater concentration of acid.

b.

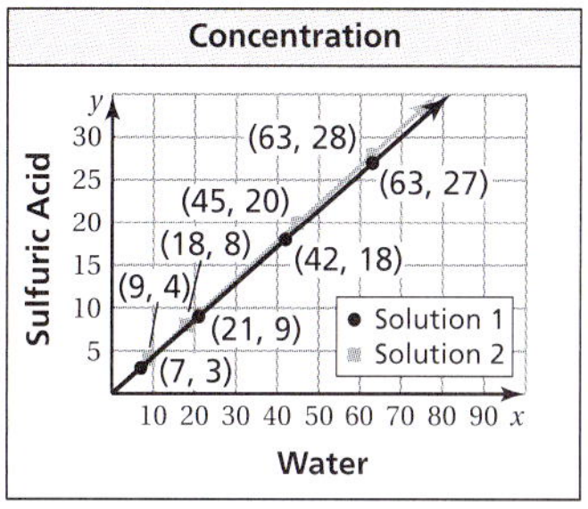

Both graphs begin at (0, 0). The graph for solution 2 is slightly steeper, so solution 2 has a greater concentration of acid.

c. *Sample answer:* A graph is preferable because the rates are visible more quickly.

17. a. *Sample answer:*

Old Mixture						
Peanuts	7	14	21	28	35	42
Almonds	4	8	12	16	20	24
Total	11	22	33	44	55	66

Old Mixture							
Peanuts	49	56	63	70	77	84	91
Almonds	28	32	36	40	44	48	52
Total	77	88	99	110	121	132	**143**

New Mixture						
Peanuts	8	16	24	32	40	48
Almonds	5	10	15	20	25	30
Total	13	26	39	52	65	78

New Mixture					
Peanuts	56	64	72	80	88
Almonds	35	40	45	50	55
Total	91	104	117	130	**143**

143 nuts

b.

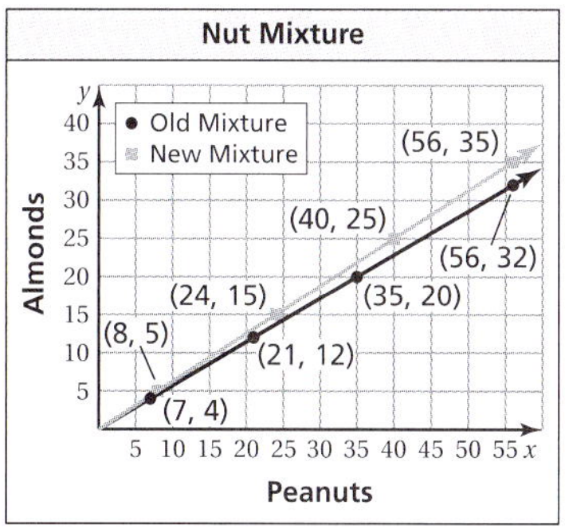

Both graphs begin at (0, 0). The graph for the new mixture is slightly steeper, so it has a greater concentration of almonds.

c. more; Almonds cost more and there are more of them in the new mixture than in the old mixture.

Section 5.5

Practice and Problem Solving

5.

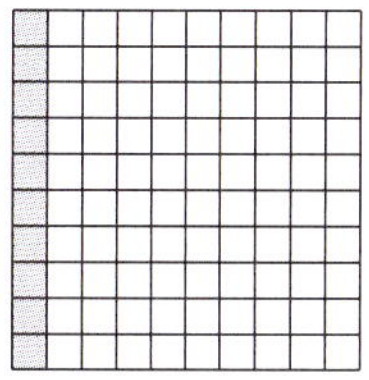

6.

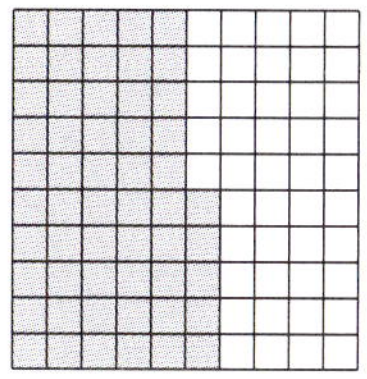

7.

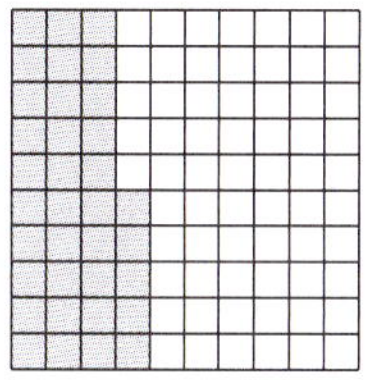

30. $\frac{3}{25}$; right-handed students; If $\frac{3}{25}$ of the students are left-handed, then $\frac{22}{25}$ of the students are right-handed and $\frac{22}{25} > \frac{3}{25}$.

38. The answer depends on the goals for each year. If the goals were the same each year, then the fundraiser raised more money this year. If this year's goal was greater than last year's goal, the fundraiser raised more money this year. If this year's goal was less than last year's goal, then the fundraiser probably did not raise more money this year. You would need to know the goal for each year.

40. $8\frac{1}{3}\%$ or $8.\overline{3}\%$; *Sample answer:* Multiply the numerator and the denominator by $\frac{25}{3}$. Then write the result as a mixed number.

Chapter 6

Record and Practice Journal Fair Game Review

16. *Sample answer:* $\frac{1}{2}$, 0.36, and 0.4

17. *Sample answer:* $\frac{2}{9}$, $\frac{1}{9}$, and 0.2

18. *Sample answer:* $2\frac{4}{5}$, 3, and 2.6

19. *Sample answer:* $\frac{3}{10}$, 0.4, and $\frac{1}{2}$

20. *Sample answer:* $\frac{2}{5}$, 0.3, and 0

21. *Sample answer:* $5\frac{9}{10}$, 6, and 10

22. $\frac{18}{25} > 0.7$

Section 6.1

Practice and Problem Solving

31. **a.** *Sample answer:* Choosing 8, the opposite is −8.

b. *Sample answer:* 8

c. The opposite of the opposite of an integer is the integer.; Yes; *Sample answer:*

Case 1:

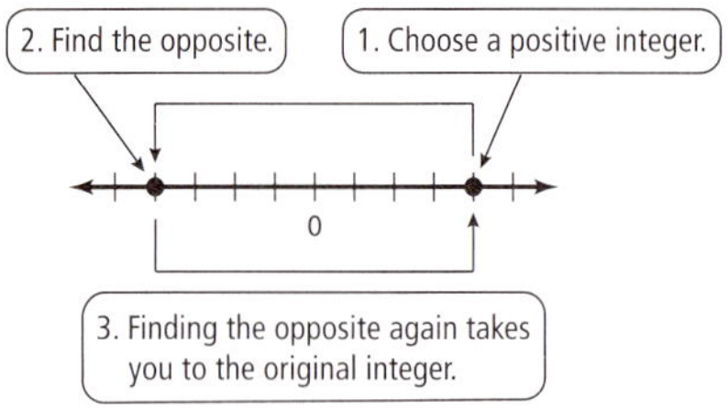

Case 2:

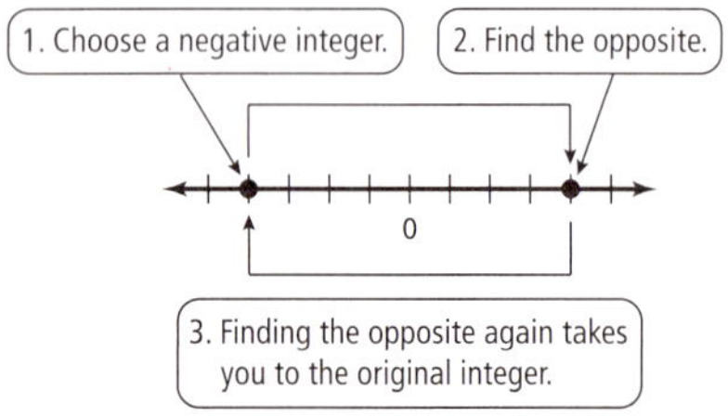

Case 3: Choose 0. The opposite of 0 is 0, so the opposite of the opposite of 0 is 0.

d. $-(-(-6))$ is the opposite of the opposite of -6; $-(-(-6)) = -6$

Section 6.2

Record and Practice Journal

1. **a.**

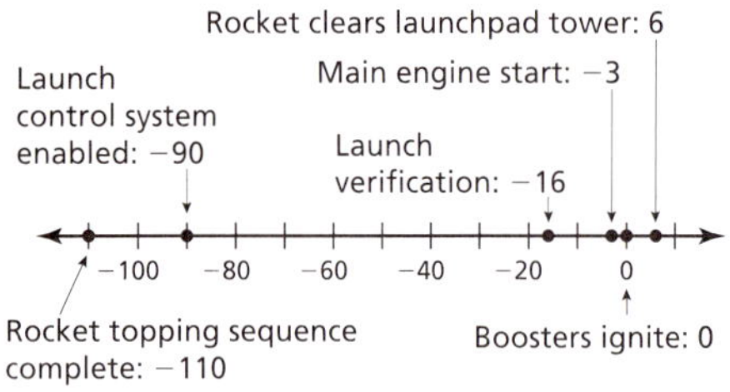

Vocabulary and Concept Check

2. **a.**

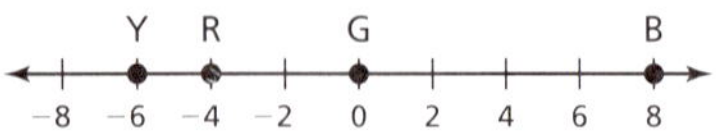

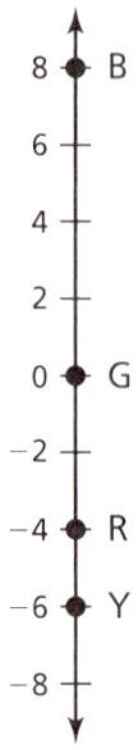

b. To order the positions from least to greatest on a horizontal number line, write the integers as they appear on the horizontal number line from left to right.

To order the positions from least to greatest on a vertical number line, write the integers as they appear on the vertical number line from bottom to top.

Practice and Problem Solving

28. **a.** Both integers are greater than or equal to 1.

b. One of the integers is −3 and the other integer is less than or equal to −3.

Section 6.4

Practice and Problem Solving

4.

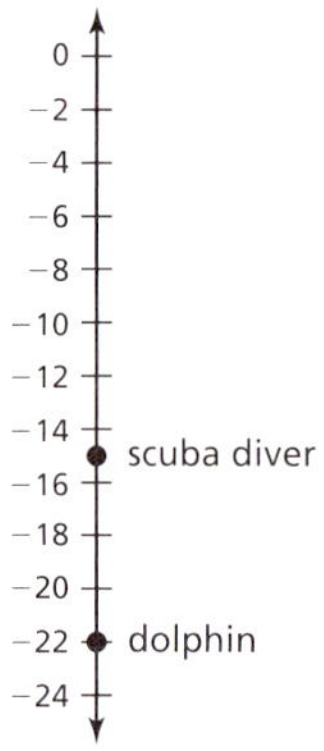

Dolphin

5.

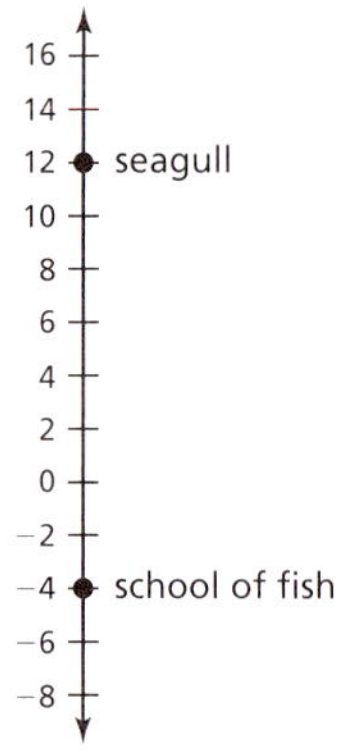

Seagull

6.

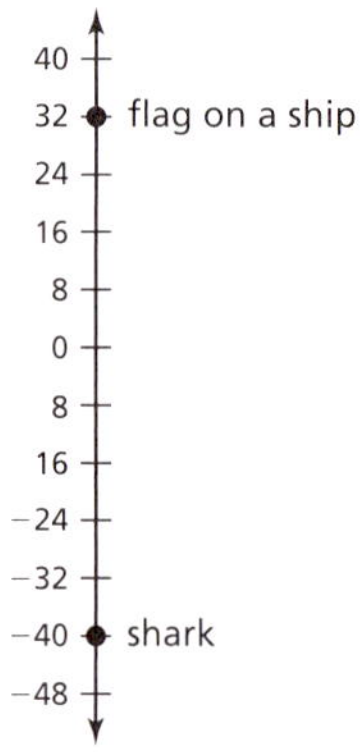

Shark

Fair Game Review

41.

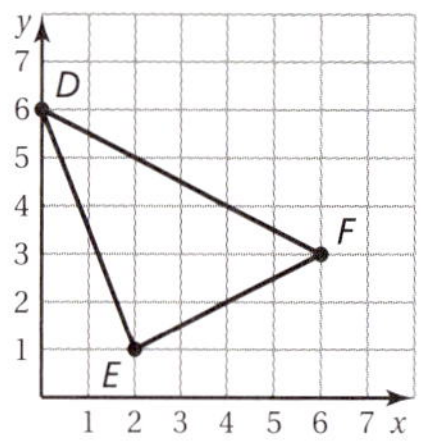

42.

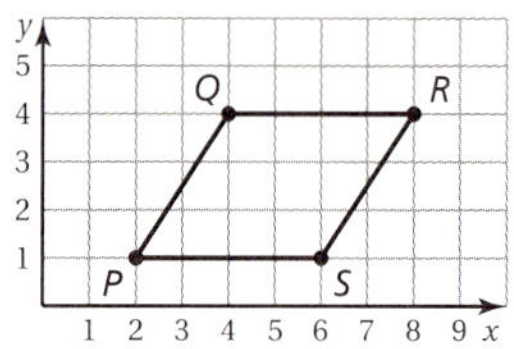

43.

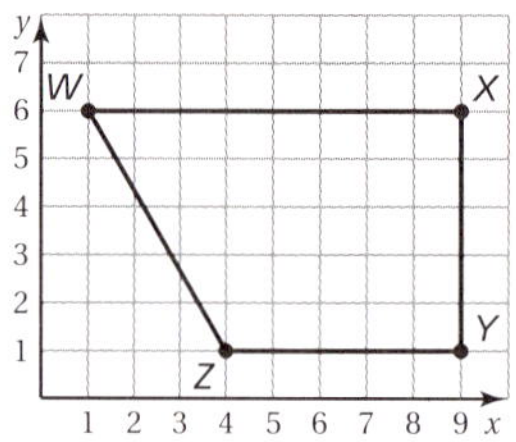

44. D

Section 6.5

Record and Practice Journal

2. a.

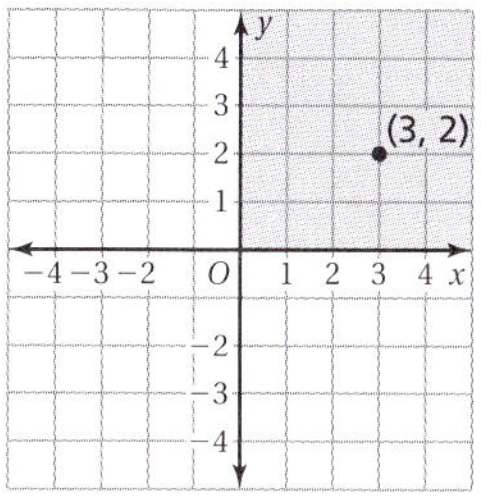

b.

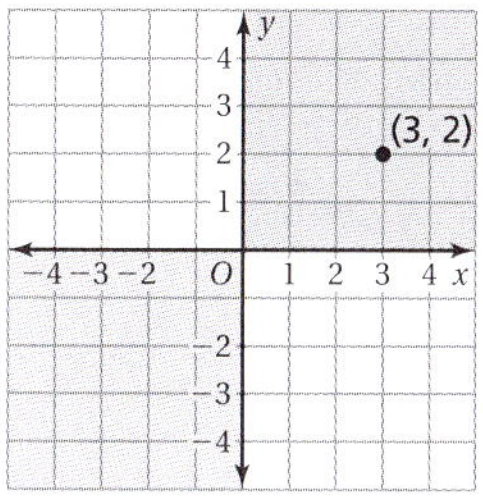

d.

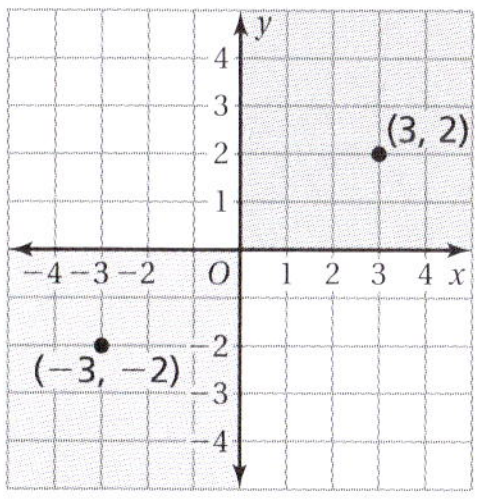

e.

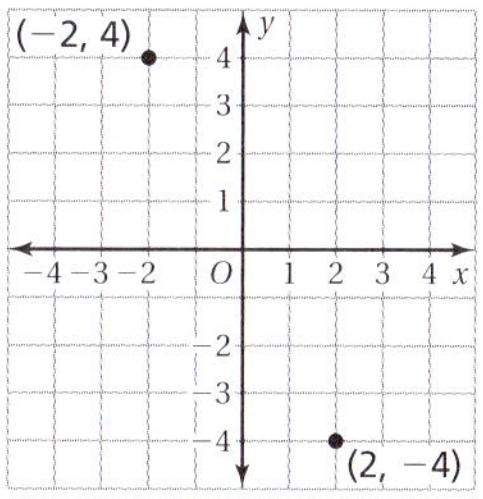

Practice and Problem Solving

28.

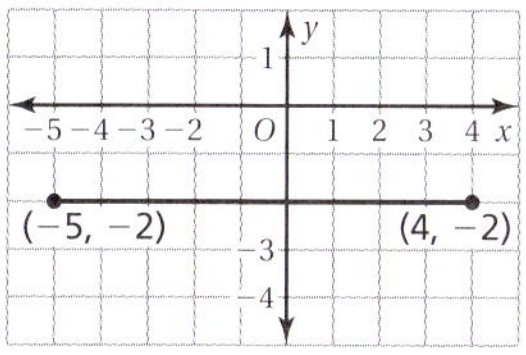

9

29.

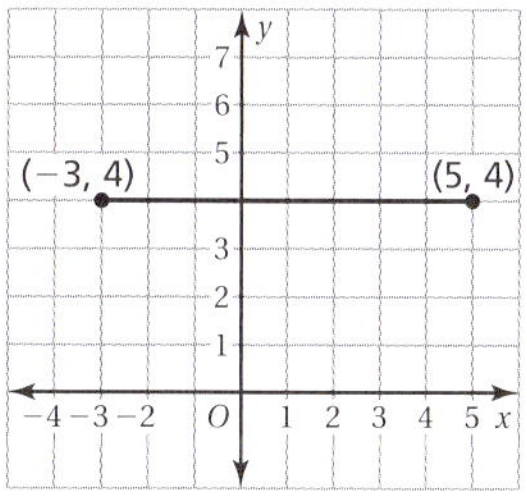

8

30.

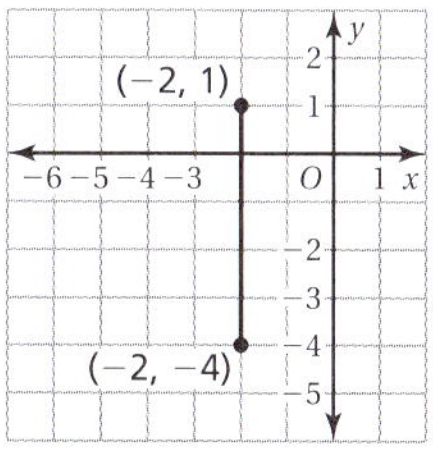

5

31. (−2, 1)

32.

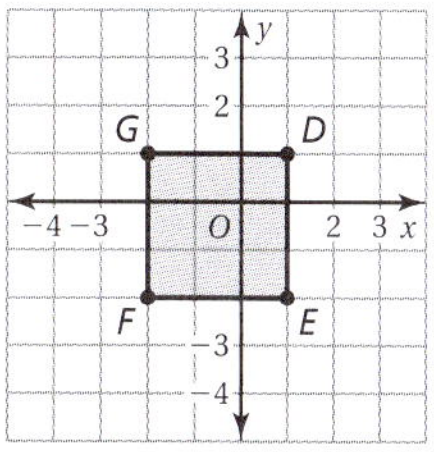

12 units; 9 units^2

33.

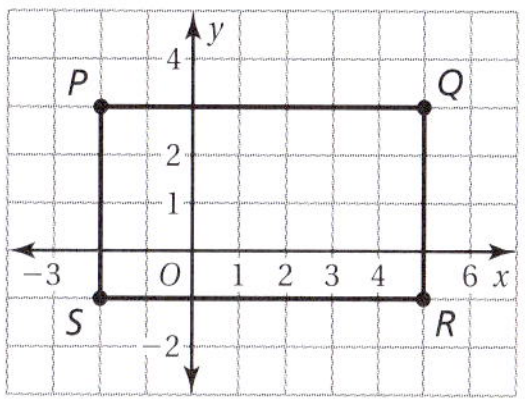

22 units; 28 units^2

34.

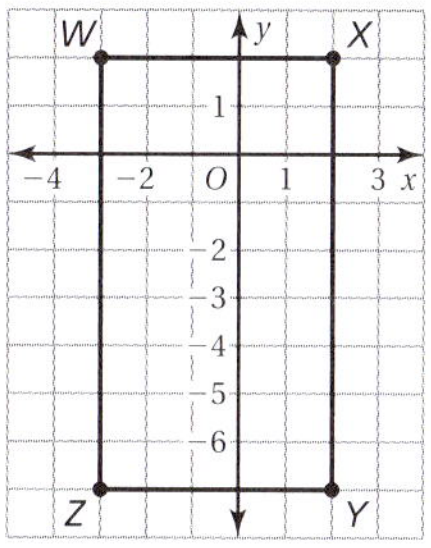

28 units; 45 units^2

35. a. about 142,000

b. 2011 and 2012

c. about 10,000

36. a.

Week	1	2	3	4	5	6	7	8	9
Miles	22	24	26	24	28	27	30	30	33

Week	10	11	12	13	14	15	16	17	18
Miles	35	38	40	40	40	36	33	24	14

b.

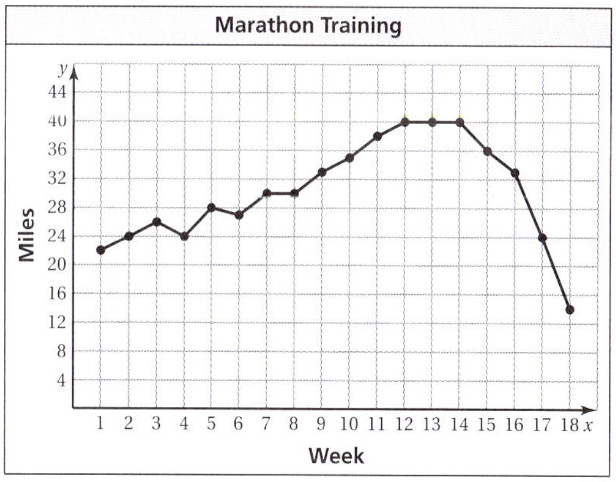

c. *Sample answer:* The number of miles per week increases for most of the first 12 weeks. There are 3 weeks in which you ran 40 miles per week. The number of miles per week decreases for the last 4 weeks.

d. *Sample answer:* The number of miles per week increases so that people get used to running longer distances. Then towards the end of the program, the number of miles decreases so that people can recover and be fully prepared for the marathon.

37. a. 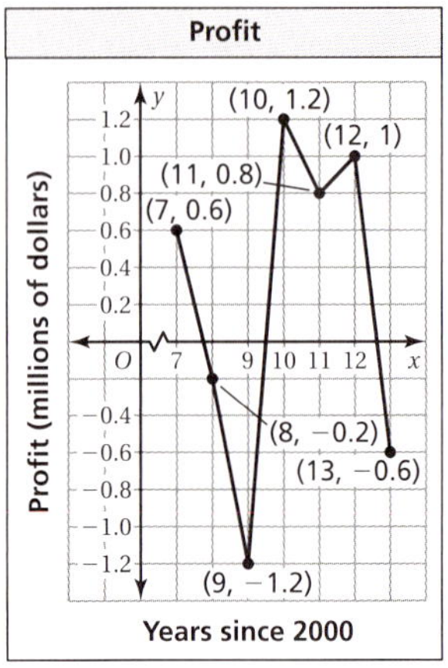

b. *Sample answer:* There are four years where the profit is positive and three years where the profit is negative. The profit decreased from 2007 to 2009. The profit increased the most from 2009 to 2010.

c. \$1.6 million

d. Because the x-axis represents the number of years since 2000, then 0 represents 2000. So, you could graph the profits for 1990 to 2006 by using x-values from -10 to 6.

38. Quadrant I

39. Quadrant III **40.** Quadrant IV

41. Quadrant I or Quadrant IV

42. Quadrant III or Quadrant IV

43. origin

44. sometimes; It is true only for (0, 0).

Extension 6.5

Practice

6. a. $(-8, 0)$ **b.** $(8, 0)$

7. a. $(2.5, -4.5)$ **b.** $(-2.5, 4.5)$

8. a. $\left(-5\frac{1}{2}, -3\right)$ **b.** $\left(5\frac{1}{2}, 3\right)$

9. $(-4, -5)$ **10.** $(1, -7)$

11. $(2, 2)$ **12.** $(-6.5, 10.5)$

13. $(3, 9)$; 18 units **14.** $(-5.75, 0)$; 11.5 units

15. a. $(-4, -5)$; $(1, -7)$; $(2, 2)$; $(-6.5, 10.5)$; yes; *Sample answer:* The order of the reflections does not matter. You are still reflecting the points in both axes.

b. *Sample answer:* Use the opposite of each coordinate.

16. *Sample answer:* Use the reflection rules on each vertex of the triangle. Then connect the reflected points to form the new triangle.

In the x-axis: $(-1, -3)$, $(-5, -3)$, $(-5, -7)$

In the y-axis: $(1, 3)$, $(5, 3)$, $(5, 7)$

Record and Practice Journal Practice

1. a. 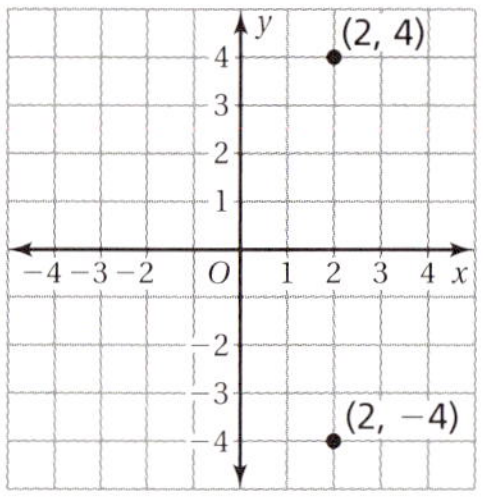

b. 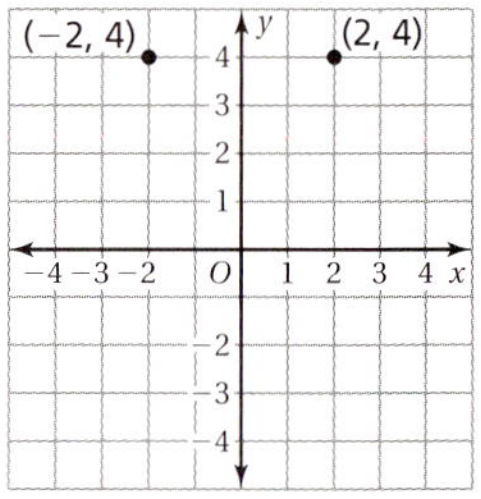

2. a. 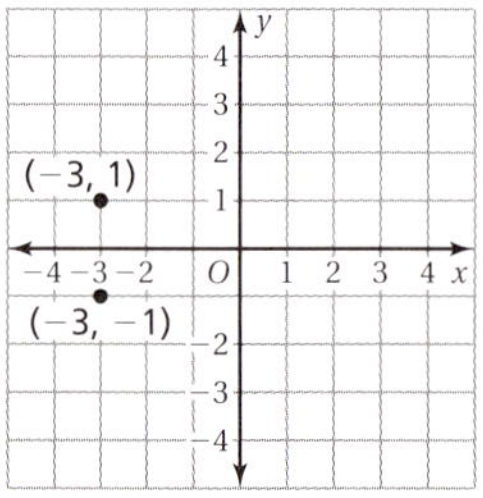

b. 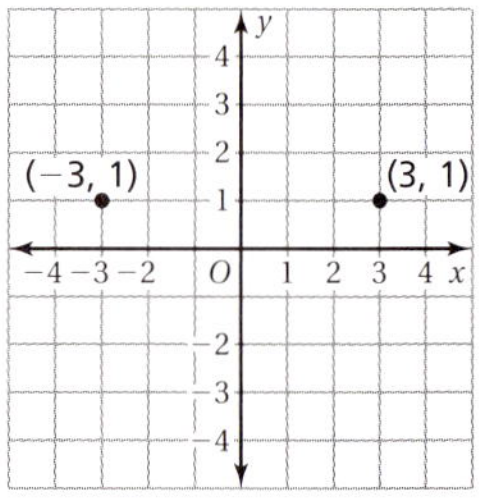

3. a. 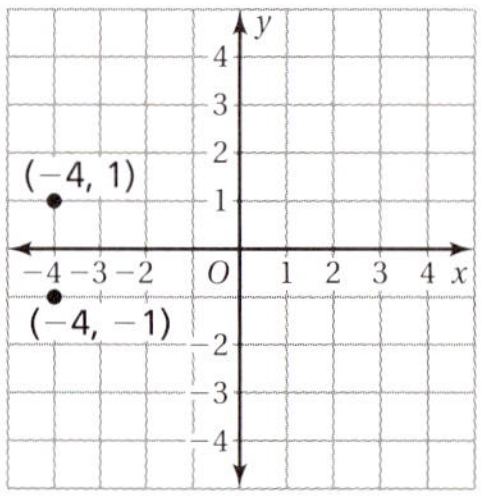

b. 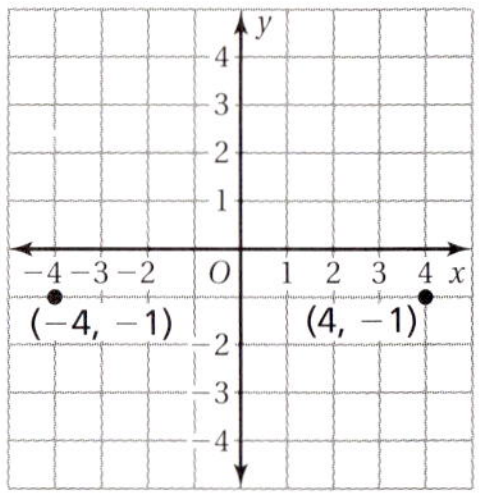

4. a. 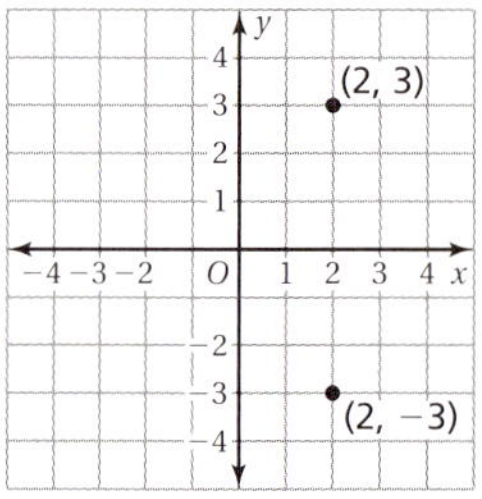

b. 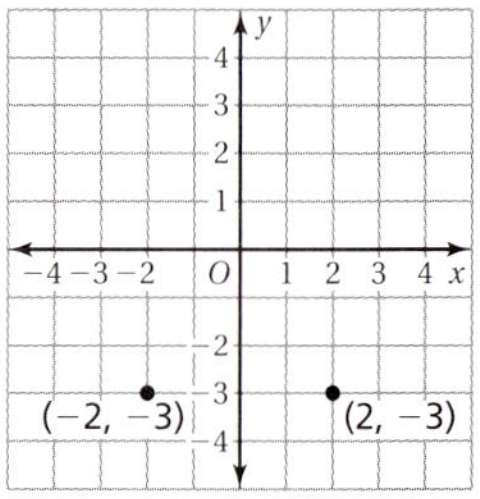

5. a. 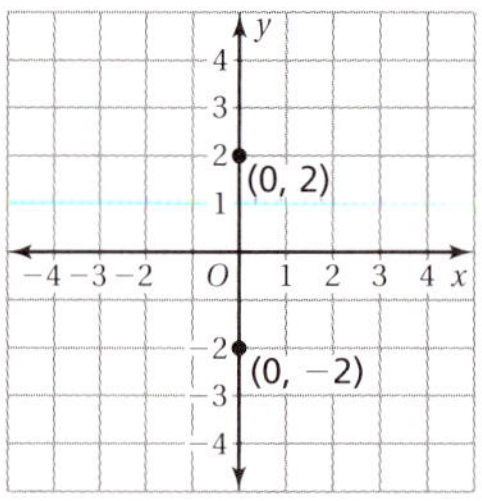

b. 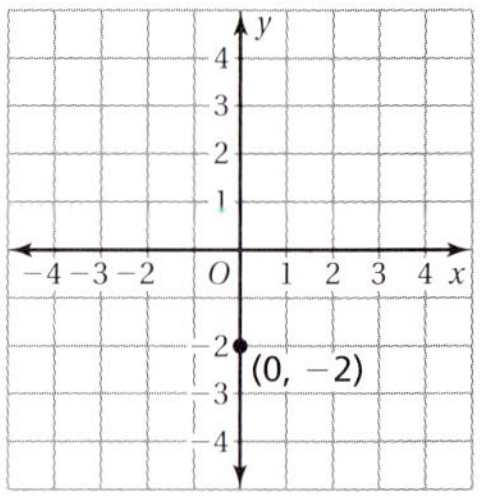

6. a. 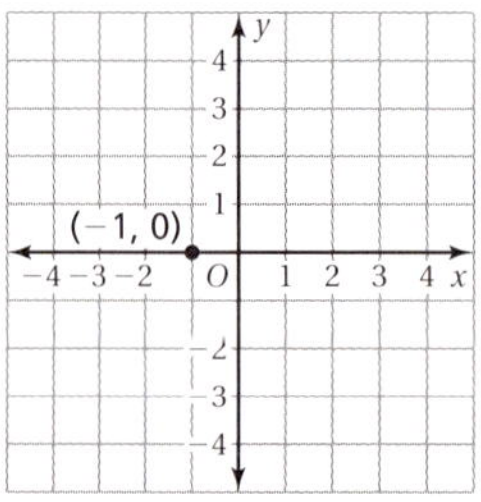

b. 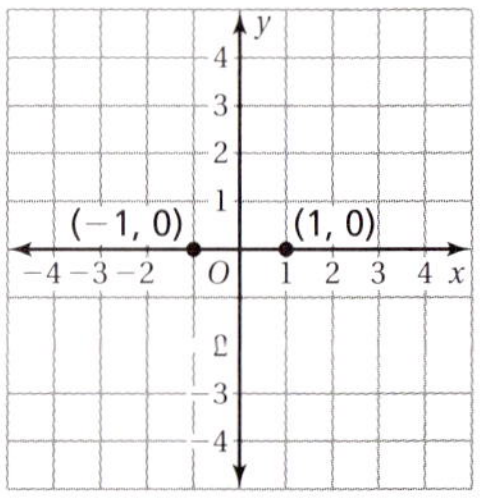

7.

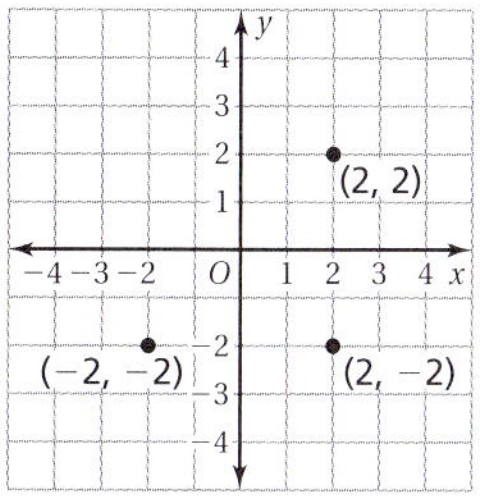

8.

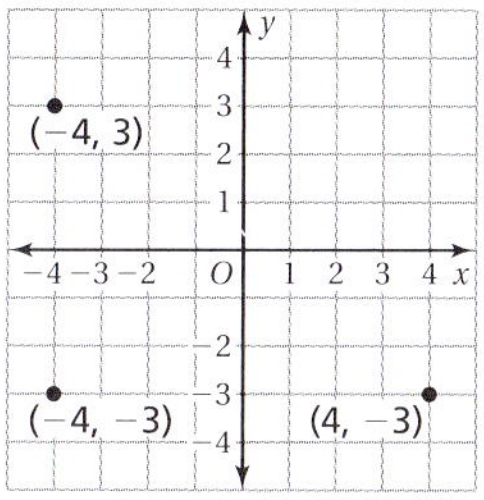

9.

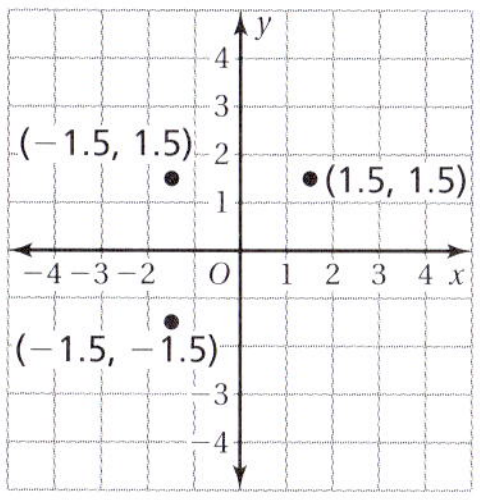

10.

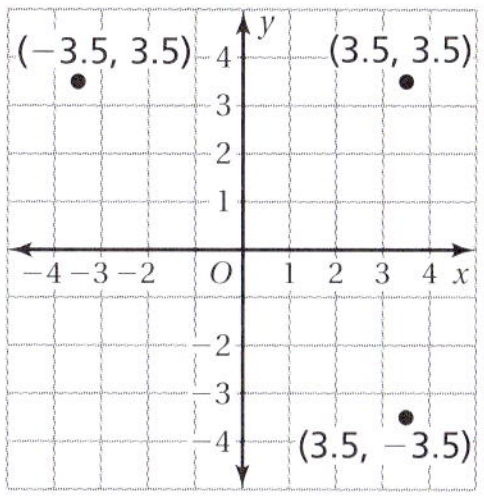

11.

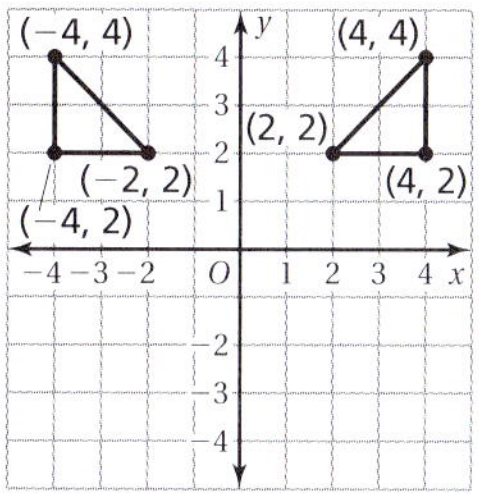

Chapter 7

Section 7.2

Record and Practice Journal

3. **a.** What number plus 1 equals 5?

   b. 4 plus what number equals 11?

   c. 8 equals what number plus 3?

   d. What number minus 9 equals 21?

   e. 13 equals what number minus 4?

Section 7.4

Record and Practice Journal

1. **b.** The horizontal distances are the same as the pattern between the values in the left column of the table. The vertical distance are the same as the pattern between the values in the right column of the table.

3. **b.** As the side length increases by 1 unit, the perimeter increases by 4 units.

Side Length	Perimeter
1	4
2	8
3	12
4	16
5	20

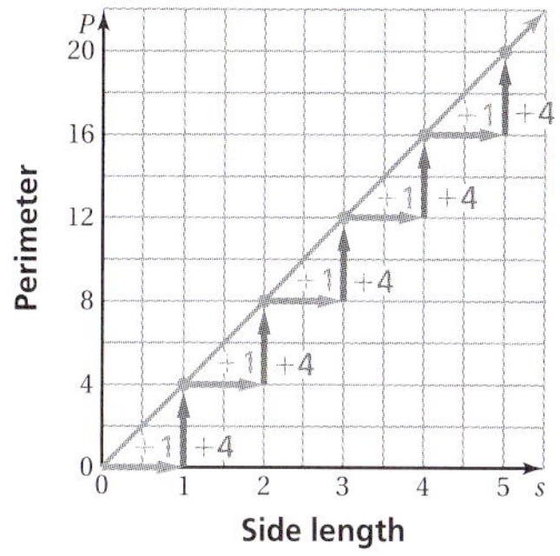

4. *Sample answer:* Identify the varying quantities. Then identify which quantity depends on which and assign variables to each quantity. Write an expression that uses the variable being depended on. Let another variable represent the value of this expression. Then write the equation using the two variables.

Practice and Problem Solving

18–21. Sample answers are given.

	Independent Variable	Dependent Variable
18.	The number of hours you study for a test	Your test score
19.	The speed you are pedaling a bike	Time it takes to stop your bike
20.	The number of minutes you use each month	Your monthly cell phone bill
21.	The number of years of education	The amount of money you earn

25. Methods to create the graph will vary.

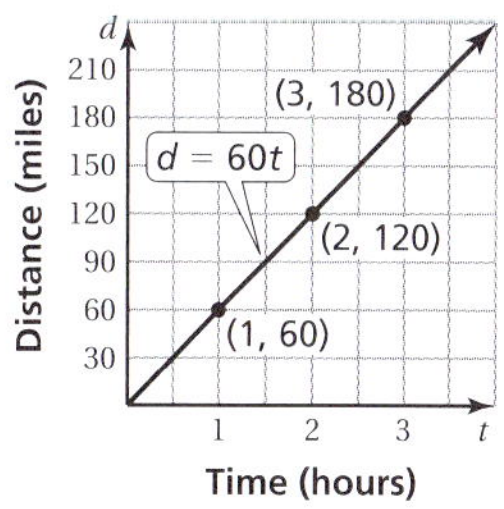

26. $d = \frac{2}{3}t$;

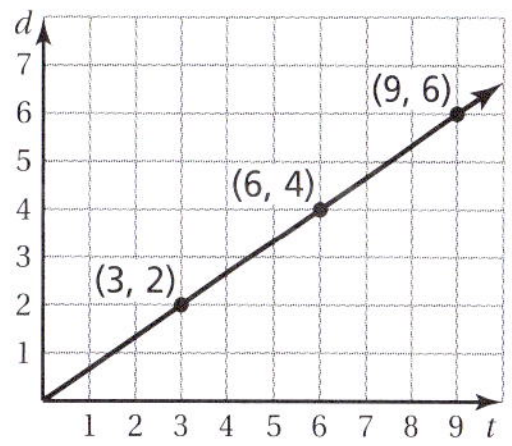

Additional Answers

27. $d = \frac{5}{6}t$;

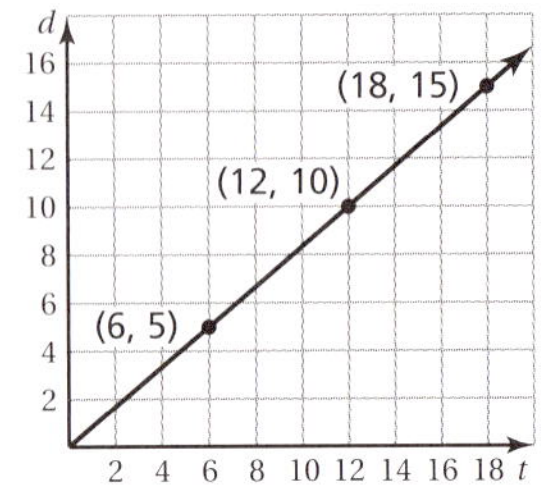

28. $d = 66t$;

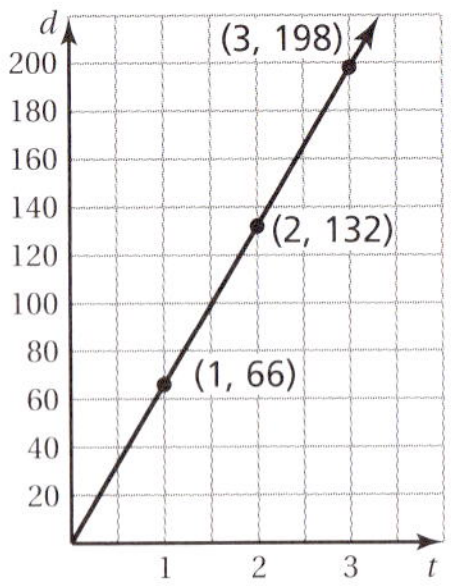

29. $d = 240t$;

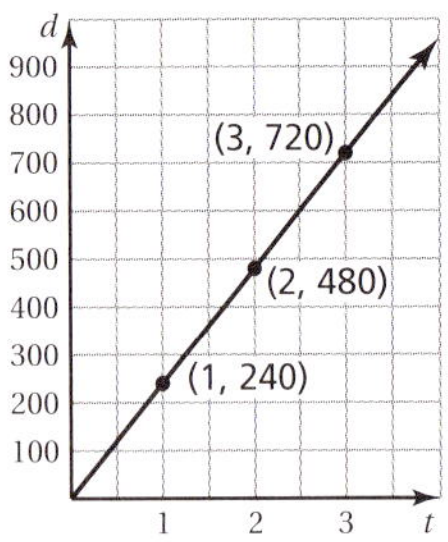

Section 7.5

Practice and Problem Solving

46. a. ;

$e \geq 0, e \leq 100$

b. 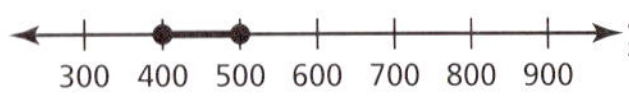;

$400 \leq e \leq 500$; The elevation is between 400 and 500. This compound inequality maintains all characteristics of the separate inequalities, $e \geq 400$ and $e \leq 500$.

c. 200 ft; *B* falls on the border of two ranges of elevation:
$100 \leq e \leq 200$
$200 \leq e \leq 300$
The only elevation that is a solution of both inequalities is 200.

Section 7.6

Practice and Problem Solving

15. $\frac{1}{4} \leq n$;

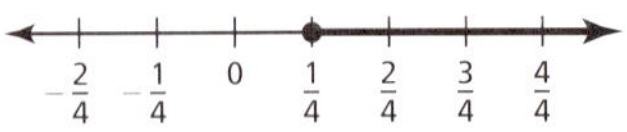

16. $b > \frac{5}{12}$;

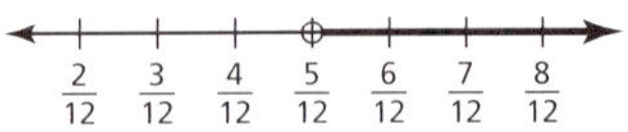

17. To solve the inequality, 9 should be added to both sides, not subtracted.

$$\begin{aligned} 28 &\geq t - 9 \\ +9 &\quad\;\; +9 \\ 37 &\geq t \end{aligned}$$

18. $22 + x \leq 40$; $x \leq 18$ lb

19. $x + 18.99 \leq 24$; $x \leq \$5.01$

26. a. $4200 + x \leq 8500$; $x \leq 4300$ lb

b. at most 2.6875 cubic yards; You can add at most 4300 pounds to your truck and each cubic yard of sand weighs 1600 pounds. So, you can haul at most $4300 \div 1600 = 2.6875$ cubic yards.

Section 7.7

Practice and Problem Solving

17. $v \leq 81$;

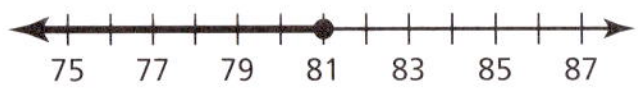

22. The right side should have been multiplied by 6.

$$\frac{x}{6} \leq 30$$

$$\frac{x}{6} \cdot 6 \leq 30 \cdot 6$$

$$x \leq 180$$

23. $8x < 168$; $x < 21$ ft

24. $\frac{x}{5} \geq 6$; $x \geq 30$ students

25. $8n < 72$; $n < 9$

26. $t \div 32 \leq 4.25$; $t \leq 136$

27. $225 \geq 12w$; $18.75 \geq w$

28. 

29. 

31. $80x > 2 \cdot 272$
$x > 6.8$ yards per play

32. A number is never less than itself.

33. *Sample answer:* the number of gallons of milk you can buy with \$20; the length of a park that has an area of at least 500 square feet

34. $\frac{5x}{50} \leq 50$ $\quad$ $\frac{5x}{50} \geq 40$
$x \leq \$500$ $\quad$ $x \geq \$400$
Selling between \$400 and \$500 will earn you between \$40 and \$50.

35. yes; $a > b$ and $x > y$

36. no; *Sample answer:* $a = 9, b = 8, x = 3, y = 2$; $a - x = 6, b - y = 6$; so, $a - x = b - y$.

37. yes; $a > b$ and $x > y$

38. no; *Sample answer:* $a = 6, b = 5, x = 10, y = 4$

$$\frac{a}{x} \overset{?}{>} \frac{y}{b}$$

$$\frac{6}{10} \overset{?}{>} \frac{4}{5}$$

$$0.6 \ngtr 0.8$$

Chapter 8

Section 8.1

Record and Practice Journal

1. b.

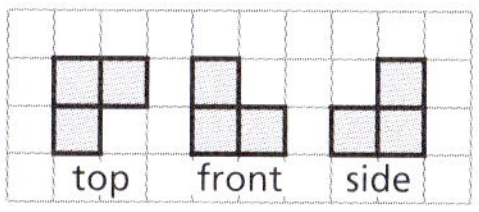

c.

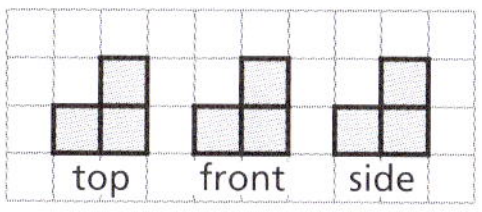

d.

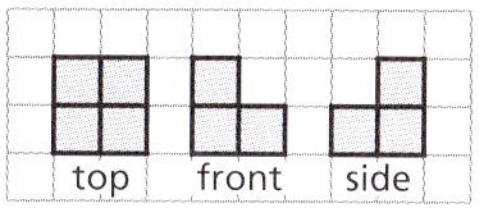

e.

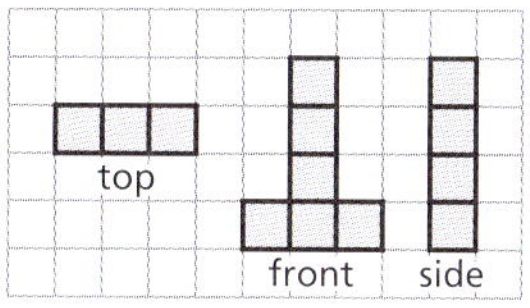

f.

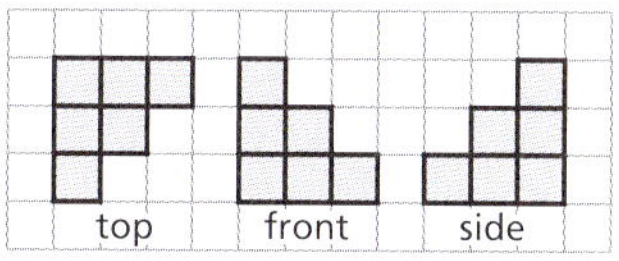

g.

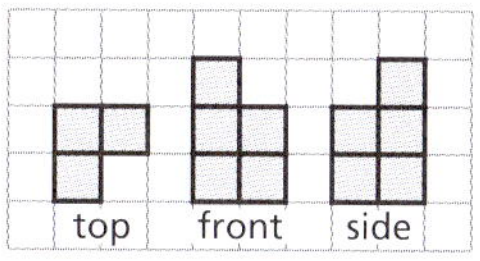

3. d. *Sample answer:* Planes are parallel if they never intersect; lines are parallel if they lie on the same plane and do not intersect; a line is parallel to a plane if they never intersect; lines are perpendicular if they meet at a right angle; planes are perpendicular if they meet at a right angle; a line is perpendicular to a plane if they meet at a right angle.

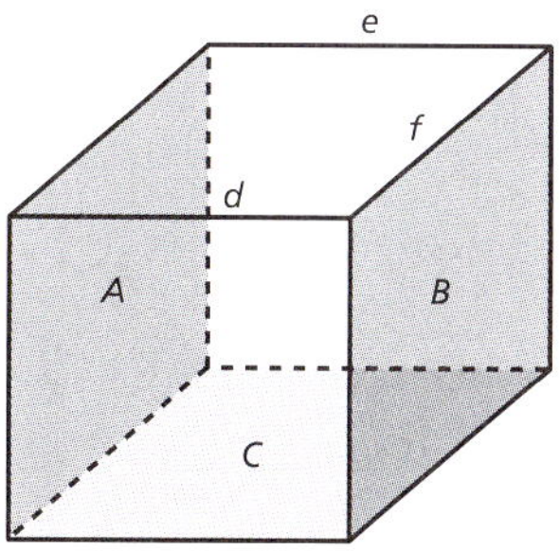

Faces *A* and *B* are parallel.
Face *C* is perpendicular to faces *A* and *B*.
Edges *d* and *e* are parallel.
Edge *f* is perpendicular to edges *d* and *e*.
Edges *d* and *e* are parallel to face *C*.
Edges *d* and *e* are perpendicular to faces *A* and *B*.

On Your Own

4. front: 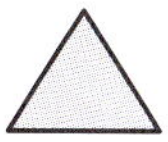 side:  top: 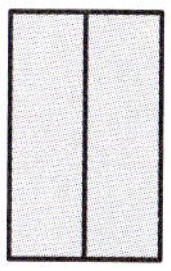

5. front:  side:  top: 

Practice and Problem Solving

13.

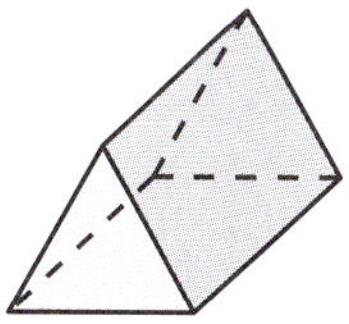

14.

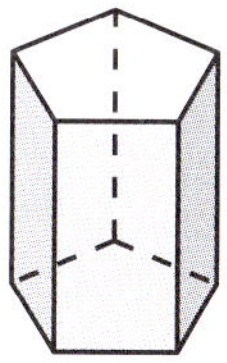

15.

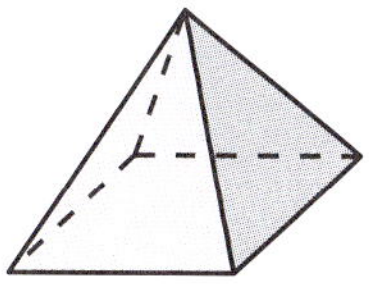

16.

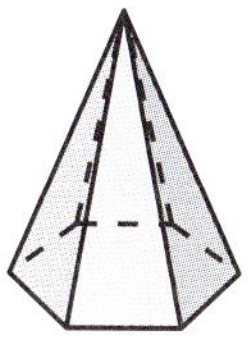

17. front:  side: 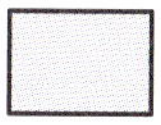 top: 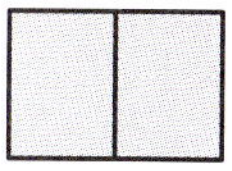

18. front: 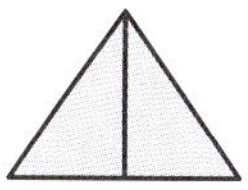 side: 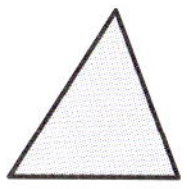

top: 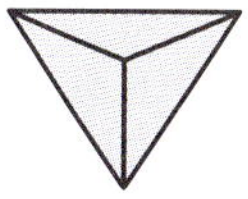

19. front: 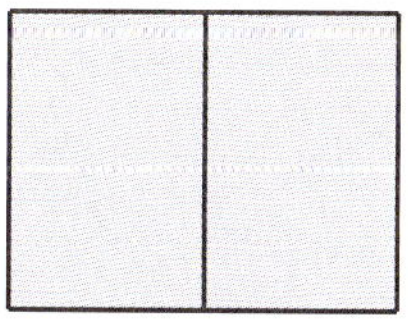

side:  top: 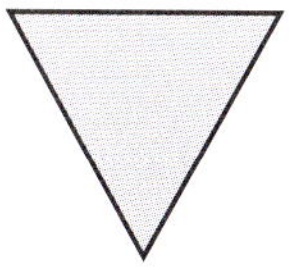

20. front: 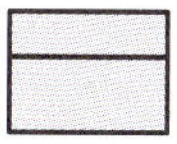 side: 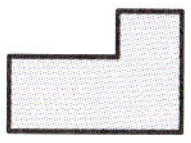 top: 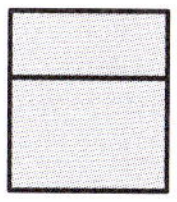

21. front:  side: 

top: 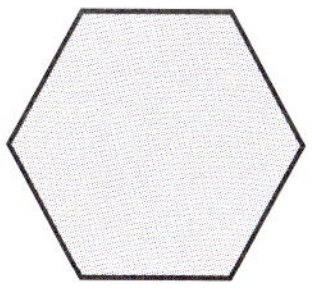

Additional Answers

22. front:  side: 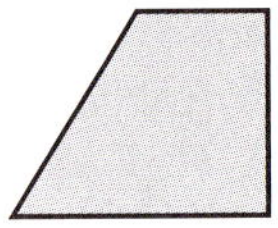

top: 

23. 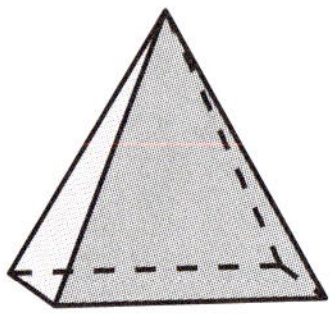

24. The Washington Monument is an *obelisk*. It consists of a pyramid sitting on top of a solid that tapers as it rises.

25. 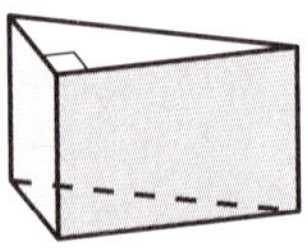

26. 

29. *Sample answer:*

a.

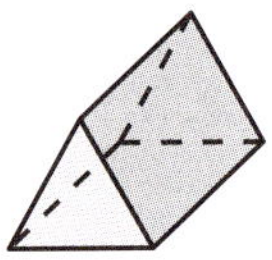

Triangular prism
6 vertices
9 edges

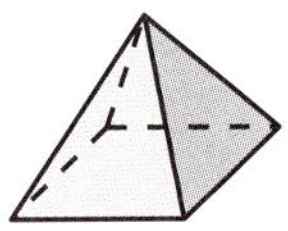

Square pyramid
5 vertices
8 edges

b. More than one solid can have the same number of faces, so knowing the number of edges and vertices can help you to draw the intended solid.

Section 8.2

Practice and Problem Solving

3. 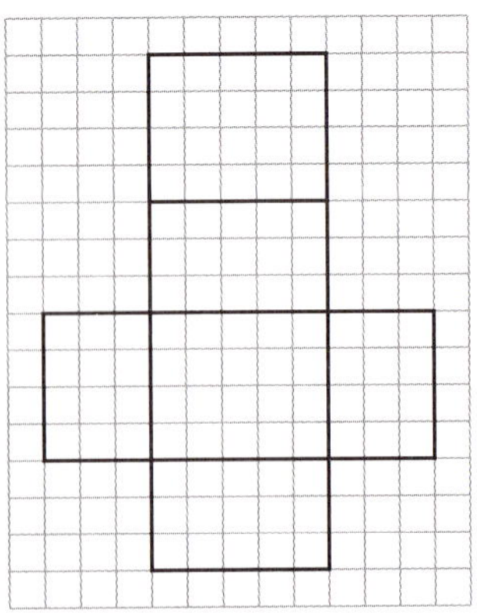

94 units^2

4. 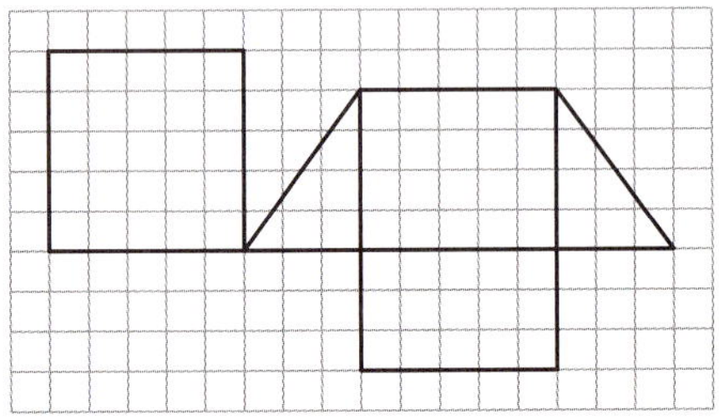

72 units^2

5. 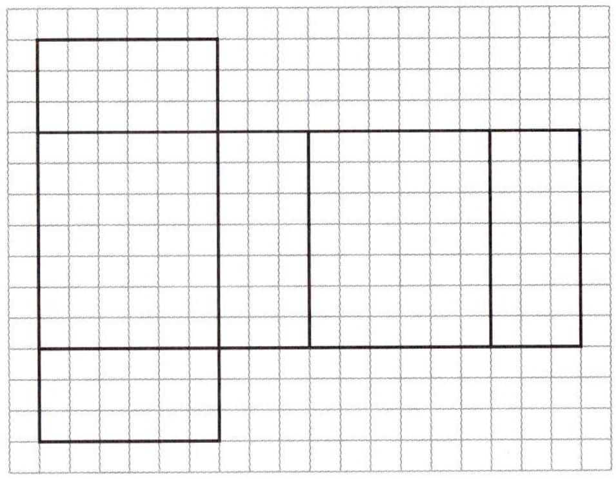

162 units^2

Section 8.3

Practice and Problem Solving

3. 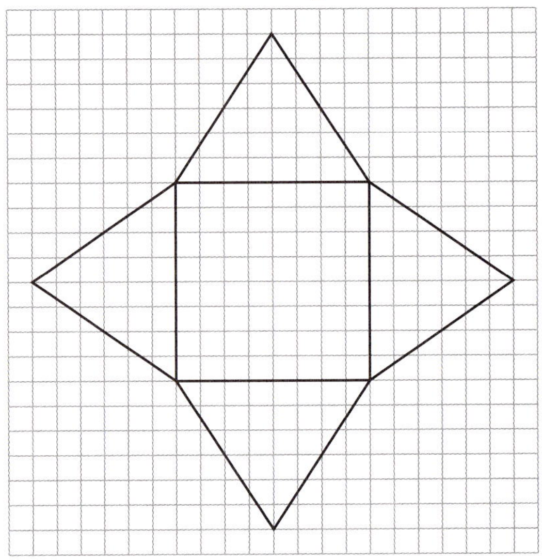

160 units^2

4. 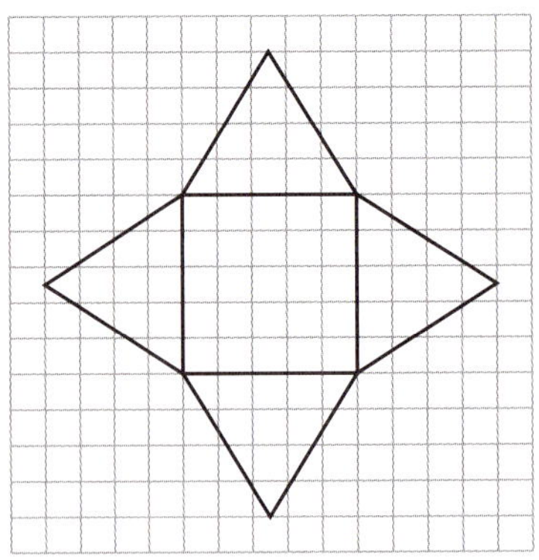

65 units^2

5.

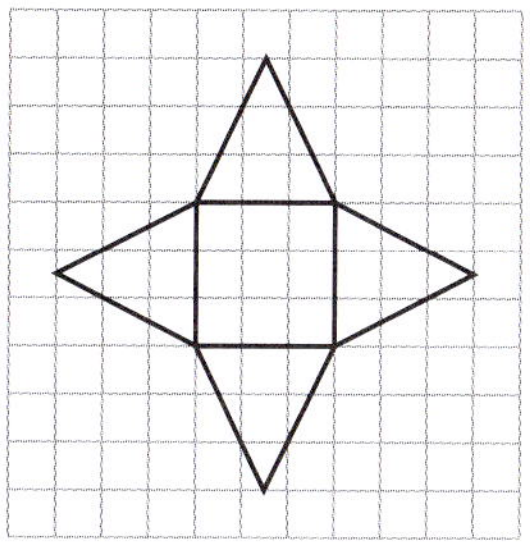

27 units^2

Section 8.4

Practice and Problem Solving

15. 1728 1-inch cubes; There are 1728 1-inch cubes in a cube with a side length of 1 foot. The area of the cube with a side length of 1 foot is 1 cubic foot, or 1728 cubic inches. So, 1 cubic foot is equal to 1728 cubic inches. You can use the conversion factors $\frac{1728\text{ in.}^3}{1\text{ ft}^3}$ and $\frac{1\text{ ft}^3}{1728\text{ in.}^3}$ to convert between cubic inches and cubic feet.

18. *Answers should include, but not limited to:*

a. Students should provide sketches of a tree house that has a surface area of at most 1400 square feet and a volume of at least 250 cubic feet. They should explain their choice of dimensions.

b. For the dimensions to be reasonable, the tree house should be able to fit people and fit in a tree.

Chapter 9

Record and Practice Journal Fair Game Review

6.

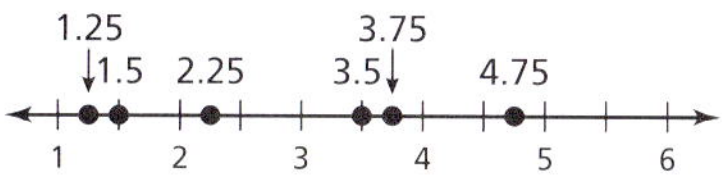

Section 9.1

On Your Own

1. a. Because you can anticipate that the ages will vary, it is a statistical question.

b.

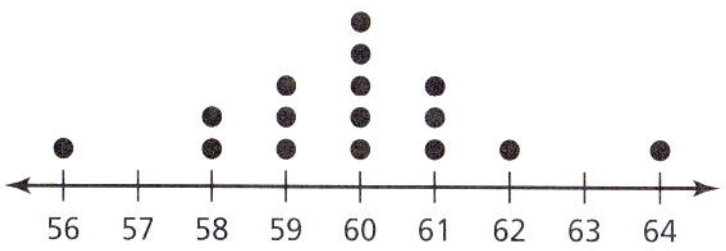

clustered around 60, peak at 60, gaps between 56 and 58 and between 62 and 64

c. Most of the people who retired early were about 60 years old.

Practice and Problem Solving

13.

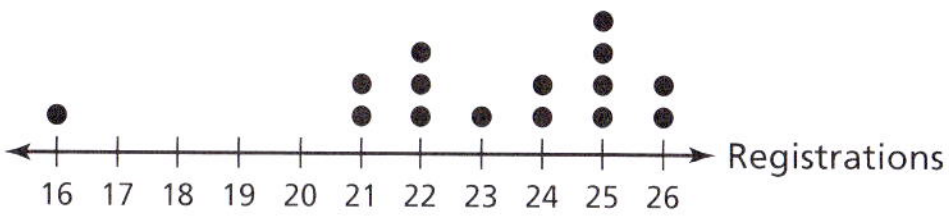

Most of the registrations are in a cluster from 21 to 26. The peak is 25. There is a gap between 16 and 21.

14.

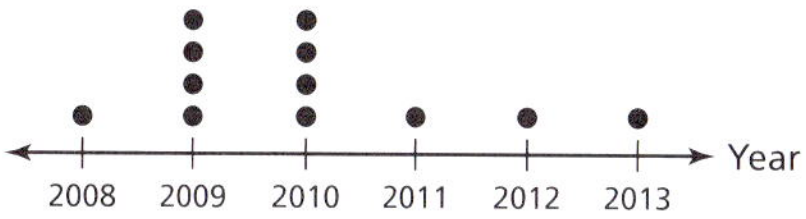

Most of the years cluster around 2009 and 2010. The peaks are 2009 and 2010. There is no gap.

15.

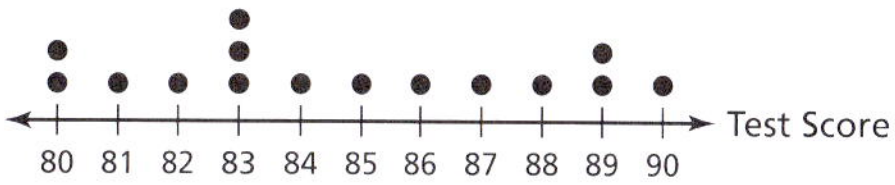

The test scores are spread out pretty evenly with no clusters or gaps. The peak is 83.

16. a. yes; It is a statistical question because you would anticipate variability in the hours spent on homework each night by students.

b.

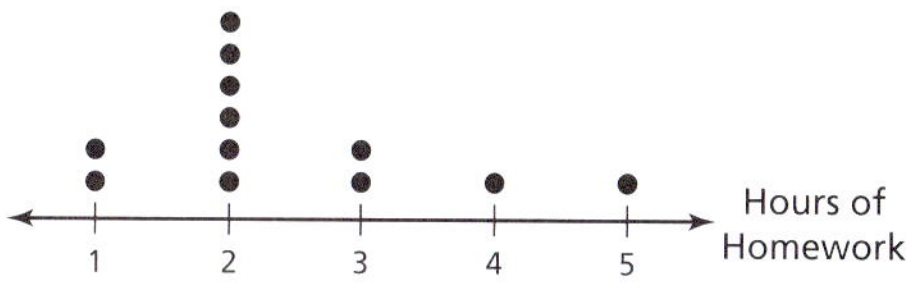

Most of the hours cluster around 2. The peak is 2. There is no gap.

c. Most students spend about 2 hours on homework during a school night.

17. a. 21 earthworms

b. *Sample answer:* Use a centimeter ruler. The units are centimeters.

c. *Sample answer:* "What is the length of an earthworm?"; The lengths are spread out pretty evenly from 15 centimeters to 28 centimeters.

18. a. 18 players

b. *Sample answer:* Use a tape measure. The units are inches.

c. *Sample answer:* "What are the heights of players on an NBA championship team?"; The heights are spread out, but most of the heights (in inches) are in the mid-to-low 80s.

19. *Sample answer:* Anemometer; miles per hour

20. *Sample answer:* Rain gauge; inches

21. *Sample answer:* Richter scale; magnitude

Additional Answers

22. *Sample answer:* 45 mi/h; Most of the data cluster around 45 and 45 miles per hour is a common speed limit.

23. *Sample answer:* 65 mi/h; Most of the data cluster around 65 and 65 miles per hour is a common speed limit.

24. "How many letters are in an alphabet?"; The original question only has one answer. The new question has many answers because there are many alphabets in the world.

Section 9.2

Practice and Problem Solving

16. 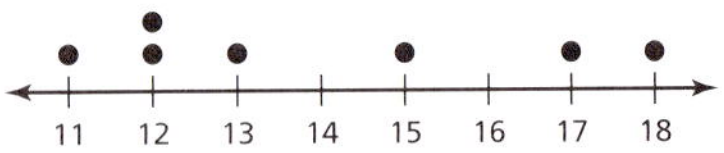

The mean is

$$\frac{11 + 13 + 17 + 15 + 12 + 18 + 12}{7} = 14.$$

Using the dot plot, the sum of the distances from the mean to points above the mean is $1 + 3 + 4 = 8$. The sum of the distances from the mean to points below the mean is $1 + 2 + 2 + 3 = 8$. These sums are equal, so the mean is the point where the data set is balanced.

18. 11 pounds; Use the mean of each collection to work backwards to find the total weight of the collection. Divide the sum of the weights of the two collections by the total number of backpacks (20) to find the mean.

Section 9.3

Practice and Problem Solving

21.

With Outlier	*Without Outlier*
mean: 48.5	mean: 53
median: 53	median: 54
mode: none	mode: none

The outlier reduces the median slightly, but reduces the mean more. There is no mode with or without the outlier.

22.

With Outlier	*Without Outlier*
mean: 103	mean: 85
median: 85	median: 85
mode: 85	mode: 85

The outlier makes the mean greater than all of the other values, but does not affect the median or mode.

23. mean: 7.61; median: 7.42; no mode

24. mean: $6\frac{3}{8}$; median: $6\frac{3}{16}$; mode: $6\frac{5}{8}$

25. **a.** mean: 94°F; median: 91°F; mode: 91°F
Sample answer: Both the median and mode are the best measures for the data, because both are very close to most of the values.

b. mean: 77°F; median: 77°F;
modes: 77°F and 78°F
Sample answer: Both the mean and median are the best measures for the data, because there are two modes.

26. *Answer should include, but is not limited to:* prices of 10 cereals with 1 outlier; unit prices calculated

a. calculation of mean, median, and mode and explanation of which is best

b. outlier identified; mean, median, and mode calculated; measure most affected described

27. 10 hours; Using the mean as the average, you would need to work 12 hours. Using the median as the average, you would need to work 10 or more hours. Using the mode as the average, you would need to work 10 hours. So, the minimum number of hours is 10 and you can use the median or mode to justify your answer.

28. *Sample answer:* A mode can be an unreliable measure of a data set because there can be no mode or more than one mode. When there is one mode, it may not be representative of the entire data set.

31. **a.** mean: \$1794; median: \$1790; mode: \$1940

b. mean: \$1883.70; median: \$1879.50; mode: \$2037; The mean, median, and mode all increased by 5%.

c. annual salaries: \$23,280, \$19,920, \$22,320, \$25,200, \$20,640, \$18,480, \$21,120, \$23,280, \$21,840, \$19,200; mean: \$21,528; median: \$21,480; mode: \$23,280; They are 12 times the mean, median, and mode of the monthly salary.

32. **a.** mean: $8x$, median: $5x$, mode: $3x$

b. yes; $23x$

Record and Practice Journal Practice

3.

	With Outlier	Without Outlier
mean	19.5	9.8
median	12	12
mode	12	12

4.

	With Outlier	Without Outlier
mean	50.75	58
median	55	55
modes	54, 55, and 62	54, 55, and 62

9.1–9.3 Quiz

1. 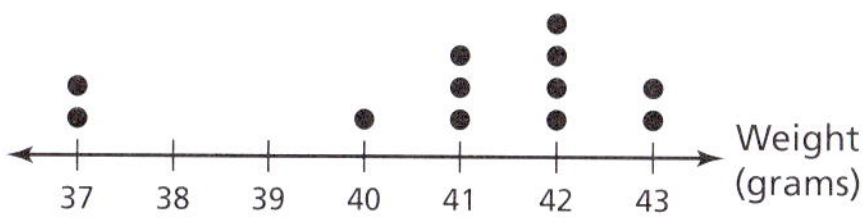

Most of the data are clustered around 42. There is a peak at 42 and a gap between 40 and 37.

2. 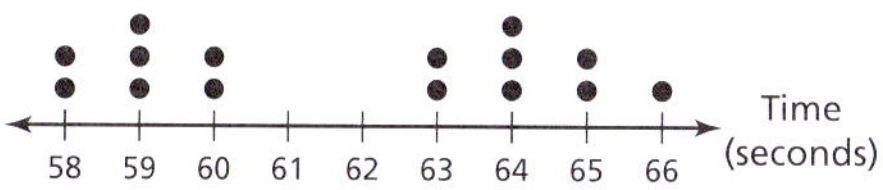

Most of the data are clustered around 59 or 64. There are peaks at 59 and 64. There is a gap between 60 and 63.

9. **a.** yes; It is a statistical question because you would anticipate variability in the hours spent on exercising each week by students.

b. 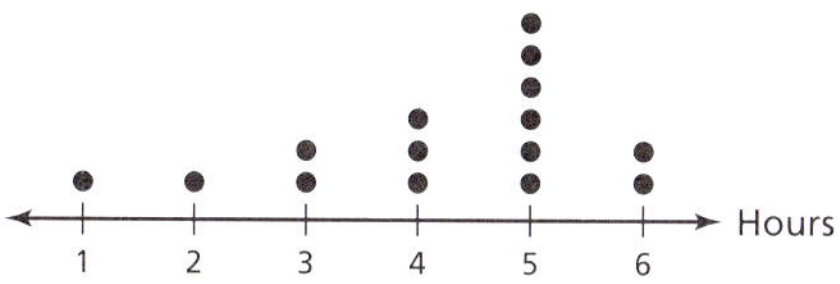

Most of the hours cluster around 5. The peak is 5. There is no gap.

c. Most students spend about 5 hours exercising during the week.

Section 9.4

Practice and Problem Solving

20. no; Two data sets can have the same range and different interquartile ranges due to outliers and/or the distribution of the data.

Example: Data set 1: 1, 5, 6, 6, 6, 7, 11; range = 10, IQR = 2
Data set 2: 1, 2, 2, 6, 10, 10, 11; range = 10, IQR = 8

21. **a.** Show A: mean = 20, median = 19.5, range = 13, IQR = 5

Show B: mean = 21, median = 20.5, range = 23, IQR = 6

The mean ages for the shows, 20 and 21, and the median ages for the shows, 19.5 and 20.5, are about the same. The interquartile ranges of ages for the shows, 5 and 6, are about the same. The range of the ages for Show A is 13 years and the range for Show B is 23 years. So, the ages for show B are more spread out.

b. Show A: The mean of the ages decreases a small amount, from 20 to $19\frac{8}{9}$. The median of the ages decreases from 19.5 to 18. The range of the ages stays at 13. The interquartile ranges of the ages increases from 5 to 6.5. Some of these values do not change by a large amount because 21 is towards the middle of the data set.

Show B: The mean of the ages decreases from 21 to $19\frac{1}{3}$. The median of the ages decreases a small amount, from 20.5 to 20. The range of the ages decreases a large amount, from 23 to 12. The interquartile ranges of the ages increases a small amount, from 6 to 6.5. Some of these values change by a large amount because 36 is an outlier of the data set.

Section 9.5

Practice and Problem Solving

12. Derek's collection: mean: 1929; median: 1930; no mode; range: 54; IQR: 48; MAD: 23.75

Paul's collection: mean: 1929; median: 1929.5; no mode; range: 15; IQR: 6; MAD: 3.5

Sample answer: The measures of center for the data sets are almost identical. But the measures of variation for Paul's coin collection are much less than the measures for Derek's coin collection. This means that the years of the coins in Paul's collection are closer together than the years of the coins in Derek's collection.

13. **a.** mean: 8.25; median: 8.5; mode: 5
range: 13; IQR: 5.5; MAD: 3

b. no; Using the interquartile range, 21 is inside the outlier boundaries.
mean: 9; median: 9; mode: 5
range: 19; IQR: 6.5; MAD: 3.5

The range is most affected by including this value. The mode stays the same. The mean, median, IQR, and MAD all increased slightly.

15. monthly amounts of water used in a home; *Sample answer:* The amount of rainfall that falls in a city during a month usually ranges from 0 to 5–6 inches. The monthly amounts of water used in a home are much greater numbers that will have more variation from month to month.

16. *Sample answer:* The range only uses two data values from a set and is greatly affected by outliers. The interquartile range ignores outliers but only uses a few data values from a set. When calculating the mean absolute deviation of a data set, you use all of the values in its calculation.

9.4–9.5 Quiz

8. a. School A: mean = 12.2, median = 9, range = 30, IQR = 4

School B: mean = 12.7, median = 11, range = 29, IQR = 8

The mean experience for the schools, 12.2 and 12.7, and the median experience for the schools, 9 and 11, are about the same. The interquartile range of experiences for School B is twice that of School A. The range of the experiences for School A is 30 years and the range for School B is 29 years.

b. School A: The mean of the experiences increases a small amount, from 12.2 to about 12.3. The median of the experiences decreases from 9 to 8. The range of the experiences stays at 30. The interquartile ranges of the experiences increases from 4 to 9.5. Some of these values do not change by a large amount because 11 is towards the middle of the data set.

School B: The mean of the experiences decreases from 12.7 to about 10.4. The median of the experiences decreases a small amount, from 11 to 10. The range of the experiences decreases a large amount, from 29 to 16. The interquartile ranges of the experiences decreases a small amount, from 8 to 7. Some of these values change by a large amount because 33 is an outlier of the data set.

Chapter 9 Review

1.

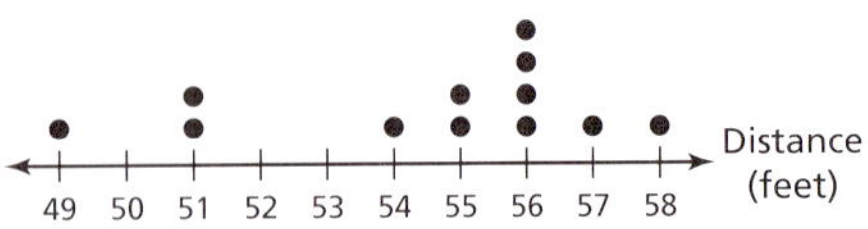

Most of the data are clustered around 56. There is a peak at 56. There are gaps between 49 and 51 and between 51 and 54.

2.

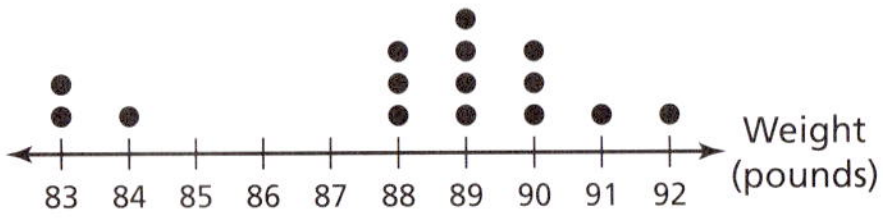

Most of the data are clustered around 89. There is peak at 89. There is a gap between 84 and 88.

Chapter 9 Test

1.

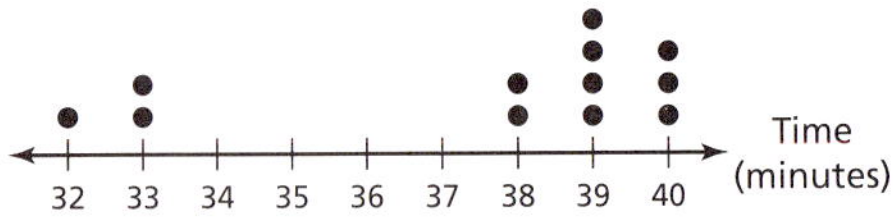

The data are clustered around 39. There is a peak at 39. There is a gap between 33 and 38.

2.

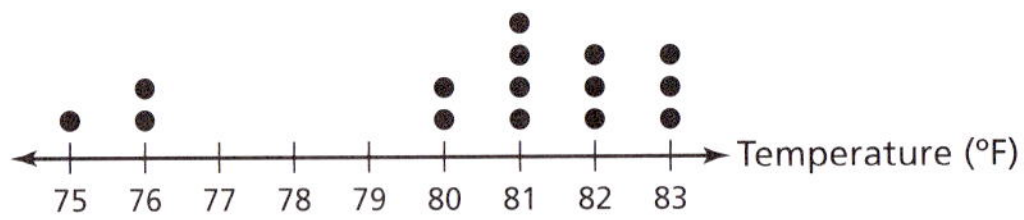

The data are clustered around 81. There is a peak at 81. There is a gap between 76 and 80.

3. mean: 6.4; median: 7; mode: 7; range: 10

14. Greg's hours: mean: 11.25; median: 10.5; modes: 9, 12; range: 18; IQR: 7.5; MAD: 4.5

Tom's hours: mean: 15; median: 15; modes: 12, 15, 18; range: 6; IQR: 4; MAD: 1.75

Sample answer: The measures of center for Tom's hours are greater than the measures of center for Greg's hours. The measures of variation for Tom's hours are less than the measures of variation for Greg's hours.

Chapter 10

Section 10.1

Record and Practice Journal

1. c.

Stem	Leaf
2	1
3	1
4	0 4 5 5 8 8 9 9
5	0 2 4 4 6 6 6 6 9
6	0 0 3

2. a.

Leaf	Stem	Leaf
	2	1
	3	1
7 7 6 3 2	4	0 4 5 5 8 8 9 9
6 6 5 5 5 4 4 4 2 1 1 1	5	0 2 4 4 6 6 6 6 9
9 4 2 1 0	6	0 0 3

Practice and Problem Solving

6. **Test Scores**

Stem	Leaf
6	2 5 8 9
7	
8	0 1 2 5 7 7 8
9	0 1 5 7

Key: 8 | 1 = 81%

7. **Points Scored**

Stem	Leaf
3	8
4	2 2 3 3 5
5	0 1 6 8 8
6	
7	0 1 1 5

Key: 3 | 8 = 38 points

8. **Bikes Sold**

Stem	Leaf
7	8 9
8	1 6
9	6 9 9
10	0 5 8
11	2 5

Key: 11 | 2 = 112 bikes

9. **Minutes in Line**

Stem	Leaf
1	6 9
2	0 2 6 7 9
3	1 1 6 8
4	0

Key: 4 | 0 = 4.0 minutes

10. Stems that fall in the range should still be shown even if there are no leaves for that stem.

Stem	Leaf
2	5 6
3	
4	2 4 7
5	0 1 5 5

Key: 4 | 2 = 42

11. **Weights**

Stem	Leaf
0	8
1	2 5 7 8
2	4 4
3	1

Key: 2 | 4 = 24 pounds

Most of the weights are in the middle.

16. The stem-and-leaf plot on the left has the lesser mean absolute deviation because it has more values closer to the mean than the stem-and-leaf plot on the right.

17. a. 6.4; The daily high temperatures differ from the mean daily high temperature by an average of 6.4 degrees.

b. Because the mean absolute deviation increases, most of the data values for the rest of the month must be further from the mean than 6.4 degrees. So, most of the data values for the rest of the month are either less than 71.6 degrees or greater than 84.4 degrees.

Section 10.2

Record and Practice Journal

2. b.

Tally Chart		
Interval	**Tally**	**Total**
10.0–11.9		0
12.0–13.9	III	3
14.0–15.9	I	1
16.0–17.9	II	2
18.0–19.9	~~IIII~~ I	6
20.0–21.9	IIII	4
22.0–23.9	II	2
24.0–25.9	II	2

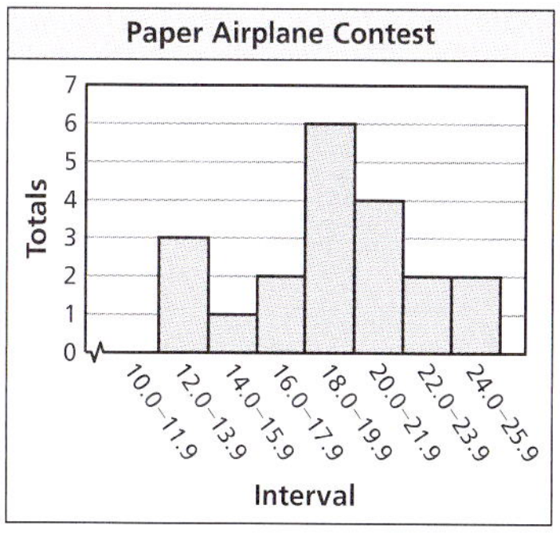

Practice and Problem Solving

4. *Sample answer:*

Interval	Tally	Total
0–9	||	2
10–19	~~||||~~ |||	8
20–29	||	2

5. *Sample answer:*

Interval	Tally	Total
20–29	|	1
30–39	|||	3
40–49	~~||||~~ ||	7
50–59	||||	4

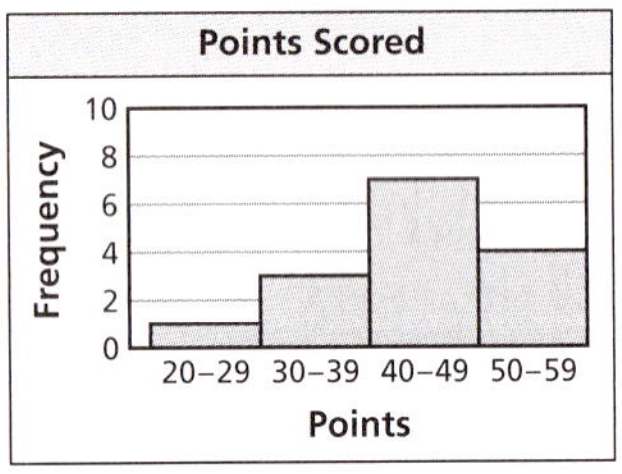

6.

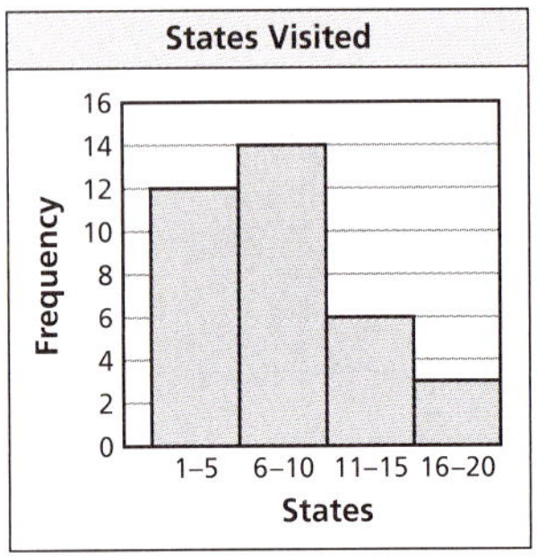

15. a. yes; The stem-and-leaf plot shows that 10 pounds is a data value.

b. no; Both displays show that 11 residents produced between 20 and 29 pounds of garbage.

Record and Practice Journal Practice

1.

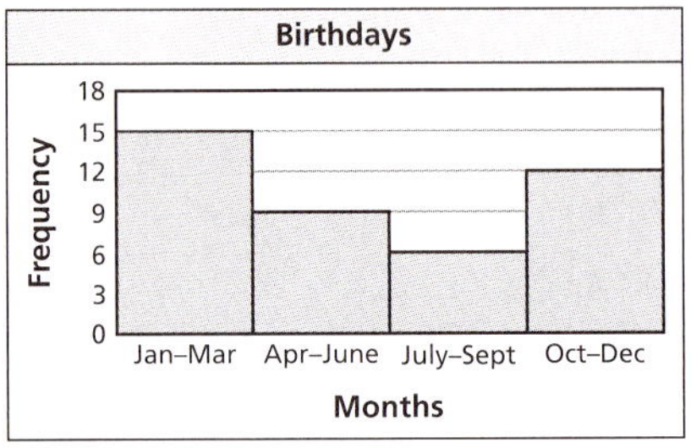

2.

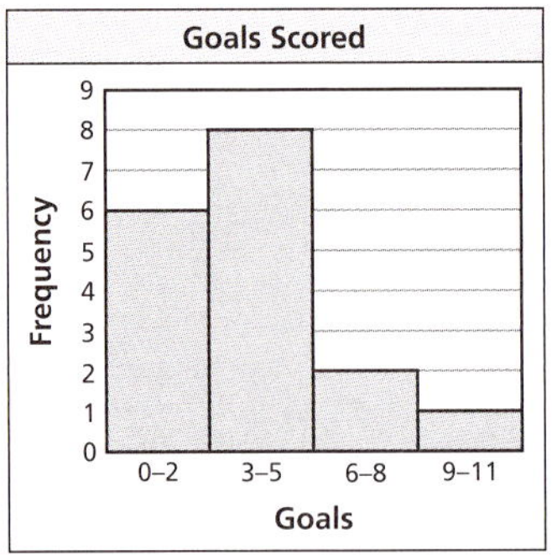

3.

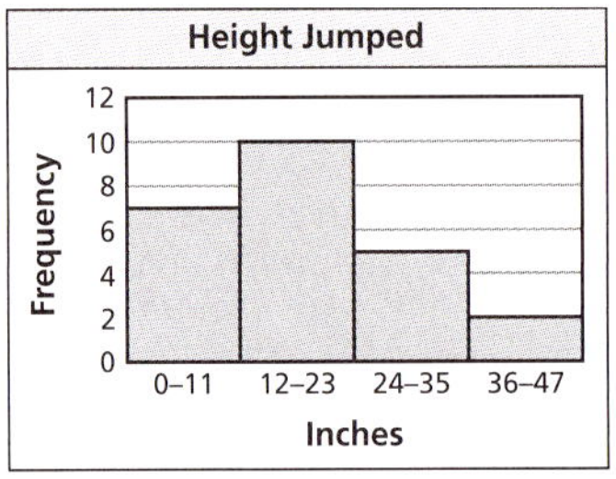

4.

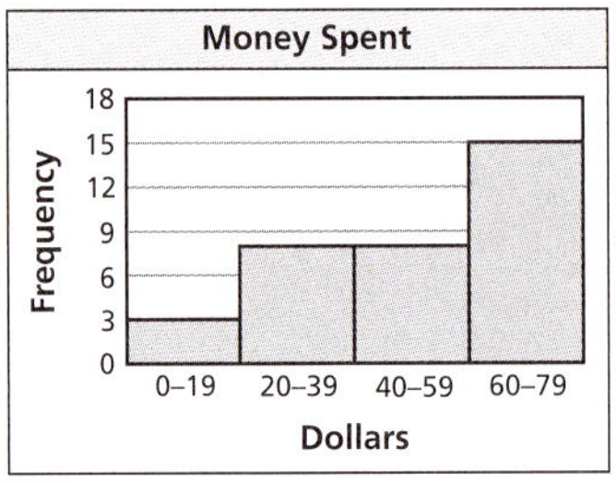

10.1–10.2 Quiz

5.

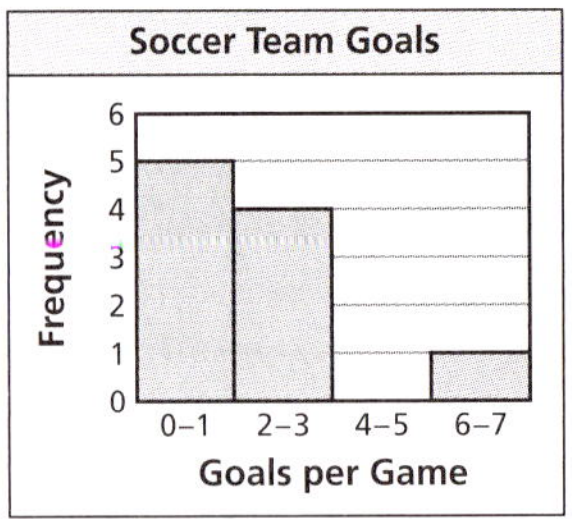

6.

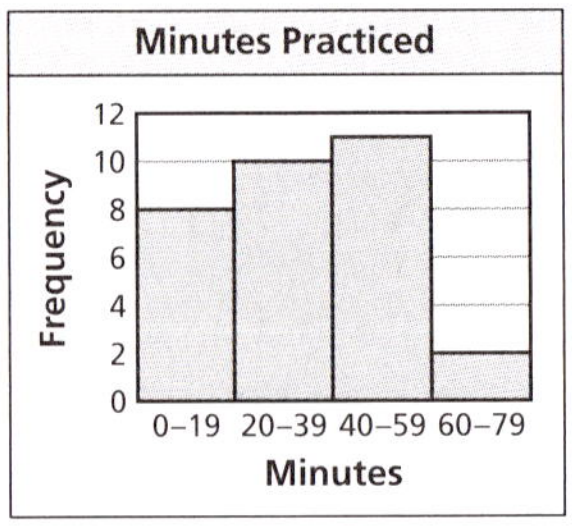

7. 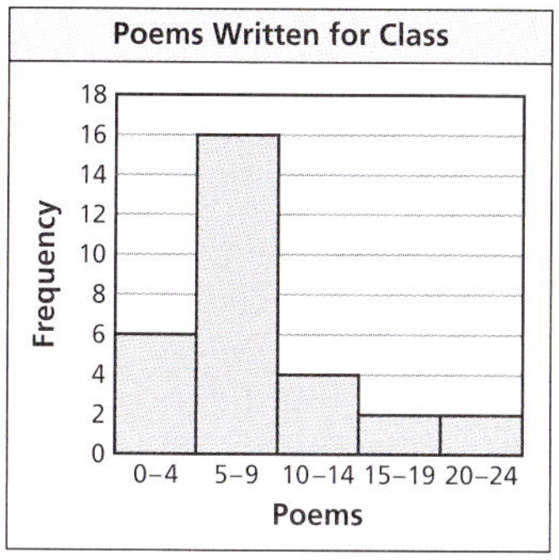

8. **Weights of Packages**

Stem	Leaf
0	6 7
1	2 3 5 6 8
2	2 5

Key: 0 | 6 = 6 ounces

Most of the data values are in the middle.

9. **a.** 2–3

b. 12 games

c. 25%

10. mean: 12; median: 11; mode: 8; range: 14; IQR: 9

Section 10.3

On Your Own

2. **a.** 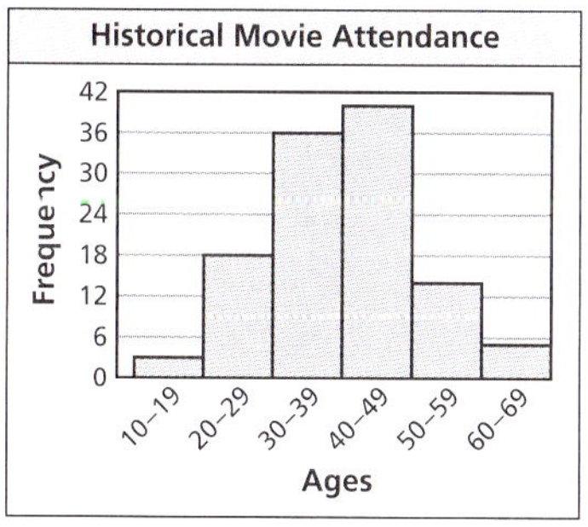

symmetric distribution

b. *Sample answer:* The historical movie has the oldest audience.

Practice and Problem Solving

3. 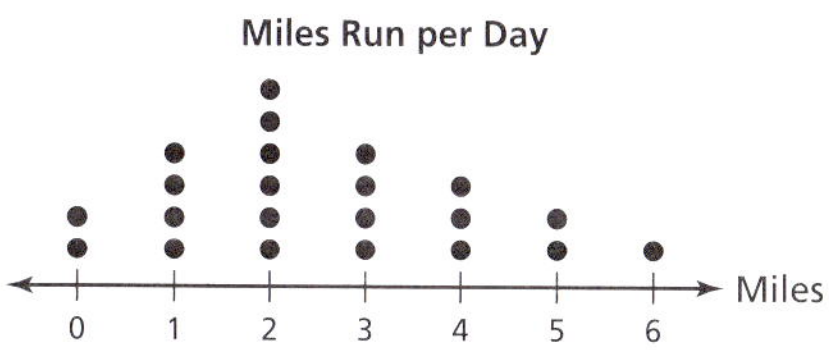

skewed

4. 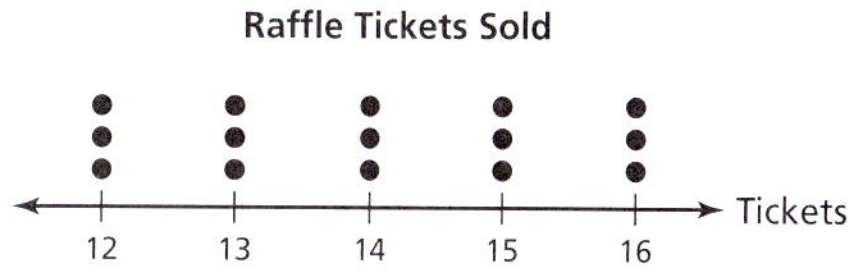

flat

11. no; Distributions can have any shape.

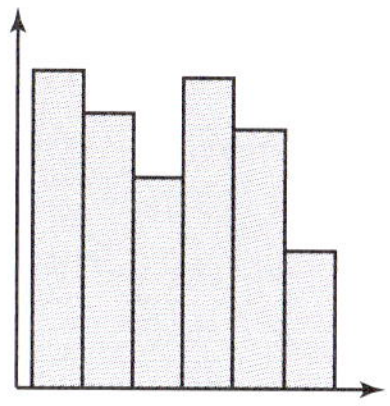

12. yes; *Sample answer:* Use interval lengths that represent what each stem represents. When the stem-and-leaf plot has most of its values on the bottom, the distribution is skewed left because when the data is placed in a dot plot or histogram, most of the data values will be on the right. When the stem-and-leaf plot is a mirror image on the top and bottom for the numbers of values, the distribution is symmetric because when the data is placed in a dot plot or histogram, the data on the left will be a mirror image of the data on the right. When the stem-and-leaf plot has most of its values on the top, the distribution is skewed right because when the data is placed in a dot plot or histogram, most of the data values will be on the left.

13. **a.** 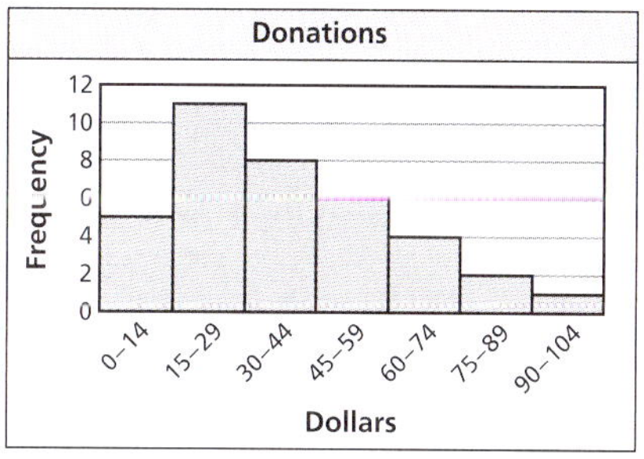

skewed right

b.

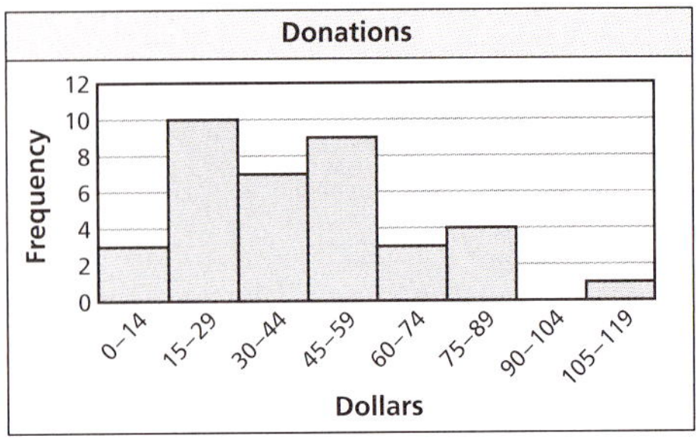

Both distributions are skewed right. The original donation distribution is more skewed right than the distribution when the increases are added to the donations. Some of the data values moved into different intervals when $5 is added to each donation, which is why the distributions are not exactly the same.

Record and Practice Journal Practice

5.

Extension 10.3

Practice

1. median and interquartile range; median: $32; IQR: $6
2. mean and mean absolute deviation; mean: 5 hours; MAD: 1.2 hours
3. no; You do not know the actual values in the data set. You can approximate the mean and MAD but your answers will not be exact.
4. **a.**

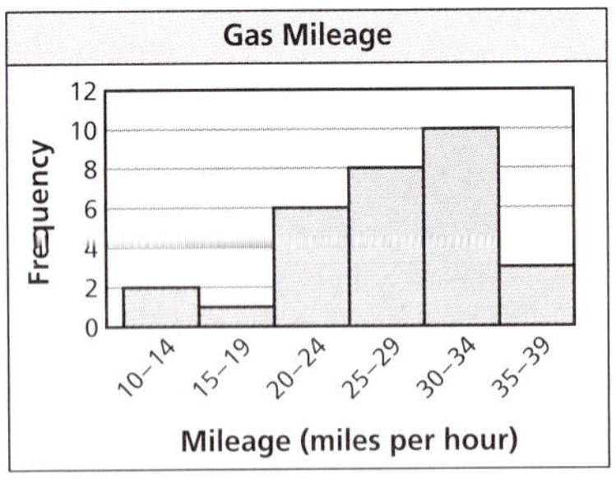

The distribution is skewed left. So, the median and interquartile range are the most appropriate measures.

b. *Sample answers:* Most of the vehicles made by the company get more than 25 miles per gallon. You can estimate that the median of the data set is in the interval 25–29.

5. *Sample answer:*

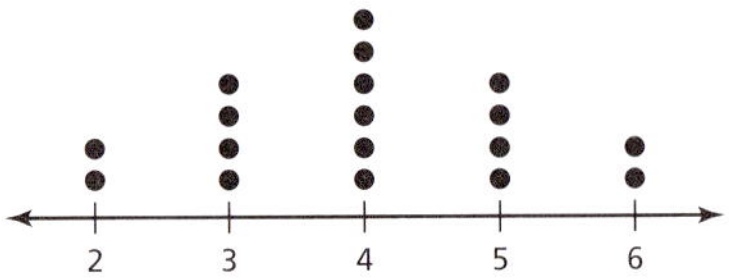

Record and Practice Journal Practice

2. The distribution is symmetric.; mean and mean absolute deviation; mean: 5 hours; mean absolute deviation: $\frac{8}{9}$ hour
3. The distribution is skewed right.; median and interquartile range; median: 8 downloads; interquartile range: 4 downloads
4. The distribution is symmetric.; mean and mean absolute deviation; mean: 5 in.; mean absolute deviation: 1.25 in.
5. **a.**

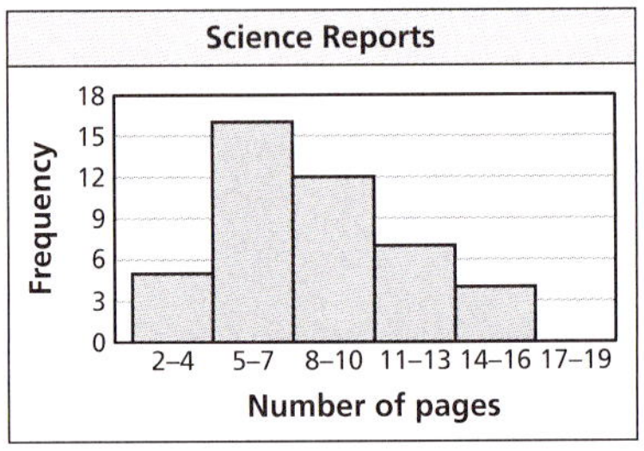

b. The distribution is skewed right. So, the median and interquartile range are the most appropriate measures.

Section 10.4

Practice and Problem Solving

5.

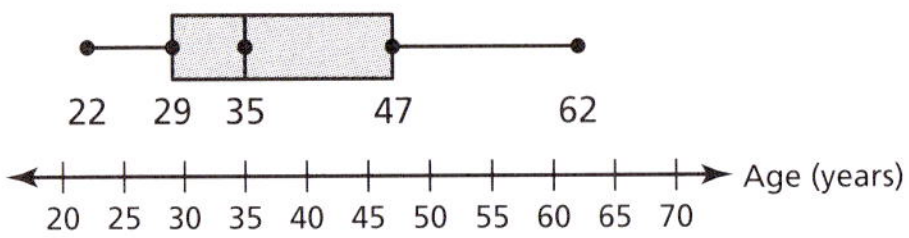

6.

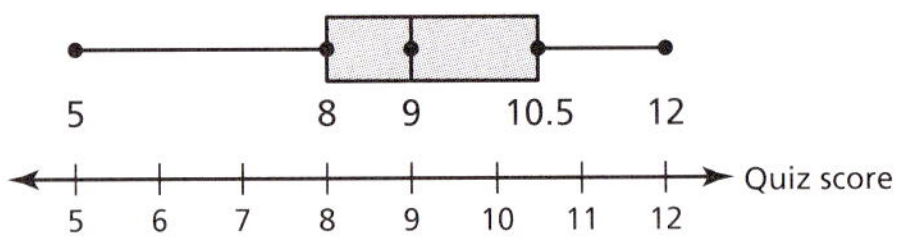

7.

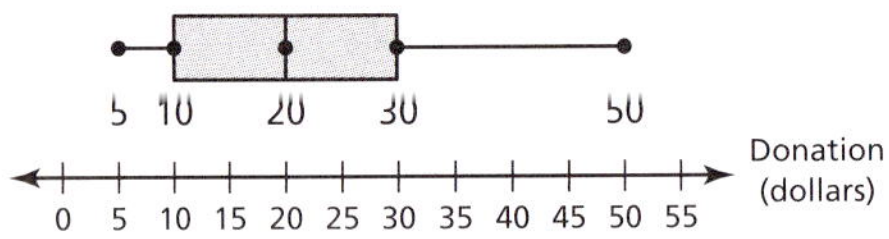

8.

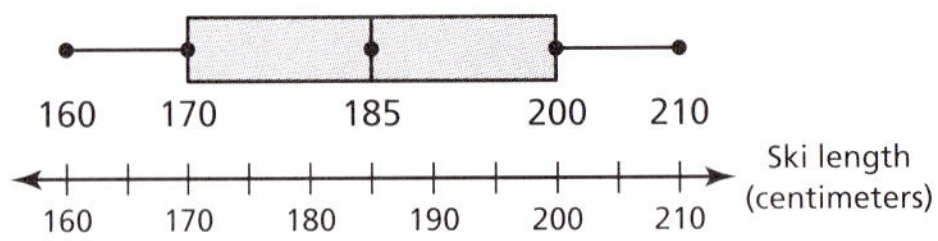

9. The data should be ordered before finding the five-number summary.

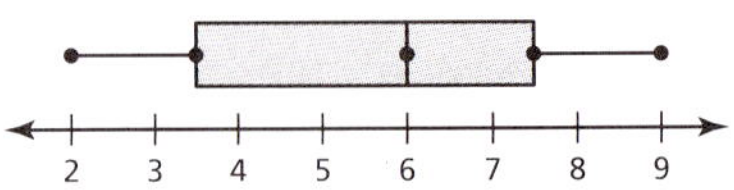

10. 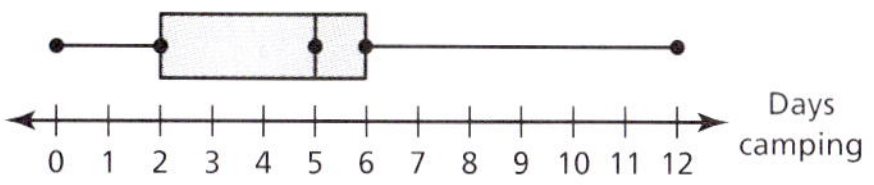

12

18. 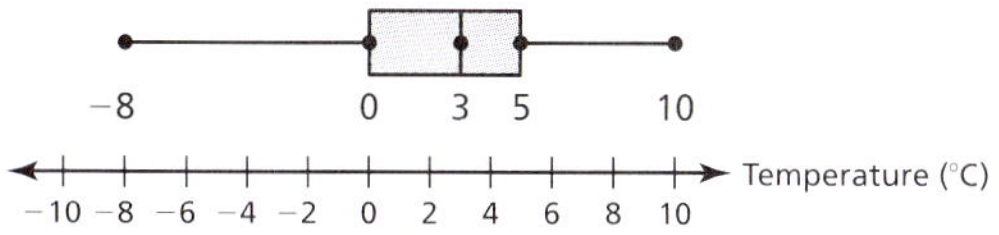

19. 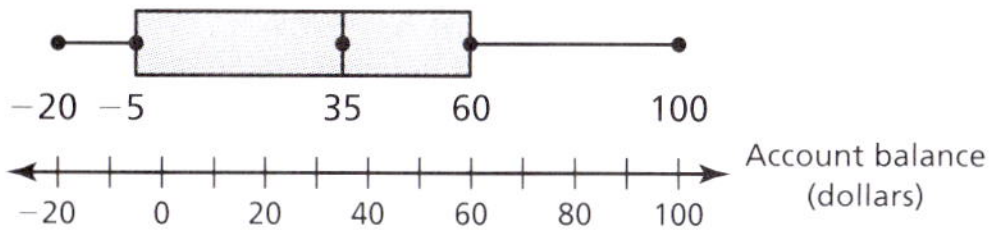

20. After removing the outlier: the left and right whiskers are shorter, the length of the box is greater, and the median and third quartile increased. In this case, the outlier had no affect on the first quartile or greatest value. See the new box-and-whisker plot below.

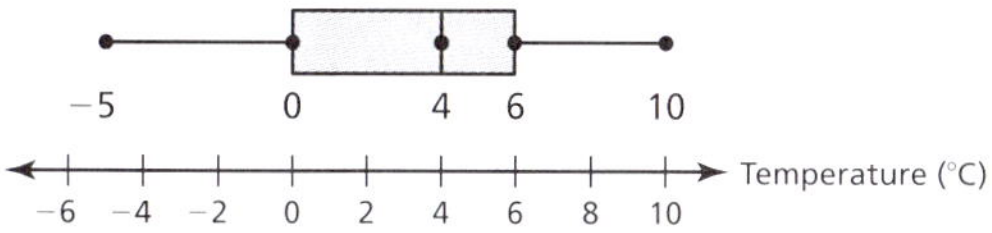

24. *Sample answer:* Each whisker is one-third the length of the box. So, draw a histogram with 5 intervals. The lower and upper intervals (whiskers) contain 25% of the data. The median is in the middle of the box, so the middle 50% of the data is distributed symmetrically.

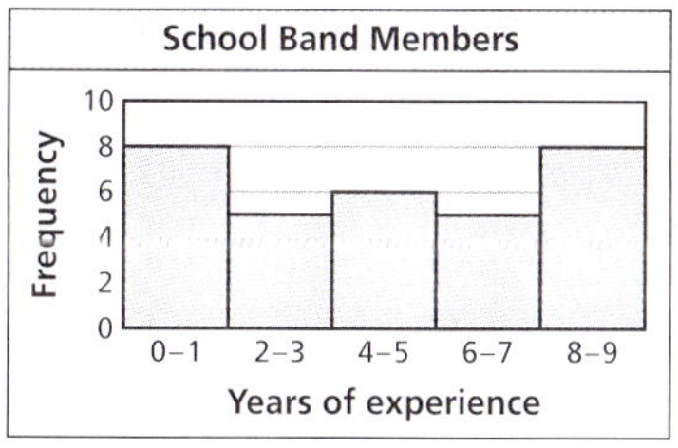

25. **a.** Team 1; There is less variability in the data.

b. 24 games

c. Team 1; In 75% of the games, Team 1 scored 6 runs or more. However, Team 2 scored 6 runs or less in 75% of the games.

d. Team 1; *Sample answer:* By looking at the shapes of the distributions, you can see that the majority of the data for Team 1 is greater than the majority of the data for Team 2.

26. *Sample answer:* box-and-whisker plot; A stem-and-leaf plot of a large data set would be impractical. The size of the display would be massive and difficult to use. While a histogram can show the variability of a data set, it is not as accurate and obvious as a box-and-whisker plot due to intervals and bar size.

Chapter 10 Test

7. 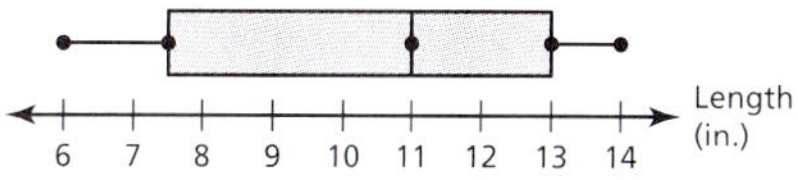

8. 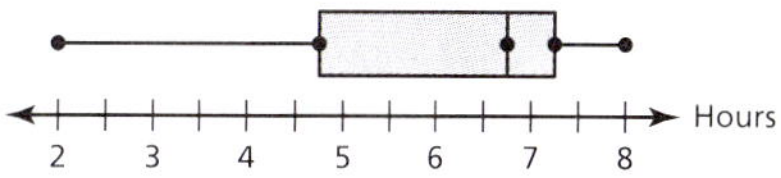

9. **a.** Brand A Range: 3.5 hours
Brand B Range: 2.7 hours

b. It is probably safe to say that Brand A is better. But, by unlucky sampling, its mean might be only slightly better than the mean for Brand B.

Additional Answers

Photo Credits

Cover
Pavelk/Shutterstock.com, Samuel Acosta/Shutterstock.com, valdis torms/Shutterstock.com

Front matter
i Pavelk/Shutterstock.com, Samuel Acosta/Shutterstock.com, valdis torms/Shutterstock.com; **iv** Big Ideas Learning, LLC; **vi** *top* ©iStockphoto.com/ALEAIMAGE, ©iStockphoto.com/Ann Marie Kurtz; *bottom* ©iStockphoto.com/Jane norton; **vii** *top* ©iStockphoto.com/ALEAIMAGE, ©iStockphoto.com/Ann Marie Kurtz; *bottom* wavebreakmedia ltd/Shutterstock.com; **viii** *top* stephan kerkhofs/Shutterstock.com, Cigdem Sean Cooper/Shutterstock.com, ©iStockphoto.com/Andreas Gradin; *bottom* Odua Images/Shutterstock.com; **ix** *top* ©iStockphoto.com/sumnersgraphicsinc, ©iStockphoto.com/Ann Marie Kurtz; *bottom* James Flint/Shutterstock.com; **x** *top* ©iStockphoto.com/Lisa Thornberg, ©iStockphoto.com/Ann Marie Kurtz; *bottom* william casey/Shutterstock.com; **xi** *top* ©iStockphoto.com/Jonathan Larsen; *bottom* Edyta Pawlowska/Shutterstock.com; **xii** *top* Varina and Jay Patel/Shutterstock.com, ©iStockphoto.com/Ann Marie Kurtz; *bottom* PETER CLOSE/Shutterstock.com; **xiii** *top* ©iStockphoto/Michael Flippo, ©iStockphoto.com/Ann Marie Kurtz; *bottom* ©iStockphoto.com/ranplett; **xiv** *top* Chiyacat/Shutterstock.com, Zoom Team/Shutterstock.com; *bottom* ©iStockphoto.com/Noraznen Azit; **xv** *top* ©iStockphoto.com/Alistair Cotton; *bottom* ©iStockphoto.com/Thomas Perkins; **xxx** *top left* AVAVA/Shutterstock.com; *bottom right* Tomasz/Shutterstock.com, Nemida/Shutterstock.com; **xxxi** *top left* Konstantin Chagin/Shutterstock.com; *top right* Roger Jegg - Fotodesign-Jegg.de/Shutterstock.com; **xxxviii** Ljupco Smokovski/Shutterstock.com

Chapter 1
1 ©iStockphoto.com/ALEAIMAGE, ©iStockphoto.com/Ann Marie Kurtz; **2** *top right* S.Dashkevych/Shutterstock.com; *bottom right* auremar/Shutterstock.com; **5** yxm2008/Shutterstock.com; **6** ©iStockphoto.com/Santino Ambrogio; **8** Jiang Dao Hua/Shutterstock.com; **9** *top right* Africa Studio/Shutterstock.com; *center left* Podriv Ustoev/Shutterstock.com; **10** Danomyte/Shutterstock.com; **11** Nicole Gordine/Shutterstock.com; **13** vonDUCK /Shutterstock.com; **15** ©iStockphoto.com/WestLight; **20** ©iStockphoto.com/clu; **21** ©iStockphoto.com/mladn61; **23** *center right* Wendy Nero/Shutterstock.com; *bottom right* ©iStockphoto.com/craftvision; **24** James Hoenstine/Shutterstock.com; **26** farbeffekte/Shutterstock.com; **29** Michael Mitchell/Shutterstock.com; **33** jmatzick/Shutterstock.com; **34** Glenda M. Powers/Shutterstock.com; **35** Africa Studio/Shutterstock.com; **39** Alinochka/Shutterstock.com; **40** Orla/Shutterstock.com; **41** yuyangc/Shutterstock.com; **44** *center left* Mike Flippo/Shutterstock.com; *bottom right* Nitr/Shutterstock.com; **48** beboy/Shutterstock.com

Chapter 2
52 ©iStockphoto.com/ALEAIMAGE, ©iStockphoto.com/Ann Marie Kurtz; **59** *center left* fivespots/Shutterstock.com; *center right* bluehand/Shutterstock.com; **68** tele52/Shutterstock.com, lithian/Shutterstock.com, Galina Barskaya/Shutterstock.com; **T-69** ©iStockphoto.com/Michael Plumb; **69** *center left* ©iStockphoto.com/Michael Plumb; *center right* g215/Shutterstock.com; **73** ©iStockphoto.com/Karin Lau; **81** Vacclav/Shutterstock.com; **82** Perednlankina/Shutterstock.com; **83** SSSCCC/Shutterstock.com; **89** Luba V Nel/Shutterstock.com; **90** KENCKOphotography/Shutterstock.com; **97** ©iStockphoto.com/bananahuman; **99** ©iStockphoto.com/Janis Litavnieks; **104** ©iStockphoto.com/graham heywood

Chapter 3
108 stephan kerkhofs/Shutterstock.com, Cigdem Sean Cooper/Shutterstock.com, ©iStockphoto.com/Andreas Gradin; **121** Kateryna Larina/Shutterstock.com; **125** Vivid Pixels/Shutterstock.com; **138** Pal Teravagimov/Shutterstock.com; **141** Inhabitant/Shutterstock.com; **142** Fotofermer/Shutterstock.com; **143** Helder Almeida/Shutterstock.com; **146** *center right* Aptyp_koK/Shutterstock.com; *bottom left* Andy Cash/Shutterstock.com

Chapter 4
150 ©iStockphoto.com/sumnersgraphicsinc, ©iStockphoto.com/Ann Marie Kurtz; **162** Terrance Emerson/Shutterstock.com; **163** ©iStockphoto.com/bamse009; **171** Kharidehal Abhirama Ashwin/Shutterstock.com; **177** Christian Musat/Shutterstock.com; **179** Brandon Seidel/Shutterstock.com; **184** *top right* ©iStockphoto.com/AlexMax; *bottom right* U.S. Geological Survey

Chapter 5
188 ©iStockphoto.com/Lisa Thornberg, ©iStockphoto.com/Ann Marie Kurtz; **190** Elnur/Shutterstock.com; **192** Vladimir Wrangel/Shutterstock.com; **193** Lepas/Shutterstock.com; **196** *top* foxie/Shutterstock.com; *center right* Constantinos/Shutterstock.com; **197** Danny Smythe/Shutterstock.com; **199** South 12th Photography/Shutterstock.com; **200** Petr Malyshev/Shutterstock.com; **202** ©iStockphoto.com/otakumania; **203** Anneka/Shutterstock.com; **205** vita khorzhevska/Shutterstock.com; **206** © HO/Reuters/Corbis; **207** Alex Staroseltsev/Shutterstock.com; **208** *top right* ©iStockphoto.com/Gord Horne; *bottom right* leonid_tit/Shutterstock.com; **209** *top left* ©iStockphoto.com/ljpat; *bottom right* ©iStockphoto.com/Birgitte Magnus; **210** Mike Flippo/Shutterstock.com; **212** *top right* ultimathule/Shutterstock.com; *top left* Christopher Kolaczan/Shutterstock.com; **213** ©iStockphoto.com/Bart Wolczyk; **214** Catalin Petolea/Shutterstock.com, Kaspri/Shutterstock.com; **215** R. Gino Santa Maria/Shutterstock.com; **225** Spasiblo/Shutterstock.com; **230** Thank You/Shutterstock.com; **236** Derek Wong / CC-BY-SA-3.0; **237** ©iStockphoto.com/Paul Tessier; **238** *bottom left* Kitch Bain/Shutterstock.com; *bottom* ©iStockphoto.com/Ermin Gutenberger; **242** Joe Gough/Shutterstock.com

Chapter 6
246 ©iStockphoto.com/Jonathan Larsen; **251** ©iStockphoto.com/Egor Mopanko; **253** haveseen/Shutterstock.com; **254** *top right* NASA/Kim Shiflett; *bottom* NASA; **255** NASA; **259** ©iStockphoto.com/Andrew Penner; **260** NASA; **261** *left* Andrey Armyagov/Shutterstock.com; *center* ©iStockphoto.com/Island Effects; *right* U.S. Navy photo by Photographer's Mate 2nd Class Prince Hughes III; **264** ©iStockphoto.com/jclegg, ©iStockphoto.com/spxChrome, ©iStockphoto.com/Laura Eisenberg; **265** A'lya/Shutterstock.com; **267** *top right* NicolasMcComber/Shutterstock.com; *bottom* ©iStockphoto.com/Zuki; **269** *left* ©iStockphoto.com/james steidl; *right* U.S. Navy photo by Photographers Mate 2nd Class Michael Sandberg; **271** Shane W Thompson/Shutterstock.com; **278** Christophe Testi/Shutterstock.com

Chapter 7

292 Varina and Jay Patel/Shutterstock.com, ©iStockphoto.com/Ann Marie Kurtz; **299** *top right* ©iStockphoto.com/Joop Snijder, ©iStockphoto.com/Michael MacFadden, ©iStockphoto.com/Steve Goodwin; *center left* ©iStockphoto.com/Kenneth C. Zirkel; **304** jocic/Shutterstock.com, Marko Poplasen/Shutterstock.com; **306** *top left* ©iStockphoto.com/Jeremy Wee; *top right* ©iStockphoto.com/Jan Will; **307** ©iStockphoto.com/Keith Reicher; **309** ©iStockphoto.com/Leo Blanchette; **312** ©iStockphoto.com/Christopher Futcher; **313** ©iStockphoto.com/Eric Isselée; **314** mangostock/Shutterstock.com; **315** Ilya Andriyanov/Shutterstock.com; **316** violetkaipa/Shutterstock.com; **318** kokandr/Shutterstock.com; **319** discpicture/Shutterstock.com; **320** ©iStockphoto.com/Mutlu Kurtbas; **324** *first* ©iStockphoto.com/Studio-Annika; *second* ©iStockphoto.com/nicholas belton; *third* ©iStockphoto.com/Robert Dant; **328** NASA/Johns Hopkins University Applied Physics Laboratory; **330** ©iStockphoto.com/George Peters; **331** ©iStockphoto.com/Anthony Ladd; **332** *top right* CLS Design/Shutterstock.com; *bottom right* Liquid Productions, LLC /Shutterstock.com; **337** *top left* ©iStockphoto.com/o-che; *center left* ©iStockphoto.com/sunygraphics; **338** *top right* Dennis Owusu-Ansah/Shutterstock.com; *bottom right* Julia Zakharova /Shutterstock.com; **341** Nathan Till/Shutterstock.com; **342** ©iStockphoto.com/rami ben ami; **343** Karin Hildebrand Lau/Shutterstock.com; **344** ©iStockphoto.com/Algimantas Balezentis; **348** ©iStockphoto.com/Jani Bryson

Chapter 8

352 ©iStockphoto.com/Michael Flippo, ©iStockphoto.com/Ann Marie Kurtz; **358** *Exercise 17* ©iStockphoto.com/Rich Koele; *Exercise 21* design56/Shutterstock.com; **359** *top right* ©iStockphoto.com/Hedda Gjerpen; *center* ©iStockphoto.com/rzdeb; **363** Niki Crucillo/Shutterstock.com; **370** PeterG/Shutterstock.com; **372** Itana/Shutterstock.com; **373** Tupungato/Shutterstock.com; **377** *top left* ©iStockphoto.com/William Britten; *center left* Denis Barbulat/Shutterstock.com; **379** *center left* ©iStockphoto.com/Jill Chen; *center right* ©iStockphoto.com/LongHa2006

Chapter 9

388 Chiyacat/Shutterstock.com, Zoom Team/Shutterstock.com; **390** LeventeGyori/Shutterstock.com; **391** *center left and right* Nattika/Shutterstock.com; **392** Eric Isselée/Shutterstock.com; **393** *top right* Iznogood/Shutterstock.com; *bottom left* AISPIX by Image Source/Shutterstock.com; **395** Laralova/Shutterstock.com; **396** Rob Byron/Shutterstock.com; **397** Denis Vrublevski/Shutterstock.com; **399** ©iStockphoto.com/Eric Isselée; **402** *top right* Hein Nouwens/Shutterstock.com; *bottom* ivelly/Shutterstock.com; **403** ©iStockphoto.com/Andrew Rich; **409** *top right* ©iStockphoto.com/suemack; *top left* ©iStockphoto.com/muratkoc; **415** Charlie Hutton/Shutterstock.com; **416** Talvi/Shutterstock.com; **417** ©iStockphoto.com/Jason Lugo; **420** Danny Smythe/Shutterstock.com; **421** *top left* Ganko/Shutterstock.com; *top right* Mark Herreid/Shutterstock.com; **422** alarich/Shutterstock.com; **423** tab62/Shutterstock.com; **424** Racheal Grazias/Shutterstock.com; **426** Jan Martin Will/Shutterstock.com; **427** Kitch Bain/Shutterstock.com

Chapter 10

432 ©iStockphoto.com/Alistair Cotton; **434** Elzbieta Szpak/Shutterstock.com; **435** ©CORBIS; **437** ©iStockphoto.com/Pekka Nikonen; **438** ©iStockphoto.com/Mehmet Salih Guler; **442** stockshoppe/Shutterstock.com; **443** *top right* ©iStockphoto.com/susaro; *bottom left* Arman Zhenikeyev/Shutterstock.com; **444** Tomasz Trojanowski/Shutterstock.com; **447** ©iStockphoto.com/Eric Isselée; **449** ©iStockphoto.com/vincent chien chow chine; **450** Sergey Mironov/Shutterstock.com; **451** fresher/Shutterstock.com; **453** mmaxer/Shutterstock.com; **457** Lightspring/Shutterstock.com; **459** *first* windu/Shutterstock.com; *second* motorolka/Shutterstock.com; *third* Preto Perola/Shutterstock.com; *fourth* nikkytok/Shutterstock.com; *center right* Apollofoto/Shutterstock.com; **460** ©iStockphoto.com/rusm; **461** Sebastian Knight/Shutterstock.com; **462** Garbuzov/Shutterstock.com; **464** *center left* Ffooter/Shutterstock.com; *bottom right* zhuda/Shutterstock.com; **465** Rob Marmion/Shutterstock.com; **469** Nikola Bilic/Shutterstock.com; **470** Monkey Business Images/Shutterstock.com

Appendix A

A0 *background* ©iStockphoto.com/Björn Kindler; *top left and bottom* ©iStockphoto.com/Ralf Hettler; **A1** *top right* Emmer, Michele, ed., The Visual Mind: Art and Mathematics, Plate 2, © 1993 Massachusetts Institute of Technology, by permission of The MIT Press.; *bottom left* ©iStockphoto.com/Andrew Cribb; *bottom right* ©iStockphoto.com/Liz Leyden; **A4** *top right* ©iStockphoto.com/Ralf Hettler; *center right* ©iStockphoto.com/Ragnarocks; *bottom left* ©iStockphoto.com/Linda Steward; *bottom right* ©iStockphoto.com/rackermann; **A5** *top and bottom right* ©iStockphoto.com/rackermann; *bottom left* ©iStockphoto.com/Ralf Hettler; **A6** *top right* Stannered, DTR; *bottom left* Emmer, Michele, ed., The Visual Mind: Art and Mathematics, Plate 2, © 1993 Massachusetts Institute of Technology, by permission of The MIT Press.; **A7** *top right* Elena Borodynkina/Shutterstock.com; *center right* ©iStockphoto.com/Andrew Cribb; *bottom left* ©iStockphoto.com/smokyme; *bottom right* Sculpture by Vladimir Bulatov; **A8** *top right* ©iStockphoto.com/Dan Van Oss; *bottom left* ©iStockphoto.com/Tomasz Tulik; **A9** *top right* ©iStockphoto.com/timoph; *bottom left* ©iStockphoto.com/Liz Leyden; *bottom right* ©iStockphoto.com/Pauline S Mills

Cartoon illustrations Tyler Stout

Common Core State Standards

Kindergarten

Counting and Cardinality	– Count to 100 by Ones and Tens; Compare Numbers
Operations and Algebraic Thinking	– Understand and Model Addition and Subtraction
Number and Operations in Base Ten	– Work with Numbers 11–19 to Gain Foundations for Place Value
Measurement and Data	– Describe and Compare Measurable Attributes; Classify Objects into Categories
Geometry	– Identify and Describe Shapes

Grade 1

Operations and Algebraic Thinking	– Represent and Solve Addition and Subtraction Problems
Number and Operations in Base Ten	– Understand Place Value for Two-Digit Numbers; Use Place Value and Properties to Add and Subtract
Measurement and Data	– Measure Lengths Indirectly; Write and Tell Time; Represent and Interpret Data
Geometry	– Draw Shapes; Partition Circles and Rectangles into Two and Four Equal Shares

Grade 2

Operations and Algebraic Thinking	– Solve One- and Two-Step Problems Involving Addition and Subtraction; Build a Foundation for Multiplication
Number and Operations in Base Ten	– Understand Place Value for Three-Digit Numbers; Use Place Value and Properties to Add and Subtract
Measurement and Data	– Measure and Estimate Lengths in Standard Units; Work with Time and Money
Geometry	– Draw and Identify Shapes; Partition Circles and Rectangles into Two, Three, and Four Equal Shares

Grade 3

Operations and Algebraic Thinking	– Represent and Solve Problems Involving Multiplication and Division; Solve Two-Step Problems Involving Four Operations
Number and Operations in Base Ten	– Round Whole Numbers; Add, Subtract, and Multiply Multi-Digit Whole Numbers
Number and Operations—Fractions	– Understand Fractions as Numbers
Measurement and Data	– Solve Time, Liquid Volume, and Mass Problems; Understand Perimeter and Area
Geometry	– Reason with Shapes and Their Attributes

Grade 4

Operations and Algebraic Thinking	– Use the Four Operations with Whole Numbers to Solve Problems; Understand Factors and Multiples
Number and Operations in Base Ten	– Generalize Place Value Understanding; Perform Multi-Digit Arithmetic
Number and Operations—Fractions	– Build Fractions from Unit Fractions; Understand Decimal Notation for Fractions
Measurement and Data	– Convert Measurements; Understand and Measure Angles
Geometry	– Draw and Identify Lines and Angles; Classify Shapes

Grade 5

Operations and Algebraic Thinking	– Write and Interpret Numerical Expressions
Number and Operations in Base Ten	– Perform Operations with Multi-Digit Numbers and Decimals to Hundredths
Number and Operations—Fractions	– Add, Subtract, Multiply, and Divide Fractions
Measurement and Data	– Convert Measurements within a Measurement System; Understand Volume
Geometry	– Graph Points in the First Quadrant of the Coordinate Plane; Classify Two-Dimensional Figures

Mathematics Reference Sheet

Conversions

U.S. Customary
1 foot = 12 inches
1 yard = 3 feet
1 mile = 5280 feet
1 acre ≈ 43,560 square feet
1 cup = 8 fluid ounces
1 pint = 2 cups
1 quart = 2 pints
1 gallon = 4 quarts
1 gallon = 231 cubic inches
1 pound = 16 ounces
1 ton = 2000 pounds
1 cubic foot ≈ 7.5 gallons

U.S. Customary to Metric
1 inch = 2.54 centimeters
1 foot ≈ 0.3 meter
1 mile ≈ 1.61 kilometers
1 quart ≈ 0.95 liter
1 gallon ≈ 3.79 liters
1 cup ≈ 237 milliliters
1 pound ≈ 0.45 kilogram
1 ounce ≈ 28.3 grams
1 gallon ≈ 3785 cubic centimeters

Time
1 minute = 60 seconds
1 hour = 60 minutes
1 hour = 3600 seconds
1 year = 52 weeks

Temperature
$C = \frac{5}{9}(F - 32)$

$F = \frac{9}{5}C + 32$

Metric
1 centimeter = 10 millimeters
1 meter = 100 centimeters
1 kilometer = 1000 meters
1 liter = 1000 milliliters
1 kiloliter = 1000 liters
1 milliliter = 1 cubic centimeter
1 liter = 1000 cubic centimeters
1 cubic millimeter = 0.001 milliliter
1 gram = 1000 milligrams
1 kilogram = 1000 grams

Metric to U.S. Customary
1 centimeter ≈ 0.39 inch
1 meter ≈ 3.28 feet
1 kilometer ≈ 0.62 mile
1 liter ≈ 1.06 quarts
1 liter ≈ 0.26 gallon
1 kilogram ≈ 2.2 pounds
1 gram ≈ 0.035 ounce
1 cubic meter ≈ 264 gallons

Number Properties

Commutative Properties of Addition and Multiplication

$a + b = b + a$

$a \cdot b = b \cdot a$

Associative Properties of Addition and Multiplication

$(a + b) + c = a + (b + c)$

$(a \cdot b) \cdot c = a \cdot (b \cdot c)$

Addition Property of Zero

$a + 0 = a$

Multiplication Properties of Zero and One

$a \cdot 0 = 0$

$a \cdot 1 = a$

Distributive Property:

$a(b + c) = ab + ac$

$a(b - c) = ab - ac$

Properties of Equality

Addition Property of Equality

If $a = b$, then $a + c = b + c$.

Subtraction Property of Equality

If $a = b$, then $a - c = b - c$.

Multiplication Property of Equality

If $a = b$, then $a \cdot c = b \cdot c$.

Multiplicative Inverse Property

$n \cdot \frac{1}{n} = \frac{1}{n} \cdot n = 1, n \neq 0$

Division Property of Equality

If $a = b$, then $a \div c = b \div c, c \neq 0$.